Lecture Notes in Artificial Intelligence 5097

Edited by R. Goebel, J. Siekmann, and W. Wahlster

Subseries of Lecture Notes in Computer Science

Leszek Rutkowski   Ryszard Tadeusiewicz
Lotfi A. Zadeh   Jacek M. Zurada (Eds.)

Artificial Intelligence and Soft Computing – ICAISC 2008

9th International Conference
Zakopane, Poland, June 22-26, 2008
Proceedings

Series Editors

Randy Goebel, University of Alberta, Edmonton, Canada
Jörg Siekmann, University of Saarland, Saarbrücken, Germany
Wolfgang Wahlster, DFKI and University of Saarland, Saarbrücken, Germany

Volume Editors

Leszek Rutkowski
Częstochowa University of Technology
Armii Krajowej 36, 42-200 Częstochowa, Poland
E-mail: lrutko@kik.pcz.czest.pl

Ryszard Tadeusiewicz
AGH University of Science and Technology
Mickiewicza 30, 30-059 Kraków, Poland
E-mail: rtad@agh.edu.pl

Lotfi A. Zadeh
University of California, Berkeley
Department of Electrical Engineering and Computer Sciences
Berkeley Initiative in Soft Computing (BISC)
Berkeley, CA 94720-1776, USA
E-mail: zadeh@cs.berkeley.edu

Jacek M. Zurada
University of Louisville
Computational Intelligence Laboratory
405 Lutz Hall, Louisville, KY 40292, USA
E-mail: jacek.zurada@louisville.edu

Library of Congress Control Number: 2008929519

CR Subject Classification (1998): I.2, F.4.1, F.1, F.2, I.4

LNCS Sublibrary: SL 7 – Artificial Intelligence

ISSN   0302-9743
ISBN-10   3-540-69572-9 Springer Berlin Heidelberg New York
ISBN-13   978-3-540-69572-1 Springer Berlin Heidelberg New York

This work is subject to copyright. All rights are reserved, whether the whole or part of the material is concerned, specifically the rights of translation, reprinting, re-use of illustrations, recitation, broadcasting, reproduction on microfilms or in any other way, and storage in data banks. Duplication of this publication or parts thereof is permitted only under the provisions of the German Copyright Law of September 9, 1965, in its current version, and permission for use must always be obtained from Springer. Violations are liable to prosecution under the German Copyright Law.

Springer is a part of Springer Science+Business Media

springer.com

© Springer-Verlag Berlin Heidelberg 2008
Printed in Germany

Typesetting: Camera-ready by author, data conversion by Scientific Publishing Services, Chennai, India
Printed on acid-free paper   SPIN: 12320011   06/3180   5 4 3 2 1 0

Preface

This volume constitutes the proceedings of the 9th Conference on Artificial Intelligence and Soft Computing, ICAISC 2008, held in Zakopane, Poland during June 22–26, 2008. The conference was organized by the Polish Neural Network Society in cooperation with the Academy of Humanities and Economics in Łódź, the Department of Computer Engineering at the Czestochowa University of Technology, and the IEEE Computational Intelligence Society - Poland Chapter. The previous conferences took place in Kule (1994), Szczyrk (1996), Kule (1997) and Zakopane (1999, 2000, 2002, 2004, 2006) and attracted a large number of papers and internationally recognized speakers: Lotfi A. Zadeh, Shun-ichi Amari, Daniel Amit, Piero P. Bonissone, Zdzislaw Bubnicki, Andrzej Cichocki, Wlodzislaw Duch, Jerzy Grzymala-Busse, Kaoru Hirota, Janusz Kacprzyk, Laszlo T. Koczy, Soo-Young Lee, Robert Marks, Evangelia Micheli-Tzanakou, Erkki Oja, Witold Pedrycz, Sarunas Raudys, Enrique Ruspini, Jörg Siekmann, Roman Slowinski, Ryszard Tadeusiewicz, Shiro Usui, Ronald Y. Yager, Syozo Yasui and Jacek Zurada. The aim of this conference is to build a bridge between traditional artificial intelligence techniques and recently developed soft computing techniques. It was pointed out by Lotfi A. Zadeh that "soft computing (SC) is a coalition of methodologies which are oriented toward the conception and design of information/intelligent systems. The principal members of the coalition are: fuzzy logic (FL), neurocomputing (NC), evolutionary computing (EC), probabilistic computing (PC), chaotic computing (CC), and machine learning (ML). The constituent methodologies of SC are, for the most part, complementary and synergistic rather than competitive." This volume presents both traditional artificial intelligence methods and soft computing techniques. Our goal is to bring together scientists representing both traditional artificial intelligence approach and soft computing techniques. The volume is divided into eight parts:

- Neural Networks and Their Applications
- Fuzzy Systems and Their Applications
- Evolutionary Algorithms and Their Applications
- Classification, Rule Discovery and Clustering
- Image Analysis, Speech and Robotics
- Bioinformatics and Medical Applications
- Various Problems of Artificial Intelligence
- Agent Systems

The conference attracted a total of 320 submissions from 39 countries and after the review process, 116 papers were accepted for publication in this volume. I would like to thank our participants, invited speakers and reviewers of the papers for their scientific and personal contribution to the conference. The following reviewers were very helpful in reviewing the papers:

R. Adamczak
R. Angryk
J. Arabas
W. Bartkiewicz
Ł. Bartczuk
A. Bielecki
L. Bobrowski
P. Boguś
T. Burczyński
K. Cetnarowicz
W. Cholewa
R. Choraś
R. Cierniak
P. Ciskowski
B. Cyganek
J. de la Rosa
J. Dolinsky
L. Dutkiewicz
P. Dziwiński
D. Elizondo
M. Gabryel
A. Galuszka
K. Grąbczewski
J. Grzymala-Busse
P. Hajek
Z. Hasiewicz
Y. Hayashi
F. Herrera
A. Horzyk
E. Jamro
A. Janczak
N. Jankowski
W. Kamiński
O. Kaynak
V. Kecman

E. Kerre
F. Klawonn
J. Kluska
P. Korohoda
M. Korytkowski
M. Korzeń
J. Kościelny
L. Kotulski
Z. Kowalczuk
J. Kozlak
M. Kretowski
B. Kryzhanovsky
A. Krzyzak
E. Kucharska
M. Kurzyński
J. Łęski
L. Magdalena
W. Malina
J. Mańdziuk
A. Materka
J. Mendel
R. Mesiar
Z. Michalewicz
A. Naud
E. Nawarecki
A. Niewiadomski
A. Obuchowicz
M. Ogiela
P. Orantek
S. Osowski
A. Owczarek
K. Patan
W. Pedrycz
A. Pieczyński
A. Piegat

E. Rafajłowicz
G. Russ
N. Sano
R. Scherer
H. Schwefel
R. Setiono
A. Skowron
E. Skubalska-Rafajłowicz
K. Slot
R. Słowiński
T. Smolinski
C. Smutnicki
P. Strumiłło
M. Studniarski
P. Szczepaniak
E. Szmidt
M. Szpyrka
P. Szymak
P. Śliwiński
J. Świątek
R. Tadeusiewicz
H. Takagi
Y. Tiumentsev
V. Torra
M. Wagenknecht
T. Walkowiak
S. Wiak
B. Wilamowski
M. Witczak
M. Wygralak
R. Yager
S. Zadrożny
J. Zieliński

Finally, I thank my co-workers Łukasz Bartczuk, Agnieszka Cpałka, Piotr Dziwiński, Marcin Gabryel, Marcin Korytkowski and the Conference Secretary Rafał Scherer, for their enormous efforts to make the conference a very successful event. Moreover, I would like to acknowledge the work of Marcin Korytkowski, who designed the Internet submission system.

June 2008

Leszek Rutkowski

Table of Contents

I Neural Networks and Their Applications

Input Signals Normalization in Kohonen Neural Networks 3
Andrzej Bielecki, Marzena Bielecka, and Anna Chmielowiec

Parallel Realisation of the Recurrent RTRN Neural Network
Learning 11
Jarosław Bilski and Jacek Smoląg

Stable Learning Algorithm of Global Neural Network for Identification
of Dynamic Complex Systems 17
Jarosław Drapała, Jerzy Świątek, and Krzysztof Brzostowski

The Influence of Training Data Availability Time on Effectiveness of
ANN Adaptation Process 28
Ewa Dudek-Dyduch and Adrian Horzyk

WWW-Newsgroup-Document Clustering by Means of Dynamic
Self-organizing Neural Networks 40
Marian B. Gorzałczany and Filip Rudziński

Municipal Creditworthiness Modelling by Kohonen's Self-organizing
Feature Maps and LVQ Neural Networks 52
Petr Hájek and Vladimír Olej

Fast and Robust Way of Learning the Fourier Series Neural Networks
on the Basis of Multidimensional Discrete Fourier Transform 62
Krzysztof Halawa

Accuracy Improvement of Neural Network State Variable Estimator in
Induction Motor Drive 71
Jerzy Jelonkiewicz and Andrzej Przybył

Ensemble of Dipolar Neural Networks in Application to Survival
Data 78
Małgorzata Krętowska

Binary Optimization: On the Probability of a Local Minimum Detection
in Random Search 89
Boris Kryzhanovsky and Vladimir Kryzhanovsky

Nonlinear Function Learning Using Radial Basis Function Networks:
Convergence and Rates 101
Adam Krzyżak and Dominik Schäfer

Efficient Predictive Control Integrated with Economic Optimisation
Based on Neural Models 111
  Maciej Ławryńczuk and Piotr Tatjewski

Model-Based Fault Detection and Isolation Using Locally Recurrent
Neural Networks 123
  Piotr Przystałka

Neural Network in Fast Adaptive Fourier Descriptor Based Leaves
Classification 135
  Dariusz Puchala and Mykhaylo Yatsymirskyy

Improving the Efficiency of Counting Defects by Learning RBF Nets
with MAD Loss 146
  Ewaryst Rafajłowicz

Robust MCD-Based Backpropagation Learning Algorithm 154
  Andrzej Rusiecki

Some Issues on Intrusion Detection in Web Applications 164
  Jaroslaw Skaruz and Franciszek Seredynski

Neural Network Device for Reliability and Functional Analysis of
Discrete Transport System 175
  Tomasz Walkowiak and Jacek Mazurkiewicz

Maximum of Marginal Likelihood Criterion instead of Cross-Validation
for Designing of Artificial Neural Networks 186
  Zenon Waszczyszyn and Marek Słoński

II Fuzzy Systems and Their Applications

Type-2 Fuzzy Decision Trees 197
  Łukasz Bartczuk and Danuta Rutkowska

An Application of Weighted Triangular Norms to Complexity Reduction
of Neuro-fuzzy Systems 207
  Krzysztof Cpalka and Leszek Rutkowski

Real-Time Road Signs Tracking with the Fuzzy Continuously Adaptive
Mean Shift Algorithm 217
  Bogusław Cyganek

A New Method for Decision Making in the Intuitionistic Fuzzy
Setting 229
  Ludmila Dymova, Izabela Róg, and Pavel Sevastjanov

Linguistic Summarization of Time Series Using Fuzzy Logic with
Linguistic Quantifiers: A Truth and Specificity Based Approach 241
   Janusz Kacprzyk and Anna Wilbik

TS Fuzzy Rule-Based Systems with Polynomial Membership
Functions 253
   Jacek Kluska

From Ensemble of Fuzzy Classifiers to Single Fuzzy Rule Base
Classifier 265
   Marcin Korytkowski, Leszek Rutkowski, and Rafał Scherer

Efficient Fuzzy Predictive Economic Set–Point Optimizer 273
   Piotr M. Marusak

Imprecision Measures for Type-2 Fuzzy Sets: Applications to Linguistic
Summarization of Databases 285
   Adam Niewiadomski

Assessing the Non-technical Service Aspects by Using Fuzzy Methods . . . 295
   Andrzej Pieczyński and Silva Robak

Adaptation of Rules in the Fuzzy Control System Using the Arithmetic
of Ordered Fuzzy Numbers 306
   Piotr Prokopowicz

Regression Modeling with Fuzzy Relations 317
   Rafał Scherer

A New Approach to Creating Multisegment Fuzzy Systems 324
   Artur Starczewski

On Defuzzification of Interval Type-2 Fuzzy Sets 333
   Janusz T. Starczewski

Combining Basic Probability Assignments for Fuzzy Focal Elements 341
   Ewa Straszecka

Using Intuitionistic Fuzzy Sets in Text Categorization 351
   Eulalia Szmidt and Janusz Kacprzyk

III   Evolutionary Algorithms and Their Applications

Improving Evolutionary Algorithms with Scouting: High–Dimensional
Problems 365
   *Konstantinos Bousmalis, Jeffrey O. Pfaffmann, and
   Gillian M. Hayes*

Memetic Algorithm Based on a Constraint Satisfaction Technique for
VRPTW.. 376
    Marco A. Cruz-Chávez, Ocotlán Díaz-Parra,
    David Juárez-Romero, and Martín G. Martínez-Rangel

Agent-Based Co-Operative Co-Evolutionary Algorithm for
Multi-Objective Optimization 388
    Rafał Dreżewski and Leszek Siwik

Evolutionary Methods for Designing Neuro-fuzzy Modular Systems
Combined by Bagging Algorithm.................................. 398
    Marcin Gabryel and Leszek Rutkowski

Evolutionary Methods to Create Interpretable Modular System 405
    Marcin Korytkowski, Marcin Gabryel, Leszek Rutkowski, and
    Stanislaw Drozda

Fractal Dimension of Trajectory as Invariant of Genetic Algorithms 414
    Stefan Kotowski, Witold Kosiński, Zbigniew Michalewicz,
    Jakub Nowicki, and Bartosz Przepiórkiewicz

Global Induction of Decision Trees: From Parallel Implementation to
Distributed Evolution 426
    Marek Krętowski and Piotr Popczyński

Particle Swarm Optimization with Variable Population Size 438
    Laura Lanzarini, Victoria Leza, and Armando De Giusti

Genetic Algorithm as a Tool for Stock Market Modelling 450
    Urszula Markowska-Kaczmar, Halina Kwasnicka, and
    Marcin Szczepkowski

Robustness of Isotropic Stable Mutations in a General Search Space 460
    Przemysław Prętki and Andrzej Obuchowicz

Ant Colony Optimization: A Leading Algorithm in Future Optimization
of Petroleum Engineering Processes 469
    Fatemeh Razavi and Farhang Jalali-Farahani

Design and Multi-Objective Optimization of Combinational
Digital Circuits Using Evolutionary Algorithm with Multi-Layer
Chromosomes 479
    Adam Słowik and Michał Białko

On Convergence of a Simple Genetic Algorithm..................... 489
    Jolanta Socała and Witold Kosiński

Tuning Quantum Multi-Swarm Optimization for Dynamic Tasks 499
    Krzysztof Trojanowski

IV Classification, Rule Discovery and Clustering

Ensembling Classifiers Using Unsupervised Learning 513
    Marek Bundzel and Peter Sinčák

A Framework for Adaptive and Integrated Classification 522
    Ireneusz Czarnowski and Piotr Jędrzejowicz

Solving Regression by Learning an Ensemble of Decision Rules 533
    Krzysztof Dembczyński, Wojciech Kotłowski, and Roman Słowiński

Meta-learning with Machine Generators and Complexity Controlled
Exploration 545
    Krzysztof Grąbczewski and Norbert Jankowski

Assessing the Quality of Rules with a New Monotonic Interestingness
Measure Z 556
    Salvatore Greco, Roman Słowiński, and Izabela Szczęch

A Comparison of Methods for Learning of Highly Non-separable
Problems 566
    Marek Grochowski and Włodzisław Duch

Towards Heterogeneous Similarity Function Learning for the k-Nearest
Neighbors Classification 578
    Karol Grudziński

Hough Transform in Music Tunes Recognition Systems 588
    Maciej Hrebień and Józef Korbicz

Maximal Margin Estimation with Perceptron-Like Algorithm 597
    Marcin Korzeń and Przemysław Klęsk

Classes of Kernels for Hit Definition in Compound Screening 609
    Karol Kozak and Katarzyna Stapor

The GA-Based Bayes-Optimal Feature Extraction Procedure Applied
to the Supervised Pattern Recognition 620
    Marek Kurzynski and Aleksander Rewak

Hierarchical SVM Classification for Localization in Multilevel Sensor
Networks 632
    Jerzy Martyna

Comparison of Shannon, Renyi and Tsallis Entropy Used in Decision
Trees 643
    Tomasz Maszczyk and Włodzisław Duch

Information Theory Inspired Weighted Immune Classification
Algorithm.. 652
  Maciej Morkowski and Robert Nowicki

Bayes' Rule, Principle of Indifference, and Safe Distribution 661
  Andrzej Piegat and Marek Landowski

MAD Loss in Pattern Recognition and RBF Learning 671
  Ewaryst Rafajłowicz and Ewa Skubalska-Rafajłowicz

Parallel Ant Miner 2 681
  Omid Roozmand, Kamran Zamanifar

Object-Oriented Software Systems Restructuring through Clustering ... 693
  Gabriela Şerban and István-Gergely Czibula

Data Clustering with Semi-binary Nonnegative Matrix Factorization.... 705
  Rafal Zdunek

An Efficient Association Rule Mining Algorithm for Classification 717
  A. Zemirline, L. Lecornu, B. Solaiman, and A. Ech-cherif

Comparison of Feature Reduction Methods in the Text Recognition
Task 729
  Jerzy Sas and Andrzej Zolnierek

V Image Analysis, Speech and Robotics

Robot Simulation of Sensory Integration Dysfunction in Autism with
Dynamic Neural Fields Model 741
  Winai Chonnaparamutt and Emilia I. Barakova

A Novel Approach to Image Reconstruction Problem from Fan-Beam
Projections Using Recurrent Neural Network 752
  Robert Cierniak

Segmentation of Ultrasound Imaging by Fuzzy Fusion: Application to
Venous Thrombosis 762
  Mounir Dhibi and Renaud Debon

Optical Flow Based Velocity Field Control 771
  *Leonardo Fermín, Wilfredis Medina-Meléndez, Juan C. Grieco, and
  Gerardo Fernández-López*

Detection of Phoneme Boundaries Using Spiking Neurons 782
  Gábor Gosztolya and László Tóth

Geometric Structure Filtering Using Coupled Diffusion Process and
CNN-Based Approach 794
    Bartosz Jablonski

MARCoPlan: MultiAgent Remote Control for Robot Motion
Planning... 806
    Sonia Kefi, Ines Barhoumi, Ilhem Kallel, and Adel M. Alimi

A Hybrid Method of User Identification with Use Independent Speech
and Facial Asymmetry... 818
    Mariusz Kubanek and Szymon Rydzek

Multilayer Perceptrons for Bio-inspired Friction Estimation............. 828
    Rosana Matuk Herrera

Color Image Watermarking and Self-recovery Based on Independent
Component Analysis 839
    Hanane Mirza, Hien Thai, and Zensho Nakao

A Fuzzy Rule-Based System with Ontology for Summarization of
Multi-camera Event Sequences.. 850
    Han-Saem Park and Sung-Bae Cho

A New Approach to Interactive Visual Search with RBF Networks
Based on Preference Modelling 861
    Paweł Rotter and Andrzej M.J. Skulimowski

An Adaptive Fast Transform Based Image Compression................ 874
    Kamil Stokfiszewski and Piotr S. Szczepaniak

Emotion Recognition with Poincare Mapping of Voiced-Speech
Segments of Utterances 886
    Krzysztof Ślot, Jaroslaw Cichosz, and Lukasz Bronakowski

Effectiveness of Simultaneous Behavior by Interactive Robot........... 896
    Masahiko Taguchi, Kentaro Ishii, and Michita Imai

VI   Bioinformatics and Medical Applications

Quality-Driven Continuous Adaptiation of ECG Interpretation in a
Distributed Surveillance System 909
    Piotr Augustyniak

Detection of Eyes Position Based on Electrooculography Signal
Analysis 919
    Robert Czabański, Tomasz Przybyła, and Tomasz Pander

An NLP-Based 3D Scene Generation System for Children with Autism
or Mental Retardation 929
   Yılmaz Kılıçaslan, Özlem Uçar, and Edip Serdar Güner

On Using Energy Signatures in Protein Structure Similarity
Searching 939
   Bożena Małysiak, Alina Momot, Stanisław Kozielski, and
   Dariusz Mrozek

SYMBIOS: A Semantic Pervasive Services Platform for Biomedical
Information Integration 951
   Myriam Mencke, Ismael Rivera, Juan Miguel Gómez,
   Giner Alor-Hernandez, Rubén Posada-Gómez, and Ying Liu

The PCR Primer Design as a Metaheuristic Search Process 963
   L. Montera and M.C. Nicoletti

On Differential Stroke Diagnosis by Neuro-fuzzy Structures 974
   Krzysztof Cpałka, Olga Rebrova, Tomasz Gałkowski, and
   Leszek Rutkowski

Novel Quantitative Method for Spleen's Morphometry in
Splenomegally 981
   Tomasz Sołtysiński

VII Various Problems of Artificial Intelligence

Parallel Single-Thread Strategies in Scheduling 995
   Wojciech Bożejko, Jarosław Pempera, and Adam Smutnicki

Artificial Immune System for Short-Term Electric Load Forecasting 1007
   Grzegorz Dudek

Ant Focused Crawling Algorithm 1018
   Piotr Dziwiński and Danuta Rutkowska

Towards Refinement of Clinical Evidence Using General Logics 1029
   Patrik Eklund, Robert Helgesson, and Helena Lindgren

An Empirical Analysis of the Impact of Prioritised Sweeping on the
DynaQ's Performance 1041
   Marek Grześ and Daniel Kudenko

Heuristic Algorithms for Solving Uncertain Routing-Scheduling
Problem 1052
   Jerzy Józefczyk and Michał Markowski

Life Story Generation Using Mobile Context and Petri Net 1064
   Young-seol Lee and Sung-Bae Cho

On the Minima of Bethe Free Energy in Gaussian Distributions 1075
    Yu Nishiyama and Sumio Watanabe

An Application of Causality for Representing and Providing Formal
Explanations about the Behavior of the Threshold Accepting
Algorithm.. 1087
    *Joaquín Pérez, Laura Cruz, Rodolfo Pazos, Vanesa Landero,
    Gerardo Reyes, Héctor Fraire, and Juan Frausto*

A Method for Evaluation of Compromise in Multiple Criteria
Problems 1099
    Henryk Piech and Pawel Figat

Neural Networks as Prediction Models for Water Intake in Water
Supply System.. 1109
    Izabela Rojek

Financial Prediction with Neuro-fuzzy Systems 1120
    Agata Pokropińska and Rafał Scherer

Selected Cognitive Categorization Systems 1127
    Ryszard Tadeusiewicz and Lidia Ogiela

Predictive Control for Artificial Intelligence in Computer Games 1137
    Paweł Wawrzyński, Jarosław Arabas, and Paweł Cichosz

Building a Model for Time Reduction of Steel Scrap Meltdown in the
Electric Arc Furnace (EAF): General Strategy with a Comparison of
Feature Selection Methods 1149
    Tadeusz Wieczorek, Marcin Blachnik, and Krystian Mączka

Epoch-Incremental Queue-Dyna Algorithm 1160
    Roman Zajdel

VIII Agent Systems

Utilizing Open Travel Alliance-Based Ontology of Golf in an
Agent-Based Travel Support System.................................... 1173
    Agnieszka Cieślik, Maria Ganzha, and Marcin Paprzycki

Collaborative Recommendations Using Bayesian Networks and
Linguistic Modelling... 1185
    Luis M. de Campos, Juan M. Fernández-Luna, and Juan F. Huete

Autonomous Parsing of Behavior in a Multi-agent Setting 1198
    Dieter Vanderelst and Emilia Barakova

On Resource Profiling and Matching in an Agent-Based Virtual
Organization 1210
  *Grzegorz Frąckowiak, Maria Ganzha, Maciej Gawinecki,
  Marcin Paprzycki, Michał Szymczak, Myon-Woong Park, and
  Yo-Sub Han*

Knowledge Technologies-Based Multi-Agent System for Semantic Web
Services Environments 1222
  *Francisco García-Sánchez, Rodrigo Martínez-Béjar,
  Rafael Valencia-García, and Jesualdo T. Fernández-Breis*

Distributed Graphs Transformed by Multiagent System 1234
  Leszek Kotulski

Multi-agent Logics with Interacting Agents Based on Linear Temporal
Logic: Deciding Algorithms 1243
  Vladimir Rybakov

On Multi Agent Coordination in the Presence of Incomplete
Information 1254
  Krzysztof Skrzypczyk

Author Index 1267

Part I

Neural Networks and Their Applications

Input Signals Normalization in Kohonen Neural Networks

Andrzej Bielecki[1], Marzena Bielecka[2], and Anna Chmielowiec[3]

[1] Institute of Computer Science,
Faculty of Mathematics and Computer Science,
Jagiellonian University,
Nawojki 11, 30-072 Kraków, Poland
[2] Department of Geoinformatics and Applied Computer Science,
Faculty of Gelogy, Geophysics and Environmental Protection,
AGH University of Science and Technology,
Al. Mickiewicza 30, 30-059, Kraków, Poland
[3] Department of Computer Design and Graphics,
Faculty of Physics, Astronomy and Applied Computer Science,
Jagiellonian University,
Reymonta 4, 30-059 Kraków, Poland
bielecki@softlab.ii.uj.edu.pl,
bielecka@agh.edu.pl,
chmielaaa@gmail.com

Abstract. In this paper a Kohonen self-organizing competitive algorithm is considered. A formal approach to classification problem, basing on equivalence relations, is proposed. The Kohonen neural networks are considered as classifying systems. The main topic of this paper is proposal of applying stereographic projection as an input signals normalization procedure. Both theoretical justification is discussed and results of experiments are presented. It turns out that the introduced normalization procedure is effective.

1 Introduction

Classification is one of the basic approach in data processing having a great practical meaning and being a specific method of information reduction - see [23], chapter 7. It is performed by classifiers basing on classifying algorithms. In such algorithms very often data describing features which are taken into consider by the classifying algorithm must be preprocessed. Input signals normalization is an example of such preprocessing.

Self-organizing Kohonen neural networks ([2], [10], [12], [15], section 3.4, [17], chapter 8, [18], section 10.2, [22], [24] section 3.8, [29], section 7.2), based on competitive-type learning processes, are used as classifiers. In this context input signals normalization was studied intensively. Using mathematical considerations and numerical simulations it was shown that the original dot product measure is applicable without input normalization when the dimension of the input space is high i.e. greater than twelve - see [5]. When adding a feature of biological neurons (accommodation) to the algorithm, the network converges with normalization as well. However in this sort of artificial neural networks (ANNs, for abbreviation) normalization of both input signals and neural weights plays a crucial role at least in some kinds of learning algorithms - see [17], section 8.1.2 and [24], section 5.2. In this paper the stereographic projection (see [1], section 2.1, [8], section 2.3 and [9], section 2.4) is proposed as a normalization procedure of Kohonen input signals.

The paper is organized in the following way. In section 2 formal approach to classification problem, basing on equivalence relations, is proposed. The Kohonen neural networks are considered as classifying systems in section 3. In the next section the input signals normalization procedures are discussed. Application of stereographic projection as an input signals normalization procedure is proposed and justified. Results of experiments are presented in section 5. It turns out that the introduced normalization procedure is effective.

2 Classification - Theoretical Foundations

In order to introduce a classification algorithm, a set of interesting features characterizing recognized objects should be specified and patterns are assigned to the same class if they are situated sufficiently closed each to other in the space of specified features. Formally, assume that the space (X, ϱ) of specified features, being a space of signals given onto a classifier input, is a metric space with a measure, say μ. The class definition formally corresponds to specification an equivalence relation REL, i.e relation which is reflexive, symmetric and transitive on the set $X \setminus Z$, where $\mu(Z) = 0$. It is necessary to introduce a set Z because, in practice, there are usually input signals which can not be classified according to the introduced criteria. Consider as an example the representatives method. Assume that n classes $C^{(1)}, ..., C^{(n)}$ and the family of their representatives sets $\left\{ \left\{ c_{k_1}^{(1)} \right\}_{k_1 \in \{1,...,n_1\}}, ..., \left\{ c_{k_m}^{(n)} \right\}_{k_m \in \{1,...,n_m\}} \right\}$ have been constituted. This means that the class $C^{(i)}$ has n_i various representatives. There are such kinds of the method that each class has a single representative. An element x is assigned to this class to which belongs a representative being most closed to x. If for an input signal x there are at least two representatives of different classes having the same minimal distance from x, i.e. $\varrho\left(x, c_{k_i}^{(i)}\right) = \varrho\left(x, c_{k_j}^{(j)}\right)$, $i \neq j$, $i, j \in \{1, ..., n\}$, $k_i \in \{1, ..., n_i\}$, $k_j \in \{1, ..., n_j\}$ and $\varrho\left(x, c_{k_i}^{(i)}\right) \leq \varrho\left(x, c_{k_t}^{(t)}\right)$, $t \notin \{i, j\}$, $t \in \{1, ..., n\}$, $k_t \in \{1, ..., n_t\}$ then x can not be assigned to any class which means that $x \in Z$. Furthermore the classifying algorithm is well defined if the set Z is small in comparison with X which means that the measure of Z is equal to zero. In particular, sometimes $Z = \emptyset$. The equivalence class of $x \in X \setminus Z$ is denoted by $[x]$ and is defined in the following way: $[x] := \{y \in X \setminus Z : xRELy\}$. The following describes very basic properties of equivalence classes - see [20], section 7.1:

(a) $x \in [x]$,
(b) $[x_1] = [x_2]$ if and only if $x_1 REL x_2$,
(c) if $[x_1] \neq [x_2]$ then $[x_1] \cap [x_2] = \emptyset$.

The simple implication of the specified properties is that

$$X \setminus Z = \bigcup_{x \in X \setminus Z} [x]. \tag{1}$$

The item (c) and (1) means that equivalence relation founds division of the set $X \setminus Z$. On the other hand, every division of a set constitutes an equivalence relation on it. Referring once more to the representatives method this means that in this method the relation REL is constituted in such a way that $xRELy$ if they are assigned to the same class.

The classifying algorithm realizes canonical mapping

$$F: X \setminus Z \to (X \setminus Z)/_{REL},$$

where $(X \setminus Z)/_{REL}$ is a quotient set.

3 Kohonen Neural Networks as Classifiers

Since 1943, when the first cybernetic neuron model was founded ([16]), a lot of types of ANNs were introduced. Most of them were originated in biological archetypes. Among others, biological visual systems were studied and their models were worked out ([25], section 6.3 [28]). Self-organizing processes were considered in the context of retinal mappings modelling - see [10], [11] and, so called, Kohonen ANN was introduced and then developed (see [12], [13], [14]) as a model of retinal processing. It should be also mentioned that theoretical aspects of Kohonen networks are studied as well ([6], [7], [21], [26]) but obtained results are very partial - see [2], section 5.6.

Kohonen networks are widely used as classifying systems. In this case input signal normalization is often demanded - see [17], section 8.1.2, [24], section 5.3, [29], section 7.2. According to the approach outlined in section 2 assume that input signals $\boldsymbol{x} \in \mathbb{R}^n$. Let X be a space of normalized input signals $\hat{x}$. When a learning process is over, the weights vectors of trained neurons play a role of a classes representatives.

Referring to biological roots of Kohonen networks it should be mentioned that thought retinal maps are usually constituted by excitation of groups of neurons there are some evidence that, sometimes, a pattern can be represented by a single neuron activation as in Kohonen nets - see [4], [19].

4 Input Signals Normalization

A normalization procedure corresponds to founding a mapping

$$F: \mathbb{R}^n \ni \boldsymbol{x} \to \hat{x} \in \mathbb{R}^n, \text{ where } \|\hat{x}\| = 1.$$

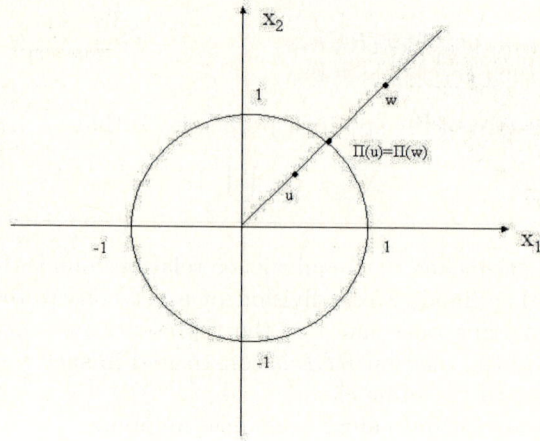

Fig. 1. Simple projection of $\mathbb{R}^2$

The most commonly used normalization is done according to the formula $\hat{x} = \frac{x}{\|x\|}$. This formula defines projection

$$\Pi : \mathbb{R}^n \setminus \{\mathbf{0}\} \to \mathcal{S}^{n-1},$$

call it a simple projection, of $\mathbb{R}^n \setminus \{\mathbf{0}\}$ onto $(n-1)$-dimensional sphere $\mathcal{S}^{n-1}$ - see Fig.1.

A simple projection has crucial drawbacks. First of all dimension is reduced. Secondly, the projection is not defined on the whole space - the mapping is undefined for $\mathbf{0}$. Furthermore, the space $\mathbb{R}^n$ having an infinite measure is projected onto a sphere having a finite measure. Additionally, the projection is not a injective mapping - if two points, say u and w, lie on the same radial line then $\Pi(u) = \Pi(w)$ - see Fig.1. Referring to the considered problem this means that if two data clusters are situated along the same redial direction then, after normalization, they can not be separated even they are well separated before normalization. Therefore, this method should be used only in such cases if it is *a priori* known that clusters in input signal space are situated in various radial directions.

Therefore sometimes is applied normalization which do not reduce the input signals space dimension ([17], section 8.1.2). The stereographic projection

$$S : \mathbb{R}^n \to \mathcal{S}^n$$

is an example of such mapping. Geometric interpretation of the stereographic projection is visualized in Fig.2 for a two-dimensional case - see [1], pages 25-26 and [9], page 36.

Stereographic projection is given explicitly by algebraic formulae for each natural n - see [8], page 73. Let $P = (x_1, ..., x_n)$. Then $S(P) = \tilde{P} = (\tilde{x}_1, ..., \tilde{x}_{n+1})$ is given as

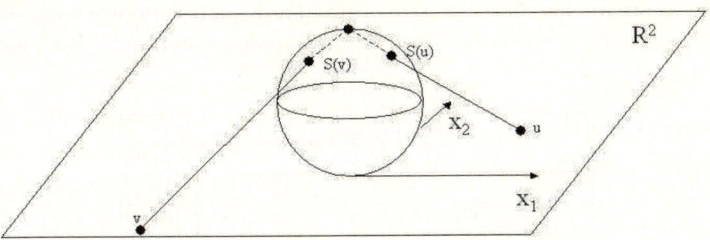

Fig. 2. Stereographic projection of $\mathbb{R}^2$

$$\tilde{x}_i = \frac{4x_i}{4+s} \text{ for } i = 1, ..., n;$$
$$\tilde{x}_{n+1} = \frac{s-4}{4+s},$$

where $s := \sum_{i=1}^n x_i^2$.

As it has been already mentioned, stereographic projection preserves the transformed space dimension and is defined on the whole $\mathbb{R}^n$. Furthermore it is an injective mapping i.e. if $u \neq v$, $u, v \in \mathbb{R}^n$ then $S(u) \neq S(v)$. However, it transforms a space having infinite measure into space having finite measure. This implied, among others, that points being far from each other in $\mathbb{R}^n$ can be closed each to other on $\mathcal{S}^n$. Therefore, two clusters which are well separated in $\mathbb{R}^n$ can be hardly separated after normalization. However such case can only take place if the clusters are far from the coordinate system origin - then they are transformed near to the north pole of the sphere. Since, in practice, norms of transformed vectors are limited, the minimal distance between clusters after signal normalization can be estimated.

5 Results

In order to test the effectiveness of the proposed normalization procedure based on the stereographic projection $\mathcal{S}$ on a certain standard testing data, the data basis IRIS was used. In this basis there are 150 data describing geometric size of three iris species: *setosa*, *versicolor* and *virginica*. For each species 50 flowers were measured and each flower is described by four numbers - width and length of sepals and petals. Thirty examples for each species were selected randomly as a Kohonen ANN learning set. Sixty remaining examples were used as testing set. The classification were performed several times for various learning sets. Two types of Kohonen nets were tested - the first one consisted of 81 neurons and the second one of 625 neurons. The ANNs were learnt using winner takes most algorithm and in both ones the learning step varied from 0.25 to 0.1 being decreased linearly. In each case three ANNs were tested parallely for the same starting parameters: the network without normalization, with simple normalization and with normalization using the stereographic projection. The mean values of correctly recognized cases in each epoch are presented in Tables 1 and 2.

Table 1. Percentage classification correctness for Kohonen network consisting of 81 neurons. (*W.n. - without normalization, Pro. - simple projection, S.pro - stereographic projection.*)

Epoch number	Learning set			Testing set		
	W.n.	Pro.	S.pro.	W.n.	Pro.	S.pro.
1	-	-	-	84	83	78
2	83	94	89	84	92	81
3	85	94	92	82	92	85
4	87	94	92	87	91	86
5	83	90	92	82	87	87
6	86	96	95	82	94	85
7	89	97	96	87	93	88
8	89	96	96	83	92	92

Table 2. Percentage classification correctness for Kohonen network consisting of 625 neurons. (*W.n. - without normalization, Pro. - simple projection, S.pro - stereographic projection.*)

Epoch number	Learning set			Testing set		
	W.n.	Pro.	S.pro.	W.n.	Pro.	S.pro.
1	-	-	-	66	57	68
2	93	93	93	81	79	81
3	95	98	97	84	80	81
4	96	99	99	84	82	84
5	95	100	99	87	83	85
6	96	100	99	85	83	86
7	97	100	99	83	83	87
8	97	100	99	87	83	90

It turned out that for input signals normalized using stereographic projection Kohonen network works significantly better than for two other types, i.e. without normalization and with simple normalization. First of all it can be observed that number of correctly classified examples in testing sets were significantly greater for stereographic normalization that for that ones that had not been normalized - 92% versus 83% and 90% versus 87% for smaller and greater ANNs respectively. Furthermore, for the networks consisting of 625 neurons the stereographic projection turned to be significantly better than simple projection - 90% correctly recognized examples for learning set versus 83%. Secondly, the learning process of a network with stereographic normalization is stable - number of correctly recognized samples increases for subsequent epochs whereas for networks without normalization strong oscillations can be observed - see Tables 1 and 2, testing set. Oscillations can be also observed for simple normalized signals - see Table 1, testing set.

6 Concluding Remarks

Presentation of a theoretical framework for classification algorithm realized by a Kohonen network and testing certain normalization procedure was the aim of this paper. It has been justified that if at least two clusters are situated along the same radial direction then, after applying the widely used simple projection, the clustering algorithm can not divide the data set properly into classes because the clusters are projected onto the same region of the sphere. Therefore the stereographic projection has been proposed and tested. It turned out that after stereographic projection classification is the most accurately and the learning process is stable. The following stages can be proposed as continuation of the presented studies:

1. testing the efficacy of the presented approach for other metrics on $\mathbb{R}^n$ (not only Euclidean);
2. testing modification of stereographic projection in order to obtain more uniform cluster distribution on the sphere knowing the maximal norm of considered vectors in $\mathbb{R}^n$;
3. testing the presented approach for Kohonen networks based on other neuron models - for instance Waxman model - see [27] and [3] as a survey paper;
4. studies concerning structure and measure of the set Z (see section 2) in order to investigate which signal can not be classified for a given normalization and learning algorithm.

References

1. Auslander, L., MacKenzie, R.E.: Introduction to Differentiable Manifolds. McGraw-Hill, New York (1963)
2. Baran, J.: Self-organizing mappings. In: Rutkowski, L. (ed.) Wydawnictwo Politechniki Częstochowskiej, Częstochowa, pp. 55–81 (1996) (in Polish)
3. Bielecki, A.: Mathematical foundations of artificial neural networks. Matematyka Stosowana 4(45), 25–55 (2003) (in Polish)
4. Connor, C.E.: Friends and grandmothers. Nature 435, 1036–1037 (2005)
5. Demartines, P., Blayo, F.: Kohonen self-organizing map: is the normalization necessary? Complex Systems 6, 105–123 (1992)
6. Erwin, E., Obermayer, K., Schulten, K.L.: Self-organizing maps: stationary states, metastability and convergence rate. Biological Cybernetics 67, 35–45 (1992)
7. Erwin, E., Obermayer, K., Schulten, K.L.: Self-organizing maps: ordering, convergence properties and energy functions. Biological Cybernetics 67, 47–55 (1992)
8. Gancarzewicz, J.: Differential Geometry, PWN, Warszawa (1987) (in Polish)
9. Gancarzewicz, J., Opozda, B.: Introduction to Differential Geometry. Jagiellonian University Press, Kraków (2003) (in Polish)
10. Kohonen, T.: Self-organizing formation of topologically correct feature maps. Biological Cybernetics 43, 59–69 (1982)
11. Kohonen, T.: Analysis of a simple self-organizing process. Biological Cybernetics 44, 135–140 (1982)
12. Kohonen, T.: Self-Organization and Associative Memory. Information Sciences, vol. 8. Springer, Berlin (1984)

13. Kohonen, T.: Adaptive, associative and self-organizing functions in neural computing. Appl. Optics 26, 4910–4918 (1987)
14. Kohonen, T.: The self-organizing maps. Proc. IEEE 78, 1464–1480 (1990)
15. Korbicz, J., Obuchowicz, A., Uciński, D.: Artificial Neural Networks - Foundations and Applications, Akademicka Oficyna Wydawnicza, Warszawa (1994) (in Polish)
16. McCoulloch, W.S., Pitts, W.H.: A logical calculus of the ideas immanent in nervous activity. Bulletin of Mathematical Biophysics 5, 115–133 (1943)
17. Osowski, S.: Neural Networks - an Algorithmic Approach, WNT, Warszawa (1996) (in Polish)
18. Osowski, S.: Neural Networks for Information Processing, Oficyna Wydawnicza Politechniki Warszawskiej, Warszawa (2006) (in Polish)
19. Quiroga, R.Q., Reddy, L.R., Kreiman, G., Koch, C., Fried, I.: Invariant visual representation by single neurons in the human brain. Nature 435, 1002–1007 (2005)
20. Rasiowa, H.: Introduction to Contemporary Mathematics, PWN, Warszawa (1998) (in Polish)
21. Ritter, H., Schulten, K.: Convergence properties of Kohonen's topology conserving maps: fluctuations, stability and dimension selection. Biological Cybernetics 60, 59–71 (1989)
22. Skubalska-Rafajłowicz, E.: Self-organizing neural networks. In: Duch, W., Korbicz, J., Rutkowski, L., Tadeusiewicz, R. (eds.) Biocybernetics and Biomedical Engineering, Nałęcz M. - the series editor, Akademicka Oficyna Wydawnicza EXIT, Warszawa, vol. 6, pp. 179–226 (2000) (in Polish)
23. Sobczak, W., Malina, W.: Methods of selection and information reduction, WNT, Warszawa (1978) (in Polish)
24. Tadeusiewicz, R.: Neural Networks, Akademicka Oficyna Wydawnicza, Warszawa (1993) (in Polish)
25. Tadeusiewicz, R.: Biocybernetics Problems, PWN, Warszawa (1994) (in Polish)
26. Tolat, V.V.: An analysis of Kohonen's self-organizing maps using a system of energy functions. Biological Cybernetics 64, 155–164 (1990)
27. Waxman, S.G.: Regional differentation of the axon: A review with special reference to the concept of the multiplex neuron. Brain Research 47, 269–288 (1972)
28. Wilshaw, D.J., Malsburg, C.: How pattern neural connections can be set up by a self-organization. Proc. Royal Society of London B-194, 431–445 (1976)
29. Żurada, J., Barski, M., Jędruch, W.: Artificial Neural Networks, PWN, Warszawa (1992) (in Polish)

Parallel Realisation of the Recurrent RTRN Neural Network Learning

Jarosław Bilski and Jacek Smoląg

Department of Computer Engineering, Częstochowa University of Technology,
Częstochowa, Poland
{Jaroslaw.Bilski,Jacek.Smolag}@kik.pcz.pl

Abstract. In this paper we present a parallel realisation of Real-Time Recurrent Network (RTRN) learning algorithm. We introduce the cuboid architecture to parallelise computation of learning algorithms. Parallel neural network structures are explicitly presented and the performance discussion is included.

1 Introduction

The RTRN network is an example of recurrent neural networks. It was proposed by Willams and Zipser in [6]. During the last decade many researchers have studied dynamical neural networks [2], [3] . Classical computer implementations of neural networks learning algorithms are serial and require high computational load. Therefore high performance parallel structure was discussed in several papers, eg. [1], [4], [5]. In this paper a new idea of the parallel realisation of the RTRN learning algorithm is described. The performance of this new architectures is very promising and useful for RTRN neural networks. A single iteration of the parallel architecture requires less computation cycles than a serial implementation. The final part of this work explains efficiency of the proposed architecture. The structure of the RTRN network is shown in Fig. 1.

The network contains N inputs and K neurons but only M from them create the network output. Moreover, all signals from outputs of neurons are connected as network inputs through unit time delay z^{-1}. Therefore the network input vector

$$[1, x_1(t), ..., x_N(t), x_{N+1}(t), ..., x_{N+K}(t)]^T \qquad (1)$$

in the RTRN network takes the form

$$[1, x_1(t), ..., x_N(t), y_1(t-1), ..., y_K(t-1)]^T \qquad (2)$$

In the recall phase the network is described by

$$s_i = \sum_{k=0}^{N+K} w_{ik} x_k \qquad (3)$$

$$y_i(t) = f(s_i(t)) \qquad (4)$$

We denote by $\mathbf{w}_i = [w_{i0}, ...w_{i,N+K}]$ vector of weights of the $i-th$ neuron and $\mathbf{W} = [\mathbf{w}_1, ..., \mathbf{w}_K]$.

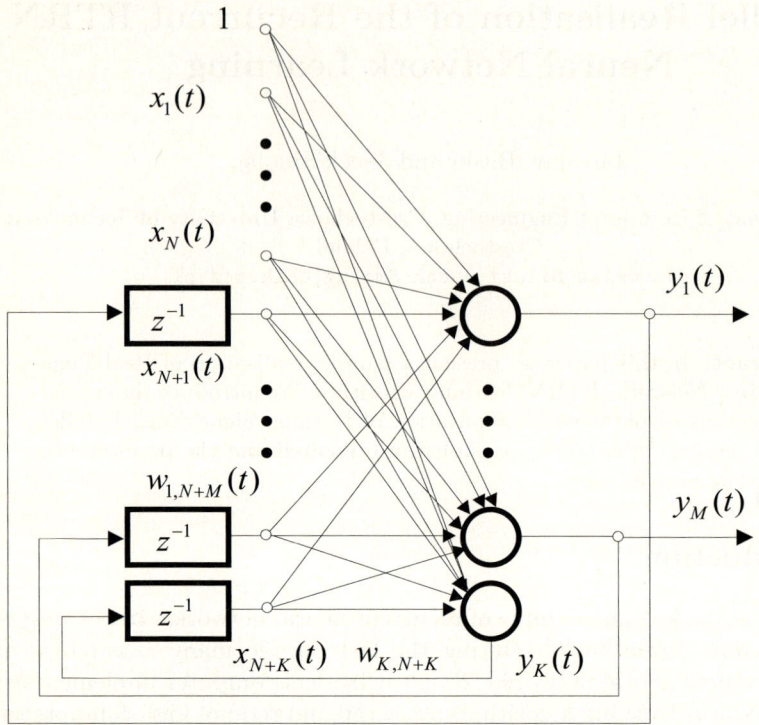

Fig. 1. Structure of the RTRN network

To learn the RTRN neural network we minimise the following criterion

$$J(t) = \frac{1}{2}\sum_{i=1}^{K}(\varepsilon_i(t))^2 = \frac{1}{2}\sum_{i=1}^{K}[d_i(t) - f(\mathbf{x}(t)\mathbf{w}_i(t))]^2 \qquad (5)$$

where $\varepsilon_i(t)$ is defined as

$$\varepsilon_i(t) = \begin{cases} d_i(t) - y_i(t) & for\ i = 1...M \\ 0 & for\ i > M \end{cases} \qquad (6)$$

By minimising equation (5) we obtain

$$\frac{\partial J(t)}{\partial \mathbf{W}} = -\sum_{i=1}^{K}\varepsilon_i(t)\frac{dy_i(t)}{d\mathbf{W}} \qquad (7)$$

and

$$\frac{dy_i(t)}{dw_{\alpha\beta}} = f'_i(t)\sum_{j=0}^{N+K}\frac{d(w_{ij}x_j(t))}{dw_{\alpha\beta}} \qquad (8)$$

where $w_{\alpha\beta}$ is any weight from $\mathbf{W}$ matrix. After calculating the derivative $\frac{d(w_{ij}x_j(t))}{dw_{\alpha\beta}}$

$$\frac{dy_i(t)}{dw_{\alpha\beta}} = f'_i(t)\left(\delta_{i\alpha}x_\beta + \sum_{j=N+1}^{N+K} w_{ij}\frac{dx_j(t)}{dw_{\alpha\beta}}\right) =$$
$$= f'_i(t)\left(\delta_{i\alpha}x_\beta + \sum_{j=N+1}^{N+K} w_{ij}\frac{dy_{j-N}(t-1)}{dw_{\alpha\beta}}\right) = \quad (9)$$
$$= f'_i(t)\left(\delta_{i\alpha}x_\beta + \sum_{j=1}^{K} w_{ij+N}\frac{dy_j(t-1)}{dw_{\alpha\beta}}\right)$$

the weights can be updated as follows

$$w_{\alpha\beta}(t+1) = w_{\alpha\beta}(t) - \eta \sum_{i=1}^{K} \varepsilon_i(t)\frac{dy_i(t)}{dw_{\alpha\beta}} \quad (10)$$

2 Parallel Realisation

The main idea of the parallel realisation of the RTRN learning algorithm is based on the cuboid computational matrix (see Fig. 2). The matrix contains processor elements realising operations given in equations (9) and (10). The cuboid matrix dimensions are $K \times K \times (K + N + 1)$. For clarity in all figures dimension K is set to 3. Any element of the cuboid matrix for indexes i and $\alpha\beta$ is obtained by

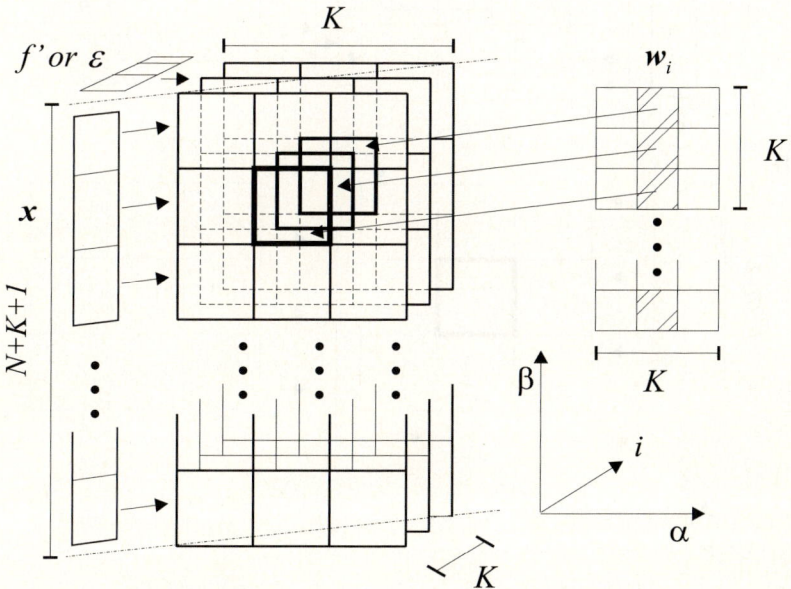

Fig. 2. Idea of the cuboid computation matrix

multiplication of vector of cuboid matrix elements with index $\alpha\beta$ by the $i-th$ weight subvector (see Eqn. 9). At the some time the whole vector of cuboid elements for index $\alpha\beta$ can be multiplied by the weight submatrix $K \times K$. In Fig. 3 accurate data flow is presented. The weight submatrix is entered parallely to all elements of the cuboid matrix. Simultaneously signals $\mathbf{f'}$ and $\mathbf{x}$ are delivered. Figure 4b shows horizontal section of the cuboid matrix. Note that weights and first derivatives lead to all elements but input x_β leads only to elements on the main diagonal. The sum in equation (9) is computed in K steps and its subtotal circles between $i-th$ elements with index $\alpha\beta$. All elements are computed at the

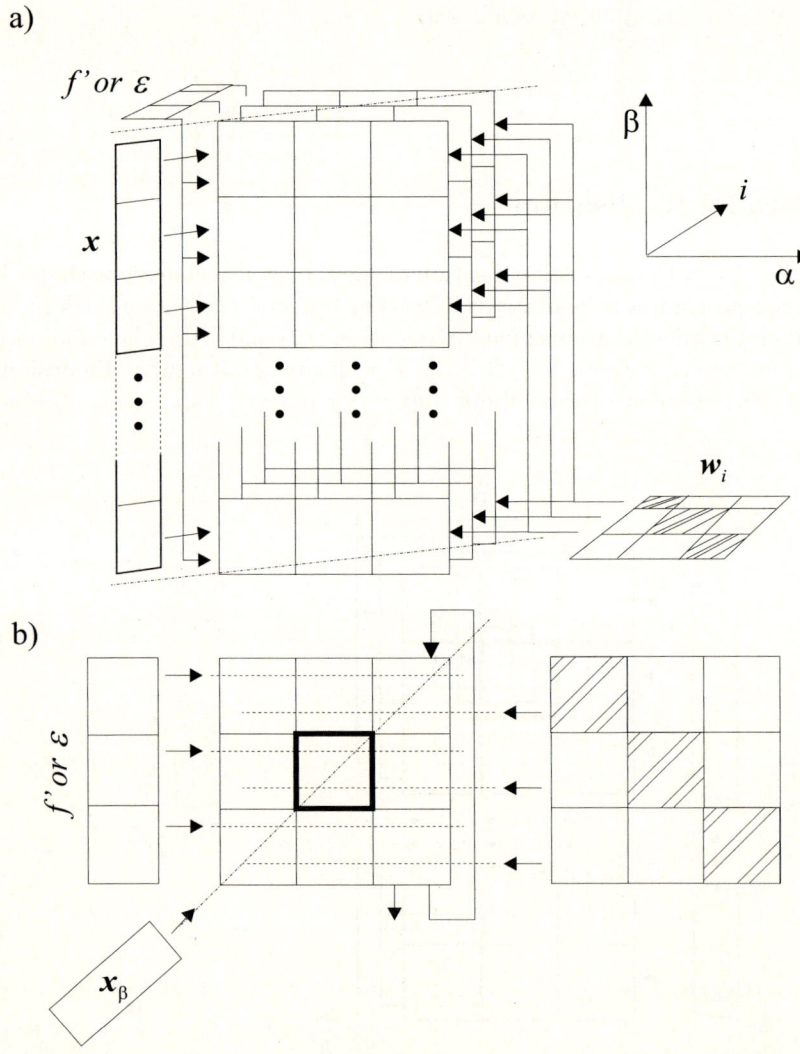

Fig. 3. Data flow in a) computational cuboid, b) one layer for $\beta = const$

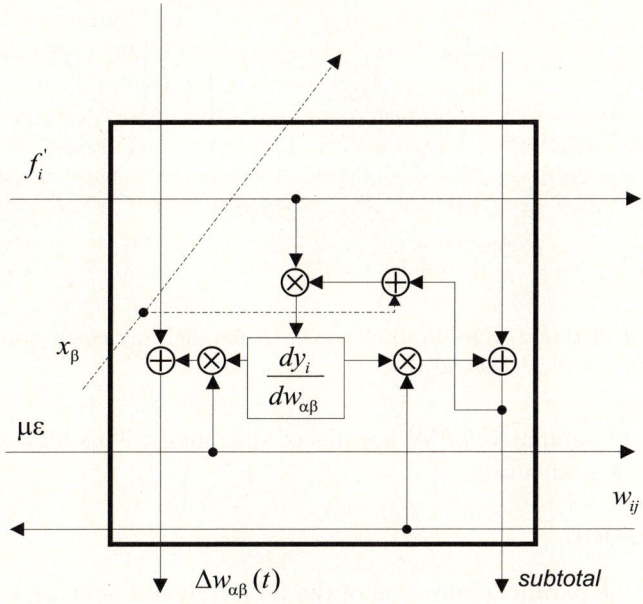

Fig. 4. Single computational element of cuboid matrix

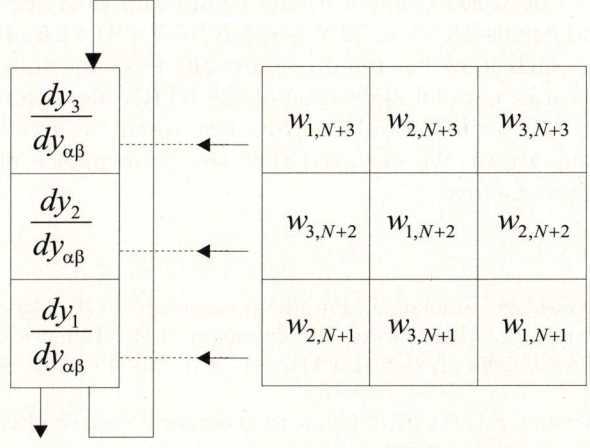

Fig. 5. Weight distribution in learning time

same time. Figure 4 shows structure of one processor element. It computes in K steps new derivative $\frac{d(w_{ij}x_j(t))}{dw_{\alpha\beta}}$ and in the next K steps values of weights' changes described by equation (9). The dotted lines exist only in elements of the main diagonal. Figure 5 depicts distribution of weights during learning phase. Note

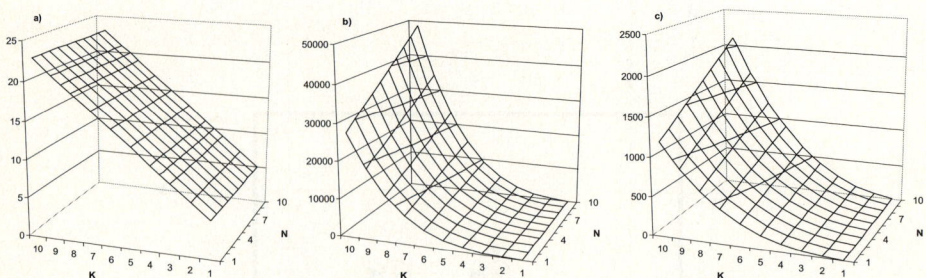

Fig. 6. Number of times cycles in a) classical, b) parallel implementation and c) performance factor parallel/classical

that columns of submatrix of **W** are placed diagonally. This allows to compute all elements simultaneously.

3 Conclusion

In this paper the parallel realisation of the RTRN neural network was proposed. We suppose that all multiplications and additions operations take the same time unit. We can compare computational performance of the RTRN parallel implementation with sequential architectures up to 10 inputs (N) and 10 neurons (K) in the neural network. Computational complexity of RTRN learning is of order $\mathcal{O}(K^4)$ and equals $2K^4 + K^3(2N+5) + K^2(3N+4) + 2K$. In the proposed parallel solution each iteration requires only $2K + 3$ time units (see Fig. 6). Performance factor of parallel realisation of the RTRN algorithm achieves 2100 for 10 inputs and 10 neurons and it grows fast when the number of network inputs or neurons grows. We observed that the performance of the proposed solution is very satisfactory.

References

1. Bilski, J., Litwiński, S., Smoląg, J.: Parallel Realisation of QR Algorithm for Neural Networks Learning. In: Rutkowski, L., Siekmann, J.H., Tadeusiewicz, R., Zadeh, L.A. (eds.) ICAISC 2004. LNCS (LNAI), vol. 3070, pp. 158–165. Springer, Heidelberg (2004)
2. Kolen, J.F., Kremer, S.C.: A Field Guide to Dynamical Recurrent Neural Networks. IEEE Press, Los Alamitos (2001)
3. Korbicz, J., Patan, K., Obuchowicz, A.: Dynamic neural networks for process modelling in fault detection and isolation. Int. J. Appl. Math. Comput. Sci. 9(3), 519–546 (1999)
4. Smoląg, J., Bilski, J.: A systolic array for fast learning of neural networks. In: Proc. of V Conf. Neural Networks and Soft Computing, Zakopane, pp. 754–758 (2000)
5. Smoląg, J., Rutkowski, L., Bilski, J.: Systolic array for neural networks. In: Proc. of IV Conf. Neural Networks and Their Applications, Zakopane, pp. 487–497 (1999)
6. Williams, R., Zipser, D.: A learning algorithm for continually running fully recurrent neural networks. Neural Computation, 270–280 (1989)

Stable Learning Algorithm of Global Neural Network for Identification of Dynamic Complex Systems

Jarosław Drapała, Jerzy Świątek, and Krzysztof Brzostowski

Institute of Information Science and Engineering, Wrocław University of Technology,
Wyb. Wyspiańskiego 27, 50-370 Wrocław, Poland
{jaroslaw.drapala,jerzy.swiatek,krzysztof.brzostowski}@pwr.wroc.pl
http://www.iit.pwr.wroc.pl/~drapala/

Abstract. Novel convergence properties of identification algorithm for complex input-output systems, which uses recurrent neural networks, are derived. By the term "complex system" we understand a system containing interconnected sub processes (elementary processes), which can operate separately. Each element of the complex system is modeled by a multi-input, multi-output neural network. A model of the whole system is obtained by composing all neural networks into one global network. Stable learning algorithm of such a neural network is proposed. We derived sufficient condition of stability using the second Lyapunov method and proved that algorithm is stable even if stability conditions for some individual neural networks are not satisfied.

Keywords: complex systems, identification, modeling, recurrent neural networks, stability.

1 Introduction

The design of computer aided management systems for a complex plant, design of computer control systems, modeling of complex systems of a different nature (technical and biological plants, management and production processes), generate new tasks of complex system identification, [2]. In such a system, we can distinguish a sub processes (elementary processes) with some inputs and outputs, which can operate separately and connections between these elements are given (system's structure). Modeling of complex systems involves modeling of each subsystem separately, taking at the same time into account that there are connections between them, [11].

Naturally, there are many possible structures of complex systems. In this work we focus attention on important case of a complex system, namely a system of cascade structure, composed of M interconnected dynamic input-output plants $O^{(m)}$, $m = 1, 2, \ldots, M$. Output of m-th plant is at the same time input of $(m + 1)$-th plant. The reason for investigations of such a system is that all complex systems without feedbacks can be transferred to this structure, [12].

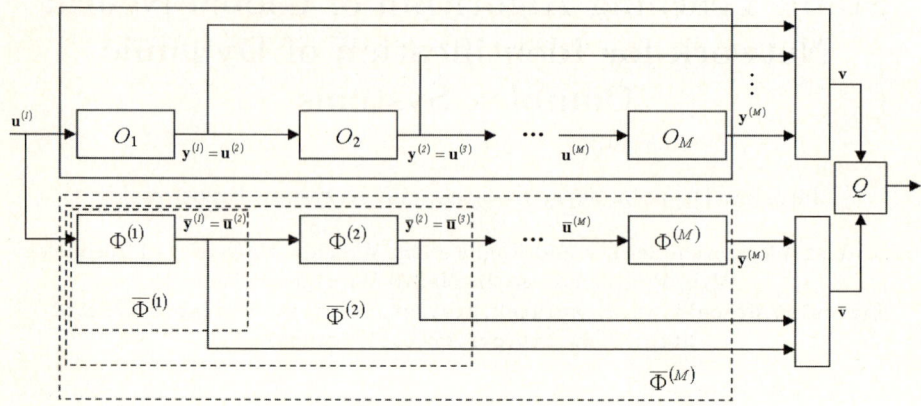

Fig. 1. Complex system with cascade structure and appropriate global model

Let the model of the complex system be M recurrent neural networks $\Phi^{(m)}$ ([3], [6]), connected into cascade structure, meaning that outputs $\overline{\mathbf{y}}^{(m)}$ of m-th network are inputs $\mathbf{u}^{(m+1)}$ of $(m+1)$-th network (see Fig. 1), where Q stands for modeling quality).

Every m-th neural network contains one hidden layer with H_m dynamic non-linear "neurons" and an output layer with L_m static linear neurons. The number of inputs is S_m and the number of outputs is L_m (see Fig. 2). Note that $S_m = L_{m-1}$. We will refer to single networks as *local networks* and to system of connected networks as *global network*.

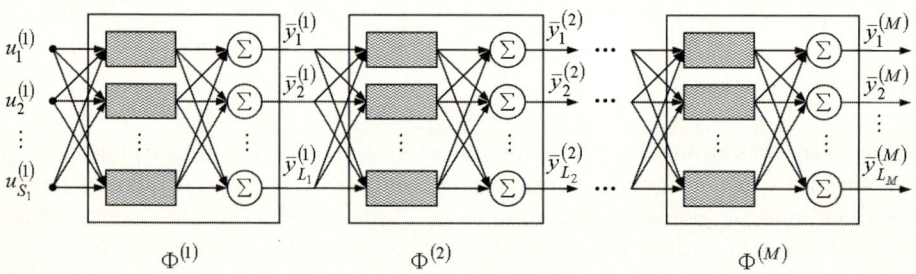

Fig. 2. Global neural network composed of M recurrent networks

Difference equations describing l-th output of m-th neural network are:

$$\overline{y}_l^{(m)}(n) = \sum_{h=0}^{H_m} \overline{w}_{hl}^{(m)} g_h^{(m)}(n), \qquad g_h^{(m)}(n) = \sigma(\chi_h^{(m)}(n)), \qquad (1)$$

$$\chi_h^{(m)}(n) = \sum_{s=0}^{S_m} \widetilde{w}_{sh}^{(m)} u_s^{(m)}(n) + \widehat{w}_h^{(m)} g_h^{(m)}(n-1), \qquad (2)$$

where $u_0^{(m)}(n) = g_0^{(m)}(n) = 1$ at each time step n; $g_h^{(m)}(n)$ and $\overline{y}_l^{(m)}(n)$ are outputs of nonlinear neuron (in the hidden layer) and linear neuron (in the output layer) respectively; $\widetilde{w}_{sh}^{(m)}$, $\widehat{w}_h^{(m)}$ and $\overline{w}_{hl}^{(m)}$ are weight parameters of connection between s-th input of h-th nonlinear neuronu, it's feedback loop and it's connection with l-th linear neuron of output layer; σ is sigmoid function:

$$\sigma(\chi^{(m)}(n)) = \frac{1 - e^{-\chi^{(m)}(n)}}{1 + e^{-\chi^{(m)}(n)}}. \tag{3}$$

In order to determine parameters of neural networks, measurements of input time series $u_s^{(m)}(n)$ and corresponding output time series $y_l^{(m)}(n)$ of each element of a complex system should be collected. Quality of model is evaluated by using global identification performance index Q:

$$Q = \frac{1}{2} \sum_{n=1}^{N} \sum_{m=1}^{M} \beta_m \sum_{l=1}^{L_m} \left[e_l^{(m)}(n) \right]^2, \quad e^{(m)}(n) = y^{(m)}(n) - \overline{y}^{(m)}(n), \tag{4}$$

where β_m are coefficients taken from interval $[0, 1]$ and their task is to put emphasis on the modeling quality of selected signals. However, we assume that weights adjustments are performed on-line, after each n-th time step, so learning algorithm makes the use of following performance index:

$$Q(n) = \frac{1}{2} \sum_{m=1}^{M} \beta_m \sum_{l=1}^{L_m} \left[e_l^{(m)}(n) \right]^2. \tag{5}$$

Neural network with weights determined in such a way, that the performance index (5) is as small as possible, is globally optimal model of the complex system with cascade structure.

2 Learning of Global Network

In order to minimize performance index (5) we propose modification of backpropagation procedure for recurrent neural networks ([8], [9], [10], [13]). Backpropagation algorithm is based on gradient descent optimization method, [1]. According to this method, globally optimal parameters of m-th neural network ($m = 1, 2, \ldots, M$) can be found by performing iterations described by following formula:

$$\mathbf{w}^{(m)}(n+1) = \mathbf{w}^{(m)}(n) - \eta \frac{\partial Q(n)}{\partial \mathbf{w}^{(m)}} \bigg|_{\mathbf{w}^{(m)} = \mathbf{w}^{(m)}(n)}, \tag{6}$$

where η is a correction coefficient. Initial solution $\mathbf{w}^{(m)}(0)$ may be chosen arbitrary. Derivatives for gradient descent procedure have general form:

$$\frac{\partial Q(n)}{\partial \mathbf{w}^{(m)}} = -\sum_{r=m}^{M} \beta_r \sum_{l=1}^{L_r} e_l^{(r)}(n) \frac{\partial \overline{y}_l^{(r)}(n)}{\partial \mathbf{w}^{(m)}} \bigg|_{\mathbf{w}^{(m)} = \mathbf{w}^{(m)}(n)}. \tag{7}$$

It can be shown, that "local" derivatives (for which $r = m$ holds), with respect $\overline{w}_{ij}^{(m)}$ and $\widetilde{w}_{ij}^{(m)}$, are:

$$\frac{\partial \overline{y}_l^{(m)}(n)}{\partial \overline{w}_{ij}^{(m)}} = \begin{cases} g_i^{(m)}(n) & \text{for } j = l \\ 0 & \text{for } j \neq l \end{cases}, \quad (8)$$

$$\frac{\partial \overline{y}_l^{(m)}(n)}{\partial \widetilde{w}_{ij}^{(m)}} = \overline{w}_{jl}^{(m)} \frac{\partial g_j^{(m)}(n)}{\partial \widetilde{w}_{ij}^{(m)}}, \quad (9)$$

$$\frac{\partial g_j^{(m)}(n)}{\partial \widetilde{w}_{ij}^{(m)}} = \Lambda_{ij}^{(m)}(n) = \sigma'\bigl(\chi_j^{(m)}(n)\bigr)\left[u_i^{(m)}(n) + \widehat{w}_j^{(m)} \Lambda_{ij}^{(m)}(n-1)\right], \quad (10)$$

where $\Lambda_{ij}^{(m)}(0) = 0$; $l = 1, 2, \ldots, L_m$, $i = 1, 2, \ldots, S_m$, $j = 1, 2, \ldots, H_m$. Recurrent term $\Lambda_{ij}^{(m)}(n)$ can be as well represented in the iterative form (more convenient for stability analysis):

$$\Lambda_{ij}^{(m)}(n) = \sum_{p=0}^{n-1} \left[\widehat{w}_j^{(m)}\right]^p u_i^{(m)}(n-p) \prod_{q=n-p}^{n} \sigma'\bigl(\chi_j^{(m)}(q)\bigr). \quad (11)$$

"Local" derivatives with respect feedback connection weights $\widehat{w}_j^{(m)}$ have form:

$$\frac{\partial \overline{y}_l^{(m)}(n)}{\partial \widehat{w}_j^{(m)}} = \overline{w}_{jl}^{(m)} \frac{\partial g_j^{(m)}(n)}{\partial \widehat{w}_j^{(m)}}, \quad (12)$$

$$\frac{\partial g_j^{(m)}(n)}{\partial \widehat{w}_j^{(m)}} = \Gamma_j^{(m)}(n) = \sigma'\bigl(\chi_j^{(m)}(n)\bigr)\left[g_j^{(m)}(n-1) + \widehat{w}_j^{(m)} \Gamma_j^{(m)}(n-1)\right], \quad (13)$$

where $\Gamma_j^{(m)}(0) = 0$. Recurrent term $\Gamma_j^{(m)}(n)$ also can be represented in the iterative form:

$$\Gamma_j^{(m)}(n) = \sum_{p=0}^{n-2} \left[\widehat{w}_j^{(m)}\right]^p g_j^{(m)}(n-p-1) \prod_{q=n-p}^{n} \sigma'\bigl(\chi_j^{(m)}(q)\bigr). \quad (14)$$

"Global" derivatives (for which $r > m$ holds) for arbitrary taken weight $w^{(m)}$ are:

$$\frac{\partial \overline{y}_l^{(r)}(n)}{\partial w^{(m)}} = \sum_{h=1}^{H_r} \overline{w}_{hl}^{(r)} \frac{\partial g_h^{(r)}(n)}{\partial w^{(m)}} = \sum_{h=1}^{H_r} \overline{w}_{hl}^{(r)} G_h^{(r,m)}(n), \quad (15)$$

$$G_h^{(r,m)}(n) = \sigma'\bigl(\chi_h^{(r)}(n)\bigr)\left[\sum_{s=0}^{S_r} \widetilde{w}_{sh}^{(r)} \frac{\partial \overline{y}_s^{(r-1)}(n)}{\partial \widetilde{w}^{(m)}} + \widehat{w}_h^{(r)} G_h^{(r,m)}(n-1)\right], \quad (16)$$

where $G_h^{(r,m)}(0) = 0$. Recurrent evaluations of $G_h^{(r,m)}(n)$ go deep as far as $r > m$. For $r = m$ equations (8)-(14) are applied. As before, recurrent term $G_h^{(r,m)}(n)$ also can be represented in the iterative form:

$$G_h^{(r,m)}(n) = \sum_{p=0}^{n-1} \left[\widehat{w}_h^{(r)}\right]^p \left\{ \prod_{q=n-p}^{n} \sigma'(\chi_h^{(m)}(q)) \right\} \sum_{s=0}^{S_r} \widetilde{w}_{sh}^{(r)} \frac{\partial \overline{y}_s^{(r-1)}(n-p)}{\partial w^{(m)}}. \quad (17)$$

At the first glance recurrent formulas for derivatives calculations suggest its high demands for computational power. But closer look at the formulas reveals that when we reserve additional memory for all recurrent terms, the number of computations decreases significantly.

Dynamic behavior of learning procedure depends strongly on value of correction coefficient η in (6). Learning process is stable as long as η is appropriately chosen at each time step. In the next sections we provide stability analysis using second Lyapunov method, [7].

3 Stability of Learning Algorithm for Local Networks

Let us assume that all elements of the complex system are stable plants, which means that absolute values of all feedback weights $\widehat{w}_j^{(m)}$ shouldn't be greater than 1. Further discussion will be also clear if we restrict to SISO systems ($L_m = S_m = 1$ for all m), see Fig. 3. All conclusions drawn in this paper can be strictly generalized to multivariate systems, but we want to keep complexity of notation on a reasonable level. As stated before, learning algorithm (6) finds suboptimal (from a point of view of performance index (5)) solution only when values of coefficient η are selected properly at each time step n. Otherwise it diverges or oscillates ([8], [9], [10]). We use in learning procedure many correction coefficients, different for each single neural network and for each layer, [3]. Let $\overline{\eta}^{(m)}(n)$, $\widetilde{\eta}^{(m)}(n)$, $\widehat{\eta}^{(m)}(n)$ denote m-th neural network's learning coefficients for output linear unit, hidden nonlinear neurons and feedback connections, respectively. In order for algorithm to end successfully, we need to establish such a set of values of learning coefficients $\eta(n)$ for which a sequence of successive values

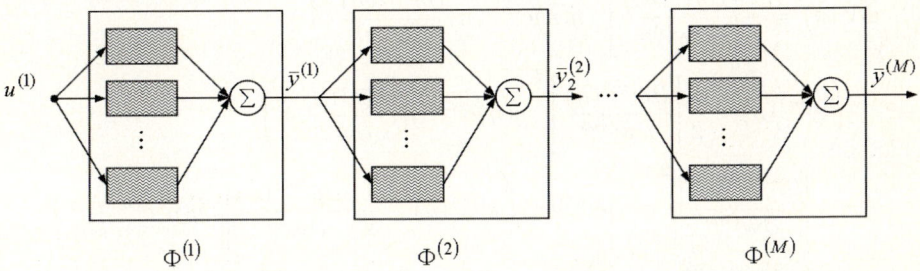

Fig. 3. Global neural network composed of M SISO recurrent networks

of global neural network parameters is convergent. We start analysis from single neural network, not taking into the account that it is connected with other networks (m is fixed).

A candidate for Lyapunov function is:

$$V(n) = \frac{1}{2}\left[e^{(m)}(n)\right]^2. \tag{18}$$

Learning algorithm is stable when first-order difference of the Lyapunov function (18):

$$\Delta V(n) = V(n+1) - V(n) = \frac{1}{2}\left(\left[e^{(m)}(n+1)\right]^2 - \left[e^{(m)}(n)\right]^2\right) \tag{19}$$

is negative, [7]. If we denote $\Delta e^{(m)}(n) = e^{(m)}(n+1) - e^{(m)}(n)$ and use first-order Taylor's expansion of error function:

$$e^{(m)}(n+1) = e^{(m)}(n) + \Delta e^{(m)}(n) = e^{(m)}(n) + \left[\frac{\partial e^{(m)}(n)}{\partial \mathbf{w}^{(m)}}\right]^T \Delta \mathbf{w}^{(m)}, \tag{20}$$

where $\Delta \mathbf{w}^{(m)}$ denotes a vector of changes of arbitrary taken weights in neural network $\Phi^{(m)}$, then we can write:

$$\Delta V(n) = \Delta e^{(m)}(n)\left(e^{(m)}(n) + \frac{1}{2}\Delta e^{(m)}(n)\right)$$

$$= \left[\frac{\partial e^{(m)}(n)}{\partial \mathbf{w}^{(m)}}\right]^T \Delta \mathbf{w}^{(m)} \left(e^{(m)}(n) + \frac{1}{2}\left[\frac{\partial e^{(m)}(n)}{\partial \mathbf{w}^{(m)}}\right]^T \Delta \mathbf{w}^{(m)}\right).$$

Owing to (7) and (15) for one output signal (β_m in a single network case can be omitted), we have:

$$\Delta \mathbf{w}^{(m)}(n) = -\eta(n)\frac{\partial Q(n)}{\partial \mathbf{w}^{(m)}} = \eta e^{(m)}(n)\frac{\partial \overline{y}^{(m)}(n)}{\partial \mathbf{w}^{(m)}}. \tag{21}$$

Taking $\frac{\partial e^{(m)}(n)}{\partial \mathbf{w}^{(m)}} = -\frac{\partial \overline{y}^{(m)}(n)}{\partial \mathbf{w}^{(m)}}$, we obtain:

$$\Delta V(n) = -\left[\frac{\partial \overline{y}^{(m)}(n)}{\partial \mathbf{w}^{(m)}}\right]^T \eta(n)e^{(m)}(n)\frac{\partial \overline{y}^{(m)}(n)}{\partial \mathbf{w}^{(m)}}.$$

$$\cdot\left(e^{(m)}(n) - \frac{1}{2}\left[\frac{\partial \overline{y}^{(m)}(n)}{\partial \mathbf{w}^{(m)}}\right]^T \eta(n)e^{(m)}(n)\frac{\partial \overline{y}^{(m)}(n)}{\partial \mathbf{w}^{(m)}}\right)$$

$$= -\left\|\frac{\partial \overline{y}^{(m)}(n)}{\partial \mathbf{w}^{(m)}}\right\|^2 \eta(n)e^{(m)}(n)\left(e^{(m)}(n) - \frac{1}{2}\left\|\frac{\partial \overline{y}^{(m)}(n)}{\partial \mathbf{w}^{(m)}}\right\|^2 \eta(n)e^{(m)}(n)\right)$$

$$= \left\|\frac{\partial \overline{y}^{(m)}(n)}{\partial \mathbf{w}^{(m)}}\right\|^2 \eta(n)\left[e^{(m)}(n)\right]^2 \left(\frac{1}{2}\eta(n)\left\|\frac{\partial \overline{y}^{(m)}(n)}{\partial \mathbf{w}^{(m)}}\right\|^2 - 1\right). \tag{22}$$

The learning algorithm is convergent as long as $\Delta V(n) < 0$, so the learning parameter η should satisfy the general condition:

$$0 < \eta(n) < \frac{2}{\left\|\dfrac{\partial \overline{y}^{(m)}(n)}{\partial \mathbf{w}^{(m)}}\right\|^2}. \tag{23}$$

Depending on the type of weights, the general stability condition has different forms. Let us evaluate it first for learning coefficient $\overline{\eta}^{(m)}$ associated with linear output layer (single neuron in this case):

$$\frac{\partial \overline{y}^{(m)}(n)}{\partial \overline{\mathbf{w}}^{(m)}} = \mathbf{g}^{(m)}(n), \quad \overline{\mathbf{w}}^{(m)} = \begin{bmatrix} \overline{w}_0^{(m)} & \overline{w}_1^{(m)} & \dots & \overline{w}_{H_m}^{(m)} \end{bmatrix}^T, \tag{24}$$

$$\mathbf{g}^{(m)}(n) = \begin{bmatrix} 1 & g_1^{(m)}(n) & \dots & g_{H_m}^{(m)}(n) \end{bmatrix}^T =$$
$$= \begin{bmatrix} 1 & \sigma(\chi_1^{(m)}(n)) & \dots & \sigma(\chi_1^{(m)}(n)) \end{bmatrix}^T. \tag{25}$$

From (3) it follows, that $-1 \leq \sigma(\chi_i^{(m)}(n)) \leq 1$ so in consequence $\|\mathbf{g}^{(m)}(n)\|^2 \leq H+1$. Hence the value of $\overline{\eta}^{(m)}$ is bounded by following inequality:

$$0 < \overline{\eta}^{(m)}(n) < \frac{2}{H_m + 1}, \tag{26}$$

which means that the greater number of neurons in hidden layer, the smaller permissible value of $\overline{\eta}^{(m)}$. In order to find interval for $\widehat{\eta}^{(m)}$, we have to evaluate (23) using (12) and (14). It is necessary to find an upper bound of the recurrent term $\Gamma_h^{(m)}(n)$. Equation (14) is convenient for this purpose:

$$\left|\Gamma_h^{(m)}(n)\right| = \sum_{p=0}^{n-2} \left|\widehat{w}_h^{(m)}\right|^p \left|g_h^{(m)}(n-p-1)\right| \prod_{q=n-p}^{n} \left|\sigma'(\chi_h^{(m)}(q))\right|. \tag{27}$$

We assumed that we deal with stable plants, which means that $|\widehat{w}_h^{(m)}| \leq 1$. Knowing also that $0 \leq \sigma(\chi_h^{(m)}(n)) \leq 0.5$ from (3), we can write:

$$\left|\Gamma_h^{(m)}(n)\right| \leq \sum_{p=0}^{n-2} \left(\frac{1}{2}\right)^{p+1} = 1. \tag{28}$$

and as a consequence of (12):

$$\left|\frac{\partial \overline{y}^{(m)}(n)}{\partial \widehat{w}_h^{(m)}}\right| = \left|\overline{w}_h^{(m)}\right|\left|\Gamma_h^{(m)}(n)\right| \leq \left|\overline{w}_h^{(m)}\right|, \tag{29}$$

which results in:

$$\left\|\frac{\partial \overline{y}^{(m)}(n)}{\partial \widehat{\mathbf{w}}_h^{(m)}}\right\|^2 \leq \left\|\begin{bmatrix}\overline{w}_1^{(m)} \\ \vdots \\ \overline{w}_{H_m}^{(m)}\end{bmatrix}\right\|^2 \leq \max_h \left[\overline{w}_h^{(m)}\right]^2 \cdot \left\|\begin{bmatrix}1 \\ \vdots \\ 1\end{bmatrix}\right\|^2 =$$

$$= \max_h \left[\overline{w}_h^{(m)}\right]^2 \cdot H_m. \qquad (30)$$

Finally, inequalities to be satisfied by $\widehat{\eta}^{(m)}$ are:

$$0 < \widehat{\eta}^{(m)}(n) < \frac{2}{\max_h \left[\overline{w}_h^{(m)}\right]^2 \cdot H_m}. \qquad (31)$$

By analogy we find interval for $\widetilde{\eta}^{(m)}$, evaluating (23) by use of (9) oraz (11). An upper bound of the term $\Lambda_{sh}^{(m)}(n)$ is:

$$\left|\Lambda_{sh}^{(m)}(n)\right| = \sum_{p=0}^{n-1} \left|\widehat{w}_h^{(m)}\right|^p \left|u_s^{(m)}(n-p)\right| \prod_{q=n-p}^{n} \left|\sigma'(\chi_h^{(m)}(q))\right|, \qquad (32)$$

$$\left|\Lambda_{sh}^{(m)}(n)\right| \leq \sum_{p=0}^{n-1} \left|u_s^{(m)}(n-p)\right| \left(\frac{1}{2}\right)^{p+1} \leq \max_h \left|u_s^{(m)}(n)\right|. \qquad (33)$$

From above equation and from (9) it follows that:

$$\left|\frac{\partial \overline{y}^{(m)}(n)}{\partial \widetilde{w}_{sh}^{(m)}}\right| = \left|\overline{w}_h^{(m)}\right| \cdot \left|\Lambda_{sh}^{(m)}(n)\right|, \qquad (34)$$

which results in:

$$\left\|\frac{\partial \overline{y}^{(m)}(n)}{\partial \widetilde{\mathbf{w}}_h^{(m)}}\right\|^2 \leq \max_{n,h} \left[\overline{w}_h^{(m)}\right]^2 (H_m + 1) \max_{n,s} \left[u_s^{(m)}(n)\right]^2 (S_m + 1), \qquad (35)$$

and finally we obtain inequality to be satisfied by $\widetilde{\eta}^{(m)}$ in order for learning algorithm to be stable:

$$0 < \widetilde{\eta}^{(m)}(n) < \frac{2}{\max_h \left[\overline{w}_h^{(m)}\right]^2 \max_s \left[u_s^{(m)}(n)\right]^2 (H_m + 1)(S_m + 1)}. \qquad (36)$$

4 Stability of Learning Algorithm for Global Network

Reasoning scheme presented in previous section can be repeated for global neural network. This time a candidate for Lyapunov function is:

$$V(n) = \frac{1}{2} \sum_{m=1}^{M} \beta_m \left[e^{(m)}(n)\right]^2, \qquad (37)$$

and its first-order difference can be evaluated in similar way than before:

$$\Delta V(n) = \frac{1}{2}\left(\sum_{m=1}^{M} \beta_m \left[e^{(m)}(n+1)\right]^2 - \sum_{m=1}^{M} \beta_m \left[e^{(m)}(n)\right]^2\right)$$

$$= \sum_{m=1}^{M} \beta_m \Delta e^{(m)}(n) \left(e^{(m)}(n) + \frac{1}{2}\Delta e^{(m)}(n)\right). \tag{38}$$

Putting equations:

$$\Delta \mathbf{w}^{(m)} = \eta(n) \sum_{r=m}^{M} \beta_r e^{(r)}(n) \frac{\partial \overline{y}^{(r)}(n)}{\partial \mathbf{w}^{(m)}}, \qquad \frac{\partial e^{(r)}(n)}{\partial \mathbf{w}^{(m)}} = -\frac{\partial \overline{y}^{(r)}(n)}{\partial \mathbf{w}^{(m)}} \tag{39}$$

to (38), we obtain:

$$\Delta V(n) = \sum_{m=1}^{M} \beta_m \left[-\frac{\partial \overline{y}^{(m)}(n)}{\partial \mathbf{w}^{(m)}}\right]^T \eta(n) \sum_{r=m}^{M} \beta_r e^{(r)}(n) \frac{\partial \overline{y}^{(r)}(n)}{\partial \mathbf{w}^{(m)}} \cdot$$

$$\cdot \left(e^{(m)}(n) - \frac{1}{2}\left[\frac{\partial \overline{y}^{(m)}(n)}{\partial \mathbf{w}^{(m)}}\right]^T \eta(n) \sum_{r=m}^{M} \beta_r e^{(r)}(n) \frac{\partial \overline{y}^{(r)}(n)}{\partial \mathbf{w}^{(m)}}\right)$$

$$= -\eta(n) \sum_{m=1}^{M} \beta_m \left\{ e^{(m)}(n) \sum_{r=m}^{M} \beta_r e^{(r)}(n) \left[\frac{\partial \overline{y}^{(m)}(n)}{\partial \mathbf{w}^{(m)}}\right]^T \left[\frac{\partial \overline{y}^{(r)}(n)}{\partial \mathbf{w}^{(m)}}\right] + \right.$$

$$\left. - \frac{\eta(n)}{2}\left(\sum_{r=m}^{M} \beta_r e^{(r)}(n) \left[\frac{\partial \overline{y}^{(m)}(n)}{\partial \mathbf{w}^{(m)}}\right]^T \left[\frac{\partial \overline{y}^{(r)}(n)}{\partial \mathbf{w}^{(m)}}\right]\right)^2\right\} = -\eta(n) \sum_{m=1}^{M} \beta_m \alpha_m.$$

Learning algorithm is stable when the term:

$$\sum_{m=1}^{M} \beta_m \alpha_m = \boldsymbol{\beta}^T \boldsymbol{\alpha} \tag{40}$$

is positive.

Requirement (40) has straight geometrical interpretation in the M-dimensional euclidean space. For a given vector (point) $\boldsymbol{\beta}$ condition (40) gives such a set of points $\boldsymbol{\alpha}$, for which scalar product $\boldsymbol{\beta}^T \boldsymbol{\alpha}$ is nonnegative. Since position of $\boldsymbol{\beta}$ depends on the choice of value η, only for points satisfying described relation learning algorithm is stable. Boundary of described set is $(M-1)$-dimensional plane (line in two-dimensional case) passing through origin and orthogonal to the vector $\boldsymbol{\beta}$. Moreover, vectors (points) $\boldsymbol{\beta}$ and $\boldsymbol{\alpha}$ must lie on the same side of the boundary line, position of which is determine by components of vector $\boldsymbol{\beta}$. According to the fact, that $\beta_m \in [0,1]$, point $\boldsymbol{\beta}$ lie inside appropriate M-dimensional unit cube. An example for two dimensions (two neural networks) is illustrated in 4. Choice of a certain point $\boldsymbol{\alpha}$ implies proper value of learning coefficient $\eta^{(m)}(n)$.

Important and very interesting observation is that it is not necessary for all α_m to be positive, which means that learning algorithm operating on all neural

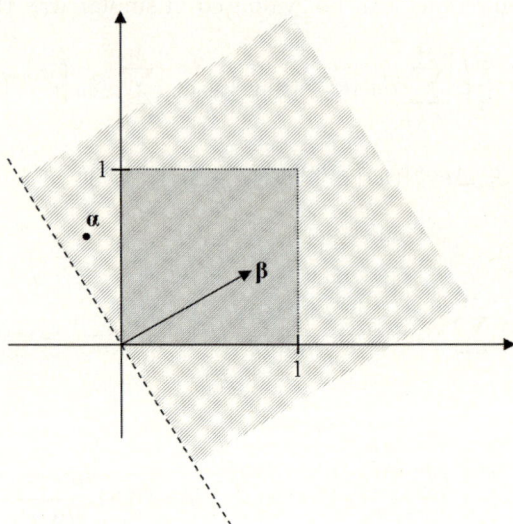

Fig. 4. Two-dimensional example of selecting $\boldsymbol{\alpha}$

networks can be stable, even if at the same time stability condition is not fulfilled from some of the networks "point of view". This result is specific for a complex systems modeling.

Naturally, we can restrict to α_m having all coordinates positive. Sufficient condition of stability of learning algorithm is then represented in the form below:

$$\eta^{(m)}(n) < \frac{2e^{(m)}(n)}{\sum_{r=m}^{M} \beta_r e^{(r)}(n) \left[\frac{\partial \overline{y}^{(m)}(n)}{\partial \mathbf{w}^{(m)}}\right]^T \left[\frac{\partial \overline{y}^{(r)}(n)}{\partial \mathbf{w}^{(m)}}\right]}. \tag{41}$$

5 Final Remarks

In complex systems many novel problems arise. Discussion given in this paper revealed new distinctive properties of learning algorithm for dynamic complex systems modeling. The learning algorithm has been based on the gradient descent method, but we suppose that similar results would be obtained for other type of optimization methods.

It should be emphasized that the stability conditions obtained in this work concern *local stability*, which means that starting from different initial points we could obtain different suboptimal solutions.

Another important issue is associated with the choice of optimal values (in the sense of speed of learning process) of learning coefficients, not only feasible. Unique analytical solution exists for single neural networks, [3], [6], but for interconnected networks this problem remains unsolved.

References

1. Bazaraa, M.S., Sherali, H.D., Shetty, C.M.: Nonlinear Programming - Theory and Algorithms. Wiley-Interscience, A John Wiley & Sons Inc, Hoboken, New Jersey (2006)
2. Bubnicki, Z.: Identification of Control Plants. (in polish) PWN, Warsaw (1980)
3. Chao-Chee, K., Kwang, Y.L.: Diagonal Recurrent Neural Networks for Dynamic Systems Control. IEEE Transactions on Neural Networks 6(1), 144–155 (1995)
4. Drapała, J., Świątek, J.: Algorithm of Recurrent Multilayer Perceptrons Learning for Global Modeling of Complex Systems. In: Proc. of 16th International Conference on Systems Science ICSS 2007, Wrocław University of Technology, Wrocław, Poland, pp. 351–358 (2007)
5. Drapała, J., Świątek, J.: Global and Local Approach to Complex Systems Modeling Using Dynamic Neural Networks– Analogy with Multiagent Systems. In: Apolloni, B., Howlett, R.J., Jain, L. (eds.) KES 2007, Part II. LNCS (LNAI), vol. 4693, pp. 279–286. Springer, Heidelberg (2007)
6. Gupta, M.M., Jin, L., Homma, N.: Static and Dynamic Neural Networks - From Fundamentals to Advanced Theory. IEEE Press, Wiley-Interscience, A John Wiley & Sons Inc (2003)
7. Kaczorek, T., Dzieliński, A., Dąbrowski, W., Łopatka, R.: The Basics of Control Theory (in polish). WNT, Warsaw (2006)
8. Narendra, K.S., Parthasarathy, K.: Identification and Control of Dynamical Systems Using Neural Networks. IEEE Transactions on Neural Networks 1(1), 4–27 (1990)
9. Nelles, O.: Nonlinear System Identification - From Classical Approaches to Neural Networks and Fuzzy Models. Springer, Heidelberg (2001)
10. Parlos, A.G., Chong, K.T., Atiya, A.M.: Application of the Recurrent Multilayer Perceptron in Modeling Complex Process Dynamics. IEEE Transactions on Neural Networks 5(2), 255–266 (1994)
11. Świątek, J.: Global and Local Modeling of Complex Input-Output Systems. In: Proc. of 16th International Conference on Systems Engineering ICSE 2003, Coventry University, England, pp. 669–671 (2003)
12. Świątek, J.: Global Identification of Complex Systems with Cascade Structure. In: Rutkowski, L., Siekmann, J.H., Tadeusiewicz, R., Zadeh, L.A. (eds.) ICAISC 2004. LNCS (LNAI), vol. 3070, pp. 990–995. Springer, Heidelberg (2004)
13. Wen, Y.: Nonlinear System Identification using Discrete-Time Recurrent Neural Networks with Stable Learning Algorithms. International Journal of Information Sciences 158, 131–147 (2004)

The Influence of Training Data Availability Time on Effectiveness of ANN Adaptation Process

Ewa Dudek-Dyduch[1] and Adrian Horzyk[2]

[1] University of Science and Technology, Department of Automatics
Mickiewicza Av. 30, 30-059 Cracow, Poland
edd@ia.agh.edu.pl
[2] University of Science and Technology, Department of Automatics
Mickiewicza Av. 30, 30-059 Cracow, Poland
horzyk@agh.edu.pl
http://home.agh.edu.pl/~horzyk

Abstract. In the paper the new approach to create artificial neural networks (ANNs) is proposed. ANN's are inspired by natural neural networks (NNNs) that receive data in time still tuning themselves. In opposite to them ANNs usually work on the training data (TD) acquired in the past and are totally available at the beginning of the adaptation process. Because of this the adaptation methods of the ANNs can be sometimes more effective than the natural training process observed in the NNNs. This paper presents the ability of ANNs to adapt more effectively than NNNs do if only all TD are known before the beginning of the adaptation process. The design and adaptation process of the proposed ANNs is divided into two stages. First, analyze or examining the set of TD. Second, the construction of neural network topology and weights computation. In the paper, two kinds of ANNs which use the proposed construction strategy are presented. The first kind of network is used for classification tasks and the second kind for feature extraction.

1 Introduction

The training process of NNNs is very specific and linked to the needs of the biological "hardware". Natural brains have to solve many different behavioral and other problems during their lives. The brain structure and training ability suit very good to temporary acquisition of the TD. Moreover, the natural ability to forget the minor and reinforce the major information is very important and lets to manage the limited capacity of a brain memory and its computational ability. The NNNs are able to correct the inner representation of the surrounding world during life searching for still better model of it [13]. This feature makes them possible to survive and adapt to a still changing situation in the world. Furthermore, the knowledge can be naturally selected and remembered in time taking into account the various and changing importance of individual events, things, thoughts etc. The NNNs can be concluded as the best developed to adapt to the many times changing TD that are acquired in a long period of time. In contrast, we look forward that the ANNs should be usually created and trained on a basis of a given TD acquired in the past. TD does usually not change during training process. The time-expensive training process necessary for adapting and tuning of NNNs can be replaced with other adaptation or construction algorithms that can compute a NN structure and its parameters more quickly. Such solutions can be even more effective than the obtained by means of traditional training [1],[2],[11],[15],[16] if only all TD are available at the beginning of the adaptation process. The complete information about all TD enables more precise tuning thanks to more global view on the TD what is a great advantage because construction of a mathematical model can take into account all specific information about all TD. This produces good generalization results of a NN constructed and adapted in this way [3],[7],[8],[13],[14]. This paper discusses advantages and disadvantages of different types of adaptation processes of ANNs and points out essential differences between two general types of adaptation mechanisms depending on the time when the TD are available. The paper describes the effectiveness of some ANN construction and adaptation if only the TD are available before the adaptation process begins. The ANN construction methods usually consist of two stages: The first one consists in analysis and examination of the input data. It is done from point of view of the specific application for which the network is constructed. The second one consists in construction of network topology and computation of its weights parameters. The mentioned thesis is illustrated by means of some examples.

2 Neural Network Global Adaptation Process

NNNs are equipped with the ability to learn. The TD can change, be contradict, have same missing values, distorted and not correctly or precisely understood. Such data can lead to misunderstanding, misclassification, mistakes, incorrect representation or associations etc. Because of this NNNs are able to forget the minor and mistaken information and to correct the inner model of the experienced surrounding world. Real data change during the time, so the inner model of the data (transformed into the network topology and other parameters) should change as well. Such changes are possible unless the network supplies the mechanisms of modifying the topology and the parameters. These mechanisms are partially known today and used to construct ANN solutions. One of the fundamental differences between NNNs and ANNs is in correctness and availability of the TD. The NNNs many times proceed untrue, incorrect, not fully or fuzzy defined data and have to change their inner models many times. On the other hand, the ANNs usually proceed true, correct and sufficiently precise data and do not need to permanently correct the inner model of the data. Unfortunately, little methods and neural models use this extra information. Many times, the data are treated in the same way independently on the time of their availability. The training process of NNNs ably drive to still better and better solutions, but the limited ability to achieve the optimum is caused by the temporary logging of the TD. In contrast, ANNs are usually designated to problems that are well-defined by some data but there is the lack of mathematical model of the solved problem. Sometimes the complexity of problem forces the use of other approximate techniques that deliver solutions in reasonable time. Global adaptation processes are available for some class of problems [2],[3],[9],[12] enabling computation of the NNs parameters and topology. In order to compare and estimate importance of the given TD and their features, they have to be available at the beginning of the adaptation process. The global estimation of TD makes possible to adapt properly the network by use of the gathered information [7],[9]. Sometimes there is possible to compute directly the networks parameters and topology without time-consuming adaptation process. Such global examination of all TD always simplifies the adaptation process and diminishes computational cost. Global information about all TD makes sometimes possible to differentiate the importance of the features or even to omit some of them if they are not representative for the given problem. The feature optimization reduces the input space dimension of the TD.

3   Self-optimizing Neural Networks

The papers [12],[14] describe ontogenic Self-Optimizing Neural Networks (SONNs) that uses data availability for global optimization processes. The method can globally estimate the importance of each input feature of all TD and adapt the network structure and weights reflecting these estimations. The importance of each k-th input feature for each n-th training sample u_n of m-th class C^m is globally measured by the use of the defined so called global discrimination coefficients (1):

$$\forall_{m\in\{1,...,M\}} \forall_{u^n \in C^m} \forall_{n\in\{1,...Q\}} \forall_{k\in\{1,...K\}} :$$
$$d_{k+}^n = \begin{cases} \frac{\hat{P}_k^m}{(M-1)\cdot Q^m} \sum_{h=1 \& h\neq m}^{M} \left(1 - \frac{\hat{P}_k^h}{Q^h}\right) & if \ u_k^n >= 0 \ \& \ u^n \in C^m \\ 0 & if \ u_k^n < 0 \ \& \ u^n \in C^m \end{cases}$$
$$d_{k-}^n = \begin{cases} \frac{\hat{N}_k^m}{(M-1)\cdot Q^m} \sum_{h=1 \& h\neq m}^{M} \left(1 - \frac{\hat{N}_k^h}{Q^h}\right) & if \ u_k^n <= 0 \ \& \ u^n \in C^m \\ 0 & if \ u_k^n > 0 \ \& \ u^n \in C^m \end{cases} \quad (1)$$

where M denotes a number of classes in the TD, Q is the number of TD, K is the number of input features, $\hat{u}_k^n$ is the probable k-th feature value for n-th training sample, and $x_k^n, y_k^n, \hat{P}_k^m, \hat{N}_k^m, Q^m$ are defined by the following formulas:

$$\forall_{m\in\{1,...,M\}} \forall_{u^n \in C^m} \forall_{n\in\{1,...Q\}} \forall_{k\in\{1,...K\}} \ \hat{u}_k^n = \begin{cases} \hat{u}_k^{n+} = x_k^n & if \ x_k^n > 0 \\ \hat{u}_k^{n-} = -y_k^n & if \ y_k^n > 0 \end{cases} \quad (2)$$

$$\forall_{m\in\{1,...,M\}} \forall_{k\in\{1,...,K\}} : \hat{P}_k^m = \sum_{u^n \in C^m} x_k^n \ \wedge \ \hat{N}_k^m = \sum_{u^n \in C^m} y_k^n \quad (3)$$

$$\forall_{m\in\{1,...,M\}} \ Q^m = \left\| \left\{ u^n \in U \bigcap C^m : n \in \{1,...Q\} \right\} \right\| \quad (4)$$

$$x_k^n = \begin{cases} 1 & if \ u_k^n = +1 \\ \frac{P_k^m}{P_k^m + N_k^m} & if \ u_k^n = 0 \\ 0 & if \ u_k^n = -1 \end{cases} \quad (5)$$

$$y_k^n = \begin{cases} 1 & if\ u_k^n = -1 \\ \frac{N_k^m}{P_k^m + N_k^m} & if\ u_k^n = 0 \\ 0 & if\ u_k^n = +1 \end{cases} \quad (6)$$

$$\forall_{m \in \{1,...,M\}} \forall_{k \in \{1,...,K\}} : P_k^m = \sum_{u_k^n \in \{u \in U \vee C^m : u_k^n > 0,\ n \in \{1,...,Q\}\}} u_k^n \quad (7)$$

$$\forall_{m \in \{1,...,M\}} \forall_{k \in \{1,...,K\}} : N_k^m = \sum_{u_k^n \in \{u \in U \vee C^m : u_k^n < 0,\ n \in \{1,...,Q\}\}} -u_k^n \quad (8)$$

The two discrimination coefficients (1) qualify the discrimination property of each k-th feature for the n-th training sample of the m-th class.

The defined discrimination coefficients (1) have the following properties:

- they are insensitive for class representation by different quantity of TD,
- each feature of each training sample is globally estimated,
- all missing values (unknown features) of TD are automatically probabilistically estimated by formulas (2), (5) and (6).

The SONN topology optimization process is so designed to compute the minimal network structure using only these features that have maximal discrimination properties for each class. The SONN-2 topology is created in a specific process of TD division and features aggregation. Each division produces a single neuron that represents a subgroup of TD that meets a division criterion. TD represented by a neuron always deal at least one same value obligatory discrimination coefficient. They are computed in such a way to maximize a quantity of aggregated same obligatory discrimination coefficients by single connections. Neurons produce a simple weighted sum output but the weight of inter-neuron connection is so computed to reflect the sum of all discrimination coefficients values represented by all input connections of this neuron. This make possible to keep the influence on classification of all inputs at a level determined by the values of their corresponding discrimination coefficients. This guarantees appropriate influence on a resultant classification and provides very good generalization.

The SONN topology optimization process is based on discrimination coefficients (1) computed for all training data samples as well as for some subgroups

Table 1. Comparison of classification results

COMPARISON OF RESULTS		Inputs		Training Data (4475 samples)				Test Data
		Dimension	Time	Average Error	Class 1	Class 2	Incorrect	Average Error
GHOST MINER 3.0	SVM	121 / 121	25 s	0.0000000	1090	3385	0	0.0241162
	IncNet Bi-central	121 / 121	2 hours	0.0017877	1082	3385	8	0.3241980
	FSM	121 / 121	18 s	0.0004469	1088	3385	2	0.3118660
	SSV Tree	4 / 121	5 s	0.0006704	1087	3385	3	0.0320636
	k-NN	121 / 121	35 s	0.0000000	1090	3385	0	0.0117841
Own implemen-	SONN-1	32 / 121	24 s	0.0000000	1090	3385	0	0.0594086
	SONN-2	14 / 121	13 s	0.0000000	1090	3385	0	0.0264254

of them. In order to find the optimal SONN topology there is necessary to create only these connections which are necessary to classify correctly and unambiguously all training data using only the input features with maximal discrimination properties [13],[14]. SONN weights parameters and a SONN topology are computed simultaneously during construction process of the network [12],[14]. Such strategy makes us possible to precisely assign to each feature representing any subgroup of TD the accurate and optimal weight value arising from its discrimination properties computed in a global view of all TD. They are fundamental in decision processes and classification. The SONN classification results have been compared with other AI methods shown in the table 1. The table 1 shows the results of classification for the "mushrooms-agaricus-lepiota" data from the ML-Repository. The results of comparisons shows that SONNs are not only very fast [12],[14] but also achieve very good generalization properties in comparison to other popular AI classification methods. The SONNs have many interesting features that can be compared to biological (natural) neural networks and different neural processes in biological brains [13]. There is not many other AI methods that directly use the information about all TD to compute each individual weight optimally fitted to global information about TD.

Training data and relations between trainings samples of the same class can be very different. The described SONN methodology does not find dependencies between different features of the data but uses their given values to probabilistically estimate their discriminative properties (1) individually. The SONN topology puts together the most important information discriminative properties of different features constructing the uniform classification model for the given TD. The SONN is built up after the most important, well-differentiating and discriminating features of all TD samples. The SONN generalization property reflects the most important and characteristic information of individual classes for any given TD set.

4 ANNs for Feature Extraction

Another new approach to creating ANNs is described and analyzed in the papers [5] and [3]. The approach is appropriated for ANNs of the feed-forward type. The paper [5] describes synthesis of ANN that indicates distribution of relative maxima of one-dimensional input signal $f(x)$, where x is a spatial coordinate. The way of synthesis is quite different from the ones usually presented in literature, namely it is not based on learning algorithms. In similar way one can make synthesis of a networks analyzing other features of an input signal, e.g. in [3] there is presented synthesis of the ANN that indicates distribution of relative extremes. This approach is appropriate for some kind of signal analysis and especially for feature extraction. It is based on the fact that a lot of features can be defined with use of relations expressed by means of linear combinations of values of signal in particular points and by means of linear combinations of differences of these values. If the features can be defined with use of relations expressed by means of linear combination of differences (any order) then the couplings between network elements (receptors or neurons) and weight coefficients may be design on a basis of formula given in [3]. A lot of such problems exist in image analysis and processing and especially in the feature extraction area. Thus the first step to design the network is to define the interesting features by means of difference calculus relations. Then the extension of the approach for two dimensional ANNs that perform image analysis and processing has been worked out. In [4] the synthesis of the ANN that perform edge detection is described and in [8] the synthesis of the ANN that indicates maxima and ridges in images is presented. It must be pointed out, however, that some parameters of the networks, especially amplification and parameter of transfer function of the first layer, as well as number of neurons in the layer must be adapted to the class of input images. Another words the class of input signals must be known *a priori* in order to a designed ANN work properly.

To illustrate our consideration let us present an architecture of two-dimensional networks that indicates ridges and relative maxima in input images. The presented net is built of one-dimensional networks the outputs of which are the inputs for the last, common for them, two-dimensional layer (figure 1). The particular networks indicate relative maxima of one-dimensional signals. A structure and parameters of this one-dimensional net are discussed briefly in a further part of the paper. If an input function $F(x,y)$ has maximum in the point (x_i, y_k) then there exists maximum in this point for each intersection crossing through this point. A ridge is a coherent set of points such that there may exist only one direction in which the intersection in the point has no maximum. (There may be maxima that lay on the ridge.) These facts induce us to apply the one-dimensional ANNs indicating relative maxima. The proposed system consists of two neural subsystems. Each of subsystems is built of parallel one-dimensional ANNs. The networks of the first subsystem indicate relative maxima for parallel intersections of the functions $F(x,y)$ in the fixed points y_i, i.e. for $F(x, y_i)$, $y_i = y_0 + i\Delta y, i = 0, 1, ...n$. The intersections are parallel to the axis x. The networks of the second subsystems indicate maxima for parallel intersections of the functions $F(x,y)$ in the fixed points $x_{i,}$, i.e. for $F(x_i, y)$, $x_i = x_0 + i\Delta x, i = 0, 1, ..n$. The intersections are parallel to the axis y. The networks of the subsystems are called $X(i)$-net and $Y(i)$-net respectively. Outputs of X-net and Y-net are connected in such a way that k-th output of $X(i)$-net is summed with i-th output of $Y(k)$-net. The presented architecture of the network is only one solution. Another two dimensional network based on hexagonal structure is given in [5].

One dimensional network. The structure of a one-dimensional network that indicates relative maxima is given if fig. (figure 2). The zero (input) layer is built from receptors, while the upper layers consist of neurons. The main task of the input layer is to measure the values of an input signal $f(x)$ in discrete points x_i (figure 2) The generalized input signal of a neuron, denoted as e, is a weighted sum of output signals of the elements connected with it in the direction in which information is transmitted. A transfer function of a neuron $h(e)$ is presented in the figure 1a and 1b. For the sake of simplicity of presentation, we assume the transfer function presented in figure 1b. and given by formula (9).

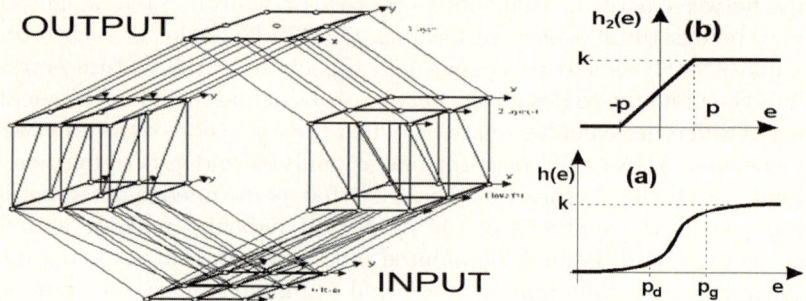

Fig. 1. Structure of the two-dimensional network indicating maxima and ridges with the transfer functions of a neuron, e-input signal (a), $h(e)$ output signal (b)

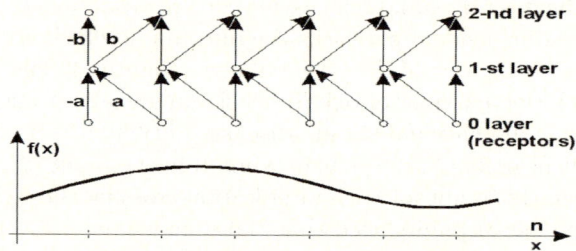

Fig. 2. Structure of the network

$$h(e) = \begin{cases} 0 & for \ e < -p_d \\ \frac{k(e+p_d)}{p_g - p_d} & for \ -p_d \leq e \leq p_d \\ 0 & for \ e > p_d \end{cases} \qquad (9)$$

where p_d, p_g, k are parameters of positive values that denote, respectively, threshold point, saturation point and value of saturation (it is assumed usually that $k = 1$). The formula is a little bit different from the one used in literature [1] but it is more convenient for our aim. For the further consideration we assume that neurons of one layer have the same transfer function parameters and homogeneous couplings. It is also assumed that each neuron obtains signals from the elements (neurons or receptors) of the preceding layer only. Such networks are named homogeneous layer feed-forward neural network (HLFNN). For such networks the following holds:

$$e_j(n) = \sum_{-q_j}^{q_j} w_j(i) f_{j-1}(n+i) \qquad (10)$$

$$f_j(n) = h_j(e_j(n)) \qquad (11)$$

where: n - number of an element in the considered layer, $e_j(n)$ - input function of the j-th layer, $f_j(n)$ - output function of the j-th layer, $w_j(i)$ - function of weight distribution of the j-th layer (defining the weight coefficient for the coupling between the n-th element of this layer and the $n+i$-th element of the $j-1$-th layer), q_j - maximal range of the function $w_j(i)$ (if $|i| > q$, then $w(i) = 0$). For such a model, the natural role of neuron couplings consists in some weighted averaging of the signals that reach the neuron. The role, however, may be determined more precisely. If a weight distribution function is given by the following formula:

$$w(i) = \begin{cases} (-1)^{r-i} \binom{r}{i} & for\ i \in [0,r] \\ 0 & for\ i \notin [0,r] \end{cases} \quad (12)$$

where i, r - integer numbers, then the input signal e is equal to the r-th difference of the given layer output function:

$$e(n) = \sum_{i=-\infty}^{+\infty} w(i) f(n+i) = \Delta^r f(n) \quad (13)$$

The zero (input) layer is built from receptors, while the upper two layers consists of neurons. As it was mentioned earlier, the main task of the input layer is to measure the values of an input signal $f(x)$ in discrete points x_i. Let d denote the length of the shortest interval $[x_1, x_2]$ such that $f(x)$ is exactly monotone in it and $f(x)$ is not exactly monotone in intervals $[x_1 - e, x_2]$ and $[x_1, x_2 + e]$ for $e > 0$. It is assumed that the distance between the receptors Δx is constant and $\Delta x \leq d/2$. Thus, the network analyses the discrete functions $f_0(n) = f(x_n)$, $x_n = x_0 + n\Delta x$. The weight distribution function of the first layer $w_1(i)$ and the transfer function $h_1(n)$ are of the form:

$$w_1(i) = \begin{cases} -a & for\ i = 0 \\ a & for\ i = 1 \\ 0 & i \neq 0, i \neq 1 \end{cases} \quad (14)$$

where parameter $a > 0$

$$h_1(e) = \begin{cases} 0 & for\ e < -p \\ \frac{k(e+p)}{2p} & for\ -p \leq e \leq p \\ k & for\ e > p \end{cases} \quad (15)$$

where p and k are parameters of positive values (figure 3a). The weight distribution function of the second layer $w_2(i)$ is of the form:

$$w_2(i) = \begin{cases} b & for\ i = 0 \\ -b & for\ i = 1 \\ 0 & i \neq 0, i \neq 1 \end{cases} \quad (16)$$

where parameter $b > 0$

The parameters of the second layer transfer function fullfil the following relations: $bk/3 \leq p_d < bk/2$ and $p_g + p_d = bk$ (p_g, p_d, k denote, respectively, threshold point, saturation point and value of saturation, figure 3b).

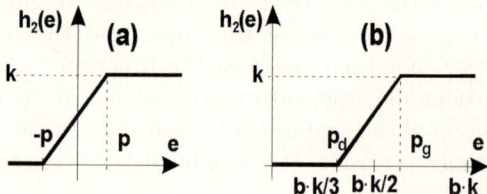

Fig. 3. Transfer function for the first and second layer neurons

It has been proven [5] that if for a relative maximum $f(x)$ the following conditions occur:

$$\forall q \in [0,1) : f(x^0 - q\Delta x) - f(x^0 - (1+q)\Delta x) \geq \frac{p}{a} \\ \wedge f(x^0 + (2-q)\Delta x) - f(x^0 + (1-q)\Delta x) \leq -\frac{p}{a} \quad (17)$$

where x^0 is the proper maximum coordinate or any point that belongs to v^* for non proper maximum, then the maximum can be located by the network

In order to the two-dimensional network works correctly, two assumptions must be held. Firstly, the relation (17) must hold for any relative maximum and any intersection of the function $F(x,y)$ (i.e. for the one-dimensional function). Thus instead of formula (17) we have:

$$\forall q \in [0,1), \forall \varphi \in [0, 2\pi) : \\ \frac{p}{a} \leq F(x^0 - q\cos(\varphi)\Delta, y^0 - q\sin(\varphi)\Delta) \\ -F(x^0 - (1+q)\cos(\varphi)\Delta, y^0 - (1+q)\sin(\varphi)\Delta) \quad (18)$$

where (x^0, y^0) is the proper maximum coordinate, or point that belong to a ridge, φ denotes an angle that determines a direction, Δ is the distance between the neighboring receptors in a direction determined by an angle φ. Secondly, a special relation between receptor distance and the shortest interval of exact monotonicity d must hold. The distance between the neighboring receptors Δ in any direction determined by a fixed angle φ is: $\Delta \leq d/2$. If the class of images represented by functions $F(x,y)$ is given one may choose number of neurons and amplifications in such a way that all points of ridges and maxima are indicated and all disturbances are omitted.

Adaptation algorithm and experiments. It results from (18) that it depends on value of p/a as well as discretization step whether the maximum is "noticed" by the neural net. Let us assume that $p = 1$ and the discretization steps in both directions are the same, i.e $\Delta x = \Delta y = \Delta$. It easy to see that the less Δ the better precision of locating maxima is. On the other hand, reduction Δ causes that the value of the first difference $\Delta f_0(n, m)$ in any direction is also less and some maxima as well as points of ridges can be not noticed by the net. Increase of Δ acts contrary to it provided $\Delta \leq d/2$. If $\Delta \geq d/2$, some maxima and points of ridges can also be not indicated. The role of amplification a is similar: the greater value of a causes that the relation (39) may by "easier" fulfilled and even

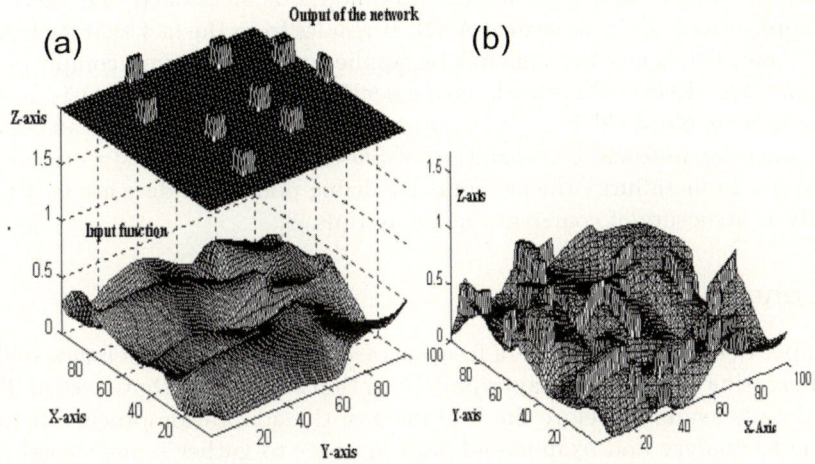

Fig. 4. The network performance presentation (a) with ridges indication (b)

Table 2. Results of adjusting experiments

No	a_{end}	Distance between sensors	No of maxims	No of adjusting cycles	No	a_{end}	Distance between sensors	No of maxims	No of adjusting cycles
1	350	1	8	35	6	50	6	12	5
2	67	6	12	61	7	40	7	14	6
3	35	6	11	29	8	11	12	14	4
4	9	9	7	4	9	34	7	8	5
5	50	7	8	5	10	40	5	8	6

"flat" maximum can be indicated. Obviously, it is not good solution when there is to many sensors (too small Δ) and very big amplification at the same time. Thus, a role of adjusting module consists in choosing best value of parameters a and Δ.

Experiment has been carried out for the class of input function G. generated as follows. First, the so called based function $F_b(x, y)$ was created with use of polynomial interpolation procedure (Lagrange'a interpolation). The interpolation nodes were generated in random way. The next functions are obtained as a sum of the basic function $F_b(x, y)$ and some random signal that intensity is of about 25% of the basic function. At the first stage of the experiments, the network was trained for 20 functions. It was done with use of the formula:

$$a_{i+1} = c_a * a \quad where \quad c_a \in [1.1, 1.6], i - iteration \qquad (19)$$

$$\Delta_{i+1} = c * \Delta_i \quad where \quad 0.5 < c < 1 \qquad (20)$$

Then the network was examined and adjusted for the other functions of the class. Only parameter a was changed: $a_{i+1} = a_i + \Delta a_i$. Results of adjusting are given in the table 2. and in the figures 4. It should be mentioned that there are a lot of applications of the described ANN. It results from the fact that such ANNs work in parallel. Thus they should be applied when real-time computing are necessary. The described network have essential application in analysis of fringe sample images obtained from interferometry measurements. Such an analysis is necessary for automatic recognition of 3D objects based on structured light method [6]. In metallurgy the networks localizing points of ridges are very useful to analysis structure of coarse-grained materials [8].

5 Conclusions

The paper deals with synthesis of feed-forward ANNs. It presents a new philosophy of creating ANNs. The paper pays attention to the fact that a set of TD or input data is available before the beginning of the adaptation process. It makes possible to analyze and examine all data in order to gather same special information that may be very useful for designing the ANNs. Obviously, this information should be utilized for the ANNs synthesis. Thanks to it, ANNs may be created much faster producing better generalization. This idea is illustrated by means of two types of ANNs: First presented type of ANNs is based on statistical analysis of all TD and probabilistic estimation of discrimination properties of TD features. The global view of data enables us to computed global discriminative coefficients that carry information about their features and can be used to optimal selection of the most general features representing classes. Thanks the global selection [12],[14] of the most representative features the SONN always returns the neural model that generalizes as well as good the quality of the given training data and how much they representative are. The SONNs build the neural model not only on the basis of the globally most discriminative features but also transforms the values of their discriminative properties into weight values. The second type of ANNs performs image processing. Consecutive steps of synthesis of ANN that indicates distribution of relative maxima and ridges in images was briefly described. The way of synthesis is quite different from the ones usually presented in literature, namely it is not based on learning algorithms. In similar way one can make synthesis of a network analyzing other features. There are two main advantages of this approach: First, it enables us to omit the time-consuming training process. Secondly, the way in which the network solves a given task can be easily understood. According to many papers, among them [10], a major disadvantage of neural networks is the lack of understanding of how they actually solve a given cognitive task. Because of that, others authors also consider possibilities "designing" neural networks instead of "learning" them [9].

References

1. Cichocki, A., Unbehauen, R.: Neural networks for optimization and signal processing. J. Wiley and Sons, New York (1993)
2. Duch, W., Korbicz, J., Rutkowski, L., Tadeusiewicz, R. (eds.): Biocybernetics and Biomedical Engineering. EXIT, Warszawa (2000)

3. Dudek-Dyduch, E.: Synthesis of feed forward neural network indicating extremes. Int. Journal System Analysis - Modelling - Simulation 24, 135–151 (1996)
4. Dudek-Dyduch, E., Bartman, J., Gomka, Z.: Detection of Edges using homogeneous neural network. In: Proc. of 7th Int. Conf. on Engineering Computer Graphics and Descriptive Geometry, Cracow, vol. 1, pp. 295–299 (1996)
5. Dudek-Dyduch, E., Dyduch, T.: On feedforward homogeneous layer neural nets. In: Proc. of the Second Conf. Neural Networks and Their Applications, Szczyrk, vol. 1, pp. 144–152 (1996)
6. Dudek-Dyduch, E., Dyduch, T.: Application of neural networks in 3D object recognition system. Int. Journal of Sample Recognition and Artificial Intelligence 12(4), 491–504 (1998)
7. Dudek-Dyduch, E., Horzyk, A.: Analytical Synthesis of Neural Networks for Selected Classes of Problems. In: Bubnicki, Z., Grzech, A. (eds.) Knowledge Engineering and Experts Systems, OWPN, Wroclaw, pp. 194–206 (2003)
8. Dudek-Dyduch, E., Tadeusiewicz, R.: Neural network indicating maxima and ridges in two-dimensional signal. In: Bulsari, A.B., Kallio, S. (eds.) Engineering Applications of ANN, Proc of Int. Conf. EANN 1995, Helsinki, pp. 485–488 (1995)
9. Dudek-Dyduch, E., Gomuka, Z., Pkala, R.: Adjusting parameters of two-dimensional neural network. In: Proc. of the Second Conf. Neural Networks and Their Applications, Szczyrk, vol. 1 (1996)
10. Dyduch, T., Dudek-Dyduch, E.: Application of neural networks to fringe samples analysis. In: Bulsari, A.B., Kallio, S. (eds.) Solving Engineering Problems with Neural Networks, Proc of EANN 1996, London, vol. 2, pp. 671–674 (1996)
11. Fiesler, E., Beale, R. (eds.): Handbook of Neural Computation. IOP Publishing Ltd. and Oxford University Press, Bristol and New York (1997)
12. Horzyk, A.: A New Extension of Self-optimizing Neural Networks for Topology Optimization. In: Duch, W., Kacprzyk, J., Oja, E., Zadrożny, S. (eds.) ICANN 2005. LNCS, vol. 3696, pp. 415–420. Springer, Heidelberg (2005)
13. Horzyk, A., Tadeusiewicz, R.: Comparison of Plasticity of Self-Optimizing Neural Networks and Natural Neural Networks. In: Mira, J., Alvarez, J.R. (eds.) Proc. of ICANN 2005, pp. 156–165. Springer, Heidelberg (2005)
14. Horzyk, A., Tadeusiewicz, R.: Self-Optimizing Neural Networks. In: Yin, F.-L., Wang, J., Guo, C. (eds.) ISNN 2004. LNCS, vol. 3173, pp. 150–155. Springer, Heidelberg (2004)
15. Jankowski, N.: Ontogenic neural networks, EXIT, Warszawa (2003)
16. Kacprzak, T., lot, K.: Cellular Neural Networks, PWN, Warszawa - d (1995)

WWW-Newsgroup-Document Clustering by Means of Dynamic Self-organizing Neural Networks

Marian B. Gorzałczany and Filip Rudziński

Department of Electrical and Computer Engineering
Kielce University of Technology
Al. 1000-lecia P.P. 7, 25-314 Kielce, Poland
{m.b.gorzalczany,f.rudzinski}@tu.kielce.pl

Abstract. The paper presents a clustering technique based on dynamic self-organizing neural networks and its application to a large-scale and highly multidimensional WWW-newsgroup-document clustering problem. The collection of 19 997 documents (e-mail messages of different *Usenet-News* newsgroups) available at WWW server of the School of Computer Science, Carnegie Mellon University (www.cs.cmu.edu/TextLearning/datasets.html) has been the subject of clustering. A broad comparative analysis with nine alternative clustering techniques has also been carried out demonstrating the superiority of the proposed approach in the considered problem.

1 Introduction

The rapidly increasing volume of electronically available World-Wide-Web resources makes more and more important the issue of helping users to locate and access relevant information as well as to organize it in an intelligible way. Since text- and hypertext documents belong to the most important and available online WWW resources, text processing techniques play the central role in this field. In turn, among them, thematic WWW-document clustering techniques (thematic text clustering techniques) are of special interest. In general, given a collection of WWW documents, the task of document clustering is to group documents together in such a way that the documents within each cluster are as "similar" as possible to each other and as "dissimilar" as possible from those of the other clusters.

This paper presents a clustering technique based on dynamic self-organizing neural networks (introduced by the same authors in [7], [8] and some earlier papers such as [5], [6]) and its application to a large-scale and highly multidimensional WWW-newsgroup-document clustering problem. First, the paper presents the concept of a dynamic self-organizing neural network for WWW-document clustering. Then, a Vector-Space-Model representation of WWW documents is outlined as well as some approaches to its dimensionality reduction are briefly presented. In turn, the application of the proposed technique to clustering of the collection of 19 997 documents (e-mail messages of different *Usenet-News* newsgroups) available at WWW server of the School of Computer Science, Carnegie Mellon University (www.cs.cmu.edu/ TextLearning/datasets.html) is presented. Finally, a broad comparative analysis with several alternative document clustering techniques is carried out.

2 The Concept of a Dynamic Self-organizing Neural Network for WWW-Document Clustering

A dynamic self-organizing neural network is a generalization of the conventional self-organizing neural network with one-dimensional neighbourhood. Consider the latter case of the network that has n inputs $x_1, x_2, \ldots, x_n$ and consists of m neurons arranged in a chain; their outputs are $y_1, y_2, \ldots, y_m$, where $y_j = \sum_{i=1}^{n} w_{ji} x_i$, $j = 1, 2, \ldots, m$ and w_{ji} are weights connecting the output of j-th neuron with i-th input of the network. Using vector notation ($\boldsymbol{x} = (x_1, x_2, \ldots, x_n)^T$, $\boldsymbol{w}_j = (w_{j1}, w_{j2}, \ldots, w_{jn})^T$), $y_j = \boldsymbol{w}_j^T \boldsymbol{x}$. The learning data consists of L input vectors $\boldsymbol{x}_l$ ($l = 1, 2, \ldots, L$). The first stage of any Winner-Takes-Most (WTM) learning algorithm that can be applied to the considered network, consists in determining the neuron $j_{\boldsymbol{x}}$ winning in the competition of neurons when learning vector $\boldsymbol{x}_l$ is presented to the network. Assuming the normalization of learning vectors, the winning neuron $j_{\boldsymbol{x}}$ is selected such that

$$d(\boldsymbol{x}_l, \boldsymbol{w}_{j_{\boldsymbol{x}}}) = \min_{j=1,2,\ldots,m} d(\boldsymbol{x}_l, \boldsymbol{w}_j), \tag{1}$$

where $d(\boldsymbol{x}_l, \boldsymbol{w}_j)$ is a distance measure between $\boldsymbol{x}_l$ and $\boldsymbol{w}_j$; throughout this paper, a distance measure d based on the cosine similarity function S (most often used for determining similarity of text documents [1]) will be applied:

$$d(\boldsymbol{x}_l, \boldsymbol{w}_j) = 1 - S(\boldsymbol{x}_l, \boldsymbol{w}_j) = 1 - \frac{\boldsymbol{x}_l^T \boldsymbol{w}_j}{\|\boldsymbol{x}_l\| \|\boldsymbol{w}_j\|} = 1 - \frac{\sum_{i=1}^{n}(x_{li} w_{ji})}{\sqrt{\sum_{i=1}^{n} x_{li}^2 \sum_{i=1}^{n} w_{ji}^2}}, \tag{2}$$

($\|.\|$ are Euclidean norms).

The WTM learning rule can be formulated as follows:

$$\boldsymbol{w}_j(k+1) = \boldsymbol{w}_j(k) + \eta_j(k) N(j, j_{\boldsymbol{x}}, k)[\boldsymbol{x}(k) - \boldsymbol{w}_j(k)], \tag{3}$$

where k is the iteration number, $\eta_j(k)$ is the learning coefficient, and $N(j, j_{\boldsymbol{x}}, k)$ is the neighbourhood function. In this paper, the Gaussian-type neighbourhood function will be used:

$$N(j, j_{\boldsymbol{x}}, k) = e^{-\frac{(j-j_{\boldsymbol{x}})^2}{2\lambda^2(k)}}, \tag{4}$$

where $\lambda(k)$ is the "radius" of neighbourhood (the width of the Gaussian "bell").

After each learning epoch, five successive operations are activated (under some conditions) [7]: 1) the removal of single, low-active neurons, 2) the disconnection of a neuron chain, 3) the removal of short neuron sub-chains, 4) the insertion of additional neurons into the neighbourhood of high-active neurons, and 5) the reconnection of two selected sub-chains of neurons.

The operations nos. 1, 3, and 4 are the components of the mechanism for automatic adjustment of the number of neurons in the chain, whereas the operations nos. 2 and 5 govern the disconnection and reconnection mechanisms, respectively. Based on experimental investigations, the following conditions for particular operations have been formulated (numberings of conditions and operations are the same). Possible operation takes place between neuron no. i and neuron no. $i+1$; $i \in \{1, 2, \ldots, r-1\}$, where r is the number of neurons in the original neuron chain or a given sub-chain.

Condition 1: $win_i < \beta_1$, where win_i is the number of wins of i-th neuron and β_1 is experimentally selected parameter (usually, for complex multidimensional WWW-document clustering, β_1 assumes the value around 50). This condition allows to remove single neuron whose activity (measured by the number of its wins) is below an assumed level represented by parameter β_1.

Condition 2: $d_{i,i+1} > \alpha_1 \frac{\sum_{j=1}^{r-1} d_{j,j+1}}{r}$, where $d_{i,i+1}$ is the distance between the neurons no. i and no. $i+1$ (see [1] for details) and α_1 is experimentally selected parameter (usually, for complex problems $\alpha_1 \in [1, 10]$). This condition prevents the excessive disconnection of the neuron chain or sub-chain by allowing to disconnect only relatively remote neurons.

Condition 3: $r_S < \beta_2$, where r_S is the number of neurons in sub-chain S and β_2 is experimentally selected parameter (usually $\beta_2 \in \{3, 4\}$). This condition allows to remove r_S-element neuron sub-chain S that length is shorter than assumed acceptable value β_2.

The operation of the insertion of additional neurons into the neighbourhood of high-active neurons in order to take over some of their activities covers 3 cases denoted by 4a, 4b, and 4c, respectively.

Condition 4a (the insertion of new neuron (n) between two neighbouring high-active neurons no. i and no. $i+1$): IF $win_i > \beta_3$ AND $win_{i+1} > \beta_3$ THEN weight vector $\boldsymbol{w}_{(n)}$ of new neuron (n) is calculated as follows: $\boldsymbol{w}_{(n)} = \frac{\boldsymbol{w}_i + \boldsymbol{w}_{i+1}}{2}$, where win_i, win_{i+1} are as in Condition 1 and β_3 is experimentally selected parameter (usually β_3 is comparable to β_1 that governs Condition 1).

Conditions 4b (the replacement of high-active neuron no. i - accompanied by low-active neurons no. $i-1$ and no. $i+1$ - by two new neurons: (n) and $(n+1)$): IF $win_i > \beta_3$ AND $win_{i-1} < \beta_3$ AND $win_{i+1} < \beta_3$ THEN weight vectors $\boldsymbol{w}_{(n)}$ and $\boldsymbol{w}_{(n+1)}$ of new neurons (n) and $(n+1)$ are calculated as follows: $\boldsymbol{w}_{(n)} = \frac{\boldsymbol{w}_{i-1} + \boldsymbol{w}_i}{2}$ and $\boldsymbol{w}_{(n+1)} = \frac{\boldsymbol{w}_i + \boldsymbol{w}_{i+1}}{2}$ (β_3 - as in Condition 4a).

Condition 4c (the insertion of new neuron in the neighbourhood of an end-chain high-active neuron accompanied by low-active neighbour; r-th neuron case will be considered; 1st neuron case is analogous): IF $win_r > \beta_3$ AND $win_{r-1} < \beta_3$ THEN weight vector $\boldsymbol{w}_{r+1}$ of new neuron $(r+1)$ is calculated as follows: $\boldsymbol{w}_{r+1} = \boldsymbol{w}_r + \frac{\boldsymbol{w}_r - \boldsymbol{w}_{r-1}}{d_{r,r-1}} d_{avr}$, where $d_{avr} = \frac{1}{r-1} \sum_{j=1}^{r-1} d_{j,j+1}$ (β_3 - as in Condition 4a and $d_{j,j+1}$ - as in Condition 2).

Condition 5: $d_{e_{S1},e_{S2}} < \alpha_2[\frac{1}{2}(\frac{\sum_{j=1}^{r_{S1}-1} d_{j,j+1}}{r_{S1}} + \frac{\sum_{j=1}^{r_{S2}-1} d_{j,j+1}}{r_{S2}})]$, where $S1$ and $S2$ are two sub-chains (containing r_{S1} and r_{S2} neurons, respectively) whose appropriate ends $e_{S1} \in \{1, r_{S1}\}$ and $e_{S2} \in \{1, r_{S2}\}$ are closest to each other; sub-chains $S1$ and $S2$ are the candidates for the connection by combining their ends e_{S1} and e_{S2} ($d_{j,j+1}$ - as in Condition 2, α_2 - experimentally selected parameter (usually α_2 is comparable to α_1 that governs Condition 2). This condition allows to connect two sub-chains not only with closest ends but also with relatively close to each other neighbouring neurons that correspond to compact pieces of the same cluster of data.

3 Vector Space Model of WWW Documents - An Outline [8]

Consider a collection of L WWW documents. In the Vector Space Model (VSM) [3, 4, 9, 12, 13], every document in the collection is represented by vector $\boldsymbol{x}_l = (x_{l1}, x_{l2}, \ldots, x_{ln})^T$ ($l = 1, 2, \ldots, L$). Component x_{li} ($i = 1, 2, \ldots, n$) of such a vector represents i-th key word or term that occurs in l-th document. The value of x_{li} depends on the degree of relationship between i-th term and l-th document. Among various schemes for measuring this relationship (very often referred to as term weighting), three are the most popular: a) binary term-weighting: $x_{li} = 1$ when i-th term occurs in l-th document and $x_{li} = 0$ otherwise, b) *tf*-weighting (*tf* stands for *term frequency*): $x_{li} = tf_{li}$ where tf_{li} denotes how many times i-th term occurs in l-th document, and c) *tf-idf*-weighting (*tf-idf* stands for *term frequency - inverse document frequency*): $x_{li} = tf_{li} \log(L/df_i)$ where tf_{li} is the term frequency as in *tf*-weighting, df_i denotes the number of documents in which i-th term appears, and L is the total number of documents in the collection. In this paper *tf*-weighting will be applied. Once the way of determining x_{li} is selected, the Vector Space Model can be formulated in a matrix form:

$$VSM_{(n \times L)} = \boldsymbol{X}_{(n \times L)} = [\boldsymbol{x}_l]_{l=1,2,\ldots,L} = [x_{li}]_{l=1,2,\ldots,L;\ i=1,2,\ldots,n}^T \qquad (5)$$

where index ($n \times L$) represents its dimensionality.

The VSM-dimensionality-reduction issues are of essential significance as far as practical usage of VSMs is concerned. There are two main classes of techniques for VSM-dimensionality reduction [11]: a) feature selection methods, and b) feature transformation methods. Among the techniques that can be included into the afore-mentioned class a) are: filtering, stemming, stop-word removal, and the proper feature-selection methods sorting terms and then eliminating some of them on the basis of some numerical measures computed from the considered collection of documents. The first three techniques sometimes are classified as text-preprocessing methods, however - since they significantly contribute to VSM-dimensionality reduction - here, for simplicity, they have been included into afore-mentioned class a).

During filtering (and tokenization) special characters, such as %, #, $, etc., are removed from the original text as well as word- and sentence boundaries are identified in it. As a result of that, initial $VSM_{(n_{ini} \times L)}$ is obtained where n_{ini} is the number of different words isolated from all documents.

During stemming all words in initial model are replaced by their respective stems (a stem is a portion of a word left after removing its suffixes and prefixes). As a result of that, $VSM_{(n_{stem} \times L)}$ is obtained where $n_{stem} < n_{ini}$.

During stop-word removal (removing words from a so-called stop list), words that on their own do not have identifiable meanings and therefore are of little use in various text processing tasks are eliminated from the model. As a result of that, $VSM_{(n_{stpl} \times L)}$ is obtained where $n_{stpl} < n_{stem}$.

Feature selection methods usually operated on term quality $q_i, i = 1, 2, \ldots, n_{stpl}$ defined for each term occurring in the latest VSM. Terms characterized by $q_i < q_{tres}$ where q_{tres} is a pre-defined threshold value are removed from the model. In this paper, the document-frequency-based method will be used to determine q_i, that is $q_i = df_i$ where df_i is the number of documents in which i-th term occurs. As a result of that, final $VSM_{(n_{fin} \times L)}$ is obtained where $n_{fin} < n_{stpl}$.

4 Application to Complex, Multidimensional WWW-Newsgroup-Document Clustering Problem

The proposed clustering technique based on the dynamic self-organizing neural networks will now be applied to real-life WWW-newsgroup-document clustering problem, that is, to clustering of the multidimensional, large-scale collection of 19 997 documents (e-mail messages of different *Usenet-News* newsgroups) available at WWW server of the School of Computer Science, Carnegie Mellon University (www.cs.cmu.edu/ TextLearning/datasets.html). Henceforward, the collection will be called 20_*newsgroups*. The considered collection is partitioned (nearly) evenly across 20 different newsgroups, each corresponding to a different topic. It is worth emphasizing that some of the newsgroups are very closely related to each other, while others are highly unrelated (see the subsequent part of the paper). Since the assignments of documents to newsgroups are known here, it allows us for direct verification of the results obtained. Obviously, the knowledge about the newsgroup assignments by no means will be used by the clustering system (it works in a fully unsupervised way).

The process of dimensionality reduction of the initial VSM for 20_*newsgroups* document collection has been presented in Table 1 with the use of notations introduced in Section 3 (additionally, in square brackets, the overall numbers of occurrences of all terms in all documents of the collection are presented). For the considered document collection, two final numerical models (identified in Table 1 as "*Small*" and "*Large*" data sets) have been obtained. For this purpose two values of threshold parameter q_{tres} have been considered: $q_{tres} = 1\,000$ - to get the model of reduced dimensionality ("*Small*"-type data sets) and $q_{tres} = 400$ - to get the model of higher dimensionality but also of higher accuracy ("*Large*"-type data sets).

Table 1. The dimensionality reduction of the initial VSM for $20\_newsgroups$ document collection

VSM	Dimensionality of VSM for $20\_newsgroups$ document collection:	
$VSM_{(n_{ini} \times L)}$	$(n_{ini} \times L) = (122\,005 \times 19\,997)$ $[2\,644\,002]$	
$VSM_{(n_{stem} \times L)}$	$(n_{stem} \times L) = (99\,072 \times 19\,997)$ $[2\,526\,731]$	
$VSM_{(n_{stpl} \times L)}$	$(n_{stpl} \times L) = (98\,599 \times 19\,997)$ $[1\,677\,316]$	
$VSM_{(n_{fin} \times L)}$	$20\_newsgroups\,"Small"$ $(q_{tres} = 1\,000)$ $(n_{fin} \times L) = (232 \times 19\,997)$ $[461\,512]$	$20\_newsgroups\,"Large"$ $(q_{tres} = 400)$ $(n_{fin} \times L) = (725 \times 19\,997)$ $[524\,321]$

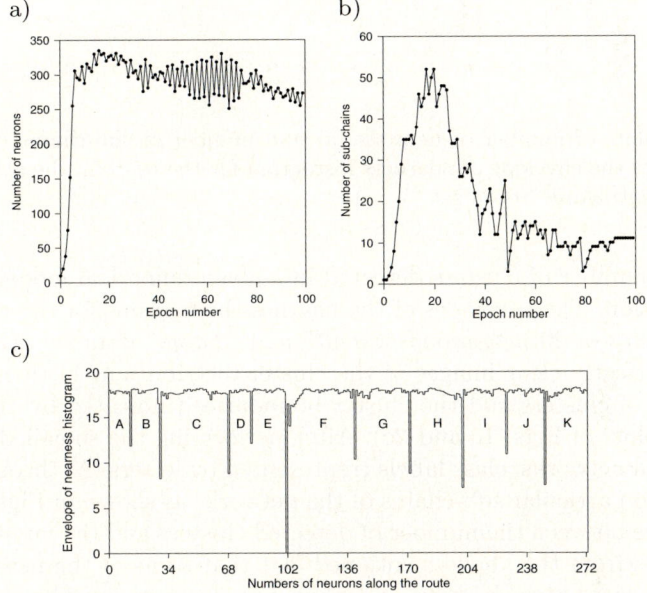

Fig. 1. The plots of number of neurons (a) and number of sub-chains (b) vs. epoch number, and c) the envelope of nearness histogram for the route in the attribute space of $20\_newsgroups\,"Small"$ data set

Figs. 1 and 2 present the performance of the proposed clustering technique for $20\_newsgroups\,"Small"$ and $"Large"$ numerical models of $20\_newsgroups$ collection of documents. As the learning progresses, both systems adjust the overall numbers of neurons in their networks (Figs. 1a and 2a) that finally are equal to 273 and 251, respectively, and the numbers of sub-chains (Figs. 1b and 2b) finally achieving the values equal to 11 in both cases; the number of sub-chains is

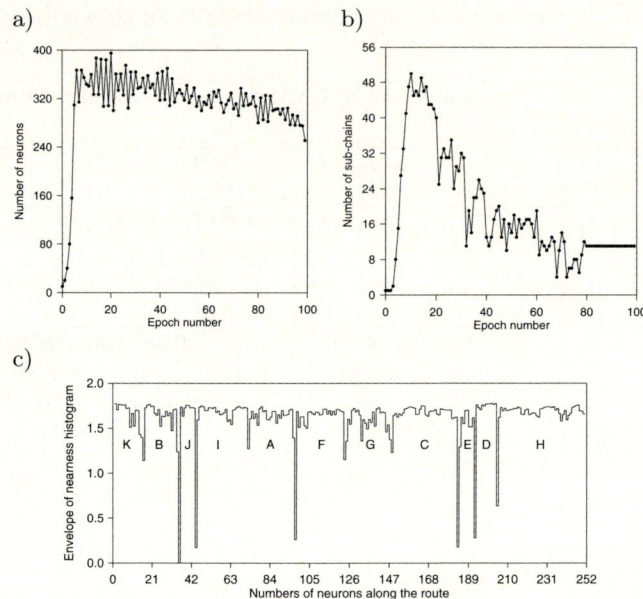

Fig. 2. The plots of number of neurons (a) and number of sub-chains (b) vs. epoch number, and c) the envelope of nearness histogram for the route in the attribute space of 20_newsgroups"Large" data set

equal to the number of clusters detected in a given numerical model of the document collection. The envelopes of the nearness histograms for the routes in the attribute spaces of 20_newsgroups"Small" and "Large" data sets (Figs. 1c and 2c) reveal perfectly clear images of the cluster distributions in them, including the numbers of clusters and the cluster boundaries (indicated by 10 local minima on the plots of Figs. 1c and 2c). After performing the so-called calibration of both neural networks, class labels (represented by letters 'A' through 'K') can be assigned to particular sub-chains of the networks as shown in Figs. 1c and 2c. The difference between the number of detected clusters and the number of newsgroups results from the above-mentioned fact that some of the newsgroups are very closely related to each other, and, thus, they are perceived by the clustering system as one cluster (details on the calibration of both neural networks are presented below). It is worth mentioning that both systems detect the same number of clusters, which confirms the internal consistency of the proposed approach.

Since the newsgroup assignments are known in the original collection of documents, a direct verification of the obtained results is also possible. Detailed numerical results of the clustering and network calibration have been collected in Tables 2 and 3 for 20_newsgroups"Small" and "Large" data sets, respectively. These tables show, for every newsgroup, numbers of its documents that have been assigned by the clustering system to particular neuron sub-chains (classes)

Table 2. Clustering results for 20_newsgroups"Small" data set

Newsgroup name	Number of decisions for sub-chain (class) labelled:										NCD[1]	NWD[2]	PCD[3] [%]
	A	B	C	D	E	F	G	H	I	J	K			
misc.forsale	5	15	12	23	105	4	39	2	**729**	12	54	729	721	72.90
comp.graphics	65	11	123	0	141	14	0	**612**	8	6	11	612	388	61.20
comp.windows.x	9	36	101	12	47	31	45	**633**	22	35	29	633	367	63.30
comp.os.ms-windows.misc	9	0	122	7	56	21	23	**586**	15	158	3	586	414	58.60
comp.sys.mac.hardware	60	12	**632**	4	15	25	28	22	78	87	37	632	368	63.20
comp.sys.ibm.pc.hardware	11	34	**599**	6	67	25	6	25	90	131	6	599	401	59.90
sci.electronics	39	36	258	2	**391**	44	58	18	38	116	0	391	609	39.10
rec.motocycles	8	**602**	36	41	146	0	0	30	67	29	41	602	398	60.20
rec.autos	15	**411**	198	21	84	11	47	61	84	43	25	411	589	41.10
talk.politics.mideast	43	0	105	20	53	130	**487**	72	21	67	2	487	513	48.70
talk.politics.guns	21	13	17	21	76	225	**583**	0	19	9	16	583	417	58.30
talk.politics.misc	0	0	99	13	9	340	**394**	1	60	81	3	394	606	39.40
rec.sport.hockey	11	3	181	17	15	19	59	26	37	54	**578**	578	422	57.80
rec.sport.baseball	0	11	137	23	27	48	35	0	21	17	**681**	681	319	68.10
sci.crypt	0	9	48	**512**	92	17	110	11	60	106	35	512	488	51.20
sci.space	**538**	5	33	89	61	25	61	29	34	35	90	538	462	53.80
sci.med	42	15	36	0	8	23	50	17	304	**429**	76	429	571	42.90
alt.atheism	0	5	54	47	5	**533**	111	32	70	120	23	533	467	53.30
talk.religion.misc	5	21	27	79	41	**562**	93	11	56	20	85	562	438	56.20
soc.religion.christian	6	2	34	15	30	**735**	39	41	25	47	23	735	262	73.72
ALL:	887	1241	2852	952	1469	2832	2268	2229	1838	1602	1827	11227	8770	56.14

[1] NCD = Number of correct decisions, [2] NWD = Number of wrong decisions,
[3] PCD = Percentage of correct decisions.

labelled by letters 'A' through 'K'. The biggest number of documents (denoted in Tables 2 and 3 by boldface) of a given newsgroup assigns that newsgroup to appropriate sub-chain (class).

The detailed results of Tables 2 and 3 have been summarized in Tables 4 and 5. Table 4 presents the overall numbers of documents that have been assigned to particular sub-chains for both "Small" and "Large" data sets. Additionally, numbers of neurons belonging to particular sub-chains of the overall chains (see also Figs. 1c and 2c for both data sets have been included. In turn, Table 5 presents the assignments of particular newsgroups to successive sub-chains (classes). It is worth emphasizing that the proposed clustering technique not only detects the same number of clusters in both "Small" and "Large" data sets (it was already mentioned earlier in this section) but also assigns the same newsgroups to appropriate clusters in both data sets. It is another confirmation of the internal consistency of the proposed clustering technique.

An important issue of the accuracy of the proposed technique will be considered in the framework of a broad comparative analysis with several alternative

Table 3. Clustering results for 20_newsgroups"Large" data set

Newsgroup name	Number of decisions for sub-chain (class) labelled:											NCD[1]	NWD[2]	PCD[3] [%]
	K	B	J	I	A	F	G	C	E	D	H			
misc.forsale	5	71	0	**794**	12	2	33	59	4	16	4	794	206	79.40
comp.graphics	11	43	30	19	22	4	5	152	5	7	**702**	702	298	70.20
comp.windows.x	1	51	45	20	6	50	23	120	4	30	**650**	650	350	65.00
comp.os.ms-windows.misc	6	10	21	80	16	19	21	211	14	11	**591**	591	409	59.10
comp.sys.mac.hardware	8	39	11	42	4	42	11	**644**	10	0	189	644	356	64.40
comp.sys.ibm.pc.hardware	4	54	15	86	29	15	0	**613**	18	6	160	613	387	61.30
sci.electronics	4	40	24	34	14	4	25	132	**405**	32	286	405	595	40.50
rec.motocycles	2	**623**	55	114	33	5	2	13	47	71	35	623	377	62.30
rec.autos	15	**432**	14	168	1	50	16	138	12	4	150	432	568	43.20
talk.politics.mideast	8	21	18	45	0	198	**503**	24	3	7	173	503	497	50.30
talk.politics.guns	18	1	11	23	5	261	**609**	21	27	19	5	609	391	60.90
talk.politics.misc	3	12	5	56	17	358	**412**	51	10	6	70	412	588	41.20
rec.sport.hockey	**601**	4	9	50	14	70	38	170	13	11	20	601	399	60.10
rec.sport.baseball	**690**	29	0	25	23	35	22	141	18	14	3	690	310	69.00
sci.crypt	8	21	3	112	10	85	11	134	1	**523**	92	523	477	52.30
sci.space	17	14	4	41	**543**	70	28	23	57	1	202	543	457	54.30
sci.med	6	16	**437**	174	4	6	50	34	11	5	257	437	563	43.70
alt.atheism	13	25	3	53	7	**547**	54	69	15	73	141	547	453	54.70
talk.religion.misc	13	22	1	125	23	**579**	98	23	34	77	5	579	421	57.90
soc.religion.christian	1	16	7	32	15	**798**	28	19	23	26	32	798	199	80.04
ALL:	1434	1544	713	2093	798	3198	1989	2791	731	939	3767	11680	8317	58.41

[1]NCD = Number of correct decisions, [2]NWD = Number of wrong decisions,
[3]PCD = Percentage of correct decisions.

clustering methods applied to 20_newsgroups"Small" and "Large" data sets as well as to some modifications of the original 20_newsgroups document collection. In Part I of Table 6 the results of comparative analysis with three alternative approaches (they are listed under Part I of Table 6) applied to "Small" and "Large" data sets are presented. In order to carry out the clustering of the 20_newsgroups"Small" and "Large" data sets with the use of the afore-mentioned techniques, WEKA (Waikato Environment for Knowledge Analysis) application that implements them has been used. The WEKA application as well as details on the clustering techniques can be found on WWW site of the University of Waikato, New Zealand (www.cs.waikato.ac.nz/ml/weka).

In order to extend the comparative-analysis aspects of this paper, in Part II of Table 6 the results of the clustering of some modifications of the original 20_newsgroups document collection that are reported in the literature have been included. This time the operation of the dynamic self-organizing neural network clustering technique has been compared with seven alternative approaches that are listed under Part II of Table 6.

Table 4. Sub-chain (class) labels, numbers of documents assigned to particular sub-chains and numbers of neurons in the chains for 20_newsgroups"Small" (a) and 20_newsgroups"Large" (b) data sets

a)

Sub-chain (class) label	Number of documents assigned to sub-chain	Number of neuron in the chain
A	887	1 – 12
B	1 241	13 – 29
C	2 852	30 – 68
D	952	69 – 81
E	1 469	82 – 101
F	2 832	102 – 140
G	2 268	141 – 171
H	2 229	172 – 201
I	1 838	202 – 226
J	1 602	227 – 248
K	1 827	249 – 273

b)

Sub-chain (class) label	Number of documents assigned to sub-chain	Number of neuron in the chain
K	1 434	1 – 16
B	1 544	17 – 35
J	713	36 – 44
I	2 093	45 – 72
A	798	73 – 82
F	3 198	83 – 123
G	1 989	124 – 148
C	2 791	149 – 183
E	731	184 – 192
D	939	193 – 204
H	3 767	205 – 251

Table 5. Assignments of particular newsgroups to sub-chains (classes) for 20_newsgroups"Small" (a) and 20_newsgroups"Large" (b) data sets

a)

Sub-chain label	Name(s) of newsgroup(s) assigned to sub-chain
A	sci.space
B	rec.motorcycles rec.autos
C	comp.sys.mac.hardware comp.sys.ibm.pc.hardware
D	sci.crypt
E	sci.electronics
F	alt.atheism talk.religion.misc soc.religion.chrystian
G	talk.politics.mideast talk.politics.guns talk.politics.misc
H	comp.graphics comp.windows.x comp.os.ms-windows.misc
I	misc.forsale
J	sci.med
K	rec.sport.hockey rec.sport.baseball

b)

Sub-chain label	Name(s) of newsgroup(s) assigned to sub-chain
K	rec.sport.hockey rec.sport.baseball
B	rec.motorcycles rec.autos
J	sci.med
I	misc.forsale
A	sci.space
F	alt.atheism talk.religion.misc soc.religion.chrystian
G	talk.politics.mideast talk.politics.guns talk.politics.misc
C	comp.sys.mac.hardware comp.sys.ibm.pc.hardware
E	sci.electronics
D	sci.crypt
H	comp.graphics comp.windows.x comp.os.ms-windows.misc

Table 6. Results of comparative analysis for 20_newsgroups numerical models

Part I	Clustering method	Percentage of correct decisions	
		20_newsgroups"Small" (dimensionality of VSM: $(n \times L) = (232 \times 19\,997))$	20_newsgroups"Large" (dimensionality of VSM: $(n \times L) = (725 \times 19\,997)))$
	DSONN	56.14%	58.41%
	EM	47.52%	49.12%
	FFTA	27.60%	33.98%
	k-means	42.59%	48.12%

DSONN = Dynamic Self-Organizing Neural Network, EM = Expectation Maximization method, FFTA = Farthest First Traversal Algorithm

Part II	Clustering method	Modifications of 20_newsgroups document collection	
		Dimensionality $(n \times L)$ of VSM	Percentage of correct decisions
	COS	$28\,101 \times \sim 19\,216$	38.28%
	BOW	$28\,101 \times \sim 19\,216$	33.97%
	sIB	$2\,000 \times 17\,446$	57.50%
	sL1	$2\,000 \times 17\,446$	15.30%
	sKL	$2\,000 \times 17\,446$	28.80%
	k-means	$2\,000 \times 17\,446$	53.40%
	sk-means	$2\,000 \times 17\,446$	54.10%

COS = COncept Space representation for paragraphs method [2], BOW = simple Bag-Of-Words characterization paragraphs method [2], sIB = sequential Information Bottleneck approach [10], sL1 and sKL = variations of sIB approach [10], sk-means = sequential k-means (presented with k-means algorithm in [10]).

Taking into account the results that have been reported in this paper, it is clear that the clustering technique based on the dynamic self-organizing neural networks is a powerful tool for large-scale and highly multidimensional cluster-analysis problems such as WWW-newsgroup-document clustering. It provides better or much better accuracy of clustering than other alternative techniques applied in this field. Moreover, it is extremely important that the proposed technique automatically determines (adjusts in the course of learning) the number of clusters in a given data set. All the alternative approaches can operate under the condition that the number of clusters is set in advance.

5 Conclusions

The application of the clustering technique based on the dynamic self-organizing neural networks (introduced by the same authors in [7], [8] and some earlier papers) to the large-scale and highly multidimensional WWW-newsgroup-document clustering task has been reported in this paper. The collection of 19 997 documents (e-mail messages of different *Usenet-News* newsgroups) available at WWW server of the School of Computer Science, Carnegie Mellon University (www.cs.cmu.edu/ TextLearning/datasets.html) has been the subject of clustering. A broad comparative analysis with nine alternative clustering techniques has also been carried out demonstrating the superiority of the proposed approach in the considered task. Especially, it is worth emphasizing the ability of the proposed technique to automatically determine the number of clusters in the considered data set and high accuracy of clustering. The proposed technique has already been successfully applied to the clustering of other WWW-document collection [8] as well as several multidimensional data sets [7].

References

1. Berry, M.W.: Survey of Text Mining. Springer, New York (2004)
2. Caillet, M., Pessiot, J., Amini, M., Gallinari, P.: Unsupervised Learning with Term Clustering For Thematic Segmentation of Texts. In: Proc. of RIAO 2004 (Recherche d'Information Assiste par Ordinateur), Toulouse, France (2004)
3. Chakrabarti, S.: Mining the Web: Analysis of Hypertext and Semi Structured Data. Morgan Kaufmann Publishers, San Francisco (2002)
4. Franke, J., Nakhaeizadeh, G., Renz, I. (eds.): Text Mining: Theoretical Aspects and Applications. Physica Verlag/Springer, Heidelberg (2003)
5. Gorzałczany, M.B., Rudziński, F.: Application of Genetic Algorithms and Kohonen Networks to Cluster Analysis. In: Rutkowski, L., Siekmann, J.H., Tadeusiewicz, R., Zadeh, L.A. (eds.) ICAISC 2004. LNCS (LNAI), vol. 3070, pp. 556–561. Springer, Heidelberg (2004)
6. Gorzałczany, M.B., Rudziński, F.: Modified Kohonen Networks for Complex Cluster-Analysis Problems. In: Rutkowski, L., Siekmann, J., Tadeusiewicz, R., Zadeh, L.A. (eds.) ICAISC 2004. LNCS (LNAI), vol. 3070, pp. 562–567. Springer, Heidelberg (2004)
7. Gorzałczany, M.B., Rudziński, F.: Cluster Analysis Via Dynamic Self-organizing Neural Networks. In: Rutkowski, L., Tadeusiewicz, R., Zadeh, L.A., Żurada, J.M. (eds.) ICAISC 2006. LNCS (LNAI), vol. 4029, pp. 593–602. Springer, Heidelberg (2006)
8. Gorzałczany, M.B., Rudziński, F.: Application of dynamic self-organizing neural networks to WWW-document clustering, International Journal of Information Technology and Intelligent Computing, 1(1), 89-101 (2006) (also presented at 8th Int. Conference on Artificial Intelligence and Soft Computing ICAISC 2006, Zakopane)
9. Salton, G., McGill, M.J.: Introduction to Modern Information Retrieval. McGraw-Hill Book Co., New York (1983)
10. Slonim, N., Friedman, N., Tishby, N.: Unsupervised Document Classification using Sequential Informaiton Maximization. In: Proc. of the Twenty-Fifth Annual International ACM SIGIR Conference, Tampere, Finland, pp. 129–136 (2002)
11. Tang, B., Shepherd, M., Milios, E., Heywood, M.I.: Comparing and combining dimension reduction techniques for efficient text clustering. In: Proc. of Int. Workshop on Feature Selection and Data Mining, Newport Beach (2005)
12. Weiss, S., Indurkhya, N., Zhang, T., Damerau, F.: Text Mining: Predictive Methods for Analyzing Unstructured Information. Springer, New York (2004)
13. Zanasi, A. (ed.): Text Mining and its Applications to Intelligence, CRM and Knowledge Management. WIT Press, Southampton (2005)

Municipal Creditworthiness Modelling by Kohonen's Self-organizing Feature Maps and LVQ Neural Networks

Petr Hájek and Vladimír Olej

Institute of System Engineering and Informatics
Faculty of Economics and Administration
University of Pardubice
Studentská 84, 532 10 Pardubice
Czech Republic
`Petr.Hajek@upce.cz, Vladimir.Olej@upce.cz`

Abstract. The paper presents the design of municipal creditworthiness parameters. Further, a model is designed based on Learning Vector Quantization neural networks for municipal creditworthiness classification. The model is composed of Kohonen's Self-organizing Feature Maps (unsupervised learning) whose outputs represent the input of the Learning Vector Quantization neural networks (supervised learning).

Keywords: Municipal creditworthiness parameters, Kohonen's Self-organizing Feature Maps, Learning Vector Quantization neural networks, classification.

1 Introduction

Municipal creditworthiness [1], [2], [3] and [4] is the ability of a municipality to meet its short-term and long-term financial obligations. Its evaluation is based on factors (parameters) relevant to the assessed objects. Municipal creditworthiness evaluation is currently being realized by methods combining mathematical-statistical methods and expert opinion-scoring systems, rating, rating based models and default models [4]. An expert designs the parameters and the weights of the scoring systems. The creditworthiness is calculated with mathematical-statistical methods. Methods based on financial analysis [5], municipal budget and municipal economic environment [6] belong to the scoring systems. They are easy to calculate and understand. Nevertheless, they fail to classify correctly and do not work with expert knowledge. Municipal rating [7] is an independent expert evaluation based on complex analysis of all known municipal creditworthiness parameters but it is rather subjective. Rating based models [8] and [9] are intended to simulate the municipal creditworthiness evaluation process of rating agencies. Therefore, the class $\omega_{i,j}$ assigned to municipalities by the rating agencies represents the output variable. Municipalities are classified into classes $\omega_{i,j}$ with the objective of both high classification accuracy and key municipal rating parameters detection. However, low classification accuracy has been obtained so far due to the selection of inappropriate parameters or methods. Low number of municipal defaults makes the application of the default models difficult to carry out. The output of the introduced methods is represented either by a score (scoring systems) or by an assignment of the i-th object $o_i \in O$, $O=\{o_1, o_2, \ldots, o_i, \ldots, o_n\}$ to the j-th class $\omega_{i,j} \in \Omega$, $\Omega = \{\omega_{1,j}, \omega_{2,j}, \ldots, \omega_{i,j}, \ldots, \omega_{n,j}\}$.

Therefore, the methods capable of processing and learning the expert knowledge, enabling their user to generalize and properly interpret, have proved to be most suitable for municipal creditworthiness modelling. For example, hierarchical structures of fuzzy inference systems [2] and [3], unsupervised methods [1], [3], [4] and neuro-fuzzy systems [4] are suitable for municipal creditworthiness evaluation. Neural networks [10] seem to be appropriate due to their ability to learn, generalize and model non-linear relations. Municipal creditworthiness evaluation is considered a problem of classification, which can be realized by various models of neural networks. Classification can be realized by supervised methods (if classes $\omega_{i,j} \in \Omega$ are known) or unsupervised methods (if classes $\omega_{i,j} \in \Omega$ are not known).

The paper presents the design of municipal creditworthiness parameters. Only those parameters were selected which show low correlation dependences. Therefore, data matrix **P** is designed where vectors $\mathbf{p}_i$ characterize municipalities $o_i \in O$. Further, the paper presents the basic concepts of the Kohonen's Self-organizing Feature Maps (KSOFM) and Learning Vector Quantization (LVQ) neural networks. The contribution of the paper lies in the model design for municipal creditworthiness evaluation. The model realizes the advantages of both the unsupervised methods (combination of the KSOFM and K-means algorithm) and supervised methods (LVQ neural networks). The final part of the paper includes the analysis of the results, comparison to other classification methods and the presentation of the classification into classes $\omega_{i,j} \in \Omega$.

2 Municipal Creditworthiness Parameters Design

In [4] common categories of parameters are mentioned namely economic, debt, financial and administrative categories. Economic parameters affect long-term municipal creditworthiness. The municipalities with more diversified economy and favourable social and economic conditions are better prepared for the economic recession. Debt parameters include the size and structure of the debt. Financial parameters inform about the budget implementation. The design of parameters [1] and [4], based on previous correlation analysis and recommendations of notable experts, can be realized as presented in Table 1. The parameters x_3 and x_4 are defined in the r-th year and parameters x_5 to x_{12} as the average value of the r-th and (r-1)th years. Based on the presented facts, the following data matrix **P** can be designed

$$\mathbf{P} = \begin{array}{c|ccccc|c} & x_1 & \cdots & x_k & \cdots & x_m & \omega \\ \hline o_1 & x_{1,1} & \cdots & x_{1,k} & \cdots & x_{1,m} & \omega_{1,j} \\ \cdots & \cdots & \cdots & \cdots & \cdots & \cdots & \cdots \\ o_i & x_{i,1} & \cdots & x_{i,k} & \cdots & x_{i,m} & \omega_{i,j} \\ \cdots & \cdots & \cdots & \cdots & \cdots & \cdots & \cdots \\ o_n & x_{n,1} & \cdots & x_{n,k} & \cdots & x_{n,m} & \omega_{n,j} \end{array},$$

where:
- $o_i \in O$, $O=\{o_1, o_2, \ldots, o_i, \ldots, o_n\}$ are objects (municipalities),
- x_k is the k-th parameter,
- $x_{i,k}$ is the value of the parameter x_k for the i-th object $o_i \in O$,
- $\omega_{i,j}$ is the j-th class assigned to the i-th object $o_i \in O$,
- $\mathbf{p}_i = (x_{i,1}, x_{i,2}, \ldots, x_{i,k}, \ldots, x_{i,m})$ is the i-th pattern,
- $\mathbf{x} = (x_1, x_2, \ldots, x_k, \ldots, x_m)$ is the parameters vector.

Table 1. Municipal creditworthiness parameters design

	Parameters
Economic	$x_1 = PO_r$, PO_r is population in the r-th year.
	$x_2 = PO_r/PO_{r-s}$, PO_{r-s} is population in the year r-s, and s is the selected time.
	$x_3 = U$, U is the unemployment rate in a municipality.
	$x_4 = \sum_{i=1}^{e}(EP_i/TEP)^2$, EP_i is the employed population of the municipality in the i-th economic sector, i=1,2, … ,e, TEP is the total number of employed population, e is the number of the economic sector.
Debt	$x_5 = DS/PR$, $x_5 \in <0,1>$, DS is debt service, PR are periodical revenues.
	$x_6 = TD/PO$, TD is a total debt.
	$x_7 = STD/TD$, $x_7 \in <0,1>$, STD is short-term debt.
Financial	$x_8 = PR/CE$, $x_8 \in R^+$, CE are current expenditures.
	$x_9 = OR/TR$, $x_9 \in <0,1>$, OR are own revenues, TR are total revenues.
	$x_{10} = KE/TE$, $x_{10} \in <0,1>$, KE are capital expenditures, TE are total expenditures.
	$x_{11} = CR/TR$, $x_{11} \in <0,1>$, CR are capital revenues.
	$x_{12} = LA/PO$, [Czech Crowns], LA is the size of the municipal liquid assets.

3 Model Design for the Classification of Municipalities

The model realizes municipal creditworthiness modelling. Data pre-processing makes the suitable economic interpretation of results possible. The KSOFM assign municipalities to clusters. Subsequently, the clusters are labelled with classes $\omega_{i,j} \in \Omega$. The outputs from the KSOFM are used as the inputs of the LVQ neural networks. Municipal creditworthiness modelling represents a classification problem. It is generally possible to define it this way:

Let $F(\mathbf{x})$ be a function defined on a set A, which assigns picture $\hat{x}$ (the value of the function from a set B) to each element $\mathbf{x} \in A$, $\hat{x} = F(\mathbf{x}) \in B$, $F: A \rightarrow B$. The problem defined this way is possible to model by supervised methods (if classes $\omega_{i,j} \in \Omega$ of the objects are known) or by unsupervised methods (if classes $\omega_{i,j} \in \Omega$ are not known).

Table 2. Descriptions of classes $\omega_{i,j} \in \Omega$

Class $\omega_{i,j}$, j=1,2, ... ,7	Description
$\omega_{i,1}$	High ability of a municipality to meet its financial obligation. Very favorable economic conditions, low debt and excellent budget implementation.
$\omega_{i,2}$	Very good ability of a municipality to meet its financial obligation.
$\omega_{i,3}$	Good ability of a municipality to meet its financial obligation.
$\omega_{i,4}$	A municipality with stable economy, medium debt and good budget implementation.
$\omega_{i,5}$	Municipality meets its financial obligation only under favorable economic conditions.
$\omega_{i,6}$	A municipality meets its financial obligations with difficulty, the municipality is highly indebted.
$\omega_{i,7}$	Inability of a municipality to meet its financial obligation.

Several Czech municipalities have the class $\omega_{i,j} \in \Omega$ assigned by specialized agencies [4] (micro-region Pardubice, n=452). However, the descriptions of classes $\omega_{i,j} \in \Omega$ are known (Table 2).

Therefore, it is suitable to realize the modelling of municipal creditworthiness by unsupervised methods. Data pre-processing is carried out by means of data standardization. Thereby, the dependency on units is eliminated. Based on the analysis presented in [1] and [3], the combination of KSOFM and K-means algorithm is a suitable unsupervised method for municipal creditworthiness modelling. The LVQ neural networks use its results as the inputs in the model presented in Fig. 1. Frequencies f of municipalities in classes in both training and testing set are based on the method [11].

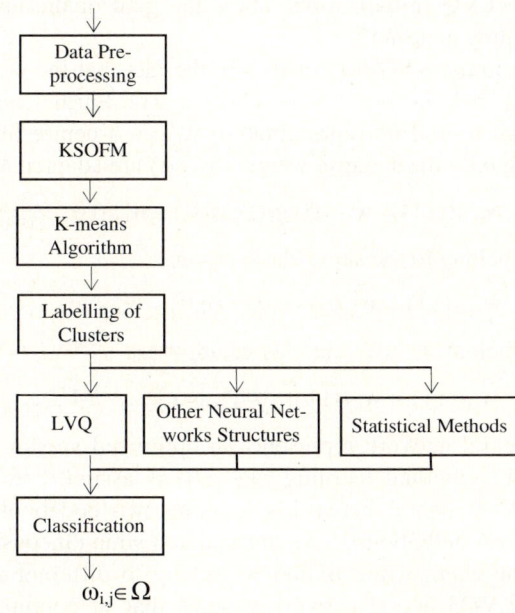

Fig. 1. Model for classification of municipalities into classes $\omega_{i,j} \in \Omega$

The KSOFMs [12] are based on competitive learning strategy. The input layer serves the distribution of the input patterns $\mathbf{p}_i = (x_{i,1}, x_{i,2}, \ldots, x_{i,k}, \ldots, x_{i,m})$. The neurons in the competitive layer serve as the representatives $\mathbf{w}_j \in R^n$, where $j=1,2,\ldots,M$ (Codebook Vectors). First, the Euclidean distances are computed between pattern $\mathbf{p}_i$ and weights $\mathbf{w}_{i,j}$ of all neurons in the competitive layer, where j goes over k neurons of the competitive layer, $\mathbf{p}_i$ is the i-th pattern and $\mathbf{w}_{i,j}$ are synapse weights. The index of the winning neuron j* (Best Matching Unit, (BMU)) is chosen, for which the Euclidean distance to the given pattern $\mathbf{p}_i$ is minimum [12]

$$j^* = \arg\min_j \left\{ \|\mathbf{p}_i - \mathbf{w}_{i,j}\| \right\}. \tag{1}$$

Synapse weights $\mathbf{w}_{i,j^*}$ of this neuron are adapted in order to approximate the i-th pattern $\mathbf{p}_i$. After the BMUs are found, the adaptation of synapse weights $\mathbf{w}_{i,j}$ follows. The principle of the sequential learning algorithm [12] is the fact, that the synapse weights $\mathbf{w}_{i,j^*}$ of the BMU and its topological neighbours (defined by a neighbourhood function $h(j^*,j)$) move towards the actual input vector $\mathbf{p}_i$ according to the relation

$$\mathbf{w}_{i,j}(t+1) = \mathbf{w}_{i,j}(t) + g(t) \times h(j^*, j) \times (\mathbf{p}_i(t) - \mathbf{w}_{i,j}(t)), \tag{2}$$

where $g(t) \in (0,1)$ is the learning rate. In [12] there are presented several versions of learning which refers to the structures of LVQ1, LVQ2, LVQ3 and OLVQ1 (Optimized Learning Vector Quantization) neural networks. They differ in the process of searching for the optimum boundaries between classes $\omega_{i,j} \in \Omega$. The LVQ neural networks are the supervised versions of the KSOFM. Let there is a LVQ1 neural network and the known number of classes $\omega_{i,j} \in \Omega$. Classes $\omega_{i,j}$ are assigned to all patterns $\mathbf{p}_i$ in the process of the LVQ initialization. Then, the goal of the learning process is the finding of the winning neuron j*.

The difference to the KSOFM consists in the fact that the process of learning finishes if $\mathbf{p}_i$ and $\mathbf{w}_{i,j^*}$ belong to the same class $\omega_{i,j} \in \Omega$. Further, let the input vector $\mathbf{p}_i$ belong to the class ω_p and its representative $\mathbf{w}_{i,j^*}$ is a centre of the class ω_q. In the process of learning only the synapse weights $\mathbf{w}_{i,j^*}(t)$ are adapted as follows

$$\mathbf{w}_{i,j^*}(t+1) = \mathbf{w}_{i,j^*}(t) + g(t) \times (\mathbf{p}_i(t) - \mathbf{w}_{i,j^*}(t)), \tag{3}$$

if $\mathbf{p}_i(t)$ and $\mathbf{w}_{i,j^*}(t)$ belong to the same class, $\omega_q = \omega_p$,

$$\mathbf{w}_{i,j^*}(t+1) = \mathbf{w}_{i,j^*}(t) - g(t) \times (\mathbf{p}_i(t) - \mathbf{w}_{i,j^*}(t)), \tag{4}$$

if $\mathbf{p}_i(t)$ and $\mathbf{w}_{i,j^*}(t)$ belong to different classes, $\omega_q \neq \omega_p$,

$$\mathbf{w}_{i,j}(t+1) = \mathbf{w}_{i,j}(t) \text{ for } j \neq j^*, j=1,2,\ldots,M. \tag{5}$$

The OLVQ1 neural network represents an optimized version of the LVQ1 neural network where an individual learning rate $g_i(t)$ is assigned to each $\mathbf{w}_i$. Within the process of the LVQ2 neural network's learning two codebook vectors $\mathbf{w}_i$ and $\mathbf{w}_j$, which are the nearest neighbours to $\mathbf{p}_i$ are updated simultaneously. Vectors $\mathbf{p}_i$ and $\mathbf{w}_j$ belong to the same class, while $\mathbf{p}_i$ and $\mathbf{w}_i$ belong to different classes. The learning algorithm of the LVQ3 neural network ensures that $\mathbf{w}_i$ continue approximating the class distributions.

4 Analysis of the Results

The input parameters of the designed KSOFM are based on a number of experiments and are specified in Table 3, where λ(t) is the size of the neighbourhood in time t. Using the KSOFM as such can detect the data structure. The K-means algorithm can be applied to the adapted KSOFM in order to find clusters as presented in Fig. 2.

Table 3. Input parameters of the KSOFM

Parameter	Initialization	h(j*,j)	Initial λ(t)	Final λ(t)	g(t)	Epochs
Value	Linear	Bubble	10	1	0.01	10000

The K-means algorithm belongs to the non-hierarchical algorithms of cluster analysis, where patterns $\mathbf{p}_i$ are assigned to clusters $\varphi_1, \varphi_2, \ldots, \varphi_z$. Interpretation of clusters is realized by the values of parameters $x_1, x_2, \ldots, x_m$, m=12, for individual representatives $\mathbf{w}_j$ (Fig. 3). The clusters interpretation makes it possible to label the clusters $\varphi_1, \varphi_2, \ldots, \varphi_z$, z=7, by the classes $\omega_{i,j} \in \Omega$, j=7, based on descriptions presented in Table 2.

Fig. 2. Clustering of the KSOFM by K-means algorithm

We designed number of the KSOFMs structures with various input parameters in the process of modelling. The specific characteristic of the KSOFM lies in the fact that it makes possible to realize the representation, which preserves the topology and characteristics of the training set. For this purpose, the neurons are ordered in a regular, mostly two-dimensional or one-dimensional structure. This structure represents the output space, where the distance of neurons is computed as the Euclidean distance of their vectors coordinates. The projection preserving the topology of the adapted KSOFM has the following important feature. Any pair of patterns $\mathbf{p}_u$ and $\mathbf{p}_v$, which are nearby in the input space, evokes the responses of the KSOFM neurons, which are also nearby in the output space. The input parameters of the LVQ neural networks' structure are presented in Table 4, where Eveninit is an equal initialization, Propinit is a proportional initialization, α is the number of codebook vectors, NN is the number of neighbours used in the K-Nearest Neighbour (KNN) classification, $\beta \in (0,1)$ is the width of the window and $\delta \in (0,1)$ is the stabilizing constant factor. Again, we designed loads of the LVQ1, LVQ2, LVQ3 and OLVQ1 structures with various input parameters. Finally, we obtained the best results with the input parameters presented in Table 4.

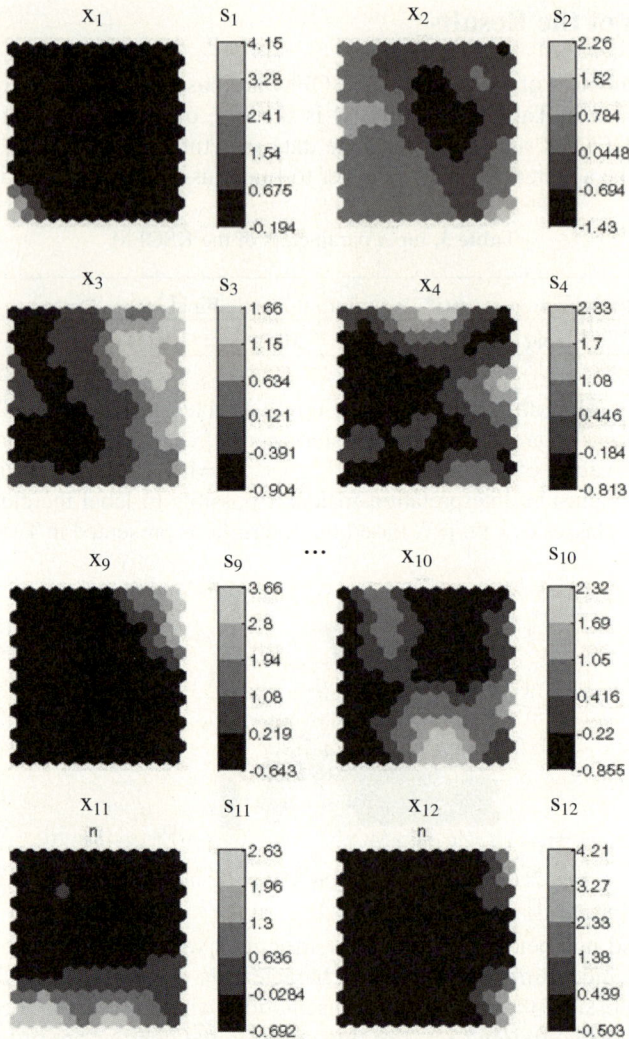

Fig. 3. Values of the parameters vectors **x** for individual representatives, $x_1, x_2, \ldots, x_k, \ldots, x_m$, m=12, are parameters of rating, s_k is a scale of the k-th parameter's standardized values

Table 4. Input parameters of the LVQ neural networks

Structure	Initialization	α	NN	g(t)	β	δ	Epochs
LVQ1	Propinit	200	5	0.05	-	-	10000
LVQ2	Eveninit	200	5	0.05	0.3	-	100
LVQ3	Eveninit	200	5	0.05	0.3	0.1	1000
OLVQ1	Propinit	200	5	-	-	-	1000

The frequencies f of municipalities in classes $\omega_{i,j} \in \Omega$ for the LVQ1 and OLVQ1 neural networks are presented in Fig. 4.

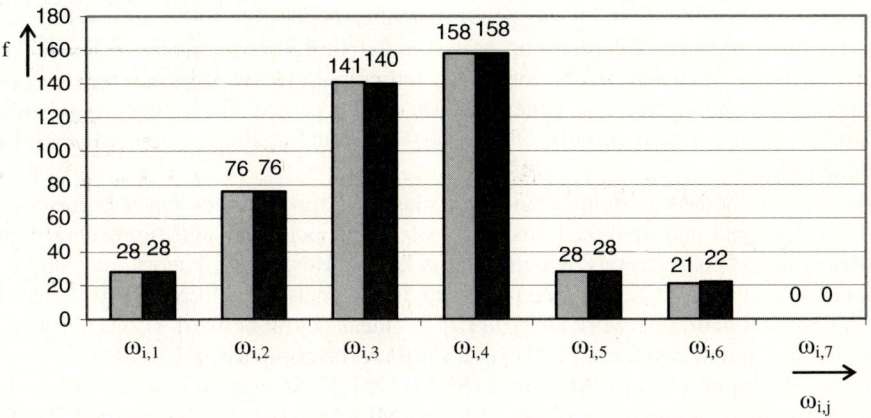

Fig. 4. Frequencies f of municipalities in classes $\omega_{i,j} \in \Omega$ for the LVQ1 (gray) and OLVQ1 (black) neural network

The results are similar. However, neither the LVQ1 nor the OLVQ1 neural network classified municipalities into the class $\omega_{i,j} \in \Omega$, j=7. The LVQ1 and OLVQ1 neural networks have the best results of all the LVQ neural networks concerning the testing set, see Table 5. The LVQ1 neural network has the maximum classification accuracy ξ_{max}=92.92[%], the average classification accuracy ξ_a=91.33[%] and the standard deviation SD=0.97[%]. We did not obtain better results even after the application of the LVQ2 and LVQ3 training algorithms on the results of the LVQ1 neural network.

Table 5. Classification accuracy ξ [%] on testing data by the LVQ neural networks

	LVQ1	LVQ2	LVQ3	OLVQ1
ξ_{max}[%]	92.92	92.04	92.04	92.92
ξ_a[%]	91.33	89.91	90.09	90.44
SD[%]	0.97	1.61	1.45	1.45

The LVQ neural networks adjust the synapse weights **w** in order to minimize the number of misclassifications coming from the classes $\omega_{i,j} \in \Omega$ overlap. The classification problem works with the set of input patterns $\mathbf{p}_i$ assigned to one of the classes $\omega_{i,j} \in \Omega$. The classifier chooses one of the classes $\omega_{i,j} \in \Omega$ for the given pattern $\mathbf{p}_i$. The classification of the pattern $\mathbf{p}_i$ is based on the label of its nearest synapse weight **w**, which represents an assignment to a class $\omega_{i,j} \in \Omega$. Contrary to the problem of vector quantization, it is not important which of the neurons is the winner. What matters is the fact that the winner should belong to one of the neurons representing the correct class $\omega_{i,j} \in \Omega$. The questionable situations can rise exactly in the areas where the classes neighbour.

The results of the designed model show the possibility of evaluating municipal creditworthiness of the given municipalities in years to come. The visualization of the municipal creditworthiness by the KSOFM makes it possible to monitor the municipalities' structure and the relations between the designed parameters. Further, the model presents an easier conception of the municipal creditworthiness for the public administration managers. The generalization of the gained knowledge (LVQ neural networks) makes it also possible to classify the municipalities not involved in the training process.

In Table 6, there is a comparison of the classification accuracy ξ [%] on testing set to other designed and analyzed structures of neural networks and representatives of statistical models. Concretely, we used an Adaptive Resonance Theory and Map field (ARTMAP) [13], a standard Feed-forward neural network (FFNN) [10], a Radial Basis Function neural network (RBF) [10], a Linear neural network (LNN) [10] and a Probabilistic neural network (PNN) [14]. Further, the comparison is realized to statistical models Support Vector Machines (SVM) [15], K-Nearest Neighbour (KNN) [16] and Multinomial Logistic Regression Model (MLRM) [16]. As presented in Table 6, the MLRM shows the worst results on testing set with the maximum classification accuracy ξ_{max}=86.73[%], the average classification accuracy ξ_a=81.42[%] and the standard deviation SD=5.31[%]. The RBF represents excellent results with the maximum classification accuracy ξ_{max}=94.69[%], the average classification accuracy ξ_a=89.93[%] and the standard deviation SD=2.88[%].

Table 6. Classification accuracy ξ [%] on testing data by other structures

	ARTMAP	FFNN	RBF	LNN	PNN	SVM	KNN	MLRM
ξ_{max}[%]	93.36	92.04	94.69	85.84	85.84	91.15	90.27	86.73
ξ_a[%]	90.34	90.56	89.93	84.60	83.34	89.76	87.46	81.42
SD[%]	3.81	1.33	2.88	0.79	1.89	1.83	3.38	5.31

5 Conclusion

The paper presents the design of municipal creditworthiness parameters. Further, the model design realizes municipal creditworthiness evaluation. First, the data are pre-processed. Next, the KSOFM realizes modelling of municipal creditworthiness. The previous analysis of unsupervised methods (KSOFM, ART-type neural networks, cluster analysis and fuzzy cluster analysis) [17] showed that the KSOFM is the most suitable one for municipal creditworthiness modelling. Namely, the visualization of clusters by the KSOFM makes the proper economic interpretation of results possible. Clustering quality indexes [18] are favourable and outlying objects do not affect the results of the KSOFM in addition. The outputs of the KSOFM are used as the inputs of the LVQ neural networks. The LVQ neural networks structures were designed and studied for the classification of municipalities into classes $\omega_{i,j} \in \Omega$ due to its high maximum classification accuracy ξ_{max}[%] and average classification accuracy ξ_a[%] with a low standard deviation SD[%]. Classification by the LVQ neural networks was carried out in LVQ_PAK, other models of neural networks in MATLAB 7.1 and statistical models in Weka 3.4.

Acknowledgements. This work was supported by the scientific research of Czech Science Foundation, under Grant No: 402/08/0849 with title Model of Sustainable Regional Development Management.

References

1. Olej, V., Hájek, P.: Modelling of Municipal Rating by Unsupervised Methods. In: WSEAS Transactions on Systems, vol. 6(7), pp. 1679–1686. WSEAS Press (2006)
2. Olej, V., Hájek, P.: Hierarchical Structure of Fuzzy Inference Systems Design for Municipal Creditworthiness Modelling. In: WSEAS Transactions on Systems and Control, vol. 2(2), pp. 162–169. WSEAS Press (2007)
3. Hájek, P., Olej, V.: Municipal Creditworthiness Modelling by means of Fuzzy Inference Systems and Neural Networks. In: 4th International Conference on Information Systems and Technology Management, TECSI-FEA USP, Sao Paulo, Brazil, pp. 586–608 (2007)
4. Hájek, P.: Municipal Creditworthiness Modelling by Computational Intelligence Methods. Ph.D. Thesis, University of Pardubice (2006)
5. Mead, D.M.: Assessing the Financial Condition of Public School Districts. Selected Papers in School Finance, National Center for Education Statistics, Washington DC (2001)
6. Mercer, T.A.: Financial Condition Index for Nova Scotia Municipalities. Government Finance Review 12(5), 36–39 (1996)
7. Miller, G.J.: Handbook of Debt Management. Marcel Dekker, New York (2003)
8. Serve, S.: Assessment of Local Financial Risk: The Determinants of the Rating of European Local Authorities-An Empirical Study Over the Period 1995-1998. EFMA Lugano Meetings, Lugano (2001)
9. Ammar, S., Duncombe, W., Hou, Y., Jump, B., Wright, R.H.: Using Fuzzy Rule-Based Systems to Evaluate Overall Financial Performance of Governments: An Enhancement to the Bond Rating Process. Public Budgeting and Finance 21(4), 91–110 (2001)
10. Haykin, S.S.: Neural Networks: A Comprehensive Foundation. Prentice-Hall, Upper Saddle River (1999)
11. Kvasnička, V., et al.: Introduction to Neural Networks. Iris, Bratislava (1997) (in Slovak)
12. Kohonen, T.: Self-Organizing Maps. Springer, New York (2001)
13. Carpenter, G.A., Grossberg, S., Reynolds, J.H.: ARTMAP: Supervised Real-Time Learning and Classification of Nonstationary Data by a Self-organizing Neural Network. Neural Networks 4(5), 565–588 (1991)
14. Speckt, D.F.: Probabilistic Neural Networks. Neural Networks 3(1), 109–118 (1990)
15. Cristianini, N., Shawe-Taylor, J.: Introduction to Support Vector Machines and Other Kernel-based Learning Methods. Cambridge University Press, Cambridge (2000)
16. Bishop, C.M.: Pattern Recognition and Machine Learning. Springer, Cambridge (2006)
17. Hájek, P., Olej, V.: Municipal Creditworthiness Modelling by Clustering Methods. In: Margaritis, Illiadis (eds.) 10th International Conference on Engineering Applications of Neural Networks, EANN 2007, Thessaloniky, Greece, pp. 168–177 (2007)
18. Stein, B., Meyer zu Eissen, S., Wissbrock, F.: On Cluster Validity and the Information Need of Users. In: International Conference on Artificial Intelligence and Applications (AIA 2003), Benalmádena, Spain, pp. 216–221 (2003)

Fast and Robust Way of Learning the Fourier Series Neural Networks on the Basis of Multidimensional Discrete Fourier Transform

Krzysztof Halawa

Institute of Computer Engineering, Control and Robotics,
Wrocław University of Technology,
Wybrzeże Wyspiańskiego 27, 50-370 Wrocław, Poland
`krzysztof.halawa@pwr.wroc.pl`

Abstract. The calculation method for weights of orthogonal Fourier series neural networks on the grounds of multidimensional discrete Fourier transform is presented. The method proposed represents high speed of operation and outlier robustness. It allows easy reduction of network structure following its training process. The paper presents also the ways of applying the method to modelling of dynamic controlled systems. It is very easy to prepare a program which would allow to use the procedure proposed.

1 Introduction

Fourier Series Neural Networks (FSNNs) represent one type of Orthogonal Neural Networks (ONN). A few works are covering this subject. Most of them appeared in the last decade. The FSNNs were presented, inter alia, in papers of [15], [12], [13] and [10]. These networks have many essential advantages including the following:

- fast convergence of the learning process with gradient descent algorithm caused by an absence of local minima for many popular cost functions, e.g. for mean squared error (MSE),
- relation between the number of inputs and outputs on one side and the maximum number of orthogonal neurons on the other side is well known,
- output is a linear function of weights, which are changed during a process of learning,
- initial values of weights are not especially important,

Inputs of FSNN are connected with neurons. These connections are associated with weights. Their values are constant and they do not change during network training. Signal from i-th input, multiplied by the aforementioned weights, stimulates $N_i/2 - 1$ neurons which activation function is the sine, and the same number of neurons which activation function is cosine, where N_i is a natural even number. The values of these weights equal to $1, 2, \ldots, N_i/2 - 1$. The multiplying nodes calculate the products of all possible combinations of outputs from the neurons and the bias equal to one. In case of Multiple-Input Single-Output Network (MISO), these nodes activate one linear neuron. Its weights change during network training process. The network output is equal to the sum calculated by this neuron. Construction of MISO Network is shown in Figure 1. The value of network output is given by the relation

$$y = [w_1, w_2, \ldots, w_Z] \cdot \left(\begin{bmatrix} 1 \\ \sin(x_1) \\ \cos(x_1) \\ \sin(2x_1) \\ \cos(2x_1) \\ \vdots \\ \sin((N_1/2-1)x_1) \\ \cos((N_1/2-1)x_1) \end{bmatrix} \otimes \begin{bmatrix} 1 \\ \sin(x_2) \\ \cos(x_2) \\ \sin(2x_2) \\ \cos(2x_2) \\ \vdots \\ \sin((N_2/2-1)x_2) \\ \cos((N_2/2-1)x_2) \end{bmatrix} \otimes \ldots \otimes \begin{bmatrix} 1 \\ \sin(x_S) \\ \cos(x_S) \\ \sin(2x_S) \\ \cos(2x_S) \\ \vdots \\ \sin((N_S/2-1)x_S) \\ \cos((N_S/2-1)x_S) \end{bmatrix} \right) \quad (1)$$

where $\otimes$ denotes Kronecker product, S is the number of network inputs, $Z = \prod_{i=1}^{S}(N_i - 1)$, $w_1, w_2, \ldots, w_Z$ are the weights of linear neuron, $x_1, x_2, \ldots, x_S$ are network inputs, y is the network output.

The comparison of learning speeds between the two-inputs-and-one-output FSNN having 49 harmonic neurons, which was trained with a gradient descent algorithm and the 2-18-1 sigmoidal network, which was trained with the Levenberg-Marquard, method is depicted in Fig. 2. The networks were trained to represent the function $f(x_1, x_2) = \tanh(0.8x_1 + x_2 - 6) - \tanh(1.25x_1 + 1.5x_2 - 9)$, where tanh means the hyperbolic tangent. The set of learning data was composed of the pairs $\{y_m, [x_{1m}, x_{2m}]^T\}_{m=1}^{1000}$, where $y_m = f(x_1, x_2) + 0.5\varepsilon_m$, x_{1m}, x_{2m} were the realizations of random variable with uniform distribution over $[0, 2\pi)$, ε_m were random variables having the standard normal distribution. The results of learning were evaluated on the testing set including 2,500 elements. It was generated in the analogous way as that for the training set. The initial values of FSNN weights were equal to 1. Sigmoidal neural networks required much more learning epochs to reach the same MSE value.

The paper presents an effective way allowing to determine the weights $w_1, w_2, \ldots, w_Z$. It allows to reduce the effect of outliers. The method makes use of S-dimensional discrete Fourier transform (DFT), which shall be calculated by means of fast Fourier transform (FFT) [2], [14]. Similar application of FFT for single-input FSNNs was given in [11]. The remaining contents of this

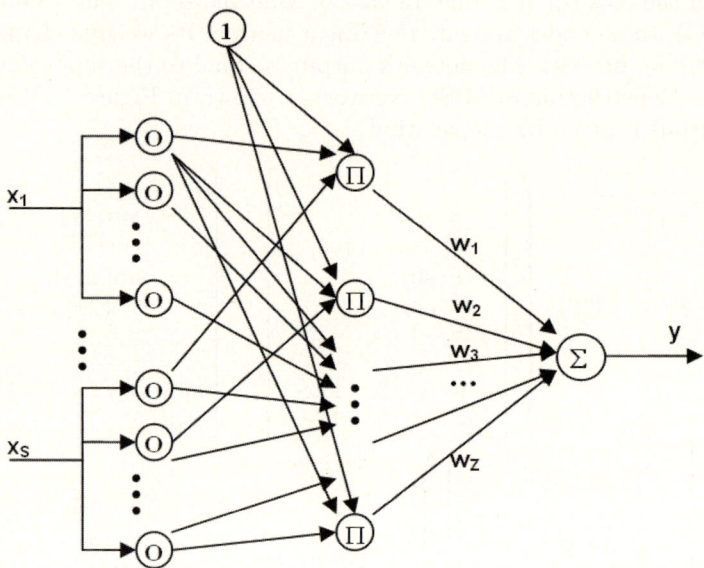

Fig. 1. A structure of MISO FSNNs (O - represents orthogonal neurons, Π - denotes multiplying nodes, Σ - is the linear neuron, 1 - denotes the bias)

paper is arranged as follows: Section 2 describes how to determine the weights using the S-dimensional DFT. Three ways of using the proposed algorithm when input data are not evenly distributed are also shown there. Section 3 deals with computational complexity of the method proposed.

2 Calculation of Weights on the Basis of S-Dimensional DFT

Below, there is the way how to determine the weights $w_1, w_2, \ldots, w_Z$ basing on the S-dimensional Fourier transform, so as the network approximate the function $f : \mathcal{R}^S \to \mathcal{R}$ which is piecewise continuous and which meets Dirichlet conditions. Let's $U_m \in \mathcal{R}^S$, $m = 1, 2, \ldots, \prod_{i=1}^{S} N_i$ are all points evenly distributed in the hypercube, the i-th Cartesian coordinates of which may take the values $\left\{0, 1 \cdot \frac{2\pi}{N_i}, 2 \cdot \frac{2\pi}{N_i}, \ldots, (N_i - 1) \cdot \frac{2\pi}{N_i}\right\}$. The method presented may be directly applied when for each U_m, there is the set M_m of the values $y_{m,1}, \ldots, y_{m,M_m}$, where

$y_{m,a} = f(U_m) + \varepsilon_{m,a}$,
$a = 1, \ldots, M_m$,
M_m are any natural numbers ≥ 1,
$\varepsilon_{m,a}$ are independent and identically distributed (i.i.d.) random variables with zero expected value and finite variance.

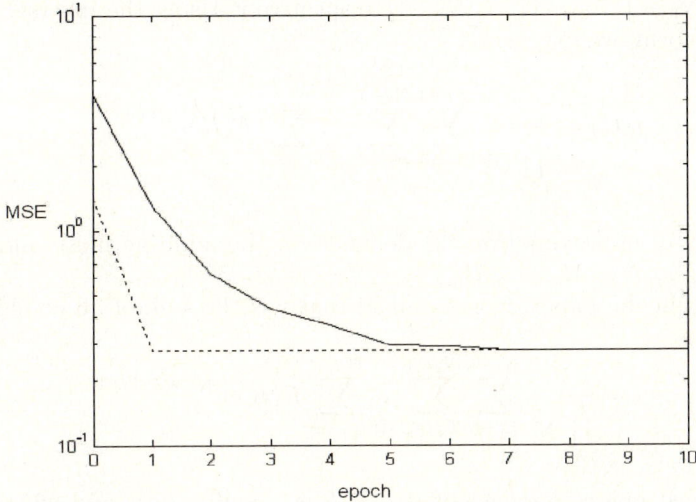

Fig. 2. Learning speed comparison for FSNN (dotted line) and sigmoidal 2-18-1 network (solid line) which was trained by the Levenberg-Marquardt method

When the vectors U_m are not evenly distributed, the necessary modifications to be made are outlined at the end of this section. Such situation is mainly the case during passive experiment and while modelling dynamic controlled system. The first step of the method consists in calculating, for each U_m, the arithmetic mean $\overline{f}(U_m) = (y_{m,1} + \cdots +, y_{m,M_m})/M_m$. In a case when among the values $y_{m,1}, \ldots, y_{m,M_m}$ there may appear outliers, it is proposed that $\overline{f}(U_m)$ equals to $\mathrm{med}(y_{m,1}, \ldots, y_{m,M_m})$, where med stands for the median. We may also use the Winsorized mean for $\overline{f}(U_m)$. For this purposes, for each m, we sort out the values $y_{m,1}, \ldots, y_{m,M_m}$ in ascending or descending order. Then, we reject some highest and smallest values and calculate the arithmetic mean for the remaining values. It is suggested to determine the median or the Winsorized mean first. Then, we calculate the arithmetic mean of those $y_{m,1}, y_{m,2}, \ldots, y_{m,M_m}$, which satisfy the condition: $y_{m,1} - c < w_m < y_{m,1} + c, \ldots, y_{m,M_m} - c < w_m < y_{m,M_m} + c$ where w_m is the median or the Winsorized mean and c is a positive constant. We may use a priori knowledge when choosing c. In this manner all data except for outliers can be used.

The S-dimensional DFT calculated from $\overline{f}$ is defined as follows

$$F(K) = \sum_{u_1=0}^{N_1-1} \sum_{u_2=0}^{N_2-1} \cdots \sum_{u_S=0}^{N_S-1} \overline{f}(U) e^{-j2\pi(k_1 u_1/N_1 + k_2 u_2/N_2 + \ldots + k_S u_S/N_S)}, \qquad (2)$$

where $j = \sqrt{-1}$, $K = [k_1, k_2, \ldots, k_S]^T$, $U = [u_1 \cdot 2\pi/N_1, u_2 \cdot 2\pi/N_2, \ldots, u_S \cdot 2\pi/N_S]^T$, $u_1, u_2, \ldots, u_S$ and $k_1, k_2, \ldots, k_S$ are non-negative integers, less than or equal to $N_1 - 1, N_2 - 1, \ldots, N_S - 1$ respectively. Using the inverse, discrete Fourier transform, we may write

$$f(U) \approx \frac{1}{\prod\limits_{i=1}^{S} N_i} \sum_{k_1=0}^{N_1-1} \sum_{k_2=0}^{N_2-1} \cdots \sum_{k_S=0}^{N_S-1} F(K) e^{jU^T K}. \tag{3}$$

The accuracy of approximation (3) depends on the additive noise and on the numbers $N_1, \ldots, N_S, M_1, \ldots, M_m$.

Further on in the paper, it is assumed that r is the sum of all combinations of S-fold sums

$$\frac{1}{\prod\limits_{i=1}^{S} N_i} \sum_{k_1=0}^{N_1-1} \sum_{k_2=0}^{N_2-1} \cdots \sum_{k_S=0}^{N_S-1} F(K) e^{jU^T K},$$

wherein at least one component of vector K is equal to zero and wherein sum symbols corresponding to zero-components of vector K are passed over. Equation (3) may be presented as

$$f(U) \approx \frac{1}{\prod\limits_{i=1}^{S} N_i} \sum_{k_1=0}^{N_1/2-1} \sum_{k_2=0}^{N_2/2-1} \cdots \sum_{k_S=0}^{N_S/2-1} F(\Psi_1) e^{jU^T \Psi_1} + F(\Psi_2) e^{jU^T \Psi_2}$$
$$+ \cdots + F(\Psi_{2^S}) e^{jU^T \Psi_{2^S}} \tag{4}$$

where the vectors Ψ_n are created as follows:
a) $n-1$ is written down in binary code as the S-bit number,
b) if the i-th digit of the binary number equals to one, the i-th element of the vector Ψ_n is $N_i - k_i$, otherwise it equals to k_i.
Example $S = 4, n = 3, n-1$ in binary notation is 0010, $\Psi_n = [k_1, k_2, N_3-k_3, k_4]^T$.

Let's notice that

$$e^{-j2\pi((N_1-k_1)u_1/N_1+(N_2-k_2)u_2/N_2+\cdots+(N_S-k_S)u_S/N_S)}$$
$$= \left(e^{-j2\pi(k_1 u_1/N_1+k_2 u_2/N_2+\cdots+k_S u_S/N_S)}\right)^*,$$
$$e^{-j2\pi(k_1 u_1/N_1+(N_2-k_2)u_2/N_2+\cdots+(N_S-k_S)u_S/N_S)}$$
$$= \left(e^{-j2\pi((N_1-k_1)u_1/N_1+k_2 u_2/N_2+\cdots+k_S u_S/N_S)}\right)^*, \tag{5}$$
$$\cdots,$$

where * represents the conjugate number.
Coefficients of DFT feature the property of symmetry, i.e.

$$F(N_1 - k_1, k_2, \ldots, k_S) = F^*(k_1, N_2 - k_2, \ldots, N_S - k_S),$$
$$F(N_1 - k_1, N_2 - k_2, \ldots, k_S) = F^*(k_1, k_2, \ldots, N_S - k_S), \tag{6}$$
$$\cdots$$

Making use the observation (5), the symmetry property of FFT coefficients and the relation $(ab)^* = (a^*)(b^*)$, the equation (4) may be transformed to the form

$$f(U) \approx r + \frac{2}{\prod_{i=1}^{S} N_i} \cdot \sum_{k_1=1}^{N_1/2-1} \sum_{k_2=1}^{N_2/2-1} \cdots \sum_{k_S=1}^{N_S/2-1} Re\{F(k_1, k_2, \ldots, k_S)\}$$
$$\cdot \cos(p_1 k_1 u_1 + p_2 k_2 u_2 + \cdots + p_S k_S u_S)$$
$$- Im\{F(k_1, k_2, \ldots, k_S)\} \sin(p_1 k_1 u_1 + p_2 k_2 u_2 + \cdots + p_S k_S u_S) \qquad (7)$$
$$+ Re\{F(k_1, k_2, \ldots, N_S - k_S)\}$$
$$\cdot \cos(p_1 k_1 u_1 + p_2 k_2 u_2 + \cdots + p_{S-1} k_{S-1} u_{S-1} - p_S k_S u_S)$$
$$- Im\{F(k_1, k_2, \ldots, N_S - k_S)\}$$
$$\cdot \sin(p_1 k_1 u_1 + p_2 k_2 u_2 + \cdots + p_{S-1} k_{S-1} u_{S-1} - p_S k_S u_S)$$
$$+ \cdots$$
$$+ Re\{F(k_1, N_2 - k_2, \ldots, N_S - k_S)\} \cos(p_1 k_1 u_1 - p_2 k_2 u_2 - \cdots - p_S k_S u_S)$$
$$- Im\{F(k_1, N_2 - k_2, \ldots, N_S - k_S)\} \sin(p_1 k_1 u_1 - p_2 k_2 u_2 - \cdots - p_S k_S u_S).$$

where $p_i = 2\pi/N_i$, $i = 1, 2, \ldots, S$. On the basis of (7), the weights may be effectively calculated independently of the value of S. It is enough to express the sine and cosine functions, whose arguments are the weighted sums of several inputs in (7), by means of the sum of the products of sine and cosine functions whose arguments depends on individual inputs only. Then, the weights will be equal to the sums of coefficients standing before identical functions having the same arguments.

The method presented may not be directly used when the data are not placed in the considered points of the hypercube. Such a situation exists during passive experiments. It also takes place during modelling of dynamic controlled systems when we use the values of system outputs from previous time moments. In such a case, one of the following methods is proposed:

Method 1. Making use of the separability feature of a multi-dimensional FFT, we modify the way of calculating single-dimensional FFTs in those dimensions where the data are not equidistantly distributed. For instance, we may use Nonuniform Fast Fourier Transform Algorithm (NUFT) which complexity is $O(Nlog_2 N)$. NUFT was presented, inter alia, in [4], [7].

Method 2. We divide the hypercube with input data into smaller separable hypercubicoids. All data belonging to specific hypercubicoid subject to averaging. Then, the multi-dimensional FFT is calculated on the basis of this data. This method requires the data set to be adequately size.

Method 3. The hypercube is divided as in the previous method. For each hypercubicoid, we calculate the coefficient of hypersurface approximating f over it using the least squares method [5], [1]. For approximation, we may use, for instance, a hyperplane or a hyperparabola. It is recommended not to use complex hypersurfaces as the number of parameters to be determined rises equipotentially with S. Then, the values of hypersurface in the considered points are calculated. They are used to determine the multi-dimensional FFT. This method requires more calculation steps than the Method 2, however it allows to reach better results. To determine the parameters of hypersurface, we need to make $O(N_h \cdot n_h + n_h^3)$ operations, where N_h is the number of data used to calculate the n_h coefficients of the hypersurface.

It is worth noticing that Methods 2 and 3 ensure good properties of network generalization, even for $M_1 = 1, M_2 = 1, \ldots$ without reducing its size after training is completed. It results from the fact that the number of generated hypercuboids is much less than the number of data items. It is possible to use several methods at the same time for various dimensions. When the values of network wages are calculated, we may easily reduce the network size. This way we may improve network generalization properties. To this purpose, it is proposed to remove some multiplying nodes and associated connections. Let it be $\phi_i = |w_i| \cdot (\sqrt{\pi})^{S-\alpha} \cdot (\sqrt{2\pi})^\alpha$, $i = 1, \ldots, Z$ where α denotes the number of neurons connected to inputs of the multiplying node which is matched with the weight w_i. The least mean square error will be reached when we remove the multiplying nodes corresponding to the least values of ϕ_i. Upon sorting out ϕ_i, the weights related to the least values of ϕ_i shall be removed until a good trade-off is reached between the bias and variance of the neural model attained.

3   Computational Complexity

For the sake of conciseness of a notation, it was assumed in this section that $N_1 = N_2 = \cdots = N_S = N$, where N is any natural number. We can easily generalize the following considerations for the case when $N_1, N_2, \ldots, N_S$ may differ each other.

Computational complexity of the FFT algorithm calculating N-point single-dimension DFT is $O(N \log_2 N)$. It is recommended to use FFT algorithms intended for real data, e.g. the 2N-Point Real FFT. The S-dimensional DFT may be found by calculating SN^{S-1} single-dimension DFTs. Computational complexity of determining the S-dimensional DFT is $O(SN^S \log_2 N)$. In order to determine, on its basis, the $(N-1)^S$ weights, we need to find the sum of 2^{S-1} real numbers for each weight. Thus, the total computational complexity is $O\left(N^S \log_2 N + (2N)^S\right)$. It is much less than that when weights are determined using the least square method. To calculate the weight using scalar products $\langle \overline{f}, g_k \rangle$ where g_k is the signal from the output of k-th multiplying node, $k = 1, \ldots, (N-1)^S$, it would be necessary to make as much as $O(N^{2S})$ multiplications and additions of complex numbers. Single-dimension FFT algorithms are the most effective when data size may be expressed in a form of a short product of prime numbers. The best situation is when N is a power of 2. The papers provides no description of well known procedures for other data sizes, such as, for instance, complementing the input data with zeros. If they are used, relevant modifications shall be made to equation (7).

4 Summary

The main attributes of the method shown here are its low computational complexity and potential to reduce outliers effects. It may be used in modelling of dynamic controlled systems. Software calculating FFT is easily accessible. In order to find a multi-dimensional FFT, its is enough to calculate single-dimension FTTs repeatedly. The drawback of ONN is that when data size grows, the number of weights rises exponentially. Dimension reduction methods may be used to reduce the input data size. In this context, a combination of self-organizing neural networks, PCA or ICA types [8] and ONNs seems to be worthy of special interest. An example of PCA network finding one principal component only is the neural network composed of a single linear neuron. Its weights may be selected according to the normalized Hebb's rule which is called the Oji rule. A single-layer network composed of k linear neurons is used to estimate k principal components. Good results are achieved when training is made using Sanger rule. To find independent components [6], a recurrent network with linear neurons may be applied. Its training is recommended with the algorithm of A.Cichocki which is effective even for large diversification of amplitudes for individual signals [8]. If the network is trained to approximate on hypercube with side length of 2π, then, according to the experiment designing theory [9], the smallest volumes of multi-dimensional confidence intervals for weights under estimation will be attained when the data are evenly distributed on this hypercube. This statement stems from the fact that in order a multi-dimensional design be D-optimal on this hypercube, it must be composed of the product of D-optimal designs in each dimension. In case of trigonometric regression, the designs are D-optimal in particular dimensions when their points are equidistant each other by and all weights of the design are equal to $1/N$. The square root of the volume of confidence hyperellipsoid of parameter estimations is proportional to the determinant of the design information inverse matrix. If the disturbance variance is constant over the whole designing area (in the whole hypercube), the D-optimal design is at the same time the G-optimal design. The D-optimality of experiment design is, apart from other advantages of the algorithm proposed, its essential asset.

Acknowledgments

This paper was supported by a grant from the Ministry of Science and High Education.

References

1. Björck, A.: Numerical Methods for Least Squares Problems. SIAM, Amsterdam (1996)
2. Brigham, E.: Fast Fourier Transform and Its Applications. Prentice-Hall, New York (1988)
3. Chu, E., George, A.: Inside the FFT Black Box: Serial and Parallel Fast Fourier Transform Algorithms. CRC Press, Boca Raton (2000)

4. Dutt, A., Rokhlin., V.: Fast Fourier Transforms for Nonequispaced Data. SIAM J. Sci. Comp. 14, 1368–1393 (1993)
5. Groß, J.: Linear Regression. Springer, London (2003)
6. Hyvrinen, A., Karhunen, J., Oja, E.: Independent Component Analysis. Wiley-Interscience, New York (2001)
7. Liu, Q.H., Nyguen, N.: An accurate algorithm for nonuniform fast Fourier transforms. IEEE Microwave Guided Wave Lett 8(1), 18–20 (1997)
8. Osowski S.: Neural Networks for Data Processing. Publishing House of Warsaw Technical University, Warsaw (2006) (In Polish)
9. Rafajowicz, E.: Experiment designing algorithms with implementations in MATHEMATICA environment. PLJ, Warsaw (1996) (In Polish)
10. Rafajowicz, E., Pawlak, M.: On Function Recovery by Neural Networks Based on Orthogonal Expansions. In: Proc. 2nd World Congress of Nonlinear Analysis, Nonlinear Analysis, Theory and Applications, vol. 30(3), pp. 1343–1354. Pergamon Press (1997)
11. Rafajowicz, E., Skubalska-Rafajowicz, E.: FFT in Calculating Nonparametric Regression Estimate Based on Trigonometrical Series. Appl. Math. and Comp. Sci 3(4), 713–720 (1993)
12. Sher, C.F., Tseng, C.S., Chen, C.S.: Properties and Performance of Orthogonal Neural Network in Function Approximation. Int. J. of Intelligent Syst. 16(12), 1377–1392 (2001)
13. Tseng, C.S., Chen, C.S.: Performance Comparison between the Training Method and the Numerical Method of the Orthogonal Neural Network in Function Approximation. Int. J. of Intelligent Syst. 19(12), 1257–1275 (2004)
14. Walker, J.S.: Fast Fourier Transforms. CRC Press, Boca Raton (1996)
15. Zhu, C., Shukla, D., Paul, F.W.: Orthogonal Functions for System Identification and Control. In: Leondes, C.T. (ed.) Neural Network Systems Techniques and Applications. Control and Dynamic Systems, vol. 7, pp. 1–73. Academic Press, San Diego (2002)

Accuracy Improvement of Neural Network State Variable Estimator in Induction Motor Drive

Jerzy Jelonkiewicz and Andrzej Przybył

Częstochowa University of Technology
Dąbrowskiego 69, 42-200 Częstochowa, Poland
{jerzy.jelonkiewicz,andrzej.przybyl}@kik.pcz.pl

Abstract. Some accuracy improvement of the neural network (NN) estimator is proposed in the paper. The estimator approximates stator current components in the rotor flux reference frame. Two approaches are considered: data mining with GMDH algorithm and gradual training of the NN in the desired frequency range. In both cases the accuracy of the estimator is significantly improved. Provided tests confirmed this feature and encourage to implement such an estimator it in a sensorless vector controlled induction motor drive.

1 Introduction

In many vector controlled induction motor drives applications rotor speed measurement or estimation is not necessary for adjustment of its operating point effectively. Such a case relates to traction drives where torque value is rather used than a vehicle speed. Moreover the absence of the rotor speed measurement increases robustness and reliability of the drive. Therefore control strategies that use other than rotor speed state variables are especially welcome. For the vector control strategy, the most needed state variable of the motor is the rotor flux in terms of its phase angle and amplitude as it allows any of advanced control schemes to be implemented.

Among various techniques to estimate rotor flux, the Model Reference Adaptive System approach looks to be the most promising as it offers higher accuracy due to closed-loop operation [1]. However considered scheme suffers from variety of drawbacks like: integration drift, sensibility to noise present in the measured signals or parameters variation of the motor. On the other hand, in a control circuit, much useful are stator current components in the rotor flux reference frame which can be easily calculated knowing rotor flux position and amplitude. Such a scheme, presented in [2], simplifies and improves estimation part of the control circuit, but requires rotor flux position detection which involves all the above mentioned drawbacks related to this task. Therefore NN estimator of stator current components in the rotor flux reference frame seems to be an interesting solution as it employs completely different approach to the problem. The intention of the paper is to find out how artificial intelligence tools are useful in the stator current components estimation in the rotor reference frame and how accurate is the estimator in the whole frequency and load range and feasible in the vector control strategy. The idea of such an estimator although mentioned in [4] has not been investigated widely. Only in [4,6,7] some results of the neural network estimator were presented but its accuracy has not been evaluated enough to confirm its usefulness in the vector control scheme. It looks like there is no possibility to find a compromise in network weights adjustment, when implemented in the straight-forward way, to fulfil high accuracy estimation of the required state variables in the whole input frequency range. Therefore the paper investigates some possibility to improve the accuracy of the estimator particularly in the low frequency range. Firstly some knowledge extraction from available input signals is presented which improves the accuracy of the estimator. Unfortunately, this modification of the input vector selection is not sufficient in the low input frequency range. Then another idea is proposed which assumes some division of the input frequency into sub-ranges with independent, separately trained NN higher accuracy estimators. Finally, the latter idea has been aggregated to create one neural network estimator with hidden layer weights defined by a polynomial function. Due to estimated stator current components the vector controlled induction motor drive is considerably simplified, while offering sensorless operation.

2 MRAS-Based Induction Motor State Variables Estimator

Considered MRAS (Model Reference Adaptive System) estimator uses reference model based on the following equations of the rotor flux components in the stationary reference frame [4]:

$$\Psi_{rd} = \frac{L_r}{L_m}[\int(u_{sD} - R_s i_{sD})dt - L_s' i_{sD}] \tag{1}$$

$$\Psi_{rq} = \frac{L_r}{L_m}[\int(u_{sQ} - R_s i_{sQ})dt - L_s' i_{sQ}] \tag{2}$$

where L_m, L_r, L_s' are: magnetizing inductance, rotor self inductance and stator transient inductance.

However adaptive model is defined as:

$$\hat{\Psi}_{rd} = \frac{1}{T_r}\int(L_m i_{sD} - \hat{\Psi}_{rd} - \omega_r T_r \hat{\Psi}_{rq})dt \tag{3}$$

$$\hat{\Psi}_{rq} = \frac{1}{T_r}\int(L_m i_{sQ} - \hat{\Psi}_{rq} + \omega_r T_r \hat{\Psi}_{rd})dt \tag{4}$$

where T_r is rotor time constant.

When both models are connected in well known closed loop system, they estimate rotor flux and motor speed. Unfortunately, due to pure integrators that exist in both models, the system fails to work properly for the sake of offsets and measurement inaccuracies in the sampled voltages and currents. This problem can be corrected with the help of a low pass filter to approximate the integrator but it fails at low speed range due to unavoidable phase shift at these frequencies. Other solution, proposed in [3] is based on drift and dc-offset compensator using feedback integrator. This idea, although effective, cures not origin of the problem but final effect of the estimator performance.

Solution proposed in [2] is based on the above mentioned MRAS estimator with slight modification. As opposed from the classical way of obtaining tuning signal where both: amplitude and phase of rotor flux components are compared, in the modified estimator only the phase angle is used to produce stator current components in the rotor flux reference frame (i_{xy}). These components then form a tuning signal which PI controller converts to the rotor speed. It is expected that proposed estimator should have better features than the classical one as used phase angle of the rotor flux is less susceptible to the integration drift. The idea of the estimator shows Fig. 1.

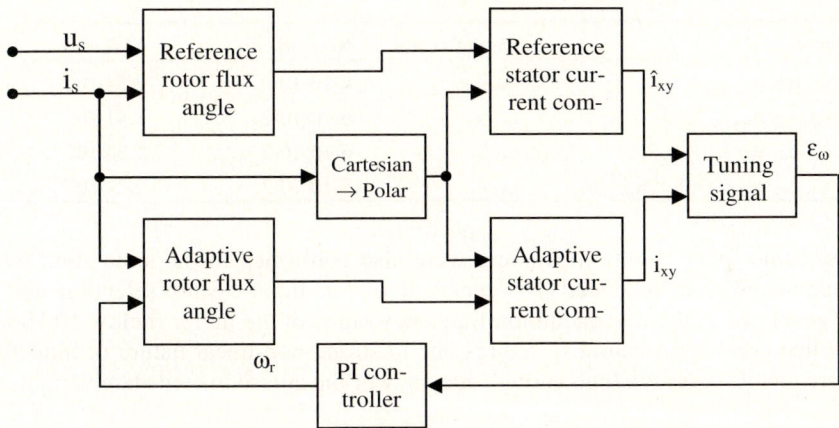

Fig. 1. Stator current components MRAS estimator

Expected features of the modified estimator have been partly confirmed. The estimator presents fine accuracy in the whole input frequency range but some sensitivity to rotor flux angle calculation error due to the integration drift is still on [2]. Encountered problems with classical MRAS-type estimator turned authors attention to a promising estimation technique using neural networks structures.

3 Neural Network Estimator Based on Stator Voltage and Current Components

As some authors reported an excellent performance of the NN estimators [6,7], first and straightforward approach to the problem was to consider only various combinations of stator voltage and current components and their delayed values as input vector. Previous authors experience with NN state variables estimators, confirmed in other papers [7], indicates that it is difficult to get a good performance with one hidden layer. Number of neurons in each hidden layer has been optimised and only for comparison purposes was set to 10. For the tests in the Matlab-Simulink environment the induction motor with the following parameters was considered:

P_n=1100 W, U_n=380 V, R_s=6.88 Ω, R_r=6.35 Ω, L_{ls}=34·10^{-3} H, L_{lr}=34·10^{-3} H, L_m=450·10^{-3} H, rated torque - T_n=7,6 Nm

Data set for training and tests purposes covered frequency range 5-50 Hz and there were 360 samples in each operating point, being a combination of the frequencies: 5, 7.5, 10, 15, 20, 30, 40, 50 Hz and three load torque levels of the motor (no load, half and full load). Obtained results show table 1.

Table 1. Results of the training of the stator current components in the rotor flux reference frame estimator

Input vector	Output vector	Network structure	MSE
$i_{sd}, i_{sq}, u_{sd}, u_{sq}$	i_{sx}, i_{sy}	4-10-10-2	2.34·10^{-4}
$i_{sd}, i_{sq}, u_{sd}, u_{sq}, i_{sd}^{-1}, i_{sq}^{-1}$	i_{sx}, i_{sy}	6-10-10-2	1.41·10^{-4}
$i_{sd}, i_{sq}, u_{sd}, u_{sq}, u_{sd}^{-1}, u_{sq}^{-1}$	i_{sx}, i_{sy}	6-10-10-2	1.38·10^{-4}
$i_{sd}, i_{sq}, u_{sd}, u_{sq}, i_{sd}^{-1}, i_{sq}^{-1}, u_{sd}^{-1}, u_{sq}^{-1}$	i_{sx}, i_{sy}	8-10-10-2	2.24·10^{-4}

These rather poor results in training were also confirmed in the simulation tests. The training error is practically independent of the input vector selection and is mostly generated in the low excitation frequency range of the motor (below 10 Hz). It appears that considered neural structures due to strong non-linear nature of induction motor are unable to assure high enough accuracy of the approximated state variables.

3.1 NN Estimator Accuracy Improvement with Data Mining in GMDH Algorithm

These unsatisfactory results lent authors to search for another configuration of the input vector which comes from data mining algorithm that reflects data mutation and exchange of the input signals [5]. The main assumption of this algorithm was to give up the idea of the deduction approach based on the expert knowledge. Second element of this theory was the evolution of the polynomial, that describes the system, from the elementary structure to the combination of the simple partial models. For the rate of the polynomial equal 2 the output function is described as :

$$y = f_1(u_1) + f_2(u_2) + f_{12}(u_1, u_2) \tag{5}$$

where f_1, f_2, f_{12} are polynomial functions and u_1, u_2 are input signals

Functions f_1, f_2 are general description of data mutation while f_{12} expresses general description of data exchange. When this relates to the NN estimator with 2 inputs the transfers function can be defined as:

$$y = a_0 + a_1 u_1 + a_2 u_2 + a_{11} u_1^2 + a_{22} u_2^2 + a_{12} u_1 u_2 \tag{6}$$

The function (6) reflects data mutation and exchange of the input signals.

According to the above defined function, additional components of the input vector were included: module of the stator current, square of the stator current module.

Table 2. Results of the training of the stator current components in the rotor flux reference frame estimator

Input vector	Output vector	Network structure	MSE
$i_{sd}, i_{sq}, u_{sd}, u_{sq}$	i_{sx}, i_{sy}	4-10-10-2	$2.34 \cdot 10^{-4}$
$i_{sd}, i_{sq}, u_{sd}, u_{sq}, |i_s|$	i_{sx}, i_{sy}	5-10-10-2	$1.13 \cdot 10^{-4}$
$i_{sd}, i_{sq}, u_{sd}, u_{sq}, |i_s|^2$	i_{sx}, i_{sy}	5-10-10-2	$6.49 \cdot 10^{-3}$
$i_{sd}, i_{sq}, u_{sd}, u_{sq}, |i_s|,|i_s|^2$	i_{sx}, i_{sy}	6-10-10-2	$1.10 \cdot 10^{-4}$
$i_{sd}, i_{sq}, |i_s|,|i_s|^2$	i_{sx}, i_{sy}	4-10-10-2	$2.10 \cdot 10^{-2}$

Contents of the Table 2 indicate that presence of the stator current module and the square of the current module slightly improves performance of the estimator. However the square of the current module when added alone spoils the training results but when has accompany of the current module helps to achieve better accuracy of the estimator [8]. The last row was added to show how important is the presence of the stator voltage components in the input vector. Unfortunately, this structure, enriched by additional input signals, was unable to follow reference stator components in the low input frequency range.

3.2 NN Estimator Trained for Sub-ranges of the Input Frequency

Results of the tests presented in the previous sections prove that there is a contradiction in weight adjustment of the neurons in the network for the whole frequency range. This observation led to the idea to create the parallel neural structures that could cover frequency sub-ranges. Table 3 confirms that parallel NN for selected frequencies can reach much higher accuracy but this solution has problematic feasibility as switching between networks seems to be troublesome. Therefore another solution was considered with only one network that is trained for selected frequencies, starting from 50 Hz down to 5 Hz (gradual training). This approach creates set of weights for selected frequencies that can then define polynomial functions to adjust weights for only one network at operating frequency. As seen in the Table 3 this approach assures similar accuracy of the network as for the first version.

Table 3. Results of training for parallel NN and gradual training

Frequency [Hz]	50	40	30	20	15	10	7	5
MSE (parallel)	$5.6 \cdot 10^{-9}$	$4.4 \cdot 10^{-10}$	$8.9 \cdot 10^{-9}$	$1.1 \cdot 10^{-7}$	$2.7 \cdot 10^{-6}$	$3.9 \cdot 10^{-6}$	$6.1 \cdot 10^{-6}$	$1.1 \cdot 10^{-6}$
MSE (gradual tr.)	$1.1 \cdot 10^{-10}$	$9.8 \cdot 10^{-11}$	$7.0 \cdot 10^{-10}$	$5.2 \cdot 10^{-7}$	$1.2 \cdot 10^{-5}$	$3.5 \cdot 10^{-6}$	$4.4 \cdot 10^{-7}$	$4.8 \cdot 10^{-7}$

Example of the weight changes for the second layer of the network and flux component of the stator current (Fig. 2) indicates that the weights change much in the frequency range 5-20 Hz, while in the range over 20 Hz are rather constant.

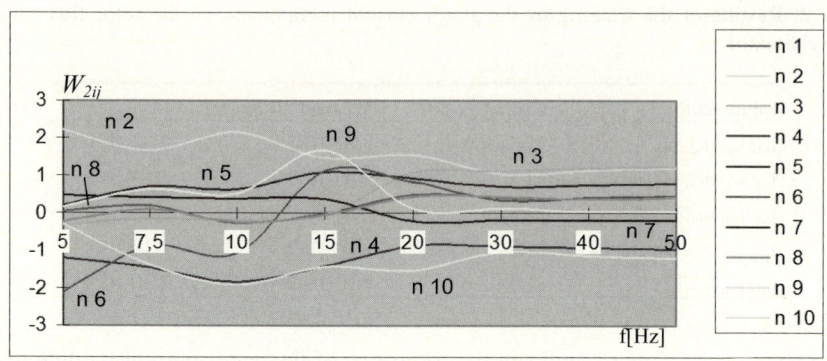

Fig. 2. Weight changes for the second layer of the "flux" stator current component

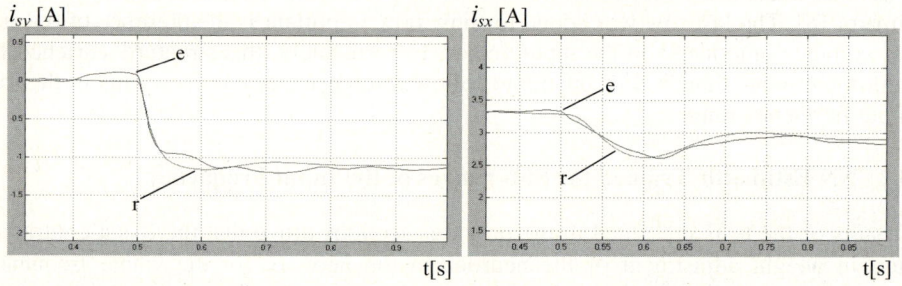

Fig. 3. NN estimation of "torque" (left) and "flux" (right) components of stator current (marked as e) with reference (r) for f=5 Hz and step load at 0.5 s

Therefore, to minimize approximation error, a polynomial function was applied only in the range 5-20 Hz. Test of proposed neural network with adjusted weights for f=5 Hz (the worst case) confirms its quite high accuracy in the whole frequency range (Fig. 3).

4 Implementation in a DSP System

Proposed NN estimator of the stator current components in the rotor reference frame was the basis of the control scheme. It plays crucial role in the control structure and significantly simplifies the estimation process of the "flux and "torque" components of the stator current. The main blocks of the structure are generated with a help of fuzzy –neural generators. Proposed new structure belongs to field oriented methods, which allows the internal parameters of the machine to be fully controlled even in transient states. The calculations of the implemented in ADDU 21161L system NN estimator take 21.36 µs which allows other parts of the structure to be implemented in a short enough time. These parts include blocks that are responsible for keeping the best relationship between i_{sx} and i_{sy} in terms of motor efficiency, especially important in the electric vehicle drive.

5 Conclusions

The paper considers NN estimator of the "flux" and "torque" components of stator current. When applied in the straightforward way its accuracy in the low frequency range is not acceptable. Proposed solution to improve its performance seems to be successful as achieved accuracy is much higher in the considered frequency range. Provided simulation and real tests confirm encouraging features of the estimator. It plays crucial role in the efficiency optimal control structure for electric vehicle. It is believed that its features can be extended to motor parameter changes resistance and noise immunity to get real robust and reliable estimator that can be used in considered and other applications.

References

1. Ohyama, K., Asher, G.M., Sumner, M.: Comparison of the practical performance and operating limits of sensorless induction motor drive using a closed loop flux observer and a full order flux observer. In: Proc. EPE 1999, Lausanne, on CD (1999)
2. Jelonkiewicz, J.: Modified MRAS estimator in sensorless vector control of induction motor. In: XII Symposium PPEE 2007, Wisla 2007, pp. 305–308 (2007)
3. Sumner, M., Spiteri Staines, C., Gao, Q., Asher, G.: Sensorless Speed Operation of Cage Induc-tion Motor using Zero Drift Feedback Integration with MRAS Observer. In: Proc. EPE 2005, Dresden, on CD (2005)
4. Vas, P.: Artificial–Intelligence-Based Electrical Machines and Drives. In: Monographs in Electrical and Electronic Engineering nr 45. Oxford University Press, Oxford (1999)
5. Korbicz J.W, Rutkowski L., Tadeusiewicz R.: Biocybernetyka i Inzynieria Biomedyczna 2000 Tom 6, Sieci Neuronowe, PAN, Akademicka Oficyna Wydawnicza EXIT, Warszawa 2000, pp. 227-255 (2000)
6. Kuchar, M., Branstetter, P., Kaduch, M.: ANN-based speed estimator for induction motor. In: Proc. EPE-PEMC 2004, Riga, on CD (2004)
7. Grzesiak, L., Ufnalski, B.: DTC drive with ANN-based stator flux estimator. In: Proc. EPE 2005, Dresden, on CD (2005)
8. Jelonkiewicz, J., Przybyl, A.: Knowledge extraction from data for neural network state variables estimators in induction motor. In: SENE 2005, Lodz 2005, pp. 211–216 (2005)

Ensemble of Dipolar Neural Networks in Application to Survival Data

Małgorzata Krętowska

Faculty of Computer Science
Białystok Technical University
Wiejska 45a, 15-351 Białystok, Poland
mmac@wi.pb.edu.pl

Abstract. In the paper the ensemble of dipolar neural networks (EDNN) for analysis of survival data is proposed. The tool is build on the base of the learning sets, which contain the data from clinical studies following patients response for a given treatment. Such datasets may contain incomplete (censored) information on patients failure times. The proposed method is able to cope with censored observations and as the result returns the aggregated Kaplan-Meier survival function. The prediction ability of the received tool as well as the significance of individual features is verified by the Brier score, $\tilde{D}_{S,x}$ and $\hat{D}_x$ measures of predictive accuracy.

1 Introduction

The main objective of regression methods is to predict the value of dependent variable y by using the set of independent features, that would be observed in the future. The regression models are usually built by minimization the sum of squares of differences between empirical (y) and theoretical $(\hat{y})$ values over all the observations from the learning set. The problem arises when the dataset does not contain the exact values of y. Such a situation is very common in survival data, in which the time of a given failure (i.e. death, disease relapse) is under investigation. The lack of knowledge of exact failure times is caused, on the one hand, by unpredictable failures being the results of other, not investigated diseases or accidents, on the other hand, by the end of follow-up time. The follow-up time in clinical trials, in which the patients response for a given treatment is studying, is determined in advance. If the failure did not occur before the end of follow-up time, the observation is cut exactly at this time. In such *censored cases* we only know that the failure time is not less than their follow-up time.

In figure 1, two described above situations are presented. Assuming that the follow-up time is a one year interval, from 01.01.2005 to 31.12.2005, the patients are included into the study just after they underwent a given treatment, often surgery. The beginning of the treatment is the starting point of their follow-up $t = 0$ (Fig. 1b). As we can see in figure 1a for patients A and D the failure occurred during the follow-up time, patient C was lost to follow-up before 31.12.2005, and patient B was observed to the end of follow-up time and during

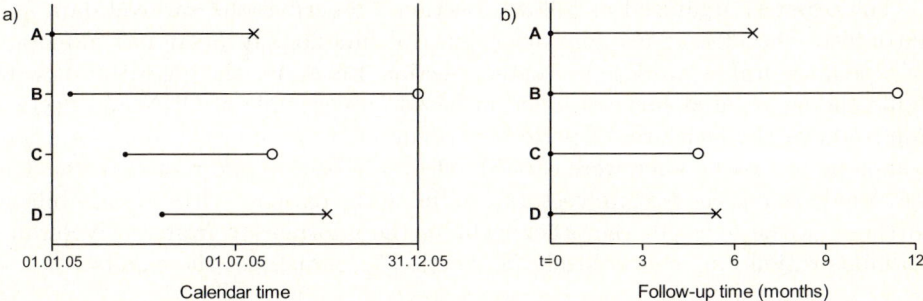

Fig. 1. Clinical trial in two points of reference: a) calendar time, b) follow-up time; x means uncensored observation, o - censored observation

this time the failure did not occur. Therefore, observations B and C are censored - the exact failure time is unknown for them.

Since the survival data is to a large extent censored, the crucial point of methods for failure time prognosis, is using the information from censored cases. The well known and widely used tool is Cox's proportional hazards model [3]. In its basic form it assumes, among other things, that covariates are independent of time and of themselves as well. If the conditions are impossible to fulfill, other, non-parametric techniques are adopted to the problem. The most common are regression trees and artificial neural networks. Recently, also methods concerning the use of random forests, bagging and boosting techniques in prognosis of survival time appear. Their application allows receiving the tool unaffected by small changes in dataset, what is particularly important in discovering the risk factors. Hothorn et al. [6] proposes boosting survival trees to create aggregated survival function. Krętowska [12] developed the approach by using the dipolar regression trees instead of the structure proposed in [6]. The technique proposed by Ridgeway [13] allows minimizing the partial likelihood function (boosting Cox's proportional hazard model). The Hothorn et al. [7] developed two approaches for censored data: random forest and gradient boosting. Breiman [2] provided the software that allows induction of the random forest for censored data.

In the paper the ensemble of dipolar neural networks (EDNN) is proposed. The individual DNN [11] is build by minimization a dipolar criterion function [1] and, by appropriate formation of the function, is able to cope with censored data. As the result of DNN a set of Kaplan-Meier survival functions is received. Each function represents the survival of an individual subgroup of observations, which is characterized by similar survival experience. A new patient may be classified to the appropriate subgroup, without exact prediction of his own failure time. The proposed algorithm enables receiving the aggregated Kaplan-Meier survival function [6] precisely for analyzed patient and predicts his survival time as the median value. The predictive ability of the proposed technique as well as the significance of individual features is assessed by using measures which were developed to cope with censored data: the Brier score [5], indirect and direct estimator of absolute predictive error and explained variation ([15,14]).

The paper is organized as follows. Section 2 describes the survival data and introduces the idea of Kaplan-Meier survival function. In Section 3 induction of dipolar neural network is presented. Section 4 contains the algorithm how to build the aggregated survival function based on ensemble of DNN and Section 5 introduces the measures of predictive accuracy. Experimental results are presented in Section 6. They were carried out on the base of two real datasets. The first one contains the feature vectors describing the patients with primary biliary cirrhosis of the liver [4], the other includes the information from the Veteran's Administration lung cancer study [8]. Section 7 summarizes the results.

2 Introduction to Survival Data

Let T^0 denotes the true survival time and C denotes the true censoring time with distribution functions F and G respectively. We observe random variable $O = (T, \Delta, \mathbf{X})$, where $T = \min(T^0, C)$ is the time to event, $\Delta = I(T \leq C)$ is a censoring indicator and $\mathbf{X} = (X_1, ..., X_N)$ denotes the set of N covariates from a sample space χ. We have learning sample $L = (\mathbf{x}_i, t_i, \delta_i)$, $i = 1, 2, ..., n$, where $\mathbf{x}_i$ is N-dimensional covariates vector, t_i - survival time and δ_i - failure indicator, which is equal to 0 for censored cases and 1 for uncensored ones.

The distribution of random variable T may be described by the marginal probability of survival up to a time $t > 0$ ($S(t) = P(T > t)$). The estimation of survival function $S(t)$ may be done by using the Kaplan-Meier product limit estimator [9], which is calculated on the base of learning sample L and is denoted by $\hat{S}(t)$:

$$\hat{S}(t) = \prod_{j|t_{(j)} \leq t} \left(\frac{m_j - d_j}{m_j} \right) \quad (1)$$

where $t_{(1)} < t_{(2)} < ... < t_{(D)}$ are distinct, ordered survival times from the learning sample L, in which the event of interest occurred, d_j is the number of events at time $t_{(j)}$ and m_j is the number of patients at risk at $t_{(j)}$ (i.e., the number of patients who are alive at $t_{(j)}$ or experience the event of interest at $t_{(j)}$).

The 'patients specific' survival probability function is given by $S(t|\mathbf{x}) = P(T > t|\mathbf{X} = \mathbf{x})$. The conditional survival probability function for the new patient with covariates vector $\mathbf{x}_{new}$ is denoted by $\hat{S}(t|\mathbf{x}_{new})$.

3 Dipolar Neural Network - DNN

A dipolar neural network model, considered in the paper, was proposed by Krętowska and Bobrowski [11]. The network consists of two layers: input and output layer. The output layer is build from neurons with binary activation function:

$$z = f(\mathbf{x}, \mathbf{w}) = \begin{cases} 1 & \text{if } \mathbf{w}^T \mathbf{x} \geq \theta \\ 0 & \text{if } \mathbf{w}^T \mathbf{x} < \theta \end{cases} \quad (2)$$

where **x** is a feature vector, **w** - a weight vector and θ is a threshold. From the geometrical point of view a neuron divides a feature space into two subspaces by using hyperplane $H(\mathbf{w}, \theta) = \{\mathbf{x} : \mathbf{w}^T \mathbf{x} = \theta\}$. If the vector **x** is situated on the positive side of the hyperplane the neuron is activated and $z = 1$. The layer of L binary neurons divided the N-dimensional feature space into disjoint regions - *active fields* (AF). Each region is represented by L-dimensional output vector: $\mathbf{z} = [z_1, z_2, \ldots, z_L]^T$, where $z_i \in \{0, 1\}$.

In case of survival analysis, the aim of learning algorithm is to receive such active fields which would contain observations with similar failure times. Then, on the base of Kaplan-Meier survival functions $(S_i(t))$ connected with individual active fields (AF_i), the survival probability at a given time t may be predicted and compared among all the subgroups (Fig. 2).

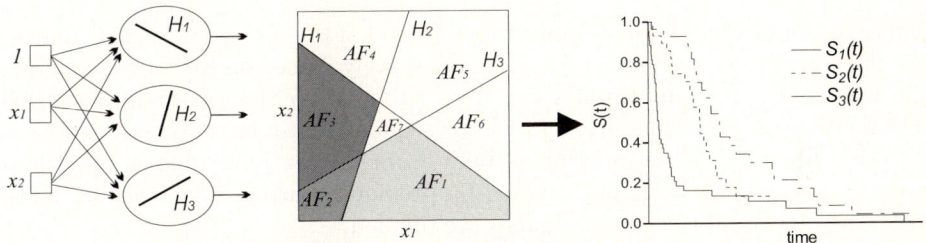

Fig. 2. Working of individual DNN

The described above objective of neural network learning procedure is realized by minimization of dipolar criterion function [1] that is built on the base of dipoles. Dipoles - pairs of feature vectors - are formed according to the following rules:

1. a pair of feature vectors $(\mathbf{x}_i, \mathbf{x}_j)$ forms the pure dipole, if
   - $\delta_i = \delta_j = 1$ and $|t_i - t_j| < \eta$
2. a pair of feature vectors $(\mathbf{x}_i, \mathbf{x}_j)$ forms the mixed dipole, if
   - $\delta_i = \delta_j = 1$ and $|t_i - t_j| > \zeta$
   - $(\delta_i = 0, \delta_j = 1$ and $t_i - t_j > \zeta)$ or $(\delta_i = 1, \delta_j = 0$ and $t_j - t_i > \zeta)$

where η and ζ are equal to quartiles of absolute values of differences between uncensored survival times. Based on earlier experiments, parameter η was fixed as 0.2 quartile and ζ - 0.6.

As we can see, the pure dipoles are formed between feature vectors for which the difference between failure times is small, and mixed dipoles between pairs with distant failure times. In the latter case we can use the information from censored cases.

Two types of piece-wise linear and convex (CPL) penalty functions $\varphi_j^+(\mathbf{v})$ and $\varphi_j^-(\mathbf{v})$ are defined:

$$\varphi_j^+(\mathbf{v}) = \begin{cases} \delta_j - <\mathbf{v}, \mathbf{y}_j> & \text{if } <\mathbf{v}, \mathbf{y}_j> \leq \delta_j \\ 0 & \text{if } <\mathbf{v}, \mathbf{y}_j> > \delta_j \end{cases} \quad (3)$$

$$\varphi_j^-(\mathbf{v}) = \begin{cases} \delta_j + <\mathbf{v}, \mathbf{y}_j> & \text{if } <\mathbf{v}, \mathbf{y}_j> \geq -\delta_j \\ 0 & \text{if } <\mathbf{v}, \mathbf{y}_j> < -\delta_j \end{cases} \quad (4)$$

where δ_j is a margin ($\delta_j = 1$, for each j), $\mathbf{y}_j = [1, x_1, \ldots, x_N]^T$ is an augmented covariate vector and $\mathbf{v} = [-\theta, w_1, \ldots, w_N]^T$ is an augmented weight vector. Each mixed dipole $(\mathbf{y}_i, \mathbf{y}_j)$, which should be divided, is associated with a function $\varphi_{ij}^m(\mathbf{v})$ being a sum of two functions with opposite signs ($\varphi_{ij}^m(\mathbf{v}) = \varphi_j^+(\mathbf{v}) + \varphi_i^-(\mathbf{v})$ or $\varphi_{ij}^m(\mathbf{v}) = \varphi_j^-(\mathbf{v}) + \varphi_i^+(\mathbf{v})$). For pure dipoles, which should stay undivided, we associate a function $\varphi_{ij}^p(\mathbf{v})$ ($\varphi_{ij}^p(\mathbf{v}) = \varphi_j^+(\mathbf{v}) + \varphi_i^+(\mathbf{v})$ or $\varphi_{ij}^c(\mathbf{v}) = \varphi_j^-(\mathbf{v}) + \varphi_i^-(\mathbf{v})$). A dipolar criterion function is a sum of penalty functions associated with each dipole:

$$\Psi_d(\mathbf{v}) = \sum_{(j,i) \in I_p} \alpha_{ij} \varphi_{ij}^p(\mathbf{v}) + \sum_{(j,i) \in I_m} \alpha_{ij} \varphi_{ij}^m(\mathbf{v}) \quad (5)$$

where α_{ij} determines relative importance (price) of the dipole $(\mathbf{y}_i, \mathbf{y}_j)$, I_p and I_m are the sets of pure and mixed dipoles, respectively. Based on earlier experiments the value of α_{ij} for pure dipoles was fixed as 1 and for the mixed ones as 1000. The neurons weight values are obtained by sequential minimization of the dipolar criterion functions. The function is built from all the pure dipoles and those mixed dipoles which were not divided by previous neurons. The learning phase is finished when all the mixed dipoles are divided.

To improve the generalization ability of the network the second phase of learning procedure - optimization - is applied. The optimization phase consists of two steps. The first step is aimed at distinguishing and enlargement of prototypes(i.e. active fields which contain the largest number of feature vectors $\mathbf{x}$) and the other at reduction of redundant neurons. More detailed description can be find in [11].

4 Ensembles of DNN

Ensemble of dipolar neural networks (EDNN) is a set of DNN_i, $(i = 1, 2, \ldots, k)$, generated on base of k learning samples $(L_1, L_2, \ldots, L_k)$ drawn with replacement from the given sample L. As the result of each DNN_i, the set of active fields $SAF_i = \{AF_i^1; AF_i^2; \ldots, AF_i^{k_i}\}$ is received. Each active field AF_i^j contains the subset of observations from the learning sample L_i. Having a new covariate vector $\mathbf{x}_{new}$, each DNN_i, $i = 1, 2, \ldots, k$ returns the active field $AF_i(\mathbf{x}_{new})$, which the new observation belongs to. Let $L_i(\mathbf{x}_{new})$ denotes the set of observation covered by active field $AF_i(\mathbf{x}_{new})$. Having k sets $L_i(\mathbf{x}_{new})$, aggregated sample $L_A(\mathbf{x}_{new})$ is built [6]:

$$L_A(\mathbf{x}_{new}) = \{L_1(\mathbf{x}_{new}); L_2(\mathbf{x}_{new}); \ldots; L_k(\mathbf{x}_{new})\}$$

The aggregated conditional Kaplan-Meier survival function, calculated on the base of set $L_A(\mathbf{x}_{new})$ can be referred to as $\hat{S}_A(t|\mathbf{x}_{new})$.

To summarize the above considerations, the algorithm leading to receive the aggregated survival function is as follows:

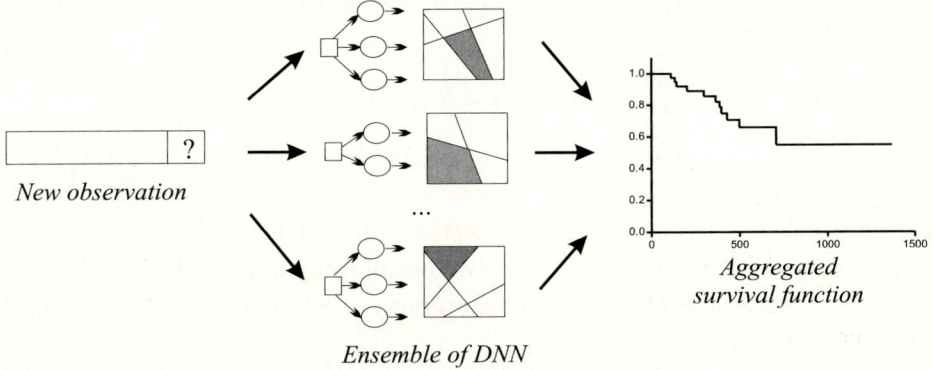

Fig. 3. EDNN in prediction of survival time for a new observation

1. Draw k bootstrap samples $(L_1, L_2, \ldots, L_k)$ of size n with replacement from L
2. Induction of dipolar neural network DNN_i based on each bootstrap sample L_i, $i = 1, 2, \ldots, k$
3. Build aggregated sample $L_A(\mathbf{x}_{new}) = \{L_1(\mathbf{x}_{new}); L_2(\mathbf{x}_{new}), \ldots, L_k(\mathbf{x}_{new})\}$
4. Compute the Kaplan-Meier aggregated survival function for a new observation $\mathbf{x}_{new}$: $\hat{S}_A(t|\mathbf{x}_{new})$ (Fig. 3).

5 Evaluation of Prediction Ability

Beside the problems concerning the use of censored data in the process of building the prediction tool, the question how to evaluate the prediction ability of received models appears. The lack of exact failure times for a part of data causes that the classical measures based on difference between empirical and theoretical values can not be used. Instead of them, other, censoring oriented, measures are proposed.

One of them is the Brier score introduced by Graf *at al.* [5]. The Brier score as a function of time is defined by

$$BS(t) = \tfrac{1}{n} \sum_{i=1}^{N} (\hat{S}(t|\mathbf{x}_i)^2 I(t_i \leq t \wedge \delta_i = 1) \hat{G}(t_i)^{-1} +$$
$$(1 - \hat{S}(t|\mathbf{x}_i))^2 I(t_i > t) \hat{G}(t)^{-1}) \qquad (6)$$

where $\hat{G}(t)$ denotes the Kaplan-Meier estimator of the censoring distribution. It is calculated on the base of observations $(t_i, 1 - \delta_i)$. $I(condition)$ is equal to 1 if the condition is fulfilled, 0 otherwise. The BS equal to 0 means the best prediction.

The Brier score belongs to direct estimators of prediction ability, because it uses the information explicitly from the data. Another direct approach is proposed by Schemper and Henderson [15]. The predictive accuracy (without covariates), expressed by absolute predictive error (APE), at each distinct failure time $t_{(j)}$ is defined as:

$$\hat{M}(t_{(j)}) = \tfrac{1}{n}\sum_{i=1}^{n}\left[I(t_i > t_{(j)})(1-\hat{S}(t_{(j)})) + \delta_i I(t_i \leq t_{(j)})\hat{S}(t_{(j)}) + \right.$$
$$\left. (1-\delta_i)I(t_i \leq t_{(j)})\left\{(1-\hat{S}(t_{(j)}))\tfrac{\hat{S}(t_{(j)})}{\hat{S}(t_i)} + \hat{S}(t_{(j)})(1-\tfrac{\hat{S}(t_{(j)})}{\hat{S}(t_i)})\right\}\right] \quad (7)$$

The measure with covariates ($\hat{M}(t_{(j)}|\mathbf{x})$) is obtained by replacing $\hat{S}(t_{(j)})$ by $\hat{S}(t_{(j)}|\mathbf{x})$ and $\hat{S}(t_i)$ by $\hat{S}(t_i|\mathbf{x})$. To receive overall estimators of APE with ($\hat{D}_x$) and without covariates ($\hat{D}$) the weighed averages of estimators over failure times are calculated:

$$\hat{D} = w^{-1}\sum_{j}\hat{G}(t_{(j)})^{-1}d_j\hat{M}(t_{(j)}) \quad (8)$$

$$\hat{D}_\mathbf{x} = w^{-1}\sum_{j}\hat{G}(t_{(j)})^{-1}d_j\hat{M}(t_{(j)}|\mathbf{x}) \quad (9)$$

where $w = \sum_{j}\hat{G}(t_{(j)})^{-1}d_j$, d_j is the number of events at time $t_{(j)}$ and $\hat{G}(t)$ denotes the Kaplan-Meier estimator of the censoring distribution (see equation 6).

The indirect estimation of predictive accuracy was proposed by Schemper [14]. In the approach the estimates (without $\tilde{M}(t_{(j)})$ and with covariates $\tilde{M}(t_{(j)}|\mathbf{x})$) are defined by

$$\tilde{M}(t_{(j)}) = 2\hat{S}(t_{(j)})(1-\hat{S}(t_{(j)})) \quad (10)$$

$$\tilde{M}(t_{(j)}|\mathbf{x}) = 2n^{-1}\sum_{i}\hat{S}(t_{(j)}|\mathbf{x}_i)(1-\hat{S}(t_{(j)})|\mathbf{x}_i) \quad (11)$$

The overall estimators of predictive accuracy with ($\tilde{D}_{S,\mathbf{x}}$) and without ($\tilde{D}_S$) covariates are calculated similarly to the estimators $\hat{D}_\mathbf{x}$ and $\hat{D}$. The only change is replacing $\hat{M}(t_{(j)})$ and $\hat{M}(t_{(j)}|\mathbf{x})$ by $\tilde{M}(t_{(j)})$ and $\tilde{M}(t_{(j)}|\mathbf{x})$ respectively.

Based on the above overall estimators of absolute predictive error, explained variation can be defined as:

$$\tilde{V}_S = \frac{\tilde{D}_S - \tilde{D}_{S,\mathbf{x}}}{\tilde{D}_S} \quad (12)$$

and

$$\hat{V} = \frac{\hat{D} - \hat{D}_\mathbf{x}}{\hat{D}} \quad (13)$$

6 Experimental Results

The analysis was conducted on the base on two datasets. The first one is from the Mayo Clinic trial in primary biliary cirrhosis (PBC) of the liver conducted between 1974 and 1984 [4]. 312 patients participated in the randomized trial. Survival time was taken as a number of days between registration and death, transplantation or study analysis time in July 1986. Patients are described by the following features: age(AGE), sex, presence of edema, logarithm of serum bilirubin [mg/dl] ($LOGBILL$), albumin [gm/dl] ($ALBUMIN$), logarithm of prothrombin time [seconds], histologic stage of disease. Dataset contains 60 per cent of censored observations.

All the experiments were performed using the ensemble of 200 DNN. The measures of predictive accuracy were calculated on the base of learning sample L. To calculate the aggregated survival function for a given example $\mathbf{x}$ from the learning set L, only such DNN_i ($i = 1, 2, \ldots, 200$) were taken into consideration, for which $\mathbf{x}$ was not belonged to the learning set L_i (i.e. $\mathbf{x}$ did not participate in the learning process of the DNN_i).

Table 1. Measures of predictive accuracy for PBC dataset

Model	BS (12years)	Indirect APE/ Explained variation	Direct APE/ Explained variation
K-M Estimator	0.23	0.37	0.37
Ensemble of DNN			
all covariates	0.17	0.29/0.22	0.27/0.26
AGE	0.22	0.36/0.036	0.36/0.038
$LOGBILL$	0.17	0.28/0.25	0.28/0.25
$ALBUMIN$	0.22	0.33/0.11	0.33/0.12

In table 1 the Brier score as well as direct and indirect estimators of absolute prediction error (APE) and explained variation for /emphPBC dataset are presented. The absolute prediction error without covariates is equivalent to the error for K-M estimator and is equal to 0.37. The indirect and direct estimators of the predition errors for ensemble of DNN with all the covariates are $\tilde{D}_{S,\mathbf{x}} = 0.29$ and $\hat{D}_\mathbf{x} = 0.27$ respectively. It means that the knowledge of the prognostic factors reduces the absolute error of prediction of survival probability in the first 12 years after registration by 0.1 (or 0.08 in direct approach). The variation explained by the model is equal to 22 per cent in indirect approach and 26 per cent according to direct approach. More detailed analysis of individual covariates shows that the logarithm of serum bilirubin is the most important prognostic factor with $\tilde{D}_{S,\mathbf{x}} = \hat{D}_\mathbf{x} = 0.28$ and $\tilde{V}_S = \hat{V} = 0.25$. The influence of age and albumin for prediction of survival probability is less important. Similar conclusions can be draw from the analysis of the Brier score. The BS after 12 years of follow-up for Kaplan-Meier estimator is equal to 0.23. Similar values are for AGE and ALBUMIN ($BS = 0.22$). The BS(12 years) is smaller and equals to 0.17 in two cases: for the ensemble with all the covariates and for the model with $LOGBILL$ only.

In figure 4 we can see Kaplan-Meier survival functions received for three different values of AGE and LOGBILL, together with predicted failure times (median values). Estimated failure times for LOGBILL (Fig. 4a) equal to 0.7 and 1.5 are 2796 and 1427 [days] respectively. The median value for LOGBILL equal to -0.5 is greater than 4000 days. We can say that greater values of LOGBILL are

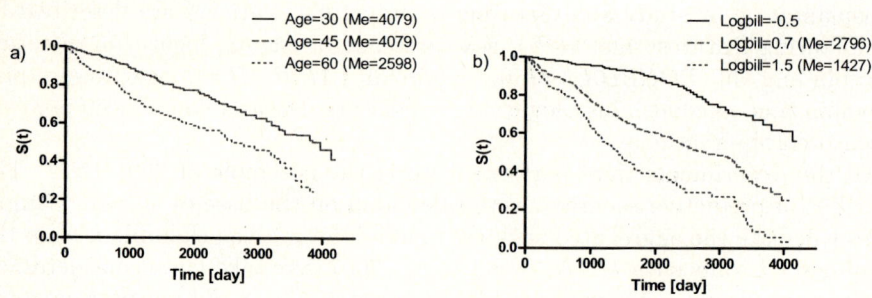

Fig. 4. Kaplan-Meier survival functions received for *PBC* dataset for different values of AGE (a) and LOGBILL (b)

connected with worse survival prediction. The failure times for three different values of AGE (30, 45, 60) are 4079, 4079 and 2598 respectively.

The other analyzed dataset contains the information from the Veteran's Administration (*VA*) lung cancer study [8]. In this trial, male patients with advanced inoperable tumors were randomized to either standard (69 subjects) or test chemotherapy (68 subjects). Only 9 subjects from 137 were censored. Information on cell type (0 - squamous, 1 - small, 2 - adeno, 3 - large) - CELL TYPE, prior therapy, performance status at baseline (Karnofsky rating - KPS), disease duration in months (TIME) and age in years at randomization (AGE), was available.

Table 2. Measures of predictive accuracy for *VA lung cancer* data

Model	BS (100 days)	Indirect $APE/$ Explained variation	Direct $APE/$ Explained variation
K-M Estimator	0.24	0.335	0.335
Ensemble of DNN all covariates	0.18	0.3/0.11	0.29/0.14
AGE	0.24	0.32/0.034	0.33/0.013
$CELL\ TYPE$	0.24	0.33/0.002	0.33/0.006
KPS	0.19	0.3/0.11	0.29/0.13
$TIME$	0.24	0.33/0.003	0.33/0.003

The measures of predictive accuracy for *VA lung cancer* data was shown in table 2. The unconditional absolute predictive error is 0.335. The ensemble of DNN, built on the base of all the covariates, reduces the error by 0.035 or 0.045 for indirect and direct approach respectively. The variance explained by the model is equal to 11 (14) per cent. The most important prognostic factor is KPS with the error equal to 0.3. Explained variation is 11 (13) per cent. Other variables have the marginal influence on the prediction of survival probability. Taking into account the values of Brier score after the first 100 days of follow-up the best prediction ability have the model built on the base of all the covariates

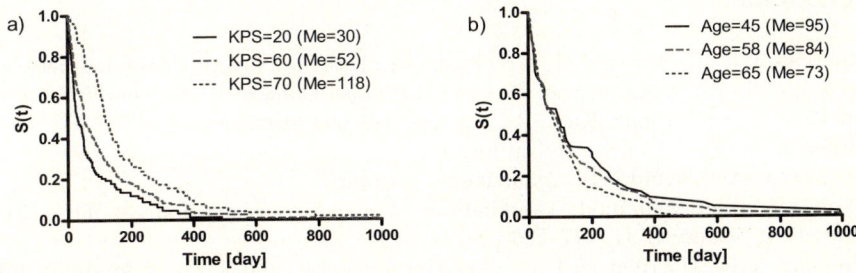

Fig. 5. Kaplan-Meier survival functions received for *VA lung cancer* data for different values of KPS (a) and AGE (b)

($BS = 0.18$). Similar value is received for the KPS feature ($BS = 0.19$) what makes it the strongest prognostic factor. For other covariates the BS value do not differ from the value for K-M estimator.

In figure 5 we can see Kaplan-Meier survival functions received for three different values of AGE and KPS together with predicted failure times. Estimated failure times for KPS (Fig. 5a) equal to 20, 60 and 70 are 30, 52 and 118 [days] respectively. We can noticed that greater values of LOGBILL are connected with better survival prediction. The failure times for three different values of AGE (45, 58, 65) are 95, 84 and 73. The functions received for KPS are more diverse than the estimators obtained for different values of AGE.

7 Conclusions

In the paper the ensemble of dipolar neural networks for prediction of survival time is proposed. The method is able to cope with censored observations, for which the exact failure time is unknown. The method, based on results of individual DNNs, produces the aggregated Kaplan-Meier survival function for a new patient described by **x**. The unknown failure time for **x** may be estimated by median value of the received function.

The prediction ability of the model was verified by several measures, such as the Brier score and direct and indirect estimators of absolute predictive errors: $\tilde{D}_{S,x}$, $\hat{D}_x$. The direct comparison of the received assessments is rather difficult. We can only noticed, that all the measures distinguished the same risk factors - the features that influence the survival the most: Karnofsky rating in case of *VA lung cancer* data and serum bilirubin for *PBC* dataset. The results were confirmed by graphical representation of survival function for different feature values.

Acknowledgements. This work was supported by the grant W/WI/4/05 from Białystok Technical University.

References

1. Bobrowski, L., Krętowska, M., Krętowski, M.: Design of neural classifying networks by using dipolar criterions. In: Proc. of the Third Conference on Neural Networks and Their Applications, Kule, Poland, pp. 689–694 (1997)
2. Breiman, L.: How to use survival forest, http://stat-www.berkeley.edu/users/breiman
3. Cox, D.R.: Regression models and life tables (with discussion). Journal of the Royal Statistical Society B 34, 187–220 (1972)
4. Fleming, T.R., Harrington, D.P.: Counting Processes and Survival Analysis. John Wiley & Sons, Inc., Chichester (1991)
5. Graf, E., Schmoor, C., Sauerbrei, W., Schumacher, M.: Assessment and comparison of prognostic classification schemes for survival data. Statistics in Medicine 18, 2529–2545 (1999)
6. Hothorn, T., Lausen, B., Benner, A., Radespiel-Troger, M.: Bagging survival trees. Statistics in Medicine 23, 77–91 (2004)
7. Hothorn, T., Buhlmann, P., Dudoit, S., Molinaro, A.M., van der Laan, M.J.: Survival ensembles. Berkeley Division of Biostatistics Working Paper Series, vol. 174 (2005), http://www.bepress.com/ucbbiostat/paper174
8. Kalbfleisch, J.D., Prentice, R.L.: The statistical analysis of failure time data. John Wiley & Sons, New York (1980)
9. Kaplan, E.L., Meier, P.: Nonparametric estimation from incomplete observations. Journal of the American Statistical Association 5, 457–481 (1958)
10. Krętowska, M.: Dipolar regression trees in survival analysis. Biocybernetics and biomedical engineering 24(3), 25–33 (2004)
11. Krętowska, M., Bobrowski, L.: Artificial Neural Networks in Identifying Areas with Homogeneous Survival Time. In: Rutkowski, L., Siekmann, J.H., Tadeusiewicz, R., Zadeh, L.A. (eds.) ICAISC 2004. LNCS (LNAI), vol. 3070, pp. 1008–1013. Springer, Heidelberg (2004)
12. Krętowska, M.: Random Forest of Dipolar Trees for Survival Prediction. In: Rutkowski, L., Tadeusiewicz, R., Zadeh, L.A., Żurada, J.M. (eds.) ICAISC 2006. LNCS (LNAI), vol. 4029, pp. 909–918. Springer, Heidelberg (2006)
13. Ridgeway, G.: The state of boosting. Computing Science and Statistics 31, 1722–1731 (1999)
14. Schemper, M.: Predictive accuracy and explained variation. Statistics in Medicine 22, 2299–2308 (2003)
15. Schemper, M., Henderson, R.: Predictive accuracy and explained variation in Cox regression. Biometrics 56, 249–255 (2000)

Binary Optimization: On the Probability of a Local Minimum Detection in Random Search

Boris Kryzhanovsky and Vladimir Kryzhanovsky

Center of Optical Neural Technologies,
SR Institute of System Analysis Russian Academy of Sciences,
44/2 Vavilov Str, Moscow 119333, Russia
kryzhanov@mail.ru

Abstract. The problem of binary optimization of a quadratic functional is discussed. By analyzing the generalized Hopfield model we obtain expressions describing the relationship between the depth of a local minimum and the size of the basin of attraction. Based on this, we present the probability of finding a local minimum as a function of the depth of the minimum. Such a relation can be used in optimization applications: it allows one, basing on a series of already found minima, to estimate the probability of finding a deeper minimum, and to decide in favor of or against further running the program. The iterative algorithm that allows us to represent any symmetric $N \times N$ matrix as a weighted Hebbian series of bipolar vectors with a given accuracy is proposed. It so proves that all conclusions about neural networks and optimization algorithms that are based on Hebbian matrices are true for any other type of matrix. The theory is in a good agreement with experimental results.

1 Introduction

Usually a neural system of associative memory is considered as a system performing a recognition or retrieval task. However it can also be considered as a system that solves an optimization problem: the network is expected to find a configuration minimizes an energy function [1]. This property of a neural network can be used to solve different NP-complete problems. A conventional approach consists in finding such an architecture and parameters of a neural network, at which the objective function or cost function represents the neural network energy. Successful application of neural networks to the traveling salesman problem [2] had initiated extensive investigations of neural network approaches for the graph bipartition problem [3], neural network optimization of the image processing [4] and many other applications. This subfield of the neural network theory is developing rapidly at the moment [5]-[16].

The aforementioned investigations have the same common feature: the overwhelming majority of neural network optimization algorithms contain the Hopfield model in their core, and the optimization process is reduced to finding the global minimum of some quadratic functional (the energy) constructed on a given $N \times N$ matrix in an N-dimensional configuration space [5,7,13,16]. The standard neural network approach to such a problem consists in a random search of an optimal solution. The procedure consists of two stages. During the first stage the neural network is initialized at random, and during the second stage the neural network relaxes into one of the possible stable states, i.e. it optimizes the energy value. Since the sought result is unknown and the search is done at random, the neural network is to be initialized many times in order to find as deep an energy minimum as possible. But the question about the reasonable number of such random starts and whether the result of the search can be regarded as successful always remains open.

In this paper we have obtained expressions that have demonstrated the relationship between the depth of a local minimum of energy and the size of the basin of attraction (Sections 3-4) like as it was made in [13,14]. Based on these expressions, we presented the probability of finding a local minimum as a function of the depth of the minimum (Section 5). Such a relation can be used in optimization applications: it allows one, based on a series of already found minima, to estimate the probability of finding a deeper minimum, and to decide in favor of or against further running of the program. Our expressions are obtained from the analysis of generalized Hopfield model, namely, of a neural network with Hebbian matrix. They are however valid for any matrices, because any kind of symmetric matrix can be represented as a Hebbian one, constructed on arbitrary number of patterns (see Section 6). A good agreement between our theory and experiment is obtained.

2 Description of the Model

Let us consider Hopfield model, i.e. a system of N Ising spins-neurons $s_i = \pm 1$, $i = 1, 2, ..., N$. A state of such a neural network can be characterized by a configuration $S = (s_1, s_2, ..., s_N)$. Here we consider a generalized model, in which the connection matrix:

$$T_{ij} = \sum_{m=1}^{M} r_m s_i^{(m)} s_j^{(m)}, \quad \sum r_m^2 = 1 \qquad (1)$$

is constructed following Hebbian rule on M binary N-dimensional patterns $\mathbf{S}_m = (s_1^{(m)}, s_2^{(m)}, ..., s_N^{(m)})$, $m = \overline{1, M}$. The diagonal matrix elements are equal to zero ($T_{ii} = 0$). The generalization consists in the fact, that each pattern $\mathbf{S}_m$ is added to the matrix T_{ij} with its statistical weight r_m. We normalize the statistical weights to simplify the expressions without loss of generality. Such a slight modification of the model turns out to be essential, since in contrast to the conventional model it allows one to describe a neural network with a non-degenerate spectrum of minima.

The energy of the neural network is given by the expression:

$$E = -\sum_{i,j=1}^{N} s_i T_{ij} s_j \qquad (2)$$

and its (asynchronous) dynamics consist in the following. Let $\mathbf{S}$ be an initial state of the network. Then the local field $h_i = -\partial E/\partial s_i$, which acts on a randomly chosen i-th spin, can be calculated, and the energy of the spin in this field $\varepsilon_i = -s_i h_i$ can be determined. If the direction of the spin coincides with the direction of the local field ($\varepsilon_i < 0$), then its state is stable, and in the subsequent moment ($t + 1$) its state will undergo no changes. In the opposite case ($\varepsilon_i > 0$) the state of the spin is unstable and it flips along the direction of the local field, so that $s_i(t+1) = -s_i(t)$ with the energy $\varepsilon_i(t+1) < 0$. Such a procedure is to be sequentially applied to all the spins of the neural network. Each spin flip is accompanied by a lowering of the neural network energy. It means that after a finite number of steps the network will relax to a stable state, which corresponds to a local energy minimum.

3 Basin of Attraction

Let us examine at which conditions the pattern $\mathbf{S}_m$ embedded in the matrix (1) will be a stable point, at which the energy E of the system reaches its local minimum E_m. In order to obtain correct estimates we consider the asymptotic limit $N \to \infty$. We determine the basin of attraction of a pattern $\mathbf{S}_m$ as a set of the points of N-dimensional space, from which the neural network relaxes into the configuration $\mathbf{S}_m$. Let us try to estimate the size of this basin. Let the initial state of the network $\mathbf{S}$ be located in a vicinity of the pattern $\mathbf{S}_m$. Then the probability of the network convergation into the point $\mathbf{S}_m$ is given by the expression:

$$\Pr = \left(\frac{1 + erf z}{2}\right)^N, \quad z = \frac{r_m \sqrt{N}}{\sqrt{2(1 - r_m^2)}} \left(1 - \frac{2n}{N}\right). \tag{3}$$

Here $erf z$ is the error function of the variable z and n is Hemming distance between $\mathbf{S}_m$ and $\mathbf{S}$. The expression (3) can be obtained with the help of the methods of probability theory, repeating the well-known calculation [17] for conventional Hopfield model.

It follows from (3) that the probability of the convergence to the point $\mathbf{S}_m$ asymptotically tends to 1 ($\Pr \to 1$ for $N \to \infty$) in the case $n < n_m$, where:

$$n_m = \frac{N}{2}\left(1 - \frac{r_0 \sqrt{1 - r_m^2}}{r_m \sqrt{1 - r_0^2}}\right), \quad r_0 = \sqrt{2 \ln N / N}. \tag{4}$$

In the opposite case ($n > n_m$) we have $\Pr \to 0$. It means that the quantity n_m can be regarded as the radius of the basin of attraction of the local minimum E_m. Accordingly, the basin of attraction have to be determined as the set of the points of the configuration space close to $\mathbf{S}_m$, for which $n \leq n_m$.

It follows from (4) that the radius of basin of attraction tends to zero when $r_m \to r_0$ (Fig.1). It means that the patterns added to the matrix (1), whose statistical weight is smaller than r_0, simply do not form local minima. Local minima exist only in those points $\mathbf{S}_m$, whose statistical weight is relatively large: $r_m > r_0$.

4 Depth of Local Minimum

From analysis of Eq. (2) it follows that the depth of a local minimum (energy E_m) can be represented in the form:

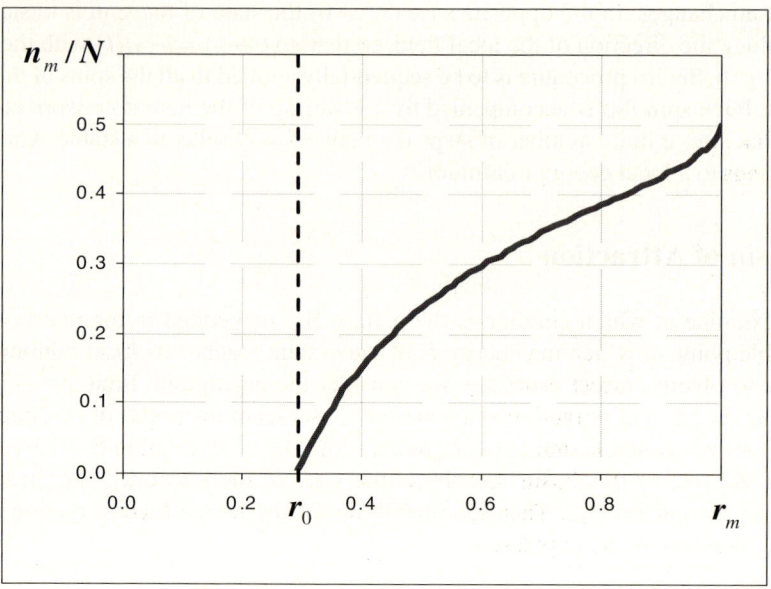

Fig. 1. A typical dependence of the width of the basin of attraction n_m on the statistical weight of the pattern r_m

$$E_m = -r_m N^2 \qquad (5)$$

with the accuracy up to an insignificant fluctuation of the order of σ_m:

$$\sigma_m = N\sqrt{1 - r_m^2}. \qquad (6)$$

Then, taking into account Eqs. (4) and (5), one can easily obtain the following expression:

$$E_m = E_0 \frac{1}{\sqrt{(1 - 2n_m/N)^2 + E_0^2/N^4}}, \quad E_0 = -N\sqrt{2N \ln N}. \qquad (7)$$

This expression describes a relationship between the depth of the local minimum and the size of its basin of attraction. One can see that the wider the basin of attraction, the deeper the local minimum and vice versa: the deeper the minimum, the wider its basin of attraction.

The quantity E_0 introduced in (7) characterizes simultaneously two parameters of the neural network. First, it determines the half-width of the Lorentzian distribution (7). Second, it follows from (7) that $-N^2 \leq E_m \leq E_0$, i.e. E_0 is the upper boundary of the local minimum spectrum and characterizes the minimal possible depth of the local minimum. These results are in a good agreement with the results of computer experiments aimed to check whether there is a local minimum at the point $\mathbf{S}_m$ or not. The results of one of these experiments ($N = 500$, $M = 25$) are shown in Fig.2. One can see a good linear dependence of the energy of the local minimum on the value of the statistical weight of the pattern. Note that the overwhelming number of the experimental points corresponding to the local minima are situated in the right lower quadrant, where

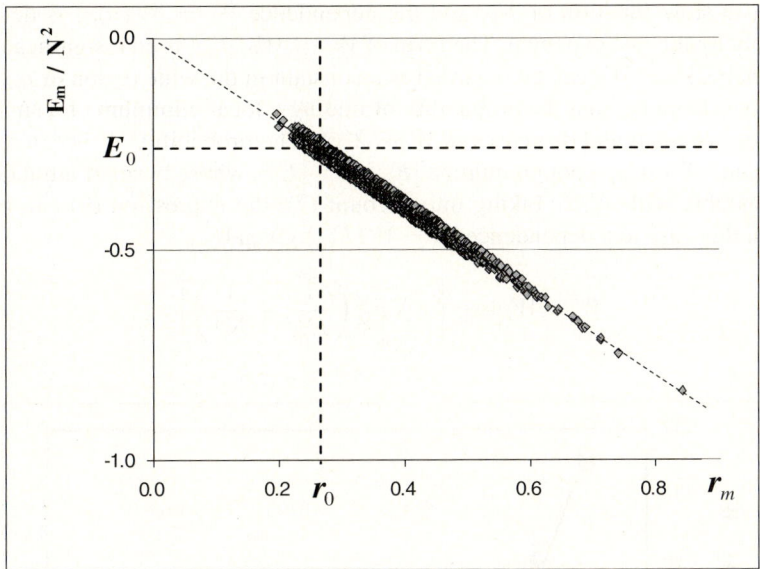

Fig. 2. The dependence of the energy E_m of a local minimum on the statistical weight of the pattern r_m

$r_m > r_0$ and $E_m < E_0$. One can also see from Fig.2 that, in accordance with (6), the dispersion of the energies of the minima decreases with the increase of the statistical weight.

5 The Probability of Finding the Minimum

Let us find the probability W of finding a local minimum E_m at a random search. By definition, this probability coincides with the probability for a randomly chosen initial configuration to get to the basin of attraction of the pattern $\mathbf{S}_m$. Consequently, the quantity $W = W(n_m)$ is the number of points in a sphere of a radius n_m, reduced to the total number of the points in the N-dimensional space:

$$W = 2^{-N} \sum_{n=1}^{n_m} \binom{N}{n}. \tag{8}$$

Equations (7) and (8) define implicitly a connection between the depth of the local minimum and the probability of its finding. Applying asymptotical Stirling expansion to the binomial coefficients (under the condition $n_m \gg 1$) and passing from summation to integration one can represent (8) as

$$W = W_0 \, e^{-Nh}, \tag{9}$$

where h is generalized Shannon function

$$h = \frac{n_m}{N} \ln \frac{n_m}{N} + \left(1 - \frac{n_m}{N}\right) \ln \left(1 - \frac{n_m}{N}\right) + \ln 2. \tag{10}$$

Here W_0 is slow function of E_m and the dependence $W = W(n_m)$ is determined completely by the fast exponent. The form of $W_0 = W_0(n_m)$ is an insignificant for the further analysis and W_0 can be regarded as a constant in the wide region of n_m.

It follows from (9) that the probability of finding a local minimum of a small depth ($E_m \sim E_0$) is small and decreases as $W \sim 2^{-N}$. The probability W becomes visibly non-zero only for deep enough minima $|E_m| \gg |E_0|$, whose basin of attraction sizes are comparable with $N/2$. Taking into account (7), the expression (9) can be transformed in this case to a dependence $W = W(E_m)$ given by

$$W = W_0 \exp\left[-NE_0^2 \left(\frac{1}{E_m^2} - \frac{1}{N^4}\right)\right]. \tag{11}$$

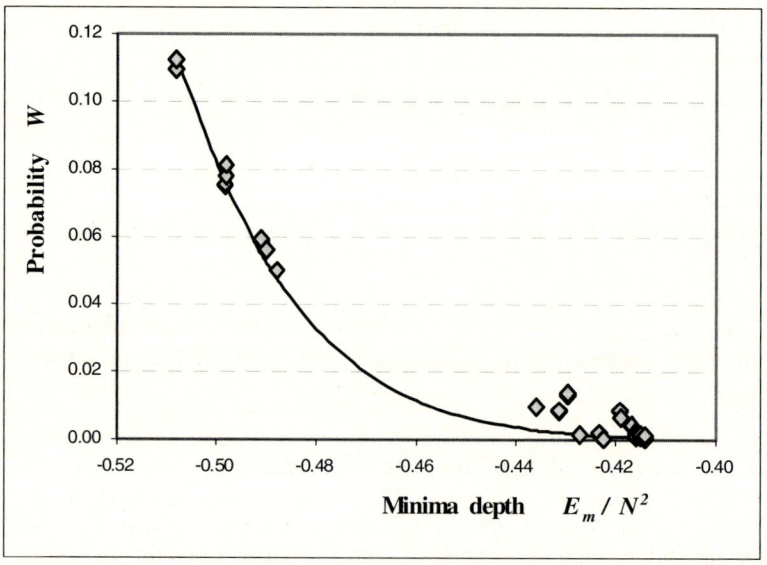

Fig. 3. The dependence of the probability W to find a local minimum in random search on its depth E_m: theory - solid line, experiment - points

It follows from (9) that the probability to find a minimum increases with the increase of its depth. This dependence "the deeper minimum → the larger the basin of attraction → the larger the probability to get to this minimum" is confirmed by the results of numerous experiments. In Fig.3 the solid line is computed from Eq.(9), and the points correspond to the experiment (Hebbian matrix with a small loading parameter $M/N \le 0.1$). One can see that a good agreement is achieved first of all for the deepest minima, which correspond to the patterns $\mathbf{S}_m$ (the energy interval $E_m \le -0.49N^2$ in Fig.3). The experimentally found minima of small depth (the points in the region $E_m > -0.44N^2$) are the so-called "chimeras". In standard Hopfield model ($r_m \equiv 1/\sqrt{M}$) they appear at relatively large loading parameter $M/N > 0.05$. In the more general case, which we consider here, they can appear also earlier. The reasons leading to their appearance are well examined with the help of the methods of statistical physics in [18], where it was shown that the chimeras appear as a consequence of interference of the minima of $\mathbf{S}_m$. At a small loading parameter the chimeras are separated from the minima of $\mathbf{S}_m$ by an energy gap clearly seen in Fig.3.

6 The Iterative Algorithm

Now we shall describe the iterative algorithm which should help us to represent any symmetric matrix $\hat{\mathbf{T}}$ as a weighted Hebbian series (1) in stationary points of the Hopfield network with a given accuracy ε_M.

Let $\hat{\mathbf{T}}_k$ be a matrix

$$\hat{\mathbf{T}}_k = \hat{\mathbf{T}}_{k-1} - r_k \mathbf{S}_k^+ \mathbf{S} = \hat{\mathbf{T}} - \sum_{m=1}^{k} r_m \mathbf{S}_m^+ \mathbf{S}_m, \tag{12}$$

whose elements can be written as

$$T_{k,ij} = T_{ij} - \sum_{m=1}^{k} r_m s_i^{(m)} s_j^{(m)}. \tag{13}$$

Let us determine the error function ε_k after k iterations as:

$$\varepsilon_k = \sum_{i=1}^{N} \sum_{j=1}^{N} T_{k,ij}^2. \tag{14}$$

We stipulate that $\varepsilon_k \to 0$ with k. If after k iterations ($k \geq M$) the error becomes smaller than a predefined value ε_M, then we can say that we have solved the problem: expression (1) is true to a given accuracy.

The iterative algorithm is as follows. We generate a random configuration vector $\mathbf{S}_k$ after k iterations. Taking (13) into consideration, we can rewrite (14) as

$$\varepsilon_k = \sum_{i=1}^{N} \sum_{j=1}^{N} \left(T_{k-1,ij} - r_k s_{ki} s_{kj}\right)^2, \tag{15}$$

or as

$$\varepsilon_k = \varepsilon_{k-1} + 2r_k E_k + r_k^2 N^2, \tag{16}$$

where

$$E_k = -\sum_{i=1}^{N} \sum_{j=1}^{N} T_{k-1,ij} s_{ki} s_{kj}. \tag{17}$$

It follows from (16) that the minimum of error ε_k can be reached if the statistical weight is defined to be

$$r_k = -E_k/N^2. \tag{18}$$

Then the error minimum can be written as

$$\varepsilon_k = \varepsilon_{k-1} - r_k^2 N^2. \tag{19}$$

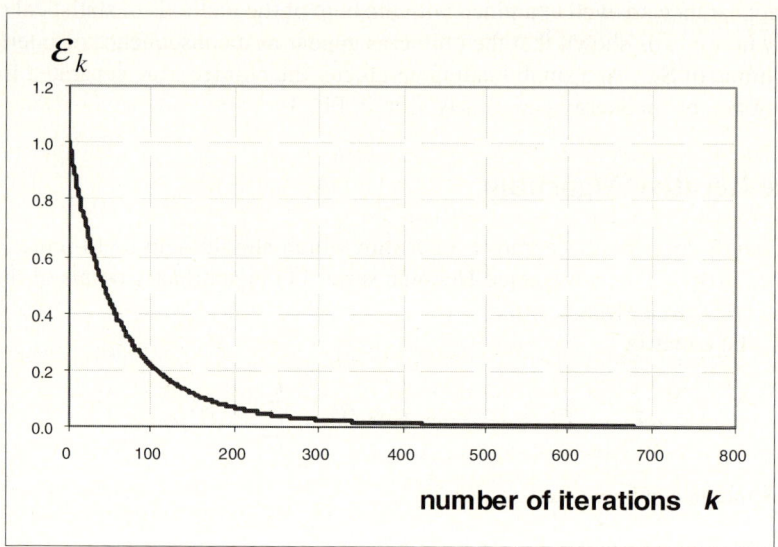

Fig. 4. The dependence of the error function value ε_k on the number of iterations k

It is seen from (19) that ε_k can be made still smaller if we increase statistical weight r_k. To do this, we should start the standard spin optimization [2] of quadratic functional E_k, in other words, we should optimize the energy of the Hopfield network (17) built around matrix $\hat{\mathbf{T}}_{k-1}$. Vector $\mathbf{S}_k$ is replaced by $\mathbf{S}_k^{(opt)}$ which corresponds to the extremum of functional E_k. The greatest (in absolute magnitude) value of statistical weight r_k and smallest value of error function ε_k are reached at point $\mathbf{S}_k^{(opt)}$. Further, taking (12) into account, we use vector $\mathbf{S}_k = \mathbf{S}_k^{(opt)}$ and corresponding statistical weight r_k to build matrix $\hat{\mathbf{T}}_k = \hat{\mathbf{T}}_{k-1} - r_k \mathbf{S}_k^+ \mathbf{S}_k$ and go to the next iteration.

Now let us show that the above algorithm converges after a finite number of iterations. If we take into consideration the relation between the energy of the extremum and pattern weight ($E_k = -r_k N^2$), we can rewrite (19) as

$$\varepsilon_k = \varepsilon_k(1 - r_k^2). \tag{20}$$

Expression (20) can be in turn rewritten as

$$\varepsilon_k = \varepsilon_0 \prod_{m=1}^{k} (1 - r_m^2), \tag{21}$$

where ε_0 is the initial value of the error function, which is equal to the sum of squares of elements of initial matrix $\mathbf{T}$. Earlier we found the limitation on the statistical weight ($r_m > r_0$) of the pattern corresponding to the extremum of the energy functional:

$$r_k^2 \geq \frac{2 \ln N}{N}. \tag{22}$$

From (20)-(22) it follows that

$$\varepsilon_k \sim \varepsilon_0 \exp\left(-\frac{2k \ln N}{N}\right). \quad (23)$$

Expression (23) allows us to estimate the number M of iterations necessary to provide the given accuracy of series (1). One can easily notice that the number of iteration is finite. Computer simulations prove the conclusion (see Fig.4).

7 Discussion

Our analysis shows that the properties of the generalized model are described by two parameters r_0 and E_0. The first determines the minimal value of the statistical weight at which the pattern forms a local minimum. The second one is the minimal depth of the local minima. It is important that these parameters are independent from the number of embedded patterns M.

Now we are able to formulate a heuristic approach of finding the global minimum of the functional (2) for any given matrix (not necessarily Hebbian one). The idea is to use the expression (11) with unknown parameters W_0 and E_0. To do this one starts the procedure of the random search and finds some minima. Using the obtained data, one determines typical value of E_0 and the fitting parameter W_0 for the given matrix. Substituting these values into (11) one can estimate the probability of finding an unknown deeper minimum E_m (if it exists) and decide in favor or against (if the estimate is a pessimistic one) the further running of the program. This approach was tested with Hebbian matrices at relatively large values of the loading parameter ($M/N \geq 0.2 \div 10$). The result of one of the experiments is shown in Fig.5. In this experiment with the aid of the found minima (the points A) the parameters W_0 and E_0 were calculated, and the dependence (solid line) was found. After repeating the procedure of the random search over and over again ($\sim 10^5$ random starts) other minima (points B) and the precise probabilities of getting into them were found. One can see that although some dispersion is present, the predicted values in the order of magnitude are in a good agreement with the precise probabilities.

In conclusion we stress once again that any given symmetric matrix can be performed in the form of Hebbian matrix (1) constructed on an arbitrary number of patterns (for instance, $M \to \infty$) with arbitrary statistical weights. It means that the dependence "the deeper minimum $\leftrightarrow$ the larger the basin of attraction $\leftrightarrow$ the larger the probability to get to this minimum" as well as all other results obtained in this paper are valid for all kinds of matrices. To prove this dependence, we have generated random matrices, with uniformly distributed elements on [-1,1] segment. The results of a local minima search on one of such matrices are shown in Fig. 6. More than 10^5 of local minima were found (in the region $-0.145 \leq E_m/N^2 \leq -0.095$) in 10^6 random starts, and most of them concentrated in central part of spectrum ($E_m/N^2 \sim -0.12$). Despite of such a complex view of the spectrum of minima, the deepest minimum is found with maximum probability ($W = 0.017$ for $E_m = -0.145N^2$). The same perfect accordance of the theory and the experimental results are also obtained in the case of random matrices, the elements of which are subjected to the Gaussian distribution with a zero mean.

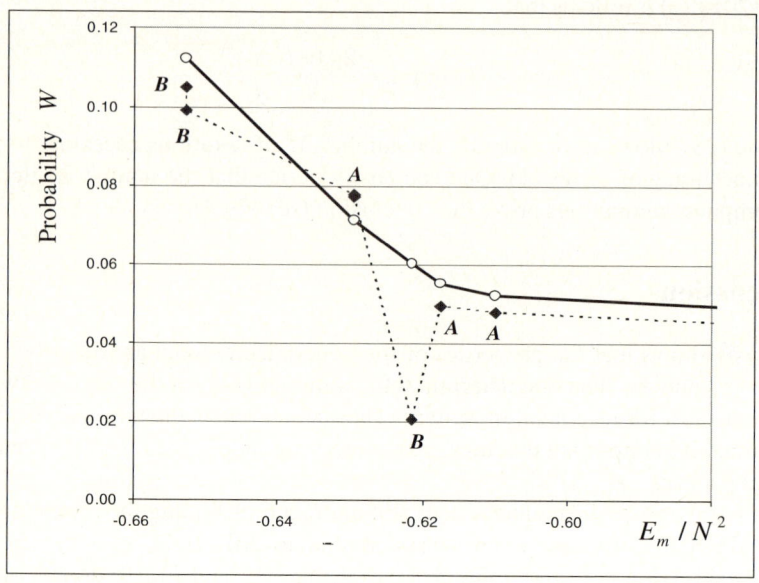

Fig. 5. The comparison of the predicted probabilities (solid line) and the experimentally found values (points connected with the dashed line)

We see that the iterative algorithm allows a finite number of iterations to represent any symmetric $N \times N$ matrix as a weighed Hebbian series of configuration vectors (1) with a given accuracy. It so proves that all conclusions about neural networks and optimization algorithms that are based on Hebbian matrices are true for any other type of matrix. In short, the whole iterative procedure looks as follows:

1. A random configuration vector $\mathbf{S}_k$ is generated in the k-th iteration step.
2. The vector is $\mathbf{S}_k$ used as an initial configuration for minimizing (maximizing) energy functional E_k built around matrix $\hat{\mathbf{T}}_{k-1}$: a conventional neural-net method of descent (ascent) over the landscape is used to find the extremum point $\mathbf{S}_k = \mathbf{S}_k^{(opt)}$ of energy E_k.
3. Formula (18) is used to compute the optimal value of statistical weight r_k and to build, according to (12), matrix $\hat{\mathbf{T}}_k = \hat{\mathbf{T}}_{k-1} - r_k \mathbf{S}_k^+ \mathbf{S}_k$;
4. The $(k+1)$-th iteration is then performed. The procedure is repeated until error ε_k becomes smaller than ε_M.

It should be pointed out again that error ε_k and functional E_k do not necessarily reach the minimum concurrently. Point $\mathbf{S}_k = \mathbf{S}_k^{(opt)}$ can correspond either to the minimum or to the maximum of the energy, depending on what we try to find, the minimum or maximum of the energy functional. The weight (18) can be either positive or negative, correspondingly. The choice of a particular algorithm of energy functional optimization (seeking the minimum or maximum) is up to the user. If the goal is the expansion in the vectors corresponding to the minimum of the functional, the dynamics that makes E_k decrease should be used. If we want to have the expansion in the vectors corresponding to the maximum of the functional, the dynamics that makes E_k grow is the right choice.

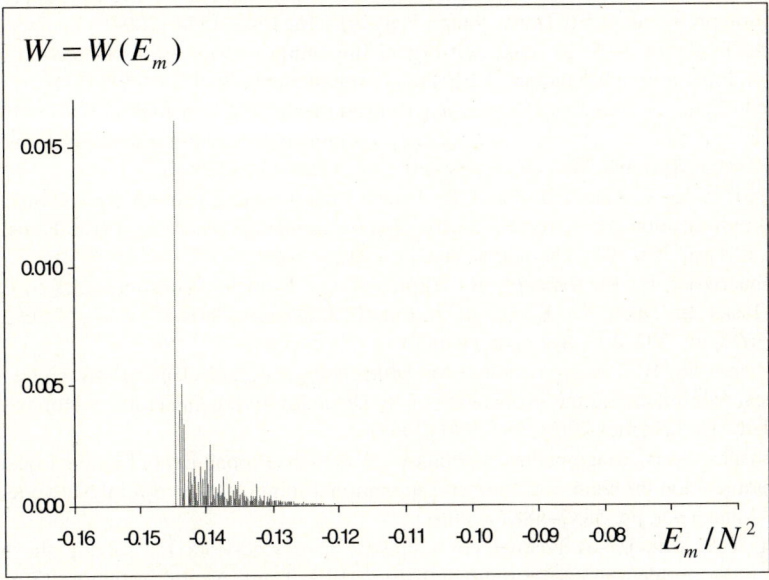

Fig. 6. Probability of finding a minimum with energy E_m in the case of a random matrix with a quasi-continuous type of spectrum

If we want to speed up the iteration process, of two procedures (seeking the minimum or maximum) we should choose the one that provides the largest magnitude of $|E_k|$.

Acknowledgment

The work supported by Russian Basic Research Foundation grant No. 06-01-00109.

References

1. Hopfield, J.J.: Neural Networks and physical systems with emergent collective computational abilities. Proc. Nat. Acad. Sci.USA. 79, 2554–2558 (1982)
2. Hopfield, J.J., Tank, D.W.: Neural computation of decisions in optimization problems. Biological Cybernetics 52, 141–152 (1985)
3. Fu, Y., Anderson, P.W.: Application of statistical mechanics to NP-complete problems in combinatorial optimization. Journal of Physics A 19, 1605–1620 (1986)
4. Poggio, T., Girosi, F.: Regularization algorithms for learning that are equivalent to multilayer networks. Science 247, 978–982 (1990)
5. Smith, K.A.: Neural Networks for Combinatorial Optimization: A Review of More Than a Decade of Research. INFORMS Journal on Computing 11(1), 15–34 (1999)
6. Joya, G., Atencia, M., Sandoval, F.: Hopfield Neural Networks for Optimization: Study of the Different Dynamics. Neurocomputing 43, 219–237 (2002)
7. Hartmann, A.K., Rieger, H.: New Optimization Algorithms in Physics. Wiley-VCH, Berlin (2004)

8. Tang, H., Tan, K.C., Yi, Z.: A columnar competitive model for solving combinatorial optimization problems. IEEE Trans. Neural Networks 15, 1568–1574 (2004)
9. Kwok, T., Smith, K.A.: A noisy self-organizing neural network with bifurcation dynamics for combinatorial optimization. IEEE Trans. Neural Networks 15, 84–98 (2004)
10. Salcedo-Sanz, S., Santiago-Mozos, R., Bousono-Calzon, C.: A hybrid Hopfield network-simulated annealing approach for frequency assignment in satellite communications systems. IEEE Trans. Systems, Man and Cybernetics 34, 1108–1116 (2004)
11. Wang, L.P., Li, S., Tian, F.Y., Fu, X.J.: A noisy chaotic neural network for solving combinatorial optimization problems: Stochastic chaotic simulated annealing. IEEE Trans. System, Man, Cybern, Part B - Cybernetics 34, 2119–2125 (2004)
12. Kryzhanovsky, B., Magomedov, B.: Application of Domain Neural Network to Optimization Tasks. In: Duch, W., Kacprzyk, J., Oja, E., Zadrożny, S. (eds.) ICANN 2005. LNCS, vol. 3697, pp. 397–403. Springer, Heidelberg (2005)
13. Kryzhanovsky, B.V., Magomedov, B.M., Mikaelyan, A.L.: A Relation Between the Depth of a Local Minimum and the Probability of Its Detection in the Generalized Hopfield Model. Doklady Mathematics 72(3), 986–990 (2005)
14. Kryzhanovsky, B., Magomedov, B., Fonarev, A.: On the Probability of Finding Local Minima in Optimization Problems. In: Proc. of International Joint Conf. on Neural Networks IJCNN-2006 Vancouver, pp. 5882–5887 (2006)
15. Wang, L.P., Shi, H.: A gradual noisy chaotic neural network for solving the broadcast scheduling problem in packet radio networks. IEEE Trans. Neural Networks 17, 989–1000 (2006)
16. Litinskii, L.B., Magomedov, B.M.: Global Minimization of a Quadratic Functional: Neural Networks Approach. Pattern Recognition and Image Analysis 15(1), 80–82 (2005)
17. Perez-Vincente, C.J.: Finite capacity of sparce-coding model. Europhys. Lett. 10, 627–631 (1989)
18. Amit, D.J., Gutfreund, H., Sompolinsky, H.: Spin-glass models of neural networks. Physical Review A 32, 1007–1018 (1985)

Nonlinear Function Learning Using Radial Basis Function Networks: Convergence and Rates

Adam Krzyżak[1,3,*] and Dominik Schäfer[2]

[1] Department of Computer Science and Software Engineering,
Concordia University,
Montreal, Canada H3G 1M8
krzyzak@cs.concordia.ca
[2] Department of Mathematcs
Stuttgart University
D-70569 Stuttgart, Germany
schaefdk@mathematik.uni-stuttgart.de
[3] Institute of Control Engineering
Technical University of Szczecin
70-313 Szczecin, Poland

Abstract. We apply normalized RBF networks to the problem of learning nonlinear regression functions. The parameters of the networks are learned by empirical risk minimization and complexity regularization. We study convergence of the RBF networks for various radial kernels as the number of training samples increases. The rates of convergence are also examined.

Keywords: Normalized radial basis function networks, convergence, rates of convergence.

1 Introduction

The most popular feedforward neural network architectures include multilayer perceptrons (MLP), standard radial basis function (RBF) networks and normalized radial basis function (NRBF) networks. They have been applied applicable to various tasks, such as estimation, classification, data smoothing and prediction, just to mention a few of them. Accounts of these methods can be found in, e.g., Anthony and Bartlett [1], Barron [2], Cybenko [4], Devroye et al. [5], Györfi et al. [12] and Hornik et al. [14] for multilayer perceptrons; for RBF networks see, e.g., Girosi and Anzellotti [9], Györfi et al. [12], Krzyżak et al. [16], Krzyżak and Linder [17], Krzyżak and Schäfer [19], Moody and Darken [21], Park and Sandberg [22,23] and Ripley [26].

---

[*] Part of this research was carried out when the first author was with the Technical University of Szczecin on sabbatical leave from Concordia University. The second author is presently at Würtembergische Versicherung AG, Stuttgart, Germany. This research has been supported by the grant from Natural Sciences and Engineering Research Council of Canada.

RBF networks have been shown to be the solution of the regularization problem in function estimation with certain standard smoothness functionals used as stabilizers (see Girosi [8], Girosi *et al.* [10], and the references therein). Universal convergence of RBF nets in function estimation and classification has been proven by Krzyżak *et al.* [16]. Approximation error convergence rates for RBF networks have been studied by Girosi and Anzellotti [9].

In the paper we study normalized RBF networks and apply them to nonlinear functions learning. Let $\mathcal{F}(k, \ell, L, R, B)$ denote *normalized radial basis function (NRBF) networks* with one hidden layer of k nodes ($k \in \mathbb{N}, L \geq \ell \geq 0, R, B > 0$), where ℓ, L, R, B are network parameters defined below. The class $\mathcal{F}$ contains the functions of the form

$$f(x) = \frac{\sum_{i=1}^{k} w_i K\left((x - c_i)^T A_i (x - c_i)\right)}{\sum_{i=1}^{k} K\left((x - c_i)^T A_i (x - c_i)\right)}$$

$$=: \frac{\sum_{i=1}^{k} w_i K_{c_i, A_i}(x)}{\sum_{i=1}^{k} K_{c_i, A_i}(x)}. \tag{1}$$

Parameters of the network satisfy the following conditions:

(i) Kernel $K : \mathbb{R}_0^+ \to \mathbb{R}^+$ is a left-continuous, decreasing function, the so-called *kernel*.
(ii) Vectors $c_1, ..., c_k \in \mathbb{R}^d$ are the so-called *center vectors* with $\|c_i\| \leq R$ for all $i = 1, ..., k$.
(iii) Matrices $A_1, ..., A_k$ are symmetric, positive definite, real $d \times d$-matrices each of which satisfies the eigenvalue inequalities $\ell \leq \lambda_{min}(A_i) \leq \lambda_{max}(A_i) \leq L$. Here, $\lambda_{min}(A_i)$ and $\lambda_{max}(A_i)$ are the minimal and the maximal eigenvalue of A_i, respectively. A_i specifies the shape of the *receptive field* about the center c_i.
(iv) Scalars $w_1, ..., w_k \in \mathbb{R}$ are the bounded *weights* satisfying $|w_i| \leq B$ for all $i = 1, ..., k$.

Throughout we use the convention $0/0 = 0$. In the paper the kernel is fixed whereas network parameters $w_i, c_i, A_i, i = 1 ..., k$ are learned from the data. Common choices for the kernel are (we implicitly assume they satisfy (i)):

- *Window type kernels.* These are kernels for which some $\delta > 0$ exists such that $K(v) \notin (0, \delta)$ for all $v \in \mathbb{R}_0^+$. The classical naive kernel $K(v) = \mathbf{1}_{[0,1]}(v)$ is a member of this class.
- *Non-window type kernels with compact support.* These comprise all kernels with support of the form $[0, s]$ which are right-continuous in s. For example, for $K(v) = \max\{1 - v, 0\}$, $K(x^T x)$ is the Epanechnikov kernel.
- *Kernels with non-compact support,* i.e. $K(v) > 0$ for all $v \in \mathbb{R}_0^+$. The best known example of this class is $K(v) = \exp(-v)$. Then $K(x^T x)$ is the classical Gaussian kernel. With $K(v) = \exp(-\sqrt{v})$ we get exponential kernel $K(x^T x)$.

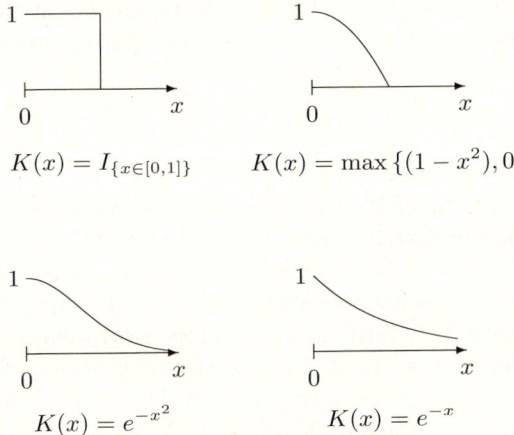

Fig. 1. Window, truncated parabolic, Gaussian, and exponential kernels

The typical kernels are (see [12]):

- $K(x) = I_{\{x \in [0,1]\}}$ (window);
- $K(x) = \max\{(1 - x^2), 0\}$ (truncated parabolic);
- $K(x) = e^{-x^2}$ (Gaussian); and
- $K(x) = e^{-x}$ (exponential).
- $K(x) = \frac{1}{\sqrt{x^2+1}}$ (inverse multiquadratic).

NRBF networks were introduced by Moody and Darken [21] and Specht [28] by normalizing standard RBF networks defined by

$$f(x) = \sum_{i=1}^{k} w_i K((x - c_i)^T A_i (x - c_i)) + w_0. \qquad (2)$$

They naturally occur in various applications, such as smoothing splines (cf. Duchon [6]), interpolation with multiquadrics, shifted surface splines or thin-plate splines (see Girosi et al. [10]), and regression analysis (see, e.g., Györfi et al. [12] and Krzyżak et al. [16]). In [16] the authors consider RBF regression estimates and classifiers based on minimizing the empirical risk and prove universal consistency results. The rate of convergence of RBF-based regression estimates using complexity regularization have been studied by Krzyżak and Linder [17]. Practical and theoretical merits and demerits of the normalization leading from RBFs to NRBFs are discussed by Shorten and Murray-Smith [27] and some references therein.

In this paper we apply NRBFs in nonlinear function (nonlinear regression) learning from random measurements. Formally, suppose that we are given the random variables X and Y taking values in $\mathbb{R}^d$ and $\mathbb{R}$, respectively. The task is to find a nonlinear function $f : \mathbb{R}^d \to \mathbb{R}$ that is a good approximation of Y in the mean squared error sense (L_2 theory). In particular, if $\mathbf{E}|Y|^2 < \infty$, we aim to find a measurable function r minimizing the L_2-risk, that is,

$$J^* = \inf_{f:\mathbb{R}^d \to \mathbb{R} \text{ measurable}} \mathbf{E}|f(X) - Y|^2 = \mathbf{E}|r(X) - Y|^2.$$

The solution of this minimization problem is given by the regression function $r(x) = \mathbf{E}[Y|X = x]$. In order to determine the regression function we need some information about distribution of the data (X, Y). However, nature rarely divulges any knowledge of distributions in practice. Thus we need to rely exclusively on the training sequences to to extract r. In this paper we assume no knowledge of the data distributions so we adopt **nonparametric** approach.

In order to estimate r without making any assumption about the distribution of (X, Y), we assume that a training set $D_n := \{(X_1, Y_1), ..., (X_n, Y_n)\}$ of independent, identically distributed copies of (X, Y) is given, where D_n is independent of (X, Y). The method of **empirical risk minimization** obtains an estimate $\hat{f}_n$ of r by selecting the parameter vector which minimizes the residual sum of squares over a suitable class $\mathcal{F}_n$ of functions. In other words, based on the training sequence, we choose an estimator $\hat{f}_n \in \mathcal{F}_n$, such that $\hat{f}_n$ minimizes the *empirical* L_2-risk (mean residual sum of squares)

$$J_n(f) = \frac{1}{n} \sum_{j=1}^{n} |f(X_j) - Y_j|^2,$$

that is,

$$J_n(\hat{f}_n) \leq J_n(f) \text{ for all } f \in \mathcal{F}_n$$

or

$$\hat{f}_n(\cdot) = \arg\min_{f \in \mathcal{F}_n} \frac{1}{n} \sum_{i=1}^{n} |f(X_i) - Y_i|^2. \tag{3}$$

The performance of the regression estimate $\hat{f}_n$ is measured by

$$J(\hat{f}_n) = \mathbf{E}[|\hat{f}_n(X) - Y|^2 | D_n].$$

In this framework a regression estimator $\hat{f}_n$ is called **strongly consistent** if it asymptotically attains the minimal L_2-risk J^* in the **P**-a.s. (probability limit) sense, i.e., if

$$J(\hat{f}_n) - J^* \to 0 \quad \mathbf{P}-a.s. \quad \text{(in probability)} \quad \text{as } n \to \infty. \tag{4}$$

Observe that $J(\hat{f}_n) - J^* \to 0$ if and only if

$$\mathbf{E}[|\hat{f}_n(X) - Y|^2 | D_n] - \mathbf{E}|r(X) - Y|^2 = \mathbf{E}[|\hat{f}_n(X) - r(X)|^2 | D_n] \to 0,$$

which is the usual notion of L_2-consistency for regression function estimates (cf. Györfi et al. [12]).

It is clear that the choice of optimal network parameters is determined by the need of balancing between two quantities, into which the error can be decomposed

$$J(\hat{f}_n) - J^* = \left(\inf_{f \in \mathcal{F}_n} J(f) - J^* \right) + \left(J(\hat{f}_n) - \inf_{f \in \mathcal{F}_n} J(f) \right) =: A_n + E_n$$

where A_n is an **approximation error** and E_n is an **estimation error**.

The idea of empirical risk minimization has been extensively used in literature. When the minimization is carried out over exceedingly rich (complex) families $\mathcal{F}_n$ of candidate functions, the resulting estimate usually overfits the data, i.e., it is not likely to perform well for new data that is independent of the training set. Different measures of complexity of $\mathcal{F}_n$ have been used for different purposes, but they are all related to the cardinality of a finite subset representing the whole family in a certain sense. Examples are metric entropy (Kolmogorov and Tihomirov [15]), VC-dimension (Vapnik [29]) and random covering numbers (Pollard [24]). Based on these measures, asymptotic properties of the method of empirical risk minimization were studied among others by Vapnik [29] and Haussler [13]. The class $\mathcal{F}_n$ of candidate functions should clearly allow the statistician to find good approximations for a multitude of target functions. Therefore, one generally needs to increase the size of the candidate family commensurate with the size of the training set. However, a good tradeoff should also be maintained between the complexity of the candidate family and the training data size to avoid overfitting. The idea of using nested candidate classes which grow in a *controlled* manner with the size of the training data is Grenander's method of sieves [11]. This approach was successfully applied to pattern recognition by Devroye et al. [5], to regression estimation by Györfi et al. [12] and Lugosi and Zeger [20], and by Faragó and Lugosi [7] in the neural network framework. It has also been used to study consistency of NRBF regression estimates in the limited framework by Krzyżak and Niemann [18] and Xu et al. [30].

In this paper we also investigate the problem of determining the optimal number of hidden nodes k and the rate of convergence via the **complexity regularization principle** (CRP). CRP for the learning problem was introduced by Vapnik [29] and fully developed by Barron [3] (see also Lugosi and Zeger [20] and Devroye *et al.* [5]). It enables the learning algorithm to choose k automatically. Complexity regularization penalizes the large candidate classes, which are bound to have small approximation error, in favor of the smaller ones. One form of this method, the minimum description length principle [25] uses as the penalty the length of a binary code describing the class.

In this paper we study strong universal convergence of the nonlinear function estimates derived from NRBF networks when *all* network parameters (including the centers) are trained by the empirical risk minimization. We also extend the results of [19] by obtaining the rates of convergence of the mean squared errors for the network with parameters trained by the empirical risk minimization and with number of hidden neurons optimized by complexity regularization. These results are obtained by applying the modern tools such as covering numbers, VC dimensions, and bu utilizing their connections with each other to feed-forward

NRBF networks, see [12]. Convergence of the learning algorithm is discussed in Section II and the rates of convergence in Section III.

2 Convergence of the Learning Algorithm

We only consider the case of non-window type kernels with compact support and kernels with non-compact support. The convergence results follow from [19]. Let the parameters $m_n \in \mathbb{N}, L_n, R_n$ and $B_n > 0$ tend to ∞ as $n \to \infty$. Note that the $\mathcal{F}(m_n, 0, L_n, R_n, B_n)$ are not nested as n increases. We therefore consider the nested models

$$\mathcal{F}_n := \bigcup_{k=1}^{m_n} \mathcal{F}(k, 0, L_n, R_n, B_n).$$

Suppose i.i.d. observations $D_n := \{(X_1, Y_1),...,(X_n, Y_n)\}$ are available. Let the estimate $\hat{f}_n \in \mathcal{F}_n$ of the regression function $r(\cdot) = \mathbf{E}[Y|X = \cdot]$ be chosen by empirical risk minimization, i.e.,

$$\hat{f}_n(\cdot) = \arg\min_{f \in \mathcal{F}_n} \frac{1}{n} \sum_{i=1}^n |f(X_i) - Y_i|^2. \qquad (5)$$

For kernels with non-compact support we have the following result:

Theorem 1. *Suppose $m_n, L_n, R_n, B_n \to \infty$ $(n \to \infty)$ in such a way that for kernel K with unbounded support*

$$\frac{B_n^4 m_n}{n} \log \frac{B_n^2}{K(4R_n^2 L_n)} \to 0.$$

and

$$B_n^4 \leq \frac{n}{(1+\beta)\log n} \qquad (6)$$

for some $\beta > 0$ and all sufficiently large n. Then the estimate $\hat{f}_n$ defined by (5) converges strongly for any distribution of (X, Y) with $\mathbf{E}Y^2 < \infty$ and $\|X\| \leq Q < \infty$.

Non-window type kernels with bounded support $[0, s]$ require special care. In the case of kernels of unbounded support we are in a position to control the size of the denominator in (1) by assuming $K(v) \notin (0, \delta)$ and the receptive field condition (iii) in Section I, respectively. For non-window type kernels with bounded support, the denominator can attain very small values, the ratio in (1) may becoming unstable. Thus we propose the following slightly modified learning procedure.

For $\delta > 0$ and $f \in \mathcal{F}(k, \ell, L, R, B)$ let $f^{(\delta)}$ be the NRBF with the same parameters as f, except for the kernel $K(v)$, which is replaced by the window type kernel

$$K^{(\delta)}(v) = \begin{cases} \delta & \text{if } K(v) \in (0, \delta] \text{ or } v = s, \\ K(v) & \text{otherwise.} \end{cases}$$

Now fix positive sequences $\delta_n, \ell_n \to 0$ $(n \to \infty)$ and choose the estimate $\hat{f}_n(\cdot)$ from $\mathcal{F}_n := \bigcup_{k=1}^{k_n} \mathcal{F}(k, \ell_n, L_n, R_n, B_n)$ such as to satisfy

$$\hat{f}_n(\cdot) = \arg \min_{f \in \mathcal{F}_n} \frac{1}{n} \sum_{i=1}^{n} |f^{(\delta_n)}(X_i) - Y_i|^2. \tag{7}$$

For suitable choice of the sequences δ_n and ℓ_n the following convergence result:

Theorem 2. *Let the kernel K be a non-window type kernel with bounded support $[0, s]$. Suppose $k_n, L_n, R_n, B_n \to \infty$ and $\ell_n, \delta_n \to 0$ $(n \to \infty)$ in such a way that*

$$\frac{B_n^4 k_n}{n} \log \frac{B_n^2 k_n}{\delta_n} \to 0 \tag{8}$$

$$\frac{B_n^4 k_n}{\ell_n^{d/2}} \left(s^{d/2} - (K^{-1}(\delta_n))^{d/2} \right) \to 0, \tag{9}$$

and let (6) hold with $K^{-1}(\delta) := \sup\{x : K(x) \geq \delta\}$, $\sup \emptyset := 0$. Then the estimate $\hat{f}_n$ defined by (7) converges strongly for any distribution of (X, Y) where X has a bounded density w.r.t. the Lebesgue measure and $\mathbf{E}Y^2 < \infty$.

The proofs of the Theorems rely on upper bounds on the covering number of the class of NRBFs and on their properties [19].

3 Rate of Convergence of the Learning Algorithm

We apply the principle of complexity regularization to establish the rate of convergence of NRBF networks regression function learning algorithm. This approach enables us to automatically adapt to the smoothness of the regression function and to adapt the structure of the network (the number of the hidden neurons) to the data.

We begin with a measure of complexity of a class of functions. Let $\mathcal{F}$ be a class of real-valued functions on $\mathbb{R}^d$. Let $x_1^n = (x_1, ..., x_n) \in \mathbb{R}^{dn}$ and $\epsilon > 0$. We say a class $\mathcal{G}$ of real-valued functions on $\mathbb{R}^d$ is an ϵ-cover of $\mathcal{F}$ in x_1^n, if for each $f \in \mathcal{F}$ there exists a $g \in \mathcal{G}$ such that

$$\frac{1}{n} \sum_{j=1}^{n} |f(x_j) - g(x_j)| \leq \epsilon.$$

The *covering number* $\mathbb{N}(\epsilon, \mathcal{F}, x_1^n)$ is the smallest integer m such that an ϵ-cover $\mathcal{G}$ of $\mathcal{F}$ in x_1^n exists with cardinality $|\mathcal{G}| = m$. Covering numbers are useful in studying convergence of algorithms learned from random data (see [24], [5], [12] for their properties and applications). Assume that output weights in NRBF networks in $\mathcal{F}_k$ satisfy $\sum_{i=1}^{k} |w_i| \leq \beta_n$.

The learning algorithm will be defined in two steps.

1. Let
$$m_{n,k} = \arg\min_{f \in \mathcal{F}_k} \frac{1}{n} \sum_{i=1}^{n} |f(X_i) - Y_i|^2.$$
Hence $m_{n,k}$ minimizes the empirical L_2 risk for n training samples over $\mathcal{F}_k$. (We assume the existence of such a minimizing function for each k and n.)
2. Define the complexity penalty of the kth class for n training samples as any nonnegative number $pen_n(k)$ satisfying

$$pen_n(k) \geq 2568 \frac{\beta_n^4}{n} \cdot (\log \mathbb{N}(1/n, \mathcal{F}_{n,k}) + t_k), \tag{10}$$

where $\mathbb{N}(\epsilon, \mathcal{F}_{n,k})$ is almost sure uniform upper bound on the random covering numbers $\mathbb{N}(\epsilon, \mathcal{F}_{n,k}, X_1^n)$ and the nonnegative constants $t_k \in \mathbb{R}_+$ satisfy Kraft's inequality $\sum_{k=1}^{\infty} e^{-t_k} \leq 1$. The coefficients t_k may be chosen as $t_k = 2 \log k + t_0$ with $t_0 \geq \log\left(\sum_{k \geq 1} k^{-2}\right)$.

The penalized empirical L_2 risk is defined for each $f \in \mathcal{F}_k$ as

$$\frac{1}{n} \sum_{i=1}^{n} |f(X_i) - Y_i|^2 + pen_n(k).$$

Our estimate f_n is then defined as the $m_{n,k}$ minimizing the penalized empirical risk over all classes

$$f_n = m_{n,k^*}, \tag{11}$$

where
$$k^* = \arg\min_{k \geq 1} \left(\frac{1}{n} \sum_{i=1}^{n} |m_{n,k}(X_i) - Y_i|^2 + pen_n(k) \right).$$

The next theorem presents the rate of convergence in the class of convex closure $\overline{\mathcal{F}}$ of $\mathcal{F} = \bigcup_k \mathcal{F}_k$ in $L_2(\mu)$.

Theorem 3. *Let $1 \leq L < \infty, n \in \mathbb{N}$, and let $L \leq \beta_n < \infty$. Suppose, furthermore, that $|Y| \leq L < \infty$ a.s. Let K be a non-window type kernel with compact support or a kernel with non-compact support. Assume that the penalty satisfies (10) for some t_k such that $\sum_{k=1}^{\infty} e^{-t_k} \leq 1$. If $r \in \overline{\mathcal{F}}$ then the NRBF regression learning algorithm with parameters learned by the complexity regularization (11) satisfies for n sufficiently large*

$$\mathbf{E} \int |f_n(x) - r(x)|^2 \mu(dx) = O\left(\beta_n^2 \left(\frac{\log(\beta_n n)}{n}\right)^{1/2}\right). \tag{12}$$

Note that if $\beta_n < const < \infty$ then

$$pen(k) = O\left(\frac{k \log n}{n}\right)$$

and the rate in (12) becomes

$$\mathbf{E}\int |f_n(x) - r(x)|^2 \mu(dx) = O\left(\sqrt{\frac{\log n}{n}}\right). \tag{13}$$

For the Gaussian kernel the rate in (13) is valid in the Sobolev space of regression functions r whose weak derivatives up to order $2m > d$ are in $L^1(R^d)$ [9]. Precise characterization of $\overline{\mathcal{F}}$ for the kernels considered in this paper is an interesting open problem. A similar problem for RBF regression estimates has been considered in [12, chapter 17.3].

References

1. Anthony, M., Bartlett, P.L.: Neural Network Learning: Theoretical Foundations. Cambridge University Press, Cambridge (1999)
2. Barron, A.R.: Universal approximation bounds for superpositions of a sigmoidal function. IEEE Trans. on Information Theory 39, 930–945 (1993)
3. Barron, A.R.: Approximation and estimation bounds for artificial neural networks. Machine Learning 14, 115–133 (1994)
4. Cybenko, G.: Approximations by superpositions of sigmoidal functions. Mathematics of Control, Signals, and Systems 2, 303–314 (1989)
5. Devroye, L., Györfi, L., Lugosi, G.: Probabilistic Theory of Pattern Recognition. Springer, New York (1996)
6. Duchon, J.: Sur l'erreur d'interpolation des fonctions de plusieurs variables par les D^m-splines. RAIRO Anal. Numér. 12(4), 325–334 (1978)
7. Faragó, A., Lugosi, G.: Strong universal consistency of neural network classifiers. IEEE Trans. on Information Theory 39, 1146–1151 (1993)
8. Girosi, F.: Regularization theory, radial basis functions and networks. In: Cherkassky, V., Friedman, J.H., Wechsler, H. (eds.) From Statistics to Neural Networks. Theory and Pattern recognition Applications, pp. 166–187. Springer, Berlin (1992)
9. Girosi, F., Anzellotti, G.: Rates of convergence for radial basis functions and neural networks. In: Mammone, R.J. (ed.) Artificial Neural Networks for Speech and Vision, pp. 97–113. Chapman and Hall, London (1993)
10. Girosi, F., Jones, M., Poggio, T.: Regularization theory and neural network architectures. Neural Computation 7, 219–267 (1995)
11. Grenander, U.: Abstract Inference. Wiley, New York (1981)
12. Györfi, L., Kohler, M., Krzyżak, A., Walk, H.: A DistributionFree Theory of Nonparametric Regression. Springer, New York (2002)
13. Haussler, D.: Decision theoretic generalizations of the PAC model for neural net and other learning applications. Information and Computation 100, 78–150 (1992)
14. Hornik, K., Stinchocombe, S., White, H.: Multilayer feed-forward networks are universal approximators. Neural Networks 2, 359–366 (1989)
15. Kolmogorov, A.N., Tihomirov, V.M.: ϵ-entropy and ϵ-capacity of sets in function spaces. Translations of the American Mathematical Society 17, 277–364 (1961)
16. Krzyżak, A., Linder, T., Lugosi, G.: Nonparametric estimation and classification using radial basis function nets and empirical risk minimization. IEEE Trans. Neural Networks 7(2), 475–487 (1996)

17. Krzyżak, A., Linder, T.: Radial basis function networks and complexity regularization in function learning. IEEE Trans. Neural Networks 9(2), 247–256 (1998)
18. Krzyżak, A., Niemann, H.: Convergence and rates of convergence of radial basis functions networks in function learning. Nonlinear Analysis 47, 281–292 (2001)
19. Krzyżak, A., Schäfer, D.: Nonparametric regression estimation by normalized radial basis function networks. IEEE Transactions on Information Theory 51(3), 1003–1010 (2005)
20. Lugosi, G., Zeger, K.: Nonparametric estimation via empirical risk minimization. IEEE Trans. on Information Theory 41, 677–687 (1995)
21. Moody, J., Darken, J.: Fast learning in networks of locally-tuned processing units. Neural Computation 1, 281–294 (1989)
22. Park, J., Sandberg, I.W.: Universal approximation using Radial-Basis-Function networks. Neural Computation 3, 246–257 (1991)
23. Park, J., Sandberg, I.W.: Approximation and Radial-Basis-Function networks. Neural Computation 5, 305–316 (1993)
24. Pollard, D.: Convergence of Stochastic Processes. Springer, New York (1984)
25. Rissanen, J.: A universal prior for integers and estimation by minimum description length. Annals of Statistics 11, 416–431 (1983)
26. Ripley, B.D.: Pattern Recognition and Neural Networks. Cambridge University Press, Cambridge (1996)
27. Shorten, R., Murray-Smith, R.: Side effects of normalising radial basis function networks. International Journal of Neural Systems 7, 167–179 (1996)
28. Specht, D.F.: Probabilistic neural networks. Neural Networks 3, 109–118 (1990)
29. Vapnik, V.N.: Estimation of Dependences Based on Empirical Data, 2nd edn. Springer, New York (1999)
30. Xu, L., Krzyżak, A., Yuille, A.L.: On radial basis function nets and kernel regression: approximation ability, convergence rate and receptive field size. Neural Networks 7, 609–628 (1994)

Efficient Predictive Control Integrated with Economic Optimisation Based on Neural Models

Maciej Ławryńczuk and Piotr Tatjewski

Institute of Control and Computation Engineering, Warsaw University of Technology
ul. Nowowiejska 15/19, 00-665 Warsaw, Poland
Tel. +48 22 234-76-73
M.Lawrynczuk@ia.pw.edu.pl, P.Tatjewski@ia.pw.edu.pl

Abstract. This paper presents a predictive control scheme integrated with economic optimisation. Two neural models are used: a dynamic one (for the control subproblem) and a steady-state one (for the economic optimisation subproblem). The algorithm is computationally efficient because it needs solving on-line only one quadratic programming problem. Unlike the classical control system structure, the necessity of repeating two nonlinear optimisation problems at each sampling instant is avoided.

1 Introduction

Model Predictive Control (MPC) is recognised as the only advanced control technique which has been very successful in large-scale industrial applications, for example distillation columns or polymerisation reactors [10,13,14]. It is mainly because MPC algorithms have a unique ability to take into account constraints imposed on process inputs (manipulated variables) and outputs (controlled variables) which decide on quality, economic efficiency and safety of production. Moreover, MPC is very efficient in multivariable process control.

MPC closely cooperates with economic optimisation in order to maximise production profit [1,2,3,6,7,8,14]. Typically, the classical multilayer control system structure is used [3,14]. The regulatory control layer keeps the process at given operating points and the optimisation layer calculates these set-points. Because in the economic optimisation problem a comprehensive nonlinear steady-state model of the process is used, it is solved reasonably less frequently than the MPC controller executes. Provided that dynamics of disturbances is much slower than the dynamics of the process, such an approach gives satisfactory results. In many practical cases, however, dynamics of disturbances is comparable with the process dynamics. Very often disturbances, for example flow rates, properties of feed and energy streams etc., vary significantly and not much slower than the dynamics of the controlled process. In such cases operation in the classical structure with low frequency of economic optimisation can result in a significant loss of economic effectiveness [14]. Ideally, it would be best to perform full nonlinear economic optimisation as frequently as the MPC executes. Because of high computational complexity and hence low reliability, such an approach has limited applicability and is rarely implemented on-line. In practice, the MPC algorithm is often supplemented with additional steady-state target optimisation [6,7,8,13,14]. The operating point determined by the nonlinear economic optimisation layer activated infrequently is recalculated as frequently as the MPC executes. Unfortunately, three optimisation problems have to be solved on-line.

This paper presents an efficient predictive control algorithm integrated with economic optimisation. It is based on the suboptimal MPC scheme with Nonlinear Prediction and Linearisation (MPC-NPL) [9,14,15]. The integrated algorithm is computationally efficient because it requires solving on-line only a quadratic programming problem. Unlike the classical control system structure, the necessity of repeating two nonlinear optimisation problems (i.e. economic optimisation and the MPC tasks) at each sampling instant on-line is avoided.

A dynamic neural model is used in the control subproblem, a steady-state neural model is used in the economic optimisation subproblem. Fundamental models, although potentially very precise, are usually not suitable for on-line control and optimisation because they are very complicated and may lead to numerical problems (ill-conditioning, stiffness, etc.). Neural models [4,5,12] are used because they have excellent approximation abilities, small number of parameters and simple structure. Problems typical of fundamental models are not encountered because neural models directly describe input-output relations of process variables, complicated systems of algebraic and differential equations do not have to be solved on-line.

2 Standard Control System Structure

Structure of the standard multilayer control system is depicted in Fig. 1 (left). The objective of economic optimisation (named local steady-state optimisation) is to maximise the production profit and satisfy the constraints, which determine safety and quality of production. Typically, the economic optimisation layer solves the following problem (for simplicity of presentation Single-Input Single-Output (SISO) process is assumed)

$$\min_{u^s} \{J_E(k) = c_u u^s - c_y y^s\}$$
$$u^{\min} \leq u^s \leq u^{\max}$$
$$y^{\min} \leq y^s \leq y^{\max} \tag{1}$$
$$y^s = f^s(u^s, h^s)$$

where u is the input of the process (manipulated variable), y is the output (controlled variable) and h is the measured (or estimated) disturbance, the superscript 's' refers to the steady-state. The function $f^s : \Re^2 \longrightarrow \Re \in C^1$ denotes a comprehensive steady-state process model. The quantities c_u, c_y represent prices resulting from economic considerations, $u^{\min}$, $u^{\max}$, $y^{\min}$, $y^{\max}$ denote constraints imposed on input and output variables, respectively.

In MPC algorithms [10,14] at each consecutive sampling instant k a set of future control increments is calculated

$$\Delta \boldsymbol{u}(k) = [\Delta u(k|k) \; \Delta u(k+1|k) \ldots \Delta u(k+N_u-1|k)]^T \tag{2}$$

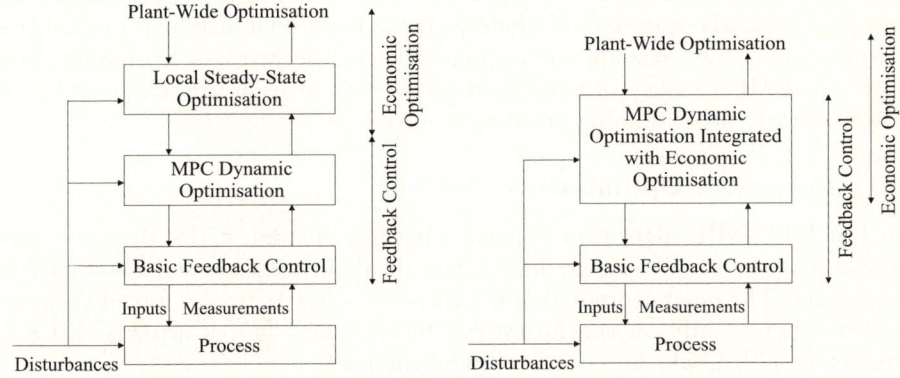

Fig. 1. Standard control structure (*left*) and the integrated structure (*right*)

Only the first element of the determined sequence (2) is applied to the process, the control law is $u(k) = \Delta u(k|k) + u(k-1)$. At the next sampling instant, $k+1$, the prediction is shifted one step forward and the whole procedure is repeated.

Let u_{eo}^s denote the optimal solution to the economic optimisation problem (1). Using the nonlinear steady-state model $y^s = f^s(u^s, h^s)$, the value y_{eo}^s corresponding to u_{eo}^s is calculated. The quantity y_{eo}^s is then passed as the desired set-point $(y^s = y_{eo}^s)$ to the MPC optimisation problem

$$\min_{\Delta \boldsymbol{u}(k)} \{J_{MPC}(k) = \sum_{p=1}^{N} \mu_p (y^s - \hat{y}(k+p|k))^2 + \sum_{p=0}^{N_u-1} \lambda_p (\Delta u(k+p|k))^2\}$$

$$u^{\min} \leq u(k+p|k) \leq u^{\max}, \quad p = 0, \ldots, N_u - 1$$

$$-\Delta u^{\max} \leq \Delta u(k+p|k) \leq \Delta u^{\max}, \quad p = 0, \ldots, N_u - 1 \quad (3)$$

$$y^{\min} \leq \hat{y}(k+p|k) \leq y^{\max}, \quad p = 1, \ldots, N$$

N and N_u are prediction and control horizons, respectively, $\mu_p \geq 0$, $\lambda_p > 0$.

If the process is nonlinear it is reasonable to use two nonlinear models: a steady-state model in economic optimisation (1) and a dynamic model in MPC optimisation (3). As a results, in the standard multilayer control system structure two nonlinear optimisation problems have to be solved on-line. Although economic optimisation can be performed infrequently, MPC optimisation has to be solved at each sampling instant.

3 Efficient Predictive Control Integrated with Economic Optimisation Based on Neural Models

Overall structure of the integrated control system is depicted in Fig. 1 (right) whereas its detailed configuration is shown in Fig. 2. It is based on the suboptimal MPC algorithm with Nonlinear Prediction and Linearisation (MPC-NPL)

[9,14,15]. The algorithm uses a dynamic neural model whereas in economic optimisation a steady-state neural model is used. Both neural models are linearised on-line taking into account the current state of the process. Economic optimisation and MPC tasks are integrated, as a result at each sampling instant only one quadratic programming problem is solved.

3.1 Integrated Optimisation Problem

In the MPC-NPL algorithm at each sampling instant k the dynamic neural model is used on-line twice: to determine a local linearisation and a nonlinear free trajectory. The output prediction is expressed as the sum of a forced trajectory, which depends only on the future (on future input moves $\Delta \boldsymbol{u}(k)$) and a free trajectory $\boldsymbol{y}^0(k)$, which depends only on the past

$$\hat{\boldsymbol{y}}(k) = \boldsymbol{G}(k)\Delta \boldsymbol{u} + \boldsymbol{y}^0(k)(k) \tag{4}$$

where

$$\hat{\boldsymbol{y}}(k) = [\hat{y}(k+1|k) \ldots \hat{y}(k+N|k)]^T \tag{5}$$

$$\boldsymbol{y}^0(k) = [y^0(k+1|k) \ldots y^0(k+N|k)]^T \tag{6}$$

The dynamic matrix $\boldsymbol{G}(k)$ of dimensionality $N \times N_u$ is comprised of step-response coefficients of the linearised model

$$\boldsymbol{G}(k) = \begin{bmatrix} s_1(k) & 0 & \ldots & 0 \\ s_2(k) & s_1(k) & \ldots & 0 \\ \vdots & \vdots & \ddots & \vdots \\ s_N(k) & s_{N-1}(k) & \ldots & s_{N-N_u+1}(k) \end{bmatrix} \tag{7}$$

Linearisation of the neural dynamic model and calculation of the nonlinear free trajectory is detailed in the following subsections.

Integrating economic optimisation (1) with MPC optimisation (3), using the superposition principle (4) and on-line linearisation of the steady-state neural model, the integrated optimisation problem is

$$\min_{\Delta \boldsymbol{u}(k),\, \boldsymbol{\varepsilon}^{\min},\, \boldsymbol{\varepsilon}^{\max},\, u^s} \{ J_{MPC}(k) + \gamma J_E(k) = \left\| \boldsymbol{y}^s(k) - \boldsymbol{G}(k)\Delta \boldsymbol{u}(k) - \boldsymbol{y}^0(k) \right\|_{\boldsymbol{M}}^2$$
$$+ \left\| \Delta \boldsymbol{u}(k) \right\|_{\boldsymbol{\Lambda}}^2 + \rho^{\min} \left\| \boldsymbol{\varepsilon}^{\min} \right\|^2 + \rho^{\max} \left\| \boldsymbol{\varepsilon}^{\max} \right\|^2 + \gamma(c_u u^s - c_y y^s) \}$$

$$\boldsymbol{u}^{\min} \leq \boldsymbol{J}\Delta \boldsymbol{u}(k) + \boldsymbol{u}^{k-1}(k) \leq \boldsymbol{u}^{\max}$$

$$-\Delta \boldsymbol{u}^{\max} \leq \Delta \boldsymbol{u}(k) \leq \Delta \boldsymbol{u}^{\max}$$

$$\boldsymbol{y}^{\min} - \boldsymbol{\varepsilon}^{\min} \leq \boldsymbol{G}(k)\Delta \boldsymbol{u}(k) + \boldsymbol{y}^0(k) \leq \boldsymbol{y}^{\max} + \boldsymbol{\varepsilon}^{\max}$$

$$\boldsymbol{\varepsilon}^{\min} \geq 0, \quad \boldsymbol{\varepsilon}^{\max} \geq 0$$

$$u^{\min} \leq u^s \leq u^{\max}$$

$$y^{\min} \leq y^s \leq y^{\max}$$

$$y^s = f^s(u^s, h^s)|_{u^s=u(k-1),\, h^s=h(k)} + \boldsymbol{H}(k)(u^s - u(k-1))$$

$$\tag{8}$$

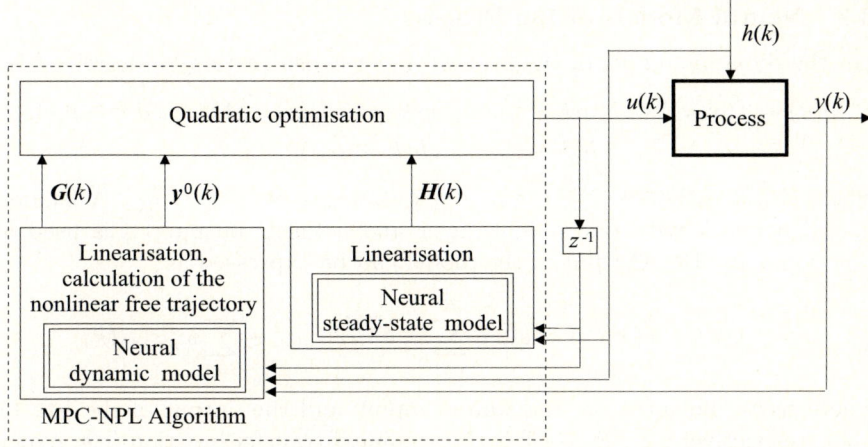

Fig. 2. Detailed configuration of the integrated structure

where vectors of length N are

$$\boldsymbol{y}^s(k) = [y^s \ldots y^s]^T, \quad \boldsymbol{y}^{\min} = \left[y^{\min} \ldots y^{\min}\right]^T, \quad \boldsymbol{y}^{\max} = [y^{\max} \ldots y^{\max}]^T \quad (9)$$

vectors of length N_u are

$$\boldsymbol{u}^{\min} = \left[u^{\min} \ldots u^{\min}\right]^T, \quad \Delta\boldsymbol{u}^{\max} = [\Delta u^{\max} \ldots \Delta u^{\max}]^T \quad (10)$$

$$\boldsymbol{u}^{\max} = [u^{\max} \ldots u^{\max}]^T, \quad \boldsymbol{u}^{k-1}(k) = [u(k-1) \ldots u(k-1)]^T \quad (11)$$

$\boldsymbol{M} = diag(\mu_1, \ldots, \mu_N)$, $\boldsymbol{\Lambda} = diag(\lambda_0, \ldots, \lambda_{N_u-1})$, $\gamma > 0$ and $\boldsymbol{J}$ is the all ones lower triangular matrix of dimensionality $N_u \times N_u$. In order to cope with infeasibility problems, output constraints are softened by means of slack variables [10,14] (vectors $\boldsymbol{\varepsilon}^{\min}$ and $\boldsymbol{\varepsilon}^{\max}$ of length N), $\rho^{\min}$, $\rho^{\max} > 0$.

In the integrated optimisation problem (8) the steady-state neural model is linearised on-line taking into account the current state of the process determined by $u(k-1)$ and $h(k)$, $\boldsymbol{H}(k)$ is the derivative of the nonlinear steady-state model

$$\boldsymbol{H}(k) = \left.\frac{df^s(u^s, h^s)}{du^s}\right|_{u^s=u(k-1),\ h^s=h(k)} \quad (12)$$

Detailed configuration of the integrated structure is depicted in Fig. 2. All things considered, at each sampling instant k the following steps are repeated:

1. Linearisation of the steady-state neural model: obtain $\boldsymbol{H}(k)$.
2. Linearisation of the dynamic neural model: obtain the matrix $\boldsymbol{G}(k)$.
3. Find the nonlinear free trajectory $\boldsymbol{y}^0(k)$ using the dynamic neural model.
4. Solve the quadratic programming problem (8).
5. Apply $u(k) = \Delta u(k|k) + u(k-1)$.
6. Set $k := k+1$, go to step 1.

3.2 Neural Models of the Process

Let the dynamic model of the process under consideration be described by

$$y(k) = f(\boldsymbol{x}(k)) = f(u(k-\tau), \ldots, u(k-n_B), y(k-1), \ldots, y(k-n_A), \quad (13)$$
$$h(k-\tau_h), \ldots, h(k-n_C))$$

where $f : \Re^{n_A+n_B+n_C-\tau-\tau_h+2} \longrightarrow \Re \in C^1$, $\tau \leq n_B$, $\tau_h \leq n_C$. A feedforward neural network with one hidden layer and a linear output [4] is used as the function f in (13). Output of the model can be expressed as

$$y(k) = f(\boldsymbol{x}(k)) = w_0^2 + \sum_{i=1}^{K} w_i^2 v_i(k) = w_0^2 + \sum_{i=1}^{K} w_i^2 \varphi(z_i(k)) \quad (14)$$

where $z_i(k)$ and $v_i(k)$ are the sum of inputs and the output of the i-th hidden node, respectively, $\varphi : \Re \longrightarrow \Re$ is the nonlinear transfer function (e.g. hyperbolic tangent), K is the number of hidden nodes. From (13) one has

$$z_i(k) = w_{i,0}^1 + \sum_{j=1}^{I_u} w_{i,j}^1 u(k-\tau+1-j) + \sum_{j=1}^{n_A} w_{i,I_u+j}^1 y(k-j) \quad (15)$$
$$+ \sum_{j=1}^{I_h} w_{i,I_u+n_A+j}^1 h(k-\tau_h+1-j)$$

Weights of the network are denoted by $w_{i,j}^1$, $i = 1, \ldots, K$, $j = 0, \ldots, n_A + n_B + n_C - \tau - \tau_h + 2$, and w_i^2, $i = 0, \ldots, K$, for the first and the second layer, respectively, $I_u = n_B - \tau + 1$, $I_h = n_C - \tau_h + 1$.

The second feedforward neural network with one hidden layer and a linear output is used as the steady-state model $y^s = f^s(u^s, h^s)$

$$y^s = f^s(u^s, h^s) = w_0^{2s} + \sum_{i=1}^{K^s} w_i^{2s} v_i^s = w_0^{2s} + \sum_{i=1}^{K^s} w_i^{2s} \varphi(z_i^s) \quad (16)$$

where

$$z_i^s = w_{i,0}^{1s} + w_{i,1}^{1s} u^s + w_{i,2}^{1s} h^s \quad (17)$$

Weights of the second network are denoted by $w_{i,j}^{1s}$, $i = 1, \ldots, K^s$, $j = 0, 1, 2$, and w_i^{2s}, $i = 0, \ldots, K^s$, for the first and the second layer, respectively.

3.3 On-Line Linearisation of Neural Models

Taking into account the structure of the dynamic neural model described by (14) and (15), coefficients of the linearised model

$$y(k) = f(\bar{\boldsymbol{x}}(k)) + \sum_{l=1}^{n_B} b_l(\bar{\boldsymbol{x}}(k))(u(k-l) - \bar{u}(k-l)) \quad (18)$$
$$- \sum_{l=1}^{n_A} a_l(\bar{\boldsymbol{x}}(k))(y(k-l) - \bar{y}(k-l))$$

are calculated on-line from

$$a_l(\bar{\boldsymbol{x}}(k)) = -\frac{\partial g(\bar{\boldsymbol{x}}(k))}{\partial y(k-l)} = -\sum_{i=1}^{K} w_i^2 \frac{d\varphi(z_i(\bar{\boldsymbol{x}}(k)))}{dz_i(\bar{\boldsymbol{x}}(k))} w_{i,I_u+l}^1 \quad l=1,\ldots,n_A \quad (19)$$

$$b_l(\bar{\boldsymbol{x}}(k)) = \begin{cases} 0 & l=1,\ldots,\tau-1 \\ \dfrac{\partial g(\bar{\boldsymbol{x}}(k))}{\partial u(k-l)} = \sum_{i=1}^{K} w_i^2 \dfrac{d\varphi(z_i(\bar{\boldsymbol{x}}(k)))}{dz_i(\bar{\boldsymbol{x}}(k))} w_{i,l-\tau+1}^1 & l=\tau,\ldots,n_B \end{cases} \quad (20)$$

If hyperbolic tangent is used as the function φ, $\frac{d\varphi(z_i(\bar{\boldsymbol{x}}(k)))}{dz_i(\bar{\boldsymbol{x}}(k))} = 1 - \tanh^2(z_i(\bar{\boldsymbol{x}}(k)))$. The linearisation point is the vector composed of past input and output signal values corresponding to the arguments of the nonlinear model (13)

$$\bar{\boldsymbol{x}}(k) = [\bar{u}(k-\tau)\ldots\bar{u}(k-n_B)\ \bar{y}(k-1)\ldots\bar{y}(k-n_A) \\ \bar{h}(k-\tau_h)\ldots\bar{h}(k-n_C)]^T \quad (21)$$

Step-response coefficients $s_j(k)$ of the linearised model are

$$s_j(k) = \sum_{i=1}^{\min(j,n_B)} b_i(k) - \sum_{i=1}^{\min(j-1,n_A)} a_i(k) s_{j-i}(k) \quad (22)$$

Taking into account structure of the steady-state neural model described by (16) and (17), one has

$$\boldsymbol{H}(k) = \sum_{i=1}^{K^s} w_i^{2s} \left.\frac{d\varphi(z_i^s)}{dz_i^s}\right|_{u^s=u(k-1),\ h^s=h(k)} w_{i,1}^{1s} \quad (23)$$

If hyperbolic tangent is used as the function φ, $\frac{d\varphi(z_i^s)}{dz_i^s} = 1 - \tanh^2(z_i^s)$.

3.4 Calculation of the Nonlinear Free Trajectory

The nonlinear free trajectory $y^0(k+p|k)$, $p=1,\ldots,N$, is calculated on-line

$$y^0(k+p|k) = w_0^2 + \sum_{i=1}^{K} w_i^2 \varphi(z_i^0(k+p|k)) + d(k) \quad (24)$$

The "DMC type" disturbance model is used [10,14]. The unmeasured disturbance $d(k)$ is assumed to be constant over the prediction horizon

$$d(k) = y(k) - y(k|k-1) = y(k) - \left(w_0^2 + \sum_{i=1}^{K} w_i^2 v_i(k)\right) \quad (25)$$

where $y(k)$ is measured while $y(k|k-1)$ is calculated from the model (14). The quantities $z_i^0(k+p|k)$ are determined from (15) assuming no changes in

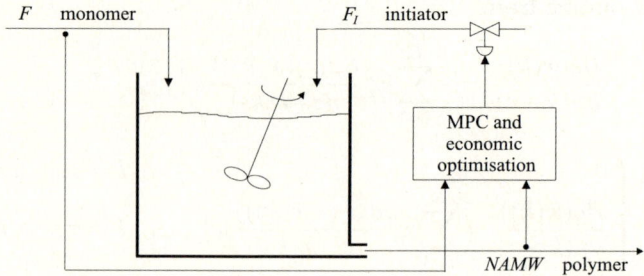

Fig. 3. Polymerisation reactor control system structure

the control signal from the sampling instant k onwards and replacing predicted output signals from $k+1$ by corresponding values of the free trajectory

$$z_i^0(k+p|k) = w_{i,0}^1 + \sum_{j=1}^{I_{uf}(p)} w_{i,j}^1 u(k-1) + \sum_{j=I_{uf}(p)+1}^{I_u} w_{i,j}^1 u(k-\tau+1-j+p)$$

$$+ \sum_{j=1}^{I_{yp}(p)} w_{i,I_u+j}^1 y^0(k-j+p|k) + \sum_{j=I_{yp}(p)+1}^{n_A} w_{i,I_u+j}^1 y(k-j+p)$$

$$+ \sum_{j=1}^{I_{hf}(p)} w_{i,I_u+n_A+j}^1 h(k-\tau_h+1-j+p|k)$$

$$+ \sum_{j=I_{hf}(p)+1}^{I_h} w_{i,I_u+n_A+j}^1 h(k-\tau_h+1-j+p) \tag{26}$$

where $I_{uf}(p) = \max(\min(p-\tau+1, I_u), 0)$, $I_{yp}(p) = \min(p-1, n_A)$, $I_{hf}(p) = \max(\min(p-\tau_h, I_h), 0)$. Typically, future values of the measured disturbance over the prediction horizon are not known at k, $h(k+p|k) = h(k)$ for $p \geq 1$.

4 Simulation Results

The process under consideration is a polymerisation reaction taking place in a jacketed continuous stirred tank reactor [11] depicted in Fig. 3. The reaction is the free-radical polymerisation of methyl methacrylate with azo-bis-isobutyronitrile as initiator and toluene as solvent. The output $NAMW$ (Number Average Molecular Weight) is controlled by manipulating the inlet initiator flow rate F_I, flow rate F of the monomer is the measured disturbance ($u = F_I$, $y = NAMW$, $h = F$). Both steady-state and dynamic properties of the process are nonlinear. Hence, it is justified to use nonlinear neural models for economic optimisation and control.

Three models of the process are used. The fundamental model [11] is used as the real process during simulations. An identification procedure is carried out, two neural models are obtained, namely a dynamic one ($K = 6$) and a steady-state one ($K^s = 4$) which are next used for MPC and economic optimisation. Parameters of MPC are: $N = 10$, $Nu = 3$, $\mu_p = 1$, $\lambda_p = 0.2$.

To maximise the production rate the economic performance function is

$$J_E(k) = -F_I^s \qquad (27)$$

The following constraints are imposed on manipulated and controlled variables

$$F_I^{\min} \leq F_I \leq F_I^{\max}, \qquad NAMW^{\min} \leq NAMW \qquad (28)$$

where $F_I^{\min} = 0.0035 \ m^3/h$, $F_I^{\max} = 0.033566 \ m^3/h$. For the first part of the simulation ($k = 1, \ldots, 100$) $NAMW^{\min} = 22500 \ kg/kmol$, for the second part ($k = 101, \ldots, 200$) $NAMW^{\min} = 20000 \ kg/kmol$. The scenario of disturbance changes is

$$F(k) = 2 - 1.6(\sin(0.008k) - \sin(0.08)) \qquad (29)$$

At first two versions of the classical multilayer structure are compared:

a) the multilayer structure with nonlinear economic optimisation repeated 30 times less frequently than the nonlinear MPC controller executes,
b) the "ideal" multilayer structure with nonlinear economic optimisation repeated as frequently as the nonlinear MPC controller executes.

In the first case nonlinear MPC optimisation is repeated at each sampling instant on-line whereas nonlinear economic optimisation every 30th sampling instant, in the second case two nonlinear optimisation problems are solved at each sampling instant. Simulation results are depicted in Fig. 4. In the first case the frequency of nonlinear economic optimisation is low. It means that changes in the disturbance (the monomer flow rate F) and changes in the output constraint are taken into account every 30th sampling instant, the calculated operating point is constant for long periods. The economic performance index calculated for the whole simulation horizon is $J_E = \sum_{k=1}^{200} J_E(k) = \sum_{k=1}^{200}(-F_I) = -4.7396$.

In the second case economic optimisation repeated as frequently as the MPC controller executes takes into account changes in the disturbance F and in the output constraint, a new optimal steady-state operating point is calculated for each sampling instant. The economic performance index improves to $J_E = -4.9730$. Because of high computational burden applicability of the "ideal" multilayer structure is limited in practice. Since the output constraint (28) is implemented as soft in the MPC-NPL algorithm, it is temporarily violated. Inefficiency of infrequent economic optimisation in multilayer structure applied to the polymerisation reactor is also discussed in [7]. To improve economic efficiency of the multilayer structure steady-state target calculation can be used which at each sampling instant recalculates the optimal operating point. Unfortunately, in such a case three optimisation problems must be solved on-line (i.e. nonlinear economic optimisation, steady-state target calculation and MPC optimisation).

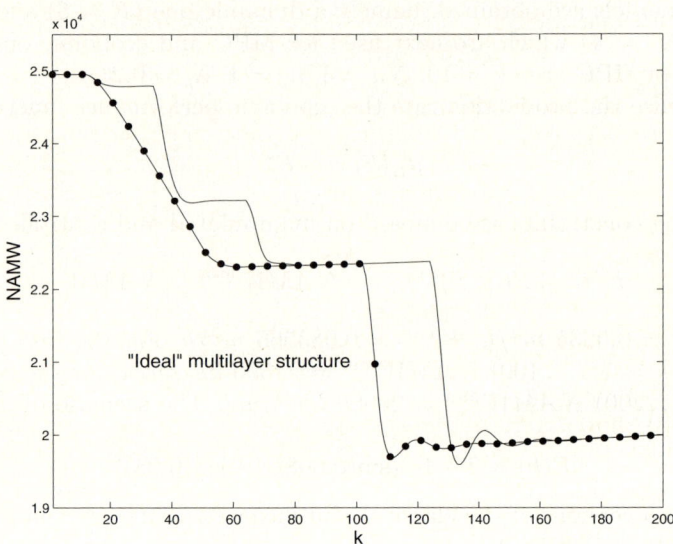

Fig. 4. Simulation results of the multilayer structure with nonlinear economic optimisation repeated 30 times less frequently than the nonlinear MPC controller executes (*solid line*) and the "ideal" multilayer structure with nonlinear economic optimisation repeated as frequently as the nonlinear MPC controller executes (*solid-pointed line*)

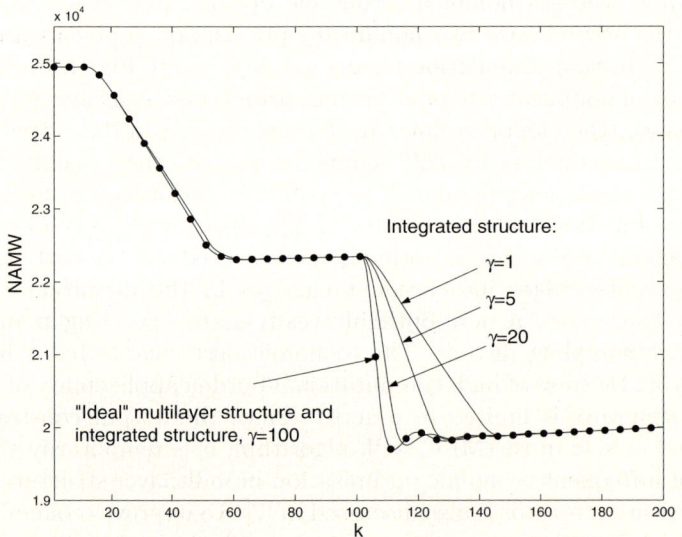

Fig. 5. Simulation results of the "ideal" multilayer structure with nonlinear economic optimisation repeated as frequently as the nonlinear MPC controller executes (*solid-pointed line*) and the integrated structure with quadratic programming (*solid line*)

Table 1. Economic performance index J_E in integrated and multilayer structures

Structure	γ	J_E
Multilayer, low frequency of economic optimisation	–	–4.7396
Integrated	1	–4.8529
Integrated	5	–4.9101
Integrated	10	–4.9345
Integrated	20	–4.9548
Integrated	50	–4.9713
Integrated	100	**–4.9730**
"Ideal" multilayer, high frequency of economic optimisation	–	**–4.9730**

In order to show advantages of the integrated structure two structures are compared:

a) the "ideal" multilayer structure with nonlinear economic optimisation repeated as frequently as the nonlinear MPC controller executes,
b) the integrated structure.

In the first case two nonlinear optimisation problems are solved on-line at each sampling instant, in the second case only one quadratic programming problem. Simulation results are depicted in Fig. 5. The bigger the value of γ, the closer the obtained response of the integrated structure to that of the "ideal" multilayer structure. For $\gamma = 100$ the trajectory of the process in the integrated structure is practically the same as in the "ideal" but unrealistic case when two nonlinear optimisation problems are solved at each sampling instant on-line (at economic optimisation and MPC layers). Table 1 shows economic performance index J_E in all considered structures.

5 Conclusions

The main advantages of the predictive control algorithm integrated with economic optimisation based on neural models presented in the paper are economic efficiency, computational efficiency and easy implementation. Thanks to on-line linearisation of nonlinear neural models of the process the algorithm requires solving on-line only one quadratic programming problem which can be done within a foreseeable time limit. At the same time, economic results obtained in the integrated structure are very close to that obtained in the classical control system structure in which two nonlinear optimisation problems must be solved at each sampling instant. Such optimisation tasks are computationally demanding, may terminate in local minima and are susceptible to numerical problems.

The described algorithm uses two neural models of the process. A dynamic neural model is used in the MPC optimisation subproblem, a steady-state neural model is used in the economic optimisation subproblem. Neural models, rather than complicated fundamental models who are usually not suitable for on-line control and optimisation are recommended. Neural models are used because they have excellent approximation abilities, small number of parameters and simple structure. Moreover, unlike fundamental models comprised of complicated systems of differential and algebraic equations which have to be solved on-line, neural models directly describe input-output relations of process variables.

Acknowledgement. This work was supported by Polish national budget funds 2007-2009 for science as a research project.

References

1. Blevins, T.L., Mcmillan, G.K., Wojsznis, M.W.: Advanced control unleashed. ISA (2003)
2. Brdys, M., Tatjewski, P.: Iterative algorithms for multilayer optimizing control. Imperial College Press, London (2005)
3. Findeisen, W.M., Bailey, F.N., Brdyś, M., Malinowski, K., Tatjewski, P., Woźniak, A.: control and coordination in hierarchical systems. J. Wiley and Sons, New York (1980)
4. Haykin, S.: Neural networks – a comprehensive foundation. Prentice-Hall, Englewood Cliffs (1999)
5. Hussain, M.A.: Review of the applications of neural networks in chemical process control – simulation and online implmementation. Artificial Intelligence in Engineering 13, 55–68 (1999)
6. Kassmann, D.E., Badgwell, T.A., Hawkins, R.B.: Robust steady-state target calculation for model predictive control. AIChE Journal 46, 1007–1024 (2000)
7. Ławryńczuk, M.: Neural Models in Computationally Efficient Predictive Control Cooperating with Economic Optimisation. In: de Sá, J.M., Alexandre, L.A., Duch, W., Mandic, D. (eds.) ICANN 2007. LNCS, vol. 4669, pp. 650–659. Springer, Heidelberg (2007)
8. Ławryńczuk, M., Marusak, M., Tatjewski, P.: Multilayer and integrated structures for predictive control and economic optimisation. In: Proceedings of the 11th IFAC/IFORS/IMACS/IFIP Symposium on Large Scale Systems: Theory and Applications, LSS 2007, Gdańsk, Poland (2007), CD-ROM, paper 60
9. Ławryńczuk, M.: A family of model predictive control algorithms with artificial neural networks. International Journal of Applied Mathematics and Computer Science 17, 217–232 (2007)
10. Maciejowski, J.M.: Predictive control with constraints. Prentice-Hall, Harlow (2002)
11. Maner, B.R., Doyle, F.J., Ogunnaike, B.A., Pearson, R.K.: Nonlinear model predictive control of a simulated multivariable polymerization reactor using second-order Volterra models. Automatica 32, 1285–1301 (1996)
12. Nørgaard, M., Ravn, O., Poulsen, N.K., Hansen, L.K.: Neural networks for modelling and control of dynamic systems. Springer, London (2000)
13. Qin, S.J., Badgwell, T.A.: A survey of industrial model predictive control technology. Control Engineering Practice 11, 733–764 (2003)
14. Tatjewski, P.: Advanced control of industrial processes, Structures and algorithms. Springer, London (2007)
15. Tatjewski, P., Ławryńczuk, M.: Soft computing in model-based predictive control. Int. Journal of Applied Mathematics and Computer Science 16, 101–120 (2006)

Model-Based Fault Detection and Isolation Using Locally Recurrent Neural Networks*

Piotr Przystałka

Silesian University of Technology, Department of Fundamentals of Machinery Design,
18A Konarskiego str., 44-100 Gliwice, Poland

Abstract. The increasing complexity of technological processes implemented in present industrial installations causes serious problems in the modern control system design and analysis. Chemical refineries, electrical furnaces, water treatments and other industrial plants are complex systems and in some cases cannot be precisely described by classical mathematical models. On the other hand, modern industrial systems are subject to faults in their components. Due to these facts, fault-tolerant control design using soft computing methods is gaining more and more attention in recent years. In this paper, the model-based approach to fault detection and isolation using locally recurrent neural networks is presented. The paper contains a numerical example that illustrates the performance of the proposed locally recurrent neural network with respect to other well-known neural structures.

1 Introduction

The development of artificial neural networks has attracted considerable research interest over the past decades. A comprehensive review of neural networks applications in areas of technical diagnostics is given by R. Tadeusiewicz in the paper [19]. Especially, the author indicates that neural networks should be also employed in such applications in an unusual way. In his opinion, it is necessary to determine all potential advantages of this means. Other important applications of neural networks in fault-tolerant control systems may be also found in [6,7,8,22]. Taking into account approaches that are related to the model-based schemes, two of the most common ways to fault detection and isolation using neural computing methods are roughly as follows: (a) two networks for state estimation: the first network as a model of a process for residual generation and the second one for residual evaluation, (b) many networks for state estimation: there is a bank of neural models for generating residuals and a neuroclassifier for their evaluating.

Obviously, there are more than one hundred neural network structures proposed altogether by researchers from varying standpoints [11,16], but in this paper only a locally recurrent net is considered in detail. The typical locally

---

* The research presented in the paper has been partially supported by the Ministry of Science and Higher Education under Grant No. N N514 3412 33.

recurrent architecture is achieved by introducing dynamic neural units into the structure of a feed-forward neural network. Such networks are well-known to be locally recurrent globally feed-forward networks. One can read in [7,18,21], that this topology may be classified somewhere in-between the feed-forward and globally recurrent architecture. Some of the most important strategies for developing dynamic elementary processors and locally recurrent topologies can be also found in [2,7]. For instance, in the paper [10] authors used the generalized Frasconi-Gori-Soda (FGS for short) neuron [4] as a basic element of the dynamic fuzzy neural network. The model comprises a set of fuzzy rules in the form suggested by Takagi, Sugeno and Kang [5,20] where the rule sub-models are implemented by locally recurrent neural networks with generalized FGS neurons. The second example is the chaotic neuron [12] that may be applied as a basic element for networks with higher-order information processing capabilities [1,17].

Generally, neural networks that are composed of such neurons are very often used to obtain better approximation of spatio-temporal data. They are able to represent deterministic dynamic behaviors of complex systems. Nevertheless, in cases when a behavior of the system is non-deterministic such networks require additional formalities. Therefore, there is the need to elaborate much more general neuron models that might be used for modeling both deterministic and stochastic objects simultaneously.

2 Theoretical Background

There are numerous ways to improve the ability of artificial neural networks to model real-world systems. One of the most common approaches is to develop dynamic neural units by introducing internal feedback connections. The general idea is not new and was considered in several publications, for instance in [2,3,7]. However, some further developments are introduced in this paper. It can be stated that it brings an artificial neuron closer and closer to the biological model.

2.1 Dynamic Elementary Processor

As discussed earlier in the previous papers of the author [14,15], a dynamic behavior (deterministic and stochastic) is embedded in the neuron by introducing linear dynamic systems (LDS$^{\mathcal{A}/\mathcal{F}}$) into its structure (see Fig. 1). Thus, the behavior of the dynamic neural unit under consideration is described by the following equations:

$$\xi_1(k) = \sum_{i=1}^{R} w_i u_i(k) + w_{(R+1)} \xi_3(k), \qquad (1)$$

where $w_1, w_2, \cdots, w_{R+1}$ are external and feedback input weights, $u_i(k)$ is the i-th external input of the unit, $\xi_3(k)$ is the output state of the neuron.

Using the backward shift operator q^{-1} the internal state of the unit at the discrete time k is formulated as follows

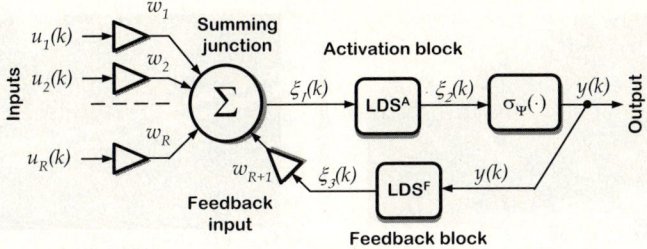

Fig. 1. Dynamic neural unit with linear dynamic systems (LDS) embedded in the activation and feedback blocks

$$A_{\mathcal{A}}\left(q^{-1}\right)\xi_2(k) = B_{\mathcal{A}}\left(q^{-1}\right)\xi_1(k) + C_{\mathcal{A}}\left(q^{-1}\right)\phi_{\mathcal{A}}(k), \quad (2)$$

and in the same way, the output state of the neuron is represented by

$$A_{\mathcal{F}}\left(q^{-1}\right)\xi_3(k) = B_{\mathcal{F}}\left(q^{-1}\right)y(k) + C_{\mathcal{F}}\left(q^{-1}\right)\phi_{\mathcal{F}}(k), \quad (3)$$

with

$$A_{\mathcal{S}}(q^{-1}) = 1 + a_1^{\mathcal{S}}q^{-1} + a_2^{\mathcal{S}}q^{-2} + \cdots + a_{A_{\mathcal{S}}}^{\mathcal{S}}q^{-A_{\mathcal{S}}},$$
$$B_{\mathcal{S}}(q^{-1}) = b_0^{\mathcal{S}} + b_1^{\mathcal{S}}q^{-1} + \cdots + b_{B_{\mathcal{S}}}^{\mathcal{S}}q^{-B_{\mathcal{S}}},$$
$$C_{\mathcal{S}}(q^{-1}) = 1 + c_1^{\mathcal{S}}q^{-1} + c_2^{\mathcal{S}}q^{-2} + \cdots + c_{C_{\mathcal{S}}}^{\mathcal{S}}q^{-C_{\mathcal{S}}},$$

where $A_{\mathcal{S}}(q^{-1}), B_{\mathcal{S}}(q^{-1}), C_{\mathcal{S}}(q^{-1})$ are polynomials of the delay operator q^{-1}, $\mathcal{S} = \{\mathcal{A}, \mathcal{F}\}$ denotes a suitably system in the activation or feedback block, $(A_{\mathcal{S}}, B_{\mathcal{S}}, C_{\mathcal{S}})$ represents the structure of the embedded system $\mathcal{S}$, term $\phi_{\mathcal{S}}$ is a random process with the mean value $\mu_{\mathcal{S}}$ and the variance $\sigma_{\mathcal{S}}^2$. Its output is calculated by using:

- hyperbolic tangent function $\sigma_\psi(\xi_2) = \frac{2}{1+\exp(-\beta\xi_2)} + b - 1$, or
- linear function $\sigma_\psi(\xi_2) = a\xi_2 + b$,

where a, β, b are parameters of the output function.

Dynamic properties

Figure 2 shows the largest Lyapunov exponent of the dynamic neural unit (Eq. 1-3) as a function of the input signal $u(k)$ and the refractory parameter a_1^A. It indicates a chaotic character of its behavior and is achieved for the following values of parameters: $w_1 = -1$, $w_2 = 1$, $\beta = 5.85$, $b = 1$, $b_0^A = -1$, $b_0^F = 0$, $b_1^F = -1$. The Lyapunov exponent is computationally useful for verifying chaotic characteristic. Its positive value is a signature of chaos.

2.2 Locally Recurrent Neural Network

Locally recurrent neural networks are black-box type models that have found wide-spread use in modeling tasks. The topology of the network (described

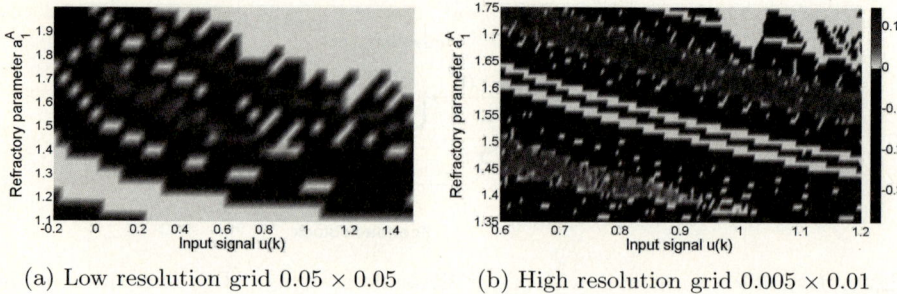

(a) Low resolution grid 0.05 × 0.05   (b) High resolution grid 0.005 × 0.01

Fig. 2. Filled contour plots of Lyapunov exponents

in Fig. 3) which has been used in this part of the author's research consists of two or three layers.

In the first layer there are simple static neurons with a non-linear output function (in the case when there are only two layers than the first one is called hidden and has dynamic units). The second hidden layer includes dynamic neurons with a non-linear output function. The last layer consists of simple static units but the output function is linear.

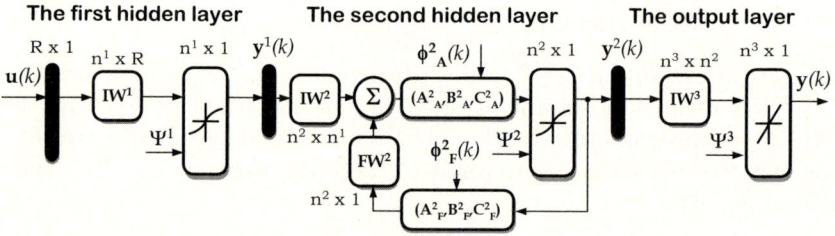

Fig. 3. General structure of locally recurrent neural network

The neural network presented in Fig. 3 can be described in a matrix form. The matrix with external input weights of the i-th layer is represented as follows

$$\mathbf{IW}^i = \begin{bmatrix} w^i_{11} & w^i_{12} & \cdots & w^i_{1R} \\ w^i_{21} & w^i_{22} & \cdots & w^i_{2R} \\ \vdots & \vdots & \ddots & \vdots \\ w^i_{n^i 1} & w^i_{n^i 2} & \cdots & w^i_{n^i R} \end{bmatrix},$$

and the vector of feedback weights

$$\mathbf{FW}^i = \begin{bmatrix} w^i_{1(R+1)} & w^i_{2(R+1)} & \cdots & w^i_{n^i(R+1)} \end{bmatrix}^T,$$

An associative input activation of neurons of the i-th layer is given by the following expression

$$\boldsymbol{\xi}^i_1(k) = \mathbf{IW}^i \mathbf{p}^i(k) + \mathbf{FW}^i \circ \boldsymbol{\xi}^i_3(k), \qquad (4)$$

Internal states of neurons in the activation block of the i-th layer can be written using a polynomial notation in a vector form

$$\mathbf{A}_{\mathcal{A}}^{i}(q^{-1}) \circ \boldsymbol{\xi}_{2}^{i}(k) = \mathbf{B}_{\mathcal{A}}^{i}(q^{-1}) \circ \boldsymbol{\xi}_{1}^{i}(k) + \mathbf{C}_{\mathcal{A}}^{i}(q^{-1}) \circ \boldsymbol{\phi}_{\mathcal{A}}^{i}(k), \qquad (5)$$

and also for the feedback block

$$\mathbf{A}_{\mathcal{F}}^{i}(q^{-1}) \circ \boldsymbol{\xi}_{3}^{i}(k) = \mathbf{B}_{\mathcal{F}}^{i}(q^{-1}) \circ \mathbf{y}^{i}(k) + \mathbf{C}_{\mathcal{F}}^{i}(q^{-1}) \circ \boldsymbol{\phi}_{\mathcal{F}}^{i}(k), \qquad (6)$$

where the vector representation of polynomials is as follows

$$\mathbf{A}_{\mathcal{S}}^{i}(q^{-1}) = \left[A_{\mathcal{S}1}^{i}(q^{-1}) \; A_{\mathcal{S}2}^{i}(q^{-1}) \; \cdots \; A_{\mathcal{S}n^{i}}^{i}(q^{-1}) \right]^{T},$$

$$\mathbf{B}_{\mathcal{S}}^{i}(q^{-1}) = \left[B_{\mathcal{S}1}^{i}(q^{-1}) \; B_{\mathcal{S}2}^{i}(q^{-1}) \; \cdots \; B_{\mathcal{S}n^{i}}^{i}(q^{-1}) \right]^{T},$$

$$\mathbf{C}_{\mathcal{S}}^{i}(q^{-1}) = \left[C_{\mathcal{S}1}^{i}(q^{-1}) \; C_{\mathcal{S}2}^{i}(q^{-1}) \; \cdots \; C_{\mathcal{S}n^{i}}^{i}(q^{-1}) \right]^{T}.$$

Finally, output of the i-th layer is computed using non-linear transform operator

$$\mathbf{y}^{i}(k) = \boldsymbol{\sigma}^{i}\left(\boldsymbol{\xi}_{2}^{i}(k), \boldsymbol{\Psi}^{i}\right). \qquad (7)$$

The exemplary structure of the network with two hidden layers where dynamic units are only included in the second hidden layer is written as follows

$$R - n^{1} - n^{2} {\binom{A_{\mathcal{A}}^{2}, B_{\mathcal{A}}^{2}, C_{\mathcal{A}}^{2}}{A_{\mathcal{F}}^{2}, B_{\mathcal{F}}^{2}, C_{\mathcal{F}}^{2}}} - n^{3},$$

where R denotes the number of inputs, n^1 and n^2 are the numbers of hidden neurons, $\binom{A_{\mathcal{A}}^{2}, B_{\mathcal{A}}^{2}, C_{\mathcal{A}}^{2}}{A_{\mathcal{F}}^{2}, B_{\mathcal{F}}^{2}, C_{\mathcal{F}}^{2}}$ represents the structures of dynamic systems embedded in dynamic units, n^3 is the number of outputs.

2.3 Hybrid Learning Algorithm

The main objective of training process is to adjust the elements of the vector

$$\boldsymbol{\Theta} = \left[w_{11}^{1} w_{12}^{1} \cdots a_{11}^{1\mathcal{A}} a_{12}^{1\mathcal{A}} \cdots b_{10}^{1\mathcal{A}} b_{11}^{1\mathcal{A}} \cdots c_{11}^{1\mathcal{A}} c_{12}^{1\mathcal{A}} \cdots a_{11}^{1\mathcal{F}} a_{12}^{1\mathcal{F}} \cdots \cdots \psi_{11}^{1} \psi_{12}^{1} \cdots \right],$$

in such a way so as to minimize some loss function

$$E(\boldsymbol{\Theta}) = \beta \sum_{k=1}^{K} \|\mathbf{y}(k) - \tilde{\mathbf{y}}(k)\|^{2}, \qquad (8)$$

where $\mathbf{y}(k)$ is the actual response of the network, $\tilde{\mathbf{y}}(k)$ is the desired output of the network, and K is the number of learning patterns. For this purpose, the EA-LM hybrid scheme is used (Evolutionary Algorithm and Levenberg-Marquardt method). EA is able to reach the region near an optimum point relatively quickly. Unfortunately it can require many network simulations to achieve convergence. Therefore EA is running for small number of generations in order to come near an optimum point. Then the solution obtained by EA is used as an starting point for the LM method that is faster and more efficient for local search. Next, the LM method explores the search space along the line containing current point $\mathbf{\Theta}_n$, parallel to the search direction

$$\mathbf{\Theta}_{n+1} = \mathbf{\Theta}_n + \alpha_n \mathbf{d}_n \quad (9)$$

$$\mathbf{d}_n = \left(\mathbf{J}_n^T \mathbf{J}_n + \lambda_n \mathbf{I}\right)^{-1} \nabla E_n, \quad (10)$$

where the Jacobian ($\mathbf{J}$) or gradient (∇E_n) information is derived using a numerical differentiation method, and scalar λ controls both the magnitude and direction ($\alpha_n = 1$). The fitness and cost function during training stage for the EA and the LM method were chosen as the sum of squares errors ($\|\cdot\|$ denotes Euclidean norm and $\beta = 1$). More details are omitted here but can be found in the papers [14,15].

2.4 Architecture Selection

To obtain good learning and generalization abilities the optimal structure of a network should be found. There are many techniques, for instance, pruning algorithms, evolutionary algorithms, direct search methods and others that can be used for solving this problem. In this paper, a few heuristics are used. These rules were prepared basing on information in the literature. Some of them are enumerated below.

Heuristic rule 1: Recurrent neural network may be used to uniformly approximate a discrete-time state-space trajectory which is produced by either a dynamic system or a continuous-time function to any degree of precision [9].

Heuristic rule 2: Locally recurrent neural network with two hidden layers is able to approximate a state-space trajectory produced by any Lipschitz continuous function with arbitrary accuracy [13].

Heuristic rule 3: The number of training patterns should be considerably larger than Vapnik-Chervonenkis's $VCdim$ measure that can be estimated as follows

$$2\left(\frac{L}{2}\right) R \leq VCdim \leq 2N_w \left(1 + \log N_n\right),$$

where L is the number of neurons in the hidden layer, R is the dimension of an input vector, N_w is the total number of network parameters and N_n is the total number of neurons in the network [11].

Heuristic rules are only used in the first step to determine an initial structure of the network. Next, it is possible to change complexity of the network during the process of optimization of neural models by increasing/reducing the number of layers, the number of units, changing structures of dynamic systems embedded in the units.

3 The Benchmark Problem

Let us consider the three coupled tanks depicted in Fig. 4. These tanks are connected by pipes where flows are controlled by different valves. Two identical pumps P_1 and P_2 cause liquid flows q_1 and q_2 into tanks T_1 and T_2 where inputs $U_1(t)$ and $U_2(t)$ describe pumps velocities.

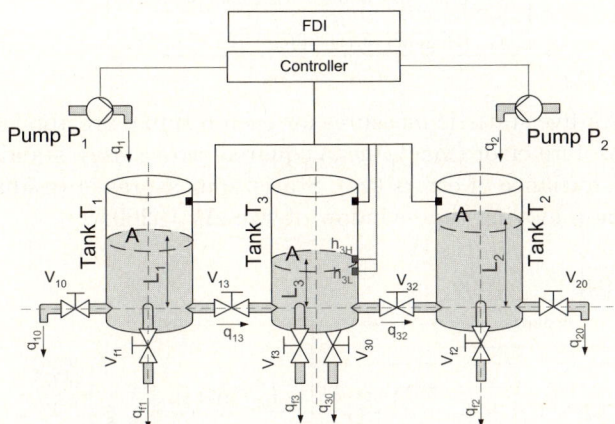

Fig. 4. The three-tank system considered in the paper

The measured signals are: streams of the medium q_1, q_2 that flow into the first and second tank, control signals U_1 and U_2, levels in tanks L_1, L_2, L_3 and, additionally, from tank T_3 discrete signals h_{3L} and h_{3H} from two capacitive proximity switches signalling whether the medium level in the tank is above or below the position of the sensor. The main aim of the control system is to keep the water level in the tank T_3 constant, while a water requirement q_{30} is changed randomly with an uniform distribution.

For this example many different types of faults like clogs and leakages may be acquired (Tab. 1). In this paper two fault scenarios are considered: the first case - only one fault occurs, and the second one - two faults may occur in the same time. Process faults f_1, f_2 and f_3 are experimented: undesirable leakages from tanks appear after 33 min. for the next 33 min. Faults f_7 and f_6 are realized by closing valves V_{13} and V_{32} in the middle of the simulation. Sensor faults are created by subtraction a 30% signal level from their output on the time window as in previous cases.

3.1 Fault Detection and Isolation System

In the proposed FDI system (Fig. 5), different neural network structures are examined. The group of three neural models ($NN1_1$, $NN1_2$, $NN1_3$) is used for residual generation, whereas the neural classifier NN2 is used for mapping the space of statistic features of residuals into the space of faults. In the features

Table 1. Set of faults for a three-tank system

Fault	Fault description
f_0	nominal conditions
f_1	undesirable leakage from the tank T_1
f_2	undesirable leakage from the tank T_2
f_3	undesirable leakage from the tank T_3
f_4	fault of the measuring channel 1
f_5	fault of the measuring channel 2
f_6	fault of the measuring channel 3
f_7	clogging of the valve V_{13}
f_8	clogging of the valve V_{32}

estimation block five statistic measures for each residual are applied: mean error (me), mean absolute error (mae), mean squared error (mse), standard deviation of errors (sde), variance of errors (ve). Statistic measures of residual signals are computed using a moving time-window of size $\Delta k = 200$.

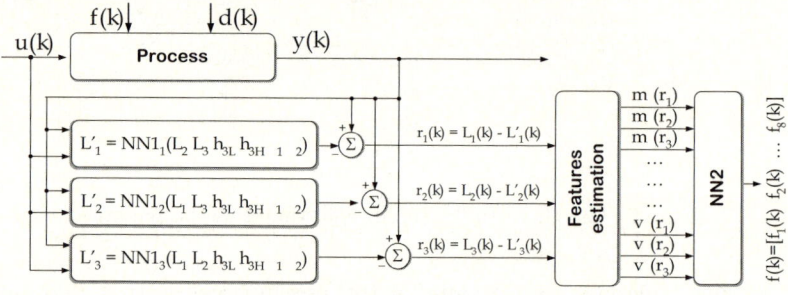

Fig. 5. Neural model-based fault detection and isolation

Three classes of neural networks were considered for residual generation block. Each of them was trained using suitable faultless data (7000 samples). Their structures were determined basing on heuristic rules discussed in Section 2.4. The processing error for the training set was defined as the sum of squares errors. On the other hand, the error for the testing set was chosen as mean absolute percentage error (mape for short). In the training stage, networks parameters were adjusted for a maximum number of epochs or until the goal error was less than 10^{-2}. It leaded to decrease mape below 2% for each neural model. In Fig. 6, the modeling results for the L_1 submodel are shown. The output of the process is marked by the solid line and the output of its neural model by the dashed one.

Locally recurrent neural networks LRNN $(6 - 5_{(0)}^{(-,2,-)} - 1)$

The notation in brackets means that the network consists of three processing layers with six inputs, five non-linear dynamic neurons and one linear output neuron. Only one hidden layer with hyperbolic tangent output function was enough to identify $NN1_i$ models accurately (mape $< 1\%$). Neural models were trained by means of the EA-LM hybrid scheme (5 generations of the EA and 15 iterations of the LM).

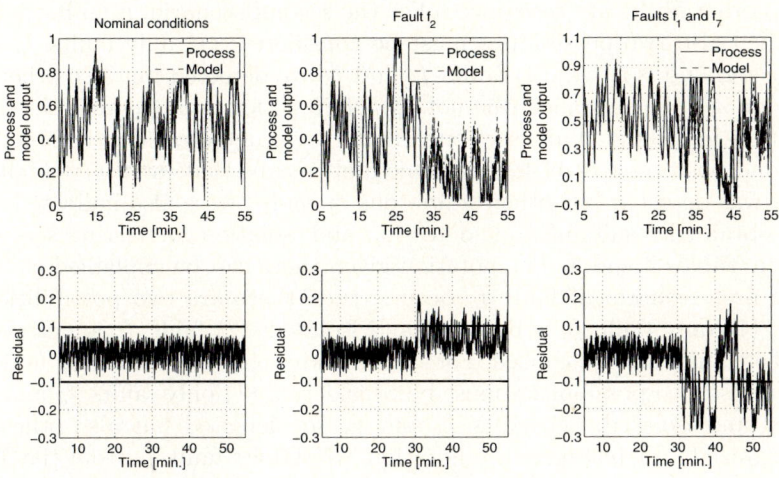

Fig. 6. Responses of the process (solid line) and LRNN model (dashed line) for nominal conditions and fault situations in the case of the testing sets

Focused time-delay neural networks FTDNN $(6^{[012]} - 5 - 1)$
These structures are very similar to the previous ones, however, there are no dynamic units included in the hidden layer. They consist of feedforward networks with tapped delay lines at the input. The training process was carried out using the same learning-pattern set, but only 20 iterations of the LM were realized. For this type of neural models mape < 2% were obtained.

NARX networks $(6 - 5^{[01]}_{[12]} - 1)$
These structures have also one hidden layer and moreover the output signal is connected to the input of the network by two delay lines q^{-1} and q^{-2}. Each neural network was trained using Bayesian regularization backpropagation method (20 iterations). The modeling results were similar to the previous case.

Fault detection and isolation were realized by means of the multilayer perceptron network. Due to the fact that for the first scenario only one fault might occur, the structure $15 - 12 - 15 - 8$ with hyperbolic tangent output functions in hidden layers and linear function in output one was sufficient for mapping the space of statistical features of residuals into the space of faults. Training patterns were obtained for no-fault and fault states (256 examples). Scaled conjugate gradient backpropagation method was used for updating network parameters (the maximum number of epochs was equal 10000). The second scenario needed 35 neurons in the second hidden layer and the number of epochs had to be set on 20000 iterations. The same training method was used, except 960 patterns were needed in order to adjust network parameters accurately.

The obtained results of fault detection and isolation for testing sets are included in Tables 2 and 3. The notation given there can be explained as follows. For the first scenario, if fault f_a occurs, then there are two possibilities considered: (1) only fault f_a is detected and isolated; (2) fault f_a is detected but also other faults are recognized. For the second scenario, if faults f_a and f_b occur then also two possibilities must be considered: (1) only faults f_a and f_b are detected and isolated; (2) faults f_a and f_b are detected but also other faults are recognized. For instance, the notation 1|75.00 for fault f_1 from the Table 2 means that in 75.00% of all cases only fault f_1 was detected. On the other hand, the second position 2|79.17 for the same fault means that in 79.17% of all cases fault f_1 was detected but other faults (one or more) were also indicated.

The obtained results of fault detection and isolation for testing sets are included in Tables 2 and 3. The notation given there can be explained as follows. For the first scenario, if fault f_a occurs, then there are two possibilities considered: (1) only fault f_a is detected and isolated; (2) fault f_a is detected but also other faults are recognized. For the second scenario, if faults f_a and f_b occur then also two possibilities must be considered: (1) only faults f_a and f_b are detected and isolated; (2) faults f_a and f_b are detected but also other faults are recognized. For instance, the notation 1|75.00 for fault f_1 from the Table 2

Table 2. Results of fault detection and isolation for the first scenario

FDI scheme	FAULT								
NN1-NN2	f_0	f_1	f_2	f_3	f_4	f_5	f_6	f_7	f_8	Avg
LRNN-MLP	1|98.44	1|75.00	1|72.92	1|62.50	1|97.92	1|95.83	1|97.92	1|100	1|100	1|88.95
		2|79.17	2|79.17	2|62.50	2|97.92	2|97.92	2|97.92	2|100	2|100	2|90.34
FTDNN-MLP	1|99.21	1|100	1|93.75	1|100	1|93.75	1|87.50	1|56.25	1|100	1|100	1|92.27
		2|100	2|93.75	2|100	2|100	2|100	2|56.25	2|100	2|100	2|94.43
NARX-MLP	1|96.61	1|0.00	1|25.00	1|29.17	1|91.67	1|4.17	1|22.92	1|97.92	1|100	1|51.04
		2|0.00	2|29.17	2|33.33	2|97.92	2|10.42	2|25.00	2|100	2|100	2|51.74

Table 3. Results of fault detection and isolation for the second scenario

FDI scheme	FAULT									
NN1-NN2	f_0	$f_1 \oplus f_2$	$f_1 \oplus f_3$	$f_1 \oplus f_4$	$f_1 \oplus f_5$	$f_1 \oplus f_6$	$f_1 \oplus f_7$	$f_1 \oplus f_8$	$f_2 \oplus f_3$	$f_2 \oplus f_4$
LRNN-MLP	1|98.83	1|70.83	1|89.58	1|62.50	1|75.00	1|62.50	1|64.58	1|95.83	1|64.58	1|52.08
		2|72.92	2|91.67	2|70.83	2|77.08	2|70.83	2|75.00	2|95.83	2|75.00	2|72.92
	$f_2 \oplus f_5$	$f_2 \oplus f_6$	$f_2 \oplus f_7$	$f_2 \oplus f_8$	$f_3 \oplus f_4$	$f_3 \oplus f_5$	$f_3 \oplus f_6$	$f_3 \oplus f_7$	$f_3 \oplus f_8$	$f_4 \oplus f_5$
	1|50.00	1|37.50	1|60.42	1|95.83	1|52.08	1|70.83	1|27.08	1|39.58	1|91.67	1|66.67
	2|70.83	2|54.17	2|70.83	2|97.92	2|62.50	2|75.00	2|35.42	2|39.58	2|91.67	2|72.92
	$f_4 \oplus f_6$	$f_4 \oplus f_7$	$f_4 \oplus f_8$	$f_5 \oplus f_6$	$f_5 \oplus f_7$	$f_5 \oplus f_8$	$f_6 \oplus f_7$	$f_6 \oplus f_8$	$f_7 \oplus f_8$	Avg
	1|87.50	1|100	1|97.92	1|45.83	1|47.92	1|97.92	1|97.92	1|93.75	1|93.75	1|72.02
	2|87.50	2|100	2|97.92	2|60.42	2|50.00	2|100	2|100	2|97.92	2|93.75	2|77.89
FTDNN-MLP	f_0	$f_1 \oplus f_2$	$f_1 \oplus f_3$	$f_1 \oplus f_4$	$f_1 \oplus f_5$	$f_1 \oplus f_6$	$f_1 \oplus f_7$	$f_1 \oplus f_8$	$f_2 \oplus f_3$	$f_2 \oplus f_4$
	1|91.28	1|25.50	1|0.00	1|37.50	1|31.25	1|18.75	1|12.50	1|31.25	1|6.25	1|18.75
		2|23.00	2|0.00	2|62.50	2|62.50	2|43.75	2|18.75	2|68.75	2|6.25	2|68.75
	$f_2 \oplus f_5$	$f_2 \oplus f_6$	$f_2 \oplus f_7$	$f_2 \oplus f_8$	$f_3 \oplus f_4$	$f_3 \oplus f_5$	$f_3 \oplus f_6$	$f_3 \oplus f_7$	$f_3 \oplus f_8$	$f_4 \oplus f_5$
	1|0.00	1|0.00	1|6.25	1|81.25	1|12.50	1|37.50	1|0.00	1|6.25	1|50.00	1|18.75
	2|6.25	2|18.75	2|25.00	2|87.50	2|18.75	2|62.50	2|0.00	2|31.25	2|81.25	2|56.25
	$f_4 \oplus f_6$	$f_4 \oplus f_7$	$f_4 \oplus f_8$	$f_5 \oplus f_6$	$f_5 \oplus f_7$	$f_5 \oplus f_8$	$f_6 \oplus f_7$	$f_6 \oplus f_8$	$f_7 \oplus f_8$	Avg
	1|62.50	1|62.50	1|93.75	1|18.75	1|31.25	1|81.25	1|81.25	1|93.75	1|62.50	1|36.93
	2|62.50	2|93.75	2|93.75	2|50.00	2|37.50	2|87.50	2|93.75	2|100	2|62.50	2|52.56
NARX-MLP	f_0	$f_1 \oplus f_2$	$f_1 \oplus f_3$	$f_1 \oplus f_4$	$f_1 \oplus f_5$	$f_1 \oplus f_6$	$f_1 \oplus f_7$	$f_1 \oplus f_8$	$f_2 \oplus f_3$	$f_2 \oplus f_4$
	1|96.88	1|31.25	1|12.50	1|25.00	1|4.17	1|20.83	1|14.58	1|93.75	1|18.75	1|29.17
		2|50.00	2|18.75	2|37.50	2|20.83	2|43.75	2|47.92	2|97.92	2|22.92	2|54.17
	$f_2 \oplus f_5$	$f_2 \oplus f_6$	$f_2 \oplus f_7$	$f_2 \oplus f_8$	$f_3 \oplus f_4$	$f_3 \oplus f_5$	$f_3 \oplus f_6$	$f_3 \oplus f_7$	$f_3 \oplus f_8$	$f_4 \oplus f_5$
	1|25.00	1|25.00	1|41.67	1|91.67	1|39.58	1|20.83	1|14.58	1|43.75	1|85.42	1|58.33
	2|31.25	2|35.42	2|47.92	2|93.75	2|52.08	2|29.17	2|27.08	2|47.92	2|87.50	2|68.75
	$f_4 \oplus f_6$	$f_4 \oplus f_7$	$f_4 \oplus f_8$	$f_5 \oplus f_6$	$f_5 \oplus f_7$	$f_5 \oplus f_8$	$f_6 \oplus f_7$	$f_6 \oplus f_8$	$f_7 \oplus f_8$	Avg
	1|68.75	1|97.92	1|97.92	1|8.33	1|54.17	1|93.75	1|95.83	1|93.75	1|91.67	1|51.54
	2|81.25	2|100	2|97.92	2|18.75	2|58.33	2|93.75	2|95.83	2|93.75	2|93.75	2|60.27

means that in 75.00% of all cases only fault f_1 was detected. On the other hand, the second position 2|79.17 for the same fault means that in 79.17% of all cases fault f_1 was detected but other faults (one or more) were also indicated.

4 Conclusion

In this paper, the model-based approach to fault detection and isolation in the case of the three-tank benchmark problem using neural computing methods was presented. Two fault scenarios were considered: the first case - only one fault was occurred, and the second one - two faults were experimented in the same time. It was shown that in this domain, locally recurrent neural networks lead to improvement of the performance of FDI systems. It can be stated that LRNN-MLP, FTDNN-MLP and NARX-MLP schemes provide the fault detection efficiency greater than 90% for both fault scenarios. Nevertheless, fault isolation by means of the LRNN-MLP scheme is more effective, especially for the second scenario where two faults occur in the same time. Finally, it can be assumed that much more complicated neuron models should be continuously developed to determine all potential advantages of this technique.

References

1. Aihara, K., Takabe, T., Toyoda, M.: Chaotic neural networks. Phys. Lett. A 144(6,7), 333–340 (1990)
2. Ayoubi, M.: Nonlinear dynamic systems identification with dynamic neural networks for fault diagnosis in technical processes. Humans, Information and Technology 3, 2120–2125 (1994)
3. Duch, W., Korbicz, J., Rutkowski, L., Tadeusiewicz, R.: Neural networks. Biocybernetics and Biomedical Engineering, vol. 6. Academic Publishing House EXIT, Warsaw (2000) (in polish)
4. Frasconi, P., Gori, M., Soda, G.: Local feedback multilayered networks. Neural Computing 4, 120–130 (1992)
5. Kang, G., Sugeno, M.: Fuzzy modeling. Trans. Society Instrument Control Engineers 23(6), 106–108 (1987)
6. Korbicz, J.: Robust fault detection using analytical and soft computing methods. Bulletin of the Polish Academy of Sciences: Technical Sciences 54(1), 75–88 (2006)
7. Korbicz, J., Kościelny, J.M., Kowalczuk, Z., Cholewa, W.: Fault Diagnosis. Models, Artificial Intelligence, Applications. Springer, Heidelberg (2004)
8. Kościelny, J.M.: Diagnostics of Automated Industrial Processes. Academic Publishing House EXIT, Warsaw (2001) (in polish)
9. Jin, L., Nikiforuk, P., Gupta, M.: Approximation of discrete-time state-space trajectories using dynamic recurrent neural networks. In: Automatic Control, July 1995, vol. 40, pp. 1266–1270. IEEE, Los Alamitos (1995)
10. Mastorocostas, P.A., Theocharis, J.B.: A recurrent fuzzy-neural model for dynamic system identification. Systems, Man and Cybernetics 32, 176–190 (2002)
11. Osowski, S.: Artificial neural network in algorithmic depiction. Wydawnictwo Naukowo-Techniczne, Warsaw (1996) (in Polish)
12. Pasemann, F.: A simple chaotic neuron. Physica D 104, 205–211 (1997)

13. Patan, K.: Approximation of state-space trajectories by locally recurrent globally feed-forward neural networks. Neural networks (November 2007), doi:10.1016/j.neunet.2007.10.004
14. Przystałka, P.: Heuristic modeling using recurrent neural networks: simulated and real-data experiments. Computer Assisted Mechanics and Engineering Sciences 14(4), 715–727 (2007)
15. Przystałka, P.: Hybrid learning algorithm for locally recurrent neural networks. In: Korbicz, J., Patan, K., Kowal, M. (eds.) Fault diagnosis and fault tolerant control, pp. 255–262. Academic Publishing House EXIT, Warsaw (2007)
16. Rutkowska, D., Piliński, M., Rutkowski, L.: Neural networks, genetic algorithms and fuzzy systems. Wydawnictwo Naukowe PWN, Warsaw (1997) (in Polish)
17. Hee-Kim, S., Park, W.-W.: Convergence analysis of chaotic dynamic neuron. In: Neural Networks, Proceedings of the International Joint Conference, July 2003, vol. 2, pp. 858–863 (2003)
18. Sinha, K., Gupta, M., Rao, H.: Dynamic neural networks: An overview. In: Proceedings of IEEE International Conference on Industrial Technology, vol. 1, pp. 491–496 (2000)
19. Tadeusiewicz, R.: Neural networks as a little used diagnostic tool. In: Proceedings of II International Congress of Technical Diagnostics, Warsaw, vol. 1, pp. 81–94 (2000) (in Polish)
20. Takagi, T., Sugeno, M.: Fuzzy identification of systems and its application to modeling and control. IEEE Trans. Syst., Man, Cybern. SMC-15, 116–132 (1983)
21. Tsoi, A.C., Back, A.D.: Locally recurrent globally feed-forward networks: a critical review of architectures. IEEE Trans. Neural Networks 5, 229–239 (1994)
22. Witczak, M.: Advances in model-based fault diagnosis with evolutionary algorithms and neural networks. International Journal of Applied Mathematics and Computer Science 16(1), 85–99 (2006)

Neural Network in Fast Adaptive Fourier Descriptor Based Leaves Classification

Dariusz Puchala[1] and Mykhaylo Yatsymirskyy[2]

[1] Institute of Computer Science, Technical University of Lodz, Lodz, Poland
dpuchala@ics.p.lodz.pl
[2] Institute of Computer Science, Technical University of Lodz, Lodz, Poland
jacym@ics.p.lodz.pl

Abstract. In this paper the results in leaves classification with non-parametrized one nearest neighbor and multilayer perceptron classifiers are presented. The feature vectors are composed of Fourier descriptors that are calculated for leaves contours with fast adaptive Fourier transform algorithm. An application of fast adaptive algorithm results in new fast adaptive Fourier descriptors.

Experimental results prove that the fast adaptive Fourier transform algorithm significantly accelerates the process of descriptors calculation and enables almost eightfold reduction in the number of contour data with no effect on classification performance. Moreover the neural network classifier gives higher accuracies of classification in comparison to the minimum distance one nearest neighbor classifier.

1 Introduction

The methods of automatic classification of two-dimensional objects find extensive applications in such fields as: medical and technical diagnosis, topography and analysis of aerial and satellite images, military systems, astronomy, forensic science, identification of persons in biometrical authorization systems, and many more [1]-[6]. The Fourier descriptors (FD) based methods, where FD are calculated for the contours of classified objects, belong to the group of methods most frequently exploited in the tasks of two-dimensional objects classification. The fundamental advantage of FD lies in good extraction of object features and ease in receiving object descriptions that are invariant to such transformations as: shifting, scaling and rotations of classified objects [7]-[10]. Some exemplary applications of FD include: shape analysis of mammographic calcifications [1] and measurements of acutance of breast tumors [2] in mammographic images, automatic classification of teeth in X-ray dental images [3], identification of military ground vehicles [4], recognition of handwritten characters [5] or classification of topographic shapes [6].

In this paper the authors present the results of experiments in leaves shape classification based on Fourier descriptors that are calculated for the contours of classified objects. For contour representation the centroid distance signature is selected [10]. The classification process exploits both: a non-parametrized one nearest neighbor (1-NN) and multilayer perceptron (MP) [11] classifiers.

In order to reduce the computational cost and an amount of object contour data required for calculation of FD, beside the fast Fourier transform radix-2 algorithm (FFT) [12], also the fast adaptive Fourier transform algorithm (AFFT) [13] is exploited. The AFFT algorithm enables automatic selection of such number of signal samples that are sufficient to calculate signal spectrum with the required accuracy, where an error of spectrum calculation can be expressed in: maximum difference (MD), mean square error (MSE) or peak signal to noise ratio (PSNR) metrics. In the tasks of shape classification it corresponds with selection of such number of contour points that allow for calculation of FD with some acceptable error expressed in one of the mentioned metrics. Moreover all of the calculations within AFFT algorithm are executed in accordance with the structures of fast time decimated Fourier transform algorithms of the computational complexity $\mathcal{O}(Nlog_2 N)$.

The application of the fast adaptive Fourier transform algorithm to the task of calculation of FD results in construction of new descriptors which are henceforth referred to as the fast adaptive Fourier descriptors (FAFD).

In the last section of this paper the results of experiments involving the image database [14] of leaves of three different tree species are presented. In all of the considered cases Fourier descriptors are calculated both by the means of fast and fast adaptive Fourier transform algorithms. The results obtained with use of 1-NN and MP classifiers for several descriptors lengths, i.e.: 8, 16, 32 and 64 coefficients are compared against each other with respect to the classification accuracies. Moreover the comparative study of computational times of FD calculation both with use of FFT and AFFT algorithms, together with the mean numbers of contour points required for calculation of fast adaptive Fourier descriptors for the selected values of acceptable error in MSE metric, are also presented.

The study of performance of leaves contour classification with 1-NN and MP classifiers and Fourier descriptors calculated exclusively with fast radix-2 algorithm can be found in previously published paper [15].

2 Theoretical Background

2.1 The Fourier Descriptors

A closed and continuous contour of a two-dimensional object can be described by continuous and periodic (with period T) function $z(t)$ of one real variable $t \in [0,T]$, where $z(t)$ takes the complex values $z(t) = x(t) + iy(t)$. The real and imaginary parts of $z(t)$ at moments t are the spacial x and y coordinates of contour points. In practice, in order to represent the contour of a classified object the centroid distance [10] signature $r(t)$ is frequently exploited. That signature in its continuous form can be written as the following real-valued function

$$r(t) = \sqrt{(x(t) - x_c)^2 + (y(t) - y_c)^2}, \qquad (1)$$

where (x_c, y_c) is the central point with coordinates calculated as the mean values of horizontal and vertical coordinates of all contour points. Then the Fourier descriptors for contour represented by $r(t)$ signature can be calculated in accordance with the formula of integral Fourier transform

$$R(k) = \frac{1}{T}\int_0^T r(t)e^{-i\frac{2\pi}{T}kt}, \qquad (2)$$

where k is an integer parameter and $k \geq 0$. However in the practical solutions of object classification mainly the small numbers M (e.g. $M = 8, 16, 32$ or 64 [10]) of low-frequency $R(k)$ coefficients are taken into consideration.

In practice of digital signal processing instead of a continuous contour its discrete form in the shape of finite set of discrete samples $(x(n), y(n))$ for parameter $n = 0, 1, ..., N-1$ is known, where N describes the maximum available number of contour samples. In this case formulas (1) and (2) take their discrete forms of: the signature

$$r(n) = \sqrt{(x(n) - x_c)^2 + (y(n) - y_c)^2}$$

for $n = 0, 1, ..., N-1$, where the central point coordinates (x_c, y_c) can be calculated with use of the following formula

$$x_c = \frac{1}{N}\sum_{n=0}^{N-1} x(n), y_c = \frac{1}{N}\sum_{n=0}^{N-1} y(n),$$

and the discrete Fourier transform (DFT) [16] defined as

$$R_N(k) = DFT_N\{x(n)\} = \frac{1}{N}\sum_{n=0}^{N-1} x(n)e^{-i\frac{2\pi}{N}kn} \qquad (3)$$

for $k = 0, 1, ..., M-1$. In the general case the following relation $R_N(k) \approx R(k)$ takes place, where an approximate equality is the direct consequence of contour sampling that results in duplication of contour spectrum $R(k)$ at integer multiplies of the discretization frequency (i.e. *aliasing*) [16]. The actual error of numerical calculation of $R(k)$ for $k = 0, 1, ..., M-1$ coefficients by the discrete Fourier transform (3) can be expressed in MSE metric with use of the following formula

$$\epsilon_{MSE}^A = \frac{1}{M}\sum_{k=0}^{M-1} \|R(k) - R_N(k)\|^2.$$

In order to obtain Fourier descriptors that are invariant to the following affine object transformations such as: shifting, scaling and rotations, the descriptors calculated with equation (3) must be further on normalized [10]. In case of the centroid distance signature the simplest form of the normalization formula can be written as follows

$$\hat{R}_N(k) = \frac{R_N(k)}{\|R_N(1)\|} \qquad (4)$$

for $k = 0, 1, ..., M - 1$. Then an invariance to object shifting ensures the choice of centroid distance signature, next an application of formula (4) to $R(k)$ coefficients gives invariance to object scaling, and by neglecting the phase component of $R(k)$ coefficients an invariance to possible object rotations can be achieved. Hence in this paper the following form $\{\left\|\hat{R}_N(k)\right\|\}$ for $k = 0, 1, ..., M - 1$ of normalized Fourier descriptors is adopted.

2.2 Fast Radix-2 Fourier Transform Algorithm

The process of calculation of Fourier descriptors directly with use of formula (3) can be characterized by the computational complexity of $\mathcal{O}(MN)$. However when the following inequality $M > log_2 N$ holds, then in order to accelerate the process of calculation of Fourier descriptors it is possible to employ fast algorithm of discrete Fourier transform with $\mathcal{O}(Nlog_2 N)$ complexity. In this paper the authors exploited the fast time decimated Cooley-Tukey radix-2 algorithm. In accordance with the paper [12] its decomposition formula can be written in the following form

$$R_N(k) = R_1(k) + R_2(k)e^{-i\frac{2\pi}{N}k}, \tag{5}$$

where $R_1(k)$ and $R_2(k)$ are two $N/2$-point discrete Fourier transforms operating upon $r(n)$ elements with even and odd n indexes respectively. Then by recursive application of formula (5) the reduction in the number of arithmetical operations to $\mathcal{O}(Nlog_2 N)$ level can be achieved, provided that N is an integer power of 2.

2.3 Fast Adaptive Fourier Transform

The fast adaptive Fourier transform algorithm described in paper [13] enables automatic selection of the number of $2N_1$ signal samples that are sufficient to calculate the required spectrum band with an acceptable error ϵ. The value of an actual error of numerical calculation of signal spectrum at subsequent stages of AFFT algorithm can be estimated on the basis of two transforms calculated upon the sets of $2N_1$ and N_1 samples. These estimations can be expressed in one of the following metrics: MD, MSE and PSNR [13]. Whenever the estimated value of an actual error is smaller than ϵ, then AFFT algorithm stops the adaptation process with the number of $2N_1$ samples. Otherwise the number of samples is doubled and another stage of adaptation begins. For calculation of Fourier transform coefficients at consecutive stages of AFFT algorithm the decomposition formula of fast time decimated algorithm can be exploited, e.g. of radix-2 or "split-radix 2/4" algorithms with the computational complexity $\mathcal{O}(Nlog_2 N)$. However in comparison with fast algorithms here the additional reduction of computations can be achieved whenever the estimated value of an actual error gets smaller than ϵ and the following inequality $2N_1 < N$ holds.

In the considered task of object contour classification AFFT algorithm is applied to automatic selection of such number of $2N_1$ contour points that are sufficient for calculation of $R_{2N_1}(k)$ for $k = 0, 1, ..., M - 1$ descriptors with an acceptable error ϵ. The estimated values of actual errors are expressed in MSE metric and at each stage of AFFT algorithm for calculation of $R_{2N_1}(k)$ coefficients the decomposition formula of FFT radix-2 algorithm is adopted. Then the estimated value ϵ_{MSE}^{E} of an actual error ϵ_{MSE}^{A} expressed at $2N_1$-point stage in MSE metric can be calculated with use of the following formula (see paper [13])

$$\epsilon_{MSE}^{E} = \frac{1}{M} \sum_{k=0}^{M-1} \|R_{2N_1}(k) - R_{N_1}(k)\|^2 / 9.$$

Hence in the considered case the stop criterion for AFFT algorithm takes the form of the following inequality $\epsilon_{MSE}^{E} \leq \epsilon$.

An application of fast adaptive Fourier transform for calculation of FD results in constructing new Fourier descriptors called fast adaptive Fourier descriptors.

3 The Database of Classified Objects

The experiments involved the database [14] of 186 photographic images of leaves belonging to three different tree species. The photographs of leaves were taken at different angles and depicted objects at different scales. In order to extract the contours of classified objects leaves images were submitted to filtering and analysis process which can be divided into the following subsequent phases: phase of preliminary filtering and image improvement, phase of search for leaf shape and the phase of contour retrieval and contour tracking.

In the first phase all images were submitted to preliminary filtering process in order to reduce the noise level and improve the brightness and image contrast.

The purpose of the second phase was at first to represent the color components of image pixels in HSV (*Hue Saturation Value*) space, in which color representation gives simplicity in separating such image areas that can be characterized by the green color of an appropriate brightness and hue, i.e. leaves shapes. And at the following step the coordinates of the central point of image area with the most dense concentration of points potentially falling into the leaf shape were calculated. Then in order to separate the leaf shape from other contents the image was filtered with the Robert's filter [17]. Next starting from the central point all possible empty spaces in leaf structures were filled, and the shape retrieved in such manner was submitted to the sequence of edge smoothing morphology-based operations, namely dilation and erosion [18], [19].

The last phase consisted in: leaf contour retrieval with Robert's filter, contour tracking with use of an algorithm for the search of inner edges of leaf in clockwise direction, and in selection of the given number of contour points that are equidistantly distributed along the whole contour line. In result the set of N points of leaf contour was finally retrieved. In the considered case the number of $N = 1024$ points were selected. In Figure 1 the sequence of operations performed at all phases of contours retrieval process is demonstrated.

From all of 186 available images the objects belonging to three classes were selected. These objects constituted the test and training sets respectively. The

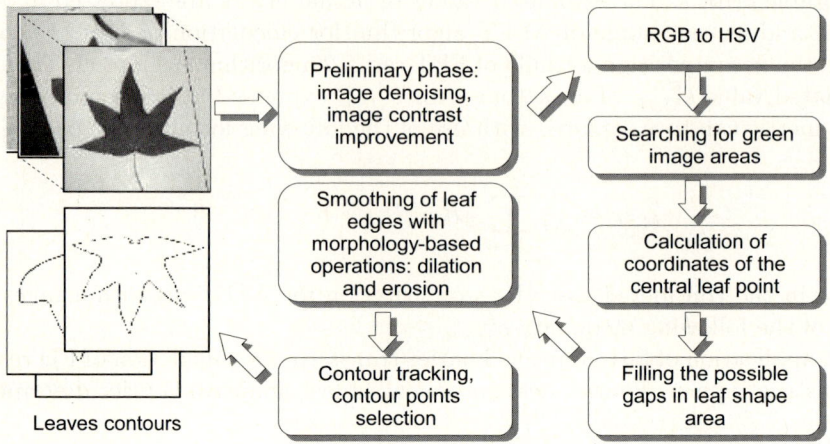

Fig. 1. The process of retrieval of leaves contours from images

test set was additionally enlarged with representatives created as transformations of existing objects, including: rotations (by 45, 135, 225, 315 angles), horizontal and vertical flips, scaling (the scaling factor was randomly taken within the range of 1.5 to 2.0), and by modifications of contour points coordinates with additive noise of normal distribution and 1/12 variance. As the result the training set composed of 30 objects and the test set including 1872 objects were retrieved.

4 Description of the Research Method

For all object contours that were retrieved during the process described in the previous section the Fourier descriptors of four different lengths: $M = 8, 16, 32$ and 64 were calculated. In order to calculate FD both the fast algorithm of radix-2 type and the fast adaptive Fourier transform algorithm were employed. Next FD were normalized in accordance with the formula (4) and the phase information of $\hat{R}(k)$ coefficients was neglected in order to obtain invariance to possible objects rotations. In result the feature vectors for all objects belonging to the test and training sets were obtained.

Into the task of objects classification a non-parametrized one nearest neighbor and multilayer perceptron classifiers were incorporated. The multilayer perceptron was composed of two hidden layers: the first one with 16, and the second output one with three sigmoidal and biased neurons [11].

All objects falling into the test set were classified. In case of 1-NN classifier the distances between FD of to-be-classified objects and objects belonging to the training set were expressed in Euclidean metric. The neural MP classifier was trained on objects belonging to the training set with use of a back-propagation learning technique [11].

In the further part of this paper by an accuracy of classification it is understood the proportional ratio of the number of objects properly classified to the number of all objects in the test set, i.e.

$$\text{Accuracy} = \frac{\text{Number of objects properly classified}}{\text{The whole number of objects in the test set}} \cdot 100\%.$$

5 Experimental Results

Several experiments were performed in order to obtain the practical verification of an effectiveness of AFFT algorithm in the tasks: of reduction of the number of contour data, and decreasing the computational times of calculation of Fourier descriptors. The second aim was to test the practical effectiveness of classification schemes constructed on the basis of AFFT algorithm with MP and 1-NN classifiers. Hence the received results are presented in the form of classification accuracies and the mean values of contour points required by AFFT algorithm for calculation of FD: at several values of acceptable ϵ errors, several descriptors lengths of $M = 8, 16, 32$ and 64 coefficients, and for two considered types of classifiers, namely MP (see Table 1) and 1-NN (see Table 2). In the second part of this section the comparative study of computational times of FD calculation with FFT and AFFT algorithms for the selected lengths M of descriptors and for MP (see Table 3) and 1-NN (see Table 4) classifiers are also presented.

In the following paragraphs by FFD we mark those Fourier descriptors that are calculated by means of FFT algorithm. The selected values of acceptable errors ϵ are computed as $\epsilon = 1/(2^n - 1)$ for several values of integer parameter n. Moreover in Tables 1 and 2 by the underlined fonts we highlight those values of n starting from which the accuracy of classification with FAFD is stabilized and equal to the one obtained with FFD.

It can be seen from results presented in Table 1 that the value of acceptable error of FAFD calculation that is required to obtain the same accuracies as with FFD is getting smaller with the growth of descriptors lengths M. In consequence

Table 1. Accuracies of classification and the mean values of FAFD contour points for different acceptable error ϵ values and MP classifier

	Descr. length $M = 8$		Descr. length $M = 16$		Descr. length $M = 32$		Descr. length $M = 64$	
n	Accuracy [%]	Mean no. of points	Accuracy [%]	Mean no. of points	Accuracy [%]	Mean no. of points	Accuracy [%]	Mean no. of points
4	93.935	32	95.000	64	95.694	64	96.111	128
8	93.889	64	95.324	128	96.111	128	96.111	128
12	93.935	256	95.463	256	96.111	256	96.111	256
16	93.981	512	95.463	512	96.111	1024	96.111	1024
20	93.981	1024	95.463	1024	96.111	1024	96.111	1024
24	93.981	1024	95.463	1024	96.111	1024	96.111	1024
28	93.981	1024	95.463	1024	96.111	1024	96.111	1024
32	93.981	1024	95.463	1024	96.111	1024	96.111	1024

the mean number of contour points required by AFFT algorithm also decreased. For example with $M = 8$ the mean number of contour points (for $n = 16$ i.e. $\epsilon = 1/(2^{16} - 1)$) equaled half of the total number of all $N = 1024$ points, and for the length of $M = 64$ it equaled $1/8$ (for $n = 4$ i.e. $\epsilon = 1/(2^4 - 1)$) of N value. The identical results to those presented for the case of $M = 64$ were also obtained for $M = 32$ (for $n = 8$ i.e. $\epsilon = 1/(2^8 - 1)$). And such descriptor length ($M = 32$) is most frequently used in the practice of object classification (see papers [7]-[10]).

Table 2. Accuracies of classification and the mean values of FAFD contour points in function of an acceptable error ϵ for 1-NN classifier

	Descr. length $M = 8$		Descr. length $M = 16$		Descr. length $M = 32$		Descr. length $M = 64$	
n	Accuracy [%]	Mean no. of points	Accuracy [%]	Mean no. of points	Accuracy [%]	Mean no. of points	Accuracy [%]	Mean no. of points
4	94.000	32	94.720	64	95.890	64	95.940	128
8	93.330	64	94.670	128	94.670	128	95.830	128
12	93.330	256	94.720	256	_95.333_	_256_	_96.000_	_256_
16	93.330	512	_94.780_	_512_	95.333	1024	96.000	1024
20	_93.280_	_1024_	94.780	1024	95.333	1024	96.000	1024
24	93.280	1024	94.780	1024	95.333	1024	96.000	1024
28	93.280	1024	94.780	1024	95.333	1024	96.000	1024
32	93.280	1024	94.780	1024	96.111	1024	96.000	1024

The results obtained with 1-NN (see Table 2) classifier indicate for almost twofold larger mean values of contour points required for FAFD to generate accuracies of classification that are stabilized and identical to those obtained with FFD. Moreover the classification accuracies for all considered lengths M are about 0.1 to 0.8 percent worse than those obtained with the classification scheme that utilizes neural MP classifier.

Further analysis of results presented in Table 3 indicates significantly smaller mean times of calculation of FAFD in comparison to the mean times required for calculation of FFD for all tested lengths: $M = 8, 16, 32$ and 64. The ratio of

Table 3. Mean computational times of calculation[1] of FAFD (measured for the first value of n with stabilized FAFD accuracies) and FFD for different lengths M of descriptors and MP classifier

Descriptors length	$M = 8$	$M = 16$	$M = 32$	$M = 64$
Mean FAFD times [ms]	3.0803	1.0676	0.5753	0.5753
Mean FFD times [ms]			4.5715	

---

[1] The experiments were performed under the control of Windows XP operating system on computer with AMD Athlon(tm) XP 2000+ processor and 512MB of memory.

Table 4. Mean computational times of calculation of FAFD (measured for the first value of n with stabilized FAFD accuracies) and FFD for different descriptors lengths M and 1-NN classifier

Descriptors length	$M = 8$	$M = 16$	$M = 32$	$M = 64$
Mean FAFD times [ms]	6.4012	3.2503	1.4628	1.4628
Mean FFD times [ms]		4.5715		

FFD to FAFD mean times was smallest for $M = 8$ and equaled about 1.5. Its highest value of about 7.9 was reached for the lengths of $M = 32, 64$. Due to the fact that FFD scheme always requires to calculate the full N-point Fourier transformation then for all considered values of M there is given only one value of the mean time of FFD calculation.

For 1-NN classifier (see Table 4) the obtained ratios of mean times of FFD to FAFD calculation are about two times worse than the ones received with MP classifier. Such results arise as the consequence of higher precision of adaptive calculation of FD that is required by 1-NN classifier in all of the tested cases.

6 Summary and Conclusions

From an analysis of experimental results it can be seen that the proposed fast adaptive Fourier descriptors gave even eightfold reduction in the number of contour points (for the scheme with neural MP classifier and descriptor lengths $M = 32$ and 64) and almost eightfold decrease in computational times in comparison with the scheme that utilizes fast radix-2 algorithm. Slightly lower ratios of computational times of FAFD to FFD than the corresponding ratios of mean values of contour points rise as the direct consequences of additional computations required by AFFT algorithm to calculate estimations of actual errors at consecutive stages of an adaptation process.

Hence it can be stated that the proposed fast adaptive Fourier descriptors meet their objective and can be applied to practical classification solutions. An application of fast adaptive descriptors results not only in significant reduction of practical times of classification but they also contribute to decrease in sizes of contour databases, and at the same time enable significant reduction of disk memory capacities and transmission bandwidths required for storage and transfer of data of classified objects.

The experiments also proved the advantage of multilayer perceptron classifier over 1-NN minimum distance one that was manifested in better classification accuracies and smaller sensitivity to numerical error of adaptive calculation of Fourier descriptors. Moreover for neural classifier all knowledge about the classified objects is encapsulated within the weights of neurons so the time required for classification of one object equals to the time of feed-forward propagation of object features vector.

The proposed fast adaptive Fourier descriptors are calculated in accordance with the structure of fast Fourier transform algorithm of radix-2 type, and hence can be efficiently implemented with parallel computers.

References

1. Shen, L., Rangayyan, R.M., Desautels, J.E.L.: Application of Shape Analysis to Mammographic Calcifications. IEEE Trans. on Medical Imaging 13(2), 263–274 (1994)
2. Rangayyan, R.M., El-Faramawy, N.M., Desautels, J.E.L., Alim, O.A.: Measures of Acutance and Shape for Classification of Breast Tumors. IEEE Trans. on Medical Imaging 16, 799–810 (1997)
3. Mahoor, M.H., Abdel-Mottaleb, M.: Automatic Classification of Teeth in Bitewing Dental Images. In: ICIP International Conference On Image Processing, pp. 3475–3478 (2004)
4. Sun, S.-G., Park, J., Park, Y.W.: Identification of Military Ground Vehicles by Feature Information Fusion in FLIR Images. In: Proc. 3rd International Symposium on Image and Signal Processing and Analysis, pp. 871–876 (2003)
5. Chung, Y.Y., Wong, M.T.: Handwritten Character Recognition by Fourier Descriptors and Neural Network. IEEE TENCON - Speech and Image Technologies for Computing and Telecommunications, 391–394 (1997)
6. Keyes, L., Winstanley, A.: Fourier Descriptors as a General Classification Tool for Topographic Shapes. In: Proc. Irish Machine Vision and Image Processing Conference, pp. 193–203 (1999)
7. Ezer, N., Anarim, E., Sankur, B.: A Comparative Study of Moment Invariants and Fourier Descriptors in Planar Shape Recognition. In: Proc. 7th Mediterranean Electrotechnical Conference (1994)
8. Kauppinen, H., Seppänen, T., Pietikäinen, M.: An Experimental Comparison of Autoregressive and Fourier-Based Descriptors in 2D Shape Classification. Trans. on Pattern Analysis and Machine Intelligence 17(2), 201–207 (1995)
9. Osowski, S., Nghia, D.D.: Fourier and Wavelet Descriptors for Shape Recognition Using Neural Networks - A Comparative Study. Pattern Recognition 35(9) (2002)
10. Zhang, D., Lu, G.: A Comparative Study of Fourier Descriptors for Shape Recognition and Retrieval. In: Proc. 5th Asian Conference On Computer Vision (2002)
11. Rutkowski, L.: Methods and Techniques of Artificial Intelligence. PWN Warsaw (2005) (In Polish)
12. Lyons, R.G.: Understanding Digital Signal Processing. Addison Wesley Longman, Inc. (1997)
13. Puchala, D., Yatsymirskyy, M.: Fast Adaptive Algorithm for Fourier Transform. In: Proc. of International Conference on Signals and Electronic Systems, pp. 183–185 (2006)
14. http://www.vision.caltech.edu/html-files/archive.html
15. Wilczura, M., Puchala, D.: Neural networks in leaves classification. In: XV-th Conference on Networks and Informatic Systems, pp. 175–178 (2007) (In Polish)
16. Cooley, J.W., Lewis, P.A., Welch, P.D.: Application of the Fast Fourier Transform to Computation of Fourier Integrals, Fourier Series, and Convolution Integrals. IEEE Trans. on Audio and Electroacoustics AU-15(2), 79–84 (1967)

17. Pavlidis, T.: Algorithms for Graphics and Image Processing. WNT Warsaw (1987) (In Polish)
18. Malina, W., Smiatacz, M.: Methods of Digital Image Processing. EXIT Warsaw (2005) (In Polish)
19. Tadeusiewicz, R., Korohoda, P.: Computer Analysis and Processing of Images. FPT (1997) (In Polish)

Improving the Efficiency of Counting Defects by Learning RBF Nets with MAD Loss

Ewaryst Rafajłowicz*

Institute of Computer Engineering, Control and Robotics, Wrocław University of Technology, Wybrzeże Wyspiańskiego 27, 50 370 Wrocław, Poland
ewaryst.rafajlowicz@pwr.wroc.pl

Abstract. The method of using a lateral histogram for evaluating the number of holes (e.g., defects) from images is known to be fast but rather inaccurate. Our aim is to propose a method of improving its performance by learning, but keeping the speed of the original method. This task is accomplished by considering a multiclass pattern recognition problem with linearly ordered labels and a loss function, which measures absolute deviations of decisions from true classes.

1 Introduction

Our aim in this paper is twofold. Firstly, we propose a method of improving the lateral histogram method (see [3], [2]), which is a well-known method of counting defects or other objects from images. This improvement is based on learning a radial basis functions (RBF) neural net with minimum absolute deviation (MAD) loss, which was introduced in [8]. The second aim is to propose a method for learning such nets and to illustrate its performance by an example[1]. By the applying MAD RBF net we do not lose too much from the speed of the lateral histogram method, since the learning of RBF net is made off line, while its decisions can be precomputed and tabulated.

The tasks of counting items from images include counting defects in continuous or discrete production of goods, counting cancer cells, estimating a traffic on a road and many others. Such tasks are frequently divided into a number of steps with separated phases of detection and recognition of objects, which are later counted. The approach proposed in this paper allows for the consideration and/or evaluation such tasks as one entity.

The above examples lead to considering problems with loss functions, which measure an absolute value of the distance between true and estimated number of items.

We remark that that other approaches to detecting and enhancing defects were recently proposed (see [14], [9] and the bibliography therein).

---

* This work was supported by the Ministry of Science and Higher Education under a research and development grant ranging from 2007 to 2009.
[1] As an example, the unexpectedly difficult task of counting holes in a slice of Swiss cheese was selected.

2   The Lateral Histogram Method

An appealing feature of the classical lateral histogram method (LHM) is its simplicity and speed of calculations, which makes it a good candidate for low level vision methods, applicable for on line image processing, e.g., in industrial quality monitoring. The only drawback is in the low accuracy of this method, providing too low a number of holes (defects) detected or too many nonexistent items detected (false alarms).

To recall briefly LHM, consider a gray scale image with pixel values scaled to the unit interval

$$y_{kl}, \quad k = 1, 2, \ldots, K, \, l = 1, 2, \ldots, L, \quad (1)$$

where $y_{kl} \in [0, 1]$ is a gray level of (k, l) pixel, while K and L denote the image width and height, respectively. As usual, 0 corresponds to black and 1 white colors.

Consider the following sums along the x and y axes

$$S_H(l) = \sum_{k=1}^{K} y_{kl}, \quad l = 1, 2, \ldots, L \quad (2)$$

and

$$S_V(k) = \sum_{l=1}^{L} y_{kl}, \quad k = 1, 2, \ldots, K. \quad (3)$$

Assuming that the holes (defects) are represented as being black or dark gray, one can expect that $S_H(l)$ and $S_V(k)$ have local minima located at the positions corresponding to projections of the holes into x and y axes. Thus, it suffices to locate local minima of these sequences in order to count the number of holes. In our experiments l^* was treated as local minimum of $S_H(l)$ if the following conditions were simultaneously fulfilled

$$S_H(l^*) < S_H(l^* + 1) < S_H(l^* + 2), \, S_H(l^*) > S_H(l^* - 1) > S_H(l^* - 2).$$

Local minima for $S_V(k)$ were detected analogously. It remains to point out how to convert the number of local minima in the horizontal direction n_H, say, and the number of local minima in the vertical direction, n_V, say, into one decision. One can use their mean, but taking into account that frequently larger holes shade into smaller ones, we suggest using

$$\max\{n_H, n_V\}. \quad (4)$$

In Figure 1 the result of applying LHM is shown. One can easily recognize the sources of errors in counting holes. The main one is in that one hole can overshadow several others and this feature is built into the method. As a partial remedy we propose adding a neural net, which is trained to reduce the consequences of overshadowing (see next sections). Alternatively, one could consider

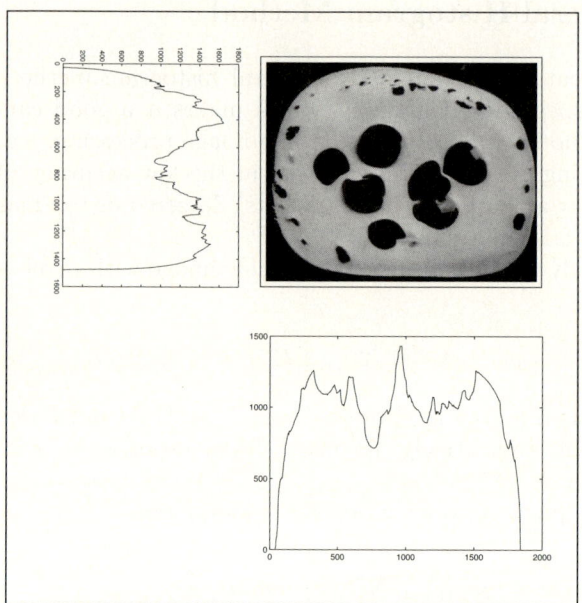

Fig. 1. An example of applying the lateral histogram method to a slice of Swiss cheese

several other projections under different angles (in the vein of the Radon transform) or to estimate how deep are local maxima in comparison to the expected sizes of holes, but these approaches introduce too much computational of a burden. The second source of errors is in detecting local minima. Small local minima can result from inhomogeneities in lighting, dust etc.

3 Minimum Absolute Deviations Decisions

Let $X \in R^d$ be a random vector. Pair (X, i) is a random vector representing a pattern X and its correct classification $i \in \mathcal{I}$, which is unknown for a new pattern X to be classified, while $\mathcal{I}$. Probability densities $f(x|i)$, describe the conditional p.d.f. of X, provided that it was drawn from i-th class. A loss $L(i, j)$, if a pattern from i-th class is classified to j-th class, is assumed to be $L(i, j) = |i-j|, i, j \in \mathcal{I}$.

A decision function $\Psi(X)$, which specifies a label of the class for X and such that it minimizes the expected loss given by:

$$R(\Psi) = E_X \left[\sum_{i=1}^{I} |i - \Psi(X)| P(i|X) \right], \qquad (5)$$

where E_X denotes the expectation w.r.t. X, while $P(i|X)$ is the a posteriori probability that observed pattern X comes from i-th class. Our aim is to minimize the risk $R(\Psi)$, provided that the minimizer $\Psi^*(x)$, say, is a measurable function.

According to the Bayes rule, $P(i|X = x)$ is given by

$$P(i|X=x) = \frac{f(x|i)\,q(i)}{f(x)}, \quad f(x) \stackrel{\text{def}}{=} \sum_{l=1}^{I} f(x|l)\,q(l), \tag{6}$$

which allows us to express the a posteriori probabilities in terms of class densities and a priori probabilities, which is easier to estimate from the learning sequence.

In [8] it was proved that a MAD optimal recognizer has the form:

$$\Psi^*(x) = \text{MED}[P(i|X = x), \, i \in \mathcal{I}], \tag{7}$$

where MED[] denotes the median of a discrete probability distribution, which is indicated in the brackets.

4 Learning a RBF Net for Improving the Lateral Histogram Method

We usually cannot apply (7) directly, since $f(x|i)$'s are unknown, but we can estimate these functions from the learning sequence $(X(n), i(n), n = 1, 2, \ldots, N$ with the aid of (6) (plug-in decision rules (see [5], [4])). We shall select estimators of $f(x|i)$ from the class of radial basis functions. We refer the reader to [6], [7], [10], [11],[12],[13], [15] for various aspects of training RBF nets.

Denote by $K(t) \geq 0$, $t \in R$ a kernel of RBF's, which is such that $\int_{-\infty}^{\infty} K(t) = 1$, $\int_{-\infty}^{\infty} t\,K(t) = 0$. Let be the set $\mathcal{I}(i)$ of those numbers of observations, for which the corresponding labels indicate i-th class, i.e., $i(n) = i$. Define $n(i) = \text{Card}(\mathcal{I}(i))$. For estimating a priori probability that a pattern comes from class i it suffices to set $\hat{q}(i) = n(i)/N$, $i = 1, 2, \ldots, I$ as the estimator of $q(i)$.

In general, learning would require a large amount of observations in order to estimate all the class densities $\hat{f}(x|i) = 1, 2, \ldots, I$. Collecting a sufficient amount of data is possible when I is relatively small, but for counting dozens of holes we have to impose some constraints on the class of learning problems. To this end we specify which features are used for training our classifier to count holes. For simplicity of exposition we select only two features, namely, the number of local minima detected in $S_H(l)$ and $S_V(k)$. Denote these numbers as κ_H and κ_V, respectively. Thus, our feature vectors have the form $x = (\kappa_H, \kappa_V) \in R^2$ and elements of the learning sequence $(X(n), i(n))$ consist of 2D vectors and the corresponding labels

$$((\kappa_H(n), \kappa_V(n)), i(n)), \quad n = 1, 2, \ldots, N. \tag{8}$$

It should be stressed that we take two features for simplicity of formulas. The learning procedure described below can be applied for more than two features. For example, instead of using only two projections on the orthogonal axes one can use projections in many other directions, or to subdivide the image into non-overlapping blocks and to take sums along their axes.

A key assumption that allows us to use moderate data sets for learning is the following: every class density $f(x|i)$ depends only on the differences $\kappa_H - i$ and $\kappa_V - i$, i.e., on the difference between a "true" class label i and number of local minima along horizontal and vertical directions, respectively. In other words, we assume that there exists a bivariate pdf, g say, such that

$$f(x|i) = f((\kappa_H, \kappa_V)|i) = g(\kappa_H - i, \kappa_V - i), \quad i = 1, 2, \ldots, I. \tag{9}$$

As estimator $\hat{g}$ of g we take the mixture of Gaussian distributions, i.e.,

$$\hat{g}(\kappa_H - i, \kappa_V - i) = \sum_{j=1}^{J} w(j) \, k_{ij}, \tag{10}$$

where $w(j)$'s are weights, $h(j)$'s are smoothing parameters, while (neglecting arguments for brevity)

$$k_{ij} \stackrel{\text{def}}{=} K\left(\frac{\|(\kappa_H - i - c_H(j), \kappa_V - i - c_V(j))\|_j}{h(j)}\right).$$

By $c_V(j)$ and $c_H(j)$ we have denoted coordinates of RBF centers in the vertical and the horizontal directions, respectively. Below we discuss the choice of parameters in (10). Once they are selected, we use it in the decision process as follows. A posteriori probability of class j, calculated for $x = (\kappa_H, \kappa_V)$ and denoted as $\hat{P}(j|(\kappa_H, \kappa_V))$ is estimated as follows:

$$\hat{P}(j|(\kappa_H, \kappa_V)) = \frac{\hat{q}_j \, \hat{g}(\kappa_H - j, \kappa_V - j)}{\sum_{j=1}^{I} \hat{q}_j \, \hat{g}(\kappa_H - j, \kappa_V - j)}, \tag{11}$$

$j = 1, 2, \ldots, I$. When a new pair of lateral histograms arrive, we calculate the number of their local minima (κ_H, κ_V) and a decision i is taken, if i is the first integer for which

$$\sum_{j=1}^{i} \hat{q}_j \, \hat{g}(\kappa_H - j, \kappa_V - j) \geq \frac{1}{2} \sum_{j=1}^{I} \hat{q}_j \, \hat{g}(\kappa_H - j, \kappa_V - j), \tag{12}$$

where $\hat{q}_j$ are the estimates of a priori probabilities for each class, i.e., $\hat{q}_j = n_j/N$, where n_j is the number of examples in the learning sequence, which has j holes.

In the experiments reported in the next section we take $K(t)$ as a univariate Gaussian pdf, while as the norms $\|\tau\|_j$ of row vectors $\tau \in R^2$ we propose to select

$$\|\tau\|_j = \left(\tau \, \hat{C}_j^{-1} \, \tau^T\right)^{\frac{1}{2}}, \quad j = 1, 2, \ldots, J, \tag{13}$$

where $\hat{C}_j$ is the standard estimator of the covariance matrix, corresponding to observations in j-th receptive field.

Now, it remains to propose a way of selecting receptive fields and other parameters in (10). To this end, firstly estimate the covariance matrix from all the feature vectors

$$(\kappa_H(n), \kappa_V(n)), \quad n = 1, 2, \ldots, N. \tag{14}$$

Denote this matrix by $\hat{C}$. Then, perform the principal component analysis (PCA) of $\hat{C}$. Denote by λ_1 and λ_2 the eigenvalues and by $\bar{v}_1$ and $\bar{v}_2$ the corresponding eigenvectors of $\hat{C}$. Consider these two cases:

1) If one direction dominates the other, $\lambda_1 >> \lambda_2$, say, then select J centers of RBF's along $\bar{v}_1$ direction only.
2) If λ_1 and λ_2 are comparable, then select J_1 in $\bar{v}_1$ direction and J_2 centers in $\bar{v}_2$ direction, $J_1 + J_2 = J$.

In both cases one can use univariate clustering algorithms.

Now, we attach observations (14) to centers $(c_H(j), c_V(j))$, $j = 1, 2, \ldots, J$, according to the nearest neighbor rule. This step results in partitioning the set of all observations into J disjoint subsets[2], denoted further as $\mathcal{X}_j$, $j = 1, 2, \ldots, J$. This partition allows us to select the rest of the parameters as follows:

a) Calculate the mean vectors from $\mathcal{X}_j$ and use them to update the positions of centers $(c_H(j), c_V(j))$. Recalculate the contents of $\mathcal{X}_j$, if changes are essential.
b) Calculate $\hat{C}_j$, i.e., estimates of covariance matrices for each set $\mathcal{X}_j$ and use them to define the norms in (13).
c) Calculate weights as follows:

$$w(j) = \text{Card}(\mathcal{X}_j)/N, \quad j = 1, 2, \ldots, J. \tag{15}$$

The above selection of parameters has the advantage that parameters are clearly interpretable in the statistical sense. It remains to discuss the choice of smoothing parameters $h(j)$. A good starting point is to put all $h(j) = 1$, since $\hat{C}_j$'s are used to define $||.||_j$. One can use the well known cross-validation technique for their tuning, taking into account the danger of overfitting.

5 Counting Holes in a Swiss Cheese

As an example for illustrating the performance of the proposed method we have chosen counting holes in slices of Swiss cheese. The reason for this choice is not only in the availability of experimental material, but also in the difficulties of applying the original lateral histogram method in this case (see Fig. 1, where holes of different sizes and shapes shade into each other).

The methodology of learning was the following:

1) We have to select an algorithm of counting holes (almost) exactly for comparisons. Fig. 1 shows that we can not count them "by hand", since the number of holes in one slice frequently exceeds forty and even patient person can make mistakes. For this reason we have selected a method of counting holes, which is based on binarization by thresholding, region growing and labelling. This classical approach is too time-consuming for on-line applications, but more accurate than the lateral histogram that we are trying to improve. Using this technique we have counted holes in 45 slices of Swiss cheese. The results of counting were

---
[2] The ties are broken according to the smallest index rule.

Table 1. Left panel – comparison of lateral histogram (left column) and improved decisions on testing data. Right panel – contour plot of density estimate $\hat{g}$.

Dec. (4)	Our Dec.	"True"
8.	23	25.
9.	22	21.
7.	22	17.
7.	19	6.
4.	18	14.
2.	17	16.
3.	16	18.
11.	21	31.
7.	19	14.
5.	19	17.

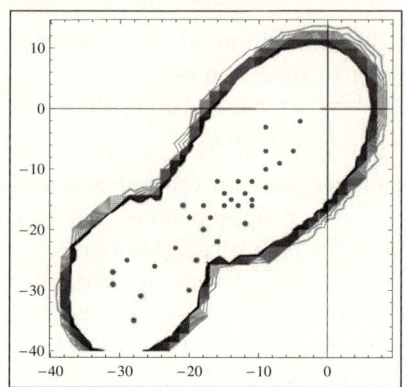

divided into two groups: the one used for learning (34 slices), which are treated as true label classes $i(n)$ in (8)) and the group for testing (11 slices).

2) For each slice calculate the number of local minima in the horizontal and vertical lateral histograms, i.e., $(\kappa_H(n), \kappa_V(n))$, $n = 1, 2, \ldots, N$.

3) Prepare the learning sequence. According to our assumption (9), the learning sequence has the form:

$$((\kappa_H(n) - i(n), \kappa_V(n) - i(n)), i(n)), \quad n = 1, 2, \ldots, N. \tag{16}$$

Note however, that for testing and applying the recognizer, we cannot use (16). Instead of $i(n)$ we then put i, treated as the decision variable.

4) Calculate parameters of density $\hat{g}((\kappa_H - i, \kappa_V - i)$ and use it for deciding on the number of holes according to (12).

The contour plot of density $\hat{g}$, which was obtained according to the above steps of learning is shown in Fig. 1 (right panel). After the learning phase, the RBF net performance was tested on a cheese slice, which was not used for learning. In order to assess the improvement of the lateral histogram method provided by our learning procedure, we take formula (4) as the classical way of decision making and rule (12) as an improved decision. The results of applying these rules is shown in Table 1. Its left column contains the results of applying the rule (4), while in the middle column one can observe the improvement introduced by the proposed approach. The overall improvement, as measured by the sum of absolute differences, is the reduction of this criterion from 118 for the lateral histogram to 45, which is obtained by learning.

6 Concluding Remarks

The method for improving the accuracy of the lateral histogram approach to counting holes (defects, non-regular shapes) by learning a specific class of RBF networks was proposed. One can easily extend this method to problems of counting from more than two dimensional projections. Furthermore, the method can be adopted to features which do not necessarily arise from counting.

References

1. Bishop, C.: Neural Networks for Pattern Recognition. Oxford University Press, Oxford (1995)
2. Davies, E.R.: Lateral histograms for efficient object location: speed versus ambiguity. Pattern Recogn. Lett. 6, 189–198 (1987)
3. Davies, E.R.: Machine Vision: Theory, Algorithms, Practicalities, 3rd edn. Morgan Kaufmann, San Francisco (2005)
4. Devroye, L., Györfi, L.: Nonparametric Density Estimation. The L_1 View. Wiley, New York (1985)
5. Devroye, L., Györfi, L., Lugosi, G.: Probabilistic Theory of Pattern Recognition. Springer, New York (1996)
6. Karayiannis, N.B., Randolph-Gips, M.M.: On the Construction and Training of Reformulated Radial Basis Function Neural Networks. IEEE Trans. on Neural Networks 14, 835–846 (2003)
7. Krzyżak, A., Skubalska-Rafajłowicz, E.: Combining Space-Filling Curves and Radial Basis Function Networks. In: Rutkowski, L., Siekmann, J.H., Tadeusiewicz, R., Zadeh, L.A. (eds.) ICAISC 2004. LNCS (LNAI), vol. 3070, pp. 229–234. Springer, Heidelberg (2004)
8. Rafajłowicz, E., Skubalska-Rafajłowicz, E.: MAD loss in pattern recognition and RBF learning. In: ICAISC 2008 (to appear, 2008)
9. Rafajłowicz, E., Pawla, M., Steland, A.: Nonlinear Image Filtering and Reconstruction: A Unified Approach Based on Vertically Weighted Regression. Int. J. Apll. Math. Comp. Sci (to appear, 2008)
10. Skubalska-Rafajłowicz, E.: Pattern Recognition Algorithms Based on Space-Filling Curves and Orthogonal Expansions. IEEE Trans. Information Theory 47, 1915–1927 (2001)
11. Skubalska-Rafajłowicz, E., Krzyżak, A.: Fast k-NN Classification Rule Using Metric on Space-Filling Curves. In: Proceedings of the 13th International Conference on Pattern Recognition, Vienna, vol. 2, pp. 121–125 (1996)
12. Skubalska-Rafajłowicz, E.: Data Compression for Pattern Recognition Based on Space-Filling Curve Pseudo-Inverse Mapping. Nonlinear Analysis, Theory, Methods and Applications 47, 315–326
13. Skubalska-Rafajłowicz, E.: RBF Neural Network for Probability Density Function Estimation and Detecting Changes in Multivariate Processes. In: Rutkowski, L., Tadeusiewicz, R., Zadeh, L.A., Żurada, J.M. (eds.) ICAISC 2006. LNCS (LNAI), vol. 4029, pp. 133–141. Springer, Heidelberg (2006)
14. Skubalska-Rafajłowicz, E.: Local correlation and entropy maps as tools for detecting defects in industrial images. Int. J. Apll. Math. Comp. Sci (to appear, 2008)
15. Xu, L., Krzyżak, A., Yuille, A.: On Radial Basis Function Nets and Kernel Regression: Statistical Consistency, Convergence Rates and Receptive Field Size. Neural Networks 4, 609–628 (1994)

Robust MCD-Based Backpropagation Learning Algorithm

Andrzej Rusiecki

Wroclaw University of Technology, Wroclaw, Poland
andrzej.rusiecki@pwr.wroc.pl

Abstract. Training data containing outliers are often a problem for supervised neural networks learning methods that may not always come up with acceptable performance. In this paper a new, robust to outliers learning algorithm, employing the concept of initial data analysis by the MCD (minimum covariance determinant) estimator, is proposed. Results of implementation and simulation of nets trained with the new algorithm and the traditional backpropagation (BP) algorithm and robust Lmls are presented and compared. The better performance and robustness against outliers for the new method are demonstrated.

1 Introduction

One of the main advantages of using feedforward neural networks (FFNs) is that they usually do not require exact mathematical knowledge about the input-output dependencies of modelled systems. These networks, considered to be universal as well as model-free approximators [1], find their application in areas such as pattern recognition, function approximation, or signal and image processing. They are trained by minimizing some kind of an error function, chosen to fit training data as close as possible. Although this method, very sufficient when no information of the modelled systems is available, is effective and easy to use, its performance depends strongly on the training data quality. In other words, this learning scheme is reliable only when the training set is not corrupted by the large noise, so when outliers and gross errors appear, the network builds a model that can be very inaccurate.

The explanation for this phenomenon seems to be clearly evident: the most common, backpropagation learning algorithm (in fact, its criterion function) is based on the least mean squares (LS) method, which is optimal for the normal error distribution. In the LS the square of the residuals is minimized to achieve the final fitting between system outputs and the training patterns, and when the errors are generated from a distribution different from Gaussian, the method may loose its efficiency.

1.1 Gross Errors and Outliers

In most real-world cases, however, the assumption that the errors are normal and iid simply doesn't hold. Data obtained from the environment are usually affected by noise of unknown form or different types of gross errors. In routine data the quantity of such outliers ranges from 1 to 10% [2]. They may be results of measurement errors, long-tailed noise, or be caused by human mistakes. It is enough to say that during obtaining the data from the environment and preprocessing them there is always a risk of gross errors.

Intuitively we can define an outlier as an observation that significantly deviates from the bulk of data. As one may notice this definition doesn't help in classifying an outlier as a gross error or a meaningful and important observation. The problem of outliers lies in the field of interest of a separate branch of statistics, called robust statistics [2,3]. Robust statistical methods are developed to act well when the true underlying model deviates from the assumed parametric model. Ideally, they should be efficient and reliable for the observations that are very close to the real values and simultaneously for the observations containing larger deviations and outliers. Unfortunately, they are not easily applicable to the FFN's learning algorithms. In this paper we propose a new robust learning algorithm based on the MCD robust estimator. This algorithm combines robust statistical method with the backpropagation learning algorithm.

1.2 Robust NN Learning Algorithms

The most popular FFNs learning algorithm, based on the backpropagation [16], minimizes the mean squared error (mse) and doesn't take into account the problem of outliers. This is why a couple robust learning algorithms have been recently proposed. Generally, all of them take advantage of the idea of robust estimators [3], replacing the mse with a loss error function of such a shape that the impact of outliers may be, in certain conditions, reduced without changing the BP strategy.

Chen and Jain [5] proposed the Hampel's hyperbolic tangent as a new error criterion, with the scale estimator β that defines the interval supposed to contain only clean data, depending on the assumed quantity of outliers or current errors values. This idea was combined with the annealing concept by Chunag and Su [6]. They applied the annealing scheme to decrease the value of β, whereas Liano [10] introduced the logistic error function Lmls (*Least Mean Log Squares*) derived from the assumption of the errors generated with the Cauchy distribution. Pernia-Espinoza et al. [13] presented an error function based on tau-estimates. Approaches based on the adaptive learning rate [17] and LTS [18] estimator were also proposed. Some authors [7,8] tried to apply robust learning algorithms also to radial basis function networks.

All these algorithms need some special assumptions to be hold. Moreover, most of them exploit the same basic idea and they are designed to perform well only for certain data types. The approach described in this paper is different because it introduces some kind of data analysis to reduce the impact of outliers to the network training process.

2 Robust MCD Algorithm

2.1 MCD Estimator

To the training data preprocessing we used the MCD estimator (*Minimum Covariance Determinant*), proposed in [14] and described in details in [11]. Applying this estimator one is able to find robust estimates of the data center and the scatter matrix. In this method we look in the data set $\theta_i = (\theta_{i1}, \ldots, \theta_{in})$ for $h > N/2$ out of N observations whose classical covariance matrix has the lowest possible determinant (obviously $n < h$). For the found subset we calculate classical estimates: the average of h points $\hat{\mu}$ and their covariance matrix $\hat{\Sigma}$. Using such results we can obtain the clean data set center and scatter matrix and divide the training data into the clean set and potential outliers.

Similarly to the well-known LTS (*Least Trimmed Squares*) estimator we get the highest resistance towards gross errors achieved by taking $h = [(N+n+1)/2]$, which unfortunately makes the estimator gaussian efficiency rather poor.

The main disadvantage of using the MCD estimator is its computation by the mean of minimization the covariance matrix for all the data subsets containing h elements. It would naively require an exhaustive investigation of all the possible h-element choices out of N. Luckily, a dedicated fast algorithm was proposed in [15]. It uses a resampling strategy and avoids such complete enumeration of given subsets.

After the proper data subset was found and the estimators $\hat{\mu}$ and $\hat{\Sigma}$ were calculated, we can start identifying outliers by estimating distances from the clean data center for each observation. It can be done by so-called robust distance [12] defined as:

$$RD_i = \sqrt{(\theta_i - \hat{\mu})^T \hat{\Sigma}^{-1}(\theta_i - \hat{\mu})}, \qquad (1)$$

The robust distance is a simple robustification of the Mahalonobis distance:

$$MD_i = \sqrt{(\theta_i - \bar{\theta})^T S^{-1}(\theta_i - \bar{\theta})}, \qquad (2)$$

where S is the covariance matrix of full data set. Once robust distances for all the observations are calculated, we can identify outliers as points with robust distance exceeding certain threshold value. In this way the data can be divided into two subsets: one of them containing only clean data, and the second one containing observations suspected to be gross errors.

2.2 Robust MCD Algorithm

Such procedure can be, with certain modifications, applied also in the field of artificial neural networks to build a new learning algorithm robust against gross errors. In this paper we propose a novel approach that combines training data analysis with the MCD estimator and a new kind of error criterion. In this algorithm, before the network training process begins, special weight coefficients for each training pattern are calculated. These coefficients are then used to create a function minimized in the algorithm.

We assume that all the training data are stored in a matrix γ composed with network inputs x and corresponding targets t as follows:

$$\gamma_i = \begin{bmatrix} x_{i1} & x_{i2} & \ldots & x_{ip} & t_{i1} & t_{i2} & \ldots & t_{iq} \end{bmatrix}, \qquad (3)$$

where $i = 1 \ldots N$.

For the matrix γ MCD estimator can be calculated. Then for each vector consisting of $(p+q)$ elements (in the most cases $q = 1$) and obtained from the MCD parameters, we calculate robust distance given by equation (1). Such procedure should help us in detecting gross errors in the input (leverage points) as well as in the output vector (outliers).

Unfortunately, a new problem appears: the MCD estimator helps in dividing data into two subsets but it doesn't guarantee that outliers are in fact gross errors. The observations suspected to be errors can be potentially very important and meaningful. This is why we cannot exclude them from the training process.

For such reasons weight coefficients need to be introduced. In this approach we propose to forward the information, about observations suspected to be errors, to the training algorithm by the MCD-based weight coefficients defined as:

$$u_i = \min\{1, \frac{\chi^2_{0.975, p+q}}{RD_i^2}\}, \qquad (4)$$

where the value of χ^2 statistics is chosen under the assumption of normality with $h/N = 0.975$. These coefficients are used then to minimize an error criterion that becomes a weighted average of the form:

$$E = \frac{1}{N} \sum_{i=1}^{N} u_i \rho(r_i), \qquad (5)$$

where ρ may be a simple square function or one of the "robust" error functions based on the one-step M-estimators.

2.3 Derivation of Robust MCD Algorithm

Let us consider, for simplicity, a simple feedforward neural network with one hidden layer. The net is trained on a data set consisting of N training pairs $\{(\boldsymbol{x}_1, \boldsymbol{t}_1), (\boldsymbol{x}_2, \boldsymbol{t}_2), \ldots, (\boldsymbol{x}_N, \boldsymbol{t}_N)\}$, where $\boldsymbol{x}_i \in R^p$ and $\boldsymbol{t}_i \in R^q$. For the given input vector $\boldsymbol{x}_i = (x_{i1}, x_{i2}, \ldots, x_{ip})^T$, the output of the jth neuron of the hidden layer may be obtained as:

$$z_{ij} = f_1(\sum_{k=1}^{p} w_{jk} x_{ik} - b_j) = f_1(inp_{ij}), \quad \text{for } j = 1, 2, \ldots, l, \qquad (6)$$

where $f_1(\cdot)$ is the activation function of the hidden layer, w_{jk} is the weight between the kth net input and jth neuron, and b_j is the bias of the jth neuron.

With such assumptions the output vector of the network $\boldsymbol{y}_i = (y_{i1}, y_{i2}, \ldots, y_{iq})^T$ is given as:

$$y_{iv} = f_2(\sum_{j=1}^{l} w'_{vj} z_{ij} - b'_v) = f_2(inp_{iv}), \quad \text{for } v = 1, 2, \ldots, q. \tag{7}$$

Here $f_2(\cdot)$ denotes the activation function, w'_{vj} is the weight between the vth neuron of the output layer and the jth neuron of the hidden layer, and b'_v is the bias of the vth neuron of the output layer. We can write new error criterion, based on the MCD estimator as in equation (5).

If we assume for simplicity that the weights are updated according to the gradient-descent learning algorithm (in our implementation the conjugate gradient method was applied [4]), then to each weight is added (α denotes a learning coefficient):

$$\Delta w_{jk} = -\alpha \frac{\partial E}{\partial w_{jk}} = -\alpha \frac{\partial \sum_{i=1}^{N} u_i \rho(r_i)}{\partial r_i} \frac{\partial r_i}{\partial w_{jk}}, \tag{8}$$

$$\Delta w'_{vj} = -\alpha \frac{\partial E}{\partial w'_{vj}} = -\alpha \frac{\partial \sum_{i=1}^{N} u_i \rho(r_i)}{\partial r_i} \frac{\partial r_i}{\partial w'_{vj}}, \tag{9}$$

where

$$\frac{\partial r_i}{\partial w_{jk}} = f'_2(inp_{iv}) w'_{vj} f'_1(inp_{ij}) x_{ik}, \tag{10}$$

and

$$\frac{\partial r_i}{\partial w'_{vj}} = f'_2(inp_{iv}) z_{ij}. \tag{11}$$

Such procedure can be easily extended to any kind of gradient-based learning algorithm.

For the MCD-based robust learning algorithm all the steps can be then written as follows:

1. Set the network initial parameters.
2. Calculate the MCD estimator for the training set.
3. Calculate robust distance for each training pattern and corresponding weight coefficients.
4. Calculate network errors based on the modified error function (backpropagation). If the stopping criterion, stop the algorithm.
5. Make one learning step and go to point 4.

3 Simulation Results

3.1 Testing Methodology

To test the MCD-based algorithm we used many different testing cases. In this paper we present only a few of many results obtained for the function approximation tasks. The first function to be approximated was:

$$y = |x|^{-2/3}, \tag{12}$$

proposed in [5], the second one was a two-dimensional spiral given as:

$$\begin{cases} x = \sin y \\ z = \cos y \end{cases} \tag{13}$$

Data types. To simulate real data containing noise and outliers we used different models, defined as follows:

- Clean data without noise and outliers;
- Data corrupted with the Gross Error Model: $F = (1-\delta)G + \delta H$, where F is the error distribution, $G \sim N(0.0, 0.1)$ and $H \sim N(0.0, 10.0)$ are Gaussin noise and outliers and occur with probability $1-\delta$ and δ (data Type 1);
- Data with high value random outliers (Type 2), proposed in [13] of the form $F = (1-\delta)G + \delta(H_1 + H_2 + H_3 + H_4)$, where:
  - $H_1 \sim N(15, 2)$,
  - $H_2 \sim N(-20, 3)$,
  - $H_3 \sim N(30, 1.5)$,
  - $H_4 \sim N(-12, 4)$.
- Data with outliers generated from the Gross Error Model, injected into the input vector x_i (Type 3).

We compared performances of the traditional backpropagation algorithm (BP), robust LMLS algorithm and out novel MCD-based algorithm. To solve the testing problem a simple three-layer network with one or two inputs (depending on a problem), one output and ten hidden sigmoid neurons was applied. Each tested method was run 500 times for each task, then a mean MSE for the networks learnt with the given algorithm was calculated. To obtain the mean MSE of hundreds trails the nets were simulated on the clean data generated as points lying on the approximated curves. To minimize the error criterion the conjugate gradient method [4,9] was used.

Mean simulation results were are presented in tables and the exemplary network responses for the testing data are shown in figures.

3.2 Results

Looking at the table 1 one may notice that for the clean training data sets all of the tested algorithms perform well and the mean error obtained for the MCD method is only slightly higher than for the others. The situation changes for the data containing gross errors - both robust algorithms act properly while traditional BP method achieves very poor error level with values even ten times larger than the best here, MCD algorithm. The results for this method are 10-20% better than for the robust Lmls algorithm.

For the data sets containing high value outliers, we can see that two robust algorithms completely outperform classical method. However, none of them can be considered as the best one in this case.

Fig. 1. Typical normalized histogram of outliers, Data Type 1, $\delta = 0.2$, outliers values normalized on [0,10]

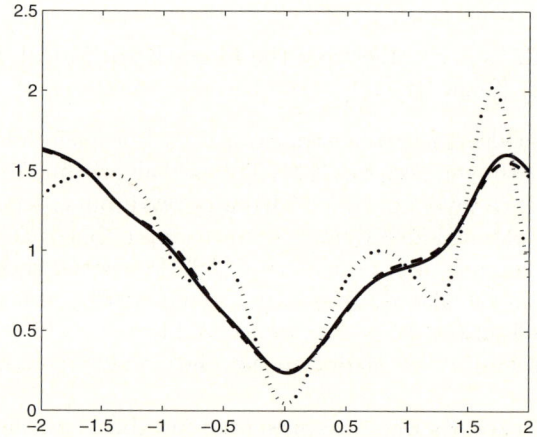

Fig. 2. Simulation results for the network trained to approximate one dimensional function (data Type 1): backpropagation algorithm (*dotted line*), LMLS alg. (*solid line*), MCD alg. (*dashed line*)

The results obtained for the second task are slightly different. After analysing table 2 we can notice that for the data with errors generated with the gross error model, our novel MCD-based algorithm is definitely the best. Its error level is only 50% of that obtained for the robust Lmls method. In the case of the data Type 2, all the methods act in the same way, as in the one-dimensional function approximation task. For the outliers injected into the input vector robust MCD algorithm is once again the best.

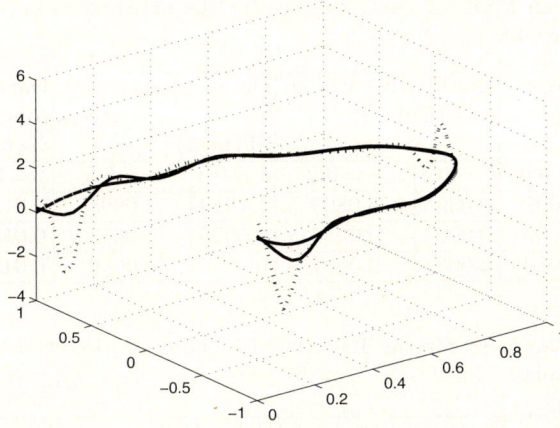

Fig. 3. Simulation results for the network trained to approximate two-dimensional spiral (data Type 2): backpropagation algorithm (*dotted line*), LMLS alg. (*solid line*), MCD alg. (*dashed line*)

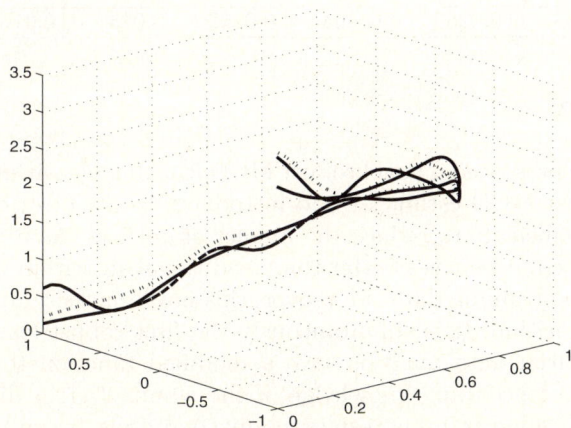

Fig. 4. Simulation results for the network trained to approximate two-dimensional spiral (data Type 3): backpropagation algorithm (*dotted line*), LMLS alg. (*solid line*), MCD alg. (*dashed line*)

To summarise, one may notice that networks trained with the robust MCD-based algorithm achieved the most often lowest errors. This new algorithm outperforms not only traditional learning method but also robust Lmls algorithm. In the few cases when the Lmls method trains network that achieved lowest error (esp. data Type 2), the new MCD algorithm presents very similar performance.

Table 1. The mean MSE for the 100 trials for the networks trained to approximate function of one variable

	Clean Data	Data with gross errors (Type 1)		Data with high value outliers (Type 2)		Data with gross errors in the input vector (Type 3)	
Algorithm	$\delta = 0.0$	$\delta = 0.1$	$\delta = 0.2$	$\delta = 0.1$	$\delta = 0.2$	$\delta = 0.1$	$\delta = 0.2$
BP	0.0007	0.0398	0.0809	1.7929	4.0996	0.0140	0.0180
LMLS	0.0007	0.0061	0.0088	0.0050	0.0053	0.0151	0.0177
MCD	0.0010	0.0051	0.0071	0.0049	0.0069	0.0138	0.0381

Table 2. The mean MSE for the 100 trials for the networks trained to approximate two-dimensional spiral

	Clean Data	Data with gross errors (Type 1)		Data with high value outliers (Type 2)		Data with gross errors in the input vector (Type 3))	
Algorithm	$\delta = 0.0$	$\delta = 0.1$	$\delta = 0.2$	$\delta = 0.1$	$\delta = 0.2$	$\delta = 0.1$	$\delta = 0.2$
BP	0.0003	0.3967	0.7722	23.6145	36.6818	0.0177	0.0206
LMLS	0.0004	0.0584	0.1442	0.0140	0.0326	0.0173	0.0235
MCD	0.0004	0.0253	0.0464	0.0202	0.0231	0.0161	0.0163

4 Summary

In this paper we presented a novel robust MCD learning algorithm. It is based on the idea of robust MCD estimator and introduces to the error criterion special data weights coefficients to reduce the impact of outliers. As it was experimentally demonstrated, it behaves better than traditional algorithm when gross error are present in the training data. Moreover, this novel method often outperforms robust Lmls algorithm. It is simultaneously the first robust learning algorithm that employs initial data analysis save a modified function to be minimized. Because it doesn't need any knowledge of the assumed error distribution, and ensures relatively good training results in any conditions, it can be considered as simple and effective mean to increase learning performance on the contaminated data sets. This robust algorithm can be easily adapted to many types of neural networks training methods.

References

1. Hornik, K., Stinchconbe, M., White, H.: Multilayer feedforward networks are universal approximators. Neural Networks 2, 359–366 (1989)
2. Hampel, F.R., Ronchetti, E.M., Rousseeuw, P.J., Stahel, W.A.: Robust Statistics the Approach Based on Influence Functions. John Wiley & Sons, New York (1986)
3. Huber, P.J.: Robust Statistics. Wiley, New York (1981)
4. Charalambous, C.: Conjugate gradient algorithm for efficient training of artificial neural networks. IEE Proceedings-G 139(3), 301–310 (1992)

5. Chen, D.S., Jain, R.C.: A robust back propagation learning algorithm for function approximation. IEEE Transactions on Neural Networks 5, 467–479 (1994)
6. Chuang, C., Su, S., Hsiao, C.: The Annealing Robust Backpropagation (ARBP) Learning Algorithm. IEEE Transactions on Neural Networks 11, 1067–1076 (2000)
7. Chuang, C.C., Jeng, J.T., Lin, P.T.: Annealing robust radial basis function networks for function approximation with outliers. Neurocomputing 56, 123–139 (2004)
8. David, V., Sanchez, A.: Robustization of a learning method for RBF networks. Neurocomputing 9, 85–94 (1995)
9. Hagan, M.T., Demuth, H.B., Beale, M.H.: Neural Network Design. PWS Publishing, Boston (1996)
10. Liano, K.: Robust error measure for supervised neural network learning with outliers. IEEE Transactions on Neural Networks 7, 246–250 (1996)
11. Mili, L., Cheniae, M., Vichare, N.S., Rousseeuw, P.J.: Robust State Estimation Based on Projection Statistics. IEEE Transactions on Power Systems 11(2) (May 1996)
12. Olive, D.J., Hawkins, D.M.: Robustifying Robust Estimators, N.Y (2007)
13. Pernia-Espinoza, A.V., Ordieres-Mere, J.B., Martinez-de-Pison, F.J., Gonzalez-Marcos, A.: TAO-robust backpropagation learning algorithm. Neural Networks 18, 191–204 (2005)
14. Rousseeuw, P.J.: Least median of squares regression. Journal of the American Statistical Association 79, 871–880 (1984)
15. Rousseeuw, P.J., Van Driessen, K.: A Fast Algorithm for the Minimum Covariance Determinant Estimator. Technometrics 41, 212–223 (1999)
16. Rumelhart, D.E., Hinton, G.E., Williams, R.J.: Learning internal representations by error propagation. In: Parallel Distribution Processing: Explorations in the Microstructures of Cignition. MIT Press, Cambridge (1986)
17. Rusiecki, A.L.: Robust Learning Algorithm with the Variable Learning Rate, In: ICAISC 2006, Artificial Intelligence and Soft Computing, pp. 83-90, Warszawa 2006 (2006)
18. Rusiecki, A.L.: Robust LTS Backpropagation Learning Algorithm. In: Sandoval, F., Prieto, A.G., Cabestany, J., Graña, M. (eds.) IWANN 2007. LNCS, vol. 4507, pp. 102–109. Springer, Heidelberg (2007)

Some Issues on Intrusion Detection in Web Applications

Jaroslaw Skaruz[1] and Franciszek Seredynski[2,3]

[1] Institute of Computer Science, University of Podlasie,
Sienkiewicza 51, 08-110 Siedlce, Poland
`jaroslaw.skaruz@ap.siedlce.pl`
[2] Polish-Japanese Institute of Information Technology,
Koszykowa 86, 02-008 Warsaw
[3] Institute of Computer Science, Polish Academy of Sciences,
Ordona 21, 01-237 Warsaw, Poland
`sered@ipipan.waw.pl`

Abstract. In the paper we present a new approach based on application of neural networks to detect SQL attacks. SQL attacks are those attacks that take the advantage of using SQL statements to be performed. The problem of detection of this class of attacks is transformed to time series prediction problem. SQL queries are used as a source of events in a protected environment. To differentiate between normal SQL queries and those sent by an attacker, we divide SQL statements into tokens and pass them to our detection system, which predicts the next token, taking into account previously seen tokens. In the learning phase tokens are passed to a recurrent neural network (RNN) trained by backpropagation through time (BPTT) algorithm. Then, two coefficients of the rule are evaluated. The rule is used to interpret RNN output. In the testing phase RNN with the rule is examined against attacks and legal data to find out how evaluated rule affects efficiency of detecting attacks. All experiments were conducted on Jordan network. Experimental results show the relationship between the rule and a length of SQL queries.

1 Introduction

Large number of Web applications, especially those deployed for companies to e-business purpose involve data integrity and confidentiality. Such applications are written in script languages like PHP embedded in HTML allowing to establish connection to databases, retrieving data and putting them in WWW site. Security violations consist in not authorized access and modification of data in the database. SQL is one of languages used to manage data in databases. Its statements can be ones of sources of events for potential attacks.

In the literature there are some approaches to intrusion detection in Web applications. In [1] the authors developed anomaly-based system that learns the profiles of the normal database access performed by web-based applications using a number of different models. A profile is a set of models, to which parts of SQL statement are fed to in order to train the set of models or to generate an anomaly score. During training phase models are built based on training data and anomaly score is calculated. For each model, the maximum of anomaly score is stored and used to set an anomaly threshold. During detection phase, for each SQL query anomaly score is calculated. If it exceeds the maximum of anomaly score evaluated during training phase, the query is considered to be anomalous. Decreasing false positive alerts involves creating models for custom data types for each application to which this system is applied.

Besides that work, there are some other works on detecting attacks on a Web server which constitutes a part of infrastructure for Web applications. In [2] a detection system correlates the server-side programs referenced by clients queries with the parameters contained in these queries. It is similar approach to detection to the previous work. The system analyzes HTTP requests and builds data model based on the attribute length of requests, attribute character distribution, structural inference and attribute order. In a detection phase built model is used for comparing requests of clients.

In [3] logs of Web server are analyzed to look for security violations. However, the proposed system is prone to high rates of false alarm. To decrease it, some site-specific available information should be taken into account which is not portable.

In this work we present a new approach to intrusion detection in Web applications. Rather than building profiles of normal behavior we focus on a sequence of tokens within SQL statements observed during normal use of an application. RNN is used to encode a stream of such SQL statements.

The paper is organized as follows. The next section discusses SQL attacks. In section 3 we describe the architecture of Jordan network. Section 4 shows training and testing data used for experiments. Next, section 5 contains experimental results. Last section summarizes results.

2 SQL Attacks

2.1 SQL Injection

SQL injection attack consists in such a manipulation of an application communicating with a database, that it allows a user to gain access or to allow it to modify data for which it has not privileges. To perform an attack in the most cases Web forms are used to inject part of SQL query. Typing SQL keywords and control signs an intruder is able to change the structure of SQL query developed by a Web designer. If variables used in SQL query are under control of a user, he can modify SQL query which will cause change of its meaning. Consider an example of a poor quality code written in PHP presented below.

```
$connection=mysql_connect();
mysql_select_db("test");
$user=$HTTP_GET_VARS['username'];
$pass=$HTTP_GET_VARS['password'];
```

```
$query="select * from users where
   login='$user' and password='$pass'";
$result=mysql_query($query);
if(mysql_num_rows($result)==1)
   echo "authorization successful"
else
   echo "authorization failed";
```

The code is responsible for authorizing users. User data typed in a Web form are assigned to variables *user* and *pass* and then passed to the SQL statement. If retrieved data include one row it means that a user filled in the form login and password the same as stored in the database. Because data sent by a Web form are not analyzed, a user is free to inject any strings. For example, an intruder can type: " ' *or 1=1* -" in the login field leaving the password field empty. The structure of SQL query will be changed as presented below.

```
$query="select * from users where login
='' or 1=1 --' and password=''";
```

Two dashes comment the following text. Boolean expression *1=1* is always true and as a result user will be logged with privileges of the first user stored in the table *users*.

2.2 Proposed Approach

The way we detect intruders can be easily transformed to time series prediction problem. According to [5] a time series is a sequence of data collected from some system by sampling a system property, usually at regular time intervals. One of the goal of the analysis of time series is to forecast the next value in the sequence based on values occurred in the past. The problem can be more precisely formulated as follows:

$$s_{t-2}, s_{t-1}, s_t \longrightarrow s_{t+1}, \qquad (1)$$

where s is any signal, which depends on a solved problem and t is a current moment of time. Given s_{t-2}, s_{t-1}, s_t, we want to predict s_{t+1}. In the problem of detection SQL attacks, each SQL statement is divided into some signals, which we further call tokens. The idea of detecting SQL attacks is based on their key feature. SQL injection attacks involve modification of SQL statement, which lead to the fact, that the sequence of tokens extracted from a modified SQL statement is different than the sequence of tokens derived from a legal SQL statement. For example, let S means recorded SQL statement and T_1, T_2, T_3, T_4, T_5 tokens of this SQL statement. The original sequence of tokens is as follows:

$$T_1, T_2, T_3, T_4, T_5. \qquad (2)$$

If an intruder performs an attack, the form of SQL statement changes. Transformation of the modified statement to tokens results in different tokens than these shown in eq.(2). The example of a sequence of tokens related to modified SQL query is as follows:

$$T_1, T_2, T_{mod3}, T_{mod4}, T_{mod5}. \tag{3}$$

Tokens number 3, 4, 5 are modified due to an intruder activity. We assume that intrusion detection system trained on original SQL statements is able to predict the next token based on the tokens from the past. If the token T_1 occurs, the system should predict token T_2, next token T_3 is expected. In the case of attacks token T_{mod3} occurs which is different than T_3, which means that an attack is performed.

Various techniques have been used to analyze time series [6,7]. Besides statistical methods, RNNs have been widely used for that problem. In our study presented in this paper we use Jordan network.

3 Recurrent Neural Networks

3.1 General Issues

Application of neural networks to solving any problem involves three steps. The first is training, during which weights of network connections are changed. Network output is compared to training data and the network error is evaluated. In the second step the network is verified. Values of connections weights are constant and the network is checked if its output is the same as in the training phase. The last step is generalization. The network output is evaluated for such data, which were not used for training the network. Good generalization is a desirable feature of all networks because it means that the network is prepared for processing data, which may occur in the future.

In comparison to feedforward neural networks, RNN have feedback connections which provide dynamics. When they process information, output neurons signal depends on input and activation of neurons in the previous steps of training RNN.

3.2 RNN Architectures

In comparison to feedforward neural network, Jordan network has context layer containing the same number of neurons as the output layer. Input signal for context layer neurons comes from the output layer. Moreover, Jordan network has an additional feedback connection in the context layer. Each recurrent connections have fixed weight equals to 1.0. Network was trained by BPTT and the related equations are presented below:

$$x(k) = [x_1(k), ..., x_N(k), v_1(k-1), ..., v_M(k-1)], \tag{4}$$

$$u_j(k) = \sum_{i=1}^{N+M} w_{ij}^{(1)} x_i(k), v_j(k) = f(u_j(k)), \tag{5}$$

$$g_j(k) = \sum_{i=1}^{K} w_{ij}^{(2)} v_i(k), y_j(k) = f(g_j(k)), \tag{6}$$

$$E(k) = 0.5 \sum_{i=1}^{M} [y_i(k) - d_i(k)]^2, \tag{7}$$

$$\delta_i^{(o)}(k) = [y_i(k) - d_i(k)] f^{'}(g_i(k)), \delta_i^{(h)}(k) = f^{'}(u_i(k)) \sum_{j=1}^{M} \delta_j^{(o)}(k) w_{ij}^{(2)}, \tag{8}$$

$$w_{ij}(k+1)^{(2)} = w_{ij}(k)^{(2)} + \sum_{k=1}^{sql-length} [v_i(k) \delta_j^{(o)}(k)], \tag{9}$$

$$w_{ij}(k+1)^{(1)} = w_{ij}(k)^{(1)} + \sum_{k=1}^{sql-length} [x_i(k) \delta_j^{(h)}(k)]. \tag{10}$$

In the equations (4)-(10), N, K, M stand for the size of the input, hidden and output layers, respectively. x(k) is an input vector, $u_j(k)$ and $g_j(k)$ are input signals provided to the hidden and output layer neurons. Next, $v_j(k)$ and $y_j(k)$ stand for the activations of the neurons in the hidden and output layer at time k, respectively. The equation (7) shows how RNN error is computed, while neurons error in the output and hidden layers are evaluated according to (8). Finally, in the last step values of weights are changed using formulas (9) for the output layer and (10) for the hidden layer.

3.3 Training

The training process of RNN is performed as follows. The two subsequent tokens of the SQL statement become input of a network. Activations of all neurons are computed. Next, an error of each neuron is calculated. These steps are repeated until all tokens have been presented to the network. Next, all weights are evaluated and activation of the context layer neurons is set to 0. For each input data, training data at the output layer are shifted so that RNN should predict the next token in the sequence of all tokens.

In this work, the following tokens are considered: keywords of SQL language, numbers and strings. We used the collection of SQL statements to define 36 distinct tokens. Each token has an index and the real number between 0.1 and 0.9, which reflects input coding of these statements according to table 1. The table shows selected tokens, indexes and the coding real values. The indexes are used for preparation of input data for neural networks. The index e.g. of a keyword *UPDATE* is 9. The index 2 points to a keyword *FROM*. The token with index 1 relates to *SELECT*.

The number of neurons at the input layer is constant and equals to 2. Network has 37 neurons in the output layer. 36 neurons correspond to each token but the neuron 37 is included to indicate that just processing input data vector is the last within a SQL query. Training data, which are compared to the output of

Table 1. A part of a list of tokens, indexes and their coding values

token	index	coding value
SELECT	1	0.1222
FROM	2	0.1444
...	...	...
...	...	...
UPDATE	9	0.3
...	...	...
number	35	0.8777
string	36	0.9

the network have value either equal to 0.1 or 0.9. If a neuron number i in the output layer has small value then it means that the next processing token can not have index i. On the other hand, if output neuron number i has value of 0.9, then the next token in a sequence should have index equal to i. Below, there is an example of a SQL query:

$$SELECT \quad name \quad FROM \quad users. \tag{11}$$

At the beginning, the SQL statement is divided into tokens. The indexes of tokens are: 1, 36, 2 and 36 (see table 1). Because there are two input neurons, we have 3 subsequent pairs (1,36), (36,2), (2,36) of training data, each pair coded by real values in the input layer:

$$0.1222 \quad 0.9 \tag{12}$$

$$0.9 \quad 0.1444, \tag{13}$$

$$0.1444 \quad 0.9. \tag{14}$$

The first two tokens that have appeared are 1 and 36. As the consequence, in the first step of training, the output signal of two neurons in the input layer are 0.1222 and 0.9. The next token in a sequence has index equal to 2. This token should be predicted by the network in the result of the input pair (1,36). It means that only neuron number 2 at the output layer should have the high value. In the next step of training, 2nd and 3rd tokens are presented to the network, which means that output of neurons is set according to tokens 36 and 2. Fourth token should be predicted so neuron number 36 in the output of RNN should have the high value. Finally, input data are 0.1444 and 0.9. In that moment weights of RNN are updated and the next SQL statement is considered.

Training the network in such a way ensures that it will posses prediction capability. While network output depends on the previous input vectors, processing the set of tokens related to the next SQL query can not be dependent on tokens of the previous SQL statement.

4 Training and Testing Data

All experiments were conducted using synthetic data collected from a SQL statements generator. The generator takes randomly a keyword from selected subset of SQL keywords, data types and mathematical operators to build a valid SQL query. Since the generator was developed on the basis of the grammar of SQL language, each generated SQL query is correct. We generated 3000000 SQL statements. Next, the identical statements were deleted. Finally, our data set contained thousands of free of attack SQL queries. The set of all SQL queries was divided into 21 subsets, each containing SQL statements of different length, from 9 to 29 tokens. Data with attacks were produced in the similar way to that without attacks. Using available knowledge about SQL attacks, we defined their characteristic parts. Next, these parts of SQL queries were inserted randomly to the generated query in such a way that it provides grammatical correctness of these new statements.

5 Experimental Results

In the first phase of experimental study we evaluated the best parameters of RNN and learning algorithm. For the following values of parameters the error of the networks was minimal. For the Jordan network *tanh* function was chosen for the hidden layer and sigmoidal function for the output layer. The number of neurons in the hidden layer equals to 38 neurons. In all cases η (training coefficient) is set to 0.2 and α value (used in momentum) equals to 0.1.

While output signal of a neuron in the output layer can be only high or low, for each input data there is a binary vector as the output of the network. If a token is well predicted by the network, it means that there is no error in the vector. Otherwise, if at any position within the vector there is 0 rather than 1 or 1 rather than 0, it means that the network predicted wrong token. Experimental results presented in [12] show that there is a great difference in terms of the number of errors and the number of vectors containing any error for cases, in which attack and legal activity were executed. To distinguish between an attack and a legitimate SQL statement we defined classification rule: an attack occurred if the average number of errors for each output vector is not less than coefficient 1 and coefficient 2 is greater than the number of output vectors that include any error [12]. Initial experiments showed that appropriate values of these coefficients were accordingly 2.0 and 80%. The problem with that rule is that assumed values of coefficients can not be used generally. They depend on the length of SQL query. The objective of this research is to show the relationship between the values of coefficients and length of SQL query. Also, we present in what extent the number of false alarms depend on the length of SQL query. Moreover, we show how the number of SQL statements used for setting the coefficients of the rule affects efficiency of intrusion detection.

5.1 Experiment Description

The scenario of the experiment is presented below:

```
Initialization: divide set of SQL queries on 21 subsets
(each one contains different length queries)
FOR each subset
    train a network
        FOR k=1 to 10
            FOR i=1 to 9
                choose randomly i/10 queries
                examine the network against chosen queries
            END
        END
    evaluate coefficients of the rule
END
```

For each data subset we had one network. After the network was trained using data with attacks, it was examined against attacks and legitimate SQL queries. Next, based on the network output for legal and illegal SQL statements, we evaluated two coefficients of the rule so that the number of false alarms is minimal.

5.2 Results

The rule which was evaluated during the experiment, was used as a classification threshold for RNN output. For testing purpose, SQL statements, which were not used in the training and during defining the rule were applied. Figure 2 presents the values of coefficients of the rule for each length of SQL query considered in this work. It can be seen that the dependence between coefficients and the length of SQL queries is similar to linear. The efficiency of RNN is limited by the number of training data, especially if large neural net architecture is used.

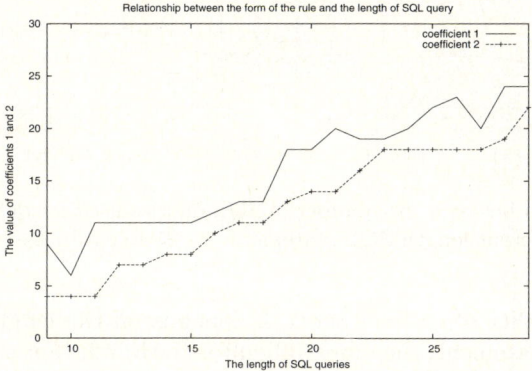

Fig. 1. The form of the rule for SQL queries

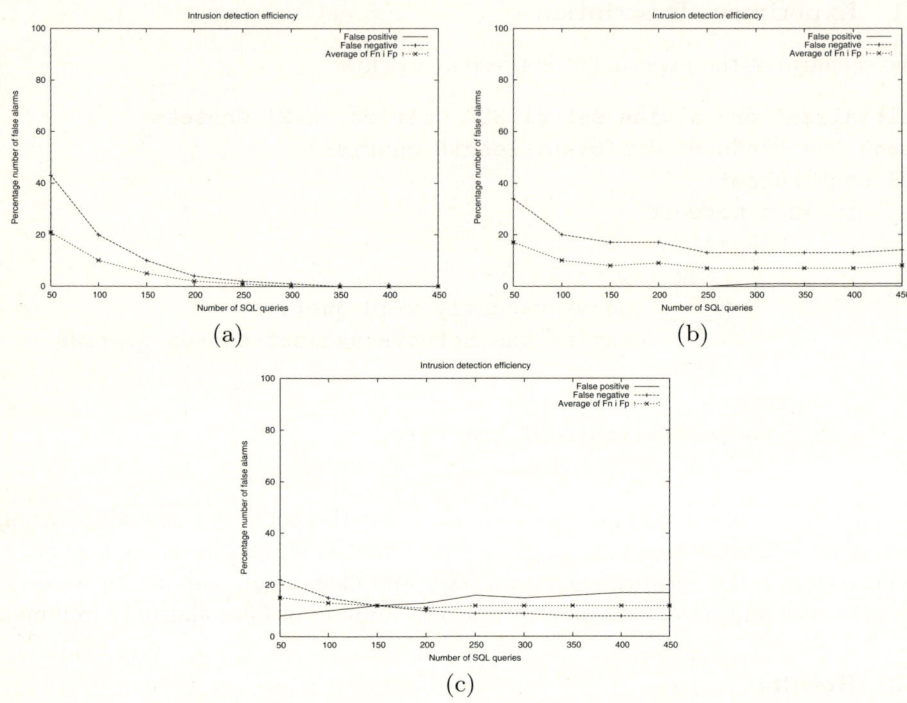

Fig. 2. Relationship between the number of SQL queries used for defining the rule and false alarms for different length SQL statements a) 10 token b) 15 tokens c) 20 tokens

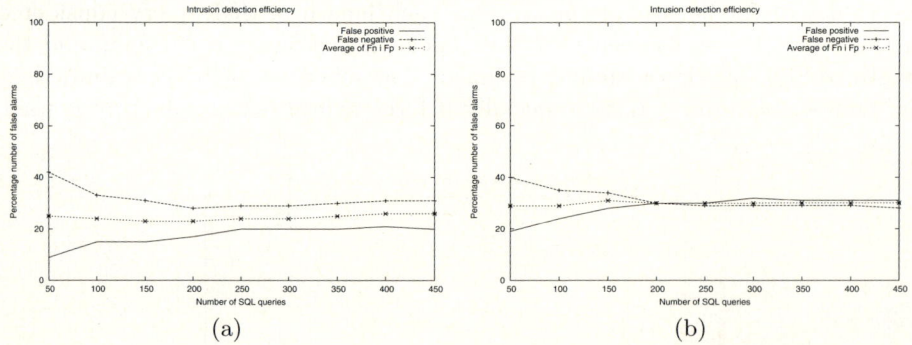

Fig. 3. Relationship between the number of SQL queries used for defining the rule and false alarms for different length SQL statements a) 25 token b) 29 tokens

Moreover, the quality of trained network depends on the length of SQL query. The longer SQL statement, the more difficult is to train it. For such cases, output vectors of RNN include more errors and there is little difference in the output network when attacks and legal SQL queries are presented to the network. That is the reason for which greater values of coefficients must be used to discriminate between attacks and legal activity. From the figure 2 it can be concluded, what is the optimal number of SQL queries that must be used to define the rule. For queries constituted from 10 and 15 tokens the optimal value of data equals to 250. The networks were trained efficiently using such SQL statements. It is easy to see that for good quality networks increasing the number of data used for defining the rule improves results in terms of the number of false alarms. If the length of query increases, it is more difficult to train the network completely. For SQL queries including 20 tokens, the optimal number of data used during setting the rule decreased to 200. This feature of decreasing repeats also for longer SQL statements, which is presented in figure 3a and 3b.

6 Conclusions

In the paper we have presented a new approach to detecting SQL-based attacks. The problem of detection was transformed to time series prediction problem and Jordan network was examined to show its potential use for such a class of attacks. Despite the fact that large architecture of RNN was used, the network is able to predict sequences of the length up to twenty nine tokens with acceptable error margin.

Finally, the classification rule was defined and validated experimentally. The relationship between the coefficients of the rule and the length of SQL query was established. We also showed how the number of SQL queries used for setting the coefficients affects the number of false alarms. The results presented in this work are helpful especially when this approach to intrusion detection is deployed in real environment, where there is no any clue about the optimal number of data, which allow to define the rule without decreasing the efficiency of intrusion detection.

References

1. Valeur, F., Mutz, D., Vigna, G.: A Learning-Based Approach to the Detection of SQL Attacks. In: Julisch, K., Krügel, C. (eds.) DIMVA 2005. LNCS, vol. 3548, Springer, Heidelberg (2005)
2. Kruegel, C., Vigna, G.: Anomaly Detection of Web-based Attacks. In: Proceedings of the 10th ACM Conference on Computer and Communication Security (CCS 2003), pp. 251–261 (2003)
3. Almgren, M., Debar, H., Dacier, M.: A lightweight tool for detecting web server attacks. In: Proceedings of the ISOC Symposium on Network and Distributed Systems Security (2000)
4. Tan, K.M.C., Killourhy, K.S., Maxion, R.A.: Undermining an Anomaly-Based Intrusion Detection System Using Common Exploits. In: Proceedings of the 5th International Symposium on Recent Advances in Intrusion Detection, pp. 54–73 (2002)
5. Nunn, I., White, T.: The Application of Antigenic Search Techniques to Time Series Forecasting. In: Proceedings of the Genetic and Evolutionary Computation Conference (GECCO), USA (2005)

6. Kendall, M., Ord, J.: "Time Series", 3rd edn. (1999)
7. Pollock, D.: A Handbook of Time-Series Analysis, Signal Processing and Dynamics. Academic Press, London (1999)
8. Lin, T., Horne, B.G., Tino, P., Giles, C.L.: Learning long-term dependencies in NARX recurrent neural networks. IEEE Transactions on Neural Networks, 1329 (1996)
9. Drake, P.R., Miller, K.A.: Improved Self-Feedback Gain in the Context Layer of a Modified Elman Neural Network. Mathematical and Computer Modelling of Dynamical Systems, 307–311 (2002)
10. http://www.insecure.org/sploits_all.html
11. http://phpnuke.org/
12. Skaruz, J., Seredyński, F., Bouvry, P.: Tracing SQL Attacks via Neural Networks. In: Parallel Processing and Applied Mathematics. LNCS, vol. 4967, Springer, Heidelberg (2007)

Neural Network Device for Reliability and Functional Analysis of Discrete Transport System

Tomasz Walkowiak and Jacek Mazurkiewicz

Institute of Computer Engineering, Control and Robotics, Wroclaw University of Technology, ul. Janiszewskiego 11/17, 50-372 Wroclaw, Poland
`Tomasz.Walkowiak@pwr.wroc.pl, Jacek.Mazurkiewicz@pwr.wroc.pl`

Abstract. This paper describes an approach of combining Monte Carlo simulation and neural nets. The approach is applied to model transport systems, with the accurate but computationally expensive Monte Carlo simulation used to train a neural net. Once trained the neural net can efficiently provide functional analysis and reliability predictions. No restriction on the system structure and on any kind of distribution is the main advantage of the proposed approach. The paper presents exemplar decision problem solved by proposed approach.

1 Introduction

Modern transport systems often have a complex network of connections. From the reliability point of view [2] the systems are characterized by a very complex structure. The main issue of reliability considerations is to model the influence of these faults at a satisfactory level of detail. This analysis can only be done if there is a formal model of the transport logistics, i.e. there are deterministic or probabilistic rules on how the transport is redirected in every possible combination of connection faults and congestion [11]. The classical models used for reliability analysis are mainly based on Markov or Semi-Markov processes [2] which are idealized and it is hard to reconcile them with practice. The typical structures with reliability focused analysis are not complicated and use very strict assumptions related to the life or repair time and random variables distributions of the analyzed system elements. Of course, there are trial experiments of a use for reliability analysis more complex than serial-parallel or k of n systems, but the number of states increases and generates calculation problems. The classical reliability models are not able to support the time-depend analysis, which seems to be very important in some practical applications. The proposed solution is to use a time event simulation with Monte Carlo analysis [1],[5],[9] to train a neural net. Once trained, the neural net can efficiently provide functional analysis and reliability predictions. One advantage of this approach is that it supports the computation of any point wise parameters. However, it also supports estimating the distributions of times when the system assumes a particular state or set of states.

2 Discrete Transport System (DTS)

The basic entities of the system are as follows: store-houses of tradesperson, roads, vehicles, trans-shipping points and store-houses of addressee and the commodities transported. An example system is shown in Fig. 1. The media transported in the system are called commodities. The commodities are taken from store-houses of tradesperson and transported by vehicles to trans-shipping points. Other vehicles transport commodities from trans-shipping points to next trans-shipping points or to final store-houses of addressees. Moreover, in time of transportation vehicles dedicated to commodities could fail and then they are repaired [6],[7]. Different commodities are characterized by common attributes which can be used for their mutual comparison. The presented analysis uses the capacity (volume) of commodities as such attribute. The following assumptions related to the commodities are taken: it is possible to transport n different kinds of commodities in the system and each kind of commodity is measured by its capacity. A road is an ordered pair of system elements. The first element must be a store-house of tradesperson or trans-shipping point, the second element must be a trans-shipping point or store-house of addressee. Moreover, each road is described by following parameters: length, the number of vehicle maintenance crews (at a given time only one vehicle could be maintained by a single crew) and the number of vehicles moving on the road. The number of maintain crews ought to be understand as the number of vehicles which can be on a single road maintained simultaneously. A single vehicle transports commodities from the start to end point of a single road, after which the empty vehicle returns and the whole cycle is repeated. Our vehicle model makes the following assumptions. Each vehicle can transport only one kind of commodity at a time. Vehicles are universal - are able to transport different kinds of commodity. Moreover, the vehicle is described by following parameters: capacity, mean speed of journey (both when hauling the commodity and when empty), journey time (described by its distribution parameters), time to vehicle failure (also described an distribution), time of vehicle maintenance (described by distribution). The choice of distribution for the random variables is flexible provided that we know both a method and the parameters needed to generate random numbers with that distribution. The store-house of tradesperson is the source of commodities. It can be only a start point of the road. Each store-house of tradesperson is an infinity source of a single kind of commodity.

The trans-shipping point can be used as a start or end point of a single road. This is a transition part of the system which is able to store the commodity. The trans-shipping point is described by following parameters: global capacity C, initial state described by capacity vector of commodities stored when the system observation begins, delivery matrix D. This matrix defines which road is chosen when each kind of commodity leaves the shipping point (1 means that a given commodity is delivered to a given road). On contradictory to previously described systems ([4],[6],[7]) in this case a commodity could be routed to more then one road (direction). The dimensions of the delivery matrix are: number of commodities x number of output roads.

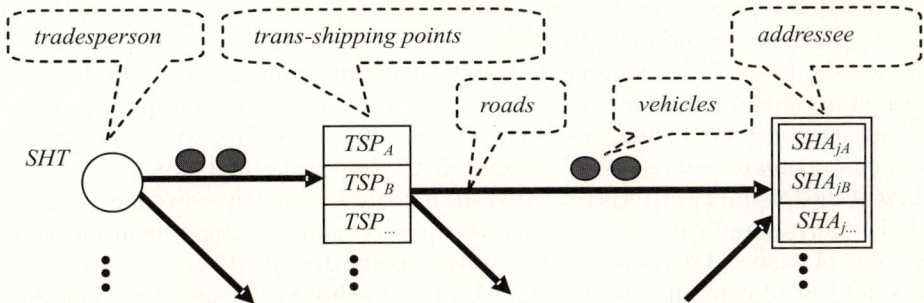

Fig. 1. Exemplar model of Discrete Transport System

Input algorithm: only one vehicle can be unloaded at a time, if the vehicle can be unloaded the commodity is stored in the trans-shipping point, if not - the vehicle is waiting in the input queue, there is only one input queue serviced by FIFO algorithm.

Output algorithm: only one vehicle can be loaded at a time, if the vehicle can be loaded, i.e. if the proper commodity is presented in the trans-shipping point, a commodity which could be routed to a given road, the state of a given commodity in the trans-shipping point is reduced, if not - the vehicle is waiting in the output queue; each output road has its own FIFO queue.

The store-house of addressee can be used only as the end point of a single road. The main task of this component of the system is to store the commodity as long as the medium is spent by recipient. The store-house of addressee is described by following parameters: global capacity C, initial state described as for the trans-shipping point, function or rule which describes how each kind of commodity is spent by recipients. Input algorithm is exactly the same as for the trans-shipping point. Output algorithm can be described as: stochastic process, continuous deterministic or discrete deterministic one. The model assumes that the capacity of the commodity can not be less than zero, "no commodity state" - is generated when there is a lack of required kind of commodity.

3 System Structure

The simulation program generates a description of all changes in the system during simulation (with all events). It is a base for calculation of any functional and reliability measures. The most valuable results of statistical analysis are: time percentage when the vehicle is present in each state, time percentage when the store-house of addressee is present in each state, mean time when the store-house of addressee is empty - this way we can say if "no commodity state" is prolonged or only momentary. We also propose a quantile calculation of time when the store-house of addressee is empty. This is the answer if "no commodity state" situation sometimes lasts significantly longer than the mean time of empty store-house. Moreover, it is possible to observe the influence of changes related to single parameter or a set of parameters - vehicle repair time for example - for other system characteristics - as vehicle utilization level, or commodity accessible in store-houses. The calculated reliability and functional measures could be a base of developing economic measures [8],[9],[12]. Such layered approach allows a high level, economic analysis of the system. It is necessary to check different variants of maintenance organization and to choose the less expensive among them if the reliability criteria are satisfied. It could be done by subsequent Monte-Carlo analysis and calculation of the required economic or functional measures for a set of analyzed parameters. The system model described in previous sections is a subject of computer simulation. A special software package for simulation of the discrete transport system has been developed. The transport system is described in the specially designed script language (with syntax similar to XML) [4]. It is an input for simulator program (written in C++) performing Monte-Carlo simulation [1],[5]. Monte Carlo simulation has an advantage in that it does not constrain the system structure or kinds of distributions used [4]. However, it requires proper data pre-processing, enough time to realize the calculations and efficient calculation engine.

4 Proposed Approach

The problem of speeding up functional and reliability analysis of discrete transport system we propose to solve by hybrid system using simulation and neural nets. In many tasks, i.e. in decision systems, there is a need to give an answer in a short time. However Monte-Carlo simulation requires quite a lot of time to realize calculations for a given set of system parameters. To solve this problem we have proposed a use of artificial neural networks [14]. The use of neural network is motivated by its universal approximation capability [3]. Knowing that most of output system parameters are continues we can expect that neural network can approximate any unknown function based on a set of examples. The time needed to get an output from learn neural network is very short. Solution generated by net seems to be satisfactory [14], because we do not need very precise results - time is the most important attribute of the solution. The neural network partly substitutes the simulation process. The neural net module is added to developed simulation software. The aim of this module is to generate an answer how to select the best system parameters (i.e. the maintenance agreements - the average time of vehicle repair) based on the achieved system functional parameters (i.e. the average time of "no commodity" in the store-house of addressee). The process of data analysis will be as follows:

1. set the input parameters for model of discrete transport system;
2. give a range of an analyzed free parameter (parameters);
3. perform initial Monte-Carlo analysis for a few parameters from a given range
   - calculate all required functional and reliability parameters;
4. build a neural network approximation tool: Multilayer Perceptron; the inputs to the network are analyzed free parameters; the outputs are functional and reliability measures;

5. build the answer about the maintenance agreement based on the output of the neural network and the proper economic measures;
6. communicate with a user: play with functional and reliability data, goto 4. If more accurate analysis of economic parameter in a function of free parameter is required goto 3 - perform more Monte-Carlo analysis.

5 Case Study: Reliability Parameters

The experiment was realized for a simple parallel system (see Fig. 2.), which is composed of two identical elements. Thus there are three reliability states: both elements are working; one element is working, another is failed; both elements are failed. We assume that, time to failure for both elements has an exponential distribution. Time of repair consists of two parts: fixed time of maintenance crew coming and time to repair - normal distribution. Also, failures of the elements are independent. Of course structure of the proposed system is rather trivial, but easily expendable. On the other hand, the distribution of repair time is not so trivial. In general, such example is very good to demonstrate the idea of reliability analysis supported by neural network and can be easily adopted for different real tasks. For a case study we focused on estimating the mean time over one year of being in one of three system stages. The reliability analysis was realized using Monte Carlo simulation. Experiments were performed for given data:

- data related to failures: mean time to failure: 10, 15, 20, 25, 30, 35, 40, 45, 50, 55, 60 days (these values were taken with an assumption about the exponential distribution of time to failure);
- data related to repairs: repair crew coming time: 0.1, 0.3, 0.5, 0.7, 0.9, 1.1, 1.3, 1.5, 1.7, 1.9, 2.1 days (the additional assumption is that the maintenance crew coming time is fixed), mean time to repair: 0.1, 1, 2, 4, 6, 8, 10 days.

These values were taken with an assumption about the normal distribution of time to repair. Standard deviation of repair time is the last data. We took the following values of this parameter: 0 - fixed time to repair, 0.5, 1.5, 2, 2.5 days. Additionally, all data where standard deviation of repair time was larger two times then mean time to repair where removed. The presented data gave in summary: 2904 performed experiments. For each set of parameters 20000 Monte Carlo simulation where performed. The Multilayer Perceptron was used to estimate the mean time over one year of being in one of three system stages. The network has four input neurons: corresponding to each of system parameters: mean time to failure, crew coming time, mean time to repair and standard deviation of time to repair. Three outputs correspond to mean time over one year of being in one of three system stages (both elements are working; one element is working, another is failed; both elements are failed.

Outputs of the network where normalized to (0,1) range - divided by 356 (duration of year in days). Moreover, the network has single hidden layer and a sigmoidal transfer function. Number of neurons in the hidden layer was set to

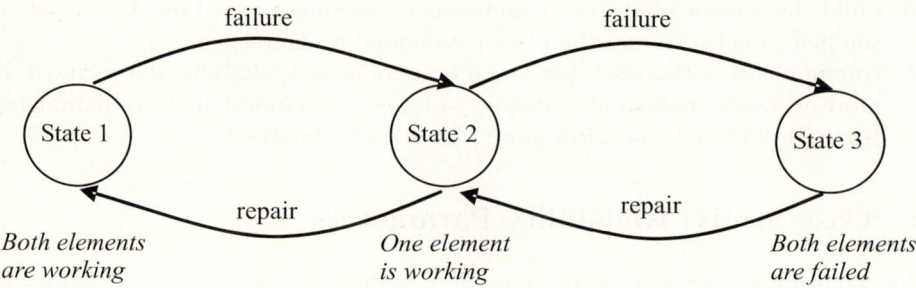

Fig. 2. Structure of an analyzed system for reliability parameters

20 by a set of preliminary experiments. The network was trained by Levenberg-Marquardt [3] algorithm using MATLAB package. Input data were divided in random way into two equally sized sets: training and testing. Three different kinds of experimental distance between resulting and original function have been used during testing procedure:

$$L_1 = \frac{1}{N} \sum_{i=1}^{N} |y(x_i) - \hat{y}(x_i)| \qquad (1)$$

$$L_2 = \sqrt{\frac{1}{N} \sum_{i=1}^{N} (y(x_i) - \hat{y}(x_i))^2} \qquad (2)$$

$$L_\infty = \max_{i=1,\ldots,N} |y(x_i) - \hat{y}(x_i)| \qquad (3)$$

where: $\hat{y}(x_i)$ - network output, $y(x_i)$ - desired output (from Monte Carlo simulation), N - number of examples. Tests where performed also for different number of hidden neurons: 4, 6, 8, 10, 12, 14, 16, 18 and 20.

5.1 Results

For each set of parameters 20000 Monte Carlo simulation where performed. Resulting values and data range of them are as follow:

T1 - average time when both objects are working in one year: 35.85 - 362.56 days,

T2 - average time when only one object is working in one year: 2.42 - 165.62 days,

T3 - average time when both objects failed in one year: 0.01 - 178.85 days.

Results related to the tests of number of neurons in hidden layer are presented in Fig. 3. The network with 20 hidden neurons performs in the best way. To compare the neural network performance two additional tests were taken. One was focussed on Monte Carlo data stability. All simulations were performed one

State number	Distance Type	Number of neurons in hidden layer								
		4	6	8	10	12	14	16	18	20
1	L_1	2.56	0.34	0.25	0.22	0.32	0.15	0.16	0.16	0.17
	L_2	3.55	0.55	0.31	0.29	0.40	0.20	0.23	0.21	0.22
	L_∞	34.74	6.79	1.26	1.20	1.91	1.08	2.45	0.92	0.93
2	L_1	2.02	0.46	0.28	0.21	0.25	0.16	0.15	0.14	0.14
	L_2	3.39	0.63	0.38	0.28	0.35	0.21	0.23	0.18	0.18
	L_∞	30.92	5.13	1.63	1.27	2.87	1.20	3.28	1.22	0.64
3	L_1	1.07	0.26	0.19	0.13	0.22	0.10	0.09	0.11	0.09
	L_2	1.53	0.35	0.24	0.17	0.28	0.14	0.13	0.14	0.13
	L_∞	7.08	2.40	1.20	0.83	0.97	0.68	0.64	0.80	0.65

Fig. 3. Errors (for different type of distances) in days for different number of hidden neurons for testing set

time more. The distances for different metrics between these two simulations are presented in Fig. 4. The second test was focussed on a classical method of time calculation of being in a given state based on stationary Markov model (with assumption of all exponential time distributions) [2]. Achieved results were compared with Monte Carlo simulation (Fig. 4). Looking at results (Fig. 4, it is clear that neural network gives much better results then Markov approach. Moreover, received performance (in meaning of a result accuracy) of the neural network method is very close to the Monte Carlo method stability and probably very close to it is possible optimal performance.

6 Case Study: Functional Parameters

To show possibilities of the proposed model and developed software we have analyzed an exemplar transport network presented in Fig. 5. The network consists of two store-houses of tradesperson (each one producing its own commodity, marked as A and B), one trans-shipping point (with one storehouse for both commodities) and two store-houses of addressee (each one with one storehouse). The commodities are spent by each recipient. The process is continuous deterministic, the amount of consumption in time unit is marked by u with subscripts corresponding to store-houses of addressee and commodity id. It's exemplar values are presented in Fig. 5. Having lengths of the roads (see Fig. 5), the amount of commodity consumption in time unit for each store-house of addressee, the capacity of each vehicle (15), vehicle speed (50 and 75 in empty return journey) the number of vehicles for each road could be easy calculated. We have take into account some redundancy due to the fact of car failure (we assumed that the time between failures is 2000 time units) what results in following number of

State number	Distance Type	Errors in days between Monte Carlo simulation results and		
		other MC simulation	best neural network	Markov model
1	L_1	0.18	0.17	8.28
	L_2	0.23	0.22	10.58
	L_∞	1.01	0.93	23.83
2	L_1	0.13	0.14	6.24
	L_2	0.17	0.18	7.51
	L_∞	0.75	0.64	14.54
3	L_1	0.11	0.09	2.04
	L_2	0.15	0.13	3.34
	L_∞	0.75	0.65	10.84

Fig. 4. Errors (for different type of distances) in days for the best neural network, the Markov stationary model and the other Monte Carlo simulation

vehicles: road one $n_1=40$, road two $n_2=12$, road three $n_3=18(A)+6(B)=24$ and road four $n_4=16(A)+8(B)=24$. The analysis time T was equal to 20000.

We have analyzed maintains and service level agreement (SLA) dependency. From one side the transport network operator has to fulfill some service level agreement, i.e. have to deliver commodity in such way that a "no commodity state" is lower then a given stated level. Therefore the analyzed functional measure was a summary time of "no commodity state" during the analyzed time period. It could be only done if a proper maintenance agreement is signed. Therefore the argument of analyzed dependency was an average time of repair of vehicles. We assumed that we have four separated maintenance agreement, one for each for each road (roads 1 and 2 with one maintains crew, and 3 and 4 with two maintains crews). Also the exponential distribution of repair time was assumed. Therefore, we have four free parameters with values spanning from 1 to 1200. The system was simulated in 1500 points. For each repair time values set the simulation was repeated 25 times to allow to get some information of summary time of "no commodity" distribution. Two measures were calculated: average time of summary of "no commodity state" and its 4% quantile (i.e. the value of summary "no commodity" time that with probability 96% could be not higher).

The achieved date from simulation was divided randomly into two sets: learning and testing. We have used the Multilayer Perceptron architecture with 4 input neurons which correspond to repair time for each road, 10 hidden layer neurons and 2 output neurons corresponding to calculated measures (average time of summary of "no commodity state" and its 4% quantile).

The number of neurons in the hidden layer was chosen experimentally. Such network produced best results and higher numbers did not give any improvement. The tan-sigmoid was used as a transfer function in hidden layer and log-sigmoid output layer. Besides that, the output values have been weighted due to the fact the log-sigmoid has values between 0 and 1. The network presented above was trained using the Levenberg-Marquardt algorithm [3].

The achieved results, the mean of absolute value of difference between neural network results (multiplied by time range: 20 000) and results from Monte-Carlo simulation, for testing data set was 364 time units and 397 respectively for an average time of summary of "no commodity state" and its 4% quantile. It is in a range of 1-2% of the analyzed transport system time. We have also tested the simulation answer stability, i.e. the difference between two different runs of Monte-Carlo simulations (25 of them each time) for both functional measures (average time of summary of "no commodity state" and its 5% quantile) was 387 time units in an average. Therefore, the neural networks outputs are on the same level of accuracy as Monte-Carlo simulation since it was used for training the neural network the results could not be better. Whereas, there is no comparison between calculation time since the calculation of neural network outputs is negligible compared to Monte-Carlo simulation.

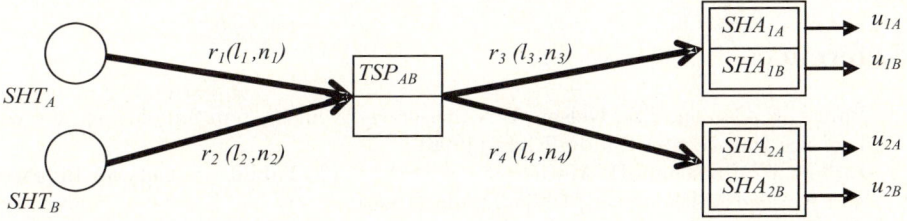

Fig. 5. Structure of case study discrete time system (parameters: $l_1=120$, $l_2=90$, $l_3=90$, $l_4=120$, $u_{1A}=60$, $u_{1B}=20$, $u_{2A}=40$, $u_{2B}=20$)

7 Conclusion

Results of functional and reliability analysis of exemplar discrete transport system are very promising. The best neural network could estimate the time of being in a given state with an error much smaller then one day in one year period. The time needed to achieve the answer is very fast. Therefore, the network could be used in any software package supporting decision process based of a reliability analysis. Time necessary for whole neural network training is less (in average 4 times) then time necessary for a single training vector preparation (run of 25 simulations for a single set of free parameters). An error related to the network answer - when the already trained network is tested by the input data which are not used during training - is in the range of disperse related to results of simulation. Of course there is an important aspect of avoiding over fitting or under training by neural network. At this stage of work it was done manually by observing the global error in function of training epochs and stopping training when the curve stops to decrease. The other interesting aspect of presented approach is the scalability projections. Increasing the number of modeled vehicles or system elements increases the Monte Carlo simulation time significantly. In case of training time of neural network (classification time is negligible) increasing a number of simulated entities has not direct influence. However, if one wants to analyze more sophisticated relation between input parameters and output measures, i.e. increases the number of input parameters, it results in an increase of input neurons, therefore needs a larger number of training data and results in a longer training time. The network solution is not free of problems. The main disadvantage is that the particular net is correctly fixed only to a single structure of the system. For each model different neural network must be trained. The overall structure of the net could be the same but a new set of weights must be estimated. Future work is planned on checking the extrapolation features of the neural network. We are going to analyze the answer of the network for input data with range outside the training set.

Acknowledgment. Work reported in this paper was sponsored by a grant No. 4 T12C 058 30, (years: 2006-2009) from the Polish Committee for Scientific Research (KBN).

References

1. Banks, J., Carson, J.S., Nelson, B.N.: Discrete-Event System Simulation, 2nd edn. Prentice Hall, Upper Saddle River (1996)
2. Barlow, R., Proschan, F.: Mathematical Theory of Reliability. Society for Industrial and Applied Mathematics, Philadelphia (1996)
3. Bischop, C.: Neural Networks for Pattern Recognition. Clarendon Press, Oxford (1996)
4. Caban, D., Walkowiak, T.: Computer Simulation of Discrete Transport System. XXX Winter School of Reliability, 93–103 (2002)
5. Fishman: Monte Carlo: Concepts, Algorithms, and Applications. Springer, New York (1996)
6. Jarnicki, J., Mazurkiewicz, J., Zamojski, W.: Model of Discrete Transport System. XXX Winter School of Reliability, 149–157 (2002)
7. Kaplon, K., Mazurkiewicz, J., Walkowiak, T.: Economic Analysis of Discrete Transport Systems. Risk Decision and Policy 8(3), 179–190 (2003)
8. Mazurkiewicz, J., Walkowiak, T.: Fuzzy Economic Analysis of Simulated Discrete Transport System. In: Rutkowski, L., Siekmann, J.H., Tadeusiewicz, R., Zadeh, L.A. (eds.) ICAISC 2004. LNCS (LNAI), vol. 3070, pp. 1161–1167. Springer, Heidelberg (2004)
9. Podofillini, L., Zio, E., Marella, M.: A multi-state Monte Carlo Simulation Model of a Railway Network System. In: Kolowrocki, K. (ed.) Advances in Safety and Reliability, European Safety and Reliability Conference - ESREL 2005, pp. 1567–1575. Taylor & Francis Group, London (2005)
10. Pozsgai, P., Bertsche, B.: Modeling and Simulation of the Operational Availability and Costs of Complex Systems - a Case Study. In: Kolowrocki, K. (ed.) Advances in Safety and Reliability, European Safety and Reliability Conference - ESREL 2005, pp. 1597–1605. Taylor & Francis Group, London (2005)
11. Sanso, B., Milot, L.: Performability of a Congested Urban-Transportation Network when Accident Information is Available. Transportation Science 33 (1999)

12. Walkowiak, T., Mazurkiewicz, J.: Fuzzy Approach to Economic Analysis of Dispatcher Driven Discrete Transport Systems. In: DepCoS-RELCOMEX 2006 International Conference, pp. 366–373. IEEE Press, Poland (2006)
13. Walkowiak, T., Mazurkiewicz, J.: Genetic Approach to Modeling of a Dispatcher in Discrete Transport Systems. In: Rutkowski, L., Tadeusiewicz, R., Zadeh, L.A., Żurada, J.M. (eds.) ICAISC 2006. LNCS (LNAI), vol. 4029, pp. 479–488. Springer, Heidelberg (2006)
14. Walkowiak, T., Mazurkiewicz, J.: Hybrid Approach to Reliability and Functional Analysis of Discrete Transport System. In: Bubak, M., van Albada, G.D., Sloot, P.M.A., Dongarra, J. (eds.) ICCS 2004. LNCS, vol. 3037, pp. 236–243. Springer, Heidelberg (2004)

Maximum of Marginal Likelihood Criterion instead of Cross-Validation for Designing of Artificial Neural Networks

Zenon Waszczyszyn[1] and Marek Słoński[2]

[1] Rzeszów University of Technology, Rzeszów, Poland
zewasz@prz.edu.pl
[2] Cracow University of Technology, Kraków, Poland
mslonski@L5.pk.edu.pl

Abstract. The cross-validation method is commonly applied in the design of Artificial Neural Networks (ANNs). In the paper the design of ANN is related to searching for an optimal value of the regularization coefficient or the number of neurons in the hidden layer of network. Instead of the cross-validation procedure, the Maximum of Marginal Likelihood (MML) criterion, taken from Bayesian approach, can be used. The MML criterion, applied to searching for the optimal values of design parameters of neural networks, is illustrated on two examples. The obtained results enable us to formulate conclusions that the MML criterion can be used instead of the cross-validation method (especially for small data sets), since it permits the design of ANNs without formulation of a validation set of patterns.

Keywords: neural network design, cross-validation, marginal likelihood, Bayesian inference.

1 Problem Description

What is an important problem in designing of artificial neural networks (ANN) is the estimation of optimal network parameters with respect to some selected criterions. This problem can be illustrated on an example of Feed-forward Layered Neural Network (FLNN) with one hidden layer and a scalar output, Fig. 1a.

In Fig. 1b the error curves for this network are presented. For a defined number of neurons in the hidden layer H, the network is trained using the training data set $\mathcal{L} = \{\mathbf{x}^n, t^n\}_{n=1}^{L}$. The trained network is then verified using a validation data set $\mathcal{V} = \{\mathbf{x}^n, t^n\}_{n=1}^{V}$. It is assumed that $\mathcal{L}$ and $\mathcal{V}$ are subsets of a known data set $\mathcal{P}$, consisting of patterns of pairs $\{\mathbf{x}^n, t^n\}_{n=1}^{P}$. From set $\mathcal{P}$ we can randomly select L training patterns and the remaining patterns $V = P - L$ are used as the validation set, where $\mathcal{P} = \mathcal{L} \cup \mathcal{V}$, $\mathcal{L} \cap \mathcal{V} = \emptyset$. Changing the values of parameter H, we obtain learning error curves $E_\mathcal{D}(H; N)$, where $\mathcal{D}$ is a set $\mathcal{L}$ or $\mathcal{V}$ with number of elements $N = L$ or V. We apply the least square error measure

$$E_\mathcal{D}(\mathbf{w}) = \frac{1}{2} \sum_{n=1}^{N} \{t^n - y(\mathbf{x}^n; \mathbf{w})\}^2, \qquad (1)$$

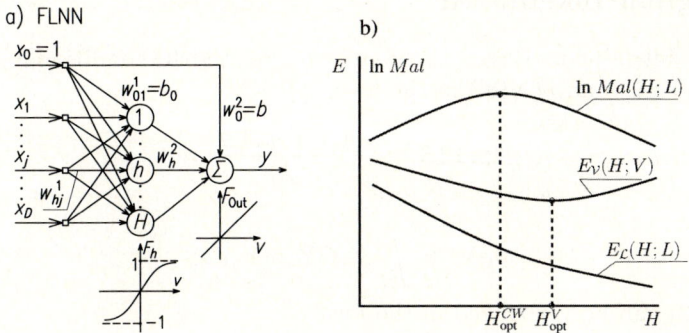

Fig. 1. a) Feed-forward Layered Neural Network (FLNN); b) training error $E_\mathcal{L}(H;L)$, validation error $E_\mathcal{V}(H;V)$ and a ln curve of marginal likelihood $\ln Mal(H;L)$

where $\mathbf{w} = \{w_i\}_{i=1}^{W}$ is a vector of generalized weights of the network (the synaptic weights w_{hj}^1, w_h^2 and the constant terms b_0, b).

Let us consider the Feed-forward Layered Neural Network (FLNN). It is important to note that increasing number H of neurons in the hidden layer (in general, increasing the model complexity), the training error curve $E_\mathcal{L}(H;L)$ monotonically decreases. The validation error curve $E_\mathcal{V}(H;V)$ has a different form, because it has the minimum at H_{opt}^V which is an estimate of the optimal number H of neurons in the hidden layer, see Fig. 1b.

The algorithm of searching for H_{opt}^V, described above, is known as cross-validation. This method is commonly used in problems of neural network designing when set $\mathcal{P}$ is large and the subsets $\mathcal{L}$ and $\mathcal{V}$ are statistically representative, see [1,2]. When the set $\mathcal{P}$ is small, the cross-validation algorithm may give not a representative estimation of H_{opt}^V.

The problems of cross-validation approach expressed above can be omitted using Bayesian inference and the marginal likelihood method, see [3,4] and references in [5]. In this method, a relation $Mal(H;L)$ is formulated and the corresponding curve is shown in Fig. 1b. It is important to note that this curve has the maximum which is called the maximum of marginal likelihood (MML). Thus, the curve of marginal likelihood can be applied for the selection of the optimal value of H_{opt}^{CW}. The criterion MML can play a crucial role since on the set of training patterns it is possible to optimize the neural model complexity only with respect to the training patterns $\mathcal{P}$ (without a validation set of patterns). This conclusion can be invaluable especially for a small set of patterns $\mathcal{P}$.

In subsequent sections we first define marginal likelihood function Mal. Next we show an application of MML method to the problem of estimation of the optimal value of regularization coefficient $\lambda_{\text{opt}}^{CW}$ for a neural network with radial basis functions (RBFN) and the optimal number of neurons in the hidden layer H_{opt}^{CW} for a two-layer neural network (FLNN).

2 Marginal Likelihood

We start by introducing Bayes' theorem which relates the conditional probability densities $p(Y|X)$ and $p(X|Y)$ in the form, see [5]

$$p(Y|X) = \frac{p(X|Y)p(Y)}{p(X)}, \tag{2}$$

where

$$p(X) = \int_{\mathcal{R}} p(X|Y)p(Y)dY. \tag{3}$$

Theorem (2) can be verbalized in the form

$$posterior = \frac{likelihood \times prior}{marginal\ likelihood}. \tag{4}$$

Eqs (2) and (3) can be specified in the form used in subsequent chapters, see [4]:

$$p(\mathbf{w}|\mathcal{D}, \alpha, \sigma^2) = \frac{p(\mathcal{D}|\mathbf{w}, \sigma^2)p(\mathbf{w}|\alpha)}{p(\mathcal{D}|\alpha, \sigma^2)} \tag{5}$$

$$p(\mathcal{D}|\alpha, \sigma^2) = \int p(\mathcal{D}|\mathbf{w}, \sigma^2)p(\mathbf{w}|\alpha)d\mathbf{w}, \tag{6}$$

where probability distribution (6) is defined as marginal likelihood, noted in short *Mal*.

In Eqs (5) and (6) there are data set and weights vector

$$\mathcal{D} = \{\mathbf{x}^n, t^n\}_{n=1}^{N}, \quad \mathbf{w} = \{w_i\}_{i=1}^{W}, \tag{7}$$

and hyperparameters α and σ^2, which are in a generalized error function for neural network $F(\mathbf{w}_{\text{MAP}})$

$$F(\mathbf{w}_{\text{MAP}}) = \frac{1}{\sigma^2} E_{\mathcal{D}}(\mathbf{w}_{\text{MAP}}) + \alpha E_W(\mathbf{w}_{\text{MAP}}). \tag{8}$$

In Eq (8) the subscript MAP corresponds to the Maximum A Posteriori, which is a solution for the network weights $\mathbf{w}_{\text{MAP}}$ using also a penalty function $E_W(\mathbf{w})$. Probability distribution in (6) can be approximated using a Gaussian distribution in the form

$$p(\mathcal{D}|\alpha, \sigma^2) = \left(\frac{1}{2\pi\sigma^2}\right)^{N/2} \left(\frac{\alpha}{2\pi}\right)^{W/2} \int \exp(-F(\mathbf{w}))d\mathbf{w}, \tag{9}$$

where $F(\mathbf{w})$ is an extended error function

$$F(\mathbf{w}) = F(\mathbf{w}_{\text{MAP}}) + \frac{1}{2}(\mathbf{w} - \mathbf{w}_{\text{MAP}})^{\text{T}}\mathbf{A}(\mathbf{w} - \mathbf{w}_{\text{MAP}}). \tag{10}$$

A matrix $\mathbf{A}$, used in (10) which is based on the Hessian matrix $\mathbf{H}$

$$\mathbf{A} = \frac{1}{\sigma^2}\mathbf{H} + \alpha\mathbf{I}, \quad \mathbf{H} = \nabla\nabla E_{\mathcal{D}} \approx \sum_n \mathbf{b}_n \mathbf{b}_n^{\text{T}}, \tag{11}$$

where $\mathbf{b}_n = \nabla y^n$.

Taking logarithm of equation (10) we obtain the following equation

$$\ln p(\mathcal{D}|\alpha, \sigma^2) = -F(\mathbf{w}_{\text{MAP}}) - \frac{1}{2}\ln|\mathbf{A}| + \frac{W}{2}\ln\alpha + \frac{N}{2}\ln 1/\sigma^2 - \frac{N}{2}\ln(2\pi). \quad (12)$$

In the next section Eq (12) is used for the estimation of the optimal value of regularization parameter λ for a neural network with radial basis functions and the estimation of the optimal number of hidden neurons H for a two-layer neural network.

3 Applications of Maximum of Marginal Likelihood

3.1 Estimation of Regularization Parameter

Let us analyze a simple example of estimation of the regularization coefficient $\lambda = \sigma^2 \alpha$, for a known value of variance of the learning target data $\sigma^2 = const$ in the generalized error function

$$F(\mathbf{w}_{\text{MAP}}) = E_\mathcal{D}(\mathbf{w}_{\text{MAP}}) + \lambda E_W(\mathbf{w}_{\text{MAP}}). \quad (13)$$

We apply a neural network with the exponential ('Gaussian') radial basis functions (RBFN), for which an explicit regression function can read

$$y(\mathbf{x}; \mathbf{w}) = b + \sum_{k=1}^{K} w_k \phi_k(\mathbf{x}; \mathbf{c}_k, \sigma^2). \quad (14)$$

In the discussed example we assume RBFN network to be an interpolation network, cf. Fig. 2, for which radial basis centres are located at learning data points $\mathbf{c}_k = \mathbf{x}_k$ and in this case the design matrix $\boldsymbol{\Phi}$ has the form

$$\boldsymbol{\Phi}_{(L \times L)} = [\Phi_{nk}] \quad \text{for} \quad k = 0, \ldots, K = L - 1; \quad n = 1, \ldots, L, \quad (15)$$

where:

$$\Phi_{nk} = \phi_k(\mathbf{x}^n) = \exp\left(-\frac{1}{2s^2}(\mathbf{x} - \mathbf{x}^n)^{\text{T}}(\mathbf{x} - \mathbf{x}^n)\right), \quad \phi_0(\mathbf{x}^n) = 1. \quad (16)$$

For RBFN network we can analytically compute matrix $\mathbf{A}$ and the network weights vector $\mathbf{w}_{\text{MAP}}$

$$\mathbf{A} = \frac{1}{\sigma^2}\boldsymbol{\Phi}^{\text{T}}\boldsymbol{\Phi} + \alpha\mathbf{I}, \quad \mathbf{w}_{\text{MAP}} = \frac{1}{\sigma^2}\mathbf{A}^{-1}\boldsymbol{\Phi}^{\text{T}}\mathbf{t}. \quad (17)$$

The number of radial basis functions determines the number of weights $W = K + 1$, because in (16) the additional parameter is a the bias term $w_0 = b$.

Eq (17) allows us to compute the value of the logarithm of the marginal likelihood function $\ln Mal = \ln p(\mathcal{D}|\lambda)$ for a known number of patterns N, the number of basis functions $K = N - 1$, weights $W = K + 1 = L$ and known regularization parameter λ.

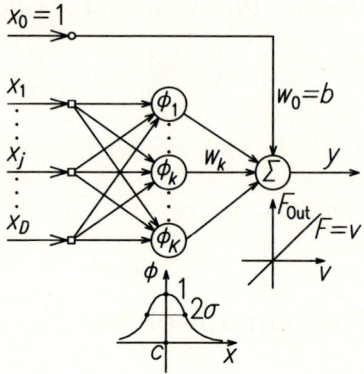

Fig. 2. Neural network with radial basis functions (RBFN)

The computations were carried out for an example taken from [4]. The $N = 15$ learning patterns were generated from a known function $h(x) = \sin 2\pi x$ for $x \in [0, 1]$ with added Gaussian noise $\epsilon = \mathcal{N}(0, \sigma^2)$ with zero mean and variance $\sigma^2 = 0.05$. The data points x^n were uniformly spaced within the interval. Also validation patterns were generated for the estimation of the regularization parameter $\lambda = \sigma^2 \alpha$. The computations were carried out for RBFN model with $K = 14$ Gaussian radial basis functions (16) for $s = 0.2$. The optimal value of regularization parameter is $\ln \lambda^V_{\text{opt}} \approx 1.66$.

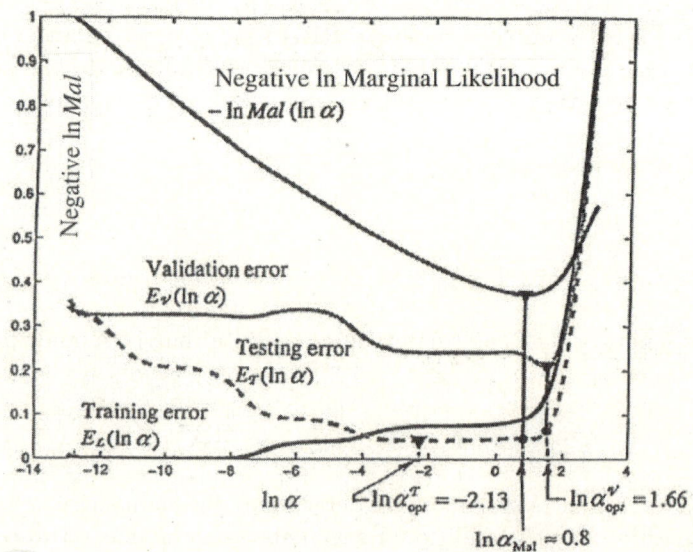

Fig. 3. Plots of error functions for learning $E_\mathcal{L}(\ln \lambda; L)$, validation $E_V(\ln \lambda; V)$, testing $E_T(\ln \lambda; T)$ and the curve of negative log of marginal likelihood $-\ln Mal(\ln \alpha; L)$

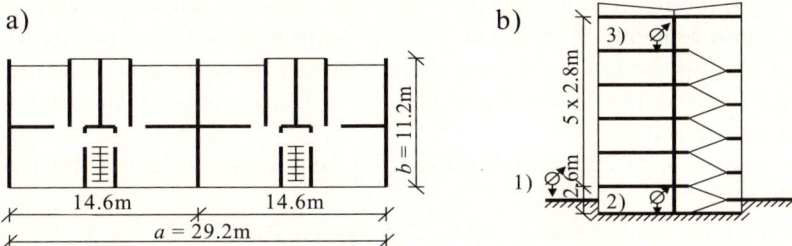

Fig. 4. Prefabricated five-storey building of WK-70 type: a) Plan, b) Sectional elevation and points of vibration measurements

Table 1. Errors and statistical parameters for different input variables and different architectures of MLNNs

Net	FLNN	NNP	MSE×10³		ARE[%]		σ(P)	r(P)
			L	T	L	T		
1	4-4-1	25	0.08	0.32	3.0	5.8	0.011	0.964
2	4-3-1	19	0.09	0.30	3.3	5.4	0.011	0.965
3	4-3-1	19	0.05	0.45	2.7	5.3	0.010	0.966

In Fig. 3 the plots of scaled error functions for learning and validation $E_{\mathcal{L}}(\ln \lambda; L)$ and $E_{\mathcal{V}}(\ln \lambda; V)$ are presented. The testing errors were also computed with respect to the known function $h(x) = \sin 2\pi x$ based on point distances $|y(\mathbf{x}; \mathbf{w}_{\text{MAP}}) - h(x^n)|$. The learning function is monotonically decreasing. The validation function has a minimum for $\ln \lambda^{\mathcal{V}}_{\text{opt}} \approx 1.66$. The testing function also has the "true" minimum for $\ln \lambda^{\mathcal{T}}_{\text{opt}} \approx -2.33$ and has significantly smaller testing error value $E_{\mathcal{T}}(\ln \lambda^{\mathcal{T}}_{\text{opt}}; T)$ than validation error value $E_{\mathcal{V}}(\ln \lambda^{\mathcal{V}}_{\text{opt}}; V)$.

In Fig. 3 plot of the curve of negative log of marginal likelihood $- \ln Mal(\ln \alpha; L)$ which has the minimum for $\ln \lambda^{Mal}_{\text{opt}} = 0.8$ is presented. This value is smaller than $\ln \lambda^{\mathcal{V}}_{\text{opt}}$ and closer to $\ln \lambda^{\mathcal{T}}_{\text{opt}}$.

3.2 Simulation of Fundamental Periods of Natural Vibration

Dynamic properties of structures are characterized by natural periods, vibrations dampings and mode shapes of natural vibrations. In recent years a great deal of attention has been paid at the Institute of Structural Mechanics of the Cracow University of Technology to the analysis of vibrations of prefabricated buildings subjected to so-called paraseismic excitations, see [6,7] for bibliographic information.

We concentrate only on the fundamental periods of natural vibrations which were used in simple expert systems developed for the evaluation of the technical state of buildings subjected to mining tremors and explosions in nearby mines or quarries, cf. [8]. We focus on a group of 14 monitored, prefabricated buildings of different types. In Fig. 4 plan and sectional elevation of a prefabricated five-storey building of WK-70 type is depicted together with three points of vibration measurements [8]. Natural vibrations are excited by propagated surface seismic waves so the full-scale measured accelerograms were used to compute 31 target patterns as the T_1 [sec.] periods of vibrations applying the Fast Fourier transformation procedures.

On the base of [6,7], the following input vector $\mathbf{x}_{(4\times 1)}$ and scalar output y were adopted

$$\mathbf{x}_{(4\times 1)} = \{C_z, b, s, r\}, \quad y = T_1, \tag{18}$$

where:
C_z – ratio of vertical unit base pressure for elastic strains; b – building dimension along the vibration direction (longitudinal or transversal); $s = \sum_i EI_i/a$, $r = \sum_i GA_i/a$ – equivalent building and shear stiffnesses of the i-th internal walls in the building plan, cf. Fig. 4a.

In Table 1 the results of neural approximation with respect to the following errors measures and statistical parameters are shown:

$$MSE(N) = \frac{1}{N}\sum_{n=1}^{N}\{t^n - y(\mathbf{x}^n; \mathbf{w})\}^2,$$

$$ARE(N) = \frac{1}{N}\sum_{n=1}^{N} |1 - y(\mathbf{x}^n; \mathbf{w})/t^n| \times 100\%, \tag{19}$$

$$\sigma(N) = RMSE(N) = \sqrt{MSE(N)},$$

$$r = \frac{\text{cov}(\mathbf{t}, \mathbf{y})}{\sigma_t \sigma_y}, \quad \text{where } \mathbf{t} = \{t^n\}_{n=1}^{N},\ \mathbf{y} = \{y(\mathbf{x}^n; \mathbf{w})\}_{n=1}^{N}.$$

In Fig. 5 the curves $\ln Mal(H; L)$ computed for $L = 31, 26, 19$ learning patterns respectively are shown and also learning $MSE(H; \mathbf{w}_{\text{MAP}}, L = 31)$ and validation $MSE(H; \mathbf{w}_{\text{MAP}}, V = 12)$ curves are drawn. The Maximum A Posteriori (MAP) solution for weights vector $\mathbf{w}_{\text{MAP}}$ was computed using 350 iterations (epochs) of scaled conjugate gradient method implemented in toolbox Netlab for MATLAB, cf. [9]. The computations were carried out for the regularization parameter $\lambda = 10^{-11}$. From the plots we can conclude that the optimal number of neurons equals $H_{\text{opt}}^{Mal} = 3$ for the network FLNN: 4-3-1 with the total number of network parameters $NNP = 19$.

In Table 1 results obtained for the neural networks 1 and 2 are compared. For the network 1 with 4 neurons in the hidden layer results were obtained using $L = 26$ learning patterns and $T = 5$ testing patterns and averaging over 100 random splits of $P = 31$ patterns into the learning and testing sets, cf. [7]. The network 2 with only 3 hidden neurons was trained using also $L = 26$ learning patterns and tested on 5 testing patterns (in [6] patterns numbered 5, 10, 16, 18, 27). It is important to note that the optimal and simpler network with 3 hidden units gives results comparable with the network with 4 hidden neurons.

The results presented in the last row were obtained for the network 3 with 3 neurons in hidden layer and after averaging over 10 random selections of initial

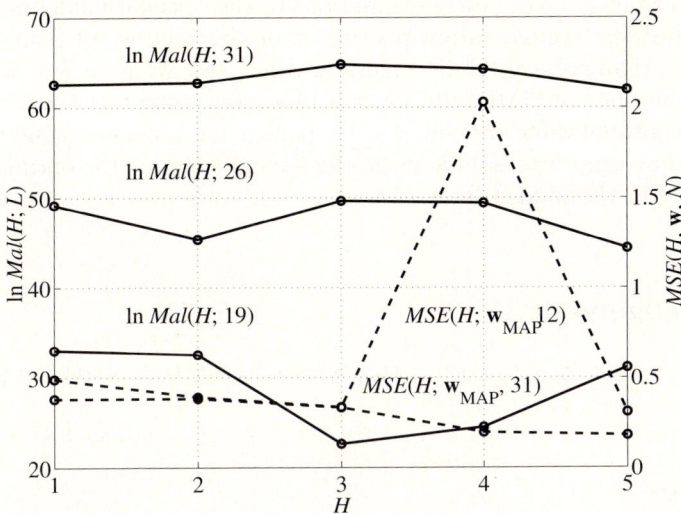

Fig. 5. Marginal likelihood curves $\ln Mal(H;L)$ computed for $L = 31, 26, 19$ learning patterns and learning MSE error curve $MSE(H; \mathbf{w}_{\text{MAP}}, L = 31)$ and validation MSE error curve $MSE(H; \mathbf{w}_{\text{MAP}}, V = 12)$

weights. Data set was randomly split into $L = 26$ learning patterns and $T = 5$ testing patterns. The averaged results are very close to the results for network 1 with 4 hidden units.

The split and size of the learning data set affect the marginal likelihood curve $\ln Mal(H;L)$ and also learning and testing curves. In Fig. 5 these curves are shown for two randomly selected data sets with $L = 26$ and $L = 19$ patterns. It is visible that the optimal number of hidden neurons $H_{\text{opt}}^{Mal}(L)$ is different and for $L = 19$ it can be $H_{\text{opt}}^{Mal} = 1$ or 2.

In the same Fig. 5 the validation curve is drawn for a randomly selected validation set with $V = 12$ patterns. This curve gives $H_{\text{opt}}^{V} = 3$, but also small validation errors are for $H = 1, 2$. The networks were trained by the set of $L = 19$ patterns using the generalized error function (13).

4 Final Remarks and Conclusions

In the paper Maximum of Marginal Likelihood (MML) criterion is presented which is based on the maximization of logarithm of marginal likelihood function $\ln Mal$, where $Mal = p(D|\alpha, \beta)$ is the marginal probability (normalizing term in Bayes' rule) for data set D given values of hyperparameters α and β. The following conclusions can be expressed:

i) MML criterion can replace the minimum of validation error criterion (cross-validation). MML criterion can be applied to any data sets, in particular to a learning set without the need of using the validation set.

ii) MML criterion gives better estimations of the optimal model metaparameters (value of the regularization parameter or the number of neurons in the hidden layer) than cross-validation method which can gives fuzzy results.

iii) The efficiency of MML criterion was presented using two examples. In the first one the optimal value of regularization parameter λ of a neural network with radial basis functions was estimated. In the second example the optimal number of neurons H in the hidden layer of feed-forward two-layer neural network was computed.

Acknowledgement

The support of this work by the Polish Grant MNiSW N N506 1814 33 is acknowledged.

References

1. Haykin, S.: Neural Networks: A Comprehensive Introduction. Prentice-Hall, Englewood Cliffs (1999)
2. Waszczyszyn, Z. (ed.): Neural Networks in the Analysis and Design of Structures. CISM Courses and Lectures, vol. 404. Springer, Wien-New York (1999)
3. Bishop, C.M.: Neural Networks for Pattern Recognition. Oxford University Press, Oxford (1995)
4. Tipping, M.E.: Bayesian Inference: An Introduction to Principles and Practice in Machine Learning. In: Bousquet, O., von Luxburg, U., Rätsch, G. (eds.) Machine Learning 2003. LNCS (LNAI), vol. 3176, pp. 41–62. Springer, Heidelberg (2004)
5. Bishop, C.M.: Pattern Recognition and Machine Learning. Springer, Heidelberg (2006)
6. Kuźniar, K.: Analysis of vibrations of medium height buildings subjected to mining tremors with application of neural networks (in Polish). Cracow University of Technology (2004)
7. Kuźniar, K., Waszczyszyn, Z.: Neural networks for the simulation and identification of building subjected to paraseismic excitations. In: Lagaros, N.D., Tsompanakis, Y. (eds.) Intelligent Computational Paradigms in Earthquake Engineering. Idea Group Publishing (2007)
8. Ciesielski, R., Kuźniar, K., Maciag, E., Tatara, T.: Empirical formulae for fundamental natural periods of buildings with load bearing walls. Archives of Civil Engineering 38, 199–291 (1992)
9. Nabney, I.T.: Netlab: Algorithms for Pattern Recognition. Springer, London (2002)

Part II

Fuzzy Systems and Their Applications

Type-2 Fuzzy Decision Trees*

Łukasz Bartczuk and Danuta Rutkowska

Department of Computer Engineering
Częstochowa University of Technology, Poland
{bartczuk,drutko}@kik.pcz.czest.pl

Abstract. This paper presents type-2 fuzzy decision trees (T2FDTs) that employ type-2 fuzzy sets as values of attributes. A modified fuzzy double clustering algorithm is proposed as a method for generating type-2 fuzzy sets. This method allows to create T2FDTs that are easy to interpret and understand. To illustrate performace of the proposed T2FDTs and in order to compare them with results obtained for type-1 fuzzy decision trees (T1FDTs), two benchmark data sets, available on the internet, have been used.

1 Introduction

In the last fifty years many methods have been proposed for data analysis and classification. Among others, there are decision trees [3],[4],[14],[18], as well as neural networks and fuzzy systems; see e.g. [5],[10],[11],[17],[21],[22], [23],[24],[28]. In the past, many algorithms to create such structures were introduced. The most popular are the well-known ID3 algorithm and its successor C4.5, proposed by Quinlan [18]-[20]. These algorithms operate on crisp (symbolic or numerical) data.

In the real world, data are often imprecise or uncertain. To analyse and process such data, fuzzy decision trees have been proposed [1],[3],[14],[15], as a combination of classical decision trees with fuzzy sets and fuzzy logic introduced by Zadeh [26]. Many algorithms have been employed to create and optimize fuzzy decision trees [1],[3],[14],[15]. From perspective of our work, the paper written by Janikow [14], that describes a generalized version of the ID3 algorithm, is especially interesting.

The most important property of fuzzy sets – that differs them from classical sets – is membership value interval [0,1] rather than {0,1} set [10],[22],[24],[26]. The fuzzy logic can be used to give a formal description of such concepts like e.g. "high", "near", "warm". Fuzzy sets are characterized by membership functions of various shapes. The shape may be crucial for performace of a classifier. The membership functions can be defined by experts or determined by means

---

* This work was partly supported by the Foundation for Polish Science (Professorial Grant 2005-2008) and the Polish State Committee for Scientific Research (Grant N518 035 31/3292), Special Research Project 2006-2009, Polish-Singapore Research Project 2008-2010, Research Project 2008-2010).

of a special algorithm based on data. In both cases, additional levels of uncertainty may appear. This is especially important when fuzzy sets are used as a description of some linguistic terms collected from many experts. In such case, like Mendel wrote, "words can mean different things to different people" [17].

To take into account the additional level of uncertainty, the so-called type-2 fuzzy sets can be used [4]. They are generalization of classical fuzzy sets in which membership grades are defined as fuzzy sets on [0,1] interval. As a special, simpler case, the type-2 fuzzy sets with an interval membership grade may be applied.

In this paper, further generalization of decision trees is considered. The Fuzzy-ID3 algorithm, proposed by Janikow [14], is adopted to work with type-2 fuzzy sets in order to take advantage of the additional level of uncertainty.

This paper is organized as follows: the next section contains a short introduction to type-2 fuzzy set theory. The third section describes type-2 fuzzy decision trees and the algorithm proposed in order to create the trees. In Section 4, an inference method that uses the type-2 tree structures is presented. Section 5 depicts an algorithm to obtain type-2 fuzzy sets that are easy to read and interpret. Experimental results are illustrated in Section 6, and final conclusions are included in Section 7.

2 Type-2 Fuzzy Sets

Type-2 fuzzy sets have been proposed by Zadeh [27] as a generalization of classic (type-1) fuzzy sets. Let X be a universe of discourse. A type-2 fuzzy set $\tilde{A}$ is defined as a collection of ordered pairs $(x, \mu_{\tilde{A}}(x))$, where $x \in X$ and $\mu_{\tilde{A}}(x)$ is a fuzzy grade of membership defined on [0,1] interval as follows [17],[24],[27]:

$$\mu_{\tilde{A}}(x) = \int f_x(u)/u \qquad (1)$$

where $u \in [0,1]$ is a primary and $f_x(u)$ is a secondary membership degree. According to this definition, the fuzzy grade of membership can be considered as a fuzzy number or a fuzzy interval.

The interval type-2 fuzzy set is a special case of a type-2 fuzzy set in which the secondary membership grade is always equal to one: $\mu_{\tilde{A}}(x) = \int_{u \in [0,1]} 1/u$ [17],[24]. This kind of fuzzy sets allows to reduce all operations performed on type-2 fuzzy sets, which – in a general case – can have very high computational cost. To simplify the notation, the interval fuzzy membership grade can be represented by minimum and maximum points in which function $f_x(u)$ has non-zero values $\mu_{\tilde{A}}(x) = \left[\underline{\mu_{\tilde{A}}(x)}, \overline{\mu_{\tilde{A}}(x)}\right]$; see [17].

The main operation performed on type-2 fuzzy sets is the intersection operation. It is defined with the so-called extended T-norm [17], [24] that can be expressed as follows:

$$\mu_{\tilde{A}}(x) \stackrel{\tilde{T}}{*} \mu_{\tilde{B}}(x) = \left[\underline{\mu_{\tilde{A}}(x)} \stackrel{\tilde{T}}{*} \underline{\mu_{\tilde{B}}(x)}, \overline{\mu_{\tilde{A}}(x)} \stackrel{\tilde{T}}{*} \overline{\mu_{\tilde{B}}(x)}\right] \qquad (2)$$

where T is the classic T-norm, e.g. the minimum function.

Another important concept associated with type-2 fuzzy sets, is the extended fuzzy relation [17],[22],[24]. In this paper, it is assumed that the fuzzy relation can be defined with any extended T-norm.

3 Type-2 Fuzzy Decision Trees

Let us assume that E is a data set and A is an attribute set. Every object included in data set E can be considered as a pair $e_i = (\mathbf{x}_i, y_i)$ where $\mathbf{x}_i = [x_1, \ldots, x_n]$ is a n-dimensional vector of input values, and y_i is a label or class (element of decision attribute A^D). Each attribute $A^k \in A$, $k = 1, \ldots, |A|$ describes a feature of an object, and in the case of the considered decision trees, it contains $|A^k|$ values a_m^k, $m = 1, \ldots, |A^k|$, which may be characterized by a symbol, crisp set, fuzzy set or type-2 fuzzy set.

Crisp decision trees, as well as fuzzy decision trees, are created in the top-down manner by recursive procedures. In every internal node of the tree, the data set is partitioned according to values of the chosen attribute.

One of the algorithms, employed in order to build fuzzy decision trees, was proposed by Janikow [14]. This algorithm is a direct extension of the very popular ID3 method introduced by Quinlan [18]. In paper [4] authors propose another extension of the ID3 algorithm that allows to create the so-called Type-2 Fuzzy Decision Trees (T2FDTs). These structures constitute a combination of classical decision trees with type-2 fuzzy sets.

The proposed algorithm to create T2FDTs is based on the Fuzzy-ID3 method introduced by Janikow [14], but every operation performed in this algorithm is extended, so it can operate on fuzzy values.

Before presentation of the proposed algorithm, let us make the following assumptions:

1. For any attribute, its values are defined by interval type-2 fuzzy sets.
2. $\mu_N(e_i)$ denotes the fuzzy membership degree of object e_i in node N. This value can be viewed as fuzzy membership degree in multidimensional type-2 fuzzy set – defined using the intersection operation of all interval type-2 fuzzy sets that are included in the path from the root node to N node.
3. The object is included in subset $E^N \subset E$ if $\underline{\mu_N(e_i)} > 0$.
4. The value $\mu_{\tilde{a}_j^D}(y_i)$ equals $[1,1]$ if $y_i = a_j^D$ and $[0,0]$ otherwise.

Having these assumptions, now let us present the proposed algorithm, in the form of the following recursive procedure, with two arguments, i.e. data set and attribute set:

Algorithm. T2FDT(data set E, attribute set A)

(1) Create node N and assign data set E to it
(2) If at least one of the stop conditions is met, convert node N to a leaf and assign to it the decision for which the examples frequency given by Equation (4), reaches the biggest value.

(3) In other case
   (a) Compute examples count P_j^N, examples frequency Q_j^N, and total examples count P^N, defined by the following formulas, respectively:

$$P_j^N = \left[\underline{P_j^N}, \overline{P_j^N}\right] = \left[\sum_{i=1}^{|E|} \underline{\mu_N}(e_i), \sum_{i=1}^{|E|} \overline{\mu_N}(e_i)\right] \quad (3)$$

$$Q_j^N = \left[\underline{Q_j^N}, \overline{Q_j^N}\right] = \left[\frac{\underline{P_j^N}}{\underline{P_j^N} + \sum_{\substack{o=1\\o\neq j}}^{|A^D|} \overline{P_j^N}}, \frac{\overline{P_j^N}}{\overline{P_j^N} + \sum_{\substack{o=1\\o\neq j}}^{|A^D|} \underline{P_j^N}}\right] \quad (4)$$

$$P^N = \left[\underline{P^N}, \overline{P^N}\right] = \left[\sum_{j=1}^{|A^D|} \underline{P_j^N}, \sum_{j=1}^{|A^D|} \overline{P_j^N}\right] \quad (5)$$

   (b) Compute the information content (entropy) I^N. In order to complete this task, the minimum and maximum values of the following function must be determined

$$Y(q_1, \ldots, q_{|A^D|}) = -\sum_{j=1}^{|A^D|} q_j \log_2 q_j \quad (6)$$

   where $q_j \in Q_j^N$ and $\sum_{j=1}^{|A^D|} q_j = 1$. In paper [4] two heuristics algorithms have been introduced in order to obtain these values.

   (c) For each attribute, not included in the path from the root node to N node, the weigted entropy and information gain must be computed. In the Fuzzy-ID3 algorithm, the weighted entropy is calculated by use of the following formula

$$I^{N|A^k} = \sum_{m=1}^{|A^k|} P^{N|\tilde{a}_m^k} I^{N|\tilde{a}_m^k} \Big/ \sum_{m=1}^{|A^k|} P^{N|\tilde{a}_m^k} \quad (7)$$

   where $P^{N|\tilde{a}_m^k}$ and $I^{N|\tilde{a}_m^k}$ denote the total examples count and the information content in the child node that contains value $\tilde{a}_m^k$, respectively. In the proposed algorithm, values $P^{N|\tilde{a}_m^k}$ and $I^{N|\tilde{a}_m^k}$ are intervals. Therefore, in order to compute the value of Equation (7), that is also an interval, the Karnik-Mendel algorithm [14] must be applied.
   The information gain can be computed as follows:

$$G^{A^k} = \left[\underline{G^{A^k}}, \overline{G^{A^k}}\right] = \left[\underline{I^N} - \overline{I^{N|A^k}}, \overline{I^N} - \underline{I^{N|A^k}}\right] \quad (8)$$

(d) The attribute with the highest value of the information gain must be chosen. Because this is an interval value, as a comparison operation, one of the methods employed to compare fuzzy numbers may be employed [1],[7],[25].

(e) For each value of the attribute with the highest information gain value, create new data set $E^{\tilde{a}_m^k}$ that contains those elements of set E for which $\mu_N(e_i) \overset{T}{*} \mu_{\tilde{a}_m^k}(e_i) > 0$, and proceed the T2FDT($E^{\tilde{a}_m^k}, A$) algorithm, in order to create a subtree for the attribute value $\tilde{a}_m^s$.

(4) return node N.

The stop conditions, mentioned in step (2), determine that N node should be further split or converted to a leaf. In the latter case, the decision for which the examples frequency reaches the biggest value is assigned to this leaf. The most popular stopping criterions are

- all atributes are included in the path from the root node to N node.
- all objects in data set E^N have the same decision label
- the maximum number of nodes have to be reached
- the information gain for any attribute does not reach the minimum value.

4 Inference Process Using Type-2 Fuzzy Decision Tree

The complete method which uses decision trees should contain at least two elements. First of them is the decision tree creation procedure presented in the previous section. The second is an inference process which is used to classify a new object. Every decision tree can be considered as a set of decision rules where each rule corresponds exactly to one leaf of the tree, and may be written as follows:

$$R^l : \text{IF } x_1 \text{ IS } \tilde{a}_l^1 \text{ AND} \ldots \text{AND } x_s \text{ IS } \tilde{a}_l^s \text{ THEN } y \text{ IS } \tilde{a}_l^D; \quad l = 1, \ldots, |L| \quad (9)$$

where $|L|$ denotes a number of leaves (rules) and s is the length of the rule; $1 \leq s \leq |A|$.

In fuzzy rule decision systems, a degree of fulfillment of antecedent part of the rule, denoted by $\mu_{A^l}(\mathbf{x})$, may be computed using any kind of the T-norm or in the case of type-2 fuzzy rule, decision systems employ the extendend T-norm.

Every rule represents a fuzzy relation that allows to determine the degree of satisfaction of the consequents. Like in the previous section, for simplicity, the fuzzy relation will be defined using the extended T-norm:

$$\mu_{R^l}(y) = \mu_{A^l}(\mathbf{x}) \overset{\tilde{T}}{*} \mu_{\tilde{a}_l^D}(y) = \left[\underline{\mu_{A^l}(\mathbf{x})} \overset{T}{*} \underline{\mu_{\tilde{a}_l^D}(y)}, \overline{\mu_{A^l}(\mathbf{x})} \overset{T}{*} \overline{\mu_{\tilde{a}_l^D}(y)} \right] \quad (10)$$

In the simplest inference method, a new object is assigned to the leaf for which the highest degree of satisfaction of the consequent is obtained. For simple tasks, this method allows to get a sufficient level of classification correctness.

The more ellaborated method, proposed by Janikow [14], requires that in every leaf, examples count P_κ^l must be computed. The purpose of this value is to limit the negative influence of leaves with small number of objects from the data set in the inference process. This can be considered as a weight of the rule generated by the decision tree

$$\mu_{R^l}(y) = \left[\mu_{A^l}(\mathbf{x}) \overset{\tilde{T}}{*} \mu_{\tilde{a}_l^p}(y)\right] P_\kappa^l \qquad (11)$$

However, this approach can generate supernatural fuzzy sets for which the biggest value of the fuzzy grade of membership is bigger than one. In order to prevent such a situation, the normalized examples count ψ_l can be applied.

By use of the normalized examples count ψ_l, Equation (11) can be rewritten as

$$\mu_{R^l}(y) = \left[\mu_{A^l}(\mathbf{x}) \overset{\tilde{T}}{*} \mu_{\tilde{a}_l^p}(y)\right] \psi_l = \left[\underline{\mu_{A^l}(\mathbf{x})} \overset{T}{*} \underline{\mu_{\tilde{a}_l^p}(y)\psi_l}, \overline{\mu_{A^l}(\mathbf{x})} \overset{T}{*} \overline{\mu_{\tilde{a}_l^p}(y)\psi_l}\right] \qquad (12)$$

The last step in this inference method is the defuzzyfication. In type-2 fuzzy systems, this step consists of two stages: type reduction and defuzzyfication. One of the most popular method of the type reduction is the height method in which the reduced set can be expressed by the following formula:

$$C_{\tilde{R}} = \int_{u_{C_1}^1 \in J_{y_1'}^1} \cdots \int_{u_{C_{|L|}}^{|L|} \in J_{y_{|L|}'}^{|L|}} 1 \bigg/ \left(\sum_{l=1}^{|L|} y_l' u_{C_l}^l \bigg/ \sum_{l=1}^{|L|} u_{C_l}^l\right) \qquad (13)$$

where $C_{\tilde{R}}$ is the reduced fuzzy set, $J_{y_1}^l, \ldots, J_{y_{|L|}}^{|L|}$ - the domain of fuzzy membership grade $\mu_{\tilde{R}^l}(y)$; $l = 1, \ldots, |L|$

In the case of interval type-2 fuzzy set, this value can be computed by the Karnik-Mendel iterative algorithm [17]. The result of the type reduction step is an interval fuzzy value which in turn can be defuzzified by the following expression:

$$y' = \left(\underline{C_{\tilde{R}'}} + \overline{C_{\tilde{R}'}}\right)/2 \qquad (14)$$

5 Generating Type-2 Fuzzy Partition

The algorithm presented in Section 3 assumes that every continuous attribute should be partitioned to a small number of type-2 fuzzy sets, before the tree is built.

The basic method that allows to generate a type-2 fuzzy partition is the so-called uncertain fuzzy clustering [12]. This method is an extension of the classical FCM (Fuzzy C-Means) [2] algorithm where the additional level of uncertainty is associated with the fuzzifier parameter which in case of this algorithm is an interval value and can be denoted as $\tilde{m} = [\underline{m}, \overline{m}]$. It has one serious disadvantage – creates multidimensional type-2 fuzzy sets that can have high degree of overlaping when are projected on the one-dimensional space. These fuzzy sets, as well as fuzzy decision trees and fuzzy rules that use them, can be hard to interpret. In order to produce type-1 fuzzy partition that is easy to read and understand, in paper [8] the so-called double fuzzy clustering is proposed.

The orginal method considers type-1 fuzzy sets and consists of two steps. First, the FCM algorithm is performed to create multidimensional fuzzy prototypes, and after that one-dimensional projections of the obtained prototypes are grouped together by a hierarchical clustering algorithm [8].

In our experiments, the orginal method was modified in the following way. In the first step, we used the uncertain fuzzy clustering algorithm, and in the second step the FCM algorithm was performed.

Next, the membership of all objects in one-dimensional clusters is computed for each dimension, with the follwing equations:

$$\underline{\mu}_{jk}(\mathbf{x}_i) = \begin{cases} \begin{cases} u_{ijk}^{(m)} & \text{if } u_{ijk}^{(m)} \leq u_{ijk}^{(\overline{m})} \\ u_{ijk}^{(\overline{m})} & \text{if } u_{ijk}^{(m)} > u_{ijk}^{(\overline{m})} \end{cases} & \text{if } v_{(j-1)}^k \leq x_{ik} \leq v_{(j+1)}^k \\ 0 & \text{otherwise} \end{cases} \quad (15)$$

$$\overline{\mu}_{jk}(\mathbf{x}_i) = \begin{cases} \begin{cases} u_{ijk}^{(m)} & \text{if } u_{ijk}^{(m)} \geq u_{ijk}^{(\overline{m})} \\ u_{ijk}^{(\overline{m})} & \text{if } u_{ijk}^{(m)} < u_{ijk}^{(\overline{m})} \end{cases} & \text{if } v_{(j-1)}^k \leq x_{ik} \leq v_{(j+1)}^k \\ 0 & \text{otherwise} \end{cases} \quad (16)$$

where:

$$u_{ijk}^{(m)} = \left(\sum_{l=1}^{C} \left(\frac{|v_j^k - x_{ik}|}{|v_l^k - x_{ik}|} \right)^{\frac{2}{m-1}} \right)^{-1} \quad (17)$$

$$u_{ijk}^{(\overline{m})} = \left(\sum_{l=1}^{C} \left(\frac{|v_j^k - x_{ik}|}{|v_l^k - x_{ik}|} \right)^{\frac{2}{\overline{m}-1}} \right)^{-1} \quad (18)$$

In the above equations $\tilde{\mu}_{jk}(\mathbf{x_i}) = \left[\underline{\mu}_{jk}(\mathbf{x_i}), \overline{\mu}_{jk}(\mathbf{x_i}) \right]$ denotes the membership degree of ith object in j^{th} type-2 fuzzy cluster defined for k^{th} attribute, and v_j^k denotes the prototype of jth cluster for k^{th} attribute, $i = 1, \ldots, |E|$; $k = 1, \ldots, |A|$; $j = 1, \ldots, |A^k|$. The condition $v_{(j-1)}^k \leq x_{ik} \leq v_{(j+1)}^k$ is necessary to assure the convexity of the created type-2 fuzzy sets.

The generated cluster center and membership table allow to determine the parameters of type-2 fuzzy sets with Gaussian primary membership functions used in the performed experiments.

6 Experimental Results

In this section, results of the experiments are presented. The results obtained for the proposed T2FDTs have been compared with results produced by classic fuzzy decision trees (created by Janikow's Fuzzy-ID3 algorithm). For these experiments, two benchmark data sets from UCI repository [6] have been used. These data sets concern the thyroid disease and breast cancer.

6.1 Data Description

The *breast cancer* data set, collected by the University of Wisconsin Hospitals, consists of 699 cases; where 458 cases represent the benign class and 241 cases represent the malignant class. There are 10 numerical input variables taking values from [1, 10] interval. The complete data set includes 16 instances with missing values which have been deleted.

The *thyroid* data, prepared by the James Cook University, Townsville Australia, consists of 215 cases, where 150 cases represent the normal condition, 35 cases represent the hyperthroid disease, and 30 cases – the hypothyroid disease. There are 5 continuous attributes with no missing values.

6.2 Results

The data sets described in the previous section have been split into two disjoint subsets: the training set (80% of all cases) and testing set (20% of all cases). The examples were chosen in a random way for both sets. The training set was applied in order to generate type-2 fuzzy sets as well as to create the decision trees. The testing data set was employed to check performace of the constructed tree structures.

The fuzzy sets were created by use of the method described in Section 5. In the case of the FCM algorithm, the most frequently used value of fuzzier m is 2, so we also applied this value in our experiments. For the same reason, we used this value as the lower limit of interval fuzzifier $\tilde{m}$ in the uncertain fuzzy clustering algorithm. The upper limit was changed from 3 to 7, taking values from interval [3,7] of natural numbers.

For all data sets employed, every experiment was repeated ten times. The average results are shown in Table 1, where the percentage of correct diagnosis for the testing set as well as the number of nodes in the generated trees are presented.

Table 1. Experiments results for fuzzy decision trees and type-2 fuzzy decision trees. In this table, FCM denotes the Fuzzy C-Means algorithm and UCF the Uncertain Fuzzy Clustering algorithm, respectively.

Algorithm type	Thyroid data		Breast cancer data	
	Error [%]	Number of nodes	Error [%]	Number of nodes
FCM $m = 2$	7.44	12.8	3.97	18.5
UCF $\tilde{m} = [2,3]$	3.02	13.3	2.72	16.6
UCF $\tilde{m} = [2,4]$	3.95	12.4	2.79	18.4
UCF $\tilde{m} = [2,5]$	3.48	13.5	3.01	17.9
UCF $\tilde{m} = [2,6]$	3.95	14.2	2.79	19.3
UCF $\tilde{m} = [2,7]$	4.18	12.7	3.01	19.7

As we can see, the number of nodes in all cases are comparable, but the number of incorrect classification is considerably lower for the proposed type-2 fuzzy decision trees, especially in the case of the thyroid diagnosis problem.

7 Conclusions

In this paper, an algorithm for creating type-2 fuzzy decision trees is presented. This algorithm allows to use type-2 fuzzy sets as attribute values. The type-2 fuzzy sets, that in the proposed algorithm have to be generated before the tree is created, are obtained by the modification of the double fuzzy clustering method. This method allows to get the type-2 fuzzy sets that are easy to read and interpret, and results in understandable type-2 fuzzy decision trees and type-2 fuzzy rules. In this paper, some experimental results are also presented, in comparison with results produced by classic fuzzy decision trees proposed by Janikow [14]. Based on the obtained results, we can say that the type-2 fuzzy decision trees are chracterized by lower classification error than the classic type-1 fuzzy decision trees.

References

1. Adamo, J.M.: Fuzzy decision trees. Fuzzy Sets and Systems 4, 207–219 (1980)
2. Bezdek, C.: Pattern Recognition with Fuzzy Objective Function Algorithms. Plenum Press, New York (1981)
3. Bartczuk, Ł., Rutkowska, D.: The new version of Fuzzy-ID3 algorithm. In: Rutkowski, L., Tadeusiewicz, R., Zadeh, L.A., Żurada, J.M. (eds.) ICAISC 2006. LNCS (LNAI), vol. 4029, pp. 1060–1070. Springer, Heidelberg (2006)
4. Bartczuk Ł., Rutkowska D., Fuzzy decision trees of type-2, in: Some Aspects of Computer Science. EXIT Academic Publishing House, Warsaw, Poland (2007) (in Polish)
5. Bilski, J.: The UD RLS algorithm for training feedforward neural networks. International Journal of Applied Mathematics and Computer Science 15(1), 115–123 (2005)
6. Blake, C., Keogh, E., Merz, C.: UCI repository of machine learning databases
7. Canfora, G., Troiano, L.: Fuzzy ordering of fuzzy numbers. In: Proc. Fuzz-IEEE, Budapest, Ungheria, pp. 669–674 (2004)
8. Castellano, G., Fanelli, A.M., Mencar, C.: A double-clustering approach for interpretable granulation of data. In: Proc. IEEE International Conference on Systems, Man and Cybernetics, pp. 483–487 (2002)
9. Czekalski, P.: Evolution-Fuzzy rule based system with parameterized consequences. International Journal of Applied Mathematics and Computer Science 16(3), 373–385 (2006)
10. Dubois, D., Prade, H.: Fuzzy Sets and Systems: Theory and Applications. Academic Press, San Diego (1980)
11. Haykin, S.: Neural Networks: A Comprehensive Foundation, Macmilan (1994)
12. Hwang., C., Rhee, F.C.-H.: Uncertain fuzzy clustering: interval type-2 fuzzy approach to C-Means. IEEE Transactions on Fuzzy Systems 15(1), 107–120 (2007)

13. Jager, R.: Fuzzy Logic in Control, Ph.D. Dissertation, Technische Universiteit Delft (1995)
14. Janikow, C.Z.: Fuzzy decision trees: issues and methods. IEEE Transactions on Systems, Man, and Cybernetics 28(3), 1–14 (1998)
15. Janikow, C.Z.: Exemplar learning in fuzzy decision trees. In: Proc. IEEE International Conference on Fuzzy Systems, Piscataway, NJ, pp. 1500–1505 (1996)
16. Lęski, J., Henzel, N.: A neuro-nuzzy system based on logical interpretation of if-then rules. International Journal of Applied Mathematics and Computer Science 10(4), 703–722 (2000)
17. Mendel, J.M.: Uncertain Rule-Based Fuzzy Logic Systems - Introduction and new directions. Prentice Hall PTR, Englewood Cliffs (2001)
18. Quinlan, J.R.: Induction of decision trees. Machine Learning 1, 81–106 (1986)
19. Quinlan, J.R.: C4.5: Programs for Machine Learning. Morgan Kaufmann Publishers, Inc., Los Altos (1993)
20. Quinlan, J.R.: Learning with continuous classes. In: Proc. 5th Australian Joint Conference on Artificial Intelligence, pp. 343–348. World Scientific, Singapore (1992)
21. Piegat, A.: Fuzzy Modeling and Control. Physica-Verlag (2001)
22. Rutkowska, D.: Neuro-Fuzzy Architectures and Hybrid Learning. Physica-Verlag, Springer, New York (2002)
23. Rutkowska, D., Nowicki, R.: Implication-based neuro-fuzzy architectures. International Journal of Applied Mathematic and Computer Science 10(4), 675–701 (2000)
24. Rutkowski, L.: Methods and Techniques of Artificial Intelligence, PWN, Warsaw, Poland (in Polish) (2005)
25. Yager, R.R.: Ranking fuzzy subsets over the unit interval. In: Proc. CDC pp. 1435–1437 (1978)
26. Zadeh, L.A.: Fuzzy sets. Information and Control 8, 338–353 (1965)
27. Zadeh, L.A.: The concept of a linguistic variable and its application to approximate reasoning - I. Information Sciences 8, 199–249 (1975)
28. Żurada, J.M.: Introduction to Artificial Nueral Systems. West Publishing Company (1992)

An Application of Weighted Triangular Norms to Complexity Reduction of Neuro-fuzzy Systems

Krzysztof Cpalka[1,2] and Leszek Rutkowski[1,2]

[1] Czestochowa University of Technology, Poland
Department of Computer Engineering
[2] Academy of Humanities and Economics, Poland
Department of Artificial Intelligence
lrutko@kik.pcz.czest.pl

Abstract. In the paper we develop a new method for designing and reduction of neuro-fuzzy systems. The method is based on the concept of the weighted triangular norms. In subsequent stages we reduce number of inputs, number of rules and number of antecedents. Simulation results are given.

1 Introduction

In literature various neuro-fuzzy systems have been developed (see e.g. [4]-[6], [9]-[10], [13]-[15]). They combine the natural language description of fuzzy systems and the learning properties of neural networks. Recently several algorithms have been proposed to increase interpretability and accuracy and decrease complexity of fuzzy rule-based systems. For various methods of designing fuzzy rule-based systems the reader is referred to [1]-[3], [7]-[12].

In this paper we propose a new method to design and reduce neuro-fuzzy systems. We will consider multi-input, single-output neuro-fuzzy system of the logical type (see e.g. [13], [14]), mapping $\mathbf{X} \to \mathbf{Y}$, where $\mathbf{X} \subset \mathbf{R}^n$ and $\mathbf{Y} \subset \mathbf{R}$. The fuzzy rule base of these systems consists of a collection of N fuzzy IF-THEN rules in the form

$$R^k : \left[\text{IF } x_1 \text{ is } A_1^k \text{ AND} \ldots \text{AND } x_n \text{ is } A_n^k \text{ THEN } y \text{ is } B^k \right], \qquad (1)$$

where $\mathbf{x} = [x_1, \ldots, x_n] \in \mathbf{X}$, $y \in \mathbf{Y}$, $A_1^k, A_2^k, \ldots, A_n^k$ are fuzzy sets characterized by membership functions $\mu_{A_i^k}(x_i)$, $\mathbf{A}^k = A_1^k \times A_2^k \times \ldots \times A_n^k$, and B^k are fuzzy sets characterized by membership functions $\mu_{B^k}(y)$, respectively, $k = 1, \ldots, N$, and N is the number of rules. The defuzzification is realized by the COA (centre of area) method defined by the following formula

$$\bar{y} = \sum_{r=1}^{R} \bar{y}_r^B \mu_{B'}\left(\bar{y}_r^B\right) / \sum_{r=1}^{R} \mu_{B'}\left(\bar{y}_r^B\right), \qquad (2)$$

where B' is the fuzzy set obtained from the linguistic model (1), using an appropriate fuzzy reasoning, $\bar{y}_r^B$ denotes centres of the output membership functions $\mu_{B^r}(y)$, i.e. for $r = 1, \ldots, R$,

$$\mu_{B^r}\left(\bar{y}_r^B\right) = \max_{y \in \mathbf{Y}} \{\mu_{B^r}(y)\}. \tag{3}$$

and R is the number of discretization points of the integrals in the continuous version of the COA method.

2 A Novel Approach to Neuro-fuzzy Modeling

Let us introduce weights $w_{i,k}^\tau \in [0,1]$, $k = 1, \ldots, N$, $i = 1, \ldots, n$, describing importance of antecedents and weights $w_k^{\text{agr}} \in [0,1]$, $k = 1, \ldots, N$, describing importance of rules. Now the linguistic model (1) is transformed to the following description

$$R^{(k)} : \left[\text{IF} x_1 \text{is} A_1^k \left(w_{1,k}^\tau\right) \text{AND} \ldots \text{AND} x_n \text{is} A_n^k \left(w_{n,k}^\tau\right) \text{THEN} y \text{is} B^k \right] \left(w_k^{\text{agr}}\right). \tag{4}$$

In order to incorporate weights into description of neuro-fuzzy systems (2) we proposed [14] the adjustable weighted t-norm

$$\vec{T}^* \left\{ a_{1,k}, \ldots, a_{n,k}; w_{1,k}^\tau, \ldots, w_{n,k}^\tau, p_k^\tau \right\} = \vec{T}^* \left\{ \mathbf{a}_k; \mathbf{w}_k^\tau, p_k^\tau \right\} = \\ \vec{T} \left\{ 1 - w_{1,k}^\tau \left(1 - a_{1,k}\right), \ldots, 1 - w_{n,k}^\tau \left(1 - a_{n,k}\right); p_k^\tau \right\} \tag{5}$$

to connect the antecedents in each rule, $k = 1, \ldots, N$, and the adjustable weighted t-norm

$$\vec{T}^* \left\{ a_1, \ldots, a_N; w_1^{\text{agr}}, \ldots, w_N^{\text{agr}}, p^{\text{agr}} \right\} = \vec{T}^* \left\{ \mathbf{a}; \mathbf{w}^{\text{agr}}, p^{\text{agr}} \right\} = \\ \vec{T} \left\{ 1 - w_1^{\text{agr}} \left(1 - a_1\right), \ldots, 1 - w_N^{\text{agr}} \left(1 - a_N\right); p^{\text{agr}} \right\} \tag{6}$$

to aggregate the individual rules in the logical models, respectively. In formula (5) parameters $a_{i,k}$, $i = 1, \ldots, n$, $k = 1, \ldots, N$, correspond to the values of $\mu_{A_i^k}(\bar{x}_i)$, whereas parameters a_k, $k = 1, \ldots, N$, in formula (6) correspond to the values of $\mu_{\bar{B}^k}(\bar{y}_r^B)$. It is easily seen that formula (5) can be applied to the evaluation of an importance of input linguistic values, and the weighted t-norm (6) to a selection of important rules. We use notation $\vec{T}^* \{a_1, \ldots, a_n; w_1, \ldots, w_n, p\}$ and $\vec{S}^* \{a_1, \ldots, a_n; w_1, \ldots, w_n, p\}$ for adjustable weighted triangular norms, and notation $\vec{T}^* \{a_1, \ldots, a_n; p\}$ and $\vec{S}^* \{a_1, \ldots, a_n; p\}$ for adjustable triangular norms. The hyperplanes corresponding to to them can be adjusted in the process of learning of an appropriate parameter p.

In the paper we propose the adjustable version of an S-implication

$$I_{fuzzy}\left(\mu_{A^k}(\bar{\mathbf{x}}), \mu_{B^k}(y)\right) = \vec{S}^* \left\{ 1 - \mu_{A^k}(\bar{\mathbf{x}}), \mu_{B^k}(y); p_k^I \right\}. \tag{7}$$

In our study we use Dombi families [13] of adjustable triangular norms.

Neuro-fuzzy architectures developed so far in the literature are based on the assumption that number of terms R in a formula (2) is equal to the number of rules N. In view of the above assumptions formula (2) takes the form

$$\bar{y} = f(\bar{\mathbf{x}}) = \frac{\sum_{r=1}^{R} \bar{y}^r \overset{N}{\underset{k=1}{\vec{T}^*}} \left\{ \vec{S} \left\{ 1 - \vec{T}^* \left\{ \begin{array}{c} \mu_{A_1^k}(\bar{x}_1), \ldots, \mu_{A_N^k}(\bar{x}_N); \\ w_{1,k}^\tau, \ldots, w_{n,k}^\tau, p_k^\tau \end{array} \right\}, \\ \mu_{B^k}(\bar{y}^r); p_k^I \\ p^{\mathrm{agr}} \end{array} \right\} \right\}}{\sum_{r=1}^{R} \overset{N}{\underset{k=1}{\vec{T}^*}} \left\{ \vec{S} \left\{ 1 - \vec{T}^* \left\{ \begin{array}{c} \mu_{A_1^k}(\bar{x}_1), \ldots, \mu_{A_N^k}(\bar{x}_N); \\ w_{1,k}^\tau, \ldots, w_{n,k}^\tau, p_k^\tau \end{array} \right\}, \\ \mu_{B^k}(\bar{y}^r); p_k^I \\ p^{\mathrm{agr}} \end{array} \right\} \right\}}. \quad (8)$$

The following parameters of system (8) are subject to learning by using the backpropagation method: (i)Parameters $\bar{x}_{i,k}^A$ and $\sigma_{i,k}^A$ of input fuzzy sets A_i^k, $k = 1, \ldots, N$, $i = 1, \ldots, n$, and parameters $\bar{y}_k^B$ and σ_k^B of output fuzzy sets B^k, $k = 1, \ldots, N$, (ii)Weights $w_{i,k}^\tau \in [0,1]$, $k = 1, \ldots, N$, $i = 1, \ldots, n$, of importance of antecedents and weights $w_k^{\mathrm{agr}} \in [0,1]$, $k = 1, \ldots, N$, of importance of rules, (iii)Parameters p^{agr}, p_k^I, p_k^τ, $k = 1, \ldots, N$, of adjustable triangular norms used for aggregation of rules, connections of antecedents and consequences and aggregation of antecedents, respectively, (iv)Discretization points $\bar{y}^r$, $r = 1, \ldots, R$.

3 Reduction of Neuro-fuzzy System (8)

Now we will develop new algorithms of reduction of neuro-fuzzy system (8). The algorithms are based on analysis of weights in antecedents of the rules $w_{i,k}^\tau \in [0,1]$, $i = 1, \ldots, n$, $k = 1, \ldots, N$, and weights in aggregation of the rules $w_k^{\mathrm{agr}} \in [0,1]$, $k = 1, \ldots, N$.

The flowcharts in Fig. 1.a and Fig. 1.b comprise 4 parts. First, we determine performance of the initial system (before the reduction process); for example, in the case of the classification we determine a percentage of mistakes of the system. The weights $w_i^x \in [0,1]$, $i = 1, \ldots, n$, are calculated using

$$w_i^x = \frac{1}{N} \sum_{k=1}^{N} w_{i,k}^\tau. \quad (9)$$

In consecutive stages we reduce number of discretization points, number of inputs, number of rules and number of antecedents. If, e.g. reduction of the i-th input is acceptable (i.e. it does not worsen the system accuracy determined before reduction) then that input is eliminated, otherwise we do not reduce it.

The algorithm of consecutive eliminations (CEA) is oriented to the most possible simplification of the system structure. Its idea is based on consecutive eliminations of contradicted, non-active and unimportant elements of the

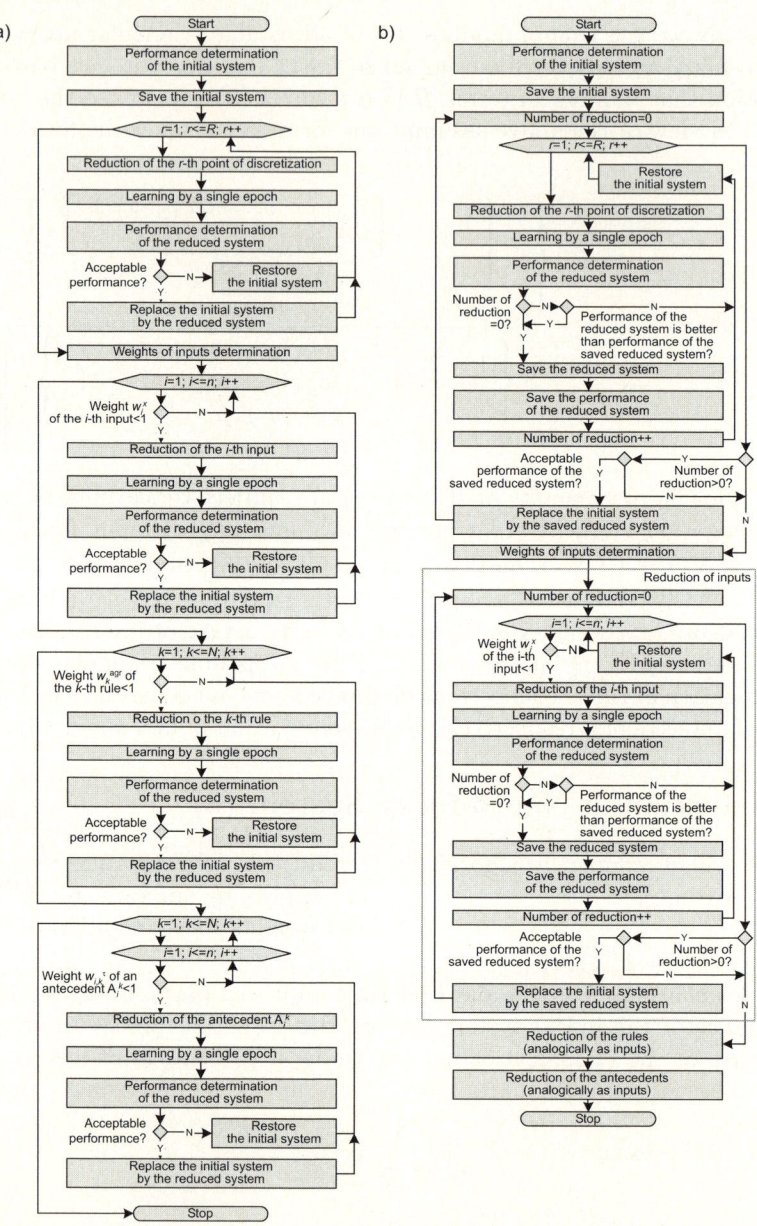

Fig. 1. Algorithms of reduction of neuro-fuzzy system (8): a) consecutive eliminations algorithm (CEA), b) algorithm of the best local eliminations (ABLE)

system starting from discretization points, and next inputs, whole rules and, finally, antecedents of rules. If a specific reduction, e.g. reduction of a concrete input, is acceptable (accuracy of the system is not worse than before the

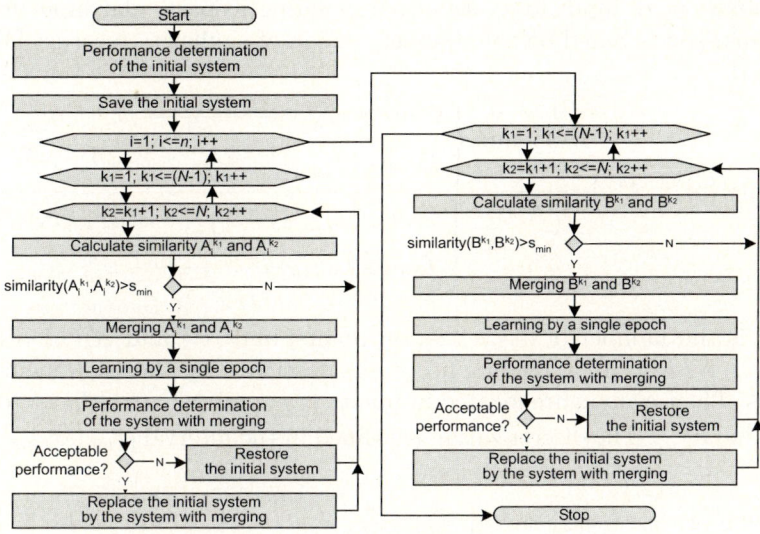

Fig. 2. Consecutive mergings algorithm (CMA)

reduction), then the reduction is accepted, otherwise it is cancelled. The flowchart of the algorithm is depicted in Fig. 1.a.

The algorithm of the best local eliminations (ABLE) is oriented to the most possible simplification of the system structure with simultaneous increase accuracy of the system. Its idea is based on finding within each group of parameters (discretization points, inputs, whole rules and antecedents of rules) such element that assures the best reduction from the view point of the system accuracy. If such an element is found, the reduction is performed and the search is repeated within the same group of parameters, otherwise another group of parameters is analyzed. This idea takes into account the fact that reduction of the next rule may have better influence on the system accuracy than reduction of the current rule. The flowchart of the algorithm is depicted in Fig. 1.b.

4 Merging of Similar Input and Output Fuzzy Sets in Neuro-fuzzy System (8)

In point 3 we eliminated in system (8) such elements as discretization points, input features, rules and antecedents. The elimination of these elements did not worsen the accuracy of system (8) and simultaneously increased its transparency and decreased computational burden. Despite of the reduction, in the linguistic model still exist similar fuzzy sets. Such fuzzy sets should be automatically detected, merged and shared by different rules.

In Fig. 2 we present the algorithm of consecutive mergings algorithm (CMM). The algorithm is initialized by the performance determination (number of correctly classified samples) of system (8) before merging. Next, we compare all combinations of input fuzzy sets corresponding to particular input features. The comparison is based on the discreet version of similarity measure [3]

$$\text{similarity}\left(A_i^{k_1}, A_i^{k_2}\right) = \frac{\sum_{j=0}^{J-1} \min \left\{ \begin{array}{l} \text{Gauss}\left(\bar{x}_{\min}+j\frac{(\bar{x}_{\max}-\bar{x}_{\min})}{J-1}; \bar{x}_{i,k_1}^A, \sigma_{i,k_1}^A\right), \\ \text{Gauss}\left(\bar{x}_{\min}+j\frac{(\bar{x}_{\max}-\bar{x}_{\min})}{J-1}; \bar{x}_{i,k_2}^A, \sigma_{i,k_2}^A\right) \end{array} \right\}}{\sum_{j=0}^{J-1} \max \left\{ \begin{array}{l} \text{Gauss}\left(\bar{x}_{\min}+j\frac{(\bar{x}_{\max}-\bar{x}_{\min})}{J-1}; \bar{x}_{i,k_1}^A, \sigma_{i,k_1}^A\right), \\ \text{Gauss}\left(\bar{x}_{\min}+j\frac{(\bar{x}_{\max}-\bar{x}_{\min})}{J-1}; \bar{x}_{i,k_2}^A, \sigma_{i,k_2}^A\right) \end{array} \right\}},$$
(10)

where J is the number of discretization points, and $A_i^{k_1}$ and $A_i^{k_2}$, $i = 1, \ldots, i$, $k_1 = 1, \ldots, N$, $k_2 = 1, \ldots, N$, are fuzzy sets described by Gaussian membership functions. Their centers are located in points $\bar{x}_{i,k_1}^A$ and $\bar{x}_{i,k_2}^A$, widths are denoted by σ_{i,k_1}^A and σ_{i,k_2}^A. The discretization is defined in the interval $[\bar{x}_{\min}, \bar{x}_{\max}]$, where

$$\bar{x}_{\min} = \min\{\bar{x}_{\min_1}, \bar{x}_{\min_2}\} = \min\left\{\bar{x}_{i,k_1}^A - \sigma_{i,k_1}^A \sqrt{-\ln(\psi)}, \bar{x}_{i,k_2}^A - \sigma_{i,k_2}^A \sqrt{-\ln(\psi)}\right\} \quad (11)$$

and

$$\bar{x}_{\max} = \max\{\bar{x}_{\max_1}, \bar{x}_{\max_2}\} = \max\left\{\bar{x}_{i,k_1}^A + \sigma_{i,k_1}^A \sqrt{-\ln(\psi)}, \bar{x}_{i,k_2}^A + \sigma_{i,k_2}^A \sqrt{-\ln(\psi)}\right\}. \quad (12)$$

In our simulations $J = 100$ and $\psi = 0.01$. Values $\bar{x}_{\min_1}$, $\bar{x}_{\min_2}$, $\bar{x}_{\max_1}$, and $\bar{x}_{\max_2}$ in formulas (11) and (12) follow from the solution of equations

$$\begin{cases} \text{Gauss}\left(\bar{x}; \bar{x}_{i,k_1}^A, \sigma_{i,k_1}^A\right) = \psi \Rightarrow \{\bar{x}_{\min_1}, \bar{x}_{\max_1}\} \\ \text{Gauss}\left(\bar{x}; \bar{x}_{i,k_2}^A, \sigma_{i,k_2}^A\right) = \psi \Rightarrow \{\bar{x}_{\min_2}, \bar{x}_{\max_2}\} \end{cases} \quad (13)$$

with respect to $\bar{x}$.

For each pair of input fuzzy sets we determine the value of similarity measure given by formula (10). If that value exceeds the threshold $s_{\min}$ (in our simulations in Section 5 we assume that $s_{\min} = 0.5$) then the input fuzzy sets are merged. More precisely, Gaussian fuzzy sets $A_i^{k_1}$ and $A_i^{k_2}$ are replaced by fuzzy set A_i^k, which is also Gaussian with the center and width given by

$$\bar{x}_{i,k}^A = \frac{\bar{x}_{i,k_1}^A \cdot w_{i,k_1}^\tau + \bar{x}_{i,k_2}^A \cdot w_{i,k_2}^\tau}{w_{i,k_1}^\tau + w_{i,k_2}^\tau} \quad (14)$$

and

$$\sigma_{i,k}^A = \frac{\sigma_{i,k_1}^A \cdot w_{i,k_1}^\tau + \sigma_{i,k_2}^A \cdot w_{i,k_2}^\tau}{w_{i,k_1}^\tau + w_{i,k_2}^\tau}. \quad (15)$$

In formulas (14) and (15) we take into account the importance of merged antecedents, described by values of weights w_{i,k_1}^τ and w_{i,k_2}^τ. The importance of the antecedent A_i^k being a result of merging antecedents $A_i^{k_1}$ and $A_i^{k_2}$ is described by

$$w_{i,k}^\tau = \frac{w_{i,k_1}^\tau + w_{i,k_2}^\tau}{2}. \qquad (16)$$

After each merging a simplified neuro-fuzzy system is trained by a single epoch and then tested. Testing enables to evaluate the influence of merging on accuracy of the simplified system. If the merging does not worsen the accuracy then the simplified system replaces the previous one. Otherwise, the merging is canceled and the initial system is restored. This procedure is performed for all combinations of antecedents corresponding to all input features.

In a similar way we merge the output fuzzy sets B^k, $k = 1, \ldots, N$. The only difference is that the centers and widths of merged fuzzy sets are given by

$$\bar{y}_k^B = \frac{\bar{y}_{k_1}^B + \bar{y}_{k_2}^B}{2} \qquad (17)$$

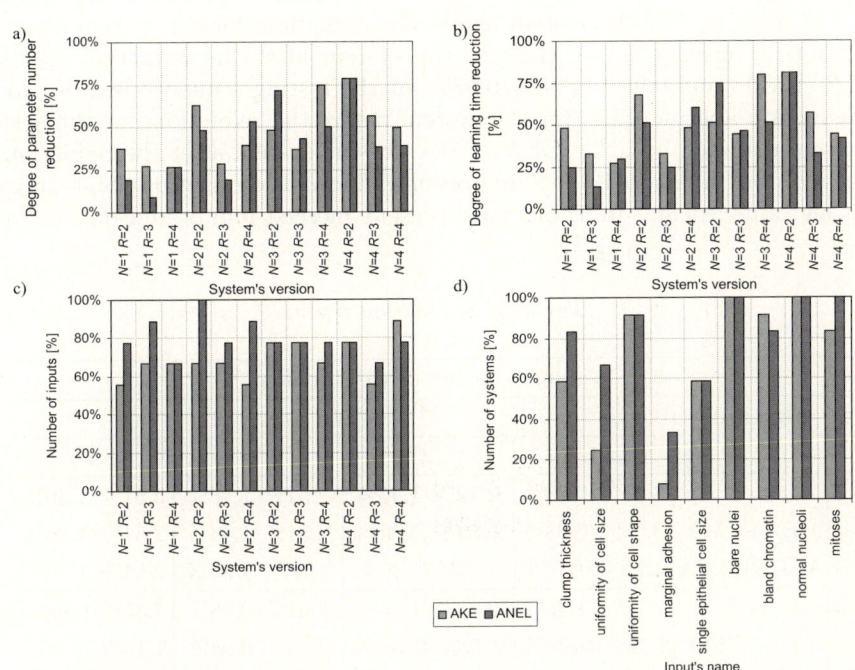

Fig. 3. Simulations results: a)degree of parameter number reduction [%], b)degree of learning time reduction [%], c)percentage of neuro-fuzzy systems having a particular input (attribute) after the reduction process, d)percentage of inputs (attributes) corresponding to a particular neuro-fuzzy system after the reduction process

and

$$\sigma_k^B = \frac{\sigma_{k_1}^B + \sigma_{k_2}^B}{2}. \tag{18}$$

The procedure described in points 3 and 4 leads to reduced (Fig. 1) and simplified (Fig. 2) fuzzy model which is less complex and more understandable than the initial system (8).

5 Simulations Results

The neuro-fuzzy system (8) is simulated on Wisconsin breast cancer problem (UCI respository of machine learning databases, Available online: *http://ftp.ics.uci.edu/pub/machine-learning-databases/*). The experimental results for the Wisconsin breast cancer problem are depicted in Table 1 and Fig. 1. In Table 1 we show the percentage of mistakes in the learning and testing sequences before and after reduction and merging, e.g. for $N = 2$ and $R = 2$ we have 3.35%/3.35%/2.72% for the learning sequence before and after reduction and merging (for the system before the reduction, for the system after the reduction by using CEA+CMA, for the system after the reduction by using ABLE+CMA) and 0.98%/0.98%/0.98% for the testing sequence before and after reduction and merging (for the system before the reduction, for the system after the reduction by using CEA+CMA, for the system after the reduction by using ABLE+CMA). In Table 2 we present reduced discretization points, inputs, rules and antecedents. In Table 3 we present merged inputs and output fuzzy sets.

Table 1. Simulation results

R	Algorithms	Wisconsin breast cancer problem			
		N			
		1	2	3	4
2	-	3.35%/0.98%	2.72%/1.46%	2.51%/1.46%	2.51%/1.46%
	CEA+CMA	3.35%/0.98%	**2.51%**/1.46%	2.51%/1.46%	2.51%/1.46%
	ABLE+CMA	**2.72%**/0.98%	2.72%/1.46%	2.51%/**0.98%**	2.51%/1.46%
3	-	2.72%/1.46%	2.72%/1.46%	2.51%/0.98%	3.35%/0.98%
	CEA+CMA	2.72%/**0.98%**	2.72%/**0.98%**	2.51%/0.98%	**3.14%**/0.98%
	ABLE+CMA	**2.51%**/1.46%	**2.51%**/**0.98%**	**2.30%**/0.98%	**2.51%**/0.98%
4	-	2.51%/1.46%	2.51%/1.46%	2.72%/1.46%	2.51%/0.98%
	CEA+CMA	2.51%/1.46%	2.51%/**0.98%**	2.51%/1.46%	2.51%/0.98%
	ABLE+CMA	2.51%/**0.98%**	**2.30%**/0.98%	2.51%/0.98%	**2.30%**/0.98%

Table 2. Simulation results

R	Algorithms	Wisconsin breast cancer problem N			
		1	2	3	4
2	CEA	$\bar{x}_2, \bar{x}_4, \bar{x}_5$	$\bar{x}_1, \bar{x}_2, \bar{x}_4, rule_2$	$\bar{x}_1, \bar{x}_4, rule_2$	$\bar{x}_1, \bar{x}_4, rule_2, rule_3, rule_4$
	ABLE	$\bar{x}_4, \bar{x}_5$	$rule_2$	$\bar{x}_2, \bar{x}_4, rule_2, rule_3$	$\bar{x}_1, \bar{x}_4, rule_2, rule_3, rule_4$
3	CEA	$\bar{x}_2, \bar{x}_4, \bar{x}_5$	$\bar{x}_2, \bar{x}_4, \bar{x}_5$	$\bar{x}_4, \bar{x}_7, A_5^1, A_2^2, A_6^2, \bar{y}^2$	$\bar{x}_2, \bar{x}_3, \bar{x}_4, \bar{x}_5, rule_2$
	ABLE	$\bar{x}_4$	$\bar{x}_2, \bar{x}_5$	$\bar{x}_4, \bar{x}_7, A_2^2, A_6^2, A_9^2, A_1^3, \bar{y}^2$	$\bar{x}_2, \bar{x}_4, \bar{x}_5$
4	CEA	$\bar{x}_1, \bar{x}_2, \bar{x}_4$	$\bar{x}_2, \bar{x}_4, \bar{x}_5, \bar{x}_9, \bar{y}^3$	$\bar{x}_1, \bar{x}_2, \bar{x}_4, rule_2, rule_3, \bar{y}^3$	$\bar{x}_2, rule_3, \bar{y}^4$
	ABLE	$\bar{x}_2, \bar{x}_4, \bar{x}_7$	$\bar{x}_2, rule_2, \bar{y}^3$	$\bar{x}_4, \bar{x}_5, A_1^2, rule_2$	$\bar{x}_4, \bar{x}_5, rule_3$

Table 3. Simulation results

R	Algorithms	Wisconsin breast cancer problem N			
		1	2	3	4
2	CEA+CMA	-	-	$B^1 + B^2$	-
	ABLE+CMA	-	-	-	-
3	CEA+CMA	-	-	$A_1^2+A_1^3, A_7^2+A_7^3, B^2+B^3$	$A_1^2+A_1^3$
	ABLE+CMA	-	-	$A_4^1+A_4^3, A_6^2+A_6^3, B^2+B^3$	$A_1^2+A_1^3+A_1^4, A_2^2+A_2^3+A_2^4$
4	CEA+CMA	-	-	-	$A_1^2+A_1^3,\ A_3^2+A_3^3, A_4^2+A_4^3,\ A_5^2+A_5^3, A_6^2+A_6^3, A_8^1+A_8^2$
	ABLE+CMA	-	-	$B^1 + B^2$	$A_1^2+A_1^3$

6 Conclusions

In the paper we described a new methods for designing and reduction of neuro-fuzzy systems. From simulations it follows that the reduction process of neuro-fuzzy structures based on adjustable weighted triangular norms do not worsen the performance of these structures. It was possible to quickly detect the inputs which can be eliminated. Our methods allows to decrease the number of parameters in neuro-fuzzy structures and consequently the learning time.

Acknowledgment

This work was partly supported by the Foundation for Polish Science (Professorial Grant 2005-2008) and Polish Ministry of Science and Higher Education (Habilitation Project 2007-2010, Special Research Project 2006-2009, Polish-Singapore Research Project 2008-2010 and Research Project 2008-2010).

References

1. Alonso, J.M., Cordon, O., Guillaume, S., Magdalena, L.: Highly Interpretable Linguistic Knowledge Bases Optimization: Genetic Tuning versus Solis-Wetts. Looking for a good interpretability-accuracy trade-off. In: Proc. of the 2007 IEEE Int. Conf. on Fuzzy Systems, pp. 1–6 (2007)
2. Amaral, T.G., Crisostomo, M.M.: An Approach to Improve the Interpretability of Neuro-Fuzzy Systems. In: Proc. of the 2006 IEEE Int. Conf. on Fuzzy Systems, pp. 1843–1850 (2006)
3. Casillas, J., Cordon, O., Herrera, F., Magdalena, L. (eds.): Interpretability Issues in Fuzzy Modeling. Springer, Heidelberg (2003)
4. Czabanski, R.: Neuro-Fuzzy Modelling Based on a Deterministic Annealing Approach. Int. J. Appl. Math. Comput. Sci. 15(4), 561–576 (2005)
5. Czogała, E., Łęski, J.: Fuzzy and Neuro-Fuzzy Intelligent Systems. Physica-Verlag, Heidelberg (2000)
6. Gorzałczany, M.: Computational Intelligence Systems and Applications: Neuro-Fuzzy and Fuzzy Neural Synergisms. Springer, Heidelberg (2002)
7. Guillaume, S.: Designing fuzzy inference systems from data: An interpretability-oriented review. IEEE Trans. Fuzzy Syst. 9(3), 426–443 (2001)
8. Kumar, M., Stoll, R., Stoll, N.: A robust design criterion for interpretable fuzzy models with uncertain data. IEEE Trans. Fuzzy Syst. 14(2), 314–328 (2006)
9. Łęski, J., Henzel, N.: A Neuro-Fuzzy System Based on Logical Interpretation of If-then Rules. Int. J. Appl. Math. Comput. Sci. 10(4), 703–722 (2000)
10. Łęski, J.: A Fuzzy If-Then Rule-Based Nonlinear Classifier. Int. J. Appl. Math. Comput. Sci. 13(2), 215–223 (2003)
11. Manley-Cooke, P., Razaz, M.: An efficient approach for reduction of membership functions and rules in fuzzy systems. In: Proc. of the 2007 IEEE Int. Conf. on Fuzzy Systems, pp. 1–6 (2007)
12. Riid, A., Rustern, E.: Interpretability of Fuzzy Systems and Its Application to Process Control. In: Proc. of the 2007 IEEE Int. Conf. on Fuzzy Systems, pp. 1–6 (2007)
13. Rutkowski, L.: Flexible Neuro-Fuzzy Systems. Kluwer Academic Publishers, Dordrecht (2004)
14. Rutkowski, L., Cpałka, K.: Flexible neuro-fuzzy systems. IEEE Trans. Neural Networks 14(3), 554–574 (2003)
15. Yager, R.R., Filev, D.P.: Essentials of fuzzy modelling and control. John Wiley & Sons, Chichester (1994)

Real-Time Road Signs Tracking with the Fuzzy Continuously Adaptive Mean Shift Algorithm

Bogusław Cyganek

AGH - University of Science and Technology,
Al. Mickiewicza 30, 30-059 Kraków, Poland
cyganek@uci.agh.edu.pl

Abstract. Tracking of multiple objects belongs to one of the fundamental tasks of computer vision. In this paper an improvement to the continuously adaptive mean shift tracking method is proposed. It consists in substitution of the probabilistic density function for the especially formed membership function. This makes possible design of tracking systems in terms of fuzzy logic. Additionally, a special data structure was developed to allow tracking of multiple objects at a time. It stores information on image regions which are active for tracking. By this it provides initial conditions for tracking in subsequent frames which also speeds up computations. The method was used and verified in an application of the road signs tracking in real time of 30 frames/s.

1 Introduction

Objects tracking is one of the key issues in computer vision. The important factors of tracking methods are accuracy, speed and ability to track multiple objects at a time. These conditions are met by the continuously adaptive mean shift method, originally proposed by Fukunaga et al. [11], and its CamShift version designed for face tracking by Bradski [1]. The method was also analyzed and extended by Cheng et al [2], Comaniciu et al. [3-4], to name a few. All of these approaches are based on estimation of the underlying probability distribution function (pdf). However, in many application a more robust approach could be developed with help of the fuzzy methods [21]. Such a modification is proposed in this paper. Instead of pdf a properly designed fuzzy membership function is used. For this purpose two methods are proposed. Both are build from the membership functions characteristic to the tracked objects. These issues are discussed in section (2) of this paper.

The method has been applied to the road sign (RS) tracking system, in which it belongs to the detection front end. RS recognition is a rapidly growing research area for automotive applications. Real-time recognition of RSs is becoming a part of the Driver Assisting Systems (DAS) built into intelligent vehicles. Their goal is to increase safety, comfort and economy of driving. Much effort has been devoted to build such systems. For example Daimler-Chrysler designed the 'thinking vehicle' system [7], Siemens a system which is able to recognize speed limits [21], etc.

There are some publications on RS detection and tracking systems. An example of this is work by Fang et al. [9]. Their system detects places with colour and shape characteristic to the RSs with help of the neural networks. Then, a fuzzy method is used for feature integration and detection. Detected objects are verified and tracked through consecutive frames based on the Kalman filter. However, the system is tuned to detect only shapes of a certain size and specific camera configuration. Also the applied neural networks are adjusted to detect only non distorted shapes.

To the best of our knowledge, all other known RS recognition systems employ a single view detection followed by a classification step without tracking. For instance, the system developed by Maldonado-Bascón et al. [16] is based on the generalization properties of the Support Vector Machine (SVM) classifiers. There are three stages of execution: HS colour segmentation, RS detection from shape with linear SVM, then the classification with Gaussian kernel SVM. The system exhibits some invariants to rotation, change of scale and position. However, the used one-versus-all SVMs with about 50 training patterns for each class make the training phase cumbersome.

The other system, proposed by Gao et al. [10], employs the CIECAM97 colour space model which is based on the human visual system. It provides colour information which is much more independent on external conditions such as lighting or weather. For shape detection the Foveal System for Traffic Signs (FOSTS), based on human behavioural model of vision, is proposed. Then the kind of a nearest-neighbour method, operating with the sign features, is employed for classification.

A solution presented by Paclik et al. [17] is based on a generic cross-correlation method. The authors propose correlation only of RS characteristic regions, instead of the whole images. The characteristic regions are obtained from a special classifier trained with real RS examples. Comparison results and discussion of different classifiers, such as the one-nearest-neighbour, Fisher and so called SIMCA, are also provided in [17].

However neither of the above three systems employs object tracking. On the other hand, a video stream can offer more information due to a strong correlation among successive frames. This ability is employed in tracking systems, as in the one presented in this paper, and allows much more reliable detection and classification of objects. Even if a sign is not reliably detected or totally missed in a frame, then suitable information can be recovered thanks to the next frames in the video stream. In this paper we propose a RSs tracking system with the continuously adaptive mean shift method adapted to use the fuzzy measure instead of pdf. The system operates in real-time of 30 frames/s.

2 The Fuzzy Continuously Adaptive Mean Shift Method

The mean shift algorithm was first proposed by Fukunaga [11] as a non-parametric method of tracing mode of a distribution, i.e. a location of the maximal probability of a searched object. For this purpose the algorithm climbs the density gradient.

Thus, the method at first necessitates assessment of the density function. However, as we will show in the next section, the method works well for models of objects defined by fuzzy rules, which not necessarily fulfils the probability conditions.

Estimation of a density function from a sample $\{x_i\}$ of M points, in an N dimensional space R^N, can be done by means of a non parametric method (so called Parzen window) with kernel K, as follows:

$$\tilde{f}_{(\delta,K)}(\mathbf{x}) = \frac{1}{M\delta^N} \sum_{j=1}^{M} K\left(\frac{\mathbf{x}-\mathbf{x}_j}{\delta}\right), \qquad (1)$$

where δ denotes size of a kernel window, called also a bandwidth. The mean shift procedure climbs the estimated density gradient by recursive computation of the mean shift vector and translation of the centre of a mean shift kernel G. Thus, the locations of the kernel G are changed at each iteration step by a mean shift vector $\mathbf{m}$, forming a sequence $\{\mathbf{y}_t\}$, as follows [3]:

$$\mathbf{m}_{(\delta,G)}(\mathbf{y}_t) = \mathbf{y}_{t+1} - \mathbf{y}_t = \frac{\sum_{j=1}^{M} \mathbf{x}_j g\left(\left\|\frac{\mathbf{y}_t-\mathbf{x}_j}{\delta}\right\|^2\right)}{\sum_{j=1}^{M} g\left(\left\|\frac{\mathbf{y}_t-\mathbf{x}_j}{\delta}\right\|^2\right)} - \mathbf{y}_t, \text{ for } t=1,2,\ldots, \qquad (2)$$

where $\mathbf{m}_{(\delta,G)}(\mathbf{y}_t)$ denotes the mean shift vector for the kernel G, which is related to the profile g as follows (c is a normalization constant):

$$G(\mathbf{x}) = cg\left(\|\mathbf{x}\|^2\right). \qquad (3)$$

The kernel K is sometimes called a shadow of G since

$$g(x) = -k'(x), \qquad (4)$$

where $k(x)$ is a profile (one-dimensional cross section) of the kernel K, i.e. $K(\mathbf{x})=k(\|\mathbf{x}\|^2)$[3]. Thus, the basic operation of the mean shift method is as follows:

1. Computation of the mean shift vector $\mathbf{m}$.
2. Shift of the kernel G by the mean shift vector $\mathbf{m}$.

It was proved that the mean shift procedure is convergent iff the kernel K has a convex and monotonically decreasing profile [3]. The problem with the mean shift formulation (2) is a choice of the bandwidth δ. An practical choice of a fixed value of δ should be a trade off between the bias and variance of the estimator $\tilde{f}$ in respect to the true pdf f, since it holds that:

$$MSE(\mathbf{x}) = E\left[\tilde{f}(\mathbf{x}) - f(\mathbf{x})\right] = Var\left(\tilde{f}(\mathbf{x})\right) + Bias^2\left(\tilde{f}(\mathbf{x})\right). \qquad (5)$$

As pointed out by Comaniciu et al. [4], δ can be adapted either in respect to the estimation point $\mathbf{x}$ (a balloon estimator) or to the sample point $\mathbf{x}_i$. The latter has a desirable property of reducing the bias. In [4] two variable bandwidth mean shift methods are proposed. The first one assumes a pilot estimate with a fixed bandwidth from which the proportionality constant is derived which is then used for estimation of δ_i. The second method is based on an assumption that density in a local neighbourhood of pixels is spherical normal with unknown mean and covariance. The very important property of those methods is that when initialized at a given local position they converge to a nearest point in which the estimator (1), with the variable bandwidth δ_i, has zero gradient. This local convergence point is a mode of pdf.

For the mean shift procedure, the task of finding an object, based on a characteristic feature **v** of this object, is formulated as finding a discrete location **y** in the candidate image for which the associated density $s(\mathbf{y},\mathbf{v})$ is the most similar to the target density $d(\mathbf{v})$. However, for this purpose the described mean shift methods rely on *static* distributions of the target and candidates. A different approach, named CamShift, was proposed by Bradski [1]. It assumes recomputation of densities in each frame (or part of it). Then the spatial moments, computed from the probability field, are used to iterate toward the mode of the (updated) probability distribution. Such an approach has many advantages for real video sequences. For a function $f(x,y)$ the moments of order (a,b) of f are given as follows:

$$m_{ab} = \sum_{(x,y)\in U} x^a y^b f(x,y), \quad c_{ab} = \sum_{(x,y)\in R} (x-\bar{x})^a (y-\bar{y})^b f(x,y), \quad (6)$$

where the point $(\bar{x}, \bar{y})$ is called a centroid. Its coordinates are as follows:

$$\bar{x} = \frac{m_{10}}{m_{00}}, \quad \bar{y} = \frac{m_{01}}{m_{00}}, \text{ assuming that } m_{00} \neq 0. \quad (7)$$

The CamShift method relies on computation of the moment of inertia M of a rigid body U, around the line with a slope φ passing through the centroid of U, in which a mass is substituted by a probability, given as follows:

$$M(\phi) = \sum_{(x,y)\in U} [(x-\bar{x})\sin\phi - (y-\bar{y})\cos\phi]^2 f(x,y). \quad (8)$$

Incorporating (6) to the above we obtain

$$M(\phi) = c_{20}\sin^2\phi - 2c_{11}\sin\phi\cos\phi + c_{02}\cos^2\phi. \quad (9)$$

The moment of inertia is minimized when $\partial M(\varphi)/\partial\varphi = 0$, which is fulfilled for

$$\tan(2\phi) = \frac{2c_{11}}{c_{20} - c_{02}}, \text{ if } c_{20} \neq c_{02}, \quad (10)$$

which leads to

$$\tan^2\phi + \frac{c_{20} - c_{02}}{c_{11}}\tan\phi - 1 = 0, \text{ if } c_{11} \neq 0. \quad (11)$$

Thus, in general we obtain two values of φ, one of which maximizes $M(\varphi)$. The value of M_{min}/M_{max} is called *the elongation ratio*. The line around which M is minimized is called *the principal axis*. It can be shown that this axis is collinear with the eigenvector corresponding to the largest eigenvector of the following tensor [15]

$$\mathbf{I} = \begin{bmatrix} c_{20} & c_{11} \\ c_{11} & c_{02} \end{bmatrix}. \tag{12}$$

An object U in an image can be approximated by a blob characterized by $\mathbf{I}$. It can be shown that such a blob can be further approximated by an ellipse with the parameters α, β, and φ, given as follows:

$$\alpha = \sqrt{\lambda_1}, \quad \beta = \sqrt{\lambda_2}, \text{ and } \phi = \frac{1}{2}\arctan\frac{2c_{11}}{c_{20}-c_{02}}, \tag{13}$$

assuming that $c_{20} \neq c_{02}$, and where $\lambda_1 \geq \lambda_2$ are eigenvalues of $\mathbf{I}$, given as follows

$$\lambda_{1,2} = \frac{1}{2}\left[(c_{02}+c_{20}) \pm \sqrt{4c_{11}^2 + (c_{02}-c_{20})^2}\right]. \tag{14}$$

However, in the CamShift method the values α, β need to be scaled by a total 'mass', given by m_{00}, to convey values which reflect the relative size of an object in the image. Moreover, from practical reasons we wish to find and track a rectangle representing the ellipse [14]. Its parameters can be obtained from (13) and (14)

$$w = 2\sqrt{\frac{\lambda_1}{m_{00}}}, \quad l = 2\sqrt{\frac{\lambda_2}{m_{00}}}. \tag{15}$$

It is interesting to observe that this way found rectangles have the same moment $\mathbf{I}$ as the object U [10]. In practice, we wish to allow the search windows to grow, so we set

$$w' = \rho w, \quad l' = \rho l, \tag{16}$$

where $1 < \rho < 2$ is a scaling parameter. Having the above definitions, the CamShift method for a *single* frame can be summarized as follows [1][14]:

1. $t = 0$; $\rho = 1.8$;
   Set up stop thresholds τ_x, τ_y, τ_w, and τ_l, as well as the initial locations and dimensions of the regions of interest R_i. Then, for each R_i perform steps 2-4 until convergence;
2. $t = t + 1$;
   Compute the probability distributions in a region R_i;
3. Estimate its centroid $(\bar{x}, \bar{y})$ and dimensions (w', l') in accordance with (16);
4. Shift R_i to a new centroid and dimensions found in the previous step;
5. Check convergence conditions: if $(|\Delta\bar{x}| = |\bar{x}(t+1) - \bar{x}(t)| < \tau_x$, and $|\Delta\bar{y}| < \tau_y$, and $|\Delta w'| < \tau_w$, and $|\Delta l'| < \tau_l)$ then stop, else go to step 2.
6. Trim the exact positions of R_i, setting $\rho = 1$ in (16).

For each frame processed with the CamShift we store final positions and sizes of the tracked regions R_i. Each new frame is initialized with these data and the above algorithm is repeated.

The robustness of the CamShift method comes from the fact that the probability values need to be evaluated *only* in the regions of interest R_i. In the worst case, and only at the first step of the algorithm, the whole image has to be checked. Such technique leads to savings in computations.

However, computing probability values for multidimensional feature vectors is usually cumbersome. Therefore, our proposition is to *replace* probability distributions with the composition of the fuzzy measures for the regions R_i. This issue is discussed in the next sections when it will be shown that this strategy works fine in practice.

3 Fuzzy Measures for Classification of Pixels

Much effort has been sacrificed for development of efficient and accurate segmentation of colour images. Recently, especially skin segmentation has gained much attention [13][18][12]. Although colours and properties of the RSs are different, results of these served us as guides in our research. In the works by Phung et. al [18] and Jones et. al [13] it was pointed out that the Bayesian classifier with the histogram technique, as well as the multilayer perceptrons perform the best, compared for example with the parametric Gaussian classifiers. They conclude also that efficient skin classification does not depend greatly on a chosen colour space. Therefore the colour space should be chosen to facilitate computations. However, it was shown that classification results obtained with all channels always outperformed feature vectors containing only the chrominance channels, e.g. HS and CrCb only.

Having two classes of *obj* and *nonObj* (such as a sign and a background) and a feature vector $\mathbf{x}$, the Bayes decision rule is given by the following inequality [8]:

$$P(obj\,|\mathbf{x}) > P(nonObj\,|\mathbf{x}), \tag{17}$$

which after considering the Bayes formula

$$P(obj\,|\mathbf{x}) = \frac{p(\mathbf{x}\,|obj)\,P(obj)}{p(\mathbf{x})}, \quad p(\mathbf{x}) = \sum_{i=1}^{N} p(\mathbf{x}\,|obj_i)\,P(obj_i) \tag{18}$$

and assuming that the evidence $p(x)$ is merely a scaling factor, leads to the equivalent

$$\frac{p(\mathbf{x}\,|objColour)}{p(\mathbf{x}\,|nonObjColour)} > \frac{P(nonObjColour)}{P(objColour)} = \pi, \tag{19}$$

where π is a certain threshold value, usually determined empirically. However, the above classification rule requires knowledge of the probability values for each possible feature (colour) value $\mathbf{x}$. This can be done by acquisition of a statistically meaningful evidence from real examples. Such information is also important for the CamShift method described in the previous section. However, direct computation of the pdf from point samples, e.g. with the Parzen window or the probabilistic neural network, requires largest amount of memory and computations. The application of the Bayes classifier with the histogram projection technique requires the smaller amount of computations and memory. Therefore we propose to use *fuzzy measures to be directly used in the CamShift method instead of probabilities*. The fuzzy measures show many desirable features, such as reduced memory and time requirements. The membership functions are derived from the point samples in the form of piecewise-linear functions. Moreover, the fuzzy membership functions are derived from the low pass filtered histograms with a suppression mechanism for values with insufficient magnitude, such as noise or outliers.

With each point P of an image let us associate a q-dimensional fuzzy measure $\mu_\xi = [\mu_1, \mu_2, \ldots, \mu_q]$ for a feature ξ. Then, in the CamShift algorithm (step 2) the probability in a region R_i is substituted by a function $g(\mu_\xi)$, defined as follows:

$$g(\mu_\xi) = \prod_{i=1}^{q} \mu_i. \qquad (20)$$

A feature attributed to a pixel in an image can be colour, shape, etc., for which we can assess its membership function, either from empirical data, a model, or even from an educated guess. Alternatively to the above we can use the following function

$$g(\mu_\xi) = \min_{1 \leq i \leq q} (\mu_i), \qquad (21)$$

which is simpler in computations. Interestingly, it is not necessary to normalize μ_ξ since a normalization is already embedded into (15).

Having proposed substitution of the probability distribution by a function of fuzzy membership functions we need to check convergence of the procedure. For this purpose we follow the proof provided in [2][1]. Since the functions (20) and (21) are nonnegative then the fuzzy CamShift climbs the gradient of g in the same way as it climbs a probability distribution function f in [2]. If g reaches its maximum, then $g' \approx 0$. The rest of the proof follows [2][1].

4 Real-Time Implementation of the Road-Signs Tracking System

The first question concerns determination of the functions (20) or (21). For RS tracking application it appears that the best results are obtained with fuzzy membership functions designed from processed histograms of the H and S channels obtained from real traffic scenes. For gathering colour information from real examples a special application was implemented (see references in [5]). It allows manual selection of the points belonging to RSs. Then the point indices and the RGB, HSI, and YCrCb values are stored. The real Polish traffic scenes were used in our experiments. Nevertheless, this application can be used for gathering colour samples of other objects as well. From data samples the histograms were built (Fig. 1) and filtered with the Savitzky-Golay filter of 4th order which preserves statistical moments of that order [19]. Such method has many desirable properties compared to a simple low-pass moving average, for instance. Finally, the piecewise-linear membership functions were obtained from the histograms. To suppress noise all values which fell below a 10% threshold of the maximum value were arbitrarily set to 0.

Table 1 presents the piecewise-linear fuzzy membership functions for different groups of the Polish RSs [20] obtained from the histograms in Fig. 1.

Table 1. Piecewise-linear fuzzy membership functions for different groups of RSs

Colour channel	Piecewise-linear membership functions – coordinates (x, y)
Red H	(0, 1) - (3, 1) - (10, 0.8) - (12, 0) - (221, 0.25) - (255, 0.25)
Red S	(50, 0.2) - (75, 0.75) - (100, 1) - (150, 1) - (220, 0.4) - (240, 0)
Green H	(85, 0.2) - (88, 0.5) - (92, 0.9) - (105, 0.9) - (110, 0.7) - (112, 0)
Green S	(60, 0.2) - (70, 0.7) - (82, 0.9) - (145, 0.8) - (170, 0)
Blue H	(125, 0) - (145, 1) - (150, 1) - (160, 0)
Blue S	(100, 0) - (145, 1) - (152, 1) - (180, 0)
Yellow H	(20, 0) - (23, 1) - (33, 1) - (39, 0)
Yellow S	(80,0) - (95, 0.22) - (115, 0.22) - (125,0) - (128,0) - (150, 0.48) – (155, 0.48) - (175, 0.18) - (200, 0.22) - (225, 1) - (249, 0.95) - (251,0)

The problem gets complicated in the case of multi object tracking which is also present in our task of RSs tracking. In such an instance we need to take into account the topological properties of the tracked objects and their corresponding membership functions. This means that a minimal separation has to be defined for objects to be treated as separate ones. This is especially severe in a first stage of tracking when objects are not recognized yet. This means that during the initial detection if the membership functions are high for all regions and no other information is provided we are not able to find occluding objects.

Selection of "areas of interest" is done after dividing an image into an arbitrary number of adjacent tiles – see Fig. 2. With each tile a bit is associated, which if set means that this tile is 'active' for subsequent CamShift procedure. Otherwise, it is either not containing any objects, or it is already occupied by a previously found object. Initially the bits for tiles are set in accordance with the membership functions μ_i computed for the whole image: If in a tile there is one or more pixels for which $\mu_i > 0$, then such a tile is set as 'active'. For instance, objects F_1 and F_2 in Fig. 2, make the underlying tiles to be set as 'inactive', so the next run of CamShift can find only new object in all tiles but those occupied for F_1 and F_2, respectively. Information stored in this data structure is used then as initial conditions for a next processed frame, and so on. In practice, for 640×480 resolution, tiles of size 16×16 pixels offer a reasonable trade off between accuracy and processing speed.

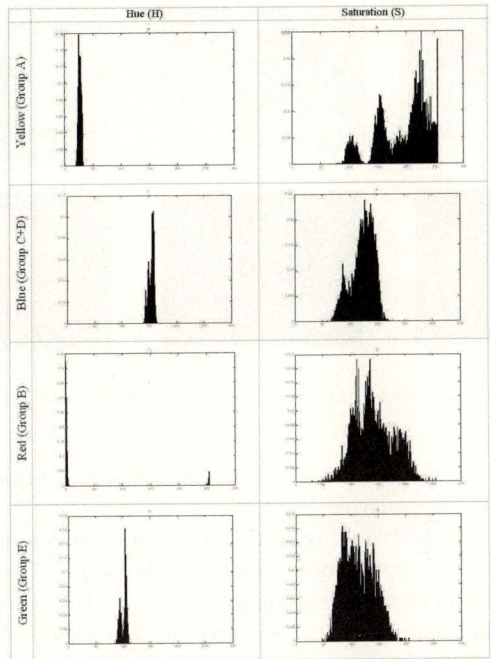

Fig. 1. Histograms of the H and S channels of the road signs from real traffic scenes

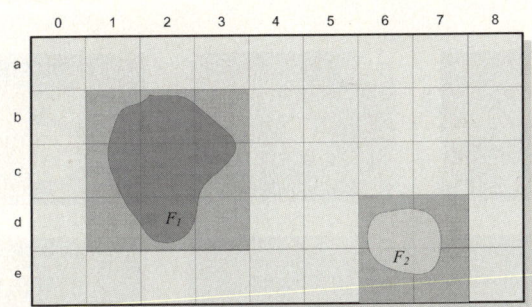

Fig. 2. Data structure for detection of multiple objects in a frame

5 Experimental Results

The experimental system consists of a setup with two Marlin-C33 cameras mounted on the adjustable tripod. The cameras are connected by the IEEE 1394 link to the laptop computer with the Intel® Duo core processor and 1GB of RAM. All software modules were written in C++ using the Microsoft® Visual IDE 6.0.

Fig. 3 presents tracking results obtained by our system. Depicted are three frames (Fig. 3a-c) from a video of a real traffic scene with tracking results (red outlined) of a single sign A-7 (inverted yellow triangle). Fig. 3d-f depict 2D fields

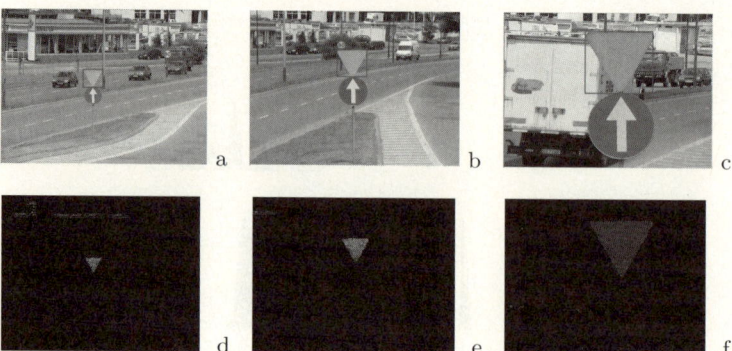

Fig. 3. Results of tracking of the A7 sign in few frames of a real traffic video

of the membership functions of the form (20), which component membership functions μ_i, are defined in Table 1.

Fig. 4 depicts results of tracking multiple objects at a time. In this case the membership functions are constructed from the H and S channels of a subset of red colour values pertinent to the signs of "A" and "B" groups. The important difference is that the function (21) is used in this case. It requires even less computations than (20) since there are used only the comparison operations, avoiding the costly multiplications. The images in the first row of Fig. 4 show tracking results of the "speed limit" (B-33) sign. The middle row depicts tracking

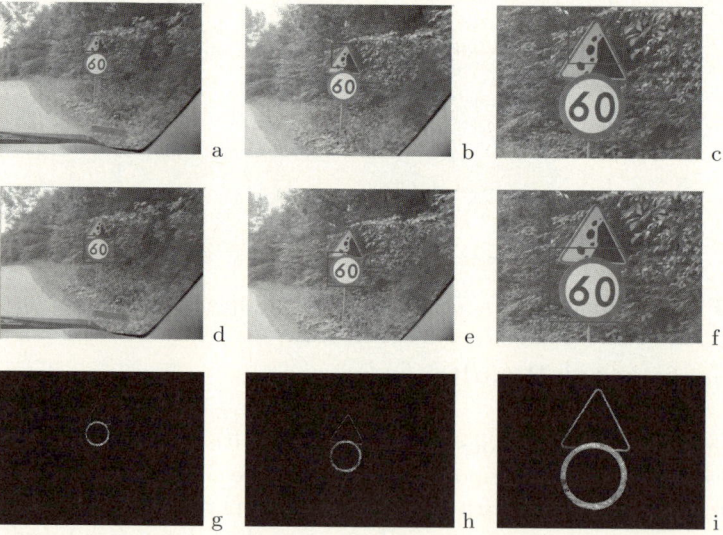

Fig. 4. Results of multi object tracking of the 60 km/h speed limit (B-33) and falling rocks (A-25) signs for frames of a real video. The minimum function (21) used in CamShift.

of the "falling rocks" sign (A-25). All detections are based on the red colour components, although the A-25 sign can be also detected from the yellow colour. This opens new possibilities of joining responses of different colour spectra for tracking which we plan to explore in a future. Fig. 4g-i visualize 2D fields of the membership functions of the form (21), for which μ_i, are also defined in Table 1. The slight up-shift of the detection windows in Fig. 4abc and Fig. 4def is due to overlap of the two signs which is well visible in Fig. 4i. Nevertheless, such a displacement can be easily fixed by a properly designed classifier. For instance the committee machine, proposed in [6], which operates on a group of deformable models can cope easily with horizontal and vertical shifts.

The system is able to process 640×480 colour frames in real time, i.e. at least 30 f/s. It is also resistant to noise and lighting variations due to noise suppression strategy in the construction of the membership functions. The two fuzzy functions (20) and (21) perform similarly, although the latter showed slightly better experimental result and a little faster computation time since it does not need the multiplications.

6 Conclusions

The paper presents a modification of the continuously adaptive mean shift tracking method which consists in replacement of the probability distribution function with the fuzzy membership functions. Such an approach has many desirable properties which allow system design in terms of fuzzy logic. For tracking the two membership functions g were proposed in the form of formulas (20) and (21). Their components μ_i are application dependant. Additionally, the specially designed support data structure registers active regions and allows tracking of multiple objects.

The method has been designed and tested in the real-time tracking system for detection of the Polish road signs. The membership functions μ_i are built from the histograms of the H and S colour components gathered from hundreds of real traffic examples. The strategy of piecewise-linear functions with small signal suppression technique showed significant noise rejection and fast computations. These have been verified by experiments showing high accuracy and real-time processing speed. Finally, although the application operates on samples specific to the Polish conditions, the method can be easily extended to the road-signs of different countries, as well as to track other objects based on their specific membership functions μ_i.

Acknowlegement

This work was supported from the Polish funds for the scientific research in 2008.

References

1. Bradski, G.: Computer Vision Face Tracking For Use in a Perceptual User Interface. Intel Technical Report (1998)
2. Cheng, Y.: Mean shift, mode seeking, and clustering. IEEE PAMI 15(6), 602–605 (1993)

3. Comaniciu, D., Meer, P.: Mean Shift: A Robust Approach Toward Feature Space Analysis. IEEE PAMI 24(5), 603–619 (2002)
4. Comaniciu, D., Ramesh, V., Meer, P.: The Variable Bandwidth Mean Shift and Data-Driven Scale Selection. In: IEEE ICCV, vol. 1, pp. 438–445 (2001)
5. Cyganek, B.: Circular Road Signs Recognition with Soft Classifiers. Integrated Computer-Aided Engineering 14(4), 323–343 (2007)
6. Cyganek, B.: Committee Machine for Road-Signs Classification. In: Rutkowski, L., Tadeusiewicz, R., Zadeh, L.A., Żurada, J.M. (eds.) ICAISC 2006. LNCS (LNAI), vol. 4029, pp. 583–592. Springer, Heidelberg (2006)
7. DaimlerChrysler: The Thinking Vehicle (2002), http://www.daimlerchrysler.com
8. Duda, R.O., Hart, P.E., Stork, D.G.: Pattern Classification. Wiley, Chichester (2001)
9. Fang, C.-Y., Chen, S.-W., Fuh, C.-S.: Road-Sign Detection and Tracking. IEEE Transactions on Vehicular Technology 52(5), 1329–1341 (2003)
10. Freeman, W.T., Tanaka, K., Ohta, J., Kyuma, K.: Computer Vision for Computer Games. In: IEEE 2nd Int. Conf. on Automatic Face and Gesture Recognition, pp. 100–105 (1996)
11. Fukunaga, K., Hostetler, L.D.: The estimation of the gradient of a density function, with application in pattern recognition. IEEE Tr. Information Theory 21, 32–40 (1975)
12. Hsu, R.-L., Abdel-Mottaleb, M., Jain, A.K.: Face Detection in Color Images. IEEE PAMI 24(5), 696–707 (2002)
13. Jones, M.J., Rehg, J.M.: Statistical Color Models with Application to Pixel-Level Human Skin Detection. In: IEEE Int. Conf. Pat. Recognition, vol. 1, pp. 1056–1059 (2000)
14. Kim, K.I., Jung, K., Kim, J.H.: Texture-Based Approach for Text Detection in Images Using Support Vector Machines and Continuously Adaptive Mean Shift Algorithm. IEEE PAMI 25(12), 1631–1639 (2003)
15. Klette, R., Rosenfeld, A.: Digital Geometry. Morgan Kaufmann, San Francisco (2004)
16. Maldonado-Bascón, S., Lafuente-Arroyo, S., Gil-Jiménez, P., Gómez-Moreno, H., López-Ferreras, F.: Road-Sign Detection and Recognition Based on Support Vector Machines. IEEE Tr. on Int. Transport. 8(2), 264–278 (2007)
17. Paclík, P., Novovièová, J., Duin, R.P.W.: Building road sign classifiers using a trainable similarity measure. IEEE Trans. on Int. Transport. 7(3), 309–321 (2006)
18. Phung, S.L., Bouzerdoum, A., Chai, D.: Skin Segmentation Using Color Pixel Classification: Analysis and Comparison. IEEE PAMI 27(1), 148–154 (2005)
19. Press, W.H., Teukolsky, S.A., Vetterling, W.T., Flannery, B.P.: Numerical Recipes in C. In: The Art of Scientific Computing. Cambridge University Press, Cambridge (1999)
20. Road Signs and Signalization. Directive of the Polish Ministry of Infrastructure, Internal Affairs and Administration (Dz. U. Nr 170, poz. 1393) (2002)
21. Rutkowski, L.: Techniques and Methods of Artificial Intelligence (in Polish), PWN (2005)
22. Siemens VDO: (2007), http://usa.siemensvdo.com/topics/adas/traffic-sign-recognition/

A New Method for Decision Making in the Intuitionistic Fuzzy Setting

Ludmila Dymova, Izabela Róg, and Pavel Sevastjanov

Institute of Comp.& Information Sci., Technical University of Czestochowa,
Dabrowskiego 73, 42-200 Czestochowa, Poland
sevast@icis.pcz.pl

Abstract. The main problem of known methods for Multiple Criteria Decision Making in the Intuitionistic Fuzzy setting is that they are generally based on the intermediate type reduction. Such approaches lead inevitable to the loss of important information. Another problem is the choice of an appropriate method for the local criteria aggregation taking into account their ranks. The aim of this paper is to present a new method which makes it possible to solve the first problem and facilitates the solution of the second one. The method is based on the Dempster-Shafer Theory (DST). It allows to solve the Multiple Criteria Decision Making problem without intermediate type reduction for different approaches to aggregation of the local criteria. The usefulness of elaborated method is illustrated with known example of Multiple Criteria Decision Making problem.

1 Introduction

Intuitionistic fuzzy set (IFS) proposed by Atanassov [1], is one of the possible generalizations of Fuzzy Sets Theory and appears to be relevant and useful in some applications. The concept of IFS is based on the simultaneous consideration of membership, μ, and non-membership, ν, of an element of a set to the set itself [1]. By definition $0 \leq \mu + \nu \leq 1$. The similar approach, the so called Vague Sets, proposed in [12] is proved to be equivalent to the IFS (see [4]). Since Vague Sets were proposed later than IFS, in this paper we will always speak of IFS. There were many papers devoted to the theoretical problems of IFS in the scientific literature (see [20] for a overview). One of the most important application of IFS is the decision making problem [6,15,17,18,19] or group decision making problem [2,3,22,23,27,28,29,30]. In the framework of IFS, the decision making problem may be formulated as follows. Let $X = \{x_1, x_2, ..., x_m\}$ be a set of alternatives, $A = \{a_1, a_2, ..., a_n\}$ be a set of local criteria, $W = \{w_1, w_2, ..., w_n\}$ be the weights of local criteria. If μ_{ij} is the degree to which x_i satisfies the criterion a_j and ν_{ij} is the degree to which x_i does not satisfy the criterion a_j then alternative x_i may be presented by its characteristics as follows: $x_i = \{(w_1, <\mu_{i1}, \nu_{i1}>), (w_2, <\mu_{i2}, \nu_{i2}>), ..., (w_n, <\mu_{in}, \nu_{in}>)\}, i = 1, ..., m$. It seems quite natural that if the alternative's attributes are intuitionistic fuzzy (IF) objects then the resulting alternative's evaluation should be an IF object as well. In general, such representation of final alternative's evaluation is not a simple task. Therefore, to avoid this problem the different real valued score functions based on the parameters μ_{ij}, ν_{ij} are usually used. In [6], the authors proposed the score function $S(x_i) = \mu(x_i) - \nu(x_i)$, where $\mu(x_i) = \min(\mu_{i1}, \mu_{i2}, ..., \mu_{in})$, $\nu(x_i) = \max(\nu_{i1}, \nu_{i2}, ..., \nu_{in})$. To take into account the weights of local criteria, the authors of [6] proposed the weighted score function $WS(x_i) = \sum_{j=1}^{n} w_j(\mu_{ij} - \nu_{ij})$. In [15], in addition to the above score function the authors introduced the so called accuracy function $H(x_i) = \mu(x_i) + \nu(x_i)$ and weighted accuracy function $T(x_i) = \sum_{j=1}^{n} w_j(\mu_{ij} + \nu_{ij})$. The score functions proposed in [19] are based on the Intuitionistic Fuzzy Point operators originating from IF triangular norm and conorm [5,9]. Different aggregating operators are usually used to calculate a final evaluation of an alternative on the base of considered real valued score functions. The most complicated formulation of IFS decision making problem has been analyzed in [17], where the weights of local criteria were IF objects too. To avoid the above mentioned problem of the representation of final alternative's evaluation in form of IF objects, the authors of [17] proposed its reduction to the linear programming task with real valued parameters. In [18], the authors proposed a method based on the aggregation of the weighted score and accuracy functions. Using this more simple method the same final results as in [17] were obtained. It is well known that the couple $<\mu(x), \nu(x)>$ can be mapped bijectively onto regular interval $[\mu(x), 1-\nu(x)]$ [1]. This interval can be presented in equivalent form $[\mu(x), \mu(x) + \pi(x)]$, where $\pi(x) = 1 - \mu(x) - \nu(x)$ is the so called intuitionistic fuzzy index. In [15], the evaluation of the alternative $x_j \in X$ with respect to the criterion $a_i \in A$ is represented by interval $[\mu_{ij}, 1-\nu_{ij}]$, where μ_{ij} indicates the degree to which x_j satisfies criterion a_i, ν_{ij} indicates the degree to which x_j does not satisfy criterion a_i. In [17], for this purpose the author propose to use intervals $[\mu_{ij}^l, \mu_{ij}^u] = [\mu_{ij}, \mu_{ij} + \pi_{ij}]$. These intervals were used in [17] only as the restrictions in the linear programming task. It is important to note that in [15,17] the semantics of the right bound, μ_{ij}^u, is not clarified. In [10], some interrelations between IFS and theories modelling imprecision such as interval valued fuzzy sets, type 2 fuzzy sets and soft sets were established. In [14], the semantics aspects of such interrelations are analyzed. Since this discussion is out of scope of this paper, we prefer to use here only the interval valued fuzzy sets interpretation of intuitionistic fuzzy sets as it seems most suitable for our purposes. In our opinion, the main problem of IF decision making is that generally the resulting alternative's evaluations should be presented in form of intuitionistic fuzzy values (or intervals). To avoid it, the different real valued score functions are usually used. Of course, such approaches provide useful, but only approximate results since any intermediate type reduction in the solution procedure leads inevitably to the loss of some information. So the next aim of this paper is to present a method allowing to obtain the solution of Multiple Criteria Decision Making ($MCDM$) problem in the intuitionistic fuzzy setting without type reduction, i.e., in such a way that the final evaluation of the alternative x_i is the interval $[\mu_i^l, \mu_i^u]$. Hence, the problem of such intervals comparison arises. The next important problem is the choice of an appropriate method for aggregation of local criteria taking into account their ranks. The weighted sum aggregation is usually used without any discussion since it seems as obvious one. On the other hand, in [34] it is shown that the choice of aggregation scheme in the decision making is a context dependent problem. The detailed analysis of the advantages and drawbacks of most popular aggregating modes can be found in [11,25], where a new method for the generalization of aggregation schemes based on the level-2 fuzzy sets methods is proposed and illustrated. On the other hand, as all known aggregation modes have their own advantages and drawbacks, it seems impossible to choose the best one especially when dealing with complicated hierarchical problem. Therefore, when dealing with a complex task characterized by a great number of local criteria, it seems reasonable to use all types of aggregations relevant to the considered problem. If the results obtained using different aggregation modes are similar, this fact may be considered as a good confirmation of their optimality. In the opposite case, an additional analysis of local criteria and their ranking should be advised. Therefore, the method for $MCDM$ in the intuitionistic fuzzy setting should allow to deal with the most popular and reputed aggregating schemes. Since some of these schemes are based on min and max operations, a special method for the comparison of alternative's evaluations represented by intuitionistic fuzzy numbers is needed. For these reasons the rest of paper is set out as follows. In Section 2, we recall the basic definitions of DST and briefly describe based on DST method which we use to get the results of interval comparison in interval form, i.e., without loss of information caused by intermediate type reduction. Section 3 is devoted to the $MCDM$ problem formulated in the spirit of proposed method. Illustrative numerical example with use of different aggregating schemes is presented as well. Finally, the concluding section summarizes the paper.

2 Interval Number Comparison Using DST Approach

In the current paper, we will use the the method for interval comparison based on DST. Due to its (DST) relative unfamiliarity we first present a brief description of some fundamentals of DST needed for the subsequent analysis.

2.1 Dempster-Shafer Theory

The origins of the Dempster-Shafer theory (DST) go back to the work by A.P. Dempster [7,8] who developed a system of upper and lower probabilities. Following this work his student G. Shafer [26] included in his 1976 book "A Mathematical Theory of Evidence" a more thorough explanation of belief functions. In [33], the authors provide a collection of articles by some of the leading researchers in this field. The close connection between DS structure and random sets is discussed in [13]. In the following, we provide an introduction to basic ideas of this theory. Assume V is a variable whose domain is a finite set X. It is important to note that variable V may be treated also as a question or proposition and X as a set of propositions or mutually exclusive hypotheses or answers [31]. A *DS* belief structure has associated with it a mapping m, called basic assignment function, from subsets of X into a unit interval, $m : 2^X \to [0,1]$ such that $m(\emptyset) = 0$, $\sum_{A \subset X} m(A) = 1$. The subsets of X for which the mapping does not assume a zero value are called focal elements. We shall denote these as A_i, for $i = 1$ to n. We note that the null set is never a focal element. In [26], Shafer introduced a number of measures associated with this structure. The measure of belief is a mapping $Bel : 2^X \to [0,1]$ such that for any subset B of X

$$Bel(B) = \sum_{i=1}^{n} m(A_i), \ A_i \subseteq B, i = 1 \text{ to } n. \tag{1}$$

With V a variable taking its value in a set X under the semantics provided by Shafer [26], *Bel(B)* is our degree of belief that the value of V lies in a set B. A second measure introduced by Shafer [26] is a measure of plausibility. The measure of plausibility associated with m is a mapping $Pl : 2^X \to [0,1]$ such that for any subset B of X.

$$Pl(B) = \sum_{i=1}^{n} m(A_i), A_i \cap B \neq \emptyset, i = 1 \text{ to } n. \tag{2}$$

The semantics associated with this measure is that *Pl(B)* is a degree of plausibility that a value of V lies in a set B. It is easy to see that $Bel(B) \leq Pl(B)$. *DS* provides an explicit measure of ignorance about an event B and its complementary $\overline{B}$ as a length of an interval $[Bel(B), Pl(B)]$ called the belief interval (*BI*). It can also be interpreted as imprecision of the "true probability" of B [26]. Alternative interpretations of the *DS* belief structure have been discussed in the literature. Most notable among these is the original framework of Dempster [7,8] and the random set viewpoint described in [13]. In both these interpretations, the *DS* structure is seen as a kind of random experiment in which m is interpreted as a probability distribution. In our case, the random set viewpoint of the *DS* structure seems as the most convenient and reflects better the nature of the intuitionistic fuzzy numbers comparison problem.

2.2 Interval Comparison without Type Reduction

As we use the interval valued fuzzy sets interpretation of intuitionistic fuzzy sets, we get the final evaluations of alternatives in form of interval. Hence, an appropriate method for such quantities comparison is needed. Moreover, if we use the *min*-type aggregation of the local criteria, we must compare the intervals in the intermediate stages of the method's realization. Therefore, to avoid the loss of information caused by intermediate type reduction, we need a method for interval comparison providing the results of comparison in interval form too.

Since we have presented such method earlier in [24] in more detail, in this section we present only its brief description.

There are only two non-trivial cases of interval locations which we call overlapping and inclusion cases (see Fig.1) deserve to be considered. Let $A = [a_1, a_2]$ and $B = [b_1, b_2]$ be independent intervals and $a \in [a_1, a_2]$, $b \in [b_1, b_2]$ be random values distributed on these intervals. As we are dealing with usual crisp intervals, the natural assumption is that the random values a and b are distributed uniformly. There are some subintervals which play an important role in our analysis. For example (see Fig. 1a), the falling of random $a \in [b_1, b_2]$, $b \in [a_1, a_2]$ into subintervals $[a_1, b_1], [b_1, a_2], [a_2, b_2]$ may be treated as a set of independent random events.

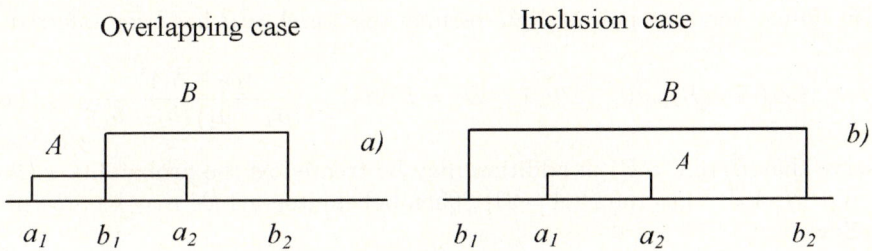

Fig. 1. The examples of interval relations

Let us consider the case of overlapping intervals (Fig.1a). Only four mutually exclusive events H_i, $i = 1$ to 4 may take place in considered situation:

$$H_1 : a \in [a_1, b_1] \& b \in [a_2, b_2], \quad H_2 : a \in [a_1, b_1] \& b \in [b_1, a_2],$$
$$H_3 : a \in [b_1, a_2] \& b \in [b_1, a_2], \quad H_4 : a \in [b_1, a_2] \& b \in [a_2, b_2]. \qquad (3)$$

For the probabilities of events H_1–H_4 from the simple geometric reasons we obtain

$$P(H_1) = \frac{b_1 - a_1}{a_2 - a_1} \frac{b_2 - a_2}{b_2 - b_1}, \quad P(H_2) = \frac{b_1 - a_1}{a_2 - a_1} \frac{a_2 - b_1}{b_2 - b_1},$$
$$P(H_3) = \frac{a_2 - b_1}{a_2 - a_1} \frac{a_2 - b_1}{b_2 - b_1}, \quad P(H_4) = \frac{a_2 - b_1}{a_2 - a_1} \frac{b_2 - a_2}{b_2 - b_1}. \qquad (4)$$

It can easily be proved that

$$P(H_1) + P(H_2) + P(H_3) + P(H_4) = 1. \qquad (5)$$

Thus, in sense of *DST* the probabilities $P(H_i)$, $i = 1$ to 4, can be used to construct a basic assignment function, m. Since in case of overlapping intervals $a_1 < b_1$ and $a_2 < b_2$, there are only two interval relations which make a sense : $A < B, A = B$. It is easy to see that events H_1, H_2 and H_4 may be considered as the "strong" evidences of $A < B$, otherwise H_3 can be treated as only the "weak" evidence of $A < B$ because it simultaneously is the witness of $A = B$. In DST's notation we obtain:

$$m(\{A < B\}) = P(H_1) + P(H_2) + P(H_4), \qquad (6)$$

$$m(\{A < B, A = B\}) = P(H_3). \qquad (7)$$

Then from (5), (6) and (7) we get

$$Bel(A < B) = m\{A < B\} = 1 - P(H_3) = 1 - \frac{(a_2 - b_1)^2}{(a_2 - a_1)(b_2 - b_1)}, \qquad (8)$$

$$Pl(A < B) = m(\{A < B\}) + m(\{A < B, A = B\}) = 1. \qquad (9)$$

In the similar way, the pair of DST estimations for $A = B$ has been inferred:

$$Bel(A = B) = 0, \quad Pl(A = B) = P(H_3) = \frac{(a_2 - b_1)^2}{(a_2 - a_1)(b_2 - b_1)}. \qquad (10)$$

Observe that $Bel(A < B)$ in addition may be treated as the probability $P(A < B)$ and $Pl(A = B)$ as the $P(A = B)$. Then belief intervals BI may be presented as follows:

$$BI(A < B) = [Bel(A < B), Pl(A < B)] = [P(A < B), 1]. \qquad (11)$$

So using DST's approach we obtain the interval estimations for the degree of interval inequality and equality. In the case of $a_1 = b_1$, $a_2 = b_2$, i.e., $A \equiv B$, from (9)–(12) we get $BI(A < B) = BI(A = B) = [0, 1]$. We introduce the degree of imprecision or ambiguity ID of interval relations (rel) as a whole:

$$ID(rel) = BI(A < B) + BI(A = B) =$$
$$= \left[1 - \frac{(a_2 - b_1)^2}{(a_2 - a_1)(b_2 - b_1)}, 1 + \frac{(a_2 - b_1)^2}{(a_2 - a_1)(b_2 - b_1)} \right].$$

It is easy to see that the length of $ID(rel)$, which may be considered as a natural real valued estimation of imprecision. It decreases along with lowering of overlapping area $a_2 - b_1$. It it is worth noting that introduced interval form of interval comparison estimations is a real embodiment of usually implicitly expressed, but pivotal requirement of interval arithmetic: the result of interval operation should be an interval too.

Let us consider the inclusion case (Fig. 1b). In this case we have three possible events:

$$H_1 : a \in [a_1, a_2] \& b \in [b_1, a_1], H_2 : a \in [a_1, a_2] \& b \in [a_1, a_2],$$
$$H_3 : a \in [a_1, a_2] \& b \in [a_2, b_2].$$

Since $b_1 \leq a_1$, in this case the relation $A > B$ may become true. For instance, there no doubts that $A > B$ if $b_1 < a_1$ and $b_2 = a_2$. We can observe the elementary evidences of events $A < B$, $A = B$, $A > B$ in this situation and we can take them into account to construct the *Bel* and *Pl* functions using nearly the same reasoning as in the case of overlapping intervals. Finally, we get

$$BI(A < B) = [Bel(A < B), Pl(A < B)] = \left[\frac{b_2 - a_2}{b_2 - b_1}, \frac{b_2 - a_1}{b_2 - b_1}\right], \quad (12)$$

$$BI(A = B) = [Bel(A = B), Pl(A = B)] = \left[0, \frac{a_2 - a_1}{b_2 - b_1}\right], \quad (13)$$

$$BI(A > B) = [Bel(A > B), Pl(A > B)] = \left[\frac{a_1 - b_1}{b_2 - b_1}, \frac{a_2 - b_1}{b_2 - b_1}\right]. \quad (14)$$

It is easy to see that in inclusion case we have $BI(A = B) < BI(A > B)$, $BI(A > B)$ at least in the "weak" sense since $BI(A = B) \cap BI(A > B) \neq \emptyset$ and/or $BI(A = B) \cap BI(A < B) \neq \emptyset$. As in the case of overlapping intervals we introduce the overall degree of imprecision or ambiguity of interval relations *rel* as follows

$$ID(rel) = BI(A < B) + BI(A = B) + BI(A > B) =$$
$$= \left[1 - \frac{a_2 - a_1}{b_2 - b_1}, 1 + 2\frac{a_2 - a_1}{b_2 - b_1}\right].$$

In contrast to the overlapping case, for degree of imprecision or ambiguity we obtain an interval which is asymmetrical in relation to 1. Indeed, the several real valued criteria may be applied in order to make a reasonable final choice when comparing intervals. Non-exhaustively, we can distinguish:

− strong preference: $B > A$ if $Bel(B > A) > Pl(A < B)$,
− weak preference: $B > A$ if $Bel(B > A) > Bel(B < A)$,
− mixed preference: $B > A$ if $MP(B > A) > MP(B < A)$,

where $MP(\cdot) = aBel(\cdot) + (1 - a)Pl(\cdot)$ with $0 \leq a \leq 1$ (the value of a reflects the risk adversity of the decision maker). Obviously, the mixed preference is the more flexible criterion.

3 MCDM Problem in the Framework of Intuitionistic/DST Approach

To make our consideration more transparent and comparable with the results obtained earlier by other authors, we will use here the example analyzed in [17],[18]: " Consider an air-condition system selection problem. Suppose there exist three air-condition systems x_1, x_2 and x_3. Denote the alternative set by $X = \{x_1, x_2, x_3\}$. Suppose three attributes a_1 (economical), a_2 (function) and a_3 (being operative) are taken into consideration in the selection problem. Denote the set of all attributes by $A = \{a_1, a_2, a_3\}$. Using statistical methods, the degrees μ_{ij} of membership and the degrees ν_{ij} of non-membership for the alternative

$x_j \in X$ with respect to the attribute $a_i \in A$ to the fuzzy concept "excellence" can be obtained, respectively. Namely,

$$((\mu_{ij}, v_{ij}))_{3\times 3} = \begin{matrix} & \begin{matrix} x_1 & x_2 & x_3 \end{matrix} \\ \begin{matrix} a_1 \\ a_2 \\ a_3 \end{matrix} & \begin{pmatrix} (0.75, 0.10) & (0.80, 0.15) & (0.40, 0.45) \\ (0.60, 0.25) & (0.68, 0.20) & (0.75, 0.05) \\ (0.80, 0.20) & (0.45, 0.50) & (0.60, 0.30) \end{pmatrix} \end{matrix} ." \quad (15)$$

"...In a similar way, the degrees ρ_i of membership and the degrees τ_i of non-membership for the three attributes $a_i \in A$ to the fuzzy concept "importance" can be obtained, respectively. Namely,

$$((\rho_i, \tau_i))_{1\times 3} = \begin{pmatrix} a_1 & a_2 & a_3 \\ (0.25, 0.25) & (0.35, 0.40) & (0.30, 0.65) \end{pmatrix} ." \quad (16)$$

Only what we change in the above example is the treatment of the set $A = \{a_1, a_2, a_3\}$. We think that in context of air-condition system selection problem, it seems more natural and convenient to treat A as a set of local criteria. Then μ_{ij} and v_{ij} are the degrees to which the alternative x_j satisfies and dissatisfies the local criterion $a_i \in A$. So the above example can be reformulated in the framework of $MCDM$. Using interval valued fuzzy sets semantics and taking into account that the couple $<\mu_{ij}, v_{ij}>$ can be mapped bijectively onto regular interval $[\mu_{ij}^l, \mu_{ij}^u]$, where $\mu_{ij}^l = \mu_{ij}$, $\mu_{ij}^u = 1 - v_{ij}$, we represent the structure (16) as follows:

$$([\mu_{ij}^l, \mu_{ij}^u])_{3\times 3} = \begin{matrix} & \begin{matrix} x_1 & x_2 & x_3 \end{matrix} \\ \begin{matrix} a_1 \\ a_2 \\ a_3 \end{matrix} & \begin{pmatrix} [0.75, 0.90] & [0.80, 0.85] & [0.40, 0.55] \\ [0.60, 0.75] & [0.68, 0.80] & [0.75, 0.95] \\ [0.80, 0.80] & [0.45, 0.50] & [0.60, 0.70] \end{pmatrix} \end{matrix} ." \quad (17)$$

Similarly, the weights of local criteria (17) can be represented by intervals

$$[w^l, w^u]_{1\times 3} = \begin{pmatrix} a_1 & a_2 & a_3 \\ [0.25, 0.75] & [0.35, 0.60] & [0.30, 0.35] \end{pmatrix}, \quad (18)$$

where $w^l = \rho_i$, $w^u = 1 - \tau_i$. Since the intervals $IA_{ij} = [\mu_{ij}^l, \mu_{ij}^u]$ and $IW_i = [w^l, w^u]$ are regular intervals, i.e., $\mu_{ij}^l \leq \mu_{ij}^u$ and $w^l \leq w^u$, the usual interval arithmetic rules [21] and interval relations introduced in Section 2 can be used to operate on them. As in the considering example we have initial data only in interval form, it seems quite natural to present the results of solution in form of intervals too. Therefore, we can present the final alternative's evaluations as $FAE(x_j) = agg(IA_{ij}, IW_i)$, where agg is used operator of aggregation. Since there are many aggregating modes proposed in the framework of $MCDM$ that are relevant in the specific conditions, we consider here only the most popular and reputed approaches: weighted sum, multiplicative mode and Yager's aggregation [32]. It is shown in [11] that these aggregating modes are often used as the atomic ones for building the more complicated methods for $MCDM$. In our case, these modes can be presented as follows:

$$FAE_{sum}(x_j) = \frac{1}{n}\sum_{i=1}^{n} IW_i IA_{ij}, \tag{19}$$

$$FAE_{prod}(x_j) = \prod_{i=1}^{n} IA_{ij}^{IW_i}, \tag{20}$$

$$FAE_{min}(x_j) = \min_{i}\{IA_{ij}^{IW_i}\}. \tag{21}$$

We have used usual interval arithmetic rules [16] to calculate the alternative's evaluations based on the aggregations (19) and (20). To obtain such evaluations in the case of aggregation mode (21), the methods for interval comparison presented in Section 2 has been used in addition. Finally, we have obtained using aggregations (19):

$$\begin{aligned} FAE_{sum}(x_1) &= [0.21, 0.47]; \\ FAE_{sum}(x_2) &= [0.19, 0.43]; \\ FAE_{sum}(x_3) &= [0.18, 0.41], \end{aligned} \tag{22}$$

using aggregation (20):

$$\begin{aligned} FAE_{prod}(x_1) &= [0.55, 0.82]; \\ FAE_{prod}(x_2) &= [0.51, 0.72]; \\ FAE_{prod}(x_3) &= [0.35, 0.76], \end{aligned} \tag{23}$$

using aggregation (21):

$$\begin{aligned} FAE_{min}(x_1) &= [0.74, 0.90]; \\ FAE_{min}(x_2) &= [0.76, 0.81]; \\ FAE_{min}(x_3) &= [0.84, 0.90]. \end{aligned} \tag{24}$$

Obviously, to select the best alternative, their final evaluations presented by corresponding intervals should be compared. Using the method described in Section 2 we get the results of such comparison in form of belief intervals as follows:

$BI_{sum}(x_1 > x_2) = [0.22, 1], BI_{sum}(x_2 > x_3) = [0.13, 1],$
$BI_{sum}(x_1 > x_3) = [0.34, 1],$
$BI_{sum}(x_1 < x_2) = 0, BI_{sum}(x_2 < x_3) = 0,$
$BI_{sum}(x_1 < x_3) = 0,$
$BI_{sum}(x_1 = x_2) = [0, 0.78], BI_{sum}(x_2 = x_3) = [0, 0.87],$
$BI_{sum}(x_1 = x_3) = [0, 0.66],$

$BI_{prod}(x_1 > x_2) = [0.49, 1], BI_{prod}(x_2 > x_3) = [0.38, 0.90],$
$BI_{prod}(x_1 > x_3) = [0.60, 1],$
$BI_{prod}(x_1 < x_2) = 0, BI_{prod}(x_2 < x_3) = [0.10, 0.62],$
$BI_{prod}(x_1 < x_3) = 0,$
$BI_{prod}(x_1 = x_2) = [0, 0.51], BI_{prod}(x_2 = x_3) = [0, 0.53],$
$BI_{prod}(x_1 = x_3) = [0, 0.40],$

$BI_{min}(x_1 > x_2) = [0.55, 0.88], BI_{min}(x_2 > x_3) = 0,$
$BI_{min}(x_1 > x_3) = [0.03, 0.40],$
$BI_{min}(x_1 < x_2) = [0.12, 0.45], BI_{min}(x_2 < x_3) = 1,$
$BI_{min}(x_1 < x_3) = [0.60, 0.97],$
$BI_{min}(x_1 = x_2) = [0, 0.33], BI_{min}(x_2 = x_3) = 0,$
$BI_{min}(x_1 = x_3) = [0, 0.37].$

To obtain the final preference on the set of comparing alternatives, the local criteria introduced in Section 2 (strong, weak and mixed preferences) could be used. Nevertheless, the structure of obtained results, (typical of the proposed method [24]) makes it possible to use the simplest, but intuitively obvious definition of interval inequality proposed by Moore [21]. In a nutshell, if A and B are intervals, we say $A < B$ if $a_1 \leq b_1$, $a_2 \leq b_2$ and at least one of the last two inequalities is strong. Using this classical definition and presented above resulting believe intervals $BI_{sum}, BI_{prod}, BI_{min}$ we get finally $x_3 \prec x_2 \prec x_1$ using the weighted sum and multiplicative modes and $x_2 \prec x_1 \prec x_3$ using the Yager's aggregation. The strengths of preferences we get using different aggregation modes are different. They can be calculated as the measures of strong, weak and mixed preferences (see Section 2). Nevertheless on the qualitative level of consideration we obtain the same results for weighted sum and multiplicative modes. As it is shown in [11,25] such coincidence of the results obtained using different types of aggregation is not usually the case. On the other hand, if the results obtained using different aggregation modes are similar, this fact may be treated as a good confirmation of the result's reliability. In the considered example of air-condition system selection problem, a small value of a local criterion can be compensated for the great values of the rest. Thus, taking into account that among considered aggregation modes only the weighted sum and multiplicative ones possess such compensative property, we can assume that the ranking $x_3 \prec x_2 \prec x_1$ is the is most preferable solution of the considered problem. In the considered example, such a reasoning excludes the results of Yager's aggregation from analysis, but generally they can be used for the generalization of the aggregation modes [11,25]. It is worthy to note that using the same example the substantially different result $x_2 \prec x_3 \prec x_1$. has been obtained in [17,18]. It is important that qualitative difference of our results from those obtained in [17,18] using type reduction of initial IFS task is observed even when we use the weighted sum aggregation as in [17]. We can explain such divergence of results only by the absence of any intermediate type reduction in the proposed method for $MCDM$ in IFS setting that makes it possible to avoid the loss of important information.

4 Conclusion

The method which allows to solve Multiple Criteria Decision Making problem in the IF setting without intermediate type reduction is presented. It is based on the treatment of Intuitionistic Fuzzy problem in context of interval valued fuzzy sets and the method for interval comparison based on DST which provides the result of interval comparison in form of believe interval. Proposed method is illustrated with known example of $MCDM$ problem in the Intuitionistic Fuzzy setting. It is shown that using new approach, the solution of $MCDM$ problem in IF setting can be obtained without any intermediate type reduction for different approaches to aggregation of local criteria. It is important that on the qualitative level of consideration we get the results quite different from those obtained in [17,18] using intermediate type reduction.

References

1. Atanassov, K.T.: Intuitionistic fuzzy sets. Fuzzy Sets and Systems 20, 87–96 (1986)
2. Atanassov, K., Pasi, G., Yager, R.: Intuitionistic fuzzy interpretations of multi-person multicriteria decision making. In: Proceedings of 2002 First International IEEE Symposium Intelligent Systems, vol. 1, pp. 115–119 (2002)
3. Atanassov, K., Pasi, G., Yager, R., Atanassova, V.: Intuitionistic fuzzy group interpretations of multi-person multicriteria decision making. In: Proceedings of the Third Conference of the European Society for Fuzzy Logic and Technology EUSFLAT'2003, Zittau, September 10-12, pp. 177–182 (2003)
4. Bustince, H., Burillo, P.: Vague sets are intuitionistic fuzzy sets. Fuzzy Sets and Systems 79, 403–405 (1996)
5. Cornelis, C., Deschrijver, G., Kerre, E.: Implication in intuitionistic fuzzy and interval-valued fuzzy set theory: Construction, classification, application. International Journal of Approximate Reasoning 35, 55–95 (2004)
6. Chen, S.M., Tan, J.M.: Handling multicriteria fuzzy decision-making problems based on vague set theory. Fuzzy Sets and Systems 67, 163–172 (1994)
7. Dempster, A.P.: Upper and lower probabilities induced by a muilti-valued mapping. Ann. Math. Stat. 38, 325–339 (1967)
8. Dempster, A.P.: A generalization of Bayesian inference (with discussion). J. Roy. Stat. Soc., Series B 30(2), 208–247 (1968)
9. Deschrijver, G., Cornelis, C., Kerre, E.: On the representation of intuitionistic fuzzy t-norms and t-conorms. IEEE Transactions on Fuzzy Systems 12(1), 45–61 (2004)
10. Deschrijver, G., Kerre, E.E.: On the position of intuitionistic fuzzy set theory in the framework of theories modelling imprecision. Information Sciences 177, 1860–1866 (2007)
11. Dimova, L., Sevastianov, P., Sevastianov, D.: MCDM in a fuzzy setting: investment projects assessment application. International Journal of Production Economics 100, 10–29 (2006)
12. Gau, W.L., Buehrer, D.J.: Vague sets. IEEE Trans. Systems Man Cybernet 23, 610–614 (1993)
13. Goodman, I.R., Nguyen, H.T.: Uncertainty Models for Knowledge-Based System. North-Holand, Amsterdam (1985)
14. Grzegorzewski, P., Mrowka, E.: Some notes on (Atanassov's) intuitionistic fuzzy sets. Fuzzy Sets and Systems 156, 492–495 (2005)
15. Honga, D.H., Choi, C.-H.: Multicriteria fuzzy decision-making problems based on vague set theory. Fuzzy Sets and Systems 114, 103–113 (2000)
16. Jaulin, L., Kieffir, M., Didrit, O., Walter, E.: Applied Interval Analysis. Springer, London (2001)
17. Li, D.-F.: Multiattribute decision making models and methods using intuitionistic fuzzy sets. Journal of Computer and System Sciences 70, 73–85 (2005)

18. Lin, L., Yuan, X.-H., Xia, Z.-Q.: Multicriteria fuzzy decision-making methods based on intuitionistic fuzzy sets. Journal of Computer and System Sciences 73, 84–88 (2007)
19. Liu, H.-W., Wang, G.-J.: Multi-criteria decision-making methods based on intuitionistic fuzzy sets. European Journal of Operational Research 179, 220–233 (2007)
20. Nikolova, M., Nikolov, M., Cornelis, C., Deschrijver, G.: Survey of the research on intuitionistic fuzzy sets. Advanced Studies in Contemporary Mathematics 4, 127–157 (2002)
21. Moore, R.E.: Interval analysis. Prentice-Hall, Englewood Cliffs (1966)
22. Pasi, G., Atanassov, K., Melo Pinto, P., Yager, R., Atanassova, V.: Multi-person multi-criteria decision making: Intuitionistic fuzzy approach and generalized net model. In: Proceedings of the 10th ISPE International Conference on Concurrent Engineering Advanced Design, Production and Management Systems, Madeira, July 26-30, pp. 1073–1078 (2003)
23. Pasi, G., Yager, Y., Atanassov, K.: Intuitionistic fuzzy graph interpretations of multi-person multi-criteria decision making:Generalized net approach. In: Proceedings of 2004 second International IEEE Conference Intelligent Systems, vol. 2, pp. 434–439 (2004)
24. Sevastjanov, P.: Numerical methods for interval and fuzzy number comparison based on the probabilistic approach and Dempster-Shafer theory. Information Sciences 177, 4645–4661 (2007)
25. Sevastjanov, P., Figat, P.: Aggregation of aggregating modes in MCDM. Synthesis of Type 2 and Level 2 fuzzy sets, Omega 35, 505–523 (2007)
26. Shafer, G.: A mathematical theory of evidence. Princeton University Press, Princeton (1976)
27. Szmidt, E., Kacprzyk, J.: Intuitionistic fuzzy sets in decision making. Notes IFS 2, 15–32 (1996)
28. Szmidt, E., Kacprzyk, J.: Remarks on some applications on intuitionistic fuzzy sets in decision making. Notes IFS 2, 22–31 (1996)
29. Szmidt, E., Kacprzyk, J.: Group decision making under intuitionistic fuzzy preference relations. In: Proc. Seventh Int. Conf. IMPU 1998, Paris, pp. 172–178 (1998)
30. Szmidt, E., Kacprzyk, J.: Applications ofintuitionistic fuzzy sets in decision making. In: Proc. Eighth Cong. EUSFLAT 1998, Pampelona, pp. 150–158 (1998)
31. Vasseur, P., Pegard, C., Mouad, E., Delahoche, L.: Perceptual organization approach based on Dempster-Shafer theory. Pattern Recognition 32, 1449–1462 (1999)
32. Yager, R.R.: Multiple objective decision-making using fuzzy sets. International Journal of Man-Machine Studies 9, 375–382 (1979)
33. Yager, R.R., Kacprzyk, J., Fedrizzi, M.: Advances in Dempster-Shafer Theory of Evidence. Wiley, New York (1994)
34. Zimmerman, H.J.: Fuzzy Sets, Decision-Making and Expert Systems. Kluwer Academic Publishers, Dordrecht (1987)

Linguistic Summarization of Time Series Using Fuzzy Logic with Linguistic Quantifiers: A Truth and Specificity Based Approach

Janusz Kacprzyk and Anna Wilbik*

Systems Research Institute, Polish Academy of Sciences
ul. Newelska 6, 01-447 Warsaw, Poland
{kacprzyk,wilbik}@ibspan.waw.pl

Abstract. We reformulate and extend our previous works (cf. Kacprzyk, Wilbik and Zadrożny [7, 8, 9, 10, 11, 12, 13, 14, 15, 16, 17]), mainly towards a more complex and realistic evaluation of results on the linguistic summarization of time series which is meant as the derivation of an linguistic quantifier driven aggregation of partial trends with respect to the dynamics of change, duration and variability. We use Zadeh's calculus of linguistically quantified propositions but, in addition to the basic criterion of a degree of truth (validity), we also use a degree of specificity to make it possible to account for a frequent case that though the degree of truth of a very general (not specific) summary is high, its usefulness may be low due to its low specificity. We show an application to the absolute performance type analysis of daily quotations of an investment fund.

1 Introduction

In spite of a huge progress in broadly perceived information technology, there still exists a huge gap between the human being and the computer ("machine"). From our point of view, this gap is mainly due to the fact that natural language is the only fully natural means of articulation and communication for a human being, and it is not the case for the existing information technology (the computer). One of promising approaches to bridge this gap would be to be able to use as much natural language as possible, to be more specific, to capture the contents and meaning of sets of data in our context. For this purpose we will employ linguistic summarization of data (bases).

A linguistic summary of a set of numerical data (database) is meant as a concise, human consistent description in (quasi)natural language that captures the very essence of the content of these data. This concept was introduced by Yager [29] and then further developed, and presented in an implementable form, by Kacprzyk and Yager [18], and Kacprzyk, Yager and Zadrożny [19], using linguistically quantified propositions dealt with via Zadeh's calculus of linguistically quantified propositions [34].

---

* Supported by the Polish Ministry of Science and Higher Education under Grant No. NN516 4309 33.

Here we deal with *time series* which are a very popular type of data, and may be exemplified by sales data, quotations of shares, etc. over a certain period of time. Traditional statistical methods employed so far, though they can provide powerful tools, are not human consistent enough because they do not rely on natural language. We explicit use natural language but we clearly do not intend to replace classic statistical analyses, but rather provide an additional tool. It is a derivative and extension of the original approach of Kacprzyk, Wilbik and Zadrożny [7, 8, 9, 10, 11, 12, 13, 14, 15, 16, 17]), mainly towards a more complex evaluation of results.

The analysis of time series data involves different elements (cf. Batyrshin and Sheremetov [2, 3]). For our purposes, first, we need to identify the consecutive parts of time series within which the data exhibit some behavioral uniformity; clearly, some variability must be neglected, depending on an assumed (degree of) granularity. Here, these consecutive parts of a time series are called trends, and are represented as straight line segments. That is, we perform first a piece wise linear approximation of a time series, i.e. a time series data is a sequence of trends. The (linguistic) summaries of a time series refer to the (linguistic) summaries of (partial) trends as meant above over the entire, or maybe long enough, time span. Such a piecewise linear approximation is derived by a modified Sklansky and Gonzalez's algorithm (cf. [27]); many other methods can be used – cf. Keogh et al. [22, 23, 24].

Then, we need to aggregate the (characteristic features of) consecutive trends over the entire time span (horizon) assumed. We follow the idea initiated by Yager [29] and then shown in an implementable way in Kacprzyk and Yager [18], and Kacprzyk, Yager and Zadrożny [19], that the most comprehensive and meaningful will be a linguistic quantifier driven aggregation resulting in linguistic summaries exemplified by *"Most trends are short"* or *"Most long trends are increasing"* which are easily derived and interpreted using Zadeh's fuzzy logic based calculus of linguistically quantified propositions. Such summaries are interpreted in terms of the number, or proportion of elements possessing a certain property related to those possessing some other property, a less restrictive one. The classic Zadeh approach, the Sugeno and Choquet integrals, the OWA operators, etc. can be employed. A new quality, and an increased generality was obtained by using Zadeh's [35] protoforms as proposed by Kacprzyk and Zadrożny [20].

In this paper we will employ the classic Zadeh's fuzzy logic based calculus of linguistically quantified propositions as in the source papers by Kacprzyk, Wilbik and Zadrożny [7, 8, 9, 10, 11, 12, 13, 14, 15, 16, 17]). However, we will use different protoforms of linguistic time series summaries which are easier comprehensible by domain experts. Moreover, in addition to the degree of truth (validity) we use a degree of specificity as the second criterion to make it possible to account for a frequent case that though the degree of truth of a very general (not specific) summary is high, its usefulness may be low due to its low specificity.

We will illustrate our analysis on a linguistic summarization of daily quotations over an eight year period of an investment (mutual) fund. We will present in detail the characteristic features of trends derived under some reasonable granulations, variability, trend duration, etc.

It should be noted that the paper is in line with some other modern approaches to linguistic summarization of time series reported in the literature. First, from a slightly more general perspective, one should refer to the SumTime project coordinated by the University of Aberdeen, an EPSRC Funded Project for Generating Summaries of Time Series Data[1]. In this project English summary descriptions of a time-series data set are sought by using advanced time series and NLG (natural language generation) technologies [28]. However, the linguistic descriptions obtained do not reflect an inherent imprecision (fuzziness) as in our approach.

2 Preprocesing

A time series is a sequence of numerical data measured at uniformly spaced moments. We identify segments as linearly increasing, stable or decreasing functions, with a variable intensity, and therefore represent a given time series as a piecewise linear function. Its consecutive segments are clearly partial trends, and a global trend concerns the entire time span of the time series, or some time window that is large enough.

There are many algorithms for extracting piecewise linear segments of time series, as, e.g., the on-line (sliding window) algorithms, bottom-up or top-down strategy, cf. the works by Keogh et al. [22, 23, 24]. We use a simple on-line algorithm [8, 9] which is a modification of Sklansky and Gonzalez one [27]. Its essence is given in Fig. 1.

In our original approach (cf. Kacprzyk, Wilbik and Zadrożny [7, 8, 9, 10, 11, 12, 13, 14, 15, 16, 17]) we have considered the following three aspects of trends in time series:

- dynamics of change,
- duration, and
- variability,

and by trends we mean here global trends, over the entire time series (or some, probably large, part of it), not partial trends extracted via the segmentation algorithm used. These features of trends are the most straightforward and intuitively appealing, and will also be used in this paper.

Dynamics of change means the speed of change of the particular (consecutive) values of time series. It can be described by the slope of a line representing the (partial) trend, (cf. any angle η from the interval $\langle \gamma, \beta \rangle$ in Fig. 1(a)). To quantify the dynamics of change we may use the interval of possible angles $\eta \in \langle -90; 90 \rangle$ but, for practical reasons, we use a fuzzy granulation, for instance as in Fig. 2. Some granulation methods are given by Batyrshin et al. [1, 2]. We map a single value α (or the interval corresponding to the grey area in Fig. 1(b)) into a best matching fuzzy set (linguistic label) using some measure of a distance or similarity, cf. the book by Cross and Sudkamp [5].

---

[1] cf. www.csd.abdn.ac.uk/research/sumtime/

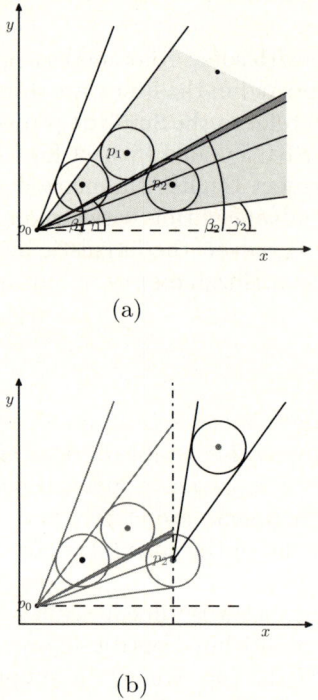

(a)

(b)

Fig. 1. An illustration of the algorithm for the uniform ε-approximation

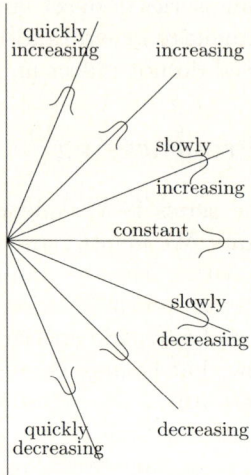

Fig. 2. A visual representation of angle granules defining the dynamics of change

Duration describes the length of a single (partial) trend, meant as a linguistic variable whose linguistic value (label) may be exemplified by a "long trend" defined as a fuzzy set with a properly defined membership function.

Variability refers to how "spread out" (in the sense of values taken on) a group of data is. We use a weighted average of values obtained by some of the following statistical measures: (1) the range, (2) the interquartile range (IQR) [the third quartile (the 75th percentile) minus the first quartile (the 25th percentile)], (3) the variance, (4) the standard deviation, (5) the mean absolute deviation (MAD).

Similarly as in the case of dynamics of change, we find for a given value of variability obtained a best matching fuzzy set (linguistic label) using, e.g., some measure of a distance or similarity.Again, the measure of variability is treated as a linguistic variable and its value is expressed using linguistic terms (labels) modeled by fuzzy sets defined by the user.

3 Linguistic Data Summaries

In Yager's [29] approach to linguistic data summarization (cf. rather Kacprzyk and Yager [18], and Kacprzyk, Yager and Zadrożny [19] for a more realistic and implementable version) we have: (1) $Y = \{y_1, \ldots, y_n\}$ is a set of objects (records) in a database, e.g., the set of workers; and (2) $A = \{A_1, \ldots, A_m\}$ is a set of attributes characterizing objects from Y, e.g., salary, age, etc. in a database of workers, and $A_j(y_i)$ denotes a value of attribute A_j for object y_i.

A linguistic summary of a data set consists of:
– a summarizer P, i.e. an attribute with a linguistic value (fuzzy predicate) defined on the domain of attribute A_j (e.g. "low" for attribute "salary");
– a quantity in agreement Q, i.e. a linguistic quantifier (e.g. most);
– truth (validity) $\mathcal{T}$ of the summary, i.e. a number from $[0, 1]$ assessing the truth (validity) of the summary;
– and, optionally, a qualifier R, i.e. another attribute together with a linguistic value defined on the domain of attribute A_k determining a (fuzzy subset) of Y (e.g. "young" for attribute "age"),
and can be exemplified by

$$\mathcal{T}(\text{most of employees earn low salary}) = 0.7 \qquad (1)$$

or, in an extended form including a qualifier (e.g. young), by

$$\mathcal{T}(\text{most of young employees earn low salary}) = 0.9 \qquad (2)$$

Thus, the core of a linguistic summary is a *linguistically quantified proposition* in the sense of Zadeh [34] which, for (1) and (2), respectively, may be written as:

$$Qy\text{'s are } P \qquad (3)$$

$$QRy\text{'s are } P \qquad (4)$$

Then, $\mathcal{T} \in [0, 1]$, i.e., the truth (validity) of a linguistic summary, directly corresponds to the truth value of (3) or (4). This may be calculated by using either the original Zadeh's fuzzy logic based calculus of linguistically quantified propositions (cf. [34]) yielding, respectively:

$$\mathcal{T}(Qy\text{'s are } P) = \mu_Q \left(\frac{1}{n} \sum_{i=1}^{n} \mu_P(y_i) \right) \qquad (5)$$

$$\mathcal{T}(QRy\text{'s are } P) = \mu_Q \left(\frac{\sum_{i=1}^{n}(\mu_R(y_i) \wedge \mu_P(y_i))}{\sum_{i=1}^{n} \mu_R(y_i)} \right) \qquad (6)$$

where $\wedge$ is the minimum operation (more generally it can be another appropriate operation, notably a *t*-norm), and Q is a fuzzy set representing the linguistic quantifier in the sense of Zadeh [34], i.e. $\mu_Q : [0, 1] \longrightarrow [0, 1]$, $\mu_Q(x) \in [0, 1]$.

We consider basically *regular non-decreasing monotone* quantifiers such that:

$$\mu(0) = 0, \quad \mu(1) = 1 \qquad (7)$$
$$x_1 \leq x_2 \Rightarrow \mu_Q(x_1) \leq \mu_Q(x_2) \qquad (8)$$

which can be exemplified by "most" given as:

$$\mu_Q(x) = \begin{cases} 1 & \text{for } x \geq 0.8 \\ 2x - 0.6 & \text{for } 0.3 < x < 0.8 \\ 0 & \text{for } x \leq 0.3 \end{cases} \qquad (9)$$

One can also use other methods to calculate $\mathcal{T}$, notably the OWA (ordered weighted aveaging) operators (cf. Yager [30, 31], Yager and Kacprzyk [33]), and the Sugeno and Choquet integrals (cf. Bosc and Lietard [4] or Grabisch [6]).

4 Protoforms of Linguistic Trend Summaries

Kacprzyk and Zadrożny [20] showed that Zadeh's [35] concept of a protoform is convenient for dealing with linguistic summaries. This approach is also employed here.

A protoform is meant as an abstract prototype (template) of a linguistically quantified proposition. For time series summaries, the use of protoforms was proposed by Kacprzyk, Wilbik and Zadrożny [7, 8, 9, 10, 11, 12, 13, 14, 15, 16, 17]. In this paper we use different protoforms of time series summaries which are more clear to the practitioners:

– for a short form:

$$\text{Among all segments, } Q \text{ are } P \qquad (10)$$

– for an extended form:

$$\text{Among all } R \text{ segments, } Q \text{ are } P \qquad (11)$$

5 Quality Measures

In Kacprzyk, Wilbik and Zadrożny's [7, 8, 9, 10, 11, 12, 13, 14, 15, 16, 17], where the new approach to the linguistic summarization of time series was proposed, the basic quality criterion was the truth value as proposed by Yager in his source paper [29].

However, then Kacprzyk and Yager [18] and Kacprzyk, Yager and Zadrożny [19] proposed other measures, notably: (1) a truth value, (2) a degree of imprecision, (3) a degree of (non)specificity, (4) a degree of fuzziness, (5) a degree of covering (support), (6) a degree of appropriateness, and (7) the length of the summary. Here, we will use, first, the traditional truth value, and then – as a second criterion – the degree of specificity.

The truth value (a degree of truth) is the basic criterion and is calculated for the simple and extended forms of the summary, as, respectively:

$$\mathcal{T}(\text{Among all } Y, \ Q \text{ are } P) = \mu_Q \left(\frac{1}{n} \sum_{i=1}^{n} \mu_P(y_i) \right) \qquad (12)$$

$$\mathcal{T}(\text{Among all } RY, \ Q \text{ are } P) = \mu_Q \left(\frac{\sum_{i=1}^{n} \mu_R(y_i) \wedge \mu_P(y_i)}{\sum_{i=1}^{n} \mu_R(y_i)} \right) \qquad (13)$$

There are a few possibilities of calculating the nonspecificity of a fuzzy set. One is the standard nonspecificity measure from fuzzy sets theory, using the so called Hartley function (cf. Klir and Wierman [25] or Klir and Yuan [26]). For a finite, nonempty (crisp) set, A, we measure this amount using a function from the class of functions $U(A) = c \log_b |A|$, where $|A|$ denotes the cardinality of A, b and c are positive constants, $b, c \geq 1$ (usually, $b = 2$ and $c = 1$). This function is applicable to finite sets only but it can be modified for infinite sets of $\mathbb{R}$ as: $U(A) = \log[1 + \mu(A)]$, where $\mu(A)$ is the measure if A defined by the Lebesque integral of the characteristic function of A. When $A = [a, b]$, than $\mu(A) = b - a$ and $U([a, b]) = \log[1 + b - a]$.

For any nonempty fuzzy set A defined on a finite universal set X, the function $U(A)$ has the form

$$U(A) = \frac{1}{h(A)} \int_0^{h(A)} \log_2 |A^\alpha| d\alpha, \tag{14}$$

where $|A^\alpha|$ is the cardinality of the α-cut of A and $h(A)$ – the height of A.

If a nonempty fuzzy set is defined in $\mathbb{R}$ and the α-cuts are infinitive sets (e.g., intervals of real numbers), then:

$$U(A) = \frac{1}{h(A)} \int_0^{h(A)} \log[1 + \mu(A^\alpha)] d\alpha, \tag{15}$$

and for other solutions, cf. Yager, Ford and Canas [32].

Here we compute the degree of nonspecificity as:

$$m(A) = \frac{\int_X \mu_A(x) dx}{range(X)} \tag{16}$$

where $range(X)$ is the length of interval of minimal and maximal possible value of X. For convenience, the value of specificity is normalized.

In most applications, both the fuzzy predicates P and R are of a simplified, atomic form referring to just one attribute. They can be extended to cover more sophisticated summaries involving some confluence of various attribute values as, e.g., "slowly decreasing and short" trends. This can be done by using t-norms (here, the minimum or product) for conjunction and a corresponding s-norm (here, the maximum or probabilistic sum) for disjunction.

Then the degrees of specificity of "Among all Y, Q are P" and "Among all RY, Q are P" are, respectively:

$$d_s(\text{"Among all } Y, Q \text{ are } P\text{"}) = 1 - (m(P) \wedge m(Q)) \tag{17}$$

$$d_s(\text{"Among all } RY, Q \text{ are } P\text{"}) = 1 - (m(P) \wedge m(R) \wedge m(Q)) \tag{18}$$

where $U(P)$ is the degree of nonspecificity of the summarizer P, given by (16), $U(R)$ is the degree of nonspecificity of the qualifier R, $U(Q)$ is the degree of nonspecificity of the quantifier Q, and $\wedge$ is a t-norm (minimum or product).

6 Numerical Experiments

Our method was tested on data on quotations of an investment (mutual) fund that invests at most 50% of assets in shares listed at the Warsaw Stock Exchange. Data shown in Figure 3 were collected from April 1998 until July 2007 with the value of one share equal to PLN 10.00 in the beginning of the period to PLN 55.27 at the end of the time span considered (PLN stands for the Polish Zloty). The minimal value recorded was PLN 6.88 while the maximal one during this period was PLN 57.85. The biggest daily increase was equal to PLN 1.27, while the biggest daily decrease was equal to PLN 2.41. Using the Sklansky and Gonzalez

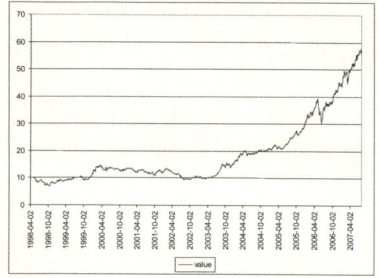

Fig. 3. Daily quotations of an investment fund in question

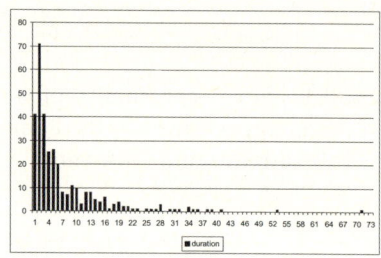

Fig. 4. Histogram of duration of trends (in the number of days)

algorithm and $\varepsilon = 0.25$ we obtained 326 extracted (partial) trends. The shortest trend took 2 days, while the longest 71 days. The histogram of the duration of trends is presented in Fig. 4. Figure 5 shows the histogram of angles which characterize the dynamics of change. The histogram of the variability of trends (in %), assumed to be – for simplicity – the interquartile range only, is presented in Fig. 6.

We will show an absolute performance type analysis that is we will just deal with the values (price quotations) of shares of the investment fund in question. We will not deal with relations between these values and some appropriate

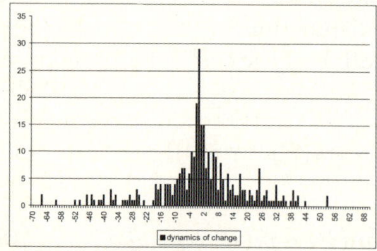

Fig. 5. Histogram of angles (in degrees) characterizing the dynamic of change

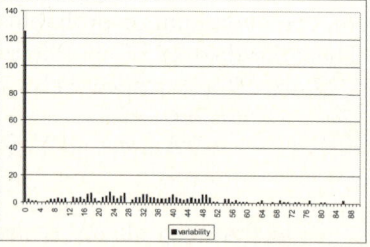

Fig. 6. Histogram of the variability (the interquartile range) of trends

Table 1.

linguistic summary	$\mathcal{T}$	d_s	α
Among all trends, most are short	0.7129	0.94	0.7697
Among all trends, most are constant	0.6318	0.861	0.6891
Among all trends, most are of a low variability	0.7028	0.7	0.7021
Among all short trends, most are of a low variability	0.8784	0.94	0.8938
Among all trends of a low variability, most are short	0.8898	0.94	0.9024
Among all increasing trends, most are of a low variability	0.8867	0.7	0.8400
Among all medium trends, at least around a half is of medium variability	1.0000	0.885	0.9713
Among all trends of a high variability, at least around a half is increasing	0.9161	0.7	0.8621
Among all decreasing trends, almost all are short	1.0000	0.94	0.985

Table 2.

linguistic summary	$\mathcal{T}$	d_s	α
Among all trends, most are constant	0.6318	0.861	0.6891
Among all trends, at least around a half is very short	1.000	0.96	0.99
Among all trends, at least around a half is of a very low variability	1.000	0.85	0.9625
Among all very short trends, most are of a very low variability	0.935	0.96	0.9413
Among all trends of a very low variability, most are very short	1.000	0.96	0.99
Among all constant trends, at least around a half are short	0.7476	0.93	0.7932
Among all trends of a very high variability, at least around a half are quickly increasing	0.8299	0.85	0.8349
Among all trends of medium length, almost all are constant	1.000	0.885	0.9713
Among all quickly decreasing trends, almost all are very short	1.000	0.96	0.99
Among all long trends, much more than a half is of a low variability	0.5519	0.8	0.6139

benchmarks exemplified by daily values of some stock market index, or an appropriate mix of indexes, daily percentual change (absolute or related to the daily percentual change of some benchmarks), etc. as is often done by professionals.

Some interesting summaries obtained, employing the classic Zadeh calculus of linguistically quantified propositions [with properly defined elements like the fuzzy linguistic quantifiers, like (9) for "most"], and for different granulations of the dynamics of change, duration and variability, are (we assume the weights for $\mathcal{T}$ and d_s equal to 0.5, and the weighted average is α):
– for 3 labels for the dynamics of change (*decreasing, constant, increasing*), the duration (*short, medium length, long*) and the variability (*low, medium, high*) are in Tab. 1.
– for 5 labels for the dynamics of change (*quickly decreasing, decreasing, constant, increasing, quickly increasing*), 5 labels for the duration (*very short, short,*

Table 3.

linguistic summary	T	d_s	α
Among all trends, almost none are of rather high variability	1.0000	0.925	0.9813
Among all trends, at least around a half are very short	1.0000	0.96	0.99
Among all trends, at least around a half are constant	1.0000	0.913	0.9783
Among all trends, at least around a half are of very low variability	0.9138	0.875	0.9041
Among all trends, at most around one third is slowly increasing	1.0000	0.945	0.9863
Among all very short trends, most are of a very low variability	0.9045	0.96	0.9184
Among all trends of a very low variability, most are very short	1.0000	0.96	0.99
Among all slowly decreasing trends, most are very short	0.9709	0.96	0.9682
Among all constant trends, at least around a half is short	0.6827	0.93	0.7445
Among all trends of medium length, almost all are constant	1.0000	0.917	0.9793
Among all quickly decreasing trends, almost all are very short	1.0000	0.96	0.99
Among all long trends, much more than a half is of a rather low variability	0.7344	0.9	0.7758

medium length, long, very long) and 5 labels for the variability (*very low, low, medium, high, very high*) are in Tab. 2.
– for 7 labels for the dynamics of change (*quickly decreasing, decreasing, slowly decreasing, constant, slowly increasing, increasing, quickly increasing*), 7 labels for the duration (*very short, short, rather short, medium length, rather long, long, very long*) and 7 labels for the variability (*very low, low, rather low, medium, rather high, high, very high*) are in Tab. 3.

It turns out that the by using two quality criteria, the degree of truth and specificity, we obtain much more insight into the essence of the time series.

7 Concluding Remarks

We reformulated and extended our previous works on linguistic summarization of time series by adding to the basic quality criterion of a degree of truth (validity), a degree of specificity. This made it possible to make a more reasonable choice between the summaries obtained as, frequently, though the degree of truth of a very general (not specific) summary may be high, its usefulness may be low. The results obtained on the analysis of the absolute performance of daily quotations of an investment fund seem very promising.

References

1. Batyrshin, I.: On granular derivatives and the solution of a granular initial value problem. International Journal Applied Mathematics and Computer Science 12(3), 403–410 (2002)
2. Batyrshin, I., Sheremetov, L.: Perception based functions in qualitative forecasting. In: Batyrshin, I., et al. (eds.) Perception-based Data Mining and Decision Making in Economics and Finance, pp. 119–134. Springer, Heidelberg (2006)

3. Batyrshin, I., Sheremetov, L.: Towards perception based time series data mining. In: Nikravesh, M., et al. (eds.) Forging New Frontiers. Fuzzy Pioneers I, pp. 217–230. Springer, Heidelberg (2007)
4. Bosc, P., Lietard, L., Pivert, O.: Quantified statements and database fuzzy queries. In: Bosc, P., Kacprzyk, J. (eds.) Fuzziness in Database Management Systems. Springer, Heidelberg (1995)
5. Cross, V., Sudkamp, T.: Similarity and Compatibility in Fuzzy Set Theory: Assessment and Applications. Springer, Heidelberg (2002)
6. Grabisch, M.: Fuzzy integral as a flexible and interpretable tool of aggregation. In: Bouchon-Meunier, B. (ed.) Aggregation and Fusion of Imperfect Information, pp. 51–72. Physica–Verlag (1998)
7. Kacprzyk, J., Wilbik, A., Zadrożny, S.: Linguistic summarization of trends: a fuzzy logic based approach. In: Proceedings of the 11th International Conference Information Processing and Management of Uncertainty in Knowledge-based Systems, pp. 2166–2172 (2006)
8. Kacprzyk, J., Wilbik, A., Zadrożny, S.: Linguistic summaries of time series via a quantifier based aggregation using the Sugeno integral. In: Proceedings of 2006 IEEE World Congress on Computational Intelligence, pp. 3610–3616. IEEE Press, Los Alamitos (2006)
9. Kacprzyk, J., Wilbik, A., Zadrożny, S.: On some types of linguistic summaries of time series. In: Proceedings of the 3rd International IEEE Conference Intelligent Systems, pp. 373–378. IEEE Press, Los Alamitos (2006)
10. Kacprzyk, J., Wilbik, A., Zadrożny, S.: A linguistic quantifier based aggregation for a human consistent summarization of time series. In: Lawry, J., et al. (eds.) Soft Methods for Integrated Uncertainty Modelling, pp. 186–190. Springer, Heidelberg (2006)
11. Kacprzyk, J., Wilbik, A., Zadrożny, S.: Capturing the essence of a dynamic behavior of sequences of numerical data using elements of a quasi-natural language. In: Proceedings of the 2006 IEEE International Conference on Systems, Man, and Cybernetics, pp. 3365–3370. IEEE Press, Los Alamitos (2006)
12. Kacprzyk, J., Wilbik, A., Zadrożny, S.: Linguistic Summarization of Time Series by Using the Choquet Integral. In: Melin, P., Castillo, O., Aguilar, L.T., Kacprzyk, J., Pedrycz, W. (eds.) IFSA 2007. LNCS (LNAI), vol. 4529, pp. 284–294. Springer, Heidelberg (2007)
13. Kacprzyk, J., Wilbik, A., Zadrożny, S.: Linguistic summarization of time series under different granulation of describing features. In: Kryszkiewicz, M., Peters, J.F., Rybinski, H., Skowron, A. (eds.) RSEISP 2007. LNCS (LNAI), vol. 4585, pp. 230–240. Springer, Heidelberg (2007)
14. Kacprzyk, J., Wilbik, A., Zadrożny, S.: Analysis of time series via their linguistic summarization: the use of the Sugeno integral. In: de Macedo Mourelle, L., et al. (eds.) Proceedings of the 7th International Conference on Intelligent Systems Design and Applications-ISDA 2007, pp. 262–267. IEEE Press, Los Alamitos (2007)
15. Kacprzyk, J., Wilbik, A., Zadrożny, S.: Linguistic Summaries of Time Series via an OWA Operator Based Aggregation of Partial Trends. In: Proceedings of the FUZZ-IEEE 2007 IEEE International Conference on Fuzzy Systems, pp. 467–472. IEEE Press, Los Alamitos (2007)
16. Kacprzyk, J., Wilbik, A., Zadrożny, S.: Mining time series data via linguistic summaries of trends by using a modified Sugeno integral based aggregation. In: Proceedings of IEEE Symposium on Computational Intelligence and Data Mining, pp. 742–749. IEEE Press, Los Alamitos (2007)

17. Kacprzyk, J., Wilbik, A., Zadrożny, S.: On Linguistic Summaries of Time Series via a Quantifier Based Aggregation Using the Sugeno Integral. In: Castillo, O., et al. (eds.) Hybrid Intelligent Systems Analysis and Design, pp. 421–439. Springer, Heidelberg (2007)
18. Kacprzyk, J., Yager, R.R.: Linguistic summaries of data using fuzzy logic. International Journal of General Systems 30, 133–154 (2001)
19. Kacprzyk, J., Yager, R.R., Zadrożny, S.: A fuzzy logic based approach to linguistic summaries of databases. International Journal of Applied Mathematics and Computer Science 10, 813–834 (2000)
20. Kacprzyk, J., Zadrożny, S.: Linguistic database summaries and their protoforms: toward natural language based knowledge discovery tools. Information Sciences 173, 281–304 (2005)
21. Kacprzyk, J., Zadrożny, S.: Fuzzy linguistic data summaries as a human consistent, user adaptable solution to data mining. In: Gabrys, B., et al. (eds.) Do Smart Adaptive Systems Exist? pp. 321–339. Springer, Heidelberg (2005)
22. Keogh, E., Pazzani, M.: An enhanced representation of time series which allows fast and accurate classification, clustering and relevance feedback. In: Proceedings of the 4th International Conference on Knowledge Discovery and Data Mining, pp. 239–241 (1998)
23. Keogh, E., Chu, S., Hart, D., Pazzani, M.: An Online Algorithm for Segmenting Time Series. In: Proceedings of IEEE International Conference on Data Mining, pp. 289–296 (2001)
24. Keogh, E., Chu, S., Hart, D., Pazzani, M.: Segmenting Time Series: A Survey and Novel Approach. In: Last, M., et al. (eds.) Data Mining in Time Series Databases. World Scientific Publishing, Singapore (2004)
25. Klir, G.J., Wierman, M.J.: Uncertainty-Based Information. In: Elements of Generalized Information Theory. Physica-Verlag (1999)
26. Klir, G.J., Yuan, B.: Fuzzy Stes and Fuzzy Logic, Theory and Applications. Prentice-Hall, Englewood Cliffs (1995)
27. Sklansky, J., Gonzalez, V.: Fast polygonal approximation of digitized curves. Pattern Recognition 12(5), 327–331 (1980)
28. Sripada, S., Reiter, E., Davy, I.: SumTime-Mousam: Configurable Marine Weather Forecast Generator. Expert Update 6(3), 4–10 (2003)
29. Yager, R.R.: A new approach to the summarization of data. Information Sciences 28, 69–86 (1982)
30. Yager, R.R.: On ordered weighted averaging aggregation operators in multicriteria decision making. IEEE Transactions on Systems, Man and Cybernetics SMC-18, 183–190 (1988)
31. Yager, R.R.: Quantifier guided aggregation using OWA operators. International Journal of Intelligent Systems 11, 49–73 (1996)
32. Yager, R.R., Ford, K.M., Canas, A.J.: An Approach to the linguistic summarization of data. In: Bouchon-Meunier, B., Zadeh, L.A., Yager, R.R. (eds.) IPMU 1990. LNCS, vol. 521, pp. 456–468. Springer, Heidelberg (1991)
33. Yager, R.R., Kacprzyk, J.: The Ordered Weighted Averaging Operators: Theory and Applications. Kluwer, Boston (1997)
34. Zadeh, L.A.: A computational approach to fuzzy quantifiers in natural languages. Computers and Mathematics with Applications 9, 149–184 (1983)
35. Zadeh, L.A.: A prototype-centered approach to adding deduction capabilities tosearch engines – the concept of a protoform. In: Proceedings of the Annual Meeting of the North American Fuzzy Information Processing Society (NAFIPS 2002), pp. 523–525 (2002)

TS Fuzzy Rule-Based Systems with Polynomial Membership Functions

Jacek Kluska

Faculty of Electrical and Computer Engineering
Rzeszow University of Technology
35-959 Rzeszow, W. Pola 2, Poland
jacklu@prz.rzeszow.pl

Abstract. The work presents some results concerning analytical modeling using the Takagi-Sugeno fuzzy rule-based system, which can be used for exact fuzzy modeling of some class of conventional systems. A special attention was paid to the so called P2-TS systems, which use the polynomial membership functions of the second degree. Theorems provide necessary and sufficient conditions for transformation of fuzzy rules into the crisp model of the system and vice-versa.

1 Introduction

Fuzzy rule-based systems have been successfully used for modeling many real processes such as fuzzy controllers and dynamical plants. However, they are often constructed via the "knowledge engineering approach" as opposed to the mathematical approaches. Including Takagi-Sugeno models [7], the fuzzy "If-then" rule-based systems have been treated and used as black-boxes in the sense that their analytical structures are unknown. Design and analysis of such (inherently nonlinear) systems is very difficult and mostly involves trial-and error methods. There is a need to have a better connection between fuzzy systems and their classical counterparts. Availability of the structure information especially in the form of analytical model may lead to much more effective analysis and design techniques that require less trial-and-error effort. From the control theory point of view, the development of these techniques should utilize the well-developed conventional control theory, as fuzzy control problems have already been translated into nonlinear control problems [9], [10]. In this spirit, an exact relationship between fuzzy models and their classical counterparts has been presented in several works [2], [3], [4], [8], [9], [10].

In [6] the so called transformation lemma developed in [3] was used for exact fuzzy modeling of some class of conventional systems. The results were based on author's theorems, which provide necessary and sufficient conditions for the fuzzy rules transformation into the conventional (crisp) model and vice-versa. In the cited papers linear membership functions of fuzzy sets were used. Such fuzzy sets have been often used in the engineering practice; they lead to the so called P1-TS rule-based systems [3]. However, the class of functions which can be obtained by the P1-TS systems is limited. Therefore in this work we will be interested in using polynomial membership functions of the degree higher than one.

Many researchers are interested in using polynomials in the context of universal approximation of continuous functions, by constructing fuzzy or neurofuzzy processing units (see e.g. [5], [11]). By such approximation, an appropriate numerical input-output data set is required and the approximation is performed with some degree of accuracy. In contrast to this approach we do not require the input-output data. By using highly interpretable zero-order TS rule-based system [1], we will be able to exactly model of some subclass of the Kolmogorov-Gabor polynomials by using the membership functions of fuzzy sets as the second degree polynomials.

2 TS Systems with Two Polynomial Membership Functions for Every Input

It would appear that by using nonlinear membership functions, one can get sufficiently large class of functions, with which the TS rule-based system is equivalent. This conjecture is something illusory, if we increase complexity of membership functions of fuzzy sets only, while preserving the number of fuzzy sets assigned for crisp inputs of the system. The number of fuzzy sets is very important, because it determines the number of consequents of the rules; thus, it constrains the class of functions performed by the zero-order TS rule-based systems. This fact will be exemplified below.

Remark 1. Suppose the inputs of a TS system are $z_k \in [-\alpha_k, \beta_k]$, ($k = 1, 2, \ldots, n$), and every input has assigned two complementary membership functions, say $N_k(z_k)$ and $P_k(z_k) = 1 - N_k(z_k)$. If all membership functions are polynomials of degree d, then

(1) the crisp output $f(z_1, \ldots, z_n)$ of this system is the following multivariate polynomial

$$f(z_1, \ldots, z_n) = \sum_{(p_1, \ldots, p_n) \in \{0,1,2,\ldots,d\}^n} \theta_{p_1, \ldots, p_n} z_1^{p_1} z_2^{p_2} \cdots z_n^{p_n}, \qquad (1)$$

where $\theta_{p_1,\ldots,p_n} \in \mathbb{R}$ in the case of the zero-order TS system,
(2) every multilinear function of type (1), can be exactly expressed by the fuzzy "If-then" rules if, and only if the degree of polynomials is $d = 1$,
(3) not every nonlinear function of type (1) can be unambiguously expressed by the fuzzy "If-then" rules, when the degree $d > 1$.

Proof. (1) First observe that system's output S is a linear combination of 2^n polynomials in the form " $\prod_{k=1}^{n} \left(a_{d,k} z_k^d + \ldots + a_{1,k} z_k + a_{0,k} \right)$ ". Thus, the output S is in the form (1), indeed. (2) For two fuzzy sets for every input (N_k and P_k), there are 2^n consequents of the rules, which are free design parameters. The polynomial of degree d is described by $(d+1)$ parameters. Thus, the number of functions (1), which are structurally different one from another, is $(d+1)^n$, and it is equal to the number of different consequents of the rules if, and only if $(d+1)^n = 2^n$. In this case we apply Theorem which has been proved in [3]. (3) For $d \geq 2$ we have $(d+1)^n > 2^n$. Thus, not every nonlinear function (1) can be exactly expressed by TS system; this finishes the proof of Remark 1. □

Now we will consider an example, which is of twofold goal. Firstly we will give an additional proof of Remark 1 for the second degree polynomial ($d = 2$). Secondly we will show that by using some nonlinear bijection for the crisp input x of the TS system with two linear membership functions, we can obtain its nonlinear output.

Example 1. Let us consider the zero-order TS system with the input x and the output S. We define a nonlinear mapping between the original input $x \in [-\alpha, \beta]$ and an ancillary variable $z \in [-\alpha, \beta]$, in the form of the second order polynomial

$$z(x) = x + m\frac{(x+\alpha)(x-\beta)}{\alpha+\beta}, \qquad (2)$$

where m is a parameter. We assume that $0 \neq |m| < 1$, because (2) is a bijection $z : [-\alpha, \beta] \to [-\alpha, \beta]$ if, and only if, $|m| < 1$, and we omit the trivial case $z = x$. If the membership functions are linear:

$$N(z) = (\alpha+\beta)^{-1}(\beta-z), \qquad P(z) = 1 - N(z),$$

then from two fuzzy rules:

$$\left.\begin{array}{l} R_1 : \text{If } z \text{ is } N, \text{ then } S = q_1, \\ R_2 : \text{If } z \text{ is } P, \text{ then } S = q_2, \end{array}\right\}$$

we obtain the system's output

$$S(x) = \frac{q_1 N(z) + q_2 P(z)}{N(z) + P(z)} = Ax^2 + Bx + C, \qquad (3)$$

where

$$A = m\frac{q_2 - q_1}{(\alpha+\beta)^2},$$

$$B = \frac{(\alpha+\beta+m(\alpha-\beta))(q_2-q_1)}{(\alpha+\beta)^2},$$

$$C = \frac{q_1\beta(\alpha+\beta+m\alpha) + q_2(\alpha\beta - m\alpha\beta + \alpha^2)}{(\alpha+\beta)^2}.$$

Thus, independently of the consequents of the rules (q_1 and q_2), the system's output is restricted to the following class of functions as second degree polynomials

$$S(x) = Ax^2 + \left(\frac{\alpha+\beta}{m} + \alpha - \beta\right)Ax + C, \qquad x \in [-\alpha, \beta], \qquad (4)$$

where $A, C \in \mathbb{R}$, by $1 > |m| \neq 0$. This means that there are "many", but not all second degree polynomials, which can be exactly represented by the rule-based system. For example, by the fixed interval $[-\alpha, \beta]$, we are not able to formulate such two fuzzy rules, that the rule-based system would be equivalent with the following polynomial

$$f(x) = Ax^2 + A(\alpha - \beta) x + C, \qquad x \in [-\alpha, \beta], \tag{5}$$

where $A, C \in \mathbb{R}$. This is because there is no m such, that $0 \neq |m| < 1$ and $\left(\dfrac{\alpha + \beta}{m} + \alpha - \beta \right) A = (\alpha - \beta) A$, for any real α, β and A. In other words, the function (5) is not from the class of functions defined by (4). This example shows by contradiction that the second part of Remark 1 is true.

The zero-order rule-based TS systems in which the membership functions of input variables are polynomials of the degree d will be called "Pd-TS systems". A special attention will be paid to P2-TS systems.

3 Normalized Membership Functions for P2-TS Systems

From the preceding Section we know that it is not possible to obtain every second degree polynomial by using the TS systems, in which only two complementary membership functions as second degree polynomials are defined. However, we will prove further on that three membership functions as the second degree polynomials suffice to model any second degree polynomial function. Such membership functions defining the fuzzy sets for input variables will be defined below.

In the interval $[-\alpha, \beta]$ we define three membership functions of fuzzy sets, say $N(z)$, $Z(z)$ and $P(z)$, which are the second degree polynomials and satisfy the following additional conditions:

1. $N : [-\alpha, \beta] \to [0, 1]$ is a monotonic function with *negative slope*, i.e.
   $dN(z)/dz < 0$ for $z \in [-\alpha, \beta]$, which satisfies two boundary conditions:
   (a) $N(-\alpha) = 1$,
   (b) $N(\beta) = 0$.
2. $P : [-\alpha, \beta] \to [0, 1]$ is the monotonic function with *positive slope*, i.e.
   $dP(z)/dz > 0$ for $z \in [-\alpha, \beta]$, symmetric to the function N with respect to the interval centre $\sigma \in [-\alpha, \beta]$:

$$\sigma = \frac{-\alpha + \beta}{2}. \tag{6}$$

3. $Z : [-\alpha, \beta] \to [0, 1]$ is the function which reaches *zero slope* in σ, i.e. $dZ(\sigma)/dz = 0$.
4. The functions N, Z and P satisfy the *normalization condition*

$$N(z) + Z(z) + P(z) = 1, \qquad \forall\, z \in [-\alpha, \beta]. \tag{7}$$

One can prove that the functions N, Z and P meeting the above needs can be expressed as follows

$$N(z) = \frac{(\alpha + \beta - \lambda(z + \alpha))(\beta - z)}{(\alpha + \beta)^2}, \tag{8}$$

$$Z(z) = 2\lambda \frac{(\beta - z)(z + \alpha)}{(\alpha + \beta)^2}, \tag{9}$$

$$P(z) = \frac{(\alpha + \beta + \lambda(z - \beta))(z + \alpha)}{(\alpha + \beta)^2}, \tag{10}$$

where the parameter λ satisfies the condition

$$0 < \lambda \leq 1. \tag{11}$$

We do not allow $\lambda = 0$, since in such case $Z(z) = 0$ for all z, and there would be two nonzero membership functions only: $N(z)$ and $P(z)$. In other words, by $\lambda = 0$, the class of P2-TS systems reduces to the P1-TS systems. Observe that N and P are normal fuzzy sets but Z is not. The cores of the fuzzy sets N, Z and P are three characteristic points of the universe of discourse: "$-\alpha$", "σ" and "β", respectively.

The membership functions N, Z and P have clear linguistic interpretation in any case of boundaries "$-\alpha$" and "β" as real numbers:

1. If $-\alpha < \beta < 0$, then N can be interpreted as *negative big*, Z - *negative medium*, and P - *negative small*,
2. If $-\alpha < \beta = 0$, then N can be interpreted as *negative*, Z - *negative small*, and P - *negative zero*,
3. If $-\alpha < 0 < \beta$, then N can be interpreted as *negative*, Z - *zero*, and P - *positive*,
4. If $0 = -\alpha < \beta$, then N can be interpreted as *positive zero*, Z - *positive small*, and P - *positive*,
5. If $0 < -\alpha < \beta$, then N can be interpreted as *positive small*, Z - *positive medium*, and P - *positive big*.

The linguistic terms can be substituted by the others depending on the context or specific application. The rule-based TS systems with the above membership functions we will call P2-TS systems for short.

4 SISO P2-TS System

Now we will consider P2-TS system with single input $z \in [-\alpha, \beta]$ and single output S. The rule-base structure is as follows

$$\left.\begin{aligned} R_1 &: \text{If } z \text{ is } N, \text{ then } S = q_0, \\ R_2 &: \text{If } z \text{ is } Z, \text{ then } S = q_1, \\ R_3 &: \text{If } z \text{ is } P, \text{ then } S = q_2. \end{aligned}\right\} \tag{12}$$

The system's output as a function of the input variable z is given by

$$S(z) = N(z)q_0 + Z(z)q_1 + P(z)q_2 = [N(z), Z(z), P(z)]\mathbf{q}, \qquad (13)$$

where $\mathbf{q} = [q_0, q_1, q_2]^T$, and N, Z and P are defined in (8)-(10). By $\mathbf{s}$ we denote the vector containing values of system's output in the cores of the fuzzy sets N, Z and P, respectively

$$\mathbf{s} = [S(-\alpha), S(\sigma), S(\beta)]^T.$$

It can be expressed equivalently by

$$\mathbf{s} = \mathbf{Rq}, \qquad (14)$$

where the matrix $\mathbf{R}$ contains the membership degrees in the cores of the fuzzy sets

$$\mathbf{R} = \begin{bmatrix} N(-\alpha) & Z(-\alpha) & P(-\alpha) \\ N(\sigma) & Z(\sigma) & P(\sigma) \\ N(\beta) & Z(\beta) & P(\beta) \end{bmatrix} = \begin{bmatrix} 1 & 0 & 0 \\ (2-\lambda)/4 & \lambda/2 & (2-\lambda)/4 \\ 0 & 0 & 1 \end{bmatrix}. \qquad (15)$$

Observe that $S(-\alpha) = q_0$ and $S(\beta) = q_2$. However, the consequent of the fuzzy rule R_2 in (12) is q_1, but

$$S(\sigma) = q_1' = \frac{2-\lambda}{4}q_0 + \frac{\lambda}{2}q_1 + \frac{2-\lambda}{4}q_2 \neq q_1,$$

and there is no such $\lambda \in (0,1]$ for which q_1' would be equal to q_1. The maximal influence of the rule consequent q_1 for the crisp output q_1' one obtains for maximal value of the parameter λ. Therefore we prefer to use $\lambda = 1$.

Now we introduce a *generator* for the SISO P2-TS system as follows

$$\mathbf{g}(z) = \begin{bmatrix} 1 \\ z \\ z^2 \end{bmatrix}. \qquad (16)$$

According to the Remark 1, the function $f(z)$ with which the rule-based system (12) is equivalent, is of the form

$$f(z) = \mathbf{g}^T(z)\boldsymbol{\theta}, \qquad (17)$$

where $\boldsymbol{\theta} = [\theta_0, \theta_1, \theta_2]^T$. The equality $S(z) = f(z)$ must be satisfied for $z \in [-\alpha, \beta]$, particularly for all three characteristic points from the set $\{-\alpha, \sigma, \beta\} \subset [-\alpha, \beta]$. Thus,

$$\mathbf{s} = \begin{bmatrix} f(-\alpha) \\ f(\sigma) \\ f(\beta) \end{bmatrix} = \begin{bmatrix} \mathbf{g}^T(-\alpha) \\ \mathbf{g}^T(\sigma) \\ \mathbf{g}^T(\beta) \end{bmatrix} \begin{bmatrix} \theta_0 \\ \theta_1 \\ \theta_2 \end{bmatrix} = \boldsymbol{\Gamma}^T\boldsymbol{\theta},$$

must be satisfied, where the matrix $\boldsymbol{\Gamma}$ is concatenation of the values of the generator (16) in the points $-\alpha$, σ, and β, respectively, i.e.

$$\boldsymbol{\Gamma} = [\mathbf{g}(-\alpha), \mathbf{g}(\sigma), \mathbf{g}(\beta)].$$

Thus, we obtain an exact relationship between consequents $\mathbf{q}$ of the rules (12) and parameters $\boldsymbol{\theta}$ of the system's function (17) as follows

$$\mathbf{R}\mathbf{q} = \boldsymbol{\Gamma}^T \boldsymbol{\theta}.$$

Thus,

$$\mathbf{q} = \mathbf{R}^{-1} \boldsymbol{\Gamma}^T \boldsymbol{\theta} = \boldsymbol{\Omega}^T \boldsymbol{\theta}, \qquad (18)$$

where the *fundamental matrix* for SISO P2-TS system is defined by

$$\boldsymbol{\Omega} = \boldsymbol{\Gamma}\left(\mathbf{R}^T\right)^{-1} = \begin{bmatrix} 1 & 1 & 1 \\ -\alpha & \sigma & \beta \\ \alpha^2 & \left(\alpha^2 + \beta^2\right)/2 - (\alpha+\beta)^2/(2\lambda) & \beta^2 \end{bmatrix}, \qquad (19)$$

where $0 < \lambda \leq 1$.

5 MISO P2-TS System with Two and More Inputs

In this Section we will investigate P2-TS systems with the inputs $z_1, \ldots, z_n$. For such systems, in order to define three membership functions N_k, Z_k and P_k as the functions of variables z_k, $(k = 1, 2, \ldots, n)$, we can choose individual parameter values $\lambda_1, \lambda_2, \ldots, \lambda_n$ for the particular inputs. The membership functions take the following general form

$$N_k(z_k) = \frac{(\alpha_k + \beta_k - \lambda_k(z_k + \alpha_k))(\beta_k - z_k)}{(\alpha_k + \beta_k)^2}, \qquad (20)$$

$$Z_k(z_k) = 2\lambda_k \frac{(\beta_k - z_k)(z_k + \alpha_k)}{(\alpha_k + \beta_k)^2}, \qquad (21)$$

$$P_k(z_k) = \frac{(\alpha_k + \beta_k + \lambda_k(z_k - \beta_k))(z_k + \alpha_k)}{(\alpha_k + \beta_k)^2}, \qquad (22)$$

where $\lambda_k \in (0, 1]$, $(k = 1, \ldots, n)$. If there are no contraindications, we prefer to assume in practice the same value $\lambda_k = 1$ for all variables.

Let M_n be a crisp set of 3^n characteristic points as n-dimensional vectors

$$M_n = \{-\alpha_1, \sigma_1, \beta_1\} \times \{-\alpha_2, \sigma_2, \beta_2\} \times \ldots \times \{-\alpha_n, \sigma_n, \beta_n\} \subset D^n. \qquad (23)$$

The set of characteristic points for P2-TS system includes all vertices of the hypercube $D^n = \times_{k=1}^n [-\alpha_k, \beta_k]$. We order M_n as follows. For every n-dimensional vector $(\gamma_1, \ldots, \gamma_n)$ as an element of the set M_n we define the corresponding index v according to the following bijection

$$v = 1 + \sum_{i=1}^{n} 3^{i-1} p_i, \qquad (24)$$

where

$$p_i = \begin{cases} 0 \Leftrightarrow \gamma_i = -\alpha_i \\ 1 \Leftrightarrow \gamma_i = \sigma_i \\ 2 \Leftrightarrow \gamma_i = \beta_i \end{cases}, \quad i = 1, \ldots, n. \qquad (25)$$

Thus, every element of the set M_n has an index. If $(\gamma'_1, \ldots, \gamma'_n) \in M_n$ and $(\gamma''_1, \ldots, \gamma''_n) \in M_n$, and the sign "$\prec$" denotes the ordering relation, then

$$(\gamma'_1, \ldots, \gamma'_n) \prec (\gamma''_1, \ldots, \gamma''_n) \quad \Leftrightarrow \quad v_{\gamma'_1, \ldots, \gamma'_n} < v_{\gamma''_1, \ldots, \gamma''_n}. \qquad (26)$$

- For $n = 1$ we have $v_{-\alpha} = 1 < v_\sigma = 2 < v_\beta = 3$ and therefore $-\alpha \prec \sigma \prec \beta$.
- For $n = 2$ the inequalities between indices are
$v_{-\alpha_1, -\alpha_2} = 1 < v_{\sigma_1, -\alpha_2} = 2 < v_{\beta_1, -\alpha_2} = 3 < v_{-\alpha_1, \sigma_2} = 4 < v_{\sigma_1, \sigma_2} = 5 < v_{\beta_1, \sigma_2} = 6 < v_{-\alpha_1, \beta_2} = 7 < v_{\sigma_1, \beta_2} = 8 < v_{\beta_1, \beta_2} = 9$.
Thus, the members of M_2 are ordered as follows
$(-\alpha_1, -\alpha_2) \prec (\sigma_1, -\alpha_2) \prec (\beta_1, -\alpha_2) \prec (-\alpha_1, \sigma_2) \prec (\sigma_1, \sigma_2) \prec (\beta_1, \sigma_2) \prec (-\alpha_1, \beta_2) \prec (\sigma_1, \beta_2) \prec (\beta_1, \beta_2)$.

The process of ordering the set M_n is unambiguous and simple for any number of system's inputs.

Finally, for the MISO P2-TS system with the inputs $z_1, \ldots, z_k$ let us introduce a *generator*

$$\mathbf{g}_0 = 1,$$

$$\mathbf{g}_{k+1}(z_1, \ldots, z_{k+1}) = \begin{bmatrix} \mathbf{g}_k(z_1, \ldots, z_k) \\ z_{k+1} \mathbf{g}_k(z_1, \ldots, z_k) \\ z_{k+1}^2 \mathbf{g}_k(z_1, \ldots, z_k) \end{bmatrix} \in \mathbb{R}^{3^{k+1}}, \quad k = 0, 1, 2, \ldots, n-1, \qquad (27)$$

which is of a great importance for such systems.

For $n = 2$ the rule-base structure is as follows

$$\left.\begin{aligned}
R_1 &: \text{If } z_1 \text{ is } N_1 \text{ and } z_2 \text{ is } N_2, \text{ then } S = q_{00}, \\
R_2 &: \text{If } z_1 \text{ is } Z_1 \text{ and } z_2 \text{ is } N_2, \text{ then } S = q_{10}, \\
R_3 &: \text{If } z_1 \text{ is } P_1 \text{ and } z_2 \text{ is } N_2, \text{ then } S = q_{20}, \\
R_4 &: \text{If } z_1 \text{ is } N_1 \text{ and } z_2 \text{ is } Z_2, \text{ then } S = q_{01}, \\
R_5 &: \text{If } z_1 \text{ is } Z_1 \text{ and } z_2 \text{ is } Z_2, \text{ then } S = q_{11}, \\
R_6 &: \text{If } z_1 \text{ is } P_1 \text{ and } z_2 \text{ is } Z_2, \text{ then } S = q_{21}, \\
R_7 &: \text{If } z_1 \text{ is } N_1 \text{ and } z_2 \text{ is } P_2, \text{ then } S = q_{02}, \\
R_8 &: \text{If } z_1 \text{ is } Z_1 \text{ and } z_2 \text{ is } P_2, \text{ then } S = q_{12}, \\
R_9 &: \text{If } z_1 \text{ is } P_1 \text{ and } z_2 \text{ is } P_2, \text{ then } S = q_{22},
\end{aligned}\right\} \qquad (28)$$

and according to (27), the generator is given by

$$\mathbf{g}_2(z_1, z_2) = \left[1, z_1, z_1^2, z_2, z_1 z_2, z_1^2 z_2, z_2^2, z_1 z_2^2, z_1^2 z_2^2\right]^T. \tag{29}$$

The crisp output of the system can be expressed as a scalar product of two vectors

$$S(z_1, z_2) = [N_1 N_2, Z_1 N_2, P_1 N_2, N_1 Z_2, Z_1 Z_2, P_1 Z_2, N_1 P_2, Z_1 P_2, P_1 P_2]\, \mathbf{q}, \tag{30}$$

where $N_k = N_k(z_k)$, $Z_k = Z_k(z_k)$ and $P_k = P_k(z_k)$ for $k = 1, 2$ are the membership functions defined by (20)-(22), and the vector $\mathbf{q}$ consists of the conclusions of the rules (28)

$$\mathbf{q} = [q_{00}, q_{10}, q_{20}, q_{01}, q_{11}, q_{21}, q_{02}, q_{12}, q_{22}]^T. \tag{31}$$

On the other hand, according to Remark 1 we have

$$S(\mathbf{z}) = \mathbf{g}_2^T(\mathbf{z})\,\boldsymbol{\theta}, \qquad \mathbf{z} \in D^2,$$

where $\boldsymbol{\theta} = [\theta_{00}, \theta_{10}, \theta_{20}, \theta_{01}, \theta_{11}, \theta_{21}, \theta_{02}, \theta_{12}, \theta_{22}]^T$ and $\mathbf{g}_2(\mathbf{z})$ is given by (29).

For $n = 3$ the rule base consists of 27 rules. Its abbreviated structure is as follows

$$\left.\begin{aligned}
&R_1 : \text{If } z_1 \text{ is } N_1 \text{ and } z_2 \text{ is } N_2 \text{ and } z_3 \text{ is } N_3, \text{ then } S = q_{000}, \\
&R_2 : \text{If } z_1 \text{ is } Z_1 \text{ and } z_2 \text{ is } N_2 \text{ and } z_3 \text{ is } N_3, \text{ then } S = q_{100}, \\
&R_3 : \text{If } z_1 \text{ is } P_1 \text{ and } z_2 \text{ is } N_2 \text{ and } z_3 \text{ is } N_3, \text{ then } S = q_{200}, \\
&R_4 : \text{If } z_1 \text{ is } N_1 \text{ and } z_2 \text{ is } Z_2 \text{ and } z_3 \text{ is } N_3, \text{ then } S = q_{010}, \\
&R_5 : \text{If } z_1 \text{ is } Z_1 \text{ and } z_2 \text{ is } Z_2 \text{ and } z_3 \text{ is } N_3, \text{ then } S = q_{110}, \\
&R_6 : \text{If } z_1 \text{ is } P_1 \text{ and } z_2 \text{ is } Z_2 \text{ and } z_3 \text{ is } N_3, \text{ then } S = q_{210}, \\
&R_7 : \text{If } z_1 \text{ is } N_1 \text{ and } z_2 \text{ is } P_2 \text{ and } z_3 \text{ is } N_3, \text{ then } S = q_{020}, \\
&R_8 : \text{If } z_1 \text{ is } Z_1 \text{ and } z_2 \text{ is } P_2 \text{ and } z_3 \text{ is } N_3, \text{ then } S = q_{120}, \\
&R_9 : \text{If } z_1 \text{ is } P_1 \text{ and } z_2 \text{ is } P_2 \text{ and } z_3 \text{ is } N_3, \text{ then } S = q_{220}, \\
&R_{10} : \text{If } z_1 \text{ is } N_1 \text{ and } z_2 \text{ is } N_2 \text{ and } z_3 \text{ is } Z_3, \text{ then } S = q_{001}, \\
&\qquad\qquad\qquad\qquad\qquad\vdots \\
&R_{27} : \text{If } z_1 \text{ is } P_1 \text{ and } z_2 \text{ is } P_2 \text{ and } z_3 \text{ is } P_3, \text{ then } S = q_{222},
\end{aligned}\right\} \tag{32}$$

and the generator

$$\begin{aligned}
\mathbf{g}_3(z_1, z_2, z_3) = [&1, z_1, z_1^2, z_2, z_1 z_2, z_1^2 z_2, z_2^2, z_1 z_2^2, z_1^2 z_2^2, \\
& z_3, z_1 z_3, z_1^2 z_3, z_2 z_3, z_1 z_2 z_3, z_1^2 z_2 z_3, \\
& z_2^2 z_3, z_1 z_2^2 z_3, z_1^2 z_2^2 z_3, z_3^2, z_1 z_3^2, z_1^2 z_3^2, z_2 z_3^2, \\
& z_1 z_2 z_3^2, z_1^2 z_2 z_3^2, z_2^2 z_3^2, z_1 z_2^2 z_3^2, z_1^2 z_2^2 z_3^2]^T.
\end{aligned} \tag{33}$$

The output of three-inputs P2-TS system can be expressed and computed in the same way as for two-inputs system - this is rather simple task, but the equations are large for the number of inputs $n \geq 3$.

6 The Fundamental Matrix for the MISO P2-TS System

Similarly to SISO P2-TS systems, for the MISO P2-TS systems, the same equations as in (18) hold, namely $\mathbf{q} = \mathbf{R}^{-1}\boldsymbol{\Gamma}^T\boldsymbol{\theta} = \boldsymbol{\Omega}^T\boldsymbol{\theta}$, where

- the vector $\mathbf{q}$ contains the consequents of the "If-then" rules,
- $\boldsymbol{\theta}$ is the vector of parameters of the crisp function (1), with which the MISO P2-TS system is equivalent,
- the meaning of matrices $\mathbf{R}$ and $\boldsymbol{\Gamma}$ is the same as in Section 4, after some generalization for MISO systems,
- the matrix

$$\boldsymbol{\Omega} = \boldsymbol{\Gamma}\left(\mathbf{R}^{-1}\right)^T \tag{34}$$

we will call the *fundamental matrix for P2-TS system*.

Both $\boldsymbol{\Omega}$, and its inverse are important, because they enable one to establish an exact relationship between the consequents $\mathbf{q}$ of the "If-then" rules and the parameters $\boldsymbol{\theta}$ of the crisp function (1), with which the rule-based system is equivalent. Therefore our goal in this Section is to give a procedure of how to compute the fundamental matrix and its inverse in general case.

Now we formulate the following

Theorem 1. *The fundamental matrix of the MISO P2-TS system with the inputs $[z_1, \ldots, z_k]^T \in D^k$, and the membership functions of fuzzy sets for the inputs defined by (20)-(22), can be computed recursively*

$$\boldsymbol{\Omega}_0 = 1,$$

$$\boldsymbol{\Omega}_k = \begin{bmatrix} 1 & 1 & 1 \\ -\alpha_k & \sigma_k & \beta_k \\ \alpha_k^2 & \frac{1}{2}\left(\alpha_k^2 + \beta_k^2 - \frac{(\alpha_k + \beta_k)^2}{\lambda_k}\right) & \beta_k^2 \end{bmatrix} \otimes \boldsymbol{\Omega}_{k-1}, \tag{35}$$

for $k = 1, \ldots, n$, where $\lambda_k \in (0, 1]$ is the parameter of membership functions and $\otimes$ denotes the Kronecker product of matrices.

Proof. Because of limited space we omit the proof of Theorem 1. □

Example 2. Our goal is to obtain the fuzzy rules for P2-TS system which exactly model the following nonlinear function

$$f(z_1, z_2) = 2z_1^2 z_2^2 + 2z_1^2 z_2 - z_1 z_2^2 + z_1^2 - 5z_1 z_2 + 3z_2^2 - 4z_1 + 6z_2 \tag{36}$$

for $(z_1, z_2) \in D^2 = [-12.8040, 16.2860] \times [-6.8844, 5.2029]$. We assume that the first input z_1 of the TS system has assigned the fuzzy sets N_1, Z_1 and P_1, whereas the second one - the fuzzy sets N_2, Z_2 and P_2. The membership functions are defined by (20)-(22), with the parameters $\lambda_1 = \lambda_2 = 1$, and boundaries of the intervals $\alpha_1 = 12.8040$, $\beta_1 = 16.2860$, $\alpha_2 = 6.8844$, and $\beta_2 = 5.2029$. The cores of fuzzy sets Z_1 and Z_2 are $\sigma_1 = 1.7410$ and $\sigma_2 = -0.8407$, respectively. Observe that the function (36) can be written equivalently as

$$f(z_1, z_2) = \boldsymbol{\theta}^T \mathbf{g}_2(z_1, z_2) = [0, -4, 1, 6, -5, 2, 3, -1, 2] \mathbf{g}_2(z_1, z_2),$$

where the generator $\mathbf{g}_2(z_1, z_2)$ is given by (29). For $\lambda_1 = \lambda_2 = 1$ after computations we obtain

$$\boldsymbol{\Omega}_2^T = \begin{bmatrix} 1 & -12.80 & 163.94 & -6.884 & 88.15 & -1128.6 & 47.39 & -606.85 & 7770.0 \\ 1 & 1.74 & -208.53 & -6.884 & -11.99 & 1435.6 & 47.39 & 82.515 & -9883.1 \\ 1 & 16.29 & 265.23 & -6.884 & -112.1 & -1826.0 & 47.39 & 771.87 & 12571. \\ 1 & -12.80 & 163.94 & -0.841 & 10.76 & -137.83 & -35.82 & 458.62 & -5872.2 \\ 1 & 1.74 & -208.53 & -0.841 & -1.464 & 175.31 & -35.82 & -62.361 & 7469.2 \\ 1 & 16.29 & 265.23 & -0.841 & -13.69 & -222.98 & -35.82 & -583.35 & -9500.4 \\ 1 & -12.80 & 163.94 & 5.203 & -66.62 & 852.98 & 27.07 & -346.61 & 4437.9 \\ 1 & 1.74 & -208.53 & 5.203 & 9.058 & -1084.9 & 27.07 & 47.129 & -5644.8 \\ 1 & 16.29 & 265.23 & 5.203 & 84.73 & 1380.0 & 27.07 & 440.86 & 7179.9 \end{bmatrix}.$$

Since $S_2 = \mathbf{g}_2^T(z_1, z_2) \left(\boldsymbol{\Omega}_2^T\right)^{-1} \mathbf{q}_2 = f(z_1, z_2) = \boldsymbol{\theta}^T \mathbf{g}_2(z_1, z_2)$ and $\mathbf{q}_2 = \boldsymbol{\Omega}_2^T \boldsymbol{\theta}$, the system of fuzzy rules for the 2-iputs-1-output P2-TS system is as follows

$$\left.\begin{aligned}
R_1 &: \text{If } z_1 \text{ is } N_1 \text{ and } z_2 \text{ is } N_2, \text{ then } S = 13764.9420, \\
R_2 &: \text{If } z_1 \text{ is } Z_1 \text{ and } z_2 \text{ is } N_2, \text{ then } S = -17032.2043, \\
R_3 &: \text{If } z_1 \text{ is } P_1 \text{ and } z_2 \text{ is } N_2, \text{ then } S = 21579.2316, \\
R_4 &: \text{If } z_1 \text{ is } N_1 \text{ and } z_2 \text{ is } Z_2, \text{ then } S = -12429.8971, \\
R_5 &: \text{If } z_1 \text{ is } Z_1 \text{ and } z_2 \text{ is } Z_2, \text{ then } S = 15030.6206, \\
R_6 &: \text{If } z_1 \text{ is } P_1 \text{ and } z_2 \text{ is } Z_2, \text{ then } S = -18707.3075, \\
R_7 &: \text{If } z_1 \text{ is } N_1 \text{ and } z_2 \text{ is } P_2, \text{ then } S = 11589.1320, \\
R_8 &: \text{If } z_1 \text{ is } Z_1 \text{ and } z_2 \text{ is } P_2, \text{ then } S = -13655.0266, \\
R_9 &: \text{If } z_1 \text{ is } P_1 \text{ and } z_2 \text{ is } P_2, \text{ then } S = 16567.7977.
\end{aligned}\right\} \quad (37)$$

One can check that the above rule-based system models exactly the function (36), because the expression $\mathbf{g}_2^T(\mathbf{z}) \left(\boldsymbol{\Omega}_2^T\right)^{-1} \mathbf{q}_2$ results in the same polynomial as in (36) for all $\mathbf{z} \in D^2$.

7 Conclusions

It was shown that it is not possible to obtain any second degree polynomial by using the TS systems, in which only two complementary membership functions in the form of second degree polynomials for the inputs are defined. However, three membership functions as the second degree polynomials suffice to model any second degree polynomial function. Three fuzzy sets for input variables, which satisfy natural requirements were defined. They have clear linguistic interpretation and therefore the whole set of fuzzy rules is highly interpretable.

Important notions of the generator and the fundamental matrix for the P2-TS system, which uses the second degree polynomials as membership functions, were defined. In this way the notions of the generator and fundamental matrix introduced formerly for the so called P1-TS models were extended. Using our approach we have a detailed insight into the TS fuzzy rule-based system with quadratic membership functions. The methodology is completely analytic and practically useful. Necessary and sufficient conditions, which have been formulated of how to transform some class of nonlinear systems into the fuzzy rules and vice-versa, deliver an exact relationship between fuzzy models and their classical counterparts. The results can be easily applied for some class of static or dynamic systems. The proposed method can be easily extended to multi-input-multi-output systems.

References

1. Cassillas, J., Cordón, O., Herrera, F., Magdalena, L. (eds.): Interpretability Issues in Fuzzy Modeling. Studies in Fuzziness and Soft Computing, vol. 128. Springer, Heidelberg (2003)
2. Galichet, S., Boukezzoula, R., Foulloy, L.: Explicit analytical formulation and exact inversion of decomposable fuzzy systems with singleton consequents. Fuzzy Sets a. Syst. 146, 421–436 (2004)
3. Kluska, J.: New results in Analytical Modeling Using Takagi-Sugeno Expert System. In: Hryniewicz, O., Kacprzyk, J., Koronacki, J., Wierzchoń, S.T. (eds.) Issues in Intelligent Systems Paradigms, pp. 79–94. EXIT (2005)
4. Kluska, J.: Exact fuzzy modeling of conventional control systems. In: Proc. of the 12th Zittau Fuzzy Colloquium, Wissenschaftliche Berichte, Heft 84/2005, No 2090-2131, September 21-23, pp. 113–125 (2005)
5. Park, B.-J., Pedrycz, W., Oh, S.-K.: Fuzzy polynomial neurons as neurofuzzy processing units. Neural Computing a. Applications 15(3–4), 310–327 (2006)
6. Kluska, J.: Transformation Lemma on Analytical Modeling Via Takagi-Sugeno Fuzzy System and Its Applications. In: Rutkowski, L., Tadeusiewicz, R., Zadeh, L.A., Żurada, J.M. (eds.) ICAISC 2006. LNCS (LNAI), vol. 4029, pp. 230–239. Springer, Heidelberg (2006)
7. Takagi, T., Sugeno, M.: Fuzzy identification of systems and its applications to modeling and control. IEEE Trans. Systems, Man, Cybern. SMC-15(1), 116–132 (1985)
8. Ying, H.: Fuzzy control and modeling. In: Analytical foundations and applications. IEEE Press, New York (2000)
9. Ying, H.: Conditions for Analytically Determining General Fuzzy Controllers of Mamdani Type to be Nonlinear, Piecewise Linear or Linear. Soft Computing 9, 606–616 (2005)
10. Ying, H.: Deriving Analytical Input–Output Relationship for Fuzzy Controllers Using Arbitrary Input Fuzzy Sets and Zadeh Fuzzy AND Operator". IEEE Trans. Fuzzy Syst. 14(5), 654–662 (2006)
11. Zarandi, M.H.F., Türksen, I.B., Sobhani, J., Ramezanianpour, A.A.: Fuzzy polynomial neural networks for approximation of the compressive strength of concrete. Applied Soft Comp. 8, 488–498 (2008)

From Ensemble of Fuzzy Classifiers to Single Fuzzy Rule Base Classifier[*]

Marcin Korytkowski[1,2], Leszek Rutkowski[1,3], and Rafał Scherer[1,3]

[1] Department of Computer Engineering, Częstochowa University of Technology
al. Armii Krajowej 36, 42-200 Częstochowa, Poland
http://kik.pcz.pl
[2] Olsztyn Academy of Computer Science and Management
ul. Artyleryjska 3c, 10-165 Olsztyn, Poland
http://www.owsiiz.edu.pl/
[3] Department of Artificial Intelligence, Academy of Humanities and Economics in Lodz, ul. Rewolucji 1905 nr 64, Łódź, Poland
marcink@kik.pcz.czest.pl, lrutko@kik.pcz.czest.pl, rafal@ieee.org
http://www.wshe.lodz.pl

Abstract. Neuro-fuzzy systems show very good performance and the knowledge comprised within their structure is easily interpretable. To further improve their accuracy they can be combined into ensembles. In the paper we combine specially modified Mamdani neuro-fuzzy systems into an AdaBoost ensemble. The proposed modification improves the interpretability of knowledge by allowing merging the subsystems rule bases into one knowledge base. Simulations on two benchmarks shows excellent performance of the modified neuro-fuzzy systems.

1 Introduction

Over the years, numerous classification methods were developed [7] based on neural networks [1], fuzzy systems [6][9][11], support vector machines, rough sets [17] and other soft computing techniques. These methods do not need the information about the probability model. Fuzzy classifiers are frequently used thanks to their ability to use knowledge in the form of intelligible IF-THEN fuzzy rules. They fall into one of the following categories [15], depending on the connective between the antecedent and the consequent in fuzzy rules:

(i) Takagi-Sugeno method - consequents are functions of inputs,
(ii) Mamdani-type reasoning method - consequents and antecedents are related by the min operator or generally by a t-norm,
(iii) Logical-type reasoning method - consequents and antecedents are related by fuzzy implications, e.g. binary, Łukasiewicz, Zadeh etc.

---

[*] This work was partly supported by the Foundation for Polish Science (Professorial Grant 2005-2008) and the Polish Ministry of Science and Higher Education (Habilitation Project 2007-2010 Nr N N516 1155 33, Special Research Project 2006-2009, Polish-Singapore Research Project 2008-2010, Research Project 2008-2010).

In the paper we use Mamdani-type neuro-fuzzy systems as subsystems of the boosting ensemble. Classifiers can be combined to improve accuracy [10]. By combining intelligent learning systems, the model robustness and accuracy is nearly always improved, comparing to single-model solutions. Popular methods are bagging and boosting which are meta-algorithms for learning different classifiers. They assign weights to learning samples according to their performance on earlier classifiers in the ensemble. Thus subsystems are trained with different datasets created from the base dataset.

2 Merging Mamdani NFS

To build the ensemble we use the AdaBoost algorithm [5][12][16] which is one of the most popular boosting method. The algorithm assigns weights to learning samples according to their performance on earlier classifiers in the ensemble. Thus subsystems are trained with different datasets created from the base dataset. To obtain the final decision in a boosting ensemble the following formula is used to combine all the hypothesis responses

$$f(\mathbf{x}) = \sum_{t=1}^{T} c_t h_t(\mathbf{x}) ,\qquad(1)$$

where $h_t(\mathbf{x})$ is the response of the hypothesis t on the basis of feature vector $\mathbf{x} = [x_1...x_n]$, and the coefficient c_t value is computed on the basis of the classifier error and can be interpreted as the measure of classification accuracy of the given classifier. Moreover the following assumptions should be met

$$\sum_{t=1}^{T} c_t = 1 .\qquad(2)$$

The output of the single Mamdani neuro-fuzzy system, shown in Fig. 1, is defined

$$h_t = \frac{\sum_{r=1}^{N_t} \bar{y}_t^r \cdot \tau_t^r}{\sum_{r=1}^{N_t} \tau_t^r} ,\qquad(3)$$

where $\tau_t^r = \underset{i=1}{\overset{n}{T}} \left(\mu_{A_i^r}(\bar{x}_i) \right)$ is the activity level of the rule $r = 1,...,N_t$ of the classifier $t = 1,...,T$. Structure depicted by (3) is shown in Fig. 1 (index t is omitted for clarity). After putting (3) into (1) we obtain the final output of the classifier ensemble

$$f(\mathbf{x}) = \sum_{t=1}^{T} c_t \frac{\sum_{r=1}^{N_t} \bar{y}_t^r \cdot \tau_t^r}{\sum_{r=1}^{N_t} \tau_t^r} .\qquad(4)$$

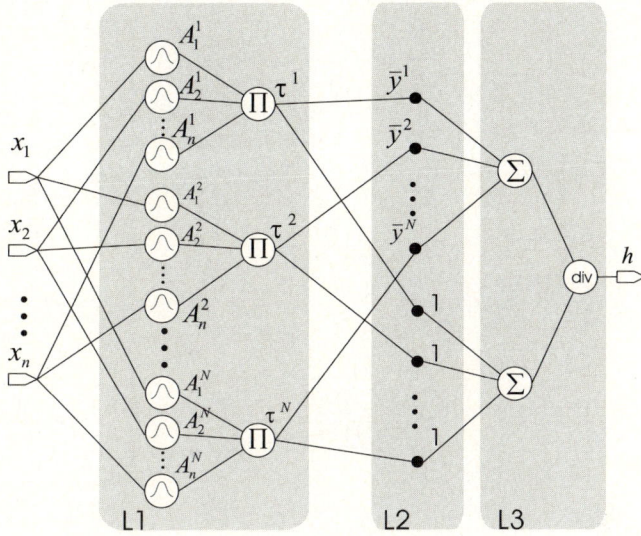

Fig. 1. Single Mamdani neuro-fuzzy system

Formula (4) is the weighted sum of all hypothesis outcomes. It is not possible to merge all rule bases coming from members of the ensemble, because (4) can not be reduced to the lowest common denominator. Let us observe that in the denominator of (4) there is sum of activity level of rules in a single neuro-fuzzy system. Thus if we want to treat formula (4) as one neuro-fuzzy system, it should be transformed so that it has the sum of activity level of all rules in the ensemble. The solution to the problem is assuming the following for every module constituting the ensemble

$$\sum_{r=1}^{N_t} \tau_t^r = 1 \ \forall t = 1, ..., T \ . \tag{5}$$

Then we obtain the following formula for the output of the whole ensemble

$$f(\mathbf{x}) = \sum_{t=1}^{T} \left(c_t \sum_{r=1}^{N_t} \left(\bar{y}_t^r \cdot \tau_t^r \right) \right) \ . \tag{6}$$

The assumption (5) guarantee that in case of creating an ensemble of several such modified neuro-fuzzy systems, summary activity level of one system does not dominate other subsystems. Unfortunately there not exists a method for designing neuro-fuzzy systems such that the sum of the activity levels $\sum_{r=1}^{N} \tau_r$ for a single system equals one. One possibility is such selection of fuzzy system parameters during learning that assumption (5) is satisfied. Considered neuro-fuzzy systems should be trained to satisfy two assumptions

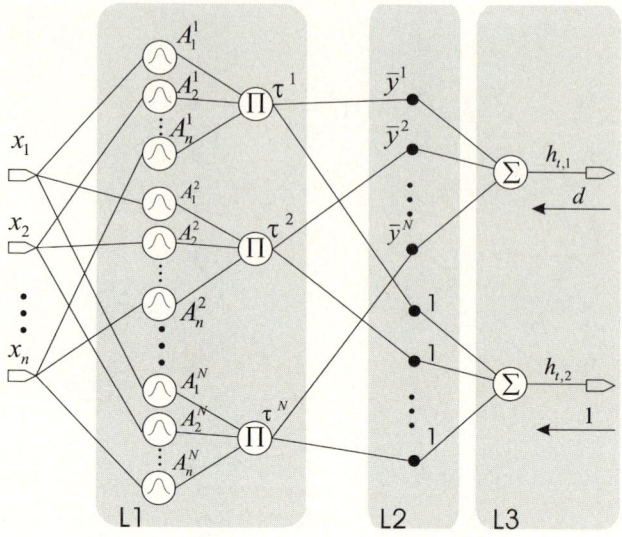

Fig. 2. Single modified Mamdani neuro-fuzzy system

$$\begin{aligned}&1)\ h_t(\mathbf{x}^q) = d^q\ \forall q = 1, ..., M\ ,\\&2)\ \sum_{r=1}^{N_t} \tau_t^r = 1\ \forall t = 1, ..., T\ .\end{aligned} \quad (7)$$

To satisfy (7) we modify the neuro-fuzzy structure in Fig. 1 to make it take the form depicted in Fig. 2. We removed the last layer performing the division thus the system has two outputs. The error on the first output will be computed taking into account desired output from learning data. Desired signal on the second output is constant and equals 1. The sum of activity levels can equal 1 in case of using triangular membership functions, but in case of Gaussian functions it is not possible to fulfill the second condition in (7). This is caused by the range of Gaussian function which do not take the exact zero value. It is not the serious problem in real life applications as we need only to approximately fulfill the first condition in (7). As we assume the fulfillment of the first condition in (7), we can remove after learning from the structure elements responsible for denominator in (3), as the denominator equals 1. The modified structure is in the Fig. 3. After learning consecutive structures according to the proposed idea, we can build the modular system presenting in Fig. 4. In such system the rules can be arranged in arbitrary order. We do not have to remember from which submodule the rule comes from. The obtained ensemble of neuro-fuzzy systems can be interpreted as a regular single neuro-fuzzy system, where coefficients c_t can be interpreted as weights of fuzzy rules. Important ability of the system is its possible fine tuning. Yet in order to make the structure acting like a neuro-fuzzy system during learning, we have to add the removed elements responsible for denominator in (3). Alternatively we can use again the modification proposed in this section for learning. System will be similar to that from the Fig. 2.

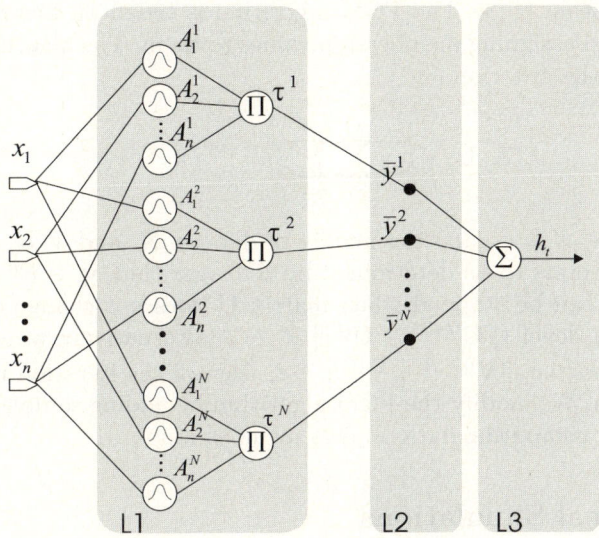

Fig. 3. Single Mamdani neuro-fuzzy system after learning. The second output is removed as it always equals one.

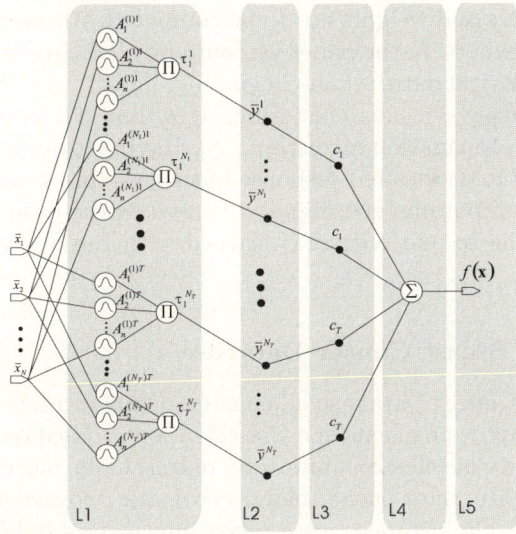

Fig. 4. Ensemble of modified Mamdani neuro-fuzzy systems combined after learning

3 FCM Modification for Boosting Initialization

One of several advantages of fuzzy systems is the possibility of interpretation of the knowledge. Thanks to this we can also initialize the parameters of the fuzzy system. A common method for such initialization is the use of the Fuzzy

C-Means algorithm [2][3]. The FCM algorithm is based on fuzzy clustering of data vectors and assigning membership values to them. The algorithm is derived by minimizing the criterion [8]

$$J\left(\mathbf{X};\mathbf{U};\mathbf{V}\right) = \sum_{i=1}^{C}\sum_{q=1}^{M}\left(\mu_{iq}\right)^{m}\left\|\mathbf{x}_{q}-\mathbf{v}_{i}\right\|_{\mathbf{A}}^{2} \qquad (8)$$

and $\mathbf{U} = [\mu_{iq}] \in Z$ is a partition matrix of dataset $\mathbf{X}$, and $\mathbf{V} = [\mathbf{v}_1, \mathbf{v}_2, ..., \mathbf{v}_C]$ is a vector of centers to be determined by the algorithm, $\mathbf{v}_i \in R^n$, $i = 1, \ldots, C$. The algorithm can be stopped when matrix $\mathbf{U}$ does not change or the change is below certain level ($\left\|\mathbf{U}^{(l+1)} - \mathbf{U}^{(l)}\right\| < \varepsilon$) . Alternatively we can check the change of centers, i.e. $\left\|\mathbf{V}^{(l+1)} - \mathbf{V}^{(l)}\right\| < \varepsilon$, where l the iteration number in the FCM algorithm. We modify the FCM algorithm by adding sample weight D_t^q to the formula for computing data objects memberships.

4 Numerical Simulations

In this section we describe learning of the neuro-fuzzy systems and simulation examples. Relational neuro-fuzzy system knowledge is expressed in the form of fuzzy rules with certainty weights. During learning the right number of rules, fuzzy set parameters and weights are to be computed. Alternatively they can be determined by an expert. In the paper we compute all the parameters by machine learning from numerical data. At the beginning, input fuzzy sets are determined by the modified fuzzy c-means clustering algorithm. Then, all parameters are tuned by the backpropagation algorithm [18]. Having given learning data set of pair $(\bar{\mathbf{x}}, d)$ where d is the desired response of the system we can use the learning error to compute all parameters. Signal d is used as the desired response of the first output and the second output is learned with the desired response which equals 1.

4.1 Wisconsin Breast Cancer Database

Wisconsin Breast Cancer Database [4] consists of 699 instances of binary classes (benign or malignant type of cancer). Classification is based on 9 features (clump thickness, uniformity of cell size, uniformity of cell shape, marginal adhesion, single epithelial cell size, bare nuclei, bland chromatin, normal nucleoli, mitoses). From the data set 205 instances were taken into testing data and 16 instances with missing features were removed. The classification accuracy is 97,45%. Detailed errors and subsystem parameters are described in Table 1.

4.2 Glass Identification

In this section we test the efficiency of our methods on the Glass Identification problem [4]. The goal is to classify 214 instances of glass into window and non-window glass basing on 9 numeric features. We took out 43 instances for a testing

Table 1. Numerical results of Wisconsin Breast Cancer Database

	Classifier 1	Classifier 2	Classifier 3
No. of epochs	15	50	30
Coefficients c_t	0.29	0.35	0.36
Classification %	97.42	97.45	97.41

Table 2. Numerical results of Glass Identification

	Classifier 1	Classifier 2	Classifier 3	Classifier 4	Classifier 5
No. of rules	2	2	2	2	2
No. of epochs	30	30	30	50	50
Coefficients c_t	0.13	0.01	0.18	0.33	0.33
Classification %	89.34	77.23	89.8	97.34	97.53

set. We obtained 97.34% classification accuracy. We used 5 neuro-fuzzy classifiers described in Table 2. Simulations shows superior performance and faster convergence of the proposed approach comparing to those from the literature.

5 Conclusions

Neuro-fuzzy systems have numerous advantages over neural networks such as the possibility to compute their initial state and the interpretable knowledge. To improve their performance we connect several fuzzy systems into an ensemble of classifiers. The ensemble is created using the AdaBoost algorithm. The fuzzy subsystems are learned by gradient learning and initialized by modified FCM clustering algorithm. To make possible merging of rulebases coming from the subsystems, we modify them by adding special second output. Thanks to the desired signal on this output being equal 1, it is possible to remove it afterwards and merge the systems as none of them dominate over the rest. Simulations on two well known benchmarks shows very good ability to classify data. Results are comparable to the best ones from the literature.

References

1. Bishop, C.M.: Neural Networks for Pattern Recognition. Oxford University Press Inc., New York (1995)
2. Bezdek, J.C., Pal, S.K.: Fuzzy Models for Pattern Recognition. IEEE Press, New York (1992)
3. Bezdek, J., Keller, J., Krisnapuram, R., Pal, N.R.: Fuzzy Models and Algorithms for Pattern Recognition and Image Processing. Kluwer Academic Press, Norwell (1999)
4. Asuncion, A., Newman, D.J.: UCI Machine Learning Repository, University of California, School of Information and Computer Science, Irvine, CA (2007), http://www.ics.uci.edu/$\sim$mlearn/{MLR}epository.html

5. Breiman, L.: Bias, variance, and arcing classifiers, Technical Report 460, Statistics Department, University of California (1997)
6. Czogała, E., Łęski, J.: Fuzzy and Neuro Fuzzy Intelligent Systems. Physica Verlag, Heidelberg (2000)
7. Duda, R.O., Hart, P.E., Stork, D.G.: Pattern Classification, 2nd edn. Wiley, Chichester (2000)
8. Konar, A.: Computational Intelligence. Springer, Heidelberg (2005)
9. Kuncheva, L.I.: Fuzzy Classifier Design. Physica Verlag, Heidelberg (2000)
10. Kuncheva, L.I.: Combining Pattern Classifiers, Methods and Algorithms. John Wiley & Sons, Chichester (2004)
11. Łęski, J.: A Fuzzy If-Then Rule-Based Nonlinear Classifier. Int. J. Appl. Math. Comput. Sci. 13(2), 215–223 (2003)
12. Meir., R., Ratsch, G.: An Introduction to Boosting and Leveraging. Advanced Lectures on Machine Learning (2003)
13. Korytkowski, M., Rutkowski, L., Scherer, R.: On Combining Backpropagation with Boosting. In: 2006 International Joint Conference on Neural Networks, IEEE World Congress on Computational Intelligence, Vancouver, BC, Canada (2006)
14. Łęski, J.: A Fuzzy If-Then Rule-Based Nonlinear Classifier. Int. J. Appl. Math. Comput. Sci. 13(2), 215–223 (2003)
15. Rutkowski, R.: Flexible Neuro-Fuzzy Systems. Kluwer Academic Publishers, Dordrecht (2004)
16. Schapire, R.E.: A brief introduction to boosting. In: Proc. of the Sixteenth International Joint Conference on Artificial Intelligence (1999)
17. Świniarski, R.W.: Rough Sets Methods in Feature Reduction and Classification. Int. J. Appl. Math. Comput. Sci. 11(3), 565–582 (2001)
18. Wang, L.-X.: Adaptive Fuzzy Systems And Control. PTR Prentice Hall, Englewood Cliffs (1994)

Efficient Fuzzy Predictive Economic Set–Point Optimizer

Piotr M. Marusak

Institute of Control and Computation Engineering, Warsaw University of Technology,
ul. Nowowiejska 15/19, 00–665 Warszawa, Poland
P.Marusak@ia.pw.edu.pl
http://www.ia.pw.edu.pl/~pmarusak

Abstract. A fuzzy predictive set–point optimizer which uses a nonlinear, fuzzy dynamic process model is proposed in the paper. The algorithm of the optimizer is formulated in such a way that only a numerically efficient, quadratic optimization problem must be solved at each algorithm iteration. It is demonstrated, using an example of a control system of a nonlinear MIMO control plant, that application of the optimizer based on a fuzzy model instead of a linear one can result in substantial improvement of control system operation. The fuzzy control plant model, the optimizer is based on, consists of local models in the form of control plant step responses. Thus, the model is easy to obtain and the proposed optimizer easy to design.

1 Introduction

The important problem in control systems is compensation of disturbances changing quickly compared to the dynamics of the control plant. The standard multilayer control system structure is not sufficient in such a case. Thus, there were a few solutions to this problem proposed; see e.g. [5,11,12] and references in there. One of the methods, being in fact a combination of other solutions, is application of a predictive economic set–point optimizer [4]. The optimizer is responsible for generation of set–point values for basic feedback controllers and for fulfillment of all constraints present in the control system. It uses the linearization of a steady–state process model, performed at each algorithm iteration. Thus, this is a numerically efficient solution, because the algorithm is formulated in such a way that only the quadratic optimization problem is solved at each iteration [4].

Moreover, the models of control plant and of the basic controllers are used by the optimizer in order to predict the behavior of the manipulated and output variables and generate set–points in such a way that fulfillment of the constraints is assured. In order to preserve the numerical efficiency of the algorithm, the linear dynamic process model is used. However, if, due to the disturbance changes, the operating point is also changing then better results can be obtained using a nonlinear dynamic control plant model. Fuzzy predictive economic set–point optimizer proposed in the paper is based on Takagi–Sugeno (TS) dynamic fuzzy models with step responses used as local models.

The step response models are used for prediction in a DMC predictive control algorithm. The DMC algorithm was designed for and applied in the process industry in late 1970s. Now, it is practically a standard in the industrial applications; see e.g. [1,9,11]. Thanks to the form of the model it is based on the algorithm can be relatively easily designed [2,3,6,11]. The proposed set–point optimizer inherits this advantage. It is based on the fuzzy (nonlinear) control plant model, thanks to such an approach it works well in a wide range of operating point changes and offers better control performance than the optimizer based on a linear dynamic process model. Moreover, the model is used in such a way that the optimization problem solved at each iteration by the optimizer is a quadratic optimization problem. Thus, the proposed algorithm is computationally efficient. It should be also emphasized that the dynamic control plant model composed of step responses, can be obtained relatively easy as it was done during the design of the controller for the example control plant.

In the next section the idea of predictive controllers is shortly reminded and the numerical and analytical predictive control algorithms are discussed. In Sect. 3 the economic optimization task, typically solved by the economic steady–state optimization layer is reminded. The predictive economic set–point optimizer is discussed in Sect. 4. Section 5 contains description of the control system used during the tests and discussion of the results obtained during the experiments, proving very good performance offered by the proposed fuzzy optimizer. The paper is summarized in the last section.

2 Predictive Control Algorithms

2.1 Basic Idea and Numerical Formulation

The Model Predictive Control (MPC) algorithms derive future values of manipulated variables predicting future behavior of the control plant many sampling instants ahead. The values of manipulated variables are calculated in such a way that the prediction fulfills assumed criteria. Usually, the minimization of a performance function is demanded subject to the constraints put on values of manipulated and output variables [2,6,11]:

$$\min_{\Delta u} \left\{ J_{\mathrm{MPC}} = \sum_{j=1}^{n_y} \sum_{i=1}^{p} \kappa_j \left(\overline{y}_k^j - y_{k+j|k}^j \right)^2 + \sum_{m=1}^{n_u} \sum_{i=0}^{s-1} \lambda_m \left(\Delta u_{k+i|k}^m \right)^2 \right\} \quad (1)$$

subject to:

$$\Delta u_{\min} \leq \Delta u \leq \Delta u_{\max}, \quad (2)$$

$$u_{\min} \leq u \leq u_{\max}, \quad (3)$$

$$y_{\min} \leq y \leq y_{\max}, \quad (4)$$

where $\overline{y}_k^j$ is a set–point value for the jth output, $y_{k+j|k}^j$ is a value of the jth output for the $(k+i)^{\text{th}}$ sampling instant predicted at the kth sampling instant using a control plant model, $\Delta u_{k+i|k}^m$ are future changes in the manipulated variables, $\kappa_j \geq 0$ and $\lambda_m \geq 0$ are weighting coefficients for the predicted control errors of the jth output and for the changes of the mth manipulated variable, respectively, p and s denote prediction and control horizons, respectively, n_y, n_u denote number of output and manipulated variables, respectively; $\boldsymbol{y}$ is a vector of length $(p \cdot n_y)$, composed of the predicted output values $y_{k+j|k}^j$ ($\boldsymbol{y} = [\boldsymbol{y}^1, \ldots, \boldsymbol{y}^{n_y}]^T$, $\boldsymbol{y}^j = [y_{k+1|k}^j, \ldots, y_{k+p|k}^j]$), $\Delta \boldsymbol{u}$ is a vector of length $(s \cdot n_u)$, composed of the future increments of manipulated variables $\Delta u_{k+i|k}^m$ ($\Delta \boldsymbol{u} = [\Delta \boldsymbol{u}^1, \ldots, \Delta \boldsymbol{u}^{n_u}]^T$, $\Delta \boldsymbol{u}^m = [\Delta u_{k+1|k}^m, \ldots, \Delta u_{k+s-1|k}^m]$), $\boldsymbol{u}$ is a vector of length $(s \cdot n_u)$ of future values of manipulated variables ($\boldsymbol{u} = [\boldsymbol{u}^1, \ldots, \boldsymbol{u}^{n_u}]^T$, $\boldsymbol{u}^m = [u_{k+1|k}^m, \ldots, u_{k+s-1|k}^m]$), $\Delta \boldsymbol{u}_{\min}$, $\Delta \boldsymbol{u}_{\max}$, $\boldsymbol{u}_{\min}$, $\boldsymbol{u}_{\max}$, $\boldsymbol{y}_{\min}$, $\boldsymbol{y}_{\max}$ are vectors of lower and upper bounds of changes and values of the control signals and of the output variable values, respectively. As a solution to the optimization problem (1–4) the optimal vector of changes in the manipulated variables is obtained. From this vector, the $\Delta u_{k|k}^m$ elements are applied in the control system and the algorithm passes to the next iteration.

The way the predicted values of output variables $y_{k+j|k}^j$ are derived depends on the dynamic control plant model the predictive algorithm is based on. If the linear model is used then the optimization problem (1–4) is a standard quadratic programming problem.

2.2 Analytical Predictive Controllers

Using the vector notation the performance index from (1) can be rewritten in the following form:

$$J_{\text{MPC}} = (\overline{\boldsymbol{y}} - \boldsymbol{y})^T \cdot \boldsymbol{\kappa} \cdot (\overline{\boldsymbol{y}} - \boldsymbol{y}) + \Delta \boldsymbol{u}^T \cdot \boldsymbol{\lambda} \cdot \Delta \boldsymbol{u} , \tag{5}$$

where $\overline{\overline{\boldsymbol{y}}} = [\overline{\boldsymbol{y}}^1, \ldots, \overline{\boldsymbol{y}}^{n_y}]^T$, $\overline{\boldsymbol{y}}^j = [\overline{y}_k^j, \ldots, \overline{y}_k^j]$ is a vector of p elements, $\boldsymbol{\kappa} = [\kappa_1, \ldots, \kappa_{n_y}] \cdot \boldsymbol{I}$, $\boldsymbol{\kappa}_j = [\kappa_j, \ldots, \kappa_j]$ is a vector of p elements, $\boldsymbol{\lambda} = [\lambda_1, \ldots, \lambda_{n_u}] \cdot \boldsymbol{I}$, $\boldsymbol{\lambda}_m = [\lambda_m, \ldots, \lambda_m]$ is a vector of s elements.

If the prediction is performed using a linear control plant model then the vector of predicted output values $\boldsymbol{y}$ can be decomposed into the following components:

$$\boldsymbol{y} = \widetilde{\boldsymbol{y}} + \boldsymbol{A} \cdot \Delta \boldsymbol{u} , \tag{6}$$

where $\widetilde{\boldsymbol{y}}$ is called a free response of the plant and contains output values predicted using only the past control values ($\widetilde{\boldsymbol{y}} = [\widetilde{\boldsymbol{y}}^1, \ldots, \widetilde{\boldsymbol{y}}^{n_y}]^T$, $\widetilde{\boldsymbol{y}}^j = [\widetilde{y}_{k+1|k}^j, \ldots, \widetilde{y}_{k+p|k}^j]$). Thus, the values of the elements of the vector $\widetilde{\boldsymbol{y}}$ are equal to the values of the control plant outputs in the future in the situation if manipulated signals were frozen at the kth time instant. The difference between predictive control algorithms based on different types of models lies in the form of the free response and the way it is calculated.

A is a matrix composed of the control plant step response coefficients, called the dynamic matrix:

$$A = \begin{bmatrix} A_{11} & A_{12} & \cdots & A_{1n_u} \\ A_{21} & A_{22} & \cdots & A_{2n_u} \\ \vdots & \vdots & \ddots & \vdots \\ A_{n_y 1} & A_{n_y 2} & \cdots & A_{n_y n_u} \end{bmatrix}, \quad (7)$$

$$A_{jm} = \begin{bmatrix} a_1^{j,m} & 0 & \cdots & 0 & 0 \\ a_1^{j,m} & a_2^{j,m} & \cdots & 0 & 0 \\ \vdots & \vdots & \ddots & \vdots & \vdots \\ a_p^{j,m} & a_{p-1}^{j,m} & \cdots & a_{p-s+2}^{j,m} & a_{p-s+1}^{j,m} \end{bmatrix}, \quad (8)$$

where $a_n^{j,m}$ are coefficients of the control plant step response described in detail later. It can be shown that the dynamic matrix is the same in different predictive control algorithms; see e.g. [11].

The performance index (5) after usage of the prediction (6) can be written in the following form:

$$J_{\mathrm{MPC}} = \left(A \cdot \Delta u - (\overline{\overline{y}} - \widetilde{y})\right)^T \cdot \kappa \cdot \left(A \cdot \Delta u - (\overline{\overline{y}} - \widetilde{y})\right) + \Delta u^T \cdot \lambda \cdot \Delta u. \quad (9)$$

If it is minimized without taking constraints into consideration then the problem has the following analytical solution:

$$\Delta u = \left(A^T \cdot \kappa \cdot A + \lambda\right)^{-1} A^T \left(\overline{\overline{y}} - \widetilde{y}\right). \quad (10)$$

Moreover, because only the $\Delta u_{k|k}^j$ elements of the vector Δu are applied to the process at each iteration, it is possible to define the control law of the predictive controller. The form of the control law depends on the form of the model the controller is based on and, in consequence, on the form of the free response; for details see e.g. [11].

2.3 Prediction Generation Using the Step Response Model

In the case of the DMC control algorithm (such version of the analytical algorithm was used in the control system of the example control plant), a process model in the form of control plant step responses is used:

$$\hat{y}_k^j = \sum_{m=1}^{n_u} \sum_{n=1}^{p_d - 1} a_n^{j,m} \cdot \Delta u_{k-n}^m + a_{p_d}^{j,m} \cdot u_{k-p_d}^m, \quad (11)$$

where $\hat{y}_k^j$ is the jth output of the control plant model at the kth time instant, Δu_k^m is a change of the mth manipulated variable at the kth time instant, $a_n^{j,m}$

($n = 1, \ldots, p_d$) are step response coefficients of the control plant describing influence of the mth input on the jth output, p_d is equal to the number of time instants after which the coefficients of the step responses can be assumed as settled, $u^m_{k-p_d}$ is a value of the mth manipulated variable at the $(k - p_d)^{\text{th}}$ time instant. The predicted values of output variables are calculated using the following formula:

$$y^j_{k+i|k} = \hat{y}^j_{k+i} + d^j_k ,\qquad(12)$$

where $d^j_k = y^j_k - \hat{y}^j_{k-1}$ is assumed the same at each time instant in the prediction horizon (it is a DMC–type disturbance model). After transformation of (12) one obtains

$$y^j_{k+i|k} = y^j_k + \sum_{m=1}^{n_u} \left(\sum_{n=i+1}^{p_d-1} a^{j,m}_n \cdot \Delta u^m_{k-n+i} + a^{j,m}_{p_d} \cdot \sum_{n=p_d}^{p_d+i-1} \Delta u^m_{k-n+i} \right) \qquad(13)$$

$$- \sum_{m=1}^{n_u} \sum_{n=1}^{p_d-1} a^{j,m}_n \cdot \Delta u^m_{k-n} + \sum_{m=1}^{n_u} \sum_{n=p}^{i} a^{j,m}_n \cdot \Delta u^m_{k-n+i|k} .$$

It is good to notice that in (13) only the last component depends on future manipulated variable changes. The rest of the prediction is the element of the free response. Thus, rewriting the predictions (13) in a vector form one obtains (6).

3 Economic Optimization

The optimization problem solved typically in the steady–state optimization layer of the standard multilayer structure is following [11]:

$$\min_{\overline{y},\overline{u}} J_{\text{E}}(\overline{y}, \overline{u}) \qquad(14)$$

subject to:

$$\overline{u}_{\min} \leq \overline{u} \leq \overline{u}_{\max} ,\qquad(15)$$

$$\overline{y}_{\min} \leq \overline{y} \leq \overline{y}_{\max} ,\qquad(16)$$

$$\overline{y} = F\left(\overline{u}, \widetilde{d}\right) ,\qquad(17)$$

where $F : \mathbb{R}^{n_u} \times \mathbb{R}^{n_d} \to \mathbb{R}^{n_y}$, $F \in C^1$ is a nonlinear, steady–state control plant model, n_d is the number of disturbances affecting the control plant, $\overline{y}$ is a vector of length n_y of the set–point values, $\overline{u}$ is a vector of length n_u of control values corresponding to set–points $\overline{y}$, calculated using the steady–state plant model, $\widetilde{d}$ is a disturbance estimate, $\overline{u}_{\min}, \overline{u}_{\max}$ are vectors of lower and upper bounds of manipulated variables, $\overline{y}_{\min}, \overline{y}_{\max}$ are vectors of lower and upper bounds of output values, $J_{\text{E}}(\overline{y}, \overline{u})$ is a performance function. The optimal solution to the optimization problem (14–17), is passed to control algorithms as the desired set–point.

The problem above is usually a nonlinear, difficult to solve optimization problem. Thus, its solution is usually time consuming. Therefore, it is repeated less often than the action of the controllers. Such a situation may result in control action far from optimal when dynamics of disturbances is comparable with dynamics of the controlled process. A desired solution to this problem would be to increase intervention rate of the steady–state economic optimization layer, up to the case in which it would be repeated at each controller sampling instant. However, it is rather unrealistic due to computational demand needed to solve the problem (14–17). One of the solutions to the described problem is to use in the control system a predictive economic set–point optimizer. One of the advantages of such an approach is the fact that it takes care not only of set–point generation, but it does it in such a way that constraints put on manipulated and output variables can be fulfilled. Moreover, it can be applied in existing control systems in which the advanced constrained control layer is not yet applied. It may radically improve the overall control system performance.

4 Fuzzy Predictive Economic Set–Point Optimizer

The optimization problem solved at each iteration by the set–point optimizer is a modified problem (14–17) of the steady–state economic optimization [4]. The modification consists in: adding to this problem the constraints (2–4) from the optimization problem of the predictive algorithm, linearizing of the steady state process model performed at each iteration and adding model of basic feedback controllers. The optimization problem solved at each iteration by the optimizer has thus the following form:

$$\min_{\overline{y},\overline{u}} J_E\left(\overline{y},\overline{u}\right) \tag{18}$$

subject to:

$$\overline{u}_{\min} \leq \overline{u} \leq \overline{u}_{\max}, \tag{19}$$

$$\overline{y}_{\min} \leq \overline{y} \leq \overline{y}_{\max}, \tag{20}$$

$$\overline{y} = F\left(u(k-1), \widetilde{d}\right) + H(k)\left(\overline{u} - u(k-1)\right), \tag{21}$$

$$\Delta u_{\min} \leq \Delta u \leq \Delta u_{\max}, \tag{22}$$

$$u_{\min} \leq u \leq u_{\max}, \tag{23}$$

$$y_{\min} \leq y \leq y_{\max}, \tag{24}$$

$$\Delta u = R(\overline{y}, y^P, u^P), \tag{25}$$

where (21) is a linearized steady–state process model (17), $u(k-1)$ is a vector of values of manipulated variables, applied to the control plant in the last algorithm iteration, $\boldsymbol{H}(k)$ is the matrix of dimension $n_y \times n_u$ that contains partial derivatives of the function $F\left(\overline{\boldsymbol{u}}, \overline{\boldsymbol{d}}\right)$ (usually computed numerically using finite difference method, for details see e.g. [11]). The economic function $J_{\mathrm{E}}(\overline{\boldsymbol{y}}, \overline{\boldsymbol{u}})$ is usually linear in relation to its arguments. However, it may be explicitly dependent on the internal variables of the steady–state model. In such a case, it should be also linearized.

The constraints (22–24) are the same as in the predictive algorithm (see Sect. 2). The predicted values of output variables $\boldsymbol{y}$ are calculated using a control plant model. In the linear case the dependence of this prediction on changes of manipulated variables is described by (6). The algorithm proposed in the paper, however, exploits a TS fuzzy model with step responses used as the local models:

Rule f: (26)

$$\text{if } y_k^{j_y} \text{ is } B_1^{f,j_y} \text{ and } \ldots \text{ and } y_{k-n+1}^{j_y} \text{ is } B_n^{f,j_y} \text{ and}$$

$$u_k^{j_u} \text{ is } C_1^{f,j_u} \text{ and } \ldots \text{ and } u_{k-m+1}^{j_u} \text{ is } C_m^{f,j_u}$$

$$\text{then } \hat{y}_k^{j,f} = \sum_{m=1}^{n_u} \sum_{n=1}^{p_d - 1} a_n^{j,m,f} \cdot \Delta u_{k-n}^m + a_{p_d}^{j,m,f} \cdot u_{k-p_d}^m \;,$$

where $y_k^{j_y}$ is the j_y^{th} output variable value at the kth time instant, $u_k^{j_u}$ is the j_u^{th} manipulated variable value at the kth time instant, $B_1^{f,j_y}, \ldots, B_n^{f,j_y}, C_1^{f,j_u}, \ldots, C_m^{f,j_u}$ are fuzzy sets, $a_n^{j,m,f}$ are the coefficients of step responses in the fth local model, $j_y = 1, \ldots, n_y$, $j_u = 1, \ldots, n_u$, $f = 1, \ldots, l$, l is number of rules.

Thanks to such an approach the model for the algorithm can be obtained in a relatively simple way. The design process of the TS model (26) with step responses used as the local models can be very simple. Instead of obtaining local models in a complicated way, one can simply collect a few step responses of the control plant for a few operating points. The premise part of the control plant model (26) can be designed using expert knowledge, simulation experiments, fuzzy neural networks or all the mentioned techniques combined.

In order to obtain the prediction (6), used then in the optimization problem (18–25), the following steps should be made at each algorithm iteration:

1. Using the TS fuzzy model (26) and fuzzy reasoning, for current values of process variables, a linear model (a step response valid for the current values of process variables) is derived. Thus, the output values of the model are calculated using the following formula:

$$\hat{y}_k^j = \sum_{m=1}^{n_u} \sum_{n=1}^{p_d - 1} \widetilde{a}_n^{j,m} \cdot \Delta u_{k-n}^m + \widetilde{a}_{p_d}^{j,m} \cdot u_{k-p_d}^m \;, \qquad (27)$$

where $\widetilde{a}_n^{j,m} = \sum_{j=1}^{l} \widetilde{w}_j \cdot a_n^{j,m,f}$, $\widetilde{w}_f$ are the normalized weights calculated using standard fuzzy reasoning, see e.g. [10].

Next steps of the prediction generation by the algorithm are the same as in the standard DMC algorithm (Sect. 2.3):

2. The step response coefficients $\widetilde{a}_n^{j,m}$ are used to generate the dynamic matrix.
3. The step response model (27) is used to generate the free response of the plant.

It is the characteristic feature of the set–point optimizer that also a model of the basic feedback controllers (25) is used by it. It is usually a linear function of set–points $\overline{y}$, past values of the output variables y^p and past values of the manipulated variables u^p. Thanks to inclusion of the controller model (25) in the optimization problem (18–25), the constraints put on manipulated variables are taken into consideration during the set–point generation [4].

5 Simulation Experiments

A control plant is an evaporator, described thoroughly in [8], with one of output variables controlled by means of a PI controller. The diagram of the control plant is shown in Fig. 1. Output variables of the plant are: $L2$ – level of the liquid in the separator (stabilized near $\overline{L2} = 1$ m by means of the PI controller with parameters $K_\mathrm{p} = 5.6$ and $T_\mathrm{i} = 8.84$ min), $X2$ – product composition, $P2$ – pressure in the evaporator. Manipulated variables are: $F2$ – product flow (variable used to stabilize level $L2$), $P100$ – steam pressure, $F200$ – cooling water flow.

The set of 14 nonlinear equations describing the control plant is presented in [8]. These equations were used to simulate the control plant. It was assumed that the manipulated variables are constrained:

$$P100_{\min} \leq P100 \leq P100_{\max} ,$$
$$F200_{\min} \leq F200 \leq F200_{\max} ,$$
(28)

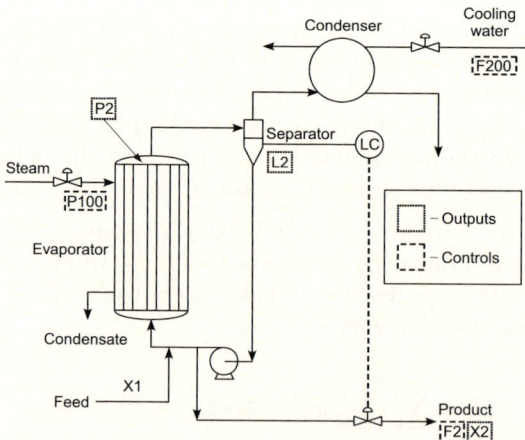

Fig. 1. Diagram of the evaporator

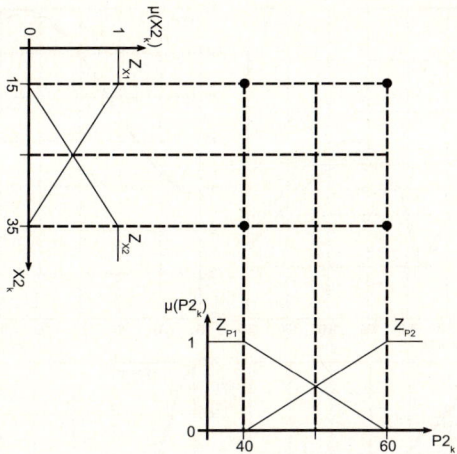

Fig. 2. Membership functions of the fuzzy TS model

where $P100_{\min} = 0$ kPa, $P100_{\max} = 400$ kPa, $F200_{\min} = 0$ kg/min, $F200_{\max} = 400$ kg/min [8].

For the control plant the DMC analytical controller was designed using step responses obtained from environs of the operating point $X2 = 25\%$, $P2 = 50.5$ kPa. It was assumed that, from the point of view of the controller, the manipulated variables are: steam pressure $P100$ and cooling water flow $F200$; the controlled variables are: product composition $X2$ and pressure in the evaporator $P2$. The following values of the parameters were assumed: $\kappa_{X2} = \kappa_{P2} = 1$, $\lambda_{P100} = \lambda_{F200} = 0.1$, $p_d = p = 100$, $s = 10$.

The TS fuzzy model used in the set–point optimizer is composed of step responses obtained from environs of four operating points ($X2 = 15\%$, $P2 = 40$ kPa; $X2 = 15\%$, $P2 = 60$ kPa; $X2 = 35\%$, $P2 = 40$ kPa; $X2 = 35\%$, $P2 = 60$ kPa). These points and the assumed membership functions are shown in Fig. 2.

It was also assumed that the product composition is constrained, i.e. the quality of the product cannot be worse than it is demanded by a customer (otherwise the product is useless):

$$X2 \geq X2_{\min} , \qquad (29)$$

where $X2_{\min} = 25\%$. Appropriate constraints put on predicted values of the output variable were included in the optimization problem (18–25) solved by the set–point optimizer. The constraint put on $\overline{X2}$ set–point value was also added:

$$\overline{X2}_{\min} + \overline{r}_{\min}^{X2} \leq \overline{X2} , \qquad (30)$$

where $\overline{X2}_{\min} = X2_{\min}$ and $\overline{r}_{\min}^{X2} = 0.3\%$ is the value shifting the constraint in order to take into account the changes in the output signal caused by controller action and modeling uncertainty.

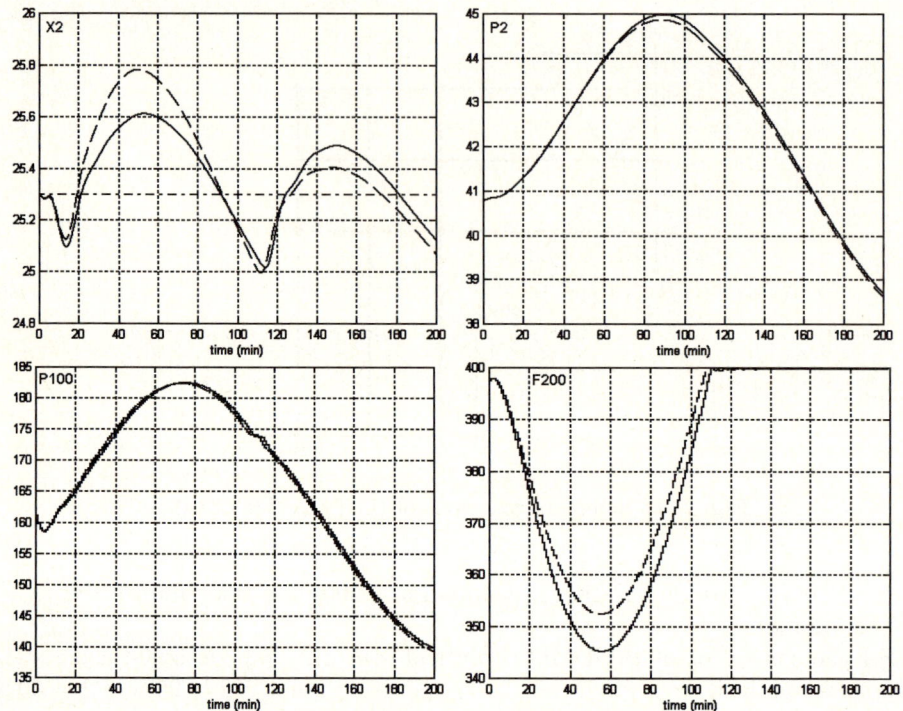

Fig. 3. Responses of the control systems with standard (dashed line) and fuzzy (solid line) algorithms to changes of the $X1$ disturbance; above – output variables $X2$ and $P2$, below – manipulated variables $P100$ and $F200$

The following economic performance index was assumed:

$$J_E = c_1 \cdot \overline{P}100 + c_2 \cdot \overline{F}2 , \qquad (31)$$

where $c_1 = 0.01$ and $c_2 = 1$ are prices: of the energy put into the process and of the product fulfilling the quality criteria, respectively.

During the example experiment it was assumed that the composition of the feed flow $X1$ is changing according to the following formula:

$$X1(t) = X1_0 + X1_a \cdot \sin(2 \cdot \pi \cdot t/T_o) , \qquad (32)$$

where $X1_0 = 5\%$, $X1_a = 0.9\%$, $T_o = 250$ min.

The obtained results are presented in Fig. 3. The operation of the control structure with the TS dynamic plant model (solid lines in Fig. 3) was compared to the operation of the structure with optimizer exploiting the linear dynamic process model (step response obtained near the operating point $X2 = 25\%$, $P2 = 50.5$ kPa). The usage of the TS model brought significant improvement of control performance. The control error values of the $X2$ variable are much smaller. Moreover, near the 110th minute of the experiment the constraint put on $X2$ output variable was violated when the optimizer with linear dynamic process model was used.

Better operation of the control system with the optimizer using fuzzy model of the process can be clearly seen in the values of economic performance index calculated as a sum of temporary values of (31). The value of the economic performance index obtained in the control system with TS model used in the optimizer was equal to $J_E = -519.6419$ whereas the value obtained in the control system with the optimizer based on the linear dynamic process model was equal to $J_E = -488.5174$ (the smaller the value the better). Thus the economic result obtained in the control system in which the fuzzy process model was used is around 6% better than the one obtained in the control system with the optimizer which used the linear model. In the case of big scale of production it is a significant difference.

It was also checked (for comparison), how application of not only fuzzy process model but also the fuzzy analytical predictive controller (and its model in the set–point optimizer) influences control system performance. The obtained result is almost the same as the one obtained in the control system with the linear analytical controller and the fuzzy process model and therefore is not shown in Fig. 3. The responses were almost the same as those marked in Fig. 3 with solid line and the value of the economic performance index was equal to $J_E = -522.6239$. The result clearly shows that very important, for control performance, is usage of a good dynamic process model and that after application of the optimizer with better (nonlinear) process model the modification of the basic feedback controller is not crucial for the control performance.

6 Summary

The results obtained during the experiments performed in the example control system with fuzzy predictive economic set–point optimizer clearly show that application of a fuzzy dynamic control plant model may significantly improve control system operation comparing to the case when linear dynamic process model is used. Application of a fuzzy analytical predictive controller practically did not have meaning for control system performance.

The optimizer, if only has a fuzzy process model at his disposal, generates the set–points in such a way that operation of a control system with a linear basic feedback controller is clearly improved. It shows the wide field of application for the set–point optimizer based on a fuzzy model. It can be used to improve operation of the existing control systems. What is also important, the proposed optimizer can be relatively easy designed because it is based on an easy to obtain TS dynamic model with step responses used as the local models. The model is used in such a way that the optimization problem, solved at each iteration by the optimizer, is the standard and easy to solve quadratic optimization problem.

Acknowledgment. This work was supported by the Polish national budget funds for science 2007–2009 as a research project.

References

1. Blevins, T.L., McMillan, G.K., Wojsznis, W.K., Brown, M.W.: Advanced Control Unleashed. ISA (2003)
2. Camacho, E.F., Bordons, C.: Model Predictive Control. Springer, Heidelberg (1999)
3. Cutler, C.R., Ramaker, B.L.: Dynamic Matrix Control – a computer control algorithm. In: Proc. Joint Automatic Control Conference, Francisco, CA, USA (1979)
4. Lawrynczuk, M., Marusak, P.M., Tatjewski, P.: Set–point optimisation and predictive constrained control for fast feedback controlled processes. In: Proc. 13th IEEE/IFAC International Conference on Methods and Models in Automation and Robotics MMAR 2007, Szczecin, Poland, pp. 357–362 (2007)
5. Lawrynczuk, M., Marusak, P.M., Tatjewski, P.: Multilayer and integrated structures for predictive control and economic optimisation. In: Proc. 11th IFAC/IFORS/IMACS/IFIP Symposium on Large Scale Systems: Theory and Applications, Gdansk, Poland (2007)
6. Maciejowski, J.M.: Predictive control with constraints. Prentice Hall, Harlow (2002)
7. Marusak, P.M.: Predictive control of nonlinear plants based on DMC technique and fuzzy modeling (in Polish). PhD Thesis, Warsaw, Poland (2002)
8. Newell, R.B., Lee, P.L.: Applied process control – a case study. Prentice Hall, London (1998)
9. Rossiter, J.A.: Model–Based Predictive Control. CRC Press, Boca Raton (2003)
10. Takagi, T., Sugeno, M.: Fuzzy identification of systems and its application to modeling and control. IEEE Trans. Systems, Man and Cybernetics 15, 116–132 (1985)
11. Tatjewski, P.: Advanced Control of Industrial Processes. In: Structures and Algorithms. Springer, London (2007)
12. Tatjewski, P., Lawrynczuk, M., Marusak, P.M.: Linking nonlinear steady-state and target set-point optimisation for model predictive control. In: Proc. International Conference Control 2006, Glasgow, Scotland (2006)

Imprecision Measures for Type-2 Fuzzy Sets: Applications to Linguistic Summarization of Databases

Adam Niewiadomski

Institute of Computer Science
Technical University of Łódź, Poland
aniewiadomski@ics.p.lodz.pl
http://edu.ics.p.lodz.pl

Abstract. The paper proposes new definitions of (im)precision measures for type-2 fuzzy sets representing linguistic terms and linguistically quantified statements. The proposed imprecision measures extend similar concepts for traditional (type-1) fuzzy sets, cf. [1,2]. Applications of those new concepts to linguistic summarization of data are proposed in the context of the problem statement of finding the best summaries.

Keywords: Type-2 Fuzzy Sets, Linguistic Summaries of Databases, Imprecision Measures, Quality Measures of Linguistic Summaries, Type-2 Linguistic Summarization, The Fuzzy Support, Cardinality-Based Imprecision Measure.

1 Representations of Linguistic Terms Using Type-2 Fuzzy Sets

1.1 Basic Definitions

The idea of a *type-2 fuzzy set* [3] extends a traditional membership function of the Zadeh fuzzy set [4]. A *type-2 membership function* is a family of type-1 (traditional) fuzzy sets in $[0,1]$, assigned to elements of a universe of discourse $\mathcal{X}$. A type-2 fuzzy set $\tilde{A}$ in $\mathcal{X}$ is defined

$$\tilde{A} = \int_{\mathcal{X}} \mu_{\tilde{A}}(x)/x \qquad (1)$$

and $\mu_{\tilde{A}}\colon \mathcal{X} \to \mathcal{F}([0,1])$ is its type-2 membership function, such that $\mu_{\tilde{A}}(x) = \int_{u \in J_x} \mu_x(u)/u$, $J_x \subseteq [0,1]$. Each *primary membership degree* u has its own *secondary membership degree*, and, moreover, many u's can be assigned to a given x. $\mu_x\colon J_x \to [0,1]$ is the *secondary membership function* – the type-1 membership function which expresses the (fuzzy) membership of x to $\tilde{A}$.

Example 1. A sample type-2 fuzzy set in a continuous $\mathcal{X}$, with triangular secondary membership functions is depicted in Fig. 1.

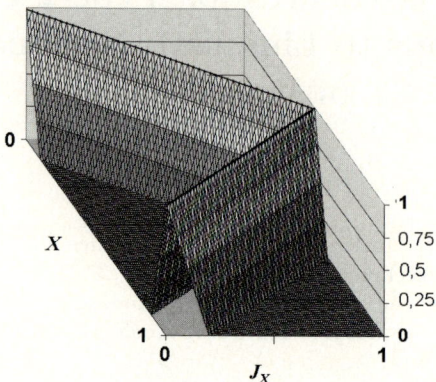

Fig. 1. A type-2 fuzzy set with triangular secondary membership functions

Example 2. Let $\mathcal{X} = \{2, 3, 4, 5\}$ be a set of scores in tests (2 – the lowest, 5 – the highest) taken into account as a discrete universe of discourse. A type-2 fuzzy set $\tilde{B}$ which models the predicate *appropriate score* is defined as follows

$$\tilde{B} = \Big\{ (0.1/0.0 + 0.7/0.1 + 1.0/0.5) / 2 + (0.3/0.1 + 0.5/0.3 + 0.9/1.0) / 3 +$$
$$+ (0.5/0.3 + 0.6/0.7 + 0.8/1.0) / 4 + (0.9/0.0 + 0.8/0.5 + 0.9/1.0) / 5 \Big\}$$

The primary membership degree for the element "2" is the fuzzy set $\{0.1/0.0 + 0.7/0.1 + 1.0/0.5\}$, in the space $J_2 = \{0.0, 0.1, 0.5\} \subset [0, 1]$, where 0.1, 0.7, and 1.0 are the secondary membership degrees. J_3, J_4, J_5 – analogously.

The set-theoretical operations on type-2 fuzzy sets are extensions of the analogous ones for type-1 fuzzy sets. Let $\tilde{A}$, $\tilde{B}$ be type-2 fuzzy sets in $\mathcal{X}$. Let t_1, t_2 be t-norms. The intersection of $\tilde{A}$ and $\tilde{B}$ is the type-2 fuzzy set $\tilde{A} \cap \tilde{B}$, the membership function of which is defined in terms of *the meet* operation "⊓" [3]:

$$\mu_{\tilde{A} \cap \tilde{B}}(x) = \mu_{\tilde{A}}(x) \sqcap \mu_{\tilde{B}}(x) = \int_{u_{\tilde{A}}} \int_{u_{\tilde{B}}} (\mu_x(u_{\tilde{A}}) \; t_1 \; \mu_x(u_{\tilde{B}})) / (u_{\tilde{A}} \; t_2 \; u_{\tilde{B}}) \quad (2)$$

where $u_{\tilde{A}}$, $u_{\tilde{B}}$ – primary membership degrees of x in $\tilde{A}$, $\tilde{B}$, respectively; $\mu_x(u_{\tilde{A}})$, $\mu_x(u_{\tilde{B}})$ – secondary membership degrees of x in $\tilde{A}$, $\tilde{B}$, respectively. (2) is applied as a model for the AND, see Sec. 3.

If common operations on a type-1 fuzzy set A and a type-2 fuzzy set $\tilde{B}$ are required, e.g. when terms represented by A and $\tilde{B}$ are composed with AND, we express A as a type-2 fuzzy set $\tilde{A}$ in which all secondary degrees are 1

$$A = \sum_{x \in \mathcal{X}} \mu_A(x)/x \; \longrightarrow \; \tilde{A} = \int_{x \in \mathcal{X}} \int_{u \in J_x} (1/u(x))/x \quad (3)$$

1.2 Cardinalities of Type-2 Fuzzy Sets

Cardinality of a crisp set A' in a finite $\mathcal{X}$ is the sum of the values characteristic function $\xi_{A'}$: $card(A') = \sum_{x \in \mathcal{X}} \xi_{A'}(x)$. The cardinality of a type-1 fuzzy set A in a finite $\mathcal{X}$ is named σ-*count*, and defined as [5]:

$$\text{card}(A) = \sum_{x \in \mathcal{X}} \mu_A(x) \tag{4}$$

One may extend this definition with respect to a continuous universe of discourse $[a,b] = \mathcal{X} \subset \mathbb{R}$. If $\mu_A(x)$ is segmentally continuous, the so-called *cardinality-like measure* of A in $\mathcal{X}$, $\text{clm}(A)$ is defined:

$$\text{clm}(A) = \int_a^b \mu_A(x)dx \tag{5}$$

The cardinality of a type-2 fuzzy set $\tilde{A}$ in a finite $\mathcal{X}$, the so-called *non-fuzzy σ-count*, assumes that a membership degree of x to $\tilde{A}$ is a fuzzy number. Hence nfσ-count($\tilde{A}$) is defined as the following a generalization of (4):

$$\text{nf}\sigma\text{-count}(\tilde{A}) = \sum_{x \in \mathcal{X}} \sup\{u \in J_x : \mu_x(u) = 1\} \tag{6}$$

Besides, in the type-2 linguistic summarization of data(bases), it is necessary to operate on type-2 fuzzy sets, the universe of discourse of which is an interval $[a,b] \subset \mathbb{R}$. Hence, because (6) produces ∞ as the result (which is the cardinality of an uncountable, so infinite, set), we define *cardinality-like measure* of $\tilde{A}$.

Definition 1. *Let $\tilde{A}$ be a type-2 fuzzy set in $[a,b] = \mathcal{X} \subset \mathbb{R}$, and the function $x: \to \sup\{u \in J_x : \mu_x(u) = 1\}$ is segmentally continuous. The* cardinality-like measure *of $\tilde{A}$, $\text{clm}(\tilde{A})$ is defined as*

$$\text{clm}(\tilde{A}) = \int_a^b \sup\{u \in J_x : \mu_x(u) = 1\}dx \tag{7}$$

1.3 Fuzzy Quantification of Type-2 Fuzzy Propositions

The forms of linguistically quantified propositions are defined in [2]. We originally generalize them with type-2 fuzzy sets as models of S_1, S_2

Definition 2. *Let $\widetilde{S}_1$, $\widetilde{S}_2$ be type-2 fuzzy sets in a finite $\mathcal{X}$ representing linguistic propositions, and Q – a type-1 fuzzy quantifier. The formulae*

$$Q \ x\text{'s are } \widetilde{S}_1 \tag{8}$$

$$Q \ x\text{'s being } \widetilde{S}_2 \text{ are } \widetilde{S}_1 \tag{9}$$

are the first (Q^I) and the second form (Q^{II}) of linguistically quantified proposition. Degrees of truth of (8) and (9) are evaluated as

$$T(\ Q \ x\text{'s are } \widetilde{S}_1\ ) = \mu_Q\left(\frac{\text{nf}\sigma\text{-count}(\widetilde{S}_1)}{M}\right) \tag{10}$$

where nfσ-count$(\widetilde{S_1})$ is a real number, see (6), $M = card(\mathcal{X})$ if Q is relative[1], or $M = 1$ if Q is absolute, and

$$T(\ Q\ x\text{'s being } \widetilde{S_2} \text{ are } \widetilde{S_1}) = \mu_Q \left(\frac{\text{nf}\sigma\text{-count}(\widetilde{S_1} \cap \widetilde{S_2})}{\text{nf}\sigma\text{-count}(\widetilde{S_2})} \right) \qquad (11)$$

where $\widetilde{S_1} \cap \widetilde{S_2}$ is given by (2).

Examples for Q^I, Q^{II}, are MANY students are ABOUT 22 and MANY of YOUNG students are ABOUT 22, respectively, where MANY=Q, about 22=$\widetilde{S_1}$, young=$\widetilde{S_2}$.

2 Imprecision Measures for Type-2 Fuzzy Sets

In this section, we define some scalar imprecision measures for type-2 fuzzy sets. They are dedicated to determine imprecision of linguistic terms represented by those sets, and used in linguistic summarization. It is worth noticing that another approach to uncertainty of type-2 fuzzy sets is presented by Wu and Mendel [6] who derive different measures, which are expressed as type-1 fuzzy sets, from The Representation Theorem [7].

2.1 The Support of a Type-2 Fuzzy Set

The *support* of a type-1 fuzzy set A in $\mathcal{X}$ is defined as $\text{supp}(A) = \{x \in \mathcal{X} : \mu_A(x) > 0\}$. However, none of the existing methods of expressing memberships in type-2 fuzzy sets extends the concept of support[2]. We assume, that the support of any fuzzy set should be a-one-type-lower-set. Due to that, we propose a new concept – *fuzzy support* – associated with a type-2 fuzzy set.

Definition 3. *Let $\widetilde{A}$ be a type-2 fuzzy set in $\mathcal{X}$. Fuzzy support of $\widetilde{A}$ is the type-1 fuzzy set* $\text{supp}(\widetilde{A}) = \{\langle x, \mu_{\text{supp}(\widetilde{A})} \rangle : x \in \mathcal{X}\}$ *where*

$$\mu_{\text{supp}(\widetilde{A})}(x) = \sup_{u \in J_x \setminus \{0\}} \mu_x(u) \qquad (12)$$

The meaning of the concept is as follows: supp($\widetilde{A}$) consists of those points of $\mathcal{X}$ are of the greatest secondary membership degree, assuming a non-zero primary membership.

Corollary 1. *For each type-1 fuzzy set A, $\mu_{\text{supp}(A)}(x) = \xi_{A_0}(x)$[3].*

Example 3. The support of the type-2 fuzzy set depicted in Fig. 1 is $|\mathcal{X}|$.

---

[1] *Relative fuzzy quantifiers* are modeled by normal and convex type-1 fuzzy sets in [0, 1], and *absolute* – by normal and convex type-1 fuzzy sets in $\mathbb{R}^+ \cup \{0\}$, cf. [2].
[2] See the concepts of *lower, upper*, or *principal membership functions* or *FOU* [3,8].
[3] A_0 is the strong 0-cut of A, $A_0 = \{x : \mu_A(x) > 0\}$, i.e. the support of A. The proof of the corollary is given in [9].

2.2 Degree of Fuzziness of a Type-2 Fuzzy Set

The degree of fuzziness is defined for a type-1 fuzzy set A as $\text{in}(A) = \frac{|\text{supp}(A)|}{|\mathcal{X}|}$ [10]. The analogous intuition can be expressed for a type-2 fuzzy set, cf. [9]:

Definition 4. *Let $\widetilde{A}$ be a type-2 fuzzy set in $\mathcal{X}$, and $\text{supp}(\widetilde{A})$ – its fuzzy support. The degree of fuzziness of $\widetilde{A}$ is the real-valued number:*

$$\text{in}(\widetilde{A}) = \frac{\text{card}\big(\text{supp}(\widetilde{A})\big)}{\text{card}(\mathcal{X})} \tag{13}$$

for a finite $\mathcal{X}$, or

$$\text{in}(\widetilde{A}) = \frac{\text{clm}\big(\text{supp}(A)\big)}{|b - a|} \tag{14}$$

for an uncountable $\mathcal{X} = [a, b] \subset \mathbb{R}$.

For each type-2 fuzzy set $\widetilde{A}$ in $\mathcal{X}$, we have:

$$0 \leq \text{in}(\widetilde{A}) \leq 1 \tag{15}$$

If $\widetilde{A}$ is a type-1 fuzzy set, then (13) is equivalent, to $\text{in}(A)$, cf. Corollary 1.

Example 4. Recall the type-2 fuzzy set $\widetilde{A}$ from Example 3. $\text{clm}(\text{supp}(\widetilde{A})) = |\mathcal{X}|$, hence $\text{in}(\widetilde{A}) = 1$.

2.3 Cardinality-Based Measure of Imprecision

This newly defined measure is based on a real-valued cardinality of a type-2 fuzzy set, in particular, in the non-fuzzy σ-count given by (6), or on the cardinality-like measure (7).

Definition 5. *Let $\widetilde{A}$ be a type-2 fuzzy set in a finite $\mathcal{X}$. The measure of imprecision called* **type-2 ratio of cardinality** *of $\widetilde{A}$ is defined as the real-valued number:*

$$\text{rc}(\widetilde{A}) = \frac{\text{nf}\sigma\text{-count}(\widetilde{A})}{\text{card}(\mathcal{X})} \tag{16}$$

If A is given in an uncountable $\mathcal{X} = [a, b] \subset \mathbb{R}$, we define

$$\text{rc}(\widetilde{A}) = \frac{\text{clm}(\widetilde{A})}{|b - a|} \tag{17}$$

For each type-2 fuzzy set $\widetilde{A}$ in $\mathcal{X}$, we have:

$$0 \leq \text{rc}(\widetilde{A}) \leq 1 \tag{18}$$

Example 5. Recall the type-2 fuzzy set $\widetilde{B}$ from Example 2. $|\mathcal{X}| = 4$, $\text{nf}\sigma\text{-count}(\widetilde{B}) = 0.5$. Thus, $\text{rc}(\widetilde{B}) = 0.125$.

For a linguistic term represented by a type-2 fuzzy set $\widetilde{A}$, the imprecision measures $\text{in}(\widetilde{A})$ and $\text{rc}(\widetilde{A})$ may appear similar. Nevertheless, they are related to different characteristics of type-2 fuzzy sets and sometimes only one of them allows to distinguish imprecision of two terms.

3 Type-2 Linguistic Summaries

In linguistic summarization of data, we frequently have to describe some values, especially, membership degrees, with words, e.g. *intensive participation, low intensity*, instead of crisp numbers, e.g. *participation to a degree of* 0.7. For this reason, at least one more measure of uncertainty should be adapted to the linguistic summarization. We definitely see this opportunity in type-2 membership functions, in particular, in secondary membership degrees and functions. Thus, we introduce the concept of *type-2 linguistic summary*, with summarizers represented by type-2 fuzzy sets [9,11].

Definition 6. *A type-2 linguistic summary of a database is a semi-natural language sentence*

$$Q\ P\ \text{are/have}\ \widetilde{S}\ [T] \tag{19}$$

where Q is the quantity pronouncement, P – the subject of the summary, and $\widetilde{S}$ – the summarizer, and T is a real number in $[0,1]$ – the degree of truth.

A sample linguistic summary is *Many of my students are excellent programmers* [0.68], where Q=*many*, P=*my students*, and $\widetilde{S}$=*excellent programmers*.

Type-2 linguistic summaries which use type-2 fuzzy sets as summarizers, are based on forms (8) and (9). Let $\mathcal{Y} = \{y_1, \ldots, y_m\}$ be a set of objects described by a table (view) in a database. Let $\mathcal{D} = \{d_1, \ldots, d_m\}$ be a table (view) the records of which are described with attributes $V_1, \ldots, V_n$ in the domains $\mathcal{X}_1, \ldots, \mathcal{X}_n$, respectively. Let the value of V_j, $j \leq n$, for the y_i object, $i \leq m$, be denoted as $V_j(y_i) \in \mathcal{X}_j$. Hence, one may denote a record d_i as $\langle V_1(y_i), \ldots, V_n(y_i) \rangle$, and – $\mu_{\widetilde{S}_j}(V_j(y_i))$ as $\mu_{\widetilde{S}_j}(d_i)$, where $\widetilde{S}_j$ is a type-2 fuzzy set in $\mathcal{X}_j$.

3.1 Summaries in the First Form

At first, we introduce the type-2 summaries based on (8). The goal is to find the degree of truth of a summary in the form of $Q\ P$ are $\widetilde{S}$. We assume that Q is represented by a type-1 fuzzy set and the real cardinality of $\widetilde{S}$ is computed via (6). The degree of truth of such a summary is a real number

$$T\left(Q\ P\ \text{are/have}\ \widetilde{S}\right) = \mu_Q\left(\frac{\sum_{i=1}^{m} \sup\{u\colon \mu_{\widetilde{S}}(d_i, u) = 1\}}{M}\right) \tag{20}$$

where $M = 1$ if Q is absolute, or $M = m$ if Q is relative. If n of type-2 fuzzy sets $\widetilde{S}_1, \ldots, \widetilde{S}_n$ represent a summarizer, $\widetilde{S} = \widetilde{S}_1$ AND $\widetilde{S}_2$ AND $\ldots$ AND $\widetilde{S}_n$, the membership function of $\widetilde{S}$ is computed as

$$\mu_{\widetilde{S}}(d_i) = \mu_{\widetilde{S}_1}(d_i) \sqcap \mu_{\widetilde{S}_2}(d_i) \sqcap \ldots \sqcap \mu_{\widetilde{S}_n}(d_i) \tag{21}$$

where all the type-1 fuzzy sets are interpreted as type-2 fuzzy sets, via (3), and the intersection (the meet) operation is given by (2).

3.2 Summaries in the Second Form

Summaries based on the second form of a linguistically quantified proposition (9) are now described. In detail, we are interested in the form

$$Q\ P\ \text{being/having}\ \widetilde{w}_g\ \text{are/have}\ \widetilde{S} \qquad (22)$$

in which $\widetilde{w}_g$ is a *qualifier* represented by a type-2 fuzzy set in $\mathcal{X}_g$, $g \in \{1,\ldots,n\}$ and $\widetilde{S}$ is a type-2 summarizer. Similarly to the method presented in [10], the use of preselected features enables producing more interesting summaries. Obviously, the qualifier w_g can be represented by more than one type-2 fuzzy set. In such a case, we use: $\widetilde{w}_g = \widetilde{w}_{g_1}\ \text{AND}\ \ldots\ \text{AND}\ \widetilde{w}_{g_x}$, in the universes of discourse $\mathcal{X}_{g_1},\ldots,\mathcal{X}_{g_x}$, respectively, where $g_1,\ldots,g_x \in \{1,\ldots,n\}$, $x \leq n$. Hence, the membership function of the qualifier represented by type-2 fuzzy sets is given

$$\mu_{\widetilde{w}_g}(d_i) = \mu_{\widetilde{w}_{g_1}}(d_i) \sqcap \ldots \sqcap \mu_{\widetilde{w}_{g_x}}(d_i) \qquad (23)$$

Hence, the degree of truth of a summary in the form of (22), is defined as

$$T\left(Q\ P\ \text{being/having}\ \widetilde{w}_g\ \text{are/have}\ \widetilde{S}\right) =$$
$$= \mu_Q \left(\frac{\sum_{i=1}^m \sup\{u\colon \mu_{\widetilde{S}}(d_i,u) = 1\} \wedge \sup\{u\colon \mu_{\widetilde{w}_g}(d_i,u) = 1\}}{\sum_{i=1}^m \sup\{u\colon \mu_{\widetilde{w}_g}(d_i,u) = 1\}} \right) \qquad (24)$$

where $\mu_{\widetilde{S}}(d_i,u)$ denotes the secondary membership degree for d_i, u. If all $\widetilde{w}_g$, $\widetilde{S}_1,\ldots,\widetilde{S}_n$ are type-1 fuzzy sets, the formula is reduced to that described in [10].

4 Quality Measures for Type-2 Summaries

The quality measures for type-2 linguistic summaries of databases have been introduced by Niewiadomski in [9]. In this paper, we present extended descriptions of those quality measures which are based on the imprecision measures for type-2 fuzzy sets presented in Sec. 2. Besides, some of the presented quality measures of databases extend the analogous ones for type-1 linguistic summaries, presented by the author in [12].

Degree of Imprecision. The degree of imprecision of a linguistic summary with a type-2 fuzzy summarizer is defined as:

$$T_2 = 1 - \left(\prod_{j=1}^n \text{in}(\widetilde{S}_j) \right)^{1/n} \qquad (25)$$

where $\text{in}(\widetilde{S}_j)$ is the degree of fuzziness of the type-2 fuzzy set $\widetilde{S}_j$ in the universe of discourse $\mathcal{X}_j$. $\text{in}(\widetilde{S}_j)$ is given by (13) if $\mathcal{X}_j$ is finite, or by (14) if $\mathcal{X}_j = [a,b] \subset \mathbb{R}$.

Degree of Type-2 Quantifier Imprecision. In case when the linguistic quantifier in a summary is represented by a type-2 fuzzy set[4] $\widetilde{Q}$ in $\mathcal{X}_{\widetilde{Q}} \subset \mathbb{R}^+ \cup \{0\}$, the quality measure of quantifier imprecision is proposed as

$$T_6 = 1 - \frac{\text{clm}(\text{supp}(\widetilde{Q}))}{|\mathcal{X}_{\widetilde{Q}}|} = 1 - \text{in}(\widetilde{Q}) \tag{26}$$

for an absolute $\widetilde{Q}$, where $\text{in}(\widetilde{Q})$ is given by (14), or simply

$$T_6 = 1 - \text{clm}(\text{supp}(\widetilde{Q})) \tag{27}$$

if $\widetilde{Q}$ is relative, i.e. $\mathcal{X}_{\widetilde{Q}} = [0, 1]$.

Degree of Type-2 Quantifier Cardinality. The quantification cardinality quality measure for a summary with a type-2 fuzzy quantifier $\widetilde{Q}$ is[5]

$$T_7 = 1 - \frac{\text{clm}(\widetilde{Q})}{|\mathcal{X}_{\widetilde{Q}}|} = 1 - \text{rc}(\widetilde{Q}) \tag{28}$$

if $\widetilde{Q}$ is absolute, where the imprecision measure $\text{rc}(\widetilde{Q})$ is given by (17), or

$$T_7 = 1 - \text{clm}(\widetilde{Q}) \tag{29}$$

if $\widetilde{Q}$ is relative, where $\text{clm}(\widetilde{Q})$ is given by (7).

Degree of Type-2 Summarizer Cardinality. Because the type-2 summarizer $\widetilde{S}$ in (19), (22) can consist of several fuzzy sets $\widetilde{S}_1, \ldots, \widetilde{S}_n$, the form of the quality measure T_8 is:

$$T_8 = 1 - \left(\prod_{j=1}^{n} \text{rc}(\widetilde{S}_j) \right)^{1/n} \tag{30}$$

where $\text{rc}(\widetilde{S}_j)$ is evaluated via (16) if $\mathcal{X}_j$ is finite, or by (17) if $\mathcal{X}_j = [a, b] \subset \mathbb{R}$.

Degree of Type-2 Qualifier Imprecision. The next quality measure, denoted as T_9, is determined by the degree of imprecision of the qualifier represented by type-2 fuzzy sets $\widetilde{w}_{g_1}, \ldots, \widetilde{w}_{g_x}$, $x \leq n$, in the universes of discourse $\mathcal{X}_{g_1}, \ldots, \mathcal{X}_{g_x}$, we propose the following form of the type-2 quality measure T_9:

$$T_9 = 1 - \left(\prod_{j=1}^{x} \text{in}(\widetilde{w}_{g_j}) \right)^{1/x} \tag{31}$$

where $\text{in}(\widetilde{w}_{g_j})$ is given by (13) if $\mathcal{X}_{g_j}$ is finite, or by (14) if $\mathcal{X}_{g_j} = [a, b] \subset \mathbb{R}$. This quality measure extends the analogous quality measure defined for summaries based on type-1 fuzzy sets, cf. [12].

---

[4] In this paper, we do not introduce type-2 fuzzy quantifiers, because it could exceed the scope of the paper. Properties of type-2 fuzzy sets representing linguistic quantifiers, and forms of related linguistically quantified statements are discussed by the author in [9]. Nevertheless, the proposed quality measure can be applied to a type-1 linguistic quantifier as well, since it is a particular case of a type-2 fuzzy quantifier.

[5] See Footnote 4.

Degree of Type-2 Qualifier Cardinality. The following quality measure directly extends the analogous quality measure T_{10} for a type-1 summary, cf. [12]. If the qualifier is represented by several type-2 fuzzy sets, we propose the following form of the type-2 quality measure T_{10}:

$$T_{10} = 1 - \left(\prod_{j=1}^{x} \mathrm{rc}(\widetilde{w}_{g_j})\right)^{1/x} \tag{32}$$

where $\mathrm{rc}(\widetilde{w}_{g_j})$ is given by (16) if $\mathcal{X}_{g_j}$ is finite, or by (17) if $\mathcal{X}_{g_j} = [a,b] \subset \mathbb{R}$.

The Optimal Type-2 Linguistic Summary. The problem statement of finding the optimal linguistic summary, is analogous to that presented in [10] for linguistic summaries based on type-1 fuzzy sets[6].

$$T = T(T_1, \ldots, T_{10}; w_1, \ldots, w_{10}) = \sum_{i=1}^{10} w_i \cdot T_i \tag{33}$$

where: $w_1 + \ldots + w_{10} = 1$.

The problem statement of finding the summary of the highest quality

$$s^* = \arg\max_{s \in \{s\}} \sum_{i=1}^{10} w_i \cdot T_i \tag{34}$$

An innovative method of automated generation of press comments on databases via finding the best summaries, is described in [9,13].

5 Conclusions

There are new concepts of imprecision measures for type-2 fuzzy sets introduced in the paper. The concepts are exemplified and applied to determine goodness of linguistic summaries of databases. Relations to existing similar methods based on type-1 fuzzy sets, are shown – the presented methods extend the older and include them as special cases.

References

1. Zadeh, L.A.: The concept of linguistic variable and its application for approximate reasoning (I). Information Sciences 8, 199–249 (1975)
2. Zadeh, L.A.: A computational approach to fuzzy quantifiers in natural languages. Computers and Maths with Applications 9, 149–184 (1983)
3. Mendel, J.M.: Uncertain Rule-Based Fuzzy Logic Systems: Introduction and New Directions. Prentice-Hall, Upper Saddle River (2001)
4. Zadeh, L.A.: Fuzzy sets. Information and Control 8, 338–353 (1965)

---

[6] Notice that in this paper we do not present all the quality measures $T_1 \div T_{10}$ for linguistic summaries, but only those in which the imprecision measures for type-2 fuzzy sets, see Sec. 2, are applied. Other quality measures for type-2 linguistic summaries are introduced in [9].

5. De Luca, A., Termini, S.: A definition of the non-probabilistic entropy in the setting of fuzzy sets theory. Information and Control 20, 301–312 (1972)
6. Wu, D., Mendel, J.M.: Uncertainty measures for interval type-2 fuzzy sets. Information Sciences, 5378–5393 (2007)
7. Mendel, J.M., John, R.I.: Type-2 fuzzy sets made simple. IEEE Transactions on Fuzzy Systems 10(2), 117–127 (2002)
8. Karnik, N.N., Mendel, J.M.: An Introduction to Type-2 Fuzzy Logic Systems. University of Southern California, Los Angeles (1998)
9. Niewiadomski, A.: A type-2 fuzzy approach to linguistic summarization of data. IEEE Trans. on Fuzzy Systems 16(1), 198–212 (2008)
10. Kacprzyk, J., Yager, R.R., Zadrożny, S.: A fuzzy logic based approach to linguistic summaries of databases. Int. J. of Applied Mathematics and Computer Sciences 10, 813–834 (2000)
11. Niewiadomski, A.: On Two Possible Roles of Type-2 Fuzzy Sets in Linguistic Summaries. In: Szczepaniak, P.S., Kacprzyk, J., Niewiadomski, A. (eds.) AWIC 2005. LNCS (LNAI), vol. 3528, pp. 341–347. Springer, Heidelberg (2005)
12. Niewiadomski, A.: Six new informativeness indices of data linguistic summaries. In: Szczepaniak, P.S., Węgrzyn-Wolska, K. (eds.) Advances in Intelligent Web Mastering, pp. 254–259. Springer, Heidelberg (2007)
13. Niewiadomski, A., Szczepaniak, P.S.: News generating based on interval type-2 linguistic summaries of databases. In: Proceedings of IPMU 2006 Conference, Paris, France, July 2-7, pp. 1324–1331 (2006)

Assessing the Non-technical Service Aspects by Using Fuzzy Methods

Andrzej Pieczyński[1] and Silva Robak[2]

[1] Institute of Control and Computation Engineering
[2] Faculty of Mathematics, Computer Science and Econometrics
The University of Zielona Góra
ul. Podgórna 50, 65-246 Zielona Góra, Poland
A.Pieczynski@issi.uz.zgora.pl,
S.Robak@wmie.uz.zgora.pl
http://www.uz.zgora.pl

Abstract. The growing expectations for service-oriented applications tailored to the customers needs comprise a definition of the service contract documents, for both, technical and non-technical service elements. The usage of the fuzzy methods and techniques allows for the customized prearrangement of such application systems, which could be developed at the costs of the standard software. In the paper the concepts associated with a service-oriented design of non-technical service aspects especially appropriate for the fuzzy description of them as the requirements destined for Service Level Agreements SLAs, are introduced. The proposed approach is demonstrated on the example of a service accessibility requirement, an important component of a contracted SLA.

1 Introduction and the Problem

The idea of Service-oriented Architecture SOA is based around the idea of software regarded as a service and the externally provided services [5]. As a precondition for such services are flexible and loosely coupled systems working in an asynchronous way. A service is defined e.g. by I. Sommerville as "an act or performance offered by one party to another. Although the process may be tied to a physical product, the performance is essentially intangible and does not normally result in ownership of any of the factors of production" [17]. Therefore a provision of a service is independent of the applications using the service, and there may be diverse stakeholders involved in a delivery, provisioning and usage of a service i.e., the service providers, the facilitators, and element provider, an aggregator and the end-user [15].

According to the SOA principles at least three roles are involved the SOA: a service provider, a service registry and the service requestors - see Fig. 1. The service provider is an organization that owns a service and is responsible for describing and publishing it. Thus, a service needs to be at first published by the service provider in a registry, or described and registered at a broker. The description of a service includes information respective to the functional properties of a service i.e., the technical information, and also the non-functional properties of a service. A registry is a directory where the service descriptions can be published and located. The requestor is a service consumer (client) that requires certain functionality offered by the provider. The primary operations in the SOA between the above mentioned roles are the publishing of a service by the service provider, finding out a service by the service consumer and binding a client to a service.

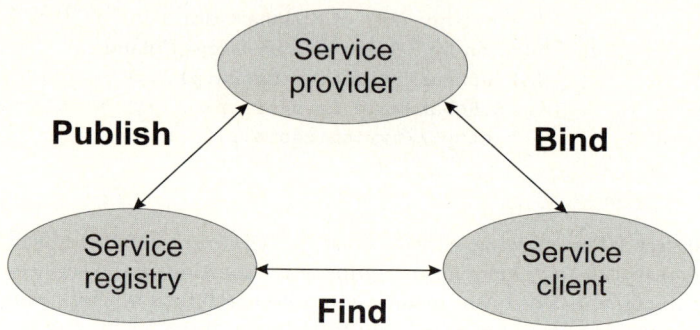

Fig. 1. Operations in the SOA [5]

A process of finding a published service consists of two steps. Firstly, the clients discover the service in the registry. Then they select a service according to some criteria like a service type, the cost of a service, and some other none-functional service's features like reliability, security, availability etc. Finally, binding of a chosen service can proceed by a direct communication between a service provider and a client, or through a broker.

The concepts introduced by the service-oriented architecture are supported by the selected technologies, mainly by the Web services platforms, which significantly expand quality of traditional distributed computing. The mentioned practices are carried out according to the accepted industry standards. There are several enterprizes vendors, and organizations involved with developing contemporary SOA specifications. The major organizations are the Word Wide Web Consortium (W3C), the Web Services Interoperability Organization (WS-I), and the Organization for the Advancement of Structured Information Standards (OASIS).

The remainder of the paper is as follows. Section 2 explains the Web services technology and the quality of service issues. Section 3 presents the approach for the application of the weighted quality requirements for Web services with partially fuzzy representation. In Section 4 we conclude our work.

2 Web Services Technology

By the definition provided by the Word Wide Web Consortium W3C a Web Service is defined as: "a software application identified by a URI, whose interfaces and bindings are capable of being defined, described, and discovered as XML artifacts. A Web service supports direct interactions with other software agents using XML-based messages exchanged via Internet-based protocols" [18].

According to the above definition the web services are the self-describing software modules that use open standards and are available over the Internet.

The Web service may be as simple as the informational services that are used for the request/response sequences or those exposing the back-end of the business applications [5]. The simple Web service are stateless i.e. they do not maintain the history of the previous actions. Examples of the simple informational services are the informational services with an access to the content like the product lists, the catalogs, and simple financial information like pricing, the stock quotes etc. Alternatively, the simple informational services are the business process services that are exposing the back-end of the business applications to the applications of the business partners that are behind the firewalls. The complex trading web services are, in comparison, the transactional Web services that may involve the multiple document exchanges what requires that they will be designed as the systems maintaining their state.

Moreover, the Web services may additionally differ in the way they expose their business logic functionality. Possible are interactive services or such that will be programmed within other applications. Thus they could work whether in an interactive way by using the web-browsers, or they may offer their functionality to other services or applications in a programmatic manner.

A collection of several related technologies is a composition of emerging specific and complementary standards. A Web services technology stack depicted in Fig. 2, and according to Papazoglou [5] includes the following layers:

- ☐ Core Layers,
- ☐ The High Level Layers,
- ☐ The Coordination and Transaction Layer,
- ☐ Value-added Service Layer.

The core layers include the common Internet transport protocols TCP/IP and HTTP, the extensible markup language XML and the packaging protocol for information exchange SOAP. As already mentioned the Web services make use of the technological infrastructure of the Internet, as well as the web servers and browsers use the HTTP protocol. The XML provides a basic format for exchanging data and SOAP is a simple XML-based messaging protocol.

The high level layers in Web services technology stack serves as a basis for the interoperability between the Web services. They include the sub-layers (see Fig. 2) with the standards for the service publication (UDDI), the service description (WSDL), the service flow (BPEL), and the service collaboration (CDL). The publication of the Web services descriptions in a registry like UDDI (Universal Description and Discovery and Integration) will enable finding the services that have been described by means of WSDL (Web Services Description

Language) specifications. The service flow sub-layer allows for the description of the execution logic of Web services that will be needed for the complex processes including multiple organizations. The Business Process Execution Language for Web services BPEL is a most representative workflow-like XML-based language destined for the purposes of orchestrating the interacting Web services. The Web Services Choreography Description Language (WS-CDL) may represent the service collaboration layer, the last sub-layer of the high level layers.

The remaining layers in the Web services technology stack include the coordination and transaction layers and the value-added service layers. Coordinating multiple services requires the existence of the adequate specifications like the WS-Coordination, and WS-Transaction for complementing the BPEL with the extensions needed by the transaction processing and other systems that would coordinate the multiple Web services.

The value-added service layers represent the additional significant aspects of the complex business interactions like security, certifications, and billings.

The enabling technology standards in the core layer constitute the backbone of the technical Web service infrastructure. The service composition and collaboration standards in the high-level layers need a support of the value-adding standards and solutions that comprise mechanisms responsible for reliability and the security concerns.

The Service Level Agreements SLAs are the instruments applied in order to ensure a delivery of the guaranteed level of performance of a service. A SLA embraces the parts such as the purpose of a service, the definitions of the

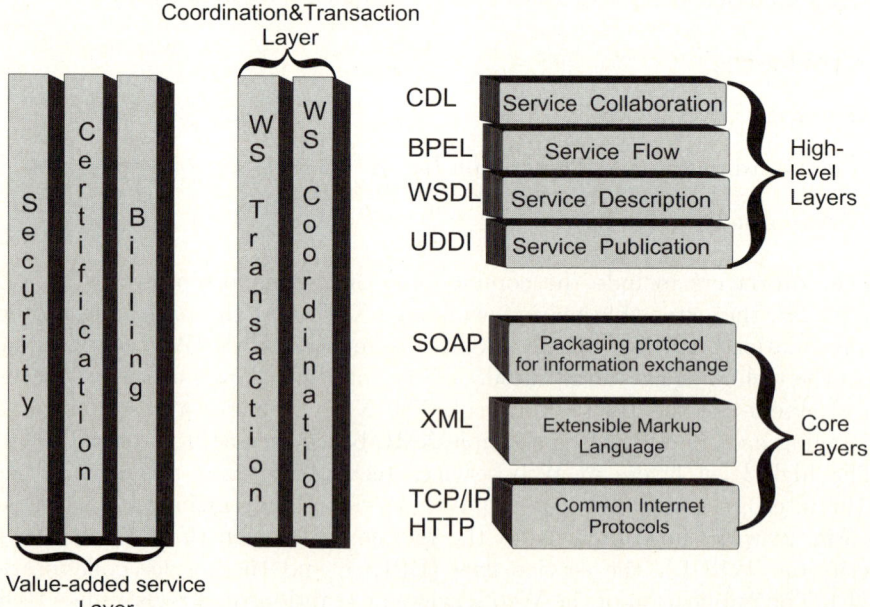

Fig. 2. Web services technology stack [5]

participants and their roles, the duration of service validity and the coverage of a service and the foreseen penalties in the case of the non-delivery of a service in an established quality. The range of the duties of a service is a key part of a SLA. It contains a description of the offered service, the constraints, the steps required for the delivery of a service and the agreed objectives of the service provider and the service consumer. Those objectives are also mostly the concerns that are associated with the quality of a service QoS.

In the next Section the issues associated with the quality of a service for the Web services will be regarded.

2.1 Web Services and Quality of Service

The quality of service is determined by a degree to which applications, systems, networks, and all other elements of the IT infrastructure support the availability of services at a required level of performance under all access and load conditions [2]. In addition to the traditional metrics the environments of the e-Business, that is a usage of the public and insecure medium as the Internet, put specific and severe demands on the organizations offering the services. For that reason delivering the quality of service over the Internet is a critical and significant challenge. The ability of a service to respond to the expected invocations and to perform them at a level adequate for the service provider and the service requestors remains a key concern in this context.

The major quality requirements of the services for supporting the business processes offered by the Web services are based on such concerns as those mentioned in [4] like:

- ☐ availability,
- ☐ accessibility,
- ☐ scalability,
- ☐ performance,
- ☐ conformance to standards,
- ☐ integrity,
- ☐ reliability,
- ☐ security,
- ☐ transactional behavior.

The availability of a service is the absence of a service downtime. It can be measured by the time to response of a service, and also by the probability determining the degree of how a service is available to the requestors (the clients). The large probability values denote that a service will be available to the requestors, as opposed to the small values which mean that a service may be not available. Another concern associated with the availability is the time-to-repair, which represents the time it takes to repair a service that has failed. The scalability denotes the ability for a consistent handling of the queries for a service despite the changing volumes of the requests for a service. Thus the availability of a service can be achieved through scalability.

The accessibility of a service denotes a degree to which the client's request for a service is handled. The high degrees mean that a service may be available to the larger amount of the clients and that they can use the service relatively easily. The performance of a service may be measured by the throughput and latency parameters. The throughput denotes the number of the queries for a service handled in a certain period of a time. The service latency may be measured by a period of time that passed from the time of a request for a service to a time that has passed to point of time of obtaining an answer.

The conformance to standards of a service guarantees a close accordance with the valid standard versions, as defined in a Service Level Agreements SLAs. The integrity of a service guarantees achieving a required functionality of a service. The reliability of a service denotes its ability to function properly and cohesive by remaining the same quality despite failures caused by the humans or machines. The further aspects of the quality of service are the security issues and the transactional behavior of a service. The security aspects include the mechanisms applied for safeguarding the authentication, the non-repudiation, the integrity and the confidentiality of a service.

Some of the quality requirements are fuzzy by their nature. Therefore in the paper we present a proposal for modeling of some the quality requirements by usage of fuzzy methods. The proposal is aimed at fuzzy modeling of the non-functional service requirements for the find operation in the SOA (see Fig. 2). In the first step of the find operation the offered services with the same interface will be discovered. In the second step the services with appropriate QoS properties will be selected. An expert system can help to retrieve an appropriate service in an automatic manner.

3 Application of the Weighted Requirements of Services with Partially Fuzzy Representation

In this Section some quality aspects of the services for the business processes represented by the complex trading Web services will be considered. To the quality attributes mentioned in the previous Section there may come some important additional concerns that should be considered. These are, for instance, the destined market segment, the competitive services offered by another enterprizes, and a period of time in which the service will be offered to the requestors. Such concerns may influence the planned availability and the accessibility of a service and hence the costs of a service in a significant way. Therefore in the example we will concentrate further on the availability and the additional aspects of the accessibility of a service.

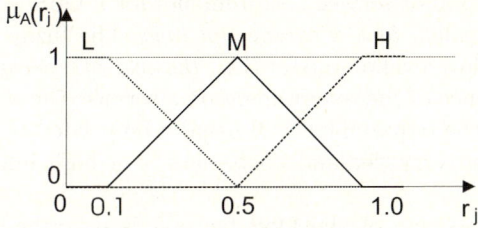

Fig. 3. Membership function used for a representation of the cost of a service and the response time

The availability denotes the absence of a service downtime. Consequently, it may be described with the Boolean representation of the situation in which a service will be whether reachable or not for the requestors. At the same time the accessibility of a service may cover the diverse values, which may in addition differ in a time. In the special periods of the time, such as a period of the day, a day in the week, a week in the month or a month of the year and according to the market segment the requestors may expect a dissimilar availability of a service during the weekends, the holidays, etc. For that reason in the further considerations the availability of a service will be defined by using the Boolean values denoting whether a service is available or not. For the accessibility requirement the three periods of time will be modeled: a day in the week, a week in the month or a month of the year.

Fuzzification operation maps a crisp point into fuzzy sets. In fuzzy representation system of the chosen requirements, two basic assumptions are to be made. The first is the type of the membership function; the other is the number and kind of distribution of fuzzy sets, necessary to perform the representation effectively. The selection of the appropriate membership functions plays an important role in any fuzzy inference system. The triangular, trapezoidal, sigmoidal, generalized bell, and Gaussian membership functions are a few examples of the membership functions that are frequently used. If we want to apply them in a fuzzy inference system, then we have further to state, how many fuzzy rules and what kind of distribution are necessary to describe the behavior of the regarded service requirements [1,16].

For the fuzzy representation of the service requirements in our system different kind of membership function are used. For fuzzy representation of the *the cost of a service* one triangular and two half trapezoidal functions have been chosen, as shown in Fig. 3.

$$\forall_{r_j \in R} \sum_{i=L}^{H} \mu_i(r_j) = 1. \qquad (1)$$

where: r_j - requirement with index j,
R - set of applied requirements,
μ - membership function,
L, M, H fuzzy sets (*low, medium, high*) used for fuzzification process.

Fuzzy representation of service requirements for a QoS is based on the three fuzzy sets that are called *low, medium and high*. The fuzzy set *low* indicates a requirement with a low level of impact (i.e., *the cost of a service* and *the response time*) or a low influence of the requirement on a service, the set *medium* describes the situation when the requirement will usually be needed by customers and the set *high* specifies the very demand customers or a high influence on a service choice decision.

The *availability*, as one of the QoS factors, is described using rectangular membership function (see Fig. 4).

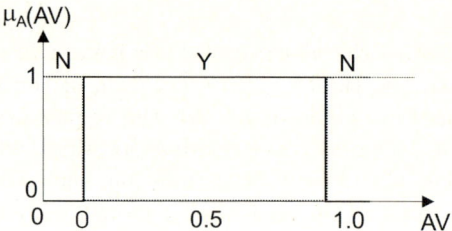

Fig. 4. Membership function used for a representation of availability

The accessibility will be strongly related with a season. We will make use of the discrete fuzzy sets for a description of the season's requirements. The membership functions applied for these aims are depicted in Fig. 5. The membership functions illustrate that the *accessibility* will be bigger Mondays than Tuesdays, at a second than at a first week in a month, and in July than in May.

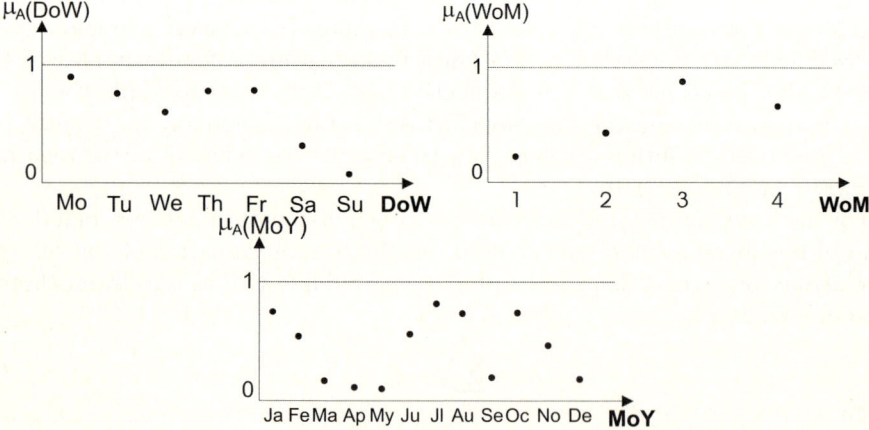

Fig. 5. Membership function used for fuzzy representation of three aspects of a accessibility at (a day of a week, a week of a month, a month of a year, where: A is an event that represents a linguistic variable: accessibility)

The next issue is the evaluation of the requirement's influence on a whole service. The influence will be described by using the suitable weights. In the example of a QoS the service's weights may be interpreted as the influence of the chosen requirement on a whole service's quality. In this case a *cost of a service* has less influence than service *availability*.

Generally speaking, some requirements for a service may be described with the fuzzy representation. In the example these are *availability, accessibility, response time* and a *cost of a service*. The weights of the described requirements are given as the constant values (1.0, 0.7, 0.4, 0.3). The weights represent the impact of the considered requirement on its parent requirements. For instance, *availability* (with the weight value 1.0) has much more influence than *accessibility* (the weight value 0.7), or *response time* (the weight value 0.4), or a *cost of a service* (the weight value 0.3). It means that the month of the year has a most impact on the season parameter of a chosen service requirement.

The importance function $\aleph(AC, WT)$, $\aleph(CS, WC)$, $\aleph(RT, WR)$ are described in the following way (2, 3, 4):

$$\aleph(AC, WT) = \sum_{i=DoW}^{MoY} \mu_i(T) wt_i, \qquad (2)$$

$$\aleph(CS, WC) = \sum_{i=L}^{H} \mu_i(CS) wc_i, \qquad (3)$$

$$\aleph(RT, WR) = \sum_{i=L}^{H} \mu_i(RT) wr_i, \qquad (4)$$

The expert knowledge may be used to define the vectors of factor importance as:

$$\mathbf{WT} = \begin{bmatrix} wt_{DoW} \\ wt_{WoM} \\ wt_{MoY} \end{bmatrix} = \begin{bmatrix} 0.7 \\ 0.4 \\ 0.3 \end{bmatrix}, \mathbf{WC} = \begin{bmatrix} wc_L \\ wc_N \\ wc_H \end{bmatrix} = \begin{bmatrix} 0.1 \\ 0.3 \\ 0.5 \end{bmatrix},$$

$$\mathbf{WR} = \begin{bmatrix} wr_L \\ wr_N \\ wr_H \end{bmatrix} = \begin{bmatrix} 0.1 \\ 0.4 \\ 0.8 \end{bmatrix}.$$

The description of the QoSs influence is defined by the function $\Phi(service)$ in the following way (5):

$$\Phi(service) = T_{NORM}(\aleph(AC, WT), \aleph(CS, WC), \aleph(RT, WR)), \qquad (5)$$

where: $T_{NORM} = prod$ [3,7].

The rule-based knowledge representation [8] of the requirements of the Web service is as following:

IF *Availability* (**AV**) **and** *Accessibility* (**AC**) **and** *Response time* (**RT**) **or** *Cost of service* (**CS**) **THEN** *Web service*;

IF *Availability* (**AV**) **and** *Response time* (**RT**) with $CF = 0.4$ **THEN** *Reference_availability* (**RAV**);
IF *Reference_availability* (**RAV**) with $CF = 0.8$ **and** *Accessibility* (**AC**) with $CF = 0.7$ **and** *Cost of service* (**CS**) with $CF = 0.3$ **THEN** *Reference_accessibility* (**RAC**);
IF *Reference_accessibility* (**RAC**) **and** $\Phi(service) \leq \delta$ **THEN** *Web service*.

An expert system can be implemented in two versions. The first one can be prepared by using the known shell expert system the EXSYS^{TM} Professional. This shell system allows to use the rule-based or tree-based knowledge representations. In this version, a tree representation should use more than 15 nodes. The user interface will be constructed on the base of a special script language, implemented in the EXSYS shell program. The other version of the expert system possible to apply is a dedicated system. The similar system has been developed at the Institute of Control and Computation Engineering at the University of Zielona Góra, Poland [6]. The both systems should base on a forward inference mechanism.

4 Conclusion

The description of a Web service contains the functional and the non-functional characteristics of a service. The functional descriptions define the operational characteristic of the service and concentrate themselves on the syntax of the messages, the configurations of the network protocols and on a delivery of the messages. The non-functional part of the description of a Web service include for instance the metering of the service costs, service quality attributes and the performance metrics. The paper addresses the problems associated with describing some aspects of quality of service for Web services for significant in the second step of the find operation in the SOA architecture. The offered software modules in form of loosely coupled Web service are growing in size and complexity and so the possibility of manual finding an appropriate service in a service registry, is in practice not sufficient.

In the paper a proposal for improvement of decision making for the second step of the find operation, i.e. the selection of a service is proposed. The approach is demonstrated on the example of availability and accessibility requirements with their partially fuzzy representation differentiated for the diverse periods of time. In the future research we shall consider some modification to the representation of fuzzy and crisp knowledge for the aims of the customization of the particular elements of the fuzzification system. These include the number of fuzzy sets, the shapes of the member functions and their arrangement, and the appropriate choice of values for the elements of weight vectors. Another open problem is the appropriate choice of fuzzy weights (importance factors), which should be tuned for specific customer profiles.

The proposed approach can be applied in the SOA's find operation for the automatic selection of an appropriate service between the discovered services.

References

1. Dubois, F., Hüllermeister, D., Prade, H.: Computer Science Today. Fuzzy Set-Based Methods in Instance-Based Reasoning. IEEE Trans. on Fuzzy Systems 10(3), 322–332 (2002)
2. IDC. Quality of service for e-Business.IDC Executive Brief, IDC #23679 (December 2000), www.iec.org/online/tutorial/sdl
3. Kosiński, W.: On fuzzy number calculus. International Journal of Applied Mathematics and Computer Sciences 16(1), 51–57 (2006)
4. Mani, A., Nagarajan, A.: Understanding the quality of service for Web services. IBM, www.ibm.com/developerworks/library/ws-quality.html
5. Papazoglou, M., Riebes, P.M.A.: e-Business - Organizational and Technical Foundations. John Wiley & Sons Ltd., Chichester (2006)
6. Pieczyński, A.: Knowledge representation in diagnostic expert systems. Lubuskie Towarzystwo Naukowe, Zielona Góra (2003) (in polish)
7. Pieczyński, A., Korbicz, J.: Application of the operation aggregation on the fuzzy object signal in detection of the damages. In: Proceedings Second Conference on the Diagnostic of Industrial Processes DPP'97, Lagow Poland, Technical University of Zielona Gora 1997 (in polish), pp. 93–98 (1997)
8. Pieczyński, A., Robak, S., Walaszek Babiszewska, A.: Features with fuzzy probability. Engineering of Computer-Based Systems ECBS 2004. In: Proceedings 11th IEEE International Conference Brno. IEEE, Los Alamitos (2004)
9. Robak S.: Contribution to the improvement of the software development process for product families. Monography, Uniwersity of Zielna Góra Press, Zielona Góra (2006)
10. Robak, S., Pieczyński, A.: Employing fuzzy logic in feature diagrams to model variability in software product-lines. In: Engineering of Computer-Based Systems ECBS 2003. Proceedings 10th IEEE International Conference, Huntsville, Alabama. IEEE, Los Alamitos (2003)
11. Robak, S., Pieczyński, A.: Employment of Fuzzy Logic in Feature Diagrams to Model Variability in Software Families. Journal of Integrated Design and Process Science 7(3), 79–94 (2003)
12. Robak, S., Pieczyński, A.: Application of Fuzzy Weighted Feature Diagrams to Model Variability in Software Families. In: Rutkowski, L., Siekmann, J.H., Tadeusiewicz, R., Zadeh, L.A. (eds.) ICAISC 2004. LNCS (LNAI), vol. 3070, pp. 370–375. Springer, Heidelberg (2004)
13. Robak, S., Pieczyński, A.: Adjusting Software-Intensive Systems Developed by Using Software Factories and Fuzzy Features. In: Rutkowski, L., Tadeusiewicz, R., Zadeh, L.A., Żurada, J.M. (eds.) ICAISC 2006. LNCS (LNAI), vol. 4029, pp. 297–305. Springer, Heidelberg (2006)
14. Robak, S., Pieczyński, A.: Fuzzy Modeling Of The Quality Of Services For Information Systems. Journal of apllied computer science (accepted, 2007)
15. Räisänen, V.: Service Modeling. In: Principles and applications. John Wiley & Sons Ltd, Chichester (2006)
16. Rutkowski, L.: Methods and techniques of Artificial Intelligence. In: Computing Intelligence. PWN, Warszawa (2006) (in polish)
17. Sommerville, I.: Software Engineering, 8th edn. Pearson Education, London (2006)
18. W3C. Web Services Architecture Requirements (October 2002), http://www.w3.org/TR/wsa-reqs

Adaptation of Rules in the Fuzzy Control System Using the Arithmetic of Ordered Fuzzy Numbers

Piotr Prokopowicz

Institute of Environmental Mechanics and Applied Computer Science
Kazimierz Wielki University, Bydgoszcz, Poland
piotrekp@ukw.edu.pl
http://www.imsis.ukw.edu.pl

Abstract. This paper describes a new look on adaptation of fuzzy rules in fuzzy control process. New idea is based on properties of the Ordered Fuzzy Numbers. The Ordered fuzzy nunbers (OFN) are a new model of fuzzy numbers, presented a few years ago [15]. Important property and advantage of the new model of fuzzy numbers is simple realization of arithmetical operations. Thanks to that we can get neutral element of adding and multiplication in the same way like in real numbers. Easy way of calculating on the Ordered Fuzzy Numbers makes possible to use them in a fuzzy control process. In the [21] new methods of processing information for a fuzzy control system were presented. These methods basing on arithmetic of the Ordered Fuzzy Numbers.

The goal of that paper is to present a way to use a good arithmetical properties of Ordered Fuzzy Numbers in the process of rules adaptation for the fuzzy control system.

Keywords: fuzzy number, arithmetic of fuzzy numbers, ordered fuzzy number, adaptation of fuzzy control.

1 Introduction

In modern complex and large-scale systems we often have to deal with imprecise data. It can have source in many aspects: different authority of people operating the system, not well-known area of knowledge, too many parameters to consider them together in short time. Effective use an imprecise data needs tools for processing them. Mathematical modelling of vagueness use a fuzzy sets and numbers as such tool.

Idea of the fuzzy sets was started in 1965 by Zadeh. Fuzzy numbers and fuzzy arithmetic (as extension of fuzzy sets) were introduced to analyze and manipulate approximate numerical values. Main ideas for fuzzy numbers are gathered under the name *convex fuzzy numbers*. The commonly accepted model of fuzzy numbers (in general compatible with convex fuzzy numbers) is one precised and set up by Dubois and Prade [3]. They proposed a restricted class of membership functions called (L, R)–numbers. The essence of such representation is that, the membership function is a particular form which is generated by two so-called shape (or spread) functions: L and R. In this context (L,R)–numbers became quite popular, because of their good interpretability and relatively easy handling for simple operation, i.e. for the fuzzy addition.

Most approaches for numerical handling of fuzzy quantities is based on the so-called extension principle. It gives a formal apparatus to carry over operations (arithmetic or algebraic) from sets to fuzzy sets. Unfortunately, the use of the extension principle results have some serious drawbacks, especially in case of performing the whole sequences of operations repeatedly (e.g. [11] [14]). A number of attempts to introduce non-standard operations on fuzzy numbers have been made [6][9][11].

The main drawbacks of the existing models of operations on fuzzy numbers (regarded as a part of the theory of fuzzy sets defined on the real axis) are: non-existing of natural neutral elements of addition and multiplication, non-existence of a unique solution X of the equation $A + X = C$, where A and C are fuzzy numbers.

To overcome these problems a new model the Ordered Fuzzy Numbers (OFN) has been defined and explored in [15][20] [18][21]. It provides a quite simple representation of non-precise information and also simple algebraic operations on them. If one performs many operations on OFNs, then the fuzziness of calculated result is limited in contrast to the operations on standard convex (als fuzzy numbers.

For constructing those numbers the concept of the membership function of a fuzzy set, introduced by Zadeh in 1965 [1] as a fundamental concept of the fuzzy (multivalued) logic, has been weakened by requiring a mere membership relations in consequent an Ordered Fuzzy Number is as an ordered pair of continuous real functions defined on the interval $[0,1]$. Four algebraic operations: addition, subtraction, multiplication and division of such fuzzy numbers have been constructed in a way that renders them into algebra.

The model of Ordered Fuzzy Numbers has interesting and important properties. Some of them are: subtracting is simply adding opposite number so $A - B = A + (-B)$ where $(-1) \cdot B = -B$, result of many operations on fuzzy number must not always be "more fuzzy", $A - A = 0$ – crisp zero, so it is neutral element of addition, operations are enough simply to programming them without very complicated methods. Moreover, the particular case of the known fuzzy numbers set of the so-called convex fuzzy numbers can be obtained as a particular subset of space of Ordered Fuzzy Numbers; for the members of this subset the arithmetic operations are compatible, to some extent, with the results for convex fuzzy numbers. In [20] [21] new propositions of operations in processing imprecise information in fuzzy control system were introduced and tested on simple simulation. The new ideas are based on arithmetical properties of Ordered Fuzzy Numbers. Those propositions shows that the Ordered Fuzzy Numbers with their "good" algebra opens new areas for work and a analysis of vagueness informations.

Further in this paper a new method for adaptation of fuzzy rules in fuzzy system will be presented. Thanks to good arithmetical properties of the Ordered Fuzzy Numbers a method is simple and intuitive.

2 Ordered Fuzzy Numbers

In the series of papers [15],[16],[17],[18],[20],[21],[22] main concepts of the idea of Ordered Fuzzy Numbers were introduced and developed. Following these papers fuzzy number A will be identified with the pair of functions defined on the interval $[0, 1]$, i.e.

2.1 Main Definition

Definition 1. *By an Ordered Fuzzy Number A we mean an ordered pair of two continuous functions*

$$A = (x_{up}, x_{down})$$

called the up-branch and the down-branch, respectively, both defined on the closed interval $[0, 1]$ with values in $\boldsymbol{R}$.

The continuity of both parts implies that their images are bounded intervals, we can call UP and $DOWN$, respectively (fig.1 a)). If we used the symbols $UP = [l_A, 1_A^-]$ and $DOWN = [1_A^+, p_A]$ to mark boundaries and add the third interval $CONST = [1_A^-, 1_A^+]$, then we have in fact three characteristic subintervals (they can be also indicated in convex fuzzy numbers) which together gives the support of fuzzy number. Notice that, in general, neither $l_A \leq 1_A^-$ nor $1_A^+ \leq p_A$ must hold (i.e. $x_{up}(1)$ does not need to be less than $x_{down}(1)$). In this way we can reach improper intervals, which have been already discussed in the framework of the extended interval arithmetic by Kaucher in [5] and called by him directed intervals, i.e. such $[n, m]$ where n may be greater than m.

In general, the functions x_{up}, x_{down} need not to be invertible, if we assume, however, that they are monotonous: x_{up} is increasing, and x_{down} is decreasing, and also $x_{up} \leq x_{down}$ (pointwise), we may define the membership function $\mu(x) = x_{up}^{-1}(x)$, if $x \in [x_{up}(0), x_{up}(1)] = [l_A, 1_A^-]$, and $\mu(x) = x_{down}^{-1}(x)$, if $x \in [x_{down}(1), x_{down}(0)] = [1_A^+, p_A]$ and $\mu(x) = 1$ when $x \in [1_A^-, 1_A^+]$.

In this way we got the membership function $\mu(x), x \in R$, which can be referred to membership function of classical fuzzy set.

When the functions x_{up} and/or x_{down} are not invertible or the second condition is not satisfied then the **membership curve** (or relation) can be defined, composed of the graphs of x_{up} and x_{down} and the line $y = 1$ over the core $\{x \in [x_{up}(1), x_{down}(1)]\}$.

It is worthwhile to point out that a class of Ordered Fuzzy Numbers (OFNs) represents the whole class of convex fuzzy numbers ([4],[11],[12],[13],[19], with continuous membership functions.

The ordered pair of two continuous functions on fig.1 c) (here just two affine functions) corresponds to a membership function of a convex fuzzy number with

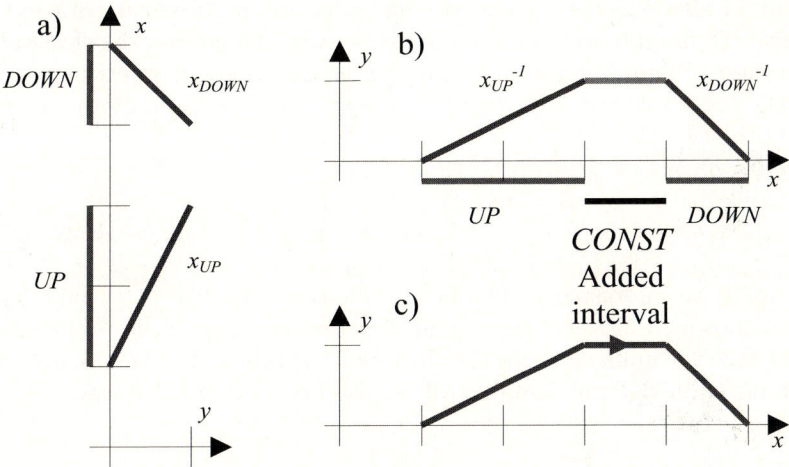

Fig. 1. a) Ordered Fuzzy Number from definition, b) Ordered Fuzzy Number as convex fuzzy number, c) Arrow to mark the order of inverted functions

an extra arrow from x_{up} to x_{down} This arrow shows that we are dealing with the ordered pair of functions. The pair of functions (x_{down}, x_{up}) determines different Ordered Fuzzy Number than the pair (x_{up}, x_{down}). Graphically the curves (x_{up}, x_{down}) and (x_{down}, x_{up}) do not differ, however they determine two different Ordered Fuzzy Numbers which are differ by the *orientation* which is marked in fig.1 c) by an arrow.

2.2 Arithmetic Operations

Now is time to define arithmetic operations to calculate on Ordered Fuzzy Numbers.

Definition 2. *Let $A = (f_A, g_A), B = (f_B, g_B)$ and $C = (f_C, g_C)$ are mathematical objects called Ordered Fuzzy Numbers. The sum $C = A + B$, subtraction $C = A - B$, product $C = A \cdot B$, and division $C = A \div B$ are defined by formula*

$$f_C(y) = f_A(y) \star f_B(y) \quad \wedge \quad g_C(y) = g_A(y) \star g_B(y) \qquad (1)$$

where "$\star$" works for "+", "−", "·", and "÷", respectively, and where $A \div B$ is defined, if the functions $|f_B|$ and $|g_B|$ are bigger than zero.

As it was already noticed in the previous section the subtraction of B is the same as addition of the opposite of B, i.e. the number $(-1) \cdot B$.

Let's analyse the example presented on fig.2. We need to calculate sum of two Ordered Fuzzy Numbers A and B:
A : $f_A(y) = y + 1 \quad g_A(y) = -y + 3$
B : $f_B(y) = 2y + 1 \quad g_B(y) = -y + 4$
We just add functions f and g separately and get result
C : $f_C(y) = f_A(y) + f_B(y) = 3y + 2 \quad g_C(y) = g_A(y) + g_B(y) = -2y + 7$.

Result in this example is exactly the same as for classical (convex) fuzzy numbers with use interval arithmethic for $\alpha - cuts$. Of course, the classical fuzzy numbers uses different notation without information about direction.

Another example (first on fig.3) concerns multiplication. Here we need to multiply the two Ordered Fuzzy Numbers A and B:
A : $f_A(y) = 0,5y + 0,5 \quad g_A(y) = -y + 2$
B : $f_B(y) = y + 2 \quad g_B(y) = -y + 5$

And again we calculate separately on functions f and g and get result
C : $f_C(y) = f_A(y) \cdot f_B(y) = 0.5y^2 + 1.5y + 1 \quad g_C(y) = g_A(y) + g_B(y) = y^2 - 7y + 10$.

Of course we interested in function's values for interval $[0,1]$ only. This example is shown in classical fuzzy numbers like view, which is proper in using Ordered Fuzzy Numbers to deal with imprecise values. For this reason all further examples in this publication will be also presented in this way.

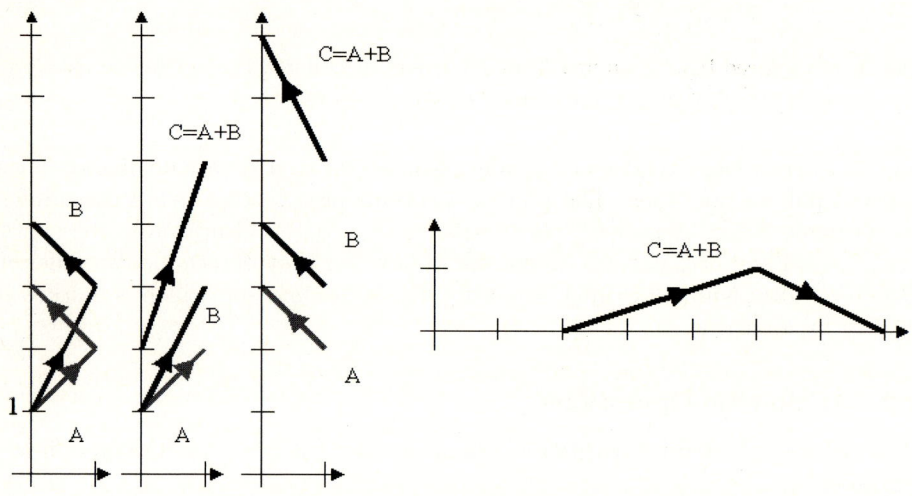

Fig. 2. Example of addition - calculations and result

3 Ordered Fuzzy Numbers in Fuzzy Control

Easy way of doing calculations on OFN also allows to use it in fuzzy control methods. We can use arithmetics to realize the individual steps in fuzzy control.

The following actions: calculating the level of activation, calculating the result of individual rule, finding a fuzzy answer for whole control process, may depend on arithmetical properties of OFNs.

3.1 Level of Activation

As first, the proposition for calculating a level of activation of the rule will be presented. We understand premise part of the rule like follows

$$\text{IF } x_1 \text{ is } A_1 \text{ AND } x_2 \text{ is } A_2 \text{ AND } ... \text{ AND } x_n \text{ is} A_n \textbf{ THEN}... \quad (2)$$

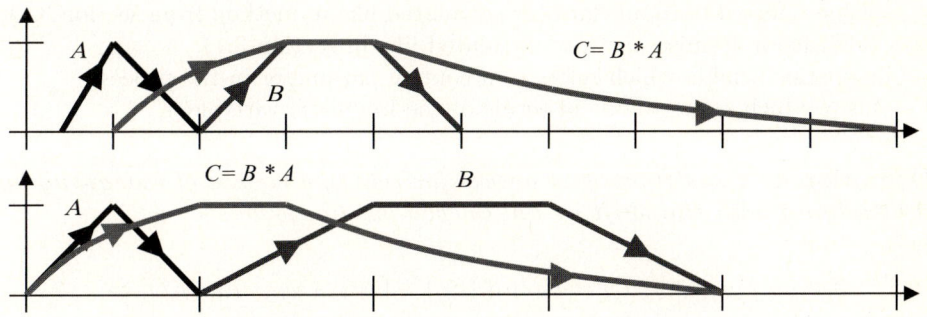

Fig. 3. Examples of multiplication

where x_i represents input parameters, A_i stands for Ordered Fuzzy Numbers and $i = 1, 2...n$ stands for amount of simply premise parts which.

Definition 3. *Calculating the level of activation a_k, for a rule like above, by the method with using **average with condition** we call receiving the result like follows:*

$$a_k = S(s) \qquad (3)$$

S - arithmetic mean of OFNs in the rule s - arithmetic mean of all input data considered in that rule, however, if exist any A_i which gives $\mu_{A_i}(x_i) = 0$ then $a_k = 0$.

S is calculated by:

$$S = \frac{\sum_{i=1}^{n} A_i}{n} \qquad (4)$$

and s by:

$$s = \frac{\sum_{i=1}^{n} x_i}{n} \qquad (5)$$

where: A_i - Ordered Fuzzy Numbers considered in that rule, x_i - consecutive input data, n - amount of Ordered Fuzzy Numbers in that rule.

We can see that calculations transforms many simple premise parts into one and as a result for calculating a level of activation, we just calculate a value of membership function of specially calculated OFN for specially calculated argument.

3.2 Answer of the Rule

We can use arithmetic of OFNs also to define a method of calculating an answer for individual rule. Such proposition is presented below.

The level of activation of any rule is real number (from interval [0,1]), so we can use it as parameter (singleton) in calculations. We will consider rule like follows:

$$\textbf{IF } x_w \text{ is } W \textbf{ THEN } y \text{ is } Y_1 \qquad (6)$$

x_w - value referred to input data (or calculated like in method from section 3.1),
W - OFN from premise part (or calculated like in section 3.1),
y - linguistic variable which refers to a control parameter in the process,
Y_1 - OFN which refers to one of terms of the linguistic variable y.

Definition 4. *Calculation of an answer for rule by a method of **reasoning by multiplying with the shift** we call calculations like follows:*
for

$$\begin{aligned}
x_w \in UP_W &\quad : Y_R = Y_1 \cdot \mu_W(x_w) + [u_{Y1} - u_{Y1} \cdot \mu_W(x_w)] \Leftrightarrow \\
&\quad \Leftrightarrow Y_R = (Y_1 - u_{Y1}) \cdot \mu_W(x_w) + l_{Y1} \\
x_w \in CONST_W &\quad : Y_R = Y_1 \cdot \mu_W(x_w), \ \mu_W(x_w) = 1 \ \Leftrightarrow \ Y_R = Y_1 \quad (7) \\
x_w \in DOWN_W &\quad : Y_R = Y_1 \cdot \mu_W(x_w) + [d_{Y1} - d_{Y1} \cdot \mu_W(x_w)] \Leftrightarrow \\
&\quad \Leftrightarrow Y_R = (Y_1 - d_{Y1}) \cdot \mu_W(x_w) + d_{Y1}
\end{aligned}$$

where $mu_W(x_w)$ - is the level of activation for this rule u_{Y1} - is UP-side border of the fuzzy number Y_1 support and d_{Y1} - DOWN-side border of support. Interesting property for this method is fact, the support of the result Y_R is included in a support of the OFN from the rule's conclusion Y_1 (fig.4).

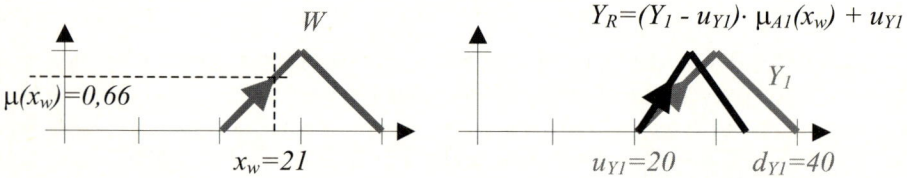

Fig. 4. An example of calculations by using a method of reasoning by multiplying with shift

This method consider which part of the membership function includes level of activation.

3.3 Fuzzy Answer of Whole Fuzzy Controller

At last, to complete processing in fuzzy control system, we can use arithmetics to determine a fuzzy answer of the whole rule base. Here we will use the idea of weighted mean where weights are the levels of activation of the considered rules:

$$Y_{res} = \frac{\sum_{i=1}^{k}(a_i \cdot Y_{Ri})}{\sum_{i=1}^{k} a_i} \quad (8)$$

where k - amount of rules, a_i - level of activation for i rule, Y_{Ri} - result for i rule.

For such calculations the result will be always OFN. We can see that, rules which were activated in level 0 have no participation in the final result.

For the final result we can use any of known methods of defuzzification i.e. *center of gravity* method, but we can also use some of new methods provided by OFN (more in [17])

4 Simple Idea of Adaptation

Important observation for all these proposed methods (for each step in fuzzy control) is a fact, we always operate on (and get in result) the Ordered Fuzzy Numbers. This is rather unique property for fuzzy control systems. In general, in classical methods, even if premise part and conlusion are described by fuzzy numbers, after reasoning we usually gets as a result fuzzy set which is not a fuzzy number. Thus is no possible to use an arithmethical operations to further processing such results.

Now is time for proposal to use a good arithmetical properties of OFNs in adaptation IF-THEN rules. It can be first step to establish (in future) simple and new self-adapting fuzzy systems based on OFNs.

Main idea presented in this paper is adopted from Artificial Neural Networks area and is based on teaching single perceptron neural network. To teach neural perceptron we use information about error, where error is difference betwen expected and real result.

For needs of this publication we consider a situation when we already calculated an error for given rule. How to do this calculation? This is a question where answer will be analysed in future work with this subject. However, we can imagine that, the error for given rule is a some kind transformation of an error for the whole fuzzy system, and this whole error is based on difference between actual reply of system and reply which is expected.

Now we can make some formal notions. We will adopt idea from adaptation perceptron to adaptation of a rule in fuzzy system based on OFNs ideas.

$$Y = Y + (\alpha \cdot \Delta Y) \qquad (9)$$

where Y - number from conclusion of the rule, ΔY - error for given rule, $\alpha \in (0, 1]$ - teaching parameter.

Let us accept that, we have as a result for considered rule an Ordered Fuzzy Number Y_R on fig.5 and we expected certain value Y_E as a result. Here, we can do a some intuitive assumptions, which can be related to a neural networks teaching methods. If ΔY is a positive number then expected answer for this rule is grater than actual, and if ΔY is minus value, then expected answer is lesser. Let us look closely into examples on fig.5 and fig.6. Here we dealing with the input values which gives us a level of activation $a = 0, 5$ for given rule, so we get as result (with use a method presented by definition 4) a number Y_R.

Fig.5 presents a situation where ΔY is positive OFN. This mean, that rule should (for better results for whole fuzzy system) generate greater OFNs as answers. If we use a teaching formula (9) with $\alpha = 1$, then we get results presented on fig.5b and this is a greater OFN as a result for the same level of activation. On fig.6 we can see the situation when ΔY is minus value, so we need lesser results for this rule (in aspect of whole fuzzy system). After applying formula (9) we indeed achieving what we expect (fig.6b).

Both examples are presenting proper results. After this analysis we can also expect correct results in other similar situations.

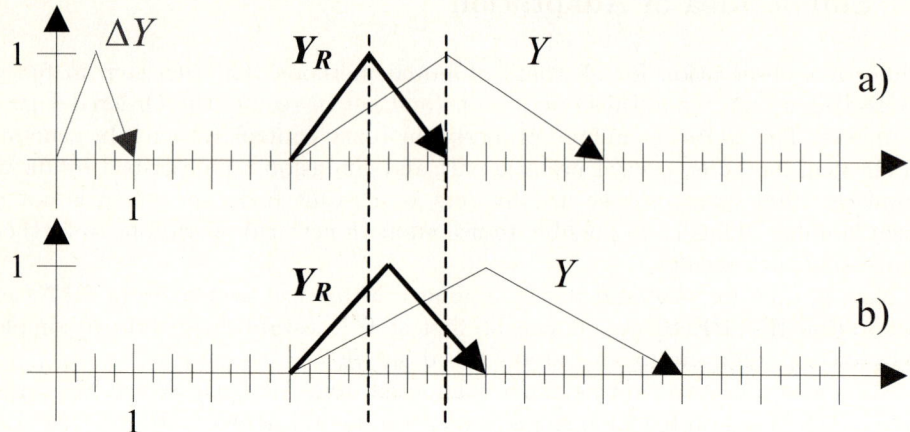

Fig. 5. Changes of conclusion in rule with positive ΔY

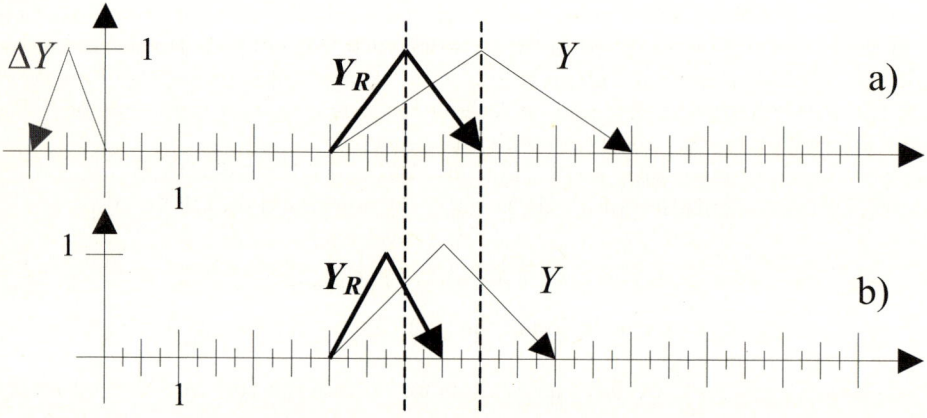

Fig. 6. Changes of conclusion in rule with negative ΔY

5 Summary

Method presented here is based on simple type (philosphy) of adaptation - learning with a teacher. In that type, in general, we compare given results with those expected and use the difference as base to establish (calculate) changes in system. These changes let us get better results next time. In consequent, many system's small changes for plus and for minus, allows to find proper (better) parameters of system, not only for a few input data-sets, but for wide class of them. Such teaching is well known for perceptron neural networks.

The lack of the good and simple arithmetic of fuzzy numbers is a barrier in using such simple methods of adaptation in the fuzzy control system. Main drawback for classical fuzzy numbers is a fact that the many operations implies more and more imprecise fuzzy number in result. However using *Ordered Fuzzy Numbers* we have no such limitations. Thanks to their good arithmetic we can apply with easy a simple and well-known (in other cases) methods in the fuzzy system adaptation.

This paper shows a way to use a new idea of OFNs to open a new areas for processing imprecise information. Complete method of adaptation for rules in fuzzy system with using OFNs need still some work. Good model for testing and analysing properties of new ideas is important step in the future work.

References

1. Zadeh, L.A.: Fuzzy sets. Information and Control 8, 338–353 (1965)
2. Zadeh, L.A.: The concept of a linguistic variable and its application to approximate reasoning, Part I. Information Sciences 8, 199–249 (1975)
3. Dubois, D., Prade, H.: Operations on fuzzy numbers. Int. J. System Science 9(6), 613–626 (1978)
4. Nguyen, H.T.: A note on the extension principle for fuzzy sets. J. Math. Anal. Appl. 64, 369–380 (1978)
5. Kaucher, E.: Interval analysis in the extended interval space IR. Computing, Suppl 2, 33–49 (1980)
6. Sanchez, E.: Solutions of fuzzy equations with extended operations. Fuzzy Sets and Systems 12, 237–248 (1984)
7. Czogała, E., Pedrycz, W.: Elements and methods of fuzzy set theory (in Polish), PWN, Warsaw, Poland (1985)
8. Yager, R.R., Filev, D.P.: Basics of modelling and fuzzy control (in Polish), WNT, Warszawa (1995)
9. Klir, G.J.: Fuzzy arithmetic with requisite constraints. Fuzzy Sets and Systems 91(2), 165–175 (1997)
10. Chen, G., Pham, T.T.: Introduction to Fuzzy Sets, Fuzzy Logic, and Fuzzy Control Systems. CRC press LLC, United States (2001)
11. Wagenknecht, M.: On the approximate treatment of fuzzy arithmetics by inclusion, linear regression and information content estimation. In: Chojcan, J., Łęski, J. (eds.) Fuzzy Sets and their Applications, pp. 291–310. Publishing House of Silesian University of Technology, Gliwice (in polish)
12. Guanrong, C., Tat, P.T.: Fuzzy Sets, Fuzzy Logic, and Fuzzy Control Systems. CRS Press, Boca Raton, London, New York, Washington, D.C (2001)
13. Drewniak, J.: Fuzzy numbers (In Polish). In: Chojcan, J., Łęski, J. (eds.) Fuzzy Sets and their Applications, WPŚ, Gliwice, Poland, pp. 103–129 (2001)
14. Wagenknecht, M., Hampel, R., Schneider, V.: Computational aspects of fuzzy arithmetic based on archimedean *t*-norms. Fuzzy Sets and Systems 123(1), 49–62 (2001)
15. Kosiński, W., Prokopowicz, P.P., Ślezak, D.: Ordered Fuzzy Number. Bulletin of the Polish Academy of Sciences, Ser. Sci. Math. 51(3), 327–338 (2003)
16. Kosiński, W., Prokopowicz, P.: Algebra of fuzzy numbers (in Polish), Applied Mathathics. Mathathics for the Society 5 (46), 37–63 (2004)
17. Kosiński, W.: On Defuzzyfication of Ordered Fuzzy Numbers. In: Rutkowski, L., Siekmann, J.H., Tadeusiewicz, R., Zadeh, L.A. (eds.) ICAISC 2004. LNCS (LNAI), vol. 3070, pp. 326–331. Springer, Heidelberg (2004)

18. Koleśnik, R., Prokopowicz, P., Kosiński, W.: Fuzzy Calculator – Useful Tool for Programming with Fuzzy Algebra. In: Rutkowski, L., Siekmann, J.H., Tadeusiewicz, R., Zadeh, L.A. (eds.) ICAISC 2004. LNCS (LNAI), vol. 3070, pp. 320–325. Springer, Heidelberg (2004)
19. Buckley James, J., Eslami, E.: An Introduction to Fuzzy Logic and Fuzzy Sets. Physica-Verlag, A Springer-Verlag Company, Heidelberg (2005)
20. Prokopowicz, P.: Methods based on the Ordered Fuzzy Numbers used in fuzzy control. In: Proc. Of the Fifth International Workshop on Robot Motion and Control – RoMoCo 2005, Dymaczewo, Poland, June 2005, pp. 349–354 (2005)
21. Prokopowicz, P.: Using Ordered Fuzzy Numbers Arithmetic, in Fuzzy Control in Artificial Intelligence and Soft Computing. In: Cader, A., Rutkowski, L., Tadeusiewicz, R., Zurada, J. (eds.) Proc. of the 8th International Conference on Artificial Intelligence and Soft Computing Zakopane, Poland, June 25-29, pp. 156–162. Academic Publishing House EXIT, Warsaw (2006)
22. Kosiński, W., Prokopowicz, P.: Fuzziness - Representation of Dynamic Changes, Using Ordered Fuzzy NumbeOrdered Fuzzy Numbers Arithmetic, New Dimensions in Fuzzy Logic nd Related Technologies, In: Proc. of the 5th EUSFLAT Conference, Ostrava, Czech Republic, September 11-14, 2007, Martin Stepnicka, Vilem Novak, Ulrich Bodenhofer(eds), University of Ostrava, vol. I, pp. 449-456 (2007)

Regression Modeling with Fuzzy Relations*

Rafał Scherer

Department of Computer Engineering, Częstochowa University of Technology
al. Armii Krajowej 36, 42-200 Częstochowa, Poland
http://kik.pcz.pl
Department of Artificial Intelligence, Academy of Humanities and Economics in
Lodz, ul. Rewolucji 1905 nr 64, Łódź, Poland
rafal@ieee.org
http://www.wshe.lodz.pl

Abstract. In the paper relational neuro-fuzzy systems are described with additional fuzzy relation connecting input and output linguistic fuzzy terms. Thanks to this the fuzzy rules have more complicated structure and can be better suited the task. Fuzzy clustering and relational equations are used to obtain the initial set of fuzzy rules and systems are then learned by the backpropagation algorithm. Simulations shows excellent performance of the modified neuro-fuzzy systems.

1 Introduction

Computational intelligence methods are used for modeling regression problems [4]. They have various advantages and drawbacks. Fuzzy models [1][5][10][11][14] are frequently used because knowledge in the form of fuzzy rules is easily understandable. Generally they fall into one of the following categories [16], depending on the connective between the antecedent and the consequent in fuzzy rules:

(i) Takagi-Sugeno method - consequents are functions of inputs,
(ii) Mamdani-type reasoning method - consequents and antecedents are related by the min operator or generally by a t-norm,
(iii) Logical-type reasoning method - consequents and antecedents are related by fuzzy implications, e.g. binary, Łukasiewicz, Zadeh etc.

Another approach, rarely studied in the literature, is based on fuzzy relations connecting input and output fuzzy linguistic values (see e.g. [8][14][17][18]). It allows setting fuzzy linguistic values in advance and fine-tuning model mapping by changing relation elements. They were used in some areas, e.g. to classification [19] and control [8]. In this paper we present possibilities of applications the new relational neuro-fuzzy structures to various nonlinear regression task. The systems are trained by relational equations, fuzzy clustering and gradient methods. The systems are described in Section 2. Learning of the systems is in Section 3. Applications of the relational NFS are in Section 4.

---
* This work was partly supported by the Foundation for Polish Science (Professorial Grant 2005-2008) and the Polish Ministry of Science and Higher Education (Habilitation Project 2007-2010 Nr N N516 1155 33, Special Research Project 2006-2009, Polish-Singapore Research Project 2008-2010, Research Project 2008-2010).

2 Fuzzy Relational Systems

Fuzzy relational models can be regarded as a generalization of linguistic fuzzy systems, where each rule has more than one linguistic value defined on the same output variable, in their consequents. Fuzzy rules in SISO relational model have the following form

$$R^k : \text{IF } x \text{ is } A^k \text{ THEN} \\ y \text{ is } B^1 \; (r_{k1}), y \text{ is } B^m \; (r_{km}), \ldots, y \text{ is } B^M \; (r_{kM}) \tag{1}$$

where r_{km} is a weight, responsible for the strength of connection between input and output fuzzy sets. Relational fuzzy systems store associations between the input and the output linguistic values in the form of a discrete fuzzy relation.

Definition 1. Binary fuzzy relation R is a fuzzy set defined in Cartesian product $\mathbf{X} \times \mathbf{Y}$ of two universes of discourse $\mathbf{X}$ and $\mathbf{Y}$

$$R : \mathbf{X} \times \mathbf{Y} \to \{0, 1\} \; . \tag{2}$$

Fuzzy set R maps each element in $\mathbf{X} \times \mathbf{Y}$ to a membership grade which is a measure of how strong the two elements are related.

In a general case of a multi-input multi-output system (MIMO), the relation R is a multidimensional matrix containing degree of connection for every possible combination of input and output fuzzy sets. In a multi-input single-output (MISO), there are N inputs x_n, and a single output. Every input variable x_n has a set A_n of K_n linguistic values A_n^k, $k = 1, \ldots, K_n$

$$A_n = \left\{ A_n^1, A_n^2, \ldots, A_n^{K_n} \right\} . \tag{3}$$

Output variable y has a set of M linguistic values B^m with membership functions $\mu_{B^m}(y)$, for $m = 1, \ldots, M$

$$B = \left\{ B^1, B^2, \ldots, B^M \right\} . \tag{4}$$

For certain MIMO and MISO systems the dimensionality of R becomes quite high and it is very hard to estimate elements of R. Sometimes training data are not enough large at all. For this reason we consider a fuzzy system with multidimensional input linguistic values. Then, we have only one set A of fuzzy linguistic values

$$A = \left\{ A^1, A^2, \ldots, A^K \right\}, \tag{5}$$

thus the relational matrix is only two-dimensional in the MISO case. Sets A and B are related to each other with a certain degree by the $K \times M$ relation matrix

$$R = \begin{bmatrix} r_{11} & r_{11} & \cdots & r_{1M} \\ r_{21} & r_{22} & \cdots & r_{2M} \\ \vdots & \vdots & r_{km} & \vdots \\ r_{K1} & r_{K2} & \cdots & r_{KM} \end{bmatrix} . \tag{6}$$

Having given vector $\bar{A}$ of K membership values $\mu_{A^k}(\bar{\mathbf{x}})$ for a crisp observed input value $\bar{\mathbf{x}}$, vector $\bar{B}$ of M crisp memberships μ_m is obtained through a fuzzy relational composition

$$\bar{B} = \bar{A} \circ R, \qquad (7)$$

implemented element-wise by a generalized form of sup-min composition, i.e. s-t composition

$$\mu_m = \underset{k=1}{\overset{K}{\mathrm{S}}} \left[\mathrm{T}\left(\mu_{A^k}(\bar{\mathbf{x}}), r_{km}\right) \right]. \qquad (8)$$

The crisp output of the relational system is computed by the weighted mean

$$\bar{y} = \frac{\sum_{m=1}^{M} \left\{ \bar{y}^m \, \mathrm{S}_{k=1}^{K} \left[\mathrm{T}\left(\mu_{A^k}(\bar{\mathbf{x}}), r_{km}\right) \right] \right\}}{\sum_{m=1}^{M} \mathrm{S}_{k=1}^{K} \left[\mathrm{T}\left(\mu_{A^k}(\bar{\mathbf{x}}), r_{km}\right) \right]}, \qquad (9)$$

where $\bar{y}^m$ is a centre of gravity (centroid) of the fuzzy set B^m. The exemplary neuro-fuzzy structure of the relational system is depicted in Fig. 1. The first layer of the system consists of K multidimensional fuzzy membership functions. The second layer is responsible for s-t composition of membership degrees from previous layer and KM crisp numbers from the fuzzy relation. Finally, the third layer realizes center average defuzzification. Depicting the system as a net structure allows learning or fine-tuning system parameters through the backpropagation algorithm. S-norm in (9) can be replaced by OWA operators [22][23] which further extends the versatility of the relational neuro-fuzzy system.

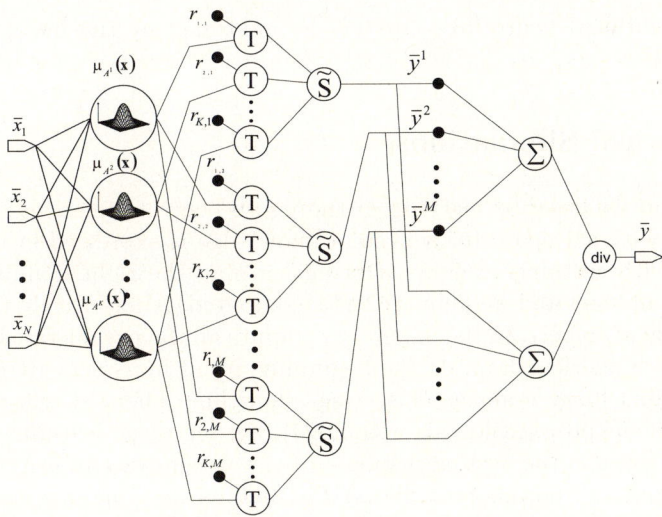

Fig. 1. Neuro-fuzzy relational system

3 Learning

Knowledge in fuzzy systems is interpretable and we can set their parameters in advance. In the simulations the Fuzzy C-Means algorithm [2][3] is used to set the input and output fuzzy sets in the cluster centers in the input and output space. After setting the fuzzy linguistic values relation R is computed. One of possible ways to obtain R is computing its elements from training data by solving fuzzy relational equations [15] using φ-composition operator and aggregating partial relations R_d

$$R = \bigcap_{d=1}^{D} R_d = \bigcap_{d=1}^{D} \left(\bar{A}_d \varphi \bar{B}_d \right), \qquad (10)$$

where $\bar{A}_d$ is the vector of membership values for the d-th input data vector, $\bar{B}_d$ is the vector of membership values for the d-th output data vector and output fuzzy sets. Each partial relation R_d corresponds to one of D training data vectors. The φ-composition operator used here is a fuzzy logical implication. For example, taking Reichenbach implication, elements of the R matrix for the k-th training vector are obtained

$$(r_{km})_d = 1 - \mu_{A^k}(\mathbf{x}^d) + \mu_{A^k}(\mathbf{x}^d)\mu_{B^m}(y^d) . \qquad (11)$$

Similarly to the fuzzy reasoning, we can use a T-norm instead of the φ-composition operator but then aggregation of partial relations is carried out by an S-norm operator

$$R = \underset{d=1}{\overset{D}{S}} R_d = \underset{d=1}{\overset{D}{S}} \left(T\left(\bar{A}_d, \bar{B}_d\right) \right) . \qquad (12)$$

Initialized relational neuro-fuzzy system is fine-tuned by the backpropagation algorithm [21].

4 Numerical Simulations

In this section we describe learning of the neuro-fuzzy systems and simulation examples. Relational neuro-fuzzy system knowledge is expressed in the form of fuzzy rules with certainty weights. During learning the right number of rules, fuzzy set parameters and weights are to be computed. Alternatively they can be determined by an expert. In the paper we compute all the parameters by machine learning from numerical data. At the beginning, input fuzzy sets are determined by the modified fuzzy c-means clustering algorithm. Then, all parameters are tuned by the backpropagation algorithm [21]. Having given learning data set of pair $(\bar{x}, d)$ where d is the desired response of the system we can use the learning error to compute all parameters. Signal d is used as the desired response of the first output and the second output is learned with the desired response which equals 1.

4.1 Chemical Plant Problem

In this problem the goal is to set monomer flow rate in a chemical plant on the basis of five input values [20]. In the experiment only three inputs are used (monomer concentration, change of monomer concentration and monomer flow rate). The dataset has 70 vectors. plant produces polymers by polymerisating some monomers. Since the start-up of the plant is very complicated, men have to perform the manual operations at the plant. Three continuous inputs are chosen for controlling the system: monomer concentration, change of monomer concentration and monomer flow rate. The output is the set point for the monomer flow rate. The rootm mean square error obtained by the relational system with three input fuzzy sets and three output fuzzy sets was 0.0044.

4.2 Box and Jenkins Furnace Data

As an simulation example we used Box and Jenkins furnace data [6]. The relational neuro-fuzzy system is learned to predict the time series, on the basis of 290-element data set. We use two inputs y(t-1), u(t-4) to predict y(t). The result is depicted in Fig. 2. Simulations shows superior performance and faster convergence of the proposed approach comparing to those from the literature.

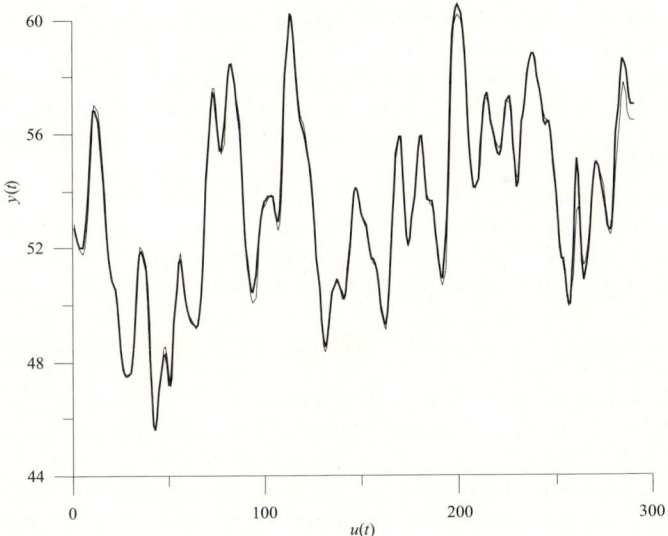

Fig. 2. Box-Jenkins furnace data (thick line), fuzzy relational system output (thin line)

5 Conclusions

In the paper we use new fuzzy systems to regression modeling. Relational neuro-fuzzy systems described in the paper uses a fuzzy relation to connect input and output linguistic fuzzy terms. Thanks to this the fuzzy rules have more complicated structure and can be better suited the task. Initial set of rules can be picked by an expert or by machine learning methods. In the paper we use fuzzy clustering and relational equations to obtain the initial set of fuzzy rules. Then the system is fine-tuned by the backpropagation algorithm. Simulations shows superior performance and faster convergence of the proposed approach comparing to those from the literature.

References

1. Babuska, R.: Fuzzy Modeling For Control. Kluwer Academic Press, Boston (1998)
2. Bezdek, J.C., Pal, S.K.: Fuzzy Models for Pattern Recognition. IEEE Press, New York (1992)
3. Bezdek, J., Keller, J., Krisnapuram, R., Pal, N.R.: Fuzzy Models and Algorithms for Pattern Recognition and Image Processing. Kluwer Academic Press, Norwell (1999)
4. Bilski, J.: The UD RLS Algorithm for Training Feedforward Neural Networks. Int. J. Appl. Math. Comput. Sci. 15(1), 115–123 (2005)
5. Bishop, C.M.: Neural Networks for Pattern Recognition. Oxford University Press Inc., New York (1995)
6. Box, G.E.P., Jenkins, G.M.: Time Series Analysis, Forecasting and Control, San Francisco, Holden Day (1970)
7. Asuncion, A., Newman, D.J.: UCI Machine Learning Repository, University of California, School of Information and Computer Science, Irvine, CA (2007), http://www.ics.uci.edu/~mlearn/MLRepository.html
8. Branco, P.J.C., Dente, J.A.: A Fuzzy Relational identification Algorithm and its Application to Predict the Behaviour of a Motor Drive System. Fuzzy Sets and Systems 109, 343–354 (2000)
9. Ischibuchi, H., Nakashima, T.: Effect of Rule Weights in Fuzzy Rule-Based Classification Systems. IEEE Transactions on Fuzzy Systems 9(4), 506–515 (2001)
10. Jang, R.J.-S., Sun, C.-T., Mizutani, E.: Neuro-Fuzzy and Soft Computing, A Computational Approach to Learning and Machine Intelligence. Prentice Hall, Upper Saddle River (1997)
11. Nauck, D., Klawon, F., Kruse, R.: Foundations of Neuro - Fuzzy Systems. John Wiley, Chichester (1997)
12. Nauck, D., Kruse, R.: How the Learning of Rule Weights Affects the Interpretability of Fuzzy Systems. In: Proceedings of 1998 IEEE World Congress on Computational Intelligence, FUZZ-IEEE, Alaska, pp. 1235–1240 (1998)
13. Nozaki, K., Ishibuchi, H., Tanaka, K.: A simple but powerful heuristic method for generating fuzzy rules from numerical data. Fuzzy Sets and Systems 86, 251–270 (1995)
14. Pedrycz, W.: Fuzzy Control and Fuzzy Systems. Research Studies Press, London (1989)
15. Pedrycz, W., Gomide, F.: An Introduction to Fuzzy Sets, Analysis and Design. The MIT Press, Cambridge (1998)
16. Rutkowski, L.: Flexible Neuro Fuzzy Systems. Kluwer Academic Publishers, Dordrecht (2004)
17. Scherer, R., Rutkowski, L.: Relational Equations Initializing Neuro-Fuzzy System. In: 10th Zittau Fuzzy Colloquium, Zittau, Germany (2002)

18. Scherer, R., Rutkowski, L.: Neuro-Fuzzy Relational Systems. In: International Conference on Fuzzy Systems and Knowledge Discovery, Singapore, November 18-22 (2002)
19. Setness, M., Babuska, R.: Fuzzy Relational Classifier Trained by Fuzzy Clustering. IEEE Transactions on Systems, Man and Cybernetics - Part B: Cybernetics 29(5), 619–625 (1999)
20. Sugeno, M., Yasukawa, T.: A Fuzzy-Logic-Based Approach to Qualitative Modeling. IEEE Transactions on Fuzzy Systems 1(1), 7–31 (1993)
21. Wang, L.-X.: Adaptive Fuzzy Systems And Control. PTR Prentice Hall, Englewood Cliffs (1994)
22. Yager, R.R., Filev, D.P.: Essentials of Fuzzy Modeling and Control. John Wiley & Sons Inc., New York (1994)
23. Yager, R.R., Filev, D.P.: On a Flexible Structure for Fuzzy Systems Models. In: Yager, R.R., Zadeh, L.A. (eds.) Fuzzy Sets, Neural Networks, and Soft Computing, pp. 1–28. Van Nostrand Reinhold, New York (1994)

A New Approach to Creating Multisegment Fuzzy Systems

Artur Starczewski

Department of Computer Engineering, Częstochowa University of Technology
Al. Armii Krajowej 36, 42-200 Częstochowa, Poland
starcz@kik.pcz.czest.pl
http://kik.pcz.pl

Abstract. Presented paper shows a new approach to creating a fuzzy system based on an exclusive use of clustering algorithms, which determine the value of necessary parameters. The applied multisegment fuzzy system functions as a classifier. Each segment makes an independent fuzzy system with a defined knowledge base and uses singleton fuzzification, as well as fuzzy inference with product operation as the Cartesian product and well-matched membership functions. Defuzzification method is not used. Only the rule-firing level must be analysed and its value suffices to determine the class. The use of clustering algorithms has allowed a qualification of the number of rules in the base of fuzzy rules for each independent segment, as well as a specification of the centers of fuzzy sets used in the given rules. The calculated parameters have proved precise, so that no additional methods have been applied to correct their values. This procedure greatly simplifies the creation of a fuzzy system. The constructed fuzzy system has been tested on medical data that come from the Internet. In the future, those systems may help doctors with their everyday work.

1 Introduction

When dealing with models of difficult real problems many researchers have recently applied "soft computing" techniques, especially neural networks and fuzzy systems [2][5][9]. Fuzzy-neural systems are often called fuzzy-neural networks and, similarly to neural networks, demonstrate a learning ability [4][8][11]. At present more complex structures are applied to many difficult problems. The structures consist of many separate independent structures, for instance: neural networks, fuzzy systems and neuro-fuzzy systems, which perform a particular task together. In this way multisegment structures are created. This paper aims to describe a multisegment fuzzy system used as a classifier in medical diagnostics, for instance in experimental diagnosing of heart diseases, breast cancer and thyroid diseases. Since the selection of suitable system parameters is based on using a clustering algorithm [3], the paper further presents a detailed description of the applied algorithm and its application at the particular stages of creating the multisegment system.

2 Description of the Multisegment System

The fuzzy system of the multisegment architecture contains fuzzy structures called segments (see Fig.1). A similar structure of the system has been presented in the articles [10][12]. As regards the classification problem, each segments recognizes only one class of data. If we enter an input vector belonging to the class c, then the vector will be recognized by the c segment. The output value of this segment is then different from zero. For the remaining segments the output signals equal zero. The segments work simultaneously and each receives an input vector and decides on a suitable class of the vector. The parameters of individual fuzzy systems are chosen in the way that an input vector can only be recognized by one fuzzy system, while the remaining segments give the answer 0. If all segments answer 0, then the vector is assigned to the so-called "additional" class.

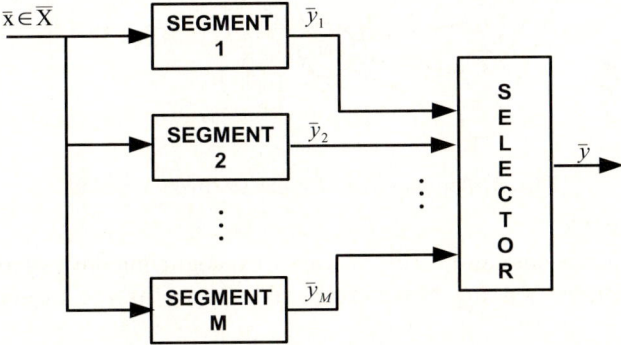

Fig. 1. The multisegment system

In the structure presented in Fig.1, the input vector is the attribute vector. This vector can characterize, for instance, a medical illness and it is passed on to all the segments, that is the fuzzy networks, which are associated with individual classes. Let M equal the number of all classes in learning data, and $c = 1, ..., M$. For instance, if an input vector belongs to class $c = 1$, then the value output of the first fuzzy network is more than zero. Then the remaining segments of this structure have the output value equal 0. All output values of individual segments are passed on to the selector. The individual inputs of the selector are numbered in turn, according to the class numbers of the learning data, and the non-zero value is the answer of the whole system. The role of the selector is only to choose the non-zero value. A detailed description of fuzzy networks constituting such a structure is given below.

2.1 Fuzzy Networks Applied to the Multisegment Structure

The idea of the multisegment system is to decompose a multi-class classification problem to simpler tasks. Each fuzzy network used in the system is associated

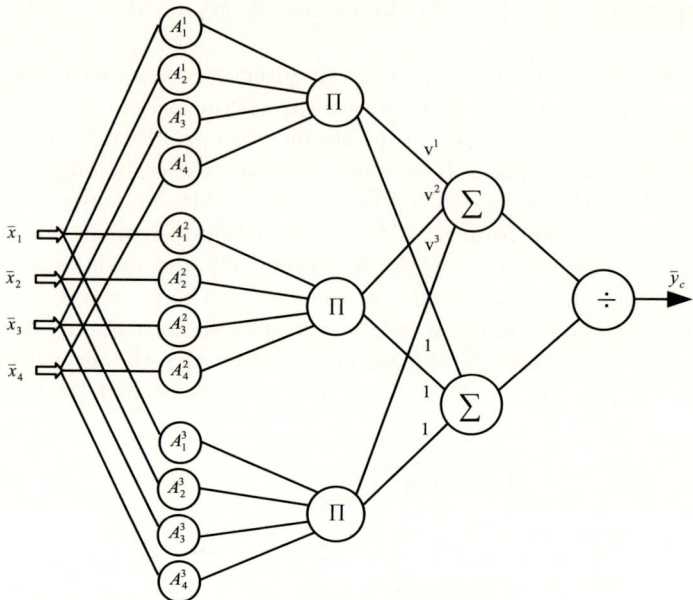

Fig. 2. An example of a classic fuzzy system

with one individual class and is responsible for recognizing data vectors belonging to this class. Before getting into details, it is important to present the classic fuzzy system.

Let $\bar{\mathbf{x}} = [\bar{x}_1, ..., \bar{x}_n]$ and $\bar{y}_c$ be an input vector and an output of the fuzzy network, and let us denote $x_1, ..., x_n$ and y_c as linguistic variables associated with the input vector and output, respectively. The fuzzy rules IF - THEN can be represented as follows:

$$\text{IF} \quad x_1 \text{ is } A_1^k \quad \text{AND} \ldots \text{AND} \quad x_n \text{ is } A_n^k \quad \text{THEN} \quad y_c \text{ is } B_c^k \tag{1}$$

where A_i^k, $k = 1, ..., w$, $i = 1, ..., n$ are fuzzy sets that define values of the linguistic variables, which correspond with the particular components of the $\bar{\mathbf{x}}$ vector. Fuzzy B_c^k sets define the membership to the c class and w denotes the number of rules in the fuzzy base. A precise qualification of A_i^k parameters and w is critical for the correct functioning of the network, that is a segment in the created structure. In many fuzzy networks or neuro-fuzzy ones, the Gaussian membership functions are employed and are represented as follows:

$$\mu_{A_i^k} = \exp\left[-\frac{(\bar{x}_i - v_{ik})^2}{\sigma_{ik}^2}\right] \tag{2}$$

where $i = 1, ..., n$. The values v_{ik} in formula (2) are the centers of the Gaussian membership function, and σ_{ik} are their widths. These parameters are usually initialized, and then exactly determined by means of, for instance, gradient methods. The classic fuzzy system [11] with singleton fuzzification, fuzzy inference with product operation as the Cartesian product, well-matched membership functions and the center-average defuzzification method can be expressed as follows:

$$\bar{y}_c = \frac{\sum_{k=1}^{w} \nu_{B_c^k} \tau_k}{\sum_{k=1}^{w} \tau_k} \tag{3}$$

where

$$\tau_k = \prod_{i=1}^{n} \mu_{A_i^k}(\bar{x}_i) \tag{4}$$

is the rule-firing level of the k-rule, and $\nu_{B_c^k}$ is the center of the B_c^k membership function of a fuzzy set. This fuzzy network is illustrated in Fig.2. The network consists of four layers, the first two of which represent the IF part of the rules base, whereas the other two layers refer to the THEN part and perform defuzzification. In the first layer there are $k \times n$ fuzzy sets, where $k = 3$ and $n = 4$. The membership function can be defined by the formula (2). The second layer has the k-elements, which perform the multiplication operation. The third and the fourth layer are responsible for the process of defuzzification.

The multisegment system presented in this paper contains fuzzy networks, which consists of four layers, as shown in Fig.3. The parameters values of individual networks are derived from the clustering algorithm, and carefully calculated. The networks are similar to the classic fuzzy system (see Fig.2). The first layer refers to antecedent fuzzy sets (see 1). The elements of the second layer produce values of the rule-firing level (see 4). In the fuzzy networks the Gaussian membership function has not been used. Much better results have been obtained by using membership functions that have a hipercuboid shape in multidimensional space. The membership function can be represented as shown below:

$$\mu_{A_i^k}(\bar{x}_i) = \begin{cases} 1 & 0 \leq \bar{x}_i \leq a_{ik} \\ 0 & otherwise \end{cases} \tag{5}$$

where $i = 1, ..., n$ and vector components $\mathbf{a_k} = [a_{1k}, ..., a_{nk}]^T$ determine the lengths of the sides. The key factor to the application of this function in multidimensional space is a precise calculation of the $\mathbf{a_k}$ vector, which uses a clustering algorithm. A detailed description of this algorithm is presented further on in this paper. This algorithm has made it possible not only to determine the suitable groups of data, but also to calculate their centers (prototype vectors) $\mathbf{v_k} = [v_{1k}, ..., v_{nk}]^T$, where $k = 1, ..., w$. In case of a hypercuboid, the $\mathbf{v_k}$ vector and the $\mathbf{a_k}$ vector determine their sides. The membership function can be expressed as follows:

$$\mu_{A_i^k}(\bar{x}_i) = \begin{cases} 1 & v_{ik} - a_{ik} \leq \bar{x}_i \leq v_{ik} + a_{ik} \\ 0 & otherwise \end{cases} \tag{6}$$

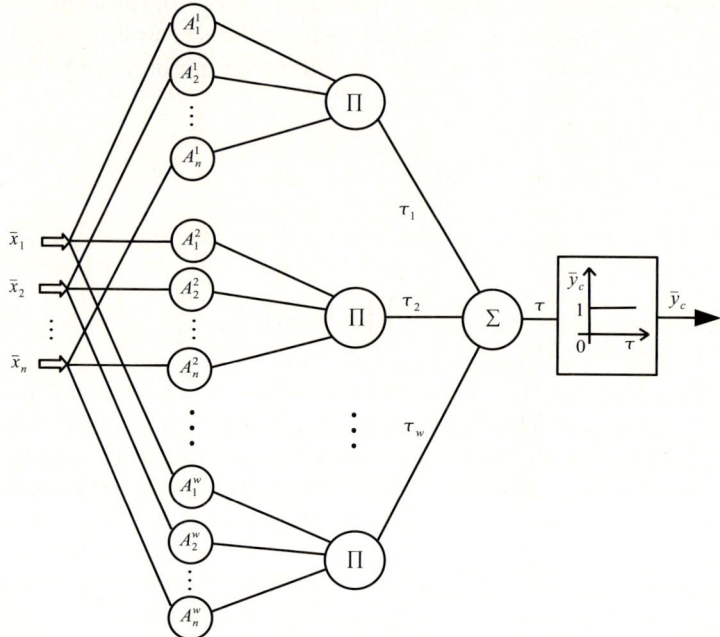

Fig. 3. The fuzzy network applied to the multisegment structure

The $\mathbf{a_k}$ vector is calculated after the individual groups of the data and their centers are specified. Then the distances from the center of a group, that is from v_{ik}, where $i = 1, ..., n$, are calculated for the elements belonging to the k group. The maximum distance for i-dimension is the searched value. Let us mark L_k as a number of the elements in the k group. We can now record the vectors that belong to the group k as $\bar{\mathbf{x}}_k^l = \left[\bar{x}_{1k}^l, ..., \bar{x}_{nk}^l\right]^T$, where $l = 1, ..., L_k$. And so, the searched value can be expressed as follows:

$$a_{ik} = \max_{1 \leq l \leq L_k} \left\| \bar{x}_{ik}^l - v_{ik} \right\| \tag{7}$$

When all the components of the $\mathbf{a}_k$ vector are identical, we obtain a hypercube. The third layer performs the summation operation. The last layer contains the element which realizes the $\bar{y}_c(\tau)$ function. The function can be expressed as follows:

$$\bar{y}_c(\tau) = \begin{cases} 1 & \tau > 0 \\ 0 & \tau \leq 0 \end{cases} \tag{8}$$

As it has been mentioned earlier, a clustering algorithm has made it possible to divide data into groups as well as to calculate their centers. Let vectors belong to the c classes, where $c = 1, ..., M$. For example, if the first network is created (Segment 1), only the vectors belonging to the class ($c = 1$) are clustered. The number of rules in the fuzzy base of the created network equals the number of the received groups of data. The partition of the learning data has been made within the classes that are in the data file. Thanks to that, in the fuzzy base of the rules we do not need to apply any methods to qualify the connections of the antecedent part (IF part) with the conclusion part (THEN part) of the rule.

3 Clustering Algorithm

There are many methods for clustering in learning data based on crisp, fuzzy or probabilistic models. It is worth emphasizing that the most widely used clustering algorithm is fuzzy c-means (FCM). However,the FCM performance is good when clusters are the same size and shape. Since the function parameters have to be a very precise calculated, the new algorithm is proposed. The applied clustering algorithm makes possible the calculation of required parameters of fuzzy networks. The number of groups corresponds to the number of rules in the fuzzy rules base. The group centers (prototype vectors) determine membership functions of fuzzy sets. Along with the term "group" the word "cluster" is used synonymously. Let $\bar{X} = \{\bar{x}_1, ..., \bar{x}_z\} \subset R^n$ be the learning data, and $\bar{x}_j$ its element, where $j = 1, ..., z$. The learning data are usually scaled. Let $V = \{v_1, ..., v_w\} \subset R^n$ be a cluster set from $\bar{X}$, and v_k its element, where $k = 1, ..., w$. The proposed algorithm involves the following steps:

1. fixing a constant value r, which represents the cluster radius and is determined on the basis of a distance analysis of the learning vectors $\bar{x}_j$ from the $\bar{X}$ center. The value is usually defined as 0.25 of the maximum distance from the $\bar{X}$ center (the center is simple averaging of the $\bar{X}$).
2. setting the initial number of the clusters $w = 1$ as well as the index $j = 1$.
3. creating a new w cluster and its center $v_w = \bar{x}_j$.
4. searching for $\bar{x}_j$, where $j = 1, ...z$, which does not belong to cluster k, where $k = 1, ..., w$. If such an $\bar{x}_j$ exists, calculating the Euclidean distance between $\bar{x}_j$ and the element v_w and checking the inequation:

$$\|\bar{x}_j - v_w\| \leq r \qquad (9)$$

If the inequation (9) is satisfied, the next step is including vector $\bar{x}_j$ into the w cluster and modifying the v_w cluster center using the following formula:

$$v_w = \frac{1}{L_w}\left(\sum_{l=1}^{L_w} \bar{x}_w^l\right) \qquad (10)$$

where L_w is a number of the elements in the w group. If the inequality (9) is not satisfied, the next stage is repeating the step 4 until the whole $\bar{X}$ set has been searched through. If there are still some vectors which don't belong to any k cluster, going to step 5 is necessary, or otherwise finishing the algorithm.

5. increasing the number of clusters $w = w + 1$ and searching the first $\bar{x}_j$, $j = 1, ..., z$, which does not belong to any k cluster and returning to step 3.

The algorithm is finished when all the vectors from $\bar{\mathbf{X}}$ have their cluster (or group) membership determined. The choice of the r parameter is of paramount importance, as it specifies the size (and the quantity) of the created groups. Of course, the results of the clustering depend in given algorithm on the order of the data. Therefore, the clustering is repeated for the random order of the data. The general principle has been to receive the smallest possible number of clusters with the fuzzy networks working properly.

4 Experimental Results

Thyroid disease diagnosis. The thyroid medical database is available on the Internet [7]. It consists of learning data and suitably prepared testing data. The learning data contains case studies of 3772 patients, and the testing data comprises 3428 cases. For each patient 21 attributes have been determined. The last attribute (number 21) is the diagnosis, which includes three classes: normal (not hypothyroid), hyperfunction, and subnormal functioning. These classes are expressed by integers 1,2,3. The other attributes are different, for example there are 6 attributes having continuous values from 0 to 1 and others having binary values. For this medical data a multisegment system has been created on the basis of 123 rules. For the learning data, it has correctly classified 98% of the cases, but for testing data it has correctly classified 93% of all the cases.

Breast cancer diagnosis. This paper uses the medical data obtained from the University of Wisconsin Hospitals, Madison, Wisconsin, USA, published by Mangasarin and Wolberg [1][6]. The breast cancer database is available on the Internet[7] and contains case studies of 683 patients, for whom 10 basic attributes have been determined, so as to represent different items of medical information. The last attribute (number 10) is the diagnosis reached by the doctor, i.e. the type of the cancer (benign - 2, malignant - 4). The remaining attributes, from 1 to 9, assume values from 1 to 10. All the attributes include: clump thickness, uniformity of cell size, uniformity of cell shape, marginal adhesion, single epithelial cell size, bare nuclei, bland chromatin, normal nucleoli, mitoses and the doctor's diagnosis. For this medical data two multisegment systems have been created, one based on 41 fuzzy rules, and the other based on only 8 rules. The first one has correctly classified all the medical cases, and the other one has produced 96% correct answers. Since there is no suitably prepared testing data on this database, the classifier has only been checked for the learning data.

Heart disease diagnosis. The paper quotes the data based on medical examination of 297 patients performed at the clinical hospital of Cleveland, Ohio, USA, which is available on the Internet [7]. Several attributes have been distinguished to represent the diagnosis, i.e. in case of a negative diagnosis, the last attribute value equals 0, otherwise the value can equal 1, 2, 3, or 4, which all refer to a heart disease. Usually only the basic classification is made, and it is either negative or positive (value 0 or 1). 13 main attributes have been determined for each patient, with the attribute number 14 designating a diagnosis. The other attributes contain different medical information, indispensable for establishing a diagnosis. The attributes include e.g.: age (in years), sex (1 – male, 0 – female), chest pain type (four different types – values 1, 2, 3 or 4), resting blood pressure (in mm Hg), serum cholestoral (in mg/dl), fasting blood sugar (1 – if greater than 120 mg/dl, and 0 – otherwise), resting electrocardiographic results (three states – values 0, 1, 2), maximum heart rate achieved, exercise induced angina (1 – yes, 0 – no), fluoroscope test, lasting changes, and a doctor's diagnosis. Again, on the basis of this medical data, two multisegment systems have been created with the application of 77 and 56 rules, respectively. The first one has correctly classified all the cases, and the other one 96% of the cases. Since there has been no testing data in the database, classification has been tested only with the learning data.

5 Conclusions

As the above-mentioned examples prove, the received results confirm a good performance of the fuzzy system. Of course, lack of testing sets for breast cancer and heart disease makes it difficult to confirm the effectiveness of the system. Therefore, the 10-fold cross validation using the testing date is planned in future investigations. It is worth emphasizing, that a good clasifier applied to the thyroid disease diagnosis should produce more then 92% correct answers, so the proposed multisegment system fulfils the requirement. Moreover, learning sets often contain incomplete data, thus the initial analysis of their attributes and deletion of incomplete data or selection of the most essential attributes are significant. Actually, in this method there is no implementation of a thorough analysis of learning data attributes. Making a multisegment system with a clustering algorithm will obviously greatly accelerate the process of its creation. Furthermore, different memberships functions can be applied, for instance the hypercuboid or the hypercube functions have facilitated an exact partition of multidimensional space for learning data. Moreover, this clustering algorithm has identified such things as: a number of rules in a fuzzy rules base, parameters of membership function and a connection of the antecedent part (IF part) with the conclusion part (THEN part) of the rules (it isn't necessary to apply any additional methods). The most important factor which conditions a good performance of the fuzzy system is certainly the clustering algorithm. To sum up, the suggested method requires further research, but the received results are already promising.

References

1. Bennett, K.P., Mangasarian, O.L.: Robust linear programming discrimination of two linearly inseparable sets. In: Optimization Methods and Software, vol. 1, pp. 23–34. Gordon & Breach Science Publishers (1992)
2. Bishop, C.M.: Neural Networks for Pattern Recognition. Oxford University Press, Oxford (1995)

3. Bezdek, J.C.: Pattern Recognition with Fuzzy Objective Function Algorithms. Plenum Press, New York (1981)
4. Jang, R.J.S., Sun, C.T., Mizutani, E.: Neuro-Fuzzy and Soft Computing. A Computational Approach to Learning and Machine Intelligence. Prentice Hall, Upper Saddle River (1997)
5. Kuncheva, L.I.: Fuzzy Classifier Design. A Springer-Verlag Company, Heidelberg (2000)
6. Mangasarian, O.L., Wolberg, W.H.: Cancer diagnosis via linear programming. SIAM News 23(5), 1–18 (1990)
7. Mertez, C.J., Murphy, P.M.: UCI repository of machine learning databases, http://www.ics.uci.edu/pub/machine-learning-databases
8. Nauck, D., Klawonn, F., Kruse, R.: Foundations of Neuro-Fuzzy Systems. John Wiley & Sons, Chichester (1997)
9. Pedrycz, W.: Fuzzy Control and Fuzzy Systems. Research Studies Press, London (1989)
10. Rutkowska, D., Starczewski, A.: Fuzzy inference neural networks and their applications to medical diagnosis. In: Szczepaniak, P.S., Lisboa, P.J.G., Kacprzyk, J. (eds.) Fuzzy Systems in Medicine, pp. 503–518. Physica-Verlag, A Springer Verlag Company, Heidelberg (2000)
11. Rutkowska, D.: Neuro-Fuzzy Architectures and Hybrid Learning. Physica-Verlag, A Springer Verlag Company, Heidelberg (2002)
12. Starczewski, A., Rutkowska, D.: New hierarchical structure of neuro-fuzzy systems. In: Proc. 5th Conference on Neural Networks and Soft Computing, Zakopane, Poland, pp. 383–388 (2000)

On Defuzzification of Interval Type-2 Fuzzy Sets*

Janusz T. Starczewski

Department of Computer Engineering, Czestochowa University of Technology,
Al. Armii Krajowej 36, 42-200 Czestochowa, Poland
jasio@kik.pcz.czest.pl

Abstract. In the paper the defuzzification of interval type-2 fuzzy sets is studied. The standard K-M type-reduction method and the uncertainty bounds approximate type-reduction are compared with the classical defuzzification of averaged type-1 fuzzy sets.

1 Introduction

Recently, the exhaustive use of interval type-2 fuzzy logic systems has been made in many areas of decision making (see eg. [1]). Type-2 fuzzy logic is supported on fuzzy sets of type-2, which are characterized by fuzzy subsets of the unit interval $[0, 1]$ in the place of membership grades.

The essential parts of type-2 fuzzy logic systems are the inference block and the two-step defuzzification. While there exist efficient methods for the intersection of interval type-2 fuzzy sets in inference, the defuzzification of interval type-2 fuzzy sets needs expensive computations. The type-2 defuzzification is divided into two steps of transformation: type-reduction and classical type-1 defuzzification. The term "type-reduction" signifies the defuzzification operation (usually center of area or height) extended with the use of the Extension Principle to deal with fuzzy subsets of $[0, 1]$ instead of crisp membership grades.

In literature, the standard method is a Karnik-Mendel (K-M) type-reduction procedure [2]. The calculation of this iterative procedure is a major computational problem. An alternative approximated method for non-iterative type-reduction based on uncertainty bounds has been developed by Wu and Mendel [3]. The functional comparison of these methods is confronted with the type-1 defuzzification of averaged type-1 fuzzy sets basing on our previous observation [4].

2 Interval Type-2 Fuzzy Logic Systems

Type-2 fuzzy sets are characterized by membership grades which are fuzzy subsets of the unit interval $[0, 1]$. More precisely, the fuzzy set of type-2, $\tilde{A}$, in the

---

* This work was partly supported by the Foundation for Polish Science (Professorial Grant 2005–2008) and Polish Ministry of Science and Higher Education (Habilitation Project N N516 372234 2008–2011, Special Research Project 2006–2009, Polish-Singapore Research Project 2008–2010, Research Project 2008–2010).

L. Rutkowski et al. (Eds.): ICAISC 2008, LNAI 5097, pp. 333–340, 2008.
© Springer-Verlag Berlin Heidelberg 2008

real line $\mathbb{R}$, is a vague collection of elements characterized by membership function $\mu_{\tilde{A}}\colon \mathbb{R} \to \mathcal{F}([0,1])$, where $\mathcal{F}([0,1])$ is a set of all classical fuzzy sets in the unit interval $[0,1]$. Each $x \in \mathbb{R}$ is associated with a secondary membership function $f_x \in \mathcal{F}([0,1])$, which is a mapping $f_x\colon [0,1] \to [0,1]$. In this paper, the interval type-2 fuzzy sets are considered, thus secondary memberships are expressed by normal rectangular functions.

It has become a standard that the type-2 FLS consists of:

- a type-2 fuzzy rule base,
- an inference engine modified to deal with type-2 fuzzy sets,
- a type reducer,
- a type-1 defuzzifier,
- and optionally a type-2 fuzzifier of inputs.

In the following we describe a simple type-2 FLS which does not have a fuzzification and whose rule base is composed of type-2 fuzzy antecedents. The rule base is formed by K rules

$$\tilde{R}^k : \text{IF } x_1 \text{ is } \tilde{A}_{k,1} \text{ and } x_{k,2} \text{ is } \tilde{A}_{k,2} \text{ and } \cdots \text{ and } x_N \text{ is } \tilde{A}_{k,N} \text{ THEN } y \text{ is } y_k,$$

where x_n is the n-th input variable, $\tilde{A}_{k,n}$ is the n-th antecedent fuzzy set of type-2, and y_k is the k-th rule consequent, $n = 1, \ldots, N$, $k = 1, \ldots, K$.

Assuming no fuzzification and singleton consequents y_k, the inference engine produces the type-2 fuzzy conclusion according to the following formula

$$\tilde{h}_k(y_k) = \tilde{T}_{n=1}^{N} \mu_{\tilde{A}_{k,n}}(x'_n). \tag{1}$$

Obviously, for interval type-2 fuzzy quantities the above extended t-norm is calculated separately by classical t-norms for upper and lower bounds of interval type-2 sets

$$\underline{h}_k(y_k) = T_{n=1}^{N} \underline{\mu}_{\tilde{A}_{k,n}}(x'_n) \tag{2}$$

$$\overline{h}_k(y_k) = T_{n=1}^{N} \overline{\mu}_{\tilde{A}_{k,n}}(x'_n) \tag{3}$$

Now we need to defuzzify type-2 fuzzy conclusions into a type-1 fuzzy set, which can be finally defuzzified into a crisp output value. In literature, the common approach to defuzzify an interval type-2 fuzzy set is by the extension principle applied to center of area or height defuzzifications. The extended operation is called the type-reduction. The following considerations have a general significance when either K denotes the number of rules for height type-reduction or K is a discretization density for centroid type-reduction.

Assume that consequents are reordered in the following way

$$y_1 < y_2 < \ldots, y_K$$

The type-reduced set for an interval type-2 FLS is an interval type-1 fuzzy set, i.e., $Y(\mathbf{x}') = [y_{\min}, y_{\max}]$. The bounds for the height type-reduction are expressed by

$$y_{\min} = \frac{\sum_{k=1}^{L-1} \overline{h}_k y_k + \sum_{k=L}^{K} \underline{h}_k y_k}{\sum_{k=1}^{L-1} \overline{h}_k + \sum_{k=L}^{K} \underline{h}_k} \quad (4)$$

$$y_{\max} = \frac{\sum_{k=1}^{R} \underline{h}_k y_k + \sum_{k=R+1}^{K} \overline{h}_k y_k}{\sum_{k=1}^{R} \underline{h}_k + \sum_{k=R+1}^{K} \overline{h}_k} \quad (5)$$

where L and R may be determined by the Karnik-Mendel (K-M) iterative type-reduction or the Simplex method. The calculation of L and R is major computational problem. For that purpose two reliable methods will be given in subsections.

The overall output of the interval type-2 fuzzy logic system is expressed by

$$y' = \frac{y_{\min} + y_{\max}}{2} \quad (6)$$

2.1 K-M Type-Reduction

The following Karnik-Mendel type-reduction algorithm has been brought to a standard [5].

Let the consequent values be ordered in the following way $y_1 < y_2 < \ldots < y_K$.

1. calculate principal output $y_{\lim pr}$ as an average of y_k weighted by mean membership grades $\left(\underline{h}_k + \overline{h}_k\right)/2$,
2. set the initial values $y_{\min} = y_{\max} = y_{\lim pr}$,
3. for each $k = 1, 2, \ldots, K$, if $y_k > y_{\max}$ then $\overline{\overline{h}}_k = \overline{h}_k$, otherwise $\overline{\overline{h}}_k = \underline{h}_k$,
4. find nearest $y_{\lim nrst} = \min_{k=1,\ldots,K} y_k : y_k > y_{\max}$,
5. calculate $y_{\max}$ as an average of y_k weighted by new grades $\overline{\overline{h}}_k$,
6. if $y_{\max} \leq y_{\lim nrst}$ calculation is completed, else go to step 3,
7. for each $k = 1, 2, \ldots, K$, if $y_k < y_{\min}$ then $\underline{\underline{h}}_k = \overline{h}_k$, otherwise $\underline{\underline{h}}_k = \underline{h}_k$,
8. find nearest $y_{\lim nrst} = \max_{k=1,\ldots,K} y_k : y_k < y_{\min}$,
9. calculate $y_{\min}$ as an average of y_k weighted by new grades $\underline{\underline{h}}_k$,
10. if $y_{\min} \geq y_{\lim nrst}$ calculation is completed, else go to step 7.

2.2 Uncertainty Bounds Approximate Type-Reduction

Instead of strict bounds $y_{\min}$ and $y_{\max}$, we can employ inner- and outer-bound sets for a type reduced set, described in [3]. Thus $y_{\min}$ and $y_{\max}$ can be determined by

$$y_{\min} = \frac{\underline{y}_{\min} + \overline{y}_{\min}}{2}, \quad (7)$$

$$y_{\max} = \frac{\underline{y}_{\max} + \overline{y}_{\max}}{2}, \quad (8)$$

where

$$\overline{y}_{\min} = \min\left(\frac{\sum_{p=1}^{P} y_p \underline{w}_p}{\sum_{p=1}^{P} \underline{w}_p}, \frac{\sum_{p=1}^{P} y_p \overline{w}_p}{\sum_{p=1}^{P} \overline{w}_p}\right),$$

$$\underline{y}_{\min} = \overline{y}_{\min} - \frac{\sum_{p=1}^{P} (\overline{w}_p - \underline{w}_p)}{\sum_{p=1}^{P} \overline{w}_p \sum_{p=1}^{P} \underline{w}_p} \cdot \frac{\sum_{p=1}^{P} (y_p - y_1) \underline{w}_p \sum_{p=1}^{P} (y_P - y_p) \overline{w}_p}{\sum_{p=1}^{P} (y_p - y_1) \underline{w}_p + \sum_{p=1}^{P} (y_P - y_p) \overline{w}_p},$$

$$\underline{y}_{\max} = \max\left(\frac{\sum_{p=1}^{P} y_p \underline{w}_p}{\sum_{p=1}^{P} \underline{w}_p}, \frac{\sum_{p=1}^{P} y_p \overline{w}_p}{\sum_{p=1}^{P} \overline{w}_p}\right),$$

$$\overline{y}_{\max} = \underline{y}_{\max} + \frac{\sum_{p=1}^{P} (\overline{w}_p - \underline{w}_p)}{\sum_{p=1}^{P} \overline{w}_p \sum_{p=1}^{P} \underline{w}_p} \cdot \frac{\sum_{p=1}^{P} (y_p - y_1) \overline{w}_p \sum_{p=1}^{P} (y_P - y_p) \underline{w}_p}{\sum_{p=1}^{P} (y_p - y_1)_p \overline{w}_p + \sum_{p=1}^{P} (y_P - y_p) \underline{w}_p}.$$

3 Defuzzification of Type-2 Fuzzy Sets

3.1 Equivalence between Type-2 and Type-1 Defuzzifications for Two Singletons with Equal Intervals of Uncertainty

Recall our previous statement [4] that the height type-reduction may be reduced to the height defuzzification, if only two rules are fired and intervals of firing fuzzy grades are equal. Suppose that all intervals of firing fuzzy grades are equal to δ,

$$\overline{h}_k - \underline{h}_k = \delta, \ \forall k = 1, 2.$$

For two active rules, in (4) and (5), $L = 2$ and $R = 1$. Combining these assumptions in (6), we obtain the overall output of the interval type-2 FLS,

$$y' = \frac{1}{2}\left(\frac{\overline{h}_1 y_1 + \underline{h}_2 y_2}{\overline{h}_1 + \underline{h}_2} + \frac{\underline{h}_1 y_1 + \overline{h}_2 y_2}{\underline{h}_1 + \overline{h}_2}\right) \quad (9)$$

$$= \frac{\left(\overline{h}_1 - \frac{1}{2}\delta\right) y_1 + \left(\overline{h}_2 - \frac{1}{2}\delta\right) y_2}{\overline{h}_1 + \overline{h}_2 - \delta}. \quad (10)$$

Thus for $\overline{h}_1 - \frac{1}{2}\delta = h_1$ and $\overline{h}_2 - \frac{1}{2}\delta = h_2$ the output realizes the classical height type defuzzification of a type-1 FLS,

$$y' = \frac{h^1 y^1 + h^2 y^2}{h^1 + h^2} = y^{T1}. \quad (11)$$

The illustration of this fact is presented in Fig. 1. This case demonstrates working conditions, when the type-2 approach to defuzzification is less applicable than type-1 fuzzy logic. For example, the uniform uncertainty of memberships in a trained interval type-2 fuzzy logic system acts a supposition that this type-2 fuzzy logic system may be reduced to the corresponding type-1 system.

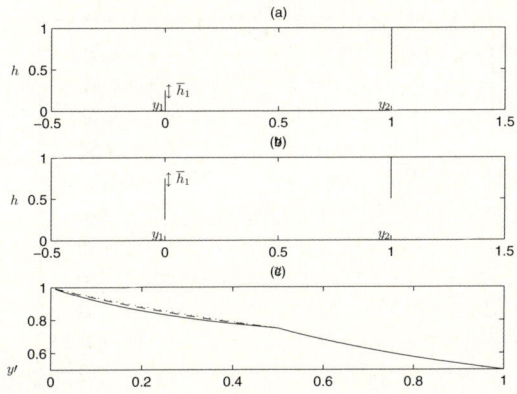

Fig. 1. Equivalence between type-2 and type-1 defuzzifications for two singletons with equal intervals of uncertainty; K-M type-reduction (solid), type-1 defuzzification (dotted) and uncertainty bounds approximate type-reduction (dashed)

3.2 Specificity of the Type-2 Approach for Two Singleton Consequents

If we know the case when the interval type-reduction may be substituted by the classical type-1 defuzzification, it is worth exploring specific cases when the type-reduction cannot be substituted by type-1 defuzzification. One of these specific cases, presented in Fig. 2, is when the height type-reduction is used. In this case the membership uncertainty of y_2 covers all unit interval, i.e. $[\underline{h}_2, \overline{h}_2] = [0, 1]$, while the upper membership of y_1 varies up to 1. Obviously, the K-M type-reduction leads directly to 0.5 value independently of $\overline{h}_1$. From the other hand,

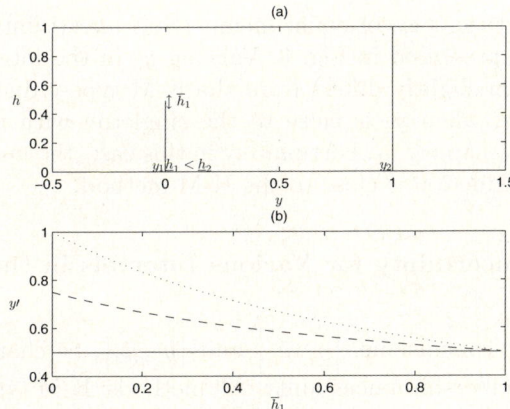

Fig. 2. Specificity of type-2 approach for two singleton consequents; K-M type-reduction (solid), type-1 defuzzification (dotted) and uncertainty bounds approximate type-reduction (dashed)

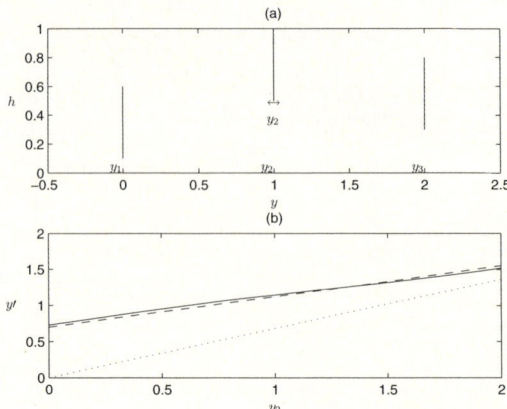

Fig. 3. Constant uncertainty for three singleton consequents; K-M type-reduction (solid), type-1 defuzzification (dotted) and uncertainty bounds approximate type-reduction (dashed)

the classical defuzzification with average memberships produces the output value between 0.5 and 1 when $\underline{h}_1 = \overline{h}_1$ are small enough but greater than $\underline{h}_2$. The uncertainty bounds approximate type-reduction places the output value just between the outputs of K-M method and type-1 defuzzification.

This case indicates premises to apply type-2 fuzzy logic with no possibility of successful application of type-1 fuzzy logic systems. When some rules are more certain while others rules are much more uncertain, the choice of interval type-2 fuzzy logic in such tasks is well-grounded.

3.3 Constant Uncertainty for Three Singleton Consequents

Now we may extend the case of equal membership uncertainties for three singleton consequents, presented in Fig. 3. Varying y_2 in the interval (y_1, y_3), the type-1 defuzzification slightly differs from the K-M type-reduction. The difference is much bigger, when y_2 is close to the singleton with more contrasting membership interval, namely y_1. Fortunately in this case, the uncertainty bounds approximate type-reduction is close to the K-M method.

3.4 Constant Uncertainty for Various Intervals in the Output Domain

Let the two interval conclusions, $\left[\underline{y}_1, \overline{y}_1\right]$ and $\left[\underline{y}_2, \overline{y}_2\right]$, be characterized by the same interval of membership uncertainty. All methods: K-M type-reduction, the uncertainty bounds approximate type-reduction and type-1 defuzzification are almost identical in a whole range of intervals width, what can be seen in Fig. 4.

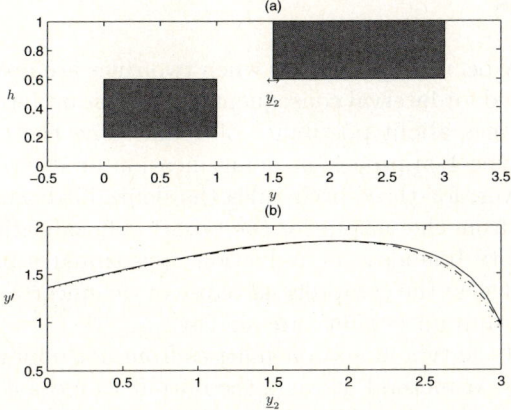

Fig. 4. Constant uncertainty for various intervals in output domain; K-M type-reduction (solid), type-1 defuzzification (dotted) and uncertainty bounds approximate type-reduction (dashed)

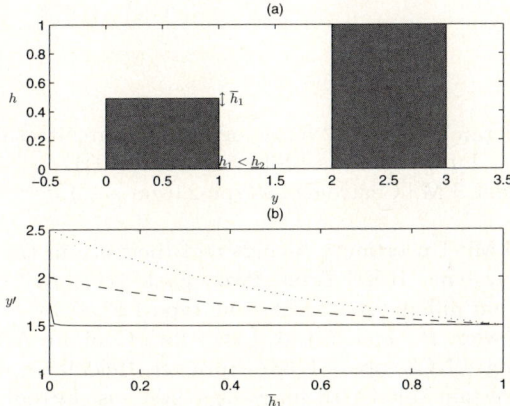

Fig. 5. Specificity of type-2 approach in centroid-type defuzzification; K-M type-reduction (solid), type-1 defuzzification (dotted) and uncertainty bounds approximate type-reduction (dashed)

3.5 Specificity of the Type-2 Approach for Centroid Defuzzification

It can be expected that the most distinct situation when the type-reduction is not close to the type-1 defuzzification is when the membership uncertainty of y_2 is $[\underline{h}_2, \overline{h}_2] = [0, 1]$ and the upper membership $\overline{h}_1$ varies in $(\underline{h}_2, 1]$. In Fig. 5 it can be seen that type-1 defuzzification gives the result varying in $[1.5, 2.5]$ while the K-M type-reduction gives 1.5 independently of $\overline{h}_1$. As in the two singleton case, the uncertainty bounds approximate type-reduction gives the output value between the outputs of K-M method and type-1 defuzzification.

4 Conclusions

The most frequently occurring cases are when two rules are fired. In these cases, both for singleton and for interval consequents, the constant intervals of membership uncertainty brings about possibility of abandoning the type-2 fuzzy logic and applying the type-1 approach in defuzzification or even in a whole inference system. However, for three fired rules the defuzzified output for the K-M method is different from the output for the type-1 defuzzification. In this occurrence, the uncertainty bounds type-reduction approximates precisely the K-M method. Thus we suggest the comprehensive use of the uncertainty bounds when intervals of membership uncertainty are similar.

The specificity of the type-2 approach arises from not uniform uncertainty of firing grades. The K-M method attracts the output to more certain rules (with smaller uncertainty interval), which is not the merit of type-1 defuzzification. Nevertheless, in such cases, the approximate uncertainty bounds type-reduction gives average results between the standard K-M method and the type-1 approach. Thus, we do not recommend the uncertainty bounds for cases of not uniform uncertainty.

References

1. Hagras, H.A.: A hierarchical type-2 fuzzy logic control architecture for autonomous robots. IEEE Trans. Fuzzy Systems 12(4), 524–539 (2004)
2. Karnik, N.N., Mendel, J.M.: Centroid of a type-2 fuzzy set. Information Sciences 132, 195–220 (2001)
3. Wu, H., Mendel, J.M.: Uncertainty bounds and their use in the design of interval type-2 fuzzy logic systems. IEEE Trans. Fuzzy Syst. 10(5), 622–639 (2002)
4. Starczewski, J.: What differs type-2 FLS from type-1 FLS? In: Rutkowski, L., Siekmann, J., Tadeusiewicz, R., Zadeh, L.A. (eds.) Int'l Conf. on Artificial Intelligence and Soft Computing. LNCS, pp. 381–387. Springer, Heidelberg (2004)
5. Mendel, J.M.: Uncertain rule-based fuzzy logic systems: Introduction and new directions 2001. Prentice Hall PTR, Upper Saddle River (2001)

Combining Basic Probability Assignments for Fuzzy Focal Elements

Ewa Straszecka

Silesian University of Technology
Institute of Electronics, 16 Akademicka St., 44 -100 Gliwice, Poland
ewa.straszecka@polsl.pl

Abstract. The Dempster-Shafer theory is convenient for implementation in models of medical diagnosis as it neglects dependence of symptoms. Yet, combination of two basic probability assignments that is defined in the theory is often criticized. The paper shows opportunities of combining that are created when the Dempster-Shafer theory is extended for fuzzy focal elements. The proposed method can help to avoid several disadvantages of the classical combination.

1 Introduction

The Dempster-Shafer theory, unlike the classical probability, does not consider dependence of focal elements in the definition of the probability assignment. Therefore, it is very convenient for heuristic inference modeling. Yet, the Dempster's combination [2] is criticized [1] as disadvantageous for implementations. In the combination formula two basic probability assignments for the same set of focal elements are considered. If a focal element has no common part with any other element and it is recognized as the most important in one basic probability assignment (has the greatest basic probability), as well as it is insignificant in the other (with the basic probability equal to zero), then the element is eliminated from the combined assignment. An elimination of a focal element should not be totally avoided, as contradictive information my disturb inference. Yet, it may happen that a focal element is disregarded in one of the assignments because of lack of information. In the latter situation the elimination is not justified. Another characteristic feature of the classical combination is an intensification of differences. When similar basic probability assignments are combined, then the resulting assignment values related to various focal elements may be extremely divergent. The combination of assignments is often essential for implementations of the Dempster-Shafer theory in medical diagnosis support. If the basic probability values are weights of symptoms, then it is inevitable to combine such weights e.g. found from training data and given by an experienced physician. Sometimes it is also valuable to combine the weights established for different populations (two training data sets). New suggestions about the combination of assignments can be made if the Dempster-Shafer theory is extended for fuzzy focal elements [5]. In case of fuzzy focal elements, not only assignments, but also membership functions can be combined. Let us assume that we want to use knowledge gathered during one population treatment in another population therapy. Symptoms of a disease will certainly enclose laboratory test results. It is obvious that membership functions of the tests are different for dissimilar populations, e.g. populations of different races or distant countries. This can't be neglected in diagnosis support. Thus, a combination of membership functions is important for inference.

The present contribution proposes a method of combining basic probability assignments when they are defined for fuzzy focal elements. The goal is to eliminate a focal element when it is contradictive for two assignments, even if its basic probability values are greater than zero in the combined assignments. Simultaneously, focal elements valuable for a diagnosis should be preserved without an excess of differences in their weights.

2 Fuzzy Focal Elements in Combining

The basic probability assignment (BPA) in the Dempster-Shafer theory is defined in the following way [2]:

$$m(f) = 0, \quad \sum_{a \in A} m(a) = 1 , \qquad (1)$$

where A is a set of focal elements and f stands for false. The classical Dempster's formula of combination is [2]:

$$m(a) = \frac{\sum_{a_i \cap a_j = a} m_1(a_i) m_2(a_j)}{\sum_{a_i \cap a_j \neq f} m_1(a_i) m_2(a_j)} , \qquad (2)$$

where m_1 and m_2 are the BPAs. When the theory is extended for fuzzy focal elements [5] then different ways of combination become possible. The fuzzy focal element a that is defined for the x variable, is represented by its membership function $\mu_a(x)$. For instance, if we choose to describe a result of the x laboratory test as 'low' or 'high', then a^1 = 'low laboratory test' and a^2 = 'high laboratory test' are focal elements, and their membership functions are e.g.:

$$\mu_{a^1}(x) = \begin{cases} 1 & x \leq x_l \\ max(0, \frac{x - x_h}{x_l - x_h}) & x > x_l \end{cases} \quad \mu_{a^2}(x) = \begin{cases} max(0, \frac{x - x_L}{x_h - x_l}) & x \leq x_h \\ 1 & x > x_h \end{cases} ,$$

where $[x_l, x_h]$ is the norm interval of the test. Usually, each of the focal elements is assigned to a different diagnosis, hence a^1 is a focal element of the D_1 diagnosis and a^2 is related to the D_2. Obviously, each diagnosis is based on multiple symptoms, not on a single test. The symptoms make a set of focal elements that is individual for the diagnosis. A focal element that refers to one medical parameter, is a single focal element. A focal element that is related to several parameters will be called a complex element. Membership of a complex element is the minimum of memberships of the included elements, i.e.:

$$\mu_{a_i^l}(x_1, \ldots, x_n) = \bigwedge_{j=1,\ldots,n} \mu_{a_{ij}^l}(x_j) , \qquad (3)$$

where a_i^l is the $i-th$ focal element of the $l-th$ diagnosis that consists of n elements, each characterized by the membership function $\mu_{a_{ij}^l}$ defined on the x_j domain. When the BPA is calculated from data, the membership of the single or the complex element is considered for an observation, i.e. for the selected value of x which is x^*. Then, for each focal element the membership becomes a number η_i in the $[0, 1]$ interval:

$$\eta_i^l = \mu_{a_i^l}(x_1^*, \ldots, x_n^*) = \bigwedge_{j=1,\ldots,n} \mu_{a_{ij}^l}(x_j^*) \ . \tag{4}$$

Specifically, for the single focal element $n = 1$. If training data is the only source of information, the BPA can be found as a normalized frequency of occurrence of patient cases in a diagnosis [6]. For instance, for the diagnosis D_1 it is counted for how many x_k^* cases $\eta_k^1 = \mu_{a^1}(x_k^*) > 0$. If we repeat counting procedure for the set of focal elements, i.e. all symptoms of the D_1, and next normalize the obtained numbers, the BPA is determined. The membership functions of the focal elements can be defined on the basis of the same data [6]. Combining the BPAs and the membership functions can join information from different sources. In case of fuzzy focal elements three types of combining can be suggested:

- classical combination (2) of two BPAs, for instance found from training data and expert's knowledge or from two databases;
- combination of membership functions, by finding minimum of the functions for the adequate focal elements:

$$a = a_i \cap a_j \Rightarrow \mu_a(x) = \min_x \left(\mu_{a_i}(x), \mu_{a_j}(x)\right) \ . \tag{5}$$

The μ_{a_i}, μ_{a_j} membership functions may originate from data or from expert's experience. Once the combined function is found, the BPA can be determined as the normalized frequency of occurrence with the application of the μ_a as characteristic for the a element.
- combining both membership functions and BPA values, according to the formula:

$$a : \{x|\mu_a(x) > 0\}; \ a_i : \{x|\mu_i(x) > 0\}; \ a_j : \{x|\mu_j(x) > 0\} \ ,$$

$$m(a) = \frac{\sum_{\mu_i \wedge \mu_j = \mu_a} m_1(a_i) m_2(a_j)}{\sum_{\mu_i \wedge \mu_j > 0} m_1(a_i) m_2(a_j)} \ . \tag{6}$$

Again data or an expert can be sources of information of the BPAs and membership functions included in the combination.

The method of BPA calculation is crucial, as values of the assignment are used to determine belief and plausibility measures, which for fuzzy focal elements can be defined as:

$$Bel(D_l) = \sum_{\substack{a_i \in A_l \\ \eta_i^l > 0}} m(a_i), \tag{7}$$

$$Pl(D_l) = \sum_{\substack{a_i \in A_l \\ \eta_i^l > 0}} m(a_i), \tag{8}$$

The combinations (5), (6) that are proposed in the present study may diminish the limitations of the classical Dempster's formula [1] and preserve its positive features.

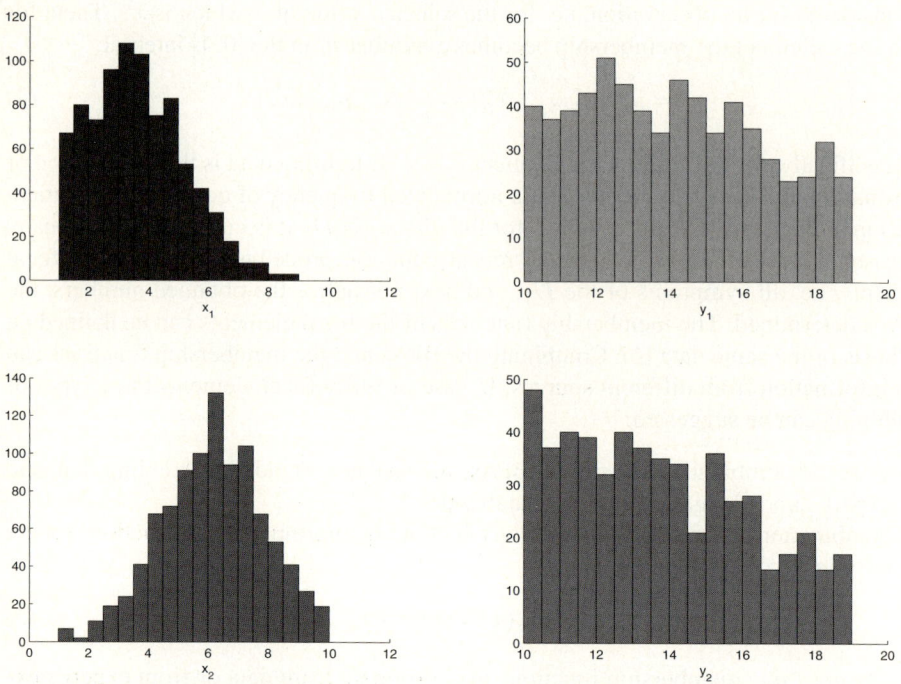

Fig. 1. Histograms of training data generated for two diagnoses

Let us observe performance of the three combinations (2), (5) and (6) on a simple example.

Example 1. Let us consider training data for two diagnoses (D_1 and D_2) each consisting of 1000 two-variable cases (stored in couples of the 1000-element vectors $\{\mathbf{x}_1, \mathbf{y}_1\}$ and $\{\mathbf{x}_2, \mathbf{y}_2\}$, respectively). The data are simulated by the normal distribution generator of the following parameters: for D_1 and the variable x - $N(3.2, 2.1)$, for D_2 and x - $N(6.2, 1.8)$, for D_1 and the variable y - $N(12, 4)$, for D_2 and y - $N(10, 4.8)$. Histograms of the values are presented in the fig.1. The x and y variables are strongly correlated (correlation coefficient is about 0.95) both for the diagnosis D_1 and D_2. From the histograms it is obvious that the variable y is useless for differentiation between the diagnoses. Let us try to eliminate the variable from inference by means of a combination. Membership functions that are defined for the data by means of the method described in [6], i.e. the membership functions $\mu_1(x)$, $\mu_1(y)$ for the diagnosis D_1 and $\mu_2(x)$, $\mu_2(y)$ for D_2, are presented in the fig.2a, b. The following fuzzy focal elements should be defined:

$a_1^1 \equiv$ "symptom X is μ_{1x}",
$a_2^1 \equiv$ "symptom Y is μ_{1y}",
$a_3^1 \equiv$ "symptom X is μ_{1x} and symptom Y is μ_{1y}",
$a_1^2 \equiv$ "symptom X is μ_{2x}",

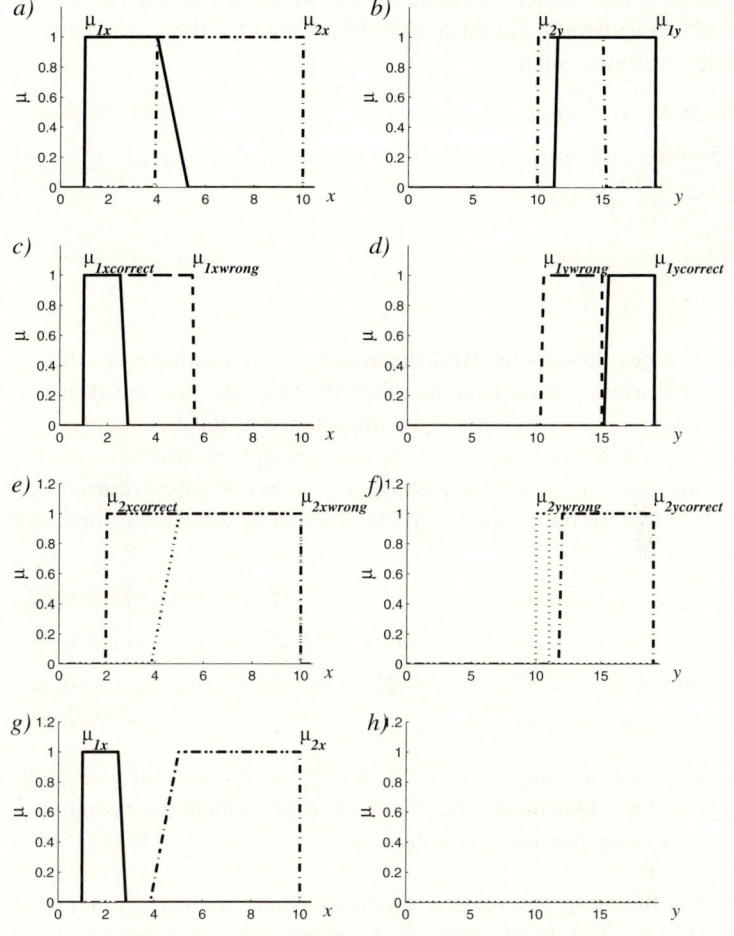

Fig. 2. Membership functions for training data

$a_2^2 \equiv$ "symptom Y is μ_{2y}",
$a_3^2 \equiv$ "symptom X is μ_{2x} and symptom Y is μ_{2y}",

where μ_{1x} denotes a linguistic value that is defined by the membership function $\mu_1(x)$, etc. The focal element a_i^j is regarded with the D_j diagnosis. Let us assume that $m(a_i^j) = 1/3$, $i = 1, 2, 3$, $j = 1, 2$. Classification of a case to the diagnosis depends on the Bel value (7). The case is recognized as an instance of D_1 if $Bel(D_1) > Bel(D_2)$ and vice versa. If belief values are equal, then an error occurs. It may happen that a case from the D_1 training set is misclassified as an instance of D_2. If BPA values are the same for all focal elements, then cases that fall into overlapping intervals of membership functions can be wrongly classified. For instance, intervals in which the membership functions overlap are: the $[3.92, 5.29]$ interval for $\mu_1(x)$ and $\mu_2(x)$ or $[11.27, 15.3]$ for $\mu_1(y)$ and $\mu_2(y)$, respectively. In consequence, after the first classification, training data can be divided for two subsets of correctly and wrongly classified cases. The classical combining (2) considerable diminishes the BPA value for the a_3^l element, but does not eliminate the y variable, since:

$$\alpha = m_{1ct}(a_1^l)m_{1wg}(a_1^l) + m_{1ct}(a_1^l)m_{1wg}(a_3^l) + m_{1wg}(a_1^l)m_{1ct}(a_3^l) \ ,$$

$$\beta = m_{1ct}(a_2^l)m_{1wg}(a_2^l) + m_{1ct}(a_2^l)m_{1wg}(a_3^l) + m_{1wg}(a_2^l)m_{1ct}(a_3^l) \ ,$$

$$\gamma = m_{1ct}(a_3^l)m_{1wg}(a_3^l) \ ,$$

$$m(a_1^l) = \frac{\alpha}{\alpha + \beta + \gamma} \ , \quad m(a_2^l) = \frac{\beta}{\alpha + \beta + \gamma} \ , \quad m(a_3^l) = \frac{\gamma}{\alpha + \beta + \gamma} \ ,$$

$$l = 1, 2,$$

where the 'ct' index denotes the BPA that results from training cases that are correctly identified in the primary classification, while the 'wg' denotes the BPA from wrongly classified cases. In case we assume equal importance of focal elements: $m_{1ct}(a_i^l) = 1/3$ and $m_{1wg}(a_i^l) = 1/3$, then $m(a_1^l) = m(a_2^l) = 0.43$, $m(a_3^l) = 0.14$. When the BPA calculation is based on frequency of case occurrence, then for the generated cases and for the membership functions in fig.2a, b the following BPA values are obtained:

$$m_{1ct}(a_1^1) = 0.33, \quad m_{1ct}(a_2^1) \ \ = 0.55, \quad m_{1ct}(a_3^1) = 0.12,$$
$$m_{1wg}(a_1^1) = 0.33, \quad m_{1wg}(a_2^1) \ = 0.28, \quad m_{1wg}(a_3^1) = 0.39,$$
$$m_{2ct}(a_1^2) = 0.41, \quad m_{2ct}(a_2^2) \ \ = 0.57, \quad m_{2ct}(a_3^2) = 0.02,$$
$$m_{2wg}(a_1^2) = 0.32, \quad m_{2wg}(a_2^2) \ = 0.29, \quad m_{2wg}(a_3^2) = 0.39.$$

Then $m(a_1^1) = 0.38$, $m(a_2^1) = 0.55$, $m(a_3^1) = 0.06$, $m(a_1^2) = 0.42$, $m(a_2^2) = 0.56$, $m(a_3^2) = 0.01$. This means that the third focal element has been practically eliminated from reasoning, but the y variable is even more important than x, so the diagnosis cannot be correct.

Membership functions that result from the division for correctly and wrongly classified cases are in fig.2c-f. If minimum of the membership functions for each diagnosis is found (see fig.2g,h), the variable y is deleted from inference. Now, if BPAs are calculated with the resulting functions, then $m(a_2^l) = m(a_3^l) = 0$. It means, that the reasoning now is based on the single focal element a_1^l, for which $m(a_1^l) = 1$ for each diagnosis and belief in each of the diagnoses of every case will be either equal 1 or 0. This unusual situation is a logical output of an elimination of one out of the two reasoning variables. Surely, this will not happen in real diagnostic problems, when reasoning is never limited to such small number of variables. Still, reasoning based on one variable can be in this case certainly more consistent.

In the example the classical combining (2) both for equal values of the BPA and the BPA based on frequency of occurrence of fuzzy focal elements, are both inadequate. Yet, when the (5) formula is used the meaningless variable can be eliminated from reasoning. If fuzzy and crisp symptoms are simultaneously included in a diagnosis, exclusively the (5) formula cannot be used for combination. Nevertheless, it can be supposed that formulas (5) and (6) used simultaneously for combination may help to tune the BPA and to prune excessive rules.

3 Tests

Data from two populations can be used in the (5) and (6) combinations instead of wrongly and correctly classified cases. One training data set may include cases of two populations or several disease syndromes of the same disease. In such situation two training subsets are identified after the primary classification. Then, it is more adequate to call the cases as 'typical and 'difficult' instead of wrong and correct. Performance of the proposed method for the BPA calculation when training data are the only source of information has been tested for the database *ftp.ics.uci.edu /pub/machine-learning-databases/thyroid-disease*, files *ann.**. The characteristic features of the database are: considerable correlation among variables and small a number of fuzzy variables in comparison to crisp variables. The data include 6 numerical and 15 binary variables with 3 diagnoses assigned. Training/test files have the following number of cases: D_1 - 191/177, D_2 - 93/73, D_3 - 3488/3178. Authors of the database demand 92% correctness of classification. Reference methods [3], [4] that introduce neural networks and genetic algorithms, report an error smaller than 2%. Still, the authors of the papers admit that such a classification lack generalization. The error of 8% is characteristic for the population from which the data originate. Thus, the present solution aimed at the error of classification just smaller than 8%. Simultaneously, the set of rules for each of the three diagnoses had to be found. Single focal elements have been defined for all 21 variables. Complex focal elements have comprised variables for which dependence have been statistically confirmed by the χ^2 test and Q-Kendall measure [7]. Each complex focal element has included at most 5 symptoms. Complex elements with numerous symptoms (up to 15) as well as combinations of statistically independent variables have been also tested. Yet, it turned out that the BPA values for focal elements including more than 5 symptoms have been equal or close to 0. Collections of dependent variables have been investigated to find focal elements with greater than 0 frequency of occurrence among training data. Eventually, 295 focal elements have been created. In every run of the algorithm 3 trapezoidal membership functions, corresponding to the three diagnoses, have been defined for every numerical variable. During calculations the following definitions of errors have been used:

- the error of the $l - th$ diagnosis:

$$\epsilon_l = \frac{n_{lwrong}}{n_l} , \qquad (9)$$

where n_{lwrong} is the number of wrongly classified cases out of the n_l-element set of data;
- the global error for N ($N = 3$) considered diagnoses

$$\epsilon = \frac{\sum_{l=1}^{N} n_{lwrong}}{\sum_{l=1}^{N} n_l} , \qquad (10)$$

- the minimal error that is achieved for the same set of focal elements for each diagnosis, but different BPA values and (generally) different membership functions. The most adequate values and functions selected individually for each diagnosis result in ϵ_{lmin}. The ϵ_{lmin} is the minimal value of ϵ_l error (9) found in an iterative process. During the iterations different divisions for typical and difficult cases have been tried that have resulted in various membership functions and the BPAs. Thus, the minimal error:

$$\epsilon_{min} = \frac{\sum_{l=1}^{N} \epsilon_{lmin} n_l}{N} ,\qquad(11)$$

Membership functions and the BPAs have been found individually for each diagnosis by means of training data. Afterwards, the same data has been classified by an appropriate comparison of Bel values. Cases that have been correctly classified created the set of 'typical cases', while the wrongly classified have been gathered in the 'difficult cases' set. Now, membership functions and BPAs have been found separately for typical and difficult cases, and next combined. The procedure has been repeated to achieve minimum of each of the errors (9)-(11).

4 Results

The membership functions found after combining functions of typical and difficult training cases are in the fig.3. Each row of diagrams resembles one variable that is denoted at the axis of the rightmost diagram. Each column gathers membership functions that have been found as the most adequate for the classification of one diagnosis. Obviously, for each variable the collection of three membership functions is constructed. The membership functions for the diagnosis D_1 are denoted by solid lines, for D_2 by dashed lines and for D_3 by dotted lines. It means that the last diagram (in the right-hand side lower corner) illustrates membership functions of the variable x_6 that the most correctly classify cases of the diagnosis 3, i.e. results in the least ϵ_3 error (9). This minimal value of the error is denoted as ϵ_{3min}. For the ϵ_{3min} error the membership function of diagnosis D_1 should be like that drawn in the solid line in this diagram, etc. Tests have shown that membership functions which are the best for individual diagnosis differ. For instance, membership functions for the variable x_1, the diagnosis D_1 and the ϵ_{1min} are different from the functions D_1 found for the ϵ_{2min} and ϵ_{3min} (see the first row of diagrams in the fig.3). On the contrary, membership functions of D_3 for x_1 determined for ϵ_{1min}, ϵ_{2min} and ϵ_{3min} are almost the same. Variations in the membership functions shapes result from change in division of cases for typical and difficult. If very different membership functions are obtained as the best for individual diagnoses, the variable is of little value as a classification feature. Thus, the variable x_1 is practically useless for a differentiation between D_1 and D_2 as well as between D_1 and D_3. Yet, we are able to distinguish between D_2 and D_3 by means of this feature. Such an analysis may help to find features 'universal' for the classification (e.g. x_5, x_6 in fig.3) and the other that may help to identify diagnosis in particular cases (for instance x_1, x_2, x_3 and x_4 when the diagnosis 1 is absolutely excluded). If all the variables are obligatory used as focal elements, the values of the BPA, which differ for individual diagnoses, may diminish importance of variables of less discrimination force in the diagnosis. Different numbers of iterations have been necessary to find ϵ_{lmin}: 23 for D_3, 131 for D_1 and 204 for D_2. Thus, relatively small number of iterations is sufficient to find acceptable weights for symptoms. Consequently, time of calculations is short. This seems to be a considerable advantage in comparison to neural networks and genetic algorithms that have been

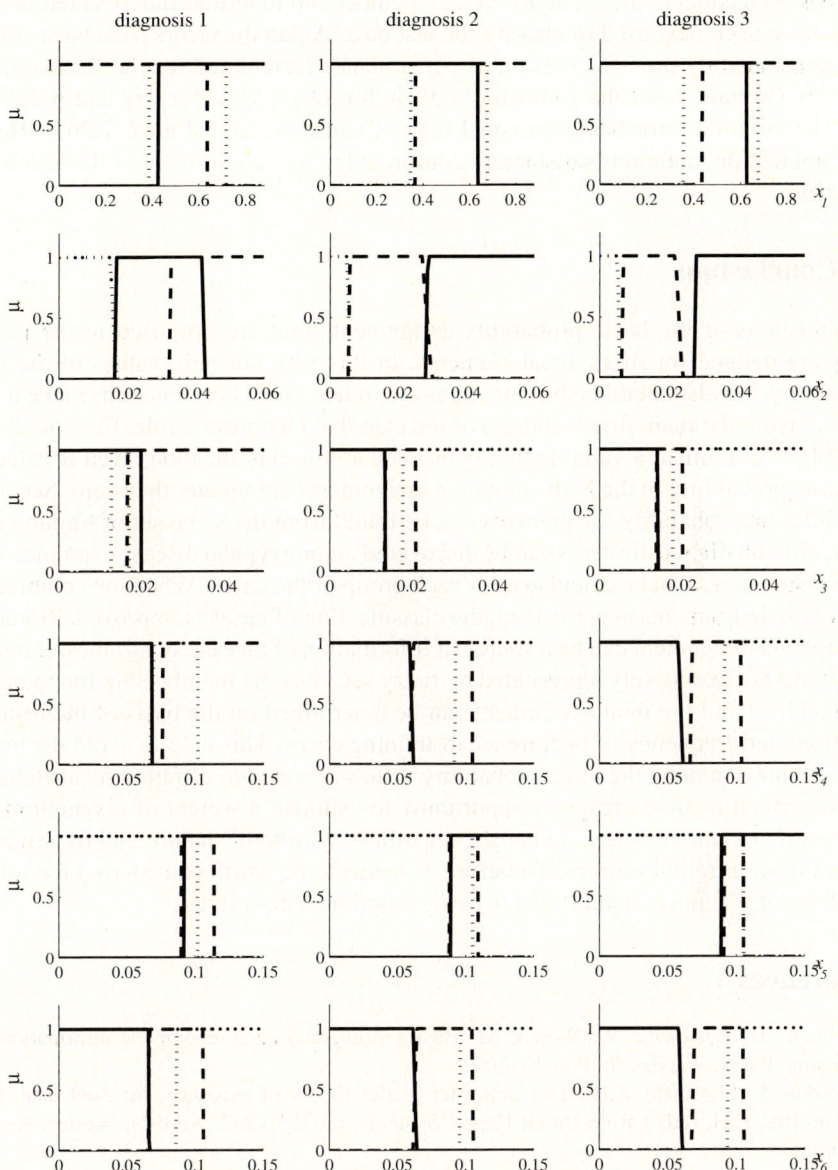

Fig. 3. Membership functions found from training data that are the best for individual diagnoses (result in ϵ_{lmin} errors). In the diagrams membership functions are denoted by solid lines for D_1, dashed lines for D_2 and dotted lines for D_3.

previously used to solve the prblem [3], [4]. At the stage of training the BPA combining resulted in the following minimal values of ϵ_{lmin} for diagnoses: 0.00% for D_1, 3.14% for D_2 and 3.90% for D_3. The minimal error (11) has equaled 3.76%. The global error

(10) has been equal 6.50%. The BPAs and membership functions that resulted in these errors have been next used to classify the test data. Again the errors have been smaller than demanded by the 8% criterion of performance formulated for the database. The ϵ_l errors (9) have been the following: 0.00% for D_1, 6.78% for D_2 and 6.42% for D_3. The minimal error has been equal 6.30%, while the global error 7.26%. Hence, the combination method is satisfactory conformed to the requirement of the acceptable 8% error.

5 Conclusions

Combinations of the basic probability assignments that are proposed in the present study are defined for fuzzy focal elements. In this way not only values of the basic probability, but also membership functions are joined. Such combinations make it possible to avoid the main disadvantages of the classical Dempster's rule. First of all, it is possible to eliminiate a variable that is inadequate for classification, even if values of the basic probability in the both combined assignments are greater than zero. Secondly, two different probability assignments can be found from the same set of training data. Thus, difficult diagnostic cases can be discovered among typical disease examples. Separate assignments can be calculated for each group of the cases. When the combination of the two assignments is performed, the classification of cases is improved. Moreover, the separate assignment can be a source of information of disease syndromes. If disease symptoms are exclusively represented by fuzzy sets, just the membership functions can be combined and the final assignment can be determined on the basis of the resulting functions and frequency of occurrence of training cases. This help to avoid the unnecessary contrast among the basic probability values assigned to different focal elements. The described method create an opportunity to estimate a weight of a symptom or a collection of symptoms in a diagnosis, regardless symptom's nature and to join information from different sources. Therefore, it seems to be worth considering for solving problems of diagnosis support that is partly based on data-mining.

References

1. Bolc, L., Borodziewicz, W., Wojcik, M.: Basics of uncertain and incomplete information processing. PWN, Warsaw (in Polish) (1991)
2. Gordon, J., Shortliffe, E.H.: The Dempster-Shafer theory of evidence. In: Buchanan, B.G., Shortliffe, E.H. (eds.) Rule-Based Expert Systems, pp. 272–292. Addison-Wesley, Reading (1984)
3. Schiffman, W., Joost, M., Werner, R.: Synthesis and performance analysis of multilayer neural network architectures, ftp archive.cis.ohio-state.edu, technical report (1992)
4. Schiffman, W., Joost, M., Werner, R.: Optomization of backpropagation algorithm for training multilayer perceptrons, ftp archive.cis.ohio-state.edu, technical report (1994)
5. Straszecka, E.: An interpretation of focal elements as fuzzy sets. Int. J. of Intelligent Systems 18, 821–835 (2003)
6. Straszecka, E.: Combining uncertainty and imprecision in models of medical diagnosis. Information Sciences 176, 3026–3059 (2006)
7. Tadeusiewicz, R., Izworski, A., Majewski, J.: Biomerty. AGH, Cracow (in Polish) (1993)

Using Intuitionistic Fuzzy Sets in Text Categorization

Eulalia Szmidt and Janusz Kacprzyk

Systems Research Institute, Polish Academy of Sciences
ul. Newelska 6, 01–447 Warsaw, Poland
{szmidt, kacprzyk}@ibspan.waw.pl

Abstract. We address some crucial problem associated with text categorization, a local feature selection. It seems that intuitionistic fuzzy sets can be an effective and efficient tool making it possible to assess each term (from a feature set for each category) from a point of view of both its indicative and non-indicative ability. It is important especially for high dimensional problems to improve text filtering via a confident rejection of non-relevant documents. Moreover, we indicate that intuitionistic fuzzy sets are a good tool for the classification of imbalanced and overlapping classes, a commonly encountered case in text categorization.

1 Introduction

Text categorization (classification) boils down to an automatic assignment of documents to a predefined set of categories (classes), cf. Sebastiani [19] for a good exposition. Since the early 1990's the effectiveness of classifiers has been considerably improved which is strongly related to a rapid development of machine learning methods exemplified by the nearest neighbor classifiers, Bayesian classifiers, decision trees, and support vector machines. For example, the support vector machines (Joachims [10]) used to text classification provide a big improvement as compared to other, weaker methods [19]. Among many problems involved in text categorization, feature selection plays a particular role as it is in a sense a point of departure.

A document is a sequence of words, and is usually represented as an array of words. The set of all words (terms) of a training set is called a vocabulary or a feature set. So a document can be represented by a binary vector with 1 if the document contains the feature word or 0 otherwise. Not all words can be used as feature words (terms), for instance auxiliary verbs, articles, conjunctions, etc. Useless words are removed in preprocessing (cf. Sebastiani [19]) which also involves stemming, the removal of words with the same stem (like "rain", "raining", "rains") and replacing them with one word ("rain"). But even after the elimination of useless words and stemming, the set of feature words is still too large for many learning algorithms. We need a feature selection techniques to further reduce the feature set (cf. Forman [7]).

Different feature selection methods are known (cf. Forman [7], Galavotti et al. [8], Torkkolla [43], Sebastiani [19], Soucy and Mineau [21], Sousa et al. [20],

Yang and Pedersen [45]). Scoring of individual words can be performed using some measures like document frequency, information gain, mutual information, chi-square, correlation coefficient, odds ratio, term strength, etc. Forman [7] presented a comparison of 12 measures on well known training sets. All feature scoring methods yield a ranking of the features and then select the top scoring features. In other words, only the features highly indicative in terms of membership for each category are single out, with no account of those which are highly indicative in terms of non-membership.

When training a binary classifier per category, all the documents in the training corpus belonging to this category are relevant while all those belonging to all other categories are irrelevant. Often there is an overwhelming majority of irrelevant training documents particularly when there is a large collection of categories with a small number of documents assigned. This is a typical imbalanced data problem that attracts much attention and is a particular challenge for classification algorithms, cf. Kubat at al. [12], Fawcett and Provost [6], Japkowicz [9], Lewis and Catlett [13], Mladenic and Grobelnik [15].

To address the imbalanced data problem, in text categorization, the so-called "query zone" and "category zone" were introduced to select a subset of non-relevant documents as the irrelevant training data. These methods try to obtain more balanced relevant and irrelevant training documents via up-sampling and down-sampling. However, in the case of overlapping classes even the artificially obtained balance does not solve the problem (some documents may appear as valid examples in both classes). This problem is still open, cf. Zhang and Mani [49], Chawla et al. [4], Visa [44]) for some proposals.

A strong impact of the commonly used local feature selection on the imbalance data problems can be explained by using the scoring feature methods (e.g., the correlation coefficient, odds ratio, gss coefficient) making use of the positive features only. For the imbalanced data problem where irrelevant documents dominate, the irrelevant documents are subjects of misclassification. How to confidently reject the irrelevant documents becomes very important.

In this paper we advocate the use of Atanassov's intuitionistic fuzzy sets which, by making to possible to account for not only to membership values (a "positive" information) but to non-membership values (a negative "information"), and hesitation margins (Section 2), may be useful both in feature selection and dealing with imbalanced data sets.

2 A Brief Introduction to Intuitionistic Fuzzy Sets

One of the possible generalizations of a fuzzy set in X (Zadeh [47]), given by

$$A^{'} = \{< x, \mu_{A'}(x) > | x \in X\} \tag{1}$$

where $\mu_{A'}(x) \in [0,1]$ is the membership function of the fuzzy set $A^{'}$, is an intuitionistic fuzzy set introduced by Atanassov [1], [2], [3], A, is given by

$$A = \{< x, \mu_A(x), \nu_A(x) > | x \in X\} \tag{2}$$

where: $\mu_A : X \to [0,1]$ and $\nu_A : X \to [0,1]$ such that

$$0 \leq \mu_A(x) + \nu_A(x) \leq 1 \tag{3}$$

and $\mu_A(x)$, $\nu_A(x) \in [0,1]$ denote a degree of membership and a degree of non-membership of $x \in A$, respectively.

In this paper we will use the intuitionistic fuzzy sets in Atanassov's sense, or A-IFSs for short.

The theory of intuitionistic fuzzy set (IFS) is based both on extensions of corresponding definitions of fuzzy sets objects and definitions of new objects and their properties (Atanassov [1,2,3]). We will use them to solve our problem. However, we will not deal with some recent discussion on the name "intuitionistic fuzzy sets" (cf. Dubois, Gottwald, Hájek, Kacprzyk, Prade H [5]) because it is irrelevant for our discussion.

Obviously, each fuzzy set may be represented by the following intuitionistic fuzzy set: $A = \{< x, \mu_{A'}(x), 1 - \mu_{A'}(x) > | x \in X\}$.

For each intuitionistic fuzzy set in X, we will call

$$\pi_A(x) = 1 - \mu_A(x) - \nu_A(x) \tag{4}$$

an *intuitionistic fuzzy index* (or a *hesitation margin*) of $x \in A$ and, it expresses a lack of knowledge of whether x belongs to A or not (cf. Atanassov [3]). It is obvious that $0 \leq \pi_A(x) \leq 1$, for each $x \in X$. The hesitation margin turns out important in theory and applications, notably while considering distances (cf. Szmidt and Kacprzyk [25], [28], [37], entropy (cf. Szmidt and Kacprzyk [30], [39], similarity (cf. Szmidt and Kacprzyk [40]), etc. i.e. measures of a crucial importance for our purposes. For some applications that may be relevant, group decision making, negotiations, etc. may be cited – cf. Szmidt and Kacprzyk [24], [26], [27], [29], [31], [33], [32], [34], [38].

The use of intuitionistic fuzzy sets instead of fuzzy sets means to have another degree of freedom (non-memberships), i.e. an additional tools to represent imperfect knowledge. We will show that this may be useful in feature selection.

3 Feature Selection

The most common form of document representation in text classification is known as a "bag of words". Most representations of this type treat each word, or a set of words, found in the corpus as a feature. Documents are represented as vectors, which are lists of words (terms) that are contained in the corpus. A binary vector representation is usually used to render the presence or absence of a word in the document. The sentence "A mountain is high" is represented using a binary vector representation defined over the set of words (terms): {mountain, deep, high, lake} as: {1, 0, 1, 0}.

Next, a document is processed (cf. Sebastiani [19]), and after removing terms irrelevant for classification, the most frequently occurring terms are placed in the feature set to create the document vectors, i.e. feature selection is performed, a key task in text classification.

Table 1. Notation

	Category c_i	Category $\neg c_i$
Term t_k	A	B
Term $\neg t_k$	C	D

To assign automatically predefined category labels to new text documents it is necessary to have a model of connections between a term t_k and a category c_i. In this paper a local feature selection is considered, i.e. for each predefined category a set of terms is chosen for classification. The feature selection measures discussed in the literature are the functions $f(t_k, c_i)$ with a term t_k and a category c_i as the parameters. The values of the functions define relationships between the term and the category.

The mathematical definitions of the most often used functions $f(t_k, c_i)$ are recalled in Section 3.1. In Table 1 the abbreviations used in the formulas are explained. Probabilities are interpreted on the event space of documents (e.g., $P(\neg t_k, c_i)$ denotes the probability that for a random document x, term t_k does not occur in x and x belongs to category c_i), and are estimated by counting occurences in the training set. All functions are specified locally to a specific category c_i. The meaning of the symbols in Table 1 is:

A – the number of times a term t_k and a category c_i co-occur,
B – the number of times a term t_k occurs without a category c_i,
C – the number of times a category c_i occurs without a term t_k,
D – the number of times neither a term t_k nor a category c_i occurs.

In further considerations N denotes the total number of documents.

To evaluate the value of a term t_k in a "global" sense (cf. Sebastiani [19]), either the sum, the weighted sum, or the maximum of the category-specific values $f(t_k, c_i)$, respectively, is usually used:

$$f_{sum}(t_k) = \sum_{i=1}^{|C|} f(t_k, c_i) \quad (5)$$

$$f_{wsum}(t_k) = \sum_{i=1}^{|C|} P(c_i) f(t_k, c_i) \quad (6)$$

$$f_{max}(t_k) = \max_{i=1}^{|C|} f(t_k, c_i) \quad (7)$$

As to the choice of $f(t_k, c_i)$, Sebastiani [19] indicates that the best results were for odds ratio (OR) - Section 3.1, corelation coefficient (CC) - Section 3.1 , GSS coefficient (GSS) - Section 3.1, and chi-square (χ^2) - Section 3.1. The following result is given in [19]: $\{OR, CC, GSS\} > \chi^2$ where ">" means "perform better than". It is the reason why we recall only these function, OR, CC, GSS, χ^2, in Section 3.1 as the most effective and most often used ones.

3.1 The Feature Selection Measures

Chi-square (CHI). The chi-square used by Yang [46] to measure a lack of independence between a term t_k and a category c_i. It is given as:

$$\chi^2 = \frac{N[P(t_k, c_i)P(\neg t_k, \neg c_i) - P(t_k, \neg c_i)P(\neg t_k, c_i)]^2}{P(t_k)P(\neg t_k)P(c_i)P(\neg c_i)} \approx$$
$$\approx \frac{N(AD - CB)^2}{(A+C)(B+D)(A+B)(C+D)} \tag{8}$$

χ^2 is equal zero if t_k and c_i are independent; it is normalized.

Correlation Coefficient (CC). The correlation coefficient $CC(t_k, c_i)$ of term t_k and category c_i (cf. Sebastiani [19]) is given as:

$$CC(t_k, c_i) = \frac{\sqrt{N}[P(t_k, c_i)P(\neg t_k, \neg c_i) - P(t_k, \neg c_i)P(\neg t_k, c_i)]}{\sqrt{P(t_k)P(\neg t_k)P(c_i)P(\neg c_i)}} \approx$$
$$\approx \frac{\sqrt{N}(AD - CB)}{\sqrt{(A+C)(B+D)(A+B)(C+D)}} \tag{9}$$

It is a variant of the CHI measure where $CC^2 = \chi^2$. The positive values of CC correspond to features indicative of membership, while negative values indicate those indicative of non-membership of term t_k to category c_i. The greater (smaller) the positive (negative) values of CC are, the stronger the term will be to indicate the membership (non-membership) of a term to a category. The standard CC based local feature selection method selects the terms with the highest CC values as features as the terms coming from irrelevant texts to a category are useless.

On the other hand, CHI is non-negative which means that its values indicate the membership or non-membership of a term to the same category. As a result, ambiguous features will be ranked lower. As opposed to CC, CHI considers the terms coming from both the relevant and non-relevant texts.

Odds ratio (OR). Odds ratio (OR) was proposed by van Rijsbergen et al. [17]:

$$OR(t_k, c_i) = \frac{P(t_k|c_i) \cdot [1 - P(t_k, \neg c_i]}{[1 - P(t_k|c_i)] \cdot P(t_k|\neg c_i)} \approx \frac{AD}{CB} \tag{10}$$

The odds ratio is always greater than or equal to zero. The values greater than 1 correspond to features indicating the membership of term t_k, while the values less than 1 correspond to features indicating the non-membership to a category c_i. In other words, odds ratio equal 1 indicates that a term t_k is equally likely in relevant and irrelevant documents. Odds ratio less than 1 indicates that a term t_k is less likely in relevant documents. If odds of relevant class approaches zero, the odds ratio approaches zero. If odds of an irrelevant class approaches zero, the odds ratio approaches positive infinity.

Example 1. Suppose that a term t_k indicates correctly 90 among 100 documents in a relevant class. So the odds of a proper classification of a document to relevant class while using term t_k are $90:10 = 9:1$. In the irrelevant class term t_k occurs in 20% of the documents. So the odds of proper classification of an irrelevant document to irrelevant class are 20 to 80, i.e., $1/4 = 0.25$. As $9/0.25 = 36$, so the odds ratio (10) is 36, showing that term t_k far better indicates documents from the relevant class.

This example also shows how odds ratio overestimates the relative power of a term. In this example the correct classification of a relevant document is 4.5 (0.9/0.2) times more possible than the correct classification of an irrelevant document but have the 36 times higher odds ratio.

GSS Coefficient. The GSS coefficient proposed by Galavotti et al. [8] is another variant of χ^2 given as:

$$GSS(t_k, c_i) = P(t_k, c_i) \cdot P(\neg t_k, \neg c_i) - P(t_k, \neg c_i) \cdot P(\neg t_k, c_i) \approx$$
$$\approx \frac{AD - CB}{N^2} \qquad (11)$$

Similarly as for CC, the positive values correspond to the features indicating the membership of a term to a category, whereas negative values indicate those of non-membership.

3.2 Problems with Feature Selection

Albeit the measures presented above, $f(t_k, c_i)$, are used most often, it is known from the experiments (cf. Forman [7]) that there is no measure that always performs better than all others. A combination of 2 measures [7] is often performed.

It is worth stressing again that all the feature scoring methods conclude by ranking the top scoring features. Negative features are not included in the feature set.

Joachims [10] has demonstrated that there are rather a few irrelevant features in text classification using the Reuters-21578 ACQ category in which he first ranks features according to their binary information gain. Joachims [10] then orders the features according to their rank and uses the features with the lowest information gain for document representation. He then trains a naive Bayes classifier using the document representation that is considered to be the "worst", showing that the induced classifier has a performance that is much better than random. By doing this, Joachims [10] demonstrates that even features which are ranked lowest according to their information gain, still contain considerable information.

Certainly, the above fact does not mean that the features selection should be abandoned. But it suggests that negative features should be taken into account, too.

In fact the local feature selection can be viewed globally as the two category problem: the negative features (usually neglected) are from the irrelevant category whereas positive features are from the relevant category. It seems that in such a two category problem both kinds of features (positive and negative) should be taken into account. The negative features found in a document indicate that the document does not belong to a category which makes it possible to eliminate effectively the irrelevant documents.

Unfortunately, the most often used feature selection measures (Section 3.1) and the fact that only the top scoring features are selected, result in not taking into account the negative features. The χ^2, combining both the positive and negative features, is an exception. On the other hand, as it has already been pointed out, just the χ^2 (Sebastiani [19]) gives worse results than other feature selection measures presented in Section 3.1. The result does not contradict the fact that the negative features should be taken into account. As χ^2 combines implicitly the terms most positive and most negative, the ratio between the negative and positive features is not optimal and can not be changed.

The following example shows the importance of taking into account irrelevant features.

Example 2. The list of selected terms $\{t_1, \ldots, t_9\}$ is scored by CC: 9.5, 8.9, 8.7, 4.5, 1, -1, -5, -5.7, -5.8, respectively. If the feature set contains 6 terms, t_1 through t_6 will be selected (due to the rule that top scored features are considered). Suppose a new document containing t_5, t_8 and t_9 is considered. The system will classify the document as relevant although it is obviously irrelevant. On the other hand, if we pay enough attention to negative features, we may use 6 following terms: t_1 through t_3 and t_7 through t_9, and hence the new document is obviously classified as irrelevant.

Because text classification is typically associated with a large number of categories that can overlap, documents are viewed as either belonging to a class (positive example) of interest or not belonging to such a class (negative example).

For the multi class problem, text classification systems use individual classifiers to recognize individual categories. A classifier is trained to recognize a document belonging to a class, or not. For an N class problem, a text classification system would consist of N classifiers trained to recognize N categories. Each classifier in the system is trained independently. In effect text classification may be viewed as a collection of binary classifiers. Viewing text classification as a collection of two class problems results in having many times more negative examples than positive examples. Typically, the number of positive examples for a given class is a few hundred, while the number of negative examples is thousands or even tens of thousands. The Reuters-21578 data set (http://www.research.att.com/lewis) consists of the average category having less than 250 examples, while the total number of negative examples exceeds 10000. It means that we face the imbalanced data problem.

Many approaches has been employed to address the imbalanced data problem. Essentially, these methods try to obtain more balanced relevant and irrelevant training documents via up-sampling and down-sampling. Unfortunately, in the case of overlapping classes even the artificially obtained balance does not solve the problem (some documents may appear as valid examples in both classes).

As the problem is still open, the new methods are investigated and trying to be improved (Zhang and Mani [49], Chawla et al. [4], Visa [44]).

As it has been shown in Example 2, the commonly used methods of feature selections (using the positive features only) may lead to misclassification of irrelevant documents. It may be even worse for the imbalanced data with dominating irrelevant documents.

We propose in this paper a new approach for feature selection, namely an approach making use of the intuitionistic fuzzy sets.

4 Intuitionistic Fuzzy Sets in Feature Selection and Text Classification

Our previous considerations have pointed out that text classification (often seen as a two-class task) may be more efficient by including in addition to highly indicative terms of a target class also highly indicative terms of the other class.

Now we will show that the natural properties of intuitionistic fuzzy sets may help not only to take into account both positive and negative features but to assess more carefully their real indicating potential.

Example 3. The problem consists in classifying documents as relevant and irrelevant. Five different terms $\{t_k\}$, $k = 1, \ldots, 5$ are taking into account. The data describing relative frequencies of terms in relevant (denoted: +) documents and irrelevant (denoted: -) documents are respectively

- relative frequencies $p^+(k)$ for terms: $k = 1, \ldots, 5$ in the relevant documents

$$p^+(t_1) = 0.25, \quad p^+(t_2) = 0.256, \quad p^+(t_3) = 0.234,$$
$$p^+(t_4) = 0.16, \quad p^+(t_5) = 0.1 \quad (12)$$

- relative frequencies $p^-(k)$ for terms: $k = 1, \ldots, 5$ in the irrelevant documents

$$p^-(t_1) = 0., \quad p^-(t_2) = 0., \quad p^-(t_3) = 0.068,$$
$$p^-(t_4) = 0.33, \quad p^-(t_5) = 0.602 \quad (13)$$

Using the algorithm describing how to convert a relative frequency distribution into an intuitionistic fuzzy set (details in Szmidt and Baldwin [22], [23]) we obtain:

- strength of the terms indicating their membership $\mu(t_k)$ into the target class (relevant documents):

$$\mu(t_1) = 1., \quad \mu(t_2) = 1., \quad \mu(t_3) = 0.795,$$
$$\mu(t_4) = 0.273, \quad \mu(t_5) = 0. \quad (14)$$

- strength of the terms indicating their non-membership $\nu(t_k)$ into the target class (relevant documents), i.e. strength of the terms indicating the irrelevant documents:

$$\nu(t_1) = 0., \quad \nu(t_2) = 0., \quad \nu(t_3) = 0.039,$$
$$\nu(t_4) = 0.263, \quad \nu(t_5) = 0.497. \quad (15)$$

- hesitation margins $\pi(t_k)$ showing lack of knowledge if k-th term indicates a relevant or irrelevant class:

$$\pi(t_1) = 0., \qquad \pi(t_2) = 0., \qquad \pi(t_3) = 0.166,$$
$$\pi(t_4) = 0.464, \qquad \pi(t_5) = 0.503 \qquad (16)$$

This way starting from relative frequencies we have obtained the values μ (14), ν (15), and π (16) characterizing the counterpart intuitionistic fuzzy set. One

Table 2. Odds ratios for data in Example 3

term	t_1	t_2	t_3	t_4	t_5
odds ratio	∞	∞	3.96	0.39	0.07

could ask what is the advantage of the intuitionistic fuzzy description (14)–(16) over the frequency description (12)–(13) or Odds ratio (OR) scoring (Section 3.1, Table 2).

Both frequency description and Odds ratio (OR) rank the terms, and the ranking is strongly related to the frequent presence of a term in the relevant documents. The odds ratio assigns the lowest score to the terms that are frequent in irrelevant documents. But in some cases we might want to use such "very negative" terms to help recognizing that some documents containing them are not relevant in spite the other terms in it. But if we use the odds ratio scoring, we could come to conclusion that besides highly indicative terms: first, second, possibly third as well (Table 2), a good choice could be the fifth term which points out irrelevant documents. But if we look more closely at the characteristics of the terms expressed via intuitionistic fuzzy description (14)–(16), we may draw more conclusions. Namely, for sure the first and second terms are highly indicative for the target class as their membership values equal 1 mean that the terms never appear in the other class (irrelevant documents), and their presence means that the document belongs for sure to the target class (relevant documents). But the fifth term which seems promising in terms of frequencies (13) and odds ratio (Table 2) for indicating irrelevant class turns out disputable with the non-membership value equal 0.497 (15) and the hesitation margin equal 0.503 (16). It means that lack of knowledge (0.503) concerning this term is bigger than its indicating power of the irrelevant class (0.497). This conclusion is possible while employing intuitionistic fuzzy sets, and impossible otherwise.

As we have already mentioned, another text classification problem is caused by the existence of imbalanced classes (the target class is usually considerably smaller than the other class). It is well known (cf. Kubat at al. [11], [12], Fawcett and Provost [6], Japkowicz [9], Lewis and Catlett [13], Mladenic and Grobelnik [15]) that typical classifiers prove to be an inadequate learners on data sets that are imbalanced.

The use of the intuitionistic fuzzy sets makes it possible to cope with the problems of imbalanced classes as shown by Szmidt and Kukier [41], [42]. The intuitionistic fuzzy index is the parameter making it possible to control the power of seeing relevant and irrelevant classes. The results obtained confirm [41], [42] that using the intuitionistic fuzzy sets makes it possible to "see" better the relevant, considerably smaller classes.

5 Conclusions

We have discussed some problems with text classification paying especially attention to feature selection and the important problem of imbalanced classes.

It seems that the natural properties of the intuitionistic fuzzy sets make it possible to use the advantages of the intuitionistic fuzzy description over other approaches to feature selection.

Intuitionistic fuzzy sets may be also a prospective tool for text classification because of their ability of dealing with imbalanced classes.

References

1. Atanassov, K.: Intuitionistic Fuzzy Sets. VII ITKR Session. Sofia (Deposed in Centr. Sci.-Techn. Library of Bulg. Acad. of Sci., 1697/84) (in Bulgarian) (1983)
2. Atanassov, K.: Intuitionistic Fuzzy Sets. Fuzzy Sets and Systems 20, 87–96 (1986)
3. Atanassov, K.: Intuitionistic Fuzzy Sets: Theory and Applications. Springer, Heidelberg (1999)
4. Chawla, N., Bowyer, K., Hall, L., Kegelmeyer, W.: Smote: synthetic minority oversampling technigue. Artificial Intelligence Research 16, 321–357 (2002)
5. Dubois, D., Gottwald, S., Hajek, P., Kacprzyk, J., Prade, H.: Terminological difficulties in fuzzy set theory - the case of "Intuitionistic Fuzzy Sets". Fuzzy Sets and Systems 156, 496–499 (2005)
6. Fawcett, T., Provost, F.: Adaptive Fraud Detection. Data Mining and Knowledge Discovery 3(1), 291–316 (1997)
7. Forman, G.: An experimental study of feature selection metrics for text categorization. Journal of Machine Learning Research 3, 1289–1305 (2003)
8. Galavotti, L., Sebastiani, F., Simi, M.: Experiments on the use of feature selection and negative evidence in automated text categorization. In: 4th European Conf. on Research and Advanced Technology for Digital Libraries ECDL 2000, pp. 59–68 (2000)
9. Japkowicz, N.: Class Imbalances: Are we Focusing on the Right Issue? In: Workshop on Learning from Imbalanced Data II, ICML, Washington (2003)
10. Joachims, T.: Text categorization with support vector machines: lerning with many relevant features. In: European Conf. on machine Learning (ECML), pp. 137–142. Springer, Berlin (1998)
11. Kubat, M., Holte, R., Matwin, S.: Learning when negative examples abound. In: van Someren, M., Widmer, G. (eds.) ECML 1997. LNCS, vol. 1224, pp. 146–153. Springer, Heidelberg (1997)
12. Kubat, M., Holte, R., Matwin, S.: Machine Learning for the Detection of Oil Spills in Satellite Radar Images. Machine Learning 30, 195–215 (1998)
13. Lewis, D., Catlett, J.: Heterogeneous Uncertainty Sampling for Supervised Learning. In: Proc. 11th Conf. on Machine Learning, pp. 148–156 (1994)

14. Lingras, P., Butz, C.J.: Precision and Recall in Rough Support Vector Machines. In: 2007 IEEE Int. Conf. on Granular Computing, pp. 654–658 (2007)
15. Mladenic, D., Grobelnik, M.: Feature Selection for Unbalanced Class Distribution and Naive Bayes. In: 16th Int. Conf. on Machine Learning, pp. 258–267 (1999)
16. Mladenic, D., Grobelnik, M.: Feature selection on hierarchy of web documents. Decision Support Systems 35, 45–87 (2003)
17. Van Rijsbergen, C.J.: Information retrieval, 2nd edn. Butterworths, London (1979)
18. Van Rijsbergen, C.J., Harper, D.J., Porter, M.F.: The selection of good search terms. Information Processing and Management 17, 77–91 (1981)
19. Sebastiani, F.: Machine Learning in Automated Text Categorizaton. ACM Coputing Surveys 34(1), 1–47 (2002)
20. Sousa, P., Pimentao, J., Santos, B., Moura-Pires, F.: Feture selection algorithms to improve documents classification performance. LNAI 2663, pp. 288–296 (2003)
21. Soucy, P., Mineau, G.: Feature Selection Strategies for Text Categorization. In: Xiang, Y., Chaib-draa, B. (eds.) Canadian AI 2003. LNCS (LNAI), vol. 2671, pp. 505–509. Springer, Heidelberg (2003)
22. Szmidt, E., Baldwin, J.: Assigning the parameters for Intuitionistic Fuzzy Sets. Notes on IFSs 11(6), 1–12 (2005)
23. Szmidt, E., Baldwin, J.: Intuitionistic fuzzy set functions, mass assignment theory, possibility theory and histograms. In: Proc. of 2006 IEEE World Congress on Computational Intelligence, Vancouver, Canada, pp. 234–243, Omnipress (IEEE Catalog Number: 06CH37726D; ISBN: 0-7803-9489-5) (2006)
24. Szmidt, E., Kacprzyk, J.: Remarks on some applications of intuitionistic fuzzy sets in decision making. Notes on IFS 2(3), 22–31 (1996c)
25. Szmidt, E., Kacprzyk, J.: On measuring distances between intuitionistic fuzzy sets. Notes on IFS 3(4), 1–13 (1997)
26. Szmidt, E., Kacprzyk, J.: Group Decision Making under Intuitionistic Fuzzy Preference Relations. In: IPMU 1998, Paris, La Sorbonne, pp. 172–178 (1998a)
27. Szmidt, E., Kacprzyk, J.: Applications of Intuitionistic Fuzzy Sets in Decision Making. In: EUSFLAT 1999, pp. 150–158 (1998b)
28. Szmidt, E., Kacprzyk, J.: Distances between intuitionistic fuzzy sets. Fuzzy Sets and Systems 114(3), 505–518 (2000)
29. Szmidt, E., Kacprzyk, J.: On Measures on Consensus Under Intuitionistic Fuzzy Relations. In: IPMU 2000, pp. 1454–1461 (2000)
30. Szmidt, E., Kacprzyk, J.: Entropy for intuitionistic fuzzy sets. Fuzzy Sets and Systems 118(3), 467–477 (2001)
31. Szmidt, E., Kacprzyk, J.: Analysis of Consensus under Intuitionistic Fuzzy Preferences. In: Proc. Int. Conf. in Fuzzy Logic and Technology. De Montfort Univ. Leicester, pp. 79–82 (2001)
32. Szmidt, E., Kacprzyk, J.: An Intuitionistic Fuzzy Set Based Approach to Intelligent Data Analysis (an application to medical diagnosis). In: Abraham, A., Jain, L., Kacprzyk, J. (eds.) Recent Advances in Intelligent Paradigms and and Applications, pp. 57–70. Springer, Heidelberg (2002)
33. Szmidt, E., Kacprzyk, J.: Analysis of Agreement in a Group of Experts via Distances Between Intuitionistic Fuzzy Preferences. In: Proc. 9th Int. Conf. IPMU 2002, Annecy, France, pp. 1859–1865 (2002)
34. Szmidt, E., Kacprzyk, J.,, J.: An Intuitionistic Fuzzy Set Based Approach to Intelligent Data Analysis (an application to medical diagnosis). In: Abraham, A., Jain, L., Kacprzyk, J. (eds.) Recent Advances in Intelligent Paradigms and Applications, pp. 57–70. Springer, Heidelberg (2002b)

35. Szmidt, E., Kacprzyk, J.: Evaluation of Agreement in a Group of Experts via Distances Between Intuitionistic Fuzzy Sets. In: Proc. IS 2002 – Int. IEEE Symposium: Intelligent Systems, Varna, IEEE Catalog Number 02EX499, pp. 166–170 (2002c)
36. Szmidt, E., Kacprzyk, J.: A New Concept of a Similarity Measure for Intuitionistic Fuzzy Sets and Its Use in Group Decision Making. In: Torra, V., Narukawa, Y., Miyamoto, S. (eds.) MDAI 2005. LNCS (LNAI), vol. 3558, pp. 272–282. Springer, Heidelberg (2005)
37. Szmidt, E., Kacprzyk, J.: Distances Between Intuitionistic Fuzzy Sets: Straightforward Approaches may not work. In: 3rd International IEEE Conference Intelligent Systems IS06, London, May 2006, pp. 716–721 (2006)
38. Szmidt, E., Kacprzyk, J.: An Application of Intuitionistic Fuzzy Set Similarity Measures to a Multi-criteria Decision Making Problem. In: Rutkowski, L., Tadeusiewicz, R., Zadeh, L.A., Żurada, J.M. (eds.) ICAISC 2006. LNCS (LNAI), vol. 4029, pp. 314–323. Springer, Heidelberg (2006)
39. Szmidt, E., Kacprzyk, J.: Some Problems with Entropy Measures for the Atanassov Intuitionistic Fuzzy Sets. In: Masulli, F., Mitra, S., Pasi, G. (eds.) WILF 2007. LNCS (LNAI), vol. 4578, pp. 291–297. Springer, Heidelberg (2007)
40. Szmidt, E., Kacprzyk, J.: A New Similarity Measure for Intuitionistic Fuzzy Sets: Straightforward Approaches not work. In: 2007 IEEE Conf. on Fuzzy Sytems, pp. 481–486 (2007a); IEEE Catalog Number: 07CH37904C,ISBN: 1-4244-1210-2
41. Szmidt, E., Kukier, M.: Classification of Imbalanced and Overlapping Classes using Intuitionistic Fuzzy Sets. In: 3rd International IEEE Conference on Intelligent Systems IS 2006, pp. 722–727 (2006)
42. Szmidt, E., Kukier, M.: A New Approach to Classification of Imbalanced Classes via Atanassov's Intuitionistic Fuzzy Sets. In: Wang, H.-F. (ed.) Intelligent Data Analysis: Developing New Methodologies Through Pattern Discovery and Recovery (in press)
43. Torkkola, K.: Discriminative features for text document classification. In: Int. Conf. on Pattern Recognition, Canada (2002)
44. Visa, S., Ralescu, A.: Experiments in guided class rebalance based on class structure. In: 15th Midwest Artificial Intelligence and Cognitive Science Conference, Dayton, USA, pp. 8–14 (2004)
45. Yang, Y., Pedersen, J.: A comparative study on feature selection in text categorization. In: Fisher Jr., D.H. (ed.) The 14th Int. Conf. on Machine Learning, pp. 412–420. Morgan Kaufmann, San Francisco (1997)
46. Yang, Y.: An evaluation of statistical approach to text categorization. Journal of Information retrieval 1(1/2), 67–88 (1999)
47. Zadeh, L.A.: Fuzzy sets. Information and Control 8, 338–353 (1965)
48. Zadrozny, S., Kacprzyk, J.: Computing with words for text processing: An approach to the text categorization. Information Sciences 176, 415–437 (2006)
49. Zhang, J., Mani, J.: knn approach to unbalanced data distributions: A case study involving information extraction. In: Proceedings of the ICML-2003 Workshop: Learning with Imbalanced Data Sets II, pp. 42–48 (2003)

Part III

Evolutionary Algorithms and Their Applications

Improving Evolutionary Algorithms with Scouting: High–Dimensional Problems

Konstantinos Bousmalis[1], Jeffrey O. Pfaffmann[2], and Gillian M. Hayes[3]

[1] School of Informatics, The University of Edinburgh, Edinburgh, UK
K.Bousmalis@sms.ed.ac.uk
[2] Department of Computer Science, Lafayette College, Easton, PA 18042, USA
pfaffmaj@cs.lafayette.edu
[3] Institute of Perception, Action and Behavior(IPAB), School of Informatics, The University of Edinburgh, Edinburgh, UK
gmh@inf.ed.ac.uk

Abstract. Evolutionary Algorithms (EAs) are common optimization techniques based on the concept of Darwinian evolution. During the search for the global optimum of a search space, a traditional EA will often become trapped in a local optimum. The Scouting-Inspired Evolutionary Algorithms (SEAs) are a recently–introduced family of EAs that use a cross–generational memory mechanism to overcome this problem and discover solutions of higher fitness. The merit of the SEAs has been established in previous work with a number of two and three-dimensional test cases and a variety of configurations. In this paper, we will present two approaches to using SEAs to solve high–dimensional problems. The first one involves the use of Locality Sensitive Hashing (LSH) for the repository of individuals, whereas the second approach entails the use of scouting–driven mutation at a certain rate, the Scouting Rate. We will show that an SEA significantly improves the equivalent simple EA configuration with higher–dimensional problems in an expeditious manner.

1 Introduction

1.1 Introduction to Evolutionary Algorithms

A simple Evolutionary Algorithm (EA) is an optimization technique that discovers satisfactory solutions to a given problem by evolving populations of candidate solutions over time. Natural evolution is simulated on this initial population by (a) assigning a measure of merit (fitness value) to each individual solution via a fitness function; (b) selecting a number of individuals for "parenthood" via a selection scheme; and (c) using these selected "parents" to create the next generation of solutions via a number of genetic operators, which usually include crossover (exchange of genes) and/or mutation (variation of one or more genes). This cycle of fitness assignment, selection and reproduction continues for a set number of generations or until a certain threshold of fitness has been reached.

1.2 Scouting-Inspired Evolutionary Algorithm: Previous Work

The essence of scouting is a cross-generational mechanism that considers past state space samples to optimize future sample generation for efficient state space exploration. [4, 10, 11, 14, 15] Scouting does this by estimating a current sample fitness, using a weighted-average of the k-Nearest Neighbors (k-NN) calculation of previous samples, and computing the difference from the actual fitness. This estimate-actual fitness difference is analogous to a "surprise-level", with a larger surprise generating a smaller variance in the generation of the next sample. A Scouting-Inspired Evolutionary Algorithm (SEA) relies on the repository of past samples to also direct the search for fitter solutions towards areas that are more "surprising," or areas that have been insufficiently explored. [4, 15] Thus, this behavior is an indirect solution to the problem of premature convergence, since the technique will migrate to different regions of the search space once a peak is effectively characterized in the past sample database.

SEA achieves this migration by slightly altering the original Scouting Algorithm. Initally, the fitness is estimated based on the existing knowledge of the given search space. A number (k) of the geometrically nearest neighbors to the current individual is calculated and the weighted average of their fitness value is then the estimated fitness for the individual of interest, as outlined in (1):

$$fit_{estimate} = \frac{\sum_{i \in k} w_i \times fit_i}{\sum_{i \in k} w_i} ,\qquad(1)$$

where $w_i = d_i^2$, d_i is the Euclidean distance between the query and individual i, and fit_i is the fitness of individual i. The selection of k is a design decision and set to $k = 8$ in previous and original work presented in this paper. The Euclidean distance is used because it is assumed the state space is not ill-scaled; the Mahalanobis distance could be used in these cases. This estimate is then compared to the real fitness value of the individual of interest to decide how "surprising" the area sampled by the individual is.

Definition 1. *The surprise value of an individual s_{ind} is defined as the absolute difference of the estimated and actual fitness values.*

An SEA requires a mutation operator that introduces Gaussian variation to an individual's genome. This operator has the effect of increasing standard deviation when the surprise is low, and the inverse when surprise is high. In previous work [4] it has been shown that a good mapping between surprise (s_{ind}) and standard deviation (σ) is achieved by:

$$\sigma(s_{ind}) = \sigma_{max} - (s_{ind})^\gamma \times (\sigma_{max} - \sigma_{min}) ,\ 0 < \gamma \leq 1 .\qquad(2)$$

The boundaries of the standard deviation range, σ_{min} and σ_{max}, are a design decision and depend on the problem and its dimensionality. After analysis of the nature of the surprise values in [4], $\gamma = 0.01305$ was chosen for this parameter and we shall also use it throughout this paper.

1.3 SEA and the "Curse of Dimensionality"

The current scouting database implementation stores the datapoints it experiences in an unstructured manner and the k-Nearest Neighbor (k-NN) lookup is performed by sorting all points by their distance from the query datapoint. One simply has to consider a 100-dimensional problem with 100 individuals per generation for 30,000 generations; ignoring the space requirements, the current k-NN lookup (1) for the fitness estimate used in the surprise calculation makes the technique extremely slow —unusable in practice.

The goal of the work presented here is to experiment with high–dimensional problems and show that an SEA configuration does significantly improve the performance of its equivalent EA. Taking advantage of recent work on k-NN optimization techniques, [7] we will present and use an alternative implementation of the scouting database that is designed for storing a large number of high–dimensional points and optimized for k-NN lookups. We will finally introduce the concept of *Scouting Rate* and present a few examples of how it could be useful to researchers who decide to use SEAs for their optimization problems, especially high–dimensional ones.

2 Approaches to an SEA for Higher–Dimensional Problems

The obvious approach to speeding up scouting–enhanced evolutionary algorithm configurations without trading for performance is the reimplementation of the SEA database. A number of alternative storing techniques were considered, but we focused on the state-of-the-art data storage structures for solving the k-Nearest Neighbor (k-NN) problem in high dimensions. [7] These were the Metric Trees [17], which are similar to the older Ball Trees, [13] and the Locality Sensitive Hashing (LSH). [3, 6, 9].

Another approach worth considering is using the scouting effect at a certain rate, which we call *Scouting Rate*.

2.1 Solutions to the k–Nearest Neighbor Problem

Ball Trees and Metric Trees. A Ball Tree or a Metric Tree is a binary tree where each node, a *ball*, represents a subspace of the n-dimensional Euclidean space bound by a hyper-sphere. The radius of the hyper-sphere is as large as required to contain the child nodes-balls. That means that the root node contains the entirety of the space covered by the datapoints in the structure. Sibling nodes are allowed to intersect and do not need to span the entire space. Ball and Metric Trees make k-NN search particularly simple and fast, since the datapoints are spatially organized and the search reduces to the problem of looking up the neighbors of a given query. The search is a depth-first one and significantly faster than the naïve linear search, like the one used in the initial SEA implementation. [7]

Locality Sensitive Hashing (LSH). Locality Sensitive Hashing (LSH) is not a structure for solving the exact k-NN problem, but it is instead designed and considered the best approach for queries that require the approximate nearest neighbors within a neighborhood radius and a certain probability. [7] This problem is referred to as the *(R, 1-δ)-near neighbor problem*, where R is the neighborhood radius and δ the probability with which an actual neighbor is *not* reported.

The way LSH achieves a fast solution to the *(R, 1-δ)-near neighbor problem* is by hashing geometrically-close datapoints to the same "bucket". [6] The "buckets" are aligned with a prespecified resolution along the axes. The method's speed has been proved both theoretically and in practice. [3, 9] The details of this technique are complex and beyond the scope of this paper. The interested readder is referred to the concise explanation in [2].

The LSH approach and the official implementation [2] allowed for results that satisfy the goals of this research. Thus, the underlying structure provided by the Metric and Ball Trees is not required for the current paper.

2.2 Scouting Rate

Definition 2. *Scouting Rate, denoted as p_s, is the probability with which the standard deviation used during scouting-driven mutation is altered based on the fitness of the nearest neighbors of the individual of interest.*

The rationale behind using a scouting rate is that it is possible to avoid entrapment in local optima by only using scouting-driven mutation on a small portion of the population. Moreover, it is certain that using a scouting rate significantly lowers the number of k-Nearest Neighbor calculations that have to be performed during an experiment. These calculations, and not storing the individuals themselves, is what makes all traditional k-NN techniques slower and almost unusable in higher dimensions.

In this paper, we will show that a scouting rate as low as 0.5 still allows an SEA to significantly improve the equivalent EA, while cutting the cost of k-Nearest Neighbor lookups to half.

3 Implementation

3.1 Test Cases and the TCG-2 Package

As with previous work [4, 15], the Test–Case Generating Package TCG-2 [16] will be used to create a variety of fitness landscapes in order to assess how effective SEAs are in higher–dimensional domains. Previous work and especially [15] fully outline the merits of TCG-2 and the reasons for using it when experimenting with SEAs. In short, TCG-2 is a very configurable C++ package and can create a vast array of non-linear programming (NLP) tasks with different levels of complexity, modality and difficulty for an EA. In this paper we experiment with test–cases of 10, 50 and 150 peaks, peak width of σ_{peak}=0.1 and 0.2, and for

Table 1. TCG-2 parameters for the high-dimensional experiments

Number of feasible components (m)	: 5
Search space feasibility (ρ)	: 0.5
Search space complexity (c)	: 0
Active constraints at global optimum (a)	: 0
Peak decay (α)	: 0.1
Component minimum distance (d)	: 0.01
Penalty (W)	: 10

6, 8, 10 and 15 dimensions. The rest of the required parameters for TCG-2 are fixed and shown in Table 1.

3.2 The Evolutionary Algorithm Framework

The Evolutionary Algorithm framework was implemented in C/C++ and follows the guidelines provided in the first two papers on SEAs [4, 15], with the necessary changes for the goals of this research. The EA/SEA configuration used here is the EAC/SEAC, as defined in [4]. This configuration uses Roulette Wheel selection, random deletion, single–point crossover at a rate of 0.5, and random (EAC) or scouting–driven (SEAC) mutation at a rate of 0.5. Minimum standard deviation for scouting–driven mutation σ_{min} and the fixed standard deviation for the EAC σ_{EA} are both set to $\sigma_{min} = \sigma_{EA} = 0.0107$, whereas the maximum one is $\sigma_{max} = 0.4$ for fitness landscapes with narrow peaks ($\sigma_{peak} = 0.1$), and $\sigma_{max} = 0.9$ for landscapes with wide peaks ($\sigma_{peaks} = 0.2$).

The evolutionary process loops for a set number of generations—30,000 for all experiments presented here, and the population size is 20 individuals. Finally, all experiments were run 75 times, using 25 different seeds generated with dice and three different random–number generators. The latter are provided by the GNU Scientific Library [5] and are the "Mersenne Twister" [12] and the two "ranlux" algorithms [8].

3.3 Locality Sensitive Hashing and the E²LSH Package

As explained earlier, the reimplementation of the scouting database is essential to this work and preliminary investigation showed that the Locality Sensitive Hashing (LSH) approach is the most suitable one for our needs. The official LSH implementation is the E²LSH (Exact Euclidean LSH) 0.1 package by Alexandr Andoni and Piotr Indyk. [2] It is a C/C++ package and was easily incorporated into the existing SEA project.

Running experiments with E²LSH was much faster compared to the naïve scouting database without affecting the performance of the SEA. Table 2 shows the averages over 10 identical runs of each sample problem with the naïve scouting implementation, and the ones with E²LSH. The E²LSH parameters used for these timing experiments are the ones in the column titled "Higher-Dim.

Table 2. Naïve vs. E²LSH implementation times. All experiments were run with a population size of 10 individuals.

TCG-2 problem	Naïve	E²LSH
2D, 1000gens	5.8032sec ± 0.1085sec	19.3287sec ± 0.1067sec
3D, 1000gens	7.4905sec ± 0.63sec	8.0773sec ± 0.5396sec
3D, 5000gens	1min 39.4522sec ± 7.4372sec	43.5159sec ± 2.3219sec
10D, 1000gens	46.6837sec ± 7.6753sec	15.4688sec ± 1.2717sec
20D, 1000gens	61.8342sec ± 12.3005sec	17.4135sec ± 0.2978sec
40D, 1000gens	Unable to cope (Seg. Fault)	25.1240sec ± 1.0382sec
100D, 1000gens	Unable to cope (Seg. Fault)	54.3983sec ± 0.5909sec
300D, 1000gens	Unable to cope (Seg. Fault)	51.0080sec ± 1.3616sec

Table 3. The E²LSH parameters used for the experiments presented in this paper

Parameter	Higher-Dim. TCG-2	15-Dim. TCG-2
k	20	6
m	35	7
L	595	21
w	4	4

TCG-2" in Table 3 and it is very likely that there exist other such parameters that make the SEA perform even faster for the problems presented.

The speed improvement of the scouting database with E²LSH was significant when the problem dimensionality was higher than two. The time duration for the three-dimensional example for 1000 generations was almost identical for the two techniques, but it is obvious that E²LSH deals with an increasing number of points in a much more effective manner than the naïve implementation. Moreover, the E²LSH implementation performed much better in the ten and twenty-dimensional examples, cutting down the time to more than one third. Finally and most importantly, the new database implementation performed better in the 300-dimensional example than the naïve one performed in the twenty-dimensional example.

It is important however to note that the times presented in Table 2 can be affected by the local memory resources available, the parameters for the E²LSH package, the way the search is performed —the maximum standard deviation scouting mutation uses, for example— and even the task at hand.

The package includes a parameter calculator, but it requires a sample of datapoints from the problem search space. The creators advised that the parameter calculator should be invoked after a certain number of points has been experienced, at which point scouting could begin playing a role. [1] Our hope was that we could avoid this step, given speed considerations, and we attempted to use the parameter calculator for a number of point samples to get an empirical understanding of the parameters and adapt them accordingly manually. Table 3 outlines the parameters used for the experiments presented in this paper.

4 Results and Discussion

Our goal in this paper is to show that an SEA configuration improves its equivalent EA in higher–dimensional problem domains. We therefore ran the EAC and the SEAC on TCG-2 test–cases of 6, 8, 10 and 15 dimensions. The latter were varied in the number of peaks —10, 50, 100 and 150— and their width —0.1 and 0.2. We ran these with $p_s=1$ and with $p_s=0.5$ to show that even a scouting rate of 0.5 is enough to significantly improve the equivalent EA, by performing only 50% of the k-NN queries.

During experimentation with these high–dimensional problems, it became apparent that every problem has an optimal σ_{max}. A higher σ_{max} made the SEAC perform worse when dealing with narrower peaks and higher dimensions. The reason for that is the fact that changing all genes by the same amount creates individuals at a lower resolution as the number of dimensions increase —the created individuals have a larger distance among them by default. Assuming the highest value in the range given by each σ and using the fact that the range is sixfold the standard deviation (σ), the distance among individuals as a function of σ and the problem dimensionality is given by the following:

$$d(\sigma, n) = \sqrt{\left(\frac{6 \times \sigma}{2}\right)^2 \times n} \qquad (3)$$

This makes it harder to find only a few narrow peaks in a large zero–fitness plateau when using a large σ_{max}. We therefore chose to use $\sigma_{max}=0.4$ for the TCG-2 cases with $\sigma_{peak}=0.1$ and $\sigma_{max}=0.9$ for those with $\sigma_{peak}=0.2$.

All the results obtained showed an improvement of performance when scouting–driven mutation was in use. Figures 1, 2, 3 and 4 show the performance of the SEAC with $p_s = 1$ against the equivalent EAC for only the 15–dimensional cases, due to lack of space. As one can clearly see, in every case the SEAC improves significantly after a certain generation continuing to improve up to and beyond the limit of 30,000 generations, whereas the EAC reaches a plateau much earlier, in the ca. $2,000^{th}$ generation for most cases presented here.

In the cases of 10 and 50 peaks, the slope of the SEAC curve is steeper and still increases rapidly at the end of the evolutionary process, whereas in the cases of 100 and 150 peaks, the SEAC has already reached a plateau at the same point. It is important to note that in the worst case the SEAC always performs significantly better than the worst case of the EAC and sometimes better than the average case of the latter. (see Fig. 3(b) for an example)

The results obtained also showed that only a small number of scouting–driven mutations are enough to get a population out of local optima and improve the relevant typical EA configuration, as shown in Figs. 5 and 6 for four of the problems we experimented with. Experiments with even lower scouting rate — $p_s=0.02$ and 0.2— showed that even when scouting–driven mutation is applied on only 2% of the population, the SEAC still exhibited a significant improvement over the EAC in most cases. We would generally suggest the use of a low scouting rate if speed is important, and a high rate if the highest possible fitness level is

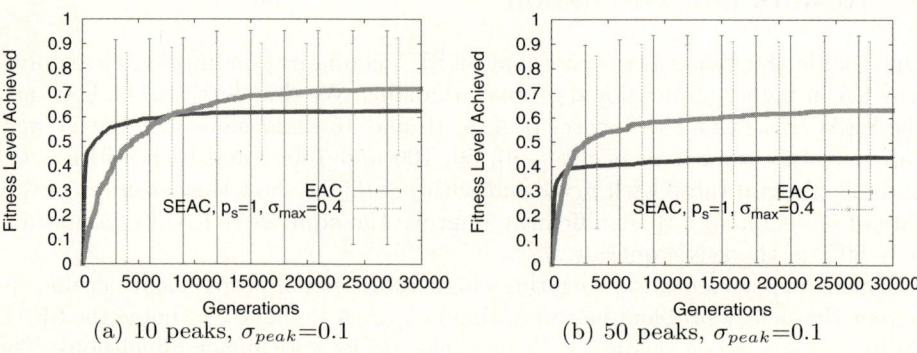

Fig. 1. Fitness level achieved per generation by EAC vs SEAC, for the 15–dimensional test cases with $\sigma_{peak} = 0.1$ for 10 and 50 peaks

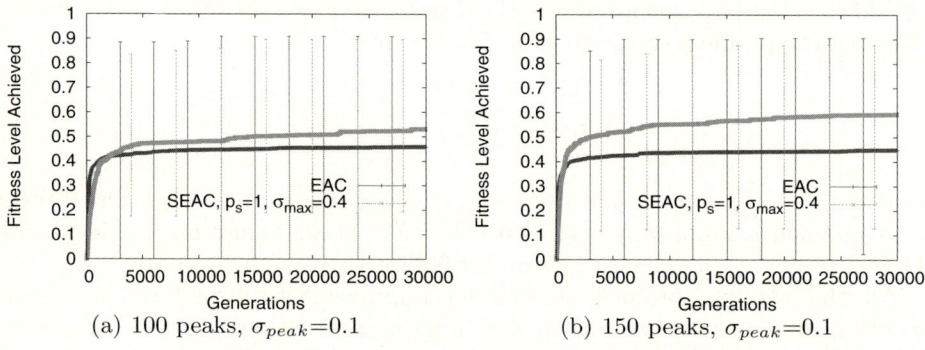

Fig. 2. Fitness level achieved per generation by EAC vs SEAC, for the 15–dimensional test cases with $\sigma_{peak} = 0.1$ for 100 and 150 peaks

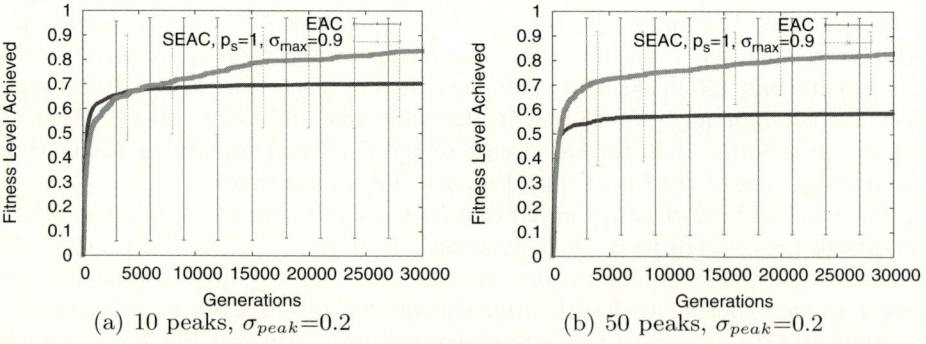

Fig. 3. Fitness level achieved per generation by EAC vs SEAC, for the 15–dimensional test cases with $\sigma_{peak} = 0.2$ for 10 and 50 peaks

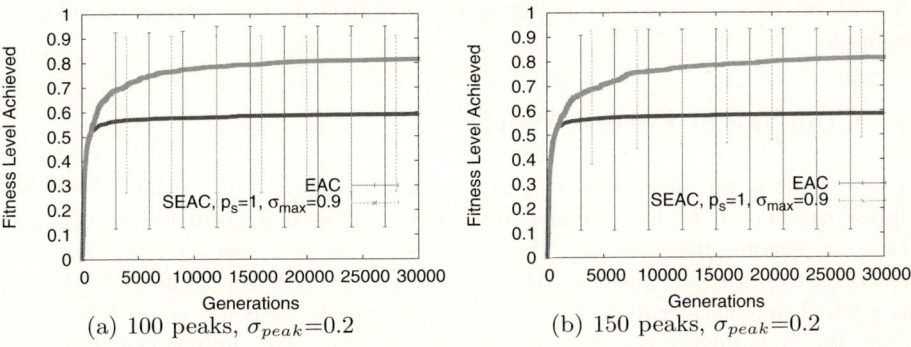

Fig. 4. Fitness level achieved per generation by EAC vs SEAC, for the 15–dimensional test cases with $\sigma_{peak} = 0.2$ for 100 and 150 peaks

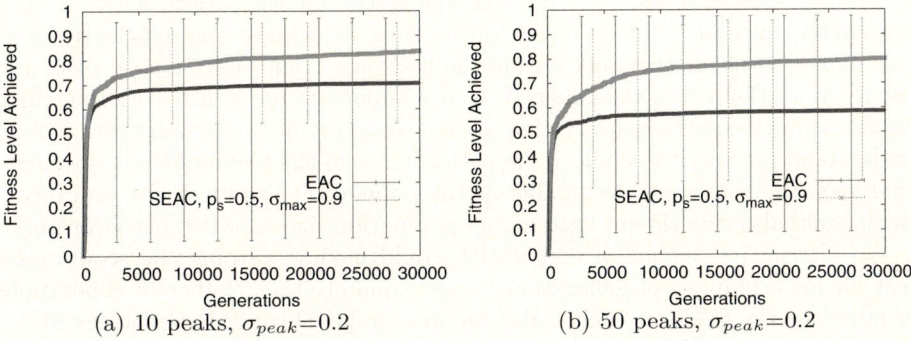

Fig. 5. Fitness level achieved per generation by EAC vs SEAC with $p_s = 0.5$, for the 15–dimensional test cases of $\sigma_{peak} = 0.2$ with 10 and 50 peaks

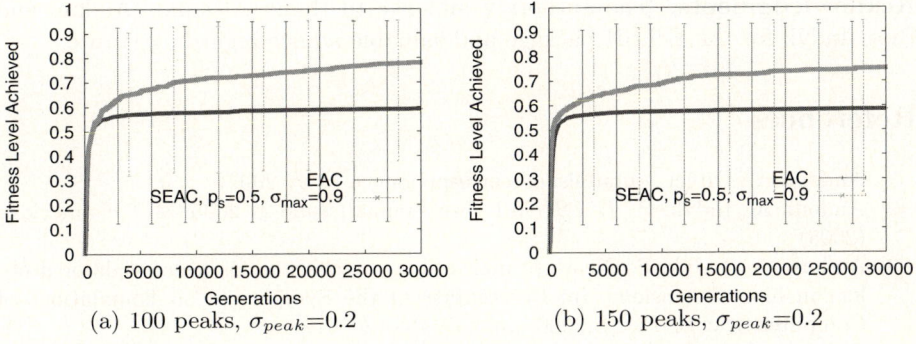

Fig. 6. Fitness level achieved per generation by EAC vs SEAC with $p_s = 0.5$, for the 15–dimensional test cases of $\sigma_{peak} = 0.2$ with 100 and 150 peaks

crucial, or if fewer passes are required, due to the cost of the fitness function for example.

5 Conclusion-Future Work

In conclusion, we have showed that an SEA performs better than the EA it improves even in higher–dimensional cases, provided a good choice of σ_{max}. This has become possible with the implementation of a new scouting database, based on the Locality Sensitive Hashing (LSH) and the use of the E²LSH package, which has accelerated the SEA by dramatically reducing the k-NN lookup time. We have introduced the concept of a scouting rate and showed that using scouting–driven mutation on only a small percentage of the individuals is sufficient to push the entire population away from local optima and towards other, fitter solutions.

The obvious next step in this research is the incorporation of the E²LSH parameter calculator. One suggestion on how to achieve this efficiently is to calculate the parameters only during the first pass of an SEA on a problem and use them for the rest of the passes. It is also important for future work to include experimentation with a variety of σ_{max} and scouting rate values, in order to fully understand the way these affect the performance of the technique. We speculate that scouting rate could be smaller as the population size increases, and future work could also experiment with higher population sizes and deduce accordingly.

An alternative avenue of exploration could include turning the scouting effect off for a number of generations, approximately the number of generations required by the EAC to reach a plateau on average. During that time the SEAC would operate with $p_s = 0$ and $\sigma_{min} = \sigma_{max}$ and attempt to populate the experience database quickly so that the SEAC would have more information and therefore be more effective when scouting does gets enabled.

Acknowledgments. The authors would like to thank Alexandr Andoni and Piotr Indyk for the E²LSH package and valuable advice regarding its use.

References

[1] Andoni, A.: Direct, unpublished correspondence (July 2007)
[2] Andoni, A., Indyk, P.: E^2LSH 0.1 User Manual, June 21 2005. MIT, Cambridge (2005)
[3] Andoni, A., Indyk, P.: Near-optimal hashing algorithms for near neighbor problem in high dimensions. In: Proceedings of the Symposium on Foundations of Computer Science–FOCS 2006, pp. 459–468 (2006)
[4] Bousmalis, K., Hayes, G.M., Pfaffmann, J.O.: Improving evolutionary algorithms with scouting. In: Main Proceedings of the 13th Portuguese Conference on Artificial Intelligence—EPIA 2007, December 3-7 2007. LNCS, vol. 4874. Springer, Heidelberg (2007)
[5] Galassi, M., Davies, J., Theiler, J., Gough, B., Jungman, G., Booth, M., Rossi, F.: GNU Scientific Library Reference Manual. Network Theory Ltd., Bristol (2003)

[6] Gionis, A., Indyk, P., Motwani, R.: Similarity search in high dimensions via hashing. In: Proceedings of the 25th International Conference on Very Large Data Bases, September 07-10 1999, pp. 518–529 (1999)

[7] Liu, T., Moore, W., Gray, A., Yang, K.: An investigation of practical approximate nearest neighbor algorithms. In: Proceedings of Neural Information Processing Systems, pp. 825–832 (2004)

[8] Lüscher, M.: A portable high-quality random number generator for lattice field theory calculations. In: Computer Physics Communications, vol. 79, pp. 1000–1110 (1994)

[9] Datar, M., Indyk, P., Immorlica, N., Mirrokni, V.: Locality-sensitive hashing scheme based on p-stable distributions. In: Proceedings of the Symposium on Computational Geometry (2004)

[10] Matsumaru, N., Colombano, S., Zauner, K.P.: Scouting enzyme behavior. In: Fogel, D.B., El-Sharkawi, M.A., Yao, X., Greenwood, G., Iba, H., Marrow, P., Shackleton, M. (eds.) 2002 World Congress on Computational Intelligence, Honolulu, Hawaii, 12-17 May 2002, pp. 19–24. IEEE, Piscataway (2002)

[11] Matsumaru, N., Centler, F., Zuner, K.P., Dittrich, P.: Self-adaptive-scouting autonomous experimentation for systems biology. In: Raidl, G.R., Cagnoni, S., Branke, J., Corne, D.W., Drechsler, R., Jin, Y., Johnson, C.G., Machado, P., Marchiori, E., Rothlauf, F., Smith, G.D., Squillero, G. (eds.) EvoWorkshops 2004. LNCS, vol. 3005, pp. 52–62. Springer, Heidelberg (2004)

[12] Matsumoto, M., Nishimura, T.: Mersenne twister: A 6234-dimensionally equidistributed uniform pseudo-random number generator. In: ACM Transactions on Modeling and Computer Simulation, vol. 8, pp. 3–30 (1998)

[13] Omohundro, S.M.: Five balltree construction algorithms, November 20 1989. International Computer Science Institute, Berkeley (1989)

[14] Pfaffmann, J.O., Zauner, K.P.: Scouting context-sensitive components. In: Keymeulen, D., Stoica, A., Lohn, J., Zebulum, R.S. (eds.) The Third NASA/DoD Workshop on Evolvable Hardware-EH 2001, 12-14 July 2001, pp. 14–20. IEEE Computer Society, Los Alamitos (2001)

[15] Pfaffmann, J.O., Bousmalis, K., Colombano, S.: A scouting-inspired evolutionary algorithm. In: Proceedings of the 2004 Congress on Evolutionary Computation-CEC 2004, Portland, OR, June 16-19, 2004, vol. 2, pp. 1706–1712 (2004)

[16] Schmidt, M., Michalewicz, Z.: Test-case generator tcg-2 for nonlinear parameter optimization. In: Deb, K., Rudolph, G., Yao, X., Lutton, E., Guervos, J.J.M., Schwefel, H.-P. (eds.) PPSN 2000. LNCS, vol. 1917, pp. 539–548. Springer, Heidelberg (2000)

[17] Uhlmann, J.K.: Satisfying general proximity/similarity queries with metric trees. In: Information Processing Letters, number 40, pp. 175–179 (November 1991)

Memetic Algorithm Based on a Constraint Satisfaction Technique for VRPTW

Marco A. Cruz-Chávez[1], Ocotlán Díaz-Parra[1],
David Juárez-Romero[1], and Martín G. Martínez-Rangel[2]

[1] CIICAp
[2] FCAeI, Autonomous University of Morelos State*
Av. Universidad 1001. Col. Chamilpa, C.P. 62110. Cuernavaca, Morelos, Mexico
{mcruz,odiazp,djuarez,mmtzr}@uaem.mx

Abstract. In this paper a Memetic Algorithm (MA) is proposed for solving the Vehicles Routing Problem with Time Windows (VRPTW) multi-objective, using a constraint satisfaction heuristic that allows pruning of the search space to direct a search towards good solutions. An evolutionary heuristic is applied in order to establish the crossover and mutation between sub-routes. The results of MA demonstrate that the use of Constraints Satisfaction Technique permits MA to work more efficiently in the VRPTW.

1 Introduction

One of the first forerunners of genetic algorithms was John Holland in 1960 [1][2].The mere structure of a GA (Genetic Algorithm) involves three types of operators: Selection, crossover and mutation[3][4].

By definition, the search carried out by a GA in the solution space of a problem is global (exploration in the search space). When global search is combined with local search (exploitation in the search space), a genetic hybrid algorithm, called a MA (Memetic Algorithm), is formed [5] This MA, because of the new characteristics contributed by the local search, is able to find better solutions than a simple GA because for each solution S obtained by the global search, the algorithm searches the neighborhood of S for the local optimum, which could turn out to be the global optimum. Researchers suggest that involving the technique of local search in GA, allows for results nearer to the global optimum to be found in combinatorial optimization problems [6][7].

In this paper a Memetic Algorithm is proposed called GA-PCP, which combines two techniques of search, local and global. For the local search, the algorithm used was one of constraint satisfaction PCP (Precedence Constraint Posting) proposed by Cheng and Smith [8]. For the global search, the simple crossover of a GA was used.

In order to prove the efficiency of the proposed algorithm, GA-PCP was applied to the Vehicles Routing Problem well-known like VRPTW (Vehicles Routing Problem with Time Windows) which is an NP-complete problem [9] [10]. The

---

* This work was supported by project 160 of the Fideicomiso SEP-UNAM, 2006-2007.

VRPTW [11] [12] [13] [14] [15] [16] is a variant of the VRP with the additional restriction of the time window associated with each client. This window defines an interval within which the client has to be assisted. The objective is to reduce the number of the vehicles, the route time sum, and the necessary wait time to provide all clients the times of attention required.

Very little research of MA exists as applied to VRPTW. In [17], a Memetic is proposed using a GA for a constraint satisfaction model of VRPTW with rescheduling and optimization of Pareto. The algorithm includes three local searches, Route-exchange, mutation and lambda-exchange. In [18] a Memetic is proposed that combines TS (Tabu Search) and GA. TS is used for its excellent local search execution capacity which allows for exploitation of the solutions space, while GA is able to diversify these local searches, allowing for the exploration of several regions in the search space. In [19], a Memetic multi-objective is proposed that incorporates three heuristics of local exploration. The first heuristic, Intra_Route, generates two different numbers based on the sequence size of the route assignment of both vehicles. This heuristic chooses two routes randomly and exchanges two nodes of each route. The second heuristic, Lambda_Interchange, assumes that two routes A and B are selected, and begins by sweeping the nodes of route A and moving the feasible nodes into B route. The third heuristic, Shortest_pf, is a modification of the shortest path first method, which tries to change the order of the nodes of a particular route and uses the optimization concept of Pareto to solve multi-objective optimization in VRPTW.

In this paper, in order to apply PCP to VRPTW, the problem was treated as a CSP (Constraint Satisfaction Problem). The constraint satisfaction works with problems that have finite domains like VRPTW, which is a discreet optimization problem. A solution to a CSP is an assignment of values to all the variables such that all restrictions of the CSP are satisfied. The most common techniques in CSP management can be organized in three groups: **systematic search techniques, inference techniques and hybrid techniques** [20].

In this work, GA-PCP uses the **hybrid search constraint satisfaction technique** for a CSP using the PCP look-ahead algorithm.

The PCP local search algorithm involves the calculation of the shortest path, partially and globally, between a pair of nodes and among all the nodes respectively, in the graph that represents the VRPTW model [21]. PCP is applied specifically to disjunctive graphs models. PCP fixes the address of each edge based on the execution of certain rules and converts the disjunctive graph into a digraph. The shortest path of the digraph represents a feasible solution to VRPTW. The representation of results that is obtained by PCP is coupled with the model of the VRPTW, which is modeled by means of a digraph in order to represent the routes, clients, demand for the client and times of attention required by the client (time window). PCP carries out a series of transformations in order to establish the address of edges in a graph, the set of transformations that is carried out to change an edge is small, since every time that it returns only a small change the address of an edge is made. PCP behaves similarly to the ramification of a tree and a bounded solution space, which carries out a local search, where each transformation is considered near. These transformations are called local transformations and the method is known as local search [22].

The result obtained in this research is that the combination of PCP with GA applied to VRPTW improves the results for several benchmarks, depending on the percentage of PCP applied to the population used in GA.

The structure of the paper is as follows; section one is the introduction, section two explains the procedure of the proposed algorithm GA-PCP for the Vehicles Routing Problem with Time Windows, section three shows the experimentation and comparison of results generated by the GA-PCP algorithm compared with the results obtained by others GA that use constraints satisfaction techniques for the Solomon benchmarks, section four presents conclusions.

2 GA-PCP Algorithm for VRPTW

Figure 1 is a general outline of the proposed algorithm called GA-PCP (Genetic Algorithm with Precedence Constraint Posting) for VRPTW, the algorithm consists of the following general steps:

Step 1. Creation of the initial population comprised of individuals with route information and individuals with Time windows information.

Step 2. Apply the tournament selection to the initial population.

Step 3. Apply crossover k

Step 4. Apply flip bit mutation.

Step 5. Construction of the following generation. With migrant quality individuals (with PCP) and individuals of the original population.

Step 6. Evaluate the fitness with objective function that is shown in equation (1), for the case of the VRPTW problem, two primordial objectives are used: the demand and attention time to each client, trying to minimize the cost implied by these two objectives.

$$\min \sum_{k \in k} \sum_{(ij) \in A} c_{ij} X_{ijk} \qquad (1)$$

In equation (1), c represents the cost of transporting of an i origin to a j destination, X represents the journey of an i origin to a j destination in a k vehicle. In order to complete the cost objective, the minimum number of vehicles assigned to each journey is searched for, while fulfilling the capacity constraint of the vehicle and the time window. For the journey, the attempt is to find the shortest distance.

Step 7. Verify whether the stop criterion is satisfied. The stop criterion is set based on the execution time, the global optimum and the generation number. If some of these criterions are satisfied, the GA-PCP execution is finished.

In order to create the next generation of the population, a certain percentage x of the population generated by the genetic phase of the GA is taken, and a certain percentage y of another migrant population generated with the PCP is taken [21]. The sum of the (x, y) percentage is 100 of the new generation to be evaluated. The creation procedure of the next generation of the population (x, y) is shown in Fig.2.

There are different types of genetic operators applied in the procedure of generation of x population for the genetic phase. One is the tournament selection method operator [13]. The crossover [13] consists of finding two points randomly in a first individual and looking for the corresponding genes to make the crossover in a second individual. This guarantees the fulfillment of one of the restrictions of the VRPTW which is not passing twice through the same node. For the mutation [13], the Flip-Bit method is used which consists of taking two genes (gen1, gen2) randomly from the same individual, with gen1 being different than gen2, and proceeding to exchange the places of gen1 and gen2.

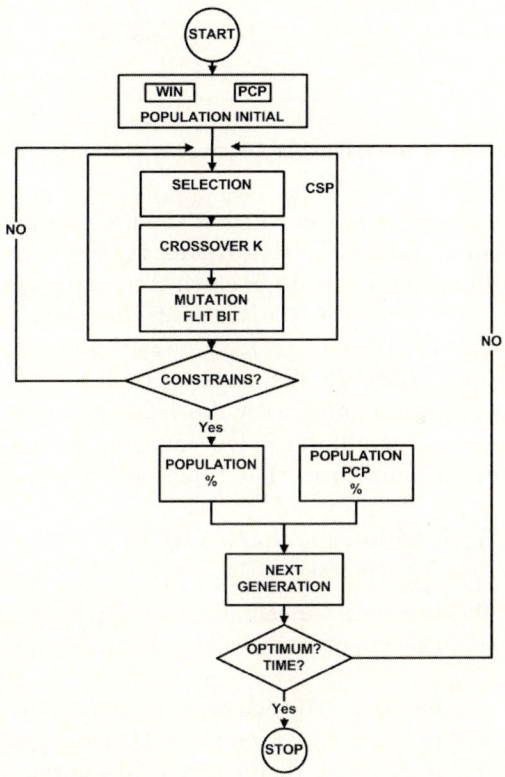

Fig. 1. Flow Diagram of GA-PCP algorithm

Within the search procedure, PCP builds the solution through Depth First using partial assignments of Ω (the set of pair of nodes i, j of the VRPTW disjunctive graph). The PCP algorithm carries out a pruning of the search space early on and provides a heuristic for the assignment of values of the $Ordering_{ij}$ variables.

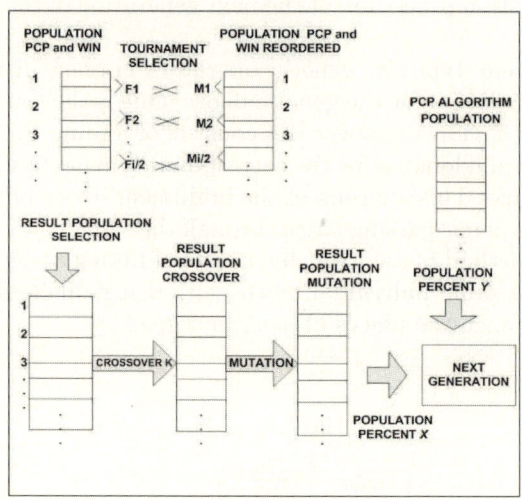

Fig. 2. Realization of the following generation of individuals in GA-PCP

PCP consists of a series of cases in which it should be true that if the shortest path sp between a pair of nodes (i,j) that represent the $Ordering_{ij}$ variable then it has a value that fulfills some of the PCP cases. According to the result obtained upon evaluating the shortest path, the value of $Ordering_{ij}$ is designated. The evaluation of sp is calculated from i to j (sp_{ij}) and from j to i (sp_{ji}).

The PCP algorithm applies the disjunctive graph model of VRPTW. PCP obtains a digraph as a result, and with the help of a greedy algorithm, the number of optimum routes is obtained that satisfies the capacity restriction of each vehicle used in optimum form that represents a feasible solution to the problem.

The PCP-VRPTW algorithm applied to VRPTW, combined with the search procedure PCP for the CSP consists of the following four steps:

Step 1.- Find the shortest path for each unordered pair of nodes sp_{ij} and sp_{ji}.

Step 2.- Classify the decision of ordination of the pairs not ordered with four cases

Case 1. If $sp_{ij} >= 0$ and $sp_{ji} < 0$ then $O_i \prec O_j$ should be selected.
Case 2. If $sp_{ji} >= 0$ and $sp_{ij} < 0$, then $O_j \prec O_i$ should be selected.
Case 3. If $sp_{ji} < 0$ and $sp_{ij} < 0$, then the partial solution is inconsistent.
Case 4. If $sp_{ji} >= 0$ and $sp_{ij} >= 0$, then no relationship of order is possible

Step 3.- Existence of cases
Does either case 1 or case 2 exist?
If one exists, go to step 4
If neither exists, go to step 1
Step 4.- Fix new precedence for unordered pairs.

The polynomial that defines the complexity in time of the proposed PCP-VRPTW algorithm for the VRPTW as a CSP is $T(n) = c_1 n^3 + c_2 n^2 + c_3 n + c_4$.

The complexity of the proposed algorithm is $O(n^3)$, where n is the number of (nodes) clients in the problem.

In order to better understand the algorithm, an example is shown of a small instance of five nodes and a vehicle with a capacity of 200 packages. The disjunctive graph model that is obtained is presented in Fig. 3.

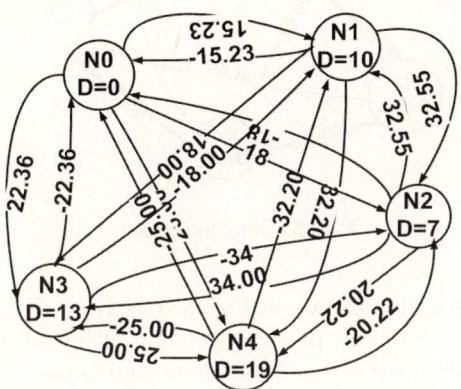

Fig. 3. Disjunctive graph

Applying the shorter path algorithm between pairs of nodes and evaluating the PCP cases, the graph shown in Fig. 4 is obtained. The resulting graph does not generate a feasible solution, this means that a route from the initial node to the final node does not exist through which each node is passed only once.

Because the resulting graph does not generate a solution, backtracking is applied in the nodes with an enter zero and exit zero, leaving fixed the nodes that have at least one entrance and one exit. The PCP algorithm is applied in order to find a route, if a feasible solution is generated, it is taken as a solution.

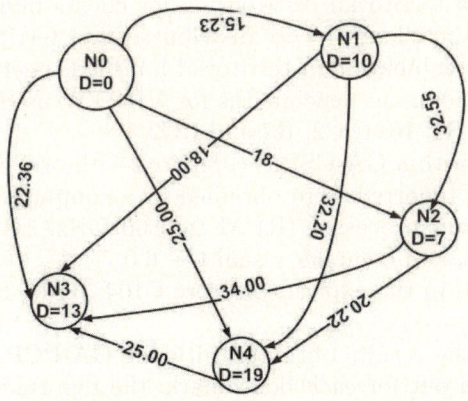

Fig. 4. Conjunctive graph

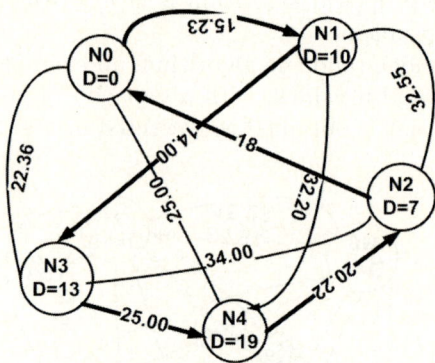

Fig. 5. Solution graph

A solution of the problem is shown in Fig. 5. Lastly, a greedy algorithm is used which divides the shortest path presented in Fig. 4 into a set of routes in order to satisfy the demand constraints of the client and capacity of the vehicles.

There are four criteria for stopping the algorithm, it is carried out: (1) for an established amount of time, (2) until the global optimum is found, (3) until within a certain range of an optimum is reached, or (4) for a certain number of generations.

3 Experimental Results

The VRPTW problems used in the experiment are taken from the Solomon benchmarks [9]. The instances for VRPTW are classified by type and by class. Two types of instances exist; type 1 manages narrow windows of time and small vehicle capacity, type 2 manages large windows of time and large vehicle capacity. Three classifications exist, C, R and RC. The C classification includes the instances that have a territorial distribution for clients bunched together. The R classification has the clients evenly distributed in a territorial area. The RC classification is the combination of territorial bunched together and distributed distribution is. The Solomon benchmarks for VRPTW used in this experimentation are types C1, R1, RC1, C2, R2 and RC2.

The proposed algorithm GA-CSP is compared with others GA that use CSP. The results that are reported were obtained in a computer with the following characteristic: Pentium processor (R) M to 1.60 GHz, 1GB RAM, operating system XP Windows, and compiler visual C+ 6.0.

The instances used in the experiment were C104, R104, RC108, C204, R208, RC208, for 25 nodes.

Table 1 presents the results obtained with the GA-PCP algorithm. Ten executions were carried out for each benchmark; the reported results include the results of the executions, the best, and medium values as well as the standard deviation. The time of each execution was one hour. Table 1 shows that for the

Table 1. Results of the GA-PCP algorithm

Benchmark	V	Best	Average	σ	Op*	V*	RE
C104-25	2	186.9	189.8	4.05	1869	3	0
R104-25	3	436.3	467.8	21.48	416.9	4	4.60
RC108-25	5	294.7	300.4	15.03	294.5	3	0.08
C204-25	2	286.0	290.9	4.51	213.1	1	34.25
R208-25	2	329.1	332.0	3.01	328.2	1	0.30
RC208-25	2	271.8	285.9	10.17	269.1	2	1.01

problems of 25 nodes, the best result is near the global optimum, for C104-25 and RC108-25 it was reached with regard to the distance, but for C204-25 the relative error was large. The results for 25 nodes show that GA-PCP works acceptably if the problem is R and/or C classification, when the clients are evenly distributed or bunched together respectively in a territorial area, and when the time windows are narrow and vehicles with a small capacity (type 1) are used, see the results for C104-25, R104-25 and RC108-25. When the problem is of R and C classification but has a big time window and the vehicles have a large capacity (type 2), the results begin decrease in quality, see the result for R208-25 and RC208-25. When the problem only has the C property but has a large time window and the vehicles have a large capacity (type 2), the results are very poor, see the result for C204-25.

The global optimum in RC108-25 is obtained with $y = 60\%$ for the population PCP of individuals in the GA-PCP. The experiments showing with when y increases from 0 to 60%, GA-PCP tends to improve the solution of RC108-25, also when y increases from 60 to 100%, GA-PCP tends to worsen the solution of RC108-25. For C204-25, with $y = 95\%$ is obtained a value near the global optimum. The experiments showing with when y increases from 0 to 100%, GA-PCP tends to improve the solution of RC204-25.

According to these the results for each instance proven in this paper, the appropriate percentage of the PCP population required in order to improve the efficiency of GA-PCP will be different and will need tuning according to the properties and type of the problem.

The following is a comparison of the results of the GA-PCP algorithm with other genetic algorithms that uses the constraints satisfaction techniques. The heuristics laboratory [23] implemented the GA used for the comparison. These comparison algorithms are the GGA (Generic Genetic Algorithm) [24], SSGA (Steady-State Genetic Algorithm) [25] and SXGA (Sexual Genetic Algorithm) [26]. The GGA algorithm uses the constraints satisfaction technique of systematic search, the SSGA algorithm uses the constraints satisfaction technique of. The SXGA algorithm uses the constraints satisfaction technique of systematic search. These genetic algorithms of the heuristics laboratory report their best results using the following tuning of their entry variables: overload penalty = 50.00, Tardiness penalty = 20.00, Route time penalty = 0.05, Travel time excess penalty = 50.00, Distance penalty = 1.00. The selection operator was tournament. The generations number and population size is the same as used for

Table 2. Comparative results of efficiency of GA-CSP vs. other algorithms that apply constraints satisfaction technique

	GA-PCP		GGA		SSGA		SXGA		
Benchmark	UB	t, sec	UB	t, sec	UB	t, sec	UB	t, sec	Op
C104-25	186.9	39.2	190.6	73.8	188.8	53.4	190.6	75.0	186.9
R104-25	436.3	45.8	417.9	75.0	417.6	0.7	418.0	75.6	416.9
RC108-25	294.7	44.9	295.4	78.0	294.9	0.6	295.4	78.0	294.5
C204-25	286.0	44.9	223.3	76.8	223.3	0.4	223.3	76.8	213.1
R208-25	329.1	43.1	329.3	75.6	329.3	0.5	329.3	76.2	328.2
RC208-25	271.8	49.9	271.6	82.8	269.5	0.6	272.0	81.6	269.1

Table 3. Comparative results of the efficacy of GA-CSP with other algorithms that apply the costraints satisfaction technique part 1

	Algorithm					Algorithm			
Results	GA-PCP	GGA	SSGA	SXGA	Results	GA-PCP	GGA	SSGA	SXGA
Problem	C104, OPTIMUM=186.9				Problem	C204, OPTIMUM=213.1			
Best*	186.9	190.6	188.8	190.6	Best*	286.0	223.3	223.3	223.3
Worst	228.2	224.4	201.0	195.6	Worst	317.6	223.4	224.7	223.4
Average	189.8	197.8	193.9	192.2	Average	290.9	223.3	223.7	223.3
σ	4.05	14.03	3.93	2.05	σ	4.51	0.05	0.48	0.04
RE*	0.00	1.98	1.02	1.98	RE*	34.25	4.78	4.79	4.78
Problem	R104, OPTIMUM=416.9				Problem	R208, OPTIMUM=328.2			
Best*	436.3	417.9	417.6	418.0	Best*	329.1	329.3	329.3	329.3
Worst	521.0	423.5	436.2	423.5	Worst	490.2	329.3	333.8	331.3
Average	467.8	419.1	422.9	418.5	Average	332.0	329.3	332.9	329.5
σ	21.48	2.32	5.72	1.74	σ	3.01	0.00	1.56	0.6198
RE*	4.60	0.24	0.17	0.25	RE*	0.30	0.34	0.34	0.34
Problem	RC108, OPTIMUM=294.5				Problem	RC208, OPTIMUM=269.1			
Best*	294.7	295.4	294.9	295.4	Best*	271.8	271.6	269.6	272.0
Worst	452.6	295.0	295.7	295.8	Worst	494.4	277.1	288.4	277.1
Average	300.4	295.4	295.7	295.5	Average	285.9	274.8	277.7	274.4
σ	15.03	0.00	0.74	0.11	σ	10.17	2.73	6.26	2.25
RE*	0.08	0.31	0.14	.31	RE*	34.25	4.78	4.79	4.78

GA-PCP, 1000 and 100 respectively. With this tuning the GGA algorithms, SSGA and SXGA were executed, giving the results presented in Table 2 and Table 3.

The tuning percentage of PCP per population depends on the problem. For the instance C104, $x = 0.1$ and $y = 0.9$. For instances R104 and C204, $x = 0.05$ and $y = 0.95$. For instances RC108, R208 and RC208, $x = 0.2$ and $y = 0.8$.

Table 3 presents the times that correspond to the time of the best solution obtained in 10 tests executed by each algorithm in each instance of VRPTW. Table 3 shows that the efficiency of GA-PCP is better than GGA and SXGA because it obtains better results with regard to the tuning of the entry parameters for each algorithm. It is observed that SSGA is better in efficacy because the times of execution are the shortest.

Table 3 presents results of 10 tests executed by each algorithm in each problem. It shows the best and worst results, the average value, the standard deviation, and the relative error. These results demonstrate that GA-PCP is competitive with these three algorithms that also use the constraints satisfaction technique. One could observe that the proposed algorithm obtains the best results in three of the six problems, that is, for 50% of the revised benchmarks.

4 Conclusions

The results reported in this research indicate that using the PCP local search algorithm in GA improves the results in VRPTW only for problems of type 1 (small window and small vehicle capacity) with C and RC classification.

The initial population is formed of feasible individuals; a randomly selected population is not used. Instead, the initial population is selected in such a way that it consists of feasible individuals that contain route information and time information. For the next generations, a certain percentage of population of PCP is worked with in order to form the total population of the following generation. It was proven that when the population is formed in great part by PCP individuals, the generated results are near the global optimum for problems of type 1. When problems of type 2 are used, it is better not to use PCP in GA.

It is demonstrated that for the revised benchmarks, the GA-PCP proposed algorithm is competitive in efficiency and efficacy to comparison algorithms used in this investigation that also apply the constraints satisfaction technique. The GA-PCP obtains the best results in 50% of the problems with competitive times of execution.

It can be seen through the results of this experiment that applying a greater percentage of PCP population improves the result of the solution of the GA.

References

1. Mitchell, M.: An Introduction to Genetic Algorithms. Massachusetts Institute of Technology Press, London (1999)
2. Holland, J.: Adaptation in Natural and Artificial Systems. The University of Michigan (1975)
3. Alvarenga, G.B., Mateus, G.R., De Tomi, G.: A genetic and set partition two-phase approach for the vehicle routing problem with time Windows. Computers & Operations Research 34(6), 1561–1584 (2007)
4. Goldberg, D.E.: Genetic Algorithms in Search, Optimization, and Machine Learning. Addison Wesley Professional, Reading (1989)
5. Krasnogor, N., Smith, J.: MAFRA a Java Memetic Algorithm Framework. Intelligent Computer System Centre University of the west of England Bristol, United Kingdom (2000)
6. Tavakkoli-Moghaddam, R., Saremi, A.R., Ziaee, M.S.: A memetic algorithm for a vehicle routing problem with backhauls. Applied Mathematics and Computation 181, 1049–1060 (2006)

7. Moscato, P.: On Evolution, Search, Optimization, Genetic Algorithms and Martial Arts: Towards Memetic Algorithms. Technical Report Caltech Concurrent Computation Program, Report. 826, California Institute of technology,Pasadena, California, USA (1989)
8. Cheng-Chung, C., Smith, S.F.: A Constraint Satisfaction Approach to Makespan Scheduling. In: Proceedings of the Third International Conference on Artificial Intelligence Planning Systems, Edinburgh, Scotland, pp. 45–52 (1996) ISBN 0-929280-97-0
9. Solomon, M.M.: Algorithms for vehicle routing and scheduling problems with time window constraints. Operations Research 35(2) (1987)
10. Garey, M.R., Johnson, D.S.: Computers and intractability, A Guide to the theory of NP-Completeness. W.H. Freeman and Company, New York (2003)
11. Toth, P., Vigo, D.: The Vehicle Routing Problem. In: Monographs on Discrete Mathematics and Applications, SIAM, Philadelphia (2001)
12. Thangiah, S.R.: Vehicle Routing with Time Windows using Genetic Algorithms. In: Chambers, L. (ed.) Application Handbook of Genetic Algorithms: New Frontiers, vol. 2, pp. 253–277. CRC Press, Boca Raton (1995)
13. Tan, K.C., Lee, L.H., Zhu, Q.L., Ou, K.: Heuristics methods for vehicle routing problem with time windows. In: Artificial Intelligence in Engineering, pp. 281–295. Elsevier, Amsterdam (2001)
14. Zhu, K.Q.: A new Algorithm for VRPTW. In: Proceedings of the International Conference on Artificial Intelligence ICAI 2000, Las Vegas. USA (2000)
15. Prins, C.: A simple and effective evolutionary algorithm for the vehicle routing problem. Computers & Operations Research 31(12) (2004) 1985-2004
16. Tan, K.C., Lee, L.H., Ou, K.: Artificial intelligence heuristics in solving vehicle routing problems with time windows constraints. Engineering Applications of Artificial Intelligence 14(6), 825–837 (2001)
17. Rhalibi, E.A., Kelleher, G.: An approach to dynamic vehicle routing, rescheduling and disruption metrics. IEEE International Conference on Systems, Man and Cybernetics 4, 3613–3618 (2003)
18. Chin, A., Kit, H., Lim, A.: A new GA approach for the vehicle routing problem. In: Proceedings 11th IEEE International Conference on Tools with Artificial Intelligence, pp. 307–310 (1999)
19. Tan, K.C., Lee, T.H., Chew, Y.H., Lee, L.H.: IEEE International Conference on Systems, Man and Cybernetics, vol. 1, pp. 1361–1366 (2003)
20. Castillo, L., Borrajo, D., Salido, M.A.: Planning, Scheduling and Constraint Satisfaction: From Theory to Practice (Frontiers in Artificial Intelligence and Applications), IOS Press, ISBN-10: 1586034847, ISBN-13: 978-1586034849. Spain (2005)
21. Cruz-Chávez, M.A., Díaz-Parra, O., Hernández, J.A., Zavala-Díaz, J.C., Martínez-Rangel, M.G.: Search Algorithm for the Constraint Satisfaction Problem of VRPTW. In: Proceeding of CERMA 2007, September 25-28, pp. 336–341. IEEE Computer Society, Los Alamitos (2007)
22. Aho, A.V., Hopcroft, J.E., Ullman, J.D.: Structure of data and algorithms. Adisson-Wesley Iberoamericana, Nueva Jersey, Nueva York, California, U.S.A (1988) (Spanish)
23. Wagner, S., Affenzeller, M.: The HeuristicLab Optimization Environment, Technical Report. Institute of Formal Models and Verification, Johannes Kepler University Linz, Austria (2004)
24. Affenzeller, M.: A Generic Evolutionary Computation Approach Based Upon Genetic Algorithms and Evolution Strategies. Journal of Systems Science 28(2), 59–72 (2002)

25. Chafekar, D., Xuan, J., Rasheed, K.: Constrained Multi-objective Optimization Using Steady State Genetic Algorithms, Computer Science Departament University of Georgia. In: Athens, Genetic and Evolutionary Computation Conference, GA 30602, USA (2003)
26. Wagner, S., Affenzeller, M.: SexualGA: Gender-Specifc Selection for Genetic Algorithms. In: Proceedings of the 9th World Multi-Conference on Systemics, Cybernetics and Informatics (WMSCI 2005), vol. 4, pp. 76–81 (2005)

Agent-Based Co-Operative Co-Evolutionary Algorithm for Multi-Objective Optimization

Rafał Dreżewski and Leszek Siwik

Department of Computer Science
AGH University of Science and Technology, Kraków, Poland
{drezew,siwik}@agh.edu.pl

Abstract. Co-evolutionary algorithms are a special type of evolutionary algorithms, in which the fitness of each individual depends on other individuals' fitness. Such algorithms are applicable in the case of problems for which the formulation of explicit fitness function is difficult or impossible. Co-evolutionary algorithms also maintain population diversity better than "classical" evolutionary algorithms. In this paper the agent-based version of co-operative co-evolutionary algorithm is presented and applied to multi-objective test problems. The proposed technique is also compared to two "classical" multi-objective evolutionary algorithms.

1 Introduction

Co-evolutionary algorithms (CEAs) [20] are a special type of *evolutionary algorithms (EAs)* [2], in which fitness of the given individual depends on fitness of other individuals existing in the population. Such algorithms have some interesting features, among others the possibility of application in the case of problems for which formulation of fitness function is difficult or impossible, maintaining the population diversity, "arms races", and so on. Co-operative co-evolutionary algorithms [21] are CEAs in which there exist several sub-populations, and each of them solves only part of the given problem—the whole solution is represented by the group of individuals composed of representants of all sub-populations. Co-operative co-evolutionary approach was applied to multi-objective optimization by Iorio and Lee [14].

Evolutionary multi-agent systems (EMAS) represent agent-based approach to evolutionary computations. In such systems the population of agents evolve—agents can reproduce, die, compete for resources, observe the environment, communicate with other agents, and make autonomously all their decisions. All these features lead to completely decentralized evolutionary processes. *Co-evolutionary multi-agent systems (CoEMAS)* additionally allow us to define interactions between species of agents. This type of systems have been already applied to multi-modal optimization [7] and multi-objective optimization [8,9].

In this paper the co-operative co-evolutionary multi-agent system for multi-objective optimization (CCoEMAS) is presented. The proposed system is evaluated with the use of test problems proposed by Zitzler ([28]) and the results are compared to those of two "classical" multi-objective evolutionary algorithms.

2 Evolutionary Multi-objective Optimization

During most real-life decision processes a lot of different (often contradictory) factors have to be considered, and the decision maker has to deal with an ambiguous situation: the solutions which optimize one criterion may prove insufficiently good considering the others. From the mathematical point of view such multi-objective (or multi-criteria) problem can be formulated as follows [1,28,26].

Let the problem variables be represented by a real-valued vector:

$$\vec{x} = [x_1, x_2, \ldots, x_N]^T \in \mathbb{R}^N \qquad (1)$$

where N is the number of variables. Then a subset of $\mathbb{R}^N$ of all possible (feasible) decision alternatives (options) can be defined by a system of:

- inequalities (constraints): $g_k(\vec{x}) \geq 0$ and $k = 1, 2, \ldots, K$,
- equalities (bounds): $h_l(\vec{x}) = 0$, $l = 1, 2, \ldots, L$

and denoted by $\mathcal{D}$. The alternatives are evaluated by a system of M functions (objectives) denoted here by vector $F = [f_1, f_2, \ldots, f_M]^T$:

$$f_m : \mathbb{R}^N \to \mathbb{R}, \quad m = 1, 2, \ldots, M \qquad (2)$$

The key issue of optimality in the Pareto sense is the *weak domination relation*. Alternative $\vec{x}^a$ is dominated by $\vec{x}^b$ (which is often denoted by $\vec{x}^b \succeq \vec{x}^a$) if and only if (assuming maximization of all the objectives):

$$\forall m \ \ f_m(\vec{x}^a) \leq f_m(\vec{x}^b) \ \text{and} \ \exists m \ \ f_m(\vec{x}^a) < f_m(\vec{x}^b) \qquad (3)$$

A solution in the Pareto sense of the multi-objective optimization problem means determination of all non-dominated (in the sense of the defined above *weak domination relation*) alternatives from the set $\mathcal{D}$, which is sometimes called a *Pareto set*:

$$\mathcal{P} = \{\vec{x} \in \mathcal{D} \ | \ \neg \exists \vec{x}^a \in \mathcal{D} \ \vec{x}^a \succeq \vec{x}\} \qquad (4)$$

At the same time the non-dominated alternatives create in criteria space a set called *Pareto frontier*:

$$\mathcal{PF} = \{\vec{y} = F(\vec{x}) \in \mathbb{R}^M \ | \ \vec{x} \in \mathcal{P}\} \qquad (5)$$

Unfortunately, when searching for the approximation of the Pareto frontier in the whole, classical computational methods often prove ineffective for many (real) decision problems. The corresponding models are too complex or the formulas applied too complicated, or it can even occur that some formulations must be rejected in the face of numerical instability of available solvers. That is why so much attention is paid to methods based on evolutionary algorithms. These methods are relatively insensitive to complexity of the problem and give the approximation of the whole Pareto frontier with controllable adequacy, which means that a solving process can be stopped by a decision maker anytime he is satisfied.

For the last 20 years a variety of evolutionary multi-criteria optimization techniques have been proposed [4,18,24,27,25]. In the Deb's typology of evolutionary multi-objective algorithms (EMOAs) firstly the elitist and non-elitist ones are distinguished[1] [5]. Each of these groups include many practically used algorithms such as:

- elitist EMOAs: Rudolph's algorithm [22], distance-based Pareto GA [19], strength Pareto EA [29], multi-objective micro GA [3], Pareto-archived evolution strategy [15], multi-objective messy GA [26], etc.
- non-elitist EMOAs: vector-optimized evolution strategy [16], random weighted GA [17], weight-based GA [11], niched-pareto GA [13], non-dominated sorting GA [23], multiple objective GA [10], distributed sharing GA [12] etc.

The main difference between these two groups of techniques consists in utilizing the so-called elite-preserving operators that give the best individuals (the elite of the population) the opportunity to be directly carried over to the next generation regardless of the actual selection mechanism used. Of course, if the algorithm finds a better solution than the one in the elite, this solution becomes a new elitist solution.

3 Co-operative Co-Evolutionary Multi-Agent System

The functioning principles of the system presented in this paper are in accordance with the general model of co-evolution in multi-agent system proposed in [6]. The formal model of the co-evolutionary multi-agent system for multi-objective optimization may be found for example in [9].

CoEMAS systems are composed of the environment, which has usually the structure of graph with every node connected with its four neighbors, and agents, which "live" within the environment. Agents can reproduce, die, and compete for limited resources. Because of decentralized nature of CoEMAS systems there is no possibility of applying selection mechanisms known from "classical" evolutionary algorithms. Resources defined within the system play the role of decentralized selection mechanism: agents need resources to perform every action (like reproduction, migration, and so on). In order to use the resources as a selection mechanism, they are transferred from "worse" to "better" agents ("better" in the sense of solution of the given problem encoded within the agent's genotype). Usually there is only one resource defined. The total amount of the resource within the system is constant, what means that there is closed circulation of the resource.

In CoEMAS systems different co-evolutionary mechanisms can be defined: predator-prey, host-parasite, co-operative interactions, sexual selection mechanism, and so on. In the system presented in this paper the co-operative interactions are used. There are several sub-populations (species) within the co-operative co-evolutionary system used in experiments (see fig.1). One criteria is assigned to each species (it is used in order to evaluate the agents), so

---

[1] Deb's typology includes also so-called *constrained EMOAs*—techniques that support handling constraints.

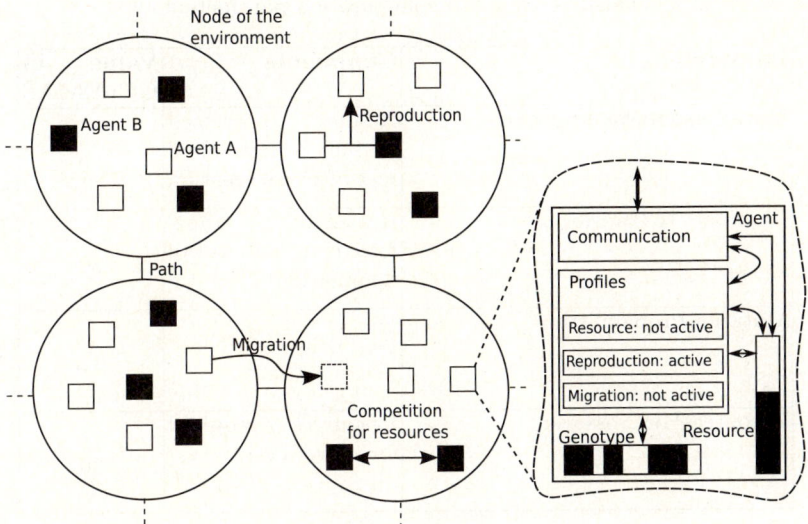

Fig. 1. Co-operative co-evolutionary multi-agent system

the number of species corresponds to the number of criteria of the given problem. Agents compete for resources only within the species—there is no competition between agents that belong to different species. Reproduction takes place when the agent has enough resources to perform it. The agent searches for a reproduction partner within the node in which it is located—the partner must come from one of the opposite species. In the system real-valued vectors are used as agents' genotypes. Mutation with self-adaptation and intermediate recombination are used as evolutionary operators [2]. Two agents produce one offspring and give it some of their resources.

The very important mechanism is the decision making process performed by the agent. Each agent has several profiles (fig. 1), which are activated in order to realize the agent's goals. Whenever there are some profiles with active goals ("active" means here that they should be realized as soon as possible) the profile with the highest priority is chosen, and its active goal is realized with the use of actions that can be performed within the given profile.

In the CCoEMAS the agent has three profiles: resource (with highest priority), reproduction, and migration. Within the resource profile there are three actions: *seek* (which is performed in order to find the agent that is dominated by the given agent and located within the same node), *get* (which gets resources from dominated agent), and *die* (which is performed when the agent is out of resources—such agent is removed from the system). Within the reproduction profile there are the following actions: *seek* (which is used to search for partner when the amount of resource is above the given level), *clone* (which clones the agent), *rec* (which performs the recombination), *mut* (which performs the mutation), and *give* (which is used in order to give the offspring some of the

Table 1. Selected configuration parameters

Parameter	Comments	Value in CCoEMAS
InitialResourcesPerSpecimen	Resources possessed initially by individual just after its creation	50
ResourcesToTransfer	Resources transferred in the case of domination	30
MutationProbability	—	0.5
ResourcesForCrossover	Resources required for reproduction	30

parent's resources). And finally, within the migration profile there is one action defined: *migr* (it is performed by the agent, which migrates to another node of the environment). The migration takes place when the agent has enough resources.

4 Experimental Results

Agent-based co-evolutionary algorithm that is the subject of considerations in the course of this paper was assessed tentatively using inter alia so called Zitzler

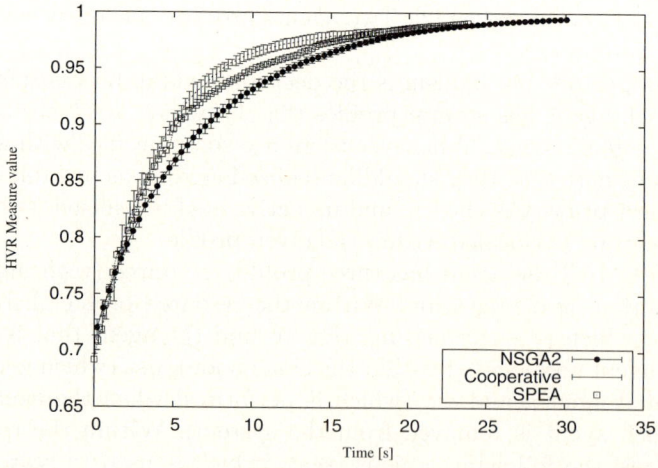

Fig. 2. HVR values obtained by CCoEMAS, SPEA2, and NSGA2 run against Zitzler's ZDT1 problem

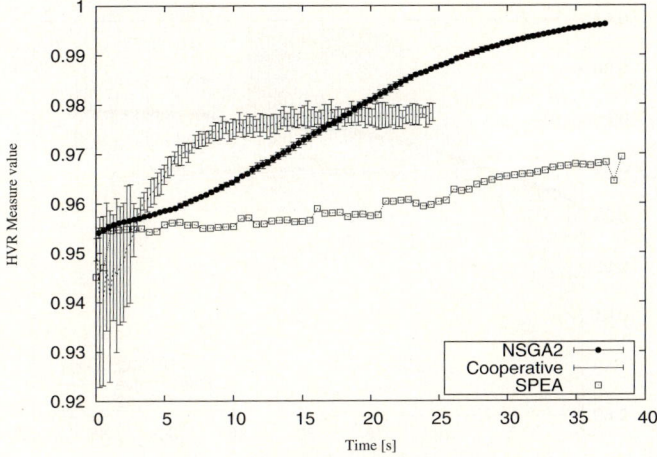

Fig. 3. HVR values obtained by CCoEMAS, SPEA2, and NSGA2 run against Zitzler's ZDT2 problem

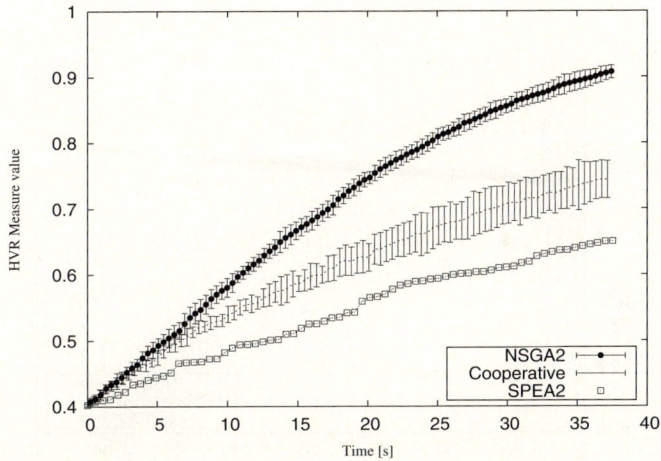

Fig. 4. HVR values obtained by CCoEMAS, SPEA2, and NSGA2 run against Zitzler's ZDT3 problem

problems—ZDT1, ZDT2, ZDT3, ZDT4 and ZDT6 (their definitions can be found in [28]). To assess proposed approach HVR measure ([5]) was used as the metrics measuring both: closeness to the model Pareto frontier and dispersing solutions over the whole frontier. The size of population of algorithm that is being assessed and benchmarking algorithms are as follows: CCoEMAS—200, NSGA2—300 and SPEA—100. In the table 1 there are presented selected values of parameters for co-operative co-evolutionary multi-agent system used during experiments.

In the figures 2–6 there are presented values of HVR measure obtained with time by co-evolutionary multi-agent system with co-operation for ZDT1 (fig. 2),

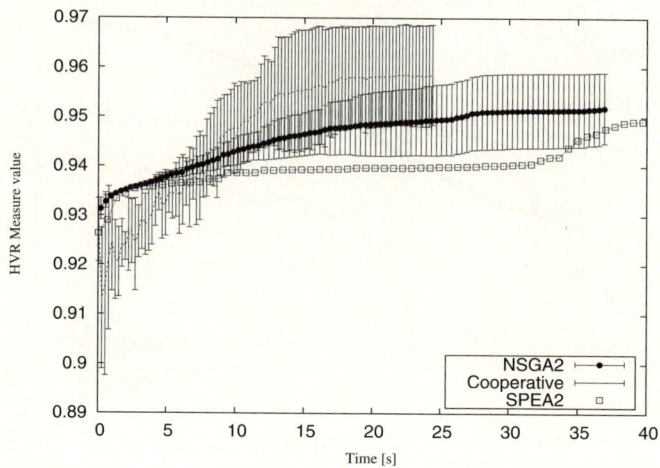

Fig. 5. HVR values obtained by CCoEMAS, SPEA2, and NSGA2 run against Zitzler's ZDT4 problem

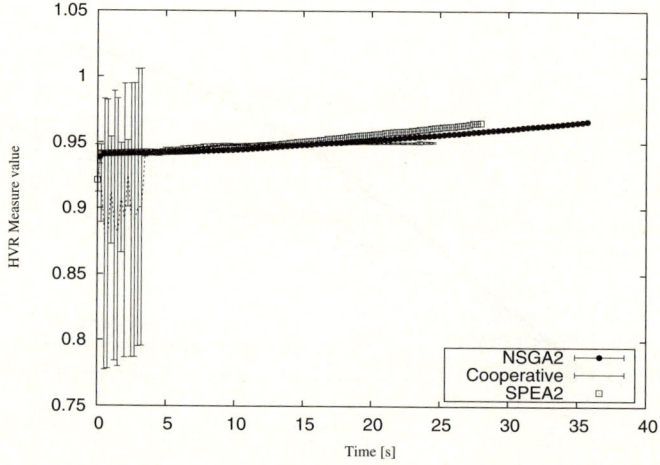

Fig. 6. HVR values obtained by CCoEMAS, SPEA2, and NSGA2 run against Zitzler's ZDT6 problem

ZDT2 (fig. 3), ZDT3 (fig. 4), ZDT4 (fig. 5) and ZDT6 (fig. 6) problems. There are also presented results obtained by NSGA2 and SPEA2 algorithms.

Co-evolutionary multi-agent system, as not so complex algorithm as NSGA2 or SPEA2, initially allows for obtaining better solutions, but with time classical algorithms—especially NSGA2—are the better alternatives. It is however worth to mention that in the case of ZDT4 problem this characteristic seems to be reversed—i.e. initially classical algorithms seem to be better alternatives, but finally co-evolutionary multi-agent system with co-operation leads for obtaining better solutions (observed as higher values of HVR metrics). Such characteristics can result from the fact that agent-based evolutionary systems are not so complex algorithms as NSGA2 or SPEA2 algorithms, so—in the consequence—from the time point of view, agent-based algorithms realize more computational steps than "classical algorithms" and they are able to obtain better solutions, but with time classical algorithms perform enough steps to obtain high-quality results.

5 Summary and Conclusions

As it can be observed in section 4 analyzing the quality of obtained solutions as the function of time, co-evolutionary multi-agent system with co-operation is very attractive alternative since initially (in our experiments during c.a. 15 seconds) solutions proposed by this algorithm were better than solutions proposed by "classical" (non agent-based) algorithms. Short explanation of such a phenomenon was given in section 4 and obviously in the function of consecutive steps classical algorithms are absolutely much more effective (but usually, it is not so important for the user/decision maker in how many steps valuable results were obtained but how fast it was done). It is enough to mention that each step of NSGA2 algorithm consists inter alia of ordering population according to the consecutive levels of domination, crowding etc.—it is obviously very effective (i.e. in each step algorithm heads effectively toward model Pareto frontier) but each step of this algorithm is complex and in the consequence time consuming. So, in the function of algorithm step—classical algorithms are more effective, also in the function of time, final results proposed by NSGA2 are slightly better than results obtained by co-evolutionary multi-agent system, but if time of obtaining valuable results is crucial—agent-based approach seems to be very attractive alternative.

The future plans include further investigation of the proposed mechanism. It could be interesting to modify the co-operation proposed in this paper in such a way that agents from different sub-populations (species) would specialize in different criteria and form aggregates (teams) composed of the representants of different species in order to solve the problem.

References

1. Abraham, A., Jain, L.C., Goldberg, R.: Evolutionary Multiobjective Optimization Theoretical Advances and Applications. Springer, Heidelberg (2005)
2. Bäck, T., Fogel, D., Michalewicz, Z. (eds.): Handbook of Evolutionary Computation. IOP Publishing and Oxford University Press (1997)
3. Coello Coello, C., Toscano, G.: Multiobjective structural optimization using a micro-genetic algorithm. Structural and Multidisciplinary Optimization 30(5), 388–403 (2005)
4. Coello Coello, C.A., Van Veldhuizen, D.A., Lamont, G.B.: Evolutionary Algorithms for Solving Multi-Objective Problems. Kluwer Academic Publishers, Dordrecht (2002)

5. Deb, K.: Multi-Objective Optimization using Evolutionary Algorithms. John Wiley & Sons, Chichester (2001)
6. Dreżewski, R.: A model of co-evolution in multi-agent system. In: Mařík, V., Müller, J.P., Pěchouček, M. (eds.) CEEMAS 2003. LNCS (LNAI), vol. 2691, pp. 314–323. Springer, Heidelberg (2003)
7. Dreżewski, R.: Co-evolutionary multi-agent system with speciation and resource sharing mechanisms. Computing and Informatics 25(4), 305–331 (2006)
8. Dreżewski, R., Siwik, L.: Co-evolutionary multi-agent system with sexual selection mechanism for multi-objective optimization. In: Proceedings of the IEEE World Congress on Computational Intelligence (WCCI 2006). IEEE, Los Alamitos (2006)
9. Dreżewski, R., Siwik, L.: Multi-objective optimization technique based on co-evolutionary interactions in multi-agent system. In: Giacobini, M. (ed.) EvoWorkshops 2007. LNCS, vol. 4448, pp. 179–188. Springer, Heidelberg (2007)
10. Fonseca, C., Fleming, P.: Genetic algorithms for multiobjective optimization: Formulation, discussion and generalization. In: Genetic Algorithms: Proceedings of the Fifth International Conference, pp. 416–423. Morgan Kaufmann, San Francisco (1993)
11. Hajela, P., Lee, E., Lin, C.: Genetic algorithms in structural topology optimization. In: Proceedings of the NATO Advanced Research Workshop on Topology Design of Structures, vol. 1, pp. 117–133 (1993)
12. Hiroyasu, T., Miki, M., Watanabe, S.: Distributed genetic algorithms with a new sharing approach. In: Proceedings of the Conference on Evolutionary Computation, vol. 1, IEEE Service Center (1999)
13. Horn, J., Nafpliotis, N., Goldberg, D.E.: A niched pareto genetic algorithm for multiobjective optimization. In: Proceedings of the First IEEE Conference on Evolutionary Computation, IEEE World Congress on Computational Intelligence, Piscataway, New Jersey, vol. 1, pp. 82–87. IEEE Service Center (1994)
14. Iorio, A., Li, X.: A cooperative coevolutionary multiobjective algorithm using non-dominated sorting. In: Deb, K., Poli, R., Banzhaf, W., Beyer, H.-G., Burke, E.K., Darwen, P.J., Dasgupta, D., Floreano, D., Foster, J.A., Harman, M., Holland, O., Lanzi, P.L., Spector, L., Tettamanzi, A., Thierens, D., Tyrrell, A.M. (eds.) GECCO 2004. LNCS, vol. 3102-3103, pp. 537–548. Springer, Heidelberg (2004)
15. Knowles, J., Corne, D.: Approximating the nondominated front using the pareto archived evolution strategy. Evolutionary Computation 8(2), 149–172 (2000)
16. Kursawe, F.: A variant of evolution strategies for vector optimization. In: Schwefel, H.-P., Männer, R. (eds.) PPSN 1990. LNCS, vol. 496, pp. 193–197. Springer, Heidelberg (1991)
17. Murata, T., Ishibuchi, H.: Moga: multi-objective genetic algorithms. In: Proceedings of the IEEE International Conference on Evolutionary Computation, November 1995, vol. 1, pp. 289–294. IEEE, IEEE Service Center (1995)
18. Osyczka, A.: Evolutionary Algorithms for Single and Multicriteria Design Optimization. Physica Verlag (2002)
19. Osyczka, A., Kundu, S.: A new method to solve generalized multicriteria optimization problems using the simple genetic algorithm. Structural and Multidisciplinary Optimization 10(2), 94–99 (1995)
20. Paredis, J.: Coevolutionary algorithms. In: Bäck, T., Fogel, D., Michalewicz, Z. (eds.) Handbook of Evolutionary Computation, 1st supplement. IOP Publishing and Oxford University Press, Oxford (1998)
21. Potter, M.A., De Jong, K.A.: Cooperative coevolution: An architecture for evolving coadapted subcomponents. Evolutionary Computation 8(1), 1–29 (2000)

22. Rudolph, G.: Evolutionary search under partially ordered finite sets. In: Sebaaly, M.F. (ed.) Proceedings of the International NAISO Congress on Information Science Innovations (ISI 2001), Dubai, U. A. E, pp. 818–822. ICSC Academic Press (2001)
23. Srinivas, N., Deb, K.: Multiobjective optimization using nondominated sorting in genetic algorithms. Evolutionary Computation 2(3), 221–248 (1994)
24. Tan, K.C., Khor, E.F., Lee, T.H.: Multiobjective Evolutionary Algorithms and Applications. Springer, Heidelberg (2005)
25. Toscano, G.: On the Use of Self-Adaptation and Elitism for Multiobjective Particle Swarm Optimization. PhD thesis, Centro de Investigacion y Estudios Avanzados del Instituto Politecnico Nacional (2005)
26. Van Veldhuizen, D.A.: Multiobjective Evolutionary Algorithms: Classifications, Analyses and New Innovations. PhD thesis, Graduate School of Engineering of the Air Force Institute of Technology Air University (1999)
27. Villalobos, M.A.: Analisis de Heuristicas de Optimizacion para Problemas Multiobjetivo. PhD thesis, Centro de Investigacion y de Estudios Avanzados del Instituto Politecnico Nacional (2005)
28. Zitzler, E.: Evolutionary algorithms for multiobjective optimization: methods and applications. PhD thesis, Swiss Federal Institute of Technology, Zurich (1999)
29. Zitzler, E., Thiele, L.: An evolutionary algorithm for multiobjective optimization: The strength pareto approach. Technical Report 43, Swiss Federal Institute of Technology, Zurich, Gloriastrasse 35, CH-8092 Zurich, Switzerland (1998)

Evolutionary Methods for Designing Neuro-fuzzy Modular Systems Combined by Bagging Algorithm*

Marcin Gabryel[1,3] and Leszek Rutkowski[1,2]

[1] Department of Computer Engineering, Częstochowa University of Technology
Al. Armii Krajowej 36, 42-200 Częstochowa, Poland
{marcing, lrutko}@kik.pcz.pl
http://kik.pcz.pl

[2] Department of Artificial Intelligence, WSHE University in Łódź
ul. Rewolucji 1905 nr 64, Łódź, Poland
http://www.wshe.lodz.pl

[3] The Professor Kotarbinski Olsztyn Academy of Computer Science and Management
ul. Artyleryjska 3c, 10-165 Olsztyn, Poland
http://www.owsiiz.edu.pl

Abstract. In this paper we present the problem of designing modular systems combined with the Bagging Algorithm. As component classifiers the Mamdani-type neuro fuzzy-systems are applied and trained using evolutionary methods. Experimental investigations presented in this paper include the classification performed by the modular system built by means of classic Bagging algorithm and its modified version which assigns evolutionary chosen weights to base classifiers.

1 Introduction

Modular system [6][12] is a group of intelligent systems which is used in order to improve accuracy. It appears that an ensemble of classifiers with component systems having not complete knowledge of the resolving problem can work far better than a single classifier. There are many ways of aggregations of outputs (answers) and many methods of designing of separate classifiers which are the parts of a modular system. In a general case the construction of a group of classifiers may consist in joining and/or generating various intelligent systems (i.e. neuro and neuro-fuzzy systems) within one system. However, an appropriate manipulation of learning data by which base systems are trained is mostly in use. This way of construction of systems contains the Bagging algorithm [3]. A Bagging is a meta-learning algorithm that allows to use any of learning algorithms as the components of a group. When adding another system to the group

---

* This work was partly supported by the Foundation for Polish Science (Professorial Grant 2005-2008) and Polish Ministry of Science and Higher Education (Special Research Project 2006-2009, Polish-Singapore Research Project 2008-2010, Research Project 2008-2010).

a new training set is created for it. As the component classifiers the neuro-fuzzy systems were used which are very popular recently [15][17]. Their great advantage is an ability to extract knowledge gained during the training and to store it in the form of fuzzy rules. In the literature neuro-fuzzy systems were trained mostly with using gradient methods [5][14][16][17]. In this paper as the alternative to the backpropagation method specificy evolutionary learning has been used based on an evolutionary strategy (μ, λ) [1][2][7] with original algorithm of initialization which takes into account the specific of a resolving problem [10].

Our paper is divided into three main sections. In the next section we will give a short description of the Mamdani-type neuro-fuzzy system, and details of evolutionary learning method. Next, we will present Bagging algorithm and its version with weights assigned to respective classifiers. In the last section the results of simulations are shown.

2 Evolutionary Learning of Neuro-fuzzy Systems

2.1 Description of Fuzzy Systems

In this paper, we consider a multi input, single output neuro fuzzy system mapping $\mathbf{X} \to \mathbf{Y}$ where $\mathbf{X} \subset \mathbf{R}^n$ and $\mathbf{Y} \subset \mathbf{R}$. The fuzzy rule base consists of a collection of N fuzzy IF-THEN rules in the form:

$$R^{(k)} : \text{IF } x_i \text{ is } A_1^k \text{ AND} \ldots \text{AND } x_n \text{ is } A_n^k \text{ THEN } y \text{ is } B^k, \tag{1}$$

where $\mathbf{x} = [x_1, \ldots, x_n]$ are input variables, n - number of inputs, y - output value, fuzzy sets $A_1^k, A_2^k, \ldots, A_n^k$ and B^k are characterized by membership functions $\mu_{A_i^k}(x_i)$ and $\mu_{B^k}(y)$, respectively, $k = 1, \ldots, N$, $i = 1, \ldots, n$. This system is based on the Mamdani-type reasoning, where antecedents and consequences in the individual rules are connected by the product t-norm. We use the most common singleton fuzzifier for mapping crisp values of input variables into fuzzy sets [17]. The defuzzification process is made by the COA (center of area) method [17]. We choose as membership functions $\mu_{A_i^k}(x_i)$ and $\mu_{B^k}(y)$ the Gaussian functions

$$\mu_{A_i^k}(x_i) = \exp\left[-\left(\frac{x_i - \overline{x}_i^k}{\sigma_i^k}\right)^2\right], \tag{2}$$

$$\mu_{B_i^k}(y) = \exp\left[-\left(\frac{y - \overline{y}^k}{\sigma^k}\right)^2\right]. \tag{3}$$

The following neuro-fuzzy system [17] will be investigated:

$$y = \frac{\sum_{r=1}^{N} \overline{y}^r \cdot \max_{1 \leq k \leq N} \left\{ \prod_{i=1}^{n} \exp\left[-\left(\frac{x_i - \overline{x}_i^k}{\sigma_i^k}\right)^2\right] \cdot \exp\left[-\left(\frac{\overline{y}^r - \overline{y}^k}{\sigma^k}\right)^2\right]\right\}}{\sum_{r=1}^{N} \max_{1 \leq k \leq N} \left\{ \prod_{i=1}^{n} \exp\left[-\left(\frac{x_i - \overline{x}_i^k}{\sigma_i^k}\right)^2\right] \cdot \exp\left[-\left(\frac{\overline{y}^r - \overline{y}^k}{\sigma^k}\right)^2\right]\right\}}. \tag{4}$$

This system has been trained using the idea of the backpropagation method [17]. In the next subsection we will shortly describe an evolutionary algorithm to train system (4).

2.2 Evolutionary Learning of Fuzzy Systems

Let $\overline{\mathbf{x}} \in \mathbf{X} \subset \mathbf{R}^n$, $y(t) \in \mathbf{Y} \subset \mathbf{R}$ and $d(t) \in \mathbf{Y} \subset \mathbf{R}$, $t = 1, ..., K$. Based on the learning sequence $((\mathbf{x}(1), d(1)), (\mathbf{x}(2), d(2)), ..., (\mathbf{x}(K), d(K)))$ we wish to determine all parameters $\overline{x}_i^k$, σ_i^k, $\overline{y}^k$, σ^k, $i = 1, ..., n$, $k = 1, ..., N$, such that

$$e(t) = \frac{1}{2}\left[f\left(\mathbf{x}(t)\right) - d(t)\right]^2 \tag{5}$$

is minimized.

We will solve the above problem by using an evolutionary strategy (μ, λ). This method consists in selection of system parameters describing shapes of membership functions. Respective parameters $\overline{x}_i^k$, σ_i^k, $\overline{y}^k$, σ^k are encoded in a chromosome as the real-value numbers. After using genetic operators and determining values of fitness functions, which reflect the effectiveness of the coded systems, the evolutionary algorithm selects the best individuals to the next generation.

The learning metod of fuzzy systems needs to initiate the primary population of individuals. It usually consists in trandom drawing of values of respective genes. In this paper a method that allows to initiate both the chromosomes coding system parameters and a chromosome preserving values of the range of mutation has been used. This method accelerates the process of optimization. In some cases the method makes possibile to complete this process in reasonable time what is not assured by the methods of evolutionary alorithms initialization which have been used untill now. The learning algorithm and the algorithm of initilaization of primary population are presented in detail in works of [9][10][11].

3 Bagging Algorithm

Bagging is a procedure for combining classifiers generated using the same training set. Bagging (bootstrap aggregating) [3][12] produces replicates of the training set $\mathbf{z}$ and trains a classifier D_k on each replicate S_k. Each classifier is applied to a test pattern $\mathbf{x}$ which is classified on a majority vote basis, ties being resolved arbitrarily.

We have a set of labels $\Omega = \{\omega_1, \omega_2, ..., \omega_C\}$, where C is the number of possible classes, labeled ω_i, $i = 1, ..., C$. We consider the ensemble of classifiers $\mathbf{D} = [D_1, ..., D_J]$, where there are J base classifiers D_k, $k = 1, ..., J$. We assume that the output of classifier D_k is $\mathbf{d}_k(\mathbf{x}) = [d_{k,1}(\mathbf{x}), ..., d_{k,C}(\mathbf{x})]^T \in \{0, 1\}^C$, where $d_{k,j} = 1$ if D_k determine that $\mathbf{x}$ belong to class ω_j, and $d_{k,j} = 0$ otherwise. The majority vote will result in an ensemble decision for class ω_k if

$$\sum_{i=1}^{J} d_{i,k}(\mathbf{x}) = \max_{j=1}^{C} \sum_{i=1}^{J} d_{i,j}(\mathbf{x}) \tag{7}$$

Table 1. The Bagging Algorithm with the majority vote

1. Initialize the parameters
   - the ensemble $\mathbf{D} = \emptyset$
   - the number of classifiers to train J
2. For $k = 1, , J$ repeat points 3-5
3. Take sample S_k from $\mathbf{Z}$
4. Build a classifier D_k using S_k as the training set
5. Add the classifier to the current ensemble $\mathbf{D} = \mathbf{D} \cup D_k$
6. Return $\mathbf{D}$ as algorithm outcome
7. Run $\mathbf{x}$ on the input $D_1, \ldots, D_J$
8. The vector $\mathbf{x}$ is a member of class ω_k, if occurs condition (6)

The Bagging algorithm is shown in table 1. We called this method as the classic Bagging algorithm.

As experiments in section 4, show the majority of classifiers, obtained during the work of the algorithm Bagging, has not been trained correctly. For this

Table 2. The Bagging Algorithm with the weighted majority vote

1. Initialize the parameters
   - the ensemble $D = \emptyset$
   - the number of classifiers to train J
   - weights $\mathbf{w} = [w_1, \ldots, w_K]$, in addition $w_i \in [0, 1]$, $\sum_{k=1}^{K} w_i = 1$
2. For $k = 1, , J$ repeat points 3-5
3. Take sample S_k from $\mathbf{Z}$
4. Build a classifier D_k using S_k as the training set
5. Add the classifier to the current ensemble $\mathbf{D} = \mathbf{D} \cup D_k$

6. Initialize the evolutionary strategy (μ, λ)
   - determine algorithm parameters μ, λ
   - initialize the population
7. Run the evolutionary strategy, the quality of individuals is determined by fitness function

$$fitness(\mathbf{X}_j) = \sum_{t=1}^{K} \left(I \left(\arg\max_{j=1}^{C} \sum_{i=1}^{J} w_i d_{i,j}(\mathbf{x}) = d(t) \right) \right) \qquad (6)$$

where $I(a = b) = \begin{cases} 1 & \text{if } a = b \\ 0 & \text{if } a \neq b \end{cases}$

8. Return $\mathbf{D}$ and $\mathbf{w}$ as algorithm outcome
9. Run $\mathbf{x}$ on the input $D_1, \ldots, D_J$
10. The vector $\mathbf{x}$ is a member of class ω_k, if occurs condition (7)

reason, we developed an algorithm, in which we replace the majority vote with the weighted majority vote. Each classifiers will have the weight, which will reflect the quality of its classification. We have introduced weights w_i, $i = 1, \ldots, J$, and rewritten equation (6) as follow: choose class label ω_k if

$$\sum_{i=1}^{J} w_i d_{i,k}(\mathbf{x}) = \max_{j=1}^{C} \sum_{i=1}^{J} w_i d_{i,j}(\mathbf{x}) \tag{8}$$

In table 2 we show algorithm Bagging with the weighted majority vote, in which weights are selected with the evolutionary strategy (μ, λ).

Firstly, weights $\mathbf{w}$ for respective classifier should be initiated. The algorithm in points 3-5 generates samples S_k, builds classifier D_k and adds it to the set of the ensemble $\mathbf{D}$. In the next step the evolutionary strategy (μ, λ) is initialized. In chromosomes all weights $\mathbf{w}$ are coded. These weights describe the importance of individual base classifiers. Fitness function is described by a formula (8). The aim of the selection of weights by evolutionary strategy is the maximization of the effectiveness of the classification. As a result, we get (in point 8) a pair of sets $\mathbf{D}$ and $\mathbf{w}$, which create the complete modular system. Testing the sample $\mathbf{x}$ by the weighted majority vote executes in points 9 and 10.

4 Simulation Result

In this section we present results of evolutionary design of modular systems. We consider classification problems taken from [19]: Monk1 problem, Glass Identification and Wine Recognition problem. To train each of base classifiers the following parameters were assumed: $\mu = 10$, $\lambda = 70$. The number of generations is 100, the number of classifiers is $J = 75$. Component classifiers are the fuzzy systems described by formula (4) where the number of rules amounts $R = 3$. The obtained results are presented in table 3. In the columns of Table 3 we present the percentage of a correct classification for the training and testing sets using classic Bagging and Bagging with weigths of classifiers. As it is easily seen the results for classic Bagging are not satisfactory. For this reason weights were assigned to the base classifiers. For the evolutionary strategy (μ, λ), which was used in this method in points 6 and 7 of Table 2, the following parameters were assumed: $\mu = 20$, $\lambda = 140$, the number of generations amounted 100. These

Table 3. The for three problems

Problem	Classic Bagging		Bagging with weigths of classifiers	
	Training	Test	Training	Test
Glass	100,0	88,89	100,0	90.48
Monk1	92,74	85.65	100.0	99.31
Wine	100.0	98.08	100.0	99.08

results show a considerable improvement of the classification using our method. It can be especially seen for the problem Monk 1.

5 Final Remarks

In our paper we have presented the method of evolutionary design of modular system which consists of the Mandani-type neuro-fuzzy systems. Base classifiers are trained using evolutionary strategies (μ, λ) with the algorithm of initialization of primary population. Simulation results showed that the modular system based on the Bagging algorithm with weights assigned to base classifiers allows to obtain better results of classification than systems based on the classic Bagging algorithm. Our method has one more advantage. It can be used for modular systems which have been in use because the process of assigning weights to respective classifiers can be done independently from the learning processes performed before.

References

1. Arabas, J.: Lectures on Evolutionary Algorithms (in Polish), WNT, Warsaw (2001)
2. Back, T.: Evolutionary Algorithms in Theory and Practice. Oxford University Press, Oxford (1996)
3. Breiman, L.: Bagging predictors. Machine Learning 26(2), 123–140 (1996)
4. Cordon, O., Herrera, F., Hoffman, F., Magdalena, L.: Genetic Fuzzy System. In: Evolutionary Tunning and Learning of Fuzzy Knowledge Bases. World Scientific, Singapore (2000)
5. Czabanski, R.: Extraction of fuzzy rules using deterministic annealing integrated with ϵ-insensitive learning. Intenrational Journal of Applied Mathematics and Computer Science 16(3), 357–372 (2006)
6. Duch, W., Korbicz, J., Rutkowski, L., Tadeusiewicz, R.: Biocybernetics and biomedical engineering (in Polish), vol. 6. Akademicka Oficyna Wydawnicza EXIT, Warszawa (2000)
7. Eiben, A.E., Smith, J.E.: Introduction to Evolutionary Computing. Springer, Heidelberg (2003)
8. Gabryel, M., Cpalka, K., Rutkowski, L.: Evolutionary strategies for learning of neuro-fuzzy systems. In: I Workshop on Genetic Fuzzy Systems, Genewa, pp. 119–123 (2005)
9. Gabryel, M., Rutkowski, L.: Evolutionary method for learning neuro-fuzzy systems with applications to medical diagnosis (in Polish), XIV Krajowa Konferencja Naukowa Biocybernetyka i Inynieria Biomedyczna, pp. 960-965, Czstochowa (2005)
10. Gabryel, M., Rutkowski, L.: Evolutionary Learning of Mamdani-Type Neuro-fuzzy Systems. In: Rutkowski, L., Tadeusiewicz, R., Zadeh, L.A., Żurada, J.M. (eds.) ICAISC 2006. LNCS (LNAI), vol. 4029, pp. 354–359. Springer, Heidelberg (2006)
11. Korytkowski, M., Gabryel, M., Rutkowski, L., Drozda, S.: Evolutionary Methods to Create Interpretable Modular System. LNCS (LNAI), vol. 5097. Springer, Heidelberg (2008)
12. Kuncheva, L.I.: Fuzzy Classifier Design. Physica Verlag, Heidelberg, New York (2000)

13. Michalewicz, Z.: Genetic Algorithms + Data Structures = Evolution Programs, 3rd edn. Springer, Heidelberg (1996)
14. Rutkowska, D., Nowicki, D.R.: Implication-Based Neuro-Fuzzy Architectures. International Journal of Applied Mathematics and Computer Science 10(4) (2000)
15. Rutkowska, D.: Neuro Fuzzy Architectures and Hybrid Learning. Springer, Heidelberg (2002)
16. Rutkowski, L.: Computational Inteligence Methods and Techniques (in Polish). PWN, Warszawa (2006)
17. Rutkowski, L.: Flexible Neuro Fuzzy Systems. Kluwer Academic Publishers, Dordrecht (2004)
18. Rutkowski, L.: Methods and Techniques of Artificial Inteligence (in Polish). Wydawnictwo Naukowe PWN, Warsaw (2005)
19. Mertz, C.J., Murphy, P.M.: UCI respository of machine learning databases, http://www.ics.uci.edu/pub/machine-learning-databases

Evolutionary Methods to Create Interpretable Modular System*

Marcin Korytkowski[1,2], Marcin Gabryel[1,2], Leszek Rutkowski[1,3], and Stanislaw Drozda[4,2]

[1] Department of Computer Engineering, Częstochowa University of Technology
Al. Armii Krajowej 36, 42-200 Częstochowa, Poland
{marcink, marcing, lrutko}@kik.pcz.pl
http://kik.pcz.pl

[2] The Professor Kotarbinski Olsztyn Academy of Computer Science and Management
ul. Artyleryjska 3c, 10-165 Olsztyn, Poland
http://www.owsiiz.edu.pl

[3] Department of Artificial Intelligence, WSHE University in Łódź
ul. Rewolucji 1905 nr 64, Łódź, Poland
http://www.wshe.lodz.pl

[4] University of Warmia and Mazury in Olsztyn
The Faculty of Mathematics and Computer Sciences
ul. Zolnierska 14, 10-561 Olsztyn, Poland
http://wmii.uwm.edu.pl

Abstract. In this paper we present an evolutionary method to create an interpretable modular system. It consists of many neuro-fuzzy structures which are merged using a very popular algorithm called AdaBoost. As the alternative to the backpropagation method to train all models a special evolutionary algorithm has been used based on the evolutionary strategy (μ, λ).

1 Introduction

Combining many classifiers is nowadays a very popular method to improve classification accuracy [8]. We can use as models many single neuro-fuzzy structures that could be separately analyzed to obtain fuzzy rules. In such case we can not obtain a single fuzzy rule base for such a modular system. In this paper we present the method which can be applied to the standard neuro-fuzzy systems to obtain new structures which joined together give one modular neuro-fuzzy system. This idea was presented previously in [7] and the backpropagation learning algorithm was applied. In our concept presented in the next sections all of models are trained using a novel evolutionary algorithm.

---

* This work was partly supported by the Foundation for Polish Science (Professorial Grant 2005-2008) and Polish Ministry of Science and Higher Education (Special Research Project 2006-2009, Polish-Singapore Research Project 2008-2010, Research Project 2008-2010).

Our paper is divided into four main sections. In the next section we will present a description of the AdaBoost algorithm. In section 3 we will present the technique for merging neuro-fuzzy classifiers. Next we will show details of the evolutionary method for learning neuro-fuzzy systems with two outputs. In the last section there are the results of simulations.

2 Adaboost Algorithm

In this section we briefly present one of the most popular algorithms used to design a modular system. The algorithm is called boosting and belongs to the AdaBoost family methods [2][9][13].

Let us denote the l-th learning vector by $\mathbf{z}_l = [x_{l1}, x_{l2}, \ldots, x_{ln}, y_l]$, $l = 1, \ldots, m$, is the number of a vector in the learning sequence, n is a dimension of input vector $\mathbf{x}_l$, and y_l is the learning class label. Weights, assigned to learning vectors, have to fulfill the following conditions

$$(i) \quad 0 < D_l < 1 \qquad (1)$$

$$(ii) \quad \sum_{l=1}^{m} D_l = 1 \qquad (2)$$

The weight D_l is the information how well classifiers learned in consecutive steps of an algorithm for a given input vector $\mathbf{x}_l$. Vector $\mathbf{D}$ for all input vectors is initialized according to the following formula

$$D_t^l = \frac{1}{m}, \text{ for } l = 1, \ldots, m, \ t = 1, \ldots, T \qquad (3)$$

where t is the number of a boosting iteration (and a number of a classifier in the ensemble). Let $\{ht(x) : t = 1, \ldots, T\}$ denotes a set of hypotheses obtained in consecutive steps t of the algorithm being described. For simplicity we present boosting algorithm to solve a binary classification (dichotomy) problem. Similarly to learning vectors weights, we assign a weight c_t for every classifier, such that

$$c_t = 1 \text{ for } t = 1, \ldots, T \qquad (4)$$

$$c_t > 0 \qquad (5)$$

Now in the AdaBoost algorithm we repeat steps 1-4 for $t = 1, \ldots, T$:

1. Create hypothesis h_t and train it with a data set with respect to a distribution d_t for input vectors.
2. Compute the classification error ϵ_t of a trained classifier h_t according to the formula

$$\epsilon_t = \sum_{l=1}^{m} D_t^l I(h_t(\mathbf{x}^l) \neq y^l) \qquad (6)$$

where l is the indicator function

$$I(a \neq b) = \begin{cases} 1 \text{ if } a \neq b \\ 0 \text{ if } a = b \end{cases} \tag{7}$$

If $\epsilon_t = 0$ or $\epsilon_t \geq 0.5$, stop the algorithm and $T = T - 1$.

3. Compute the value

$$\alpha_t = 0.5 \frac{1 - \epsilon_t}{\epsilon_t} \tag{8}$$

4. Modify weights for learning vectors according to the formula

$$D_{t+1}(\mathbf{z}_l) = D_t(\mathbf{z}_l) \exp\{-\alpha_t I(h_t(\mathbf{x}_l) = y_l))\} N_t \tag{9}$$

where N_t is a constant such that

$$\sum_{l=1}^{m} D_{t+1}(\mathbf{z}_l) = 1 \tag{10}$$

The AdaBoost algorithm is a meta-learning algorithm and does not impose the way of learning for classifiers in the ensemble. To compute the final answer of all classifiers in a modular system we have the following formula:

$$F(\mathbf{x}) = \sum_{t=1}^{T} c_t h_t(\mathbf{x}) = \sum_{t=1}^{T} c_t \frac{\sum_{r=1}^{N_t} \bar{y}_t^r \cdot \tau_t^r}{\sum_{r=1}^{N_t} \tau_t^r} \tag{11}$$

where $h_t(\mathbf{x})$ is the response of the hypothesis t on the basis of feature vector $\mathbf{x} = [x_1, \ldots, x_n]$, coefficient c_t value is computed using

$$c_t = \frac{1}{\alpha_t} \tag{12}$$

and can be interpreted as the measure of quality of performance of the given classifier h_t.

3 The Technique for Merging Neuro-fuzzy Classifiers

In paper [7] we presented our concept to join many neuro-fuzzy classifiers to obtain one neuro-fuzzy system that could be interpreted and the knowledge could be extracted in the form of fuzzy rules. Now we shortly remind this idea for Mamdani neuro-fuzzy systems [11][12]. According to [7] in the learning process we have to adapt all parameters of fuzzy sets in the system which structure is presented in Fig. 1. One can see that in this model we have two outputs - one extra output is added to the standard neuro-fuzzy structure. During learning process we want to obtain on this output a constant value which equals 1 for all vectors from the training set. This assumption can be written as follows

$$\begin{aligned} &1)\ h_t(\mathbf{x}_l) = d_l\ \forall l = 1, \ldots, m\ , \\ &2)\ \sum_{r=1}^{N_t} \tau_t^r = 1\ \forall t = 1, \ldots, T\ . \end{aligned} \tag{13}$$

After learning process, if we met conditions (13), we can simplify the structure of a single neuro-fuzzy system given in Fig. 1 to the form presented in Fig. 2. A modular system which consists of such modules has several advantages:

(i) it is possible to interpret the resulting rulebase,
(ii) the merged system can be fine-tuned, regarding boosting as a method for initial choosing of parameters,
(iii) the resulting rulebase can be further simplified and reduced. Its structure is shown in Fig. 3.

Its structure is shown in Fig. 3.

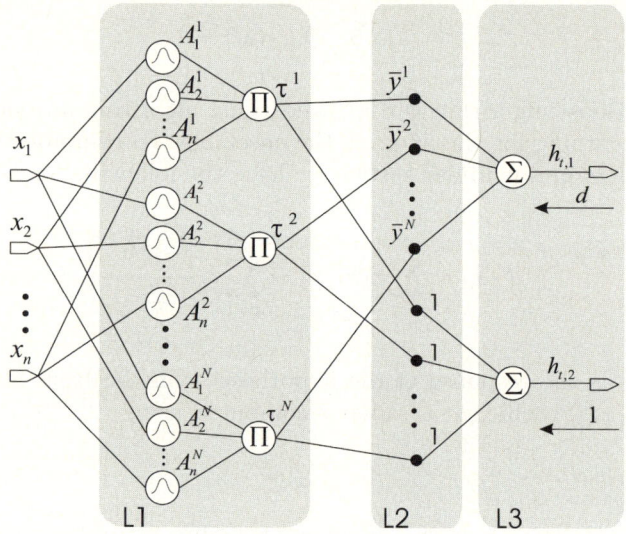

Fig. 1. Single modified Mamdani neuro-fuzzy system

4 Evolutionary Method to Learn Neuro-fuzzy Systems

In this section we present the evolutionary strategy (μ, λ) for learning Mamdani neuro-fuzzy systems of the type presented in Fig. 1. This learning algorithm is based on method described in [4], [5] and [6]. It is assumed that the chromosome of an individual is formed by a pair of real-valued vectors $(\mathbf{X}, \sigma)$. We use uncorrelated mutation [3] and we extend the standard evolution strategy by making use of a uniform recombination operator [10]. Based on the learning sequence $\mathbf{z}_l$, $l = 1, \ldots, m$ we wish to determine all parameters of neuro-fuzzy system $\overline{x}_i^r$, σ_i^r, $\overline{y}^r$, $i = 1, \ldots, n$, $r = 1, \ldots, N$, such that

$$e(l) = \frac{1}{2} \left[(h_{t,1}(\mathbf{x}_l) - d_l) + (h_{t,2}(\mathbf{x}_l) - 1) \right]^2 \tag{14}$$

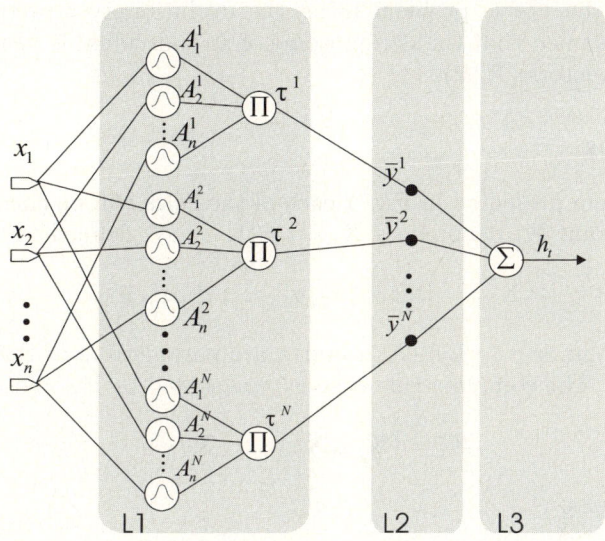

Fig. 2. Single Mamdani neuro-fuzzy system after learning. The second output is removed as it always equals one.

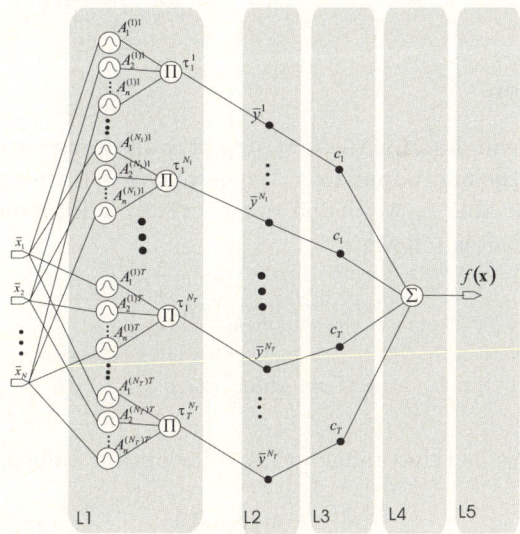

Fig. 3. Ensemble of modified Mamdani neuro-fuzzy systems combined after learning

is minimized. We choose as membership functions $\mu_{A_i^r}(x_i)$, the Gaussian functions

$$\mu_{A_i^r}(x_i) = \exp\left[-\left(\frac{x_i - \overline{x}_i^r}{\sigma_i^r}\right)^2\right], \tag{15}$$

We will solve the above problem using the evolutionary strategy (μ, λ) (see [1],[3]). It is assumed that the chromosome of an individual is formed by a pair of real-valued vectors $(\mathbf{X}, \boldsymbol{\sigma})$.

4.1 Encoding

In a fuzzy system presented in Fig. 1 each of the rules will be encoded in a part of the chromosome $\mathbf{X}_j$ denoted by $\mathbf{X}_{j,r}$, $r = 1, \ldots, N$, defined by

$$\mathbf{X}_{j,r} = (\overline{x}_1^r, \sigma_1^r, \overline{x}_2^r, \sigma_2^r, \ldots, \overline{x}_n^r, \sigma_n^r, \overline{y}^r) \tag{16}$$

where $j = 1, ..., \mu$ or $j = 1, ..., \lambda$, μ and λ are parametrers of the evolutionary strategy (μ, λ). The complete rule base is represented by chromosome $\mathbf{X}_j$

$$\mathbf{X}_j = (\mathbf{X}_{j,1}, \mathbf{X}_{j,2}, \ldots, \mathbf{X}_{j,N}) \tag{17}$$

or more in details

$$\mathbf{X}_j = \begin{matrix} (\overline{x}_1^1, \sigma_1^1, \overline{x}_2^1, \sigma_2^1, \ldots, \overline{x}_n^1, \sigma_n^1, \overline{y}^1, \\ \overline{x}_1^2, \sigma_1^2, \overline{x}_2^2, \sigma_2^2, \ldots, \overline{x}_n^2, \sigma_n^2, \overline{y}^2, \\ \vdots \\ \overline{x}_1^N, \sigma_1^N, \overline{x}_2^N, \sigma_2^N, \ldots, \overline{x}_n^N, \sigma_n^N, \overline{y}^N) \,. \end{matrix} \tag{18}$$

4.2 Initialization

In this section we will describe the first step of evolutionary strategy, i.e. process of initialization of the first population. When we have a learning sequence $(\mathbf{x}_l, d_l)$, for every input variable x_i we define range $[x_i^-, x_i^+]$. Boundary values x_i^- and x_i^+ can be computed as follows

$$\begin{aligned} x_i^- &= \min_{1 \leq l \leq m} (x_{li}), \\ x_i^+ &= \max_{1 \leq l \leq K} (x_{li}). \end{aligned} \tag{19}$$

In the same way we can find range $[d^-, d^+]$ for output values y, i.e.

$$\begin{aligned} d^- &= \min_{1 \leq l \leq K} (d_l), \\ d^+ &= \max_{1 \leq l \leq K} (d_l). \end{aligned} \tag{20}$$

On the ranges defined above uniform fuzzy partitions are made, where every region has a membership function assigned. The number of parts for every partition $[x_i^-, x_i^+]$ and $[d^-, d^+]$ is equal N. Therefore width a_i for input variables $(i = 1, \ldots, n)$ has value $a_i = \frac{x_i^+ - x_i^-}{N}$, $i = 1, ..., n$. For output variable we have

$a_{n+1} = \frac{d^+ - d^-}{N}$. Initial parameters of membership functions $\mu_{A_i^r}(x_i)$ (see (15)) take values

$$\overline{x}_i^k = x_i^- + (b_r - 0.5) \cdot a_i + N(0, \frac{a_i}{4}), \tag{21}$$

$$\sigma_i^r = \frac{a_i}{2} \tag{22}$$

where $\mathbf{b} = [b_1, ..., b_N]$ is a vector with randomly chosen components $b_r \in \{1, ..., N\}$, $b_r \neq b_p$, $r, p = 1, ..., N$, and $N(0, \frac{a_i}{4})$ is a random value from the normal distribution. Similarly, intial parameters $\overline{y}^r$, $r = 1, ..., N$ assume value

$$\overline{y}^r = d^- + (b'_r - 0.5) \cdot a_{n+1} + N(0, \frac{a_{n+1}}{4}), \tag{23}$$

where b'_r is determined analogously to b_r, $r = 1, ..., N$ and $N(0, \frac{a_{n+1}}{4}))$ is a random value from the normal distribution. Numbers b_r and b'_r, determined in the random way, give us the method of generating parameters of membership functions in the initial population.

We will present a method of initialization of the standard deviation vector $\boldsymbol{\sigma}$. This problem is not clearly described in literature and most often elements of vector $\boldsymbol{\sigma}$ are chosen experimentally. We use information which is encoded in chromosome $\mathbf{X}_j$. Therefore the vector of standard deviation $\boldsymbol{\sigma}$ will be initiated by using values of a_i and a_{n+1} as follows

$$\boldsymbol{\sigma}_j = \begin{matrix} (h \cdot a_1, h \cdot a_1, h \cdot a_2, h \cdot a_2, ..., h \cdot a_n, h \cdot a_n, h \cdot a_{n+1}, \\ \vdots \\ h \cdot a_1, h \cdot a_1, h \cdot a_2, h \cdot a_2, ..., h \cdot a_n, h \cdot a_n, h \cdot a_{n+1}) \end{matrix} \tag{24}$$

where $h = 0.1$ is a constant, which value was selected experimentally.

5 Simulation Result

We assume for learning each classifier t the following: $\mu = 10$, $\lambda = 50$, maximal number of generations: 50, we take the mean square error as a fitness function

$$fitness = \sqrt{\frac{1}{m} \sum_{l=1}^{m} ((h_{t,1}(\mathbf{x}_l) - d_l) + (h_{t,2}(\mathbf{x}_l) - 1))^2}. \tag{25}$$

In tables 1 and 2 we present results of individual classifiers and the merged system.

5.1 The Glass Identification

The Glass Identification problem [14] contains 214 instances and each instance is described by nine attributes (RI: refractive index, Na: sodium, Mg: magnesium, Al: aluminium, Si: silicon, K: potassium, Ca: calcium, Ba: barium, Fe: iron). All attributes are continuous. There are two classes: the window glass and the non-window glass. In our experiments, all sets are divided into a learning sequence (150 sets) and a testing sequence (64 sets). All individual classifiers in the ensemble have 3 rules. Detailed accuracy for all subsystems are shown in Table 1.

Table 1. The experimental results for the Glass Identification problem

Classifier no.	The classification accuracy on training set [%]	The classification accuracy on testing set [%]
1	66.22	57.14
2	74.83	73.02
3	88.07	88.89
4	90.06	82.54
5	76.16	76.20
6	92.72	93.65
All	94.03	93.66

Table 2. The experimental results for the Pima Indians Diabetes problem

Classifier no.	The classification accuracy on training set [%]	The classification accuracy on testing set [%]
1	49.44	49.13
2	50.19	48.26
3	51.12	50.87
4	49.26	47.83
All	76.21	76.52

5.2 The Pima Indian Diabets

The Pima Indians Diabetes (PID) data [14] contains two classes, eight attributes (number of times pregnant, plasma glucose concentration in an oral glucose tolerance test, diastolic blood pressure, triceps skin fold thickness, 2-hour serum insulin, body mass index, diabetes pedigree function, age). We consider 768 instances, 500 (65.1%) healthy and 268 (34.9%) diabetes cases. All patients were females at least 21 years old of Pima Indian heritage. In our experiments all sets are divided into a learning sequence (538 sets) and testing sequence (230 sets). The experimental results for the Pima Indians Diabetes problem are shown in Table 2. We obtain 76% the classification accuracy.

6 Final Remarks

In the paper we solved the problem of merging several fuzzy rule bases, resulting from boosting training, into one comprehensive fuzzy rule base. For learning we have used a special evolutionary algorithm. Having one rule-base is very convenient in terms of interpretability and possibility to reduce its size. In the future we will use our methods to create interpretable modular system where classifiers

References

1. Arabas, J.: Lectures on Evolutionary Algorithms (in Polish). WNT, Warsaw (2001)
2. Breiman, L.: Bias, variance, and arcing classifiers. Technical Report 460, Statistics Department, University of California (1997)
3. Eiben, A.E., Smith, J.E.: Introduction to Evolutionary Computing. Springer, Heidelberg (2003)
4. Gabryel, M., Cpalka, K., Rutkowski, L.: Evolutionary strategies for learning of neuro-fuzzy systems. In: I Workshop on Genetic Fuzzy Systems, Genewa, pp. 119–123 (2005)
5. Gabryel, M., Rutkowski, L.: Evolutionary method for learning neuro-fuzzy systems with applications to medical diagnosis (in Polish), XIV Krajowa Konferencja Naukowa Biocybernetyka i Inynieria Biomedyczna, pp. 960–965, Czstochowa (2005)
6. Gabryel, M., Rutkowski, L.: Evolutionary Learning of Mamdani-Type Neuro-fuzzy Systems. In: Rutkowski, L., Tadeusiewicz, R., Zadeh, L.A., Żurada, J.M. (eds.) ICAISC 2006. LNCS (LNAI), vol. 4029, pp. 354–359. Springer, Heidelberg (2006)
7. Korytkowski, M., Scherer, R., Rutkowski, L.: From Ensemble of Fuzzy Classifiers to Single Fuzzy Rule Base Classifier. LNCS (LNAI), vol. 5097. Springer, Heidelberg (2008)
8. Kuncheva, L.I.: Combining Pattern Classifiers, Methods and Algorithms. John Wiley & Sons, Chichester (2004)
9. Meir, R., Ratsch, G.: An Introduction to Boosting and Leveraging. Advanced Lectures on Machine Learning (2003)
10. Michalewicz, Z.: Genetic Algorithms + Data Structures = Evolution Programs, 3rd edn. Springer, Heidelberg (1996)
11. Rutkowski, L.: Flexible Neuro Fuzzy Systems. Kluwer Academic Publishers, Dordrecht (2004)
12. Rutkowski, L.: Computational Intelligence. Springer, Heidelberg (2008)
13. Schapire, R.E.: A brief introduction to boosting. In: Proc. of the Sixteenth International Joint Conference on Artificial Intelligence (1999)
14. Mertz, C.J., Murphy, P.M.: UCI respository of machine learning databases, http://www.ics.uci.edu/pub/machine-learning-databases

Fractal Dimension of Trajectory as Invariant of Genetic Algorithms

Stefan Kotowski[1,2], Witold Kosiński[1,3], Zbigniew Michalewicz[1,4],
Jakub Nowicki[1], and Bartosz Przepiórkiewicz[1]

[1] Faculty of Computer Science, Polish-Japanese Institute of Information Technology
02-008 Warszawa, Poland
[2] Institute of Fundamental Technological Research
IPPT PAN, 00-049 Warszawa, Poland
[3] Institute of Environmental Mechanics and Applied Computer Science
Kazimierz Wielki University, ul. Chodkiewicza 30, 85-064 Bydgoszcz, Poland
[4] Adelaide University, Australia
skot@ippt.gov.pl, wkos@pjwstk.edu.pl, zbyszek@cs.adelaide.edu.au

Abstract. Convergence properties of genetic algorithms are investigated. For them some measures are introduced. A classification procedure is proposed for genetic algorithms based on a conjecture: the entropy and the fractal dimension of trajectories produced by them are quantities that characterize the classes of the algorithms. The role of these quantities as invariants of the algorithm classes is presented. The present approach can form a new method in construction and adaptation of genetic algorithms and their optimization based on dynamical systems theory.

Keywords: genetic algorithm, entropy, fractal dimension, box-counting dimension, dynamical system.

1 Introduction

Evolutionary algorithms are the methods of optimizations which use a limited knowledge about investigated problem. On the other hand, our knowledge about the algorithm in use is often limited as well. Evolutionary algorithms are probabilistic optimisation algorithms. They are inspired by some observations made in the nature and concern the evolution of living organisms.

During their action several random operations are performed and each action generates a sequence of random variables, wich are candidate solutions. The action of EA/GA could be represented in the solution (or rather -search) space as a random trajectory.

The "no free lunch" results indicate that matching algorithms to problems give higher average performance than those applying a fixed algorithm to all problems.

In the view of these facts, the choice of the best algorithm may be correctly stated only in the context of the optimisation problem.

Then, to overcome both problems, one can ask whether there is the best genetic algorithm, which always generates an optimal solution for a sufficiently large class of problems. Moreover, one could ask whether one can select from the known operators such as crossover, mutation and selection, which are supreme over all the other operators for solving a given class of problems. Unfortunately, the answers to both questions are not positive.

These facts imply the necessity of searching particular genetic algorithms suitable to the problem at hand.

The present paper is an attempt to introduce an enlarged investigation method to the theory of genetic (evolutionary) algorithms. We aim at

1° the formulation of a new method of analysis of evolutionary algorithms regarded as dynamical processes, and

2° the development of some tools suitable for characterization of evolutionary algorithms based on the notions of the symbolic dynamics.

2 Genetic Algorithms

Let X be a space of solutions of an optimisation problem characterized by a fitness function

$$f : X \to \mathbf{R}, \qquad X \subset \mathbf{R}^l \tag{1}$$

for which a genetic algorithm will be invented. Each element $x \in X$ will be encoded in the form of a binary chromosome of the length N. The coding function

$$\varphi : X \to \{0,1\}^N = B \tag{2}$$

maps elements of X into chromosome from the B space.

Let us assume that the genetic algorithm acts on K-element populations. Each population forms a subset $[P^K]$ in the product space B^K, for the i-th generation we will use the denotation $[P_i^K]$, for the population and each element of this subset is a vector

$$P_i^K = [x_1^i, x_2^i, \ldots, x_K^i] \ . \tag{3}$$

Other elements of the same subset $[P_i^K]$ are obtained from the element (vector) P_i^K by any permutation of its components (factors). Let us notice that each component of this vector is an element of the space B, and some components can be repeated, i.e. different components of the vector could be the same element of the space B. One can say that a population is an equivalent class of points from the vector space B^K. The equivalent relation is defined by the class of all possible permutations of the set of K-th numbers $\{1, 2, ..., K\}$ that can be applied to permute components of a given point from B^K.

Let us notice that we can identify points from X with their encoded targets in B under the action of space X^K.

By a trajectory of the genetic algorithm of the duration M we mean a set

$$T = \bigcup_{i=1}^{M} [P_i^K] \ , \tag{4}$$

where M is the number of steps (generations) of the genetic algorithm which is realized.

Let p_m and p_c be the probabilities of the mutation and crossover, respectively, while p_s is the probability of selection, all independent from the generation.

Then, for such a genetic algorithm the probability of the appearance of the population $[P_{i+1}^K]$ at the generation $i+1$ after the population $[P_i^K]$ at the generation i, is the conditional probability[1].

$$\mathcal{P}(P_{i+1}^K | P_i^K, f(P_i^K), p_m, p_c, p_s) \ . \tag{5}$$

The initial population $[P_1^K]$ is generated by the use of a uniform probability distribution over the set B, i.e. each point from B has the same probability of being selected as a member (component) of $[P_1^K]$. Next populations following that one are the results of the action of the GA and, hence, may have a non-uniform probability distribution.

Let us notice that in view of our assumptions it follows from (5) that the probability of the appearance of each population depends on the previous population and does not depend on the history (i.e. on earlier population; the probabilities p_m, p_c and p_s can be regarded as parameters of the function $\mathcal{P}$).

3 Classification of Algorithms and Its Invariants

The convergence of GAs is one of the main issues of the theoretical foundations of GAs, and has been investigated by means of Markov's chains. The model of GA as a Markov's chain is relatively close to the methods known in the theory of dynamical systems.

In the analysis of GAs regarded as (stochastic) dynamical systems one can use the fact, (proven by Ornstein and Friedman [4,10]) which state that mixing Markov's chains are Bernoulli's systems and consequently, the entropy of the systems is a complete metric invariant.

Those facts enable us to classify GAs using the entropy. The systems for which the entropies have the same value are isomorphic. Hence the entropy makes it possible to classify GAs by splitting them into equivalence classes.

3.1 Isomorphism of Algorithms

The domain of research of the ergodic theory is a space with measure and mappings which preserve it. The measure space is the point set X with a measure m (when normalised to one, it is called the probability) defined on σ - algebra of its subsets $\mathcal{B}$, called measureable.

Definition 1. *The probability spaces* $(X_1, \mathcal{B}_1, m_1), (X_2, \mathcal{B}_2, m_2)$ *are said to be **isomorphic** if there exist* $M_1 \in \mathcal{B}_1, M_2 \in \mathcal{B}_2$ *with* $m_1(M_1) = 1 = m_2(M_2)$ *and an invertible measure preservimg transformation* $\phi : M_1 \to M_2$.

---

[1] Here by $f(P_i^K)$ we understand (cf. (3)) the vector–valued function $[f(x_1^i), f(x_2^i), \ldots, f(x_K^i)]$.

In order to investigate genetic algorithms and their similarity (or even more - isomorphism) we need to consider mappings defined on probability space.

Definition 2. *Suppose* $(X_1, \mathcal{B}_1, m_1), (X_2, \mathcal{B}_2, m_2)$ *are probability spaces together with measure preserving transformations* $T_1 : X_1 \to X_1, T_2 : X_2 \to X_2$. *We say that* T_1 *is **isomorphic to** T_2 if there exist* $M_1 \in \mathcal{B}_1, M_2 \in \mathcal{B}_2$ *with* $m_1(M_1) = m_2(M_2) = 1$ *such that*

- $T_1(M_1) \subseteq M1, T_2(M_2) \subseteq M_2$,
- *there is an invertible measure-preserving transformation*

$$\phi : M_1 \to M_2 \text{ with } \phi(T_1(x)) = T_2(\phi((x)) \text{ for all } x \in M_1.$$

The Bernoulli shift will play the main role in our approach [15,16].

Consider infinite strings made of k symbols from $[1, ..., k]$. Put $X = \prod_1^\infty \{1, ..., k\}$. An element x of X is denoted by $(x_1 x_2 x_3 ...)$.[2] Let a finite sequence $p_1, p_2, ..., p_k$, where for each i the number $p_i \in [0, 1]$ be such that $\sum_{i=1}^{k} p_i = 1$. For $t \geq 1$ define a cylinder set (or a block) of length n by

$$[a_1, a_2, ..., a_n]_{t,...,t+n-1} = \{x \in X : x_{t+1} = a_1, ..., x_{t+n} = a_n\} \tag{6}$$

With this denotation let us introduce the main definition

Definition 3. *Define a measure* μ *on cylinder sets by*

$$\mu([a_1, a_2, ..., a_n]_{t,...,t+n-1}) = p_{a_1}...p_{a_n} . \tag{7}$$

A probability measure on X*, again denoted by* μ*, is uniquely defined on the* σ *- algebra generated by cylinder sets. The **one-sided Bernoulli shift transformation** on* X *defined by*

$$(x_1 x_2 x_3 ...) \longmapsto (x_2 x_3 x_4 ...). \tag{8}$$

Let us notice that the shift preserves the measure μ.

Having the probability distribution (5) characterizing the mapping $\mathcal{T}_i$ from one population to another, we can define the entropy of the mapping

$$H(\mathcal{T}_i) = -\sum_{j=1}^{2^{NK}} \mathcal{P}(P_{i+1,j}^K | P_i^K, f(P_i^K), p_m, p_c, p_s)$$
$$\log \mathcal{P}(P_{i+1}^K | P_i^K, f(P_i^K), p_m, p_c, p_s) \tag{9}$$

where $[P_{i+1,j}^K]$ is a possible population from $B, j = 1, 2..., 2^{NK}$.

Entropy is the function of the probability of mutation and selection; it grows with the growing mutation probability and decreases when the selection pressure grows. Then the entropy could realize a measure of interactions between

---
[2] If $k = 2$ then x is said to be a **binary** sequence.

mutations and selection operators. The entropy value of the trajectory could be linked with computational complexity of the evolutionary algorithms.

Since the determination of the probability of the mapping T_i, as well as the entropy H_i, in an analytical way is rather difficult to be performed, we are proposing to substitute them with a fractal dimension which is related to the entropy [10] and can characterize non-deterministic features of GA. In paper [8] general statistical and topological methods of analysis of GAs have been introduced.

Lemma 1. *(Ornstein) Transformation corresponding to the mixed Markov chain is equivalent to the Bernoulli shift.*

Theorem 1. *(Ornstein) Every two Bernoulli shifts with the same entropy are isomorphic.*

Lemma 2. *(Choe) Let $X = \prod_1^\infty \{0,1\}$ be the $(p, 1-p)$ Bernoulli shift space that is regarded as the unit interval $[0,1]$ endowed with the Euclidean metric. Let X_p denote the set of all binary sequences $x \in X$ such that $X_p = \{x \in X : \lim_{n \to \infty} \frac{1}{n} \sum_{i=1}^{n} x_i = p\}$ then Hausdorff dimension of the x_p is equal to the entropy $-p \log_2 p - (1-p) \log_2(1-p)$ of the Bernoulli shift transformation. Similar results can be obtained for a Markov shift space.*

Moreover one can use Hausdorff's dimension or its approximation as an invariant of equivalence of algorithms.

4 Fractal Dimensions

To be more evident, let us recall the notion of the s-dimensional Hausdorff measure ([5]) of the subset $E \subset \mathbf{R}^l$, where $s \geq 0$. If $E \subset \bigcup_i U_i$ and diameter of U_i, denoted by $\delta(U_i)$, is less than ϵ for each i, we say that $\{U_i\}$ is an ϵ-cover of E. For $\epsilon > 0$, let us define

$$\mathcal{H}^s_\epsilon(E) = \inf \sum_{i=1}^{\infty} [\epsilon(U_i)]^s \qquad (10)$$

where the infimum is over all ϵ-covers $\{U_i\}$ of E. The limit of $\mathcal{H}^s_\epsilon$ as $\epsilon \to 0$ denoted by $\mathcal{H}^s(E)$, is the s-dimensional Hausdorff measure of E. Let us notice that in the space R^l one can prove that $\mathcal{H}^l(E) = \kappa_l \mathcal{L}^l(E)$, where $\mathcal{L}^l$ is the l-dimensional Lebesgue measure and κ_l is a ratio of volume of the l-dimensional cube to l-dimensional ball inscribed in the cube.

It is evident that $\mathcal{H}^s_\epsilon(E)$ increases as the maximal diameter ϵ of the sets U_i tends to zero, therefore, it requires to take finer and finer details, that might not be apparent in the larger scale into account. On the other hand for the Hausdorff measure the value $\mathcal{H}^s(E)$ decreases as s increases, and for large s this value becomes 0. Then the **Hausdorff dimension** of E is defined by

$$\dim(E) = \inf\{s : \mathcal{H}^s(E) = 0\}, \qquad (11)$$

and it can be verified that $\dim(E) = \sup\{s : \mathcal{H}^s(E) = \infty\}$.

In fractal geometry the Minkowski-Bouligand dimension or Minkowski dimension is a way of determining the fractal dimension of a set S in a Euclidean space $\mathbf{R}^n$, or more generally of a metric space (X, d). This dimension is also, less accurately, known as the packing dimension or the box-counting dimension. To calculate this dimension for a set S imagine this set lying on an evenly-spaced grid. Let us count how many boxes are required to cover the set. The **box-counting** dimension is calculated by observing how this number changes as we make the grid finer. Suppose that $N(\epsilon)$ is the number of boxes of the side length ϵ required to cover the set. Then the box-counting dimension is defined as:

$$\dim_{box}(S) = \lim_{\epsilon \to 0} \frac{\log N(\epsilon)}{\log(1/\epsilon)} \quad (12)$$

If the limit does not exist, hence, one must talk about the **upper box dimension** and the **lower box dimension** which correspond to the upper limit and lower limit, respectively, in the above expression. In other words, the box-counting dimension is well defined only if the upper and lower box dimensions are equal. The upper box dimension is sometimes called the **entropy dimension**, Kolmogorov dimension, Kolmogorov capacity or upper Minkowski dimension, while the lower box dimension is also called the **lower Minkowski dimension**. Both are strongly related to the more popular Hausdorff dimension. Only in very specialized applications it is important to distinguish between the three. Also, another measure of fractal dimension is the correlation dimension.

Both box-counting dimensions are finitely additive, i.e. if $\{A_1, A_2, ..., A_n\}$ is a finite collection of sets then $\dim_{box}(\{A_1 \cup A_2 \cup .... \cup A_n\}) = $ $=$ $\max\{\dim_{box}(A_1), \dim_{box}(A_2), ..., \dim_{box}(A_n)\}$.

In relation to the Hausdorff dimension the box-counting dimension is one of a number of definitions for the dimension that can be applied to fractals. For many well behaved fractals all these dimensions are equal. For example, the Hausdorff dimension, lower box dimension, and upper box dimension of the Cantor set are all equal to $\log(2)/\log(3)$. However, the definitions are not equivalent. The box dimensions and the Hausdorff dimension are related by the inequality

$$\dim_H(S) \leq \dim_{lowerbox}(S) \leq \dim_{upperbox}(S) \quad (13)$$

In general, both inequalities may be strict. The upper box dimension may be bigger than the lower box dimension if the fractal has different behaviour in different scales.

It is possible to define box dimensions using balls, with either the covering number or the packing number. The covering number $N_{covering}(\epsilon)$ is the minimal number of open balls of radius ϵ required to cover the fractal, or, in other words, such that their union contains the fractal. We can also consider the intrinsic covering number $N'_{covering}(\epsilon)$, which is defined in the same way but with the additional requirement that the centers of the open balls lie inside the set S. The packing number $N_{packing}(\epsilon)$ is the maximal number of disjoint balls of radius ϵ one can situate in such way that their centers would be inside the fractal. While $N, N_{covering}, N'_{covering}$ and $N_{packing}$ are not exactly identical, they are closely related, and give rise to identical definitions of the upper and lower box dimensions. This is easy to prove the following inequalities :

$$N'_{covering}(2\epsilon) \leq N_{covering}(\epsilon), \text{ and } N_{packing}(\epsilon) \leq N'_{covering}(\epsilon). \qquad (14)$$

These, in turn, follow with a little effort from the triangle inequality.

Sometimes it is just too hard to find the Hausdorff dimension of a set E, but it is possible for other definitions to have some restriction on the ϵ-covers considered in the definition. We recall here the most common alternative. It is the **box dimension**, introduced by Kolmogorov in 1961 (cf.[5]), and defined in the same way as Hausdorff dimension except that in the definition of measure only balls (discs) in $\mathbf{R}^l$ of the same radius ϵ are considered for covers of E. It follows that box dimension of E is always $\geq \dim(E)$. Moreover, the box dimension of the closure of E is the same as for the set E itself. (Thus, box dimension of the rational numbers is 1.) Harrison in [5] recommends the box dimension to be used only for closed sets, although even for compact sets it can differ from Hausdorff dimension and, moreover, the box dimension gives the most natural result than the measure $\mathcal{H}^s$.

5 Dimension of Trajectory

By inventing the fractal (Hausdorff) dimension the trajectory of GA's or its attractor can be investigated. Algorithms could be regarded as equivalent if they have the same computational complexity while solving the same problem. As the measure of computational complexity of genetic algorithm, we propose a product of population's size and the number of steps after which an optimal solution is reached. This measure of computational complexity of genetic algorithms joins the memory and the temporal complexity.

During the execution of genetic algorithms, a trajectory is realized and should "converge" to some attraction set. It is expected that an ideal genetic algorithm produces an optimal solution which, in the term of its trajectory, leads to an attractor which is one–element set. On the other hand, for an algorithm without selection the attractor is the whole space. Then, we could say that algorithms are equivalent when they produce similar attractors [6].

Our proposal is to use fractal dimensions to measure the similarity of attractors on the base of Lemma 2.

Definition 4. *Two genetic algorithms are* **equivalent** *if they realize trajectories with the same fractal dimension.*

Hence, instead of the entropy, the fractal dimension will be used as an indicator, or better to say – a measure of the classifications of GAs.

The transfer from the entropy to the new indicator can be made with the help of particular gauges. The gauge is the so-called **information dimension** of the trajectory defined by:

$$D_I(T) = -\lim_{\epsilon \to 0} \frac{\sum_{i=1}^{W(\epsilon)} p_i \ln p_i}{\ln(\frac{1}{\epsilon})} \qquad (15)$$

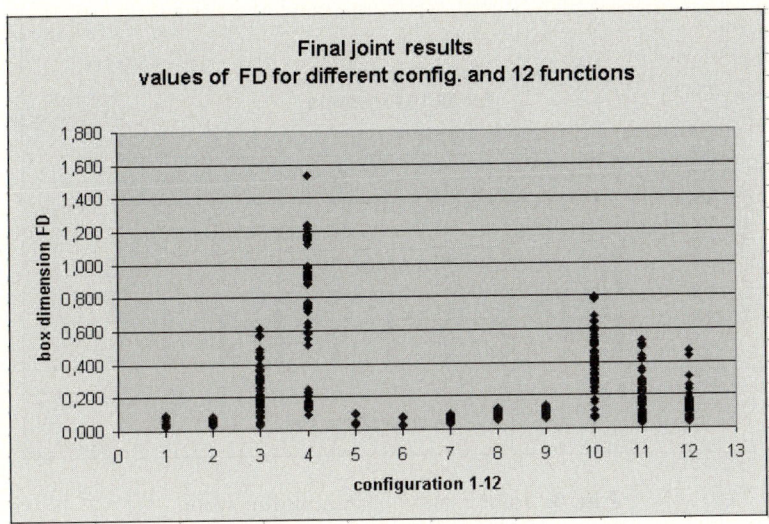

Fig. 1. Final joint results of fractal dimension

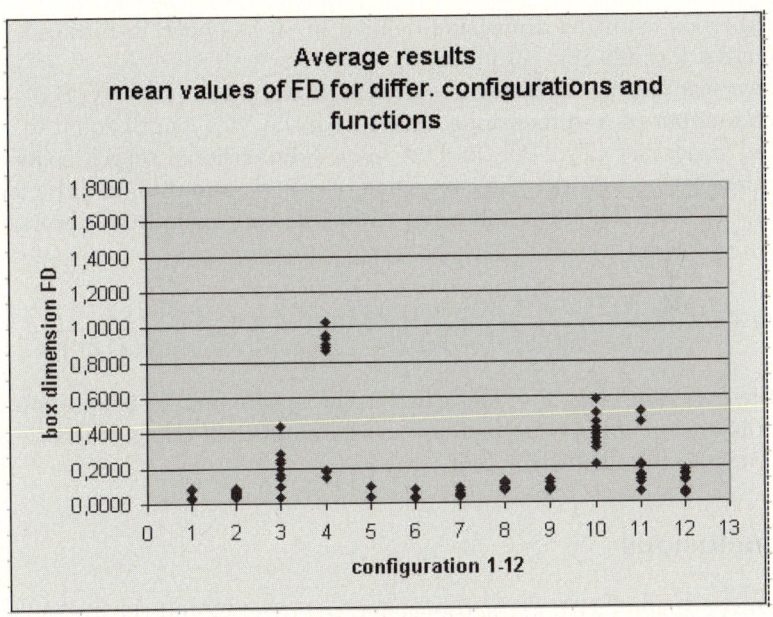

Fig. 2. Average results of fractal dimension

where $W(\epsilon)$ is the number of elements of the trajectory which are contained in a l–dimensional cube with the edge length equal to ϵ, and $p_i = \frac{N_i}{N}$ is the probability of finding of $i - th$ element, and N_i – number of points in i-th hypercube, N - number of trajectory points. In further analysis we are going

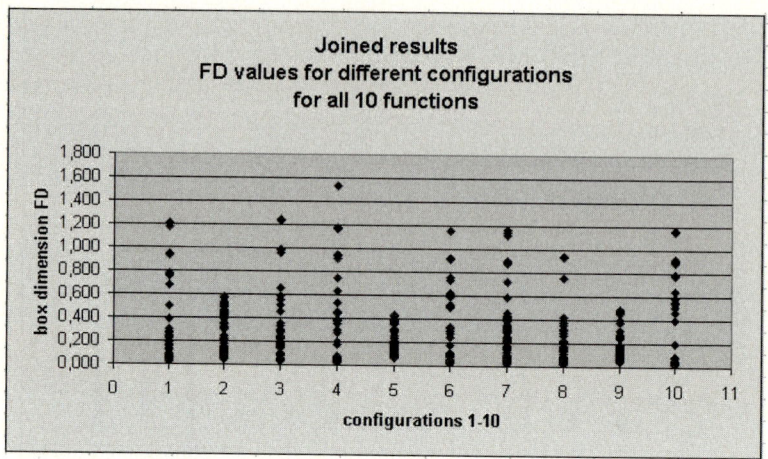

Fig. 3. Joint results of fractal dimension

to replace (15) and (12) with its approximation, namely the box or capacity dimension.

In [6] the box counting dimension defined in [3] has been introduced with its approximated formula (cf. (2) in [6]).

Here we use another approach to the approximation. Let $N(T, \epsilon)$ be the minimum number of K-dimensional cubes with the edge length equal to ϵ, that covers the trajectory $T \subset X$, and X is a l-dimensional search space. To be more evident let us consider the case when $\epsilon = 2^{-k}$ and diminish the length of cube edges by half. Then the following ratio will approximate the box counting dimension of trajectory T

$$D_c(T) \approx \frac{\log_2 N(T, 2^{-(k+1)}) - \log_2 N(T, 2^{-k})}{\log_2 2^{k+1} - \log 2^k} = \log_2 \frac{N(T, 2^{-(k+1)})}{N(T, 2^{-k})} \qquad (16)$$

due to the fact that $\log_2 x = \log_2 e \ln x$. The approximated expression (16) of the box dimension counts the increase in the number of cubes when the length of their edges is diminished by half.

6 Conclusions

The first experiments with attractors generated by GAs and the expression (16) have been performed by our co-worker in [6]. His results allow us to claim that the present approach can be useful in the GA's dynamics research.

In our paper we include new calculation results. 12 benchmark functions were used (cf. [7]) in the analysis. Experiments were performed for different dimension: 10, 15, 20 bits with operator parameters and Popsize. Then the box counting dimension was used to calculate the trajectory dimension.

As far as the analytical approach and the formal definitions of dimensions (10) and (15) are concerned their computer implementation needs additional investigations. Computer accuracy is finite, hence all limits with ϵ tending to zero will give unrealistic results. For example, if in (15) the calculation drops below the computing accuracy the expression value becomes zero or undefined. It means that we have to stop taking limit values in early stage. Hence, the questions arise: to which minimal value of ϵ the calculation should be performed and whether and how the relations with limits should be substituted with finite, non-asymptotic, expression? This, however, will be the subject of our further research.

Experimental results

The main idea of our experiments was the verification and confrontation of our theoretical considerations and conjectures with real genetic algorithms.
    On the basis of our experiments we can conclude that:

1. Selection
Change of the selection methods while preserving the other parameters does not effect the values of fractal dimension.
2. Crossover
When the number of crossover positions is changing the fractal dimension is growing with roulette selection method and is decreasing when selection is a tournament.
3. Populations
Fractal dimension is growing with the number of individuals in population.
4. Mutation probability changes have small implication on the value of fractal dimension.

   The analysis of the experimental result.
   The value of box-counting dimension of the trajectory of genetic algorithms is not random. When we use the same fitness function and the same configurations, then the box dimensions become clustered near the same value. Whole trials of the independent running attains the same values. Moreover with the different functions but the same configuration we deal with the conservation of box-counting dimension clustering.
   Average values of the box-counting dimension for the united trajectories of the algorithms from the same trial were similar to these which were calculated by averaging of the dimension of individual trajectories. This fact acknowledges the conjectures that box-counting dimension could characterize the complexity of algorithms.
   Box-counting dimension describes the way of evolution during search. Algorithms which attain the maximum in a wide loose set have bigger dimension than others which trajectories were narrow, with small differences between individuals.
   One can say that bigger box dimension characterizes more random algorithms. The main result of the experiments states that fractal dimension is the same in

the case when some boxes contains one individual as well as when these boxes contain many elements (individuals). Box dimension does not distinguish the fact that two or more elements are in the same place. They undergo counting as one element. The value of dimension should depend on the number of elements placed in each box. Our main conclusion is that good characterization is the information dimension.

Acknowledgement

The research work on the paper done by W.K., S.K. and Z.M. was supported by the KBN Project No. 3 T11 C007 28 in 2005-2007.

The authors are grateful to Dr. J. Socała for valuable discussions.

References

1. Baker, G.L., Gollub, J.P.: Chaotic Dynamics: an Introduction. Cambridge Univ. Press, Cambridge (1992)
2. Barnsley, M.F.: Lecture notes on iterated function systems. In: Denamney, R.L., Keen, L. (eds.) Chaos and Fractals. The Mathematics Behind the Computer Graphics, Proc.Symp. Appl. Math., vol. 39, pp. 127–144. American Mathematical Society, Providence, Rhode Island, (1989)
3. Falconer, K.J.: Fractal geometry. In: Math. Found. Appl., pp. 15–25. John Wiley, Chichester, Chichester (1990)
4. Friedman, N.A., Ornstein, D.S.: On isomorphisms of weak Bernoulli transformations. Adv. in Math. 5, 365–394 (1970)
5. Harrison, J.: An introduction to fractals. In: Denamney, R.L., Keen, L. (eds.) Chaos and Fractals. The Mathematics Behind the Computer Graphics, Proc. Symp. Appl. Math., American Mathematical Society, Providence, Rhode Island, vol. 39, pp. 107–126 (1989)
6. Kieś, P.: Dimension of attractors generated by a genetic algorithm. In: Proc. of Workshop Intelligent Information Systems IX, Bystra, Poland, June 12-16, pp. 40–45 (2000)
7. Michalewicz, Z.: Genetic Algorithms + Data Structures = Evolution Programs, 3rd edn. Springer, Heidelberg (1996)
8. Ossowski, A.: Statistical and topological dynamics of evolutionary algorithms. In: Proc. of Workshop Intelligent Information Systems IX, Bystra, Poland, June 12-16, pp. 94–103 (2000)
9. Ott, E.: Chaos in Dynamical Systems. Cambridge Univ. Press, Cambridge (1996)
10. Ornstein, D.S.: Ergodic theory, Randomness and Dynamical Systems. Yale Univ. Press (1974)
11. Vose, M.D.: Modelling Simple Genetic Algorithms. Evolutionary Computation 3(4), 453–472 (1996)
12. Wolpert, D.H., Macready, W.G.: No Free Lunch Theorems for Optimization. IEEE Transaction on Evolutionary Computation 1(1), 67–82 (1997), http://ic.arc.nasa.gov/people/dhw/papers/78.pdf
13. Igel, C., Toussaint, M.: A No-Free-Lunch Theorem for Non-Uniform Distributions of Target Functions. Journal of Mathematical Modelling and Algorithms 3, 313–322 (2004)

14. English, T.: No More Lunch: Analysis of Sequential Search. In: Proceedings of the 2004 IEEE Congress on Evolutionary Computation, pp. 227–234 (2004), http://BoundedTheoretics.com/CEC04.pdf
15. Szlenk, W.: An Introduction to the Theory of Smooth Dynamical Systems, PWN, Warszawa. John Wiley &Sons, Chichester (1984)
16. Choe, G.H.: Computational Ergodic Theory. Springer, Heidelberg, New York (2005)

Global Induction of Decision Trees: From Parallel Implementation to Distributed Evolution

Marek Krętowski and Piotr Popczyński

Faculty of Computer Science, Białystok Technical University
Wiejska 45a, 15-351 Białystok, Poland
mkret@wi.pb.edu.pl

Abstract. In most of data mining systems decision trees are induced in a top-down manner. This greedy method is fast but can fail for certain classification problems. As an alternative a global approach based on evolutionary algorithms (EAs) can be applied. We developed *Global Decision Tree* (GDT) system, which learns a tree structure and tests in one run of the EA. Specialized genetic operators are used, which allow the system to exchange parts of trees, generate new sub-trees, prune existing ones as well as change the node type and the tests. The system is able to induce univariate, oblique and mixed decision trees. In the paper, we investigate how the *GDT* system can profit from a parallelization on a compute cluster. Both parallel implementation and distributed version of the induction are considered and significant speedups are obtained. Preliminary experimental results show that at least for certain problems the distributed version of the *GDT* system is more accurate than its panmictic predecessor.

1 Introduction

Evolutionary algorithms [17] are methods inspired by the process of natural evolution, which are applied to solve many difficult optimization and search problems. Among others they are successfully used in various knowledge discovery systems [7]. However, it is known that the evolutionary approach is not the fastest one and a lot of effort is put into speeding it up. This issue is especially important for future data mining applications, where larger and larger datasets will be processed and analyzed.

Fortunately evolutionary techniques are naturally prone to parallelism and in most of the cases they can be efficiently implemented in distributed environments [1]. Such an implementation can be only aimed at speeding up the calculations without changing the original sequential algorithm (so called *global parallelism*) or can try to extend it by using structured populations. Among the most widely known types of structured EAs are *distributed* (*coarse-grained*) and *cellular* (*fine-grained*) algorithms. In the distributed version a population is partitioned into several independent subpopulations (islands) performing a sparse exchange of individuals, whereas in the cellular algorithm individuals evolve in overlapped small neighborhoods.

In our previous papers, *Global Decision Tree (GDT)* system was introduced for a global induction of decision trees based on evolutionary algorithms. The system was able to generate accurate and simple univariate [12], oblique [13] and mixed [14] trees. In this paper we want to investigate how the global induction of decision trees can profit from a parallelization on a compute cluster. Among many top-down (e.g. [5],[4]) and global (e.g. [10], [8], [19]) evolutionary decision tree inducers, according to our knowledge only two attempts in the framework of the fine-grained approaches were developed: *CGP/SA - (Cellular Genetic Programming/Simulated Annealing)* [6] and *GALE - (Genetic and Artificial Life Environment)* [16].

The rest of the paper is organized as follows. In the next section the *GDT* system is briefly recalled. In section 3 a parallel implementation of the global induction is presented and in section 4 a distributed version is investigated. Experimental validation of the presented approaches is performed in section 5. The paper is concluded and possible research directions are sketched in the last section.

2 GDT System

In this section, the *GDT* system is briefly presented. The system is able to generate univariate [12], oblique [13] and mixed [14] decision trees. Its general structure follows a typical framework of evolutionary algorithms [17] with an unstructured population and a generational selection.

2.1 Representation, Initialization and Termination Condition

A decision tree is a complicated tree structure, in which the number of nodes, test types and even the number of test outcomes are usually not known in advance for a given learning set. Moreover additional information, e.g. about learning vectors associated with each node, should be accessible during the induction. As a result, decision trees are not specially encoded in individuals and they are represented in their actual form.

Three test types can be used in non-terminal nodes depending on the decision tree type: two types of univariate tests for an univariate tree, only oblique tests for an oblique tree and all previously mentioned for a mixed tree.

In case of univariate tests, a test representation depends on the considered attribute type. For nominal attributes at least one attribute value is associated with each branch starting in the node, which means that an internal disjunction is implemented. For continuous-valued features typical inequality tests with two outcomes are used. In order to speed up the search process only boundary thresholds are considered as potential splits and they are calculated before the EA starts. In an oblique test with binary outcome a splitting hyperplane is represented by a fixed-size table of real values corresponding to a weight vector and a threshold. The inner product is calculated to decide where an example is routed.

Before starting the actual evolution, an initial population is created by applying a simple top-down algorithm based on a dipolar principle [11] to randomly selected sub-samples of the learning set [14].

The evolutionary induction terminates when the fitness of the best individual does not improve during a fixed number of generations (default value is equal to 1000) or the maximum number of generations (default value: 5000) is reached.

2.2 Variation Operators

We use two specialized genetic operators corresponding to the classical mutation and cross-over. Application of any operator can result in a necessity for relocation of the learning examples between tree parts rooted in the modified nodes. Additionally the local maximization of the fitness can be performed by pruning lower parts of the sub-tree on condition that it improves the value of the fitness.

Mutation operator. A mutation-like operator [14] is applied with a given probability to a tree (default value is 0.8) and it guarantees that at least one node of the selected individual is mutated. First, the type of the node (leaf or internal node) is randomly chosen with equal probability and if a mutation of a node of this type is not possible, the other node type is chosen. A ranked list of nodes of the selected type is created and a mechanism analogous to a ranking linear selection [17] is applied to decide which node will be affected.

While concerning internal nodes, the location (the level) of the node in the tree and the quality of the subtree rooted in the considered node are taken into account. Nodes on lower levels of the tree are mutated with higher probability, which promotes local changes. Nodes on the same level are sorted according to the number of misclassified objects by the subtree. As for leaves, the number of misclassified learning examples in leaves is used to put them in order, but homogenous leaves are excluded. As a result, leaves for which classification accuracies are worse are mutated with higher probability.

Modifications performed by a mutation operator depend on the tree type and the node type (i.e. if the considered node is a leaf node or an internal node). For a non-terminal node a few possibilities exist:

- A completely new test (of the approved type) can be found. A pair of objects from different classes (mixed dipole) in this node is randomly chosen and a test which separates them to distinct sub-trees is searched. In case of a univariate tree, such a test can be constructed directly for any feature with different feature values. When an oblique test is considered, the splitting hyperplane is perpendicular to the segment connecting two drawn objects and placed in a halfway position.
- The existing test can be altered by shifting the splitting threshold (continuous-valued feature), by re-grouping feature values (nominal features) or by shifting the hyperplane (oblique test). These modifications can be

purely random or can be guided by the dipolar principles [11] of splitting mixed dipoles and avoiding to split pure ones.
- A test can be replaced by another test or tests can be interchanged.
- One sub-tree can be replaced by another sub-tree from the same node.
- A node can be transformed (pruned) into a leaf.

Modifying a leaf makes sense only if it contains objects from different classes. The leaf is transformed into an internal node and a new test is chosen in the aforementioned way.

Cross-over operator. There are also several variants of exchanging information between individuals. Most of them start with a random selection of one node in each of two affected trees. In the first variant, the subtrees starting in the selected nodes are exchanged. In the second variant, which can be applied only when non-internal nodes are randomly chosen and the numbers of outcomes are equal, only tests associated with the nodes are exchanged. The third variant is also applicable with the same assumptions and branches which start from the selected nodes are exchanged in random order. There is also a variant of crossover inspired by the dipolar principles. In the internal node in the first tree a cut mixed dipole is randomly chosen and for the cross-over the node with the test splitting this dipole is selected in the second tree.

2.3 Fitness Function

A fitness function drives the evolutionary search and is very important and sensitive component of the induction. It is well-known that it is not possible to optimize directly the classification quality on unseen data. Instead, the reclassification quality measured on the learning set can be used, but this may lead to an over-fitting problem. In order to mitigate the problem a complexity term is incorporated into the fitness function. In the GDT system the fitness function is maximized and has the following form:

$$Fitness(T) = Q_{Reclass}(T) - \alpha \cdot (Comp(T) - 1.0), \qquad (1)$$

where $Q_{Reclass}(T)$ is the reclassification quality of the tree T and α is the relative importance of the classifier complexity (default value is 0.005).

In case of the univariate trees, the tree complexity $Comp(T)$ is defined as the classifier size which is equal to the number of nodes. The penalty associated with the classifier complexity increases proportionally with the tree size and prevents classifier over-specialization. Subtracting 1.0 eliminates the penalty when the tree is composed of only one leaf (in majority voting).

A little bit more elaborated definition of the tree complexity is used when oblique tests are allowed (i.e. in oblique and mixed trees), because it reflects the complexity of tests:

$$Comp(T) = |N_{leaf}(T)| + \sum_{n \in N_{int}(T)} (1 + \beta \cdot (F(n) - 1)), \qquad (2)$$

where $N_{leaf}(T)$ and $N_{int}(T)$ are sets of leaves and internal nodes respectively, $F(n)$ is the number of features used in the test associated with the node n and $\beta \in [0,1]$ is the relative importance of the test complexity (default value 0.2). The complexity of the tree is defined as a sum of the complexities of the nodes and it is assumed that for leaves and internal nodes with univariate tests the node complexity is always equal to 1.0. It can be also observed that when $\beta = 1$ the number of features included in the test is used as the node complexity, and if $\beta = 0$, then the node complexity is 1.0.

3 Parallel Implementation of Global Induction

The most costly (time consuming) operation in the typical evolutionary algorithm is evaluation of the fitness of all individuals in the population. As the fitness is calculated independently for every individual this feature gives the most straightforward way of exploiting parallelism in the evolutionary search. The population is evenly distributed to available processors (slaves) and the fitness is calculated in parallel. The remaining parts of the evolution are executed in the master processor.

In evolutionary data mining an individual usually represents a certain type of classifier and its goodness of fit is evaluated by using the learning set. When this set is large, it can be also profitable to divide the learning set into subsets and to distribute them among processors. In each computational node the partial fitness calculations are performed in parallel and finally the overall fitness is summed up in the master processor [15].

In case of the global induction of decision trees none of the aforementioned approaches can be directly applied. In the GDT system, information about the learning vectors reaching any node of decision tree is stored in that node. It allows us to apply the genetic operators efficiently and to directly obtain the fitness corresponding to the individual. In fact, the actual fitness calculation is embedded into the post mutation and cross-over processing, when the learning vectors in the affected parts of the tree (or trees) are relocated. This mechanism increases the memory complexity of the induction but significantly reduces its computational complexity. As a consequence, the most time consuming elements of the algorithm are genetic operators and they should be performed in parallel.

The general scheme of a single generation in the parallel implementation of the global induction is presented in Fig. 1. The proposed approach is based on the classical master-slave model. After the selection and the reproduction, subpopulations are sent to the slave processors where individual variation operations are performed. When all individuals in the subpopulation migrated to the given processor are processed, they are sent back to the master processor, where the rest of the algorithm is performed. It should be noticed that also an initial population of the algorithm is created in parallel by slave processors.

Migrating an individual between processors within the framework of the message-passing interface is composed of 3 steeps: packing the individual into a flat message, transferring the message between processors (sending/receiving)

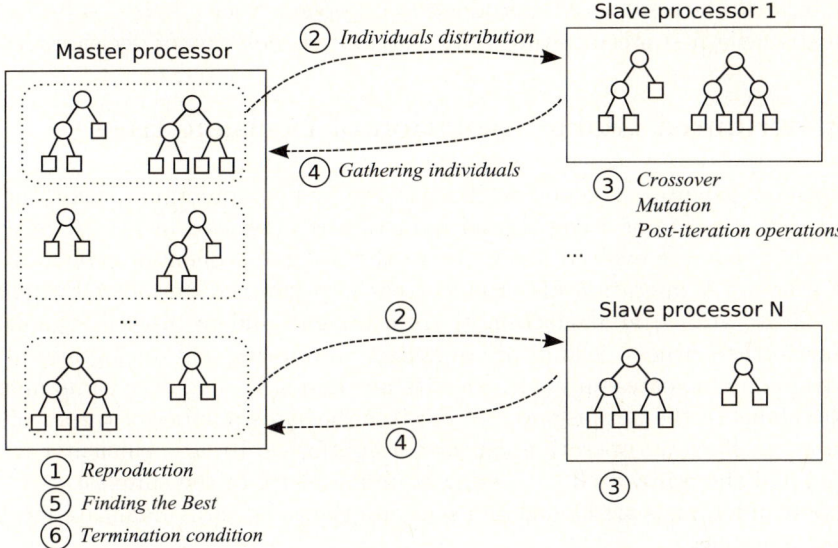

Fig. 1. Parallel implementation of the global decision tree induction based on the master-slave model

and unpacking the corresponding tree. In order to minimize the message size information about learning vectors associated with tree nodes is not stored in the message. Only the number of vectors is included in the message and this speeds up the reconstruction during the message unpacking in the target processor. It should be also noticed that in the master processor, when centralized parts of the algorithm are executed (i.e. the reproduction with elitism and the termination condition verification) packed individuals can be processed on condition that the corresponding values of the fitness are known. This way unnecessary unpacking-packing operations can be eliminated in the master processor. Additionally, a certain number of individuals from the given slave processor survive (or replicate) and in the next iteration they are scheduled to be sent back to that processor. This observation gives another possibility to eliminate unproductive calculation, however the cloned trees should be sent to different processors in order to avoid crossing with identical or very similar individuals.

In spite of all it can be expected that the computational cost associated with scattering and gathering of decision trees is high and the significant improvement in the speedup can be obtained only by reducing migrations. One of the simplest solutions consists in performing the population synchronization not in every iteration. Hence, in these iterations when the synchronization is not performed, subpopulations evolve locally. There are two elements which become different compared to the original algorithm: reproduction and crossover. In every subpopulation the best individual is searched and the reproduction is performed locally. Moreover, individuals can mate only with other individuals located in the same processor. After a few local iterations (default value: 10) a standard iteration with the centralized reproduction is applied. Such a hybrid solution can be treated as a first steep toward fully distributed evolution of decision trees.

4 Distributed Global Evolution of Decision Trees

The proposed solution is based on the classical multi-population (island) model. The general scheme of a distributed induction is presented in Fig. 2. In every dem individuals are evolved locally as in the original sequential version of the *GDT* system. A migration of selected trees is performed periodically (default value: 20 iterations) and selection of trees to send and to discard is done according to their fitness. Islands are organized into a ring and during migrations any dem communicates only with two its neighbors. It sends the same number of individuals to the next island and receives the same number of trees from its predecessor. Such an operation can be straightforwardly and efficiently implemented and the centralized processing is unnecessary. In the simplest case only local best individuals are cloned and sent and the worst individuals are replaced by incoming ones.

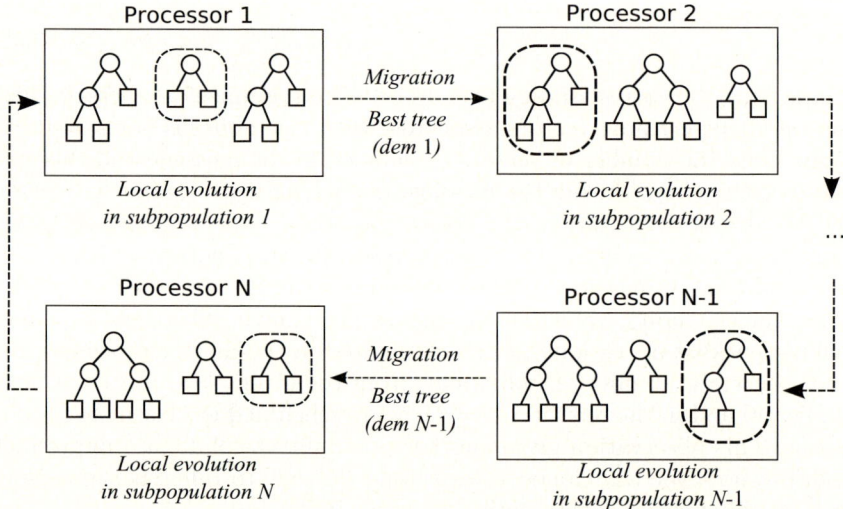

Fig. 2. Distributed evolution based on the island model (ring topology and migration of single individual)

The termination condition of the distributed evolution is verified in a centralized way after migrations. If the overall best value of the fitness does not improve during the fixed number of generations, the algorithm is finished even if the maximum number of generations is not reached. The resulting decision tree of the distributed evolution is chosen according to the fitness function among the best local individuals stored in each island.

5 Experimental Results

In this section the proposed parallel and distributed versions of global decision tree induction are experimentally verified. In order to focus and clarify the presentation only results concerning the mixed decision trees, as the most representative, are included. All presented results were obtained with a default setting of parameters from the sequential version of the *GDT* system.

In the experiments a cluster of sixteen SMP servers running Linux 2.6 and connected by an Infiniband network was used. Each server is equipped with two 64-bit Xeon 3.2GHz CPUs with 2MB L2 cache, 2GB of RAM and an Infniband 10 Gb/s HCA connected to a PCI-Express port. We used the MVAPICH version 0.9.5 [9] MPI implementation.

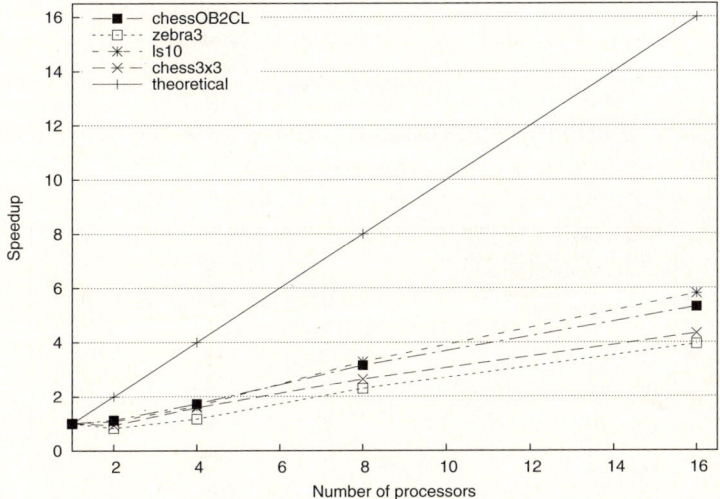

Fig. 3. Efficiency of parallel implementation: speedup measured on exemplar datasets

In Fig. 3 scalability of the parallel implementation measured on four exemplar artificial datasets[1] is presented. It can be observed that the obtained speedup is satisfactory. In order to explain this result detailed time-sharing is necessary and it is presented for one dataset in Fig. 4. It is clearly visible that the problem is caused by the overhead connected with unpacking individuals and redistributing the learning examples to tree nodes at slave processors. The time of applying the variation operators (denoted as *Operation time*) is properly reduced. It can be also noticed that the sending/receiving time (denoted as *MPI time*) is relatively small and it increases with the number of slave processors.

In the next experiment it was verified that the significant increase of speedup can be obtained by reducing migrations. In Fig. 5 execution times measured for

---

[1] For detailed description of artificial datasets used in the experiments please refer to [14].

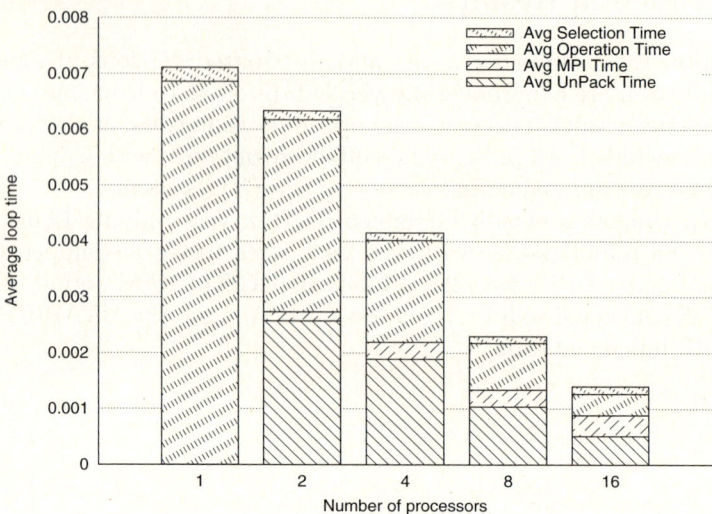

Fig. 4. Efficiency of parallel implementation: detailed time-sharing on *chessOB2CL* dataset

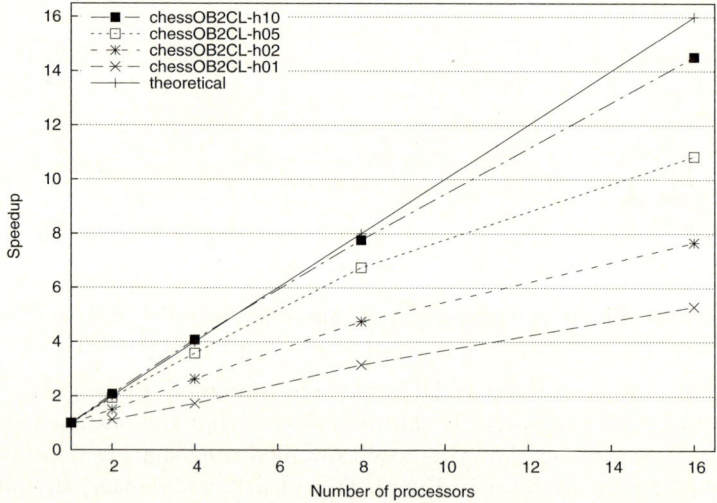

Fig. 5. Speedup of the hybrid solution measured on *chessOB2CL* dataset (with different rate of migrations)

the hybrid implementation with increasing number of local iterations (without centralized reproduction) are presented. It can be observed that sparser migrations result in the important improvement in terms of speedup (almost linear speedup obtained for the run with 10 local iterations after the population synchronization).

Table 1. Results on artificial data

Dataset	C4.5 size	C4.5 quality	OC1 size	OC1 quality	GDT size	GDT quality	dGDT size	dGDT quality
chess2x2	1	50	10.1	89.3	4	99.8	4	99.9
chess3x3	9	99.7	21.1	73.7	9	99.3	9	99.6
chessOB2CL	33	95.6	7	77.3	4.1	99.1	4	99.7
chessOB4CL	35	94.6	4.3	49.8	4	98.9	4	99.7
house	21	97.4	8.2	92.8	3.8	96.9	5	99.5
ls10	284	77.3	7.3	95.3	2	97.6	2	98.2
ls2	22	97	2	99.7	2	99.9	2	99.9
zebra1	25	95.3	3	83.5	3	99.6	3	99.9
zebra2	2	59.5	4.8	94.1	4	98.2	4	99.6
zebra3	57	91.2	8.2	24.3	8.4	97	9.3	98.4

Table 2. Results on real datasets

Dataset	C4.5 size	C4.5 quality	OC1 size	OC1 quality	GDT size	GDT quality	dGDT size	dGDT quality
balance-scale	57	77.5	5.4	90	2.6	89.5	9.3	94.8
bcw	22.8	94.7	4.7	91.2	2	96.7	2.1	96.7
bupa	44.6	64.7	5.8	65.6	3.5	68.6	32.6	64.1
glass	39	62.5	4.5	55.7	11.6	66.4	31.6	68.5
page-blocks	82.8	97	15.6	96.6	3	94.9	7.3	96.6
pima	40.6	74.6	6.5	69.6	2.2	75.4	13.6	74.4
sat	435	85.5	58.3	78.9	6.4	81.5	14.6	84.8
vehicle	138.6	72.7	21.6	66.4	8.8	65.4	42.3	70.3
waveform	107	73.5	10.5	77.4	4.2	80.5	12.3	81.1
wine	9	85	3.2	87	4.2	91.5	5	88.8

As it can be expected based on the experiments with the hybrid implementation, in case of the distributed version of the *GDT* system almost linear speedup is achieved (15.4 for 16 processors). In the final experiments, an impact of the distributed evolution on accuracy and complexity of decision tree classifiers is verified.

The distributed approach (denoted as *dGDT* in tables) is assessed on 10 artificial and 10 real life datasets and is compared to the well-known top-down univariate (*C4.5* [20]) and oblique (*OC1* [18]) decision tree systems. It is also compared to the sequential versions of the global inducer - *GDT*. All prepared artificial datasets comprise training and testing parts. In case of data from *UCI Repository* [2] for which testing data are not provided, a 10-fold stratified cross-validation was employed. The population size for *dGDT* was set to 256 giving 8 sub-populations of 32 individuals each.

In table 1 and table 2 results of experiments with artificial and real life datasets are presented. It can be observed that two global inducers performed better both in terms of accuracy and classifier complexity that its top-down counterparts on artificial datasets. It seems that $dGDT$ was even slightly better than the sequential version on these datasets. In case of real life datasets, the situation is more equilibrated. For six datasets $dGDT$ was more accurate then GDT, but for three problems the sequential version was better. It can be also noticed that the distributed inducer produced more complex trees. Compared to the top-down systems both global inducers performed well, however for certain problems they were outperformed.

6 Conclusions

In the paper, the parallelization of the global decision tree induction based on evolutionary algorithms is investigated. It was shown that the migration of the individuals is a sensitive point of the process and the cost associated with the migration is relatively high. As a result the most efficient solution is based on the multi-population model. Moreover, it was experimentally shown that at least for certain problems the distributed version of the system is more accurate than its sequential predecessor.

The GDT system is still in development and many possible directions of future improvements exist. Currently we are working on more sophisticated island topologies, where random but still decentralized and balanced migrations will be possible. We also plan to implement the global induction in the framework of shared memory multiprocessing (OpenMP), which will reduce the problem of migrations.

Acknowledgments

This work was supported by the grant W/WI/5/08 from Białystok Technical University.

References

1. Alba, E., Tomassini, M.: Parallelism and evolutionary algorithms. IEEE Transactions on Evolutionary Computation 6(5), 443–462 (2002)
2. Blake, C., Keogh, E., Merz, C.: UCI repository of machine learning databases (1998), http://www.ics.uci.edu/~mlearn/MLRepository.html
3. Breiman, L., Friedman, J., Olshen, R., Stone, C.: Classification and Regression Trees. Wadsworth Int. Group (1984)
4. Cantu-Paz, E., Kamath, C.: Inducing oblique decision trees with evolutionary algorithms. IEEE Transactions on Evolutionary Computation 7(1), 54–68 (2003)
5. Chai, B., et al.: Piecewise-linear classifiers using binary tree structure and genetic algorithm. Pattern Recognition 29(11), 1905–1917 (1996)
6. Folino, G., Pizzuti, C., Spezzano, G.: Genetic Programming and Simulated Annealing: A Hybrid Method to Evolve Decision Trees. In: Poli, R., Banzhaf, W., Langdon, W.B., Miller, J., Nordin, P., Fogarty, T.C. (eds.) EuroGP 2000. LNCS, vol. 1802, pp. 294–303. Springer, Heidelberg (2000)

7. Freitas, A.: Data Mining and Knowledge Discovery with Evolutionary Algorithms. Springer, Heidelberg (2002)
8. Fu, Z., Golden, B., Lele, S., Raghavan, S., Wasil, E.: A genetic algorithm-based approach for building accurate decision trees. INFORMS Journal on Computing 15(1), 3–22 (2003)
9. Liu, J., Wu, J., Panda, D.K.: High performance RDMA-based MPI implementation over InfiniBand. Int. Journal of Parallel Programming 32(3), 167–198 (2004)
10. Koza, J.: Concept formation and decision tree induction using genetic programming paradigm. In: Schwefel, H.-P., Männer, R. (eds.) PPSN 1990. LNCS, vol. 496, pp. 124–128. Springer, Heidelberg (1991)
11. Krętowski, M.: An Evolutionary Algorithm for Oblique Decision Tree Induction. In: Rutkowski, L., Siekmann, J.H., Tadeusiewicz, R., Zadeh, L.A. (eds.) ICAISC 2004. LNCS (LNAI), vol. 3070, pp. 432–437. Springer, Heidelberg (2004)
12. Krętowski, M., Grześ, M.: Global learning of decision trees by an evolutionary algorithm. In: Information Processing and Security Systems, pp. 401–410. Springer, Heidelberg (2005)
13. Krętowski, M., Grześ, M.: Evolutionary Learning of Linear Trees with Embedded Feature Selection. In: Rutkowski, L., Tadeusiewicz, R., Zadeh, L.A., Żurada, J.M. (eds.) ICAISC 2006. LNCS (LNAI), vol. 4029, pp. 400–409. Springer, Heidelberg (2006)
14. Krętowski, M., Grześ, M.: Evolutionary induction of mixed decision trees. International Journal of Data Warehousing and Mining 3(4), 68–82 (2007)
15. Kwedlo, W., Krętowski, M.: Learning decision rules using a distributed evolutionary algorithm. TASK Quarterly 6(3), 483–492 (2002)
16. Llora, X., Garrell, J.: Evolution of decision trees. In: Proc. of CCAI 2001, pp. 115–122. ACIA Press (2001)
17. Michalewicz, Z.: Genetic Algorithms + Data Structures = Evolution Programs, 3rd edn. Springer, Heidelberg (1996)
18. Murthy, S., Kasif, S., Salzberg, S.: A system for induction of oblique decision trees. Journal of Artificial Intelligence Research 2, 1–33 (1994)
19. Papagelis, A., Kalles, D.: Breeding decision trees using evolutionary techniques. In: Proc. of ICML 2001, pp. 393–400. Morgan Kaufmann, San Francisco (2001)
20. Quinlan, J.: C4.5: Programs for Machine Learning. Morgan Kaufmann, San Francisco (1993)

Particle Swarm Optimization with Variable Population Size

Laura Lanzarini, Victoria Leza, and Armando De Giusti

III-LIDI (Institute of Research in Computer Science LIDI)
Faculty of Computer Sciences. National University of La Plata
La Plata, Buenos Aires, Argentina

Abstract. At present, the optimization problem resolution is a topic of great interest, which has fostered the development of several computer methods forsolving them.

Particle Swarm Optimization (PSO) is a metaheuristics which has successfully been used in the resolution of a wider range of optimization problems, including neural network training and function minimization. In its original definition, PSO makes use, during the overall adaptive process, of a population made up by a fixed number of solutions.

This paper presents a new extension of PSO, called VarPSO, incorporating the concepts of age and neighborhood to allow varying the size of the population. In this way, the quality of the solution to be obtained will not be affected by the used swarms size.

The method here proposed is applied to the resolution of some complex functions, finding better results than those typically achieved using a fixed size population.

Keywords: Evolutionary Computation, Swarm Intelligence, Particle Swarm Optimization, Function Optimization.

1 Introduction

Optimization, in the sense of finding the best solution or at least an acceptable one for a given problem, is a field of critical importance in the real life. We are constantly solving optimization problems, such as for example the shortest way to get to a place or the organization of our daily duties so as to spend as little time as possible. However, when the problem to solve is extremely complex it is essential to have the computer tools to address it [9].

Due to the great importance of optimization problems, several computer methods have been developed to solve them, which can be generally classified into exact and approximate. At present, in the context of approximate solutions, research has been focused on design and application of metaheuristics, which are based on the integration of local improvement procedures and high level strategies creating efficient processes in terms of computing times and memory space. Such metaheuristics are capable of delivering a good solution, i.e., relatively closed to the optimal, by examining just a small subset of solution of the overall number. Particle Swarm Optimization (PSO) is a metaheuristics which has been successfully used in the resolution of a wider range of optimization problems, including neural network training and function minimization. In the original definition of PSO, the quantity of solutions analyzed remains fixed during the adaptive process [6] [11].

2 Objective

This paper aims at presenting a new extension of PSO, called VarPSO, incorporating the concepts of age and neighborhood to allow varying the size of the population. In this way, not only the quality of the solution to be obtained will be independent of the initial swarms size but also the commitment relationship existing between the convergence speed and the population diversity will be improved. The variation of the population size is based on a modification of the adaptive process allowing adding and/or deleting individuals in function of its capacity to solve the posed problem. This is carried out through the concept of age, which allows determining the time of permanence of each element within the population. In addition, since PSO tends to drive particles towards the explored areas with good fitness and in order to not excessively overpopulate certain solution spaces, each individuals environment is analyzed and the worst solutions of the areas with larger number of particles are eliminated.

The method here proposed, VarPSO, is applied to the resolution of some complex functions, finding better results than those typically achieved using a fixed size population.

This paper is organized as follows: Section 3 briefly describes the basic PSO algorithm; Section 4 presents the concepts necessary to carry out the variation of the swarms size; Section 5 presents in detail the proposed algorithm; Section 6 introduces the results obtained, and Section 7, the conclusions and future lines of work.

3 Swarm Particle-Based Algorithms

A Swarm Particle-based algorithm, also called Particle Swarm Optimization (PSO), is a heuristic population technique in which each individual represents a potential solution to the problem and makes its adaptation taking into account three factors: its knowledge on the environment (its fitness value), its historical background or previous experiences (its memory), and its historical background or previous experiences of its neighborhoods individuals [6]. Its objective consists in evolving its behavior so as to look like those most successful individuals within its environment. In this type of technique, each individual remains constantly moving within the search space and never dies. On its part, the population can be considered as a multi-agent system in which each individual or particle moves around the search space saving and, eventually, communicating the best solution found.

There exist various versions of PSO; the most known are gBest PSO, which uses as neighborhood criterion the overall population, and lBest PSO which, on the other hand, makes use of a small sized population [6] [11]. The size of the neighborhood impacts on the algorithm convergence neighborhood as well as on the diversity of the populations individuals. When the neighborhood size increases, the convergence of the algorithm is faster but the diversity of individuals is lower. Each particle p_i is made up by three vectors and two fitness values:

- Vector $x_i = (x_{i1}, x_{i2}, , x_{in})$ stores the current position of the particle in the search space.
- Vector $pBest_i = (p_{i1}, p_{i2}, , p_{in})$ stores the best position of the solution found by the particle up to the moment.
- Speed vector $v_i = (v_{i1}, v_{i2}, , v_{in})$ stores the gradient (direction) according to which the particle will move.
- The fitness $value fitness\_x_i$ stores the current solution capacity value (vector x_i). The fitness value $fitness_p Best_i$ stores the capacity value of the best local solution found up to the moment(vector $pBest_i$).

The position of a particle is updated as follows

$$x_i(t+1) = x_i(t) + v_i(t+1) \qquad (1)$$

As previously explained, the speed vector is modified taking into account its experience and the environments. The expression is the following:

$$v_{ij}(t+1) = w.v_i(t) + \varphi_1.rand_1.(pBest_i - x_i(t)) + \varphi_2.rand_2.(g_i - x_i(t)) \qquad (2)$$

where w represents the inertia factor [10], φ_1 and φ_2 are acceleration constants, $rand_1$ and $rand_2$ are random values belonging to the interval (0,1), and g_i represents the position of the particle with the best fitness of the environment of p_i (lBest o localbest) or the whole swarm (gBest o globalbest). Values of w, φ_1 and φ_2 are essential to assure the algorithms convergence. For more details on the selection of these values, consult [3] and [1].

4 Variable Population Size Particle Swarms

The variation of the populations size is accomplished by allowing the particle to reproduce and die during the adaptation process. This requires the definition of mechanisms regulating the corresponding insertion and elimination processes.

4.1 Life Time

One of the most important concepts of the proposed strategy is the particle's life time, since it determines the duration of its permanence within the population. Such value is expressed in quantity of iterations, which once over, the particle is removed. This value is closely related to the capacity of each particle and allows the best to remain longer in the population, influencing the behavior of the others.

In order to assess the life time of each population individual, the method of assignment by classes defined in [7] was used, since it has proved to be capable of providing good results with a smaller quantity of individuals than that applied by conventional methods. In [7], individuals of a population are grouped according to their capacity value in k classes using a winner-take-all-type competitive clustering method. Over the result of this clustering, we can apply one of the following methods:

a) Fixed life time assignment by class. The maximum time of the life time to assign is divided by the quantity of classes, k. This allows us to know the time range corresponding to each class. Within a same class, its individuals will receive a life time proportional to the class to which they belong and to the quantity of individuals in the same class, as follows: $WidthClass := MAX\_LT/k$
$TVPrev := (NumC - 1) * WidthClass$
$TVCurrent := WidthClass$
$Displacement = (fitness[i] - Class[NumC].MinFit)/$
$\qquad\qquad abs(Class[NumC].MaxFit - Class[NumC].MinFit))$
$LifeTime[i] := trunc(TVPrev + TVCurrent * Displacement)$

where

- $WidthClass$ is the life time range assigned to each class.
- $NumC$ is the number of class to which each individual belongs.
- $TVCurrent$ is the life time range of the class to which the individual belongs.
- $TVPrev$ is the life time range assigned to the classes previous to $NumC$.
- $Class[NumC].MinFit$ and $Class[NumC].MaxFit$ are the values of minimum and maximum capacity of the class to which the individual under consideration belongs.
- $Fitness[i]$ is the value of the population i-th individuals capacity.

b) Life time assignment proportional to the quantity of each classs individuals. Each class receives a life time range proportional to the quantity of elements it contains. That is, individuals belonging to numerous classes could have a wider life time range. The computation is as follows:
$TotalPrev := 0$
for $i := 1$ to $NumC - 1$ do $TotalPrev := TotalPrev + Class[i].Cant$
$TVPrev := MAX\_LT * TotalPrev/TotalIndiv$
$TVCurrent := MAX\_LT * Class[NumC].Cant/TotalIndiv$
$Displacement = (fitness[i] - Class[NumC].MinFit)/$
$\qquad\qquad abs(Class[NumC].MaxFit - Class[NumC].MinFit))$
$LifeTime[i] := trunc(TVPrev + TVCurrent * Displacement)$

where

- $Class[NumC].Cant$ represents the total quantity of individual of the closest class.
- $TotalIndiv$ is the total quantity of the individuals of the population.

These two ways of computing the individuals life time should be combined in order to achieve the proper assignment. We propose to apply assignment b) during acertain percentage of the algorithms maximum generation quantity and apply a) in the remaining. This is due to the fact that initial clustering is carried out over individuals which are not completely adapted yet and, thus, they give place to highly dissimilar sized clusterings. If we directly apply the distribution indicated in a) over these clusterings, several individuals will received similar life times, leading the algorithm to unnecessarily increase the quantity of individuals in the population.

4.2 Particle Insertion

Particle insertion has two objectives: increasing the convergence speed by incorporating individuals in the less populated areas and compensating the particle elimination caused by the fulfillment of the corresponding life times. Determining the convenient locations, within the search space, where the new individuals should be inserted is not a trivial task. In fact, it is a commitment between the optimal area identification and the new individuals insertion process speed.

The adopted solution divides the original problem in two parts: first, it seeks to determine how many particles it is necessary to incorporate so as to then establish where they should be placed within the search space.

The quantity of incorporated particles at each iteration coincides with the quantity of *isolated* individuals. An *isolated* individual is that which does not have any neighbor within a pre-established r radius. Equations 3 and 4 show the way in which such radius should be computed [2]. As it can be seen, r is computed as the average of each particles distances with their closest neighbor.

$$d_i = min\{\|x_i - x_j\|; \forall j x_i, x_j \epsilon S; x_i \neq x_j\} \qquad i = 1..n \qquad (3)$$

$$r = \frac{\sum_{i=1}^{n} d_i}{n} \qquad (4)$$

It only remains to determine the position of these new individuals. The adopted criterion was the following: the 20 % of these new particles receive position vector of the best individuals of the population, but its speed vector is random; the remaining 80% is random. In this way, part of the new individuals will begin to move from the positions that have shown better performance up to the moment, but with different directions and speeds from those of the best individuals. The remaining 80% will allow exploring other areas of the search space.

It is important to notice that the efficacy of the distance measure used in 3 will depend on the selected search space representation. If necessary, you may consult other alternatives in [4].

5 Proposed Algorithm

The algorithm begins with apopulation of N individuals generated at random within the search space and computes, for each of them, their corresponding fitness and life time.

During the process, individuals move according equations 1 and 2.The inertia used in order to update speed vectors is adjusted as follows [8]:

$$w = w_{start} - \frac{(w_{start} - w_{end})}{TotalIteraciones}.CurrentIteration \qquad (5)$$

where w_{start} is the initial value of w and w_{end} is the end value.

A high value of w at the beginning of the evolution allows the particles to make large movements placing themselves in different position of the search space. As the number of iterations advances, the value of w is reduced, allowing them to make a finer adjustment.

From the new positions within the search space, the individuals fitness value is recomputed and radius r is obtained according to equation 4.

Then, as many individuals as particles exist in the population without neighbors within this radius are created. These new individuals will have random speed vectors, within the allowed ranges. The 20% of these new particles will receive the position vectors of the best individuals of the population, and the remaining 80% will have random position vectors. For these new particles, their fitness is assessed and they are then incorporated to the population. Using the complete population, the recently incorporated individuals life time is computed.

The life time of every individual is decreased in 1, and those who have reached zero value are eliminated from the population.

The proposed algorithm makes use of elitism, reason why the best individual of each iteration is preserved. In this way, it is assured that the population will have at least one particle. This is carried out by replacing the particle with smaller fitness by the best one of the previous iteration.

Finally, the algorithm ends when one of the following conditions is met:

- The initially indicated maximum quantity of iteration is reached.
- The best fitness has not been modified during the 15% of the total iterations.

Figure 1 presents the pseudo-code of the described algorithm.

The *CreatePopulation* function receives as parameter the quantity of particles to be created and returns a swarm with random position and speed vectors within the established limits and with null life times. In order to estimate them, it is necessary to first assess each individuals fitness.

The *ComputeLifeTimes* process receives a complete swarm and only computes the life time corresponding to the particles that have null life time at the moment of the invocation. The second parameter corresponds to the type of computation that should be carried out and it values 1 for the assignment described in 4.1.a), and 2 for that described in 4.1.b).

The radius computation,according to equation 4, is carried out within the *ComputeRadius* process which receives as parameter the complete swarm and returns the quantity of new particles that should be inserted in the population. This module is the one in charge of avoiding the concentration of several particles in the same place of the search space; for this reason, it also returns the list of individuals that have really closed neighbors. Such particles are eliminated in the *SeeEnvironmnet* module, in function of their fitness and the quantity of generated sons.

```
Pop= CreatePopulation(N)
ComputeFitness(Pop);
ComputeLifeTimes(Pop,2);
w←MAXIMUMINTERTIA
while no end condition is reached do
    for i = 1 to size(S) do
        evaluate particle x_i of swarm S
        if fitness(x_i) is better than  fitness(pBest_i) then
            pBest_i ← x_i; fitness(pBest_i) ← fitness(x_i)
        end if
    end for
    Save individual with maximum fitness.
    for i = 1 to size(S) do
        Choose g_i according to neighborhood criterion used
        v_i ← w.v_i + (φ_1.rand_1.(pBest_i − x_i) + φ_2.rand_2.(g_i − x_i)
        x_i ← x_i + v_i
    end
    ComputeFitness(Pop);
    ComputeRadius(Pop, SonQuant, Sentenced);
    New = CreatePopulation(SonQuant);
    Assign 20% to these new individuals the position vectors of the best individuals of Pop.
    ComputeFitness(New);
    SeeEnvironment(Pop, Sentenced, SonQuant);
    Pop = Pop ∪New;
    if (CurrentIteration is greater than 5%of the TOTALITERATIONS))
            ComputeLifeTimes(Pop, 1);
        else ComputeLifeTimes(Pop, 2);
    end
    Deduct 1 to each particles life time
    Remove the particles with null life time
    Replace the worst individual by the saved at the beginnning of this iteration
    w ← dynamically modify the inertia
end while
Output : the best solution found
```

Fig. 1. Algorithm of the proposed VarPSO method

6 Results Obtained

VarPSO was used in order to obtain the minimum value of several functions. Thus, each particle contains in its position vector the values of the function statements. The capacity of each particle is computed as follows:

$$(c\_max - Value\_of\_the\_Particle) \qquad (6)$$

where *c_max* represents the function top limit in the interval to be optimized, and *Value_of_the_Particle* is the result from assessing the function in the corresponding particles position vector.

Next, the functions used are presented in detail. For each of them, the interval used to determine the search space is shown together with the *c_max* value used to compute the fitness.

$$F1(x,y) = x^2 + y^2 \qquad x,y \in [-1,5]; c\_max = 50$$

$$F2(x) = -x * sin(10 * \pi * x) + 1 \qquad x \in [-2,1]; c\_max = 3$$

$$F3(x,y) = 0.5 + \frac{(sin(\sqrt{x^2+y^2+4}))^2 - 0.5}{(1+0.001.(x^2+y^2))^2} \qquad x,y \in [-50,50]; c\_max = 1$$

$$F4(x_1,x_2) = \frac{1}{0.002 + \sum_{j=1}^{25} \frac{1}{50j + \sum_{i=1}^{2}(x_i - a_{ij})^6}} \qquad x_1, x_2 \in [-50,50]; c\_max = 500$$

$$F5(x_1,x_2) = \frac{1}{0.002 + \frac{1}{1+(x_1+1)^{12}+(y_1+1)^{12}})} \qquad x_1, x_2 \in [-200, 200]; c\_max = 500$$

$$F6(x_1,x_2) = \frac{1}{0.002 + \sum_{j=1}^{3} \frac{1}{50j^2 + \sum_{i=1}^{2}(x_i - b_{ij})^{12}}} \qquad x_1, x_2 \in [-50,50]; c\_max = 500$$

$$b_{ij} = \begin{pmatrix} -30 & 16 & 30 \\ -40 & -32 & 35 \end{pmatrix}$$

The six functions present a single minimum. Each of them seeks measuring a different aspect of the proposed algorithm. F1 allows analyzing the precision of the solution obtained. F2, F3, and F4 show their capacity of moving along a search space with really changing fitness values. Notice that function F4 is a modification of the De Jong's function 5. Functions F5 and F6 have been introduced to analyze the algorithm's exploratory capacity. In F5 a single gap appears, which leads to the minimum within a completely flat surface. F6 is similar, but it presents three gaps with very different depths.

For each function, 100 tests were carried out, using in each case, a maximum quantity of 500 iterations. The used cognitive and social learning values, φ_1 and φ_2, described in equation 2, were both in 0.5. Inertia values were between 0.2 and 1.5. The allowed speed range was established between -0.5 and 0.5.

The quantity of classes used for computing the individuals life time was 4. Tests were carried out with values between 2 and 10, confirming that a high value for the number of classes improves the fitness distinction among individuals, but significantly increases the population size. On the other hand, if the class quantity is very low, the swarms size could be considerably decreased. A value of 4 classes is adequate for the minimization of the afore mentioned functions. The value used as maximum life times was of 9 iterations, except in functions F5 and F6, which used a value of 12.

Table 1. Results obtained with PSO and VarPSO

Function F1	Avg.Ite.	Min. Fit.	Med. Fit.	Max. Fit.	Ini.Pop	Fin.Pop
gBest PSO	106.90	48.0544	49.2226	50.0000	20	20.0
lBest PSO	142.08	48.0373	49.2735	50.0000	20	20.0
gBestVarPSO	181.89	30.4806	45.3319	49.9996	30	33.8
lBestVarPSO	169.82	30.3254	45.0502	49.9995	30	32.8
Function F2	Avg.Ite.	Min. Fit.	Med. Fit.	Max. Fit.	Ini.Pop	Fin.Pop
gBest PSO	102.67	1.6791	2.1150	3.8466	60	60.0
lBest PSO	102.40	1.7325	2.0647	3.8456	70	70.0
gBestVarPSO	115.22	1.3443	2.3861	3.8489	40	13.1
lBestVarPSO	122.18	1.2349	2.3705	3.8489	30	17.7
Function F3	Avg.Ite.	Min. Fit.	Med. Fit.	Max. Fit.	Ini.Pop	Fin.Pop
gBest PSO	231.06	0.7288	0.9074	0.9762	70	70.0
lBest PSO	197.59	0.6130	0.8607	0.9717	60	60.0
gBestVarPSO	175.17	0.3032	0.5857	0.9942	20	53.3
lBestVarPSO	154.88	0.3025	0.5866	0.9936	40	62.6
Function F4	Avg.Ite.	Min. Fit.	Med. Fit.	Max. Fit.	Ini.Pop	Fin.Pop
gBest PSO	284.10	441.7729	445.3109	446.6185	60	60.0
lBest PSO	218.77	437.8757	443.5489	444.9094	70	70.0
gBestVarPSO	132.04	145.0876	285.1007	453.0275	30	30.7
lBestVarPSO	124.64	156.9515	299.9142	454.5333	40	55.0
Function F5	Avg.Ite.	Min. Fit.	Med. Fit.	Max. Fit.	Ini.Pop	Fin.Pop
gBest PSO	140.19	468.3641	482.4536	484.0312	70	70.0
lBest PSO	138.37	448.4032	477.8286	479.0419	70	70.0
gBestVarPSO	117.01	171.4216	280.7481	494.0287	40	16.2
lBestVarPSO	121.72	134.1485	238.5021	496.0285	30	23.8
Function F6	Avg.Ite.	Min. Fit.	Med. Fit.	Max. Fit.	Ini.Pop	Fin.Pop
gBest PSO	103.27	373.4571	391.0167	391.8489	60	60.0
lBest PSO	115.49	404.7819	440.9218	442.8571	50	50.0
gBestVarPSO	92.56	71.9364	194.3381	443.2076	40	85.7
lBestVarPSO	108.44	120.5613	228.8209	443.4110	20	52.9

Table 1 allows us to compare the results obtained when applying the algorithm proposed in this paper and the fixed population PSO algorithm. Tests were carried out with initial population size of 5, 10, 20, 30, 40, 50, 60, and 70 particles. In all the cases, the values correspond to the averages of the 100 tests carried out for each function, taking as initial population size that which allowed obtaining the best maximum average fitness.

As it can be seen, in all the cases, the solution obtained using VarPSO is superior or equal to that reached by PSO with fixed population, presenting greater population diversity. In addition, except function F1 for which, it is quite simple to reach the maximum fitness, the proposed method makes use of a inferior initial population than that of the fixed population method. If we analyze the quantity of particles displaced in average in each case, we can see that, in most of the functions, the proposed method carries out less than half of the job

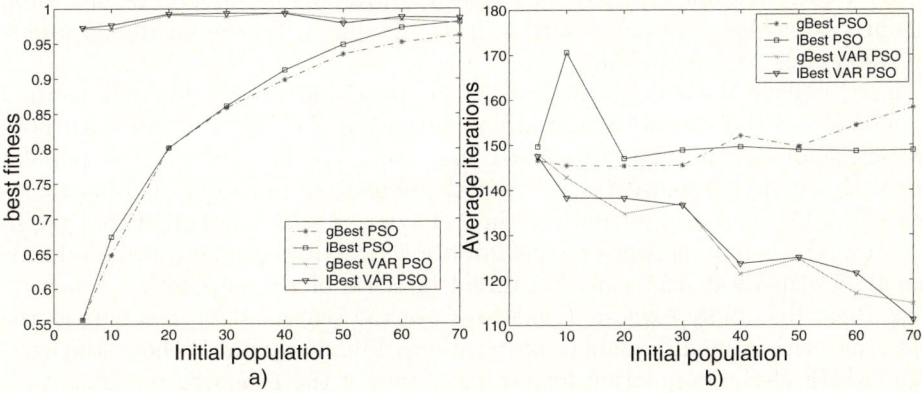

Fig. 2. a) Average maximum fitness obtained for different initial population sizes. b) Variation of the average iteration number necessary to obtain the best fitness in function of the initial population size.

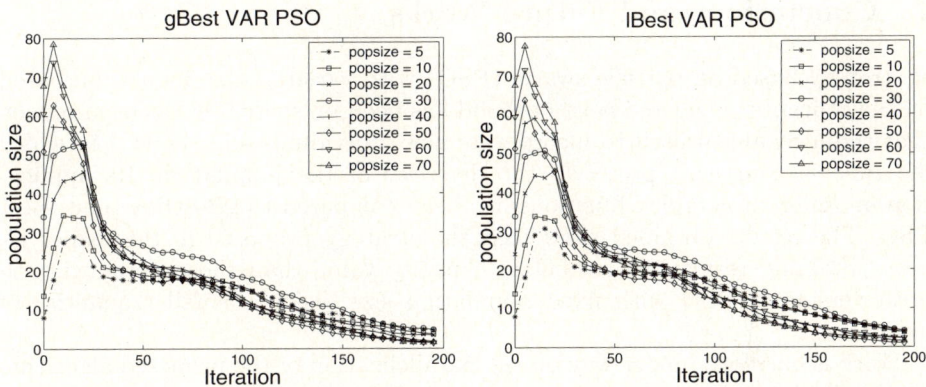

Fig. 3. Average size of the population using gBest VarPSO and lBest VarPSO. The values correspond to the first 200 iterations average of the 600 tests carried out (100 over each function).

than that carried out by fixed population solutions. This last is not verified for functions F1 and F6. In the case of F1, it is due to the functions simplicity, which contrasts with the exploratory behavior of the proposed method, and in the case of F6, it is due to the fact that the method does not prematurely converge like its fixed population peer, but goes on analyzing the search space obtaining best solutions.

Figure 2.a) shows the capacity of the proposed method to adapt itself to the searching surface, finding an almost optimal solution independent of the size of the initial population. We can also observe that fixed population methods provide proper results when the adequate initial population size is used. Each point of figure 2.a) corresponds to the maximum average fitness of the 600 tests

(100 for each function) carried out for each initial population size. In each case, the fitness has been linearly scaled to [0,1], dividing it by the maximum fitness corresponding to the function.

If we analyze the average quantity of iterations carried out for each method in function of the size of the initial population, we can see that, for fixed population solutions, it tends to slightly increase while for variable size populations, it rapidly decreases as the initial population increases. This is represented in Figure 2.b) and reflects the behavior of each method. Fixed population methods only depends on the initial particle displacement, while variable population methods carry out an initial increase of the population which allows them to rapidly explore a wider area of the search space, reaching at the optimum with a smaller number of iterations. Finally, Figure 3 shows the average growth of the population for both variants of the proposed method. As it can be seen, the behavior is rather similar in both cases and presents a growth phase, within the first 30 iterations, followed by a reduction and stabilization phase.

7 Conclusions and Future Works

A strategy based on particle swarm (PSO) with variable size population based on the concepts of age and neighborhood has been presented. It has been proven that the mechanism used to incorporate new individuals and the way in which life time is computed preserves the diversity of the population. Its application on different complex functions has been compared to gBestPSO and lBest PSO. The results obtained show that the strategy proposed in this paper allows obtaining results with a better capacity value than those obtained with both version of PSO with fixed population size, using a smaller quantity of iterations.

Work is currently under way on the parallelization of the proposed algorithm using only the information environment of each particle to determine whether this is an isolated individual or not. Also algorithm migration to cluster/grid arquitectures is analyzed. Moreover, according to the great exploration capacity shown by method proposed in this paper, we are currently working on its applications to neural networks' adaptation. In this case, each particle represents a complete feedforward neural network with fixed architecture and, along iterations, we intend to adapt the connection weights. It remains defining a similarity measure assuring the proper quantification of the proximity between two individuals in the search space.

Acknowledgment

This research was partially funded by the project CyTED "Grid Technology for Boosting Regional Development".

References

1. Van den Bergh, F.: An analysis of particle swarm optimizers. Ph.D. dissertation. Department Computer Science. University Pretoria. South Africa (2002)
2. Bird, S., Li, X.: Adaptively Choosing Niching Parameters in a PSO. In: Keijzer, M., et al. (eds.) Proceeding of Genetic and Evolutionary Computation Conference 2006 (GECCO 2006), pp. 3–9. ACM Press, New York (2006)
3. Clerc, M., Kennedy, J.: The particle swarm-explosion, stability and convergence in a multidimensional complex space. IEEE Transactions on Evolutionary Computation 6(1), 58–73 (2002)
4. Ertöz, L., Steinbach, M., Kumar, V.: A new shared nearest neighbor clustering algorithm and its applications. In: Proc. Workshop on Clustering High Dimensional Data and its Applications, Arlington, VA, USA, pp. 105–115 (2002)
5. Fernandes, C., Ramos, V., Rosa, A.: Varying the Population Size of Artificial Foraging Swarms on Time Varying Landscapes. In: Duch, W., Kacprzyk, J., Oja, E., Zadrożny, S. (eds.) ICANN 2005. LNCS, vol. 3696, pp. 311–316. Springer, Heidelberg (2005)
6. Kenedy, J., Eberhart, R.: Particle Swarm Optimization. In: Proceedings of IEEE International Conference on Neural Networks, Australia, vol. IV, pp. 1942–1948 (1995)
7. Lanzarini, L., Sanz, C., Naiouf, M., Romero, F.: Mixed alternative in the assignment by classes vs. conventional methods for calculation of individuals lifetime in GAVaPS. In: Proceedings of the 22nd International Conference on Information Technology Interfaces, 2000. ITI 2000, pp. 383–389 (2000); ISBN: 953-96769-1-6.
8. Meissner, M., Schmuker, M., Schneider, G.: Optimized Particle Swarm Optimization (OPSO) and its application to artificial neural networks training. In: BMC Bioinformatics 2006 (Published online 2006 March 10) pp. 7–125 (2006) DOI: 10.1186/1471-2105-7-125
9. José, G.N.: Algorithms based on swarms of particles for solving complex problems. University Málaga (In Spanish) (2006)
10. Shi, Y., Eberhart, R.: Parameter Selection in Particle Swarm Optimization. In: Proceedings of the 7th International Conference on Evolutionary Programming, pp. 591–600. Springer, Heidelberg (1998)
11. Shi, Y., Eberhart, R.: An empirical study of particle swarm optimization. In: Proceeding on IEEE Congress Evolutionary Computation, Washington DC, pp. 1945–1949 (1999)

Genetic Algorithm as a Tool for Stock Market Modelling

Urszula Markowska-Kaczmar, Halina Kwasnicka, and Marcin Szczepkowski

Wroclaw University of Technology, Poland
urszula.markowska-kaczmar@pwr.wroc.pl

Abstract. The paper describes the model of virtual stock market which is evolved by a genetic algorithm. The model consists of cooperating Agents that imitate behaviour of real investors. They act on the virtual market buying or selling stocks. The aim of the model is to generate stocks prices on a virtual market that are similar to real ones for a short period of time. Each Agent is described by its unique characteristics which determine his performance. The details of the model are presented in the paper. The applied genetic algorithm is generic one. Its main components such as: an individual, genetic operators and fitness function are described here, as well. The results of experiments investigating the role of genetic algorithm parameters are presented in the paper. Agent's ability to predict the quotations values are presented and analysed. Future plans referring to the further development of the system are presented at the end of the paper.

1 Introduction

Along with origins of financial markets different approaches to forecasting and modelling of market behaviour have been developed. Fundamental analysis is connected to analysis of financial results of companies, their assets and liabilities perspective of a development and a macroeconomic climate. On this basis the value of the share is calculated. Its comparison with the market value allows to make a decision about purchase or sale of the share. This approach is very helpful for longtime investments. But the problem is the choice of appropriate model of company evaluation, availability of data and elimination of irrelevant data. Technical analysis, in practice, relies in analysing share price charts and searching for a given chart pattern of price trends that is called formation. This analysis is supported by technical analysis coefficients. It is used during short- and longtime investments. Experience and good understanding mechanisms of financial market is necessary to apply it in the full extent.

Very common other approach to market prediction is an application of intelligent methods. In many research neural networks (e.g. [1]) are used to predict share prices but the basic fault of this solution is the lack of explanation of a neural network final decision.

Another popular approach is the use of multi-agent techniques. A great similarity to the real world is its advantage. The next one is a better interpretation of market behaviour in comparison to neural networks. In the literature we can find different goals of research applying agent techniques. They spread across simulation of the investor's behaviour [2] through simulation of a market [3] and trials to find the influence of different factors on the exchange rate [4] to simulation of turnover volume of real market or simulation of such a phenomenon like speculation bubbles [4]. Taking into account a goal of simulation approaches can be divided into two categories. The aim of the first group is to build a model of the market. While the second group searches for mechanism to enable higher investment return. The structure of the virtual market and the principles describing interactions between Agents are essential for successful model. In many papers, for instance in [5,3] a Market Maker exists, which role is to ensure the stock market liquidity. It is responsible for covering all bids for sales and to purchase. Various solutions are applied taking into account the number of Agents. Some authors assume that there are specified categories of Agents acting on the market [2,6].

Different assumptions about financial instruments of Agents are made. The instruments can include real stocks [7], indices of stock market [2] and foreign exchange [1]. Various techniques are applied to built Agents, for instance they can be based on neural networks [1] or LCSs (Leaning Classifying Systems) are used [2] and [6]. In the latter case Agents make their decisions on the rules operating on technical analysis indicators. In recent papers genetic algorithm is applied to learn Agents strategies of a stock, [8] and [9]. The strategy can be represented as a set of evolved rules or evolved neural networks responsible for Agent decisions informing it when to sell or to buy stocks.

In our approach a genetic algorithm is used to evolve multiagents system which models the stock market behaviour. Agents (a model of the real market) acting on the virtual market by selling and buying various stocks cause the changes of prices that mimic the real ones. The accuracy of the evolved model is defined as a difference between price movement on the real stock and virtual stock. The model is searched by genetic algorithm where one individual represents a model of a real stock market. Unique characteristics of Agent that define its behaviour on virtual stock market are evolved during artificial evolution.

The paper is organized as follows. The next section gives more detailed description of the system. Then, the basic elements of the system are presented. The subsequent section describes performed experiments and the results. The conclusion and future direction of the system enhancement finish the paper.

2 General Idea of the System

The system is composed of Virtual Market (VM) and a group of Agents that represent investors. VM has a database with real quotations of stocks. We assume that the group of Agents creates the model of real market. The goal is to find a model that behaves as real investors, i.e. their Agents by selling and buying stocks generate the prices changes similar to the real ones.

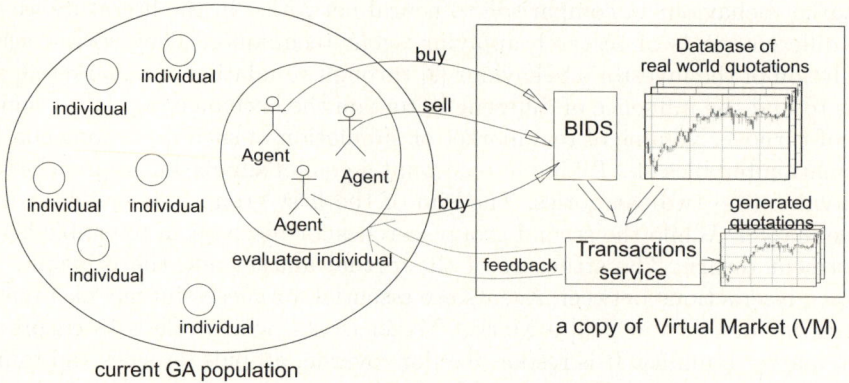

Fig. 1. The idea of the system

The role of VM is to keep track of all transactions, to clear transactions, to update stock prices and to provide Agents with necessary information about actual prices. Each of the Agents in the model has its own set of chart patterns and parameters and thus reacts in an unique way. They represent its own expectations and experience on the basis of which an Agent makes decisions about purchase or sale of shares, which are realized during transactions (Fig.1). Agent chart patterns can be divided into two groups: negative (when to supply) and positive (when to demand). The superposition of actions of all Agents in the model creates changes of prices of stocks on the virtual market.

Searching for a model can be perceived as an optimization task. Because the space of possible solutions is huge in the proposed system a genetic algorithm (GA) is used to search for a model. One individual in GA represents the set of Agents. In order to evaluate each individual, a set of Agents acts on the copy of the VM assumed number of sessions buying and selling shares. The searched model should generate quotations highly correlated with real-world data. This is the basis of individual's evaluation. After evaluation, individuals are exposed to genetic operations. Then, this process is repeated assumed number of generations. At the end, the best individual is the model of the market.

This model can be used to predict the behaviour of a real stock in the period of time beyond of training data. The future analysis of data being the basis of Agents decision will allow to find those elements that have the greatest influence on their decisions.

3 Basic Elements of the System

The simulated system consists of the elements listed below:

- Agents,
- Genetic Algorithm,
- Virtual Market.

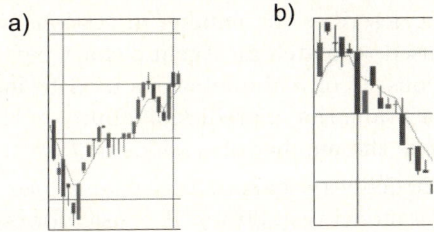

Fig. 2. The examples of data used to generate: a) - positive pattern, b) - negative pattern

An Agent represents a single investor that searches for investments satisfying its stock chart patterns. Each pattern is a sequence of quotations with the assumed number of session, chosen randomly as a part of archived stock quotations. There are two kinds of patterns – negative and positive (Fig. 2) ones. A positive chart pattern is a sequence in which the price of the last quotation is the highest one in the sequence. A negative chart pattern is a pattern in which the price of the last quotation is the lowest in the sequence.

Behaviour of an Agent is assigned by coefficients, which affect on his decision making. (Fig. 3).

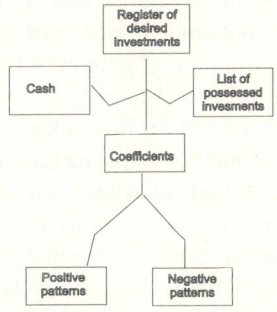

Fig. 3. The basic elements of an Agent

These coefficients are as follows:

- *Purchase Margin* – margin by which the price of a purchase bid placed by an Agent for a stock is higher than the current stock market price (e.g. 5%),
- *Sale Margin* – margin by which the price of a sale bid placed by an Agent for a stock is lower than the current stock market price (e.g. 5%),
- *Investment Evaluation Threshold* – threshold value of the correlation between an Agent's chart pattern and stock quotations, above which Agent creates a bid to buy or to sell an investment (e.g. 0.8).
- *Investment Cancellation Threshold* – threshold value of correlation between the pattern used to create an investment and the current stock quotations, below which Agent cancels an investment (e.g. 0.5).

- *Investment Span* – it describes the number of sessions with quotations of a real stock that are used to match an Agent's chart pattern while searching for investments. It consists of real quotations used to match chart patterns and the part used for evaluation a prediction ability of the model defined by *Forecast Horizon*; N is the number of sessions in *Investment Span*.
- *Forecast Horizon* – it defines a part of *Investment Span* used to evaluate a prediction ability of a model (e.g. 30%). It is used for evaluation of individuals; K defines the number of sessions used for prediction.
- *Maximum Investment Count* – it defines the number of investments a single Agent can have.
- *Common Sense Coefficient* – it defines in what extend an Agent will follow in its decision after the similarity to its chart patterns and what role will play the expected profit. A high value of this coefficient causes that an Agent will perceive stocks with a higher correlation of prices to the chart pattern as better ones.
- *Negative pattern factor* – threshold of correlation between quotation of a stock and Agent's negative pattern above which Agent will decide to resign from investment.

The next elements that characterize an Agent are: *Register of Desired Investments*, *List of Possessed Investments* and *Cash* that are in an Agent's disposal. On the basis of the difference between *Register of Desired Investments* and *List of Possessed Investments* an Agent will generate bids to purchase or to sell investments.

The first phase of an Agent activity relies in searching new investments. In its disposal are real stocks quotations in the period defined by (N-K) sessions. Each positive Agent's pattern is matched up with each stock. If such a chart pattern is strongly correlated with current stock quotations (correlation is greater than *Investment Threshold*) an Agent creates an entity called an investment which he would like to have. An investment consists of the stock and the pattern on the basis of which it was found.

Next, an Agent evaluates his current investments. If an investment doesn't behave like the Agent had wished (recent quotations of the stock associated with the investment do not behave like the rest chart pattern, associated with the investment, that was not used to create the investment) then that investment is cancelled. Investment can also be cancelled if quotations of the stock associated with that investment correlate strongly to a negative pattern the Agent has.

After evaluating the investments the Agent uses the remaining ones (those that have not been cancelled) to create a list of stocks the Agents would like to have (the list of desired stocks). This list is matched up with a list of Agent's stocks. The difference between both lists is used to create purchase and sale bids. They are realized by Virtual Market and they generate price movement. During one session Agents may repeat the process of buying and selling stocks several times. Closing price of the stock in the session is taken into account in the next session. In the same way Agents behave for K sessions. Then the model (the individual) is evaluated by comparison of the prices generated by the model in K sessions with real ones for the same period.

Genetic Algorithm (GA) evolves a set of Agents which role is to generate similar prices movement as in the real stock market. The GA used in this system is generic one. It starts with an initial population of individuals, where one individual encodes a set of Agents. Evaluation of an individual needs to create a copy of actual state of the virtual market. Agents from an evaluated individual act assumed number of sessions (K) and at the end the generated quotations on the virtual market are compared with those from the real stock market. Evaluation is done for each individual in the population. Next, on this basis, individuals are chosen to create offspring population. Then, they are subjected to genetic operations – *mutation* and *crossover*. After that, they are evaluated and the process is repeated until the stop condition arrives (assumed number of generations has elapsed).

Mutation operator is realized by replacing the number of Agents (assigned by the parameter of GA) with new ones created randomly. A *crossover* operator needs two parents and generates one offspring. The offspring is generated by taking random Agents from both of the parents. The number of Agents in an offspring is equal to an average number of Agents in both its parents rounded up.

The *fitness function* R is used to evaluate individuals. It is defined as a difference between two values (eq.1): correlation $R_{correlation}$ (eq.2) and difference $R_{difference}$ (eq.3) between stock quotations generated by a model and corresponding to them real-world quotations for K sessions. Both these values are an average of values calculated for all n stocks in the market. The value $R_{correlation}$ informs about similarity of tendency price movement on the real and virtual stocks. $R_{difference}$ informs about prices convergence on the both markets.

$$R = R_{correlation} - R_{difference} \tag{1}$$

$R_{difference}$ is calculated as an average difference between real $p_{i,j}^r$ and virtual $p_{i,j}^v$ quotations for a given period of time; j is an index of session, i – an index of stock.

$$R_{difference} = 1/n \sum_{i=1}^{n} (1/K \sum_{j=1}^{K} \frac{|p_{i,j}^r - p_{i,j}^v|}{min\{p_{i,j}^r; p_{i,j}^v\}}, \tag{2}$$

$R_{correlation}$ is formally expressed by eq. 3

$$R_{correlation} = 1/n \sum_{i=1}^{n} \frac{\sum_{j=1}^{K}(p_{i,j}^r - \overline{p_i^r})(p_{i,j}^v - \overline{p_i^v})}{s_{p^r} s_{p^v}}, \tag{3}$$

where $\overline{p_r}$ and $\overline{p_v}$ are respectively mean real price and mean virtual price of a given stock during K sessions; s_{p_r} and s_{p_v} are standard deviations of real and virtual prices. Other values are defined in the same way as in eq. 2.

The Virtual Market (VM) is a central part of the system, which joins the majority of the system elements. The role of VM is to keep track of all transactions, to clear transactions, to update stock prices and to provide Agents with necessary information about those prices. The performance of VM takes place in steps. One step corresponds to one stock session. The opening of a given session is connected with creation for each stock new quotation. The information about current stock is updated. It relies in addition new values to the sequence describing the virtual stock quotations. The volume for the current session is set to 0. Minimal and maximal prices are updated, as well.

The VM waits for bids to be placed. When all present bids from Agents have been collected, they are sorted and VM begins to search for stocks that match Agents bids. If sale and purchase bids are matched, they are resolved, and the current price of the virtual stock is updated. The price of transaction is an average of both prices – of purchase and of sale. The number of shares is the lowest number from matched pair purchase-sale. To update the account after transaction the following steps are performed:

- transfer of cash on the seller account,
- transfer of stocks for buyer,
- updating of quotation of the stock that was the subject of transaction,
- updating the number of stocks in the sale offers,
- updating the number of purchase offers.

This process is repeated until there are no matching bids. The Agents from a given individual act K number of sessions (the parameter of the system). The values of quotations generated during K sessions are then used to evaluate the individual by using the fitness function described by eq. 1.

4 Experiments

The system capability to predict real stock quotations has been verified in a series of experiments. The goal of those experiments was to test:

- an impact of genetic operators on system forecasting capability,
- forecasting capability of the system in reference to the different number of stocks existing on the VM,
- the influence of different data time spans on the forecasting capability of the system,
- the influence of the amount of Agent's capital on the system forecast capability,
- the influence of the number of chart patterns each Agent can have on the system forecast capability.

Each of those experiments was carried on different stock sets in order to avoid results being characteristic just to one stock set.

The experiments showed that the use of genetic operators has positive impact on the speed of creation a correct model. Generally, the use of the mutation operator prevented a stagnation and encouraged a diversity among the elite solutions and thus led to creation of better solutions.

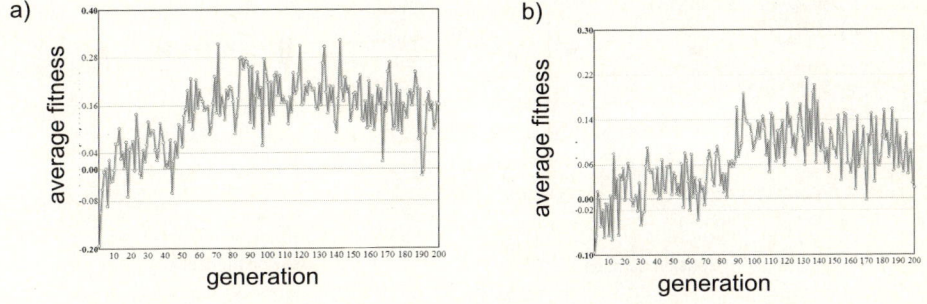

Fig. 4. Average fitness for a) two stocks, b) three stocks

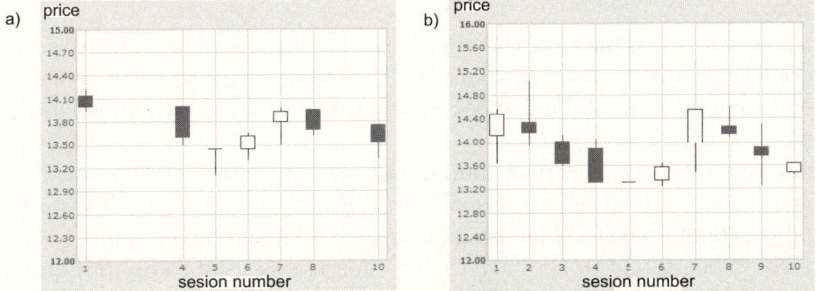

Fig. 5. Quotations generated by a) real world market and b) the model for stock *Getin*

The lesser was the number of available stocks on the VM the better was the average fitness evolved by the genetic algorithm. Typical situation is illustrated in Fig. 4, where part a) presents the average fitness value for two stocks – Electrim and Getin, b) for three stocks – Millenium, Electrim and Getin.

This conclusion is natural, because it is easier to generate a set of Agents whose actions replay quotations of one stock rather than quotations of two or more stocks. As the number of stocks present on the VM roses, the average fitness values of all populations decreases rather quickly. However, the fitness value of the best found individual does not much differ along with different stock number.

Further experiments showed that the choice of a stock set is more critical for the system forecasting capability than the choice of a specific time period. Results generated for different time periods were similar but the choice of a different stock set had quite high impact on the system forecasting capability.

The choice the initial amount of cash and the initial number of Agent's stocks is also very important. One can easily imagine a situation where all of the Agents have lots of cash, but none has stock or in reverse. In such a situation even a single transaction would not take place.

Experiments also showed that the number of chart patterns each Agent uses in his decision making process is not critical for forecasting performance. Lowering that number had no significant impact on the fitness value of created individuals.

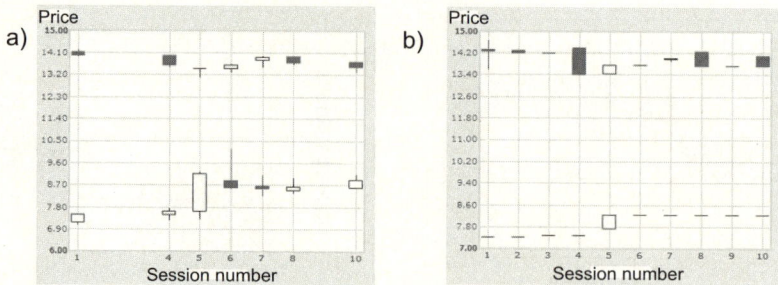

Fig. 6. Quotations generated by a) real world market and b) the model for two stocks – *Getin* – the upper chart and *Electrim* – the lower chart

It had however an impact on the speed of calculations. The algorithm used to match patterns and stocks quotations is CPU intensive and reducing the number of Agent's patterns in virtual stock reduces the time of individual evaluation greatly.

Prices generated by the best model and the corresponding quotations from real market for *Getin* stock are visualized in Fig. 5. The fitness value for this individual was equal 0.6314. The model was evolved for archived quotations in the period from 2004-02-24 to 2007-04-19 and then it was used to predict quotations from 2007-04-20 to 2007-04-20. The result is illustrated in Fig. 5b). The model composed of 71 Agents. This experiment has shown an oversight of the system, which consists in considering holidays as normal working days by the model. The quotations values of nonexisting in reality sessions were approximated on the basis of existing sessions. This problem will be solved in the nearest future. As it can be noticed, comparing the results generated by the model (after removing nonexisting quotations) with the real ones, the model is convergent to quotations of this stock on Warsaw Stock Market.

The results generated by the model for two stocks *Getin* and *Electrim* are shown in Fig. 6b). By comparison Fig. 6a) and Fig. 6b) the similarity of price values generated by the model and trends of real quotations can be noticed again. However for *Electrim* stock one anomaly can be observed. The horizontal lines represent sessions in which there were no transactions. This shows that the model was not able to match properly both stocks quotations simultaneously.

5 Conclusions and Future Plans

The results obtained during experiments are promising but in order to evaluate the proposed solution in a full extent more experimental studies are necessary. The experiments have exhibited some shortcomings that should be improved. The highest correlation between the prices generated by the model and prices from a real market was equal 0.8543 and it shows the solid ability of the system to model a trend of stock price movements. The average correlation of all created during experiments stock market models was equal 0.3924. This result takes into consideration random individuals, too. This means that the applied approach and the created virtual market have a natural gift to match the trends existing on the stock market.

The number of stocks tested in the experiments presented in this paper was not greater than 4. It is too small number to fully evaluate the proposed solution.

In the further development of the system we plan to change mutation operator. Now it is realized by introducing assumed number of new randomly created individuals. It causes very rapid changes of individuals. In the future the implemented system will be enhanced in calculation of some statistics allowing to determine the features of Agents that cause specific behaviour on the market, for instance how many of them invested in aggressive way and how many were conservative investors.

References

1. Grothmann, R.: Multi-Agent Market Modeling based on Neural Networks. PhD thesis, University of Bremen, Germany (2002)
2. Schoreels, C., Logan, B., Garibaldi, J.: Agent based genetic algorithm employing financial technical analysis for making trading decisions using historical equity market data. In: Proc. of the IEEE/WIC/ACM Int'l Conf. on Intelligent Agent Technology 2004 (IAT 2004), Beijing, China, pp. 421–424 (2004)
3. Blok, H.: On the nature of the stock market: Simulations and experiments. PhD thesis, The University of British Columbia, Canada (2001)
4. Neuberg, L., Bertels, K.: An artificial stock market. In: Proceedings of the IASTED International Conference, pp. 308–313 (2002)
5. Ankenbrand, T., Tomassini, M.: Agent based simulation of multiple financial markets. Neural Network World 4, 397–405 (1997)
6. Schoreels, C., Garibaldi, J.: Genetic algorithm evolved agent-based equity trading using technical analysis and the capital asset pricing mode. In: Proc. 6th Int'l Conf. on Recent Advances in Soft Computing 2006 (RASC 2006), pp. 194–199 (2006)
7. Schulenburg, S., Ross, P.: Advances in Learning Classifier Systems. In: Lanzi, P.L., Stolzmann, W., Wilson, S.W. (eds.) Explorations in LCS Models of Stock Trading, vol. 2321, pp. 150–179 (2002)
8. Tokinaga, S.: Modeling and analysis of agent-based artificial demand-supply market by using the genetic programming and its applications. Journal of political economy 71, 107–118 (2004)
9. Kendall, G., Su, Y.: Learning with imperfections-a multi-agent neural-genetic trading system with differing levels of social learning. In: Proceedings of the 2004 IEEE Conference on Cybernetics and Intelligent Systems, pp. 47–52 (2004)

Robustness of Isotropic Stable Mutations in a General Search Space

Przemysław Prętki and Andrzej Obuchowicz

University of Zielona Góra, Zielona Góra 65-001, Poland
{P.Pretki,A.Obuchowicz}@uz.zgora.pl

Abstract. One of the most serious problem concerning global optimization methods is their correct configuration. Usually algorithms are described by some number of external parameters for which optimal values strongly depend on the objective function. If there is a lack of knowledge on the function under consideration the optimization algorithms can by adjusted using trail-and-error method. Naturally, this kind of approach gives rise to many computational problems. Moreover, it can be applied only when a lot of function evaluations is allowed. In order to avoid trial-and-error method it is reasonable to use an optimization algorithm which is characterized by the highest degree of robustness according to the variations in its control parameters. In this paper, the robustness issue of evolutionary strategy with isotropic stable mutations is discussed. The experimental simulations are conducted with the help of special search environment - the so-called general search space.

1 Introduction and Preliminaries

The well-known No-Free-Lunch Theorem [17] states that the average performances of all optimization algorithms across all possible problems are identical. It means that a general-purpose universal optimization strategy is theoretically impossible, and the only way one strategy can outperform another is if it is specialized to the specific problem [6]. In practice, having two different algorithms, it is very difficult to mark off a class of functions in which one algorithm outperform the other. It must to be stressed that the last sentence does not concern any trivial cases e.g. it is rather indisputable that stochastic algorithms work much much worst for unimodal, differentiable and well-conditioned functions than any deterministic method that is based on gradient evaluation technique. Moreover, it is not clear how to choose adequate criteria that would make it possible to split the space of functions. Ideally the space should have parametrization which allows to evaluate the performance of the global optimization techniques in different subspaces.

In general, the performance evaluation of optimization algorithms is a challenging task. It is common practice to test some techniques on the basis of a set of popular benchmark functions. Such a situation is most frequent encountered in practice, as it is simple to implement. On the other hand it is not clear whether or not typical benchmark functions truly reflects real problems. In the light of the so-called black-box optimization [7,16], the experimental results which are focused only on several exemplary search environments are rather useless. Instead of using a small set of test functions, in this paper it is proposed to evaluate the performance and robustness of evolutionary algorithms on the basis of a special class of continuous problems. First and foremost the class is described in a stochastic manner. Probabilistic measure under consideration makes it possible to emphasize various features of search environments and compare global optimization techniques from different angles.

The paper is organized as follows. First section contains a brief introduction to the class of isotropic stable distributions. In the second section an evolutionary strategy is presented. A notion of the general search space is explained in the third section, where also its stochastic parametrization is provided. In the forth section the Statistical Learning Theory and Monte-Carlo Integration techniques are combined to evaluate performance of the algorithm. Fifth section includes the comparison of evolutionary strategy and deterministic algorithms with respect to their robustness and effectiveness. Finally the paper is summarized in conclusions.

2 Isotropic Stable Mutation

Let us start this section with a brief introduction of the concept of isotropic distributions. The class of spherically symmetric random distributions can be defined in the following manner [4]:

Definition 1. *The random vector* $\boldsymbol{Z} = (z_1, ..., z_n)^T$ *has spherically symmetric distribution iff its characteristic function is of the form* $\phi(\boldsymbol{\theta}^T \boldsymbol{\theta})$, *for some scalar function* $\phi(\cdot)$ *called a characteristic generator,*

As a matter of fact, it is not easy to generalize univariate α-stable distributions [18] to their multivariate isotropic counterparts, especially as the Definition 1 does not supply any clues how to do this. Fortunately, it turns out that there exists a special subclass of stable distributions, the so-called sub-Gaussian distributions [13], which makes it possible to obtain isotropic versions of α-stable distributions. Since, deeper discussion about this topic decidedly exceeds the scope of the paper, we confine ourselves to just quoting main facts about it.

Namely, the isotropic α-stable vector $\boldsymbol{Z}$ can be represented in the form of the following stochastic decomposition [13]:

$$\boldsymbol{Z} = A_{\alpha/2}^{1/2} \boldsymbol{G}, \quad \alpha \in (0, 2), \tag{1}$$

where $A_{\alpha/2} \sim S_{\alpha/2}\Big((\cos(\frac{\pi\alpha}{4}))^{2/\alpha}, 1, 0\Big)$ is stable random variable, and $G \sim \mathcal{N}(\boldsymbol{0}, \sigma \boldsymbol{I}_n)$. Moreover, random vector (1) possesses a uniform spectral measure [9] and characteristic function of the form [13]

$$\varphi(\boldsymbol{\theta}) = E[\exp(-j\boldsymbol{\theta}^T \boldsymbol{Z})] = \exp\big(-2^{-\alpha/2}\sigma^\alpha \|\boldsymbol{\theta}\|^\alpha\big) \tag{2}$$

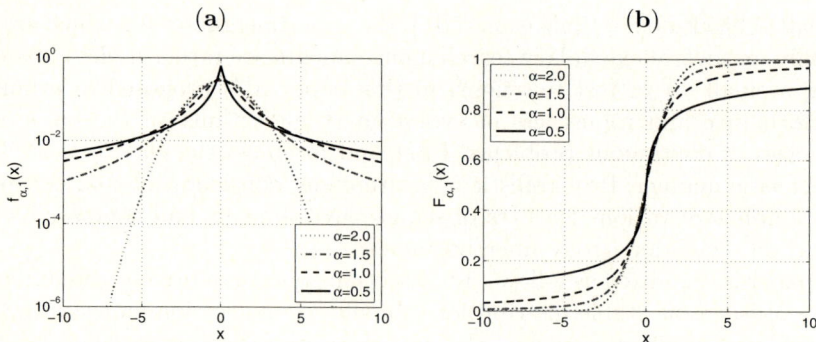

Fig. 1. Probability density functions (**a**) and cumulative density functions (**b**) of marginal distribution of standard ($\sigma = 1$) isotropic α-stable vectors

It is clear that characteristic function (2) depends only on the magnitude of the vector $\boldsymbol{\theta} = [\theta_1, \theta_2, \ldots, \theta_n]^T$, so that one can immediately affirm its consistency with Definition 1. Due to the fact that there exist effective numerical routines dedicated to generating pseudo-random numbers of univariate α-stable distribution [13], it is relatively easy to build a generator according to (1). Moreover, it can be shown that all marginal distributions of vector (1) are univariate stable distributions with the same stability index and the following characteristic function:

$$\varphi(\theta) = \exp\left(-\left(\frac{\sigma}{\sqrt{2}}\right)^\alpha |\theta|^\alpha\right) \quad (3)$$

Marginal densities and cumulative distribution functions of (1) are depicted in Fig. 1.

2.1 Evolutionary Strategy (1 + 1)ES with Stable Mutations

Evolutionary strategy $(1+1)$ES [1] is probably the simplest evolutionary algorithm. Its idea can be described in the following way: the point $\boldsymbol{x}_k$ that represents a solution at iteration k is mutated according to the formula

$$\boldsymbol{x}'_k = \boldsymbol{x}_k + \boldsymbol{Z}, \quad (4)$$

where $\boldsymbol{Z}$ is defined by (1). In order to emphasis the fact that mutation (1) is carried out on the basis of α-stable distribution, in the subsequent part of the paper the notation $(1+1)\text{ES}_\alpha$ will be used. The new solution $\boldsymbol{x}'_k$ is subjected to an evaluation procedure, and the better individual from a pair $\{\boldsymbol{x}_k, \boldsymbol{x}'_k\}$ survives to the next generation. The whole process is repeated until some stop condition is met. The selection scheme guaranties that the fitness function in successive iterations does not increase (minimization problem) or decrease (maximization). In fact it is very easy to prove the convergence of the algorithm to the arbitrary small neighborhood of a global optimum, provided that the neighborhood has non-zero Lebesgue measure [5].

3   General Search Space

Vast majority of optimization problems that are considered in engineering consists of continuous objective functions. Because of this, it is reasonable to restrict our further considerations only to the $\mathcal{C}[\mathbb{S}]$ - a class of all continuous functions defined on $\mathbb{S} \subset \mathbb{R}^n$. The well-known Stone-Weierstrass theorem [3] says that the set $\mathcal{F} = \bigcup_{i=1}^{\infty} \mathcal{F}_i$, where

$$\mathcal{F}_N = \left\{ f(\boldsymbol{x}) = \sum_{i=1}^{N} w_i \exp\left(-b_i \|\boldsymbol{x} - \boldsymbol{m}_i\|^2\right); \; w_i, b_i \in \mathbb{R}_+, \boldsymbol{m}_i \in \mathbb{S} \right\} \quad (5)$$

is dense in $\mathcal{C}[\mathbb{S}]$. This means that every $g \in \mathcal{C}[\mathbb{S}]$ can be approximated by a function $h \in \mathcal{F}$ with an arbitrary small error, i.e. for any $\varepsilon > 0$ and $g \in \mathcal{C}[\mathbb{S}]$, there is a function $f \in \mathcal{F}$ such that

$$|g(\boldsymbol{x}) - f(\boldsymbol{x})| < \varepsilon \quad (6)$$

for all $\boldsymbol{x} \in \mathbb{S}$.

In this paper it is proposed to treat the set $\mathcal{F}$ as a test environment for evolutionary strategy $(1+1)\text{ES}_\alpha$ presented in section 2.1. The choice of the class $\mathcal{F}$ is supported by a supposition that some potential conclusions can be generalized to the whole class of continuous functions $\mathcal{C}[\mathbb{S}]$. Bearing in mind mentioned before No-Free-Lunch Theorem, the average performance evaluated on $\mathcal{F}$ should by the same for every algorithm, regardless the value of stability index α. For that reason, it is desirable to establish some criteria that would allow to split the set $\mathcal{F}$ into separate subclasses. The proposed approach aims, on the one hand, at assessing effectiveness of evolutionary strategies for these subsets, and on the other hand, at assigning the best strategies to specific classes. In order to make such a classification possible a special probabilistic measure $\mu(\cdot)$ for the space $\mathcal{F}$ is established. Since the measure belongs to a parametric family, it is possible, modifying its parameters, to prefer different areas of $\mathcal{F}$. A definition of μ is further presented in the form of its marginal density functions. The number of radial functions N is identified with a Poisson random variable, for which a probability function is given by:

$$P(N = k) = \frac{\lambda_N^k \exp(-\lambda_N)}{k!} \quad (7)$$

Without lost of generality, we assume that the domain $\mathbb{S}$ is a two-dimensional hypercube $[-10, 10] \times [-10, 10]$. Moreover, locations of centers $\boldsymbol{m}_i \in \mathbb{S}$, $i = 1, \ldots, N$ are equally probable, so their density functions are equal to:

$$f(\boldsymbol{m}_i) = \frac{1}{400} \quad (8)$$

Marginal distributions of coefficients w_i and b_i belong to the exponential family:

$$f(t; \lambda) = \begin{cases} \lambda \exp(-\lambda t) & , t \geq 0 \\ 0 & , t < 0 \end{cases} \quad (9)$$

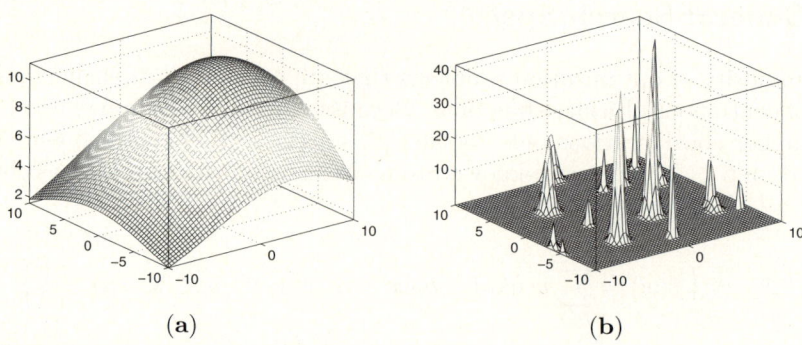

Fig. 2. Two typical search environments generated according to scheme presented in section 3. (a) - $\lambda_w = 10$, $\lambda_b = 0.01$, $\lambda_N = 0.01$, (b) - $\lambda_w = 20$, $\lambda_b = 20$, $\lambda_N = 20$.

with different expectations: $E[w_i] = \lambda_w = \frac{1}{\lambda}$ and $E[b_i] = \lambda_b = \frac{1}{\lambda}$, $i = 1, \ldots, N$. The choice of the exponential distribution is not an accidental one. It is dictated by the fact, that among all continuous probability distributions with support $[0, \infty)$ and fixed mean, the exponential distribution has the largest entropy. Since the all distributions are independent, thus the probabilistic measure μ of any set can computed as product of probabilities (7)-(9). Moreover, the measure has three parameters $\lambda_N, \lambda_w, \lambda_b$ which control intensity of sampling $\mathcal{F}$, thereby their also determine the form of benchmark environment. Two exemplary elements of $\mathcal{F}$ with measure $\mu(\cdot; \lambda_N, \lambda_w, \lambda_b)$ are depicted in Fig. 2. The fact that it has been decided to treat the set $\mathcal{F}$ in a probabilistic manner has several interesting consequences. First and foremost, if one considers any continuous function g, and an arbitrary small error $\varepsilon > 0$, there always exists a non-zero measure set $F_g \subset \mathcal{F}$, such that every element $f_g \in F_g$ is similar to $g \in \mathcal{C}[\mathbb{S}]$ [1]. Roughly speaking, it is possible (at least theoretically) to obtain almost identical function to any freely chosen continuous one. Of course, depending on the parameters of probabilistic measure $\lambda_N, \lambda_w, \lambda_b$, some of them are approximated more easily than the others. It makes it possible to provide a distinction between problems and allows to classify algorithms according to their performances.

4 Effectiveness of Stable Mutation in the General Search Space

Lets $\boldsymbol{\theta}_N = [\boldsymbol{w}^T, \boldsymbol{b}, \boldsymbol{m}_1, \ldots, \boldsymbol{m}_N] \in \Theta_N$ with $\boldsymbol{w}^T = [w_1, \ldots, w_N] \in \mathbb{R}_+^N$, $\boldsymbol{b}^T = [b_1, \ldots, b_N] \in \mathbb{R}_+^N$, stands for a vector of all parameters of a function $f \in \mathcal{F}_N$. If we denote a performance function of the algorithm $A(\boldsymbol{\xi})$ by $H_A(\boldsymbol{\xi})$, then its average effectiveness for the problems belonging to set $\mathcal{F}_N$ can be evaluated using the following formula:

$$\overline{H}_A(\boldsymbol{\xi}) = \int_{\Theta_N} H_A(\boldsymbol{\xi}, \boldsymbol{\theta}_N) d\boldsymbol{\theta}_N, \qquad (10)$$

---

[1] i.e. $\int_\mathbb{S} |g(\boldsymbol{x}) - f(\boldsymbol{x})| d\boldsymbol{x} \leq \varepsilon \int_\mathbb{S} d\boldsymbol{x}$

where $\boldsymbol{\xi}$ stands for the vector of external parameters that control behavior the algorithm A. It must to be stressed that the performance function must be bounded e.g. $0 \leq H(\cdot) \leq 1$. In order to meet this requirement the performance function $H_A(\boldsymbol{\xi})$ can be of the form (for maximization problem):

$$H_A(\boldsymbol{\xi}) = \frac{f(\boldsymbol{x}; \boldsymbol{\xi})}{f^*}, \qquad (11)$$

where f^* is global maximum (if only it is known), or some predefined threshold such that $f^* > f(\boldsymbol{x}; \boldsymbol{\xi}) \forall \boldsymbol{x} \in \mathbb{S}$, and $f(\boldsymbol{x}; \boldsymbol{\xi})$ is an estimate of global solution after a stop condition of the algorithm A is met.

Moreover, if one looks at the class $\mathcal{F}_N$ from probabilistic point of view, then it is possible to substitute average (10) by an expected performance of the algorithm A according to probabilistic measure μ, i.e.:

$$E[H_A(\boldsymbol{\xi})] = \int_\Theta H_A(\boldsymbol{\xi}, \boldsymbol{\theta}_N) \mu(d\boldsymbol{\theta}_N) \qquad (12)$$

It is a well-known fact that for a vast majority of global optimization algorithms there is a lack of general rules that would allow to achieve high effective configuration of external parameters $\boldsymbol{\xi}$. Using the expression (12), it can be done (at least theoretically) by solving the problem:

$$\boldsymbol{\xi}^* = \arg\max_{\boldsymbol{\xi} \in \Xi} E[A(\boldsymbol{\xi})] \qquad (13)$$

From practical point of view, however, it is just impossible to compute integral (12) analytically. One way out from this difficult situation is to apply Monte-Carlo integration, that approximates value (12) with the estimator:

$$\hat{E}[H_A(\boldsymbol{\xi})] = \frac{1}{N} \sum_{i=1}^{N} H_A(\boldsymbol{\xi}, \boldsymbol{\theta}_N^i), \qquad (14)$$

where particular vectors $\boldsymbol{\theta}_N^i$ are drawn independently according to probabilistic measure μ. Naturally, the result of Monte-Carlo integration is a random variable, which asymptotically approaches (12). As far as one is time limited the asymptotic behavior is of a minor significance. Decidedly more interesting are properties of the estimator (14) evaluated on basis of a finite number of samples. In spite of the fact that the distribution of (14) is generally unknown, the so-called Chernoff inequality [15] delivers a tool that allows to control the quality of (14):

$$P^k\left\{\boldsymbol{\theta}_N \in \Theta_N : \left|E[H_A(\boldsymbol{\xi})] - \hat{E}[H_A(\boldsymbol{\xi})]\right| > \varepsilon\right\} \leq 2\exp(-2\,k\,\varepsilon^2), \qquad (15)$$

where $\varepsilon > 0$. In this way, if one wants to be sure that $E[H_A(\boldsymbol{\xi})]$ belongs to the confidence interval $[\hat{E}[H_A(\boldsymbol{\xi})] - \varepsilon, \hat{E}[H_A(\boldsymbol{\xi})] + \varepsilon]$ with probability $1 - \delta$, then the estimate (14) must be evaluated with the sample of size k, where

$$k \geq \frac{1}{2\varepsilon^2} \ln\left(\frac{2}{\delta}\right) \qquad (16)$$

Since in the black-box optimization [7,16], there is no a-priori knowledge about the problem under consideration, one does not known how to choose initial scale parameters σ for passive evolutionary strategy $(1+1)\text{ES}_\alpha$. Theoretically σ can be a positive real number. In the literature [14], for a commonly applied normal distribution, it is very often recommended to fix the standard deviation σ as a half of the diameter of $\mathbb{S}$. Since the choice is of a great importance for optimization process [8,10,12], we have decided to give an opportunity to test the robustness of stable mutations with respect to scale parameter σ. That is way, the strength of mutation σ for the considered evolutionary strategies $(1+1)\text{ES}_\alpha$ is chosen to be random variable uniformly distributed in a set $(0, \text{diam}(\mathbb{S})/2)^2$.

5 Experimental Simulations

The experimental simulation concerns seven optimization algorithms: four evolutionary strategies with different stable indices i.e. $(1+1)\text{ES}_{2.0}$, $(1+1)\text{ES}_{1.5}$, $(1+1)\text{ES}_{1.0}$ and $(1+1)\text{ES}_{0.5}$, the popular quasi-Newton method, direct search polytope and modified Newton algorithm. The last three optimization techniques are completely deterministic one, and their implementations from IMSL Fortran 90 MP Library were used. The experiment was conducted in the following way. First, according to probabilistic measure presented in section 3, two-dimensional objective function was created. After that, each algorithm was started from the same randomly chosen initial point $\boldsymbol{x}_0$. Evolutionary strategies were stop when the probability of improving the quality of solution decreases below some level i.e. when the improvement has not occurred during a hundred of consecutive iterations. The stop conditions of deterministic methods are precisely described in [2]. The whole experiment consists of $k = 73778$ independent runs of each algorithm, what allowed to obtain a confidence interval with $\varepsilon = 0.005$ and level $1 - \theta = 0.95$. Table 1 contains the estimate of the expected performances (14), and the averaged number of function evaluations are presented in Tab. 2.

Remark 1. *As one can notice while analyzing results in Tab. 1, evolutionary strategies are more efficient than deterministic algorithms A_1 and A_3 whenever multimodal problems are considered. Since the algorithms take advantage of the local information in the form of the first and second-order derivatives of the objective function, it can be concluded that they find the nearest solution, not necessary the global one. Due to the fact, that results presented in the first row of the Tab. 1 can be perceived as a local optimization $\lambda_N = 0.01$ well-conditioned $\lambda_b = 0.01$ problem, it is worth noticing that heavy-tailed mutations are more efficient in such situations than their counterparts with $\alpha = 2.0$ and $\alpha = 1.5$. The results make a stand against common opinion that Gaussian mutation leads to better local convergence than e.g. Cauchy distribution [11]. Moreover, the set of isotropic stable mutations seems to be ordered according to α. Namely, in all eight scenarios it can be noticed that smaller stability indices always lead to the better performance of evolutionary strategies. For this reason, the presented results bring the application of Gaussian distribution in stochastic algorithms into question.*

---
[2] $\text{diam}(A) = \sup \|x - y\| : x, y \in A$.

Table 1. The expected performance $\hat{E}[H_A(\sigma)]$ of the algorithms obtained on the base of Monte Carlo integration (14). $\alpha = \{2.0, 1.5, 1.0, 0.5\}$ correspond to evolutionary strategy $(1+1)\text{ES}_\alpha$ and symbols A_1, A_2, A_3 stands for the quasi-Newton method and user supplied gradient, direct search polytope algorithm and modified Newton method and a user supplied Hessian, respectively.

λ_w	λ_b	λ_N	$\alpha=2.0$	$\alpha=1.5$	$\alpha=1.0$	$\alpha=0.5$	A_1	A_2	A_3
10	0.01	0.01	0.9992	0.9994	0.9999	1.0000	1.0000	1.0000	1.0000
1	1	0.1	0.9777	0.9869	0.9913	0.9975	0.7846	0.9982	0.7847
1	1	1	0.9064	0.9178	0.9289	0.9417	0.7236	0.9319	0.7247
1	1	5	0.6215	0.6462	0.6920	0.7453	0.5163	0.6798	0.5195
1	0.1	10	0.8247	0.8504	0.8913	0.9330	0.8141	0.9037	0.8064
5	5	10	0.4682	0.4929	0.5354	0.5875	0.3754	0.5051	0.3804
10	10	10	0.4356	0.4550	0.4862	0.5269	0.3505	0.4684	0.3543
20	20	20	0.3728	0.3931	0.4265	0.4672	0.2890	0.3995	0.2912

Table 2. The averaged numbers of function evaluations that were needed to obtain performances reported in Tab. 1. The results were rounded to a nearest integer number.

λ_w	λ_b	λ_N	$\alpha=2.0$	$\alpha=1.5$	$\alpha=1.0$	$\alpha=0.5$	A_1	A_2	A_3
10	0.01	0.01	352	350	291	297	30	75	141
1	1	0.1	257	259	241	273	41	84	63
1	1	1	261	265	245	275	41	84	65
1	1	5	274	279	261	295	48	86	84
1	0.1	10	295	303	274	308	47	84	149
5	5	10	250	256	245	286	43	93	68
10	10	10	239	245	239	278	42	97	61
20	20	20	231	240	233	280	42	97	59

6 Conclusions

This paper can be perceived as a first attempt at establishing a connection between an optimal optimization algorithm and a class of functions for which it obtains the best performance. First and foremost a stochastic parametrization of the set of functions is proposed. The probabilistic measure established on the set $\mathcal{F}$ allows to build general test environments for global optimization algorithms. The numerical experiments show very clearly, that heavy tailed mutations are more robust and therefore more efficient than e.g. Gaussian mutation or deterministic methods. The further research will be mainly concentrated on the searching patterns in data collected during numerical simulations. The authors place their faith in presumption that data mining method and pattern recognition procedures could help to assign some subsets of functions to particular algorithms. This problem is of a great importance in engineering computations, since the well matched optimization method allows to save a lot of time, money and various resources.

Acknowledgments

The work was partially supported by Ministry of Science and Higher Education in Poland under the grant 3T11A01530 *Evolutionary algorithms with α-stable mutation in parametric global optimization*

References

1. Beyer, H.G., Schwefel, H.P.: Evolution strategies - a comprehensive introduction. Natural Computing 1(1), 3–52 (2002)
2. Bonnans, J.F., Gilbert, J.C., Lemaréchal, C., Sagastizábal, C.: Numerical Optimization. Springer, Heidelberg (2003)
3. Cullen, H.F.: The stone-weierstrass theorem and complex stone-weierstrass theorem. In: Introduction to General Topology, pp. 286–293. Heath, Boston, MA (1968)
4. Fang, K.T., Kotz, S., KaiWang, N.: Symmetric Multivariate and Related Distributions. Chapman and Hall, London (1990)
5. Feller, W.: An Introduction to Probability Theory and Its Application. John Wiley and Sons, Inc., Chichester (1957)
6. Ho, Y.C., Pepyne, D.L.: Simple explanation of the no-free-lunch theorem and its implications. Journal of Optimization Theory and Applications (115) (2002)
7. Jones, D.R., Schonlau, M., Welch, W.J.: Efficient global optimization of expensive black-box functions. J. of Global Optimization 13(4), 455–492 (1998)
8. Karcz-Dulęba, I.: Asymptotic behaviour of a discrete dynamical system generated by a simple evolutionary process. International Journal of Applied Mathematics and Computer Science 14(1), 79–90 (2004)
9. Nolan, J.P.: Stable Distributions - Models for Heavy Tailed Data, ch. 1, Birkhäuser, Boston (in progress, 2007), academic2.american.edu/~jpnolan
10. Obuchowicz, A., Prętki, P.: Phenotypic evolution with a mutation based on symmetric α-stable distributions. International Journal of Applied Mathematics and Computer Science 14(3), 289–316 (2004)
11. Rudolph, G.: Convergence of evolutionary algorithms in general search spaces. In: Proc. ICEC 1996, Nagoya, pp. 50–54 (1996)
12. Rudolph, G.: Local convergence rates of simple evolutionary algorithms with cauchy mutations. IEEE Transactions on Evolutionary Computation 1(4), 249–258 (1997)
13. Samorodnitsky, G., Taqqu, M.S.: Stable Non-Gaussian Random Processes. Chapman and Hall, New York (1994)
14. Törn, A., Zilinskas, A.: Global optimization. Springer, New York (1989)
15. Vidyasagar, M.: Randomized algorithms for robust controller synthesis using statistical learning theory. Automatica 37(10), 1515–1528 (2001)
16. Wang, G., Goodman, E.D., Punch, W.F.: Toward the optimization of a class of black box optimization algorithms. In: ICTAI 1997: Proceedings of the 9th International Conference on Tools with Artificial Intelligence, p. 348. IEEE Computer Society, Washington, DC, USA (1997)
17. Wolpert, D., Macready, W.: The mathematics of search (1994)
18. Zolotariev, A.: One-Dimensional Stable Distributions. American Mathematical Society, Providence (1986)

Ant Colony Optimization: A Leading Algorithm in Future Optimization of Petroleum Engineering Processes

Fatemeh Razavi[1] and Farhang Jalali-Farahani[2,*]

[1] Institute of Petroleum Engineering, University of Tehran
f_s_razavi@yahoo.com
[2] Institute of Petroleum Engineering, University of Tehran
P.O. Box 11365/4563, Tehran, Iran
fjlali@ut.ac.ir

Abstract. The objective of the research presented in this paper is to investigate the application of a metaheuristic algorithm called *Ant Colony Algorithm* to petroleum engineering problems. This algorithm usually used for discrete domains, but with some modifications could be applied to continuous optimization. In this Paper, two examples with continuous and discrete parameters also known solutions and varying degrees of complexities are presented as an illustration for solving a large class of process optimization problems in petroleum engineering. Results of case studies show ability of Ant Colony Algorithm to provide fast and accurate solutions.

Keywords: Ant Colony, Optimization, History Matching, Separator, Ackley Function.

1 Introduction

Within the past decade, use of evolutionary algorithms, such as genetic algorithm, simulated annealing and more recently Ant Colony Optimization (ACO), have been considered extensively. Ant Colony Optimization was introduced as a novel nature-inspired metaheuristic method by M. Dorigo et al. [1]. ACO, using principles of communicative behavior occurring in real ant colonies, has been applied successfully to solve various combinatorial optimization problems especially discrete problems such as traveling salesman problem [2], however it has been also applied to continuous domains [3,4].

Petroleum engineers face a wide variety of parameters estimation and optimization problems. The objective in these problems is to determine the optimum values of controllable variables (decision variables) to obtain maximum profitability.

In the present study, two important petroleum engineering problems have been analyzed using a version of multi-dimensional ACO for continuous domain, Continuous ACO (CACO), proposed by V.K. Jayaraman et al. [3]. The problems have been

---

* Corresponding author.

presented with two different degrees of complexities: I) Parameters Estimation in History Matching Problem in Petroleum Reservoirs, II) Determine Optimum Number of Phase Separators and Separators Pressure in Oil Industry.

The rest of this paper is organized as follows: in section 2, the general aspects of Ant Colony Optimization are discussed. Section 3 describes the mathematical background of CACO. In section 4, the authority of CACO written code is tested. In Section 5, CACO is applied to the mentioned petroleum engineering problems and finally, section 6 is conclusion part.

2   Ant Colony Optimization: General Aspects [1]

Ant Colony algorithms were inspired by the observation of real ant colonies. Ants are social insects that live in colonies. An important and interesting behavior of ant colonies is their foraging behavior and in particular, how ants can find shortest path between food sources and their nest.

While walking from food sources to the nest and vice versa, ants deposit on the ground a substance called *pheromone*, forming a pheromone trail. Ants can smell pheromone and, when choosing their way, they tend to choose paths marked by strong pheromone concentrations. The pheromone trail allows the ants to find their way back to the food source (or to the nest). Also, other ants to find the location of the food sources found by their nest-mates can use it. Considering ants are moving on a straight line that connects a food source to their nest. It is well known that the primary means for ants to form and maintain the line is pheromone trail. Ants deposit a certain amount of pheromone while walking, and each ant probabilistically prefers to follow a direction rich in pheromone. This elementary behavior of real ants can be used to explain how they can find the shortest path that reconnects a broken line after the sudden appearance of an unexpected obstacle has interrupted the initial path. In fact once the obstacle has appeared, those ants that are just in front of the obstacle cannot continue to follow the pheromone trail and therefore they have to choose between turning right or left. In this situation we can expect half the ants to choose to turn right and the other half to turn left. A very similar situation can be found on the other side of the obstacle. It is interesting to note that those ants which choose, by chance, the shorter path around the obstacle will more rapidly reconstitute the interrupted pheromone trail compared to those which choose the longer path. Thus the shorter path will receive a greater amount of pheromone per time unit and in turn a large number of ants will choose the shorter path.

ACO, introduced by M. Dorigo et al. [1], is a metaheuristic algorithm to solve combinational optimization problems by using principles of communicative behavior occurring in ant colonies. A metaheuristic refers to a master strategy that guides and modifies other heuristics to produce solutions beyond those that are normally generated in a quest for local optimality.

As discussed before, ants can communicate information about the paths they found to food sources by marking these paths with pheromone. The pheromone trails can lead other ants to the food sources. It is an evolutionary approach where several generations of artificial ants search for good solution. Every ant of the generation builds up a solution, which is found. Ants that find a good solution mark their paths through the decision space by putting some amount of pheromone on the path. Ants of the next generation are attracted by the pheromone that they will put in the solution space near good solutions.

3 Mathematical Background (CACO)

ACO is mainly applicable to discrete optimization problems such as the traveling salesman problem. However, with some modifications it has been also applied to continuous domains [3,4].

3.1 Data Structure for Continuous ACO Algorithm

To apply ACO algorithm in continuous domain, a data structure must be used at first. The data structure applied in this work was proposed by F. Jalalinejad et al. [4], shown in Fig.1. The two dimensional $n \times m$ matrix is the search area where n is the number of variable to be optimized and m is the number of ants where $m \geq n$ [1]. In Continuous ACO, number of ants is equal to number of regions. In fact, each region is a solution in which an ant moves on it to improve the solution. In Fig.1 f_i and τ_i are $1 \times m$ vectors representing the ith region objective function and pheromone trail amount, respectively.

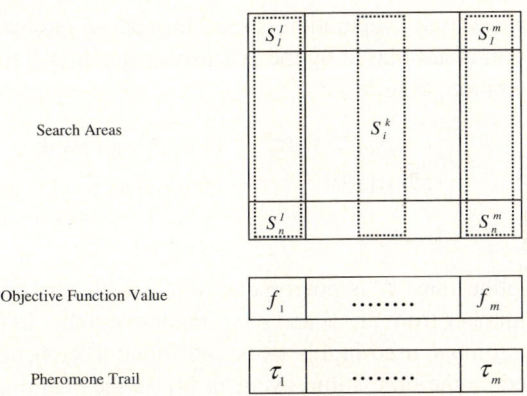

Fig. 1. Data Structure used for CACO

3.2 Continuous Ant Colony Optimization (CACO) Algorithm

The CACO algorithm formed from following steps [4]:

   a) Initialization: in this stage each region is randomly initialized for each variable in the feasible interval. Feasible interval is defined by problem constraint. The equal amount of pheromone is also placed in pheromone trail vector.
   b) Global search: the main responsibility of global ants is to search globally in order to escape local optimum traps. The global search consists of three operations: crossover,

mutation and trail diffusion. The crossover operation is done as follows: select a parent randomly and set the first variable of the child's position vector same as that of the first element of the parent's positions vector. The subsequent values of the variables of the child are set to the corresponding values of a randomly chosen parent with a probability equal to the crossover probability (CP). A cross over probability of one means that each element of the child's position vector has a different parent. A CP of zero means that the child region has all the elements same as the chosen single parent. After the crossover step, mutation is carried out by randomly adding or subtracting a value to each and every variable of the newly created region with a probability equal to a suitably defined mutation probability [3]. The mutation step size is reduced as per the relation: [4]

$$\text{Step for variable } i = (-1)^{1+\text{rand}(1)} \times \text{max.step for } i \times (1 - r^{(1-T)^b}) \quad (1)$$

where r is a random number from [0,1], T is the ratio of the current iteration number to that of the total number of iterations, and b is a positive parameter controlling the degree of non-linearity [3].

c) Local search: local ants search in a smaller region around the potential solution to improve the objective function. Different methods such as simulated annealing could be used for local search. Direct search is used in the present work to decrease the load of computation in this stage. Direct search checks nearby regions in both directions of each variable to find a better solution.

d) Pheromone update: trail evaporation is used in order to ensure that the search during the next generation is not biased by the proceeding iteration. The pheromone trail is updated after each iteration, as follows:

$$\tau_i^{(t+1)} = \begin{cases} (1-\rho)\tau_i^{(t)} + \Delta\tau_i^{(t)} & \text{Fitness improved} \\ (1-\rho)\tau_i^{(t)} & \text{Otherwise} \end{cases} \quad (2)$$

$$\Delta\tau_i^{(t)} = e(\text{rank})^f$$

where ρ is evaporation rate, τ is pheromone trail, e and f are constants which most often f is a random number from [1, 2] and e is a random number in [0.1, 1.5]. This definition for adding pheromone used in this work was found to be better than the one used before which is based on the fitness improvement [4]. In the pheromone update strategy, Eq. 2, evaporation and laying pheromone are kept in the proportional range and would prevent stagnation in the local minima.

4 Testing the Authority of CACO Code

The Continuous Ant Colony Algorithm has been written in MATLAB environment. For checking the validity and power of the code, a benchmark problem was solved with the written algorithm. This benchmark problem is Ackley function. The Ackley function is a continuous benchmark and multimodal function which is included nonlinear exponential and cosinusoidal functions. General form of this function is as follows [5]:

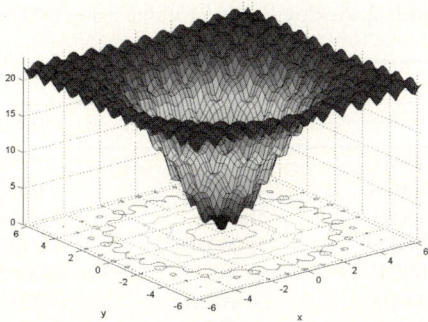

Fig. 2. Two dimensional Ackley function

$$f(x) = 20 + e - 20\exp\left[-0.2\sqrt{\frac{1}{n}\sum_{i=1}^{n}x_i^2}\right] - \exp\left[\frac{1}{n}\sum_{i=1}^{n}\cos(2\pi x_i)\right] \quad (3)$$

$-5 \leq x_i \leq 5 \quad (i = 1, 2, ..., n)$

"n" determines the dimension of Ackley function. In the above figure, two dimensional Ackley function ($n=2$) is shown. As it can be seen, this function has a symmetric function with too many local minimum which all of them are moving to one global minimum (zero) from all directions. Two, five, eight and ten dimensional Ackley function have been solved with our written code to test the power of the CACO.

First, CACO parameters must be determined. These parameters are number of ants, crossover probability, mutation probability, evaporation rate and local ants (Table 1); finally select those with the best results for objective function. Optimum value of Ackley function for each dimension is shown in Table 2. It is obvious that in all dimensions, CACO could find global minimum with excellent approximation.

Table 1. CACO Parameters with different Ackley function dimensions

Dimension	Ants	Crossover Probability	Mutation Probability	Evaporation rate	Local ants %
$n=10$	25	50	40	0.75	40
$n=8$	20	60	55	0.8	45
$n=5$	30	55	60	0.7	40
$n=2$	25	65	45	0.65	45

Table 2. Ackley Function Value using CACO

Runs	Ackley Function Dimension			
	n=2	n=5	n=8	n=10
1	0.000041	-0.000097	0.0003	0.0001
2	0.000046	0.00046	-0.0005	0.0001
3	0.000081	0.00077	0.0011	0.0001
4	-0.000019	0.00046	0.0007	-0.0002
5	-0.000035	0.00033	-0.0005	0.0003
6	0.000048	-0.00038	0.0002	0.0013
7	-0.000021	0.00061	0.0007	0.0001
8	0.00011	-0.00011	0.001	0
9	-0.00004	0.00023	-0.0001	0.0002
10	-0.0000198	-0.00036	0.0008	0.0015
Mean	0.000019	0.00014522	0.000369	0.0003444
The Best	-0.000019	-0.0000975	-0.0001	0
The Worst	0.00011	0.0007738	0.0011	0.0015
S.D.*	0.0000533	0.000426	0.000569	0.000556
C.V.**	2.805	2.9335	1.542	1.6144

* Standard Deviation
** Coefficient of Variation

5 Applying CACO to Petroleum Engineering Problems

Subsequent to checking the reliability of the algorithm, we investigate the application of CACO algorithm to the problem of history matching of petroleum reservoirs and determining some optimum parameters for phase separators in oil industry.

5.1 Parameters Estimation in History Matching Problem in Oil Industry

First, we apply CACO algorithm for a History Matching problem with known solution. In this problem the following data are available: well flow pressure versus time, fluid compressibility, well radius, well production rate, initial pressure and reservoir thickness. Moreover, the equation which is used for simulation well pressure behavior is as follows:

$$P_{wf} = P_i - \frac{q\mu}{4\pi kh}[\ln\frac{kt}{\phi\mu c r_w^2} + 0.8090] \tag{4}$$

Unknowns are porosity (ϕ) and permeability (k) values. The proper values for porosity (ϕ) and permeability (k) are those which with substituting them in the simulation equation, the resulted well flow pressure versus time would be the same as one in available data. In fact we match the new results with our data history to obtain correct values for the unknowns.

The objective function is defined as:

$$\text{Error Function} = \sqrt{\sum (\text{Calculated } P - P \text{ from graph})^2} \tag{5}$$

The control variables are porosity (ϕ) and permeability (k). Both control variables are continuous variables ($\phi_{min} < \phi < \phi_{max}$ and $k_{min} < k < k_{max}$). First, CACO parameters must be adjusted and determined. Results are shown in Table 3. Final results are shown in Table 4. In this case CACO converges to the solution after 14 iterations. For the available data, we will have:

Table 3. Ant Colony parameters for History Matching problem

Ants	Crossover Probability	Mutation Probability	Evaporation rate	Local ants %
30	55	60	0.60	45

Table 4. Unknown Parameters

Runs	Unknowns		Objective function value (Error)
	Permeability (Darcy)	Porosity %	
1	0.0979	26.4946	0.00001964
2	0.0986	26.4799	0.00004014
3	0.0971	24.4066	0.0000889
4	0.1034	25.6233	0.00000263
5	0.0983	28.5789	0.00006346
6	0.0996	27.4261	0.00009216
7	0.0988	24.2862	0.00002413
8	0.0973	25.2112	0.00009544
9	0.1006	22.2227	0.00002036
10	0.0973	24.2712	0.00008393
Mean	0.0989	25.5001	0.000053079
The Best	0.1034	25.6233	0.00000263
The Worst	0.0973	25.2112	0.00009544
S.D.	0.0019	1.8305	0.000035596
C.V.	0.0192	0.0718	0.6706

The values of permeability and porosity which were calculated before (not including CACO) are 0.101 Darcy and 25.75%, respectively and the error function value is 10^-5. Because the estimated values for two parameters are calculated simultaneously, according to the results, it can be concluded that the best values for permeability and porosity are 0.1034 and 25.6233 respectively and simultaneously (4th run) and the worst values are 0.0973 and 25.21 simultaneously (8th run).

5.2 Determine Optimum Number of Phase Separators and Separators Pressure in Oil Industry

The performance of a hydrocarbon production well is a function of several variables. Separators are production facilities that have high effect on the performance of a producing well. So, optimal design for separators is one of the most important aspects of production optimization in oil industry. An important concept is that for a finite number of separators, there is an optimal combination of separator pressures that will maximize the fractional liquids recovery [6].

At the optimal combination of separator pressures the API gravity of the crude oil will be maximized and the gas-oil ratio (GOR) will be minimized. So the objective function could be GOR/ 'API [7].

Our developed code in MATLAB environment, based on CACO, is able to obtain the most optimum number of separators, and separator pressures. The optimization of these parameters should result in improved hydrocarbon liquid production.

Each separator is modeled with flash calculation. A flash calculation takes the overall composition of a mixture and determines number of phases presents, composition and quantity of each phase at a new temperature and pressure.

The procedure of performing a flash calculation is iterative and converges when the fugacity of each component is the same in both phases.

Applying CACO to the model, optimum number of separators (discrete parameter) and separators pressure (continuous parameter) have been determined.

CACO parameters and final solution using CACO for the available data are shown in Table 5 and 6, respectively. In this case CACO converges to the solution after 11 iterations. Moreover, the unknown parameters, which were calculated before using Separator Test (not including CACO), are shown in Table 7 [7].

Comparing the results in Table 6 and 7, it is apparent that the problem solution by means of CACO algorithm (see Table 6) has good agreement with the known solution using separator test (see Table 7). For the available data, the number of separator is 1.

The two mentioned problems are multi-variable optimization problems which CACO could find the solution successfully.

Table 5. Ant Colony parameters for separator problem

Ants	Crossover Probability	Mutation Probability	Evaporation rate	Local ants %
30	60	50	0.65	40

Table 6. Unknown Parameters

Runs	Number of Separators=1 Unknown Separator Pressure	Objective Function Value (GOR/'API)
1	97.2	17.13
2	96.7	16.97
3	100.3	16.71
4	99.2	16.49
5	98.9	14.9
6	97.1	15.86
7	101.1	15.94
8	96.6	15.77
9	96.6	14.92
10	100.1	14.88
Mean	98.4	15.96
The Best	98.9	14.9
The Worst	97.2	17.13
S.D.	1.7399	0.86
C.V.	0.0177	0.0539

Table 7. Unknown parameters using Separator Test

Num. of Separator	Separator Pres. (psig)	Degree API	GOR	Objective Function (GOR/'API)
1	100	40.7	676	16.6

6 Conclusions

Continuous Ant Colony Algorithm has been successfully applied to two petroleum engineering problems. The algorithm is simple and could be applied to problems with different degrees of complexity. As a quick look on this research, some advantages of CACO can be stated as follows:

I) It optimizes with continuous or discrete parameters or a combination of both types.
II) It doesn't need any derivative information.
III) Simultaneously searches from a large number of decision variables.
IV) It can jump out local optima to find global optima.
V) Provides a list of all optimum parameters as a solution.

In conclusion, Ant Colony Approach can be proposed as a reliable and useful optimization tool in petroleum engineering problems.

References

1. Dorigo, M., Stutzle, T.: Ant Colony Optimization. MIT Press, United States (2004)
2. Dorigo, M., Gambardella, L.M.: Ant Colony System: A Cooperative Learning Approach to the Traveling Salesman Problem. IEEE Transactions on Evolutionary Computation, 53–66 (1997)
3. Jayaraman, V.K., Kulkarni, B.D., Karale, S., Shelokar, P.: Ant Colony Framework for Optimal Design and Scheduling of Batch Plants. Computer and Chemical Engineering 24, 1901–1912 (2000)
4. Jalalinejad, F., Jalali-Farahani, F., Mostoufi, N., Sotudeh-Gharebagh, R.: Ant Colony Optimization: A Leading Algorithm in Future Optimization of Chemical Process. In: 17th European Symposium on Computer Aided Process Engineering, pp. 1–6. Elsevier, Amsterdam (2007)
5. Afshar, A., Abbasi, H., Jalali, M.R.: Optimum Design of Water Conveyance Systems by Ant Colony Optimization Algorithms. In: International Journal of Civil Engineering, Iranian Society of Civil Engineers, Iran University of Science and Technology, Tehran, Iran, 1735-0522, 1–13 (2006)
6. Tavakkolian, M., Jalali-Farahani, F., Emadi, M.A.: Production Optimization Using Genetic Algorithm Approach, SPE 88901 (2004)
7. Danesh, A.: PVT and Phase Behavior of Petroleum Reservoir Fluids, Netherlands (1998)

Design and Multi-Objective Optimization of Combinational Digital Circuits Using Evolutionary Algorithm with Multi-Layer Chromosomes*

Adam Słowik and Michał Białko

Department of Electronics and Computer Science, Koszalin University of Technology,
ul. Śniadeckich 2, 75-453 Koszalin, Poland
aslowik@ie.tu.koszalin.pl

Abstract. In the paper an application of evolutionary algorithm with multi-layer chromosomes to the design and multi-objective optimization of combinational digital circuits is presented. The optimization criterions are minimizations of: number of gates, number of transistors in the circuit, and circuit propagation time. Four combinational circuits, chosen from literature, are designed, and optimized using proposed method. Results obtained using this method are compared with results obtained by other methods. The results obtained using this method are in many cases better than those obtained using other methods.

1 Introduction

The process of evolutionary circuit design is fundamentally different form traditional design process, because it is not based on designer knowledge and experience, but on the evolution process [1]. The evolutionary circuit design has less constraints than the design based on the designer knowledge and experience; the designers are not only limited by the technology in which the circuit will be produced, but also by their own habits (routines), imagination and creative thinking [1]. An application of evolutionary methods to the circuit design allows to escape from limitations characterized earlier and to obtain an access to new possibilities [1]. Many different evolutionary methods for design of digital circuits have been created until now. Among those it is possible to mention the papers: Coello [2, 11, 12], Slowik [3, 4], Miller [9], Kalganova [10], and Nilagupta [5]. In all mentioned papers evolutionary algorithms were used for search the optimal solution with one chosen criterion, such as minimization of number of gates or number of transistors, necessary for the physical realization of a chosen circuit. In the paper [4], it was noticed that a minimization of transistor number

---

* This work was supported by the Polish State Committee for Scientific Research (KBN) under Grant No. 3 T11B 025 29.

does not always lead to the minimization of the gate number in the designed circuit. So, an optimal solution with respect to the transistor number may be a non-optimal solution with respect to the gate number in the designed circuit.

In this paper an application of the evolutionary algorithm with multi-layer chromosomes to the design and multi-objective optimization of combinational digital circuits is presented. Described method is named MLCEA-MOO (*Multi Layer Chromosome Evolutionary Algorithm - Multi Objective Optimization*) and is used to the design and optimization of combinational digital circuits with respect to minimization of: number of gates, number of transistors, and propagation time of the designed circuit.

2 MLCEA-MOO Method

Because proposed MLCEA-MOO method is linking two criteria used in the earlier methods: MLCEA [3] (minimization number of gates) and MLCEA-TC [4] (minimization number of transistors), the following set of gates is used in this method: NOT, NOR, XOR, NAND, DC (direct connection of gate input with its output), OR, and AND, because this set is the sum of the sets of gates available in methods MLCEA, and MLCEA-TC. According to the work [5] it is assumed that, the NOT gate consists of 2 transistors, NOR and NAND gates are composed of 4 transistors, OR and AND gates contain 6 transistors, and the XOR gate is constructed using 12 transistors, in the case of implementation in silicon, or 16 transistors (similarly as at work [5]) in the case of realization using elementary NAND and NOT gates. Here, we assume that the XOR gate is constructed of 12 transistors. The electrical structures of used gates are presented in the following figures: NOT (Fig. 1a), AND (Fig. 1b), OR (Fig. 2a), XOR (Fig. 2b), NAND (Fig. 3a), and NOR (Fig. 3b). In Figures 1-3, *Vdd* represents voltage supply, x and y corresponds to gate inputs, and z represents gate output.

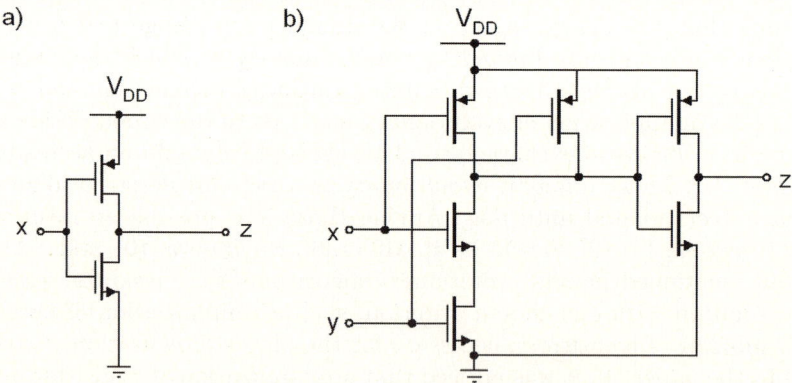

Fig. 1. Electrical structures of the gates NOT (a), and AND (b)

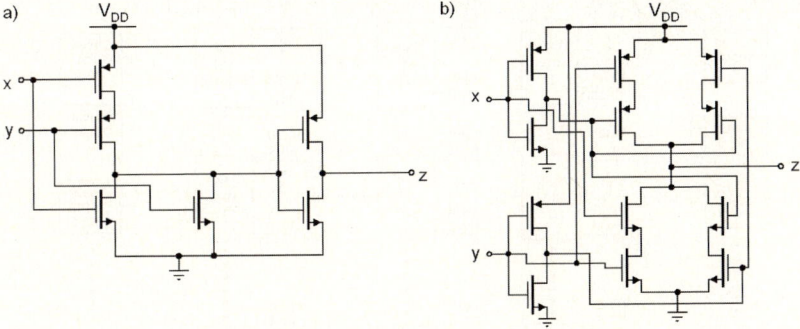

Fig. 2. Electrical structures of the gates OR (a), and XOR (b)

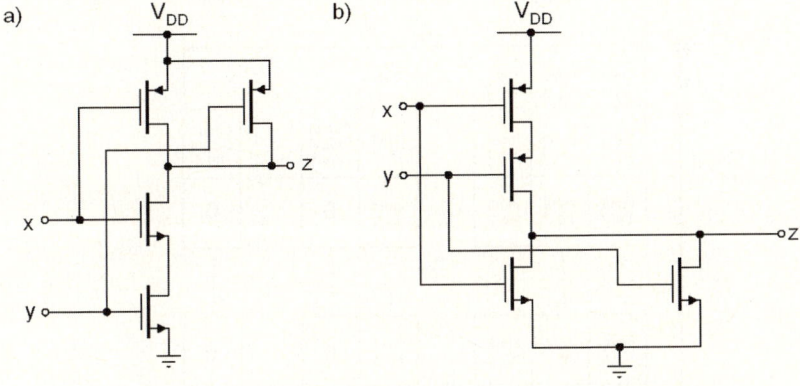

Fig. 3. Electrical structures of the gates NAND (a), and NOR (b)

In order to determine the propagation time (the third optimization criterion) we have assumed, according to work [6], that the gates possess following delays (CMOS 3.3 [V] technology):

$$\text{NOT} \Rightarrow 0.0625 \text{ [ns]}$$
$$\text{NOR} \Rightarrow 0.1560 \text{ [ns]}$$
$$\text{XOR (12 transistors)} \Rightarrow 0.2120 \text{ [ns]}$$
$$\text{NAND} \Rightarrow 0.1300 \text{ [ns]}$$
$$\text{OR} \Rightarrow 0.2160 \text{ [ns]}$$
$$\text{AND} \Rightarrow 0.2090 \text{ [ns]}$$

Presented delays are the average values of the gate switching-on delay (transition from the state of the logical zero to the state of logical one) and the gate switching-off delay (transition from the state of logical one to the state logical zero).

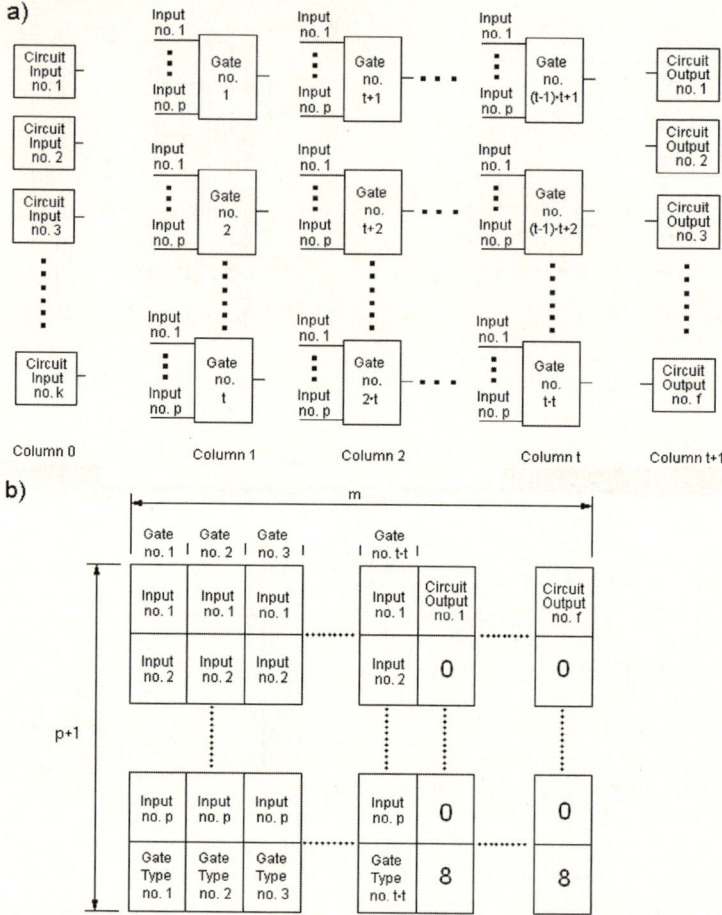

Fig. 4. The structure: pattern of gates (a), multi-layer chromosome representing this pattern (b)

The MLCEA-MOO method is operating in the two phases. At the beginning in order to create an initial population, we create a pattern (template) of the designed circuit, which structure is shown in Fig. 4a. The structure and coding of chromosomes representing the pattern are shown in Fig. 4b.

Similarly as in MLCEA method, in the section of the chromosome marked as "Input no. x" we put the number of the circuit input or the gate number from the pattern, which output is to be connected to this input; in the place "Gate Type no. x" we put one of the seven digits, which represent respectively: 1-NOT gate, 2-NAND gate, 3-XOR gate, 4-NOR gate, 5-DC, 6-AND gate, 7-OR gate. Symbol "DC" represents direct connection of a given gate input with its output. Digit "0" represents a lack of connection of a given gate input in the pattern, and digit "8" represents the circuit output. In the place "Output no. x" we put

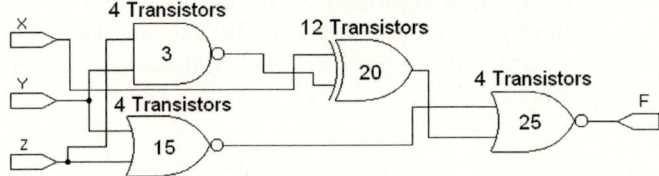

Number of gates = 4; Number of transistors = 24; Propagation time = 0.4980 [ns]

Fig. 5. Example of digital circuit fulfilling the logic function of "Circuit no. 1" from section 3 (of this paper)

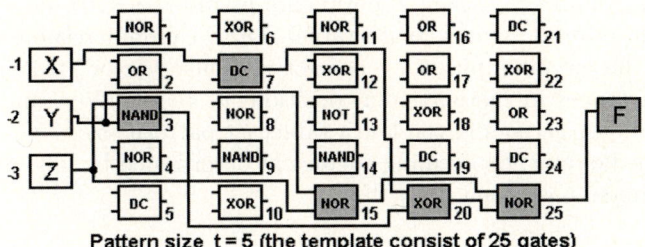

Pattern size t = 5 (the template consist of 25 gates)

Fig. 6. Pattern of gates corresponding to the circuit from Figure 5

-3	-2	-3	-1	-1	1	-1	-1	-3	3	-1	1	10	3	-2	13	-3	3	11	7	14	2	13	13	15	25
-3	-3	-2	-1	0	1	0	5	3	-3	-3	8	0	5	-3	6	9	14	0	3	0	6	19	0	20	0
4	7	2	4	5	3	5	4	2	3	4	3	1	2	4	7	7	3	5	3	5	3	7	5	4	8

Fig. 7. Multilayer chromosome corresponding to the pattern of gates from Figure 6

the pattern gate number, which output is to be connected to the circuit output. All circuit inputs are represented by negative numbers, that is, the first input is represented by the number "-1", etc.

Each individual (chromosome) in the population represents selected combinational circuit. For example in Fig. 5 the three-input circuit realizing the logic function of "Circuit no. 1" from section 3 of this paper is shown. The pattern (template) of the two-input gates, with connections between them corresponding to this circuit, is shown in Fig. 6; this pattern is represented in a population of individuals (potential solutions) by the multi-layer chromosome, shown in Fig. 7, in which each column corresponds to single gate of the template. In Fig. 6, and Fig. 7 the gates from the circuit (Fig. 5) are marked by grey color.

The pattern size t for designed circuit is determined experimentally using following formula:
$$t = MAX(NI, NO) \qquad (1)$$
where: NI - number of circuit inputs, NO - number of circuit outputs.

In the case when the computed pattern size t does not allow to find the circuit fulfilling a given truth table, then it is increased by one, and the design process is repeated. After creation the initial population each individual is evaluated using following objective function $FC1$:

$$FC1 = \sum_{i=1}^{I} fc1_i + Sc \cdot (t^2 - NG) \qquad (2)$$

where:

$$fc1_i = \begin{cases} v \cdot j, & \text{when } O(c_i) \neq C_i \\ 0, & \text{when } O(c_i) = C_i \end{cases} \qquad (3)$$

v - positive real number (during experiments assumed $v=10$), c_i - i-th vector of combination of input signals (truth table), $O(c_i)$ - circuit response for vector c_i applied to the circuit input, C_i - correct response vector (truth table), j - number of differences in the particular positions of vector $O(c_i)$ and vector C_i, I - number of input vectors in the truth table, t - pattern size, NG - number of gates in designed circuit, Sc - scaling coefficient determined by following formula (during experiments assumed $Sc = 0.1$):

$$0 < Sc < \frac{v}{t^2} \qquad (4)$$

The value of the $FC1$ function is lower when the properties of a given individual (circuit) better fit the given truth table. The function $FC1$ specifies the number of constraints that follow from the truth table, which are not fulfilled by a particular individual (first factor of $FC1$ function). In the design process we are looking for circuits that fulfill the constraints posed by the truth table, but with the highest number of gates allowed by the circuit pattern size (second factor of $FC1$ function). This treatment increases a search space in the second phase of the algorithm. The algorithm is minimizing the $FC1$ function during its operation. In the case when the circuit fulfilling the truth table is not found ($FC1 \geq v$) following operators are applied: crossover, mutation and the fan selection [8]. The crossover operator depends on cutting all layers of two randomly chosen chromosomes in one randomly chosen point, and exchange of the cut fragments between them. Due to the application of the multi-layer chromosomes, an interchange of the whole gate in the pattern is possible without damage of its internal structure (such a damage is possible in single-layer chromosomes). The mutation operator causes a random change of the gate type in the pattern (in the case of choosing to the mutation the gene belonging to the last layer of the chromosome) or change of the connections between gates in the pattern (when the gene from remaining layers is chosen to the mutation). In the case when the circuit fulfilling the truth table ($FC1 < v$) is found, the algorithm starts the second phase of its operation. Simultaneous minimization of the gate number, the transistor number in the circuit, and the circuit propagation time is the goal

Table 1. Truth tables of designed circuits

Circuit no. 1				Circuit no. 2					Circuit no. 3					Circuit no. 4						
In			O	In				O	In				O	In				O		
X	Y	Z	F	Z	W	X	Y	F	A	B	C	D	F	A_1	A_0	B_1	B_0	X_2	X_1	X_0
0	0	0	0	0	0	0	0	1	0	0	0	0	1	0	0	0	0	0	0	0
0	0	1	0	0	0	0	1	1	0	0	0	1	0	0	0	0	1	0	0	1
0	1	0	0	0	0	1	0	0	0	0	1	0	0	0	0	1	0	0	1	0
0	1	1	1	0	0	1	1	1	0	0	1	1	0	0	0	1	1	0	1	1
1	0	0	0	0	1	0	0	0	0	1	0	0	1	0	1	0	0	0	0	1
1	0	1	1	0	1	0	1	0	0	1	0	1	1	0	1	0	1	0	1	0
1	1	0	1	0	1	1	0	1	0	1	1	0	1	0	1	1	0	0	1	1
1	1	1	0	0	1	1	1	1	0	1	1	1	1	0	1	1	1	1	0	0
				1	0	0	0	1	1	0	0	0	1	1	0	0	0	0	1	0
				1	0	0	1	0	1	0	0	1	1	1	0	0	1	0	1	1
				1	0	1	0	1	1	0	1	0	0	1	0	1	0	1	0	0
				1	0	1	1	0	1	0	1	1	0	1	0	1	1	1	0	1
				1	1	0	0	0	1	1	0	0	0	1	1	0	0	0	1	1
				1	1	0	1	1	1	1	0	1	1	1	1	0	1	1	0	0
				1	1	1	0	0	1	1	1	0	0	1	1	1	0	1	0	1
				1	1	1	1	0	1	1	1	1	1	1	1	1	1	1	1	0

of the second phase of the algorithm. In this phase of the MLCEA-MOO method a similar approach as in the MOGA [7] algorithm (Multiple Objective Genetic Algorithm) was applied. In this approach a given individual fitness depends on the number of solutions dominating (better with respect to all optimization criteria) this individual in a given population. The second phase of MLCEA-MOO method consists of several steps. In the first step the fulfilling of the truth table is checked (the circuit fulfilling the truth table is an acceptable solution) for each individual (circuit), and also the number of gates (NG), number of transistors (NT), and the propagation time (PT) are determined for a given circuit. In the second step the parameter *dom* value (5) is computed for each acceptable individual; the value of this parameter corresponds to the number of acceptable individuals that dominate given circuit (individual). The value of the *dom* parameter for a given individual is computed as follows:

$$dom = \sum_{i=1}^{N} d_i \qquad (5)$$

where:

$$d_i = \begin{cases} 1, \text{ when } (NG_i < NG) \wedge (NT_i < NT) \wedge (PT_i < PT) \\ 0, \text{ otherwise} \end{cases} \qquad (6)$$

i - number of individual, N - population size, NG (NT, PT) - number of gates (number of transistors, propagation time) for a given individual, NG_i (NT_i, PT_i) - number of gates (number of transistors, propagation time) for i-th individual.

Table 2. Comparison results for designed circuits

Method	Obtained function - circuit number 1	NG	NT	PT [ns]
HD	F=(X'·Y·Z)+(X·(Y⊕Z))	6	38	0.6965
GA	F=(Z·(X+Y))⊕(X·Y)	4	30	0.6370
GT	F=(((Y·Z)'⊕X)+(Y+Z)')'	4	24	0.4980
MO	F=((Y+Z)'+(X⊕(Z·Y)')))'	4	24	0.4980
Method	Obtained function - circuit number 2	NG	NT	PT [ns]
HD	F=((Z'·X)+(Y'·W'))+((X'·Y)·(Z⊕W'))	11	66	0.7035
GA	F=(((W⊕Y)+(W·X))⊕((Z+X+Y)⊕Z))'	8	62	0.9185
GT	F=((Z⊕Y)+(X+Y)')'⊕((X·Y)'·W)'	6	40	0.6400
MO	F=(((W·X)'·Y)'·(X+Y))'⊕(Z⊕W)	6	42	0.6020
Method	Obtained function - circuit number 3	NG	NT	PT [ns]
HD	F=((A⊕B)⊕((A·D)·(B+C)))+((A+C)+D)'	9	66	0.9155
GA	F=((A⊕B)⊕A·D)+(C+(A⊕D))'	7	56	0.9115
GT	F=((A+D)'⊕(B⊕D)·((C+(B⊕D))'+((A·D)'+C)')')'	8	48	0.7100
MO	F=(((A'+D)⊕B)·((D⊕A)+C))'	6	42	0.6205
Method	Obtained function - circuit number 4	NG	NT	PT [ns]
HD	$X_0 = A_0 \oplus B_0$ $X_1 = (A_1 \oplus B_1) \cdot B_0' + ((A_1 \oplus B_1) \oplus A_0) \cdot B_0$ $X_2 = (A_1 \cdot B_1) + (A_0 \cdot B_0) \cdot (A_1 + B_1)$	12	86	0.8490
GA	$X_0 = A_0 \oplus B_0$ $X_1 = (A_0 \cdot B_0) \oplus (A_1 \oplus B_1)$ $X_2 = (A_1 \cdot B_1) + (A_0 \cdot B_0) \cdot (A_1 \oplus B_1)$	7	60	0.6370
GT	-	-	-	-
MO	$X_0 = ((A_0 \cdot B_0) + (A_0 + B_0)')'$ $X_1 = (A_1 \oplus B_1) \oplus (A_0 \cdot B_0)$ $X_2 = (((A_0 \cdot B_0) \cdot (A_1 \oplus B_1))' \cdot (A_1 \cdot B_1)')'$	8	50	0.4720

In the third step all acceptable and non-dominated solutions (for which the *dom* value is equal to 0) are remembered. All unacceptable solutions are removed from the population in the fourth step, and individuals chosen randomly from the remembered set of acceptable and non-dominated solutions are inserted in their place. In the fifth step, for each individual in the population, the values of *FC2* function are computed (the algorithm minimizes this function) according to the following formula (similarly as in the MOGA algorithm [7]):

$$FC2 = 1 + dom \qquad (7)$$

Then, the operators of the selection, crossover and mutation are performed. As the result of the MLCEA-MOO algorithm, not a single individual, but a set of acceptable and non-dominated solutions, called also the set of Pareto optimal solutions are obtained. The minimum value of the product of: number of gates, number of transistors, and circuit propagation time, is taken as a choice criterion for the single individual from obtained population.

3 Description of Experiments

Four test digital circuits were chosen (identical as in work [3, 5, 12]) in experiments; their logic functions are shown in Table 1. Symbol "In" represent inputs, and symbol "O" corresponds to circuit outputs.

During performed experiments the parameters of evolutionary algorithm were: population size 100, crossover probability 0.5, mutation probability 0.05, fan selection [8] coefficient a=0.3. The pattern size was determined experimentally according to the formula (1); for all circuits t=5 was used (the pattern consisted of 25 gates).

In Table 2 results obtained using MLCEA-MOO (marked as MO) method and other methods: Human Design (HD), Genetic Algorithm (GA) (taken from works [3, 4, 12]), GT (results obtained using method described in paper [5]) are presented. Symbols used in all Tables are as follows: "NG" - number of gates, "NT" - number of transistors, "PT" - propagation time. Character " ' " in logic function description represents negation. All circuits were designed and optimized using two-input gates.

4 Conclusion

Due to application of the MLCEA-MOO method the polyoptimal solutions, in view of many criteria can be obtained. Results obtained are better (in 10 cases for 11 possible and comparable in one case) than the results obtained using other methods. Also it is necessary to point out, that the result of the operation of the MLCEA-MOO method is not a single solution, but the whole set of non-dominated (i. e. Pareto optimal) and acceptable solutions, and an open question is a selection of a single solution (fulfilling required criterion) from this set. In the described case as a choice criterion for the selection of this single solution from the set of Pareto optimal solutions was the minimum value of the product of: numbers of gates, numbers of transistors and the circuit propagation time. However, using other selection rules it is possible to obtain other results which are polyoptimal still.

References

1. Greene, J.: Simulated Evolution and Adaptive Search in Engineering Design, Experiences at the University of Cape Town, In: 2nd Online Workshop on Soft Computing (July 1997)
2. Coello, C.A., Christiansen, A.D., Aguirre, A.H.: Use of Evolutionary Techniques to Automate the Design of Combinational Circuits. International Journal of Smart Engineering System Design (2000)
3. Slowik, A., Bialko, M.: Design and Optimization of Combinational Digital Circuits Using Modified Evolutionary Algorithm. In: Rutkowski, L., Siekmann, J.H., Tadeusiewicz, R., Zadeh, L.A. (eds.) ICAISC 2004. LNCS (LNAI), vol. 3070, pp. 468–473. Springer, Heidelberg (2004)

4. Slowik, A., Bialko, M.: Evolutionary Design and Optimization of Combinational Digital Circuits with Respect to Transistor Count. In: 4th National Conference of Electronics Technical University of Koszalin, June 2005, pp. 207–212 (2005)
5. Nilagupta, P., Ou-thong, N.: Logic Function Minimization Based On Transistor Count Using Genetic Algorithm. In: Proc. of the 3rd ICEP, Songkla, Thailand (January 2003)
6. Ercegovac, M.D., Lang, T., Moreno, J.H.: Introduction to Digital Systems. John Wiley, Chichester (1999)
7. Fonseca, C.M., Fleming, P.J.: Genetic Algorithms for Multiobjective Optimization: Formulation, Discussion and Generalization. In: Proc. of the 5th International Conference on Genetic Algorithms, pp. 416–423. Morgan Kauffman Publishers, San Francisco (1993)
8. Slowik, A., Bialko, M.: Modified Version of Roulette Selection for Evolution Algorithms – The Fan Selection. In: Rutkowski, L., Siekmann, J.H., Tadeusiewicz, R., Zadeh, L.A. (eds.) ICAISC 2004. LNCS (LNAI), vol. 3070, pp. 474–479. Springer, Heidelberg (2004)
9. Miller, J., Kalganova, T., Lipnitskaya, N., Job, D.: The Genetic Algorithm as a Discovery Engine: Strange Circuits and New Principles. In: Proc. of the AISB Symposium on Creative Evolutionary Systems (CES 1999), Edinburgh, UK (1999)
10. Kalganova, T., Miller, J.: Evolving more efficient digital circuits by allowing circuit layout and multi-objective fitness. In: Proc. of the First NASA/DoD Workshop on Evolvable Hardware, Los Alamitos, California, pp. 54–63 (1999)
11. Coello, C.A., Christiansen, A.D., Aguirre, A.H.: Automated Design of Combinational Logic Circuits using Genetic Algorithms. In: Proc. of the International Conference on Artificial Neural Nets and Genetic Algorithms, pp. 335–338 (April 1997)
12. Coello, C.A., Aguirre, A.H., Buckles, B.P.: Evolutionary Multiobjective Design of Combinational Logic Circuits. In: Proc. of the Second NASA/DoD Workshop on Evolvable Hardware, Los Alamitos, California, July 2000, pp. 161–170 (2000)

On Convergence of a Simple Genetic Algorithm

Jolanta Socała[1] and Witold Kosiński[2,3]

[1] Institute of Technology and Mathematics
Państwowa Wyższa Szkoł Zawodowa
State Higher Vocational School in Racibórz
ul. Słowackiego 55, 47-400 Racibórz
`jolanta.socala@pwsz.raciborz.edu.pl`
[2] Department of Intelligent Systems
Polish-Japanese Institute of Information Technology
ul. Koszykowa 86, 02-008 Warszawa, Poland
[3] Institute of Environmental Mechanics and Applied Computer Science
Kazimierz Wielki University
ul. Chodkiewicza 30, 85-064 Bydgoszcz, Poland
`wkos@pjwstk.edu.pl`

Abstract. The simple genetic algorithm (SGA) and its convergence analysis are main subjects of the article. The SGA is defined on a finite multi-set of potential problem solutions (individuals) together with mutation and selection operators, and appearing with some prescribed probabilities. The selection operation acts on the basis of the fitness function defined on individuals, and is fundamental for the problem considered. Generation of new population is realized by iterative actions of those operators written in the form of a transition operator acting on probability vectors. The transition operator is a Markov one. Conditions for convergence and asymptotic stability of the transition operator are formulated.

1 Introduction

Recently there has been growing interest in universal optimisation algorithms to solve complex and not well defined, often, optimisation problems. An important role play here genetic, and more general, evolutionary algorithms that are probabilistic optimisation algorithms inspired by some observations made in the nature. That algorithms perform a kind of search for good solutions in a space of "candidate solutions".

During their action a sequence of random variables, candidate solutions are generated. Wide applications of those methods in practical solutions of complex optimal problems cause a need to develop theoretical foundations for them and to investigate their convergence properties [1,5,6,8,12,13].

Genetic algorithm (GA) performs a multi-directional search by maintaining a population of potential solutions and encourages information formation and exchange between these directions due to the iterative action with some probability distributions of a composition of mutation, crossover and selection operators. If we imagine that a population is a point in the space Z of (encoded) potential solutions then the effect of one iteration of this composition is to move that population to another point. The action of that composition is a random operation on populations. In this way the action of GA can be regarded as a discrete (stochastic) dynamical system.

In the paper we use the term *population* in two meanings; in the first it is a finite multi-set (a set with elements that can repeat) of solutions, in the second it is a frequency vector composed of fractions, i.e. the ratio of the number of copies of each element $z_k \in Z$ to the total population size (*PopSize*).

In our analysis we are concerned with a limiting (terminal) probability distributions of all populations for a particular case of the simple genetic algorithm (SGA) in which the mutation follows the proportional selection and the crossover is not present. The composition of both operations is a matrix which leads to the general form of the transition operator (cf.(13)) acting on a new probability vector representing a probability distribution of appearance of all populations of the same *PopSize*.

It seems that the present result can be extended to that more general case when crossover takes place, however, the components of the corresponding transition matrix are more difficult to calculate. Hence we confine to the case when mutation and selection take place, only, and the components of the transition matrix are given explicitly by the formula (12).

The matrix appearing there turns to be Markovian and each subsequent application of SGA is the same as the subsequent composition of that matrix with itself. (cf.(14)). Thanks to the well-developed theory of Markov operators ([3,4,7,9]) new conditions for the convergence asymptotic stability of the transition operator are formulated.

2 Mathematical Preliminaries

Let
$$Z = \{z_0, ..., z_{s-1}\},$$
be the set of individuals called *chromosomes*. If one considers all binary 1-element sequences then after ordering them, one can compose a set Z with $s = 2^1$, where its typical element (chromosome) is of the form $z_j = \{0, 0, 1, 0, ..., 1, 0, 0\}$. By a *population* we understand any multi-set of r chromosomes from Z, then r is the population size: *PopSize*.

Definition 1. *By a frequency vector of population we understand the vector*

$$p = (p_0, ..., p_{s-1}) , \quad \text{where } p_k = \frac{a_k}{r} , \quad k = 0, 1, ..., s - 1, \tag{1}$$

where a_k is a number of copies of the element z_k.

The set of all possible populations (frequency vectors), with $\mathbf{N}$ - natural numbers, is

$$\Lambda = \{p \in \mathbf{R}^s : p_k \geq 0, \ p_k = \frac{d}{r}, \ d \in \mathbf{N}, \ \sum_{k=0}^{s-1} p_k = 1\ \}. \tag{2}$$

When a genetic algorithm is realized, then we act on populations, and new populations are generated. The transition between two subsequent populations is random and is realized by a probabilistic operator. If one starts with a frequency vector, a probabilistic vector can be obtained. It means that in some cases p_i cannot be rational any more. Hence the closure of the set Λ, namely

$$\overline{\Lambda} = \{x \in \mathbf{R}^s : x_k \geq 0, \text{ and } \sum_{k=0}^{s-1} x_k = 1\ \}, \tag{3}$$

is more suitable for our analysis of such random processes acting on probabilistic vectors; they are in the set $\overline{\Lambda}$.

2.1 Selection and Mutation Operators

In SGA when the proportional selection occurs the probability of appearance of the individual in the next generation (population) is equal to the quotient of the fitness of the individual to the average fitness of the population. If a fitness function $f : Z \to \mathbf{R}^+$ and population p are given then the first main genetic operator, called the *fitness proportional selection* can be defined. It is given by the probability that the element z_k will appear in the next population

$$\frac{f(z_k)p_k}{\overline{f}(p)}, \tag{4}$$

where $\overline{f}(p)$ is the *average population fitness* denoted by

$$\overline{f}(p) = \sum_{k=0}^{s-1} f(z_k)p_k. \tag{5}$$

Hence we create the matrix $\boldsymbol{S}$ of the size s, where its values on the main diagonal are

$$S_{kk} = f(z_k), \tag{6}$$

and the transition from the population p into the new one, say q, is given by

$$q = \frac{1}{\overline{f}(p)} \boldsymbol{S} p. \tag{7}$$

Additional to the matrix $\boldsymbol{S}$ describing *selection operator* [6,8,12,13] we define the matrix $\boldsymbol{U} = [U_{ij}]$ with U_{ij} as the probability of mutation of the element z_j into the element z_i, and U_{ii} - the probability of the surviving of the element z_i, as the second genetic operator: *mutation*. One requires that

$$1.\ U_{ij} \geq 0 \text{ and } 2.\ \sum_{i=0}^{s-1} U_{ij} = 1\ , \text{ for all } j. \tag{8}$$

If a parameter μ describes the probability of changing bits 0 into 1 or vice versa in the case of the binary chromosome, and if the chromosome z_i differs from z_j at c positions then

$$U_{ij} = \mu^c(1-\mu)^{l-c}. \tag{9}$$

3 Transition Operator

Let $p = (p_0, ..., p_{s-1})$ be a probabilistic vector. If we consider $p \in \overline{\Lambda}$ then transition operators should transform set $\overline{\Lambda}$ into itself. The action of the genetic algorithm at the first and at all subsequent steps is the following: if we have a given population p then we sample with returning r-elements from the set Z, and the probability of sampling the elements $z_0, ..., z_{s-1}$ is described by the vector $\mathcal{G}(p)$, where

$$\mathcal{G}(p) = \frac{1}{\overline{f}(p)} \boldsymbol{U}\boldsymbol{S}p . \tag{10}$$

This r-element vector is our new population q.

Let us denote by W the set of all possible r-element populations composed of elements selected from the set Z, where elements in the population could be repeated. This set is finite and let its cardinality be M. It can be proven that the number M is given by some combinatoric formula

$$M = \binom{s+r-1}{s-1} = \binom{s+r-1}{r} . \tag{11}$$

Let us order all populations, then we identify the set W with the list $W = \{w^1, \ldots, w^M\}$. Every $w^k, k = 1, 2, ..., M$, is some population for which we used the notation p in the previous section. According to what we wrote, the population will be identified with its frequency vector or probabilistic vector. This means that for the population $p = w^k = (w_0^k, \ldots, w_{s-1}^k)$, the number w_i^k, for $i \in \{0, \ldots, s-1\}$, denotes the probability of sampling from the population w^k the individual z_i (or the fraction of the individual z_i in the population w^k).

Let us assume that we begin our implementation of SGA from an arbitrary population $p = w^k$. In the next stage each population $w^1, \ldots, w^M$ can appear with the probability, which can be determined from our analysis. In particular, if in the next stage the population has to be q, with the position l on our list W (it means $q = w^l$), then this probability [6,10,11], denoted here by p_{lk}, is equal

$$r! \prod_{j=0}^{s-1} \frac{(\mathcal{G}(p)_j)^{rq_j}}{(rq_j)!} . \tag{12}$$

After two steps, every population $w^1, \ldots, w^M$ will appear with some probability, which is a double composition of this formula. It will be analogously in the third step and so on. Then it is well founded to analyse the probability distribution of the population's realization in the next steps.

Notice that beginning from one, specific population $p = w^j$, we will denote this particular situation of the distribution in the step zero 0, by $u = (0, \ldots, 0, 1, 0 \ldots, 0) \in \mathbf{R}^M$, where 1 stands on the j-th place.

Let us denote by

$$\Gamma = \{x \in \mathbf{R}^M : \forall k \; x_k \geq 0 \text{ oraz } ||x|| = 1\},$$

where $||x|| = x_1 + \ldots + x_M$, for $x = (x_1, \ldots, x_M)$, the set of new M-dimensional probabilistic vectors. A particular component of the vector x represents the probability of the appearance of this population from the list W of all M populations. The set Γ is composed of the all possible probability distributions for M populations. Then above described implementation transforms, at every step, the set Γ into the same.

The formula (12) gives a possibility of determining all elements of a matrix $\boldsymbol{T}$ which defines the probability distribution of appearance of populations in the next steps, if we have current probability distribution of the populations. With our choice of denotations for the populations p and q, the element (l, k) of the matrix will give transition probability from the population with the number k into the population with the number l. It is important that elements of the matrix are determined once forever, independently of the number of steps. The transition between elements of different pairs of populations is described by different entries of the fundamental *transition operator*

$$T(\cdot) : \mathbf{N} \times \Gamma \to \Gamma \tag{13}$$

which for fixed $t \in \mathbf{N}$ is the square matrix $T(t)$ of dimension M. If we consider one time step only and if $u \in \Gamma$ is the actual probabilistic vector of all population, then an operator $\boldsymbol{T}$ can be defined on the set Γ, with elements p_{lk}, $l, k = 1, 2, \ldots, M$, and then the probability distribution of all M populations in the next step is given by the formula $\boldsymbol{T}u$. By t - application of this method and using the mathematical induction, one can prove that at steps $t = 1, 2, \ldots$, the probability distribution of all M populations is given by the formula $\boldsymbol{T}^t u$. This means that $\boldsymbol{T}^t u = T(t)u = \big((T(t)u)_1, \ldots, (T(t)u)_M\big)$ is the probability distribution for M populations in the step number t, if we have begun our implementation of SGA given by $\mathcal{G}$ (cf. (10)) from the probability distribution $u = (u_1, \ldots, u_M) \in \Gamma$. The number $(T(t)u)_k$ for $k \in \{1 \ldots, M\}$ denotes the probability of appearance of the population w^k in the step of number t. By the definition $\mathcal{G}(p)$ in (10),(12) and the remarks made at the end of the previous section the transition operator $T(t)$ is linear for all natural t.

Hence the above introduced transition operator $T(t)$ is linked with the above matrix by the dependence

$$T(t) = \boldsymbol{T}^t . \tag{14}$$

Notice that though the formula (12) determining individual entries (components) of the matrix $\boldsymbol{T}$ is a population dependent, and hence nonlinear, the transition operator $T(t)$ is linear thanks to the order relation introduced in the set W of all M populations. The multi-index l, k of the component p_{lk} kills, in some sense, this nonlinearity, since it tells (i.e. it is responsible) for a pair of populations between which the transition takes place. The matrix T is a Markovian matrix, i.e. the sum of the components in one row is 1. This fact permits us to apply Theory of Markov operators to analyze the convergence of genetic algorithms [3,4,7,9].

Let $e_k \in \Gamma$ be a vector which at the k-th position has one and zeroes at the other positions, i.e. $e_k = (0, 0, ..., 0, 1, 0, 0, ..., 0)$. Then e_k describes the probability distribution in which the population w^k is attained with the probability 1.

By the notation $T(t)w^k$ we will understand

$$T(t)w^k = T(t)e_k \qquad (15)$$

which means that we begin the GA at the specific population w^k.

4 Asymptotic Stability

We shall say that from the chromosome z_a it is possible to obtain z_b in one mutation step with a positive probability if $U_{ba} > 0$. We shall say that from the chromosome z_a it is possible to get the chromosome z_b with positive probability in n-step mutation if there exists a sequence of chromosomes $z_{i_o}, ..., z_{i_n}$, such that $z_{i_o} = z_a$, $z_{i_n} = z_b$ and any z_{i_j}, for $j = 1, ..., n$, is possible to attain from $z_{i_{j-1}}$ in one step with a positive probability.

Let us formulate two conditions:

(I) For any $i, j \in \{0, ..., s-1\}$ the probability of mutation U_{ij} is positive, it means that from each chromosome one can get in a one step an arbitrary chromosome;

(II) For any $j \in \{0, ..., s-1\}$ the probability of preserving the chromosome z_j is positive, i.e. the probability $U_{jj} > 0$.

Notice that the condition $U_{jj} > 0$ for the case of binary mutation (9) will be satisfied if $0 \leq \mu < 1$.

For a given probability distribution $u = (u_1, ..., u_M) \in \Gamma$ it is easy to compute that the probability of sampling the individual z_i, for $i \in \{0, ..., s-1\}$, is equal to

$$\sum_{k=1}^{M} w_i^k \cdot u_k, \qquad (16)$$

where w_i^k is the probability of sampling from the population k the chromosome z^i, and u_k - the probability of appearance of the k-th population. By an *expected population* we call the vector from $\mathbf{R}^s$, of which i-th coordinate is given by (16). Since $u_k \geq 0$, $w_k^i \geq 0$ for $k \in \{1, ..., M\}$, $i \in \{0, ..., s-1\}$ and

$$\sum_{i=0}^{s-1}\left(\sum_{k=1}^{M} u_k \cdot w_i^k\right) = \sum_{k=1}^{M} u_k \left(\sum_{i=0}^{s-1} w_i^k\right) = \sum_{k=1}^{M} u_k = 1,$$

the vector belongs to $\overline{\Lambda}$. From (16) we obtain that the expected population is given by

$$\sum_{k=1}^{M} w^k \cdot u_k \qquad (17)$$

Obviously, it is possible that expected population could not be any possible population with r-elements.

For every $u \in \Gamma$ and for every t certain probability distribution for M populations $T(t)u$ is given. Consequently the expected population in this step is known.

By $R(t)u = \big((R(t)u)_0, \ldots, (R(t)u)_{s-1}\big)$ we denote the expected population at the step t, if we begun our experiment from the distribution $u \in \Gamma$; of course we have $R(t)u \in \overline{\Lambda}$.

Definition 2. *We will say that the model is asymptotically stable if there exists $u^* \in \Gamma$ such that:*

$$T(t)u^* = u^* \quad \text{for} \quad t = 0, 1, \ldots \qquad (18)$$

$$\lim_{t \to \infty} \|T(t)u - u^*\| = 0 \quad \text{for all } u \in \Gamma. \qquad (19)$$

Since for $k \in \{1, \ldots, M\}$ we have

$$\big|(T(t)u)_k - u_k^*\big| \leq \|T(t)u - u^*\|, \qquad (20)$$

then (19) will gives

$$\lim_{t \to \infty} (T(t)u)_k = u_k^*. \qquad (21)$$

It means that probability of appearance of the population w^k in the step number t converges to a certain fixed number u_k^* independently of the initial distribution u. It is realized in some special case, when our implementation begun at one specific population $p = w^j$.

Theorem 1. *If the model is asymptotically stable, then*

$$\lim_{t \to \infty} \|R(t)u - p^*\| = 0 \quad \text{for} \quad u \in \Gamma, \qquad (22)$$

where $p^* \in \overline{\Lambda}$ is the expected population adequate to the distribution u^*. Particularly, we have also

$$\lim_{t \to \infty} \|R(t)p - p^*\| = 0 \quad \text{for} \quad p \in W. \qquad (23)$$

PROOF. From (17) we have

$$R(t)u = \sum_{i=1}^{M} w^i \cdot (T(t)u)_i$$

and

$$p^* = \sum_{i=1}^{M} w^i \cdot u_i^*.$$

Then
$$\|R(t)u - p^*\| = \sum_{j=0}^{s-1}|\sum_{i=1}^{M} w_j^i \cdot (T(t)u)_i - \sum_{i=1}^{M} w_j^i \cdot u_i^*|$$
$$\leq \sum_{j=0}^{s-1}\sum_{i=1}^{M} w_j^i |(T(t)u)_i - u_i^*| = \|T(t)u - u^*\|.$$

On the basis of (19) it follows the equality (22). Taking into account our notation, given in (15), the formula (23) is the particular case of (22). △

From Theorem 1 it follows that if the model is asymptotically stable then the expected population stabilizes, converging to $p^* \in \overline{\Lambda}$ independently of initial conditions. This result has a fundamental meaning for the analysis of the convergence of genetic algorithms. This generalization will be the subject of our next paper.

Definition 3. *Model is pointwise asymptotically stable if there exists such a population w^j that*
$$\lim_{t \to \infty}(T(t)u)_j = 1 \text{ for } u \in \Gamma .\qquad(24)$$

Condition (24) denotes that in successive steps the probability of appearance of another population than w^j tends to zero. It is a special case of the asymptotic stability for which
$$u^* = e_j.$$
Let us formualate the conditions for the asymptotic stability.

Theorem 2. *Let us assume that there is a chromosome $z_a, a \in \{0,...,s-1\}$ such that, it can be attained from an arbitrary chromosome z_i in a finite number of steps. Let us assume that the condition (II) above is satisfied. Then the model is asymptotically stable.* △

Theorem 3. *Let us assume that both conditions (I) and (II) are satisfied. Then the model is asymptotically stable and moreover, the limiting distributions u^* and the expected population p^* satisfy the conditions:*
$$u_k^* > 0 \text{ for } k = 1, 2, ..., M \qquad(25)$$
$$p_i^* > 0 \text{ for } i = 0, 1, 2, ..., s-1.\qquad(26)$$
△

Notice that the condition (25) means that each r-th element population will repeat in our implementation with a nonzero frequency. The second condition (26) means that in the limiting expected population all chromosomes will appear.

Corollary 1. *If the condition (II) is satisfied then model is pointwise asymptotically stable if and only if there exists exactly one chromosome z_a with such a property that it is possible to attain it from any chromosome in a finite number of* steps with a positive probability. In this situation the population w^j is exclusively composed of the chromosomes z_a and

$$T(t)w^j = w^j \qquad (27)$$

holds. Moreover, the probability of appearance of other population than w^j tends to zero in the step number t with a geometrical rate, i.e. there exists $\lambda \in (0,1)$, $D \in \mathbf{R}^+$ such that

$$\sum_{i=1,\ i\neq j}^{M} (T(t)u)_i \leq D \cdot \lambda^t \ .\triangle \qquad (28)$$

The proofs of our theorems and auxiliary corollary are stated in original articles [11,2]. There one can find examples proving the necessity of conditions (I) and (II).

Numbers λ and D could be determined for a specific model. It will be the subject of the next articles. Formula (24) means, that even really we realize with descripted by us algorithm, then population w^j practically becomes after finite number of steps. By the formula (27) follows, that from population w^j we receives w^j with probability equal 1. Then, if w^j becomes once, then from this moment, we shall permanently populations w^j.

Theorem 2 states that the convergence to one population could occur only under specific assumptions. This justifies the investigation of asymptotic stability that in Definition 3.

Definition 4. *By an attainable chromosome we denote $z_a \in Z$ such that it is possible to attain it from any other chromosome in a finite number of steps with a positive probability. Let us denote by Z^* the set of all z_a with this property.*

Theorem 4. *Model is asymptotically stable if and only if $Z^* \neq \emptyset$.* $\triangle$

Theorem 5. *Let us assume that the model is asymptotically stable. Then the next relationship holds:*

(war) $u_k^* > 0$ if and only if the population w^k is exclusively composed of chromosomes belonging to the set Z^*. $\triangle$

In SGA with a positive mutation probability, it is possible to attain any individual (chromosome) from any other individual. Then there is more than one chromosome which is possible to attain from any other in a finite number of steps with a positive probability. Hence, by Theorem 2, it is impossible to get the population composed exclusively of one type of chromosome.

Acknowledgement. The research work on the paper was done by W.K in the framework of the KBN Project (Minister of Higher Education and Science) No. 3 T11 C007 28. Authors thanks Professor Zbigniew Michalewicz and Mr. Stefan Kotowski for the inspiration and discussions.

References

1. Kieś, P., Michalewicz, Z.: Foundations of Genetic Algorithms (in Polish). Matematyka Stosowana. Matematyka dla Społeczeństwa, Applied Mathematics. Mathematics for Society 1(44), 68–91 (2000)
2. Kotowski, S., Socała, J., Kosiński, W., Michalewicz, Z.: Markovian model of simple genetic algorithms and its asymptotic behaviour (under preparation)
3. Lasota, A.: Asymptotic properties of Markov operators semigroups(in Polish). Matematyka Stosowana. Matematyka dla Społeczeństwa, Applied Mathematics. Mathematics for Society 3(46), 39–51 (2002)
4. Lasota, A., Yorke, J.A.: Exact dynamical systems and the Frobenius–Perron operator. Trans. Amer. Math. Soc. 273, 375–384 (1982)
5. Michalewicz, Z.: Genetic Algorithms + Data Structures = Evolution Programs, 3rd edn. Springer, Heidelberg (1996)
6. Rowe, J.E.: The dynamical system models of the simple genetic algorithm. In: Kallel, L., Naudts, B., Rogers, A. (eds.) Theoretical Aspects of Evolutionary Computing, pp. 31–57. Springer, Heidelberg (2001)
7. Rudnicki, R.: On asymptotic stability and sweeping for Markov operators. Bull. Polish Acad. Sci. Math. 43, 245–262 (1995)
8. Schaefer, R.: Foundations of genetic global optimization (in Polish). In: Podstawy genetycznej optymalizacji globalnej, Jagiellonian University Press, Cracow (2002)
9. Socała, J.: Asymptotic behaviour of the iterates of nonnegative operators on a Banach lattice. Ann. Polon. Math. 68(1), 1–16 (1998)
10. Socała, J., Kosiński, W., Kotowski, S.: On asymptotic behaviour of a simple genetic algorithms (in Polish). Matematyka Stosowana. Matematyka dla Społeczeństwa, Applied Mathematics. Mathematics for Society 47, 70–86 (2005)
11. Socała, J., Kosiński, W.: Zastosowanie metody funkcji dolnej do badania zbieżności algorytmów genetycznych (in Polish). Matematyka Stosowana. Matematyka dla Społeczeństwa, Applied Mathematics. Mathematics for Society 8(49), 23–36 (2007)
12. Vose, M.D.: The Simple Genetic Algorithm: Foundation and Theory. MIT Press, Cambridge (1999)
13. Vose, M.D.: Modelling Simple Genetic Algorithms. Evolutionary Computation 3(4), 453–472 (1996)

Tuning Quantum Multi-Swarm Optimization for Dynamic Tasks

Krzysztof Trojanowski

Institute of Computer Science, Polish Academy of Sciences
Ordona 21, 01-237 Warsaw, Poland
trojanow@ipipan.waw.pl

Abstract. Heuristic approaches already proved their efficiency for the cases where real-world problems dynamically change in time and there is no effective way of prediction of the changes. Among them a mixed multi-swarm optimization (mSO) is regarded as the most efficient. The approach is a hybrid solution and it is based on two types of particle swarm optimization (PSO): pure PSO and quantum swarm optimization (QSO). Both types are applied in a set of simultaneously working subswarms. In spite of the fact that there appeared a series of publications discussing properties of this approach the motion mechanism of quantum particles was just briefly studied, and there is still some research to do. This paper presents the results of our research on this subject. The novelty is based on a new type of distributions of particles in a quantum cloud. Obtained results allow to derive some guidelines of an effective tuning of the mechanism of distribution in the quantum cloud and show that further improvement of mSO is possible.

1 Introduction

Heuristic optimization in dynamic environments has been a subject of growing interest in the last decade. In this case the requested output of the optimization algorithm is a continuous adaptation of the proposed best found solution to the current state of the fitness landscape. This type of problems appears in the real world very often. It could be a problem of efficient finding the shortest path from one point to another in the crowded streets of a big city as well as a task of the continuous control of the parameters of the engine to keep it working the most efficiently i.e. in the optimal conditions for varying requests of the user.

Starting with evolutionary approach the list of metaheuristics employed to solve this type of problems is steadily increased. Among them a significant and wide stream of publications on this subject is concerned with particle swarm optimization. The particle swarm optimization is based on the ideas of swarm intelligence and stigmergy [11,1]. Briefly speaking the heuristic is inspired by observations of swarms of living beings in the nature and especially their amazing skills of self-organization present in their movement dynamics. The heuristic is based on a set of rules of communication between participants of the swarm, which exchange information about location of e.g. food or any other objects being a subject of interest of the swarm.

Applications of particle swarm to non-stationary optimization tasks were already studied and presented in publications. Especially multi-swarm with mixed types of particles proved its usefulness and outperformed other metaheuristics. Approaches with varying number of subs-warms [2] as well as approaches with adaptive number of species in the swarm [12,17,13] have been researched. The results of the research showed the key role of appropriate sub-swarms management during the process of search. In this paper multi-swarm approach is also a subject of interest, however, the main goal is concerned rather with studying properties of the particles rules of movement in the multi-swarm. Here we focus on the properties of one of the classes of particles – quantum particles. The main idea of this type of particles is based on atom analogy where a single swarm represents a single atom. The model of movement of the quantum particles is closely related to the movement of the atom which is governed by quantum rules. In this paper we study two strategies of reallocation of quantum particles. The experimental results showed that the proposed approaches improved efficiency of the algorithm. For the tests the Moving Peaks Benchmark generator [6] was selected. In MPB we optimize in real-valued parametrical search space and the fitness landscape is built of a set of unimodal functions individually controlled by the parameters allowing to create different types of changes.

The paper is organized as follows. In Section 2 a brief description of the optimization algorithm is presented. Section 3 includes some details of the selected testing environment and the applied measure. Section 4 shows the results of experiments performed with the first of the tested strategies of reallocation of quantum particles while Section 5 – the results for the second one. Section 6 concludes the presented research.

2 Quantum Multi-swarm

A simple scheme of the particle swarm optimization algorithm is given in Algorithm 1. A PSO optimizer is equipped with a set of particles $\mathbf{x}_i$ where $i \in [1, \ldots N]$. Each of the particles represents a solution in an n-dimensional real valued search space. For the search space a fitness function $f(\cdot)$ is defined which is used to evaluate the quality of the solutions. A particle $\mathbf{y}_i$ represents the best solution found by the i-th particle (called particle attractor), and a particle $\mathbf{y}^*$ – the best solution found by the swarm (called swarm attractor).

Search properties of the PSO scheme presented in Algorithm 1 are represented by the step "update location and velocity". In this step there are two main actions performed: first the velocity of each of the particles is updated and then all the particles change their location in the search space according to the new values of velocity vectors and the kinematic laws. Formally for every iteration t of the search process every i-th coordinate of the vector of velocity $\mathbf{v}$ as well as the coordinate of the location $\mathbf{x}$ undergo the following transformation [10]:

$$\begin{aligned} v_i^{t+1} &= \chi(v_i^t + c_1 r_1^t (y_i^t - x_i^t) + c_2 r_2^t (y_i^{*t} - x_i^t)) \, , \\ x_i^{t+1} &= x_i^t + v_i^{t+1} \, , \end{aligned} \quad (1)$$

Algorithm 1. the particle swarm optimization

1: Create and initialize the swarm
2: **repeat**
3:     **for** $i = 1$ to N **do**
4:         **if** $f(\mathbf{x}_i) > f(\mathbf{y}_i)$ **then**
5:             $\mathbf{y}_i = \mathbf{x}_i$
6:         **end if**
7:         **if** $f(\mathbf{y}_i) > f(\mathbf{y}^*)$ **then**
8:             $\mathbf{y}^* = \mathbf{y}_i$
9:         **end if**
10:    **end for**
11:    update location and velocity of all the particles
12: **until** stop condition is satisfied

where χ is a constriction factor and $\chi < 1$ and c_1 and c_2 control the attraction to the best found personal and global solutions, respectively.

The basic idea presented in Algorithm 1 has been extended for non-stationary optimization tasks. One of the first significant changes in this scheme was introduction of multi-swarm. In the approach presented in this paper the number of sub-swarms is constant during the process of search. Each of them is treated as an independent self-governing population which is not influenced by any of the neighbors, however, there are mechanisms which periodically perform some action based on the information about the state of search of the entire swarm [4].

The first of the mechanisms called *exclusion* is used to ensure the appropriate distribution of the sub-swarms over the entire search space. It eliminates sub-swarms which are located too close to each other assuming that most probably they occupy the same optimum. In this case one of them is eliminated and a new one is generated from scratch. Any two sub-swarms are too close if for the best solutions from the compared two sub-swarms the distance is closer than the defined threshold ρ.

Except for exclusion yet another mechanism of sub-swarms' management called *anti-convergence* was proposed in [4], which protects against convergence of sub-swarms. However in preliminary tests performed with this mechanism very small or even none influence on the quality of obtained results was observed and thus eventually this mechanism was omitted in the implementation of mSO presented in this paper.

The last of the sub-swarms' management mechanisms described in [4] is the hybridization based on the employment of mixed sub-swarms. While the location of the so-called neutral particles is evaluated according to classic formulas as discussed above, another ones are treated as quantum particles which change their location according to the analogy with quantum dynamics of the particles. All the particles in such a mixed sub-swarm share the information about the current best position and the best position ever found by the sub-swarm.

Idea of quantum particle proposed by Blackwell and Branke in [3] originates from the quantum model of atom where the trajectories of electrons are described as quantum clouds. Adaptation of this idea to the movement of the particles in the PSO model is simple: instead of using kinematics laws for evaluation of a distance traveled by the particle with a constant velocity in a period of time a new position of the particle is randomly generated inside a cloud of the given range r_{cloud} surrounding the particle. In this model the particle's speed becomes irrelevant because every location inside the cloud has equal probability to be chosen as a new location of the particle.

2.1 Quantum Particle Movement Rules

The proposed concept of quantum particles has been extended in our research. Since the model by Blackwell and Branke assumes the uniform distribution of the set of possible new location candidates over the cloud's space it was interesting to verify another types of distributions. A class of new distributions was proposed where the uniform distribution of the set of candidates is wrapped by another distribution $g(\cdot)$ as defined in equations (2):

$$\begin{cases} g(d) = (1 - d/r_{cloud})^\kappa & \text{iff } k > 0 \ , \\ g(d) = (d/r_{cloud})^\kappa & \text{otherwise} \ , \end{cases} \quad (2)$$

where d represents the distance between the current location and a candidate for a location of the particle and κ is the parameter of the distribution.

The way of generation of m-dimensional random variate **x** uniformly distributed inside a hyper-sphere with a center located in the point of origin is presented in Algorithm 2. For clarity of the pseudo-code it is assumed that the particle is located in the point of origin of the coordinate system. However it is easy to use for a particle located anywhere in the domain simply by adding each of its coordinates to the respective coordinates calculated in Algorithm 2.

Algorithm 2. Generation of m-dimensional random variate **x** uniformly distributed in a hyper-sphere with radius 1

1: Generate m independent normal random variates: $X = (X_1, \ldots, X_m)$
2: **for** $j = 1$ to m **do**
3: $\quad x_j = X_j / \|X\|$
4: **end for**
5: Generate exponential random variate $Z = Y^{1/m}$ where Y is a uniformly distributed random variate on $(0, 1)$
6: **return** $\mathbf{x} = (Zx_1, \ldots, Zx_m)$

To generate non-uniformly distributed candidates for a new location in the cloud the von Neumann's acceptance-rejection method was applied as presented in Algorithm 3. As in previous procedures just for simplicity of the pseudo-code this one also assumes that the particle is in the point of origin.

Figure 1 presents samples of possible distributions of new location candidates inside the 2-dimensional quantum cloud generated with the procedure given in Algorithm 3. In Figure 1 there are eight distributions of candidates for eight different values of κ: $\kappa = -2, -1, 0, 1, 2, 3, 4,$ and 5.

Algorithm 3. generation of a new location of the quantum particle with use of the von Neumann's acceptance-rejection method

1: **repeat**
2:     Generate **x** as an m-dimensional uniformly distributed random variate in a hyper-sphere with radius r_{cloud} (*with use of Algorithm 2*)
3:     compute the distance d from the center of the hyper-sphere to **x**: $d = \sqrt{(x_1)^2 + \ldots + (x_m)^2}$
4:     generate Y – the uniformly distributed random variate on $(0, 1)$
5:     **if** $k > 0$ **then**
6:         **if** $(Y \leq (1 - d/r_{cloud})^\kappa)$ **then**
7:             **break**
8:         **end if**
9:     **else**
10:         **if** $(Y \leq (d/r_{cloud})^\kappa)$ **then**
11:             **break**
12:         **end if**
13:     **end if**
14: **until** false
15: **return** $\mathbf{x} = (x_1, \ldots, x_m)$

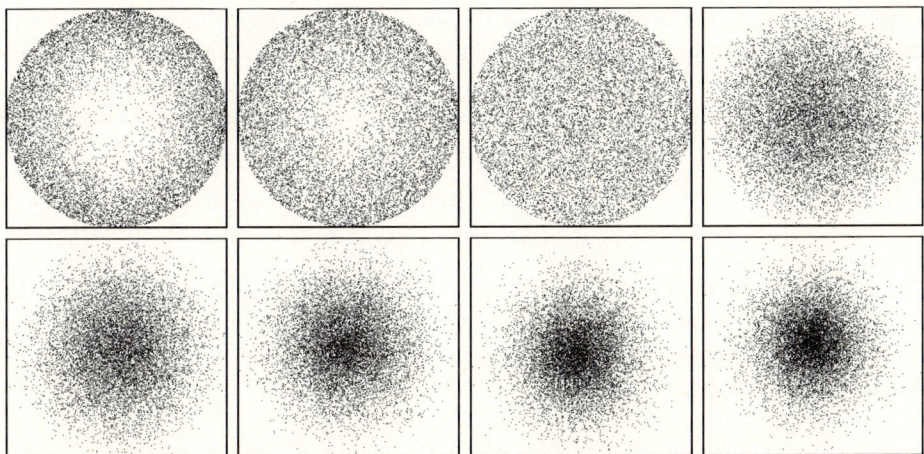

Fig. 1. Distribution of the new location candidates in the quantum cloud in 2-dimensional search space for $k \in \langle -2, -1, 0, 1 \rangle$ in the first row and for $k \in \langle 2, 3, 4, 5 \rangle$ in the second row

2.2 Tested Algorithm

The algorithm parameters' settings applied to the experiments presented below originate from [4]. In the cited publication authors present results of experiments obtained for different configurations of swarms tested with the MPB benchmark where there are 10 moving peaks. Among many tested configurations the best results for the optimization problem with 10 moving peaks are obtained where there are 10 sub-swarms and each of them consists of five neutral particles and five quantum ones (see Table III in [4]). The total population of particles consists of 100 solutions divided equally into 10 sub-swarms. The values of pure PSO parameters are: $c_{1,2} = 2.05$ and $\chi = 0.7298$. For QSO the range of exclusion is set to 31.5 (for the best performance the value of ρ should be set close to 30. However, the precision of this parameter's setting is not crucial. In [4] the authors claim that the algorithm is not very sensitive to small changes in this parameter).

In the presented algorithm there is no strategy of detecting the appearance of change in the fitness landscape. Since our main goal was studying the properties of the distributions of quantum particles, we assumed that such an on-line detecting would just introduce yet another unnecessary bias into the obtained values of offline error and make their analysis even more difficult. Therefore a change is known to the system instantly as it appears and there is not any additional computational effort for its detection. When the change appears, all the solutions stored in both neutral and quantum particles are reevaluated and the swarm memory is forgotten. Neutral particle's attractors are overwritten by current solutions represented by these particles and sub-swarms' attractors are overwritten by the current best solutions in the sub-swarms.

We also assumed that the domain of possible solutions is limited by a set of some box constraints and only the solutions fitted in the constraints are classified as feasible. We selected a very simple procedure of immediate repairing unfeasible particles. Clearly the i-th coordinate of the solution $\mathbf{x}$ breaking its box constraints is trimmed to the exceeded limit, i.e.:

$$\texttt{if } x_i < lo_i \quad \texttt{then} \quad x_i = lo_i,$$
$$\texttt{if } x_i > hi_i \quad \texttt{then} \quad x_i = hi_i.$$

The procedure is applied in the same way to both types of the particles, the neutral and the quantum ones. In case of neutral particles the velocity vector $\mathbf{v}$ of the repaired particle stays unchanged even if it still leads the particle outside the acceptable search space.

3 Applied Measure and the Benchmark

In the performed experiments the *offline error* (briefly *oe*) measure [6,7] of obtained results was used. The offline error represents the average deviation from the optimum of the fitness of the best individual evaluated since the last change of the fitness landscape. Formally:

$$oe = \frac{1}{N_c} \sum_{j=1}^{N_c} \left(\frac{1}{N_e(j)} \sum_{i=1}^{N_e(j)} (f_j^* - f_{ji}^*) \right), \qquad (3)$$

where:

N_c is the total number of changes of the fitness landscape in the experiment, $N_e(j)$ is the number of evaluations of the solutions performed for the j-th state of the landscape, f_j^* is the value of optimal solution for the j-th landscape (i.e. between the j-th and $(j + 1)$-th change in the landscape) and f_{ji}^* is the current best found fitness value for the j-th landscape i.e. the best value found among the ones belonging to the set from f_{j1} till f_{ji} where f_{ji} is the value of the fitness function returned for its i-th call performed for the j-th landscape.

During the process of searching the offline error can be calculated in two ways: in one of them the error is evaluated from the beginning of the experiment while in another one – the value of offline error starts to be evaluated only after some number of changes in the fitness landscape. The latter way is advised as saddled with the less measurement error caused by the initial phase of the search process (please see e.g. [19] for a discussion about the possible influence of the initial phase on the quality of the results obtained for MPB). Therefore, just this way was applied in our tests.

For compatibility with experiments published by others the number of evaluations between subsequent changes of the fitness landscape equals 5000. During a single experiment the fitness landscape changes 110 times (however for the first 10 changes the error is not evaluated). Every experiment was repeated 50 times and the means are presented.

For our tests we selected the MPB generator [6]. The parameters of MPB were set exactly the same as specified in [4] in scenario 2. The fitness landscape was defined for the 5-dimensional search space with boundaries for each of dimensions set to $\langle 0; 100 \rangle$. For the search space there exist a set of 10 moving peaks which vary their height randomly within the interval $\langle 30; 70 \rangle$, width within $\langle 1; 12 \rangle$ and position by a distance of 1.

4 Results of Experiments

At the beginning of the research the plan of experiments included just one group of experiments where the influence of type od distribution as well as the size of the quantum cloud on the quality of the results was verified. In the experiments the range of the quantum cloud changed from 0.1 to 1.5 (the following values were used in experiments: 0.1, 0.2, 0.25, 0.3, 0.4, 0.5, 0.6, 0.7, 0.75, 0.8, 0.9, 1, 1.1, 1.2, 1.25, 1.3, 1.4, 1.5) and the parameter κ in the distribution procedure changed from -2 to 6 with step 0.5. For each of the configurations of the algorithm's parameters the value of offline error was obtained. The results are presented in Figure 2. The best result was $oe = 1.488$ and it was obtained for the configuration where $\kappa = 5.5$ and $r_{cloud} = 0.6$. The second best was $oe = 1.496$ (for $\kappa = 4$ and $r_{cloud} = 0.4$) and the third was $oe = 1.501$ (for $\kappa = 0.5$ and $r_{cloud} = 0.25$). For efficiency comparison lets note that the best obtained result obtained for $\kappa = 0$ i.e. for the quantum cloud with uniform distribution proposed in [4] was $oe = 1.525$ ($r_{cloud} = 0.6$).

It can be observed that the optimal configuration of parameters is closely concerned with the specific distribution of particles inside the cloud. With

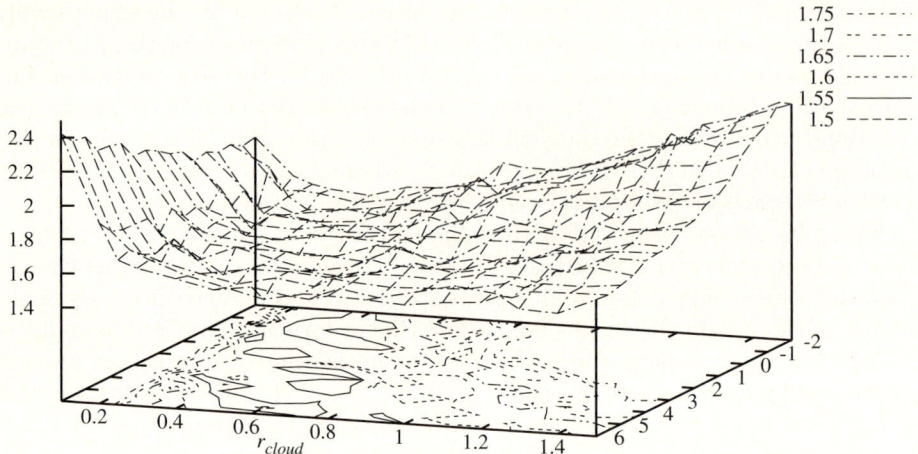

Fig. 2. Characteristics of offline error obtained by mQSO for the tested parameters: range of the quantum cloud r_{cloud} and the value of the distribution parameter κ

increasing values of κ the cloud gathers new location candidates closer to its center while the increasing value of r_{cloud} makes the cloud able to stretch the distribution over its entire volume and even repulse the candidates from the center. That is why the best results were gained for $\kappa = 5.5$ and $r_{cloud} = 0.6$, then for $\kappa = 4$ and $r_{cloud} = 0.4$ and also for $\kappa = 0.5$ and $r_{cloud} = 0.25$. This dependency is visible in the offline error landscape in Figure 2. Probably this is also a symptom of at least partial tuning of the distribution parameters to the solved problem. However, there is also visible an increasing robustness of the algorithm with growing value of κ. For higher values of κ the efficiency of the algorithm is less sensitive to the changes in the value of r_{cloud} which means that the effect of tuning to the solved problem disappears. Unfortunately the increasing computational cost of a single call of the procedure of generation of new locations in the quantum cloud for higher values of κ accompanies this profit. The growth of the cost comes from the increasing number of rejected candidates necessary to generate during the execution of the procedure for higher values of κ. The additional cost of production of the rejected candidates is the natural property of the acceptance-rejection method.

5 Alternative Rules of Quantum Particle Movement

Since the acceptance-rejection method showed to be of high computational cost for higher values of κ an alternative method of generation of new location candidates in the quantum cloud was proposed. The method consists of two phases. In the first phase a direction θ is selected i.e. a new random point **x** uniformly distributed over the hyper-sphere of the radius equal 1 surrounding the original location is found. Such a point **x** can be generated e.g. in the way as described in

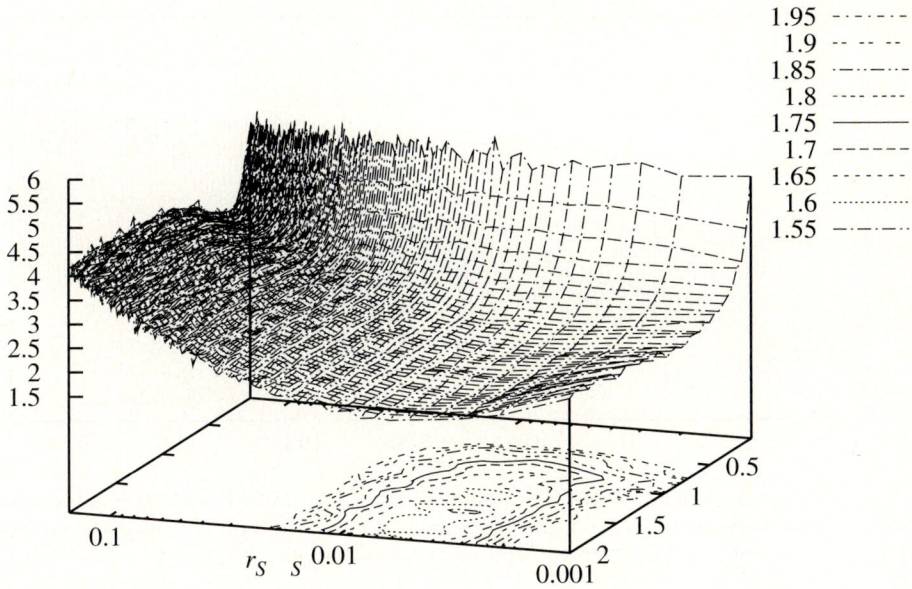

Fig. 3. Characteristics of offline error obtained by mQSO with the alternative method of generation of the location candidates for the tested parameters: $r_{S\alpha S}$ and α

the first part of Algorithm 2 (lines 2-4). The original location and the new point **x** allow to compute θ i.e. the direction of move. In the second phase a distance d from the original is calculated. The distance is an α-stable random variate and is computed as follows:

$$d = S\alpha S(0, \sigma), \quad \text{and} \tag{4}$$

$$\sigma = r_{S\alpha S} \cdot (D_w/2), \tag{5}$$

where $S\alpha S(\cdot, \cdot)$ represents α-stable symmetric distribution variate. The α-stable distribution is controlled by four parameters: stability index α ($\alpha \in \langle 0 < \alpha \leq 2 \rangle$), skewness parameter β, scale parameter σ and location parameter μ. In the symmetric version of this distribution (called $S\alpha S$, i.e. symmetric α-stable distribution) β is set to 0. In the practical implementation the Chambers-Mallows-Stuck method of generation of the α-stable symmetric random variables [9] can be used. D_w is a width of the feasible part of the domain, i.e. a distance between a lower and an upper boundary of the search space. The computation of the new location is based on the found direction θ and a distance d from the original. The two phase distribution using α-stable symmetric distribution variate is an isotropic one i.e. the location candidates are distributed equally in all directions so none of directions is distinctive in any sense. Such mechanism employing α-stable symmetric distribution was first published in [16] as the mutation operator in evolutionary algorithms. However the idea of two phase generation of new solutions in the real values n-dimensional search space is much older. Two-phase

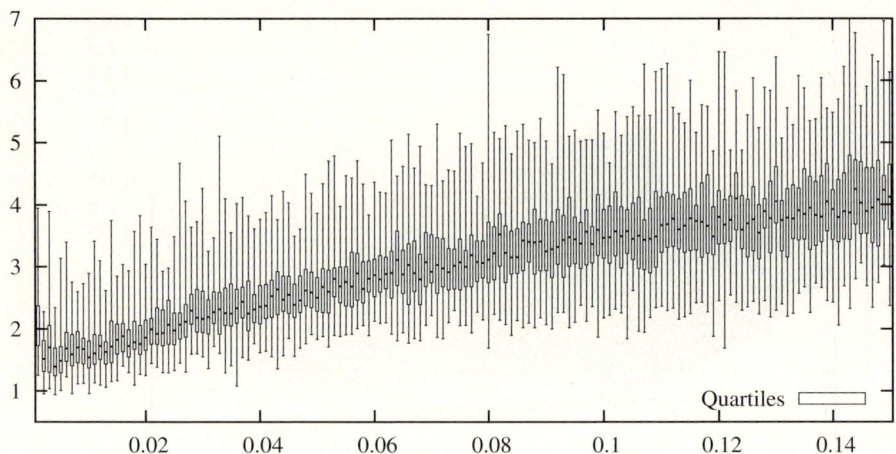

Fig. 4. Box-and-whisker diagrams of offline error for mQSO with the alternative method of generation of the location candidates for $\alpha = 1.25$ and a range of values of $r_{S\alpha S}$

mechanisms based on a direction and a distance have already been proposed for the procedure of the next candidate generation in the simulated annealing e.g. in [5,8,18].

In the experiments with the alternative method of generation of the location candidates we controlled this method with two parameters: α and $r_{S\alpha S}$. The former parameter varied from 0.05 to 2 with step 0.05 while the latter – from 0.05 to 7.5 with step 0.005. It gave 6000 configurations. The graph in Figure 3 represents obtained mean values where a boxcar smoothing (with a window of size 3) has been applied. The best result was $oe = 1.473$ (obtained for $\alpha = 1.25$), the second best – $oe = 1.479$ and the third – $oe = 1.484$. All of them were obtained for $r_{S\alpha S} = 0.004$. As in previous method of generation of the candidates the best mean values of offline error are slightly higher than the best values obtained for the quantum cloud with uniform distribution.

A box-and-whisker diagrams of offline error with all five quartiles for $\alpha = 1.25$ is presented in Figure 4. The quartiles for the best median value obtained for $r_{S\alpha S} = 0.004$ were: $Min = 0.939$, $1stQuartile = 1.248$, $Median = 1.396$, $3rdQuartile = 1.694$ and $Max = 2.344$.

6 Conclusions

In this paper two new methods of generation of the new location candidates for quantum particles are discussed. In the first of them the distribution is controlled by two parameters: r_{cloud} which is the range of the quantum cloud and κ which controls the distribution inside the cloud. The second method is also controlled by two parameters: $r_{S\alpha S}$ and α. The first parameter controls the distribution of candidates over the search domain. The second one controls the random number generator used for evaluation of the distance from the original location. For both methods and for two control parameters of each of the methods a set of configurations was experimentally tested which gave satisfactory results. The main drawback of the first of the two methods of distribution is the growth of the computational cost appearing with growth of κ. This disadvantage could be solved by finding an alternative and more computationally optimized technology of generation of candidates which gives the same proposed distribution. Fortunately the alternative method of distribution has also been proposed. The alternative method gives as good results as the first one and its computational cost is low and constant for all values of its parameters. When we look at the best obtained results, both proposed methods are comparable to the classic one proposed in [4]. However, they outperform the classic method when we compare their characteristics. Clearly the proposed methods are more robust i.e. less sensitive to lack of appropriate tuning of r_{cloud} or $r_{S\alpha S}$ parameters to the solved problem.

Finally it is necessary to stress that the results showed clearly that the way of distribution of new locations in the quantum component of the multi-swarm approach is also a key feature of the efficient dynamic optimization with mQSO.

Acknowledgements

I would like to thank Paweł Szyller from Warsaw School of Information Technology (Warsaw, Poland) for his assistance in computer simulations.

References

1. Abraham, A., Grosan, C., Ramos, V. (eds.): Stigmergic Optimization. Studies in Computational Intelligence, vol. 31. Springer, Heidelberg (2006)
2. Blackwell, T.: Particle Swarm Optimization in Dynamic Environments. In: Evolutionary Computation in Dynamic and Uncertain Environments. Studies in Computational Intelligence, vol. 51, pp. 29–49. Springer, Heidelberg (2007)
3. Blackwell, T., Branke, J.: Multi-swarm Optimization in Dynamic Environments. In: Raidl, G.R., Cagnoni, S., Branke, J., Corne, D.W., Drechsler, R., Jin, Y., Johnson, C.G., Machado, P., Marchiori, E., Rothlauf, F., Smith, G.D., Squillero, G. (eds.) EvoWorkshops 2004. LNCS, vol. 3005, pp. 489–500. Springer, Heidelberg (2004)
4. Blackwell, T., Branke, J.: Multiswarms, exclusion, and anti-convergence in dynamic environments. IEEE Transactions on Evolutionary Computation 10(4), 459–472 (2006)
5. Bohachevsky, I.O., Johnson, M.E., Stein, M.L.: Generalized simulated annealing for function optimization. Technometrics 28(3), 209–217 (1986)
6. Branke, J.: Memory enhanced evolutionary algorithm for changing optimization problems. In: Proc. of the Congress on Evolutionary Computation, vol. 3, pp. 1875–1882. IEEE Press, Piscataway (1999)
7. Branke, J.: Evolutionary Optimization in Dynamic Environments. Kluwer Academic Publishers, Dordrecht (2002)

8. Brooks, D.G., Verdini, W.A.: Computational experience with generalized simulated annealing over continuous variables. Am. J. Math. Manage. Sci. 8(3-4), 425–449 (1988)
9. Chambers, J.M., Mallows, C.L., Stuck, B.W.: A method for simulating stable random variables. J. Amer. Statist. Assoc. 71(354), 340–344 (1976)
10. Clerc, M., Kennedy, J.: The particle swarm-explosion, stability, and convergence in a multi-dimensional complex space. IEEE Trans. on Evolutionary Computation 6(1), 58–73 (2002)
11. Eberhart, R.C., Kenendy, J.: Swarm Intelligence. Morgan Kaufmann, San Francisco (2001)
12. Li, X.: Adaptively Choosing Neighbourhood Bests Using Species in a Particle Swarm Optimizer for Multimodal Function Optimization. In: Deb, K., et al. (eds.) GECCO 2004. LNCS, vol. 3102, pp. 105–116. Springer, Heidelberg (2004)
13. Li, X., Branke, J., Blackwell, T.: Particle swarm with speciation and adaptation in a dynamic environment. In: Keijzer, M., et al. (eds.) GECCO 2006: Proc. Conference on Genetic and Evolutionary Computation, pp. 51–58. ACM Press, New York (2006)
14. Morrison, R.W., De Jong, K.A.: A test problem generator for non-stationary environments. In: Proc. of the Congress on Evaluationary Computation, vol. 3, pp. 1859–1866. IEEE Press, Piscataway, NJ (1999)
15. Obuchowicz, A., Pretki, P.: Phenotypic evolution with a mutation based on symmetric α stable distributions. Int. J. Appl. Math. Comput. Sci. 14(3), 289–316 (2004)
16. Obuchowicz, A., Pretki, P.: Isotropic symmetric α-stable mutations for evolutionary algorithms. In: Corne, D., et al. (eds.) The 2005 IEEE Congress on Evolutionary Computation, pp. 404–410. IEEE Press, Los Alamitos (2005)
17. Parrot, D., Li, X.: Locating and tracking multiple dynamic optima by a particle swarm model using speciation. IEEE Trans. Evol. Comput. 10(4), 440–458 (2006)
18. Romeijn, H.E., Smith, R.L.: Simulated annealing for constrained global optimization. J. Global Optim. 5(2), 101–126 (1994)
19. Trojanowski, K.: B-cell algorithm as a parallel approach to optimization of moving peaks benchmark tasks. In: Saeed, K., Abraham, A., Mosdorf, R. (eds.) Sixth International Conference on Computer Information Systems and Industrial Management Applications (CISIM 2007), Poland, June 28-30, pp. 143–148. IEEE Computer Society Conference Publishing Services (2007)

Part IV

Classification, Rule Discovery and Clustering

Ensembling Classifiers Using Unsupervised Learning

Marek Bundzel and Peter Sinčák

Technical University of Kosice, Department of Cybernetics and Artificial Intelligence,
Letna 9, 04001, Kosice, Slovakia
{marek.bundzel,peter.sincak}@tuke.sk

Abstract. This paper describes a method for production of an ensemble of general classifiers using unsupervised learning. The method uses the 'divide and conquer' strategy. Using competitive learning the feature space is divided into subregions where the classifiers are constructed. The structure of the ensemble starts from a single member and new members are being added during the training. The growth of the ensemble is self determined until the ensemble reaches the desired accuracy. The overall response of the ensemble to an input pattern is represented by the output of a winning member for the particular pattern. The method is generic, i.e. it is not bound to a specific type of classifier and it is suitable for parallel implementation. It is possible to use the method for data mining. A basic wide margin linear classifier is used in the experiments here. Experimental results achieved on artificial and real world data are presented and compared to the results of Gaussian SVM. Parallel implementation of the method is described.

Keywords: Ensemble, Parallel, Structural Adaptation.

1 Introduction

Weak classifiers can be combined into accurate classification systems by boosting, ensembling and other methods. The method described here produces an ensemble of general classifiers. The motivation is to increase efficiency of the single classifier with respect to the learning time, accuracy, generalization performance and parallelization. .

The feature space is progressively divided into subregions which are ruled by one member of the ensemble. If the member does not meet the criteria on the accuracy in its subregion it is deleted and the subregion is further divided. The method works with general dichotomous classifiers of any type. A wide margin classifier with the functionality of a linear Support Vector Machine (SVM) was used here.

The construction of the ensemble starts with a single member. If it does not reach the desired accuracy it is deleted and new members are being introduced into the determined subregions of the feature space until the stop criteria are met. The ensemble is 'growing'. Each subregion is defined by its center found by competitive learning and the boundaries corresponding to the Voronoi tessellation build over the set of the centers. Because a linear classifier was used here the resulting decision boundary is a polygonal object. The simplicity of the polygons can be used for data mining purposes, an example will be shown. The method was extended for multiclass classification. The experiments were performed on artificial and real world data (remote sensed images). The overall behavior of the method, its accuracy and generalization performance were observed. The features useful for data analysis and knowledge retrieval were identified.

Several models for ensembling artificial neural networks have been proposed [2, 3, 4]. This work was inspired by [1]. The method for division of the feature space described in [1] (confidence output) was replaced by a new mechanism and a dynamical adaptation of the structure was introduced. The new method is also no longer dependent on simple perceptrons.

2 Division of the Feature Space into Subregions

Let us have a set of training examples

$$P = \{\bar{a}_1, \bar{a}_2, ..., \bar{a}_{Na}, \bar{b}_1, \bar{b}_2, ..., \bar{b}_{Nb}\},$$
$$\bar{a}_i, \bar{b}_j \in R^n, \quad i = 1, 2, ..., N_a, \quad j = 1, 2, ..., N_b. \quad (1)$$

where examples $\bar{a}_i$ and $\bar{b}_j$ belong to different classes. The idea is to divide P into the subsets $P_1, P_2, ..., P_{Mp} \in P$, $P = P_1 \cup P_2 \cup ... \cup P_{Mp}$ and to construct a dichotomous classifier C_k for each P_k, $k = 1, 2, ... M_p$. The subset P_k is determined by its 'center' $\bar{x}_k \in R^n$ identifying the Voronoi area containing the training examples closer to $\bar{x}_k$ than to other centers from the set $X = \{\bar{x}_1, \bar{x}_2, ..., \bar{x}_{Mp}\}$.

The algorithm starts with a single center $\bar{x}_1$ the elements of which are initialized with random numbers from the intervals delimited with minimum and maximum values of the elements of the training examples of P. $P_1 = P$ now. Classifier C_1 is constructed on P_1. The accuracy $A(C_1, P_1)$ of the classifier C_1 on the subset P_1 is calculated. Percentage accuracy or average percentage accuracy per class can be used or any other method considered suitable by the user. The user determines the minimal accuracy A_{min} to be achieved by each classifier. If $A(C_1, P_1) < A_{min}$ a new center $\bar{x}_2$ is initialized.

The final positions of the centers are determined by competitive learning. The winner takes all principle is applied in E_C epochs:

$$\forall \bar{e} \in P_1 \quad \text{repeat } E_C \text{ times}$$
$$\text{if } \|\bar{e} - \bar{x}_1(t)\| < \|\bar{e} - \bar{x}_2(t)\| \quad \text{then} \quad \bar{x}_1(t+1) = \bar{x}_1(t) + \alpha(\bar{e} - \bar{x}_1(t)) \quad (2)$$
$$\text{else} \quad \bar{x}_2(t+1) = \bar{x}_2(t) + \alpha(\bar{e} - \bar{x}_2(t))$$

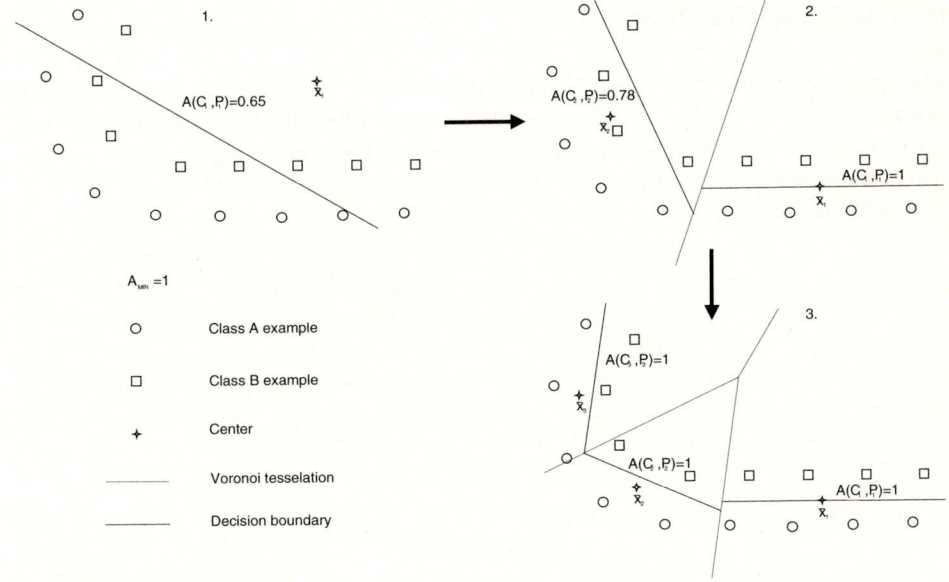

Fig. 1. Three steps of ensemble growth. The accuracy is calculated as percentage here

Now there are two centers: $\bar{x}_1$ and $\bar{x}_2$, two subsets: P_1 and P_2 and classifiers C_1 and C_2 can be constructed.

The classifiers performing worse than A_{min} are identified in every consequent training epoch. If $A(C_k, P_k) < A_{min}$ two new centers are initialized within P_k (the original center $\bar{x}_k$ is erased). Competitive learning (2) is applied on the new centers using P_k. As the result the two new centers divide P_k in two subsets and two new classifiers are trained. This process is repeated until the accuracy of all classifiers is sufficient or the maximum number of epochs is reached. Fig. 1. depicts an example of three ensemble growing epochs.

The advantage of the growing approach is that more training subsets are created in the regions where the decision boundary is complex while the separable areas remain ruled by one classifier. This decreases the complexity of the ensemble. The algorithm can separate arbitrary number of training examples – it finishes with subsets containing only single class examples in an extreme case. The entire region ruled by the classifier is then claimed for the class populating it automatically. Empty subsets are created in rare cases. A simple test identifies and removes such subsets.

By obtaining results, the center closest to the evaluated example is found. The decision of the classifier corresponding to the closest center is accepted as the overall response of the ensemble.

3 The Basic Classifier

The Optimal Linear Separation (OLS) method is an alternative to linear SVM. It represents a wide margin classifier. The method described here was introduced in [5]. OLS is applied here to construct members of the growing ensemble.

Let us again consider the sets (1). Next let us consider normal vector $\bar{v} \in R^n$, with unity norm:

$$\bar{v} = \{v_1, v_2, \ldots, v_n\}, \quad \|v\|_2 = \left[\sum_{i=1}^{n} v_i^2\right]^{1/2} = 1. \tag{3}$$

For the hyperplanes with the normal vector $\bar{v}$ we consider the next criterial function which describes the dichotomous classification separation quality, achieved with the direction choice:

$$Q_{\bar{v}} = \max\left([\min_{\bar{a} \in P}(\bar{a} \cdot \bar{v}) - \max_{\bar{b} \in P}(\bar{b} \cdot \bar{v})]; [\min_{\bar{b} \in P}(\bar{b} \cdot \bar{v}) - \max_{\bar{a} \in P}(\bar{a} \cdot \bar{v})] \right). \tag{4}$$

Then the best separation direction will be that, which maximizes this criterial function. The best separation value is given by:

$$Q = \max_{\|\bar{v}\|_2 = 1} Q_{\bar{v}}. \tag{5}$$

Then the Optimal Separation Problem can be formulated as finding the direction $\bar{v}*$ for which:

$$Q_{\bar{v}*} = \max_{\|\bar{v}\|_2 = 1} Q_{\bar{v}}. \tag{6}$$

Finding $\bar{v}*$ is a constrained optimization problem subject to constraints (3). The 'optimal' direction, which maximizes the criterial function on the set of unity vectors $\bar{v}$ is computed using a numerical iterative method.

The separation hyperplane is given by:

$$\bar{z} \cdot \bar{v}*' + g = 0, \quad z \in R^n, g \in R. \tag{7}$$

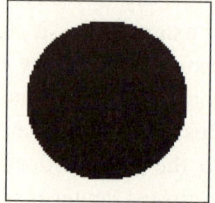 

Fig. 2. Artificial Datasets

Constant g is calculated as follows:

If $[\min_{\overline{a} \in P}(\overline{a} \cdot \overline{v}) - \max_{\overline{b} \in P}(\overline{b} \cdot \overline{v})] > [\min_{\overline{b} \in P}(\overline{b} \cdot \overline{v}) - \max_{\overline{a} \in P}(\overline{a} \cdot \overline{v})]$

then $\quad g = -\frac{1}{2}[\min_{\overline{a} \in P}(\overline{a} \cdot \overline{v}) + \max_{\overline{b} \in P}(\overline{b} \cdot \overline{v})]$ \hfill (8)

else $\quad g = -\frac{1}{2}[\min_{\overline{b} \in P}(\overline{b} \cdot \overline{v}) - \max_{\overline{a} \in P}(\overline{a} \cdot \overline{v})].$

If the data are linearly separable, the separation hyperplane will have the widest margin. Otherwise, the separation hyperplane will have the narrowest margin relative to the closest misclassified examples.

4 Extending the Method for Multiclass Classification

Implementation of multiclass classification is relatively simple here. The training set is processed into N_{class} training sets where N_{class} is the number of the classes. In every trainset the examples of one class are designated as class A and the examples of the remaining classes as class B. N_{class} ensembles are constructed for each of the processed trainsets. The centers of all ensembles live in the same space. The normal vectors of the classifiers as described in Section 3 have unity norm therefore the classifiers output Euclidian distance of the evaluated example from the separating hyperplane with the sign corresponding to the position of the evaluated example relative to the hyperplane. When obtaining results for the multiclass classification all ensembles produce the output relevant to the evaluated example. Assuming that the probability of correct classification increases with the distance from the separating hyperplane the maximum of the outputs of the ensembles is adopted as the overall result.

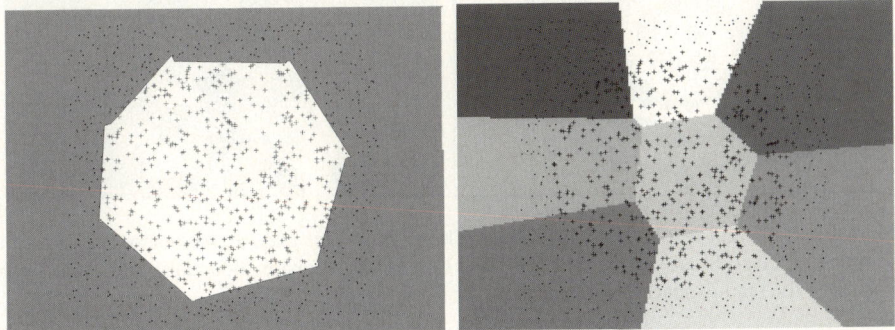

Fig. 3. Experimental results on 'circle' data. The examples of class A are marked by dots, the examples of class B are marked by crosses. Left is the decision boundary and right is the division of the feature space depicted.

5 Experiments

The experiments were performed on artificial and real world data. The purpose of the experiments was to test the performance of the method and to identify features useful for knowledge retrieval.

Fig. 2 shows the artificial datasets used. These are often referred to as 'circle in the square' and 'spiral' data. The sets contained 100 x 100 points each. 1000 points from the 'circle' set and 1250 points from the 'spiral' set were randomly drawn for training. The class misbalance in the training sets was lower than 15%.

The inputs were normalized to the interval [0,1].

Behavior of the method was also observed on multispectral, remotely sensed image data. The image containing 775x475 pixels was taken from Landsat satellite over the eastern Slovakia region. Each pixel of the image was identified by 6 spectral characteristics. The representation set was determined by a geographer and it was supported by a ground verification procedure. However, later investigation showed conflicts and noise in the representation set. The main goal was land use identification. There were seven classes of interest picked up for classification procedure. The representation set contained 6331 verified pixels, 3164 patterns was used for training. The classes were not evenly populated (74 patterns in the least populated class vs. 1174 patterns in the most populated class). According to the previous experimental analysis the features of some classes are overlapping. The experimental results on the real world data were compared to the experimental results of Gaussian SVM. Accuracy on the class, overall percentage accuracy and average accuracy per class were used to evaluate the results.

The average accuracy per class was used to calculate the growth stop criteria A_{min}.

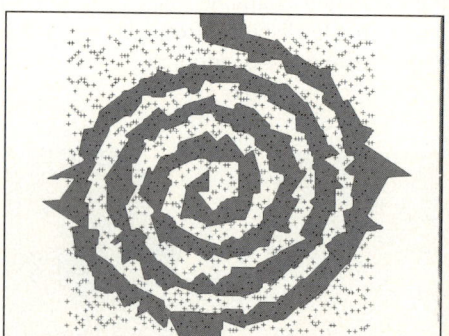 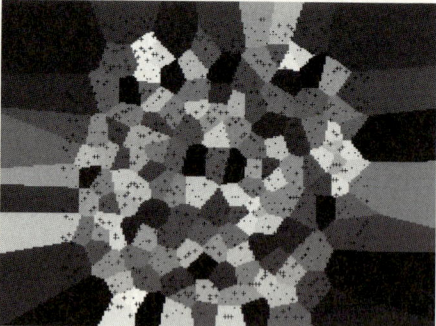

Fig. 4. Experimental results on 'spiral' data. The examples of class A are marked by dots, the examples of class B are marked by crosses. Left is the decision boundary and right is the division of the feature space depicted.

5.1 Experimental Results – Artificial Data

Fig. 1 and Fig. 2 depict the experimental results on the artificial datasets. Table 1 shows the parameters of the ensembles and the quantified experimental results.

The experimental results on the artificial data are useful for illustration. The required decision boundaries were nonlinear but the partial linear approximation was relatively successful especially on the 'circle' data. It is visible that the feature space is divided evenly which is appropriate for this data.

Please note that lowering A_{min} by the spiral data led to a significant reduction of the resulting size of the ensemble while the accuracy was almost not influenced (Table 1).

Table 1. Settings of the algorithm and experimental results on artificial datasets

Dataset	A_{min} (%)	No. of Members	Overall Accuracy Train set (%)	Average Accuracy Train set (%)	Overall Accuracy Test set (%)	Average Accuracy Test set (%)
'circle'	95	8	97.68	97.52	97.68	97.52
'spiral'	100	159	100	100	88.65	88.51
'spiral'	98	128	98.56	98.54	88.56	88.60

Table 2. Settings and experimental results on real world data. Comparison to Gaussian SVM

Ensemble Class	Urban Area	Barren Field	Bush	Agricult. Field	Meadow	Forest	Water
No. of Examples (Train)	155	483	206	902	170	1174	74
No. of Examples (Test)	156	484	206	902	171	1174	74
A_{min}	97.5	94	100	95	95	1	99
No. of Members	1	44	13	1	16	18	24
Accuracy on the Class (Train) %	98.71	90.48	99.51	99.67	88.24	99.91	52.70
Overall Accuracy (Train)	96.55 %						
Average Accuracy (Train)	89.89 %						
Accuracy on the Class (Test) %	87.18	81.48	100	84.48	53.22	99.32	47.30
Overall Accuracy (Test)	88.13 %						
Average Accuracy (Test)	79.01 %						
Gaussian SVM							
Overall Accuracy (Train)	95.83 %						
Average Accuracy (Train)	91.19 %						
Accuracy on the Class (Test) %	83.33	79.96	100	94.57	60.82	99.40	72.97
Overall Accuracy (Test)	91.60 %						
Average Accuracy (Test)	84.44 %						

5.2 Experimental Results – Real World Data

Table 2 summarizes the experimental results on real world data. Comparison to Gaussian SVM is given. Gaussian SVM was chosen because the previous experimental analysis showed its superiority over other methods (e.g. Backpropagation ANN) on given data. Comparison to similar methods is to come.

The experimental results of growing ensemble of linear classifiers are slightly inferior to Gaussian RBF. However, useful information on the data can be derived. We can see that the urban areas and agricultural fields are separated from the other examples in the training set with only one hyperplane and with high accuracy. The drop in the accuracy on the testing set signalizes these classes are not sufficiently populated. There exist examples of these classes with different spectral characteristics. The decision boundary dividing bushes from other classes is more complex (16 members were needed) but the separation is reliable. The results show barren fields characteristics are rather noisy. The most problematic classes are meadows and water. This is partially due to the underpopulation of the classes and partially due to conflicting data. The conflicts are caused by faulty ground verification and by overlapping of the spectral characteristics (e.g. meadows look like forests in the sun or infra red characteristics of bushes in the shadow can be similar to water etc. which can be seen in the confusion table). The advantage of the method is that the training parameters can be tuned for each ensemble separately while the resulting classifiers live in the same space with the same metrics.

6 Conclusion

The proposed method for growing ensembles of linear classifiers proved to be viable. The method is generic, i.e. it is not bound to certain type of classifier and extensions for nonlinear classifiers are possible. The type of the decision boundary enables to derive knowledge on the data by comparison of the numbers of members in the ensembles with the accuracy achieved on the training and the testing sets. Decomposition of a multiclass classification problem does not cause the resulting classifiers to live in different transformed spaces. The method is suitable for parallel implementation because the members of the ensemble can be optimized simultaneously.

Future work includes experimental analysis of usage of nonlinear classifiers. Quadratic surfaces are to be tested. This should cause reduction of the number of ensemble's members needed to achieve the required performance. The resulting decision boundary will be smoother. Future work includes also application of fuzzy logic on determination of the regions ruled by the classifiers. This should smooth up the transitions between the subregions. The work in progress includes development of parallel implementation of the method. The nature of the method enables scalability of the cluster of computational units. Each unit runs the same software which has 4 operational modes: waiting for subspace data, busy, sending subspace data and sending partial results to the master. The communication uses TCP protocol and the computational units can be distributed around the world.

References

1. Hartono, P., Hasimoto, S.: Ensemble of Linear Perceptrons with Confidence Level Output. In: 4th International Conference on Hybrid Intelligent Systems, Kitakiushu, Japan, December 5-8, pp. 186–191. IEEE Computer Society, Los Alamitos (2005)
2. Baxt, W.: Improving accuracy of artificial neural network using Multiple differently trained networks. Neural Computation 4, 772–780 (1992)
3. Hashem, S.: Optimal linear combination of neural networks. Neural Networks 10(4), 299–313 (1996)
4. Sharkey, A.: On combining artificial neural nets. Connection Science 9(4), 299–313 (1996)
5. Bundzel M.: Structural and Parametrical Adaptation of Artificial Neural Networks Using Principles of Support Vector Machines. Ph.D. Thesis, Technical University of Kosice, Slovakia, Faculty of Electrical Engineering and Informatics, Department of Cybernetics and Artificial Intelligence (2005)

A Framework for Adaptive and Integrated Classification

Ireneusz Czarnowski and Piotr Jędrzejowicz

Department of Information Systems, Gdynia Maritime University
Morska 83, 81-225 Gdynia, Poland
{irek@am.gdynia.pl, pj@am.gdynia.pl}

Abstract. This paper focuses on classification tasks. The goal of the paper is to propose a framework for adaptive and integrated machine classification and to investigate the effect of different adaptation and integration schemes. After having introduced several integration and adaptation schemes a framework for adaptive and integrated classification in the form of the software shell is proposed. The shell allows for integrating data pre-processing with data mining stages using population-based and A-Team techniques. The approach was validated experimentally. Experiment results have shown that integrated and adaptive classification outperforms traditional approaches.

Keywords: machine learning, data mining, integrated and adaptive classifiers, learning classifiers.

1 Introduction

The amount of data being collected in contemporary databases makes it impossible to reduce and analyze data without the use of automated analysis techniques. Knowledge discovery in databases (KDD) is the field that has evolved to provide such techniques. Frawley et.al. define knowledge discovery as "the non-trivial extraction of implicit, unknown, and potentially useful information from data" [11].

The KDD process can be viewed as an iterative sequence of the following steps [11], [27]:

- defining the problem,
- data pre-processing (data preparation),
- data mining including intelligent method selection,
- post data mining.

Data pre-processing comprises data collecting, data integration, data transformation, data cleaning, and data reduction.

There are many knowledge discovery methodologies available. Some of these techniques are generic, while others are domain-specific. Typical tasks for KDD are the identification of classes (clustering), the prediction of new, unknown objects (classification), the discovery of associations or deviations in spatial databases. Classification is understood as the process of finding a set of models (or functions) that describe and distinguish data classes or concepts, for the purpose of being able to use the model to predict the class of objects whose class label is unknown [13]. The derived model is based on the analysis of a set of training data (i.e., data objects whose class label is known).

This paper focuses on classification tasks. The goal of the paper is to propose a framework for adaptive and integrated machine classification and to investigate the effect of different adaptation and integration schemes.

Integration of both important stages of the KDD process i.e. data pre-processing (in particular data reduction) with data mining has been already recognized as an important step towards improving quality of machine learning tools (see for example: [1], [4], [5], [16], [22]). Another improvement has been achieved by introducing adaptation mechanisms as exemplified by the idea of learning classifier systems [3], [14], [24]. The paper proposes a framework allowing integration of the data reduction process with data mining through applying a population-based search with the inherent adaptation mechanism.

The paper is organized as follows: Section 2 introduces a set of integration and adaptation schemes; Section 3 provides an overview of the proposed framework for integrated and adaptive classification; Section 4 presents the results of the computation experiment carried to validate the approach; Section 5 contains conclusions and suggestions of further research.

2 Adaptation and Integration Schemes

In this Section several possible approaches to constructing integrated and adaptive classifiers are reviewed. From what has been observed in Section 1 it seems clear that an integrated adaptive classifiers should have two important features:

- Integration, at least partial, of data reduction and data mining stages.
- The existence of a positive feedback whereby more effective data reduction leads to higher classification accuracy, and in return, higher classification accuracy results in even more effective data reduction.

For the purpose of this paper data reduction is classified into following two types: feature selection and instance selection. Feature selection, also known as variable selection, feature reduction, attribute selection or variable subset selection, is the technique, commonly used in machine learning, of selecting a subset of relevant features for building robust learning models. The main aim of doing feature selection is to reduce the search space (for attributes), by selecting relevant features and then removing the remaining irrelevant features. That is, feature selection selects m features from the entire set of n features such that $m < n$. Ideally $m <<< n$.

In turn, instance selection is the technique allowing selection of representative instances in the data, based on some criterion, and removal of the remaining instances. Instance selection can be based either on sampling techniques or on search methods. In the later case instance selection is based on forming prototype instances from the actual instances, which would mimic the performance

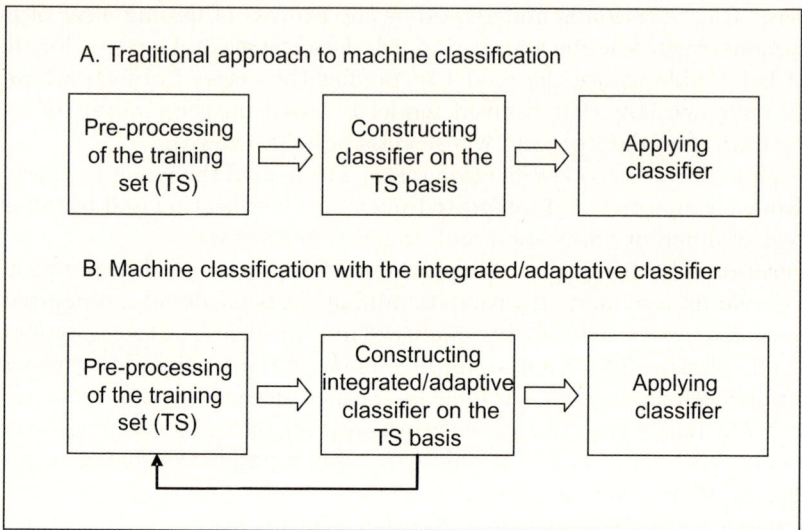

Fig. 1. Traditional versus integrated/adaptive approach to constructing machine classifiers

of these instances and then using only the prototype instances. Alternatively, instances can be selected using statistical measures (number of instances, mean or standard deviations) to replace redundant instances with their representative pseudo-instances or using support vectors to represent the entire set of instances from the data-set.

Traditional KDD process is assumed to be a sequential one in which stages or steps do not overlap. Traditional classifier possess neither integration nor adaptation features (understood as proposed at the beginning of this Section). In Fig. 1 the model of an approach to constructing traditional machine classifiers versus an approach to constructing integrated/adaptive ones is shown.

Pre-processing stage within the traditional model (scheme A in Fig. 1) may include feature and instance selection processes. Both are however based on some properties of the training dataset and are not integrated with the construction of the classifier. Training set thus produced forms approximation space, which is not adaptable. In another words approximation space used for the purpose of the classifier construction remains constant. The machine classification problem can be then defined as follows: given the approximation space in the form of the training set produced at the pre-processing stage, construct classifier which performs best from the point of view of the performance criterion used.

Situation changes with the introduction of the integrated and adaptive classifier (scheme B in Fig. 1). Some processes within the pre-processing stage are now integrated with the construction of the classifier. The approximation space becomes adaptable. The machine classification problem can be defined as: construct classifier which performs best from the point of view of the performance

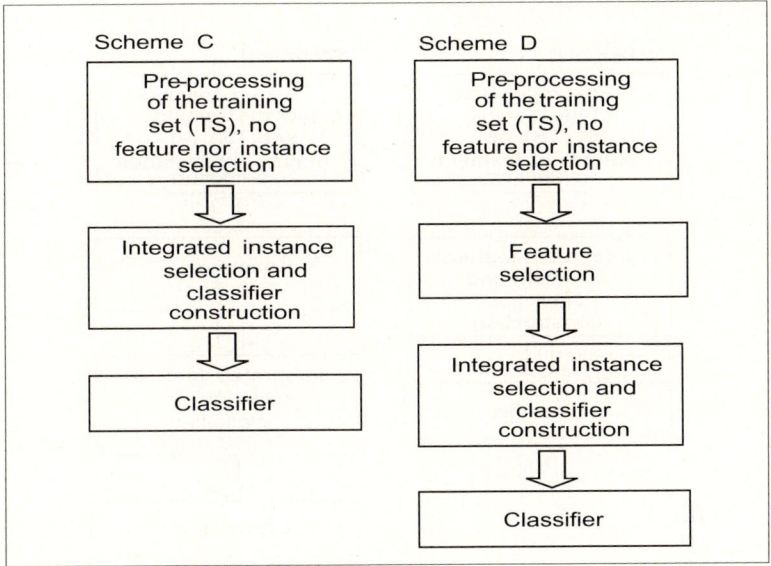

Fig. 2. Machine classification with the integrated instance selection and classifier construction without (Scheme C) and with (Scheme D) feature selection

criterion by finding the representation of the approximation space and, at the same time, deciding on the classifier features.

Integration of the pre-processing and classifier construction stages leads to a considerable extension of the decision space at the classifier construction phase. Both - feature selection and instance selection are computationally difficult (see [17], [21]) and of course the resulting problem of constructing the integrated and adaptive classifier is also computationally difficult. To deal with it there are several integration and adaptation schemes possible, as depicted in Fig. 2, 3 and 4.

Schemes C and D represent solutions where instance selection has been integrated with the classifier construction stage. This allows to construct the classifier and, at same time, to modify the approximation space obtained from the training set by changing the way instances are represented or simply by selecting instances. Modified approximation space allows to improve the classifier performance. Scheme D, additionally, provides for feature selection at the preprocessing stage.

Schemes E and F represent solutions where feature selection has been integrated with the classifier construction stage. This allows to construct the classifier and, at same time, to modify the approximation space obtained from the training set by selecting a subset of features. Modified approximation space allows to improve the classifier performance. Scheme F, additionally, provides for instance selection at the preprocessing stage.

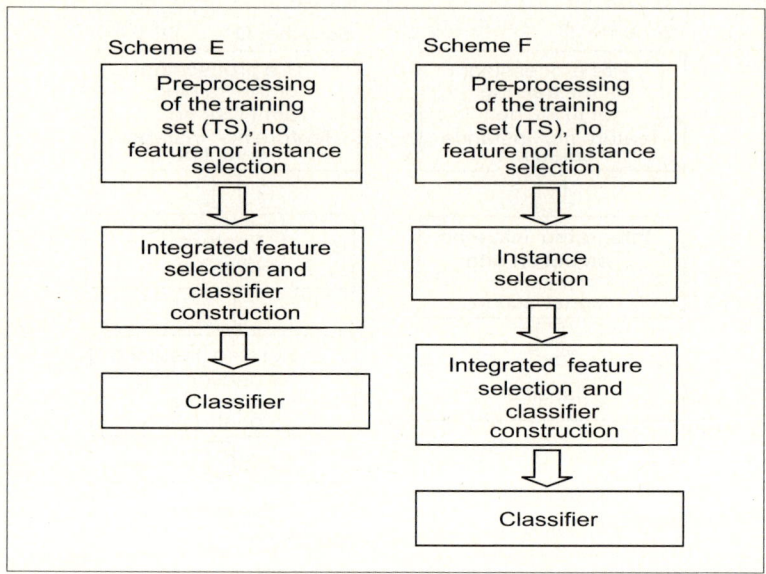

Fig. 3. Machine classification with the integrated feature selection and classifier construction without (Scheme E) and with (Scheme F) instance selection

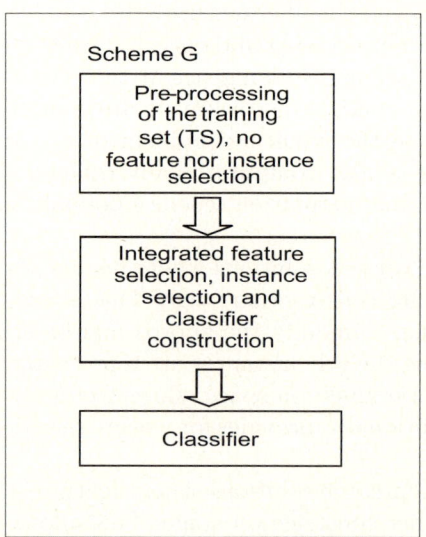

Fig. 4. Machine classification with the integrated feature selection, instance selection and classifier construction

Finally, scheme G represent solutions where both - instance and feature selection have been integrated with the classifier construction stage. Instances and features are selected iteratively while constructing the classifier.

3 Agent-Based Shell for the Integrated and Adaptive Classification

Since 1976 when Holland proposed Learning Classifier Systems [14] the idea of applying machine learning techniques which combine evolutionary computing, reinforcement learning, supervised learning and heuristics to produce adaptive systems have been gaining a growing interest from the machine intelligence research community. The framework for construction learning classifier systems proposed in this paper is based on emergence of evolutionary computation [19], local search algorithms [12], population learning algorithm [15], multiple-agent systems and the A-Team concept [23].

The proposed framework can be considered as a shell providing instance and feature reduction capabilities integrated with the classification algorithm chosen by the user. Technically, the shell has an A-Team structure where multiple agents search for the best combination of features, instances and classifier settings using local search heuristics and population based methods. The best combination of features the A-Team is searching for, is selected from the population of potential solutions which are kept in the common memory. Specialized agents try to improve solutions from the common memory by changing values of the decision variables (either selected instances or features or classifier settings). Their interactions provide for the required adaptation capabilities and for the evolution of the population of potential solutions.

In the reported approach the A-Team was designed and implemented using JADE-based A-Team (JABAT) environment. JABAT is a middleware allowing to design and implement an A-Team architecture for solving combinatorial optimization problems (for particulars of JABAT see [2]).

Main functionality of JABAT is searching for the optimum solution of a given problem instance through employing a variety of the solution improvement algorithms. The search involves a sequence of the following steps:

- Generation of the initial population of solutions.
- Application of solution improvement algorithms which draw individuals from the common memory and store them back after attempted improvement, using some user defined replacement strategy.
- Continuation of the reading-improving-replacing cycle until a stopping criterion is met.

The above functionality is realized by the two main types of classes. The first one includes *OptiAgents*, which are implementations of the improvement algorithms. The second are *SolutionManagers*, which are agents responsible for maintenance and updating of individuals in the common memory. All agents act in parallel. Each *OptiAgent* is representing a single improvement algorithm (for example tabu search, genetic algorithm, local search heuristics etc.). An *OptiAgent* has two basic behaviours defined. The first is sending around messages on readiness for action including the required number of individuals (solutions). The second is activated upon receiving a message from some *SolutionManager* containing the problem instance description and the required number of individuals. This behaviour involves improving fitness of individuals and resending the improved ones to a sender. A *SolutionManager* is brought to life for each problem instance. Its behaviour involves sending individuals to *OptiAgents* and updating the common memory.

The initial population of solutions is generated randomly. To solve the data reduction problem four types of agents representing different improvement procedures have been implemented. In each case the agent's classes are inherited from the *OptiAgent* class. The first and the second procedure aim at improving current solution through modification and exchange of the data instances grouped in clusters. The approach is based on calculating, for each instance from the original set, the value of its similarity coefficient proposed in [6], and then grouping instances into clusters consisting of instances with identical values of this coefficient, selecting the representation of instances for each cluster and removing the remaining instances, thus producing the reduced training set. The third procedure improves a solution by changing randomly content of the feature list. The fourth improvement procedure modifies a solution by changing values on both lists - of data instances and features. The modification of both lists is based on random moves. The first and the third improvement procedure use the tabu search algorithm [12], where the tabu list (the short term memory) contains moves that, for some number of iterations, are disallowed. The details of the described optimization agents can be found in [7], [8].

Quality measure (or fitness function) of a solution is the classifier accuracy calculated over the available test set using the 10 cross-validation approach. The classifier is provided by the user. In case it requires optimizing its parameters the respective *OptiAgents* must be designed and added to the available set of agents. Otherwise, the classifier is used directly to evaluate quality of solutions from the common memory.

4 Computational Experiment Results

To evaluate the performance of the proposed learning classifier framework it has been decided to carry a computational experiment with a view to compare the performance of different integration and adaptation schemes. Datasets for each problem have been obtained from [18]. They include: Cleveland heart disease (303 instances, 13 attributes, 2 classes), credit approval (690, 15, 2), Wisconsin breast cancer (699, 9, 2), and sonar problem (208, 60, 2).

Each benchmarking problem has been solved 30 times and the reported values of the quality measures have been averaged over all runs. The quality measure in all cases was the correct classification ratio calculated using the 10-cross-validation approach. The results obtained by using the: C 4.5 classifier

Table 1. Accuracy (%) of classifiers results obtained for selected integrated learning schemes

Classifier Problem	1NN	10NN	Bayes Network	WLSVM	C 4.5 (pruned)	C 4.5 (unpruned)
Scheme A (no data reduction applied)						
cancer	95.71	96.71	96.00	95.57	94.57	95.00
credit	82.46	86.38	75.36	85.22	84.93	83.19
heart	77.23	80.86	83.50	80.53	77.89	76.90
sonar	94.23	75.00	73.08	72.12	74.04	74.04
average	87.41	84.74	81.98	83.36	82.86	82.28
Scheme A' (feature selection at the pre-processing stage)						
cancer	94.57	95.00	96.00	95.14	94.43	94.43
credit	79.57	84.64	84.93	85.51	77.25	74.78
heart	71.95	80.53	80.53	80.86	79.87	79.54
sonar	79.33	81.25	61.54	73.08	72.10	71.23
average	81.35	85.35	80.75	83.65	80.91	79.99
Scheme C (integrated instance selection, no feature selection)						
cancer	96.86	97.43	96.87	90.59	97.44	98.43
credit	83.33	88.70	75.22	85.94	90.72	90.43
heart	84.00	85.67	87.33	87.00	91.21	92.42
sonar	94.23	75.00	75.00	40.38	83.65	83.65
average	89.61	86.70	83.61	75.98	90.76	91.24
Scheme D (integrated instance selection, feature selection carried at the pre-processing stage)						
cancer	94.15	96.15	96.72	74.49	95.15	95.44
credit	75.22	72.61	77.54	64.49	81.88	81.16
heart	83.33	87.00	86.00	87.00	85.00	85.67
sonar	76.32	72.31	63.56	71.23	82.02	81.72
average	82.26	82.02	80.95	74.30	86.01	86.00
Scheme F (integrated feature selection, instance selection carried at the pre-processing stage)						
cancer	89.31	94.73	75.16	73.64	95.01	86.45
credit	69.28	61.30	69.71	59.57	81.32	76.71
heart	75.04	81.33	76.00	83.00	81.67	81.33
sonar	74.02	79.32	71.65	65.86	78.43	76.43
average	76.91	79.17	73.13	70.52	84.11	80.23
Scheme G (integrated instance and feature selection)						
cancer	98.15	98.29	98.15	88.01	98.15	97.94
credit	87.68	89.57	85.22	85.51	92.61	90.87
heart	88.67	89.33	89.33	90.00	93.00	92.17
sonar	95.19	82.69	80.77	72.08	87.50	88.85
average	92.42	89.97	88.37	83.90	92.81	92.45

without pruned leaves and with pruned leaves [20], support vector machine (WLSVM) [26] and 1NN, kNN, Bayes Network implemented in WEKA library [25], are shown in Table 1. In Table 2 the results obtained by using fully integrated approach (Scheme G) are compared with the best classification results reported in the literature.

Table 2. Best results obtained by different classifiers (accuracy in %)

Approach	cancer	heart	credit	sonar
Scheme G (10NN)	98.29%	89.33%	89.57%	82.69%
Scheme G (C4.5 pruned)	98.15%	93.00%	92.61%	87.50%
Scheme G (1NN)	98.15%	88.67%	87.68%	95.19%
Naive-Bayes [9]	97.10%	83.60%	82.90%	90.40%
kNN [9]	96.6%	81.5% (1NN)	84.8% (11NN)	96.2% (1NN)
MLP+BP [9]	96.7%	81.30%	84.60%	90.40%
SVM [9]	96.90%	81.50%	-	76.90%
C 4.5 [9]	94.7%	77.80%	85.00%	76.90%

Non-integrated feature selection was carried at the pre-processing stage using the *wrapper* technique [10]. Non-integrated instance selection was based on selection of reference vectors as proposed in [6]. All integrated cases were obtained applying the proposed framework.

The experiment results have been used to perform the two-way analysis of variance. The following null hypothesis were formulated:

I. Choice of the integration and adaptation scheme does not influence the classifier performance.
II. Choice of the classifier type does not influence the classification accuracy.
III. There are no interactions between both factors (i.e. choice of the integration and adaptation scheme and choice of the classifier type).

It was established that with the degree of confidence set at 95% hypothesis II and III hold true. However hypothesis I should be rejected. In addition Tukey test confirmed that in case of the integrated approach there are no statistically significant differences between mean performances obtained using different classifiers.

5 Conclusion

The paper proposes a framework allowing for several schemes of integrating data pre-processing with data mining. Main tool of such framework is the software shell constructed as a multiple agent system. The agents work in parallel as an A-Team evolving population of solutions. Such an approach proved to produce excellent classification results using most popular classifiers. The proposed shell allows to embed any known classifier into the system with a view to achieve adaptive and integrated learning.

Computational experiment carried out confirmed that the integrated and adaptive solutions perform better then the traditional ones. It has been also confirmed that the choice of integration and adaptation scheme significantly influences performance of the classifier. On the other hand the choice of classifier does not influence the performance of integrated and adaptive classifiers.

Future research will aim at refining the available data reduction techniques, decreasing computation time required by the multiple agent system and extending the approach to distributed datasets.

References

1. Aksela, M.: Adaptive Combinations of Classifiers with Application to On-line Handwritten Character Recognition. Ph.D., Department of Computer Science and Engineering, Helsinki University of Technology, Helsinki (2007)
2. Barbucha, D., Czarnowski, I., Jędrzejowicz, P., Ratajczak-Ropel, E., Wierzbowska, I.: JADE-Based A-Team as a Tool for Implementing Population-Based Algorithms. In: Chen, Y., Abraham, A. (eds.) Proceedings of the Sixth International Conference on Intelligent Systems Design and Applications (ISDA 2006), vol. 3, pp. 144–149. IEEE Computer Society Press, Los Alamitos (2006)
3. Bull, L.: Learning Classifier Systems: A Brief Introduction. In: Bull, L. (ed.) Applications of Learning Classifier Systems (2004)
4. Bull, L., Kovacs, T.: Foundations of Learning Classifier Systems: An Introduction. Foundations of Learning Classifier Systems. Springer, Heidelberg (2005)
5. Bhanu, B., Peng, J.: Adaptive Integration Image Segmentation and Object Recognition. IEEE Trans. on Systems, Man and Cybernetics 30(4), 427–441 (2000)
6. Czarnowski, I., Jędrzejowicz, P.: An Approach to Instance Reduction in Supervised Learning. In: Coenen, F., et al. (eds.) Research and Development in Intelligent Systems XX, pp. 267–282. Springer, London (2004)
7. Czarnowski, I., Jędrzejowicz, P.: An Agent-Based Algorithm for Data Reduction. In: Bramer, M., et al. (eds.) Research and Development in Intelligent Systems XXIV and Applications and Innovations in Intelligent Systems XV, pp. 351–356. Springer, London (2008)
8. Czarnowski, I., Jędrzejowicz, P.: Data Reduction Algorithm for Machine Learning and Data Mining. In: Nguyen, N.T., et al. (eds.) IEA/AIE 2008. LNCS (LNAI), vol. 5027, pp. 276–285. Springer, Heidelberg (2008)
9. Duch, W.: Results - comparison of classification. Nicolaus Copernicus University (2002), http://www.is.umk.pl/projects/datasets.html
10. Dash, M., Liu, H.: Feature Selection for Classification. Intelligence Data Analysis 1(3), 131–156 (1997)
11. Frawley, W.J., Piatetsky-Shapiro, G., Matheus, C.: Knowledge Discovery in Databases - An Overview. In: Piatetsky-Shapiro, G., Matheus, C. (eds.) Knowledge discovery in databases. AAAI/MIT Press (1991)
12. Glover, F.: Tabu Search. Part I and II, ORSA Journal of Computing. 1 (3), Summer (1990) and 2 (1) Winter (1990)
13. Han, J., Kamber, M.: Data Mining. In: Concepts and Techniques. Academic Press, San Diego (2001)
14. Holland, J.H.: Adaptation. In: Rosen, Snell (eds.) Progress in Theoretical Biology, vol. 4, Plenum (1976)
15. Jędrzejowicz, P.: Social Learning Algorithm as a Tool for Solving Some Difficult Scheduling Problems. Foundation of Computing and Decision Sciences 24, 51–66 (1999)

16. Krawiec, K.: Konstruktywna indukcja cech we wspomaganiu decyzji na podstawie informacji obrazowej. Rozprawa doktorska. Instytut Informatyki Politechniki Poznanskiej, Poznan (2000) (in Polish)
17. Meiri, R., Zahavi, J.: Using Simulated Annealing to Optimize the Feature Selection Problem in Marketing Applications. European Journal of Operational Research 17(3), 842–858 (2006)
18. Merz, C.J., Murphy, P.M.: UCI Repository of Machine Learning Databases. University of California, Department of Information and Computer Science, Irvine, CA (1998), http://www.ics.uci.edu/~mlearn/MLRepository.html
19. Michalewicz, Z.: Genetic Algorithms + Data Structures = Evolution Programs. Springer, Berlin (1996)
20. Quinlan, J.R.: C 4.5: Programs for Machine Learning. Morgan Kaufmann, SanMateo (1992)
21. Rozsypal, A., Kubat, M.: Selecting Representative Examples and Attributes by a Genetic Algorithm. Intelligent Data Analysis 7(4), 291–304 (2003)
22. Sahel, Z., Bouchachia, A., Gabrys, B., Rogers, P.: Adaptive Mechanisms for Classification Problems with Drifting Data. In: Apolloni, B., Howlett, R.J., Jain, L. (eds.) KES 2007, Part II. LNCS (LNAI), vol. 4693, pp. 419–426. Springer, Heidelberg (2007)
23. Talukdar, S., Baerentzen, L., Gove, A., de Souza, P.: Asynchronous Teams: Cooperation Schemes for Autonomous, Computer-Based Agents. Technical Report EDRC 18-59-96, Carnegie Mellon University, Pittsburgh (1996)
24. Wilson, S.W.: Classifier Fitness Based on Accuracy. Evolutionary Computation 3(2), 149–176 (1995)
25. Witten, I.H., Frank, E.: Data Mining: Practical Machine Learning Tools and Techniques with JAVA Implementations. Morgan Kaufmann, San Francisco (2003)
26. EL-Manzalawy, Y., Honavar, V.: WLSVM: Integrating LibSVM into Weka Environment (2005), http://www.cs.iastate.edu/~yasser/wlsvm
27. Zhang, C., Zhang, S.: Association Rule Mining. LNCS (LNAI), vol. 2307, p. 243. Springer, Heidelberg (2002)

Solving Regression by Learning an Ensemble of Decision Rules

Krzysztof Dembczyński[1], Wojciech Kotłowski[1], and Roman Słowiński[1,2]

[1] Institute of Computing Science, Poznań University of Technology,
60-965 Poznań, Poland
{kdembczynski,wkotlowski,rslowinski}@cs.put.poznan.pl
[2] Systems Research Institute, Polish Academy of Sciences, 01-447 Warsaw, Poland

Abstract. We introduce a novel decision rule induction algorithm for solving the regression problem. There are only few approaches in which decision rules are applied to this type of prediction problems. The algorithm uses a single decision rule as a base classifier in the ensemble. Forward stagewise additive modeling is used in order to obtain the ensemble of decision rules. We consider two types of loss functions, the squared- and absolute-error loss, that are commonly used in regression problems. The minimization of empirical risk based on these loss functions is performed by two optimization techniques, the gradient boosting and the least angle technique. The main advantage of decision rules is their simplicity and good interpretability. The prediction model in the form of an ensemble of decision rules is powerful, which is shown by results of the experiment presented in the paper.

1 Introduction

Decision rule is a logical statement of the form: *if [condition], then [decision]*. If an object satisfies condition of the rule, then the decision is taken; otherwise no action is performed. A rule can be treated as a simple classifier that gives a constant response for the objects satisfying the condition, and abstains from the response for all other objects.

Induction of decision rules has been widely considered in the early machine learning approaches. The most popular algorithms were based on a sequential covering procedure (also known as separate-and-conquer approach) [27,7,8,18]. Apart from the sequential covering, some other approaches to rule induction exist. For instance, the apriori-based algorithms are also used for induction of predictive rules [25,34]. There are several rule-based approaches of lazy learning type, possibly combined with instance-based methods [11,19]. Other algorithms based on Boolean reasoning and mathematical programming try to select the most relevant rules – this is the case of Logical Analysis of Data [3]. Let us also notice that decision rule models are strongly associated with rough set approaches to knowledge discovery [28,32,21,33,20], where also Boolean reasoning has been applied [31]. Such a wide interest in decision rules may be explained by their simplicity and good interpretability. It seems, however, that decision trees (e.g. C4.5 [30], CART [6]) are more popular in data mining and machine learning applications.

Recently, we are able to observe again a growing interest in decision rule models. As an example, let us mention such algorithms as RuleFit [16], SLIPPER [9], Lightweight Rule Induction (LRI) [36], and ensemble of decision rules [1,2]. All these algorithms can be explained within the framework of *forward stagewise additive modeling* (FSAM) [22], a greedy procedure for minimizing a loss function on the dataset.

Let us notice that there are only few rule induction algorithms tailored to the problem of regression. An example is an extension of LRI [24], in which the decision attribute is discretized and then the problem is solved via classification rules. Another example is RuleFit, which uses FSAM framework explicitly, therefore can utilize a variety of loss functions, including those adapted to regression problems, like the squared-error loss.

However, in RuleFit the decision rules are not generated directly – trees are used as base classifiers instead. Rules are produced from each node (interior or terminal) of each resulting tree. This is set up by conjunction of conditions associated with all the edges on the path from the root of the tree to the considered node. Rule ensemble is then fitted by gradient directed regularization [15]. LRI, in turn, uses a specific reweighting schema (cumulative error), similar to Breiman's Arc-xf algorithm [5], which can also be explained in the context of loss function minimization [26]. Single rules are in the form of DNF-formulas.

The algorithm described here, called ENDER (from ENsemble of DEcision Rules), benefits from the achievements in boosting machines [17,26,12,13]. The main contribution of this paper is transmission of these achievements to the ground of a specific weak learner being a decision rule. Similarly to FSAM, our approach is stated as a greedy minimization of a loss function on the training set. However, contrary to RuleFit and LRI, the method introduced in this paper generates simple single rules (conjunctions of elementary conditions) directly, one rule in each iteration of the algorithm. Our approach is also distinguished by the fact of using the same single measure (value of the empirical risk) at all stages of the learning procedure: setting the best cuts (conditions), stopping the rule's growth and determining the response (weight) of the rule; no additional measures or tools (e.g. impurity measures, pruning procedures) are employed. Our research includes detailed analysis of the algorithm, both from the theoretical and experimental point of view. We report experiments with two types of loss functions, the squared- and the absolute-error loss, using two optimization techniques, the gradient boosting [13] and least angle technique [26].

The paper is organized as follows. In section 2, the regression problem is formulated. Section 3 presents a general framework for learning an ensemble of decision rules. Section 4 is devoted to the problem of a single rule generation. Derivation of particular algorithms for different optimization techniques and loss functions is described in section 5. Section 6 contains experimental results and comparison with other methods. The last section concludes the paper and outlines further research directions.

2 Problem Statement

In the regression problem, the aim is to predict the unknown value of an attribute y (called *decision attribute, output* or *dependent variable*) of an object using known joint values of other attributes (called *condition attributes, inputs, predictors*, or *independent variables*) $\mathbf{x} = (x_1, x_2, \ldots, x_n)$. The decision attribute is quantitative and it is assumed that $y \in \mathbb{R}$, where $\mathbb{R}$ is a set of real numbers.

The aim is to find a function $F(\mathbf{x})$ that predicts accurately value of y. The accuracy of a single prediction is measured in terms of the *loss function* $L(y, F(\mathbf{x}))$ which is the penalty for predicting $F(\mathbf{x})$ when the actual value is y. The overall accuracy of the function $F(\mathbf{x})$ is measured by the expected loss (*prediction risk*) over the joint distribution of variables $P(y, \mathbf{x})$ for the data to be predicted:

$$R(F) = E_{y\mathbf{x}} L(y, F(\mathbf{x})) = E_{\mathbf{x}}[E_{y|\mathbf{x}} L(y, F(\mathbf{x}))].$$

Therefore, the optimal (risk-minimizing) decision function (or Bayes optimal decision) is given by:

$$F^* = \arg\min_{F} E_{y\mathbf{x}} L(y, F(\mathbf{x})) = \arg\min_{F} E_{\mathbf{x}}[E_{y|\mathbf{x}} L(y, F(\mathbf{x}))]. \tag{1}$$

Since $P(y, \mathbf{x})$ is generally unknown, the learning procedure uses only a set of training examples $\{y_i, \mathbf{x}_i\}_1^N$ to construct $F(\mathbf{x})$ to be the best possible approximation of $F^*(\mathbf{x})$. Usually, this is performed by minimization of the *empirical risk*:

$$R_{\text{emp}}(F) = \frac{1}{N} \sum_{i=1}^{N} L(y_i, F(\mathbf{x}_i)),$$

where function F is chosen from a restricted family of functions.

The regression problem is solved typically by using the squared-error loss:

$$L_{\text{se}}(y, F(\mathbf{x})) = (y - F(\mathbf{x}))^2. \tag{2}$$

Bayes optimal decision for the squared-error loss has the following form:

$$F^*(\mathbf{x}) = \arg\min_{F(\mathbf{x})} E_{y|\mathbf{x}} L_{\text{se}}(y, F(\mathbf{x})) = E_{y|\mathbf{x}}(y). \tag{3}$$

It follows that minimization of the squared-error loss on the dataset can be seen as estimation of the expected value of y for a given $\mathbf{x}$. The absolute-error loss is also considered:

$$L_{\text{ae}}(y, F(\mathbf{x})) = |y - F(\mathbf{x})|. \tag{4}$$

Bayes optimal decision in this case is:

$$F^*(\mathbf{x}) = \arg\min_{F(\mathbf{x})} E_{y|\mathbf{x}} L_{\text{ae}}(y, F(\mathbf{x})) = \text{median}_{y|\mathbf{x}}(y). \tag{5}$$

Thus, minimization of the absolute-error loss on the dataset can be seen as estimation of median of y for given $\mathbf{x}$. It is often observed that the absolute-error makes the fitting procedures less sensitive to outliers, because the error grows linearly, not quadratically, with the distance between actual and predicted value.

In order to solve the regression problem, other loss functions can also be employed, for example, Huber loss [23] or ϵ-insensitive error loss, well-known from support vector regression [35]. In this paper, however, we restrict our considerations to the squared- and the absolute-error loss only.

3 Learning an Ensemble of Decision Rules

This section describes the general scheme of learning an ensemble of decision rules. We start with definition of a single rule, further we define the ensemble and outline the learning procedure.

Let X_j be a value set of attribute j, i.e. the set of all possible values of attribute j. Condition part of the rule consists of a conjunction of elementary expressions of the general form $x_j \in S_j$, where x_j is the value of object $\mathbf{x}$ on attribute j and S_j is a subset of X_j, $j \in \{1, \ldots, n\}$. We assume that in the case of ordered value sets, S_j has the form of the interval $[s_j, \infty)$ or $(-\infty, s_j]$ for some $s_j \in X_j$, so that the elementary expression takes the form $x_j \geq s_j$ or $x_j \leq s_j$. For nominal attributes, we consider elementary expression of the form $x_j = s_j$ or $x_j \neq s_j$. Let $\phi_{S_j}(\mathbf{x}) = I(x_j \in S_j)$, where $I(a)$ is an indicator function, i.e. if a is true then $I(a) = 1$, otherwise $I(a) = 0$. Let further Φ denote the set of elementary expressions constituting the conditions of the rule. Moreover, let $\Phi(\mathbf{x}) = \prod_{\phi \in \Phi} \phi_{S_j}(\mathbf{x})$ be a function that indicates whether an object satisfies condition part of the rule. We say that a rule *covers* an object, if $\Phi(\mathbf{x}) = 1$. It is easy to see that conjunction of elementary expressions defines an arbitrary axis-parallel region in the attribute space. Decision (also called response), denoted by α, is a real non-zero value assigned to the region defined by Φ. Therefore, we define a decision rule as:

$$r(\mathbf{x}) = \alpha \Phi(\mathbf{x}). \tag{6}$$

Notice that the decision rule takes only two values, $r(\mathbf{x}) \in \{\alpha, 0\}$, depending whether $\mathbf{x}$ satisfies the condition part or not.

We assume that the optimal decision function $F^*(\mathbf{x})$ is approximated by a linear combination of M decision rules:

$$F(\mathbf{x}) = \alpha_0 + \sum_{m=1}^{M} r_m(\mathbf{x}), \tag{7}$$

where α_0 is a constant value, which can be interpreted as a default rule, covering the whole attribute space. Construction of optimal combination of rules minimizing the empirical risk is a hard optimization problem. That is why we follow here FSAM [22], i.e. the rules are added one by one, greedily minimizing the loss function. We start with the default rule defined as:

$$\alpha_0 = \arg\min_{\alpha} \sum_{i=1}^{N} L(y_i, \alpha). \tag{8}$$

Algorithm 1. Ensemble of decision rules – ENDER

Input   : set of training examples $\{y_i, \mathbf{x}_i\}_1^N$,
              M – number of decision rules to be generated.
Output: default rule α_0, ensemble of decision rules $\{r_m(\mathbf{x})\}_1^M$.

$\alpha_0 = \arg\min_\alpha \sum_{i=1}^N L(y_i, \alpha)$;
$F_0(\mathbf{x}) = \alpha_0$;
for $m = 1$ **to** M **do**
$\quad r_m(\mathbf{x}) = \arg\min_{\Phi,\alpha} \left\{ \sum_{\Phi(\mathbf{x}_i)=1} L(y_i, F_{m-1}(\mathbf{x}_i) + \alpha) + \sum_{\Phi(\mathbf{x}_i)=0} L(y_i, F_{m-1}(\mathbf{x}_i)) \right\}$
$\quad F_m(\mathbf{x}) = F_{m-1}(\mathbf{x}) + \nu \cdot r_m(\mathbf{x})$;
end
$ensemble = \{r_m(\mathbf{x})\}_1^M$;

In each next iteration, the new rule is added taking into account previously generated rules. Let $F_{m-1}(\mathbf{x})$ be a prediction function after $m-1$ iterations, consisting of first $m-1$ rules and the default rule. In the m-th iteration, a decision rule is obtained from:

$$r_m(\mathbf{x}) = \arg\min_{\Phi,\alpha} \left\{ \sum_{\Phi(\mathbf{x}_i)=1} L(y_i, F_{m-1}(\mathbf{x}_i) + \alpha) + \sum_{\Phi(\mathbf{x}_i)=0} L(y_i, F_{m-1}(\mathbf{x}_i)) \right\} \quad (9)$$

Expression (9) defines the main step of the algorithm. Different solutions of this step are described in the next two sections.

It has been shown that in order to improve the accuracy of the ensemble, the base classifiers should be shrunk towards α_0 [22]. That is why we use a shrinkage parameter $\nu \in (0,1]$ when the ensemble is augmented by the newly generated rule:

$$F_m(\mathbf{x}) = F_{m-1}(\mathbf{x}) + \nu \cdot r_m(\mathbf{x}).$$

In other words, values of ν determine the degree to which previously generated rules $r_k(\mathbf{x})$, $k = 1, \ldots, m-1$, affect the generation of the successive one in the sequence, i.e., $r_m(\mathbf{x})$.

It has also been observed that training base classifiers on a subsample of the training set leads to improvement of both accuracy and computational complexity [14]. This is due to the fact that classifiers trained on the random subsamples become diversified and less correlated. We also apply this technique and take subsample of size $\eta \leq N$ randomly drawn with or without replacement from the original training set, when a single rule is generated.

The whole procedure for constructing an ensemble of decision rules is presented as Algorithm 1. We called this procedure ENDER (from ENsemble of DEcision Rules).

4 Generation of a Single Rule

Exact solution of (9) is still computationally hard. That is why, in order to generate a single rule, we proceed in two steps using a fast approximate algorithm. The general scheme of this algorithm is the following. At the m-th iteration:

1. Find Φ_m by minimizing a functional $\mathcal{L}_m(\Phi)$ in a greedy manner (the particular form of the functional depends on the loss function and minimization technique used; see next section):

$$\Phi_m = \arg\min_{\Phi} \mathcal{L}_m(\Phi). \tag{10}$$

2. Find α_m to be a solution of the following line-search problem:

$$\alpha_m = \arg\min_{\alpha} \sum_{i=1}^{N} L(y_i, F_{m-1}(\mathbf{x}_i) + \alpha \Phi_m(\mathbf{x}_i)). \tag{11}$$

In the case of the squared- and the absolute-error loss, computation of α_m is straightforward, because analytical expressions for (11) can be given (see next section).

The greedy procedure for finding Φ_m is the following:

- At the beginning, Φ_m is empty (no elementary expressions are specified),
- In the next step, an elementary expression ϕ_{S_j} is added to Φ_m that minimizes $\mathcal{L}_m(\Phi)$ (if it exists). Such expression is searched by consecutive testing of elementary expressions, attribute by attribute. For ordered attributes, let $x_j^{(1)}, x_j^{(2)}, \ldots, x_j^{(l)}$ be a sequence of ordered values on attribute j of training examples satisfying Φ_m, such that $x_j^{(i-1)} \geq x_j^{(i)}$, for $i = 2, \ldots, l$. Then, each elementary expression $x_j \geq s_j$ or $x_j \leq s_j$, for each $s_j = \frac{x_j^{(i-1)} + x_j^{(i)}}{2}$, is tested. For nominal attributes, we test each expression $x_j = s_j$ or $x_j \neq s_j$, for each value $s_j \in X_j$.
- The above step is repeated until $\mathcal{L}_m(\Phi)$ cannot be decreased.

The output of the whole procedure is the decision rule $r_m(\mathbf{x}) = \alpha_m \Phi_m(\mathbf{x})$. The details of the algorithm for two loss functions and two optimization techniques are given in the next section.

Let us underline that this procedure is very fast and proved to be efficient in computational experiments. The ordered attributes can be sorted once before generating any rule. The procedure for finding optimal Φ resembles the way the decision trees are generated. Here, we look for only one branch instead of the whole decision tree. Moreover, let us notice that minimal value of $\mathcal{L}_m(\Phi)$ is a natural stop criterion in building a single rule and we do not use any other measure (e.g. impurity measures) for choosing the optimal cuts.

5 ENDER Algorithms with Different Optimization Techniques and Loss Functions

We use two minimization techniques that determine the form of $\mathcal{L}_m(\Phi)$.

Gradient boosting [13]. In this method, the rule is fitted to the negative gradient of the loss function:

$$\tilde{y}_i = -\left.\frac{\partial L(y_i, F(\mathbf{x}))}{\partial F(\mathbf{x})}\right|_{F(\mathbf{x})=F_{m-1}(\mathbf{x}_i)}, \quad i = 1, \ldots, N. \tag{12}$$

The fitting procedure is defined by minimization of the squared-error between rule response and negative gradient:

$$r(\mathbf{x}) = \arg\min_{\Phi, \alpha} \left(\sum_{\Phi(\mathbf{x}_i)=1} (\tilde{y}_i - \alpha)^2 + \sum_{\Phi(\mathbf{x}_i)=0} \tilde{y}_i^2 \right). \tag{13}$$

The term in brackets can be solved for:

$$\alpha = \frac{\sum_{\Phi(\mathbf{x}_i)=1} \tilde{y}_i}{\sum_{i=1}^{N} \Phi(\mathbf{x}_i)}, \tag{14}$$

where $\sum_{i=1}^{N} \Phi(\mathbf{x}_i)$ is the number of objects satisfying $\Phi(\mathbf{x}_i) = 1$. Thus, α is an average of the negative gradient in the region covered by the rule. Putting (14) into (13), expanding the term in the sum for $\Phi(\mathbf{x}_i) = 1$, and performing some simple calculations, we obtain:

$$r(\mathbf{x}) = \arg\min_{\Phi} \left(\sum_{i=1}^{N} \tilde{y}_i^2 - \frac{1}{\sum_{i=1}^{N} \Phi(\mathbf{x}_i)} \left(\sum_{\Phi(\mathbf{x}_i)=1} \tilde{y}_i \right)^2 \right).$$

After removing the first term, which is constant, and taking the square root of the second term (which does not affect the minimization) we get:

$$-\frac{\left|\sum_{\Phi(\mathbf{x}_i)=1} \tilde{y}_i\right|}{\sqrt{\sum_{i=1}^{N} \Phi(\mathbf{x}_i)}} \tag{15}$$

which plays the role of $\mathcal{L}_m(\Phi)$, to be minimized with respect to Φ.

Least angle technique [26]. In this technique, the rule is also fitted to the negative gradient of the loss function. However, contrary to gradient boosting, the angle between negative gradient vector (12) and rule response vector is minimized. This can be formulated as minimization of the inner product with fixed value (norm) of the rule response:

$$r(\mathbf{x}) = \arg\min_{\Phi} \left(\sum_{\Phi(\mathbf{x}_i)=1} \alpha \tilde{y}_i \right) = |\alpha| \arg\min_{\Phi} \left(\pm \sum_{\Phi(\mathbf{x}_i)=1} \tilde{y}_i \right) \tag{16}$$

which can be expressed in the following form, independent of α:

$$r(\mathbf{x}) = \arg\min_{\Phi} -\left| \sum_{\Phi(\mathbf{x}_i)=1} \tilde{y}_i \right|. \tag{17}$$

The term $-\left|\sum_{\Phi(\mathbf{x}_i)=1} \tilde{y}_i\right|$ in (17) plays the role of $\mathcal{L}_m(\Phi)$. Notice the similarity between this term and (15). The latter is divided by the square root of the rule size, which results in more specific rules covering smaller regions in the attribute space. Thus, minimizing the angle results in more general rules.

Below, we present how the above techniques apply to the squared- and the absolute-error loss functions.

Squared-error loss. The negative gradient is in this case:

$$\tilde{y}_i = -\left.\frac{\partial L_{se}(y_i, F(\mathbf{x}_i))}{\partial F(\mathbf{x}_i)}\right|_{F(\mathbf{x}_i)=F_{m-1}(\mathbf{x}_i)} = \frac{y_i - F_{m-1}(\mathbf{x}_i)}{2}, \quad i = 1, \ldots, N. \quad (18)$$

Putting (18) to (15) or (17), one obtains Φ_m by the gradient boosting or the least angle approach, respectively.

Finally, we find α_m as a solution of the line-search (11). The optimal value is obtained by:

$$\alpha_m = \arg\min_\alpha \sum_{i=1}^N L_{se}(y_i, F_{m-1}(\mathbf{x}_i) + \alpha \Phi_m(\mathbf{x}_i)) = \frac{\sum_{\Phi(\mathbf{x}_i)=1}(y_i - F_{m-1}(\mathbf{x}_i))}{\sum_{i=1}^N \Phi(\mathbf{x}_i)}. \quad (19)$$

It is interesting that gradient boosting gives in this case an exact solution to (9) (but still we have to use the greedy procedure for finding Φ_m). This is due to the fact that putting (19) into (9), after some simple calculations, we get the following expression to be minimized

$$-\frac{\left|\sum_{\Phi(\mathbf{x}_i)=1}(y_i - F_{m-1}(\mathbf{x}_i))\right|}{\sqrt{\sum_{i=1}^N \Phi(\mathbf{x}_i)}}. \quad (20)$$

(20) is equivalent to (15) with negative gradient (18) up to the constant $\frac{1}{2}$ that does not affect the solution.

Absolute error loss. In this case, the negative gradient is:

$$\tilde{y}_i = -\left.\frac{\partial L_{ae}(y_i, F(\mathbf{x}_i))}{\partial F(\mathbf{x}_i)}\right|_{F(\mathbf{x}_i)=F_{m-1}(\mathbf{x}_i)} = \text{sgn}(y_i - F_{m-1}(\mathbf{x}_i)), \quad i = 1, \ldots, N, \quad (21)$$

and the optimal α_m is obtained by:

$$\alpha_m = \arg\min_\alpha \sum_{i=1}^N L_{ae}(y_i, F_{m-1}(\mathbf{x}_i) + \alpha \Phi_m(\mathbf{x}_i)) = \text{median}_{\Phi(\mathbf{x}_i=1)}(y_i - F_m(\mathbf{x}_i)). \quad (22)$$

For this loss function, there is no simple and exact solutions to (9) and $\mathcal{L}_m(\Phi)$ has to be determined by gradient boosting or least angle techniques.

Table 1. Experimental results on eight benchmark datasets. RMSE from 10-fold cross validation repeated 10 times and ranks (in parentheses) of the algorithms are given.

DATA SET	ENDER LS	ENDER LA	ENDER GS	ENDER GA	LR	M5P	AR w/ DS	AR w/ RT	BA w/ RT
PYRIM.	0.098(4)	0.089(2)	0.102(5)	0.09(3)	0.12(8)	0.107(6)	0.088(1)	0.13(9)	0.116(7)
CPU	55.092(3)	102.277(9)	53.399(1)	70.053(6)	68.404(5)	53.512(2)	56.297(4)	93.343(8)	71.326(7)
BOSTON	3.621(3)	4.402(8)	3.604(2)	3.831(6)	4.875(9)	3.593(1)	3.856(7)	3.807(5)	3.731(4)
ABALONE	2.203(5)	2.399(9)	2.199(4)	2.234(8)	2.218(6)	2.138(1)	2.232(7)	2.194(3)	2.148(2)
BANK	0.034(6)	0.04(9)	0.033(4)	0.034(5)	0.039(8)	0.03(1)	0.036(7)	0.031(2)	0.032(3)
COMP.	3.558(7)	4.281(8)	3.105(3)	3.394(6)	9.897(9)	3.166(4)	3.194(5)	2.843(1)	3.076(2)
CALIF.	59088(6)	62637(7)	54566(3)	56914(5)	69632(9)	55705(4)	64698(8)	48393(1)	50421(2)
CENSUS	31442(5)	36570(8)	30851(3)	32337(6)	41606(9)	31396(4)	32900(7)	30143(2)	30028(1)
AVE. RANK	4.875	7.5	3.125	5.625	7.875	2.875	5.75	3.875	3.5

6 Experimental Results

We designed an experiment to compare performance of ENDER algorithm with other regression methods. We collected eight benchmark datasets taken from http://www.liacc.up.pt/~ltorgo/Regression/DataSets.html.

We tested four types of ENDER algorithm: least angle with absolute- (LA) and squared-error (LS), and gradient boosting with absolute- (GA) and squared-error (GS). We used 200 rules, shrinkage parameter ν set to 0.5, and resampling in which 50% of training examples were drawn without replacement. These parameters were found to work well in a few previous experimental runs on several datasets. In order to perform the comparison, we chose five methods implemented in Weka [37]. These are: linear regression (LR), M5P trees [29] (M5P), additive regression [14] (AR) with decision stumps (DS) and with reduced-error pruning trees (RT), and also bagging [4] (BA) with RT. To perform the experiment in a fair manner, we ran several times these algorithms with different parameters, and the best result of each method for each dataset was reported. This is in fact a real challenge for ENDER algorithms that were applied with one set of parameters for all datasets.

We performed 10-fold cross validation that was repeated 10 times, in which we used root mean squared-error (RMSE) as a measure of performance. Table 1 reports RMSE and ranks (in parentheses) for each algorithm and dataset. In order to compare classifiers, we followed Demšar [10] and made Friedman test, which uses average ranks of each algorithm to check whether all the algorithms performed equally well (null hypothesis). Friedman statistic gave 28.83 which exceeded the critical value 15.51 for confidence level 0.05 (approximated by Chi-squared distribution). In such a case, we could reject the null hypothesis and proceed to a post-hoc analysis. We calculated the critical differences, CD= 4.332 (for confidence level 0.05), using Nemenyi test, which means that difference in average ranks of algorithms greater than 4.332 is significant. Thus, we have that M5P and ENDER GS are significantly better than linear regression and ENDER LA. However, we cannot say that others algorithms differ with each other.

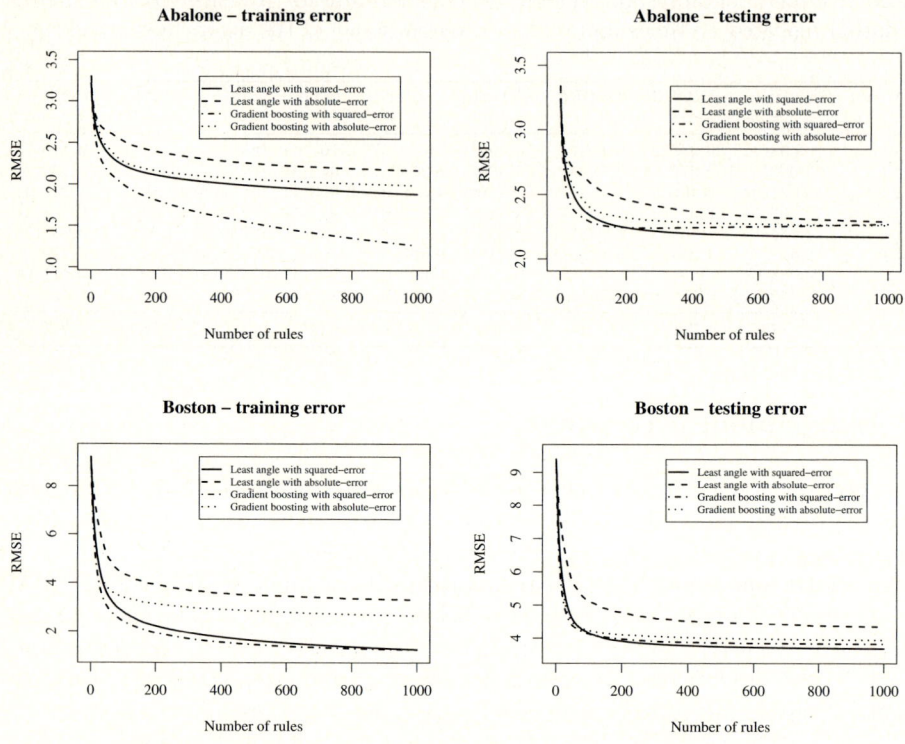

Fig. 1. Errors on Abalone and Boston datasets over 30 runs

The results show that the best variant of ENDER algorithm is the gradient boosting with the squared-error loss that was the second best in the experiment ranked after M5P. Let us remind that this approach gives an exact solution to (9). The second best is the least angle variant with the same loss function. The superiority of the squared-error loss over the absolute-error loss is due to the fact that the chosen measure was RMSE. There is, however, still a place for improvements for the ensemble of decision rules. In Figure 1 the RMSE on training and testing set for Abalone and Boston datasets is given for up to 1000 rules. The results are averaged over 30 runs. Datasets were split into training and testing set in the proportion 60% to 40%. One can observe that ENDER is quite stable and results given in Table 1 can be better for higher number of rules. One can also see that in the case of Abalone dataset, ENDER GS has mild tendency to overfit.

7 Conclusions and Future Plans

We introduced a novel algorithm for solving the regression problems that generates an ensemble of decision rules. We derived four different variants of the algorithm incorporating two optimization techniques with two loss functions. Let us underline that there are only few approaches in which decision rules are used for solving this type of prediction problems. Examples of such methods are LRI and RuleFit. Unfortunately, we still did not compare our algorithm to these methods. This is included in our future research plans. Experiment performed in this paper shows that the algorithm is competitive to other methods commonly used in regression problems. Moreover, the algorithm has an additional advantage. Its output is a set of decision rules that are simple and interpretable. However, the ensemble of 1000 rules could lose this advantage. That is why, an additional tool for post-processing of rules is desirable.

References

1. Błaszczyński, J., Dembczyński, K., Kotłowski, W., Słowiński, R., Szeląg, M.: Ensemble of decision rules. Foundations of Computing and Decision Sciences (31), 3–4 (2006)
2. Błaszczyński, J., Dembczyński, K., Kotłowski, W., Słowiński, R., Szeląg, M.: Ensembles of Decision Rules for Solving Binary Classification Problems in the Presence of Missing Values. In: Greco, S., Hata, Y., Hirano, S., Inuiguchi, M., Miyamoto, S., Nguyen, H.S., Słowiński, R. (eds.) RSCTC 2006. LNCS (LNAI), vol. 4259, pp. 224–234. Springer, Heidelberg (2006)
3. Boros, E., Hammer, P., Ibaraki, T., Kogan, A., Mayoraz, E., Muchnik, I.: An Implementation of Logical Analysis of Data. IEEE Trans. on Knowledge and Data Engineering 12, 292–306 (2000)
4. Breiman, L.: Bagging Predictors. Machine Learning 24, 123–140 (1996)
5. Breiman, L.: Arcing classifiers. Annals of Statistics 26, 801–824 (1998)
6. Breiman, L., Friedman, J., Olshen, R., Stone, C.: Classification and Regression Trees. Wadsworth (1984)
7. Clark, P., Nibbet, T.: The CN2 induction algorithm. Machine Learning 3, 261–283 (1989)
8. Cohen, W.: Fast effective rule induction. In: International Conference on Machine Learning, pp. 115–123 (1995)
9. Cohen, W., Singer, Y.: A simple, fast, and effective rule learner. In: National Conference on Artificial Intelligence, pp. 335–342 (1999)
10. Demšar, J.: Statistical comparisons of classifiers over multiple data sets. Journal of Machine Learning Research 7, 1–30 (2006)
11. Domingos, P.: Unifying instance-based and rule-based induction. Machine Learning 24, 141–168 (1996)
12. Friedman, J., Hastie, T., Tibshirani, R.: Additive logistic regression: a statistical view of boosting. Annals of Statistics 28, 337–407 (2000)
13. Friedman, J.: Greedy Function Approximation: A Gradient Boosting Machine. Annals of Statistics 29, 1189–1232 (2001)
14. Friedman, J.: Stochastic Gradient Boosting. Computational Statistics & Data Analysis 38, 367–378 (2002)
15. Friedman, J., Popescu, B.: Gradient directed regularization. Technical Report, Dept. of Statistics, Stanford University (2004)
16. Friedman, J., Popescu, B.: Predictive Learning via Rule Ensembles. Technical Report, Dept. of Statistics, Stanford University (2005)

17. Freund, Y., Schapire, R.: A decision-theoretic generalization of on-line learning and an application to boosting. J. of Comp. and System Sc. 55, 119–139 (1997)
18. Fürnkranz, J.: Separate-and-conquer rule learning. AI Review 13, 3–54 (1996)
19. Góra, G., Wojna, A.: RIONA: A New Classification System Combining Rule Induction and Instance-Based Learning. Fundamenta Informaticae 54, 369–390 (2002)
20. Greco, S., Matarazzo, B., Słowiński, R., Stefanowski, J.: An Algorithm for Induction of Decision Rules Consistent with the Dominance Principle. In: Ziarko, W., Yao, Y. (eds.) RSCTC 2000. LNCS (LNAI), vol. 2005, pp. 304–313. Springer, Heidelberg (2001)
21. Grzymala-Busse, J.: LERS — A system for learning from examples based on rough sets. In: [32], pp. 3–18. Kluwer Academic Publishers, Dordrecht (1992)
22. Hastie, T., Tibshirani, R., Friedman, J.: Elements of Statistical Learning: Data Mining, Inference, and Prediction. Springer, Heidelberg (2003)
23. Huber, P.: Robust Estimation of a Location Parameter. Annals of Mathematical Statistics 35, 73–101 (1964)
24. Indurkhya, N., Weiss, S.: Solving Regression Problems with Rule-based Ensemble Classifiers. In: ACM SIGKDD International Conference on Knowledge Discovery and Data Mining, pp. 287–292 (2001)
25. Jovanoski, V., Lavrac, N.: Classification Rule Learning with APRIORI-C. In: Brazdil, P., Jorge, A. (eds.) EPIA 2001. LNCS (LNAI), vol. 2258, pp. 44–51. Springer, Heidelberg (2001)
26. Mason, L., Baxter, J., Bartlett, P., Frean, M.: Functional gradient techniques for combining hypotheses. In: Advances in Large Margin Classifiers, pp. 33–58. MIT Press, Cambridge (1999)
27. Michalski, R.: A Theory and Methodology of Inductive Learning. In: Michalski, R., Carbonell, J., Mitchell, T. (eds.) Machine Learning: An Artificial Intelligence Approach, pp. 83–129. Tioga Publishing, Palo Alto (1983)
28. Pawlak, Z.: Rough Sets. Theoretical Aspects of Reasoning about Data. Kluwer Academic Publishers, Dordrecht (1991)
29. Quinlan, J.: Learning with continuous classes. In: Australian Joint Conference on Artificial Intelligence, pp. 343–348 (1992)
30. Quinlan, J.: C4.5: Programs for Machine Learning. Morgan Kaufmann, San Francisco (1993)
31. Skowron, A.: Extracting laws from decision tables - a rough set approach. Computational Intelligence 11, 371–388 (1995)
32. Słowiński, R. (ed.): Intelligent Decision Support. Handbook of Applications and Advances of the Rough Set Theory. Kluwer Academic Publishers, Dordrecht (1992)
33. Stefanowski, J.: On rough set based approach to induction of decision rules. In: Skowron, A., Polkowski, L. (eds.) Rough Set in Knowledge Discovering, pp. 500–529. Physica Verlag (1998)
34. Stefanowski, J., Vanderpooten, D.: Induction of decision rules in classification and discovery-oriented perspectives. Int. J. on Intelligent Systems 16, 13–27 (2001)
35. Vapnik, V.: The Nature of Statistical Learning Theory, 2nd edn. Springer, Heidelberg (1998)
36. Weiss, S., Indurkhya, N.: Lightweight rule induction. In: International Conference on Machine Learning, pp. 1135–1142 (2000)
37. Witten, I., Frank, E.: Data Mining: Practical machine learning tools and techniques, 2nd edn. Morgan Kaufmann, San Francisco (2005)

Meta-learning with Machine Generators and Complexity Controlled Exploration

Krzysztof Grąbczewski and Norbert Jankowski

Department of Informatics
Nicolaus Copernicus University
Toruń, Poland
{norbert,kgrabcze}@is.umk.pl
http://www.is.umk.pl/

Abstract. We present a novel approach to meta-learning, which is not just a ranking of methods, not just a strategy for building model committees, but an algorithm performing a search similar to what human experts do when analyzing data, solving full scope of data mining problems. The search through the space of possible solutions is driven by special mechanisms of machine generators based on meta-schemes. The approach facilitates using human experts knowledge to restrict the search space and gaining meta-knowledge in an automated manner. The conclusions help in further search and may also be passed to other meta-learners. All the functionality is included in our new general architecture for data mining, especially eligible for meta-learning tasks.

1 Introduction

Meta-learning is learning how to learn. In order to perform meta-level analysis of *learning from data* one needs a robust system for different kinds of learning with uniform management of miscellaneous learning machines and their results. Our data mining system is an implementation of a very general view of learning machines and models. Therefore it is very flexible and eligible for sophisticated meta-level analysis of learning processes.

A *learning problem* can be defined as $\mathcal{P} = \langle D, \mathcal{M} \rangle$, where $D \subseteq \mathcal{D}$ is a *learning dataset* and $\mathcal{M}$ is a *model space*.

In computational intelligence attractive models $m \in \mathcal{M}$ are determined with learning processs:

$$L^p : \mathcal{D} \to \mathcal{M}, \qquad (1)$$

where p defines the parameters of the learning machine. This view of learning encircles many different approaches of supervised and unsupervised learning including classification, approximation, clustering, finding associations etc. Such definition does not limit the concept of search to specific kinds of learning methods like neural networks or statistical algorithms, however such reduction of model space is possible in practice.

In real life problems, sensible solutions $m \in \mathcal{M}$ are usually so complex, that it is very advantageous to decompose the given problem $\mathcal{P} = \langle D, \mathcal{M} \rangle$ into subproblems:

$$[\mathcal{P}_1, \ldots, \mathcal{P}_n] \qquad (2)$$

where $\mathcal{P}_i = \langle \mathcal{D}_i, \mathcal{M}_i \rangle$. In this way, the vector of solutions of the problems $\mathcal{P}_i$ constitutes a model for the main problem $\mathcal{P}$:

$$m = [m_1, \ldots, m_n], \tag{3}$$

and the model space gets the form

$$\mathcal{M} = \mathcal{M}_1 \times \ldots \times \mathcal{M}_n. \tag{4}$$

In practice $\mathcal{M}$ is usually a subspace of $\mathcal{M}_1 \times \cdots \times \mathcal{M}_n$, because the submachines may extract not only the component needed for the main model but also some other information. It is not a problem, because in such cases, the final model is just a simple projection of the whole complex model.

The solution constructed by a decomposition is often much easier to find, because the main task gets reduced to a series of simpler tasks: model m_i solving the subproblem $\mathcal{P}_i$, is the result of a learning process

$$L_i^{p_i} : \mathcal{D}_i \to \mathcal{M}_i, \quad i = 1, \ldots, n, \tag{5}$$

where

$$\mathcal{D}_i = \prod_{k \in K_i} \mathcal{M}_k, \tag{6}$$

and $K_i \subseteq \{0, 1, \ldots, i-1\}$, $\mathcal{M}_0 = \mathcal{D}$. It means that the learning machine $L_i^{p_i}$ may take advantage of some of the models $m_1, \ldots, m_{i-1}$ learned by preceding subprocesses and of the original dataset D of the main problem $\mathcal{P}$. Naturally, also parameters $p = (p_1, \ldots, p_n)$.

So, the main learning process L^p is decomposed to the vector

$$[L_i^{p_i}, \ldots, L_n^{p_n}]. \tag{7}$$

Such decomposition is often very natural: a standardization or feature selection naturally precedes classification, a set of classifiers precedes a committee module etc. Note that, the subprocesses need not to be dependent on all preceding subprocesses, so such decomposition has natural consequences in the possibility of parallelization of problem solving.

A real life example of a learning process decomposition is presented in figure 1, where a classification committee is constructed, but member classifiers need the data transformed before they can be applied. The structure of the project directly corresponds to both the learning process and the final model decomposition. The rectangle including all small boxes but "Data", depicts the whole learning process, which, given dataset, is expected to provide a classification routine. For different kinds of analysis like testing classification accuracy etc. it must be treated as a whole, but from the formal point of view each inner rounded rectangle is a separate process solving its task and providing its model.

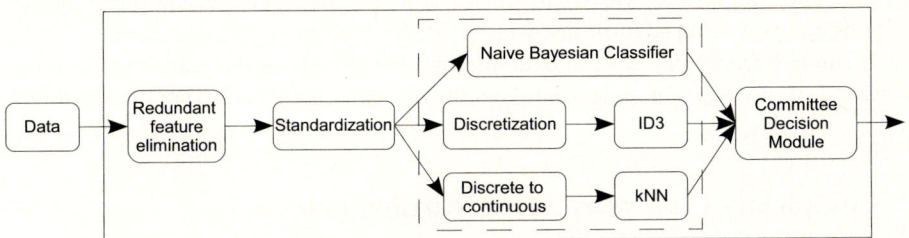

Fig. 1. An example of a DM project

Because each DM process is a directed acyclic graph, it is easy to show the composite process and composite model it corresponds to. The model of figure 1 may be decomposed as

$$\begin{aligned}
m_{rfe} &= L_{rfe}(Data) \\
m_{std} &= L_{std}(m_{rfe}) \\
m_{nbc} &= L_{nbc}(m_{std}) \\
m_{discr} &= L_{discr}(m_{std}) \\
m_{id3} &= L_{id3}(m_{discr}) \\
m_{dtc} &= L_{dtc}(m_{std}) \\
m_{knn} &= L_{knn}(m_{dtc}) \\
m_{comm} &= L_{comm}(m_{nbc}, m_{id3}, m_{knn})
\end{aligned} \quad (8)$$

The subscripts are easy to decode, when compared to the figure 1. Each of the components learns some part of the final model, which has a corresponding structure.

Such general and uniform foundations of our DM system facilitate solving problem of any kind, requiring any structural complexity, provided appropriate components. It is especially important when undertaking meta-learning challenges, where we must try many different methods, from simple ones to those of large complexity. Nontriviality of model selection is evident when browsing the results of NIPS 2003 Challenge in Feature Selection [1,2] or WCCI Performance Prediction Challenge [3] in 2006.

Some meta-learning approaches [4,5,6,7] are based on data characterization techniques (characteristics of data like the numbers of features/vectors/classes, features variances, information measures on features, also from decision trees etc.) or on *landmarking* (machines are ranked on the basis of simple machines performances before starting the more power consuming ones). Although the projects are really interesting, they still may be done in different ways or, at least, may be extended in some aspects. The whole space of possible and interesting models is not browsed so thoroughly by the mentioned projects, thereby some types of solutions can not be found with them.

In our approach the term *meta-learning* encompasses the whole complex process of model construction including adjustment of training parameters for different parts of the model hierarchy, construction of hierarchies, combining miscellaneous data transformation methods and other adaptive processes, performing model validation and complexity analysis, etc.

This article presents some details of the structure and functionality of a meta-learning machine, naturally implemented within the architecture of our recently created system.

It is an efficient algorithm, which can find many interesting solutions and is a good starting point to even better algorithms, which will be certainly created as further steps of our research, because our general data mining platform opens the gates to easy implementation of advanced meta-learning techniques, capable of gathering and exploiting meta-knowledge.

2 Complexity Controlled Meta-learning Process

The space of potential solutions is usually very huge, but it does not mean that experts should be more effective than dedicated meta-learning algorithms which search through the model space in intelligent ways. From the other side, even the most advanced expert is limited in some ways—it can be seen for instance when browsing the difference of quality of solutions, presented by experts in competitions in the area of computational intelligence.

The algorithm, presented below, can find solutions to different kinds of computational intelligence problems like classification, approximation, prediction, etc. Also, it may optimize different criteria, the selection of which, usually depends on the task which is to be solved. The solutions generated by our algorithm may be of simple or complex structure. They are searched for in a uniform process controlled with real complexity of algorithms (learning machines). Note that a single machine is not always of smaller complexity than another one of more complex structure, but composed of submachines of small complexities. The complexity based control of meta-learning processes is of highest importance, because it helps avoid some traps which could crush the whole learning process.

Given a dataset representing the problem and a goal criterion, some learning machines can find a solution (with different efficiency and accuracy) but for some others the problem may be unsolvable (for example, may encounter convergence troubles because of their stochastic behavior, typical for some neural networks). Moreover, in accordance with insolvability of the halting problem, some learning processes may be infinite. The meta-learning algorithm, we propose, deals successfully also with such cases.

Our solution to these problems was inspired by the definition of complexity by Levin [8,9]:

$$C_L(P) = \min_p \{c_L(p) : p \text{ is a program which solves } P\}, \tag{9}$$

where P is the problem to be solved and

$$c_L(p) = l(p) + \log(t(p)), \tag{10}$$

$l(p)$ is the length of the program p and $t(p)$ is the time in which p solves P.

In more advanced meta-learning the Eq. 10 may be substituted by

$$c_{NiK}(p) = [l(p) + \log(t(p))] \cdot q(p), \tag{11}$$

where $q(p)$ is a function term responsible to reflect the inverse of an estimate of reliability of p, and p denotes a learning machine (or a submachine) (the same applies to Eq. 9 when it is adapted to computational intelligence problems).

The main idea of the algorithm is to iterate in the main loop through the *programs* (*algorithms*, *learning machines*), constructed by a system of *machine generators* (described below), in the order of their complexity measured with Eq. 11 or, in a simplified version, with Eq. 10. In fact, the complexity which is used by the meta-learning algorithm to order machines, is a sum of two complexities: the first for the *learning part* and the second for the *test part*.

In general, our meta-learning algorithm may be seen as a loop of test estimation trials with a complexity control mechanism. Each generated machine is nested in the test procedure (adequate for the problem type and configured goal), then the test procedure starts and the loop supervises whether the complexity of the task does not exceed current complexity threshold.

An outline of the meta-learning algorithm may be presented as:

```
1  procedure ML;
2  while( stop_condition != true )
3  {
4     start tasks if possible
5     check the complexity of running tasks
6     analyze finished tasks
7  }
```

The following subsections describe details of subsequent parts of this algorithm.

2.1 The Stopping Condition of the Loop

As long as machines are generated by machine generators, the main loop may continue the job. However the process may be stopped for example when the goal is obtained (remember, that the goal may depend on the problem type and on our preferences). We may wish to:

- find the *best model* in given time for given dataset,
- find the *best model* satisfying a goal condition with given threshold θ,
- find the *best model* satisfying a goal condition with given threshold θ, and of as simple structure as possible,
- find several *best models* which can be used as complementary and which satisfy a goal condition with given threshold θ,
- stop when the progress of objective function (test criterion) is smaller than a given ϵ.

Also the term of *best model* (or rather of *better model*) may be defined differently (based on several concepts), however it is the simpler part of the algorithm. It is important to see that stopping criterion is not a problem itself—we just need to declare our preferences.

2.2 Starting New Tasks

Line number 4 of the procedure ML is devoted to starting new tasks. The algorithm keeps the started tasks in a special queue Q of a specified, limited size. A new task can be added only if the number of tasks in Q is smaller than the limit. The tasks in Q may run in parallel.

The tasks are constructed on the basis of machine configurations obtained from a set of generators. The procedure always gets the machine of the smallest estimated complexity, according to Eq. 11 or 10, considering all active generators (a meta-learner may change the set of machine generators up to its needs). The selected machine or rather its configuration is nested in a task which performs a test of the machine, for example in a cross-validation test. The type of the test and its parameters is also a subject to configuration. If the complexity of selected machine was not bigger than *current complexity* level, the current complexity is set to the maximum of current complexity and complexity of selected machine[1].

The outline of the procedure starting new tasks looks like:

```
languagelanguage
1  procedure  start tasks if possible;
2   while(  started tasks count  <  limit  )
3   {
4    m :=  find machine of simplest complexity in generators set
5    form new test task  t  for machine  m
6    add  t  to Q
7    current_complexity  :=  max(current_complexity, complexity_of(m))
8   }
```

The crucial role in the above symbolic code, plays the set of machine generators which is a source of machine configurations. Different machine generators may form significantly different solution spaces. Machine generators are also strongly goal–dependent (depend on the problem type and the criterion used for testing). The machine generators are asked to present or give single machines of the smallest complexity, one by one. The meta-learning procedure selects a machine of smallest complexity among the results obtained from all the generators. All of these ideas are realized very efficiently using appropriate data structures.

To enable the calculation of *machine complexity*, for each machine, there is a function estimating the complexity on the basis of the machine configuration and the inputs it gets. In the case, when the estimation is not possible, the complexity is approximated from averaged past observation of the behavior of the method on different inputs. In both cases the estimated complexity is additionally weighted to obtain similar runtime behavior of machines which declare the same (or very similar) complexity.

The goal of using a set of generators instead of a single generator was that it is simpler to define several *dedicated* generators, than a single universal one for any type of tasks. The generators may form different levels of abstractions in machines construction. They may be more or less sophisticated and produce more or less complex machines. The meta-learner may exchange results of the explorations between generators, integrating the possibilities of generators. The generators may be added or removed, during meta-learning, according to the needs of the ML procedure. They may also adjust their behavior to the knowledge collected while learning, to produce new machines, more adequate to the experience, providing lower $q(p)$ of Eq. 11.

---

[1] It can not be simply set to the complexity of selected machine because it may happen (from different reasons) that a generator generates a new machine of smaller complexity.

2.3 Complexity Control of Running Tasks

The tasks which are running, must be checked whether they don't consume more time or memory than planned. All tasks are supervised, because otherwise, some of them could use too much time and block the resources for other tasks. When the (realized/-consumed) complexity of a task exceeds the threshold calculated on the basis of the value of *current_complexity*, the task is stopped and removed from the tasks queue Q. The estimated complexity of such task is increased with a fixed factor or according to the estimated progress of the task and the task is moved to the *quarantine*. If possible, (it depends on implementation of given machine) the task state is saved (outside of memory) to be restarted from the stopping-point, when *current_complexity* gets adequately large. Thus, the quarantine plays the role of a machine generator, which stores the stopped tasks, for future use.

Similarly to the idea of machine examination in the order of increasing complexity, braking too complex processes resembles what human experts do when searching for attractive models, but here, instead of the fuzzy criterion of expert's patience we have a formal complexity-based test.

2.4 Analysis of Finished Tasks

Each finished task is removed from the task queue Q and the estimated quality of the tested machine, together with machine configuration and the results of learning, is moved to *results repository*. Partial results (current ranking of models) are available in real time (e.g. accessible from GUI).

All finished tasks help find more and more interesting solutions. Even if they do not provide very attractive solutions, they are a source of some meta-knowledge, helpful in further exploration, for example in estimation of the reliability of machines created by active generators for next generations. This information is very useful for adjustment of $q(p)$ from Eq. 11, which has crucial influence on the ordering of generated machines. For instance, if it is found, that a combination of given feature selection method works well with some classifier, we may promote such submachine structures in new machines.

2.5 Examples of Machine Generators

The simplest form of a machine generator is the one providing learning machines configuration from a predefined set. Such a generator must be capable of pointing to the simplest machine in the set. The same generator is used by our meta-learning algorithm to realize the *quarantine* for too complex machines.

The generators are free in the choice of knowledge used to generate machines. The *scheme based generator (SBG)* was designed to produce new machines using *meta-schemes*. A meta-scheme is a template which defines how to build structures of machines. Some examples of meta-schemes are presented in figures 2, 3 and 4. Meta-schemes may contain machines, placeholders for machines and connections between machines inputs and outputs. SBGs fill the meta-schemes with particular machines provided in some sets obtaining complex machines. The fact that the structure is more complex, does not imply a higher complexity of such new machine. Imagine a machine

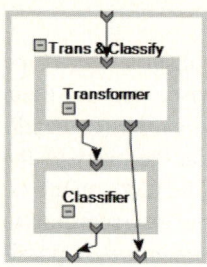 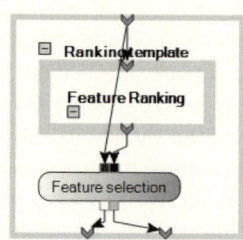 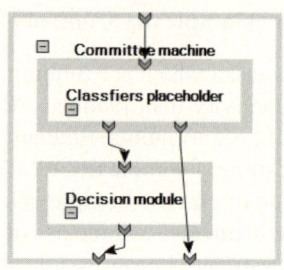

Fig. 2. A meta-scheme of data transformation and classification

Fig. 3. A meta-scheme of feature selection transformation with placeholder for ranking machine

Fig. 4. A meta-scheme of a committee machine with placeholder for a number of classifiers and decision module

composed of a feature selection and a classifier (by filling the meta-scheme of figure 2). It may happen that the complexity of the feature selection is small and the transformation leaves small amount of features in the output dataset. The classifier trained on transformed data may have much smaller complexity, because of the dimensionality reduction, and final complexity of such composite model may be significantly lower than the complexity of the same classifier, when not preceded by the feature selection machine. This is a very important feature of our algorithm, because it facilitates finding solutions, even when the base algorithms are too complex, if only some compound machines can solve the problem effectively.

The meta-scheme of figure 2 enables creating machines which consist of any dataset transformation method and any classification machine. The choice of data transformation depends on initial configuration but also on newly produced machines. Note that such compound, as a product of the meta-scheme, forms another classifier and it may be nested in another scheme. Also the transformation placeholder of this scheme may be filled directly by a data transformer or by an instance of a scheme which plays the role of dataset transformer (for example the meta-scheme of figure 3). The SBG type of generators should avoid producing tautology or nonsense (from computational intelligence point of view), however, in general, it is impossible to prevent the generators from providing unnecessary or useless (sub-)solutions.

Figure 3 presents a meta-scheme dedicated to feature selection. The role of the ranking machine (the placeholder) is to determine the importance order of features and the feature selection machine performs the selection of the top ranked features. A filled instance of that scheme may be nested in the previous meta-scheme to compose a classifier preceded by the feature selection. Each machine generator may have its own tactic for building/composing new machines. Especially the generator which composes machines from meta-schemes can be realized in a number of ways.

Figure 4 presents a general meta-scheme of a committee model. The classifiers can be inserted in the classifiers placeholder and a decision module in the other placeholder (it may be a voting/weighting/WTA or any other kind of decision module).

Another very important machine generator may be seen as a sub-meta-learning and is devoted to search for optimal (or close to optimal) configuration parameters for a given machine (including complex structures of machines). This machine generator produces a specialized test machine (*meta parameter search machine*) to search for *meta parameters*. By meta parameters of given machine we mean its configuration parameters which are declared to be searched automatically. Such parameters can be described by their types, interval of acceptable values, default values, interval of recommended values, recommended search strategy, etc. A meta parameter search machine tests given machine using one of several search strategies. The strategy should reflect behavior of the meta-parameter (linear, logarithmic, exponential or nominal). We have implemented several search strategies for 1D and 2D. The description of meta-parameters and their search methods provide a very interesting knowledge for the parameters search automation. The knowledge may be used by a machine generator to produce a series of independent machines and efficiently explore the space of possible machine configurations.

2.6 How It All Works Together

Meta-learning based on machine generators is a search process similar to what human experts do when analyzing data. The machine generators construct machines according to figures 2–4. The generated machines are validated in proper order. The simplest machines are constructed by some substitutions to the meta-scheme of figure 2. One of the simplest transformations is data standardization, another one removes useless features with the filter of invariance[2]. They fit the first placeholder in the meta-scheme. Replacing the second box by a Naive Bayesian Classifier (NBC)[3] results in the instance of the meta-scheme of one of the smallest possible complexity. Thus, NBC trained on simply filtered data is one of the first candidate validated.

Not all the instances of this meta-scheme are so simple. We can also use Principal Components Analysis (PCA) as data transformation and a version of kNN with automated adjustment of k, obtaining quite computationally complex instance of the meta-scheme. Because of its large complexity, such machine is not tested at the very beginning of the search. It may get into the queue, even behind some models of more complex structure (for example composed of a data normalization, a simple feature selector and a classifier), but with more attractive time complexity prediction.

The complexity control also facilitates withdrawal of some methods, when their adaptive processes take too much time. It is quite natural, that for example a Support Vector Machine (SVM) training may be very difficult, when run on raw data, but after some feature selection, or other data transformation, the optimization process is very fast. In such cases the SVM which has been running for some time without success, is withdrawn, and other machines are tried. Otherwise, problematic machines could block the whole meta-learning.

The recursive nature of the meta-scheme presented in figure 4 facilitates taking advantage of what has been learned in the earlier stages of the search—the most successful (and most different) methods may be easily put into a committee to obtain even better or more stable results. It is not necessary to learn everything from scratch, when we

---

[2] It removes each feature, which variance is equal to zero.
[3] Our implementation of NBC works with both nominal and continuous features.

start searching for committees, it is enough to combine the decisions of already created models, which may save a lot of time.

It is also worth to notice, that evolutionary algorithms may be very easily implemented within our framework—it is enough to implement a machine generator capable of producing next generations and define the fitness function which will serve as the meta-learning validation criterion.

Small number of simple machine generators allows us to create quite complex machines and search for optimal configuration of their components. Experts meta-knowledge used to define an adequate set of meta-schemes and the mechanism of complexity control significantly reduce the search space, while not resigning from the most attractive solutions.

Obviously, providing unreasonable machine generators (for example generating very large number of similar machines of simple structure but poorly performing) or misleading complexity estimators, may easily spoil the whole meta-learning process, so all the components of the algorithm must be carefully selected.

3 Summary

We have presented a meta-learning algorithm based on machine generators and complexity control. Using meta-schemes restricts testing to only such machine architectures, that we regard as sensible. Our definition of machine (and model) complexity, and the mechanisms for their estimation (before starting adaptive processes) and control (during their runs), facilitate testing machines in proper order. Validating candidate machines in the order of increasing complexity guarantees success in the pursuit for suboptimal models—if there is an accurate structure (compatible with the meta-schemes), then it will be found in a finite time, for the same reasons, for which the breadth first search successfully explores possibly infinite trees.

Our system supplies tools for easy meta-level activity, so that meta-knowledge may be easily extracted from data mining projects. Our algorithm collects such information to improve further search stages, for more efficient selection of committee members etc. More advanced methods for collecting, exchange and exploiting meta-knowledge will be one of our most important interests in the future.

Acknowledgements. The research was supported by the Polish Ministry of Science with a grant for years 2005–2007.

References

1. Guyon, I.: Nips 2003 workshop on feature extraction (December 2003),
   http://www.clopinet.com/isabelle/Projects/NIPS2003/
2. Guyon, I., Gunn, S., Nikravesh, M., Zadeh, L.: Feature extraction, foundations and applications. Springer, Heidelberg (2006)
3. Guyon, I.: Performance prediction challenge (July 2006),
   http://www.modelselect.inf.ethz.ch/

4. Pfahringer, B., Bensusan, H., Giraud-Carrier, C.: Meta-learning by landmarking various learning algorithms. In: Proceedings of the Seventeenth International Conference on Machine Learning, pp. 743–750. Morgan Kaufmann, San Francisco (2000)
5. Brazdil, P., Soares, C., da Costa, J.P.: Ranking learning algorithms: Using IBL and meta-learning on accuracy and time results. Machine Learning 50(3), 251–277 (2003)
6. Bensusan, H., Giraud-Carrier, C., Kennedy, C.J.: A higher-order approach to meta-learning. In: Cussens, J., Frisch, A. (eds.) Proceedings of the Work-in-Progress Track at the 10th International Conference on Inductive Logic Programming, pp. 33–42 (2000)
7. Peng, Y.H.: Falch, P., Soares, C., Brazdil, P.: Improved dataset characterisation for meta-learning. In: The 5th International Conference on Discovery Science, January 2002, pp. 141–152. Springer, Luebeck (2002)
8. Levin, L.A.: Universal sequential search problems. In: Problems of Information Transmission (translated from Problemy Peredachi Informatsii (Russian)), vol. 9 (1973)
9. Li, M., Vitányi, P.: An Introduction to Kolmogorov Complexity and Its Applications. In: Text and Monographs in Computer Science, Springer, Heidelberg (1993)

Assessing the Quality of Rules with a New Monotonic Interestingness Measure Z

Salvatore Greco[1], Roman Słowiński[2,3], and Izabela Szczęch[2]

[1] Faculty of Economics, University of Catania, Corso Italia, 55, 95129 Catania, Italy
salgreco@unict.it
[2] Institute of Computing Science Poznan University of Technology, 60-965 Poznan, Poland
{Izabela.Szczech,Roman.Slowinski}@cs.put.poznan.pl
[3] Systems Research Institute, Polish Academy of Sciences, 01-447 Warsaw, Poland

Abstract. The development of effective interestingness measures that help in interpretation and evaluation of the discovered knowledge is an active research area in data mining and machine learning. In this paper, we consider a new Bayesian confirmation measure for "*if..., then...*" rules proposed in [4]. We analyze this measure, called Z, with respect to valuable property M of monotonic dependency on the number of objects in the dataset satisfying or not the premise or the conclusion of the rule. The obtained results unveil interesting relationship between Z measure and two other simple and commonly used measures of rule support and anti-support, which leads to efficiency gains while searching for the best rules.

1 Introduction

When mining large datasets, the number of knowledge patterns, often expressed in a form of "*if..., then...*" rules, can easily be overwhelming for the human capabilities to understand them and to find the useful results. To guide the data analyst identifying valuable rules, various quantitative measures of interestingness (attractiveness measures) have been proposed and studied (e.g. support, anti-support, measures of confirmation) [9]. They all reflect some different characteristics of rules. The problem of choosing an appropriate interestingness measure for a certain application is difficult because the number and variety of measures proposed in the literature is so big. Therefore, studies analyzing theoretical properties of these measures, as well as relationships among them, is worth consideration. Moreover, there are some theoretical properties of interestingness measures which are particularly valuable for practical applications. Properties of measures also naturally group them unveiling relationships between them, and are helpful in choosing an appropriate measure for a particular application.

In this paper, we focus on a new interestingness measure, from the category of Bayesian confirmation, proposed by Crupi et al. [4] and called the Z measure. It is a measure that quantifies the degree to which the premise of a rule provides support for or against the rule's conclusion. We analyze it with respect to a valuable property M, introduced by Greco et al. [7], of monotonic dependency of the measure on the number of objects satisfying or not the premise or the conclusion of the rule. Moreover, on the basis of satisfying the property M, we draw some practical conclusions about very particular relationship between measure Z and two other simple but meaningful measures of rule support and anti-support.

The paper is organized as follows. In section 2, there are preliminaries on rules and their quantitative description. Next, in section 3, we analyze Z with respect to property M. Section 4 presents practical application of the obtained results. The paper ends with conclusions.

2 Preliminaries

Let us consider discovering rules from a sample of larger reality given in a form of a data table. Formally, a *data table* is a pair $S = (U, A)$, where U is a nonempty finite set of objects, called *universe*, and A is a nonempty finite *set of attributes*. The set V_a is the set of values of the attribute $a \in A$.

Let us associate a formal language L of logical formulas with every subset of attributes. Formulas for a subset $B \subseteq A$ are built up from attribute-value pairs (a, v), where $a \in B$ and $v \in V_a$, using logical operators $\neg$ (not), $\wedge$ (and), $\vee$ (or).

A *rule* induced from S and expressed in L is denoted by $\phi \rightarrow \psi$ (read as "if ϕ, then ψ"). It consists of antecedent ϕ and consequent ψ, being formulas expressed in L, called *premise* and *conclusion* (hypothesis or decision), respectively, and therefore it can be seen as a consequence relation between premise and conclusion (see critical discussion [7] about interpretation of rules as logical implications). The rules mined from data may be either *decision* or *association* rules, depending on whether the division of A into condition and decision attributes has been fixed or not.

2.1 Support and Anti-support Measures of Rules

One of the most popular measures used to identify frequently occurring association rules in sets of items from data table S is the *support*. Support of condition ϕ, denoted as $sup(\phi)$, is equal to the number of objects in U having property ϕ. The support of rule $\phi \rightarrow \psi$ (also simply referred to as support), denoted as $sup(\phi \rightarrow \psi)$, is the number of objects in U having property ϕ and ψ. Thus, it corresponds to statistical significance [9]. Naturally, support is a gain-type criterion, i.e. its higher values are more desirable.

Anti-support of a rule $\phi \rightarrow \psi$ (also simply referred to as anti-support), denoted as $anti - sup(\phi \rightarrow \psi)$, is equal to the number of objects in U having property ϕ but not having property ψ. Thus, anti-support is the number of counter-examples, i.e. objects for which the premise ϕ evaluates to true but which fall into a class different than ψ. Note that anti-support can also be regarded as $sup(\phi \rightarrow \neg \psi)$. Thus, it is considered as a cost-type criterion, which means that the smaller the value of anti-support, the more desirable it is.

In literature, there can also be found definitions of support and anti-support as relative values with respect to the number of objects in the whole dataset. In this paper, we will not take under consideration such interpretation of support and anti-support, however, doing so would not influence anyhow the generality of the conducted analysis and the obtained results.

2.2 Z Measure

Among commonly used interestingness measures there is a large group of Bayesian confirmation measures which quantify the degree to which the premise provides "support for or against" the conclusion [6]. Thus, formally, a measure $c(\phi \rightarrow \psi)$ can be regarded as a measure of confirmation if it satisfies the following condition:

$$c(\phi \rightarrow \psi) \begin{cases} > 0 & if \quad Pr(\psi|\phi) > Pr(\psi), \\ = 0 & if \quad Pr(\psi|\phi) = Pr(\psi), \\ < 0 & if \quad Pr(\psi|\phi) < Pr(\psi). \end{cases} \quad (1)$$

Under the "closed world assumption" adopted in inductive reasoning, and because U is a finite set, it is legitimate to estimate probabilities $Pr(\phi)$ and $Pr(\psi)$ in terms of frequencies $sup(\phi)/|U|$ and $sup(\psi)/|U|$, respectively. In consequence, we can define the conditional probability as $Pr(\psi|\phi) = Pr(\psi \wedge \phi)/Pr(\phi)$, and it can be regarded as $sup(\phi \rightarrow \psi)/sup(\phi)$. Thus, the above condition can be re-written as:

$$c(\phi \rightarrow \psi) \begin{cases} > 0 & if \quad \dfrac{sup(\phi \rightarrow \psi)}{sup(\phi)} > sup(\psi)/U, \\ = 0 & if \quad \dfrac{sup(\phi \rightarrow \psi)}{sup(\phi)} = sup(\psi)/U, \\ < 0 & if \quad \dfrac{sup(\phi \rightarrow \psi)}{sup(\phi)} > sup(\psi)/U. \end{cases} \quad (2)$$

Over the years, many authors have proposed their own definitions of particular measures that satisfy condition 2 and now the catalogue of confirmation measures proposed in the literature is quite large. Among the most commonly used ones, there are those shown in Table (1).

Crupi et al. [4] have considered the above confirmation measures from the viewpoint of classical deductive logic [2] introducing function v such that for any argument (ϕ, ψ), v assigns it the same positive value (e.g., 1) iff ϕ entails ψ, i.e. $\phi| = \psi$, an equivalent value of opposite sign (e.g., -1) iff ϕ entails the negation of ψ, i.e. $\phi| = \neg\psi$, and value 0 otherwise. The relationship between the logical implication or refutation of ψ by ϕ, and the conditional probability of ψ by ϕ requires that $v(\phi, \psi)$ and $c(\phi \rightarrow \psi)$ should always be of the same sign. However, Crupi et al. [4] also argue that any confirmation measure $c(\phi \rightarrow \psi)$ should also satisfy principle (3):

$$if \quad v(\phi_1, \psi_1) > v(\phi_2, \psi_2), \quad then \quad c(\phi_1 \rightarrow \psi_1) > c(\phi_2 \rightarrow \psi_2). \quad (3)$$

Table 1. Common confirmation measures

$D(\phi \rightarrow \psi) = \dfrac{sup(\phi \rightarrow \psi)}{sup(\phi)} - sup(\psi)$	Carnap [2]
$S(\phi \rightarrow \psi) = \dfrac{sup(\phi \rightarrow \psi)}{sup(\phi)} - \dfrac{sup(\neg\phi \rightarrow \psi)}{sup(\neg\phi)}$	Christensen [3]
$M(\phi \rightarrow \psi) = \dfrac{sup(\phi \rightarrow \psi)}{sup(\psi)} - sup(\phi)$	Mortimer [11]
$N(\phi \rightarrow \psi) = \dfrac{sup(\phi \rightarrow \psi)}{sup(\psi)} - \dfrac{sup(\phi \rightarrow \neg\psi)}{sup(\neg\psi)}$	Nozick [12]
$C(\phi \rightarrow \psi) = \dfrac{sup(\phi \rightarrow \psi)}{|U|} - \dfrac{sup(\phi)sup(\psi)}{|U|}$	Carnap [2]
$R(\phi \rightarrow \psi) = \dfrac{sup(\phi \rightarrow \psi)|U|}{sup(\phi)sup(\psi)} - 1$	Finch [5]
$G(\phi \rightarrow \psi) = 1 - \dfrac{sup(\phi \rightarrow \neg\psi)|U|}{sup(\phi)sup(\neg\psi)}$	Rips [13]

They have proved that neither of the above mentioned confirmation measures satisfies principle 3. However, their further analysis has unveiled a rather simple way to obtain a measure of confirmation that does fulfill this principle from either D, S, M, N, C, R, or G. They have normalized these measures by dividing them by the maximum they obtain in case of confirmation (i.e. when $sup(\phi \rightarrow \psi)/sup(\phi) \geq sup(\psi)/|U|$), and the absolute value of the minimum they obtain in case of disconfirmation (i.e. when $sup(\phi \rightarrow \psi)/sup(\phi) \leq sup(\psi)/|U|$). It has also been shown that those normalized confirmation measures are all equal:

$$D_{norm} = S_{norm} = M_{norm} = N_{norm} = C_{norm} = R_{norm} = G_{norm}. \qquad (4)$$

Crupi et al. have therefore proposed to call them all by one name: **Z-measure**. They have proved that Z, and all confirmation measures equivalent to it, satisfy principle 3. Thus, Z is surely a valuable tool for measuring the confirmation of decision or association rules induced from datasets. Throughout this paper let us consider Z defined as follows:

$$Z(\phi \to \psi) \begin{cases} \dfrac{\dfrac{sup(\phi \to \psi)}{sup(\phi)} - \dfrac{sup(\psi)}{|U|}}{1 - \dfrac{sup(\psi)}{|U|}} & if \quad \dfrac{sup(\phi \to \psi)}{sup(\phi)} \geq \dfrac{sup(\psi)}{|U|}, \\[2ex] \dfrac{\dfrac{sup(\phi \to \psi)}{sup(\phi)} - \dfrac{sup(\psi)}{|U|}}{\dfrac{sup(\psi)}{|U|}} & if \quad \dfrac{sup(\phi \to \psi)}{sup(\phi)} < \dfrac{sup(\psi)}{|U|}. \end{cases} \qquad (5)$$

2.3 Property M of Monotonicity

Greco, Pawlak and Słowiński have proposed in [7] property M of monotonic dependency of an interestingness measure on the number of objects satisfying or not the premise or the conclusion of a rule. Formally, an interestingness measure F satisfies the property M if:

$$F[sup(\phi \to \psi), sup(\neg\phi \to \psi), sup(\phi \to \neg\psi), sup(\neg\phi \to \neg\psi)] \qquad (6)$$

is a function non-decreasing with respect to $sup(\phi \to \psi)$ and $sup(\neg\phi \to \neg\psi)$, and non-increasing with respect to $sup(\neg\phi \to \psi)$ and $sup(\phi \to \neg\psi)$.

The property M with respect to $sup(\phi \to \psi)$ (or, analogously, with respect to $sup(\neg\phi \to \neg\psi)$) means that any evidence in which ϕ and ψ (or, analogously, neither ϕ nor ψ) hold together increases (or at least does not decrease) the credibility of the rule $\phi \to \psi$. On the other hand, the property of monotonicity with respect to $sup(\neg\phi \to \psi)$ (or, analogously, with respect to $sup(\phi \to \neg\psi)$) means that any evidence in which ϕ does not hold and ψ holds (or, analogously, ψ holds and ϕ does not hold) decreases (or at least does not increase) the credibility of the rule $\phi \to \psi$. In order to present the interpretation of property M let us use the following example used by Hempel [8]. Let us consider a rule $\phi \to \psi$:

if x is a raven then x is black.

In this case ϕ stands for the property of being a raven and ψ is the property of being black. If an attractiveness measure $I(\phi \to \psi)$ possesses the property M, then:

- the more black ravens or non-black non-ravens there will be in the dataset, the more credible will become the rule, and thus $I(\phi \to \psi)$ will obtain greater (or at least not smaller) values,
- the more black non-ravens or non-black ravens in the dataset, the less credible will become the rule and thus, $I(\phi \to \psi)$ will obtain smaller (or at least not greater) values.

2.4 Partial Preorder on Rules in Terms of Rule Support and Anti-support

Let us denote by $\preceq_{s\sim a}$ a partial preorder given by the dominance relation on a set X of rules in terms of two interestingness measures *support* and *anti-support*, i.e. given a set of rules X and two rules $r_1, r_2 \in X$, $r_1 \preceq_{s\sim a} r_2$ if and only if

$$sup(r_1) \leq sup(r_2) \quad \wedge \quad anti - sup(r_1) \geq anti - sup(r_2).$$

Recall that a *partial preorder* on a set X is a binary relation R on X that is reflexive and transitive. The partial preorder $\preceq_{s\sim a}$ can be decomposed into its asymmetric part $\prec_{s\sim a}$ and its symmetric part $\sim_{s\sim a}$ in the following manner: given a set of rules X and two rules $r_1, r_2 \in X$, $r_1 \prec_{s\sim a} r_2$ if and only if:

$$\begin{aligned} sup(r_1) \leq sup(r_2) \quad &\wedge \quad anti - sup(r_1) > anti - sup(r_2), \quad or \\ sup(r_1) < sup(r_2) \quad &\wedge \quad anti - sup(r_1) \geq anti - sup(r_2), \end{aligned} \quad (7)$$

moreover, $r_1 \sim_{s\sim a} r_2$ if and only if:

$$sup(r_1) = sup(r_2) \quad \wedge \quad anti - sup(r_1) = anti - sup(r_2). \quad (8)$$

If for a rule $r \in X$ there does not exist any rule $r' \in X$, such that $r \prec_{s\sim a} r'$, then r is said to be *non-dominated* (i.e. *Pareto-optimal*) with respect to support and anti-support. A set of all non-dominated rules with respect to these measures is also referred to as a *support–anti-support Pareto-optimal border*. In other words, it is the set of rules such that there is no other rule having greater support and smaller anti-support.

3 Analysis of Z-Measure with Respect to Property M

For the clarity of presentation, the following notation shall be used from now on:

$$a = sup(\phi \to \psi), \quad b = sup(\neg\phi \to \psi), \quad c = sup(\phi \to \neg\psi), \quad d = sup(\neg\phi \to \neg\psi),$$
$$a + c = sup(\phi), \quad a + b = sup(\psi), \quad b + d = sup(\neg\phi), \quad c + d = sup(\neg\psi),$$
$$a + b + c + d = |U|.$$

We also assume that set U is not empty, so that at least one of a, b, c or d is strictly positive. Moreover, for the sake of simplicity, we assume that any value in the denominator of any ratio is different from zero. In order to prove that a measure has the property M we need to show that it is non-decreasing with respect to a and d, and non-increasing with respect to b and c.

Theorem 1. *Measure Z has the property M.*

Proof. First, let us consider Z in case of confirmation, i.e. when: $sup(\phi \to \psi)/sup(\phi) \geq sup(\psi)/|U|$:

$$Z = \frac{\frac{a}{a+c} - \frac{a+b}{a+b+c+d}}{\frac{c+d}{a+b+c+d}}. \quad (9)$$

Through simple mathematical transformations we obtain:

$$Z = \frac{ad - bc}{(a + c)(c + d)}. \tag{10}$$

Let us verify if Z is non-decreasing with respect to a, i.e. if an increase of a by $\Delta > 0$ will not result in decrease of Z. Simple algebraic transformations show that:

$$\frac{(a + \Delta)d - bc}{(a + \Delta + c)(c + d)} - \frac{ad - bc}{(a + c)(c + d)} = \frac{cd\Delta + bc\Delta}{(a + \Delta + c)(c + d)(a + c)} \geq 0. \tag{11}$$

Thus, Z (in case of confirmation) is non-decreasing with respect to a.

Clearly, Z is also non-increasing with respect to b, as increase of b by $\Delta > 0$ will result in decrease of the numerator of (10) and therefore in decrease of Z.

Now, let us verify if Z is non-increasing with respect to c, i.e. if an increase of c by $\Delta > 0$ will not result in increase of Z. Simple algebraic transformations show that:

$$\frac{ad - b(c + \Delta)}{(a + c + \Delta)(c + \Delta + d)} - \frac{ad - bc}{(a + c)(c + d)} =$$
$$= \frac{(c\Delta + \Delta^2)(bc - ad) - ad(b\Delta + c\Delta + a\Delta + d\Delta)}{(a + c + \Delta)(c + \Delta + d)(a + c)(c + d)}. \tag{12}$$

Let us observe that: $c\Delta + \Delta^2 > 0$, $ad(b\Delta + c\Delta + a\Delta + d\Delta) > 0$, and $(bc - ad) < 0$ because we consider the case of confirmation. Thus, the numerator of (12) is negative. Since the denominator is positive, we can conclude that Z (in case of confirmation) is non-increasing with respect to c.

Finally, let us verify if Z is non-decreasing with respect to d, i.e. if an increase of d by $\Delta > 0$ will not result in decrease of Z. Simple algebraic transformations show that:

$$\frac{a(d + \Delta) - bc}{(a + c)(c + d + \Delta)} - \frac{ad - bc}{(a + c)(c + d)} = \frac{ac\Delta + bc\Delta}{(a + c)(c + d)(c + d + \Delta)} > 0. \tag{13}$$

Thus, Z (in case of confirmation) is non-decreasing with respect to d.

Since all four conditions are satisfied, the hypothesis that Z measure in case of confirmation has the property M is true. The proof that in case of disconfirmation Z has the property M is analogous.

4 Practical Application of the Results

The approach to evaluation of a set of rules with the same conclusion in terms of two interestingness measures being rule support and anti-support was proposed and presented in detail in [1]. The idea of combining those two dimensions came as a result of looking for a set of rules that would include all rules optimal with respect to any confirmation measure with the desirable property M.

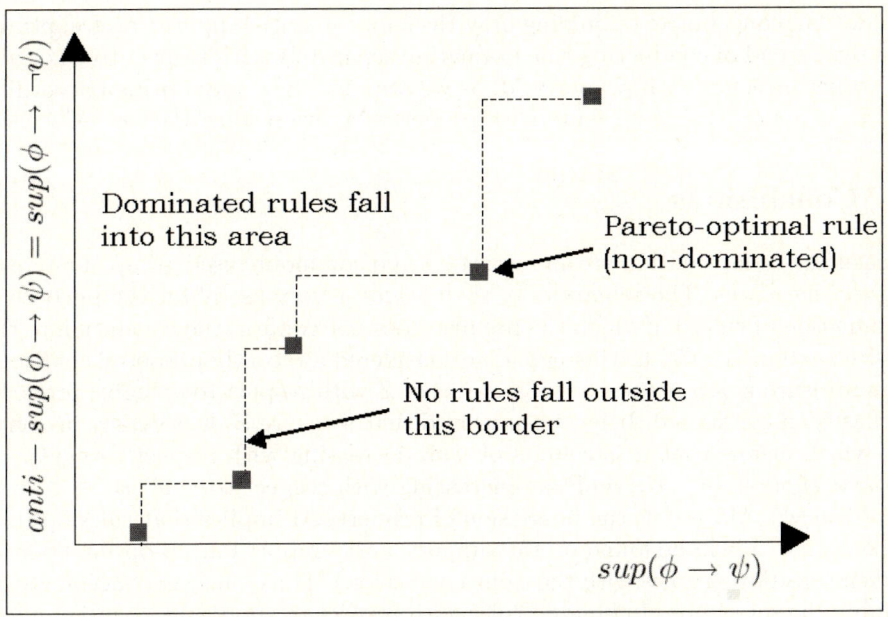

Fig. 1. Support–anti-support Pareto-optimal border

Theorem 2. *[1] When considering rules with the same conclusion, rules that are optimal with respect to any interestingness measure that has the property M must reside on the support–anti-support Pareto-optimal border.*

It means that the best rules according to any of confirmation measures with M are in the set of non-dominated rules with respect to support–anti-support. This valuable result unveils some relationships between different interestingness measures. Moreover, it allows to identify a set of rules containing most interesting (optimal) rules according to any interestingness measures with the property M simply by solving an optimized rule mining problem with respect to rule support and anti-support.

As we have proved, measure Z satisfies the property M. This result allows us to conclude that this interestingness measure is monotonically dependent on the number of objects satisfying both the premise and conclusion of the rule (or neither the premise nor the conclusion) and anti-monotonically dependent on the number of objects satisfying only the premise or only the rule's conclusion. Moreover, possession of property M means that rules optimal with respect to Z reside on the Pareto-optimal border with respect to support and anti-support (when considering rules with the same conclusion).

It is a very practical result as it allows potential efficiency gains:

– rules optimal with respect to Z can be mined from the support–anti-support Pareto-optimal set instead of searching the set of all rules,

– we can concentrate on mining only the support–anti-support Pareto-optimal set instead of conducting rule evaluation separately with respect to Z, or any other measure with property M, as we are sure that rules optimal according to Z, or any other measure with property M, are in that Pareto set.

5 Conclusions

Measures of confirmation are an important and commonly used group of interestingness measures. The semantics of their scales is very useful for the purpose of elimination of rules for which the premise does not confirm the conclusion to the desired extent [14,15]. In this paper we considered a recently proposed confirmation measure Z [4]. A theoretical analysis of Z with respect to valuable property M has been conducted. It has been proved that measure Z does satisfy property M, which means that it is a function non-decreasing with respect to $sup(\phi \to \psi)$ and $sup(\neg\phi \to \neg\psi)$, and non-increasing with respect to $sup(\neg\phi \to \psi)$ and $sup(\phi\, to \neg\psi)$. Moreover, the possession of property M implies that rules optimal according to Z will be found on the support–anti-support Pareto-optimal border (when considering rules with the same conclusion). Thus, one can concentrate on mining the set of non-dominated rules with respect to support and anti-support and be sure to obtain in that set all rules that are optimal with respect to any measure with the property M, which includes measure Z.

References

1. Brzezińska, I., Greco, S., Słowiński, R.: Mining Pareto-optimal rules with respect to support and anti-support. Engineering Applications of Artificial Intelligence 20(5), 587–600 (2007)
2. Carnap, R.: Logical Foundations of Probability, 2nd edn. University of Chicago Press, Chicago (1962)
3. Christensen, D.: Measuring confirmation. Journal of Philosophy 96, 437–461 (1999)
4. Crupi, V., Tentori, K., Gonzalez, M.: On Bayesian measures of evidential support: Theoretical and empirical issues. Philosophy of Science (to appear)
5. Finch, H.A.: Confirming Power of Observations Metricized for Decisions among Hypotheses. Philosophy of Science 27, 293-307, 391–404 (1999)
6. Fitelson, B.: Studies in Bayesian Confirmation Theory. Ph.D. Thesis, University of Wisconsin, Madison (2001)
7. Greco, S., Pawlak, Z., Słowiński, R.: Can Bayesian confirmation measures be useful for rough set decision rules? Engineering Applications of Artificial Intelligence 17, 345–361 (2004)
8. Hempel, C.G.: Studies in the logic of confirmation (I). Mind 54, 1–26 (1945)
9. Hilderman, R., Hamilton, H.: Knowledge Discovery and Measures of Interest. Kluwer Academic Publishers, Dordrecht (2001)
10. Kemeny, J., Oppenheim, P.: Degrees of factual support. Philosophy of Science 19, 307–324 (1952)
11. Mortimer, H.: The Logic of Induction. Prentice-Hall, Paramus (1988)
12. Nozick, R.: Philosophical Explanations. Clarendon Press Oxford, Oxford (1981)
13. Rips, L.J.: Two Kinds of Reasoning. Psychological Science 12, 129–134 (2001)

14. Słowiński, R., Brzezińska, I., Greco, S.: Application of Bayesian Confirmation Measures for Mining Rules from Support-Confidence Pareto-Optimal Set. In: Rutkowski, L., Tadeusiewicz, R., Zadeh, L.A., Żurada, J. (eds.) ICAISC 2006. LNCS (LNAI), vol. 4029, pp. 1018–1026. Springer, Heidelberg (2006)
15. Słowiński, R., Szczęch, I., Urbanowicz, M., Greco, S.: Mining Association Rules with Respect to Support and Anti-support-Experimental Results. In: Kryszkiewicz, M., Peters, J.F., Rybinski, H., Skowron, A. (eds.) RSEISP 2007. LNCS (LNAI), vol. 4585, pp. 534–542. Springer, Heidelberg (2007)

A Comparison of Methods for Learning of Highly Non-separable Problems

Marek Grochowski and Włodzisław Duch

Department of Informatics, Nicolaus Copernicus University, Toruń, Poland
grochu@is.umk.pl, Google: W. Duch

Abstract. Learning in cases that are almost linearly separable is easy, but for highly non-separable problems all standard machine learning methods fail. Many strategies to build adaptive systems are based on the "divide-and-conquer" principle. Constructive neural network architectures with novel training methods allow to overcome some drawbacks of standard backpropagation MLP networks. They are able to handle complex multidimensional problems in reasonable time, creating models with small number of neurons. In this paper a comparison of our new constructive *c3sep* algorithm based on *k*-separability idea with several sequential constructive learning methods is reported. Tests have been performed on parity function, 3 artificial Monks problems, and a few benchmark problems. Simple and accurate solutions have been discovered using *c3sep* algorithm even in highly non-separable cases.

1 Introduction

Multi-layer neural networks, basis set expansion networks, kernel methods and other computational intelligence methods are great tools that enable learning from data [1]. Recently the class of non-separable problems has been characterized more precisely using k-separability index [2,3] that measures the minimum number of intervals needed to separate pure clusters of data in a single projection. This index divides problems into different classes of complexity. Problems with $k = 2$ are linear separable (a single hyperplane separates data from two classes), but convergence and accuracy of standard methods decreases quickly with growing k-separability index. Parity function for n bit binary strings requires $k = n + 1$ clusters and thus is highly non-separable. Each vector is surrounded by the vectors from the opposite class, and therefore methods based on similarity (including kernel methods) do not generalize and fail, while methods based on discriminating hyperplanes may in principle solve such problems, but require $O(n^2)$ parameters and do not converge easily. Universal approximation does not entail good generalization.

Many statistical and machine learning methods fail on problems with inherent complex logic. Training feedforward networks with backpropagation-type algorithms requires specification of correct architecture for a given task. Highly non-separable problems even for relatively modest number of dimensions lead to complex network architectures. Learning in such cases suffers from high computational costs and problems with convergence. Even for a simple XOR problem convergence fails fairly often. Constructive neural techniques allow to overcome some of these drawbacks. Furthermore, simplest models should be sought to avoid overfitting and ensure good generalization. Often quite complex data may be described using a few simple rules [4], therefore using an appropriate constructive strategy should create network with small number of neurons and clear interpretation.

Several sequential constructive learning methods have been proposed [5,6,7,8,9,10,11], based on computational geometry ideas. Some of them work only for binary problems and therefore pre-processing of data based on Gray coding is necessary. However, it is interesting to see how well these algorithms will work on Boolean functions and real benchmark problems, and how do they compare to our *c3sep* algorithm. Next section gives short description of the most promising constructive learning methods, and the third section contains comparison made on artificial as well as on real benchmark problems.

2 Algorithms

All algorithms reported below may be viewed as a realization of general sequential constructive method described by Muselli in [5]. In each step of these algorithms searching for best separation between maximum possible number of vectors of the same class from the rest of the training samples is performed. Consider a separation with a hyperplane created by a threshold neuron:

$$\varphi(\boldsymbol{x}) = \begin{cases} +1 \text{ if } \boldsymbol{wx} - w_0 \geq 0 \\ -1 \text{ otherwise} \end{cases} \quad (1)$$

Let's denote by $\mathcal{Q}^+$ a set of all vectors from a single selected class, by $\mathcal{Q}^-$ the remaining training samples, and by $\mathcal{R}$ a set of training vectors for which the neuron performing separation gives an output equal to $+1$. Then each step of this algorithm could be seen as searching for a subset $\mathcal{R} \in \mathcal{Q}^+$ with maximum number of elements. After each step vectors from the set $\mathcal{R}$ are removed from the training dataset, and if some unclassified patterns still remain the search procedure is repeated. After a finite number of iterations this procedure leads to a construction of neural network that classifies all training vectors (unless there are conflicting cases, i.e. identical vectors with different labels), where each separating hyperplane corresponds to a hidden neuron in the final neural network. Weights in the output layer do not take part in the learning phase and their values can be determined from a simple algebraic equation proposed in [5]. The main difference between different learning algorithms described below lies in the search method employed for creating hidden neurons. Most of these algorithms were designed to handle datasets with binary attributes. Some of them (Irregular Partitioning, Carve, Target Switch) have extended versions for the general case of real valued features. Two methods (Sequential Window and *c3sep*) employ neurons with window-like activation function

$$\tilde{M}_i(\boldsymbol{x}; \boldsymbol{w}, a, b) = \begin{cases} 1 \text{ if } \boldsymbol{w} \cdot \boldsymbol{x} \in [a, b] \\ 0 \text{ if } \boldsymbol{w} \cdot \boldsymbol{x} \notin [a, b] \end{cases} \quad (2)$$

while the rest of them use threshold neurons, Eq. (1).

2.1 Irregular Partitioning Algorithm

This algorithm [10] starts with an empty set $\mathcal{R}$ and in each iteration one vector from $\mathcal{Q}^+$ is moved to $\mathcal{R}$. A new pattern is kept in $\mathcal{R}$ only if the $\mathcal{R}$ and $\mathcal{Q}^-$ sets are linearly separable. Quality of obtained results and computational cost of this algorithm depends on the type of methods employed to verify existence of the separating hyperplane. Although simple perceptron learning may be used, Muselli proposed to use the Thermal Perceptron Learning Rule (TPLR) [5], because it has high convergence speed. In this paper linear programming techniques are used to find an optimal solution.

2.2 Carve Algorithm

The Carve algorithm considers a convex hull generated by the elements of $\mathcal{Q}^-$. Searching for the best linear separation of points from class $\mathcal{Q}^+$ is done here by a proper traversing and rotating of hyperplanes that pass through boundary points of the convex hull. Using the general dimension gift wrapping method a near optimal solution may be found in polynomial time [6].

2.3 Target Switch Algorithm

In every step of the Target Switch Algorithm, proposed by Zollner *et al.* [11] and extended for real valued features by Campbell and Vicente [8], search for the threshold neuron that performs best separation between the set of vectors $\mathcal{R}$ from a given class $\mathcal{Q}^+$ and the rest of samples $\mathcal{Q}^-$ may be made using any algorithm that searches for linear separability (e.g. perceptron learning, TPLR or some version of linear discrimination method). If the current solution misclassifies some vectors from the set $\mathcal{Q}^-$ the vector from $\mathcal{Q}^+$ nearest to the misclassified vector from $\mathcal{Q}^-$, with the largest distance to the current hyperplane, is moved to $\mathcal{Q}^-$, and the learning process is repeated. This procedure leads to a desired solution in a finite number of iterations, leaving on one side of hyperplanes only vectors from a single class.

2.4 Oil Spot Algorithm

Data with the binary-valued features may be represented as n-dimensional hypercube. There are two conditions that must be satisfied to separate two sets $\mathcal{R}$ and $\mathcal{Q}^-$ with a hyperplane: (1) vectors in $\mathcal{R}$ are nodes of connected subgraph, and (2) two parallel edges connecting vectors from $\mathcal{R}$ and $\mathcal{Q}^-$ are always with the same orientation. This concept is used by Oil Spot algorithm to search for optimal separation of binary data with a hyperplane [9].

2.5 Sequential Window Learning

This method uses neurons with window-like transfer function of the form

$$\varphi(\boldsymbol{x}) = \begin{cases} 1 & \text{if} \quad |\boldsymbol{w}\boldsymbol{x} - w_0| \leq \delta \\ -1 & \text{otherwise} \end{cases} \tag{3}$$

For this type of neurons fast and efficient learning algorithm for binary valued data based on solution of a system of algebraic equations has been proposed by Muselli [7]. This algorithm starts with two patterns from $\mathcal{Q}^+$ in $\mathcal{R}$ and incrementally adds new vectors from $\mathcal{Q}^+$, maximizing the number of correctly classified vectors by the window-like neuron.

2.6 *c3sep* - Constructive 3-Separability Learning

The *c3sep* algorithm [3] uses traditional error minimization algorithm to train neurons with soft-windowed transfer functions, e.g. bicentral functions [12]:

$$\tilde{\varphi}(\boldsymbol{x}) = \sigma(\beta(\boldsymbol{w}\boldsymbol{x} - w_0 - \delta))(1 - \sigma(\beta(\boldsymbol{w}\boldsymbol{x} - w_0 + \delta))) \tag{4}$$

Sharp decision boundaries, like those in Eq. (2), are obtained by putting a large value of the slope β at the end of the learning procedure. Function (4) separates a cluster of vectors projected on a line defined by $\boldsymbol{w}$ with boundaries defined by parameters w_0 and δ. This transformation splits input space with two parallel hyperplanes into 3 disjoint subsets. A large and pure cluster $\mathcal{R}$ of vectors from one class is generated by minimization of the following error function:

$$E(\boldsymbol{x}) = \frac{1}{2}\sum_{\boldsymbol{x}}(y(\boldsymbol{x}) - c(\boldsymbol{x}))^2 + \lambda_1 \sum_{\boldsymbol{x}}(1 - c(\boldsymbol{x}))y(\boldsymbol{x}) - \lambda_2 \sum_{\boldsymbol{x}} c(\boldsymbol{x})y(\boldsymbol{x}) \tag{5}$$

where $c(\boldsymbol{x}) = \{0, 1\}$ denotes the label of a vector $\boldsymbol{x}$, and $y(\boldsymbol{x}) = \sum \tilde{\varphi}_i(\boldsymbol{x})$ is the actual output of a network. All weights in the output layer are fixed to 1. First term is the standard mean square error (MSE) measure, second term with λ_1 (penalty factor) increases the total error for vectors x_i from the $c(x_i) = 0$ class that fall into cluster of vectors from the opposite class 1 (it is the penalty for "contaminated" clusters), and the third term with λ_2 (reward factor) decreases the value of total error for every vector x_i from class 1 that was correctly placed inside created clusters (it is the reward for large clusters) [3]. To speed up training procedure, after each learning phase, samples that were correctly handled by previous nodes may be removed from the training data. However, larger clusters (and thus better generalization) may be obtained if these vectors are not removed and different neurons that respond to clusters of data sharing some vectors are created. The *c3sep* algorithm defines a simple feedforward network with window-line neurons in the hidden layer, and every node trying to find best 3-separable solution in the input space (projection on the line that create large cluster of vectors from one class between two clusters of vectors from the opposite class). This is the simplest extension of linear separability; other goals of learning have been discussed in [3,13].

3 Results

Results for three types of problems are reported here: Boolean functions offering systematically increasing complexity, artificial symbolic benchmark problems, and real benchmark problems.

3.1 Boolean Functions

All algorithms described above have been applied to several artificial Boolean functions. In this test no generalization is required since only the complexity and computational cost of data models that each method creates are compared, estimating the capacity of different algorithms to discover simplest correct solutions. Fig. 1 presents results of learning of constructive methods applied to the parity problems from 2 bits (equivalent to the XOR problem) to 10 bits, and Fig. 2 shows the results of learning randomly selected Boolean functions with $P(C(\boldsymbol{x}) = 1) = 0.5$ (the same function is used to train all algorithms). The most likely random function for n bits has k-separability index around $n/2$. The n dimensional parity problem can be solved by a two-layer neural network with n threshold neurons or $(n+1)/2$ window-like neurons in the hidden layer. It may also be solved by one neuron with n thresholds [2]. All results are averaged over 10 trials.

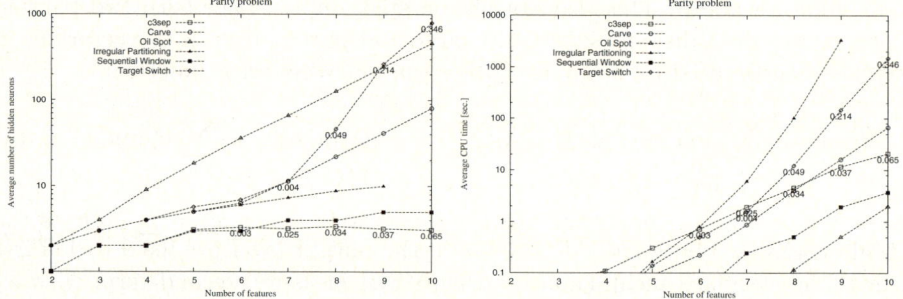

Fig. 1. Number of hidden units created and time consumed during learning of parity problems. Each result is averaged over 10 trials.

Most algorithms are able to learn all parity patterns without mistake. The *c3sep* network, using a stochastic algorithm made in some runs small errors, although repeating the training a few times perfect solution is easily found. The target switch algorithm for problems with dimension larger then $n = 7$ generate solution with rapidly growing training errors. Values of the error are placed in corresponding points of Fig. 1 and Fig. 2. Sequential window learning and irregular partitioning algorithms were able do obtain optimal solution for all dimensions.

Learning of random Boolean function is much more difficult and upper bound for the number of neurons needed for solving of this kind of functions is not known. Learning of irregular partitioning with linear programming for problems

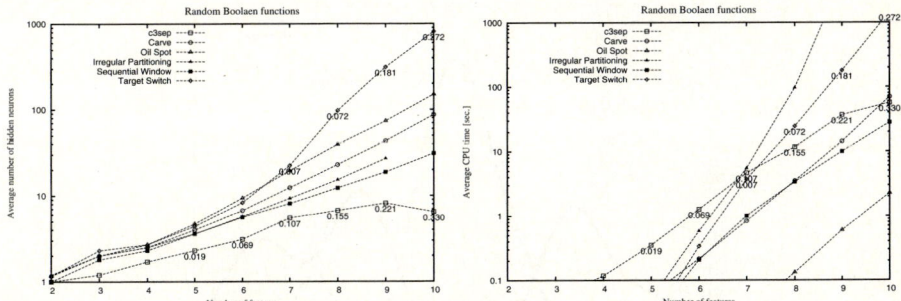

Fig. 2. Number of hidden units created and time consumed during learning of random Boolean functions. Each result is averaged over 10 trials.

with large dimension is time consuming. Oil spot algorithm is the fastest algorithm but leads to networks with high complexity. Sequential window learning gave solutions with a small number of neurons and low computational cost. The Carve algorithm can also handle all patterns in a reasonable time and with a small number of neurons. The *c3sep* network was able to create smallest architectures but the average times of computations are longer than needed by most other algorithms. This network provides near optimal solution, as not all patterns were correctly classified; it avoids small clusters that may reduce classification error but are not likely to generalize well in crossvalidation tests. Algorithms capable of exact learning of every example by creating separate node for single vectors are rarely useful.

4 Benchmark Problems

Monk's Problems. Monk's problems are three symbolic classification problems designed to test machine learning algorithms [14]. 6 symbolic attributes may be combined in 432 possible ways, and different logical rules are used to divide these cases into two groups, first group containing "monks", and the second other objects. The Monk 3 problem is intentionally corrupted by noise, therefore an optimal accuracy is here 94%. The other two problems have perfect logical solutions.

Sequential learning algorithm and the Oil Spot algorithm require binary valued features therefore proper transformation of the symbolic data must be applied first. Resulting binary dataset contains 15 inputs where a separate binary feature is associated with presence of each symbolic value. For binary valued features Hamming clustering [15] can be applied to reduce the size of training set and reduce the time of computations. This algorithm iteratively clusters together all binary strings from the same class that are close in the Hamming distance.

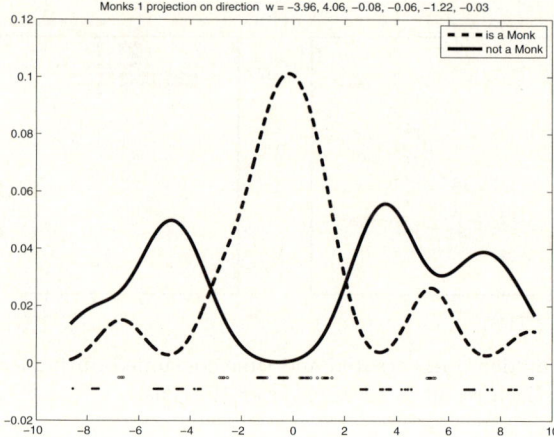

Fig. 3. Probability density for two classes projected on first direction found for the Monk1 problem

All algorithms have been trained on the 3 Monk's problems 30 times, and averaged results for generalization error, the size of created network and computational costs of training are reported in Tables 1, 2 and 3, respectively. Sequential window learning algorithm and *c3sep* network were able to achieve great accuracy with very small number of neurons. With binary features only a single neuron is sufficient to solve the problem in each case (only for Monk1 sometimes 2 neurons are created), showing that in the binary representation all these problems are 3-separable. For Monk3 problem (with binary features) the *c3sep* is the only algorithm that achieves optimal accuracy. Sequential Window algorithm produces also quite simple networks and finds very good solution with Hamming clustering, staying behind *c3sep* in both accuracy and complexity only for the Monk3 problem. The use of the Hamming clustering in most cases increased generalization powers and reduced the number of neurons in the hidden layer. CPU times for all algorithms are comparable.

An example of a projection on the first direction found by the *c3sep* algorithm for the Monk1 problem shows (Fig. 3) a large cluster in the center that can be separated using a window function. If the direction of the projection is binarized logical rules for classification may be extracted.

Benchmark Problems with Crossvalidation Tests. Boolean problems are rather distinct from typical learning problems. Several tests have been performed on a few benchmark datasets from the UCI repository [16]. The Iris dataset is perhaps the most widely used simple problem, with 3 types of Iris flowers described by 4 real valued attributes. A Glass identification problem has 9 real valued features with patterns divided into float-processed and non float-processed piece of glass. United States congressional voting record database, denoted here as Voting0 dataset, contains 12 features that record decisions of congressmans who belong to a democratic or republican party. The Voting1 dataset has been

Table 1. Accuracy of neural networks for Monk's problems. Each result is averaged over 30 trials.

	Monk's 1	Monk's 2	Monk's 3
Carve	93.93 ± 2.63	82.42 ± 2.10	88.50 ± 2.47
Irregular Partitioning	87.93 ± 4.19	83.94 ± 1.65	87.99 ± 2.06
Target Switch	93.46 ± 2.57	65.80 ± 2.64	80.38 ± 1.67
c3sep	99.69 ± 1.16	79.78 ± 1.32	87.84 ± 2.36
binary features			
Carve	86.94 ± 3.79	87.14 ± 2.89	85.46 ± 3.09
Oil Spot	86.00 ± 1.63	84.32 ± 0.99	71.84 ± 1.16
Irregular Partitioning	100.00 ± 0.00	98.15 ± 0.00	91.44 ± 1.58
Sequential Window	95.18 ± 5.78	100.00 ± 0.00	84.68 ± 5.82
Target Switch	94.84 ± 3.18	91.23 ± 5.29	90.73 ± 1.97
c3sep	92.49 ± 9.12	100.00 ± 0.00	94.41 ± 0.31
with Hamming clustering			
Carve	100.00 ± 0.00	97.50 ± 0.57	90.29 ± 2.85
Irregular Partitioning	83.84 ± 3.74	99.92 ± 0.46	92.58 ± 1.33
Oil Spot	100.00 ± 0.00	95.07 ± 0.52	90.22 ± 1.15
Sequential Window	100.00 ± 0.00	100.00 ± 0.00	91.72 ± 1.66
Target Switch	100.00 ± 0.00	98.09 ± 0.16	90.62 ± 1.00
c3sep	99.86 ± 0.76	99.38 ± 1.03	91.40 ± 2.54

Table 2. Number of hidden neurons created for Monk's problems. Each result is averaged over 30 trials.

	Monk's 1	Monk's 2	Monk's 3
Carve	3.30 ± 0.70	10.83 ± 0.99	4.90 ± 0.61
Irregular Partitioning	4.63 ± 0.76	8.77 ± 0.68	4.20 ± 0.41
Target Switch	10.67 ± 2.59	75.73 ± 4.83	12.33 ± 1.83
c3sep	2.43 ± 0.63	2.90 ± 1.94	3.87 ± 0.51
binary features			
Carve	5.30 ± 0.60	8.17 ± 0.99	5.83 ± 0.75
Oil Spot	18.67 ± 2.34	34.37 ± 1.77	31.53 ± 2.29
Irregular Partitioning	3.60 ± 0.62	2.03 ± 0.18	2.00 ± 0.00
Sequential Window	2.70 ± 0.70	1.20 ± 0.41	4.47 ± 0.51
Target Switch	3.73 ± 0.91	12.97 ± 5.04	4.03 ± 0.56
c3sep	1.63 ± 0.49	1.00 ± 0.00	1.00 ± 0.00
with Hamming clustering			
Carve	3.00 ± 0.00	5.00 ± 0.00	5.67 ± 0.61
Oil Spot	3.20 ± 0.41	5.53 ± 0.51	9.20 ± 1.06
Irregular Partitioning	3.00 ± 0.00	5.00 ± 0.00	4.37 ± 0.61
Sequential Window	2.00 ± 0.00	1.00 ± 0.00	5.10 ± 0.96
Target Switch	3.00 ± 0.00	5.00 ± 0.00	5.53 ± 0.63
c3sep	2.00 ± 0.00	1.67 ± 0.88	3.23 ± 0.43

Table 3. CPU time consumed during construction of neural networks solving Monk's problems. Each result is averaged over 30 trials.

	Monk's 1	Monk's 2	Monk's 3
Carve	0.18 ± 0.02	0.63 ± 0.06	0.21 ± 0.02
Irregular Partitioning	2.04 ± 0.16	8.94 ± 1.64	2.00 ± 0.19
Target Switch	0.53 ± 0.15	6.71 ± 0.66	0.53 ± 0.14
c3sep	0.78 ± 0.19	3.54 ± 1.81	2.36 ± 0.35
binary features			
Carve	0.72 ± 0.09	1.41 ± 0.22	0.69 ± 0.07
Oil Spot	0.18 ± 0.02	0.39 ± 0.06	0.08 ± 0.01
Irregular Partitioning	3.25 ± 0.42	6.32 ± 1.53	2.39 ± 0.35
Sequential Window	1.06 ± 0.34	0.25 ± 0.27	1.72 ± 0.22
Target Switch	0.21 ± 0.08	3.97 ± 1.46	0.06 ± 0.01
c3sep	1.19 ± 0.35	0.73 ± 0.13	1.33 ± 0.12
with Hamming clustering			
Carve	0.01 ± 0.00	0.11 ± 0.01	0.14 ± 0.02
Oil Spot	0.01 ± 0.00	0.02 ± 0.00	0.03 ± 0.00
Irregular Partitioning	0.01 ± 0.00	0.12 ± 0.01	0.11 ± 0.03
Sequential Window	0.01 ± 0.00	0.01 ± 0.00	0.05 ± 0.02
Target Switch	0.01 ± 0.00	0.05 ± 0.00	0.03 ± 0.01
c3sep	0.15 ± 0.06	0.31 ± 0.15	0.45 ± 0.06

Table 4. 30x3 CV test accuracy

	Iris	Glass	Voting0	Voting1
Carve	90.18 ± 1.58	74.17 ± 3.28	93.24 ± 1.00	87.45 ± 1.37
Irregular Partitioning	90.98 ± 2.29	72.29 ± 4.39	93.41 ± 1.13	86.96 ± 1.43
Target Switch	65.45 ± 5.05	46.76 ± 0.91	94.64 ± 0.63	88.13 ± 1.47
c3sep	95.40 ± 1.30	70.68 ± 2.97	94.38 ± 0.72	90.42 ± 1.15
binary features				
Carve	71.84 ± 3.46	62.08 ± 4.55	91.79 ± 1.22	86.77 ± 1.43
Oil Spot	75.16 ± 2.86	66.05 ± 2.41	90.93 ± 0.90	86.68 ± 1.47
Irregular Partitioning	75.53 ± 3.20	62.38 ± 3.66	92.73 ± 1.19	86.79 ± 2.20
Target Switch	84.93 ± 3.28	71.69 ± 3.42	94.66 ± 0.69	88.36 ± 0.98
c3sep	75.58 ± 6.15	60.92 ± 4.47	94.50 ± 0.89	89.78 ± 1.26
with Hamming clustering				
Carve	79.93 ± 3.16	61.92 ± 3.77	93.40 ± 0.69	86.40 ± 1.10
Oil Spot	80.84 ± 2.46	61.83 ± 3.46	92.86 ± 1.09	85.33 ± 1.41
Irregular Partitioning	79.56 ± 4.45	62.38 ± 4.89	93.05 ± 1.10	85.84 ± 1.27
Sequential Window	77.36 ± 4.71	54.18 ± 3.50	91.40 ± 1.21	82.28 ± 1.82
Target Switch	81.09 ± 3.70	62.54 ± 3.57	93.31 ± 0.79	87.02 ± 1.42
c3sep	78.58 ± 4.74	58.51 ± 3.90	83.48 ± 13.39	76.35 ± 11.62

Table 5. Average number of hidden neurons generated during 30x3 CV test

	Iris	Glass	Voting0	Voting1
Carve	5.72 ± 0.46	7.00 ± 0.50	4.99 ± 0.39	8.34 ± 0.45
Irregular Partitioning	5.49 ± 0.53	4.69 ± 0.26	2.04 ± 0.21	3.48 ± 0.30
Target Switch	22.76 ± 2.17	55.49 ± 2.38	3.69 ± 0.29	9.22 ± 0.85
c3sep	3.00 ± 0.00	1.14 ± 0.26	1.00 ± 0.00	1.02 ± 0.12
binary features				
Carve	8.02 ± 0.52	6.79 ± 0.26	5.56 ± 0.32	8.59 ± 0.46
Oil Spot	27.78 ± 1.41	21.54 ± 1.80	22.76 ± 1.39	37.32 ± 2.32
Irregular Partitioning	3.00 ± 0.00	1.00 ± 0.00	0.99 ± 0.06	2.50 ± 0.30
Target Switch	3.07 ± 0.14	1.72 ± 0.25	3.20 ± 0.26	7.46 ± 0.48
c3sep	3.30 ± 0.35	1.03 ± 0.10	1.00 ± 0.00	1.00 ± 0.00
with Hamming clustering				
Carve	7.81 ± 0.61	7.28 ± 0.46	5.07 ± 0.37	8.88 ± 0.50
Oil Spot	10.94 ± 1.13	12.17 ± 1.17	7.70 ± 0.74	15.31 ± 1.38
Irregular Partitioning	6.68 ± 0.73	5.39 ± 0.40	4.00 ± 0.48	6.19 ± 0.43
Sequential Window	9.90 ± 0.68	5.54 ± 0.40	5.46 ± 0.50	7.10 ± 0.61
Target Switch	7.02 ± 0.54	8.10 ± 0.63	4.58 ± 0.47	8.98 ± 0.57
c3sep	5.37 ± 0.79	3.38 ± 0.38	3.29 ± 0.40	4.97 ± 0.36

Table 6. Average of CPU time consumed during 30x3 CV test

	Iris	Glass	Voting0	Voting1
Carve	0.07 ± 0.00	0.40 ± 0.02	1.08 ± 0.04	1.85 ± 0.08
Irregular Partitioning	0.64 ± 0.08	4.86 ± 0.57	122.66 ± 24.86	242.48 ± 32.82
Target Switch	0.37 ± 0.05	7.12 ± 1.46	0.08 ± 0.02	1.86 ± 0.37
c3sep	2.05 ± 0.21	1.27 ± 0.17	3.49 ± 0.31	4.58 ± 0.34
binary features				
Carve	1.53 ± 0.09	4.82 ± 0.21	4.22 ± 0.20	5.42 ± 0.24
Oil Spot	0.06 ± 0.00	0.68 ± 0.06	1.74 ± 0.23	2.20 ± 0.40
Irregular Partitioning	0.39 ± 0.02	1.38 ± 0.04	49.34 ± 7.14	474.56 ± 83.94
Target Switch	0.00 ± 0.00	0.02 ± 0.00	0.15 ± 0.04	2.20 ± 0.45
c3sep	2.61 ± 0.34	4.07 ± 0.51	7.28 ± 0.96	7.98 ± 0.18
with Hamming clustering				
Carve	0.28 ± 0.03	0.49 ± 0.04	0.52 ± 0.04	1.47 ± 0.09
Oil Spot	0.04 ± 0.00	0.12 ± 0.01	0.29 ± 0.03	0.62 ± 0.05
Irregular Partitioning	0.51 ± 0.22	3.54 ± 0.21	0.81 ± 0.40	6.59 ± 1.36
Sequential Window	0.29 ± 0.05	1.11 ± 0.16	0.60 ± 0.13	2.52 ± 0.39
Target Switch	0.06 ± 0.02	0.76 ± 0.13	0.29 ± 0.04	0.98 ± 0.12
c3sep	2.29 ± 6.73	0.77 ± 0.07	0.68 ± 0.11	1.78 ± 0.25

obtained from the Voting0 by removing the most informative feature. Each input can assume values: yea, nay or missing.

As in the case of Monk's problems some algorithms require additional transformations of datasets. Real valued features were transformed to binary by employing Gray coding. Resulting Iris and Glass dataset in binary representation have 22 and 79 features, respectively. For the three valued input of Voting dataset the same transformation as in the Monk's problems has been adopted. The corresponding binary data contain 48 features for Voting0 and 45 for Voting1 datasets.

Accuracy obtained after averaging 30 repetitions of the 3-fold crossvalidation test is reported in Tab. 4. The average number of neurons and average CPU time consumed during computations is reported in Tab. 5 and Tab.6. In all these tests in most cases *c3sep* network gave very good accuracy and low variance with very small number of neurons created in the hidden layer. Most algorithms that can handle real features suffer from data binarization, particularly in the case of Iris and Glass where all features in the original data are real valued.

5 Conclusions

Constructive algorithms that may learn functions with inherently complex logics have been collected and compared on learning of Boolean functions, symbolic and real benchmark problems. These algorithms are rarely used but certainly worth more detailed study. Sequential Window algorithm works particularly well for binary problems but is not competitive for benchmark problems with real-valued features. Note that on the Boolean problems used here for testing all the off-the shelf methods that are found in data mining packages (decision trees, MLPs, SVMs and nearest neighbor methods) will completely fail. Therefore investigation of methods that can handle such data is very important.

The key element is to use non-local projections $\boldsymbol{w} \cdot \boldsymbol{x}$ that do not lead to separable distribution of data, followed by a window that covers the largest pure cluster that can be surrounded by vectors from other classes. Constructive neural networks are well suited to use this type of projections. The Support Vector Neural Training constructive algorithm [17] may also be used here; during training all vectors that are too far from the current decision border, that is excite the sigmoidal neuron too strongly or too weakly, are removed, so the final training involves only the vectors in some margin around the decision border. Another alternative algorithm may be based on modified oblique decision tree (such as OC1 [18]), with each node defining $\boldsymbol{w} \cdot \boldsymbol{x}$ projection, followed by univariate decision tree separating the data along the projection line. Several improvements of the *c3sep* algorithm are also possible, for example by using better search algorithms, second order convergent methods etc. A significant progress in handling problems that require complex logic should be expected along these lines. More tests on ambitious bioinformatics and text analysis problems using such algorithms are needed to show their real potential.

Acknowledgement. we are grateful to dr Marco Muselli (Genova) for his implementation of sequential learning algorithms.

References

1. Cherkassky, V., Mulier, F.: Learning from data. In: Adaptive and learning systems for signal processing, communications and control. John Wiley & Sons, Inc., New York (1998)
2. Duch, W.: K-Separability. In: Kollias, S., Stafylopatis, A., Duch, W., Oja, E. (eds.) ICANN 2006. LNCS, vol. 4131, pp. 188–197. Springer, Heidelberg (2006)
3. Grochowski, M., Duch, W.: Learning Highly Non-separable Boolean Functions Using Constructive Feedforward Neural Network. In: de Sá, J.M., Alexandre, L.A., Duch, W., Mandic, D. (eds.) ICANN 2007. LNCS, vol. 4668, pp. 180–189. Springer, Heidelberg (2007)
4. Duch, W., Setiono, R., Zurada, J.: Computational intelligence methods for understanding of data. Proceedings of the IEEE 92(5), 771–805 (2004)
5. Muselli, M.: Sequential constructive techniques. In: Leondes, C. (ed.) Optimization Techniques. Neural Network Systems, Techniques and Applications, vol. 2, pp. 81–144. Academic Press, San Diego (1998)
6. Young, S., Downs, T.: Improvements and extensions to the constructive algorithm carve. In: Vorbrüggen, J.C., von Seelen, W., Sendhoff, B. (eds.) ICANN 1996. LNCS, vol. 1112, pp. 513–518. Springer, Heidelberg (1996)
7. Muselli, M.: On sequential construction of binary neural networks. IEEE Transactions on Neural Networks 6(3), 678–690 (1995)
8. Campbell, C., Vicente, C.: The target switch algorithm: a constructive learning procedure for feed-forward neural networks. Neural Computations 7(6), 1245–1264 (1995)
9. Mascioli, F.M.F., Martinelli, G.: A constructive algorithm for binary neural networks: The oil-spot algorithm. IEEE Transactions on Neural Networks 6(3), 794–797 (1995)
10. Marchand, M., Golea, M.: On learning simple neural concepts: from halfspace intersections to neural decision lists. Network: Computation in Neural Systems 4, 67–85 (1993)
11. Zollner, R., Schmitz, H.J., Wünsch, F., Krey, U.: Fast generating algorithm for a general three-layer perceptron. Neural Networks 5(5), 771–777 (1992)
12. Duch, W., Jankowski, N.: Survey of neural transfer functions. Neural Computing Surveys 2, 163–213 (1999)
13. Duch, W.: Towards comprehensive foundations of computational intelligence. In: Duch, W., Mandziuk, J. (eds.) Challenges for Computational Intelligence, vol. 63, pp. 261–316. Springer, Heidelberg (2007)
14. Thurn, S.: The monk's problems: a performance comparison of different learning algorithms. Technical Report CMU-CS-91-197, Carnegie Mellon University (1991)
15. Muselli, M., Liberati, D.: Binary rule generation via hamming clustering. IEEE Transactions on Knowledge and Data Engineering 14, 1258–1268 (2002)
16. Merz, C., Murphy, P.: UCI repository of machine learning databases (1998-2004), http://www.ics.uci.edu/~mlearn/MLRepository.html
17. Duch, W.: Support Vector Neural Training. In: Duch, W., Kacprzyk, J., Oja, E., Zadrożny, S. (eds.) ICANN 2005. LNCS, vol. 3697, pp. 67–72. Springer, Heidelberg (2005)
18. Murthy, S., Kasif, S., Salzberg, S.: A system for induction of oblique decision trees. Journal of Artificial Intelligence Research 2, 1–32 (1994)

Towards Heterogeneous Similarity Function Learning for the k-Nearest Neighbors Classification

Karol Grudziński[1,2]

[1] Department of Physics,
Kazimierz Wielki University
Plac Weyssenhoffa 11, 85-072 Bydgoszcz, Poland
[2] Institute of Applied Informatics
University of Economy,
ul. Garbary 2, 85-229 Bydgoszcz, Poland
karol.grudzinski@wp.pl

Abstract. In order to classify an unseen (query) vector q with the k-Nearest Neighbors method (k-NN) one computes a similarity function between q and training vectors in a database. In the basic variant of the k-NN algorithm the predicted class of q is estimated by taking the majority class of the q's k-nearest neighbors. Various similarity functions may be applied leading to different classification results. In this paper a heterogeneous similarity function is constructed out of different 1-component metrics by minimization of the number of classification errors the system makes on a training set. The HSFL-NN system, which has been introduced in this paper, on five tested datasets has given better results on unseen samples than the plain k-NN method with the optimally selected k parameter and the optimal homogeneous similarity function.

Keywords: Machine Learning, Concept Learning, Similarity-Based Methods, k-Nearest Neighbors Data Classification.

1 Introduction

In the simplest, brute-force implementation of the k-Nearest Neighbors method [1,2,3,4,5,6,7,8,9,10] one computes a distance or more generally the similarity function between a query vector q and samples from a training set. The adaptive parameter k restricts the neighborhood of q to k training instances and the predicted class of q is estimated by taking the majority class of the q's k nearest neighbors. Various similarity functions may be used, which usually leads to a significant improvement of the generalization of this algorithm. Despite of this, many researchers restrict themselves to using only Euclidean metric in their experiments and often set the k parameter to the value of 1. This is because the information about the extensions of the basic k-NN method is scattered in many papers and these modifications are known to a rather narrow group of case-based reasoning (CBR) researchers. Another reason for common usage of only the simplest versions of the k-NN method is a lack of freely available programs realizing sophisticated versions of this algorithm. The most popular and well established public domain machine learning workbenches such as Weka [11], Yale [12] or Knime [13] include only the simplest k-NN implementations. The aim of the development of the SBL system [14] (which so far exists only in a very flaky, experimental state) is to fill this gap of freely available sophisticated k-NN programs.

The k-NN method had been developed in the late sixties and the basic version of this model belongs to the family of, so called, 'lazy learning' algorithms. In this case the learning is restricted to accessing the training data memory. The research on the k-NN method is not so commonly conducted by the machine learning community as on the other popular algorithms as neural-networks, Bayesian models or decision trees mainly because it is extremely hard to beat the plain (basic) k-NN algorithm described above. All so called 'eager learning' algorithms which involve optimization of the adaptive parameters are usually much more time consuming and the main natural classic extensions of this method have been already developed. The most common modifications of the k-NN method include the optimization of the k parameter and the selection of similarity functions. Excellent results can often be attained by optimization of attribute weights. Currently, the research concentrates on speeding up calculations by various ways of partitioning the instance memory space. This method also starts to be more and more often embedded in real-time classification systems. Nowadays the research on the k-NN improvements is nearly completely restricted to the field of case-based reasoning (CBR) where this method has been at the core of interests of the researchers from the beginning. Enormous amount of work related to the development of the extensions of the k-NN algorithm has been done in this area however the outcome of the research that has been conducted by CBR community is rather not known to the pattern recognition and machine learning researchers.

The structure of the paper is the following. In the second section a new algorithm which is called HSFL-NN (**H**eterogeneous **S**imilarity **F**unction **L**earning for the k-**N**earest **N**eighbors method) is described. Section three is devoted to the preliminary numerical experiments which have been conducted in this paper. Finally, the summary of this article is given where plans for the further developments are contained.

2 The HSFL-NN Method

Let us denote by $\mathbf{X}^i$ vectors from the test (query) set $\{\mathcal{Q}\}_{i=1}^{N}$ and by $\mathbf{Y}^j$ vectors from a training set $\{\mathcal{T}\}_{j=1}^{M}$. Let us assume that the vectors are $n+1$ dimensional and by $\mathcal{C}_m$ the class attribute is denoted where $m = 1, 2, \ldots, C$ and C is the number of classes in a problem domain. i.e.:

$$\mathbf{X}^i = (X_1^i, X_2^i, \ldots, X_n^i, \mathcal{C}_m)$$

and
$$\mathbf{Y}^j = (Y_1^j, Y_2^j, \ldots, Y_n^j, \mathcal{C}_m).$$

Formally, the plain (basic) k-NN method proceeds as follows:

1. For each test case $\mathbf{X}^i$:
   – Compute all distances $D(\mathbf{X}^i, \mathbf{Y}^j)$,
   – Find k nearest neighbors to $\mathbf{X}^i$ among $\mathbf{Y}^j$,
   – Perform classification step for $\mathbf{X}^i$.
2. Report the results.

The total similarity function can be expressed as the sum of 1-dimensional contributions:

$$D(\mathbf{X}^i, \mathbf{Y}^j) = \sum_{l=1}^{n} D_l(X_l^i, Y_l^j), \qquad l = 1, 2, \ldots, n$$

and in the standard k-NN procedure[1], we have:

$$D_1 \equiv D_2 \equiv \cdots \equiv D_n.$$

The idea behind the HSFL-NN method (**H**eterogeneous **S**imilarity **F**unction **L**earning for the k-**NN** algorithm) is very simple. Instead of using the same 1-component similarity function in every dimension, the total similarity function

$$D = \sum_{l=1}^{n} D_l$$

is built from different 1-component metrics, i.e. in general

$$D_p \not\equiv D_q \not\equiv \cdots \not\equiv D_r,$$

where $p, q, r \in \{1, 2, \ldots n\}$ and $p < q \leq r$. The 1-dimensional functions to be found, in the software program are mapped on integer numbers which take values from $\{1,2,\ldots,7\}$ since in our calculations we have used seven 1-component similarity functions in order to construct the heterogeneous distance. Thus we have n-parameter vector and every parameter may take the value from the range $[1,7]$ and is optimized on a training set with the non-gradient minimization procedure. The target function returns the number of errors the k-NN classifier makes and the learning is performed by conducting cross-validation on a training partition. The algorithm which is used to calculate the heterogeneous distance function is the following:

---

[1] In real world k-NN implementations the square root in the calculation of the Euclidean metric is usually omitted as it does not change the order of the reference vectors but only scales the value of their distance to the query instance.

```
float heterogeneous_distance(int P[ ], float X[ ], float Y[ ], int n){

    /*
    Input:

        * P[ ] - the array of optimization parameters.
        * X[ ] - an instance from a test set.
        * Y[ ] - an instance from a training set.
        * n - the number of attributes.

    Output:

        * distance: heterogeneous distance.
    */

    dis_0 = dis_1 = ... = dis_7 = 0.0;

    for(int l = 1; l <= n; l++){

        switch(P[l]){

            case 0: dis_0 += 0;
            break;
            case 1: dis_1 += Euclidean_Distance_1(X[l],Y[l]);
            break;
            case 2: dis_2 += Manhattan_Distance_1(X[l],Y[l]);
            break;
            case 3: dis_3 += Chebychev_Distance_1(X[l],Y[l]);
            break;
            case 4: dis_4 += Canberra_Distance_1(X[l],Y[l]);
            break;
            case 5: dis_5 += Cosine_Distance_1(X[l],Y[l]);
            break;
            case 6: dis_6 += Dice_Distance_1(X[l],Y[l]);
            break;
            case 7: dis_7 += Jacard_Distance_1(X[l],Y[l]);
            break;
        }
    }

    distance = dis_0 + dis_1 + ... + dis_7;

    return distance;
}
```

In our experiments the 1-component metrics of the following total homogeneous similarity functions have been taken to construct a heterogeneous similarity function:

1. Euclidean metric[1]: $D(\mathbf{X}^i, \mathbf{Y}^i) = \sum_l (X_l^i - Y_l^i)^2$
2. Manhattan metric: $D(\mathbf{X}^i, \mathbf{Y}^i) = \sum_l |X_l^i - Y_l^i|$
3. Chebychev metric: $D(\mathbf{X}^i, \mathbf{Y}^i) = max|X_l^i - Y_l^i|$
4. Canberra metric: $D(\mathbf{X}^i, \mathbf{Y}^i) = \sum_l \frac{|X_l^i - Y_l^i|}{|X_l^i + Y_l^i|}$
5. Cosine metric: $D(\mathbf{X}^i, \mathbf{Y}^i) = 1 - \frac{\langle \mathbf{X}^i | \mathbf{Y}^i \rangle}{||\mathbf{X}^i|| ||\mathbf{Y}^i||}$
6. Dice metric: $D(\mathbf{X}^i, \mathbf{Y}^i) = 1 - \frac{2\langle \mathbf{X}^i | \mathbf{Y}^i \rangle}{||\mathbf{X}^i||^2 + ||\mathbf{Y}^i||^2}$
7. Jacard metric: $D(\mathbf{X}^i, \mathbf{Y}^i) = 1 - \frac{\langle \mathbf{X}^i | \mathbf{Y}^i \rangle}{||\mathbf{X}^i||^2 + ||\mathbf{Y}^i||^2 - \langle \mathbf{X}^i | \mathbf{Y}^i \rangle}$

where $\langle \mathbf{X}^i | \mathbf{Y}^i \rangle = \sum_l X^i Y^i$ and $||\mathbf{X}^i||^2 = \langle \mathbf{X}^i | \mathbf{X}^i \rangle$.

The algorithm which is used to calculate the cost function proceeds as following:

```
int cost_function(int P[ ], float X[N][ ], float Y[M][ ], int n){

    /*
    Input:

        * P[ ] - the array of optimization parameters.
        * X[ ][ ] - the test set matrix.
        * Y[ ][ ] - the training set matrix.
        * n - the number of attributes.

    Output:

        * num_errors: number of classification errors.
    */

    num_errors = crossvalidate(int P[ ], float X[N][ ], float Y[M][ ], int n){
        /*
         *
         * Run k-NN classifier with heterogeneous similarity function.
         *
         */
    }

    return num_errors;
}
```

The idea of heterogeneous similarity functions is not a new concept as it is known in case-based reasoning as the local-global principle. Also the learning of the similarity functions is not a new concept [15]. However our approach seems to be original.

3 Numerical Experiments

The usefulness of the HSFL-NN method has been shown on five datasets. Two artificial (the well-known monk1 and monk3 [16]) and three real databases (appendicitis, nevi and Sklodowski) have been taken for our experiments. The appendicitis dataset [17] has been obtained from Professor Shalom Weiss by Professor Włodzisław Duch. The well known nevi dataset [18] has been obtained from Professor Zdzisław Hippe. Finally, the Sklodowski psychological data has been obtained from Professor Henryk Skłodowski [19].

The results of our computations are quite promising. However, it should be noted that we have managed to perform only very preliminary numerical experiments and little effort has been made to select the optimal parameters. All our computations have been conducted with our system SBL [14].

In the early stage of our calculations we used the well known simplex method for function minimization [20]. It is a local minimization procedure and it is very simple as it basically has no parameters that control a minimization process. Better results than those obtained with the simplex algorithm have been attained with the ASA minimization routine (Adaptive Simulated Annealing) [21]. We are however aware of the fact that experiments on five datasets are not enough to demonstrate the suitability of HSFL-NN in real-world applications.

Three datasets which have been taken for our experiments have a test set so the standard procedure to estimate the classification accuracy of the unseen samples has been performed in these cases. On the appendicitis and Sklodowski datasets 10-fold stratified cross-validation test has been done. All calculations have been repeated 10 times and the averages and standard deviations of the results are given.

The learning process, in most cases, has been performed by the leave one out cross validation and is denoted by (L1O). Sometimes, probably due to overfitting, n-fold stratified cross validation has been chosen for learning and it is denoted by (n-SCV).

No data preprocessing has been made and the calculations have been performed with all attributes present. HSFL-NN will probably work better if it is preceded by the attribute selection but for monk1 and nevi application of the attribute selection leads to the 100% correct classification. Thus, there is no room left for the study of the improvement of the generalization over the one that is obtained with the homogeneous metric optimization procedure.

If the missing values occurred in the data, the calculations of the corresponding 1-dimensional distances which contribute to the total similarity function have not been performed and these dimensions were skipped.

In the all experiments involving homogeneous metric optimization, first the optimal values for the k parameter have been found with the Euclidean metric set and after that the best homogeneous metric was selected by iterating over the available functions from the pool and applying the classifier with these metrics set to the training set using cross-validation. If more than one function corresponded to the peak cross-validation training accuracy, the first best metric from the pool has been selected. In case of the cross validation tests which were used for the appendicitis and the Sklodowski data, the most common optimal value found in cross-validation on the training set was selected as the final optimal setting for the k parameter. In all cases the HSFL model has been evaluated with the same number of nearest neighbors set as in case of the homogeneous metric optimization calculation. It should be noted that avoiding re-optimization of parameters for every model individually degrades the results significantly and our HSFL-NN computations certainly could have lead to much larger classification accuracies if the optimal k parameter value had been found for this model.

The results in tables are sorted with respect to the classification accuracy obtained on test partitions. The averages and standard deviations are taken over 10 runs. In the first column of a table, after the k parameter, the information about the classification model and its additional parameters is provided. What concerns the HSFL calculations, the minimization procedure is listed, as well as (in the square brackets), the limitation on the maximum number of cost function evaluations. The information about the optimal metric selected with the homogeneous metric optimization model is provided only for datasets containing the test set. In the case of the cross-validation experiments, the optimal homogeneous metrics found in every cross-validation fold usually differ from one another and listing them here makes no sense as it would only fill up the valuable space. All homogeneous metric optimization models have been trained with the leave-one-out cross-validation and this is the reason why in the listings containing results obtained on separated test sets the std. dev. of the classification accuracy value equals zero. By Cn majority committees [22,23] of n models constructed out of the system which is listed are denoted. They are used to stabilize classifiers.

The first dataset that has been used for our study is the monk1 database. The training file consists of 124 and the test set has 432 samples. There are 6 attributes. There are two classes evenly distributed. This is a relatively easy dataset to analyze and the classification accuracy close to 100% can be attained with many systems. The optimal k value is 1 and the Canberra metric is the optimal homogeneous similarity function. The calculation with these settings gives classification accuracy of 88.4%. For this data HSFL-NN improves the results significantly. The accuracy of 98.2% has been obtained with HSFL-NN using ASA minimization method with the limitation to 300 cost evaluations. We are certain that tuning the parameters that control the ASA minimization procedure better would result in attaining the accuracy which is indistinguishable from the ones obtained with the classifiers which perform best on this data. A modified HSFL-NN model which is called HSFL-AS-NN improved the results further and lowered the standard deviation of the results. The HSFL-AS-NN model uses additional (virtual) 1-component similarity function which tells to omit given attribute. This provides some sort of attribute selection which is built into HSFL-NN.

Table 1. Results for the test on monk1 data

System and Parameters	Train	Test %	± %
1-HSFL-AS-NN-ASA-[100]	(L1O)	99.8	0.8
1-HSFL-NN-ASA-[300]	(L1O)	98.2	2.0
1-HSFL-NN-ASA-[100]	(L1O)	97.3	5.3
1-NN, Best Similarity Function: Canberra	(L1O)	88.4	0.0

Monk3 is the second dataset that has been taken for our experiments. The training set consists of 122 and the test set of 432 instances. There are two classes and the distribution of them in the test set is (52.8%, 47.2%). There are 6 attributes. The optimal k value for this data equals 2 but with $k = 10$ and the best homogeneous metric selected similar results are obtained so we have taken $k = 10$ as the target value for this parameter. With these settings, running the homogeneous metric optimization procedure leads to 92.8% of classification accuracy. Again a significant but not so large improvement of the generalization ability as for the monk1 data can be noted when using HSFL-NN.

The next dataset that has been taken for our experiments is the appendicitis database. It consists of 106 cases distributed in two classes. The base rate is 80.2%. There are eight attributes including class. HSFL-NN improves the results obtained by the homogeneous metric optimization method by approximately 2%.

Table 2. Results for the test on monk3 data

System and Parameters	Train	Test %	± %
10-HSFL-NN-ASA-[100]	(L1O)	96.3	1.1
10-HSFL-NN-ASA-[300]	(L1O)	95.2	2.5
10-NN, Best Similarity Function: Manhattan	(L1O)	92.8	0.0

Table 3. Results for the 10-fold CV test on the appendicitis data

System and parameters	Train	Test %	± %
10-HSFL-NN-ASA-[300]	(3-SCV)	87.7	7.7
10-HSFL-NN-ASA-[300]-C5	(L1O)	86.5	1.5
10-HSFL-AS-NN-ASA-[300]-C5	(L1O)	86.2	1.3
10-HSFL-NN-ASA-[300]	(L1O)	85.4	10.2
10-NN, Best Similarity Function	(L1O)	84.9	1.9

Table 4. Results for the test on the nevi data

System and parameters	Train	Test %	± %
1-HSFL-AS-NN-ASA-[100]-C10	(L1O)	93.1	5.4
1-HSFL-AS-NN-ASA-[100]	(L1O)	91.5	10.5
1-HSFL-NN-ASA-[300]	(3-SCV)	87.1	2.0
1-HSFL-NN-ASA-[300]	(L1O)	86.4	1.6
1-NN, Best Similarity Function: Manhattan	(L1O)	80.8	0.0

Experiments conducted on this data have shown strong instability of the HSFL-NN method but with the application of majority committees it was possible to stabilize this model and satisfactory results have been obtained.

Another dataset from the real-world group of databases that have been taken for our experimentation is the well known nevi data. It is a fifteen-dimensional dataset (including the class attribute). The training/test set consists of 250 and 26 cases respectively. It is a relatively easy dataset to be analyzed by k-NN and 100% classification accuracy is obtained provided that classification is preceded by attribute selection. In order to study the improvement of the generalization of HSFL-NN over the homogeneous metric optimization method we have not performed attribute selection on this data. The improvement of the generalization of about 10% on the test data has been recorded.

The last real-world data we have taken is the Sklodowski psychological database. The distribution of classes is approximately (56%, 44%). There are 48 attributes including class and the data consists of 180 cases. This is an extremely hard data, base rate is obtained by most classifiers. HSFL-NN improved the results significantly over our reference model however strong instability which is typical for HSFL-NN has been observed. We had not enough time to perform the calculations with the ASA minimization procedure which would probably improve the results significantly.

Table 5. Results for the 10-fold CV test on the Sklodowski data

System and parameters	Train	Test %	± %
10-HSFL-NN-Simplex-[300]	(L1O)	65.8	4.9
10-HSFL-NN-Simplex-[300]-C5	(L1O)	63.0	3.8
10-NN, Best Similarity Function	(L1O)	57.7	2.1

4 Summary and Conclusions

Considering the fact that for the datasets under study the optimal values for the number of nearest neighbors and the best homogeneous metrics have been found, these results are already close to the best possible to be attained by the k-NN classifier that is not preceded by running the attribute selection filters. The statistical significance of the results obtained with the HSFL-NN method with respect to the reference ones has not been computed because there is no support for this sort of analysis in our SBL system. Despite of this the generalization ability of the HSFL-NN model, as reflected in our preliminary experiments, is very good and the system under study seems to outperform classic homogeneous similarity metric optimization method. Room for the potential further improvement is left by combining HSFL-NN with the attribute selection and weighting in the future. This method also may turn out to play a very important role in our SBM MetaLearning model [24,25]. The next step is to employ searching instead of minimization in order to find optimal 1-dimensional similarity functions. We believe that this would improve the results even further and lower the standard deviation of the HSFL-NN method. What may be a little bit worrying is a very high computational cost of the HSFL-NN method. It involves learning by cross-validation on the entire training data partition and this process has to be repeated sometimes a couple of hundred times in order to reach sufficient level of optimization of parameters. One of the solutions is to use data pruning, i.e. training set reduction as a preprocessing step for the classification with the HSFL-NN method. This could be done, for example, with our excellent SBL-PM-M method [26].

References

1. Fix, E., Hodges Jr., J.L.: Discriminatory analysis, nonparametric discrimination consistency properties. Technical Report 4, Randolph Filed, TX: US Air Force, School of Aviation Medicine (1951)
2. Sebestyen, G.S.: Decision-making process in pattern-recognition. The Macmillan Company, New York (1962)
3. Cover, T., Hart, P.: Nearest Neighbor Pattern Classification. IEEE Transactions on Information Theory 13, 21–27 (1967)
4. Cover, T.: Estimation by the nearest neighbor rule. IEEE Transactions on Information Theory 14, 50–55 (1968)
5. Duda, R., Hart, P.: Pattern Classification and Scene Analysis. John Wiley & Sons, Chichester (1973)

6. Aha, D.W., Kibler, D., Albert, M.K.: Instance-based learning algorithms. Machine Learning 6, 37–66 (1991)
7. Dasarathy, B.V.: Nearest Neighbor (NN) Norms: NN Pattern Classification Techniques. IEEE Computer Society Press, Los Alamitos (1991)
8. Cost, S., Salzberg, S.: A weighted nearest neighbor algorithm for learning with symbolic features. Machine Learning 10, 57–78 (1993)
9. Duch, W.: Similarity Based Methods: a general framework for classification, approximation and association. Control and Cybernetics 29(4), 1–30 (2000)
10. Grudzinski, K.: Similarity Based Methods in Application to Analysis of Scientific and Medical Data, PhD Thesis, Department of Applied Informatics, Nicholaus Copernicus University, Torun, Poland (2002)
11. Witten, I.H., Frank, E.: Data Mining: Practical machine learning tools and techniques, 2nd edn. Morgan Kaufmann, San Francisco (2005)
12. Mierswa, I., Wurst, M., Klinkenberg, R., Scholz, M., Euler, T.: YALE: Rapid Prototyping for Complex Data Mining Tasks. In: Proceedings of the 12th ACM SIGKDD International Conference on Knowledge Discovery and Data Mining (KDD 2006) (2006)
13. Knime: Konstanz Information Miner, http://www.knime.org/index.html
14. SBL, Similarity Based Learner, Software Developed by Karol Grudzinski, Nicholaus Copernicus University: 1997-2002, Kazimierz Wielki University: 2002-2008, University of Economy: 2005-2008s
15. Stahl, A.: Learning of Knowledge-Intensive Similarity Measures in Case-Based Reasoning. PhD thesis, University of Kaiserslautern, Germany
16. Mertz, C.J., Murphy, P.M.: UCI repository of machine learning databases, http://www.ics.uci.edu/pub/machine-learning-data-bases
17. Weiss, S.M., Kulikowski, C.A.: Computer Systems that Learn. Morgan Kaufmann, San Francisco (1991)
18. Duch, W., Grabczewski, K., Adamczak, R., Grudzinski, K., Hippe, Z.S.: Rules for melanoma skin cancer diagnosis. Komputerowe Systemy Rozpoznawania, KOSYR, Wrocaw 2001, pp. 59–68 (2001)
19. Hab, Sklodowski, H., Zarzadzania, W.: Spoleczna Wyzsza Szkola Przedsiebiorczosci i Zarzadzania w Lodzi
20. Nelder, J.A., Mead, R.: A simplex method for function minimization. Computer Journal 7, 308–313 (1965)
21. Ingber, L.: Adaptive simulated annealing (ASA): Lessons learned. Control and Cybernetics 25(1), 33–54 (1996)
22. Ortega, J., Koppel, M., Argamon, S.: Arbitrating Among Competing Classifiers Using Learned Referees. Knowledge and Information Systems 3, 470–490 (2001)
23. Bauer, E., Kohavi, R.: An empirical comparison of voting classification algorithms: bagging, boosting and variants. Machine Learning 36, 105–142 (1999)
24. Duch, W., Grudziński, K.: Meta-Learning: searching in the model space. In: Proceedings of the International Conference on Neural Information Processing, Shanghai, vol. I, pp. 235–240 (2001)
25. Duch, W., Grudziński, K.: Meta-learning via search combined with parameter optimization. In: Intelligent Information Systems, Sopot, Poland, 2002, Advances in Soft Computing, pp. 13–22. Physica-Verlag (Springer) (2002)
26. Grudziński, K.: SBL-PM-M: A System for Partial Memory Learning. In: Rutkowski, L., Siekmann, J.H., Tadeusiewicz, R., Zadeh, L.A. (eds.) ICAISC 2004. LNCS (LNAI), vol. 3070, pp. 586–591. Springer, Heidelberg (2004)

Hough Transform in Music Tunes Recognition Systems

Maciej Hrebień and Józef Korbicz

Institute of Control and Computation Engineering
University of Zielona Góra, ul. Podgórna 50, 65-246 Zielona Góra
{m.hrebien,j.korbicz}@issi.uz.zgora.pl
http://www.issi.uz.zgora.pl

Abstract. This paper presents a method of music tunes recognition based on adopted Hough transform. One can also find here experimental results showing the effectiveness of the presented solution. Perspectives of further work and quality improvements are also stated as a base for subsequent research.

1 Introduction

In last decade we can observe a very dynamic grow in the number of cell phones per person, especially in the so-called developing countries - new members of the EU. For example, in Poland there are nearly 36758 thou. cell phones per 38157 thou. citizens and this number is still increasing [5]. The cell phone, besides the main function, which is voice communication at a distance, is starting to be a sort of mini-centre of information. Many modern cell phones have the ability of accessing the internet, exchanging data, many of them heave photo-camera, MP3 decoder or a mini-dictaphone onboard. The computational power of cell phones per Watt is increasing what gives the opportunity to threat them as a personal mini-computers with many useful features. This features can be connected with work (business notes, meeting remainders), fun (games, photo editing) or more dedicated solutions like e.g. telemedicine [3].

Most people like music and many of them listen to it in a car, home or even during work. How many times one was wondering what is the title of the song one is currently hearing and who is the performer, is hard to count. As most of people have cell phones it would be interesting to have an operator supporting system that would be able to recognize the played song. Such a system should include facts that a given recording is short (nobody wants to hold a cell phone more than a few seconds in front of a speaker), contains an additive noise generated by surrounding (car engine, detuned radio, wind, voices of passers-by, etc.) and its quality is limited to the capabilities of the microphone acquiring the audio signal and losses in the lossy GSM compression.

According to authors knowledge there are currently two commercial music recognition systems in the western EU countries, that is Shazam in United Kingdom and Musiwave in Spain [13]. Because this systems are fully commercial it is very hard to obtain the full information of how they work, what is its *real* effectiveness and computation complexity.

In this paper a method of music tunes recognition based on adopted Hough transform is presented. One can also find here experimental results collected during research and creation of a GSM based telephony simulator. Perspectives of further work and quality improvements are also stated as a base for subsequent researches.

2 System Assumptions

The main task of the system is to recognize and identify a music tune sample with the use of a cell phone. A client calls a dedicated phone number and puts its phone near sound source for a period of a few seconds. What the client expects as a result is a short message containing the information about the tune like e.g.: who is the performer, what is the title of the song and album it comes from, when it was recorded/produced, etc. Unfortunately, the recognition task performed at the operators side is not very easy because of cell phones' hardware restrictions and the GSM lossy compression.

A human voice contains sound waves with base frequencies varying from about 300 to 3500 Hz [12]. A human ear, on the other hand, plays the role of a biological amplifier and is the most sensitive to frequencies characteristic to human voices generated by vibrating vocal cords [16]. This facts are used with success in the cell phone telephony. Thus, a given voice signal is passed through a band-pass filter with the boundary frequencies characteristic to human voice. The signal

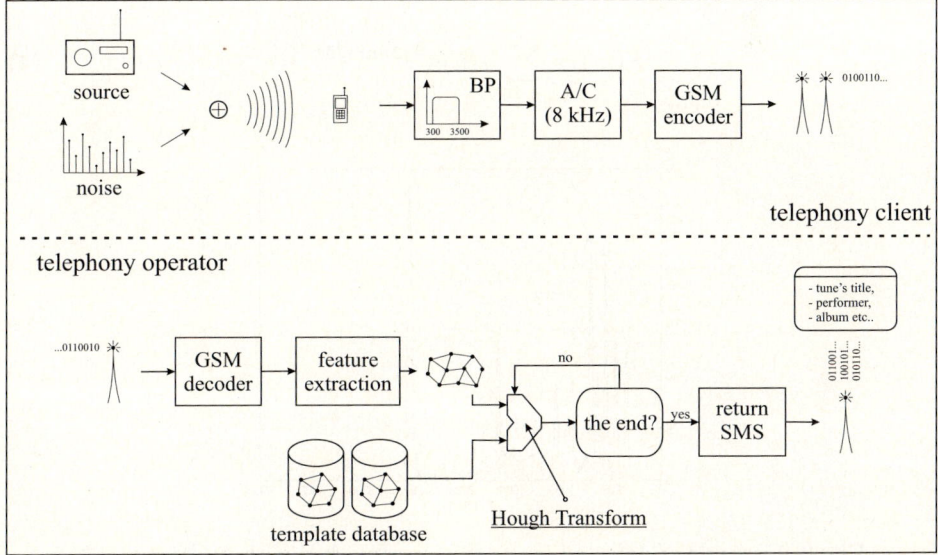

Fig. 1. Schematic of the music tunes recognition system

is then sampled with 7 ÷ 16 bits precision at 8 kHz frequency to satisfy the Nyquist's condition and lossy compressed by the GSM encoder [19].

The system schematic is given in Fig. 1.

3 Features Extraction

The presented solution is based on short-time frequency spectrum analysis. According to authors' observations, the frequency spectrum is the least sensitive to the impact of any external noise, is unique for every music tune and valuably represents its nature and dynamics.

A given tune sample is at first decompressed by the GSM decoder and then analyzed frame by frame with the overlaying and frequency leak stopping technique. For each frame a set of features is calculated. In our approach features (S) are these frequencies for those its amplitude estimated using the Fourier transform creates a local maxima in a given milliseconds of the analyzed tune:

$$S = \{\{(t_1, f_{1,1}), (t_1, f_{1,2}), \ldots, (t_1, f_{1,n})\}, \quad (1)$$
$$\{(t_2, f_{2,1}), (t_2, f_{2,2}), \ldots, (t_2, f_{2,n})\},$$
$$\ldots,$$
$$\{(t_k, f_{k,1}), (t_k, f_{k,2}), \ldots, (t_k, f_{k,n})\}\},$$

where:

$$f_{t,1\ldots n} = \arg\max_n \left(peaks\big(A(m)\big)\right), \quad (2)$$

$$A(m) = |F(m)|, \quad (3)$$

$$F(m) = \sum_{n=0}^{N-1} x(n) e^{-j2\pi nm/M}, \quad (4)$$

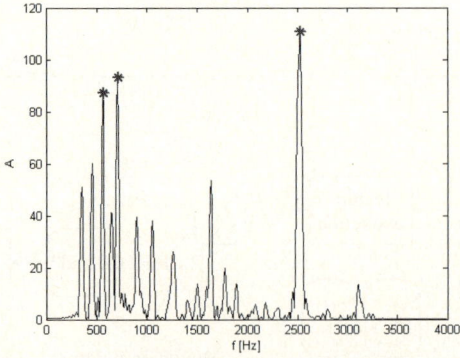

Fig. 2. Example of detected frame features (marked with "*") for $n = 3$

$$m = 1 \ldots M, \tag{5}$$

n is the number of features per frame, t is the time period and F is the frequency spectrum. In situations where the number of features in a frame is less than n the smaller group of features is also included in the S set. All empty frames (silence) are omitted.

A graphical example of feature detection is given in Fig. 2.

4 Matching

The Hough transform, originaly designed for line detection [6], was many times used in literature to find a specific shape [1,8,9,10,11] or a group of features in a larger set [7,14]. The transform is known for its quality in the presence of noise, the ability to adopt it to the detection of a given shape [2] and the accumulator based voting.

The accumulator in our approach is two-dimensional. The first dimension, which is the time, is obvious because a given tune sample is a part of a much longer music tune and the shift in time have to be detected to find the best match. The second dimension, which is the frequency, is added because a research shown that the same music tune can be played by a musician higher or lower in the sense of sound timbre what can be observed mainly in classical music. What was additionally assumed is that the tempo has to be the same to talk about tunes equality. Thus, any tempo differences are not included in the accumulator's dimensionality.

The task of music tune recognition is a problem of finding the shift parameters $(\Delta t, \Delta f)$ which added to the features of tune sample send by a client will give the higher level of fitness to the features of a template taken from early prepared database. Since the features of a tune create an irregular shape the Hough transform have to be defined in an algorithmic form presented in Fig. 3.

For each pair of features: $(t_i^A, f_i^A) \in S^A$ and $(t_j^B, f_j^B) \in S^B$ the accumulator A collects votes in those cells that the best shows the difference between considered features. Thus, a given cell of the accumulator collects:

$$A(\Delta t, \Delta f) = \sum_{i=1}^{m} \sum_{j=1}^{n} \left[\delta_2(t_i^A - t_j^B, \Delta t) \delta_2(f_i^A - f_j^B, \Delta f) \right], \tag{6}$$

where:

$$\delta_2(x_1, x_2) = \begin{cases} x_1 = x_2, 1 \\ x_1 \neq x_2, 0 \end{cases} \tag{7}$$

what gives the possibility to find the best shift parameters $(\Delta t, \Delta f)$ in the accumulator by taking the argument of the maximal value of A (Fig. 4).

As a measure of the final match (Fig. 5) sum of Euclidian distances from features of the send by a client sample to their nearest features from a template is calculated:

$$
\begin{array}{|l|}
\hline
\forall_i \forall_j A(i,j) \leftarrow 0 \\
\\
FOR\ \{t_i^A, f_i^A\} \in S^A,\ i = 1\ldots m \\
\quad FOR\ \{t_j^B, f_j^B\} \in S^B,\ j = 1\ldots n \\
\quad \{ \\
\qquad \Delta t_{ij} = t_i^A - t_j^B \\
\qquad \Delta f_{ij} = f_i^A - f_j^B \\
\\
\qquad \Delta t_{ij}, \Delta f_{ij} \leftarrow align(\Delta t_{ij}, \Delta f_{ij}) \\
\qquad A(\Delta t_{ij}, \Delta f_{ij}) \leftarrow A(\Delta t_{ij}, \Delta f_{ij}) + 1 \\
\quad \} \\
\\
\Delta t, \Delta f \leftarrow \arg\max(A) \\
\hline
\end{array}
$$

Fig. 3. Algorithmic form of the adopted Hough transform

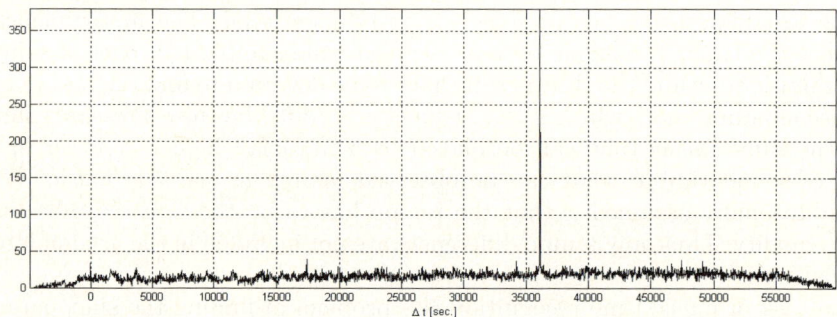

Fig. 4. Exemplary accumulator's slice for $\Delta f = 0$ Hz and low noise sample

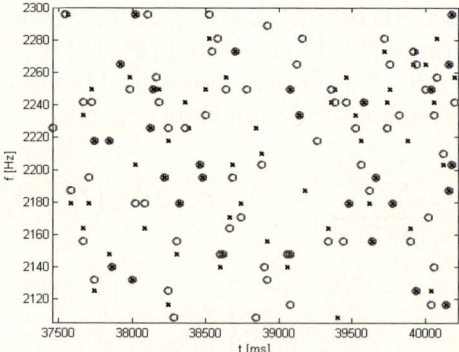

Fig. 5. Exemplary fragment of the best match with template's features marked with "×" and tested sample's features with "○" (experiment for $SNR = 0$ dB)

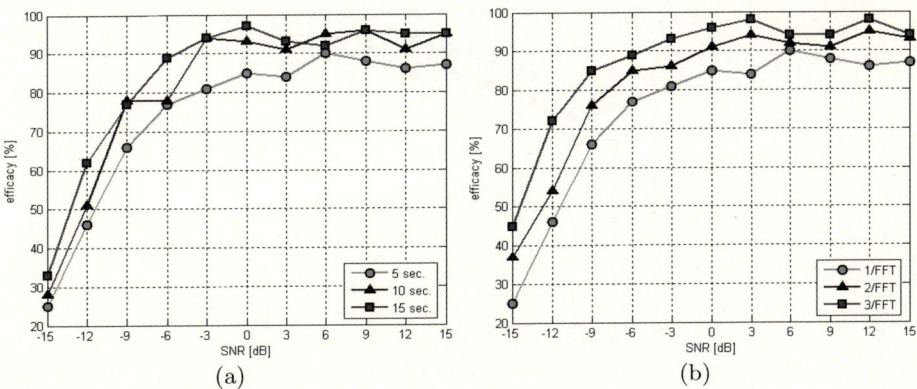

Fig. 6. Experimental results showing the influence of: (a) recording time, (b) number of features used

$$E = \sum_{i=1}^{m} \min \left(\parallel x_{i1} \parallel, \parallel x_{i2} \parallel, \ldots, \parallel x_{ij} \parallel \right), \qquad (8)$$

where:
$$x_{ij} = \left[(t_i^A + \Delta t) - t_j^B, (f_i^A + \Delta f) - f_j^B \right], \quad j = 1 \ldots n. \qquad (9)$$

The E measure can be considered as a match error, so the smaller value of E the better fitness to the template.

5 Experimental Results

To test the quality of the presented solution three experiments were performed in the presence of additive noise. The first one shows the influence of the length of a tune sample on the match quality. In this case tests were performed for 5, 10 and 15 second samples with one feature per frame (Fig. 6a). The second one shows the influence of the number of features used per frame with the length of a send sample equal to 5 seconds (Fig. 6b). The third experiment shows the influence of the GSM 6.10 compression with comparison to the same experiment performed but with the compression switched off (Fig. 7).

The database consisted of 1000 randomly selected music tunes of different genres. The test set consisted of 100 randomly generated tune samples with noise levels raging from -15 to 15 dB with 3 dB interval. All experiments were simulated with software, no telephony hardware was used.

The performed experiments show that the Hough transform adopted for sound features localization can be effectively used in music tunes recognition systems. The presented solution needs less then a quarter of second on today's machines (Athlon 64 3500+ 2.8 GHz, Pentium 4 2.2 GHz) per sample–template comparison depending on the hardware, the number of features and the length of sample used. For an average 3 minute music tune the method needs about 50 MB of

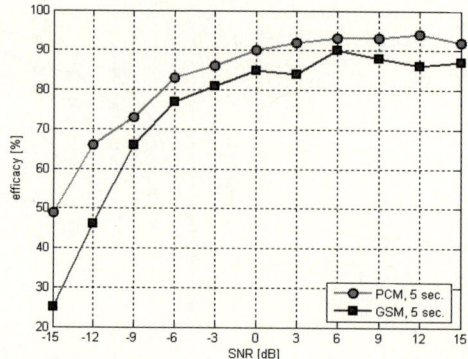

Fig. 7. GSM compression influence on match efficacy

memory for the accumulator with 7.8125 Hz and 20 milliseconds cell granularity which is not much for todays computer systems.

What we can observe is that the presented method tolerates frequency restrictions, losses in the GSM compression (8.72% on average) and noise well. Additionally, experimental results illustrating the influence of the recording time on the match quality are very similar to the ones showing the influence of the number of features used. This gives the conclusion, that if a short sample is recorded by a client a higher number of features per frame is desired for better efficacy and vice versa – if the recording is long the smaller number of features per frame can be used to speed up the search.

The global speed up of the presented solution can be achieved by a classifier that would be able to decide what kind of music a sample represents [4] or a classifier that will create a probability vector that will show what subset of the database should be checked at first [17]. The better quality of recognition for very noisy samples can be achieved, in authors opinion, by a fuzzy way [18] of giving votes in accumulator cells. High level of noise disturbs the voting process by creating a cluster of cells that share votes for one specific parameters of the shift. Since we are searching the maximal value in the accumulator there can be situations where the noise dominates the accumulator giving a false match. Votes given in a fuzzy way will enhance proper cells in the accumulator what is illustrated as an example in Fig. 8 where a vote is described by the t-class function [15]:

$$t(x; a, b, c) = \begin{cases} 0 & for & x \leq a, \\ \frac{x-a}{b-a} & for & a < x \leq b, \\ \frac{c-x}{c-b} & for & b < x \leq c, \\ 0 & for & x > c. \end{cases} \quad (10)$$

The function can be easily redefined for two-dimensional accumulators but to simplify the idea it is left without modifications. Thus, lets consider that the first vote goes to $t_1(x; 2, 5, 8)$, the second one to $t_2(x; 1, 4, 7)$ and the third one

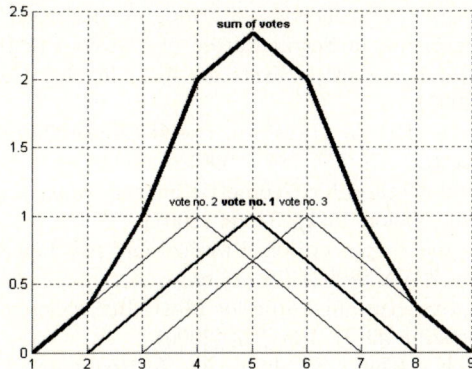

Fig. 8. Example of fuzzy voting – three votes and the sum

to $t_3(x; 3, 6, 9)$. Assuming that the second and the third vote was disturbed, the sum $(t_1 + t_2 + t_3)$ still shows the right position of parameters.

6 Conclusions

In conclusion, preliminary experiments in the Hough transform adaptation to the problem of music tunes recognition are promising. This gives the base for further research, that is solving above mentioned noise influence and global speed up problems. Performance and quality tests on much bigger database for mixed as well as one genre music tunes should also be considered in the future works.

References

1. Antiquzzamann, M.: Coarse-to-Fine Search Technique do Detect Circles in Images. Int. J. Adv. Manuf. Technol. 15, 96–102 (1999)
2. Ballard, D.: Generalizing the Hough Transform to Detect Arbitrary Shapes. Pattern Recognition 13(2), 111–122 (1981)
3. Bożewicz, D., Pałko, T., Jeżewski, M., Łabaj, P., Kupka, T., Bernyś, M.: Telemedicine Framework for Follow-Up High Risk Pergnancy. In: Proc. 11th Int. Conf. Medical Informatics and Technology, pp. 236–240 (2006)
4. Cendrowska, D.: Strict Maximum Separability of Two Finite Sets: an Algorithmic Approach. Int. J. Appl. Math. Comput. Sci. 15(2), 295–304 (2005)
5. Dmochowska, H. (ed.): Concise statistical yearbook of Poland. Statistical Publishing Establishment (2007), http://www.stat.gov.pl
6. Hough, P.: Method and Means for Recognizing Complex Patterns, U.S. Patent No. 3,069,654 (1962)
7. Hrebień, M., Korbicz, J.: Human Identification Based on Fingerprint Local Features. In: Rutkowski, L., Tadeusiewicz, R., Zadeh, L.A., Żurada, J.M. (eds.) ICAISC 2006. LNCS (LNAI), vol. 4029, pp. 796–803. Springer, Heidelberg (2006)
8. Hrebień, M., Steć, P., Nieczkowski, T., Obuchowicz, A.: Segmentation of Breast Cancer Fine Needle Biopsy Cytological Images. Int. J. Appl. Math. Comput. Sci, Special Issue (accepted, 2007)

9. Hrebień, M., Korbicz, J., Obuchowicz, A.: Hough transform (1+1) Search Strategy and Watershed Algorithm in Segmentation of Cytological Images. In: Proc. 5th Int. Conf. Computer Recognition Systems, Adv. in Soft Comput., pp. 550–557. Springer, Berlin (2007)
10. Nair, P., Saunders, A.: Hough Transform Based Ellipse Detection Algorithm. Pattern Recognition Letters 17, 777–784 (1996)
11. Olson, C.: A General Method for Geometric Feature Matching and Model Extraction. Int. J. Comput. Vision 45(1), 39–54 (2001)
12. Ozimek, E.: Sound and its perception. Physical and psychoacoustic aspects. Polish Scientific Publishers PWN (2002) (in polish)
13. Prado, B.: Finding Structure in Audio for Music Information Retrieval. IEEE Signal Processing Magazine 23(3), 126–132 (2006)
14. Ratha, N., Karu, K., Chen, S., Jain, A.: A Real-time Matching System for Large Fingerprint Databases. IEEE Trans. Pattern Analysis and Machine Intelligence 28(8), 799–813 (1996)
15. Rutkowski, L.: Artificial intelligence methods and techniques. Polish Scientific Publishers PWN (2005) (in Polish)
16. Tadeusiewicz, R.: Speech signal. Transport and Communication Publishers, Warsaw (1988) (in Polish)
17. Toth, L., Kocsor, A., Csirik, J.: On Naive Bayes in Speech Recognition. Int. J. Appl. Math. Comput. Sci 15(2), 287–294 (2005)
18. Zadeh, L.: From Computing with Numbers to Computing with Words – From Manipulation of Measurements to Manipulation of Perceptions. Int. J. Appl. Math. Comput. Sci. 12(3), 307–324 (2002)
19. Zieliński, T.: Digital signal processing. From theory to practise. Transport and Communication Publishers, Warsaw (2005) (in Polish)

Maximal Margin Estimation with Perceptron-Like Algorithm

Marcin Korzeń and Przemysław Klęsk

Faculty of Computer Science and Information Systems
Szczecin University of Technology,
ul. Żołnierska 49, 71-210, Szczecin, Poland
{mkorzen,pklesk}@wi.ps.pl

Abstract. In this paper we propose and analyse a γ-margin generalisation of the perceptron learning algorithm of Rosenblatt. The difference between the original approach and the γ-margin approach is only in the update step. We consider the behaviour of such a modified algorithm in both separable and non-separable case and also when the γ-margin is negative. We give the convergence proof of such a modified algorithm, similar to the classical proof by Novikoff. Moreover we show how to change the margin of the update step in the progress of the algorithm to obtain the maximal possible margin of separation. In application part, we show the connection of the maximal margin of separation with SVM methods.

1 Introduction

We consider a binary classification problem. Let $P = \{(x_i, d_i)\}_{i=1,\ldots,p}$ denote the set of training pairs, where $x_i \in \mathbb{R}^{n+1}$, are input vectors and $d_i \in \{-1, 1\}$ are the corresponding decisions assigned to them. We distinguish the case without a bias term using the $\prime$ symbol in the following way:

$$x_i = (1, x_{i1}, \ldots, x_{in}),\ x'_i = (x_{i1}, \ldots, x_{in}),$$
$$w = (w_0, w_1, \ldots, w_n),\ w' = (w_1, \ldots, w_n),$$

where w denotes a vector of weights (coefficients of a hyperplane).

The inner product is denoted by $\langle w, x \rangle$, the norm of a vector with the bias term by $\|x\|$, $\|w\|$ and the norm without the bias by $\|x'\|$, $\|w'\|$ respectively. Also, let $R = \max_{i=1,\ldots,p} \|x'_i\|$ and $R_0 \max_{i=1,\ldots,p} \|x'_i - \bar{x}'\|$.

Given a certain point x and a hyperplane defined by a vector w, the signed distance of x to the hyperplane is

$$d_i \langle \frac{w}{\|w'\|}, x \rangle. \tag{1}$$

Definition 1. *We say that a data set P is γ-separable (or separable with the γ-margin) when:*

$$\exists w : \|w\| = 1, \forall_{i \in \{1,\ldots,p\}}\ d_i \langle \frac{w}{\|w'\|}, x_i \rangle \geq \gamma. \tag{2}$$

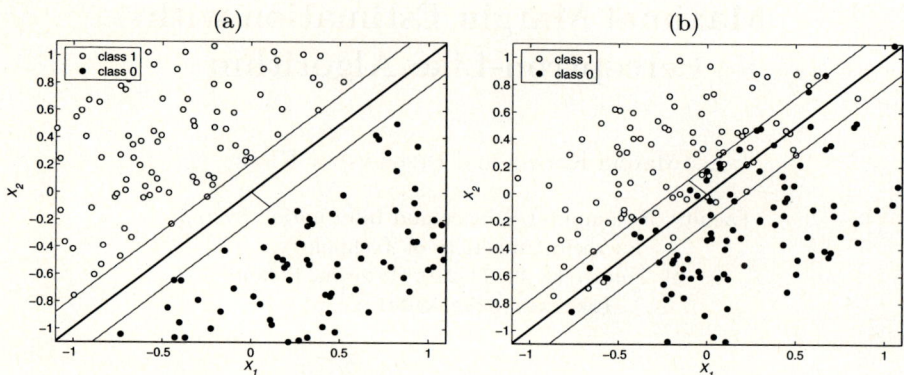

Fig. 1. Illustration of hyperplanes with a positive (a) and a negative (b) margin

In this definition γ could be either negative or positive. If there exists a vector of weights w and a number $\gamma = 0$ such that (2) is true, then we have an ordinary linearly separable case. When there exists $\gamma > 0$ we talk about a positive margin and when $\gamma < 0$ we talk about a negative margin. Fig. 1 illustrates this notion.

In the sense of definition 1 all data sets are γ-separable with a certain γ — in all cases we can take any w such that $\|w'\| = 1$ and $\gamma = -(R+1)R$. This fact is a consequence of the *Cauchy–Schwartz inequality*:

$$|d_i\langle w, x_i\rangle| \leq \|w\| \cdot \|x_i\| \leq \|w\| \cdot R \leq (\|w'\| + |w_0|)R \leq (R+1)R \qquad (3)$$
$$\text{therefore } \forall i \quad d_i\langle w, x_i\rangle \geq -(R+1)R. \qquad (4)$$

∎

It should be clear that for the practical cases $w_0 \leq R$.

We want to consider an optimal hyperplane with the maximal value of the margin. Such a hyperplane always exists, it is a consequence of the *extreme value theorem*. Consider a function $F: \{w: \|w'\| = 1\} \to \mathbb{R}^p$ defined as $F(w) = [f_1, \ldots, f_p]$, where $f_i = d_i\langle w, x_i\rangle$. Each coordinate f_i of F is a real continuous function of vector of weights w, and so $g(w) = \min_i(F(w)) = \min_i(f_i)$ is also a continuous real function defined over the compact set. Then, *extreme value theorem* implies that the function g attains its maximum, i.e. there exists w^*, $\|w^{*\prime}\| = 1$ and a number γ^* such that:

$$\gamma^* = \sup_{\substack{w \\ \|w'\|=1}} \min_{i \in \{1,\ldots,p\}} d_i\langle w, x_i\rangle,$$

and for all x_i we have:

$$d_i\langle w^*, x_i\rangle \geq \gamma^*$$

∎

Further on in this paper we name the optimal vector of weights and the value of margin corresponding to it as w^* and γ^* respectively. We will also denote $\gamma_w(x) = \langle \frac{w}{\|w'\|}, x\rangle$.

2 Perceptron Algorithm with γ-Margin Modification

Rosenblatt's perceptron algorithm does not guarantee that the final margin has a certain positive value even if the data set is separable. Different authors consider several modifications of the classical algorithm e.g.: the voting perceptron [4], the margin perceptron [2], and the maximal margin algorithms: [8] (ROMMA),[7].

In the fig. 2 the perceptron learning algorithm with the γ-margin is presented. In comparison to the classical algorithm of [10] the difference is in the step 5, where the original update condition $d_i \langle w_k, x_i \rangle \leq 0$ is replaced by the γ-margin condition. When the original perceptron algorithm stops, we know only that the final margin is greater than zero or equal zero.

Note that if we divide the vector of weights w_k by $\|w'_k\|$ (step 5), then $d_i \langle \frac{w_k}{\|w'_k\|}, x_i \rangle$ can be interpreted as the signed distance from the i-th point to the k-th step hyperplane. When that value is positive then the i-th sample lies at the correct side and when that value is negative then sample lies at the wrong side of the k-th step hyperplane. The distance is equal to $|\langle \frac{w_k}{\|w'_k\|}, x_i \rangle|$.

In the paper [2] a similar algorithm with the update condition (step 5), if $d_i \langle w, x_i \rangle \leq \gamma_{\text{update}_2}$) is presented. The only difference is the lack of norm of vector w'_k. Because of that such a γ_{update_2} cannot be interpreted as the distance to the k-th step hyperplane.

2.1 Convergence Proof for the γ-Separable Case

The algorithm in the fig. 2 has a convergence proof similar to the classical proof of Novikoff's theorem [9]. We give the following theorem:

Theorem 1. *Let w^* be the vector of weights of the optimal hyperplane for a γ^*-separable P set. If $\gamma_{\text{update}} < \gamma^*$ then the γ-margin perceptron algorithm (fig. 2) always stops after a finite number of update steps.*

The presented proof concerns only the case when the margin is positive $\gamma^* > \gamma_{\text{update}} > 0$, but the theorem seems to be true in general case for any

```
 1: function (w_k, γ_k)=GAMMAMARGINPERCEPTRON(P, γ_update)
 2:     k ← 0, w_k ← (0, ..., 0), stopCounter ← 0
 3:     while stopCounter < p do
 4:         draw next training pair (x_i, d_i) from P
 5:         if d_i⟨w_k, x_i⟩ ≤ γ_update ||w'_k|| then
 6:             w_{k+1} ← w_k + d_i x_i
 7:             stopCounter ← 0
 8:             k ← k + 1
 9:         else
10:             stopCounter ← stopCounter + 1
11:         end if
12:     end while
13: end function
```

Fig. 2. The perceptron learning algorithm with the γ-margin

$\gamma^* > \gamma_{\text{update}}$. It should be clear that if the algorithm stops then the final margin of the obtained hyperplane is greater than (or equal to) γ_{update} and obviously less than γ^*.

Without loss of generality we skip the bias term (we make $\|w^*\| = \|w^{*\prime}\|$) so that the hyperplane $\langle w^*, x \rangle = 0$ crosses the origin[1]. Resigning from this simplification doesn't bring relevant difficulties to the proof.

Proof: Similarly like in Novokoff's proof, we consider the inequalities for $\langle w_k, w^* \rangle$ and for $\|w_k\|$. Directly from the definition of the γ-separable set we have:

$$\langle w_k, w^* \rangle = \langle w_{k-1} + d_i x_i, w^* \rangle = \langle w_{k-1}, w^* \rangle + d_i \langle x_i, w^* \rangle \geq \langle w_{k-1}, w^* \rangle + \gamma^*$$

Then, because $w_0 = \mathbf{0}$ we have:

$$\langle w_k, w^* \rangle \geq k \cdot \gamma^*, \text{ for all } k = 0, 1, \ldots. \tag{5}$$

Similarly, taking into account that $\|x_i\| \leq R$ we have:

$$\|w_k\|^2 = \langle w_k, w_k \rangle = \|w_{k-1}\|^2 + 2d_i \langle w_{k-1}, x_i \rangle + \|x_i\|^2$$
$$\leq \|w_{k-1}\|^2 + 2\gamma_{\text{update}} \|w'_{k-1}\| + R^2$$
$$\leq \|w_{k-1}\|^2 + 2\gamma_{\text{update}} \|w_{k-1}\| + R^2 \tag{6}$$

Consider a sequence defined as : $a_k = \sqrt{a_{k-1}^2 + 2\gamma_{\text{update}} a_{k-1} + R^2}$, $a_0 = 0$, and the difference:

$$r_k = a_k - a_{k-1} = \frac{2\gamma_{\text{update}} a_{k-1} + R^2}{\sqrt{a_{k-1}^2 + 2\gamma_{\text{update}} a_{k-1} + R^2} + a_{k-1}}. \tag{7}$$

It is clear that $\|w_k\| \leq a_k$. Since $r_k > 0$ then a_k is a rising sequence. It is easy to check that r_k changes monotonically from R when $k = 1$ to γ_{update} when k approaches infinity, more strictly: $\lim_{k \to \infty}(a_k - a_{k-1}) = \gamma_{\text{update}}$. It means that for any $\epsilon > 0$ there exists such step k_0, that for all $k \geq k_0$ we have: $a_k - a_{k-1} < \gamma_{\text{update}} + \epsilon$. Denoting a_{k_0} by A we have that $a_k \leq a_{k-1} + \gamma_{\text{update}} + \epsilon$ for $k > k_0$ and

$$a_k \leq \begin{cases} A, & k \leq k_0; \\ A + (\gamma_{\text{update}} + \epsilon) \cdot (k - k_0), & k > k_0. \end{cases} \tag{8}$$

Taking into account that $\|w_k\| \leq a_k$ and using: the *Cauchy-Schwartz inequality*, (5) and (8) we have:

$$k \cdot \gamma^* \leq \langle w_k, w^* \rangle \leq \|w_k\| \leq a_k.$$
$$k \cdot \gamma^* \leq a_k. \tag{9}$$

Because $\gamma^* > \gamma_{\text{update}}$ then this inequality can hold true only for a finite set of numbers k, and the maximal value of k satisfying (9) is the bound for the number of update steps of γ-margin perceptron algorithm. ∎

Please note that numbers k_0 and A could be evaluated effectively, and this way we can give more strict upper bound on the number of updates.

---

[1] Every general case of P and w^* can be transformed to this simplified situation.

Convergence when the margin is negative. The theorem 1 seems to be true in general case for any $\gamma_{\text{update}} < \gamma^* < 0$ but the proof causes some difficulties. During our experiments the algorithm always satisfied the stop condition.

2.2 γ-Non–separable Case

If $\gamma^* < \gamma_{\text{update}}$ in the step 2. of the γ-margin algorithm, one cannot expect that the algorithm stops. In this case one could perform one of the following strategies:

1. to stop the algorithm after a given number k_{max} of steps,
2. to decrease γ_{update} in the progress of the algorithm,
3. to take the average of vectors of weights over all update steps and to stop the algorithm after a given number of steps,
4. to use the voting procedure (see [4]).

The first strategy is the simplest and gives the worst results. In the following section we show how one can change γ_{update} in the progress of the algorithm to obtain an estimate of the maximal margin in the data. Taking an average of vectors of weights over all update steps is very similar but not identical to the voting procedure. Such an approach (similar to the voting) is very fast and gives good results in the classification task.

3 How Do We Estimate the Maximal Margin in the Data

3.1 Motivation

Let us see what is happening to the margin of a sample when an update step (see the alg. 2 step 5) has occured. The weight vector $w_{k+1} = w_k + x_i d_i$ is updated, and

$$\langle w_{k+1}, x_i d_i \rangle = \langle w_k + x_i d_i, x_i d_i \rangle = \langle w_k, x_i d_i \rangle + \|x_i\|^2$$
$$= \gamma_{w_k}(x_i)\|w'_k\| + \|x_i\|^2 \geqslant \langle w_{k+1}, x_i d_i \rangle. \tag{10}$$

Thus, the margin of the sample after an update step is increased by $\|x_i\|^2$, which is a positive value when $x_i \neq \mathbf{0}$. In this case there are two possibilities: (1) $d_i \langle w_{k+1}, x_i \rangle > \gamma_{\text{update}} \|w'_{k+1}\|$ or (2) $d_i \langle w_{k+1}, x_i \rangle \leqslant \gamma_{\text{update}} \|w'_{k+1}\|$. The first possibility means that after the update the sample (x_i, d_i) is correctly classified. The second possibility means that the γ_{update} is too high or the algorithm did not do enough number of steps yet. In the second case we will assume that γ_{update} is too high and we will decrease γ_{update} to the value of $\frac{d_i}{\|w'_{k+1}\|} \langle w_{k+1}, x_i \rangle$.

3.2 Perceptron Algorithm with a Varying Margin

In the fig. 3 we present the perceptron learning algorithm with the simplest strategy of changing the γ_{update}. This simple strategy in many cases allows to find quite a good estimation of the maximal margin in the data. We don't give

```
 1: function (w_k, γ_k)=VARMARGINPERCEPTRON(P,γ_0)
 2:    k ← 0, w_k ← (0,...,0), stopCounter ← 0
 3:    while stopCounter < p do
 4:       if d_i⟨w_k, x_i⟩ ⩽ γ_k‖w'_k‖ then
 5:          w_{k+1} ← w_k + d_i x_i
 6:          stopCounter ← 0
 7:          if d_i⟨w_{k+1}, x_i⟩ ⩽ γ_k‖w'_{k+1}‖ then
 8:             γ_{k+1} ← d_i⟨ w_k/‖w'_k‖ , x_i⟩
 9:          else
10:             γ_{k+1} ← γ_k
11:          end if
12:          k ← k + 1
13:       else
14:          stopCounter ← stopCounter + 1
15:       end if
16:    end while
17: end function
```

Fig. 3. The Perceptron learning algorithm with a varying margin

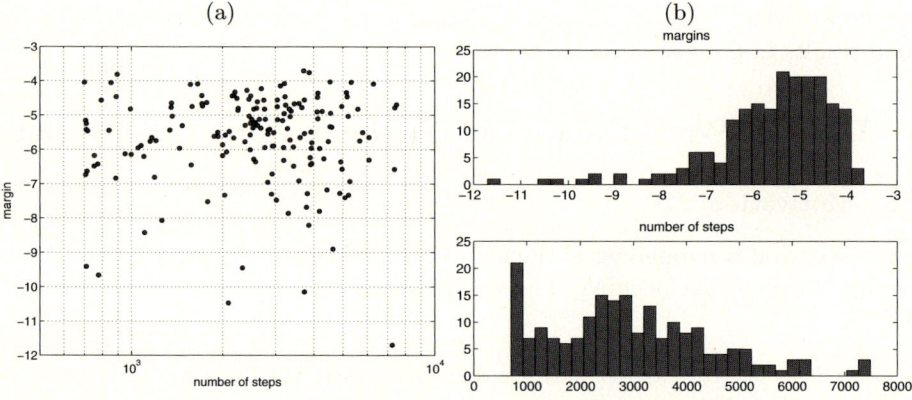

Fig. 4. 200 random runs of the algorithm 3 for the breast_w dataset from [3]. The number of steps and obtained margins are presented together in a single graph in (a) and separately as histograms in (b).

the convergence proof for this algorithm, nevertheless this algorithm is very fast. The final margin obtained by the algorithm 3 obviously is not optimal, but often it is close to the optimal one.

The figure 4 shows 200 random runs of algorithm on the `breast_w` (699 × 9) data set [3]. As we can see in most cases the number of steps is less than 3000 and the margin is greater than −5.5. The maximal margin found over 200 was equal to −3.7. Below we describe an improved version of algorithm 3 which allows to estimate the maximal margin equal to −2.6.

```
 1: function $(w_k, \gamma_k)$=SOFTVARMARGINPERCEPTRON$(P, \gamma_0, \eta, q)$
 2:     $k \leftarrow 0$, $w_k \leftarrow (0, \ldots, 0)$, stopCounter $\leftarrow 0$, impatienceCounter $\leftarrow 0$
 3:     while stopCounter $< p$ do
 4:         draw next trainig pair $(x_i, d_i)$ from $P$
 5:         if $d_i \langle w_k, x_i \rangle \leq \gamma_k \|w'_k\|$ then
 6:             $w_{k+1} \leftarrow w_k + d_i x_i$
 7:             stopCounter $\leftarrow 0$
 8:             if $d_i \langle w_{k+1}, x_i \rangle \leq \gamma_k \|w'_{k+1}\|$ then
 9:                 $\gamma_{k+1} \leftarrow \gamma_k + \eta(d_i \langle \frac{w_{k+1}}{\|w'_{k+1}\|}, x_i \rangle - \gamma_k)$
10:                 impatienceCounter $\leftarrow 0$
11:             else
12:                 impatienceCounter $\leftarrow$ impatienceCounter $+ 1$
13:                 if impatienceCounter $< q$ then
14:                     $\gamma_{k+1} \leftarrow \gamma_k$
15:                 else
16:                     $\gamma_{k+1} \leftarrow \gamma_k + \eta(d_i \langle \frac{w_k}{\|w'_k\|}, x_i \rangle - \gamma_k)$
17:                     impatienceCounter $\leftarrow 0$
18:                 end if
19:             end if
20:             $k \leftarrow k + 1$
21:         else
22:             stopCounter $\leftarrow$ stopCounter $+ 1$
23:         end if
24:     end while
25: end function
```

Fig. 5. The perceptron learning algorithm with a softly varying margin

3.3 Perceptron Algorithm with a Soft-Decrease of Margin

Hereby, we propose a modified version of the former algorithm 3, introducing two modifications: (1) a softer way of decreasing γ-margins, (2) so called *impatience counter*, which forces the decrease in the case when the algorithm performed a long sequence of weight updates but no decrease of the γ-margin.

As regards the first modification, instead of setting the new γ_{k+1} directly to the value of a new margin $d_i \langle \frac{w_{k+1}}{\|w'_{k+1}\|}, x_i \rangle$ as it was the case in the former algorithm (line 8), one can change it only partially towards this value, i.e.:

$$\gamma_{k+1} \leftarrow \gamma_k + \eta(d_i \langle \frac{w_{k+1}}{\|w'_{k+1}\|}, x_i \rangle - \gamma_k), \tag{11}$$

where $\eta \in (0, 1]$ is the fraction of this change. In particular, one can note that setting $\eta = 1$ brings us back to the former version of the algorithm. After experiments with this modificaiton, we suggest to make η dependent on the size p of the training set, e.g. $\eta = \frac{1}{p}$ or similar.

As regards the second modification, it can be viewed as a certain protection against the situation where no decrease of γ-margin takes place for many steps despite the fact that weigths are being updated. All such steps are simply counted and when the counter reaches a given treshold q, then the decrease of γ is forced. Please note however, that now one must not set the γ_{k+1} with respect to the new margin $d_i\langle\frac{w_{k+1}}{\|w'_{k+1}\|}, x_i\rangle$ (implied by new weights w_{k+1}) since this margin is greater than γ_k. This would lead to the increase not the dicrease. Therefore now one must refer to the margin implied by the former weights w_k:

$$\gamma_{k+1} \leftarrow \gamma_k + \eta(d_i\langle\frac{w_k}{\|w'_k\|}, x_i\rangle - \gamma_k). \tag{12}$$

We suggest to set the treshold q as a certain fraction of the training set size, e.g. $q = 10\%p$. The algorithm is presented in the fig. 5.

In comparison to the algorithm 3, the advantage of this algorithm is that it is more stable during the first iterations and that it allows to find a good estimate of the optimal γ^* in fewer runs or even in a single run. Of course its success may depend on the good settings of η and q.

4 Applications

Classification task. First of all we want to stress that the maximal margin in the data set is rather a geometrical notion than a statistical one. In most classification tasks, one often uses statistical criteria: the mean square error (MSE) or minimal absolute error (MAE) — both are well acknowledged for practical classification tasks. On the other hand, by using the maximal margin perceptron as a classifier, in fact we use $Q(w) = \max_i\{-\gamma_w(x_i)\}$ (including negative margin case) as a quality of classification criteria. This corresponds to uniform convergence (l_∞), whereas the MSE corresponds to l_2 convergence. Of course both criteria should be minimized.

Please note that the samples lying close to margin line are the worst misclassified samples in the training data set. The maximal margin classifier is clearly non-robust — one sample may significantly change the location of the hyperplane. On the other hand such an algorithm could be specifically applied to detect outliers in the training data set.

An illustration comparing the varying margin classifier and the SVM linear classifier is presented in the fig. 6. As one can see the execution of the varying margin algorithm leads to finding the exact maximal margin classifier (fig. 6a). Applying the SVM method and its criterion does not necessarily lead to finding the maximal margin hyperplane (fig. 6a). It depends on the suitable setting of the C parameter of SVM (the discussion on C is presented in further sections). Nevertheless, from the classification task's point of view, the hyperplane found by SVM in this case seems to be a better one. It is even more noticeable in the case when the margin in the data is negative.

The figure 6c,d concerns the non-linear separable case. We transform the data using the RBF kernels: $K(x_i, x_j) = \exp(-\frac{\|x_i-x_j\|^2}{2\sigma^2})$. We apply the varying margin perceptron algorithm to learning the linear part of classifier (fig. 6c), and we compare it with the RBF/SVM classifier (fig. 6d).

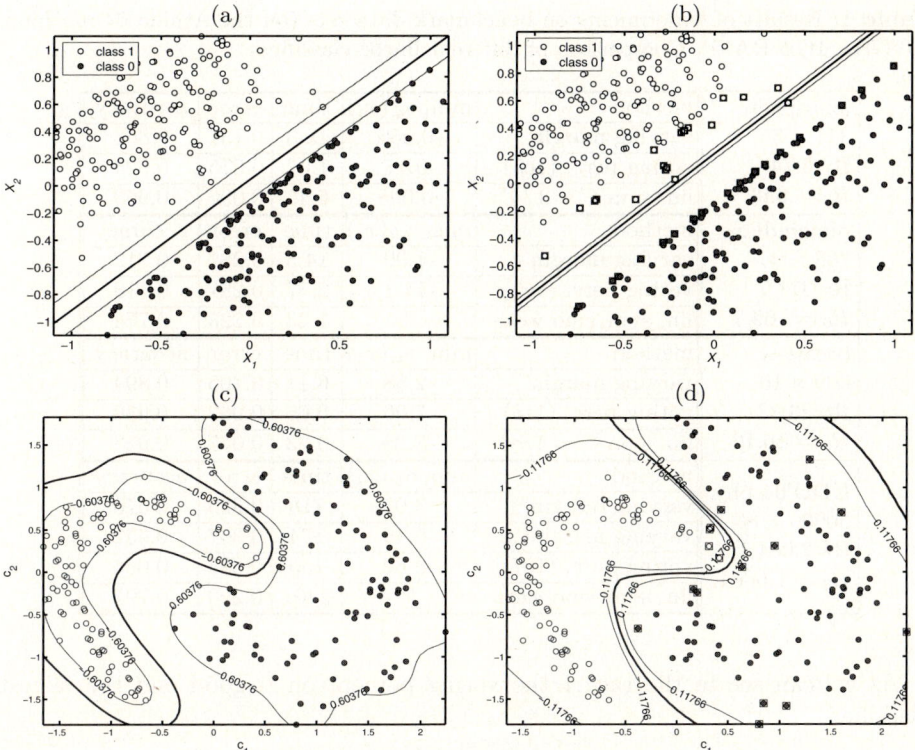

Fig. 6. Comparison of maximal margin classifier (a) and linear SVM classifier (b) in separable case. Non-linear separble case: (c) maximal margin classifier, (d) SVM solution. The Separation line and the margin lines are presented. Data points being the support vectors are marked as squared.

In the table 1 we can find comparison of different linear classifiers with maximal margin classifier, on the different data sets. As we can see in certain cases the maximal margin classifier gives good results. This depends on the kinds of noise in the data. This phenomenon can be observed e.g. in the case of *KDD cup '2004 physics* data set, see the table 1 when maximal margin classifier is quite good. In this data set the value of R is about[2] $2 \cdot 10^4$ whereas the found estimation of the maximal is about -3.02, (note that -5.49 estimation can be found significantly faster).

In fact, this is actually the worst case scenario, i.e. when most of the data points are distributed along the margin closely to the optimal hyperplane and in the same time when R is large and γ is small. This fact is known and is a basic conclusion of classical Novikoff's proof ($k \leq \frac{R^2}{\gamma^2}$). Some analysis on the convergence of the perceptron algorithm and statistical bounds on the number of steps are shown in [6].

---

[2] In all experiments the data sets were not normalised.

Table 1. Results of experiments on benchmark data sets (on the Athlon 64 machine, 2.0 GHz, 1GB RAM). Comparison of different linear classifiers.

ionosphere	method	$\min_i \gamma_w(x_i)$	time	error	accuracy
351×35	varying margin	-0.058	18.3	0.131	0.868
R=5.74	voting perc. (1e5)	-0.17	1.07	0.075	0.924
$R_0 = 6.02$	lin. svm $C = 1/\gamma^*$	-0.065	6.45	0.096	0.903
pimaindians	method	$\min_i \gamma_w(x_i)$	time	error	accuracy
768×9	varying margin	-1.99	14.3	0.287	0.712
R=871.7	voting perc. (1e5)	-11.1	2.57	0.281	0.718
$R_0 = 769.7$	lin. svm/smo weka	?	1.23	0.225	0.774
breast-w	method	$\min_i \gamma_w(x_i)$	time	error	accuracy
699×10	varying margin	-2.58	6.11	0.105	0.894
R=23.83	voting perc. (1e5)	-7.06	0.65	0.062	0.938
$R_0 = 16.12$	lin. svm $C = 1/\gamma^*$	-8.24	0.23	0.077	0.923
KDD'04 phy.	method	$\min_i \gamma_w(x_i)$	time	error	accuracy
50000×80	varying margin	-3.02	718	0.329	0.670
R=2.0e4	varying margin(2)	-5.49	174	0.367	0.632
$R_0 = 1.56e4$	voting perc. (1e7)	-8.82	156	0.335	0.664
	lin. svm/smo weka	?	2461	0.293	0.707

As we can see in the tab .1 the voting perceptron is good and the fastest classifier.

Relation to SVM methods. Recall the quadratic optimisation problem connected with SVM methods [1], [11]. We want to minimize

$$\|w'\|^2 + C \sum_{i=1,\ldots,p} \xi_i \qquad (13)$$

subject to constraints:

$$d_i (\langle w, x_i \rangle + \xi_i) \geqslant 1, i = 1, \ldots, p \qquad (14)$$
$$\xi_i \geqslant 0, i = 1, \ldots, p$$

Suppose we find the optimal solution w, ξ satisfying (13), and $\gamma = \frac{1}{\|w'\|}$ is the margin. The value $\frac{-\xi_i}{\|w'\|}$ can be interpreted as the signed distance from the margin hyperplane $\langle w, x_i \rangle = \pm 1$. Let us rewrite the goal function in the form $\|w'\|(\|w'\| + C \sum_i \frac{-\xi_i}{\|w'\|})$. Both terms will be in a comparable scale when the C value is equal to $\|w'\|$ or $\frac{1}{\gamma}$. In the case when the C value is less than $\frac{1}{\gamma}$ we can expect the greater number of support vectors, on the other hand when $C \gg \frac{1}{\gamma}$ then we only obtain the support vectors lying close to the margin lines, in this case the $\xi_i \approx 0$.

The negative margin case is more complicated. If we choose a suitably large number for C then all the samples with negative margin become support vectors.

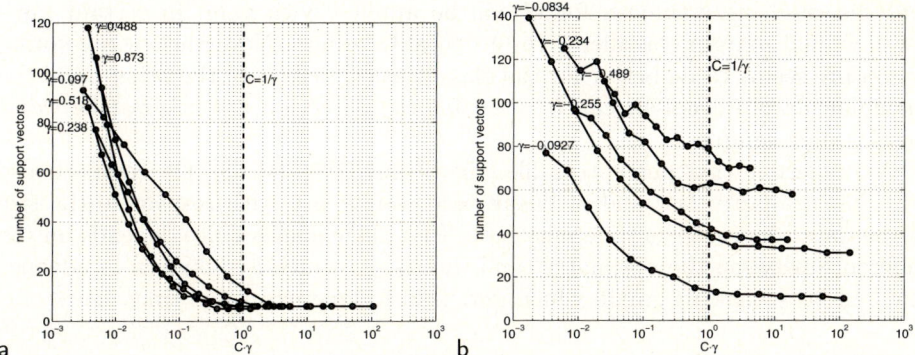

Fig. 7. Illustration of how the number of support vectors changes along with the change of C. (a) Positive margin and (b) negative margin in the data. The value of 1 on the horizontal axis means $C = 1/\gamma$ for the positive margin case and $C = 1/|\gamma|$ for the negative margin case. The all plots are unified for better understanding.

The number of support vectors depends on the distribution of the margins of samples around the optimal hyperplane.

In the fig. 7 we can see how the number of support vectors changes along with the change of C. In this example we compare different artificial data sets. The figure 7(a) concerns the case when the data set has a positive margin, and the figure 7(a) concerns a negative margin. The data sets were chosen randomly with different kinds of distributions and different margins. The number of samples was between 80 and 150 and there were 5 attributes. The maximal margin was estimated using the algorithm 5. The value of C changes from $1/\sum_{\{i:\ \langle w^*, x_i\rangle < 0\}} \langle w^*, x_i \rangle$ to $1/\gamma^{*2}$, where γ^* is the estimated maximal margin, and w^* is the obtained vector of weights.

One can observe that the number of support vectors rapidly decreases when C rises to the value of $\frac{1}{\gamma^*}$ and slowly becomes fixed when the value of C is greater than $\frac{1}{\gamma^*}$. We want to recall that some SVM learning algorithms have the time of computation depended on C (see e.g. [5]).

5 Conclusions

In the paper a γ-margin generalisation of the classical Rosenblatt's perceptron was proposed with the convergence proof. Two modifications of the proposed algorithm were also shown, where the γ value is not fixed but varying. These modifications can be successfully applied to estimate the maximal margin of separation in the data.

Despite the fact that the maximal margin in the data is rather a geometrical notion than a statistical one, one can use it to speed up the construction of known classifiers like SVM's by pre-estimating the value of C. It should be stressed that this parameter is commonly estimated by means of the cross-validation procedure which is very time-consuming. Experiments show that our algorithms and the resulting estimates for C can be applied with gain. In certain cases the maximal margin classifier gives acceptable results. Nevertheless the voting procedure seems to be better for the classification tasks. On the other hand the maximal margin classifier can be used for the detection of outliers (the worst misclassified samples).

We underline that perceptron-like algorithms ([10], [4], [8], [2], compare also with [5]) are fairly fast and of linear time complexity with respect to the size of the data set. However, one must note that their convergence depends on the difference between the optimal γ^* and the γ_{update} which is unknown in advance and can be arbitrarily small, which may lead to long execution time.

References

1. Burgers, J.C.: A tutorial on support vector machines for pattern recognition. Data Mining and Knowledge Discowery 2, 121–167 (1998)
2. Collobert, R., Bengio, S.: Links between perceptrons, mlps and svms. In: Proc. of the 21st International Conference on Machine Learning, Banff, Canada (2004)
3. Blake, C.L., Newman, D.J., Hettich, S., Merz, C.J.: UCI repository of machine learning databases (1998)
4. Freund, Y., Schapire, R.E.: Large margin classification using the perceptron algorithm. Machine Learning 37(3), 277–296 (1999)
5. Joachims, T.: Training linear SVMs in linear time. In: KDD 2006, Pennsylvania, USA (2006)
6. Klęsk, P., Korzeń, M.: Perceptron algorithm: notes on statistically expected convergence. Polish Journal of Environmental Studies 15(4C) (2006)
7. Kowalczyk, A., Smola, A.J., Wiliamson, R.C.: Kernel machines and boolean functions. Advances in Neural Information Processing Systems 14 (2002)
8. Li, Y., Long, P.M.: The relaxed online maximum margin algorithm. Machine Learning 46(1-3), 361–378 (2002)
9. Novikoff, A.B.J.: On convergence proofs on perceptrons. In: Proceedings of the Symposium on the Mathematical Theory of Automata, Polytechnic Institute of Brooklyn, vol. XII, pp. 615–622 (1962)
10. Rosenblatt, F.: Principles of neurodynamics: Perceptron and theory of brain mechanisms. Spartan Books, Washington D.C (1962)
11. Vapnik, V.N.: The Nature of Statistical Learning Theory. In: Statistics for Engineering and Information Science. Springer, New York (2000)

Classes of Kernels for Hit Definition in Compound Screening

Karol Kozak[1] and Katarzyna Stapor[2]

[1] ETH, Schafmattstr. 18,
8093 Zurich, Switzerland
kozakk@bc.biol.ethz.ch
[2] Silesian Technical University, Akademicak 16,
44-100 Gliwice, Poland
katarzyna.stapor@polsl.pl

Abstract. In this paper we analyze Support Vector Machine (SVM) algorithm to the problem of chemical compounds screening with a desired activity, definition of hits. The support vector machine transforms the input data in an (unknown) high dimensional feature space and the kernel technique is applied to calculate the inner-product of feature data.The problem of automatically tuning multiple parameters for pattern recognition SVMs using our new introduced kernel for chemical compounds is considered. This is done by simple eigen analysis method which is applied to the matrix of the same dimension as the kernel matrix to find the structure of feature data, and to find the kernel parameter accordingly. We characterize distribution of data by the principle component analysis method.

1 Introduction

Biological screening processes are used in drug discovery to screen large numbers of compounds against a biological target. By quantifying structure-activity relationship (QSAR) means a group of bioinformatics or chemoinformatics algorithms, typically using physico-chemical parameters or three-dimensional molecular fields with statistical techniques. Finding active compounds amongst those screened, QSAR algorithms are aimed at planning additional active compounds without having to screen each of them individually. Classically, QSAR models are designed by representing molecules by a large set of descriptors, i.e. by a high dimensional vector, and then applying some kind of Pattern Recognition methods. Modeling the relationship between the chemical structure, descriptors and activities of sampled compounds provides an insight into the QSAR, gives predictions for untested compounds of interest, and guides future screening. Recently, the support vector machine (SVM) has drawn much attention in QSAR due to its excellent generalization performance. One of the most important design choices for SVMs is the kernel-parameter, which implicitly defines the structure of the high dimensional feature space where a maximal margin hyperplane will be found. However, before this stage is reached in the use of SVMs, the actual kernel must be chosen, different kernels may exhibit vastly different performance. Our previous work [7] focused on graph-structured data on screening compounds and the on development of kernel methods for graph-structured data in combination with molecular descriptors, with particular emphasis on QSAR in the screening process and the prediction of biological activity of chemical compounds. We combined graph kernel based on chemical structure and Gausian kernel where we considered Burden descriptors [1] as features of high dimensional space. Model selection of the SVM involves two hyperparameters: the penalty parameter C and the kernel parameter γ used in Gaussian kernel. The selection of suitable parameters for Gaussian kernel used in proposed screening kernel is crucial for effective classification. In this paper we analyze the distribution of data in the feature space on the basis of the principal component analysis (PCA) method, and propose a simple method that predicts a good set of Gaussian kernel parameter γ quickly for a particular problem domain in chemical compounds. PCA is designed to be very simple and efficient to compute in compare to other optimization methods. Moreover, the method is completely general in that it can be used for any data classification problem.

This paper is organized as follows: Section 2 outlines the theory behind SVM classifiers with a particular emphasis on kernel functions. Next we review the basic theory of kernel functions for chemical compounds. Section 3 describes the PCA method in the feature space and the properties of feature data structure based on kernel parameter. In section 4 we experimentally evaluate our approach. The conclusion is given will in section 5.

2 Support Vector Machine Classification

The SVM is a structural risk minimization based classifier which finds an optimum decision region and is using kernel functions. Let us have a training data set $\{m_i; \omega_i\}$; $i = 1,..., l$, where $\omega_i \in \{-1,1\}$ represents the label of an arbitrary example $\omega_i \in R^d$, d being the dimension of the input space. Also, let us define a linear decision surface by the equation $f(x) = w{:}m+b = 0$. The original formulation of support vector machine algorithm seeks a linear decision surface that maximizes the margin between positive and negative examples.

SVMs exactly implement the idea of regularized risk minimization by maximizing the margin between the two classes. This can be achieved through the minimization of $\|w\|^2$ which solution is

$$w = \sum_{i=1}^{l} \alpha_i \omega_i m_i \quad (1)$$

with the constraint $\sum_i \alpha_i \omega_i = 0$. The parameters α_i are found by maximizing the dual objective [12,13]

$$L_D = \sum_i \alpha_i - \frac{1}{2} \sum_{i,j} \alpha_i \alpha_j \omega_i \omega_j m_i m_j \quad (2)$$

However, in real-life classification problems, the algorithm as stated above is unable to achieve perfect separation between two classes especially in case of noisy data. Cortes et al. in [2] slightly modified the model by adding a heuristic that accounts for accepting misclassified examples while penalizing them. Mathematically, this does not imply any major modification except that α_i must be upper bounded as $0 \leq \alpha_i \geq C$ where C is a penalization parameter also called trade-off parameter. An infinite value for C yields a classifier that seeks a well separated data. In another work, Boser et al. [16] added an important feature that allows a different insight into support vector theory. In fact, they enable these models to produce complex nonlinear boundaries in the original space. The technique consists of projecting the data into higher order spaces of possibly infinite dimension through a mapping function φ. One key aspect of the SVM model is that the data enters the above expressions only in the form of the dot product of pairs. This leads to the resolution of the second problem mentioned above, namely that of non-linearly separable data. The basic idea with SVMs is to map the training data into a higher dimensional feature space via some mapping $\varphi(m)$ and construct a separating hyperplane with maximum margin there. This yields a non-linear decision boundary in the original input space.

Kernel can deal in a uniform way with a multitude of data types and can be used to detect many types of relations in data. In kernel methods, examples are mapped into a feature space implicitly, and only the inner products of the vector representations are used when learning machines access the examples. This means that even in cases where the dimension of the vector representations is extremely high, the dimensions do not explicitly appear in the process of training and classification as long as an efficient procedure to compute the inner products is available.

We assume that m is mapped to p dimensional kernel feature space F by $\Phi(m) \in F$ $m = (m_1,\ldots,m_n)t \rightarrow \Phi(m)=(\Phi_1(m),\ldots, \Phi_p(m))^t$. Kernel methods computes inner product $K(m,m') = \langle \phi(m) \cdot \phi(m') \rangle = \phi(m)^t \cdot \phi(m')$, without referring the feature $\Phi_i(i=1,\ldots,p)$ directly. Let K be $K_{ij} = \langle \phi(m_i) \cdot \phi(m'_j) \rangle = k(m_i, m'_j)$ then K= $\Phi^t \Phi$ with $\Phi=(\Phi(x_1),\ldots,\Phi(x_l))$ (sample matrix). Typical choice for kernels is listed in table 2.1.

Clearly the structures present in screening data are essential for the more-or-less automated extraction of meaning, patterns, and regularities. Here we focus on graph-structured data on screening compound and use our application of kernel methods for graph-structured data in combination with molecular descriptors, with particular emphasis on QSAR in the screening process and the prediction of biological activity of

Table 1. Common kernels

Kernels	Formula
Linear	K(m, m')= x. m'
Sigmoid	K(m, m') = tanh(a m. m'+b)
Polynomial	K(m, m') = (1+ m. m')^d
RBF	K(m, m') = exp(- γ ‖ m - m'‖^2)
Gaussian	K(m, m') = exp(- γ ‖ m - m'‖)

chemical compounds. Because chemical compounds are often represented by the graph of their covalent bonds, pattern recognition methods in this domain must be capable of processing graphical structures with variable size. Let us assume now we have two molecules m and m', which have atoms $a_1,...,a_{|m|}$ and $a'_1,..., a'_{|m'|}$. Let us further assume we have a kernel k_{nei}, which compares a pair of atoms $(a_h, a'_{h'})$ from both molecules, including information on their neighborhoods, membership to certain structural elements and other characteristics. Then a valid graph kernel between m, m' is the optimal assignment kernel [5]:

$$k_{OA}(m,m') := \begin{cases} \max_\pi \sum_{h=1}^{|m|} k_{nei}(a_h, a'_{\pi(h)}) & if\, |m'| \geq |m| \\ \max_\pi \sum_{j=1}^{|m'|} k_{nei}(a_{\pi(h')}, a'_{h'}) & otherwise \end{cases}, \quad (3)$$

where k_{nei} is calculated based on two kernels k_{atom} and k_{bond} which compare the atom and bond features, respectively. Each atom of the smaller of both molecules is assigned to exactly one atom of the larger molecule such that the overall similarity score is maximized. To prevent larger molecules automatically achieving a higher kernel value than smaller ones, kernel is normalized (Schoelkopf & Smola, 2002) [12], i.e.

$$k_{OA}(m,m') \leftarrow \frac{k_{OA}(m,m')}{\sqrt{k_{OA}(m,m)k_{OA}(m',m')}}. \quad (4)$$

OA graph kernel, is of general use for QSAR problems, but is more focused on screening data. In developing a good quality QSAR model for these data we combined OA graph kernel and kernel where we considered Burden descriptors [1]. Giving two molecules m and m', the basic idea of our method is to construct a kernel $k(m, m')$ which measures the similarity between *m* and *m'*. Gaussian kernel was chosen because it readily produces a closed decision boundary, which is consistent with the method used to select the molecular descriptors.

$$k_{Gaus}(m,m') = \exp(-\gamma\, |m - m'|) \quad (5)$$

Calculating the OA graph kernel and at same time the Gaussian kernel with eight Burden features give us two ways of comparing molecules: looking at the chemical structure and looking at molecular activity. Having two values from these two kernels and making, at the same time an average, gives us more precise molecule similarity information.

$$K_{scr}(m,m') = \frac{K_{Gaus}(m,m') + K_{OA}(m,m')}{2}. \quad (6)$$

To improve accuracy of classification in screening data using introduced screening kernel we want also optimized parameters in Gausian kernel where we use as properties Burden descriptors [1]. The parameter γ in Gausian kernel can be determined by the *loo* bound described below. Vapnik [14] showed that the following radius margin bound holds for SVM without the bias term b.

$$loo \leq 4R^2 \|w\|^2 \tag{7}$$

where *loo* is the number of leave-one-out error, ◆ is the solution of (1) and R is the radius of the smallest sphere. The R^2 can be obtained as the objective function value of the following optimization problem:

$$\max \quad 1 - \beta^T K \beta \tag{8}$$

$$\text{subject to} \quad \sum_{i=1}^{l} \beta_i = 1, \tag{9}$$

$$\beta_i \geq 0, i = 1, 2, ..., l,$$

where $K_{ij} = K(m_i, m'_j)$

Vapnik and Chapelle [14] further extend the bound for the general case where b is present. The optimization problem [6] is needed to solve to get the radius in each iterative step for minimizing the *loo* bound. As we consider the Gaussian kernel as one part of screening kernel, we may think that data structure of the input data in the feature space changes by changing the kernel parameter. Intuitively, we get different distribution of the same input data in the feature space by changing the kernel parameter. From the error bound in equation (7), we see that R^2 and $\|w\|^2$ both depend on the structure of feature data. Therefore, the performance of the SVM depends on the distribution of the input data in the feature space. For our Gaussian kernel used with graph kernel we analyze the distribution of data in the feature space on the basis of the principal component analysis (PCA) method, and apply a simple method that predicts a good set of kernel parameters quickly.

3 PCA

Once the files have been Performing PCA in kernel space, we assume that p dimensional linear space is transformed to d dimensional space. Let n_i be an orthonormal vector of k-dimensional linear sub-space, and let $A = (n_1, n_2, ..., n_k)$. Then, $\Phi(m)$ is transformed into H subspace of space F by the linear operator A as

$$\psi(m_i) = A^T \phi(m_i) \in R^k. \tag{10}$$

Let 1 be a 1 x l matrix of all ones and $\Phi = (\Phi(m_1), \Phi(m_2), ..., \Phi(m_l))$. The covariance matrix of feature vectors Φ in the feature space F is given by

$$\sum_\phi = \frac{1}{l} \phi (I - \frac{1}{l})(I - \frac{1}{l}) \phi^T. \tag{11}$$

The covariance matrix of $\Phi(m_i)$ in the k-dimensional subspace is given by

$$\sum_\psi = A^T \sum_\phi A. \tag{12}$$

The orthonormal bases n_i lie in the span of $\Phi(m_1), \Phi(m_2),\ldots, \Phi(m_l)$. We can represent them by the linear combination of $(\Phi(m_i) - \Phi) \in F$ where $\Phi = \sum_{i=1}^{l} \phi(m_i)$ (centering the data). Then, n_i can be written as

$$n_i = \sum_{i=1}^{l} s_{ij}(\phi(m_i) - \phi), \qquad (13)$$

where $s_i = (s_{i1}, s_{i2},\ldots,s_{ik})^T$ is a coefficient vector. Letting $S = (s_1, s_2,\ldots,s_k)$,

$$A = \phi(I - \frac{1}{l})S. \qquad (14)$$

Let $G = (I - \frac{1}{l})\phi^T \phi (I - \frac{1}{l})$ where the kernel matrix is $K = \Phi^T \Phi$. The covariance matrix of feature vectors is given by

$$\sum_{\psi} = \frac{1}{l} S^T G^2 S, \qquad (15)$$

and

$$A^T A = G^T GS. \qquad (16)$$

The S can be calculated according to the PCA variance maximization criterion [8ref], [9ref], i.e., maximizing variances such that $A^T A = I$. This gives the Lagrangian $L(S;\Lambda) = (1/l)tr(S^T G^2 S) - tr(\ (S^T G^2 S - I)\ \Lambda)$ with Lagrange multiplier $\Lambda > 0$ where is a diagonal matrix. Then setting

$$\frac{\partial L}{\partial S} = \frac{2}{l} G^2 S - 2 GS \Lambda \qquad (17)$$

by zero, we obtain Karush-Kuhn-Tucker(KKT) condition

$$\frac{1}{l} G^2 S = GS\Lambda, \qquad (18)$$

$$S^T GS = I_k \qquad (19)$$

Let, i-th eigenvalue of G be μ_i with corresponding eigenvectors v_i, and $\mu_1, \mu_2,\ldots, \mu_k \geq 0$. Since G is symmetric matrix, eigenvectors constitute orthonormal set. We denote $V = (v_1, v_2,\ldots,v_k)$, and a k x k diagonal matrix M whose $M_{ii} = \mu_i$. If we consider $S = VM^{-\frac{1}{2}}$ where

$$[M^{-\frac{1}{2}}]_{i,j} = \begin{cases} \frac{1}{\sqrt{\mu_i}} & if \quad i = j \\ 0 & if \quad i \neq j \end{cases} \qquad (20)$$

And eq. (18) can be transformed to

$$\text{(left side)} = \frac{1}{l}G^2 S = \frac{1}{l}VM^2 M^{-\frac{1}{2}} \tag{21}$$

$$\text{(right side)} = GS\Lambda = VMM^{-\frac{1}{2}}\Lambda \tag{22}$$

by using $GV=VM$ and V, M are non zero matrix.
Since $GV = VM$,

$$\sum_{\psi} = \frac{1}{l}S^T G^2 S = \frac{1}{l}M^{-\frac{1}{2}}V^T G^2 VM^{-\frac{1}{2}} = \frac{1}{l}M^{-\frac{1}{2}}V^T GVMM^{-\frac{1}{2}} =$$
$$= \frac{1}{l}M^{-\frac{1}{2}}V^T VMM^{-\frac{1}{2}} = \frac{1}{l}M^{-\frac{1}{2}}IMM^{-\frac{1}{2}} = \frac{1}{l}M. \tag{23}$$

Thus eigenvalues of $\sum_{\psi}$ equal to the eigenvalues of $(1/l)G$. From the eigenvalues and corresponding eigenvectors we may get the abstraction of the data. Usually, eigenvalues determine the structure of the data. This is immaterial since we need only the positive eigenvalues of G.

3.1 Kernel Parameter Optimization

For optimization of γ parameter we used method presented by Debnath and Takahashi [9]. Eigenvalues, μ_i's, are large for some kernel parameters γ means data spread in a large area. So we may not get good performance for those γ for which eigenvalues are large enough. If eigenvalues, μ_i's are too small, however the number of positive eigenvalues are large, data lie in a very compact area. Thus the number of support vectors will be large and the support vector machine may implement a rapidly oscillating function rather than the smooth mapping. As a results, the performance will not be good. In [6], Chung et. al. showed that a good bound should avoid minima that happens at the boundary (i.e., too small and too large γ do not show good performance). In [10], Keerthi and Lin also showed that too small γ creates overfitting/underfitting problems, and too large γ creates underfitting problem. For large and small value of kernel parameter γ, our ideas follow Chung et. al. and Keerthi and Lin's idea. When the number of positive eigenvalues of G is small enough, many feature data linearly depend on other feature data. Let the k-th eigenvalue of G be μ_k with corresponding eigenvectors $\upsilon_k \in R_l$ and $\Phi'(xi) = \Phi(xi) - \Phi$.

$$\mu_k = \upsilon_k^T G \upsilon_k = \upsilon_k^T (I - \frac{1}{l})\phi^T \phi (I - \frac{1}{l})\upsilon_k = (\sum_{i=1}^{l} \upsilon_{ki}(\phi(x_i) - \phi))^T \sum_{j=1}^{l} \upsilon_{kj}(\phi(x_j) - \phi) =$$
$$= (\sum_{i=1}^{l} \upsilon_{ki}\phi(x_i))^T \sum_{j=1}^{l} \upsilon_{kj}\phi(x_j) \tag{24}$$

As the K=$\Phi^T\Phi$ is the rank one matrix when $\gamma \to \infty$, many eigenvalues of G become zero (or smaller than a small threshold value) when the value of γ is large. The μ_k is zero means $\sum_{i=1}^{l} v_{ki}\phi'(x_i) = 0$, i.e. some data are linearly depend on other data. So for large value of γ, we loose information for many data as some feature data are becoming the linear combination of others, and the kernel matrix K does not have their information. The small number of positive eigenvalues means the SVM learns with a small fraction of training data from the training set. Thus, very small number of positive eigenvalues can not show good performance. Considering all above conditions, good kernel parameters must avoid either too large or too small eigenvalues, and the G has sufficient number of positive eigenvalues. Therefore, given a set of kernel parameters we choose a subset of kernel parameters for which G has sufficient number of positive eigenvalues and the largest eigenvalue μ_1, is not too large. This will give a good set of kernel parameters. In the following section we will show our experimental results that use screening kernel with optimized parameters.

4 Experimental Evaluations

We experimentally evaluated the performance of graphical kernel where parameters are optimized in a classification algorithm and compared it against that achieved by earlier kernels on a variety of chemical compound datasets.

4.1 Datasets

We used two different public available datasets to derive a total of five different classification problems. The first dataset was obtained from the National Cancer Institute's DTP AIDS Anti-viral Screen program [3, 11]. Each compound in the dataset is evaluated for evidence of anti-HIV activity. The second dataset was obtained from the Center of Computational Drug Discovery's anthrax project at the University of Oxford [7]. The goal of this project was to discover small molecules that would bind with the heptameric protective antigen component of the anthrax toxin, and prevent it from spreading its toxic effects. For these datasets we generated 8 features, called Burden descriptors [1] (eight dimensional spaces).

4.2 Results

We tested a proposed extension of molecule graph kernel in two benchmark experiments of chemical compound classification. We use the standard SVM algorithm for binary classification described previously. The regularization factor of SVM was fixed to *C = 10*. In order to see the effect of generalization performance on the size of training data set and model complexity, experiments were carried out by varying

Table 2. Common kernels. Recognition performance comparison of K_{OA} with K_{scr} kernel and K_{scr} with optimized parameters ($K_{scr\ OP}$) for different number of training sets. Numbers represent correct classification rate [%].

Training set	OA Kernel			Kscr			Kscr OP		
	DTP	Toxic	Diff.	DTP	Toxic	Diff.	DTP	Toxic	Diff.
100	79.6	73.9	5.5	79.1	74.9	4.9	79.5	73.2	5.3
200	79.9	74.1	5.8	80.1	75.6	4.5	81.0	77.1	3.9
300	80.1	79.1	1.0	80.8	79.6	1.2	82.2	80.3	1.9
400	81.3	79.8	1.5	82.5	80.3	2.2	81.9	80.8	1.1
500	82.5	81.3	1.2	83.2	81.5	1.8	83.6	82.0	1.6

the number of training samples (100, 200, 300, 400, 500) according to a 5-fold cross validation evaluation of the generalization error.

The $K_{scr\ OP}$ shows performance improvements over the standard K_{scr} and K_{OA} kernel, for both of the (noisy) and real data sets. Moreover it is worth mentioning that K_{scr} does slightly better than K_{OA} kernel in general. We would like now to determine if optimized $K_{scr\ OP}$ kernel has an effect in SVM classification in comparison to K_{scr} and K_{OA}. A permutation test was selected as an alternative way to test for differences in our kernels in a nonparametric fashion (so we do not assume that the population has a normal distribution, or any particular distribution and, therefore, do not make distributional assumptions about the sampling distribution of the test statistics). The R package "exactRankTests" [17] was used for permutation test calculation. Table 3 lists the 9 calculation of SVM accuracy on different number of test sets and results from the test. This Table shows four columns for each pair of compared different kernel methods (both data sets), the first and second giving the classification accuracy, while the last two columns have the raw (i.e., unadjusted) t-statistic result and p-values computed by the resampling algorithm described in [15]. The permutation test based on 2000 sample replacements estimated a p-value to decide whether or not to reject the null hypothesis. The null hypotheses for this test were $H0_1$: $K_{scr\ OP} = K_{scr}$, and alternative hypothesis HA_1: $K_{scr\ OP} > K_{scr}$, additionally let's assume at a significance level $\alpha = 0.05$. The permutation test will reject the null hypothesis if the estimated P-value is less than α. More specifically, for any value of $\alpha < p$-value, fail to reject H_0, and for any value of $\alpha \geq P$-value, reject $H0$. The P-values on average for DTS AIDS and toxic data sets of 0.0527, 0.0025, 0.0426, 0.0020 indicates that the SVM with $K_{scr\ OP}$ kernel is probably not equal to K_{scr} kernel. The P-value 0.0527 between $K_{scr\ OP}$ and K_{scr} kernel for DTS AIDS data sets, indicates weak evidence against the null hypothesis. There is strong evidence that all other tests null hypothesis can be rejected. Permutation tests suggest, on average, that K_{scr} kernel for screening data is statistically significantly larger than $K_{OA\ OP}$ and RBF.

Table 3. Statistical test between $K_{scr\ OP}$ and K_{scr} kernel for different number of training subset Numbers represent correct classification rate [%], t-statistic (without permutation) and calculated p-value from permutation test. Calculated t*-statistic to the new data set with replacement 2000 times gave result in average $t^*_{Min} = 0.852$, $t^*_{Max} = 1.254$.

γ	DTP AIDS			
	Kscr	Kscr optimized parameter	t-stat	p-value
100	79.8	79.6	3.78	0.0231
150	80.1	79.9	4.12	0.0633
200	80.8	80.1	2.59	0.0712
250	82.5	81.3	3.22	0.0541
300	83.2	82.5	2.99	0.0634
350	84.2	82.6	-3.12	0.0773
400	84.9	83.6	3.02	0.0424
450	84.8	83.9	-2.89	0.0561
500	83.6	82.8	4.55	0.0233
			Average:	0.0527

5 Conclusion

We propose a method for finding good kernel parameter in graph kernel for chemical compounds only applying the eigen analysis method that is suitable for discriminative classification with unordered sets of local features. We have applied tuned gausian kernel our screening kernel to SVM-based object recognition tasks, and demonstrated recognition performance with accuracy comparable to existing kernels on screening data. Our experimental evaluation showed that parameter tuning leads to substantially better results than those obtained by existing same kernel without optimization.

References

1. Burden, F.R.: Molecular Identification Number For Substructure Searches. Journal of Chemical Information and Computer Sciences 29, 225–227 (1989)
2. Cortes, C., Vapnik, V.: Support-vector networks. Machine Learning 20(3), 273–297 (1995)
3. dtp.nci.nih.gov. Dtp aids antiviral screen dataset
4. Graham, W.R.: Virtual screening using grid computing: the screensaver project. Nature Reviews: Drug Discovery 1, 551–554 (2002)
5. Froehlich, H., Wegner, J.K., Zell, A.: QSAR Comb. Sci. 23, 311–318 (2004)
6. Chung, K.-M., Kao, W.-C., Sun, T., Wang, L.-L., Lin, C.-J.: Radious margin bounds for support vector machines with Gaussian kernel. Neural Computation 15, 2643–2681 (2003)
7. Kozak, K., Kozak, M., Stapor, K.: Kernels for Chemical Compounds in Biological Screening. In: Beliczynski, B., Dzielinski, A., Iwanowski, M., Ribeiro, B. (eds.) ICANNGA 2007. LNCS, vol. 4432, pp. 327–337. Springer, Heidelberg (2007)
8. Pearlman, R.S., Smith, K.M.: Novel software tools for chemical diversity. Perspectives in Drug Discovery and Design, 9/10/11, 339–353 (1998)

9. Debnath, R., Takahashi, H.: Generalization of kernel PCA and automatic parameter tuning. Neural Information Processing 5(3) (December 2004)
10. Keerthi, S.S., Lin, C.-J.: Asymptotic behaviours of support vector machines with Gaussian kernel. Neural Computation 15, 1667–1689 (2003)
11. Kramer, S., De Raedt, L., Helma, C.: Molecular feature mining in hiv data. In: 7th International Conference on Knowledge Discovery and Data Mining (2001)
12. Schoelkopf, B., Smola, A.J.: Learning with kernels. MIT Press, Cambridge (2002)
13. Vapnik, V.N.: Stasistical Learning Theory. Wiley, New York (1998)
14. Vapnik, V., Chapelle, O.: Bounds on error expectation for support vector machines. Neural computation 12(9), 2013–2036 (2000)
15. Westfall, P.H., Young, S.S.: Resampling-based multiple testing: Examples and methods for p-value adjustment. John Wiley & Sons, Chichester (1993)
16. Boser, Y., Guyon, I., Vapnik, V.: A training algorithm for optimal margin classifiers. In: Fifth Annual Workshop on Computational Learning Theory, Pittsburg (1992)
17. "exactRankTests": Exact Distributions for Rank and Permutation Tests, http://cran.r-project.org/src/contrib/Descriptions/exactRankTests.html.

The GA-Based Bayes-Optimal Feature Extraction Procedure Applied to the Supervised Pattern Recognition

Marek Kurzynski and Aleksander Rewak

Wroclaw University of Technology, Chair of Systems and Computer Networks,
Wyb. Wyspianskiego 27, 50-370 Wroclaw, Poland
{marek.kurzynski,aleksander.rewak}@pwr.wroc.pl

Abstract. The paper deals with the extraction of features for statistical pattern recognition. Bayes probability of correct classification is adopted as the extraction criterion. The problem with complete probabilistic information is discussed and next the Bayes-optimal feature extraction procedure for the supervised classfication is presented in detail. As method of solution of optimal feature extraction a genetic algorithm is proposed. Several computer experiments for wide spectrum of cases were made and their results demonstrating capability of proposed approach to solve feature extraction problem are presented.

1 Introduction

Feature dimension reduction has been an important and long-stading research problem in statistical pattern recognition. In general, dimension reduction can be defined as a transformation from original high-dimensional space to low-dimensional space where an accurate classifier can be constructed.

There are two main methods of dimensionality reduction ([4], [5]): *feature selection* in which we select the best possible subset of input features and *feature extraction* consisting in finding a transformation (usually linear) to a lower dimensional space. Although feature selection preserves the original physical meaning of selected features, it costs a great degree of time complexity for an exhaustive comparison if a large number of features is to be selected. In contrast, feature extraction is considered to create a new and smaller feature set by combining the original features. We shall concentrate here on feature extraction for the sake of flexibility and effectiveness [10].

There are many effective methods of feature extraction. Linear Discriminant Analysis (LDA) and Principal Component Analysis (PCA) are the two popular feature extraction algorithms [5]. Both of them extract features by projecting the feature vectors into a new feature space through a linear transformation. But they optimize the transformation matrix with different intention. PCA optimizes the transformation matrix by finding the largest variation in the original feature space. LDA pursues the largest ratio of between-class variation and within-class variation when projecting the original feature space to a derivative space. The drawback of independent feature extraction algorithm is that their optimization criteria are different from the classifier minimum classification error criterion, which may degrade the performance of clasifier. A direct way to overcome this problem is to conduct feature extraction and classification jointly with a consistent criterion. Minimum Classification Error (MCE) training algorithm provides such an integrated framework [11].

As it seems, the Bayes probability of error (or equivalently, the Bayes probability of correct classification) i.e. the lowest attainable classification error is the most appropriate criterion for feature extraction procedure [1]. Unfortunately, this criterion is very complex for mathematical treatment, therefore researches have restored to other criteria like various functions of scatter matrices (e.g. Fisher criterion) or measures related to the Bayes error (e.g. Bhattacharyya distance) [5], [4].

In this paper we formulate the optimal feature extraction problem adopting the Bayes probability of correct classification as an optimality criterion. Since this problem cannot be directly solved using analytical ways (except simple cases including for example Gaussian, triangular or uniform distributions [13]), we propose to apply genetic algorithm (GA), which is very-well known heuristic optimization procedure and has been successfully applied to a broad spectrum of optimization problems, including many pattern recognition and classification tasks [17], [18]. This concept of feature extraction via GA was applied to the case of supervised pattern recognition for which several experiments were made. Obtained results demonstrate effectiveness of proposed feature extraction procedure and its superiority over PCA method.

The contents of the paper are as follows. In section 2 we introduce necessary background and formulate the Bayes-optimal feature extraction problem. In section 3 and 4 optimization procedures for the cases of complete probabilistic information and recognition with learning are presented and discussed in detail. Section 5 describes computer experiments for wide range of cases which were carried out in order to evaluate effectiveness and quality of proposed feature extraction method.

2 Preliminaries and the Problem Statement

Let us consider the pattern recognition problem with probabilistic model. This means that n-dimensional vector of features describing recognized pattern $x = (x_1, x_2, ..., x_n)^T \in \mathcal{X} \subseteq \mathcal{R}^n$ and its class number $j \in \mathcal{M} = \{1, 2, ..., M\}$ are observed values of a pair of random variables $(\mathbf{X}, \mathbf{J})$, respectively. Its probability distribution is given by *a priori* probabilities of classes

$$p_j = P(\mathbf{J} = j), \; j \in \mathcal{M} \qquad (1)$$

and class-conditional probability density function (CPDFs) of $\mathbf{X}$

$$f_j(x) = f(x/j), \; x \in \mathcal{X}, \; j \in \mathcal{M}. \qquad (2)$$

In order to reduce dimensionality of feature space let consider linear transformation

$$y = Ax, \tag{3}$$

which maps n-dimensional input feature space $\mathcal{X}$ into m-dimensional derivative feature space $\mathcal{Y} \subseteq \mathcal{R}^m$, or - under assumption that $m < n$ - reduces dimensionality of space of object descriptors. It is obvious, that y is a vector of observed values of m-dimensional random variable $\mathbf{Y}$, which probability distribution given by CPDFs depends on mapping matrix A, viz.

$$g(y/j; A) = g_j(y; A), \quad y \in \mathcal{Y}, \ j \in \mathcal{M}. \tag{4}$$

Let introduce now a criterion function $Q(A)$ which evaluates discriminative ability of features y, i.e. Q states a measure of feature extraction mapping (3). As a criterion Q any measure can be involved which evaluates both the relevance of features based on a feature capacity to discriminate between classes or quality of a recognition algorithm used later to built the final classifier. In the further considerations the Bayes probability of correct classification will be used, namely

$$Q(A) = Pc(A) = \int_{\mathcal{Y}} \max_{j \in \mathcal{M}} \{p_j \, g_j(y; A)\} \, dy. \tag{5}$$

Then the feature extraction problem can be formulated as follows: for given *priors* (1), CPDFs (2) and reduced dimension m find the matrix A^* for which

$$Pc(A^*) = \max_A Pc(A). \tag{6}$$

3 Optimization Procedure

In order to solve (6) first we must explicitly determine CPDFs (4). Let introduce the vector $\bar{y} = (y, x_1, x_2, ..., x_{n-m})^T$ and linear transformation

$$\bar{y} = \bar{A} \, x, \tag{7}$$

where

$$\bar{A} = \begin{bmatrix} A \\ \hline I \mid 0 \end{bmatrix} \tag{8}$$

is a square matrix $n \times n$. For given y equation (7) has an unique solution given by Cramer formulas

$$x_k(y) = \mid \bar{A}_k(y) \mid \cdot \mid \bar{A} \mid^{-1}, \tag{9}$$

where $\bar{A}_k(y)$ denotes matrix with k-th column replaced with vector $\bar{y}$. Hence putting (9) into (2) and (4) we get CPDFs of $\bar{y}$ ([6]):

$$\bar{g}_j(\bar{y}; A) = J^{-1} \cdot f_j(x_1(\bar{y}), x_2(\bar{y}), \cdots, x_n(\bar{y})), \tag{10}$$

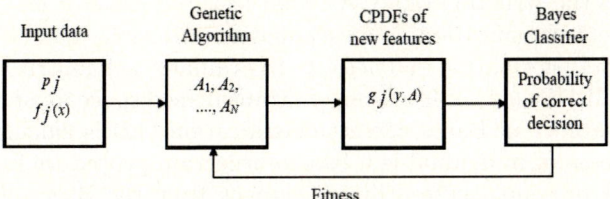

Fig. 1. GA-based Bayes-optimal feature extractor

where J is a Jacobian of mapping (7). Integrating (10) over variables $x_1, ..., x_{n-m}$ we simply get

$$g_j(y;A) = \int_{\mathcal{X}_1}\int_{\mathcal{X}_2}\cdots\int_{\mathcal{X}_{n-m}} \bar{g}_j(\bar{y};A)\, dx_1\, dx_2\, ...\, dx_{n-m}. \tag{11}$$

Formula (11) allows one to determine class-conditional density functions for the vector of features y, describing the object in a new m-dimensional space. Substituting (11) into (5) one gets a criterion defining the probability of correct classification for the objects in space $\mathcal{Y}$:

$$Q(A) = Pc(A) = \int_{\mathcal{Y}} \max_{j \in \mathcal{M}} \left\{ p_j \cdot \int_{\mathcal{X}_1} \cdots \int_{\mathcal{X}_{n-m}} J^{-1} \times \right.$$

$$\left. \times f_j(x_1(\bar{y}), \cdots, x_n(\bar{y}))\, dx_1\, ...\, dx_{n-m} \right\} dy =$$

$$= \int_{\mathcal{Y}} \max_{j \in \mathcal{M}} \left\{ p_j \cdot \int_{\mathcal{X}_1} \cdots \int_{\mathcal{X}_{n-m}} J^{-1} f_j(|\bar{A}_1(y)| \times \right.$$

$$\left. \times |\bar{A}|^{-1}, \cdots, |\bar{A}_n(y)| \cdot |\bar{A}|^{-1})\, dx_1\, ...\, dx_{n-m} \right\} dy. \tag{12}$$

Thus, the solution of the feature extraction problem (6) requires that such matrix A^* should be determined for which the Bayes probability of correct classification (12) is the maximum one.

Consequently, complex multiple integration and inversion operations must be performed on the multidimensional matrices in order to obtain optimal values of A. Although an analytical solution is possible (for low n and m values), it is complicated and time-consuming. Therefore it is proposed to use numerical procedures. For linear problem optimization (which is the case here) classic numerical algorithms are very ineffective. In a search for a global extremum they have to be started (from different starting points) many times whereby the time needed to obtain an optimal solution is very long. Thus it is only natural to use the parallel processing methodology offered by genetic algorithms, which is based on ideas borrowed from the theories of natural selection - the "survival of the fittest" ([8]).

Fig. 1 shows the structure of a GA-based feature extractor using Bayes probability of correct classification as an evaluation criterion. The GA maintains a population of transformation matrices A. To evaluate each matrix in this population, first the CPDFs (11) of features y in transformed space must be determined and next probability of Bayes correct classification (12) is calculated. This accuracy, i.e. fitness of individual is a base of selection procedure in GA. In other words, the GA presented here utilizes feedback from the Bayes classifier to the feature extraction procedure.

4 The Case of Recognition with Learning - Nonparametric Estimation

It follows from the above considerations that an analytical or numerical solution of the optimization problem is possible. But for this, one must know the class-conditional density functions and the *a priori* probabilities of the classes. In practice, such information is rarely available. All we know about the classification problem is usually contained in the so-called learning set in the space $\mathcal{X}$:

$$S_N(x) = \{(x^{(1)}, j^{(1)}), (x^{(2)}, j^{(2)}), ..., (x^{(N)}, j^{(N)})\}. \tag{13}$$

Applying for this case GA-based Bayes-optimal feature extractor from previous section, we get recursive procedure, for which one iteration dealing with determination of fitness function can be expressed in the following points:

1. For given matrix A, transform learning set to the space $\mathcal{Y}$:

$$S_N(y) = \{(y^{(1)}, j^{(1)}), (y^{(2)}, j^{(2)}), ..., (y^{(L)}, j^{(N)})\}, \tag{14}$$

where

$$y^{(k)} = A \cdot x^{(k)}; \quad k = 1, 2, ..., N. \tag{15}$$

2. Estimate from learning set (14) *a priori* probabilities $\hat{p}_j$ and CPDF's $\hat{g}_j(y; A)$. In the next experimental investigations Parzen procedure with Gaussian kernel was used for this purpose.
3. Calculate probability of correct classification with estimated probability distributions instead of the real ones (fitness function)

$$\hat{P}c(A) = \int_{\mathcal{Y}} \max_{j \in \mathcal{M}} \{\hat{p}_j \, \hat{g}_j(y; A)\} \, dy, \tag{16}$$

using procedure of numerical integration.

Alternatively, the criterion (16) can be calculated experimentally as the number of correctly classified patterns (divided by N) from (14) using empirical Bayes classifier.

5 Experimental Investigations

5.1 Description of Experiments

In order to study the performance of the proposed method of reducing feature vector dimensionality for the case of supervised pattern recognition, some computer experiments were made. Obtained results were compared with quality of known Principal Component Analysis (PCA) technique or Karhunen-Loeve transform [5], in respect of criterion (16).

Five experiments were carried out in the MatLab environment. For all experiments, in computer generated learning set (13) we adopted equal *a priori* probabilities of classes (1) and Gaussian CPDF's (2) with identical covariance matrices $\Sigma_i = I$. Detailed information on the remaining experiment parameters are presented in the Tab. 1.

5.2 Genetic Algorithm Procedure

In order to find matrix A the genetic algorithm was applied, which was proceeded as follows:

- *Coding method* - Although binary representation has been widely used for GA analysis, in recent years many researchers have been concentrated on the use of real-coded GA (RGA). It is robust, accurate and efficient approach because the floating-point representation is conceptually closest to the problems with real optimization parameters. The chromosome used by the RGA is a string of floating-point numbers (real-valued genes) of the same length as the solution vector. It means, that in our task, the elements of matrix $A = [a_{ij}]_{n \times m}$ were directly coded to the chromosome, namely:

$$C = [a_{11}, a_{12}, ..., a_{1n}, a_{21}, a_{22}, ..., a_{2n}, ..., a_{nm}] = [c_1, c_2, ..., c_L], \qquad (17)$$

where n and m are dimensions of input and derivative feature space, respectively.
- *The fitness function* - Each chromosome was evaluated based on the probability of correct classification $\hat{P}c(A)$ (16) calculated by numerical procedure in the discretized feature space $\mathcal{Y}$.

Table 1. Parameters of experimental investigations

Experiment	1	2	3	4	5
Dimension of space $\mathcal{X}$	3	10	30	30	30
Dimension of space $\mathcal{Y}$	2	4	2	4	5
Number of classes	3	6	9	9	9
Size of learning set	300	600	900	900	900
Number of GA runs	5	5	5	3	1
Number of points in the numerical procedure of $\hat{P}c$ calculation	900	10,000	100	10,000	100,000

- *Initialization* - GA needs an initial individual population to carry out parallel multidirectional search of optimal solution. The RGA starts with constructing an initial population of individuals generated randomly within the search space. Each gene of chromosome was a random number uniformly distributed in $[0, 1]$. The size of population was set to 20.
- *Selection* - The probability of selecting a specific individual can be calculated by using the individuals fitness and the sum of population fitness. In this research a roulette wheel approach was applied. Additionally, an elitism policy, wherein the best individual from the current generation is copied directly to the next generation, was also used for faster convergence.
- *Crossover* - The crossover process defines how genes from the parents have been passed to the offspring. In experiments both arithmetic and directional crossover [9] were used and in each generation one of them was selected randomly. If C_p^1 and C_p^2 are parent chromosomes, then we get offspring chromosomes C_c^1 and C_c^2 according to the following formulas (α is a random number uniformly distributed in $[0, 1]$):

$$C_c^1 = \alpha \cdot C_p^1 + (1 - \alpha) \cdot C_p^2, \quad C_c^2 = \alpha \cdot C_p^2 + (1 - \alpha) \cdot C_p^1, \qquad (18)$$

for arithmetic crossover,

$$C_c^1 = \alpha \cdot (C_p^1 - C_p^2) + C_p^1, \quad C_c^2 = \alpha \cdot (C_p^2 - C_p^1) + C_p^2, \qquad (19)$$

for directional crossover.

These crossover procedures represent a promising way for introducing a correct exploration/exploitation balance in order to avoid premature convergence and reach approximate final solution.
- *Mutation* - Mutation is carried out by randomly perturbing genes of chromosomes after crossover. The mutation used in experiments is the random (Gaussian) procedure ([15]), i.e.:
  - an individual (in offspring population) is randomly selected,
  - a mutation site is randomly fixed in the interval $[1, L]$,
  - the selected gene (value) c_i is replaced by $\bar{c}_i$, randomly generated with Gaussian $N(0, 1)$ distribution.

  The probability of mutation was equal to 0.01.
- *Stop procedure* - evolution process was terminated after 20 and 50 iterations. In fact, the fitness value usually converged within this latter value. Fig. 2-5. show the fitness change against generation number in one run of RGA.

5.3 Results and Discussion

Results of experiments, i.e. empirical probabilities of correct classifications after feature extraction procedure for proposed methods and for PCA technique are presented in Tab. 2 and 3. Furthermore, in Fig. 2-5 are depicted examples of the course of the fitness value versus number of iteration in one GA run.

Table 2. Results of experiment 1 (NGA - number of GA iterations, a_{ik} - values of matrix A elements, $\hat{P}c$ - empirical probability of Bayes correct classification)

GA run	1	2	3	4	5	PCA method
$\hat{P}c$	0.9942	0.9966	0.9909	0.9960	0.9963	0.9911
a_{11}	0.561	0.647	0.518	0.854	0.251	0.671
a_{12}	-0.135	3.287	0.214	-0.788	2.227	-0.410
a_{13}	0.768	0.255	0.317	0.519	0.247	0.616
a_{21}	0.138	1.550	0.176	0.205	0.317	0.354
a_{22}	0.933	-1.033	0.957	1.110	-0.504	0.909
a_{23}	0.119	0.698	0.037	-0.030	0.360	0.219
NGA	11	20	6	7	13	-

Table 3. Results of experiments 2 - 5

GA run	1	2	3	4	5	PCA method
Experiment 2						
NGA	20	20	20	20	20	-
$\hat{P}c$	0.983	0.991	0.984	0.994	0.987	0.992
NGA	50	50	50	50	50	-
$\hat{P}c$	0.996	0.998	0.995	0.997	0.996	0.992
Experiment 3						
NGA	20	20	20	20	20	-
$\hat{P}c$	0.733	0.706	0.731	0.726	0.692	0.689
NGA	50	50	50	50	50	-
$\hat{P}c$	0.794	0.814	0.826	0.834	0.46	0.689
Experiment 4						
NGA	20	20	20			-
$\hat{P}c$	0.871	0.905	0.856			0.948
NGA	50	50	50			-
$\hat{P}c$	0.952	0.969	0.958			0.948
Experiment 5						
NGA	20					-
$\hat{P}c$	0.918					0.974
NGA	50					-
$\hat{P}c$	0.972					0.974

These results imply the following conclusions:

1. Except the experiment 5, results for proposed method are always better than results for PCA technique.

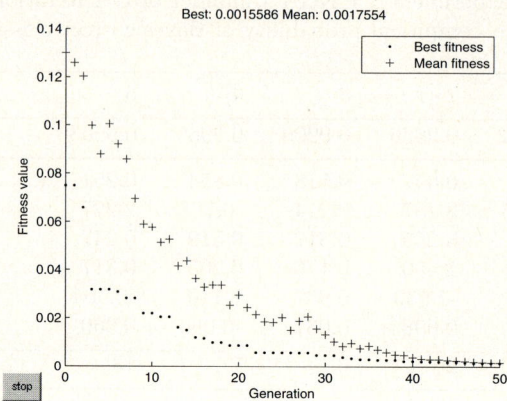

Fig. 2. The example of the course of the fitness value vs. number of generation for experiment 2

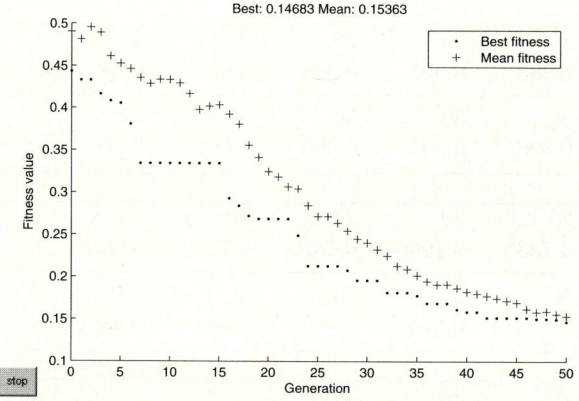

Fig. 3. The example of the course of the fitness value vs. number of generation for experiment 3

2. In the experiment 5, the greater number of GA iterations should lead to the better result. Unfortunately, due to the dimension of the feature space $\mathcal{Y}$, one iteration of GA run needs over 20 mins of PC time. This is also reason why in this experiment only one run of GA was carried out.
3. There are no essential difference between different runs of GA for the same experiment.
4. Although in the experiment 1, matrices A resulting from proposed approach and PCA method are quite different there are no serious difference between quality of both algorithms.

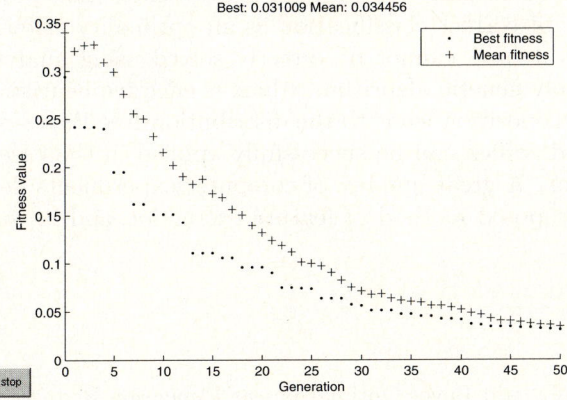

Fig. 4. The example of the course of the fitness value vs. number of generation for experiment 4

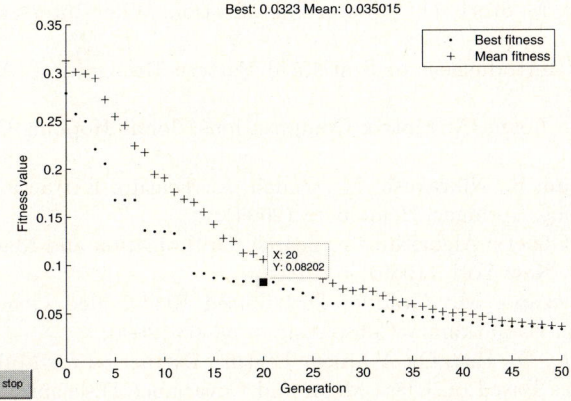

Fig. 5. The course of the fitness value vs. number of generation for experiment 5

6 Conclusions

Feature extraction is an important task in any pattern recognition problem. For this purpose, it is sensible to use linear feature extraction which is considered as a linear mapping of data from a high to a low-dimensional space, where class separability is approximately preserved. Construction of linear transformation is based on minimization (maximization) of proper criterion in the transformed space. In other words, in order to define a linear transformation one should determine the values of the transformation matrix components as a solution of an appropriate optimization problem.

In this paper we formulate the optimal feature extraction problem with the Bayes probability of correct classification as an optimality criterion. Since this problem, in general case, cannot be directly solved using analytical methods, we propose to apply genetic algorithm, which is effective heuristic optimization procedure. This proposition leads to the distribution-free Bayes-optimal feature extraction method, which can be successfully applied in the case of supervised pattern recognition. A great number of computer experiments demonstrate the effectiveness of proposed method of feature extraction and its superiority over the PCA method.

References

1. Buturovic, L.: Toward Bayes-Optimal Linear Dimension Reduction. IEEE Trans. on PAMI 16, 420–424 (1994)
2. Choi, E., Lee, C.: Feature Extraction Based on the Bhattacharyya Distance. Pattern Recognition 36, 1703–1709 (2002)
3. Devroye, L., Gyorfi, P., Lugossi, G.: A Probabilistic Theory of Pattern Recognition. Springer, New York (1996)
4. Duda, R., Hart, P., Stork, D.: Pattern Classification. Wiley-Interscience, New York (2001)
5. Fukunaga, K.: Introduciton to Statistical Pattern Recognition. Academic Press, London (1990)
6. Golub, G., Van Loan, C.: Matrix Computations. Johns Hopkins University Press (1996)
7. Guyon, I., Gunn, S., Nikravesh, M., Zadeh, L.: Feature Extraction. Foundations and Applications. Springer, Heidelberg (2004)
8. Goldberg, D.: Genetic Algorithms in Search, Optimization and Machine Learning. Adison-Wesley, New York (1989)
9. Herrera, F., Lozano, M.: Gradual Distributed Real-Coded Genetic Algorithm. IEEE Trans. on Evolutionary Computing 4, 43–63 (2000)
10. Hsieh, P., Wang, D., Hsu, C.: A Linear Feature Extraction for Multiclass Classification Problems Based on Class Mean and Covariance Discriminant Information. IEEE Trans. on PAMI 28, 223–235 (2006)
11. Kubota, S., Mizutani, H., Yoshiaki, K.: A Discriminative Learning Criterion for the Overall Optimization of Error and Reject. In: Proc. 16th Int. Conf. on Pattern Recognition, vol. 4, pp. 498–502 (2002)
12. Kuo, B., Landgrebe, D.: A Robust Classification Procedure Based on Mixture Classifiers and Nonparametric Weighted Feature Extraction. IEEE Trans. on GRS 40, 2486–2494 (2002)
13. Puchala, E., Kurzynski, M.W., Rewak, A.: The Bayes-Optimal Feature Extraction Procedure for Pattern Recognition Using Genetic Algorithm. In: Kollias, S., Stafylopatis, A., Duch, W., Oja, E. (eds.) ICANN 2006. LNCS, vol. 4131, pp. 21–30. Springer, Heidelberg (2006)
14. Loog, M., Duin, R., Haeb-Umbach, R.: Multiclass Linear Dimension Reduction by Meighted Pairwise Fisher Criteria. IEEE Trans. on PAMI 23, 762–766 (2001)
15. Michalewicz, Z.: Genetic Algorithms + Data Structure = Evolution Programs. Springer, New York (1996)

16. Park, H., Park, C., Pardalos, P.: Comparitive Study of Linear and Nonlinear Feature Extraction Methods - Technical Report. Minneapolis (2004)
17. Raymer, M., Punch, W., et al.: Dimensionality Reduction Using Genetic Algorithms. IEEE Trans. on EC 4, 164–168 (2002)
18. Rovithakis, G., Maniadakis, M., Zervakis, M.: A Hybrid Neural Network and Genetic Algorithm Approach to Optimizing Feature Extraction for Signal Classification. IEEE Trans. on SMC 34, 695–702 (2004)

Hierarchical SVM Classification for Localization in Multilevel Sensor Networks

Jerzy Martyna

Institute of Computer Science, Jagiellonian University, ul. Nawojki 11,
30-072 Cracow, Poland
martyna@softlab.ii.uj.edu.pl

Abstract. We show that the localization problem for multilevel wireless sensor networks (WSNs) can be solved as a pattern recognition with the use of the Support Vector Machines (SVM) method. In this paper, we propose a novel hierarchical classification method that generalizes the SVM learning and that is based on discriminant functions structured in such a way that it contains the class hierarchy. We study a version of this solution, which uses a hierarchical SVM classifier. We present experimental results the hierarchical SVM classifier for localization in multilevel WSNs.

1 Introduction

In multilevel wireless sensor networks (WSNs), the collected data must be transmitted from the monitored region called a sensor field to a clusterhead sensor and next to a sink. This often requires the information about the physical localization of sensors. The localization belongs to the first duty of all applications used in WSNs. In the traditional localization approach signal propagation models [2] are used. In these models the distance between a pair of points is estimated. The accuracy of signal propagation models is limited by the ranging errors. As an alternative to measuring distance between nodes angles can be measured. Thus, in order to measure angles the directional antennas should be used. However, the localization based on angulation is expensive and inappropriate for sensor nodes.

Artificial intelligence methods for the solutions of the localization problem in WSNs are given in the paper [1], [10]. In the first of them, a kernel-based learning algorithm for defining the matrix of signal strengrts received by the sensor is used. In the paper by Abhishek [1] the localization problem in WSNs is viewed as a classification problem that generalized Support Vector Machine (SVM) learning. The output of the SVM was mapped onto probability. With the use of the Laplace Eigenmaps and an appropriate Affine Transformation a good estimate of node position was obtained.

We recall that the SVM method has been introduced by Cortes and Vapnik [4] and [15]. Briefly, the SVM method maps input data onto a high dimensional hypersurface where it may become linearly separable. Thus, the SVM method allows us to minimize the structural task whereas the previous techniques are based on the minimization of the number of misclassified points on the training set.

It is worth noting that the SVM methods have been explored recently in many pattern recognition problems (see Miteéran, 2003) [9], face authentication (see Jonsson, 1999) [5], and feature selection (see Barzilay, 1999) [3]. A modified version of the SVM called the least squares SVM (LS-SVM) was proposed by Suykens and Vandewolle (1999) [13] and Suykens et al. (2002) [14]. In the LS-SVM method a set of linear equations was used instead of a quadratic programming problem. Currently, many solutions based on the LS-SVM method are being implemented in low-cost VLSI chips.

The main goal of this paper is to introduce a novel hierarchical classification method that generalizes the SVM learning. This approach incorporates knowledge about multilevel relationships for an efficient training algorithm. However, the presented method is not restricted to the zero-one classification loss, but is able to incorporate a specific sense classification. This means that the given method satisfies all the required demands.

The hierarchical approach has been used many times before and has led to a number of problem solutions, such as a document classification [7], text categorization [17], a building hierarchical classifiers [16].

The paper is organized as follows. We begin with multilevel WSNs. Further, the formulation of a hierarchical SVM classification for the localization in multilevel WSNs is described. We then present details of the learning algorithm and its implementation. An evaluation of our algorithm with a simulated WSN is given in section 5. Conclusions are given in section 6.

2 Multilevel Wireless Sensor Networks

Multilevel wireless sensor networks are obtained with the help of the hierarchical approach at first. Thus, we have the partitioning of all auxiliary nodes into several clusters or local groups. These nodes form the first layer of hierarchy in the WSN. The "controllers" of such groups form the second layer of hierarchy in the WSN and are referred to as clusterheads. All clusterheads are natural places to aggregate and compress traffic converging from many sensors to a single node. Moreover, all clusterheads constitute a higher layer of hierarchy in WSN. They provide the medium for communication between clusters called a backbone network. With the use of the backbone network data collected from the sensor field are sent to the sink (see Fig 1).

Formally, a multilevel WSN provides $G = (V, E)$ the identification of a set of subsets of nodes V_i, $i = 1, \ldots, n$ such that $\bigcup_{i=1,\ldots,n} V_i = V$. Some properties must be satisfied:

1) The partitioning of V into clusters does not mandate anything about the internal structure of a cluster. One option would be to assign the same number of nodes to clusters. In the second option the summarized energy in the batteries in all the clusters should be balanced.

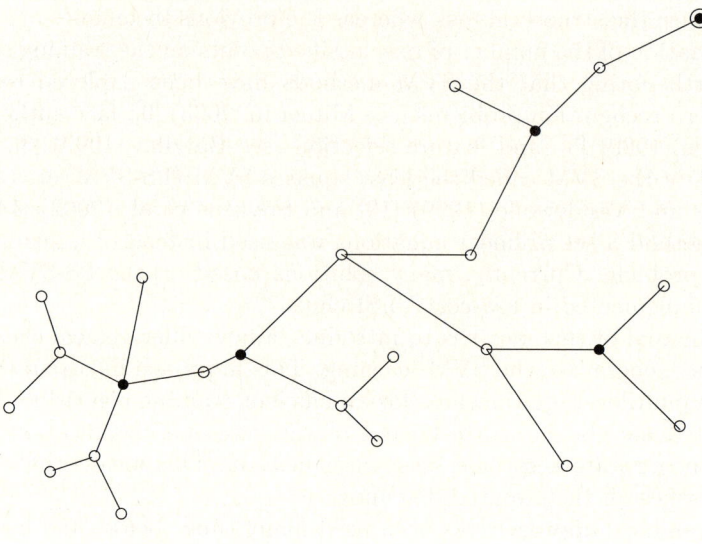

○ - *ordinary sensor*
● - *clusterhead*
◉ - *sink*

Fig. 1. An example of clustered layers in sensor network

2) For each set V_i there is a unique node c_i referred to as a clusterhead that represents the set and can take some tasks such as data collection, data aggregation, etc.
3) Clusterheads form an independent subset C, $C \subset V$ such that no two nodes in C are joined by an edge in $E - \forall c_1, c_2 \in C : (c_1, c_2) \notin E$.
4) None of the clusters should not overlap. It means that those nodes that are adjacent to two clusterheads must be assigned unambiguously to the clusters.
5) In many situations the two-level hierarchy is sufficient. Nevertheless, it is possible to consider a multi-level hierarchy of WSNs.

3 Hierarchical SVM Classification for the Localization in WSNs

In this section, we will present a hierarchical SVM classification for the localization in WSNs.

Let $\{(x_i, y_i)\}_{i=1}^n$ be a set of n labeled coordinates of cluster i location. Here x_i denotes a vector representation for the i-th location. Each label y_i denotes a unique category encoded as an integer, $y_i \in Y \equiv \{1, \ldots, q\}$, where q is the total number of possible cluster locations.

We introduce the discriminant function F as

$$F(x, y; \mathbf{w}) \equiv \langle \mathbf{w}, \Phi(x, y) \rangle \tag{1}$$

where $\mathbf{w}$ is the stacked vector of all weights, namely $\mathbf{w} = (w_1, \ldots, w_q)$, and $\Phi(x, y) = \Lambda(y) \otimes x$. Here $\otimes$ denotes a tensor product, i.e. $\Phi(x, y) \in \Re^{d \cdot s}$ is a vector containing all the products of coefficients from the first and second vector argument. Writing out $\Phi(x, y)$ one gets

$$\Phi(x, y) = \begin{pmatrix} \lambda_1(y) \cdot x \\ \lambda_2(y) \cdot x \\ \vdots \\ \lambda_s(y) \cdot x \end{pmatrix} \tag{2}$$

and for $\lambda_r(y) = \delta_{ry}$ this simply reduces to

$$\Phi(x, y) = \begin{pmatrix} \vdots \\ 0 \\ x \\ 0 \\ \vdots \end{pmatrix} \leftarrow y\text{-th position} \tag{3}$$

An exemplary taxonomy with 7 possible locations of sensor clusters equal to 12 is shown in Fig. 2a. The decomposition of the discriminant function for localization is given in Fig. 2b.

Equation (1) can be rewritten as an additive superposition of a linear discrimination as follows

$$F(x, y : \mathbf{w}) = \sum_{r=1}^{s} \lambda_r(\mathbf{w}) \langle w_r, \mathbf{x} \rangle \tag{4}$$

where $w_r \in \Re^d$ is the weight vector associated with the r-th class attribute.

With the use of Eq. (4) for defining the margin, it becomes a linear constraint

$$\langle \delta \Phi_i(y), \mathbf{w} \rangle \geq 1 - \xi_i, \text{ for } \forall_i, \, y \neq y_i, \, \xi_i \geq 0 \tag{5}$$

where $\delta \Phi_i(y) \equiv \Phi(x_i, y_i) - \Phi(x_i, y)$.

We suppose that a set of labeled clusters $\{(x_i, y_i)\}_{i=1}^n$ is given. Thus, $y_i \in \{-1, 1\}^q$ and $y_{ir} = 1$ encode the fact that cluster x_i belongs to the r-th location. Then we can formulate the Lagrangian for this problem

$$\mathcal{L}(\mathbf{w}, \xi, \alpha, \psi) = \frac{1}{2} w^2 + C \sum_{i=1}^{n} \sum_{r=1}^{q} \xi_{ir} - \sum_{i=1}^{n} \xi_i \psi_i \tag{6}$$

$$- \sum_{i=1}^{n} \sum_{y \neq y_i} \alpha_{iy} (\langle \delta \phi_i(y), \mathbf{w} \rangle - 1 + \xi_i) \tag{7}$$

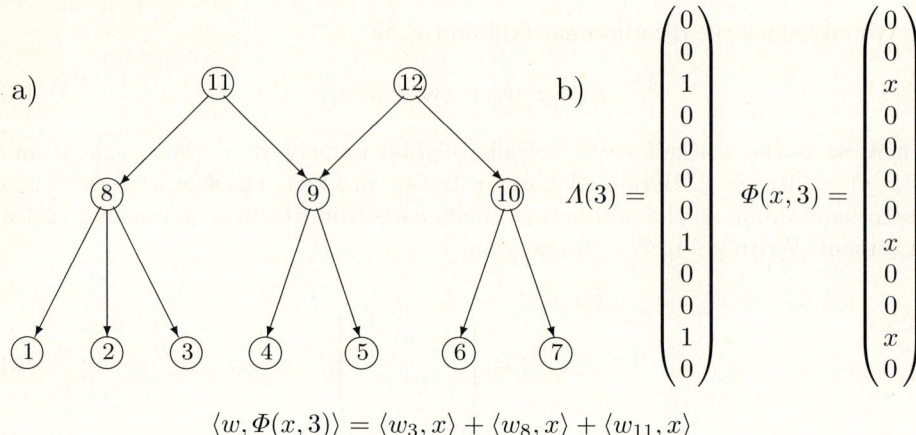

$$\langle w, \Phi(x,3) \rangle = \langle w_3, x \rangle + \langle w_8, x \rangle + \langle w_{11}, x \rangle$$

Fig. 2. An examplary taxonomy with 7 possible locations of 12 clusters (a) and the discriminant function for location 3 (b)

with conditions for optimality

$$\frac{\partial \mathcal{L}}{\partial \mathbf{w}} = 0 \rightarrow \mathbf{w} = \sum_{i=1}^{n} \sum_{y \neq y_i} \alpha_{iy} \delta \phi_i(y) \qquad (8)$$

$$\frac{\partial \mathcal{L}}{\partial \xi} = 0 \rightarrow \psi_i = C - \sum_{y \neq y_i} \alpha_{iy} \qquad (9)$$

where the α_{iy} values are the Lagrange multipliers, which can be positive or negative due to equality constraints, ψ_i is an additional slack variable for $i = 1, \ldots, n$.

Since $\psi_i \geq 0$, we obtain the condition for $\forall i$, namely

$$\sum_{y \neq y_i} \alpha_{iy} \leq C_i \qquad (10)$$

With the help of the optimality equation for w and the exploitation Eq. (9) we obtain the objective

$$\theta(\alpha) = \sum_{i=1}^{n} \sum_{y \neq y_i} \alpha_{iy} - \frac{1}{2} \sum_{i=1}^{n} \sum_{j=1}^{n} \sum_{r=1}^{q} \sum_{y \neq y_i} \sum_{y \neq y_j} \alpha_{iy} \alpha_{jy'} \langle \delta \phi_i(y), \delta \phi_j(j) \rangle \qquad (11)$$

The solution of the dual QP is characterized by

$$\alpha^\star = \arg\max_{\alpha} \theta(\alpha) \text{ for } \forall i, r : \alpha_{iy} \geq 0, \sum_{y \neq y_i} \leq C \text{ and } \xi_{ir} \geq 0$$

$$\text{and } y_{ir} \langle \Phi(x_i, r), \mathbf{w} \rangle \geq 1 - \xi_{ir} \qquad (12)$$

We can notice that

$$\langle \Phi(x_i, y_i), \Phi(x_j, y') \rangle = \langle \Lambda(y), \Lambda(y') \rangle \langle x_i, y_j \rangle \qquad (13)$$

and thus from the definition of Φ we can obtain

$$\langle \delta\Phi_i(y), \delta\Phi_j(y') \rangle = \langle \Lambda(y_i) - \Lambda(y), \Lambda(y_i) - \Lambda(y) \rangle \langle x_i, x_j \rangle \tag{14}$$

It is obvious that the inner products $\langle x_i, x_j \rangle$ can be replaced by any kernel function $K(x_i, x_j)$ as in the standard SVM classification.

4 The Training of a Hierarchical SVM Classifier for Localization in Multilevel Sensor Networks

In this section, we give the method used for the training of a hierarchical SVM classifier for localization in multilevel WSNs.

The traditional approach to the training of SVM classifiers is centralized. Thus, a direct communication between each sensor and the clusterhead sensor is difficult. Firstly, the clusterhead may be far away from the given sensor node thus a direct communication between them both can be energy costly. The secondly, with increased number of nodes in one location, multiple sensors may obtain the same sample vectors, and thus, may be redundant in terms of their usefulnes for determining a separatle sample.

Therefore, we propose a use of a distributed algorithm where the final estimation is obtained through a sequence of incremental steps concerning a place at a given cluster. The applied distributed algorithm for the training of the SVM estimates at cluster i the previous estimation of cluster $i-1$. Thus, all the sample vectors are measured by the sensors belonging to cluster i. This method is referred to as an incremental learning with the SVM and was first described in the paper by Ruping [11].

Now, we can give the training algorithm of our SVM classifier in multilevel WSNs as follows:

Step 1. Determine the support vectors using the hierarchical SVM. As a result the separating line which gives the maximal margin is obtained.
Step 2. As a result of ignoring the first cluster of sensor nodes in a multilevel WSN, we train our hierarchical SVM classifier by determining additional support vectors needed for the invariant condition. Thus two virtual vectors are obtained.
Step 3. We construct a new hierarchical SVM classifier for the given patterns (virtual and real). As a result a new line with a smaller margin but in accordance with the actual knowledge is obtained.
Step 4. If the learning of a hierarchical SVM is more costly than the error on the new sample, then the training algorithm is stopped. Otherwise, go to **Step 1**.

As the cost function we used the value of energy consumed for transmitting its measurements to the clusterhead during the training of a hierarchical SVM up to this stage of the training algorithm.

5 Numerical Experiments

In this section, we compare the standard method for cluster localization in WSNs with our hierarchical SVM classifier.

In order to localize clusters in WSNs we used one of the most popular technique for positioning, namely the multilateration method [12], [6]. In this method the distance between two clusterhead nodes or the angle in triangle is calculated by means of elementary geometry. This information can be used to derive the information about clusterhead positions. We assume that the clusterhead node is usually applied in the center of cluster. In order to minimize the number of hops each clusterhead node should be placed in the center of cluster.

We used three measures to evaluate the performance of the hierarchical SVM classifiers. The first of them, called a *precision for a localization* measures the percentage of a correct localization among all the localizations. The second one, called *accuracy* gives the percentage of correct localizations among all the localizations that should be assigned to this localization. Both measures are as follows

$$precision = \frac{TP}{TP + FN} \times 100\% \qquad (15)$$

$$accuracy = \frac{TP + TN}{TP + FN + TN + FP} \times 100\% \qquad (16)$$

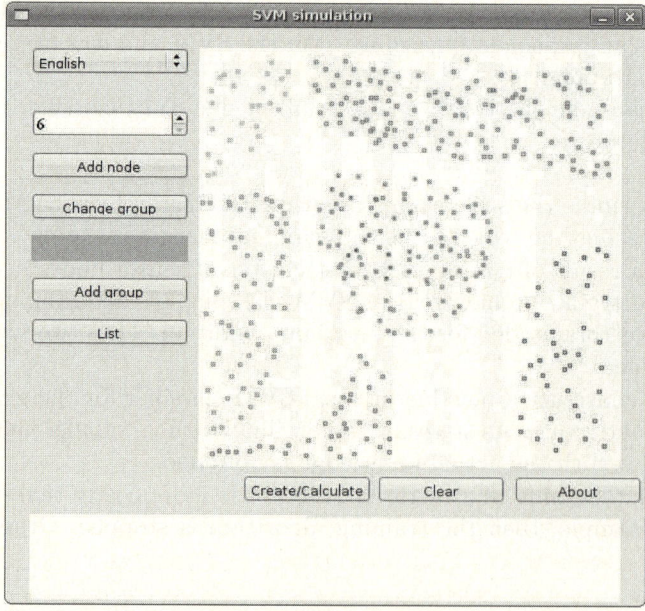

Fig. 3. The process of the cluster localization with the use of the LS-SVM method

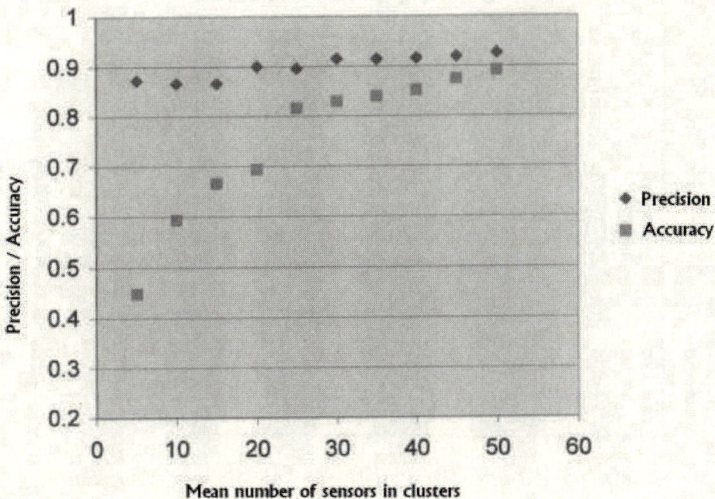

Fig. 4. Precision and accuracy depending on the number of sensors in clusters

where TP stands for true positive, TN true negative, FP for false positive and FN for false negative localization. For example, when a random given cluster localization is concerned, TP represents the cluster localization classified as good, TN represents the cluster localization in a false way.

Otherwise, when a random given localization is excluded FP represents misclassified cluster localization and FN represents a localization misclassified as a false localization given.

Additionally, we introduce a new measure for hierarchical SVM classifiers here, namely the *parent accuracy*. Parent accuracy gives accuracy at the level of the cluster's parent. Let $\mathcal{Z} = \{z : z \in Z \wedge \exists y \in \mathcal{Y}, \text{ so that } parent(y) = z\}$. For instance, $parent(y) = z$ means that z is y's immediate parent. Then, our parent accuracy is given by

$$parent\ accuracy(f(x)) = \frac{1}{n} \cdot accuracy \sum_{i=1}^{n} parent(y_i) \times 100\% \qquad (17)$$

Our parent accuracy defines the percentage of the parent accuracy. When accuracy is the same for different localizations, the higher parent accuracy is assigned to this localization which is farther away from the correct localization.

The process of the cluster localization for an exemplary data with six possible locations of sensor clusters is shown in Fig. 3. The data used for the cluster localization consisted of 1120 sessions, of which 555 were used for training and the remaining 565 for the evaluation of the hierarchical SVM performance. Using the LIBSVM library [8], we obtained among others, the three measures of our SVM classifier. The graph in Fig. 4 shows the obtained precision and accuracy

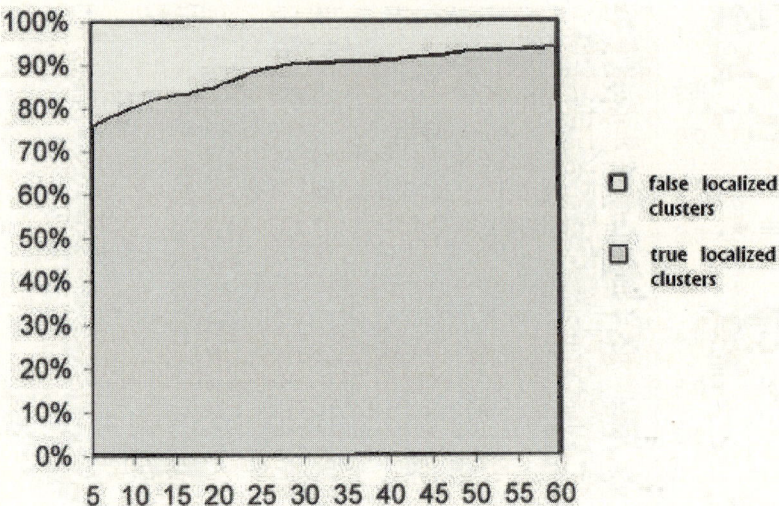

Fig. 5. Percentage of true and false localized clusters depending on the number of sensors in clusters

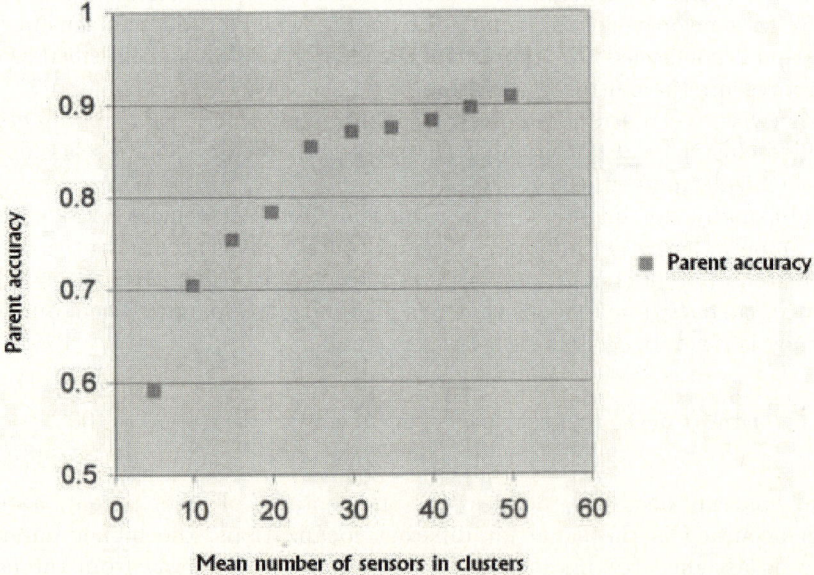

Fig. 6. Parent accuracy depending on the mean number of sensors in clusters

depending on the mean number of sensors in clusters. The percentage of true and false localized clusters depending on the mean number of sensors in clusters is given in Fig. 5. Figure 6 shows the parent accuracy measure depending on the mean number of sensors in clusters used in localizations.

We have observed that the increase of the number of sensors in clusters used for cluster localization improves both measures. It is appreciable yet for the number of sensors in clusters equal to 25.

6 Conclusions

We have proposed a new hierarchical SVM classifier for localization in multilevel WSNs. In our method, we used a standard vector representation for the i-th cluster as a localization representation. Each cluster was assigned to only one category. With the help of the hierarchical SVM classifier built this way, we have found the localization of each clusters in multilevel WSNs. We have derived a training algorithm for our hierarchical SVM classifiers. The obtained results have confirmed the competitiveness of our approach.

References

1. Abhishek, V.: Localization in Ad Hoc Sensor Network: A Machine Learning Based Approach, CS229 Project Report (Fall 2005),
   http://www.stanford.edu/class/cs229/proj2005/
2. Bahl, P., Padmanabhan, V.N.: RADAR: An In-building RF-based User Localization and Tracking System. In: IEEE INFOCOM 2000, pp. 775–784 (2000)
3. Barzilay, O., Brailovsky, V.L.: On Domain Knowledge and Feature Selection Using a Support Vector Machine. Pattern Recognition Letters 20, 475–484 (1999)
4. Cortes, C., Vapnik, V.N.: Support Vector Networks. Machine Learning 20, 273–297 (1995)
5. Jonsson, K., et al.: Support Vector Machines for Face Autentication. In: Pridmore, T., Ellman, D. (eds.) British Machine Vision Conference, London, pp. 543–553 (1999)
6. Karl, H., Willig, A.: Protocols and Architectures for Wireless Sensor Networks. John Wiley and Sons, Hoboken (2005)
7. Koller, D., Sahami, M.: Hierarchically Classifying Documents Using Very Few Words. In: Proc. of the 14th Int. Conf. Machine Learning (ICML) (1997)
8. LIBSVM - A Library for Support Vector Machines (2007),
   http://www.csie.ntu.edu.tw/cjlin/libsvm/
9. Mitéran, J., Bouillant, S., Bourennane, E.: SVM Approximation for Real-Time Image Segmentation by Using an Improved Hyperrectangles-based Method. Real-Time Imaging 9, 179–188 (2003)
10. Ngueyen, X., et al.: A Kernel-based Learning Approach to Ad Hoc Sensor Network Localization. ACM Trans. on Sensor Networks 1(1), 134–152 (2005)
11. Ruping, S.: Incremental Learning with Support Vector Machines. In: Proc. IEEE Int. Conf. on Data Mining, San Jose, CA, USA, November 2001, pp. 641–642 (2001)
12. Savvides, A., Han, C.-C., Srivastava, M.: Dynamic Fine-Grained Localization in Ad Hoc Networks of Sensors. In: Proc. of the 7th Annual International Conference on Mobile Computing and Networking, pp. 166–179. ACM Press, New York (2001)
13. Suykens, J.A.K., Vandewalle, J.: Least Squares Support Vector Machine Classifiers. Neural Processing Letters 9(3), 293–300 (1999)

14. Suykens, J.A.K., van Gestel, T., de Brabanter, J., de Moor, B., Vandewalle, J.: Least Squares Support Vector Machines. World Scientific, Singapore (2002)
15. Vapnik, N.V.: Statistical Learning Theory. John Wiley and Sons, Chichester (1998)
16. Wang, K., Zhou, S., Liew, S.C.: Building Hierarchical Classifiers Using Class Proximity. In: Atkinson, M.P., Orlowska, M.E., Valduriez, P., Zdonik, S.B., Brodie, M.L. (eds.) Proc. of VLDB 1999, 25th Int. Conf. on Very Large Data Bases, pp. 363–374. Morgan Kaufman Publishers, San Francisco (1999)
17. Weigend, A.S., Wiener, E.D., Pedersen, J.O.: Exploiting Hierarchy in Text Categorization. Information Retrieval 1(3), 193–216 (1999)

Comparison of Shannon, Renyi and Tsallis Entropy Used in Decision Trees

Tomasz Maszczyk and Włodzisław Duch

Department of Informatics, Nicolaus Copernicus University
Grudziądzka 5, 87-100 Toruń, Poland
{tmaszczyk,wduch}@is.umk.pl
http://www.is.umk.pl

Abstract. Shannon entropy used in standard top-down decision trees does not guarantee the best generalization. Split criteria based on generalized entropies offer different compromise between purity of nodes and overall information gain. Modified C4.5 decision trees based on Tsallis and Renyi entropies have been tested on several high-dimensional microarray datasets with interesting results. This approach may be used in any decision tree and information selection algorithm.

Keywords: Decision rules, entropy, information theory, information selection, decision trees.

1 Introduction

Decision tree algorithms are still the foundation of most large data mining packages, offering easy and computationally efficient way to extract simple decision rules [1]. They should always be used as a reference, with more complex classification models justified only if they give significant improvement. Trees are based on recursive partitioning of data and unlike most learning systems they use different sets of features in different parts of the feature space, automatically performing local feature selection. This is an important and unique property of general divide-and-conquer algorithms that has not been paid much attention. The hidden nodes in the first neural layer weight the inputs in a different way, calculating specific projections of the input data on a line defined by the weights $g_k = \boldsymbol{W}^{(k)} \cdot \boldsymbol{X}$. This is still non-local feature that captures information from a whole sector of the input space, not from the localized region. On the other hand localized radial basis functions capture only the local information around some reference points. Recursive partitioning in decision trees is capable of capturing local information in some dimensions and non-local in others. This is a desirable property that may be used in neural algorithms based on localized projected data [2].

The C4.5 algorithm to generate trees [3] is still the basis of the most popular approach in this field. Tests for partitioning data in C4.5 decision trees are based on the concept of information entropy and applied to each feature $x_1, x_2, ...$ individually. Such tests create two nodes that should on the one hand contain data that are as pure as possible (i.e. belong to a single class), and on the other hand increase overall separability of data. Tests that are based directly on indices measuring accuracy are optimal from Bayesian point of view, but are not so accurate as those based on information theory that may be evaluated with greater precision [4]. Choice of the test is always a hidden compromise in how much weight is put on the purity of samples in one or both nodes and the total gain achieved by partitioning of data. It is therefore worthwhile to test other types of entropies that may be used as tests. Essentially the same reasoning may be used in applications of entropy-based indices in feature selection. For some data features that can help to distinguish rare cases are important, but standard approaches may rank them quite low.

In the next section properties of Shannon, Renyi and Tsallis entropies are described. As an example of application three microarray datasets are analyzed in the third section. This type of application is especially interesting for decision trees because of the high dimensionality of microarray data, the need to identify important genes and find simple decision rules. More sophisticated learning systems do not seem to achieve significantly higher accuracy. Conclusions are given in section four.

2 Theoretical Framework

Entropy is the measure of disorder in physical systems, or an amount of information that may be gained by observations of disordered systems. Claude Shannon defined a formal measure of entropy, called Shannon entropy[5]:

$$S = -\sum_{i=1}^{n} p_i \log_2 p_i \qquad (1)$$

where p_i is the probability of occurrence of an event (feature value) x_i being an element of the event (feature) X that can take values $\{x_1...x_n\}$. The Shannon entropy is a decreasing function of a scattering of random variable, and is maximal when all the outcomes are equally likely.

Shannon entropy is may be used globally, for the whole data, or locally, to evaluate entropy of probability density distributions around some points. This notion of entropy can be generalized to provide additional information about the importance of specific events, for example outliers or rare events. Comparing entropy of two distributions, corresponding for example to two features, Shannon entropy assumes implicite certain tradeoff between contributions from the tails and the main mass of this distribution. It should be worthwhile to control this tradeoff explicitly, as in many cases it may be important to distinguish weak signal overlapping with much stronger one. Entropy measures that depend on powers of probability, $\sum_{i=1}^{n} p(x_i)^\alpha$, provide such control. If α has large positive value this measure is more sensitive to events that occur often, while for large negative α it is more sensitive to the events which happen seldom.

Constantino Tsallis [6] and Alfred Renyi [7] both proposed generalized entropies that for $\alpha = 1$ reduce to the Shannon entropy. The Renyi entropy is defined as [7]:

$$I_\alpha = \frac{1}{1-\alpha} \log \left(\sum_{i=1}^{n} p_i^\alpha \right) \qquad (2)$$

It has similar properties as the Shannon entropy:

- it is additive
- it has maximum $= \ln(n)$ for $p_i = 1/n$

but it contains additional parameter α which can be used to make it more or less sensitive to the shape of probability distributions.

Tsallis defined his entropy as:

$$S_\alpha = \frac{1}{\alpha - 1} \left(1 - \sum_{i=1}^{n} p_i^\alpha \right) \qquad (3)$$

Figures 1-3 shows illustration and comparison of Renyi, Tsallis and Shannon entropies for two probabilities p_1 and p_2 where $p_1 = 1 - p_2$.

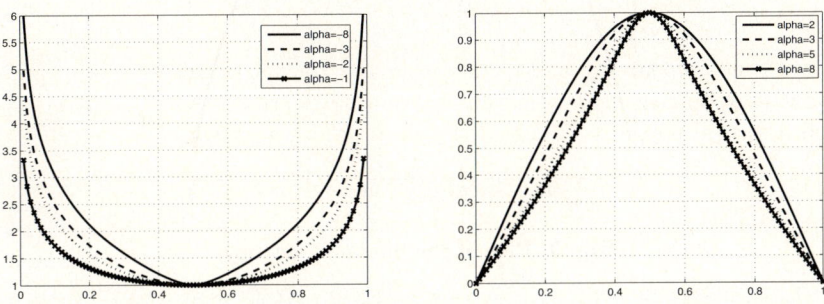

Fig. 1. Plots of the Renyi entropy for several negative and positive values of α

The modification of the standard C4.5 algorithm has been done by simply replacing the Shannon measure with one of the two other entropies, as the goal here was to evaluate their influence on the properties of decision trees. This means that the final split criterion is based on the gain ratio: a test on attribute A that partitions the data D in two branches with D_t and D_f data, with a set of classes *omega* has gain value:

$$G(\omega, A|D) = H(\omega|D) - \frac{|D_t|}{|D|} H(\omega|D_t) - \frac{|D_f|}{|D|} H(\omega|D_f) \qquad (4)$$

where $|D|$ is the number of elements in the D set and $H(\omega|S)$ is one of the 3 entropies considered here: Shannon, Renyi or Tsallis. Parameter α has clearly an influence on what type of splits are going to be created, with preference for negative values of α given to rare events or longer tails of probability distribution.

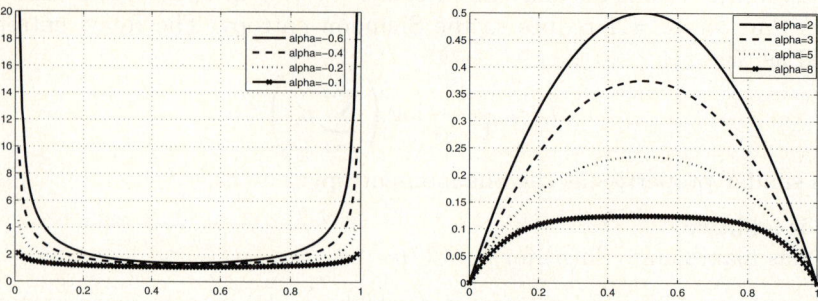

Fig. 2. Plots of the Tsallis entropy for several negative and positive values of α

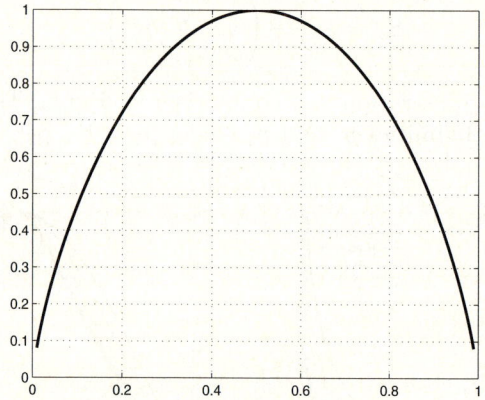

Fig. 3. Plot of the Shannon entropy

3 Empirical Study

To evaluate the usefulness of Renyi and Tsallis entropy measures in decision trees the classical C4.5 algorithm has been modified and applied first to artificial data to verify its usefulness. The results were encouraging, therefore experiments on three data sets of gene expression profiles were carried out. Such data are characterized by large number of features and very small number of samples. In such situations several features may by pure statistical chance seem to be quite informative and allow for good generalization. Therefore it is deceivingly simple to reach high accuracy on such data, although it is very difficult to classify these data reliably. Only the simplest models may avoid overfitting, therefore decision trees providing simple rules may have an advantage over other models. A summary of these data sets is presented in Table 1, all data were downloaded from the Kent Ridge Bio-Medical Data Set Repository http://sdmc.lit.org.sg/GEDatasets/Datasets.html. Short description of these datasets follows:

1. Leukemia: training dataset consists of 38 bone marrow samples (27 ALL and 11 AML), over 7129 probes from 6817 human genes. Also 34 samples testing data is provided, with 20 ALL and 14 AML cases.
2. Colon Tumor: data contains 62 samples collected from the colon cancer patients. Among them, 40 tumor biopsies are from tumors (labeled as "negative") and 22 normal (labeled as "positive") biopsies are from healthy parts of the colons of the same patients. Two thousand out of around 6500 genes were pre-selected by the authors of the experiment based on the confidence in the measured expression levels.
3. Diffuse Large B-cell Lymphoma (DLBCL) is the most common subtype of non-Hodgkins lymphoma. The data contains gene expression data from distinct types of such cells. There are 47 samples, 24 of them are from "germinal centre B-like" group while 23 are "activated B-like" group. Each sample is described by 4026 genes.

Table 1. Summary of data sets

Title	#Genes	#Samples	#Samples per class		Source
Colon cancer	2000	62	40 tumor	22 normal	Alon at all (1999) [8]
DLBCL	4026	47	24 GCB	23 AB	Alizadeh at all (2000) [9]
Leukemia	7129	72	47 ALL	25 AML	Golub at all (1999) [10]

4 Experiment and Results

For each data set the standard C4.5 decision tree is used 10 times, each time averaging the 10-fold crossvalidation mode, followed on the same partitioning of data by the modified algorithms based on Tsallis and Renyi entropies for different values of parameter α. Results are collected in Tables 2-7, with accuracies and standard deviations for each dataset. The best values of α parameter differ in each case and can be easily determined from crossvalidation. Although overall results may not improve accuracy for different classes they may strongly differ in sensitivity and specificity, as can be seen in the tables below.

Table 2. Accuracy on Colon cancer data set; Shannon and $\alpha = 1$ results are identical

Entropy	Alpha							
	-1.5	-0.9	-0.5	-0.1	0.1	0.3	0.5	0.7
Renyi	64.6±0.2	64.6±0.2	64.6±0.2	77.3±4.1	75.4±2.1	77.7±3.3	77.8±4.7	79.1±2.6
Tsallis	64.6±0.2	64.6±0.2	64.6±0.2	64.6±0.2	77.3±3.7	75.4±4.0	74.4±4.3	71.3±5.4
	Alpha							
	0.9	1.1	1.3	1.5	2.0	3.0	4.0	5.0
Renyi	78.8±4.4	82.1±4.2	82.8±4.0	82.9±2.5	84.0±3.9	79.4±3.0	80.8±3.1	78.9±2.2
Tsallis	73.0±3.4	74.9±1.8	73.4±2.4	71.1±4.0	70.2±3.9	73.9±4.4	72.8±3.6	71.1±4.4
Shannon	81.2±3.7							

Table 3. Accuracy per class on Colon cancer data set; Shannon and $\alpha = 1$ results are identical

Entropy	Class	Alpha						
		-1.5	-0.5	-0.1	0.1	0.3	0.5	0.7
Renyi	1	0.0±0.0	0.0±0.0	59.7±4.7	58.7±6.8	60.8±6.4	63.2±7.3	66.0±6.5
	2	100±0.0	100±0.0	87.2±4.1	84.7±2.4	87.2±2.8	85.8±4.0	86.2±3.9
Tsallis	1	0.0±0.0	0.0±0.0	0.0±0.0	58.2±5.1	59.8±9.3	59.8±5.2	50.7±10.2
	2	100±0.0	100±0.0	100± 0.0	87.6±4.8	83.9±4.3	82.8±4.4	82.6±3.9
	Class	Alpha						
		0.9	1.1	1.5	2.0	3.0	4.0	5.0
Renyi	1	65.8±6.5	70.0±7.2	67.3±5.4	69.2±6.6	58.5±2.9	61.0±3.8	58.7±3.5
	2	85.8±4.6	88.8±4.5	91.5±2.1	92.1±2.9	90.7±4.3	91.6±3.7	90.1±3.6
Tsallis	1	55.7±7.7	58.5±4.9	58.3±8.2	53.2±7.6	67.2±9.2	60.5±7.4	60.0±10.4
	2	82.8±3.4	84.2±2.2	78.9±4.6	80.0±3.7	77.7±6.0	79.7±5.3	77.3±3.4
Shannon	1	69.5±4.2						
	2	87.7±4.8						

Table 4. Accuracy on DLBCL data set

Entropy	Alpha							
	-1.5	-0.9	-0.5	-0.1	0.1	0.3	0.5	0.7
Renyi	46.0±4.2	46.0±4.2	46.0±4.2	69.9±5.2	71.8±5.4	70.7±5.4	70.5±5.0	73.0±4.9
Tsallis	52.4±6.8	52.4±6.8	52.4±6.8	52.4±6.8	71.1±5.6	69.8±5.2	72.4±6.0	79.9±5.0
	Alpha							
	0.9	1.1	1.3	1.5	2.0	3.0	4.0	5.0
Renyi	76.5±6.7	81.0±6.2	81.0±4.8	80.5±5.0	79.3±5.1	79.5±5.6	75.9±7.2	69.7±6.3
Tsallis	81.3±4.7	82.0±4.3	81.8±5.2	80.8±6.5	81.5±5.7	78.8±6.9	81.8±4.1	80.5±4.0
Shannon	78.5±4.8							

Results depend quite clearly on the α coefficient, with $\alpha = 1$ always reproducing the Shannon entropy results. The optimal coefficient should be determined through crossvalidation. For the Colon dataset (Tab. 2-3) peak accuracy is achieved for Renyi entropy with $\alpha = 2$, with specificity (accuracy of the second class) significantly higher than for the Shannon case, and with smaller variance. Tsallis entropy seems to be very sensitive and does not improve over Shannon value around $\alpha = 1$. For DLBCL both Renyi and Tsallis entropies with α in the range $1.1 - 1.3$ give the best results, improving both specificity and sensitivity of the Shannon measure (Tab. 4-5). For the Leukemia data best Renyi result for $\alpha = -0.1$, around 88.5 ± 2.4 is significantly better than Shannon's $81.4 \pm 4.1\%$, improving both the sensitivity and specificity and decreasing variance; results with $\alpha = 3$ are also very good. Tsallis results for quite large $\alpha = 5$ are even better.

Table 5. Accuracy per class on DLBCL data set

Entropy	Class	Alpha						
		-1.5	-0.5	-0.1	0.1	0.3	0.5	0.7
Renyi	1	90.0±10.5	90.0±10.5	72.3±8.4	75.5±10.6	74.3±11.2	76.3±8.7	79.7±7.1
	2	10.0±10.5	10.0±10.5	65.5±11.8	66.7±10.5	65.7±10.9	62.5±8.5	65.3±5.9
Tsallis	1	64.8±10.6	64.8±10.6	64.8±10.6	74.5±12.1	73.3±10.7	80.2±8.5	85.8±7.2
	2	41.8±11.0	41.8±11.0	41.8±11.0	65.7±10.6	65.2±12.0	64.8±9.1	74.7±9.8
	Class	Alpha						
		0.9	1.1	1.3	1.5	2.0	3.0	5.0
Renyi	1	82.7±8.4	85.5±8.1	86.5±5.8	85.2±5.8	84.8±6.4	84.2±5.3	68.0±12.0
	2	70.2±11.1	77.3±9.0	77.0±7.4	77.3±7.6	74.7±8.1	75.3±9.0	69.3±4.7
Tsallis	1	88.2±6.3	88.2±5.7	86.2±5.3	85.2±6.4	84.7±5.7	83.2±8.8	87.3±5.2
	2	76.0±7.5	77.3±5.3	78.7±6.9	77.8±9.3	80.0±6.9	76.3±6.4	75.3±4.9
Shannon	1	84.8±7.0						
	2	72.7±8.7						

Table 6. Accuracy on Leukemia data set

Entropy	Alpha						
	-1.5	-0.5	-0.1	0.1	0.3	0.5	0.7
Renyi	65.4 ±0.4	65.4 ±0.4	88.5 ±2.4	85.6 ±3.9	84.6 ±3.8	82.4 ±4.6	82.0 ±4.6
Tsallis	65.4 ±0.4	65.4 ±0.4	65.4 ±0.4	83.5 ±4.4	84.8 ±4.2	84.3 ±3.5	82.3 ±3.9
	Alpha						
	0.9	1.1	1.3	1.5	2.0	3.0	5.0
Renyi	80.5±3.8	81.5±3.5	82.2±3.5	82.4±2.6	85.3±2.8	86.1±2.8	83.8±2.0
Tsallis	82.5±4.4	81.5±2.9	82.3±1.1	83.3±1.4	82.2±2.5	86.5±2.7	87.5±3.6
Shannon	81.4±4.1						

Table 7. Accuracy per class on Leukemia data set

Entropy	Class	Alpha						
		-1.5	-0.5	-0.1	0.1	0.3	0.5	0.7
Renyi	1	100±0.0	100±0.0	89.2±1.6	89.4±3.2	88.7±2.5	86.2±3.2	84.8±4.0
	2	0.0±0.0	0.0±0.0	86.6±5.4	77.7±9.9	75.8±10.5	74.3±10.5	76.2±10.6
Tsallis	1	100±0.0	100±0.0	100±0.0	88.3±3.6	88.7±2.9	89.0±3.3	85.4±3.2
	2	0.0±0.0	0.0±0.0	0.0±0.0	73.5±10.1	76.8±10.3	74.8±9.3	76.6±10.3
	Class	Alpha						
		0.9	1.3	1.5	2.0	3.0	4.0	5.0
Renyi	1	85.0±3.9	84.7±4.6	85.2±3.7	88.6±4.5	90.2±3.4	83.9±3.9	86.8±2.0
	2	71.3±11.9	77.4±5.0	76.7±5.7	79.1±4.6	78.3±3.7	80.3±7.1	78.2±5.9
Tsallis	1	84.5±3.9	86.1±3.4	87.4±4.0	85.0±3.4	90.0±3.8	89.1±3.0	91.3±3.5
	2	78.1±11.8	74.4±7.5	75.2±7.8	77.9±7.0	80.2±6.7	84.3±6.2	80.3±6.5
Shannon	1	83.8±5.3						
	2	76.6±5.7						

Below one of the decision trees generated on the Leukemia data set with Renyi's entropy and $\alpha = 3$ is presented:

```
g760 >  588 : AML
g760 <= 588 :
|    c1926 <= 134 : ALL
|    c1926 >  134 : AML
```

Rules extracted from this tree are:

```
Rule 1:
        g760 > 588
    --> class AML   [93.0%]

Rule 2:
        g1926 > 134
    --> class AML   [89.1%]

Rule 3:
        g760 <= 588
        g1926 <= 134
    --> class ALL   [93.9%]
```

The trees are very small and should provide good generalization for small datasets, avoiding overfitting that other methods may suffer from. However, for gene expression data this may not be sufficient as the results are not stable against small perturbation of data (all learning methods suffer from this problem, see [11]). Therefore in practical applications a better approach is to define approximate coverings of redundant features and replace such groups of features with their linear combinations to reduce their dimensionality, aggregating information about similar genes (Biesiada and Duch, in preparation). For the demonstration of efficiency of non-standard entropy measures this is not so important.

5 Conclusions

Information Theoretic Learning (ITL) has been used to train neural networks for blind source separation [12], definition of new error functions in neural networks [13], classification with labeled and unlabeled data, feature extraction and other applications [14]. The quadratic Renyi's error entropy has been used to minimize the average information content of the error signal for supervised adaptive system training. However, the use of non-Shannon entropies in extraction of logical rules using decision trees has not yet been attempted. The importance of decision trees in large-scale data mining knowledge extraction applications and the simplicity of this approach encouraged us to explore this possibility.

First experiments with modified split criterion of the C4.5 decision trees were presented here. Theoretical results and tests on artificial data show that this can be useful particularly for datasets with one or more small classes. Additional parameter α that may be easily adapted using crossvalidation tests gives possibility to tune the tree to the discrimination of classes of different size. This algorithm makes it more attractive than standard approach based on Shannon entropy that does not allow for exploration of the tradeoff between the probability of different classes and the overall information gain. This opens a way to applications in many types of decision trees, encouraging also modification of split criteria that are not based on entropy to account for the same tradeoff. An explicit formula for such tradeoff may be defined, aimed at separation of nodes of high purity at the expense of lower overall information gain. This as well as applications of non-standard entropies to the text classification data, where small classes are very important, remain to be explored. Bearing in mind the results and the simiplicity of the approach presented here the use of non-standard entropies may be highly recommended.

References

1. Duch, W., Setiono, R., Zurada, J.: Computational intelligence methods for understanding of data. Proceedings of the IEEE 92(5), 771–805 (2004)
2. Duch, W., Grochowski, M.: Learning Highly Non-separable Boolean Functions Using Constructive Feedforward Neural Network. In: de Sá, J.M., Alexandre, L.A., Duch, W., Mandic, D. (eds.) ICANN 2007. LNCS, vol. 4668, pp. 180–189. Springer, Heidelberg (2007)
3. Quinlan, J.: C 4.5: Programs for machine learning. Morgan Kaufmann, San Mateo (1993)
4. Duch, W.: Filter methods. In: Guyon, I., Gunn, S., Nikravesh, M., Zadeh, L. (eds.) Feature extraction, foundations and applications, pp. 89–118. Physica Verlag, Springer, Heidelberg (2006)
5. Shannon, C., Weaver, W.: The Mathematical Theory of Communication. University of Illinois Press, Urbana (1964)
6. Tsallis, C., Mendes, R., Plastino, A.: The role of constraints within generalized nonextensive statistics. Physica 261A, 534–554 (1998)
7. Renyi, A.: Probability Theory. North-Holland, Amsterdam (1970)
8. Alon, U.: Broad patterns of gene expression revealed by clustering analysis of tumor and normal colon tissues probed by oligonucleotide arrays. PNAS 96, 745–750 (1999)
9. Alizadeh, A.: Distinct types of diffuse large b-cell lymphoma identified by gene expression profiling. Nature 403, 503–511 (2000)
10. Golub, T.: Molecular classification of cancer: Class discovery and class prediction by gene expression monitoring. Science 286, 531–537 (1999)
11. Duch, W.: Towards comprehensive foundations of computational intelligence. In: Duch, W., Mandziuk, J. (eds.) Challenges for Computational Intelligence, vol. 63, pp. 261–316. Springer, Heidelberg (2007)
12. Hild, K., Erdogmus, D., Principe, J.: Blind source separation using renyi's mutual information. IEEE Signal Processing Letters 8, 174–176 (2001)
13. Erdogmus, D., Principe, J.: Generalized information potential criterion for adaptive system training. IEEE Trans. on Neural Networks 13, 1035–1044 (2002)
14. Hild, K., Erdogmus, D., Torkkola, K., Principe, J.: Feature extraction using information-theoretic learning. IEEE Trans. on Pattern Analysis and Machine Intelligence 28, 1385–1392 (2006)

Information Theory Inspired Weighted Immune Classification Algorithm*

Maciej Morkowski[1] and Robert Nowicki[1,2]

[1] Department of Computer Engineering, Częstochowa University of Technology
Al. Armii Krajowej 36, 42-200 Częstochowa, Poland
maciejm@kik.pcz.czest.pl, rnowicki@kik.pcz.czest.pl
http://kik.pcz.pl
[2] Department of Artificial Intelligence
Academy of Humanities and Economics in Łódź
ul. Rewolucji 1905 nr 64, Łódź, Poland
http://www.wshe.lodz.pl

Abstract. This article presents an example of a handwritten numbers classifier based on the immune system. We study mutual relations between the system operation parameters, as well as new mechanisms introduced in order to make the system work faster. To achieve the goal the weights inspired by the information theory have been inserted to the immune system.

1 Introduction

Leukocytes (white blood cells) are blood components specialised in defending the organism against harmful microbes. There are two major types of leukocytes. Granulocytes, which are generated in the bone marrow, detect and absorb bacteria and dead cells. The number of such cells increases in the case of allergies and parasite infections. The other type, the granulocytes, includes lymphocytes (which produce antibodies and attack bacteria and viruses directly) and macrophages (which destroy bacteria via phagocytises). Both specific and non-specific defence mechanisms function in all vertebrates.

Specific immunity consists of humoral immunity and cell-mediated immunity. Cell-mediated immunity relies on lymphocytes attacking pathogens directly, while humoral immunity relies on the production of specific antibodies by the lymphocytes in order to eliminate pathogens.

The main types of lymphocytes are the B-lymphocytes and the T-lymphocytes. The T-lymphocytes originate from hematopoietic stem cells in the bone marrow and mature in the thymus. There, they become immunologically competent, that

---

* This work was partly supported by the Foundation for Polish Science (Professorial Grant 2005-2008) and Polish Ministry of Science and Higher Education (Habilitation Project 2008-2010, Special Research Project 2006-2009, Polish-Singapore Research Project 2008-2010, Research Project 2008-2010).

is, they develop the ability to perform the immune response – a reaction to specific antigens. The T-lymphocytes are involved in the cell-mediated immune response. The ability to detect pathogens depends on the protein receptors on the T cell's surface. We can distinguish three classes of T-lymphocytes, the killer T cells, which eliminate pathogens, the helper T cells, which activate and amplify the immune response, and the suppressor T cells, which moderate the response.

The B-lymphocytes are involved in the humoral immune response. Each B cell has an ability to bind specific antigens. After it recognises an antigen, a lymphocyte begins to divide, cloning itself, and secrete antibodies into the bloodstream.

The immune system's ability to discern infected cells from the healthy ones is based on a group of protein markers (antigens) called the MHC (major histocompatibility complex). The foreign antigen creates, by means of surface contact, a complex with the macrophage's MHC antigens. Antibodies on the surface of the B cell react to that complex as receptors. Competent B cells bind themselves to the MHC-antigen complex on the macrophage's surface. As an effect of the interaction between the macrophage and the helper T cell, the receptors of which respond to the presented antigen, the macrophage releases interleukine-1, which in turn activates helper T cells and stimulate them to detect the B cells already bound to the MHC-antigen complex and attach themselves to the complex. After they attach, the T cells activate the B cells they attached to by secreting suitable compounds. Stimulated B cells begin to grow in size and divide via mitosis, producing numerous identical clones. Some B cells transform into plasmatic cells producing antibodies, others become immunological memory cells, which enable the organism to respond immediately in case of a future infection by the same pathogen.

What is worth noting is that the antibodies produced by the B cell clones are not identical. This is because only parts of the antibody-coding genes, and not the whole genes, are inherited during the cloning process. Their shuffling and juxtaposition take effect only in the lymphocytes' maturation. Another source of the diversity of the antibodies are the mutations of the aforementioned DNA segments. Thanks to this, the antibodies are able to react to specific amino acid sequences on the antigens surfaces. Those sequences are called epitopes [17].

2 Artificial Immune System

Farmer, Packard and Perelson, who in their work [7] presented the ideas behind the functioning of the immune system, had decided to eschew many important elements, such as the presence of the T cells and macrophages, while creating its model. They had concentrated on the most vital elements, so that their system could easily be simulated by a computer [5], [8], [9].

The most important simplification was the decision to present epitopes and their matching amino acid sequences, called paratopes, as binary chains. It was also assumed that each antigen has only one epitope.

The basic element conditioning the actions of the artificial immune system is the stimulation of the m_{ij} lymphocyte by the antigen. Marking the epitope

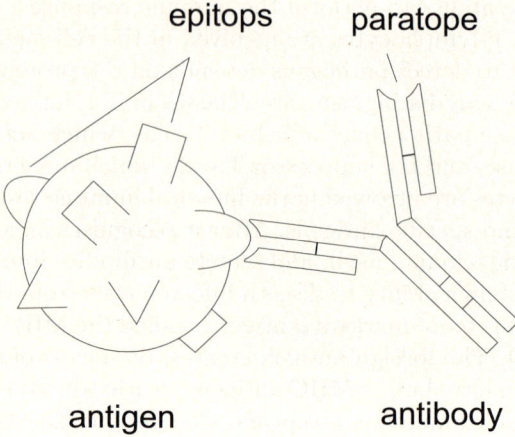

Fig. 1. Antigen-antibody complex

chain length as l_e, the paratope chain length as l_p, the value of the n-th bit in the i-th epitope as $e_i(n)$, the value of the n-th bit in the j-th epitope as $p_j(n)$, the similarity threshold as s and the maximum shift value as k, we form the equation:

$$m_{ij} = \sum_k G\left(\sum_n |e_i(n-k) - p_j(n)| - s + 1\right), \qquad (1)$$

where $G(x) = x$ for $x > 0$ and $G(x) = 0$ in all other cases.

The aforementioned authors based the system dynamics process on the fact that antibodies have paratropes as well, thus the system is constantly active, thanks to the lymphocytes being stimulated by the antibodies on the surfaces of other lymphocytes. This takes place even if no external antigens influence the system. This, however, is not compatible with a system, in which an immune response is conditioned by the stimulation of the helper T cell, able to recognise the cells of its own organism. It is also necessary to add that inactive B cells are suppressed and that the organism generates new lymphocytes all the time, even as many as 10^7 a day [10].

3 The Classification System

All numbers, subjected to the identification of the classifier immune system, were written on a grid of 32 x 32 points. Each point could exist in two states: black or white. Then, each grid was divided into 64 non-overlapping squares of 4 x 4 points each. The number of black fields in subsequent squares is the classifier input data vector. The learning sequence is stored in a text file. Each line contains an input vector comprising of 64 integers in the $<0; 16>$ range and a digit defining the classifier correct answer [1], [2], [3], [4].

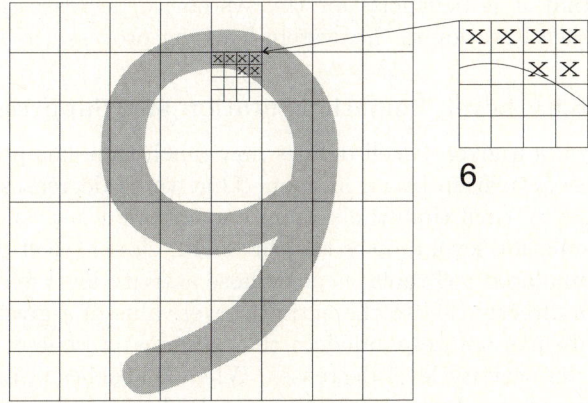

Fig. 2. Representation of a number

3.1 The Representation of Lymphocytes and Antigens

Each lymphocyte is represented by a paratope — a chain of 64 numbers in the < 0; 16 > bracket, a value — a numerical equivalent of the antigen class, and stimulation the system's knowledge of how much time has passed since the antigen's last stimulation. Antigens, whose stimulation falls to 0, are suppressed.

Antigens are represented by the paratopes counterparts, called in this case epitopes, and a numerical value describing their affiliation with a certain class.

3.2 The Similarity Level, Initial Lymphocyte Population, Stimulation Threshold

As the paratopes and epitopes are represented in a non-binary way, the way of similarity level calculation had to be changed. It is now the difference between the values of the epitope and the paratope in respective positions. The maximum dissimilarity value equals 1024.

In the case of random generation of the starting epitopes, we encounter the problem of high dissimilarity level between the lymphocytes and the antigens. It becomes necessary to generate the initial population based on the division of the teaching series into two subseries. The first subseries lets us determine the initial population while the other remains the actual teaching series. Thanks to this, we avoid the situation when the lymphocytes are alternately stimulated by different-class antigens, which leads to the system constantly getting out of order instead of learning.

For research purposes, it is good to set each class-related lymphocyte value at -1, which lets us determine, after teaching the system, how many lymphocytes were not stimulated, or stimulated rarely enough to be suppressed. We can thus determine the immune system initial stimulation threshold. If it is too low, many lymphocytes are inactive and do not improve the system functionality. It has been observed that it is beneficial for the system to decrease the stimulation threshold with each iteration of the system learning process.

3.3 The Activity Level, Somatic Mutation and Suppression

The activity level of a given B cell defines how much time has passed since the lymphocyte's last activation by an antigen. That value decreases for every not reacting lymphocyte, each time the system encounters a new antigen. When a lymphocyte stimulation level crosses the threshold, clonal selection takes place. Clones of the stimulated cell replace cells whose activity level fell to 0.

Suppression is directly tied to the activity level value of a given lymphocyte. If the lymphocyte has not responded to the antigens presented to the system for a long time, its activity level decreases. When it reaches 0, and the system demands new space for cells created in the clonal selection process, lymphocytes with 0 activity level will be replaced with the newly-created clones.

4 The Learning Algorithm

The process of teaching consists in presenting the system with the full set of antigens n times. The similarity level of the system lymphocytes is checked for each presented antigen. The activity value of lymphocytes whose similarity level is below the stimulation threshold is lowered. When a lymphocyte crosses the stimulation threshold, clonal selection takes place. A stimulated lymphocyte activity level is set to maximum. Its clones replace cells with activity level 0 and then mutate. Subsequent lymphocytes are then checked. Finally, at the end of a cycle, the stimulation threshold lowers.

```
Read antigen population from file
Generate initial lymphocyte population
Teaching cycles number > 0
    for each antigen
        get next B cell
        calculate similarity level
        if similarity level crosses similarity threshold
            delete a lymphocyte with activity level = 0
            replace it with a B cell clone
            mutate B cell
        if similarity level does not cross similarity threshold
            decrease lymphocyte activity
    decrease stimulation threshold
    decrease teaching cycle number by 1
```

5 Experimental Results without Weights

Table 1 illustrates the effectiveness of classification in relation to set functioning parameters.

A higher stimulation threshold is represented by a lower numeral value.

Table 1. Effectiveness of classification

cycle number	population size	stimulation treshold	lymphocyte activity	clonal sel. lvl	locus number	treshold decrease	system lvl efficiency
1	30000	450	30	10	16	0.02	93.8787
2	30000	450	30	10	16	0.02	94.6578
1	30000	450	30	10	16	0.01	94.1569
2	30000	450	30	10	16	0.01	93.9900
3	30000	450	30	10	16	0.01	93.8787
1	20000	450	30	10	16	0.01	92.5431
1	15000	450	30	10	16	0.01	92.9327
1	10000	450	30	10	16	0.01	89.5938
1	30000	425	30	10	16	0.01	94.2126
1	30000	400	30	10	16	0.01	93.9900
1	30000	375	30	10	16	0.01	93.9343
1	30000	425	50	10	16	0.01	94.2682
1	30000	425	100	10	16	0.01	94.1013
1	30000	425	200	10	16	0.01	94.7134
1	30000	425	500	10	16	0.01	93.9343
1	30000	425	200	20	16	0.01	94.1013
1	30000	425	200	50	16	0.01	94.3795
1	30000	425	200	100	16	0.01	94.1569
1	30000	425	200	100	12	0.01	94.4908
1	30000	425	200	100	10	0.01	93.4335
2	30000	425	200	100	12	0.02	95.4368
2	20000	425	200	100	12	0.02	93.8230
2	30000	425	200	50	12	0.02	95.0473

6 Information Theory Inspired Weights

Not all fields of the analysed image matter equally during the number classification process. One of the ways to express the differences between fields are weights. The additional weights have been used to aggregation of rules and to connectives of antecedents in neuro-fuzzy systems [12], [13] as well as in other methods [6], [11], [18]. Usually, the weights come to improve the obtained results. There are many methods to get the values of the weights (e.g. a usual gradient optimization [13]).

We propose to extend equation (1) using weights in the process of paratope – epitope similarity calculation as follows

$$m_{ij} = \sum_k G \left(\sum_n |e_i(n-k) - p_j(n)| \cdot w(n-k) - s + 1 \right) . \quad (2)$$

To calculate the values of the weight we used the method inspired by Shannon theory of information [15], [16]. In this theory the entropy of message (event) A is given as follows

Table 2. Weights calculated from information theory

attribute	weight	attribute	weight	attribute	weight	attribute	weight
1	0.087	17	0.190	33	0.194	49	0.403
2	0.371	18	0.843	34	0.638	50	0.830
3	0.872	19	0.901	35	0.878	51	0.902
4	0.884	20	0.913	36	0.913	52	0.919
5	0.914	21	0.882	37	0.889	53	0.934
6	0.912	22	0.868	38	0.915	54	0.911
7	0.749	23	0.751	39	0.789	55	0.802
8	0.467	24	0.335	40	0.260	56	0.348
9	0.168	25	0.239	41	0.356	57	0.211
10	0.775	26	0.805	42	0.615	58	0.651
11	0.889	27	0.895	43	0.843	59	0.877
12	0.935	28	0.913	44	0.881	60	0.861
13	0.924	29	0.877	45	0.920	61	0.880
14	0.899	30	0.905	46	0.937	62	0.884
15	0.833	31	0.676	47	0.800	63	0.687
16	0.534	32	0.260	48	0.273	64	0.370

Table 3. Effectiveness of classification by modified system

cycle number	population size	stimulation treshold	lymphocyte activity	clonal sel. lvl	locus number	treshold decrease lvl	system efficiency
2	30000	350	200	100	12	0.02	94.9360
1	30000	350	200	100	12	0.02	94.6578
1	30000	350	200	100	12	0.01	94.4352
1	30000	325	200	100	12	0.01	94.1569
1	30000	300	100	10	16	0.01	95.1586
1	20000	300	30	10	16	0.01	95.1586
1	20000	300	30	30	16	0.01	94.9360
1	15000	300	30	50	16	0.01	94.9917
2	15000	300	100	10	16	0.01	94.7691
1	10000	300	30	50	12	0.01	93.9900

$$H = P(A) \cdot i(A) , \qquad (3)$$

where $P(A)$ is the probability than event A occurs, and $i(A)$ is the amount of information which is according to event A and is given by

$$i(A) = log_b \frac{1}{P(A)} = -log_b P(A) . \qquad (4)$$

In case of the event series $A_1, \ldots, A_Q$ the entropy is calculated as follows

$$H = \sum_{q=1}^{Q} P(A_q) \cdot i(A_q) . \qquad (5)$$

We propose to calculate the weight of n-th attribute using the equation

$$w(n) = \frac{H(n)}{Q} , \qquad (6)$$

where Q_l is the number of the records in the learning sequence and $H(n)$ is calculated as follows

$$H(n) = \sum_{q=1}^{Q} H_q(n) . \qquad (7)$$

The $H_q(n)$ values are calculated in following way

$$H_q(n) = \sum_{l=1}^{L} P(x \in \omega_l | q_n = q) \cdot (-log(P(x \in \omega_l | q_n = q))) . \qquad (8)$$

The probability $P(x \in \omega_l | q_n = q)$ is calculated based on the learning sequence.

Using the above model we calculate the relation between an element value and its affiliation with a certain class or group of classes. We record weights obtained this way in Table 2. Table 3 illustrates the effectiveness of classification in relation to set functioning parameters using the weights.

7 Conclusions

The introduction of weights enabled a decrease of the number of lymphocytes and an increase of the initial stimulation threshold. This was achieved without a decrease in the classification correctness. This hastens both the learning process and the system response to a presented pathogen.

References

1. Alpaydin, E., Kaynak, C., Alimoglu, F.: Cascading Multiple Classifiers and Representations for Optical and Pen-Based Handwritten Digit Recognition. In: IWFHR, Amsterdam, The Netherlands (September 2000)
2. Alimoglu, F., Alpaydin, E.: Combining Multiple Representations for Pen-based Handwritten Digit Recognition. ELEKTRIK: Turkish Journal of Electrical Engineering and Computer Sciences 9(1), 1–12 (2001)
3. Alpaydin, E., Alimoglu, F.: Optical Recognition of Handwritten Digits, http://www.ics.uci.edu/~mlearn/databases/optdigits/
4. Alpaydin, E., Kaynak, C.: Pen- Based Recognition of Handwritten Digits, http://www.ics.uci.edu/~mlearn/databases/pendigits/
5. de Castro, L.N., von Zuben, F.J.: Learning and optimization using the clonal selection principle. IEEE Transactions on Evolutionary Computation, Special Issue on Artifical Immune System 6, 239–251 (2002)
6. Chen, S.-M.: Weighted fuzzy reasoning using weighted fuzzy Petri nets. IEEE Transactions on Knowledge and Data Engineering 14(2), 386–397 (2002)
7. Farmer, J.D., Packard, N.H., Perelson, A.S.: The immune system, adaptation, and machine learning. Physica D 22, 187–204 (1986)

8. Garret, S.M.: How do we evaluate artificial immune systems? Evolutionary Computation 13, 145–178 (2005)
9. Hofmeyr, S.A., Forrest, S.: Architecture for an artificial immune system. Evolutionary Computation 8, 443–473 (2000)
10. Osmond, D.G.: The turn-over of B cell populations. Immunology Today 14, 34–37 (1993)
11. Rafajłowicz, E., Pawlak, M., Steland, A.: Nonlinear image processing and filtering: A unified approach based on vertically weighted regression. Int. J. Appl. Math. Comput. Sci. 18(1), 49–61 (2008)
12. Rutkowski, L.: Flexible Neuro-Fuzzy Systems. Kluwer Academic Publishers, Dordrecht (2004)
13. Rutkowski, L., Cpałka, K.: Flexible neuro-fuzzy systems. IEEE Trans. Neural Netw. 14(3), 554–574 (2003)
14. Rutkowski, L., Cpałka, K.: Designing and learning of adjustable quasi-triangular norms with applications to neuro-fuzzy systems. IEEE Trans. Fuzzy Syst. 13(1), 140–151 (2005)
15. Shannon, C.E.: A Mathematical Theory of Communication. Bell System Technical Journal 27, 379–423, 623–656 (1948)
16. Shannon, C.E., Weaver, W.: The Mathematical Theory of Communication. Univ of Illinois Press (1949)
17. Solomon, E.P., Berg, L.R., Martin, D.W.: Biology, 6th edn. Thomson Brooks/Cole (2001)
18. Yeung, D.S., Ysang, E.C.C.: A multilevel weighted fuzzy reasoning algorithm for expert systems. IEEE Transactions on Systems, Man and Cybernetics, Part A 28(2), 149–158 (1998)

Bayes' Rule, Principle of Indifference, and Safe Distribution

Andrzej Piegat[1] and Marek Landowski[1,2]

[1] Faculty of Computer Science and Information Systems,
Szczecin University of Technology, Zolnierska 49, 71-210 Szczecin, Poland
Andrzej.Piegat@wi.ps.pl, mlandowski@wi.ps.pl
[2] Quantitative Methods Institute,
Szczecin Maritime University, Waly Chrobrego 1-2, 70-500 Szczecin, Poland
m.landowski@am.szczecin.pl

Abstract. Bayes' rule is the basis of probabilistic reasoning. It enables to surmount information gaps. However, it requires the knowledge of prior distributions of probabilistic variables. If this distribution is not known then, according to the principle of indifference, the uniform distribution has to be assumed. The uniform distribution is frequently and heavily criticized. The paper presents a safe distribution of probability density that can be often used instead of the uniform distribution to surmount information gaps. According to the authors' knowledge the concept of the safe distribution is new and unknown in the literature.

1 Introduction

Human intelligence comprises many skills. The basic skill is the skill of reasoning, especially under uncertainty. One type of automatic reasoning is the probabilistic reasoning. The aim of probabilistic reasoning can be shortly defined as "to build network models to reason under uncertainty according to the laws of probability theory" [3]. These network models for reasoning are called Bayesian networks, belief networks, and probabilistic networks. An extension of Bayesian networks are called decision networks and enable to calculate decisions under uncertainty. Solving problems under uncertainty (partial lack of knowledge) is one of the most difficult aims of artificial intelligence (AI). People can solve such problems. To make AI comparable with the human intelligence, it also has to be able to solve problems under uncertainty. Problems of information gaps are being intensively investigated at present [4]. An information gap means lack of knowledge of variables values, of distributions of their probability or possibility, of variability intervals, etc. One of ways to surmount information gaps is proposed by probability theory in the form of Bayes' theorem [3] [5]. Let A and B denote two events. The conditional probability $p(A\backslash B)$ is not known. However, the inverse of conditional probability $p(B\backslash A)$ and the prior probability $p(A)$ is known. Then the unknown probability $p(A\backslash B)$ can be determined using Bayes' theorem [5], formula (1).

$$p(A\backslash B) = \alpha L(B\backslash A)p(A) \qquad (1)$$

where: α - normalizing constant, $L(B\backslash A)$ - likelihood of A given B $(= p(B\backslash A))$. The Bayes' theorem can be paraphrased as:

posterior probability = (normalizing constant) × likelihood × prior probability.

Bayes' theorem is a very valuable tool because it tells us how to update or revise beliefs, expressed in the form of probabilities, in the light of new evidence a posteriori. However, in practical applications it is sometimes difficult to determine the prior probability distribution. Let us consider the known *Monty Hall problem* [5].

We are presented with 3 doors numbered 1, 2, 3 - one of which has a prize. We choose the 1-door, which is not opened until the presenter performs an action. The presenter, who knows which door the prize is behind, and who must open a door, but is not permitted to open the door we have picked or the door with the prize, opens the 3-door and reveals that there is no prize behind it and subsequently asks if we wish to change our mind about the initial selection of 1. What is the probability that the prize is behind the 1- or 2-door?

This problem can be solved using Bayes' theorem. However, the knowledge of probability of the prize being hidden behind the particular door is necessary. Because we know nothing about preferences of the presenter (he can e.g. most frequently hid the prize behind the 3-door) we assume that, according to *the principle of indifference* [6], shortly PI, the uniform distribution of probability for particular door is:

$$p(1) = p(2) = p(3) = 1/3.$$

The above means, we have applied principle of indifference (principle of insufficient reason).

Another interesting "continuous" example is the Bertrand problem [2].

The train leaves at noon to travel a distance of 300 km. It travels at a speed of between 100 km/h and 300 km/h. What is the probability that it arrives before 2 p.m.?

To give an answer to the above question, the knowledge of the probability density distribution of the train velocity $PDD(\nu)$ is necessary. However, this distribution is not given in the problem description (information gap). Because we do not know which velocities occur more and which less frequently we use the PI and assume that the distribution is a uniform one, Fig. 1.

Using the principle of indifference the Bertrand problem can be solved. However, the PI is strongly criticized by its opponents. For instance, Yakov in [4] writes:

The uncertainty is an information gap .... We have no information, not even probabilistic. By principle of indifference elementary events, about which nothing is known, are assigned equal probabilities. Knowing a specific probability is knowledge.

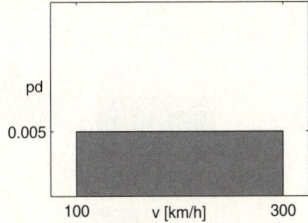

Fig. 1. Uniform distribution of probability density (PD) of the train suggested by the principle of indifference

The Yakov's argumentation is incorrect. Ofra Magidor gives in [2] a convincing contra-argumentation:

> Assume that two alternatives are equally probable if you do not have any reason not to do so. Such formulations present *action-guiding principles*. They do not claim that in the cases mentioned the probabilities are equal- they simply direct us to assume so. In these cases, the principle can at best be viewed as a *rationality principle*.

The principle of indifference helps us to make decisions if we have to make them, although our knowledge is insuffiecient.

2 Is the Uniform Prior a Rational Assumption?

If we do not know the probability distribution of the prize behind the doors in the Monty Hall problem, we cannot exclude any type or any form of the prior distribution. We have to assume that the distribution can be unimodal (of any shape) and bimodal (of any shape) and trimodal (of any shape) etc. The number of possible distributions is infinitely large because we do not know preferences of the presenter. Using *the method of decreasing granulation of elementary events and probability* (shortly: *method of decreasing granulation-DG-method*), conceived by Andrzej Piegat, one can prove that the uniform distribution is a very rational assumption because it minimizes the average absolute error in relation to all possible distributions of probability. In the Monty Hall problem the number of possible events is given a priori and equals 3 doors. Let i, $i \in \{1, 2, 3\}$, be the door number. This number can not be changed; therefore the event granularity is constant. Firstly let us assume only 4 possible values $p \in \{0, 1/3, 2/3, 1\}$ for probability. The generated distributions will be numbered by index j. At the next stages of the DG-method the number of possible probability values will be gradualy increased to infinity. At the assumed granularity there exist 10 possible distributions shown in Fig. 2.

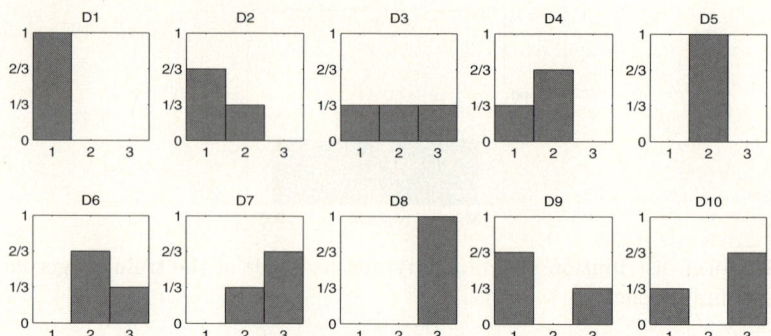

Fig. 2. Possible distributions D_j, $j = 1 - 10$, of probability $p_i \in \{0, 1/3, 2/3, 1\}$ of the prize hidden behind doors numbered by i, $i \in \{1, 2, 3\}$ (8 unimodal and 2 bimodal distributions)

Let us calculate the average probability $p_{aver}(i)$ resulting from all 10 possible distributions for each door $i \in \{1, 2, 3\}$, formula (2).

$$p_{aver}(i) = \frac{1}{10} \sum_{j=1}^{10} p_{ji}. \qquad (2)$$

The calculations result in:

$$p_{aver}(1) = p_{aver}(2) = p_{aver}(3) = \frac{1}{3}.$$

The above result means that the average probability distribution is uniform one, Fig. 3.

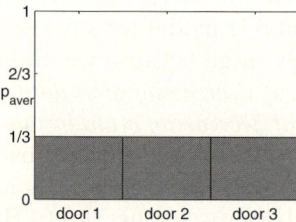

Fig. 3. The average distribution of probability $p_{aver}(i)$ of the prize being hidden behind the door $i \in \{1, 2, 3\}$

The average distribution of probability shown in Fig. 3 gives a minimal sum of the absolute and also square errors in relation to all 10 possible probability distributions shown in Fig. 2. At the first stage of the decreasing granulation only 4 discrete probability values $\{0, 1/3, 2/3, 1\}$ were allowed. At the second stage the probability granularity was decreased from $1/3$ to $1/4$. Now, 5 discrete values of probability are possible; $p_i \in \{0, 1/4, 2/4, 3/4, 1\}$. It allows a more

Table 1. 15 possible distributions of probability p_i of the prize being hidden behind 3 doors i and the average distribution of the probability at granularity 1/4 ($p_i \in \{0, 1/4, 2/4, 3/4, 1\}$)

Probability distribution j	Door i		
	1 p_{j1}	2 p_{j2}	3 p_{j3}
1	4/4	0	0
2	3/4	1/4	0
3	3/4	0	1/4
4	2/4	2/4	0
5	2/4	0	2/4
6	2/4	1/4	1/4
7	0	4/4	0
8	1/4	3/4	0
9	0	3/4	1/4
10	0	2/4	2/4
11	1/4	2/4	1/4
12	0	0	4/4
13	0	1/4	3/4
14	1/4	0	3/4
15	1/4	1/4	2/4
$\sum_{j=1}^{15} p_{ji}$	20/4	20/4	20/4
$p_{aver}(i)$	1/3	1/3	1/3

precise analysis of the Monty Hall problem. The number of possible probability distributions of the prize being hidden behind the particular door numbered i increases from 10 to 15, Table 1.

As it can be seen in Table 1 also at probability granularity 1/4 the average probability of the prize behind door i is equal for each door ($p_{aver}(1) = p_{aver}(2) = p_{aver}(3) = 1/3$).

At the third stage 6 values $\{0, 1/5, 2/5, 3/5, 4/5, 1\}$ of probability are allowed, etc.

The granularity of events (number of doors) is constant and does not change in this problem. At each stage all possible distributions are generated and the average distribution of probability is calculated. The granularity is decreased at each stage. Thus finally we approach the infinite number of possible distributions of probability. One can check that at each stage of the method the achieved average distribution of probability is the uniform one. It means that the principle of indifference suggests a rational distribution.

Thus, the analysis of the Monty Hall problem, with only 3 possible events, has been finished. In Chapter 3 the situation will be analysed, when we have some qualitative (but not quantitative) knowledge of the distribution. We know that the real distribution is a unimodal one.

3 Should We, in Case of Lack of Knowledge, Always Assume the Prior Distribution Is the Uniform One?

Lack of knowledge can be higher or lower. If we virtually know nothing about the distribution (about its modality and shape), the best assumption is the uniform distribution. However, in practical terms, we frequently know (with full certainty or with high probability) that the *distribution in question is a unimodal* one. Ilustrative examples of unimodal distributions are presented in Fig. 4.

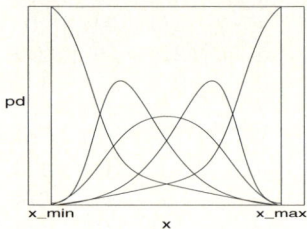

Fig. 4. Example of unimodal distributions

For instance, the distribution of probability density of the train travel time or of the train speed or of a car speed between two cities is mostly unimodal (the author investigated it experimentally for a distance of 15 km between his work place and home). Similarly the distribution of height of adult men in a country is mostly unimodal, etc. If we know that the distribution is unimodal then, using the method of decreasing granulation, we can determine the average unimodal distribution resulting from all possible unimodal distributions, which minimizes the sum of absolute errors. For the sake of simplicity, let us assume that there is a normalized variability interval [0, 1] of the stochastic variable x. At the first stage of the method of decreasing granulation 3 large intervals are assumed: $0 \leq x < 1/3$, $1/3 \leq x < 2/3$, $2/3 \leq x < 1$, and only 4 possible discrete values of probability $\{0, 1/3, 2/3, 1\}$ of events (x is in interval $i = 1$ or 2 or 3). Fig. 5 presents all possible unimodal distributions of probability (histograms) for granularity 1/3.

For each sub-interval i of variable x the average probability can be calculated with formula (3).

$$p_{aver}(i) = \frac{1}{8} \sum_{j=1}^{8} p_{ji} \qquad (3)$$

$p_{aver}(1) = 7/24$, $p_{aver}(2) = 10/24$, $p_{aver}(3) = 7/24$. The average histogram of probability and average distribution of probability density of elementary events for granularity 1/3 is shown in Fig. 6.

Now, both the probability granularity and the granularity of the variable x will be decreased to 1/4. It means that ($p_i \in \{0, 1/4, 2/4, 3/4, 1\}$) and the number of possible intervals of the variable x equals 5: $x \in \{[0, 1/4], [1/4, 2/4], [2/4, 3/4],$

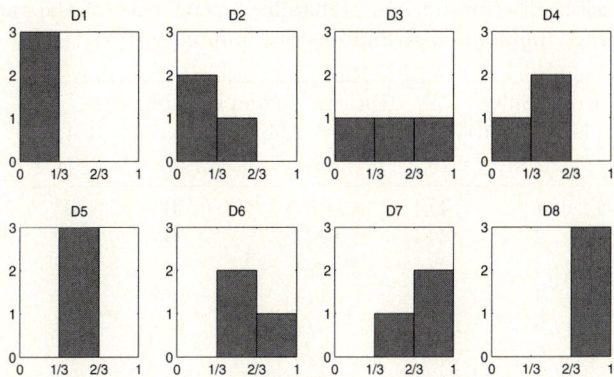

Fig. 5. All possible 8 unimodal distributions Dj of probability (histograms) of the stochastic, continuos variable x achieved at granularity 1/3 of atomic (elementary) events

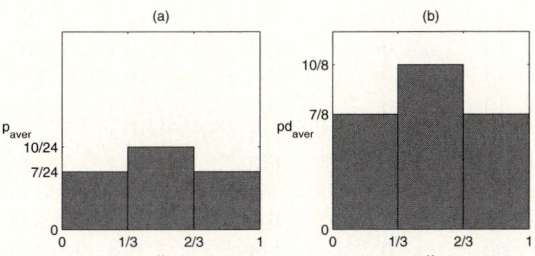

Fig. 6. The average histogram of probability (a) and average probability density (PD) (b) distribution achieved at granularity 1/3 of elementary events (i-door's number)

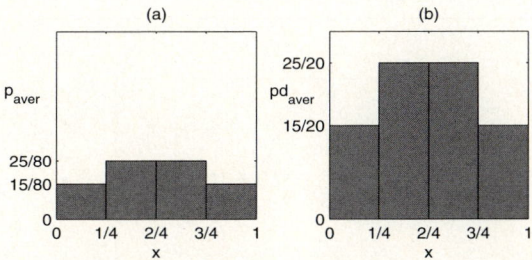

Fig. 7. The average unimodal histogram of probability and the average probability distribution achieved at granularity 1/4 of probability and of the variable x

[3/4, 1]}. Table 2 shows all 20 possible unimodal distributions of probability. The results contained in Table 2 are presented in Fig. 7.

For granularity 1/5 the number of possible unimodal distributions equals 45. The average distribution of probability density is shown in Fig. 8.

Table 2. 20 possible distributions at probability granularity of the variable x equal $1/4$, and the average unimodal distributions of probability

Probability distribution j	Interval of the variable x			
	$[0, 1/4]$ p_{j1}	$[1/4, 2/4]$ p_{j2}	$[2/4, 3/4]$ p_{j3}	$[3/4, 1]$ p_{j4}
1	4/4	0	0	0
2	3/4	1/4	0	0
3	2/4	2/4	0	0
4	2/4	1/4	1/4	0
5	1/4	1/4	1/4	1/4
6	0	4/4	0	0
7	1/4	3/4	0	0
8	0	3/4	1/4	0
9	0	2/4	2/4	0
10	1/4	2/4	1/4	0
11	0	2/4	1/4	1/4
12	0	0	4/4	0
13	0	1/4	3/4	0
14	0	0	3/4	1/4
15	0	0	2/4	2/4
16	1/4	1/4	2/4	0
17	0	1/4	2/4	1/4
18	0	0	0	4/4
19	0	0	1/4	3/4
20	0	1/4	1/4	2/4
$\sum_{j=1}^{23} p_{ji}$	15/4	25/4	25/4	15/4
$p_{aver}(i)$	15/80	25/80	25/80	15/80

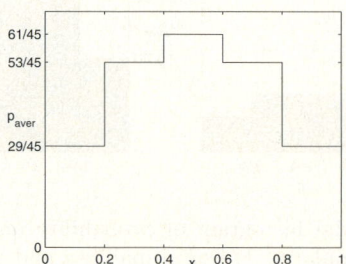

Fig. 8. The average unimodal distribution of probability density resulting from 45 possible unimodal distributions for granularity $1/5$

The number of possible unimodal distributions sharply increases with the change of granularity. Thus, it equals 191 for granularity $1/7$, 676 for granularity $1/9$, 2112 for granularity $1/11$, 5996 for granularity $1/13$, 15763 for granularity

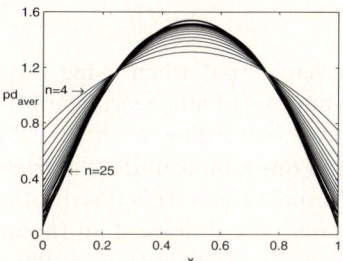

Fig. 9. The average unimodal distributions of probability density for different granularities 1/4 - 1/25 of elementary events (width of subintervals of variable x), the distributions were smoothed

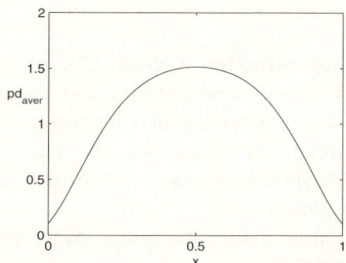

Fig. 10. The average unimodal distribution of probability density for number of subintervals $n = 25$ that can be acknowledged as precise approximation of the limiting distribution for $n \to \infty$

1/15, 646363 for granularity 1/25, and so on. Fig. 9 presents average distributions of probability density for different granularities $1/4, 1/5, 1/6, 1/7, ..., 1/25$.

As it can be observed in Fig. 9 at number n of subintervals greater than 15 (granularity $\leq 1/15$) changes of average distributions are insignificant and imperceptible. They approach the limiting distribution for $n \to \infty$. Thus, further increasing of the number of subintervals is unnecessary and one can assume that the distribution for $n = 25$ is a sufficiently close approximation of the limiting distribution for n approaching infinity. This distribution is shown in Fig. 10.

The approximation of the average distribution from Fig. 10 for $n \to \infty$ can be called *safe unimodal distribution* because it minimizes the average absolute error of all possible unimodal distributions. The safe distribution of probability density was approximated with formula (4):

$$\begin{aligned} pd_{safe}(x) = & 72.5451x^6 - 217.6353x^5 + 246.7007x^4 \\ & - 130.6759x^3 + 26.1685x^2 + 2.8969x + 0.1058 \end{aligned} \quad (4)$$

where the mean absolute error equals 0.0299.

4 Conclusions

The authors showed in the paper that when using Bayes' theorem to determine posterior probability distributions of stochastic variables we do not need to use the uniform prior distribution (according to the principle of indifference). If we know that the distribution in question is unimodal then the safe unimodal prior-distribution can be used. In most cases, this distribution will have smaller absolute error compared to the real distribution than the uniform prior-distribution. The safe distribution is the average distribution of all possible unimodal probability distributions, which can occur in the problem. The safe unimodal distribution was derived from the method of decreasing granulation of elementary events.

References

1. Kneale, W.: Probability and induction. Oxford University Press, Oxford (1952)
2. Magidor, O.: The classical theory of probability and the principle of indifference. In: 5th Annual Carnegie Mellon/University of Pittsburgh Graduate Philosophy Conference, pp. 1–17 (2003), http://www.andrew.cmu.edu/org/conference/2003
3. Russel, R., Norvig, P.: Artificial Intelligence - A Modern Approach, 2nd edn. Prentice Hall, Upper Saddle River (2003)
4. Yakov, B.H.: Info-gap decision theory-decisions under severe uncertainty, 2nd edn. Academic Press, London (2006)
5. http://en.wikipedia.org/wiki/Bayes'_theorem
6. http://en.wikipedia.org/wiki/Principle_of_indifference

MAD Loss in Pattern Recognition and RBF Learning

Ewaryst Rafajłowicz and Ewa Skubalska-Rafajłowicz*

Institute of Computer Engineering, Control and Robotics, Wrocław University of Technology, Wybrzeże Wyspiańskiego 27, 50 370 Wrocław, Poland
ewaryst.rafajlowicz@pwr.wroc.pl

Abstract. We consider a multi-class pattern recognition problem with linearly ordered labels and a loss function, which measures absolute deviations of decisions from true classes. In the bayesian setting the optimal decision rule is shown to be the median of a posteriori class probabilities. Then, we propose three approaches to constructing an empirical decision rule, based on a learning sequence. Our starting point is the Parzen-Rosenblatt kernel density estimator. The second and the third approach are based on radial bases functions (RBF) nets estimators of class densities.

1 Introduction

In [14] a multi-class pattern recognition problem was discussed, in a setting which differs from the classical, statistical pattern recognition problem in that a metric (or more generally, a topology) is defined in the space of outputs (labels). In other words, in opposite to the classical setting, in which labels of the classes are arbitrary and unordered, in [14] the family of problems with class labels that have neighbors, which are closer or further in a specified metric were considered.

In this paper we specialize the class of labels to one in which labels can be linearly ordered. In opposite to [14], where the sum of squared differences was used, here we take the minimum of absolute differences (MAD) between the output of a recognition system and a true class label as the loss function. Without further loss of the generality, we can assume that $\mathcal{I} \stackrel{\text{def}}{=} \{1, 2, \ldots, I\}$, $I \geq 2$ is the set of labels.

To motivate the above sketched problem statement consider the following examples.

1) Quality control procedures frequently classify products to more than two classes of quality. Usually, these classes can be ordered according to the increasing quality of products. It is reasonable to attach more loss when product of quality I, say, is classified as having quality III than, if it is (also incorrectly) classified to a class of quality II.

---

* This work was supported by the grant of the Polish Foundation for Science and by Ministry of Science and Higher Education under a grant ranging from 2006 to 2009.

2) In fault localization problems along, e.g., a water supply line the loss should be proportional to the difference between the fault point location and our decision.
3) Tasks of counting items or objects from images or image sequences seems to be the most representative for the class of pattern recognition problems considered in this paper. These include counting cancer cells, counting the number of people from a watchdog camera and counting cars. Such tasks are frequently divided into into a number of steps including detection and recognition of objects, which are later counted. The approach proposed in this paper allows for considering and/or evaluating such tasks as one entity. This does not mean that enhancement of low contrast defects should not be used as a preliminary step (see [16], [21] and the bibliography therein for recently proposed methods).

The above examples lead to considering pattern recognition problems with loss functions, which measure an absolute value of the distance between labels of classes (further called MAD loss for brevity).

There are three more important reasons for considering absolute difference loss functions instead of zero-one loss, which has been mainly considered in the literature so far (see [6], [3] for extensive bibliographies).

A) If the learning sequence is enlarged by even one example, then an empirical decision rule can change a decision. If we have ordering of labels, then this type of changing decisions will, probably, result in selecting a decision which is close to the previous one.
B) Multiple class problems with a large number of classes, a hundred say, are known to cause problems in practice (see, [2], [7], [9]). By using the approach proposed here we reduce them to a large extent.
C) Minimization of absolute differences is known to be a criterion which provides more robust estimation procedures in regression estimation problems. One can hope that also in pattern recognition tasks we obtain recognizers, which are less sensitive to erroneous classification of labels in a learning sequence.

An empirical version of the proposed classifier, which is deduced from MAD criterion, is based on the theoretical median of estimated class densities. We emphasise the difference between our approach and a median based approach presented in [1]. Namely, the loss function in that paper is the classical 0-1 loss function and the empirical median is introduced additionally, in order to increase robustness, while in our case the theoretical median of aposteriori probablities arises as the solution of the Bayes decision problem. As a consequence, we also obtain RBF structure, which is different from the one considered in [1]. Median based estimation in regression estimation and pattern recognition was considered in [12], but also there only 0-1 loss is discussed.

It should be mentioned that different orderings were suggested in [17], [18], but they were introduced in the pattern space, while here, we consider ordering in the space of labels.

2 Problem Statement

Consider the standard problem statement in the bayesian setting (see, e.g., [3] or [6]). Its main ingredients are defined below.

1) Let $X \in R^d$ be a random vector, representing a pattern, which is a member of one of I classes, labelled as $1, 2, \ldots, I$.
2) Pair (X, i) is a random vector representing a pattern and its correct classification i, which is unknown for a new pattern X to be classified.
3) Probability distribution of (X, i) is unknown, but we also have a learning sequence $(X^{(k)}, i^{(k)})$, $k = 1, 2, \ldots, n$ of observed patterns $X_k \in R^d$ and their correct classifications $i^{(k)} \in \{1, 2, \ldots, I\}$. We assume that $(X^{(k)}, i^{(k)})$'s are independent, identically distributed random vectors with the same probability distribution as (X, i).
4) Denote by $0 \leq q(i) \leq 1$, a priori probability that X comes from i-th class, $i = 1, 2, \ldots, I$, $\sum_{i=1}^{I} q(i) = 1$.

Remark 1. *For simplicity of the exposition assume the existence of probability densities $f(x|i)$, which describes the conditional p.d.f. of X, provided that it was drawn from i-th class.*

5) The next ingredient of the problem setting is a loss function, $L(i, j)$ say, which attaches loss $L(i, j)$ if a pattern from i-th class is classified to j-th class. In this paper we take

$$L(i, j) = |i - j|, \quad i, j \in \mathcal{I} \tag{1}$$

as the loss function for the reasons explained in the Introduction.

Problem statement – bayesian setting

The aim is to find (or to approximate from a learning sequence) a decision function $\Psi(X)$, which specifies a label of the class for X and such that it minimizes the expected loss given by:

$$R(\Psi) = E_X \left[\sum_{i=1}^{I} |i - \Psi(X)| P(i|X) \right], \tag{2}$$

where E_X denotes the expectation w.r.t. X, while $P(i|X)$ is the a posteriori probability that observed pattern X comes from i-th class. In other words, $P(i|X = x)$ is the conditional probability of the event that label i is the correct classification of a given pattern $X = x$. Our aim is to minimize the risk $R(\Psi)$, provided that the minimizer $\Psi^*(x)$, say, is a measurable function.

It is well known, that in order to minimize $R(\Psi)$ it suffices to minimize the conditional risk

$$r(\psi, x) \stackrel{\text{def}}{=} \sum_{i=1}^{I} |i - \psi| P(i|X = x) \tag{3}$$

with respect to ψ, which is a real variable taking values in the range of $\Psi(x)$, while x is treated as a parameter. According to the above statement, the optimal decision rule $\Psi^*(x)$ is obtained as

$$\Psi^*(x) = \arg\min_{\psi} r(\psi, x)$$

for all $x \in R^d$ in the range of X and it is called the Bayes classifier.

According to the Bayes rule, $P(i|X = x)$ is given by

$$P(i|X = x) = \frac{f(x|i)\, q(i)}{f(x)}, \quad i = 1, 2, \ldots, I, \quad f(x) \stackrel{\text{def}}{=} \sum_{l=1}^{I} f(x|l)\, q(l), \quad (4)$$

which allows us to express the a posteriori probabilities in terms of class densities and a priori probabilities, being easier to estimate from the learning sequence.

3 MAD Decision Rule

The minimum absolute differences decision rule $\Psi^*(x)$, which minimizes (3), can be derived by noticing that the minimization with respect to ψ is in fact minimization of a function with respect to scalar variable ψ, while x plays the role of a parameter. Thus, for fixed x we have to minimize the sum of $|i - \psi|$ with weights, which sum up to 1. This problem is the well known solution, namely, the best ψ is the median of a discrete probability distribution. In our case this result reads as follows:

$$\Psi^*(x) = \text{MED}[P(i|X = x), i \in \mathcal{I}], \quad (5)$$

where MED[] denotes the median of a discrete probability distribution, which is indicated in the brackets.

Summarizing the above considerations, we obtain.

Theorem 1. *For linearly ordered class labels the expected loss (2) is minimized by decision rule (5).*

Recall (see, e.g., [4] page 71) that for a random variable Z, say, with values z_i, and attached probabilities p_i, $i = 1, 2, \ldots I$, $\sum_{i=1}^{I} p_i = 1, 2, \ldots, I$ the median $\text{MED}[p_i, i \in \mathcal{I}]$ is any number m, for which the following conditions hold

$$P(Z \leq m) \geq \frac{1}{2} \quad \text{and} \quad P(Z \geq m) \geq \frac{1}{2}. \quad (6)$$

Note that for discrete distributions the median is frequently not unique. In our case one can expect that $\text{MED}[P(i|X = x), i \in \mathcal{I}]$ is left closed, right open interval $[j, j+1)$, say, unless either $P(j|X = x)$ or $P(j+1|X = x)$ is zero. One can resolve this difficulty either by randomization, i.e., by selecting j or $(j+1)$ at random or selecting just j as the smallest value in the median range.

From the above considerations and Thm. 1 as well as from (4), we obtain the following corollary: for a given pattern $X = x$ to be recognized, the optimal MAD decision is to select class i^*, which is the smallest integer i for which the following condition holds

$$\sum_{j=1}^{i} q_j\, f(x|j) \geq \frac{1}{2} \sum_{j=1}^{I} q_j\, f(x|j) \quad (7)$$

or, equivalently,
$$\sum_{j=1}^{i} q_j f(x|j) \geq \sum_{j=i+1}^{I} q_j f(x|j) \qquad (8)$$

Remark 2. *Its is worth mentioning that for a two class problem, i.e., when the cardinality of ($\mathcal{I}$) equals 2, decision (8) coincides with the maximum a posteriori probability, which is the optimal solution for the problem with $0-1$ loss function.*

Example 1. *To illustrate the above results in an idealized case, assume for a while that we have only one feature $x \in [0, 1]$ and that class densities have the form:*
$$f(x|i) = (I+1) B_I^{(i)}(x), \quad x \in [0, 1], \quad i = 0, 1, \ldots, I, \qquad (9)$$
where $B_I^{(i)}(x) = \binom{I}{i} x^i (1-x)^{I-i}$. For convenience, we have labelled classes from 0 to I. In (9) one can easily recognize the Bernstein polynomials, which are normalized so as to integrate to 1. We also assume that a priori class probabilities are equal to $q_i = 1/(I+1)$, $i = 0, 1, \ldots, I$. Taking into account that
$$\sum_{i=0}^{I} B_I^{(i)}(x) = 1, \quad x \in [0, 1], \qquad (10)$$
we obtain from (7) that MAD decision rule has the form: $x \in [0, 1]$ is categorized to i^* class, if it is the smallest integer i for which
$$\sum_{j=0}^{i} B_I^{(j)}(x) \geq 1/2. \qquad (11)$$

Class densities for $I = 5$ are shown in Fig. 1 (left panel). The r.h.s. panel of this figure shows the optimal decision function (solid line) in the MAD sense.

It is usually not reasonable to compare decisions, which minimize different criterions, but the popularity of the $0-1$ loss function forces us to compare it with MAD optimal decisions. The dashed line in Fig. 1 (right panel) indicates that in some regions the decisions would be different. In this example the decision, which minimizes $0-1$ loss has the form:
$$k^* = \arg\max_{j} B_I^{(j)}(x), \quad j = 0, 1, \ldots, I. \qquad (12)$$

Our next step is to propose ways of estimating either $P(i|X=x)$' in (5) or directly $\Psi^*(x)$ from the learning sequence.

4 Bank of RBF Nets for MAD Decisions

The classical way of converting a decision rule (5) into an empirical decision rule is to estimate probability density functions $f(x|i)$ from $(X^{(k)}, i^{(k)})$,

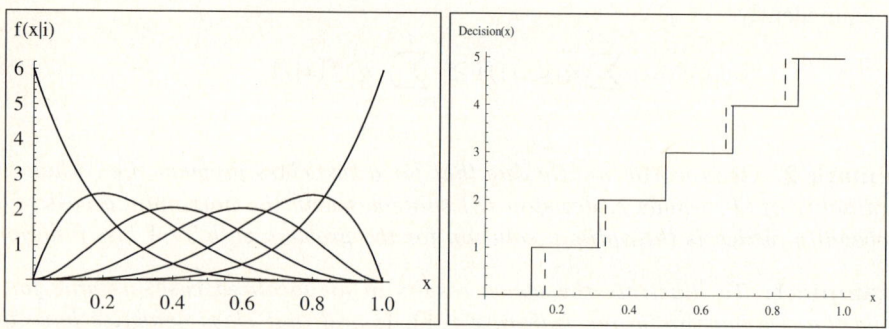

Fig. 1. The normalized Bernstein polynomials as class densities in Example 1 (left panel). MAD optimal decision function (solid line) and the optimal decision function under $0-1$ loss function (dashed line).

$k = 1, 2, \ldots, n$ with the aid of (4) (the so called plug-in decision rules (see [6])). We shall present the details of this approach in this section, while in the next one possible direct approaches will be discussed.

Estimators of $f(x|i)$ can be based on orthogonal expansions (see [17] and the bibliograhy on earlier approaches cited therein) or the well known Parzen-Rosenblatt kernel estimator, partitioning estimators and many others (see [5] and [6] for detailed discussions and extensive bibliographies). All these approaches operate directly on the whole learning sequence, which has to be stored and available for each pattern to be recognized. We shall use RBF estimators as those, which have compression of the learning sequence built-in in their construction.

We shall select estimators of $f(x|i)$ from the class of radial basis functions (see [22] for the results on their approximation abilities and [13], [20], [10], [11], [19] for recent results in related directions).

Denote by $K(t) \geq 0$, $t \in R$ a kernel of RBF's, which is such that $\int_{-\infty}^{\infty} K(t) = 1$, $\int_{-\infty}^{\infty} t \, K(t) = 0$, i.e., K fulfils the same conditions as a density function.

Let us denote by $\mathcal{I}(i)$ those number of observations k in the learning sequence $(X^{(k)}, i^{(k)})$, $k = 1, 2, \ldots, n$, for which the corresponding labels indicate i-th class, i.e., $i^{(k)} = i$. Clearly, $\sum_{i=1}^{I} \text{Card}(\mathcal{I}(i)) = n$, where Card denote the cardinality of a set. Let $n(i) = \text{Card}(\mathcal{I}(i))$ denotes the number of observations with the label i in the learning sequence.

The simplest task is estimating a priori probability that a pattern comes from class i, since it suffices to set $\hat{q}(i) = n(i)/n$, $i = 1, 2, \ldots, I$ as the estimator of $q(i)$. As estimators $\hat{f}(x|i)$ of $f(x|i)$ we take

$$\hat{f}(x|i) = \sum_{j=1}^{J(i)} w(i,j) \, K\left(\frac{||x - c(i,j)||_{ij}}{h(i)}\right), \quad i = 1, 2, \ldots, I, \qquad (13)$$

where
$c(i,j)$ is j-th center for estimating density in i-th class,
$w(i,j)$ is j-th weighting coefficient in i-class,
$h(i)$ is i-th smoothing parameter,
$J(i)$ is the number of centers and weights for estimating $f(x|i)$,
$\|.\|_{ij}$ is the norm of the form $\|x\|_j = (x^T A(i,j) x)^{\frac{1}{2}}$, where $A)i,j)$ is $d \times d$ positive definite matrix, which should be chosen during a learning process.

Summarizing, the empirical MAD decision structure is a bank of RBF nets (13) and the empirical decision block, which acts as follows: it selects the first i, for which

$$\sum_{j+1}^{i} \hat{q}_j \, \hat{f}(x|j) \geq \frac{1}{2} \sum_{j+1}^{I} \hat{q}_j \, \hat{f}(x|j). \tag{14}$$

We shall not describe the techniques of selecting the above parameters, since one can use those described in the bibliography cited in the Introduction. One such technique is demonstrated in the next section.

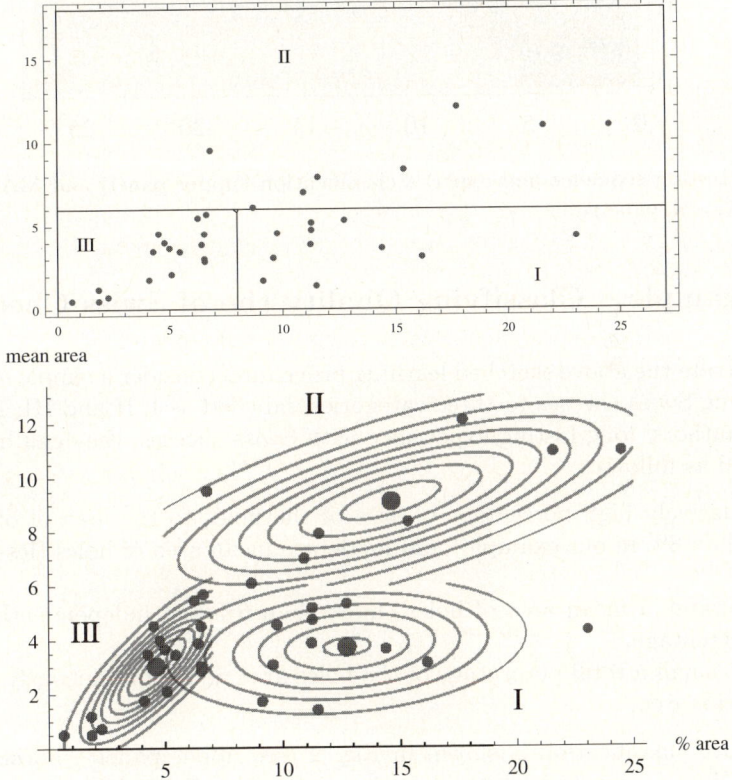

Fig. 2. Learning sequence and the expert classification (upper panel) and the estimates of class densities (lower panel) in the example of classifying Swiss cheeses

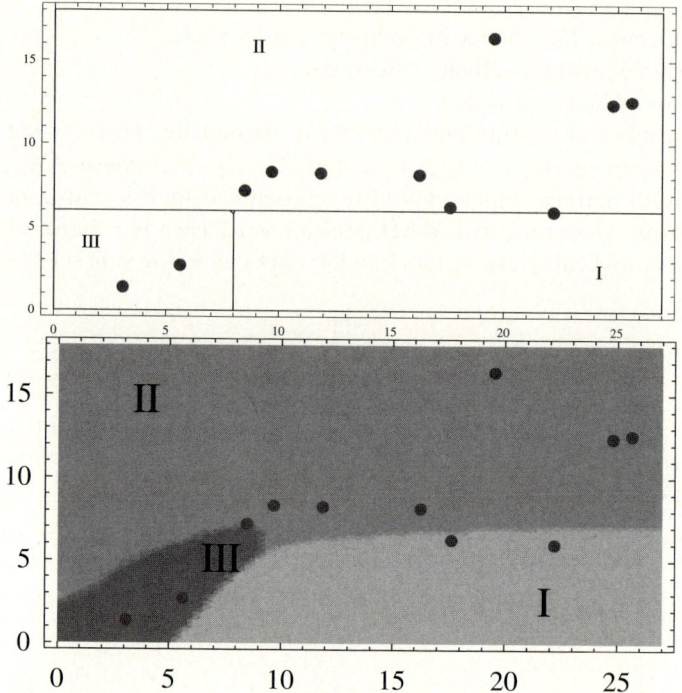

Fig. 3. Testing sequence and expert's classification (upper panel) and MAD decision regions (lower panel)

5 Example – Classifying Quality the of Swiss Cheeses

To illustrate the above sketched learning procedure, consider a simple example of classifying Swiss cheeses to three categories, labelled as I, II and III. According to the authors' long lasting experience with Swiss cheeses, they can be roughly classified as follows:

I – a relatively high percentage of area of the holes to the area of of the slice (more than 8% in our example) and moderate mean area of holes (less than 6% say),
II – too large a mean area of holes (more than 6% say, independently of their total percentage,
III – too small a total percentage of the holes area and simultaneously too small their mean area.

The above classification is shown in Fig. 2 (see upper panel – horizontal and vertical lines) together with the learning sequence (36 dots). A relatively short learning sequence forces us to confine ourselves to one element RBF's ($J(i) = 1$ in (13)). We take $K(t) = \exp(-t^2/2)$ as kernels, which are centered at mean vectors of data points from classes I, II and III, while the standard estimates of the covariance matrices of these points are taken as $A(i, j)$'s. Smoothing parameters $h(i) = 1$, $i = I, II, III$ were selected and the proportions of samples from each group were used as $w(i, j)$ weights. The contour plots of the resulting class densities estimates are shown in Fig. 2 (lower panel).

The testing sequence consisting of 11 data points and expert's classification is shown in Fig. 3 (upper panel). In the lower panel of this figure the estimated decision regions are shown. As one can notice, our MAD classifier committed only one mistake (point with coordinates (17.5, 6) was classified to class I instead of II).

6 Conclusions

The usefulness of MAD criterion in pattern recognition was discussed. It was shown that its minimization leads to a simple and interpretable decision rule. Then, an outline of the learning algorithm is proposed, using the plug-in approach and a bank of RBF nets. In addition to the example presented above, the example of applying MAD criterion is presented in [15].

In addition to the above approach one can consider two direct approaches. The first one is based on calculating the empirical median (not the theoretical one as above) from $k(n)$ nearest neighbors of X, where $k(n)$ indicates that the number of neighbours, which are taken into account is a (slowly growing) function of n. The second one, calculates the empirical median from neighbors contained in a ball centered at X and having the radius $R(n)$, where the radius is a (slowly growing) function of n.

References

1. Bors, A.G., Pitas, I.: Median Radial Basis Function Neural Network. IEEE Trans. on Neural Networks I, 1351–1364 (1996)
2. Allwein, A., Schapire, R., Singer, Y.: Reducing multiclass to binary: A unifying approach for margin classifiers. J. Machine Learning Research 1, 113–141 (2000)
3. Bishop, C.: Neural Networks for Pattern Recognition. Oxford University Press, Oxford (1995)
4. Chow, Y.S., Teicher, H.: Probability Theory. Springer, New York (1988)
5. Devroye, L., Györfi, L.: Nonparametric Density Estimation. The L_1 View. Wiley, New York (1985)
6. Devroye, L., Györfi, L., Lugosi, G.: Probabilistic Theory of Pattern Recognition. Springer, New York (1996)
7. Dietterich, T., Bakiri, G.: Solving Multiclass Learning Problems via Error-Correcting Output Codes. J. Artificial Intelligence Research 2, 263–286 (1995)
8. Greblicki, W., Pawlak, M.: Necessary and Sufficient Conditions for Bayes Risk Consistency of Recursive Kernel Classification Rule. IEEE Trans. Information Theory 33, 408–412 (1987)
9. Hastie, T., Tibshirani, R.: Classification by Pairwise Coupling. The Annals of Statistics 26, 451–471 (1998)
10. Karayiannis, N.B., Randolph-Gips, M.M.: On the Construction and Training of Reformulated Radial Basis Function Neural Networks. IEEE Trans. on Neural Networks 14, 835–846 (2003)

11. Krzyżak, A., Skubalska-Rafajłowicz, E.: Combining Space-Filling Curves and Radial Basis Function Networks. In: Rutkowski, L., Siekmann, J.H., Tadeusiewicz, R., Zadeh, L.A. (eds.) ICAISC 2004. LNCS (LNAI), vol. 3070, pp. 229–234. Springer, Heidelberg (2004)
12. Lugosi, G., Zeger, K.: Nonparametric Estimation via Empirical Risk Minimization. IEEE Trans. on Information Theory 41, 677–687 (1995)
13. Pawlak, M., Siu, D.: Classification with Noisy Features. Advances in Pattern Recognition 1451, 845–852 (1999)
14. Rafajłowicz, E.: RBF Nets in Faults Localization. In: Rutkowski, L., Tadeusiewicz, R., Zadeh, L.A., Żurada, J.M. (eds.) ICAISC 2006. LNCS (LNAI), vol. 4029, Springer, Heidelberg (2006)
15. Rafajłowicz, E.: Improving the efficiency of counting defects by learning RBF nets with MAD loss. ICAISC 2008 (to appear)
16. Rafajłowicz, E., Pawla, M., Steland, A.: Nonlinear Image Filtering and Reconstruction: A Unified Approach Based on Vertically Weighted Regression. Int. J. Apll. Math. Comp. Sci (to appear, 2008)
17. Skubalska-Rafajłowicz, E.: Pattern Recognition Algorithms Based on Space-Filling Curves and Orthogonal Expansions. IEEE Trans. Information Theory 47, 1915–1927 (2001)
18. Skubalska-Rafajłowicz, E., Krzyżak, A.: Fast k-NN Classification Rule Using Metric on Space-Filling Curves. In: Proceedings of the 13th International Conference on Pattern Recognition, Vienna, vol. 2, pp. 121–125 (1996)
19. Skubalska-Rafajłowicz, E.: Data Compression for Pattern Recognition Based on Space-Filling Curve Pseudo-Inverse Mapping. Nonlinear Analysis, Theory, Methods and Applications 47, 315–326
20. Skubalska-Rafajłowicz, E.: RBF Neural Network for Probability Density Function Estimation and Detecting Changes in Multivariate Processes. In: Rutkowski, L., Tadeusiewicz, R., Zadeh, L.A., Żurada, J.M. (eds.) ICAISC 2006. LNCS (LNAI), vol. 4029, pp. 133–141. Springer, Heidelberg (2006)
21. Skubalska-Rafajłowicz, E.: Local correlation and entropy maps as tools for detecting defects in industrial images. Int. J. Apll. Math. Comp. Sci (to appear, 2008)
22. Xu, L., Krzyżak, A., Yuille, A.: On Radial Basis Function Nets and Kernel Regression: Statistical Consistency, Convergence Rates and Receptive Field Size. Neural Networks 4, 609–628 (1994)

Parallel Ant Miner 2

Omid Roozmand[1] and Kamran Zamanifar[2]

[1] PhD Student of Software Engineering, Isfahan University, Iran
[2] Assistant Professor of Isfahan University, Iran
roozmand@eng.ui.ac.ir, K.zamanifar@yahoo.com

Abstract. In this paper, we propose a flexible parallel ant colony algorithm for classification rule discovery in the large databases. We call this algorithm Parallel Ant-Miner2. This model relies on the extension of real behavior of ants and data mining concepts. The artificial ants are firstly generated and separated into several groups. Each group is assigned a class label which is the consequent parts of the rules it should discover. Ants try to discover rules in parallel and then communicate with each other to update the pheromones in different paths. The communication methods help ants not to gather irrelevant terms of the rule. The parallel executions of ants reduce the speed of convergence and consequently make it possible to extract more new high quality rules by exploring all search space. Our experimental results show that the proposed model is more accurate than the other versions of Ant-Miner.

Keywords: Ant Colony Optimization, Classification Rule Discovery, Data Mining, Parallel Computing.

1 Introduction

Data Mining is the process of extracting useful knowledge from real-world data. In this paper we focus on classification rule discovery among the several data mining tasks. One popular category of classification model consists of classification rules, which is the model category used in this paper. In this context, the aim of the classification algorithm is to discover a set of classification rules. Classification rules have the advantage of representing knowledge at a high level of abstraction, so that they are intuitively comprehensible to the user [1].

Classification rules are often represented in the IF-THEN form, and it can be described as: IF<conditions> THEN <class>. In each rule, the <condition> part (which is called the antecedent) usually contains a logic combination of predictor attributes in the form: term1 AND term2 AND…AND term¡. And each term is a triple <attribute, operation, value>, such as <Color =Red>. The consequent part <class> shows the predicted class whose cases satisfy the <condition>. Efficiency, accuracy and simplicity become important criteria for the design of rule discovery algorithms [2].

Some classification algorithms are as follows: decision tree algorithms [3] (such as the C4.5 algorithm [4]), statistical algorithms [5], neural network approaches [6], and genetic algorithms [7]. In this paper, we study another mining paradigm: ant colony and parallel ant colony algorithms, which are efficient, easy to implement and parallel mining [2].

One algorithm for solving this task is Ant-Miner, proposed by Parpinelli and colleagues [8], which employs ant colony optimization techniques to discover classification rules. Ant-Miner has produced good results when compared with more conventional data mining algorithms, [9] and it is still a relatively recent algorithm, which motivates further research trying to improve it. Although some modifications to the Ant-Miner algorithm have already been proposed [10-12]. One of the most important algorithms for classification rules by ant colony is parallel ant-miner [2]. Also, the idea of parallel ant colony system has been applied for solving the traveling sales man problem in [13].

2 Analysis of Different Versions of Ant-Miner

In this section, we take a brief look at different versions of Ant-Miner algorithms. Subsection 2.1 describes the original Ant-Miner [8] proposed by Parpinelli and colleagues. Also, in subsections 2.2, 2.3, 2.4 and 2.5 Ant-Miner2 [10], Ant-Miner3 [11] and new version of Ant-Miner algorithm for discovering unordered rule sets proposed by James Smaldon and Alex A.Freitas are discussed [12], and parallel ant Miner [2], respectively.

2.1 Original Ant-Miner

The original Ant-Miner algorithm [8] is described in this section. We provide here just a brief overview of the algorithm; for more details the reader is referred to that reference.

2.1.1 Pheromone Initialization
All cells in the pheromone table are initialized equally to the following value:

$$\tau_{ij}(t=0) = \frac{1}{\sum_{i=1}^{a} b_i} \qquad (1)$$

Where a is the total number of attributes, b_i is the number of values in the domain of attribute i.

2.1.2 Rule Construction
Each rule in Ant-Miner contains a condition part as the antecedent and a predicted class. The condition part is a conjunction of attribute-operator-value tuples. The operator used in all experiments is "=" since in Ant-Miner2, just as in Ant-Miner, all attributes are assumed to be categorical. Let us assume a rule condition such as $term_{ij} \approx A_i = V_{ij}$, where A_i is the ith attribute and V_{ij} is the jth value in the domain of A_i. The probability, that this condition is added to the current partial rule that the ant is constructing, is given by the following Equation:

$$P_{ij}(t) = \frac{\tau_{ij}(t).\eta_{ij}}{\sum_{i=1}^{a} x_i . \sum_{j=1}^{b_i} (\tau_{ij}(t).\eta_{ij})} \quad (2)$$

η_{ij} is the value of a problem-dependent heuristic function for $term_{ij}$. The higher the value of η_{ij}, the more relevant for classification the $term_{ij}$ is, and so the higher its probability of being chosen. $\tau_{ij}(t)$ is the amount of pheromone associated with $term_{ij}$ at iteration t, corresponding to the amount of pheromone currently available in the position i,j of the path being followed by the current ant. The better the quality of the rule constructed by an ant, the higher the amount of pheromone added to the trail segments visited by the ant. x_i is set to 1 if the attribute A_i was not yet used by the current ant, or to 0 otherwise.

2.1.3 Heuristic Value

In Ant-Miner, the heuristic value is taken to be an information theoretic measure for the quality of the term to be added to the rule. The quality here is measured in terms of the entropy for preferring this term to the others, and is given by the following equations:

$$\eta_{ij} = \frac{Log(K) - InfoT_{ij}}{\sum_{i}^{a} \sum_{j}^{b_i} Log_2^{(K)} - InfoT_{ij}} \quad (3)$$

In which:

$$InfoT_{ij} = -\sum_{w=1}^{k} \left[\frac{freqT_{ij}^w}{|T_{ij}|} \right] \times Log_2 \left[\frac{freqT_{ij}^w}{|T_{ij}|} \right] \quad (4)$$

Where K is the number of classes, $|T_{ij}|$ is the total number of cases in partition T_{ij} (partition containing the cases where attribute A_i has value V_{ij}), $freqT_{ij}^w$ is the number of cases in partition T_{ij} with class w. The higher the value of $InfoT_{ij}$, the less likely that the ant will choose $term_{ij}$ to add to its partial rule.

2.1.4 Rule Pruning

Immediately after the ant completes the construction of a rule, rule pruning is undertaken to increase the comprehensibility and accuracy of the rule. After the pruning step, the rule may be assigned a different predicted class based on the majority class in the cases covered by the rule antecedent. The rule pruning procedure iteratively removes the term whose removal will cause a maximum increase in the quality of the rule. The quality of a rule is measured using the following equation:

$$Q = \left(\frac{TruePos}{TruePos + FalseNeg}\right) \times \left(\frac{TrueNeg}{FalsePost + TrueNeg}\right) \quad (5)$$

Where *TruePos* is the number of cases covered by the rule and having the same class as that predicted by the rule, *FalsePos* is the number of cases covered by the rule and having a different class from that predicted by the rule, *FalseNeg* is the number of cases that are not covered by the rule, while having the class predicted by the rule, *TrueNeg* is the number of cases that are not covered by the rule which have a different class from the class predicted by the rule [8].

2.1.5 Pheromone Update Rule

After each ant completes the construction of its rule, pheromone updating is carried out as follows:

$$\tau_{ij}(t+1) = \tau_{ij}(t) + Q.\tau_{ij}(t), \quad \forall i,j \in R \quad (6)$$

To simulate the phenomenon of pheromone evaporation in real ant colony systems, the amount of pheromone associated with each termij which does not occur in the constructed rule must be decreased,. The reduction of pheromone of an unused term is performed by dividing the value of each τ_{ij} by the summation of all τ_{ij} [8].

2.2 Ant-Miner2

In [11], a new version of ant-miner is proposed called Ant-Miner2. In this algorithm, a computable density estimation equation (7) instead of equation (3) is presented.

$$\eta_{ij} = \frac{majority\_classT_{ij}}{|T_{ij}|} \quad (7)$$

where $majority\_classT_{ij}$ is the majority class in partition T_{ij}.

2.3 Ant-Miner3

In Ant-Miner3 [12], a new equation is proposed for updating pheromone associated with each term that occurs in the constructed rule. The pheromone of unused terms is updated by normalization.

$$\tau_{ij}(t) = (1-p).\tau_{ij}(t-1) + (1-\frac{1}{1+Q}).\tau_{ij}(t-1), \quad \forall i,j \in R \quad (8)$$

where ρ is the pheromone evaporation rate, Q is quality of the constructed rule, ρ is the pheromone evaporation rate, which controls how fast the old path evaporates. This parameter controls the influence of the history on the current pheromone trail [6].

2.4 New Version of Ant-Miner Algorithm for Discovering Unordered Rule Sets

This new version of Ant-Miner presents an improvement to the original Ant-Miner so that the algorithm discovers a set of rules which do not need to be applied to test data in the order in which they were discovered. There are some modifications on High-Level algorithm, heuristic function and pheromone updating. For more information see [12].

2.5 Parallel Ant Colony Algorithm for Mining Classification Rules (Parallel Ant-Miner)

A key observation have been made in parallel ant-miner [2] is that, by executing multiple agents (ants) independently, ant colony algorithms have strong parallelism that can be exploited to support fast parallel processing. In parallel ant-miner, each processor is assigned a class label which is the consequent part of the rules it should discover. A group of ants are allocated on the processor to search for the antecedent part of the rules. In this parallel algorithm, pruning the rules is done during the process of rule construction. The innovation of this work is parallelism and extended formula for term selection. Term selection is described as follows:

Suppose there are g classes in the training set, ants are divided into g groups, each of which is responsible for mining the rules for one class. The ants of the groups will excavate the classification rules in parallel with the corresponding consequent parts. During the process of rules construction, the most crucial step is to choose $term_{ij}$. $term_{ij}$ is a rule condition of the form $A_i = V_{ij}$, where A_i is the ith attribute and V_{ij} is the jth value of the domain of A_i. The probability that $term_{ij}$ is chosen to be added to the current partial rule is given by Equation (9):

$$P_{ij}^p = \frac{\tau_{ij}^{\alpha}(t).\eta_{ij}^{\beta}(t).\lambda_{ip}^{\gamma}}{\sum_{i=1}^{a} x_i . \sum_{j=1}^{b_i} \tau_{ij}^{\alpha}(t).\eta_{ij}^{\beta}(t).\lambda_{ip}^{\gamma}} \quad (9)$$

In which, λ_{ip} is the heuristic factor for the ants in processor p to choose values of attribute A_i. τ_{ij} and η_{ij} are the pheromone and the heuristic factor of the $term_{ij}$ on processor p at iteration t. a is the total number of attributes. b_i is the number of values in the domain of the ith attribute. x_i is set to 1 if the attribute A_i was not yet used by the current ant, or to 0 otherwise. α, β and γ reflect the importance of pheromone of $term_{ij}$, heuristic function of $term_{ij}$ and heuristic function of processor p, respectively.

In the Parallel Ant-Miner algorithm, it has been applied a dynamic strategy to update the parameters α, β, and γ which determine the importance of the pheromone and heuristic functions. At the beginning of the algorithm the pheromone of each term is small. Therefore it has been considered a small value for α and high value for heuristic functions [2].

3 New Parallel Ant-Miner (Parallel Ant-Miner2)

We apply the new idea of data parallelism to Ant Colony Systems (ACS) in order to discover classification rules by improving parallel ant-miner [2]. We have applied the idea of communication between ants [13] to enable them not to gather irrelevant rules. Algorithm1 shows the High-Level pseudo code of the proposed algorithm. In this algorithm all groups work parallel (The groups have been distributed on different processors) and all ants in each group execute parallel as well (It has been implemented by multithreading in java). The key point about proposed algorithm is communication between ants of each group with the ants of their own group and with the best of other groups which is done once a rule constructed. Therefore, when an ant constructs a rule then waits to be synchronized with the other ants.

By the use of efficient communication methods between ants, the algorithm becomes able to discover more high quality rules by reduce the speed of convergence and also avoid gathering irrelevant terms of the rule and decrease the time complexity of rule pruning.

3.1 Global Pheromone Level Updating Rule

After all iterations of the most inner DO-WHILE loop of Algorithm1 one rule is constructed by the corresponding ant. The quality Q is evaluated at each iteration as well as rule pruning. The last evaluated quality at last iteration Q_{last} shows the quality of constructed rule. Therefore, global pheromone updating is updated by equation (10).

$$\tau_{ij}(t+1) = (1-\rho).\tau_{ij}(t) + \rho.(1 - \frac{1}{1+Q_{last}}).\tau_{ij}(t) \qquad (10)$$

Where ρ is the coefficient of evaporation.

3.2 Update the Pheromone Level from Communication

Two communication methods are proposed to power the ants to update the pheromone of terms as follows:

Method1: Update the pheromone level for $term_{ij}$ of processor p by using the pheromones produced by ants on same processor (communication between ants of same group). The pheromone is updated by equation (11).

$$\tau_{ijp}(t+1) = (1-\rho).\tau_{ijp}(t) + \rho.\frac{\sum \tau_{ijp}(t)}{|X|} \qquad (11)$$

Where $|X|$ is the number of ants on processor p. $\sum \tau_{ijp}(t)$, is sum of all pheromone produced by all ants on processor p for $term_{ij}$.

Algorithm 1. High-Level Description of Parallel Ant-Miner

- **Input:** TrainingSet = {all training cases};
- DiscoveredRuleList[ ][ ]; ϕ ; iteration[][]== ϕ ;
/*iteration[i][j]: Number of iteration of jth ant on i^{th} group*/.
No_of_Max_Iteration is initialized; /*Max iteratin which has been considered for each ant*/
/* DiscoveredRuleList [i][j]: shows the jth best rule discovered by ants of ith group*/
- Consider G groups and P processor (P=G);
- Each processor is assigned a class label which is the consequent parts of the rules it should discover;
- Generate N_j ants for j^{th} group and then assign each group a processor;
- BroadCast the TrainingSet to all processors;
- Initialize all trails with the same amount of pheromone;
FOR ALL GROUPS DO-PARALLEL
/*each processor works seperately, but when all ants of each group construct a rule, wait for the other groups to Communicate and synchronize*/
    FOR ALL ANTS IN EACH GROUP DO-PARALLEL
      /*ants in each group work concurrently by multithreading implementation in java*/
      WHILE (Number of uncovered cases in the training set of each Class > max_uncovered_cases)
        DO
          - Ant_i^g starts with an empty rule. /* Ant_i^g is i^{th} ant of gth group*/
          Do /* at the end of this loop one rule is constructed */
            - Ant_i^g use equation (9) to select that which term should be add to the current rule[2];
            - Prune rule R_t ; /*is derived from[2] */
          WHILE (Rule is not built)/*in each iteration one term can be added to the rule*/
          iteration[p][i]++;
          - Do global pheromone level updating rule by equation (10);
          - Wait for the ants of your own group and other groups to construct their rules;/*according to this fact that ants need to communicate with the ants of their own group and the other groups, therefore they should become synchrone with each other */
          - Update pheromone from communication with the ants of it's own group by equation (11);
          - Update pheromone from communication with the best of other groups by equation (12);
        WHILE (iteration[p][i] < No_of_Max_Iteration) OR (j < No_rules_converg)
      - Choose the best rule R_{Best}^g among all rules R_i^g constructed by all the ants;
      - Add rule R_{Best}^g to DiscoveredRuleList;
      - TrainingSet = TrainingSet - {set of cases correctly covered by R_{Best}^g};
    END WHILE
  END FOR
END FOR

Method2: Update the pheromone level for $term_{ij}$ of processor p by communication with the best of other groups. The pheromone is updated by equation (12).

$$\tau_{ij}(t+1) = (1-\rho).\tau_{ij}(t) + \rho.\frac{\sum_{p=1}^{G}\tau_{ij\_best\_p}(t)}{|G|} \qquad (12)$$

Where $|G|$ is the number of all groups (which is equal to the number of classes and processors). $\tau_{ij\_best\_p}(t)$ is the best pheromone produced by the ants on processor number p for $term_{ij}$ and consequently $\sum_{p=1}^{G}\tau_{ij\_best\_p}(t)$ is sum of the all pheromones produced by the all best ant of all processors.

4 Experimental Results

The performance of our parallel Ant-Miner (parallel ant-miner2) was evaluated using five public-domain data sets presented in Table1. The main characteristics of the data sets used in our experiment are summarized in Table1are as follows: The first column of this table gives the data set name, while the other columns indicate, respectively, the number of cases, the number of categorical attributes, the number of continuous attributes, and the number of classes of the data set.

Table 1. Number of Classes of the Data Set

Data Set	#Cases	#Categorical Attribute	#Continues Attribute	#Classes
Ljubljana breast cancer	341	9	-	2
Wisconsin breast cancer	699	-	9	2
Heart disease	315	8	5	5
dermatology	412	31	1	6
hepatitis	321	13	6	2

Recall that Parallel Ant-Miner has the following parameters.

- ✓ $|G|$ is the number of groups. Each group is assigned to discover the rules related to one class. This is equal to 17. (The number of processors $|P|$ is 17 too).
- ✓ Number of ants: This is also the maximum number of complete candidate rules constructed and pruned during an iteration of the WHILE loop of

Algorithm1, since each ant is associated with a single rule. In each iteration the best candidate rule found is considered a discovered rule. The larger numbers of ants, the more candidate rules are evaluated per iteration, but the slower the system is. We have considered totally 3400 ants. Each group contains 200 ants.

✓ Number of max iteration (No_of_Max_Iteration): Each ant for completing its search has two criteria *No_of_Max_Iteration* and *No_rules_converg*. *No_of_Max_Iteration* specifies the number of iteration which each ant can do.

✓ Maximum number of uncovered cases in the training set (*Max_uncovered_cases*): The process of rule discovery is iteratively performed until the number of training cases that are not covered by any discovered rule is smaller than this threshold. *Max_uncovered_cases* is set to 10.

✓ Number of rules used to test convergence of the ants (*No_rules_converg*): If the current ant has constructed a rule that is exactly the same as the rule constructed by the previous No_rules_converg − 1 ants, then the system concludes that the ants have converged to a single rule (path). *No_rules_converg* is set to 10.

✓ The value of ρ is set to 0.1.

Our experimental results are based on two criteria, namely the predictive accuracy of the discovered rule and the simplicity. The database is divided into ten partitions; one is used as the test set, and the others as the training set. The predictive accuracies (on the test set) of the 10 runs are then averaged and reported as the predictive accuracy of the discovered rule list. As we said before we use 3400 ant, 200 ants on each processor. Table 2 shows accuracy rates for the rule sets produced by Parallel Ant_Miner2, Parallel Ant-Miner and Ant-Miner3 for ten runs on *the Ljubljana breast cancer* datasets.

Table 2. Accuracy rate of 10 runs of Ljubljana breast cancer

Run Number	Ljubljana breast cancer		
	Parallel Ant-Miner2	Parallel Ant-Miner	Ant-Miner3
1	97.85	93.33	92.40
2	96.83	94.91	90.94
3	97.22	94.95	91.64
4	98.09	95.32	92.67
5	97.11	93.36	92.88
6	97.74	94.01	91.26
7	95.68	94.58	93.41
8	96.32	93.78	92.43
9	98.00	94.57	92.22
10	97.76	95.09	91.45

The results comparing the predictive accuracy of Parallel Ant-Miner, Ant-Miner3 and Ant-Miner is reported in table 3. As shown in this table 3, Parallel Ant-Miner2 discovered rules with a better predictive accuracy than parallel Ant-Miner and Ant-Miner3 in all data sets. In two data sets, namely Wisconsin breast cancer and dermatology, the difference in predictive accuracy between the three algorithms was quite small. In two data sets, Heart disease and hepatitis, Parallel Ant-Miner2 was significantly more accurate than parallel Ant-Miner and Ant-Miner3.

Table 3. Comparison Predictive Accuracy rate (%) of Parallel Ant-Miner2, parallel Ant-Miner and Ant-Miner3

Data Set	Parallel Ant-Miner2 Predictive accuracy (%)	Parallel Ant-Miner Predictive accuracy (%)	Ant-Miner3 Predictive Accuracy
Ljubljana breast cancer	97.26	94.39	92.13
Wisconsin breast cancer	95.62	95.23	94.50
Heart disease	74.36	68.15	65.61
dermatology	94.74	94.55	94.10
hepatitis	96.35	90.31	90.26

The results comparing the simplicity of the rule lists discovered by Parallel Ant_Miner2, Parallel Ant-Miner and Ant-Miner3 are reported in Table 4.

Table 4. Comparisons of average numbers of rules of Parallel Ant_Miner2, Parallel Ant-Miner and Ant-Miner3

Data Set	Average number of rules		
	Parallel Ant-Miner2	Parallel Ant-Miner	Ant-Miner3
Ljubljana breast cancer	5.3	5.8	6.8
Wisconsin breast cancer	6.9	7.2	7.2
Heart disease	8.6	10.5	10.8
dermatology	5.1	6.3	6.6
hepatitis	4.3	5.6	5.9

Our results in table 4, show that for all five data sets the rule discovered by Parallel Ant-Miner2 is simpler, that is, it had a smaller number of rules. In other word, Parallel Ant-Miner2 discovers the rules with high quality rather than extracting a lot of rules with low quality. In two out of five data sets the difference between the number of rules discovered by Parallel Ant-Miner2 and parallel Ant-Miner is somewhat small, as follows: *Ljubljana breast cancer* and *Wisconsin breast cancer*. However, discovered rules by Parallel Ant Miner2 are simpler in *Heart disease, dermatology* and *hepatitis*.

5 Conclusion

In this paper we propose a parallel ant colony algorithm for classification rules discovery which is improvement of [2]. In this algorithm, we divided ants into several groups that each group is responsible for discovering of rules of one class. A group of ants is allocated to each processor to search for the antecedent part of the rules. Ants search the all space in parallel to discover classification rules and then communicate with the other ants of their own group and the best of the other groups to update the pheromone of terms. Two methods of communications are considered to provide the better pheromone updating. Parallelization and communication methods cause that algorithm discovers the more high quality rules and avoid gathering irrelevant ones. Also, it reduces the speed of convergence by exploring the all search space. Our experimental results on several benchmark datasets show that our algorithm can discover classification rules with significantly better accuracy and lower redundancy than other leading methods

References

[1] Witten, I.H., Frank, E.: Data Mining: practical machine learning tools and techniques, 2nd edn. Morgan Kaufmann, San Francisco (2005)
[2] Chen, H., Chen, L., Li, T.: Parallel Ant Colony Algorithm for Mining Classification Rules. In: 2006 IEEE International Conference on Granular Computing, May 10-12, 2006, pp. 85–90 (2006)
[3] Chang, S.-Y., Lin, C.-R., Chang, C.-T.: A fuzzy diagnosis approach using dynamic fault trees. Chemical Engineering Science 57(15), 2971–2985 (2002)
[4] Quinlan, J.R.: C4.5: Programs for Machine Learning. Morgan Kaufmann, San Francisco (1993)
[5] Hsiung, J.T., Himmelblau, D.M.: Detection of leaks in a liquid-liquid heat exchanger using passive acoustic noise. Computers & Chemical Engineering 20(9), 1101–1111 (1996)
[6] Venkatasubramanian, V., Vaidyanathan, R., Yamamoto, Y.: Process fault detection and diagnosis using neural networks-I. Steady-stateeee processes. Computers & Chemical Engineering 14(7), 699–712 (1990)
[7] Romão, W., Freitas, A.A., de S. Gimenes, I.M.: Discovering interesting knowledge from a science and technology database with a genetic algorithm. Applied Soft Computing 4, 21–137 (2004)
[8] Parpinelli, R.S., Lopes, H.S., Freitas, A.A.: Data mining with an ant colony optimization algorithm. IEEE Transactions on Evolutionary Computing 6(4), 321–332 (2002)
[9] Clark, P., Boswell, R.: Rule induction with CN2: some recent improvements. In: Kodratoff, Y. (ed.) EWSL 1991. LNCS, vol. 482, pp. 151–163. Springer, Heidelberg (1991)
[10] Bo, L., Abbas, H.A., McKay, B.: Density-based Heuristic for Rule Discovery with Ant-Miner. In: The 6th Australia-Japan Joint Workshop on Intelligent and Evolutionary System, pp. 180–184 (2002)
[11] Bo, L., Abbas, H.A., McKay, B.: Classification Rule Discovery with Ant Colony Optimization. In: International Conference on Intelligent Agent Technology, 2003. IAT 2003. IEEE/WIC, October 13-16, 2003, pp. 83–88 (2003)

[12] Smaldon, J., Freitas, A.A.: A new version of the ant-miner algorithm discovering unordered rule sets. In: Genetic and Evolutionary Computation Conference (GECCO-2006), pp. 43–50. ACM Press, New York (2006)
[13] Chu, S.C., Roddick, J.F., Pan, J.S., Su, C.J.: Parallel Ant Colony System. In: Zhong, N., Raś, Z.W., Tsumoto, S., Suzuki, E. (eds.) ISMIS 2003. LNCS (LNAI), vol. 2871, pp. 279–284. Springer, Heidelberg (2003)

Object-Oriented Software Systems Restructuring through Clustering

Gabriela Şerban and István-Gergely Czibula

Department of Computer Science, Babeş-Bolyai University
1, M. Kogalniceanu Street, Cluj-Napoca, Romania
{gabis, istvanc}@cs.ubbcluj.ro

Abstract. It is well-known that maintenance and evolution represent important stages in the lifecycle of any software system (about 66% from the total cost of the software systems development). That is why in this paper we are focusing on the problem of automating an essential activity that appears in the maintenance and evolution of software systems: the problem of identifying refactorings that would improve the structure of the system. *Refactoring* is the process of improving the design of software systems, by improving their internal structure, without altering the external behavior of the code. The aim of this paper is to introduce a new clustering algorithm, *CASYR* (*Clustering Algorithm for Software Systems Restructuring*), that can be used for improving software systems design, by identifying the appropriate refactorings. The proposed approach can be useful for assisting software engineers in their daily work of refactoring software systems. We evaluate our approach on a real software system and we also provide a comparison with previous approaches.

1 Introduction

The structure of a software system has a major impact on the maintainability of the system. This structure is the subject of many changes during the system lifecycle. Improper implementations of these changes imply structure degradation that leads to costly maintenance.

Refactoring is a solution adopted by most modern software development methodologies (extreme programming and other agile methodologies), in order to keep the software structure clean and easy to maintain. Refactoring becomes an integral part of the software development cycle: developers alternate between adding new tests and functionality and refactoring the code to improve its internal consistency and clarity.

Fowler defines in [1] refactoring as "the process of changing a software system in such a way that it does not alter the external behavior of the code yet improves its internal structure. It is a disciplined way to clean up code that minimizes the chances of introducing bugs". Refactoring is viewed as a way to improve the design of the code after it has been written. Software developers have to identify parts of code having a negative impact on the system's maintainability, and to apply appropriate refactorings in order to remove the so called "bad-smells" [2].

We have previously introduced in [3] a clustering approach for identifying refactorings in order to improve the structure of software systems. To our knowledge, there is no approach in the literature that uses clustering in order to improve the class structure of a system, excepting the approach introduced in [3].

In this paper we propose a new hierarchical clustering algorithm that would help developers to identify the appropriate refactorings in a software system. Our approach takes an existing software and reassembles it using hierarchical clustering, in order to obtain a better design, suggesting the needed refactorings. Applying the proposed refactorings remains the decision of the software engineer.

The main contributions of this paper are:

- To improve the approach from [3], by defining a new *agglomerative hierarchical* clustering algorithm for identifying refactorings in order to recondition the class structure of software systems. The proposed approach can be useful for assisting software engineers in their daily work of restructuring software systems.
- To validate our approach by evaluating the obtained results on a real software system.
- To emphasize the advantages of our approach in comparison with existing approaches.

The rest of the paper is structured as follows. Section 2 presents the main aspects related to the problem of *clustering* and to the clustering approach for systems design improvement (that we have previously introduced in [3]). A new hierarchical clustering algorithm for identifying refactorings is introduced in Section 3. Section 4 provides an evaluation of our approach on a real software system. A comparison of our approach with other similar approaches is given in Section 5. Section 6 contains some conclusions of our paper and future research directions.

2 Background

2.1 Clustering

Clustering [4], also known as unsupervised classification, is a data mining activity that aims to differentiate groups (classes or clusters) inside a given set of objects, $\mathcal{O}$. The measure used for discriminating objects can be any *metric* or *semimetric* function $d : \mathcal{O} \times \mathcal{O} \longrightarrow \Re$, called *distance*. A large collection of clustering algorithms is available in the literature [4]. Most clustering algorithms are based on two popular techniques known as *partitional* and *hierarchical* clustering.

In this paper we are focusing only on *hierarchical* clustering, that is why, in the following, a short overview of the *hierarchical* clustering methods is presented.

Hierarchical clustering methods represent a major class of clustering techniques [5]. There are two types of hierarchical clustering algorithms. Given a set of n objects, the agglomerative (bottom-up) methods begin with n singletons (sets with one element), merging them until a single cluster is obtained. At each step, the most similar two clusters are chosen for merging. The divisive (top-down) methods start from one cluster containing all n objects and split it until n clusters are obtained.

The agglomerative clustering algorithms that were proposed in the literature differ in the way the two most similar clusters are calculated and the *linkage-metric* used (single, complete or average).

Single link algorithms merge the clusters whose distance between their closest patterns is the smallest. *Complete link* algorithms, on the other hand, merge the clusters whose distance between their most distant patterns is the smallest [5]. In general, complete link algorithms generate compact clusters while single link algorithms generate elongated clusters. Thus, complete link algorithms are generally more useful than single link algorithms. *Average link* algorithms merge the clusters whose average distance (the average of distances between the objects from the clusters) is the smallest.

2.2 A Clustering Approach for Refactorings Determination - *CARD*

In this subsection we briefly describe the clustering approach (*CARD*) that we have previously introduced in [3] in order to find adequate refactorings to improve the structure of software systems. *CARD* approach consists of three steps:

- **Data collection** - The existing software system is analyzed in order to extract from it the relevant entities: classes, methods, attributes and the existing relationships between them.
- **Grouping** - The set of entities extracted at the previous step are re-grouped in clusters using a clustering algorithm (as *CASYR* algorithm introduced in this paper). The goal of this step is to obtain an improved structure of the existing software system.
- **Refactorings extraction** - The newly obtained software structure is compared with the original one in order to provide a list of refactorings which transform the original structure into an improved one.

As described above, at the **Grouping** step of *CARD*, the software system S has to be re-grouped. This re-grouping can be viewed as a ***partition*** $\mathcal{K}$ of S, $\mathcal{K} = \{K_1, K_2, ..., K_p\}$. We mention that a software system S is viewed in [3] as a set $S = \{s_1, s_2, ..., s_n\}$, where s_i ($1 \leq i \leq n$) is an *entity* from the software system (it can be an *application class*, a *method* from a class or an *attribute* from a class).

In our clustering approach, the objects to be clustered are the entities from the software system S. Our focus is to group similar entities from S in order to obtain high cohesive groups (clusters). A cluster from the partition $\mathcal{K}$ obtained at the *Grouping* step is a set of entities from S and represents an application class in the new (improved) structure of the software system.

3 A New Clustering Algorithm for Software Systems Restructuring (CASYR)

In this section we introduce a new hierarchical agglomerative clustering algorithm (*CASYR*), that aims at identifying a **partition** of a software system S that corresponds to an improved structure of it. *CASYR* algorithm can be used in the **Grouping** step of *CARD*.

3.1 *CASYR* Algorithm

CASYR is based on the idea of *hierarchical agglomerative clustering*, but uses an heuristic for determining the number of clusters (the number of application classes in the improved structure of the software system) and a measure for evaluating the "quality" of a partition.

In our clustering approach, the objects to be clustered are the entities from the software system S, i.e., $\mathcal{O} = \{s_1, s_2, \ldots, s_n\}$. Our focus is to group similar entities from S in order to obtain high cohesive groups (clusters).

We will adapt the generic cohesion measure introduced in [6] that is connected with the theory of similarity and dissimilarity. In our view, this cohesion measure is the most appropriate to our goal. We will consider the dissimilarity degree between any two entities from the software system S. Consequently, we will consider the distance $d(s_i, s_j)$ between two entities s_i and s_j as expressed in Equation (1).

$$d(s_i, s_j) = \begin{cases} 1 - \frac{|p(s_i) \cap p(s_j)|}{|p(s_i) \cup p(s_j)|} & if\ p(s_i) \cap p(s_j) \neq \emptyset \\ \infty & otherwise \end{cases}, \quad (1)$$

where, for a given entity $e \in S$, $p(e)$ represents a set of relevant properties of e, defined as:

- If e is an attribute, then $p(e)$ consists of: the attribute itself, the application class where the attribute is defined, and all the methods from S that access the attribute.
- If e is a method, then $p(e)$ consists of: the method itself, the application class where the method is defined, and all the attributes from S accessed by the method.
- If e is a class, then $p(e)$ consists of: the application class itself, and all the attributes and the methods defined in the class.

We have chosen the distance between two entities as expressed in Equation (1), because it emphasizes the idea of cohesion. The authors define in [7] cohesion as "*the degree to which module components belong together*". Our distance, as defined in Equation (1), highlights the concept of cohesion, i.e., entities with low distances are cohesive, whereas entities with higher distances are less cohesive. We have given a theoretical validation for this statement in [8].

Based on the definition of distance d given in Equation (1) it can be easily proved that d is a semi-metric function, so a clustering approach can be applied.

We will consider the distance $dist(k, k')$ between two clusters $k \in \mathcal{K}$ and $k' \in \mathcal{K}$ ($k \neq k'$) given by the average link metric, as expressed in Equation (2). We mention that we use *average link* as linkage metric, because we have obtained better results with this metric.

$$dist(k, k') = \frac{1}{|k| \cdot |k'|} \cdot \sum_{e \in k, e' \in k'} d(e, e') \qquad (2)$$

In the following we will introduce an heuristic for choosing the number of clusters, i.e., the number of application classes in the restructured software system. This heuristic is particular to our problem and it will provide a good enough choice for the number of application classes in the restructured software system. In order to determine the appropriate number p of clusters, we are focusing on determining p representative entities, i.e., a representative entity for each cluster.

The main idea of *CASYR*'s heuristic for choosing the representative entities and the number p of clusters is the following:

(i) The initial number p of clusters is n (the number of entities from the software system).
(ii) The first representative entity chosen is the most "distant" entity from the set of all entities (the entity that maximizes the average distance from all other entities).
(iii) In order to choose the next representative entity we reason as follows. For each remaining entity (that was not already chosen), we compute the minimum distance ($dmin$) from the entity and the already chosen representative entities. The next representative entity is chosen as the entity e that maximizes $dmin$ and this distance is greater than a positive given threshold ($distMin$). If such an entity does not exist, it means that e is very close to all the already chosen representatives and should not be chosen as a new representative (from the software system structure point of view this means that e should belong to the same application class with an already chosen representative). In this case, the number p of clusters will be decreased.
(iv) The step (iii) will be repeatedly performed, until p representatives are chosen.

We have to notice that step (iii) described above assures, from the software system design point of view, that near entities (with respect to the given threshold $distMin$) will be merged into a single application class (cluster), instead of being distributed in different application classes (clusters).

We mention that at steps (ii) and (iii) the choice could be a non-deterministic one. In the current version of *CASYR* algorithm, if such a non-deterministic case exists, the first selection is chosen. Heuristics can be used in non-deterministic selection cases. Improvements of *CASYR* algorithm will deal with these kind of situations.

Starting from the idea that we intend to obtain high cohesive clusters and as a high cohesion between two entities from a cluster is given by a low distance (dissimilarity) between them, we will search for the clustering with the lowest overall dissimilarity. For each cluster, we define its dissimilarity as the sum of pairwise entities dissimilarity, and we seek to minimize that sum over all clusters.

Consequently, the dissimilarity of a partition $\mathcal{K} = \{K_1, K_2, \ldots, K_p\}$, $DISS(\mathcal{K})$, is defined as given in Equation (3).

$$DISS(\mathcal{K}) = \sum_{i=1}^{p} diss(K_i),\qquad(3)$$

where $diss(K_i)$ represents the dissimilarity of cluster K_i and is defined as:

$$diss(K_i) = \begin{cases} \sum_{e \in K_i,\ e' \in K_i,\ e \neq e'} d(e, e') & if\ |K_i| \neq 1 \\ \infty & otherwise \end{cases}$$

Intuitively, the *dissimilarity* of a cluster indicates the *cohesion degree* between the entities from the corresponding application class. This is due to the fact that if an entity e should belong to an application class (cluster) K_i, then it is very likely that the *distance* (Equation (1)) between e and the elements from K_i is less than the *distance* between e and all the other elements from the other application classes (clusters).

For these reasons, we intend to minimize the *dissimilarity* of a partition, in order to maximize the cohesion of the corresponding application classes from the software system. We mention that we are currently working on giving a rigorous proof for this statement.

The main steps of *CASYR* algorithm are:

- Determine the number p of clusters using the heuristic presented above. In the current implementation of *CASYR* we have chosen the value 1 for the threshold *distMin*, because distances greater than 1 are obtained only for unrelated entities (Equation (1)).
- Each entity from the software system is put in its own cluster (singleton).
- The following steps are repeated until p clusters are reached:
  - Select the two most similar clusters K_i and K_j from the current partition, i.e, the pair of clusters that minimizes the distance from Equation (2).
  - Merge the clusters K_i and K_j into a single new cluster. The number of clusters in the partition is decreased. Let us denote by $\mathcal{K}$ the obtained partition.
  - Compute the dissimilarity of partition $\mathcal{K}$ (see Equation (3)) and retain $\mathcal{K}$ if its dissimilarity is minimum.

3.2 Identified Refactorings

In this subsection we briefly discuss about the refactorings that *CASYR* algorithm is able to identify. The main refactorings identified by *CASYR* algorithm are:

1. **Move Method refactoring** [1]
   It moves a method m of a class C_1 to another class C_2 that uses the method most; the method m of class C_1 should be turned into a simple delegation, or

it should be removed completely. The bad smell motivating this refactoring is that a method uses or is used by more features of another class than the class in which it is defined [9]. This refactoring is identified by $CASYR$ algorithm by moving the method m in the cluster K_i corresponding to the application class C_2.

2. **Move Attribute refactoring** [1]
   It moves an attribute a of a class C_1 to another class C_2 that uses the attribute most. The bad smell motivating this refactoring is that an attribute is used by another class more than the class in which it is defined [9]. This refactoring is identified by $CASYR$ algorithm by moving the attribute a in the cluster K_i corresponding to the application class C_2.

3. **Inline Class refactoring** [1]
   It moves all members of a class C_1 into another class C_2 and deletes the old class. The bad smell motivating this refactoring is that a class is not doing very much [9]. This refactoring is identified by $CASYR$ algorithm by decreasing the number of elements in the partition $\mathcal{K}$. Consequently, classes C_1 and C_2 with their corresponding entities (methods and attributes) will be merged in the same cluster.

4. **Extract Class refactoring** [1]
   Creates a new class C and move some cohesive attributes and methods into the new class. The bad smell motivating this refactoring is that one class offers too much functionality that should be provided by at least two classes [9]. This refactoring is identified by $CASYR$ algorithm by increasing the number of elements in the partition $\mathcal{K}$. Consequently, a new cluster appears, corresponding to an application class in the new structure of S.

We have currently implemented the above enumerated refactorings, but $CASYR$ algorithm can also identify other refactorings, like: *Pull Up Attribute*, *Pull Down Attribute*, *Pull Up Method*, *Pull Down Method*, *Collapse Class Hierarchy*. Future improvements will deal with these situations, also.

4 Experimental Evaluation - A Real Software System

In order to evaluate $CASYR$ algorithm, we have chosen a real software system. It is DICOM [10](*Digital Imaging and Communications in Medicine*) and HL7 [11] (*Health Level 7*) compliant PACS (*Picture Archiving and Communications System*) system, facilitating medical images management, offering quick access to radiological images, and making the diagnosing process easier.

In the following, we will briefly describe the *Data Collection* step from our approach.

The analyzed system is written in Java. In order to extract from the system the data needed in the *Grouping* step of our approach (Subsection 2.2) we use ASM 3.0 [12]. ASM is a Java bytecode manipulation framework. We use this framework in order to extract the structure of the system (attributes, methods, classes and relationships between all these entities).

The analyzed application is a distributed system, currently used by hospitals in locations such as Romania, United Kingdom, South Africa, Bulgaria and the Republic of Moldova. It is a large system that consists of several subsystems in form of stand-alone and web-based applications. We have applied *CASYR* algorithm on one of the subsystems from this application.

For confidentiality reasons, we will refer the analyzed application as $\mathcal{A}$. $\mathcal{A}$ is a stand-alone Java application used by physicians in order to interpret radiological images. The application fetch clinical images from an image server (using DICOM protocol), display them, and offer various tools to manage radiological images.

Even if the application is currently used, it also continuously evolves in order to satisfy change requirements and to provide better user experience based on feedback. That is why, the developers are often faced with the need of structural and conceptual changes.

$\mathcal{A}$ consists of **1015** classes, **8639** methods and **4457** attributes.

After applying *CASYR* algorithm, a total of 88 refactorings have been suggested: 8 *Move Attribute* refactorings, 78 *Move Method* refactorings, and 2 *Inline Class* refactorings.

The obtained results have been analyzed by the developers of $\mathcal{A}$ and the following conclusions were made:

- 29.5% from the refactorings identified by *CASYR* were accepted by the developers as useful in order to improve the system.
- 22.5% from the refactorings were acceptable for the developers, but they concluded that these refactorings are not necessary in the current stage of the project.
- 48% from the refactorings were strongly rejected by the developers.

Analyzing the obtained results, based on the feedback provided by the developers, we have concluded the following:

- *CASYR* successfully identified smart GUI anti-patterns (parts of software were the presentation layer contains business logic), misplaced constants (constants used only on a subtree of a class hierarchy, but defined in some base class). These kind of weaknesses can be discovered only if the developer manually inspects all the classes, or if a bug (related to the misplaced business logic) arises. That is why automatic detection by *CASYR* of these kind of weaknesses can prevent system failure or other kind of bugs and also save a lot of manual work.
- A large number of miss-identified refactorings are due to technical issues: the use of Java anonymous inner classes, introspection, the use of dynamic proxies. These kind of technical aspects appear frequently in projects developed in JAVA. In order to correctly deal with these aspects, we have to improve only the **Data collection** step of our approach, without modifying *CASYR* algorithm.
- Another cause of miss-identified refactorings is due to the fact that the *distance* (Equation (1)) used for discriminating entities in the clustering process

take into account only two aspects of a good design: *low coupling* and *high cohesion*. It would be also important to consider other principles related to an improved design, like: *Single Responsibility Principle*, *Open-Closed Principle*, *Interface Segregation Principle*, *Common Closure Principle* [13], etc.
- Our approach is currently implemented as a stand-alone application: the user provides the .jar files containing the classes of the analyzed software system and our application displays the suggested refactorings. The developers have suggested that it would be preferable to integrate our tool with existing IDE (as a plugin), instead of a stand-alone application.

5 Related Work

In this section we present some approaches existing in the literature in the field of *refactoring*. We provide, for similar approaches, a comparison with our approach.

Even if various approaches exist in the literature in the field of *refactoring*, only very limited support exists for detecting refactorings.

Deursen et al. have approached the problem of *refactoring* in [14]. The authors illustrate the difference between refactoring test code and refactoring production code, and they describe a set of bad smells that indicate trouble in test code, and a collection of test refactorings to remove these smells.

Xing and Stroulia present in [15] an approach for detecting refactorings by analyzing the system evolution at the design level.

An approach for restructuring programs written in Java starting from a catalog of bad smells is introduced in [16]. Based on some elementary metrics, the approach in [17] aids the user in deciding what kind of refactoring should be applied.

Clustering techniques have already been applied for program restructuring. A clustering based approach for program restructuring at the functional level is presented in [18]. This approach focuses on automated support for identifying ill-structured or low cohesive functions. The paper [19] presents a quantitative approach based on clustering techniques for software architecture restructuring and reengineering as well for software architecture recovery. It focuses on system decomposition into subsystems.

A complete comparison between our approach and the approaches from [16], [17], [18] and [19] can not be provided, because of the following reasons:

- The obtained results for relevant case studies are not available. There are given only short examples indicating the obtained refactorings.
- The techniques [18] and [19] address particular refactorings: the one from [18] focuses on automatic support only for identifying ill-structured or low cohesive functions and the technique from [19] focuses on system decomposition into subsystems.

The paper [9] describes a software vizualization tool which offers support to the developers in judging which refactoring to apply. The only result provided by the authors is a short example, illustrating the *Move Method* refactoring. We have applied *CASYR* algorithm on this example and the *Move Method* refactoring indicated by the authors was determined.

A search based approach for refactoring software systems structure is proposed in [20]. The authors use an evolutionary algorithm for identifying refactorings that improve the system structure. The authors provide an list of refactorings obtained on JHotDraw [21]. In order to provide a comparison with the approach from [20], we have applied *CASYR* algorithm on JHotDraw version 5.1 [21]. JHotDraw is a Java GUI framework for technical and structured graphics, developed by Erich Gamma and Thomas Eggenschwiler, as a design exercise for using design patterns. It consists of **173** classes, **1375** methods and **475** attributes and it is well-known as a good example for the use of design patterns and as a good design.

The advantages of our approach in comparison with the approach presented in [20] are illustrated bellow:

- Our technique is deterministic, in comparison with the approach from [20]. The evolutionary algorithm from [20] is executed **10** times, in order to judge how stable are the results, while *CASYR* algorithm from our approach is executed just **once**.
- The technique from [20] reports **10** misplaced methods, while in our approach there are only **4** misplaced methods.
- The overall running time for the technique from [20] is about **300** minutes (30 minutes for one run), while *CASYR* algorithm in our approach provide the results in about **4.8** minutes. We mention that the execution was made on similar computers.
- Because the results are provided in a reasonable time, our approach can be used for assisting developers in their daily work for improving software systems.

Based on the previous analysis, *CASYR* is better than the approach from [20].

We have previously introduced in [3] a clustering approach for identifying refactorings in order to improve the structure of software systems. For this purpose, a clustering algorithm named *kRED*, was introduced.

The advantages of *CASYR* algorithm introduced in this paper in comparison with *kRED* algorithm from [3] are the following:

- The number of methods misplaced by *CASYR* (**4**) is equal to the number of methods misplaced by *kRED* (**4**).
- The number of attributes misplaced by *CASYR* (**0**) is less than the number of attributes misplaced by *kRED* (**2**).
- The running time of *CASYR* (**4.8 minutes**) is less than the running time of *kRED* (**5**). We mention that the execution was made on similar computers.
- *CASYR* algorithm, unlike *kRED* algorithm, identifies the *Extract Class* refactoring, also.

Considering the above, *CASYR* algorithm provides better results than *kRED* algorithm.

Consequently, we can conclude that $CARD$ clustering approach using $CASYR$ algorithm is better than similar approaches existing in the literature.

6 Conclusions and Future Work

We have presented in this paper a new hierarchical clustering algorithm ($CASYR$) that can be used for improving software systems design.

We have demonstrated the potential of our algorithm by applying it to a real software system. Based on the feedback provided by the developers of the system we have identified in Section 4 some potential improvements of our approach. We have also presented the advantages of our approach in comparison with existing similar approaches from the literature.

As a conclusion, the advantages of our approach for determining refactorings using clustering are:

- it is deterministic;
- it can deal with various types of refactorings;
- it can be applied for large software systems;
- it can offer support to software developers for identifying ill-structured software modules.

Further work can also be done in the following directions:

- To study the applicability of other learning techniques [22] in order to improve software systems design.
- To develop a tool (as a plugin for Eclipse) that is based on determining refactorings using $CASYR$ algorithm.
- To apply our approach in order to transform non object-oriented software into object-oriented systems.

Acknowledgements

This work was supported by the research project CRONIS No. 11-003/2007, with governmental funding.

References

1. Fowler, M.: Refactoring: Improving the Design of Existing Code. Addison-Wesley Longman Publishing Co., Inc., Boston (1999)
2. Brown, W.J., Malveau, R.C., Hays, W., McCormick, I., Mowbray, T.J.: AntiPatterns: refactoring software, architectures, and projects in crisis. John Wiley & Sons, Inc., New York (1998)
3. Czibula, I., Serban, G.: Improving systems design using a clustering approach. International Journal of Computer Science and Network Security 6, 40–49 (2006)
4. Han, J.: Data Mining: Concepts and Techniques. Morgan Kaufmann Publishers Inc., San Francisco (2005)

5. Jain, A.K., Murty, M.N., Flynn, P.J.: Data clustering: a review. ACM Computing Surveys 31, 264–323 (1999)
6. Simon, F., Loffler, S., Lewerentz, C.: Distance based cohesion measuring. In: Proceedings of the 2nd European Software Measurement Conference (FESMA), Technologisch Instituut Amsterdam, pp. 69–83 (1999)
7. Bieman, J.M., Kang, B.K.: Measuring design-level cohesion. Software Engineering 24, 111–124 (1998)
8. Serban, G., Czibula, I.: On evaluating software systems design. Studia Universitatis "Babes-Bolyai", Informatica LII, 55–66 (2007)
9. Simon, F., Steinbrückner, F., Lewerentz, C.: Metrics based refactoring. In: CSMR 2001: Proceedings of the Fifth European Conference on Software Maintenance and Reengineering, pp. 30–38. IEEE Computer Society, Washington, DC, USA (2001)
10. (Digital Imaging and Communications in Medicine), http://medical.nema.org/
11. (Health Level 7), http://www.hl7.org/
12. (ObjectWeb: Open Source Middleware), http://asm.objectweb.org/
13. DeMarco, T.: Structured analysis and system specification, pp. 529–560 (2002)
14. van Deursen, A., Moonen, L., van den Bergh, A., Kok, G.: Refactoring test code, 92–95 (2001)
15. Xing, Z., Stroulia, E.: Refactoring detection based on umldiff change-facts queries. WCRE, 263–274 (2006)
16. Dudzikan, T., Wlodka, J.: Tool-supported dicovery and refactoring of structural weakness. Master's thesis, TU Berlin, Germany (2002)
17. Tahvildari, L., Kontogiannis, K.: A metric-based approach to enhance design quality through meta-pattern transformations. In: CSMR 2003: Proceedings of the Seventh European Conference on Software Maintenance and Reengineering, pp. 183–192. IEEE Computer Society, Washington, DC, USA (2003)
18. Xu, X., Lung, C.H., Zaman, M., Srinivasan, A.: Program restructuring through clustering techniques. In: SCAM 2004: Proceedings of the Source Code Analysis and Manipulation, Fourth IEEE International Workshop on (SCAM 2004), pp. 75–84. IEEE Computer Society, Washington, DC, USA (2004)
19. Lung, C.H.: Software architecture recovery and restructuring through clustering techniques. In: ISAW 1998: Proceedings of the third international workshop on Software architecture, pp. 101–104. ACM Press, New York (1998)
20. Seng, O., Stammel, J., Burkhart, D.: Search-based determination of refactorings for improving the class structure of object-oriented systems. In: GECCO 2006: Proceedings of the 8th annual conference on Genetic and evolutionary computation, pp. 1909–1916. ACM Press, New York (2006)
21. Gamma, E. (JHotDraw Project), http://sourceforge.net/projects/jhotdraw
22. Mitchell, T.M.: Machine Learning. McGraw-Hill, New York (1997)

Data Clustering with Semi-binary Nonnegative Matrix Factorization

Rafal Zdunek

Institute of Telecommunications, Teleinformatics and Acoustics,
Wroclaw University of Technology, Wybrzeze Wyspianskiego 27, 50-370 Wroclaw,
Poland
rafal.zdunek@pwr.wroc.pl

Abstract. Recently, a considerable growth of interest in using Nonnegative Matrix Factorization (NMF) for pattern classification and data clustering has been observed. For nonnegative data (observations, data items, feature vectors) many problems of partitional clustering can be modeled in terms of a matrix factorization into two groups of vectors: the nonnegative centroid vectors and the binary vectors of cluster indicators. Hence our data partitional clustering problem boils down to a semi-binary NMF problem. Usually, NMF problems are solved with an alternating minimization of a given cost function with multiplicative algorithms. Since our NMF problem has a particular characteristics, we apply a different algorithm for updating the estimated factors than commonly-used, i.e. a binary update with simulated annealing steering. As a result, our algorithm outperforms some well-known algorithms for partitional clustering.

1 Introduction

Data clustering can be regarded as the unsupervised classification of patterns into groups (clusters) that have similar features. The grouping can be basically obtained with hierarchical or partitional techniques [1]. In the hierarchical clustering, a nested series of partitions is performed with varying dissimilarity level whereas the partitional clustering technique yields a single partition of data that optimizes (usually locally) a given clustering criterion. The former technique is usually more robust but due to its high computational complexity it is not suitable for a large set of data. For such a case, the partitional technique is more favorable, and hence there is a need to develop an algorithmic approach to partitional clustering.

In our approach, we restrict our considerations to the partitional clustering in which a whole set of data is proceeded simultaneously in one updating cycle. There are many approaches to partitional clustering: k-means, [2] and its variants [3] (ISODATA [4], dynamic clustering [5,6], Mahalanobis distance based clustering, [7], spectral k-means [8]), graph-theoretic clustering [9,10], mixture-resolving and mode-seeking algorithms [11,12], nearest neighbor clustering [13], fuzzy clustering [14,15,16], neural networks based clustering [17,18], Kohonens learning vector quantization (LVQ), self-organizing map (SOM) [19], evolutionary clustering [20, 21, 22], branch-and-bound technique [23, 24], and simulated annealing clustering [25, 26].

Recently, many researchers proposed another approach to partitional clustering of nonnegative data that involves Nonnegative Matrix Factorization (NMF) which has been popularized by Lee and Seung [27, 28]. Some examples of this aspect can be found in [29, 30, 31, 32, 33, 34, 35, 36, 37]. Ding *et al* showed in [38] the equivalence between NMF, spectral clustering and k-means clustering.

NMF became also very useful in many other applications, including blind source separation [39, 40, 41], spectra recovering [42, 43], music transcription [44], pattern recognition [27, 45], image retrieval and classification [46, 32], and many others.

In general, NMF decomposes a data matrix into a product of two lower-rank matrices (factors) with nonnegative entries. Such a decomposition is typically obtained with alternating minimization of a given objective function subject to nonnegativity constraints. The basic NMF algorithms, which were proposed by Lee and Seung [27, 28], have found a variety of applications, especially in unsupervised learning problems in which there is a strong redundancy of information and the factors to be estimated are very sparse. Nevertheless, it has been demonstrated by many researchers that in many applications the well-known multiplicative Lee-Seung algorithms give rather poor performance, and hence should be replaced with many other optimization strategies, which in consequence leads to development of new NMF algorithms. Examples include the following algorithms: projected gradient descent [47], quasi-Newton [48, 41], extended SMART [40], quadratic programming [49, 50], alternating least squares [51]. Also, there exist many modifications of the basic NMF algorithms due to the additional terms incorporated to the objective function to enforce a certain priorly known characteristics of the desired solution.

In the assumed strategy for using NMF to partitional clustering, the prior information on the estimated factors is very strong. Assuming that one of the estimated factors contains the vectors of centroids (central points of clusters), the other one should contain binary numbers that assign cluster indicators to the observation (data) vectors. Moreover, assuming non-overlapping clusters, each column vector of this factor should contain exactly one non-zero entry. Thus, our extension to NMF can be regarded as semi-binary NMF.

Taking into account such a strong prior information about the binarity, the area of feasible solutions can be considerably narrowed. Also, the updates of the binary factor can be done with binary steering algorithms that are very robust in other applications, e.g. in binary tomography [52].

Our approach to clustering with NMF considerably differs from many extensions of NMF discussed in [30, 31]. The main difference persists in the way of formulating and incorporating the prior information. In many extensions to NMF, which were proposed by the Ding's group (uni-orthogonal, bi-orthogonal, symmetric 3-factor NMF), orthogonality constraints are explicitly assumed. It has been observed by many researchers [31, 53, 42] that adopting such constraints to orthogonal NMF problems gives very good results. In our approach the orthogonality constraints are assumed implicitly since the so-defined binary matrix (factor) of cluster indicators is orthogonal. Furthermore, we expect to get better results since the binarity and orthogonality constraints are stronger than only the nonnegativity and orthogonality ones. In this respect, our approach is similar to the binary matrix factorization that has been proposed by Zhang et al [54] recently. However, the main difference between the two matrix factorization methods is that our extension concerns the semi-binary NMF (only one factor is binary), and we use a completely different algorithm for obtaining a binary solution.

The Section 2 introduces to semi-binary NMF with respect to partitional clustering. The algorithmic methodology is discussed in Section 3. The numerical results obtained for a typical example of clustering are illustrated in Section 4. Finally, the conclusions are given in Section 5.

2 Semi-binary NMF

The aim of NMF is to find such lower-rank nonnegative matrices $A \in \mathbb{R}^{I \times J}$ and $X \in \mathbb{R}^{J \times K}$ that $Y \cong AX \in \mathbb{R}^{I \times K}$, given the data matrix Y, the lower rank J, and possibly some prior knowledge on the matrices A or X. Assuming each column vector of $Y = [y_1, \ldots, y_K]$ represents a single observation (a datum point in $\mathbb{R}^I$), the cluster are disjoint (non-overlapping), and J is *a priori* known number of clusters, we can interpret the column vectors of $A = [a_1, \ldots, a_J]$ as the centroids (indicating the directions of central points of clusters in R^I) and binary values in $X = [x_{jk}]$ as indicators to the clusters. If $x_{jk} = 1$, y_k belongs to the j-th cluster; $x_{jk} = 0$, otherwise.

Let us assume the following example:

$$Y = \begin{bmatrix} 0.5847 & 0.5714 & 0.5867 & 0.7301 & 0.7309 & 0.7124 & 0.7246 \\ 0.1936 & 0.1757 & 0.1920 & 0.4193 & 0.3955 & 0.3954 & 0.4147 \end{bmatrix}. \quad (1)$$

It easy to notice that the first 3 columns should be in one cluster, and the last 4 columns in the other. Thus X should be

$$X = \begin{bmatrix} 1 & 1 & 1 & 0 & 0 & 0 & 0 \\ 0 & 0 & 0 & 1 & 1 & 1 & 1 \end{bmatrix}. \quad (2)$$

The matrix A can be computed as $A = YX^+$, where X^+ is a pseudo-inverse to X. Thus

$$A = \begin{bmatrix} 0.5809 & 0.7245 \\ 0.1871 & 0.4062 \end{bmatrix}. \quad (3)$$

The column vector a_1 constitutes the algebraic mean over the first 3 columns of Y, and a_2 – the algebraic mean over the last 4 columns of Y. Thus the column vectors in A can be interpreted as the centroids of the clusters.

It is easy to notice that $\boldsymbol{X}$ is an orthogonal matrix, i.e. $\boldsymbol{X}\boldsymbol{X}^T = \boldsymbol{I}_J$, where $\boldsymbol{I}_J$ is an identity matrix, thus the orthogonality constraints are met here intrinsically.

In the next Section, we propose the algorithm that within a finite number of iterations is convergent to a binary solution, and we use such an algorithm to update the matrix $\boldsymbol{X}$. The centroid matrix $\boldsymbol{A}$ can be updated in a typical way for NMF.

3 Algorithms

Given an objective function $D(\boldsymbol{Y}||\boldsymbol{A}\boldsymbol{X})$ that measures the misfitting between $\boldsymbol{Y}$ and $\boldsymbol{A}\boldsymbol{X}$, the NMF algorithms minimize $D(\boldsymbol{Y}||\boldsymbol{A}\boldsymbol{X})$ with the alternating optimization strategy: for $s = 1, 2, \ldots$ do

$$\boldsymbol{X}^{(s)} = \arg\min_{\boldsymbol{X} \geq 0} D(\boldsymbol{Y}||\boldsymbol{A}^{(s-1)}\boldsymbol{X}), \tag{4}$$

$$\boldsymbol{A}^{(s)} = \arg\min_{\boldsymbol{A} \geq 0} D(\boldsymbol{Y}||\boldsymbol{A}\boldsymbol{X}^{(s)}). \tag{5}$$

In general, the objective function $D(\boldsymbol{Y}||\boldsymbol{A}\boldsymbol{X})$ does not need to be the same for the updates (4) and (5). For example, it is well-known that minimization of the l_1-norm leads to a sparse solution whereas the l_2-norm rather to a smooth one. Certainly, the expected (binary) solution to the problem in the step (4) should be very sparse but this is not the case in (5).

Let

$$D_X(\boldsymbol{Y}||\boldsymbol{A}\boldsymbol{X}) = \sum_{k=1}^{K} F(\boldsymbol{x}_k) \tag{6}$$

and $D_A(\boldsymbol{Y}||\boldsymbol{A}\boldsymbol{X})$ be objective functions in the problems (4) and (5), respectively. Assuming we expect to find such a centroid from which the sum over all the distances between the centroid and the data points belonging to its cluster is minimized, $D_A(\boldsymbol{Y}||\boldsymbol{A}\boldsymbol{X})$ should be defined according to the least squares criterion. Thus

$$D_A(\boldsymbol{Y}||\boldsymbol{A}\boldsymbol{X}) = ||\boldsymbol{Y} - \boldsymbol{A}\boldsymbol{X}||_F^2, \tag{7}$$

where $||\cdot||_F$ is the Frobenious norm.

The function $F(\boldsymbol{x}_k)$ in (6) can be defined in many ways, e.g. as the l_p norm, α- or β-divergence [40], or the Green's clique energy function [55] that is also known as the logistic function.

Usually, in many NMF problems $F(\boldsymbol{x}_k)$ is expressed by the squared Euclidean distance. For sparse solutions the l_1 norm is more suitable. For a binary solution, the logistic function should be appropriate since it makes smoothing while preserving discontinuities, i.e. small entries are smoothed to zero and the predominate entries tend to ones. Thus we selected the logistic function:

$$\psi(\xi) = \beta^2 \ln \cosh\left(\frac{\xi}{\beta}\right), \tag{8}$$

where β is a scaling parameter. Thus

$$F(\boldsymbol{x}_k) = \sum_{i=1}^{I} \psi\left([\boldsymbol{y}_k - \boldsymbol{A}\boldsymbol{x}_k]_i\right). \tag{9}$$

3.1 Updates for $\boldsymbol{X}$

The function (8) is convex with respect to $\boldsymbol{x}_k \in \mathbb{R}^J$ [55] but its minimization subjected to binarity constraints is not straightforward. For the optimization task we inherited one of the strategies used for discrete tomography [56]. Let us associated $F(\boldsymbol{x}_k)$ with the Gibbs–Boltzmann distribution, i.e.

$$P_F(\boldsymbol{x}_k) = \frac{\exp\left\{-\frac{1}{T}F(\boldsymbol{x}_k)\right\}}{\sum_{\boldsymbol{x}_k \in \{0,1\}^J} \exp\left\{-\frac{1}{T}F(\boldsymbol{x}_k)\right\}}, \tag{10}$$

where T is a temperature parameter controlling the ascent towards a global maximum of $P_F(\boldsymbol{x}_k)$.

To find a binary solution to the problem $\min_{\boldsymbol{x}_k \in \{0,1\}^J} F(\boldsymbol{x}_k)$, we take advantage of the following theorem:

Theorem 1. *[57] Let $\boldsymbol{x}_k^*$ be the global maximizer of the Gibbs–Boltzmann distribution (10), then*

$$\lim_{T \to 0} \langle \boldsymbol{x}_k \rangle_{P_F} \to \boldsymbol{x}_k^*,$$

where $<\cdot>_{P_F}$ means the expectation with respect to P_F.

Proof. Assuming $\forall \boldsymbol{x}_k \neq \boldsymbol{x}_k^* : F(\boldsymbol{x}_k) \geq F(\boldsymbol{x}_k^*)$:

$$\lim_{T \to 0} \langle \boldsymbol{x}_k \rangle_{P_F} = \lim_{T \to 0} \sum_{\boldsymbol{x}_k \in \{0,1\}^J} \boldsymbol{x}_k \frac{\exp\left\{-\frac{1}{T}F(\boldsymbol{x}_k)\right\}}{\sum_{\boldsymbol{x}_k \in \{0,1\}^J} \exp\left\{-\frac{1}{T}F(\boldsymbol{x}_k)\right\}}$$

$$= \lim_{T \to 0} \frac{\sum_{\boldsymbol{x}_k \in \{0,1\}^J} \boldsymbol{x}_k \exp\left\{\frac{1}{T}(F(\boldsymbol{x}_k^*) - F(\boldsymbol{x}_k))\right\}}{\sum_{\boldsymbol{x}_k \in \{0,1\}^J} \exp\left\{\frac{1}{T}(F(\boldsymbol{x}_k^*) - F(\boldsymbol{x}_k))\right\}}$$

$$= \lim_{T \to 0} \frac{\sum_{\boldsymbol{x}_k \neq \boldsymbol{x}_k^*} \boldsymbol{x} \exp\left\{\frac{1}{T}(F(\boldsymbol{x}_k^*) - F(\boldsymbol{x}_k))\right\} + \boldsymbol{x}_k^*}{\sum_{\boldsymbol{x}_k \neq \boldsymbol{x}_k^*} \exp\left\{\frac{1}{T}(F(\boldsymbol{x}_k^*) - F(\boldsymbol{x}_k))\right\} + 1} \to \boldsymbol{x}_k^*.$$

Since for each k only one entry in $\boldsymbol{x}_k$ is equal to one, the denominator in (10) can be considerably simplified as

$$D_F = \sum_{\boldsymbol{x}_k \in \{0,1\}^J} \exp\left\{-\frac{1}{T}F(\boldsymbol{x}_k)\right\} = \left[\exp\left\{-\frac{1}{T}F(\boldsymbol{I}_J)\right\}\right]\boldsymbol{e}_J, \tag{11}$$

where $\boldsymbol{I}_J \in \mathbb{R}^{J \times J}$ is an identity matrix, and $\boldsymbol{e}_J = [1, \ldots, 1]^T \in \mathbb{R}^J$. Note that $F(\boldsymbol{I}_J)$ can further expressed as

$$F(\boldsymbol{I}_J) = \sum_{i=1}^{I} \psi\left([\boldsymbol{e}_J^T \otimes \boldsymbol{y}_k - \boldsymbol{A}]_i\right), \tag{12}$$

where $\otimes$ stands for the Kronecker product.

Having the denominator D_F computed, the Gibbs–Boltzmann distribution can be easily determined, and for $P_F(\boldsymbol{I}_J)$ we got J real numbers in the range $[0, 1]$. For $T \to \infty$, only one number goes to one whereas the others tend to zeros, because $P_F(\boldsymbol{x}_k = \boldsymbol{x}^*) \to 1$ and $P_F(\boldsymbol{x}_k \neq \boldsymbol{x}^*) \to 0$ and we know that one of the column vectors in $\boldsymbol{I}_J$ is equal to $\boldsymbol{x}_k^*$. Thus computing $P_F(\boldsymbol{I}_J)$ we should know with probability dependent on T which column vector from $\boldsymbol{I}_J$ tends to $\boldsymbol{x}_k^*$. This statement is justified by Theorem 1.

Proposition 1. *Since $\boldsymbol{A}$ is gradually updated with the alternating optimization scheme, the updates for $\boldsymbol{X}$ should be also slowly steered to a binary solution, which can be done with gradual decreasing the temperature T. Assuming the update $\boldsymbol{X} \leftarrow \langle \boldsymbol{X} \rangle_{P_F}$ according to Theorem 1, zero-values in $\boldsymbol{X}$ would appear even in the first iterative step, which for the non-convex (alternating) optimization is not recommendable. To avoid too fast descent towards a local binary solution, we propose to use the following update:*

$$\forall k : \boldsymbol{x}_k \leftarrow P_F(\boldsymbol{I}_J). \tag{13}$$

Note that $P_F(\boldsymbol{I}_J) \in [0, 1]^J$ and $\lim_{T \to 0} P_F(\boldsymbol{I}_J) \in \{0, 1\}^J$.

For simultaneous processing of all the observations, we have

$$\boldsymbol{\Psi} = \psi(\boldsymbol{Y} \otimes \boldsymbol{e}_J^T - \boldsymbol{e}_K^T \otimes \boldsymbol{A}) \in \mathbb{R}^{J \times JK}, \tag{14}$$

with $\psi(\cdot)$ defined by (8) and $\boldsymbol{e}_K = [1, \ldots, 1]^T \in \mathbb{R}^K$. Let $\text{Matrix}(\boldsymbol{v}, m, n)$ be a matricization operation that transforms a vector $\boldsymbol{v} \in \mathbb{R}^{mn}$ to a matrix $\boldsymbol{V} \in \mathbb{R}^{m \times n}$. In consequence, we have the following update rule for $\boldsymbol{X}$:

$$\boldsymbol{X} \leftarrow \tilde{\boldsymbol{R}} \oslash \left(\boldsymbol{e}_J \otimes \boldsymbol{e}_J^T \tilde{\boldsymbol{R}} \right), \tag{15}$$

where $\oslash$ denotes a component-wise division, and

$$\tilde{\boldsymbol{R}} = \text{Matrix} \left(\exp \left\{ -\frac{\boldsymbol{e}_J^T \boldsymbol{\Psi}}{T} \right\}, J, K \right).$$

To steer the updates (15) gradually to a binary approximation, we set the temperature parameter T in (15) according a typical exponential temperature schedule as in evolutionary algorithms, e.g. in simulated annealing. Hence,

$$T = \bar{T} + T_0 \exp\{-\lambda s\}, \tag{16}$$

where $\bar{T} > 0$ is the minimal temperature to avoid numerical instabilities, $T_0 > 0$ is an initial temperature, $\lambda > 0$ is a decay constant that determines how fast the temperature goes to $\bar{T}$.

3.2 Updates for $\boldsymbol{A}$

The minimization of the objective function $D_A(\boldsymbol{Y} \| \boldsymbol{A}\boldsymbol{X})$ given in (7) with respect to $\boldsymbol{A}$ and subject to nonnegativity constraints can be done with many algorithms. We selected the update rule based on the $FNMA^I$ algorithm [48] that was demonstrated to be monotonically convergent. This algorithm belongs to a class of second-order methods which in general are very efficient to solve NMF problems (see, e.g. [41, 50]).

The $FNMA^I$ algorithm can be regarded as the projected Newton method that updates only the components that do not belong to the active set which is defined as follows:

$$\mathcal{A} = \{(i,j) : a_{ij} \leq 0, [\boldsymbol{G}_A]_{ij} > 0\}, \quad (17)$$

where $\boldsymbol{G}_A = \nabla_A D_A(\boldsymbol{Y}\|\boldsymbol{A}\boldsymbol{X}) = (\boldsymbol{A}\boldsymbol{X} - \boldsymbol{Y})\boldsymbol{X}^T$ and $\boldsymbol{A} = [a_{ij}]$. According to (17) the projection gradient is defined as:

$$[\boldsymbol{G}_A^+]_{ij} = \begin{cases} [\boldsymbol{G}_A]_{ij} & \text{if } (i,j) \notin \mathcal{A} \\ 0 & \text{otherwise.} \end{cases} \quad (18)$$

Consequently, the matrix of the projected Newton-based search directions is given by

$$[\boldsymbol{D}_A^+]_{ij} = \begin{cases} [\boldsymbol{G}_A^+(\boldsymbol{X}\boldsymbol{X}^T)^{-1}]_{ij} & \text{if } (i,j) \notin \mathcal{A} \\ 0 & \text{otherwise.} \end{cases} \quad (19)$$

The projected Newton update rule has the following form:

$$\boldsymbol{A} \leftarrow [\boldsymbol{A} - \alpha \boldsymbol{D}_A^+]_+. \quad (20)$$

with the projection defined as $[\xi]_+ = \max\{\xi, 0\}$.

In the $FNMA^I$, the step-length α is heuristically selected according to the inexact search technique. In our approach, α is determined according to a different heuristic technique that is a little similar to the Armijo rule [47]. In each alternating step, α is iteratively decreased using the rule $\alpha = 2^{-p}$, for $p = 0, 1, \ldots, p_{max}$ until the condition

$$D_A(\boldsymbol{Y}\|\boldsymbol{A}(\alpha)^{(s)}\boldsymbol{X}) - D_A(\boldsymbol{Y}\|\boldsymbol{A}^{(s-1)}\boldsymbol{X}) > 0 \quad (21)$$

is held, where $\boldsymbol{A}^{(s-1)}$ is the estimate from the previous alternating step. If the condition (21) already fails for $p = 0$, $\alpha = 1$ is chosen.

This rule relaxes the risk of increasing the objective function during the alternating process. Obviously, it does not guarantee a monotonic convergence because it may happen that for some unfortunate estimate of $\boldsymbol{A}$, the update for $\boldsymbol{X}$ will raise the value of the objective function, and the next update for $\boldsymbol{A}$ will remain steady due to $\alpha \rightarrow 0$. However, if this case takes place, α always remains very small, the failure can be easily detected, and the alternating process can be repeated for a new initial guess for $\boldsymbol{A}$.

Due to intrinsic scale and permutation indeterminacies in NMF, the lengths of the column vectors in $\boldsymbol{A}$ should be additionally controlled in each alternating step to avoid a serious misclassification. This event may happen when the distance between the estimated centroid and the true one grows up, and finally this may lead to the case of zero-rows in X (misclassification). One of the ways to handle this problem is to normalize the column vectors in A to unit lengths, e.g. in the sense of the Euclidean metrics. Obviously, the normalization nearly always moves the estimated centroids from the origins but it stabilizes the estimation and relaxes the indeterminacies.

4 Experiments

To test our algorithm we used the 3D data with 3 partially overlapping clusters. The small cluster overlapping is introduced to check the behavior of the algorithms for a perturbed model (inexactly satisfied). Each cluster contains 500 points that are distributed in a 3D space according to a skew-Gaussian distribution. The corresponding means and covariances of the distributions are as follows:

$$\boldsymbol{\mu}_1 = \begin{bmatrix} 40 \\ 80 \\ 30 \end{bmatrix}, \quad \boldsymbol{\mu}_2 = \begin{bmatrix} 70 \\ 40 \\ 60 \end{bmatrix}, \quad \boldsymbol{\mu}_3 = \begin{bmatrix} 20 \\ 20 \\ 30 \end{bmatrix},$$

$$\boldsymbol{\Sigma}_1 = \begin{bmatrix} 50 & -0.2 & 0.1 \\ -0.2 & 0.1 & 0.1 \\ 0.1 & 0.1 & 1 \end{bmatrix}, \quad \boldsymbol{\Sigma}_2 = \begin{bmatrix} 50 & -5 & -1 \\ -5 & 5 & -0.5 \\ -1 & 0.5 & 1 \end{bmatrix}, \quad \boldsymbol{\Sigma}_3 = \begin{bmatrix} 2 & 0 & 0 \\ 0 & 50 & 0 \\ 0 & 0 & 30 \end{bmatrix}.$$

Obviously, all the covariance matrices are positive-definite. The mixed data are stored in the matrix $Y \in \mathbb{R}^{3 \times 1500}$. For the NMF clustering, the column vectors of Y are normalized to the unit l_2 norm. The distribution of the data in a 3D space is illustrated in Fig. 1.

The performance of the algorithm has been evaluated with the measure χ which stands for the number of misclustered observations. Since the objective functions with respect to both sets of arguments are not convex, we repeat the

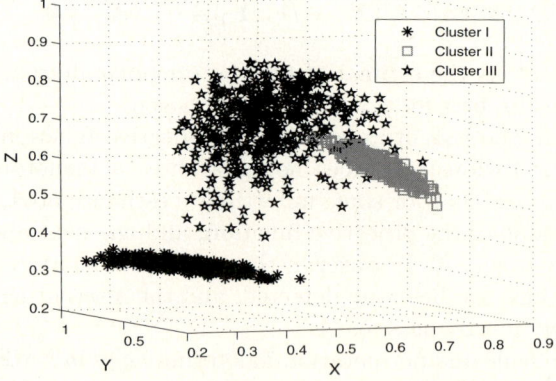

Fig. 1. Distribution of Y in 3D space

clustering with each tested algorithm 100 times, and then we evaluate the mean and standard deviation for χ.

We compare our algorithm with the uni-orthogonal NMF [31]. Additionally, the standard k-means algorithm implemented in the Statistics Toolbox in MAT-LAB 7.0 is used with the default settings. Its application is justified since our data points are distributed according to a Gaussian statistics. In our algorithm, we set heuristically $\beta = 0.05$ in (8), and the temperature parameters in (16): $\bar{T} = 0.001$, $T_0 = 1000$ and $\lambda = 0.5$. Both uni-orthogonal and our NMF algorithm are applied to the data with the normalized column vectors. The standard Lee-Seung NMF algorithm (with Frobenius norm) is applied to the data with normalization of the rows in $\boldsymbol{Y}$ to the unit row vectors. Otherwise, this algorithm completely fails. In all the NMF algorithms the matrices $\boldsymbol{A}$ and $\boldsymbol{X}$ are initialized with random nonnegative numbers generated from a uniform distribution.

The means and standard deviations (in parenthesis) of χ versus the number of alternating steps (only NMF algorithms) are given in Table 1.

Table 1. Mean-values and standard deviations (in parenthesis) of measure χ (number of misclustered vectors) versus parameter s (numbers of alternating steps) for the tested NMF algorithm. The mean and standard deviation for the k-means algorithm are 47.9 and 170.9, respectively.

Algorithm	$s = 20$	$s = 50$	$s = 100$	$s = 200$
NMF (Frobenius)	172.9 (118.7)	138.3 (118.1)	151.1 (123.1)	161.5 (116.6)
Uni-orthogonal NMF	216.5 (96.6)	213.6 (121.1)	177 (134.3)	80.3 (70.1)
Our algorithm	68 (0)	48 (0)	48 (0)	48 (0)

5 Conclusions

We proposed the algorithm for semi-binary NMF that can be used for partitional data clustering. The algorithm uses the Gibbs–Boltzmann distribution that allows us to steer the updates for the indicator matrix gradually to a binary case. The speed of the steering is determined by the exponential temperature schedule that is commonly used in the simulated annealing techniques. Our approach to data clustering implicitly assumes orthogonality constraints.

As the results, we got the algorithm that outperforms the well-known Lee-Seung algorithm and the uni-orthogonal algorithm proposed by Ding et al [31]. When all the associated parameters (temperature and scaling in logistic function) are well selected (could be roughly), our algorithm can be very stable, i.e. for each initial matrices $\boldsymbol{A}$ and $\boldsymbol{X}$ it gives the same final estimates (see Table 1). In this respect, it is much better than the k-means algorithm. Moreover, our algorithm gives better results even after 20 iterations than the uni-orthogonal NMF after 200 iterations. We believe that our algorithm can find other applications in NMF problems. However, the further research is needed to determine how to select the associate parameters optimally for real problems, and to compare the proposed algorithm with some density-based clustering methods, e.g. DBSCAN or RDBC.

References

[1] Jain, A.K., Murty, M.N., Flynn, P.J.: Data clustering: a review. ACM Compututing Surveys 31(3), 264–323 (1999)
[2] Mcqueen, J.: Some methods for classification and analysis of multivariate observations. In: Proceedings of the Fifth Berkeley Symposium on Mathematical Statistics and Probability, pp. 281–297 (1967)
[3] Anderberg, M.R.: Cluster Analysis for Applications. Monographs and Textbooks on Probability and Mathematical Statistics. Academic Press, Inc., New York (1973)
[4] Ball, G.H., Hall, D.J.: ISODATA, a novel method of data analysis and classification. Technical report, Stanford University, Stanford, CA (1965)
[5] Diday, E.: The dynamic cluster method in non-hierarchical clustering. J. Comput. Inf. Sci. 2, 61–88 (1973)
[6] Symon, M.J.: Clustering criterion and multi-variate normal mixture. Biometrics 77, 35–43 (1977)
[7] Mao, J., Jain, A.K.: A self-organizing network for hyperellipsoidal clustering (HEC). IEEE Trans. Neural Netw. 7(1), 16–29 (1996)
[8] Dhillon, I.S., Modha, D.M.: Concept decompositions for large sparse text data using clustering. Machine Learning J. 42, 143–175 (2001)
[9] Zahn, C.T.: Graph-theoretical methods for detecting and describing gestalt clusters. IEEE Trans. Comput. C-20, 68–86 (1971)
[10] Ozawa, K.: A stratificational overlapping cluster scheme. Pattern Recogn. 18, 279–286 (1985)
[11] Jain, A.K., Dubes, R.C.: Algorithms for clustering data. Prentice-Hall, Inc., Upper Saddle River (1988)
[12] Mitchell, T. (ed.): Machine Learning. McGraw Hill, Inc., New York (1997)
[13] Lu, S.Y., Fu, K.S.: A sentence-to-sentence clustering procedure for pattern analysis. IEEE Trans. Syst. Man Cybern. 8, 381–389 (1978)
[14] Zadeh, L.A.: Fuzzy sets. Inf. Control 8, 338–353 (1965)
[15] Ruspini, E.H.: A new approach to clustering. Inf. Control 15, 22–32 (1969)
[16] Bezdek, J.C.: Pattern Recognition With Fuzzy Objective Function Algorithms. Plenum Press, New York (1981)
[17] Sethi, I., Jain, A.K. (eds.): Artificial Neural Networks and Pattern Recognition: Old and New Connections. Elsevier Science Inc., New York (1991)
[18] Jain, A.K., Mao, J.: Neural networks and pattern recognition. In: Zurada, J.M., Marks II, R.J., Robinson, E.G. (eds.) Computational Intell. Imitating Life, pp. 194–212. IEEE Press, Los Alamitos (1994)
[19] Kohonen, T.: Self-organization and associative memory, 3rd edn. Springer, New York (1989)
[20] Raghavan, V.V., Birchard, K.: A clustering strategy based on a formalism of the reproductive process in natural systems. SIGIR Forum 14(2), 10–22 (1979)
[21] Special issue on evolutionary computation. In: Fogel, D.B., Fogel, L.J. (eds.) IEEE Transactions Neural Networks (1994)
[22] Jones, D., Beltramo, M.A.: Solving partitioning problems with genetic algorithms. In: Proc. of the Fourth International Conference on Genetic Algorithms, pp. 442–449. Morgan Kaufmann Publishers, San Francisco (1991)
[23] Koontz, W.L.G., Fukunaga, K., Narendra, P.M.: A branch and bound clustering algorithm. IEEE Trans. Comput. 23, 908–914 (1975)

[24] Cheng, C.H.: A branch-and-bound clustering algorithm. IEEE Trans. Syst. Man Cybern. 25(5), 895–898 (1995)
[25] Rojas, M., Santos, S.A., Sorensen, D.C.: Deterministic annealing approach to constrained clustering. IEEE Trans. Pattern Anal. Mach. Intell. 15, 785–794 (1993)
[26] Baeza-Yates, R.A.: Introduction to data structures and algorithms related to information retrieval. In: Information retrieval: data structures and algorithms, pp. 13–27. Prentice-Hall, Inc., Upper Saddle River (1992)
[27] Lee, D.D., Seung, H.S.: Learning the parts of objects by nonnegative matrix factorization. Nature 401, 788–791 (1999)
[28] Lee, D.D., Seung, H.S.: Algorithms for nonnegative matrix factorization. In: NIPS, pp. 556–562 (2000)
[29] Shahnaz, F., Berry, M., Pauca, P., Plemmons, R.: Document clustering using nonnegative matrix factorization. Inf. Process. Manage. 42(2), 373–386 (2006)
[30] Li, T., Ding, C.: The relationships among various nonnegative matrix factorization methods for clustering. In: ICDM 2006, pp. 362–371. IEEE Computer Society, Washington, DC, USA (2006)
[31] Ding, C., Li, T., Peng, W., Park, H.: Orthogonal nonnegative matrix trifactorizations for clustering. In: KDD 2006: Proceedings of the 12th ACM SIGKDD international conference on knowledge discovery and data mining, pp. 126–135. ACM Press, New York (2006)
[32] Okun, O.G.: Non-negative matrix factorization and classifiers: experimental study. In: Proc. of the Fourth IASTED International Conference on Visualization, Imaging, and Image Processing (VIIP 2004), Marbella, Spain, pp. 550–555 (2004)
[33] Zass, R., Shashua, A.: A unifying approach to hard and probabilistic clustering. In: International Conference on Computer Vision (ICCV), Beijing, China (2005)
[34] Banerjee, A., Merugu, S., Dhillon, I.S., Ghosh, J.: Clustering with Bregman divergences. In: SIAM International Conf. on Data Mining, Lake Buena Vista, Florida. SIAM, Philadelphia (2004)
[35] Carmona-Saez, P., Pascual-Marqui, R.D., Tirado, F., Carazo, J.M., Pascual-Montano, A.: Biclustering of gene expression data by non-smooth non-negative matrix factorization. BMC Bioinformatics 7(78) (2006)
[36] Cho, H., Dhillon, I.S., Guan, Y., Sra, S.: Minimum sum squared residue based coclustering of gene expression data. In: Proc. 4th SIAM International Conference on Data Mining (SDM), Florida, pp. 114–125 (2004)
[37] Wild, S.: Seeding non-negative matrix factorization with the spherical k-means clustering. M.Sc. Thesis, University of Colorado (2000)
[38] Ding, C., He, X., Simon, H.D.: On the equivalence of nonnegative matrix factorization and spectral clustering. In: Jonker, W., Petković, M. (eds.) SDM 2005. LNCS, vol. 3674, pp. 606–610. Springer, Heidelberg (2005)
[39] Cichocki, A., Zdunek, R., Amari, S.: New algorithms for non-negative matrix factorization in applications to blind source separation. In: Proc. IEEE International Conference on Acoustics, Speech, and Signal Processing, ICASSP 2006, Toulouse, France, pp. 621–624 (2006)
[40] Cichocki, A., Amari, S., Zdunek, R., Kompass, R., Hori, G., He, Z.: Extended SMART algorithms for non-negative matrix factorization. In: Rutkowski, L., Tadeusiewicz, R., Zadeh, L.A., Żurada, J.M. (eds.) ICAISC 2006. LNCS (LNAI), vol. 4029, pp. 548–562. Springer, Heidelberg (2006)
[41] Zdunek, R., Cichocki, A.: Nonnegative matrix factorization with constrained second-order optimization. Signal Processing 87, 1904–1916 (2007)

[42] Li, H., Adali, T., Wang, W., Emge, D., Cichocki, A.: Non-negative matrix factorization with orthogonality constraints and its application to Raman spectroscopy. Journal of VLSI Signal Processing 48(1-2), 83–97 (2007)
[43] Sajda, P., Du, S., Brown, T.R., Shungu, R.S.D.C., Mao, X., Parra, L.C.: Nonnegative matrix factorization for rapid recovery of constituent spectra in magnetic resonance chemical shift imaging of the brain. IEEE Trans. Medical Imaging 23(12), 1453–1465 (2004)
[44] Cho, Y.C., Choi, S.: Nonnegative features of spectro-temporal sounds for classification. Pattern Recognition Letters 26, 1327–1336 (2005)
[45] Liu, W., Zheng, N.: Non-negative matrix factorization based methods for object recognition. Pattern Recognition Letters 25(8), 893–897 (2004)
[46] Guillamet, D., Schiele, B., Vitrià, J.: Analyzing non-negative matrix factorization for image classification. In: 16th International Conference on Pattern Recognition (ICPR 2002), Quebec City, Canada, vol. 2, pp. 116–119 (2002)
[47] Lin, C.J.: Projected gradient methods for non-negative matrix factorization. Neural Computation 19(10), 2756–2779 (2007)
[48] Kim, D., Sra, S., Dhillon, I.S.: Fast Newton-type methods for the least squares nonnegative matrix approximation problem. In: Proc. 6-th SIAM International Conference on Data Mining, Minneapolis, Minnesota, USA (2007)
[49] Heiler, M., Schnörr, C.: Learning sparse representations by non-negative matrix factorization and sequential cone programming. J. Mach. Learn. Res. 7, 1385–1407 (2006)
[50] Zdunek, R., Cichocki, A.: Nonnegative matrix factorization with quadratic programming. Neurocomputing (accepted, 2008)
[51] Cichocki, A., Zdunek, R.: Regularized alternating least squares algorithms for non-negative matrix/tensor factorizations. In: Liu, D., Fei, S., Hou, Z., Zhang, H., Sun, C. (eds.) ISNN 2007. LNCS, vol. 4493, pp. 793–802. Springer, Heidelberg (2007)
[52] Herman, G.T., Kuba, A. (eds.): Discrete Tomography: Foundations, Algorithms, and Applications. Birkhauser, Boston (1999)
[53] Cao, B., Shen, D., Sun, J.T., Wang, X., Yang, Q., Chen, Z.: Detect and track latent factors with online nonnegative matrix factorization. In: Proc. the 20th International Joint Conference on Artificial Intelligence (IJCAI), Hyderabad, India, pp. 2689–2694 (2007)
[54] Zhang, Z., Li, T., Ding, C., Zhang, X.S.: Binary matrix factorization with applications. In: Proc. IEEE Intternational Conference on Data Mining (ICDM) (to appear, 2007)
[55] Green, P.J.: Bayesian reconstruction from emission tomography data using a modified EM algorithm. IEEE Trans. Medical Imaging 9, 84–93 (1990)
[56] Zdunek, R., Pralat, A.: Detection of subsurface bubbles with discrete electromagnetic geotomography. Electronic Notes in Discrete Mathematics 20, 535–553 (2005)
[57] Phillips, J.W., Leahy, R.M., Mosher, J.C.: MEG-based imaging of focal neuronal current sources. IEEE Trans. Medical Imaging 16, 248–338 (1997)

An Efficient Association Rule Mining Algorithm for Classification

A. Zemirline, L. Lecornu, B. Solaiman, and A. Ech-cherif*

ITI Department, ENST Bretagne
29285 Brest, France
{abdelhamid.zemirline,laurent.lecornu,basel.solaiman}@enst-bretagne.fr
Laboratoire Lamosi-USTO-Oran*
Algeria
a.echerif@alum.rpi.edu

Abstract. In this paper, we propose a new Association Rule Mining algorithm for Classification (ARMC). Our algorithm extracts the set of rules, specific to each class, using a fuzzy approach to select the items and does not require the user to provide thresholds. ARMC is experimentaly evaluated and compared to state of the art classification algorithms, namely CBA, PART and RIPPER. Results of experiments on standard UCI benchmarks show that our algorithm outperforms the above mentionned approaches in terms of mean accuracy.

1 Introduction

Association Rule Mining [1] is the most popular data mining task due to its numerous applications and the efficiency of the APRIORI algorithm which produces interesting relationships among items of large databases. Associative classification [8], [7] attempts to solve classical classification problems using association rule mining. Successful algorithms in this category including the well known $C4.5$ predict the class label for novel (unseen) examples .

In this paper, we propose a new Association Rule Mining algorithm for Classification (ARMC) based on the extraction of both common and exception rules, specific to each class. ARMC uses a fuzzy method to automatically select the items that generate the rules and does not require the user to provide thresholds.

ARMC proceeds by combining on one hand, the weighted voting and the decision list algorithms. On the other hand, a fuzzy method is used in order to distinguish the important rules from the less important ones for each class.

The remainder of this paper is structured as follows. In the next section the basic terminology used throughout the paper is introduced. Related research on associative classification is surveyed in section 3. The ARMC algorithm is presented in section 4. Experimental results are given in section 5. We conclude the study in section 6.

2 Preliminaries

2.1 Association Rule Mining

Let B be a database consisting of one table over n items $I = \{a_1, a_2 \cdots, a_n\}$ whose values are nominal and containing k instances.

A database instance d is said to satisfy an item set $X \subseteq \{a_1, a_2, \cdots a_n\}$ if $X \subseteq d$.

An association rule is an implication $X \Rightarrow Y$ where $X, Y \subseteq I$, $Y \neq \emptyset$ and $X \cap Y = \emptyset$.

The support of an item set X is the number of database instances d which satisfy X: $sup(X) = \frac{|\{t \in B| \ X \subseteq t\}|}{|B|}$.

The confidence of an association rule is a percentage value that shows how frequently the consequent part occurs among all the groups containing the rule antecedent part : $conf(X \rightarrow Y) = \frac{|\{t \in B | X \cup Y \subseteq t\}|}{|\{t \in B | X \subseteq t\}|}$.

2.2 Association Rules for Classification

Given a relational table B containing N cases (training examples) belonging to C classes, each one is described by an itemset and let I be the set of all items of B. Classical classification methods seek a rule or a hypothesis that predicts the label of an unseen example.

Table 1. An example of training dataset

Id	Set of items	Class	Id	Set of items	Class
1	x_1, x_4, x_9	A	8	x_1, x_4, x_9	B
2	x_1, x_4, x_{10}	A	9	x_1, x_4, x_9	B
3	x_1, x_4, x_{11}	A	10	x_1, x_5, x_{11}	B
4	x_1, x_5, x_{11}	A	11	x_2, x_6, x_{12}	B
5	x_2, x_6, x_{12}	A	12	x_2, x_6, x_{10}	B
6	x_2, x_7, x_{13}	A	13	x_2, x_6, x_{12}	B
7	x_3, x_8, x_{14}	A	14	x_2, x_7, x_{12}	B
			15	x_3, x_8, x_{13}	B

The class association rule is an implementation of the form $X \rightarrow c$, where $X \subseteq I$, and $c \in C$.

The objectives are to generate the complete set of class association rules that satisfy the minimum support as well as the minimum confidence ($minConf$) constraints and to build a classifier from the class association rule set.

The classifier is applied to classify examples. To this aim, one combines the prediction of all rules which satify the exemple: if there is only one rule, the consequent of this rule is taken to be the predicted class for the example; if there is no rule satisfying the example, then a default class is taken to be the predicted class; and if there are multiple rules satisfying the example, then their predictions must be combined. Various strategies for realizing this are discussed in section §3.3.

3 Methodology

The associative classification algorithm can be divided into two fundamental parts: association rule mining and classification. The mining of association rules is a typical data mining task that works in an unsupervised manner. A major advantage of association rules is that they are theoretically capable of revealing all interesting relationships in a database.

3.1 Current Approaches for Discovering Frequent Items and Rule Generation

Many approaches for frequent items discovery and rule generation have been proposed. In this section, we present some paradigms used in the most important association rules and associative classification methods: type algorithms *Apriori* Agrawal et *al.* [1] discover large frequent itemsets, by making multiple passes over the dataset. In the first pass, the support of individual are counted and then the frequent ones are selected. In each subsequent pass, the algorithm starts with a seed set of frequent items found in the previous pass. This process continues until no new frequent items are found. This algorithm needs to scan several times the dataset in order to produce the frequent itemsets.

The CMAR algorithm [7] adopts some properties of the FP-growth method [5]. FP-growth is a frequent pattern mining algorithm which is faster than conventional *Apriori* method. Whose advantages are: it constructs a highly compact Frequent-Pattern-Tree, which is smaller than dataset and that avoids to scan the dataset in the subsequent mining process. Also, in contrary to FOIL and CPAR methods, it avoids costly candidate generation.

The MMAC [11], [12] algorithms use the technique based on intersection methods [15]. They scan the dataset once to count the support of individual items and determine those having minimum support. They store items along with their location (rowIds) inside arrays. Then, by intersecting the rowIds of the frequent items discovered so far, can easily obtained the remaining frequent items that involve more than one attribute.

3.2 Current Approaches for Extracting Exception Rules

After performing data mining tasks as using association rule methods [6], which usually terminate with a large set of rules, there is a need to find some interesting rules that can be used by the decision maker. In [6], [9], exception rules are defined as the rules that contradict the common belief. They play an important role in making critical decision. The exception rules usually have a small cardinality as a set, and they are not known or omitted.

There are some intuitive ways [6], [9] such as multi-support or generate-and-remove. For discovering exception rules which generate the weak patterns using the traditional data mining techniques. For example, in order to find exception rules, we can introduce two support thresholds $[s_1, s_2]$ as delimiter and we search the itemsets whose support values fall into the range $[s_1, s_2]$.

Generate-and-remove is straightforward method where the rules are generated from the training data T by an induction algorithm as follows: remove T' from T that is covered by R; then generate rules R' on $T - T'$; repeat the process until no data are left. This procedure can find many weak patterns.

3.3 Current Approaches to Classification

In most associative classification methods such as CBA, MMAC, ... we use the classification approach which is called decision list algorithm. This approach takes into account just one rule in order to classify an instance. Therefore the set of class association rules is stored in a list data structure whose first rule covering the instance to be classified is used for prediction.

In CBA [8], the class association rules are stored as follows: given the rules, r_a and r_b, r_a precedes r_b if the confidence of r_a is greater than that of r_b, or if their confidences are identical, but the support of r_a is greater than that of r_b, or both the confidences and support of r_a and r_b are the same but r_a is generated earlier than r_b.

Any associative classification using this approach, has approximately these same definition rules. The resulting classifier has the form: $< r_1, r_2, ..., r_n, default\_class >$ where r_i precedes r_{i+1}. For classifying a new case, the first rule that satisfies the case classifies it. If no rule applies to the case, it takes the default class.

Another approach called the weighted vote algorithm. Uses a set of the best rules of each class for prediction, according to the following procedure: (1) select all the rules whose bodies are satisfied by the instance; (2) from the rules selected in step (1), select the best rules for each class; and (3) compare the measure of the combined effect of selected rules of each class and choose the class with the highest measure.

There are many possible ways to measure the combined effect of a group of rules [13],[7]. For example, one can compute the average of the confidence values of a group of rules or expected accuracy [13] as the predicted class. Another alternative is to use the strongest rule as a representative of the predicted class [7]. In other words, the rule with highest χ^2 value is selected. However, this method is based on a single rule for making predictions.

4 ARMC Algorithm

We present a new algorithm for associative classification, based on multi-rule classification. Our algorithm extracts rules from data subset instead of the whole dataset. Each class of the dataset has its instance subset which consists of the cases of the same class. In this approach, exception rules are discovered by all.

4.1 Item Discovery and Rule Generation

Our approach generates local rules (i.e., the rules are generated from a subset of dataset that is composed of the instances of the same class label for freqent items discovery and rule generation tasks, and employs a technique).

Based on the intersection method [15] in order to scan the dataset once and to compute confidence value of rules from confidence value of frequent single items.

It clusters the training data set T into several subsets. Each subset holds the instances of the same class. We scan each subset once in order to count the occurrence of single items. Each single item has a number of occurences for each class. For a given class, the whole number of items is used in order to build a memberships function to the given class. These memberships sort the single items in two subsets: one rare items subset and another frequent items one. We use fuzzy sets [14] subsets in order to extract the frequent items inherent in a given class.

The support threshold value for some methods is pre-fixed, in order to select the items iteration numbers exceeding this support threshold. Although this procedure is widely used by most methods it has some disadvantages as:

- If the whole number of item iterations are higher or equal than the threshold, then no frequent items are selected.
- Some interesting items are not selected as frequent items though their iteration numbers are close to the threshold.

In order to build membership functions for determination of the frequent items, we use the method introduced in [14]. This method enables the interpretation of the linguistic variable frequency for the term set $\{rare, frequent\}$. Each term is characterized by a fuzzy sets in a universe of frequency values of items for a specific class label.

Before presenting how to build these membership functions for class c_i, some notations have to be specified: use of subsets of a database instead of the entire dataset requires the adaptation of certain functions such as the *support* function which calculates the frequency of appearances of an itemset in the whole data base. The *support* function is adapted to calculate the support of an itemset in a subset of data and named *supportLocal*:

$$suppLocal_{c_i}(X \to c_i) = \frac{|\{t \in B_{c_i}|\ X \subseteq t\}|}{|B_{c_i}|}$$

where : B_{c_i} is a set of instances of the class c_i.

To define these membership functions, we denote the linguistic variable which is characterized by [14]: $(x, T(x), U)$ where

- x is the linguistic variable. In our application x is equal to the linguistic variable frequency.
- $T(x)$ is the set of terms associated with the linguistic value, in which the frequency is represented according to the following set $\{Never, Rare, Frequent, Always\}$.
- U is the universe of discourse and $U = \{s \in I|\ suppLocal_{c_i}(s \to c_i)\}$.

The terms of $T(x)$ are characterized by fuzzy subsets defined by the following membership functions [14]:

Table 2. SupportLocals calculated from training dataset of Table 2

Item	Support	$suppLocal_A$	$suppLocal_B$
x_1	7/15	4/7	3/8
x_2	6/15	2/7	4/8
x_3	2/15	1/7	1/8
x_4	5/15	3/7	2/8
x_5	3/15	1/7	2/8
x_6	3/15	1/7	2/8
x_7	2/15	1/7	1/8
x_8	2/15	1/7	1/8
x_9	2/15	1/7	1/8
x_{10}	3/15	1/7	2/8
x_{11}	3/15	2/7	1/8
x_{12}	4/15	1/7	3/8
x_{13}	2/15	1/7	1/8
x_{14}	1/15	1/7	0

- K: is the set of centroids obtained by the K-means algorithm which is applied to U for $K = \{0, s_{rare}, s_{freq}, 1\}$ (freq = frequent).
- $\mu_{c_i, rare}$: corresponds to the membership function in the linguistic term *rare*. It is built from a set of instance frequencies which belong to the class c_i.

$$\mu_{c_i, rare}(s) = \begin{cases} 1 & \text{if } s \leq s_{rare} \\ 1 - \frac{s - s_{rare}}{s_{freq} - s_{rare}} & \text{if } s_{rare} < s \leq s_{freq} \\ 0 & \text{otherwise.} \end{cases}$$

- $\mu_{c_i, freq}$: corresponds to the membership function in the linguistic term *frequent*. It is built from a set of frequency values of instances which belong to label class c_i.

$$\mu_{c_i, freq}(s) = \begin{cases} 0 & \text{if } s \leq s_{rare} \\ 1 - \frac{s - s_{freq}}{s_{freq} - s_{rare}} & \text{if } s_{rare} < s \leq s_{freq} \\ 1 & \text{otherwise.} \end{cases}$$

where : $s = suppLocal_{c_i}(X \to c_i)$ such as $X \in I$

These membership functions are able to define fuzzy subset of single items. The frequent single items are all the items in which the frequency value of membership degree of the term *frequent* is higher than that of term *rare*, i.e. $\{s \in I | \mu_{c_i, freq}(suppLocal_{c_i}(s \to ci)) > \mu_{c_i, rare}(suppLocal_{c_i}(s \to c_i))\}$. Therefore two subjective thresholds (first support $t_1^{c_i}$ and second support $t_2^{c_i}$) can be computed (Figre 1). The first support $t_1^{c_i}$ is the value that corresponds to the intersection of $\mu_{c_i, freq}$ and $\mu_{c_i, rare}$. The second support $t_2^{c_i}$ is the centroid s_{rare} ($t_2^{c_i} = s_{rare}$).

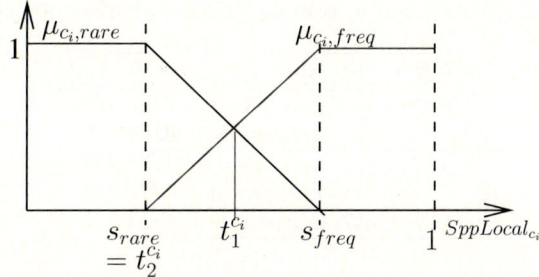

Fig. 1. Membership functions ($\mu_{c_i,freq}$, $\mu_{c_i,rare}$) and the thresholds (t_1, t_2)

According to this method, we define subjective thresholds for a specific class by selecting. All the single items whose local support value exceeds the first threshold. We call these items frequent single items. Then, to produce the rules from these items, the local support value is calculated, and the rules whose local support is greater than the first threshold $t_1^{c_i}$ are considered as common rules. The others are considered as exception rules if their local support exceeds the second threshold $t_2^{c_i}$ and they are stored respectively in common rule set and exception set of this class.

In order to illustrate the advantages of the combination of these approaches to find the exception rules, we present an example: the training dataset of Table 2, represents a set of the instances whose label class is A or B. From this table, we apply our method to produce the rules. In table 3 and 3, there are the rules extracted from the instances whose class label is A and B respectively. In these tables, the rules are grouped by the way either by the first or by the second threshold or by the generate-and-remove.

Double local support thresholds are computed by the method defined above. For the instances whose class label is A, the results obtained are $t_1^A = 0.29$ and to $t_2^A = 0.17$. Using the first threshold t_1^A, we select the set of the single frequent item $\{x_1, x_4\}$, with which, our method produces three rules: $x_1 \rightarrow A$, $x_1 x_4 \rightarrow A$ and $x_4 \rightarrow A$. The last rule is pruned because its included in the second rule and the both of them cover the same instances. Using the second threshold t_2^A, the items x_2 and x_{11} are added to the set of the frequent single item, which are combined to generate many rules, but only rules $x_1 x_{11} \rightarrow A$ and $x_2 \rightarrow A$ are saved. The other rules are pruned as: $x_2 x_4 \rightarrow A$, $x_2 x_{11} \rightarrow A$ because they do not cover any instances. The rule $x_3 x_8 x_{13} \rightarrow A$ is found by the generate-and-remove approach.

For class B, results of threshold computing gives $t_1^B = 0.307$ and $t_2^B = 0.23$ respectively. The items x_1, x_2, x_6, and x_{12} are the frequent single items. The rules generated from these items are those having the Id 1, 2, 3, and 4 of Table 4. Other rules are generated but they are pruned because, either they do not cover any instances or they are included in other rules. the second set of frequent single items are $\{x_{10}, x_4\}$ and the rules generated by adding these new frequent

Table 3. Rules found in training subset instances of class A

Approach	Id	Rule	Sup	SupL
1st threshold t_1^A	1	$x_1 \to A$	2/7	4/7
	2	$x_1 x_4 \to A$	3/14	3/7
2nd threshold t_2^B	3	$x_2 \to A$	1/7	2/7
	4	$x_1 x_{11} \to A$	1/7	2/7
gen-and-rem	5	$x_3 x_8 x_{14} \to A$	1/14	1/7

Table 4. Rules found in training subset instances of class B

Approach	Id	Rule	Sup	SupL
1st threshold t_1^B	1	$x_2 \to B$	4/15	1/2
	2	$x_1 \to B$	3/15	3/8
	3	$x_2 x_6 \to B$	3/15	3/8
	4	$x_2 x_7 \to B$	3/15	3/8
2nd threshold t_2^B	5	$x_1 x_4 \to B$	1/3	1/4
	6	$x_2 x_6 x_{12} \to B$	2/15	1/4
	7	$x_{10} \to B$	2/15	1/4
gen-and-rem	8	$x_3 x_8 x_{13} \to B$	1/15	1/8

items are the rules with Id 5, 7, 8 of Table 4. The rule $x_3 x_8 x_{13} \to B$ is found by the generate-and-remove approach.

This approach is quite effective in terms of runtime and storage because it does not rely on the traditional Apriori approach [2] of discovering frequent items requiring multiple scans. Moreover, it generates more rules than the traditional method and therefore avoids missing important ones.

4.2 Classification

Our approache builds a global model for classification by combining the two approaches presented above (3.3): the weighted vote algorithm and the decision list algorithm. Firstly, for each class c_i, we store a set of rules R_{c_i} in two sets: the set of the *common* rules to class c_i and the set of the *exception* rules to the class c_i.

Given R a rule set containing the rules for each class, R is composed of common and exception rules, and given also the instance, we use the best rules of each class for prediction, with the following procedure:
(1) Select all the rules from R whose bodies are satisfied by the instance;
(2) Compare the average local support value of the *common* rules of each class and choose the class with the highest average as the predicted class. If there are no the common rules for a specific class, the local suport of the best rule from the specific class is selected.

The following function is used for selecting and combining rules as discussed above:

$$Average_{c_i}(e) = \begin{cases} \frac{\sum_{r \in R_{c_i,com}(e)} SuppLocal_{c_i}(r)}{|R_{c_i,com}(e)|} \\ if\ R_{c_i,com}(e) \neq \emptyset. \\ Max_{r \in R_{c_i, \neg com}(e)}(SuppLocal_{c_i}(r)) \\ if\ R_{c_i,com}(e) = \emptyset\ and\ R_{c_i, \neg com}(e) \neq \emptyset. \\ 0\ else. \end{cases}$$

where:

- $R_{c_i,com}(e)$: set of common rules which cover the instance e from class c_i.
- $R_{c_i,\neg com}(e)$: set of exception rules which cover the instance e from class c_i.

We use multiple rules in prediction for the following reasons: the accurate rules cannot be precisely estimated and we cannot expect that any single rule can perfectly predict the class label of every example satisfying its body. Moreover, we use the best rule instead of using all of them because there are different number of rules for different classes and we do not want to use low rank rules in prediction when there are already enough rules to make a prediction.

4.3 ARMC Algorithm

Our algorithm proceeds in two phases: The first phase generates the common and exception rules. The last one builds the classifier.

Before presenting this algorithm, some notations have to be specified:

- B_{c_i} is the set of instances of the class c_i.
- R_{c_i} is the set of rules extracted from B_{c_i}.
- S_f is the *frequent* item set.
- S_r is the *rare* item set.
- $B_{c_i}(r)$ is set of istances covered by the rule r.
- $B_{c_i}(R_{c_i})$: is the set of instances covered by the set of rules R_{c_i}.

ARMC Algorithm
Input: Database B.
Output: The rule set for each class (R_{c_i}).

1. Scan the database B to group the instances according to class labels.
   For each set B_{c_i} :
   (a) Generation of S_f and S_r from B_{c_i}
   (b) Generation of common rules :
   For each $X_1 \subset S_f$ and $X_2 \subset S_f$
   $R_{c_i} \leftarrow\ <X_1 \cup X_2 \rightarrow c_i>$ if $suppLocal_{c_i}(X_1 \cup X_2 \rightarrow c_i) \geq t_1^{c_i}$

(c) Generation of exception rules :
   For each $X_1 \subset S_f$ and $X_2 \subset S_r$
   $R_{c_i} \leftarrow\, <X_1 \cup X_2 \rightarrow c_i>$ if $suppLocal_{c_i}(X_1 \cup X_2 \rightarrow c_i) \geq t_2^{c_i}$
   For each $X_1 \subset S_r$ and $X_2 \subset S_r$
   $R_{c_i} \leftarrow\, <X_1 \cup X_2 \rightarrow c_i>$ and $suppLocal_{c_i}(X_1 \cup X_2 \rightarrow c_i) \geq t_2^{c_i}$
(d) Purning R_{c_i} :
   Remove $X' \rightarrow c_i$ if $B_{c_i}(X' \rightarrow c_i) \subseteq B_{c_i}(X \rightarrow c_i)$ and $X \subseteq X'$
(e) $B_{c_i} \leftarrow B_{c_i} - B_{c_i}(R_{c_i})$
(f) Repeat (a), (c), (d) and (e) until $B_{c_i} = \emptyset$
2. Build the classifier.

In phase 1 of our algorithm, double local support threshold is applied, then the generate-and-remove is applied to discover the remaining rules in order to cover all the training instance by rules. This combination enables us to find more rules than the traditional methods and also to find some rules that can be to be generated by either generate-and-remove approach alone or by double support threshold approach alone.

5 Evaluation

We have evaluated the accuracy, efficiency and scalability of our algorithm. In this section, we report on our experimental results by comparing our algorithm with three popular classification techniques namely RIPPER [3], PART [4] and CBA [8] in order to evaluate the predictive power of the proposed method. As in [11], datasets from UCI Machine Learning Repository [10] are used. 10-fold cross validation is used for every dataset. In our experiments, confidence threshold was set to 3, the datasets selected from [10] were reduced by ignoring their integer and/or real attributes. The accuracy of our algorithm has been computed using the top label evaluation measure. We obtain better results with ARMC.

Table 5 shows the classification average accuracy of the classifiers extracted by RIPPER, PART, CBA, and ARMC on the 26 benchmark problems. Our algorithm outperforms the rule-learning methods in terms of accuracy rate, and the won_loss_tied records of ARMC against RIPPER, PART and CBA are 16_10_0, 15_10_1 and 18_8_0, respectively.

Table 6 compares the number of classification rules discovred by RIPPER, PART, CBA and ARMC algorithms. We can see that on average, ARMC finds many more rules than RIPPER, PART and CBA. The reason why ARMC finds more rules is that it mines rules which cover all the instances of a dataset by the generate_and_remove procedure.

Table 7 shows the classification average accuracy of the different classifiers approach on the 26 benchmark problems.

Table 5. Classification accuracy average of PART, RIPPER and ARMC

Algorithm	RIPPER	PART	CBA	ARMC
Average	83.519	83.923	83.242	85.367

Table 6. Rule number Average of PART, RIPPER and ARMC

Algorithm	RIPPER	PART	CBA	ARMC
Average	6.29	18.58	44.71	47.79

Table 7. Classification average accuracy of ARMC-1, ARMC-Multi, ARMC

Approach	ARMC-1	ARMC-Multi	ARMC
Average	84.91	82.1	85.36

The ARMC-1 denotes the result of the weighted vote algorithm when sed by our method. The ARMC-Multi denotes the result of the decision list algorithm when used by our method.

We conclude that the combination of two approaches (the weighted vote algorithm and the decision list algorithm) outperforms the weighted vote algorithm and the decision list algorithm.

6 Conclusion

A new approach for classification rules has been proposed having different features: (1) Covering all training instances and leaving no unclassified instances (2) requiring only one pass to discover rules (3) using a novel approach for building a classification model. All these features are not offered by the traditional associative classification methods. The first series of experiments on databases in UCI machine learning repository showed that ARMC is robust and highly effective for various classification tasks.

References

1. Agrawal, R., Imielinski, T., Swami, A.: Mining association rules between sets of items in large databases. In: Proc. of the 1993 ACM SIGMOD International Conference on Management of Data SIGMOD 1993, Washington, DC, pp. 207–216 (1993)
2. Agrawal, R., Srikant, R.: Fast algorithms for mining association rules. In: Bocca, J.B., Jarke, M., Zaniolo, C. (eds.) Proc. 20th Int. Conf. Very Large Data Bases, VLDB, pp. 487–499. Morgan Kaufmann, San Francisco (1994)
3. Cohen, W.: Fast effective rule induction. In: Proceedings of the 12th International Conference on Machine Learning, pp. 115–123. Morgan Kaufmann, San Francisco (1995)
4. Frank, E., Witten, I.: Generating accurate rule sets without global optimisation. In: Morgan Kaufmann Madison (ed.)Proceedings of the Fifteenth International Conference on Machine Learning, pp. 144–151 (1998)
5. Han, J., Pei, J., Mortazavi-Asl, B., Chen, Q., Dayal, U., Hsu, M.: Freespan: frequent pattern-projected sequential pattern mining. In: KDD, pp. 355–359 (2000)
6. Hussain, F., Liu, H., Suzuki, E., Lu, H.: Exception rule mining with a relative interestingness measure. In: Pacific-Asia Conference on Knowledge Discovery and Data Mining, pp. 86–97 (2000)

7. Li, W., Han, J., Pei, J.: Cmar: Accurate and efficient classification based on multiple class-association rules. In: ICDM, pp. 369–376 (2001)
8. Liu, B., Hsu, W., Ma, Y.: Integrating classification and association rule mining. In: KDD, pp. 80–86 (1998)
9. Liu, H., Lu, H., Feng, L., Hussain, F.: Efficient search of reliable exceptions. In: Zhong, N., Zhou, L. (eds.) PAKDD 1999. LNCS (LNAI), vol. 1574, pp. 194–203. Springer, Heidelberg (1999)
10. Merz, C.J., Murphy, P.M.: UCI repository of machine learning databases. Department of Information and Computer Science, University of California, Irvine (1996)
11. Thabtah, F.A., Cowling, P.I., Peng, Y.: Mmac: A new multi-class, multi-label associative classification approach. In: ICDM, pp. 217–224 (2004)
12. Thabtah, F.A., Cowling, P.I., Peng, Y.: Multiple labels associative classification. Knowl. Inf. Syst. 9(1), 109–129 (2006)
13. Yin, X., Han, J.: CPAR: Classification based on predictive association rules. In: Proceedings of 2003 SIAM International Conference on Data Mining, San Fransisco, CA (2003)
14. Zadeh, L.: The concept of a linguistic variable and its application to approximate reasoning - ii. Information Sciences (Part 2) 8(4), 301–357 (1975)
15. Zaki, M.J., Parthasarathy, S., Ogihara, M., Li, W.: New algorithms for fast discovery of association rules. Technical report, Rochester, NY, USA (1997)

Comparison of Feature Reduction Methods in the Text Recognition Task

Jerzy Sas[1] and Andrzej Zolnierek[2]

[1] Wroclaw University of Technology, Institute of Applied Informatics, Wyb. Wyspianskiego 27, 50-370 Wroclaw, Poland
jerzy.sas@pwr.wroc.pl
[2] Wroclaw University of Technology, Faculty of Electronics, Chair of Systems and Computer Networks, Wyb. Wyspianskiego 27, 50-370 Wroclaw, Poland
andrzej.zolnierek@pwr.wroc.pl

Abstract. In this paper two level model of handwritten word recognition is considered. On the first level the consecutive letters are recognized by the same classifier using preprocessed data from the optical device, while on the second level we try to recognize the whole word. From the other point of view we can treat the first level as a feature reduction level. Then, in this paper two different methods of feature reduction for handwritten word recognition algorithm are described. On the lower level different well-known in the literature methods are taken into account (for example multi layer perceptron, k-NN algorithm). The results of classification from the first level serve as a feature for the second level and two different cases are considered. The first one consist in taking into account the crisp result of classification from the first level while in the second approach we take into account the support vector of decision on this level. On the second level, in order to improve the word recognition accuracy, for both methods, the Probabilistic Character Level Language Model was applied. In this model, the assumption of first-order Markov dependence in the sequence of characters was made. Moreover, we comment the possibility of using Markov model in forward and backward directions. For both methods of feature reduction the appropriate word recognition algorithms are presented. In order to find the best solution, the Viterbi algorithm is used. A number of experiments were carried out to test the properties of the proposed methods of feature reduction. The experiment results are presented and concluded in the end of the paper.

1 Introduction

Hand printed letters classification ([1], [4], [6]) and consequently text (words) recognition is one of the most important practical application of pattern recognition theory. In this task we have to deal with two different problems of pattern recognition. The first problem consist in letter recognition, while the second one consists in the word recognition. Then we can treat the recognition task as a two-level recognition process. On the first lower level, we have to deal with the vector of preprocessed features obtained from optical device (let us denote it by y), which dimension is very large. Then, we classify the separate letter using certain classifier $\psi(y)$. The results of classification on this level, either in the form of decision about separate character or in the form of support values for all possible classes, are the input of the second level. We can treat this process as a feature reduction procedure for the second level classifier. In the first case the reduction is to one feature, which is the decision on this level $i_n \in \mathcal{C} = \{c_1, c_2, ..., c_L\}$, while in the second case the particular values of the support decision vector on the lower level are considered on the upper level as elements of the feature vector $x_n = [x_n^{(1)}, x_n^{(2)}, ..., x_n^{(L)}]$, where L is the number of letters in the alphabet. The figure 1 presents the scheme of the recognition process for both cases of the input data for the upper level (see either on the left or on the right side of the arrows and consequently use appropriate recognition algorithm for this level).

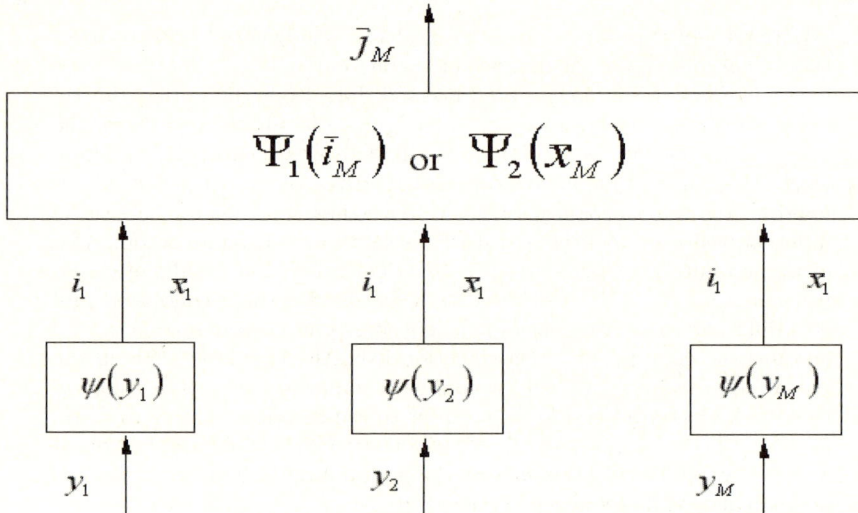

Fig. 1. Two level word classification scheme

Next, on the second (upper) level we try to recognize the whole word using another classifier, in which we take into account the sequence of the results of the first level classifier as well as the probabilistic character language model (PCLM). PCLM defines the conditional probabilities of the characters occurrence provided that the preceding or succeeding character in the word is known. It means that on this level, assuming Markov dependence in the sequence of letters and using hidden Markov model (HMM), the pattern recognition algorithms can be constructed. The details of the application of the first and the second-order Markov chain and consequently HMM for construction of the words recognition algorithm supported by bi-gram and tri-gram character models are presented in ([9],[11]). The word recognition can be done by analyzing the character sequence in forward and backward directions and then by combining the recognition results obtained for the two directions. Experiments described in [10] show that the combined bi-directional algorithm outperforms its components. Because the aim of our work is to test usefulness of different kind of input data to the classical HMM, in this paper we use only the first of the above mentioned probabilistic approaches to the problem of word recognition, i.e. first order Markov model for the forward direction is applied. For the classifiers on the second level, we used two kinds of input data (or equivalently two methods of the feature reduction) in order to compare their influence on the recognition quality. The recognition quality is determined by the average error rate of the pattern recognition algorithms for word classification.

Many other approaches to handwritten word recognition ([4], [7]) make use of the dictionary of words that can appear in the text. The approach presented here does not need the dictionary. Hence, it can be used in the cases, where the set of allowed words is not available or it is so big that its use is impractical for efficiency reasons. The recognizer based on PCLM model constructed as described in this article can be combined with the word recognizer based on the incomplete probabilistic lexicon. Experiments described in [10] prove that such combination results in further improvement of the word recognition accuracy.

The organization of the paper is as follows: after introduction, the problem statement is presented in the section 2. Next, the description and the probabilistic properties of every model of word recognition are presented. In the section 3, the algorithms for word recognition, based on two kinds of input data are described in detail. Then, in section 4, the results of empirical investigation are presented and in the end we conclude this paper.

2 Problem Statement

In this paper we treat the problem of word recognition as two-level classification problem. On the lower (character) level separate letters are being classified and results of classification serve as feature vector on the upper (word) level. Then, we can treat the first level of classification as feature reduction level. The word is hand printed, i.e it is written in block capital letters, enclosed in the separated regions (character fields) located in fixed positions on the printed form. Hence, it is easy to extract isolated images of subsequent characters and the problem of text segmentation is almost trivial and not discussed here. We also assume that the text image analyzer can perfectly distinguish between empty and nonempty character fields, so the length M of the word is fixed before start of recognition procedure. Let us consider the problem of correctly segmented word recognition. The word consists of M characters, each of them belongs to a finite alphabet $\mathcal{C} = \{c_1, c_2, ..., c_L\}$. The word image is segmented into M subimages $(y_1, y_2, ..., y_M)$ containing consecutive characters. On the character level, isolated characters $c_i, i = 1, ..., M$ are recognized independently by the same soft classifier ψ ([3]) which gets consecutive character image y as its input and evaluates the vector of support factors for all characters in the alphabet, i.e.:

$$\psi(y) = [d_1(y), d_2(y), ..., d_L(y)]^T. \qquad (1)$$

Without the loss of generality we can restrict every $d_k(y)$ within the interval $[0,1]$ and additionally achieve that $\sum_k d_k(y) = 1$. Crisp classification $i = \bar{\psi}(y), i \in \mathcal{C}$ can be easily obtained from the support vector by selecting this character, for which the support factor has the greatest value. In this paper two different approaches in preparing the input information for the upper level are considered. These approaches can be treated as two different methods of feature reduction. As the input information for the word level recognition we can use either the support vectors, i.e. results of soft classification obtained by classifier ψ for particular characters constituting the word $(\psi(y_1), \psi(y_2), ..., \psi(y_M))$ or alternately their results of crisp character classification $(i_1, i_2, ..., i_M)$. In the first case the dimension of the vector of original data y is reduced to L while in the second case to one. As it was mentioned above on the word level, i.e. on the upper level we applied HMM in which the state (letter in the word) can not be observed directly, but we obtain the results of classification of the lower level, which are now the features of the vector of observation. In such case, we need the probabilistic characteristic of a random vector, observed value of which is the feature vector on the upper level given true classification. In the second case the problem is not so complicated from computational point of view because we need the information about the confusion matrix of character recognizer $E_{L \times L}$. In this matrix, every element $e_{i,j} = p(\bar{\psi}(y_n) = c_i \mid c_j)$ is a conditional probability that decision of character recognizer is c_i while the actual character on the image y_n is c_j. For the first case we need to assume that for every $n = 1, \ldots, M$ we have the following equation:

$$x_n = [x_n^{(1)}, x_n^{(2)}, ..., x_n^{(L)}]^T = [d_1(y_n), d_2(y), ..., d_L(y)]^T. \tag{2}$$

Moreover, we assume that consecutive vectors x_n, $n = 1, ..., M$ are observed values of random vector X_n and there exist the conditional density functions $f(x_n \mid j_n)$ of random vector X_n given that $J_n = j_n, j_n \in \mathcal{C}$. Then, for the word recognition level our input information is either the sequence of independent decisions of lower level $\bar{i}_n = (i_1, i_2, ..., i_n)$ or the sequence of observed values $\bar{x}_n = (x_1, x_2, ..., x_n)$ of the sequence independent random vectors $\bar{X}_n$.

In order to use the methodology of HMM we need also to adopt the probabilistic character language model. In this paper we take into account two different PCLMs, which consist of:

- probability distribution of leading and trailing characters (i.e. characters beginning and ending words) $P_I = (p_1^I, p_2^I, ..., p_L^I)$ and $P^E = (p_1^E, p_2^E, ..., p_L^E)$, and corresponding set of probabilities:
- set A_{LxL}^S of character succession conditional probabilities, where $a_{i,i-1}^S = p_S(c_i \mid c_{i-1})$ is the probability that the next character in a word is c_i provided that the preceding character is c_{i-1},
- set A_{LxL}^P of character precedence conditional probabilities, where $a_{i,i-1}^P = p_P(c_{i-1} \mid c_i)$ is the probability that the preceding character in a word is c_{i-1} provided that the next characters is c_i.

The probability distribution of leading and trailing characters and sets of succession and precedence conditional probabilities can be obtained by simple analysis of the sufficiently big text corpus. It can be assumed that character succession and precedence probability distributions are common properties of the language in which recognized texts are written and they are domain invariant. Hence the texts in the corpus does not have to be from the area of interest. Large corpora of general texts in the language can be used to estimate sets A^S, A^P and probabilities: P_I, P_E respectively. It should be noted that in this way the need of the domain-specific dictionary typically used in handwriting recognition can be avoided.

3 Algorithms of Word Classification

On the upper level we have to deal with the classical sequential compound decision problem, in which there are dependences in the sequence of recognized patterns. Usually, the Markov dependence (first or higher order) in the sequence of classes to which these patterns belong is assumed. Generally, two attempts to this problem are possible. In the first one, we have to classify the current pattern using observations known at this moment, while in the second one we can wait for the whole sequence of observations and after then we classify the whole sequence of patterns. The first case is typical for example for medical diagnosis, but in the word recognition problem we can wait for the whole information about all characters in the word and after then we classify it. Using Bayes approach to the second case of sequential compound decision problem and assuming the first-order Markov dependence (forward or backward) our aim is to find such sequence of decisions $\bar{j}_M = j_1, j_2, ..., j_M$ for which the corresponding posterior probability is maximum. As it was presented in the problem statement we consider two kinds of input data, treated as realizations of corresponding random variables:

- the whole sequence of crisp decisions from the lower level $\bar{i}_M = (i_1, i_2, ..., i_M)$ or
- the whole sequence of observed values of their soft support vectors $\bar{x}_M = (x_1, x_2, ..., x_M)$.

Algorithm with crisp decisions on the lower level

We have to find the sequence of decisions $\bar{j}_M$ for which the posterior probability: $p(\bar{j}_M \mid \bar{i}_M)$ is maximal. Taking into account the first-order Markov dependence in the sequence of classes and analyzing only the numerator of this probability the pattern recognition algorithm on the upper level $\bar{\Psi}_1(\bar{i}_M)$ consists in finding the following sequence of letters:

$$\bar{j}_M^* = \max_{\bar{j}_M} (\prod_{\alpha=1}^{M} e_{i_\alpha, j_\alpha} \cdot P_I \cdot \prod_{\beta=2}^{M} p_S(j_\beta \mid j_{\beta-1})) \qquad (3)$$

Algorithm with soft decisions on the lower level

We have to find the sequence of decisions $\bar{j}_M$ for which the posterior probability: $p(\bar{j}_M \mid \bar{x}_M)$ is maximal. By taking into account the first-order Markov dependence in the sequence of classes and by analyzing only the numerator of this probability, we can define the pattern recognition algorithm on the upper level $\bar{\Psi}_2(\bar{x}_M)$ in such way that it consists in finding the following sequence of letters:

$$\bar{j}_M^* = \max_{\bar{j}_M} (\prod_{\alpha=1}^{M} f(x_\alpha \mid j_\alpha) \cdot P_I \cdot \prod_{\beta=2}^{M} p_S(j_\beta \mid j_{\beta-1})) \qquad (4)$$

Analyzing both problems of finding the best sequence of decisions concerning the successive characters constituting the words we can see that we have to deal with classical problem of HMM in which very efficient Viterbi procedure can be applied. The theoretical and technical details of this procedure are very well described in the literature and also can be found in our previous papers ([9], [11]), then it will not be presented here. In the next section we present experimental results obtained using this algorithm.

The analogous method can be applied to the word recognition carried out in reversed order. The procedure remains the same but the character precedence probabilities from PCLM should be used in place of succession probabilities. Also second order HMMs can be used for both forward and backward directions.

4 Experiments

In order to test the properties of the proposed approaches to the word recognition problem, series of experiments have been conducted. As the training set, a collection of images of hand printed capital letters from the Polish alphabet was used. The sample character images were extracted from the form prepared appropriately for the automatic acquisition of the training set. The content of the form was arranged so as to simplify the procedure of particular letter localization on the form. Individual alphabet letters were located in distinct block areas labeled with the requested letter. The writers filling the form were asked to write isolated character samples in the labeled areas on the form. The forms were then automatically scanned and the position of the form image was normalized so as to obtain the localization of the character fields in precisely defined areas of the image. Next the individual character samples were extracted from fixed areas of the normalized image and stored in image files. The training set consisted of the letter samples from 150 writers. Totally 5250 character samples were collected in this way.

Unfortunately, we were not able to collect sufficiently large set of hand printed words which could be used as the testing set. Therefore the word images constituting the testing set were created artificially from the individual characters coming from the test forms prepared by other 31 writers. The sets of writers whose texts were used in training and testing sets were disjoint. The word images for testing purposes were created by concatenating the character samples so as to create the words randomly drawn from the text corpus. Because the work being described here is the part of a wider project related to handwritten medical documents recognition, the text corpus was the collection of patient records stored in the medical information system. The corpus consisted of almost 16000 notes. Each note typically consists of several sentences. 3600 notes were used as the testing set while remaining ones were applied to estimate the probabilities in PCLM.

The training set was next divided into two subsets. The first subset was used to train the character classifier. The second one was applied to estimate the probabilities in the confusion matrix E or to estimate the needed values of pdf's $f(x_n \mid j_n)$.

Various character classifiers following the soft recognition paradigm were used at the character level. The best results were obtained with the classifier based on neural network (MLP). The accuracy of the classification interpreted as a crisp recognition for this classifier was 93.8%. Slightly worse results were obtained with k-NN algorithm (KNN). KNN classifier accuracy was 92.6% In order to evaluate the proposed methods in the case of low performance character classifiers we used also the classifier based on unconstrained elastic matching concept, which was trained using the training subsets of varying cardinality. In effect, three classifiers were obtained giving the accuracies: 88.4%, 80.3% and 57.3%. They will be referenced as EM_88, EM_80 and EM_57 correspondingly. All used classifiers were constructed so as to provide the support vector (1). The details of support values calculation for used classifier are presented in [8]. For EM and MLP classifiers 200-element directional features vector was extracted from the character sample image. The method of the directional feature extraction is described in [5] and [8]. The classifier based on unconstrained elastic matching used directly the binarized raster image pixel array in order to determine the similarity measures between images.

For pdf's estimation, nonparametric Parzen estimator was used ([2]). The Gaussian multivariate function was used as the Parzen window:

$$\phi(x, x_k) = \frac{1}{(\sqrt{2\pi}\sigma)^L} e^{\left(-\frac{(x-x_k)^T(x-x_k)}{2\sigma^2}\right)}, \qquad (5)$$

where x_k is the feature vector of the k-th sample in the training set for particular class and σ determines the Parzen window width. The value of σ depends on the number of samples used to estimate the probability density function. According to suggestions in [2] we applied the formula $\sigma = w_0 \sqrt{N_j}$ where N_j is the number of samples for j-th character in the alphabet and w_0 was experimentally set to 0.7.

Finally, for each character j in the alphabet its conditional probability density function $f(x \mid j)$ in tested point x is estimated as:

$$f(x \mid j) = \sum_{k=1}^{N_j} \phi(x, x_k). \qquad (6)$$

The number of samples in the set used for matrix E or pdf $f(x \mid j)$ estimation is 1250 for MLP, KMM and EM_88 classifiers. For EM_80 and EM_57 classifiers less training samples were used to train the character classifier and in result more samples left for the estimation procedure. 2600 and 4000 samples were used in the case of EM_80 and EM_57 classifiers correspondingly.

Isolated word recognition results are subject to the further text recognition stages, where n-gram language model or the language grammar is used in order to improve the text recognition accuracy. For this reason, the word classifier is expected to provide not the single most probable word but rather the set of the most likely words with corresponding likelihood measures. Therefore in the experiment carried out we evaluated the tested algorithm accuracies by considering as a success such word recognition result, where the actual word is among the K character sequences evaluated by the word recognizer as the most likely ones. The experimental results presented here were obtained for $K = 1, 3, 5, 10$. The word recognition accuracies achieved with the first order HMM and the result of character crisp recognition as scalar discrete feature used in HMM are given in Tab. 1. Analogous results obtained if the feature is the support vector provided by the soft character classifier are shown in Tab. 2

Table 1. Word recognition accuracy with the crisp character recognition result used as the feature in the upper level HMM

Top K	MLP	KNN	EN_88	EN_80	EN_57
K=1	82.1%	79.6%	71.0%	50.9%	27.9%
K=3	95.0%	94.4%	88.8%	71.2%	42.4%
K=5	97.5%	97.1%	92.9%	78.1%	48.9%
K=10	99.0%	98.9%	96.5%	84.7%	57.7%

Table 2. Word recognition accuracy with the support vector as the feature in the upper level HMM

Top K	MLP	KNN	EN_88	EN_80	EN_57
K=1	71.4%	73.4%	63.2%	57.5%	39.3%
K=3	75.4%	85.5%	76.9%	73.7%	56.2%
K=5	76.4%	89.0%	80.7%	79.3%	64.5%
K=10	77.5%	91.3%	84.9%	85.2%	67.9%

It can be observed that the method using the support vector as the observation in HMM results in higher recognition accuracy in cases where the number of samples used to estimate the pdf's is sufficiently big, i.e. in the case of EM_80 and EM_57 classifiers. In cases where the number of samples used to estimate the density functions is small, the approach based on crisp character recognition and estimation of confusion matrix is more accurate.

5 Conclusions, Further Works

In this paper, two variants of the handwritten text recognition method are presented and compared. The recognition is a two-level process, where on the lower level individual characters are recognized. In the upper level results of the character recognition and the probabilistic properties of character succession in the language are utilized. From the other point of view we can consider our approach as two methods of feature reduction for the algorithm of word recognition. Then, the novelty of this approach consists in fact that the results of the recognition on the lower level are used as the observed features of the sequential pattern recognition process on the upper level. The sequential recognition is described by hidden Markov model. In the first compared variant the index of the class (character) recognized by the crisp classifier on the lower level is used as the observation in HMM. In the alternate variant the support vector evaluated by the soft character recognizer is used as the multivariate observation.

For the character classification the soft recognizer is used. The process of conversion of the elaborated support vector to the index of the most likely class leads evidently to the loss of information. One can therefore expect that higher recognition accuracy on the upper level can be obtained if the complete support vector is used. The conducted experiments show however that it is true only in the case if sufficiently big set of samples can be used in the nonparametric pdf's estimation. In practice, the proposed method can be mainly utilized in writer-dependent handwriting recognition, where the number of correctly labeled character samples that can be used to train the recognizer is limited. Hence, the method based on crisp recognition and character recognizer confusion matrix estimation seems to be of greater practical value.

The experiments described here dealt with the case of the recognition of the word images that can be reliably segmented into the images of isolated characters. In cursive handwriting, the segmentation into characters is unreliable and typically a number of segmentation variants is created for each word image. Then the segmentation variants are recognized independently. Finally such variant and corresponding recognition is selected, for which the reliability assessment is highest. The method presented here can be also applied to the recognition of cursive handwriting. In such case the two level soft recognizer can be applied to each of the segmentation variants.

In order to investigate proposed methods in more details it should be tested in what proportions the training samples set should be divided into the part used to train the lower level character classifier and to the part used to estimate the confusion matrix or pdf's. Similar experiments should be also carried out for second order forward and backward HMM as well as for combined bi-directional model as described in [9].

References

1. Brakensiek, A., Rottland, J., Kosmala, A., Rigoll, G.: Off-Line Handwriting Recognition Using Various Hybrid Modeling Techniques and Character N-Grams. In: Proc. of the Seventh Int. Workshop on Frontiers in Handwriting Recognition, pp. 343–352 (2000)
2. Duda, R.O., Hart, P.E., Stork, D.G.: Pattern Classification, 2nd edn. Combining Classifiers: Soft Computing Solutions. John Wiley & Sons, Chichester
3. Kuncheva, L.: Combining Classifiers: Soft Computing Solutions. In: Pal, S., Pal, A. (eds.) Pattern Recognition: from Classical to Modern Approaches, pp. 427–451. World Scientific, Singapore (2001)
4. Marti, U.V., Bunke, H.: Using a Statistical Language Model to Improve the Performance of an HMM-Based Cursive Handwritting Recognition System. Int. Journ. of Pattern Recognition and Artificial Intelligence 15, 65–90 (2001)
5. Liu, C., Nakashima, K., Sako, H.: Handwritten Digit Recognition: Benchmarking of State-of-the-Art Techniques. Pattern Recognition 36, 2271–2285 (2003)
6. l-Nasan, A., Nagy, G., Veeramachaneni, S.: Handwriting recognition using position sensitive n-gram matching. In: Proc. 7th Int. Conf. on Document Analysis and Recognition, pp. 577–582 (2003)
7. Vinciarelli, A., Bengio, S., Bunke, H.: Offline Recognition of Unconstrained Handwritten Text Using HMMs and Statistical Language Models. IEEE Trans. on PAMI 26, 709–720 (2004)
8. Sas, J., Luzyna, M.: Combining Character Classifier Using Member Classifiers Assessment. In: Proc. of 5th Int. Conf. on Intelligent Systems Design and Applications, ISDA 2005, pp. 400–405. IEEE Press, Los Alamitos (2005)
9. Sas, J.: Application of Bidirectional Probabilistic Character Language Model in Handwritten Words Recognition. In: Corchado, E.S., Yin, H., Botti, V., Fyfe, C. (eds.) IDEAL 2006. LNCS, vol. 4224, pp. 679–687. Springer, Heidelberg (2006)
10. Sas, J.: Handwritten Text Recognition Using Incomplete Probabilistic Lexicon and Character Language Model. Systems Science 32(2), 45–61 (2006)
11. Sas, J., Zolnierek, A.: Handwritten Word recognition with Combined Classifier Based on Tri-grams. In: Kurzynski, M., et al. (eds.) Advances in Soft Computing. Computer Recognition Systems 2, vol. 45, pp. 477–484. Springer, Heidelberg (2007)

Part V

Image Analysis, Speech and Robotics

Part V

Image Analysis, Speech and Robotics

Robot Simulation of Sensory Integration Dysfunction in Autism with Dynamic Neural Fields Model

Winai Chonnaparamutt and Emilia I. Barakova

Eindhoven University of Technology P.O. Box 513
5600MB Eindhoven, The Netherlands
w.chonaparamutt@tue.nl, e.i.barakova@tue.nl

Abstract. This paper applies dynamic neural fields model [1,23,7] to multimodal interaction of sensory cues obtained from a mobile robot, and shows the impact of different temporal aspects of the integration to the precision of movements. We speculate that temporally uncoordinated sensory integration might be a reason for the poor motor skills of patients with autism. Accordingly, we make a simulation of orientation behavior and suggest that the results can be generalized for grasping and other movements that are performed in three dimensional space. Our experiments show that impact of temporal aspects of sensory integration on the precision of movement are concordant with behavioral studies of sensory integration dysfunction and of autism. Our simulation and the robot experiment may suggest ideas for understanding and training the motor skills of patients with sensory integration dysfunction, and autistic patients in particular, and are aimed to help design of games for behavioral training of autistic children.

1 Introduction

Movement disturbance symptoms in individuals with autism have not been considered as an important symptom for a long time. During the last decade, Leary and Hill [17] have offered a radical perspective on this subject. After thorough analysis of the bibliography on movement impairments in autism they argue that motor disorder symptoms may have a significant impact on the core characteristics of autism.

Imprecise grasping, or other motor or executive dysfunctions, observed by autistic patients are caused by a disturbance in a dynamic mechanism that involves multisensory processing and integration. Therefore we investigate how the dynamic aspects of integration of multisensory input influences forming of coherent percept, planning, and coordination of action.

Temporal multisensory integration has previously been discussed in the context of autism in [4,13], in attempts to revile and simulate the underlying biological mechanism of interaction in [10,11] and in the robotics setting in [2,25], and implicitly in many other robotics studies.

Proper modeling of the temporal integration mechanism requires a dynamic neural model. The main stream connectionist methods, like self-organizing or supervised feed-forward networks and Hopfield type recurrent networks produce static outputs, because their internal dynamics lacks feedback loops and their input space is static. Therefore they are suitable for modeling static behaviors. We are interested in a neural system that can spontaneously exhibit several dynamic behaviors, derived by the interaction between changing input and complex inner dynamics. However, for the sake of controllability and computational expense, we choose the model with least complexity needed. Schoner and colleagues [23,25,8] have adapted the dynamic neural field model of Amari [1] for controlling mobile robots and robot-manipulators. It produces smooth behavioral trajectories satisfying more than one external variable. In this model the attractor is a fixed point, but continuous attractor is approximated in sequential steps. The system goes from one attractor to the other through input-dependent variations. More complex dynamic models that have continuous attractors may suffer high computational expense.

This paper is organized as follows. Section 2 discusses the method for sensory integration used, the experiment design, and the results of a computer simulation. In Section 3, the results of robot simulation are shown. Discussion is offered in Section 4. Last, but not least, we acknowledge the people who have contributed to this work.

2 Temporal Multisensory Integration

2.1 Dynamic Neural Field Model for Multisensory Integration

The dynamic neural fields (DNF) model has been proposed as a simplified mathematical model for neural processing [1,7]. The main characteristics of this model are its inherent properties for stimulus enhancement, cooperative, and competitive interactions within and across stimuli-response representations.

Recently Erlhagen and Schoener [8] formalized the extension of the theoretical model to dynamic field theory of motor programming, explaining the way it was and could be used for robotics and behavioral modeling applications. Before and since, DNF model has been used in robotics for navigation and manipulation of objects [9,25,14], for multimodal integration [22] and imitation [21]. Applications feature biologically convincing methods that can optimize more than one behavioral goal, contradicting sensory information, or sensory-motor task that requires common representation. For instance, Iossifidis and Steinhage [14] applied the dynamic neural fields to control the end-effector's position of a redundant robot arm. Two problems were solved by this implementation: smooth end-effector trajectory is generated and obstacles are avoided. Faubel and Schoener [9] use dynamic neural fields to represent the low-level features of the object such as color, shape, and size. The fast object recognition achieved is beneficial for an interaction with a human user. Thelen et al. [26] have modeled the dynamics of the movement planning by integrating the visual input and motor memory to generate the decision for the direction of reaching.

The mathematical description of the DNF model incorporates the formation of patterns of excitation, their interaction, and their response to input stimuli. The basic equation of one dimensional homogeneous field of lateral-inhibition can be represented in the following way:

$$\tau\frac{\partial u(x,t)}{\partial t} = -u(x,t) + \int w(x-y)f[u(y)]dt + h + s(x,t) \qquad (1)$$

where

- τ is the time constant for dynamics of a neuron
- x and y is the located positions of neurons
- u is the average membrane potential of neurons located a position x at time t
- h is the resting potential
- s(x,t) is the input stimulation level at position x at time t.

An interesting for us feature of the model is that it possesses dynamical properties useful for multisensory and sensory-motor integration. We suggest that the dynamical characteristics of the model can be exploited for investigating the temporal aspects of multimodal integration. The temporal window for integration is shown to have an impact on the multisensory interaction, so we investigate the possibilities for its adaptation within the neural field model and its impact on the computational outcomes. The presentation of the sensory cues within the DNF model is in the form of Gaussian distributions. We tune the variance of these distributions according to the experimental findings, and experiment with the delay in the presentation of each cue in accordance with the realistic times of sensory processing of different modalities, and of course, following the restrictions of the experimental platform.

2.2 Experimental Setting

We intend to test the temporal aspects of multimodal interaction by grasping. Since at present we have available only a mobile robot the experiments are restricted to a two-dimensional task of reaching a target. Based on earlier findings [2] two complementary sensory cues are sufficient and necessary for reaching, as well as for precision grip of the robot. An example for complementary sensory cues are proprioception and vision. In this application, the proprioceptive or self-motion information is the angular deviation of the head direction of the robot from the initial position. Vision data are used for spotting the landmark or goal direction.

The heading direction is defined by the output potential that is generated after the integration of both cues. The robot will typically find a compromise between target direction and free of obstacles space. The DNF model would supply a smooth solution of this problem, once the model parameters are tuned for the particular application. For tuning of the parameters a computer simulation is used. One of the reasons to choose for the dynamic neural fields model for sensory interaction is that it uses a window of time to combine all sensory stimuli and make a decision accordingly. Experimental studies of sensory integration propose that there is a window of time during which the stimuli are integrated for producing a perception of a unitary sensory event [12,19,5,16,20,24]. We intended to vary the size of the temporal window of the dynamic neural field model to find out whether there is an optimal window for integration for the particular sensory cues. Since the window of integration corresponds most closely to the time constant of the neural field model, its change will produce a linear dependence. At this stage we did not found a reason to change it to more complex (nonlinear) function: Instead, the window is defined by the necessary processing time for the visual and the proprioceptive cues and from the guidelines from experimental studies.

Our hypothesis is that the delay in activation caused by each of the sensory cues may cause or contribute to imprecise motor behavior. With the following experiment we are going to test the impact of the delay in the activation caused by each of the sensory modalities. We experiment with different delay intervals.

Each cue was delayed with different time interval when a goal finding task was performed. In figure 1 are shown the response times for movement direction. The visual cue delay causes longer response time.

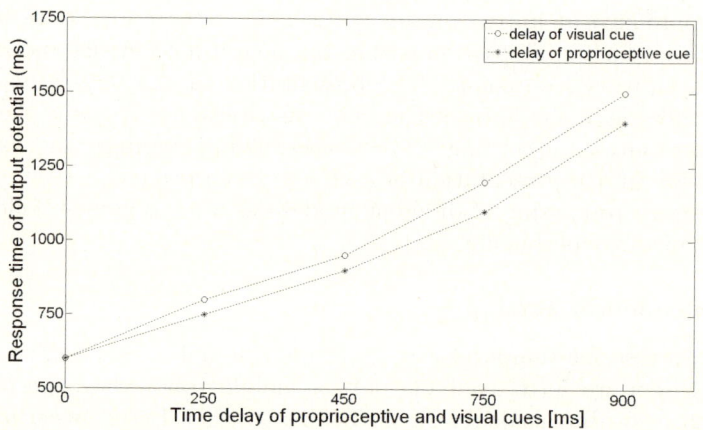

Fig. 1. The time to generate the output potential when each cue is delayed

To get further information on the delay effect of each cue, the experiment of changing heading direction for three successive steps was carried out.

Several tests with a simulated robot that performs target following task were made. In each test the target was moved so that the heading direction of the robot changes with different angles. Figure 2 depicts trajectories with a change of the heading direction correspondingly with 5 - 15 - 25 and 15 - 30 - 50 degrees. Figure 2-top shows the output potential of the second trajectory, and Figure 2-bottom shows the two trajectories in polar coordinates. Polar coordinates representation was chosen, because it corresponds to the actual movement of the robot, from its egocentric perspective. Several experiments were made to compare the effect of changing heading direction by different change of the heading direction when there was no delay, when there was a delay in the proprioceptive cue, and when there was a delay in the visual cue. Delay of the proprioceptive cue has less effect for generating the new heading direction in all experiments. The experiments differed in the sharpness of change in the heading direction, when both cues have the same period of delay, and the neural field parameters are constant for both cues.

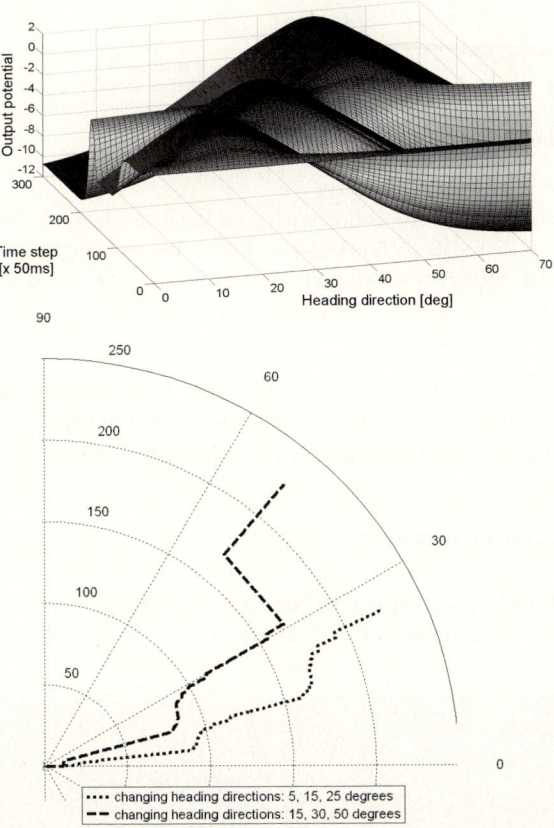

Fig. 2. Top:The output potential with HD changing with 15 - 30 - 50 degrees, bottom: the trajectories of the robot in polar coordinates with heading direction changing correspondingly with 5 - 15 - 25 , and 15 - 30 - 50 degrees

Experimental data from [27] show that although this is true in general,the precision of movements is determined differently by the visual and proprioceptive cues for depth and azimuth motion. Proprioceptive cue is more precise when the depth (distant goal) is targeted, and vision is more accurate in proximal (moment to moment) movements. To simulate this effect, the weight parameters of the neural field model were tuned to correspond to the variances for movement accuracy as found by Van Beers at al. [27]. Figure 3 shows the change of heading direction of the robot with tuned weight parameters of the neural field model

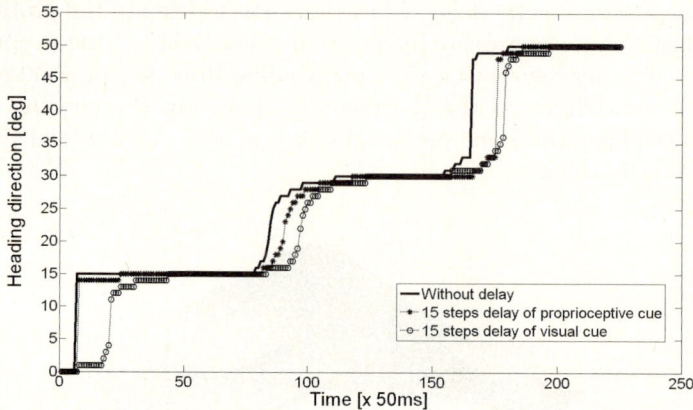

Fig. 3. Heading direction of the robot with and without delay by changing the target direction from 0 to 15 to 30 to 50 degrees. The 3 lines depict the change of heading direction by sensory integration without delays in the cues, and with delay with 15 steps of each cue.

in the cases of no delay, delay of the visual cue, and delay of the proprioceptive cue. Since this result is in consensus with the experimental studies [27,15,3], we intend to use it for grasping behaviors in robot for training of autistic children.

3 Robot Experiment

For the robot simulation an e-puck robot was used. The e-puck is a two-wheel mobile robot that was originally developed at Swiss Federal Institute of

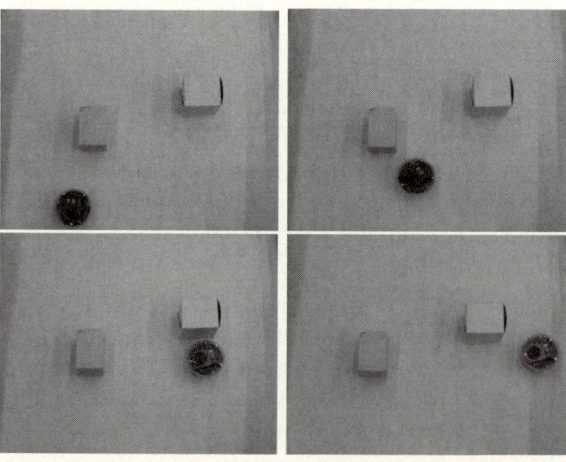

Fig. 4. The example path of the robot in an unstructured environment is shown in the sequence of pictures from left to right and top to bottom

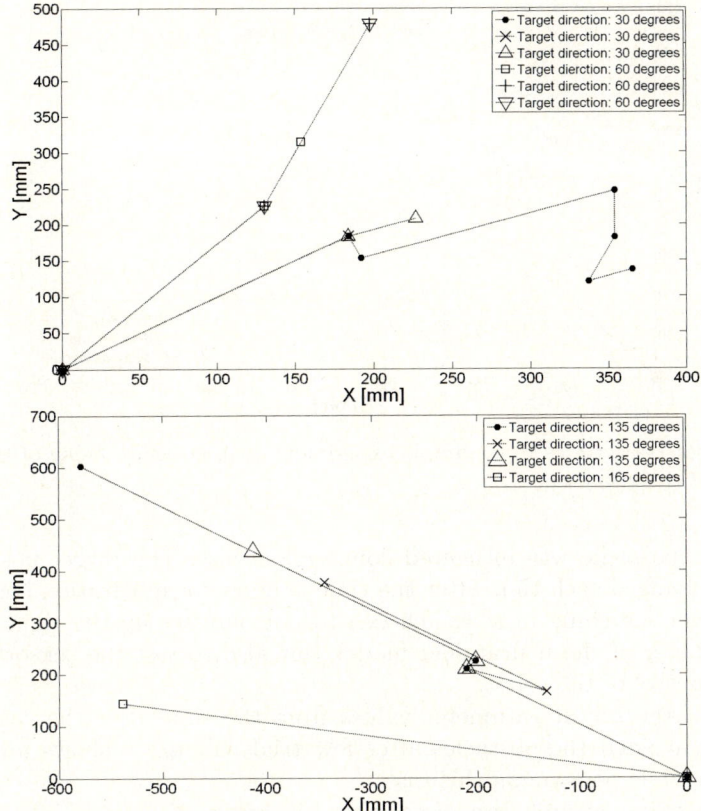

Fig. 5. Trajectories with target directions of: top - 30 and 60 degrees, bottom - 135 and 165 degrees, used to validate the model on a physical robot

Technology (EPFL) [6]. The robot is equipped with dsPIC processor. It is equipped with infrared sensors (IRs) that were used to derive the information about the turning angle of the robot. The free from obstacle space determined the possible direction of the robot for the next moment to moment movement. Vision was used to determine the target direction of the robot.

The experiment was divided into two parts: model validation and the hypothesis testing. We need to validate the model on the real robot because the DNF model parameters might differ by computer simulation and robot experiment.

3.1 Model Verification

To validate the model a task of searching for a randomly changing target was designed. The polygon shaped arena contained several objects that served as obstacles (figure 4). The heading direction was detected with respect to the initial position of the robot: the zero degree direction was chosen to be at the positive

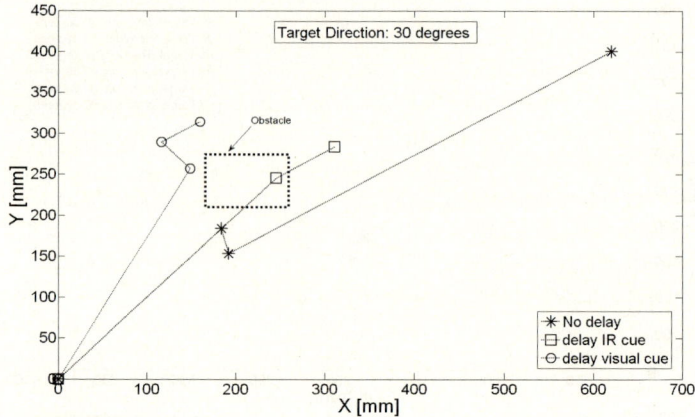

Fig. 6. Robot trajectories from sample experiment with no delay, delay of the IR cue, and delay of the visual cue

x-axis and the angle was measured counterclockwise. The target position was randomly changed each time after the time window for integration has passed.

The target searching task was chosen to not only set up the right values of each parameter of the neural field model, but also to test the sensors and the low-level control of the robot.

Based on the initial parameter values from the simulation,the robot found target and avoided the obstacles after few trials. Figure 4 shows an example robot path while performing this task.
The results from a sample test are depicted in Figure 5.

These results show movement with no added delay for the sensory cues. The time window was defined to be 550ms, since the sensory processing for this robot and control program is that large at present. The model is quite robust because e-puck can find any target direction. Moreover, the results in each target direction are similar.

The influence of each sensory cue on the output potential is tested after the experimental scenario was simplified by using only one obstacle in the arena, as shown in Figure 6. With this simplification the influence of any artifact on the outcome of the experiment is excluded. Without delay in the sensory cues, the robot can avoid the obstacle and reach the target direction in (right upper corner in Figure 6).

When the delay was added to the infrared sensor or to the visual input, the robot took different trajectories. Depending on the distance of the obstacle and the speed of the robot, changing the delays had different effects on it. Figure 6 shows three sample trajectories of the robot: without delay, with delay of the IR sensory cue, and with delay of the visual cue, respectively.

The delay of IR cue resulted in a collision between the robot and the obstacle. When the visual cue was delayed, e-puck started to move in arbitrary free direction until the visual input was received, without hitting to the obstacle.

4 Discussion

We investigated different temporal aspects of multisensory integration on the motor behavior of a robot, namely the effect of the size of the integration window and the delay of different sensory cues. DNF model was used to guide the robot movements, because it contains two parameters which, as we have shown, can simulate the effects of both temporal parameters: the influence of the interaction window and the delay in the sensory cues. The interaction window simulates the time for relaxation of system dynamics to the next fixed point, i.e. it isolates moment to moment multisensory integration. Using this property, we can delay each of the sensory cues and keep the integration within the span of the behavioral window.

The results show that delay of the proprioceptive cue has less effect on close interactions, while visual cue will have less impact on distant target finding.

The DNF model requires a certain time to generate output potential. When there were three successive changing directions the outputs were different when adding the same period of delay to each cue, single cue per trial. Implementing the model on physical robot showed that sensory integration with DNF model provides realistic behavior except for the length of the sensory integration window that has to be tuned according to the restrictions of processing capacity of this robot. The DNF model insures human-like decision making and smooth motion when different external stimuli are present. However, the unreliable sensory information can result in totally different behavioral solutions when the robot started from the same starting point in the same arena. Unrepeatable behavior may be caused by detection failure of the sensors or imprecise tuning of the parameters of the DNF model. The infrared sensors of e-puck robot are sometimes too sensitive to detect the obstacles in the environment, or sometimes they cannot detect anything when the robot stays too close to the obstacle. This results either in robot departing from the natural path or in a collision with an obstacle. To fulfill our ambition of simulating the sensory integration process by autistic people, we would need a more advanced platform, and we are in process of purchasing such. However, the obtained results with the current restrictions are very promising.

Our initial hypothesis was that a bad timing in sensory integration causes poor motor performance in children with autism. Masterton and Biederman [18] have shown that children with autism relied on proprioceptive feedback over visual feedback to modulate goal-directed motor actions, including reaching and placing objects under conditions that required adaptation to the displacement of a visual field by prisms. This finding might be indicative of a perceptual deficit resulting in poor visual control and visual sequential processing [18]. Leary and Hill [17] argued that motor deficits of autism can be not peripheral, but central to the development of children with autism and to have significant impact on the development of higher cognitive atypical behaviors that include unusual sensory or motor behaviors, in addition to social and communicative differences. We aim to extend our integration model to robot simulation of behavior of typically developing and autistic individuals, and use it for behavioral training of autistic children.

Acknowledgments. Our thanks to the Institut fr Neuroinformatik Ruhr-Universitt-Bochum for giving a high proficient discussion about the dynamic neural field theory, in particularly Christian Faubel and Dr. Ioannis Iossifidis.

References

1. Amari, S.: Dynamics of pattern formation in lateral-inhibition type neural fields. Biol. Cybern. 27(2), 77–87 (1977)
2. Barakova, E.I., Lourens, T.: Event based self-supervised temporal integration for multimodal sensor data. J. Integr. Neurosci. 4(2), 265–282 (2005)
3. Bays, P.M., Wolpert, D.M.: Computational principles of sensorimotor control that minimize uncertainty and variability. J. Physiol. 578(Pt 2), 387–396 (2007)
4. Brock, J., Brown, C.C., Boucher, J., Rippon, G.: The temporal binding deficit hypothesis of autism. Dev. Psychopathol. 14(2), 209–224 (2002)
5. Calvert, G.A., Thesen, T.: Multisensory integration: methodological approaches and emerging principles in the human brain. J. Physiol. Paris 98(1-3), 191–205 (2004)
6. EPFL education robot, http://www.e-puck.org
7. Erlhagen, W., Bicho, E.: The dynamic neural field approach to cognitive robotics. J. Neural. Eng. 3(3), R36–R54 (2006)
8. Erlhagen, W., Schner, G.: Dynamic field theory of movement preparation. Psychol. Rev. 109(3), 545–572 (2002)
9. Faubel, C., Schner, G.: Fast learning to recognize objects – dynamic fields in label-feature spaces. In: Proceedings of the 5th International Conference On Development and Learning 2006 (2006)
10. Galambos, R.: A comparison of certain gamma band (40-hz) brain rhythms in cat and man. Induced Rhythms in the Brain, 201–216 (1992)
11. Gray, C.M., Knig, P., Engel, A.K., Singer, W.: Oscillatory responses in cat visual cortex exhibit inter-columnar synchronization which reflects global stimulus properties. Nature 338(6213), 334–337 (1989)
12. Hairston, W.D., Burdette, J.H., Flowers, D.L., Wood, F.B., Wallace, M.T.: Altered temporal profile of visual-auditory multisensory interactions in dyslexia. Exp. Brain Res. 166, 474–480 (2005)
13. Iarocci, G., McDonald, J.: Sensory integration and the perceptual experience of persons with autism. J. Autism. Dev. Disord. 36(1), 77–90 (2006)
14. Iossifidis, I., Steinhage, A.: Controlling an 8 dof manipulator by means of neural fields. In: Proceedings of the IEEE Int. Conf. on Field and Service Robotics 2001 (2001)
15. Krding, K.P., Tenenbaum, J.B.: Causal inference in sensorimotor integration. In: NIPS, pp. 737–744 (2006)
16. Laurienti, P.J., Kraft, R.A., Maldjian, J.A., Burdette, J.H., Wallace, M.T.: Semantic congruence is a critical factor in multisensory behavioral performance. Exp. Brain Res. 158(4), 405–414 (2004)
17. Leary, M.R., Hill, D.A.: Moving on: autism and movement disturbance. Ment. Retard. 34(1), 39–53 (1996)

18. Masterton, B.A., Biederman, G.B.: Proprioceptive versus visual control in autistic children. J. Autism. Dev. Disord. 13(2), 141–152 (1983)
19. Mustovic, H., Scheffler, K., Salle, F.D., Esposito, F., Neuhoff, J.G., Hennig, J., Seifritz, E.: Temporal integration of sequential auditory events: silent period in sound pattern activates human planum temporale. Neuroimage 20(1), 429–434 (2003)
20. Nishida, S.: Interactions and integrations of multiple sensory channels in human brain. In: 2006 IEEE International Conference on Multimedia and Expo, pp. 509–512 (2006)
21. Sauser, E.L., Billard, A.G.: Biologically inspired multimodal integration: Interferences in a human-robot interaction game. In: 2006 IEEE/RSJ International Conference on Intelligent Robots and Systems (2006)
22. Schauer, C., Gross, H.-M.: Design and optimization of amari neural fields for early auditory-visual integration. In: Proceedings 2004 IEEE International Joint Conference on Neural Networks, vol. 4, pp. 2523–2528 (2004)
23. Schoener, G., Dose, M., Engels, C.: Dynamics of behavior: theory and applications for autonomous robot architectures. Robotics and Autonomous System 16, 213–245 (1995)
24. Senkowski, D., Talsma, D., Grigutsch, M., Herrmann, C.S., Woldorff, M.G.: Good times for multisensory integration: Effects of the precision of temporal synchrony as revealed by gamma-band oscillations. Neuropsychologia 45(3), 561–571 (2007)
25. Steinhage, A.: The dynamic approach to anthropomorphic robotics. In: Proceedings of the Fourth Portuguese Conference on Automatic Control (2000)
26. Thelen, E., Schner, G., Scheier, C., Smith, L.B.: The dynamics of embodiment: a field theory of infant perseverative reaching. Behav. Brain Sci. 24(1), 1–34, discussion 34–86 (2001)
27. van Beers, R.J., Sittig, A.C., van der Gon, J.J.D.: The precision of proprioceptive position sense. Exp. Brain Res. 122(4), 367–377 (1998)

A Novel Approach to Image Reconstruction Problem from Fan-Beam Projections Using Recurrent Neural Network

Robert Cierniak

Technical University of Czestochowa, Departament of Computer Engineering,
Armii Krajowej 36, 42-200 Czestochowa, Poland

Abstract. This paper presents a novel approach to the problem of image reconstruction from projections using recurrent neural network. The reconstruction process is performed during the minimizing of the energy function in this network. Our method is of a great practical use in reconstruction from discrete fan-beam projections. Experimental results show that the appropriately designed neural network is able to reconstruct an image with better quality than obtained from conventional algorithms.

1 Introduction

The most popular and the most widespread among tomograph methods used in medicine is x-ray computed tomography (CT). The possibility to acquire three-dimensional images of the investigated objects is realized by applying an appropriate method of projections acquisition and an appropriate image reconstruction algorithm. The key problem arising in computed tomography is image reconstruction from projections obtained from the x-ray scanner of a given geometry. There are several reconstruction methods to solve this problem, for example the most popular reconstruction algorithms using convolution and back-projection ([6], [10], [14]) and the algebraic reconstruction technique (ART) ([5]). Considering the increasing amount of soft computing algorithms used in different science disciplines, it is possible that in the foreseeable future they will occupy an important place in computerized tomography as well. The applications of neural networks in computerized tomography were presented in the past for example in [8], [9], [12]. The so-called neural algebraic approaches to reconstruction problem comparable to our algorithm are presented in papers [15], [16]. The main disadvantage of that approach to image reconstruction from projections problem is the extremely large number of variables which are used during calculations. The computational complexity of the reconstruction process is proportional in that case to the square of the image size multiplied by the number of performed projections. Unfortunately, it leads to a huge number of connections between neurons in the neural networks presented there. In this paper a new approach to the reconstruction problem will be developed based on transformation methodology. It resembles the traditional ρ-filtered layergram reconstruction method where the two-dimensional filtering is the crucial point of that approach [10].

Unfortunately, two-dimensional filtering is computationally complex. Therefore, in our approach a recurrent neural network [3] is proposed to design the reconstruction algorithm. Some authors [1], [4] applied similar neural network structures to solve another problem, namely unidimensional signal reconstruction. Our approach significantly decreases the complexity of the tomographic reconstruction problem. It means that the number of neurons in the proposed network is proportional only to the square of the image size and is independent of the resolution of the performed earlier projections. The reconstruction method presented herein, originally formulated by the author, can be applied to the fan-beam scanner geometry of the tomography device. The weights of the neural network arising in our reconstruction method will be determined in a novel way. The calculations of these weights will be carried out only once before the principal part of the reconstruction process is started. It should be underlined that our algorithm could be in easy way impemented in hardware.

2 Image Reconstruction Algorithm

Presented in this paper reconstruction neural network algorithm resembles the ρ-filtered layergram method [10]. The main difference between these two methods is a realization of the filtering. In our case the neural network is implemented instead of the two-dimensional filtering of the blurred image obtained after the back-projection operation. The scheme of the proposed reconstruction method using the Hopfield-type neural network is shown in Fig. 1, where the fan-beam geometry of collected projections is taken into consideration.

2.1 The Acquisition of the Projections

In the first step of the reconstruction algorithm a set of all fan-beams projections is collected using a scanner whose geometry is depicted in Fig.2. A value of projection depends on the depth of the shadow cast by the object onto a certain place on the screen positioned opposite the radiation source. A given ray from a fan-beam is involved in obtaining a particular projection value $p^f(\beta, \alpha^f)$ where the projection value is obtained at angle α^f and β is the angle of divergence of the ray from the symmetry-line of the fan-beam. In practice only samples $p^f(\beta_\eta, \alpha_\gamma^f)$ of the projections are measured, where $\beta_\eta = \eta \cdot \Delta_\beta$ are equiangular rays, $\eta = -(H-1)/2, \ldots, 0, \ldots, (H-1)/2$ are indexes of these rays, $\alpha_\gamma^f = \gamma \cdot \Delta_\alpha^f$ are particular angles of the x-ray source from which projections are obtained, and $\gamma = 0, \ldots, \Gamma-1$ are the indexes of these angles. For simplicity we can define the discrete projections $\hat{p}^f(\eta, \gamma) = p^f(\eta \cdot \Delta_\beta, \gamma \cdot \Delta_\alpha^f)$.

2.2 Rebinning

After getting a set of fan-beam projections we can proceed with the image reconstruction. One of the most widespread algorithms for image reconstruction from fan-beam projections is a method, which consists of rebinning (re-sorting) operation, which transform them into equivalent parallel projection data ([6], [10]).

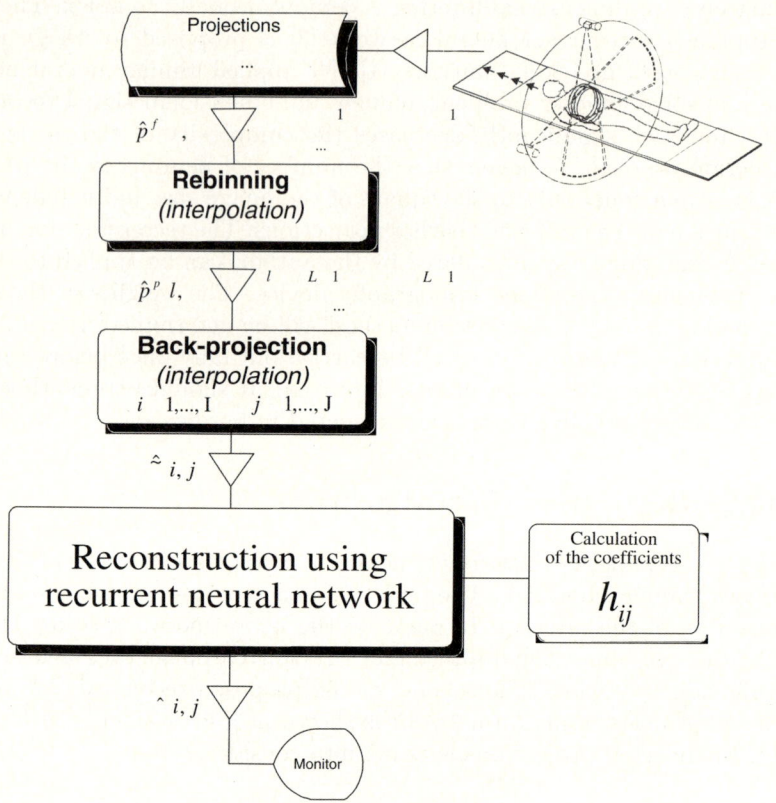

Fig. 1. A neural network image reconstruction algorithm with parallel beam geometry of the scanner

After this operation it is possible to use any algorithm destined to reconstruct an image from parallel projections. In our case it is a method using a recurrent neural network. Before we start the rebinning operation it is convenient to determine a set of parallel projections, which are needed for further signal processing. In the case of the parallel geometry of the scanner, the projection is called Radon's transformation ([13]). In continuous domain it can be expressed as follows

$$p^p(s, \alpha^p) = \int_{-\infty}^{+\infty} \int_{-\infty}^{+\infty} \mu(x,y) \cdot \delta(x\cos\alpha^p + y\sin\alpha^p - s) \, dxdy, \qquad (1)$$

where: α^p —is the angle of parallel projection; x, y —the co-ordinates of the examined object; $\Delta s = (x\cos\alpha^p + y\sin\alpha^p - s)$ —a distance from the centre of rotation to the axis of the ray falling on the projection screen; $\mu(x,y)$ —a distribution of the attenuation of X-rays in analysed cross-section. One can find following relation between parameters in both considered geometries of scanners as

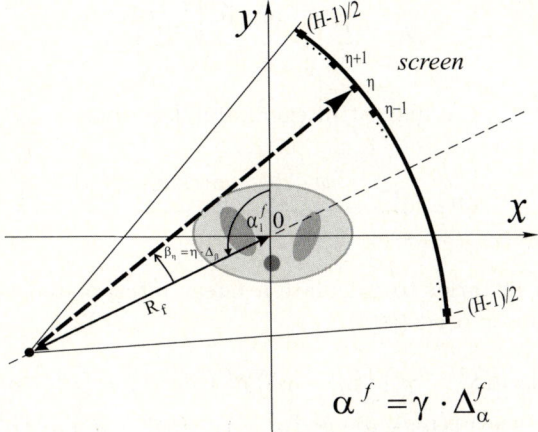

Fig. 2. A single fan-beam projection

$$p^p(s, \alpha^p) = p^f(\beta, \alpha^f) = p^f\left(\arcsin\left(\frac{\dot{s}}{R_f}\right), \alpha^p - \arcsin\left(\frac{\dot{s}}{R_f}\right)\right). \quad (2)$$

Now we determine a uniform sampling on the screen at points $l = -(L-1)/2, \ldots, 0, \ldots, (L-1)/2$, where L is an odd number of virtual detectors from the projection obtained at angle α_ψ^p. It is easy to calculate the distance between each parallel ray from the origin in space (x, y) if these detectors are symmetrically placed on the screen. The distance is given by

$$s(i, j) = l \cdot \Delta_s, \quad (3)$$

where Δ_s is a sample interval on the screen of virtual projections and could be determined as follows

$$\Delta_s = R_f \sin(\Delta_\beta). \quad (4)$$

Taking into consideration the sample of parameters s and α^p of the parallel projections, we can write $\hat{p}^p(l, \psi) \equiv p^p\left(l \cdot \Delta_s, \alpha_\psi^p\right)$. After defining the parameters of the virtual parallel projections we can start the rebinning operation. Unfortunately, in a lot of cases there is a lack of equivalences for parallel rays in the set of fan-beam projections. As a remedy we use an interpolation, in the simplest way —bilinear interpolation. In this case an estimation of the parallel projection $\hat{p}_p(l, \psi)$ can begin by identifying the neighbourhood of the fan-beam projection given by

$$\hat{p}^p(l, \psi) = p^f\left(\arcsin\left(\frac{l \cdot \Delta_s}{R_f}\right), \alpha_\psi^p - \arcsin\left(\frac{l \cdot \Delta_s}{R_f}\right)\right). \quad (5)$$

The neighbourhood is determined based on four real measures from a whole set of fan-beam projections: $\hat{p}^f(\eta^\uparrow, \gamma^\uparrow), \hat{p}^f(\eta^\uparrow, \gamma^\downarrow), \hat{p}^f(\eta^\downarrow, \gamma^\uparrow), \hat{p}^f(\eta^\downarrow, \gamma^\downarrow)$, where: $\eta^\downarrow$ is the highest integer value less than

$$\eta^p = \frac{\beta}{\Delta_\beta} = \frac{\arcsin\left(\frac{l\cdot\Delta_s}{R_f}\right)}{\Delta_\beta}. \tag{6}$$

$\eta^\uparrow = \eta^\downarrow + 1$ and $\gamma^\downarrow$ is the highest integer value less than

$$\gamma^p = \frac{\alpha^f}{\Delta_\alpha^f} = \frac{\psi\cdot\Delta_\alpha^p - \arcsin\left(\frac{l\cdot\Delta_s}{R_f}\right)}{\Delta_\alpha^f} \tag{7}$$

and $\gamma^\uparrow = \gamma^\downarrow + 1$. In order to calculate a linear interpolated value $\hat{p}^p(l,\psi)$ the following expression is used ([10])

$$\hat{p}^p(l,\psi) = \left(\gamma^\uparrow - \gamma^p\right)\left[\left(\eta^\uparrow - \eta^p\right)\hat{p}^f\left(\eta^\downarrow,\gamma^\downarrow\right) + \left(\eta^p - \eta^\downarrow\right)\hat{p}^f\left(\eta^\uparrow,\gamma^\downarrow\right)\right] \tag{8}$$
$$+ \left(\gamma^p - \gamma^\uparrow\right)\left[\left(\eta^\uparrow - \eta^p\right)\hat{p}^f\left(\eta^\downarrow,\gamma^\uparrow\right) + \left(\eta^p - \eta^\downarrow\right)\hat{p}^f\left(\eta^\uparrow,\gamma^\uparrow\right)\right]$$

In this way we obtained all the imagined parallel projections $\hat{p}^p(l,\psi)$ given on a grid $l = -(L-1)/2,\ldots,0,\ldots,(L-1)/2$, $\psi = 0,\ldots,\Psi-1$, where $\Psi = \pi/\Delta_\alpha^p$. These projections which will be used in the next steps of the reconstruction procedure.

2.3 The Back-Projection Operation

The next step in the proceeding sequence is the back-projection operation [6], [10]. In practical realization of the proposed reconstruction algorithm it is highly possible that for any given projection no ray passes through a certain point (x,y) of the image. To take this into account we can apply interpolation expressed by the equation

$$\dot{p}^p(\dot{s},\alpha^p) = \int_{-\infty}^{+\infty} p^p(s,\alpha^p)\cdot I(\dot{s}-s)ds, \tag{9}$$

where $I(\Delta s)$ is an interpolation function. Now the back-projection operation can be expressed as

$$\widetilde{\mu}(x,y) = \int_0^\pi \dot{p}^p(s,\alpha^p)d\alpha^p. \tag{10}$$

Function $\widetilde{\mu}(x,y)$ denotes a blurred image obtained after operations of projection and back-projection.

Owing to relation (1), (9) and (10) it is possible to define the obtained, after back-projection operation, image in the following way

$$\widetilde{\mu}(x,y) = \int_0^\pi \int_{-\infty}^{+\infty} \left(\int_{-\infty}^{+\infty}\int_{-\infty}^{+\infty} \mu(\ddot{x},\ddot{y})\cdot\delta(\ddot{x}\cos\alpha^p + \ddot{y}\sin\alpha^p - \dot{s})d\ddot{x}d\ddot{y}\right)\cdot I(\dot{s}-s)dsd\alpha^p. \tag{11}$$

According to the properties of the convoluation we can transform formula (11) to the form

$$\tilde{\mu}(x,y) = \int_{-\infty}^{+\infty}\int_{-\infty}^{+\infty} \mu(\ddot{x},\ddot{y}) \left(\int_0^\pi I(\ddot{x}\cos\alpha^p + \ddot{y}\sin\alpha^p - x\cos\alpha^p - y\sin\alpha^p)d\alpha^p \right) d\ddot{x}d\ddot{y}. \tag{12}$$

2.4 Discrete Reconstruction Problem

In presented method we take into consideration the discrete form of images $\mu(x,y)$ and $\tilde{\mu}(x,y)$. That means we will substitute continuous functions of images in equation (6) for their discrete equivalents $\hat{\mu}(i,j)$ and $\hat{\tilde{\mu}}(i,j)$; $i = 0, 1, \ldots, I$; $j = 0, 1, \ldots, J$, where I and J are numbers of pixels in horizontal and vertical directions, respectively. Additionally, we approximate the 2-D convolution function by two finite sums. In this way we express relation (12) in the following form

$$\hat{\tilde{\mu}}(i,j) \approx \sum_{\ddot{i}}\sum_{\ddot{j}} \hat{\mu}(i-\ddot{i}, j-\ddot{j}) \cdot h_{\ddot{i}\ddot{j}}, \tag{13}$$

where

$$h_{\ddot{i}\ddot{j}} = \Delta_\alpha^p (\Delta_s)^2 \cdot \sum_{\psi=0}^{\Psi-1} I\left(\ddot{i}\Delta_s\cos\psi\Delta_\alpha^p + \ddot{j}\Delta_s\sin\psi\Delta_\alpha^p\right). \tag{14}$$

As one can see from equation (13), the original image in a given cross-section of the object, obtained in the way described above, is equal to the amalgamation of this image and the geometrical distortion element given by (14). The number of coefficients $h_{\ddot{i}\ddot{j}}$ is equal to $I \cdot J$ and owing to expression (14) values of these coefficients can be easily calculated.

The discrete reconstruction from projections problem can be formulated as following optimisation problem [11]

$$\min_{\Omega} \left(p \cdot \sum_{\ddot{i}=1}^{I}\sum_{\ddot{j}=1}^{J} f\left(e_{\ddot{i}\ddot{j}}(\Omega)\right) \right), \tag{15}$$

where: $\Omega = [\hat{\mu}(i,j)]$—a matrix of pixels from original image; p—suitable large positive coefficient; $f(\bullet)$—penalty function and

$$e_{\ddot{i}\ddot{j}}(\Omega) = \sum_i\sum_j \hat{\mu}(i,j) \cdot h_{\ddot{i}-i, \ddot{j}-j} - \hat{\tilde{\mu}}(\ddot{i},\ddot{j}). \tag{16}$$

If a value of coefficient p tends to infinity or in other words is suitably large, then the solution of (15) tends to the optimal result. Our research has shown that the following penalty function yields the best result

$$f\left(e_{\ddot{i}\ddot{j}}(\Omega)\right) = \lambda \cdot \ln\cosh\left(\frac{e_{\ddot{i}\ddot{j}}(\Omega)}{\lambda}\right), \tag{17}$$

and derivation of (17) has the convenient form

$$f'\left(e_{ij}^{...}\left(\Omega\right)\right) = \frac{df\left(e_{ij}^{...}\left(\Omega\right)\right)}{de_{ij}^{...}\left(\Omega\right)} = \frac{1 - \exp\left(e_{ij}^{...}\left(\Omega\right)/\lambda\right)}{1 + \exp\left(e_{ij}^{...}\left(\Omega\right)/\lambda\right)}, \qquad (18)$$

where: λ—slope coefficient.

2.5 Reconstruction Process Using Recurrent Neural Network

Now we can start to formulate the energy expression

$$E^t = p \cdot \sum_{i=1}^{I} \sum_{j=1}^{J} f\left(e_{ij}^{...}\left(\Omega^t\right)\right). \qquad (19)$$

which will be minimized by the constructed neural network to realize the deconvolution task expressed by equation (13). In order to find a minimum of function (19) we calculate the derivation

$$\frac{dE^t}{dt} = p \cdot \sum_{\bar{i}=1}^{I} \sum_{\bar{j}=1}^{I} \sum_{i=1}^{I} \sum_{j=1}^{J} \frac{\partial f\left(e_{\bar{i}\bar{j}}^{...}\left(\Omega^t\right)\right)}{\partial \left(e_{\bar{i}\bar{j}}^{...}\left(\Omega^t\right)\right)} \frac{\partial \left(e_{\bar{i}\bar{j}}^{...}\left(\Omega^t\right)\right)}{\partial \hat{\mu}^t\left(i,j\right)} \frac{d\hat{\mu}^t\left(i,j\right)}{dt}. \qquad (20)$$

If we let

$$\frac{d\hat{\mu}^t\left(i,j\right)}{dt} = -p \sum_{\bar{i}=1}^{I} \sum_{\bar{j}=1}^{I} \frac{\partial f\left(e_{\bar{i}\bar{j}}^{...}\left(\Omega^t\right)\right)}{\partial \left(e_{\bar{i}\bar{j}}^{...}\left(\Omega^t\right)\right)} \frac{\partial \left(e_{\bar{i}\bar{j}}^{...}\left(\Omega^t\right)\right)}{\partial \hat{\mu}^t\left(i,j\right)} = -p \sum_{\bar{i}=1}^{I} \sum_{\bar{j}=1}^{I} f'\left(e_{\bar{i}\bar{j}}^{...}\left(\Omega\right)\right) h_{ij}, \qquad (21)$$

equation (20) takes the form

$$\frac{dE^t}{dt} = -\sum_{i=1}^{I} \sum_{j=1}^{J} \left(\frac{d\hat{\mu}^t\left(i,j\right)}{dt}\right)^2. \qquad (22)$$

The structure of the recursive neural network performing the reconstruction process consists of two layers and is depicted in Fig. 3.

3 Experimental Results

A mathematical model of the projected object, a so-called phantom, is used to obtain projections during simulations. The most common matematical phantom of head was proposed by Kak (see eg. [6]). In our experiment the size of the image was fixed at $I \times J = 129 \times 129$ pixels.

The discret approximation of the interpolation operation expressed by equation (9) takes the form

$$\hat{\hat{p}}^p(s,\psi) = \sum_{l=-L/2}^{L/2-1} \hat{p}^p(l,\psi) \cdot I(s - l\Delta_\psi^p). \qquad (23)$$

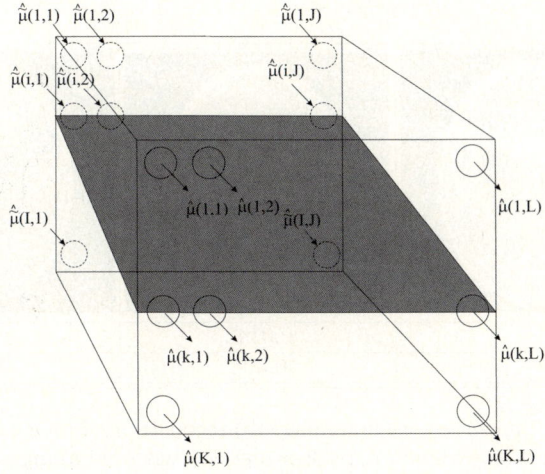

Fig. 3. The structure of the designed recurrent neural network

The interpolation function $I(\Delta s)$ can be defined for example as linear interpolation function

$$I_L(\Delta s) = \begin{cases} \frac{1}{\Delta_s}\left(1 - \frac{|\Delta s|}{\Delta_s}\right) & if\ |\Delta s| \geq \Delta_s \\ 0, & if\ |\Delta s| > \Delta_s \end{cases}, \qquad (24)$$

where $\Delta s = (i\cos\psi \Delta_\alpha^p + j\sin\psi \Delta_\alpha^p)$.

The image was next subjected to a process of reconstruction using the recurrent neural network presented in section 2. The Euler's method (see eg. [16]) was used to approximate (21) in following way

$$\mu(i,j)^{t+1} := \mu(i,j)^t + \Delta t \left(-p \sum_{\ddot{i}=1}^{I} \sum_{\ddot{j}=1}^{I} f'\left(e_{\ddot{i}\ddot{j}}(\Omega)\right) h_{\ddot{i}\ddot{j}}\right), \qquad (25)$$

where $e_{\ddot{i}\ddot{j}}$ is expressed by (16) and Δt is a sufficient small time step.

The difference between reconstructed images using the recurrent neural network algorithm described in this paper 4 and the standard convolution/backprojection method 4 is depicted below ($R_f = 110$). The quality of the reconstructed image has been evaluated in this case by error measures defined as follows

$$MSE = \frac{1}{I \cdot J} \sum_{i=1}^{I} \sum_{j=1}^{J} [\mu(i,j) - \hat{\mu}(i,j)]^2 \qquad (26)$$

and

$$ERROR = \left[\frac{\sum_{i=1}^{I}\sum_{j=1}^{J}(y(i,j) - \hat{y}(i,j))^2}{\sum_{i=1}^{I}\sum_{j=1}^{J}(y(i,j) - \bar{y}(i,j))^2}\right]^{1/2}, \qquad (27)$$

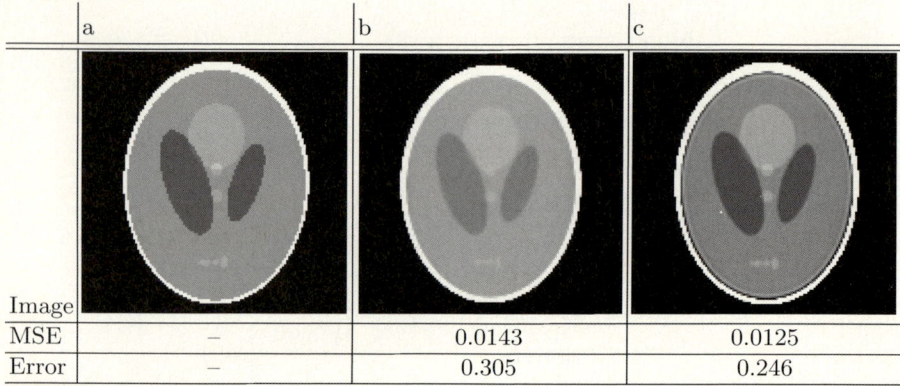

Image			
MSE	–	0.0143	0.0125
Error	–	0.305	0.246

Fig. 4. View of the images: a) original image; b) reconstructed image using standard reconstruction method (convoluation/back-projection with rebinning method); c) neural network reconstruction algorithm described in this paper

where $y(i,j)$, $\hat{y}(i,j)$ and $\bar{y}(i,j)$ are the image of the Shepp-Logan phantom, the reconstructed image and the mean of the Shepp-Logan image, respectively. These images are obtained using window described in [7]. In this case ware used following parameters of this window $C = 1.02$, $W = 0.2$.

4 Conclusions

The performed simulations demonstrated a convergence of the image reconstruction from projections algorithm based on the recurrent neural network described in this work. The reconstructed images obtained after ten thousand iterations showed in Fig.4 have better level of quality in comparison to result of standard reconstruction method. Although our procedure is time consuming, the hardware implementation of the described neural network structure could give incomparable better results than other reconstruction methods.

Acknowledgements

This work was partly supported by the Polish Ministry of Science and Higher Education (Special Research Project 2006-2009).

References

1. Cichocki, A., Unbehauen, R., Lendl, M., Weinzierl, K.: Neural networks for linear inverse problems with incomplete data especially in application to signal and image reconstruction. Neurocomputing 8, 7–41 (1995)
2. Frieden, B.R., Zoltani, C.R.: Maximum bounded entropy: application to tomographic reconstruction. Appl. Optics 24, 201–207 (1985)

3. Hopfield, J.J.: Neural networks and physical systems with emergent collective computational abilities. Proc. National Academy of Science USA 79, 2554–2558 (1982)
4. Ingman, D., Merlis, Y.: Maximum entropy signal reconstruction with neural networks. IEEE Trans. on Neural Networks 3, 195–201 (1992)
5. Jaene, B.: Digital Image Processing - Concepts, Algoritms and Scientific Applications. Springer, Heidelberg (1991)
6. Jain, A.K.: Fundamentals of Digital Image Processing. Prentice Hall, New Jersey (1989)
7. Ka, k.A.C., Slanley, M.: Principles of Computerized Tomographic Imaging. IEEE Press, New York (1988)
8. Kerr, J.P., Bartlett, E.B.: A statistically tailored neural network approach to tomographic image reconstruction. Medical Physics 22, 601–610 (1995)
9. Knoll, P., Mirzaei, S., Muellner, A., Leitha, T., Koriska, K., Koehn, H., Neumann, M.: An artificial neural net and error backpropagation to reconstruct single photon emission computerized tomography data. Medical Physics 26, 244–248 (1999)
10. Lewitt, R.M.: Reconstruction algorithms: transform methods. Proceeding of the IEEE 71, 390–408 (1983)
11. Luo, F.-L., Unbehauen, R.: Applied Neural Networks for Signal Processing. Cambridge University Press, Cambridge (1998)
12. Munlay, M.T., Floyd, C.E., Bowsher, J.E., Coleman, R.E.: An artificial neural network approach to quantitative single photon emission computed tomographic reconstruction with collimator, attenuation, and scatter compensation. Medical Physics 21, 1889–1899 (1994)
13. Radon, J.: Ueber die Bestimmung von Functionen durch ihre Integralwerte Tangs gewisser Mannigfaltigkeiten. Berichte Saechsiche Akad. Wissenschaften. Math. Phys. Klass 69, 262–277 (1917)
14. Ramachandran, G.N., Lakshminarayanan, A.V.: Three-dimensional reconstruction from radiographs and electron micrographs: II. Application of convolutions instead of Fourier transforms. Proc. Nat. Acad. Sci. 68, 2236–2240 (1971)
15. Srinivasan, V., Han, Y.K., Ong, S.H.: Image reconstruction by a Hopfield neural network. Image and Vision Computing 11, 278–282 (1993)
16. Wang, Y., Wahl, F.M.: Vector-entropy optimization-based neural-network approach to image reconstruction from projections. IEEE Transaction on Neural Networks 8, 1008–1014 (1997)

Segmentation of Ultrasound Imaging by Fuzzy Fusion: Application to Venous Thrombosis

Mounir Dhibi and Renaud Debon

ENST Bretagne, ITI, CS83818 Brest Cedex
29238 Brest cedex-3 France
{dhibi.mounir,debon.renaud}@enst-bretagne.fr

Abstract. In this work we propose a new method for ultrasound imaging segmentation in objective to help doctors and specialists to interpret anatomical structure. The proposed method is based on fuzzy fusion theory in objective to extract the venous thrombosis contour in ultrasound images acquired in vivo case. The first obtained results by optimization algorithm, adapted to our particular problem case are presented.

1 Introduction

Ultrasound (US) image segmentation plays an important role in many clinical applications to interpret anatomical structures as qualitative as quantitative. In many situations in medical imaging, and more particularly in ultrasound imaging, we try to detect, to identify and to localize automatically particular zones like organs deformations, tumors, and other problems in bi-dimensional or three dimensional images. To help expert in their diagnosis we must know various information about these regions. Also image segmentation allows identifying the object to be studied by its boundary, its shape and its intrinsic properties as the length, surface and volume [6], [13] and [7]. These geometric proprieties allow information about gravity degree of gravity of the disease to be diagnosed and the therapeutic method to be employed. Ultrasound image segmentation is a particularly important problem and one of the most difficult to study in the field of the analysis and of image processing. It lies especially with the nature of this medical image modality. Indeed, it marked by a relatively little contrast and present a rather important vagueness and a no uniform noise called the speckle. This type of noise is attributable to the ultrasound systems and results from unpredictable fluctuations in the acoustic signal due to the physical phenomena such as reflection and retro diffusion.

Venous thrombus (VT) results from the formation of blood clot in the vein. VT is a pathology vascular being able to lead to a cardiovascular complication (pulmonary embolism or vascular occlusion at the patient who sulphur of this disease. Doctor or expert use generally ultrasound imaging for the first clinical exam. The diagnosis of the thrombus is based only on a subjective evaluation with a visually image analysis by a doctor. A good diagnosis of ultrasound image of vein in vivo case depends on so much experience of a doctor as the image quality. Venous thrombosis volume quantification and automatic form recognition become necessary when, beyond the positive diagnosis of the thrombosis, the popular objective is either the characterization of the thrombus to recognize, or etiologic structure, or a precise surveillance and its maturation [5] and [8]. When studying the VT, we found that the perimeter and the volume are primordial parameters to date and predict the thrombus evolution.

The problem of VT automatic detection in US imaging is not enough aborted subject in the literature. It is a problem of detecting and localizing one or several floating shape objects, relatively with a little contrast, in the light of the blood vessel, to extract from its information shape, size, structure, and volume [10], [9] and [1]. In literature, many studies are proposed to enhance segmentation approach and to optimize results for a particular application. However, a different approach using the complementary aspects of different techniques to build processing methods, allow obtaining improved results. In US imaging, information content is so poor. Thus, its necessary to use all available information and, if possible, the complementary aspect of different methods to obtain a robust edge detection. Such a methodology has proved its efficiency in [11] where a fuzzy model assisted by a dynamic approach allows achieving the detection of VT image sequences. In [2] the ability of the Hough transform integrates a prior information to detect aorta sections in the same sequences. Dhibi and al [4], proposed an elliptic fitting method to detect the VT edges is US image. In [3], authors propose another method based on active contour technique to improve US image segmentation. But in the last method many parameters are necessary in segmentation procedure. This constraint makes the proposed method with non automatic aspect.

2 Methodology

Before segmentation step a pre-processing method is used. This method is composed of three primary steps. A VT has a structure of pseudo-cylinder, or more precisely a generalized cylinder structure. A Cartesian to polar conversion is operated in order to significantly simplify the detection stage. In the second step the sequence photometry equalization is operated. This step is followed by image filtering. This is a difficult problem in the particular case of US imaging. For instance, the filter used in this work does not depend on any noise model. The inhomogeneous diffusion [12], based on the diffusion equation (1) has been chosen in this study because of its capacity to integrate a prior knowledge via the diffusion coefficient.

$$\frac{\partial I}{\partial t} = div(k.\nabla I) \qquad (1)$$

where k is the anisotropic diffusion coefficient which dependent on the gradient magnitude and I the US gray levels image. Thereafter, the relation between the anisotropic diffusion and the error function based on the amplitude of the gradient are defined by the influence function of $\psi(x) = x.k(x)$ which is written as follows:

$$\psi(x,\sigma) = \begin{cases} \frac{1}{2}x.[1-(\frac{x}{\sigma})^2]^2 & |x \leq \sigma| \\ 0 & else \end{cases} \quad (2)$$

In literature we distinguish several forms of the diffusion coefficient. Only the coefficient suggested by Black which uses the Tukeys robust estimator is characterized by a fast descent. This type of estimator shows its robustness compared to the others one, then it permit to minimize the error between the image gradient and its approximation. The σ parameter behave on the evolution of the amplitude and on the iteration number of the diffusion process.

Fuzzy logic is an efficient mathematical tool [14], which makes easier the manipulation and the fusion of imprecise and heterogeneous concepts like position, size, and intensity and so on. In our particular case, we can define, using a prior knowledge, several primitives which denote, with a given ambiguity, that a pixel belongs to the venous contour. Relatively to [11], the following concepts are represented in terms of fuzzy images (i.e. images where grey levels correspond to membership values of the considered pixels to a given concept). Firstly, the contour concept is strong information on the presence of the thrombus in US image. A gradient operator, defined by two 3 by 3 convolution masks, similar to Sobel operator, is used to estimate contours in the data volume. After normalization, using an S-Shape function (Fig. 1), we obtain an image μ_c where grey levels pixels correspond to the membership values to the edge concept. Secondly, the intensity in US image provides that a hyper echogenicity tissue appears as brilliant in an ultrasound image; and the gray level intensity becomes an important feature to consider. An image μ_i, where each pixel level denotes the suitability of the concept being brilliant, is obtained by normalizing the original image using an S-Shape function.

Thirdly, the region parameter is used in this work in order to build the final contour and consider volume. Due to the acquisition system, a strong contrast defines two different regions, which can be easily distinguished: venous thrombosis lumen (appears in black) and tissue area (appears usually brilliant). This information is very precious for the computation of contour in belief image.

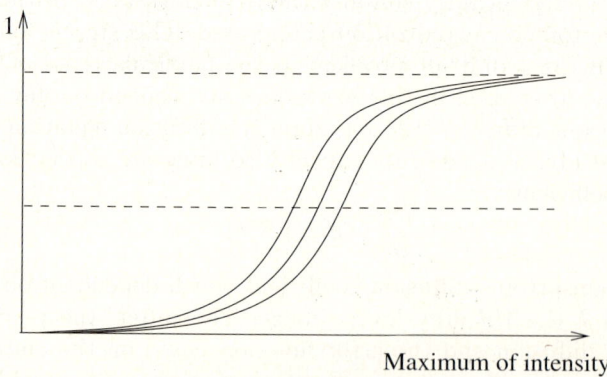

Fig. 1. The S-Shape functions of membership values

Cartesian to polar transformation eases image clustering. An image where each grey level corresponds to a membership value to VT contour is attributed (Fig. 2) with a manually initialisation which is drawled by the expert. In order to build the final contour believes volumes, all fuzzy images information have to be combined. This operation is achieved using a conjunctive operator min. the venous wall is obtained by the combination of the following images: contour, intensity and then region. The merging is given by the Eq. 3 for edge detection, as follows:

$$\mu_f = min(\mu_c, \mu_i, \mu_r) \qquad (3)$$

Two important arguments justify the choice of this kind of fusion operator:

- The principal objective is to develop an algorithm independent to the arbitrary concept of initialization which is necessary for the determinist models of optimization like the dynamics one;
- Weak information component of the US images.

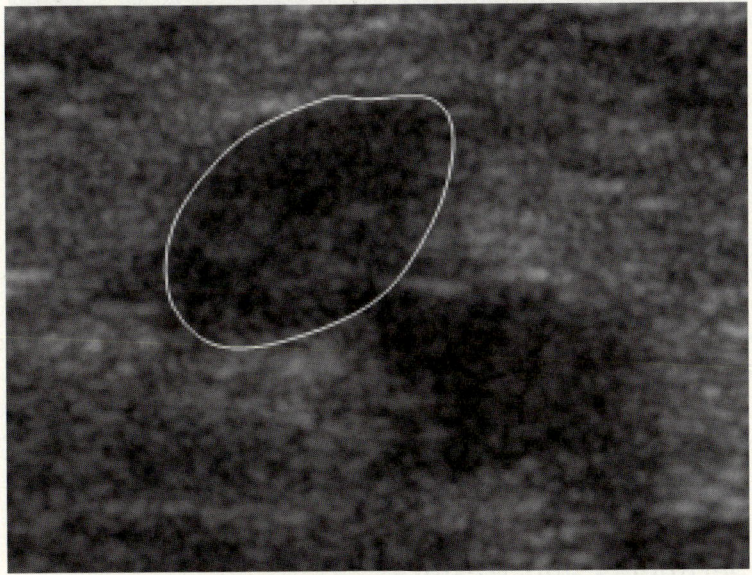

Fig. 2. Venous thrombosis in cartesian plane

After obtaining the optimal contour, the curve parameterized by these n first coefficients of Fourier transform is calculated. Then, the solution is evaluated relatively with the following criterion:

$$C = \sum_{i \in \Delta} \mu_f \qquad (4)$$

For the $(j+1)$ iteration, the contour founded in the previous one will be used in this iteration in order to estimate all contours in the US sequence. This iteratively

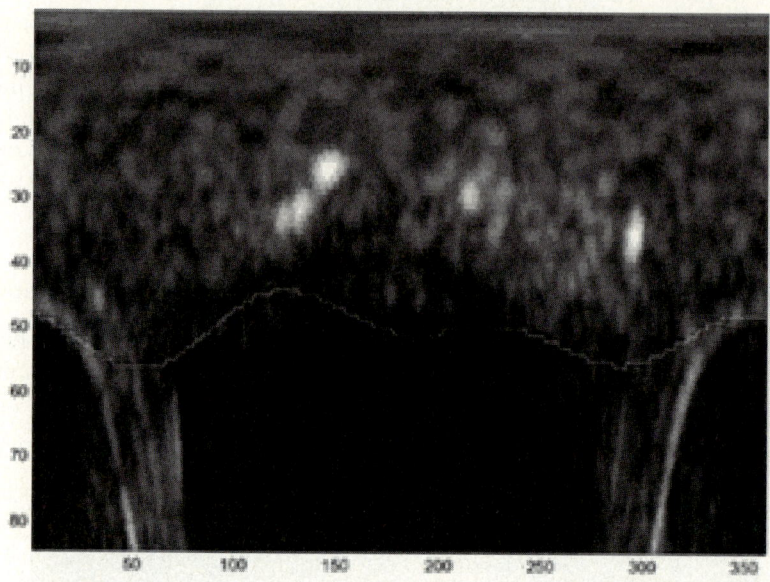

Fig. 3. Venous thrombosis after polar transformation

concept will reduce the Δ space in each iteration, i.e., with each generation only one solution C relative to Eq. 4 is selected. The following generation is made up by considering the update of the image position in function of the space of research Δ reduced to nearest of C_{final}. The last point can be compared with the reduction in the temperature, which ensures the convergence of the algorithm of simulated annealing. Finally the optimal solution, closely connected to detect contour representing thrombosis, is generated by the following equation:

$$C_{final} = argmax \sum_{i \in \Delta} \mu_f \qquad (5)$$

3 Experiments and Results

An acquisition system of the US slices consists of a computer (Authentic AMD 1700Mhz), an electromagnetic locator system, a probe (type SDII210 whose frequency varies between 7 and 12MHz) and a video acquisition card (Matrox Meteor II). We use a system tracking of the position of each US slice instantly in order to retrieve the bi-dimensional data. The user sweeps the studied region with the probe. Throughout the movement, the system records the US images and the electromagnetic sensor positions (translations and rotations).

This section presents results obtained from real sequences acquired. We can see two original images issued from two different sequences, respectively from two different patients (Fig. 4.a and Fig. 5.a). The result of contour detection is

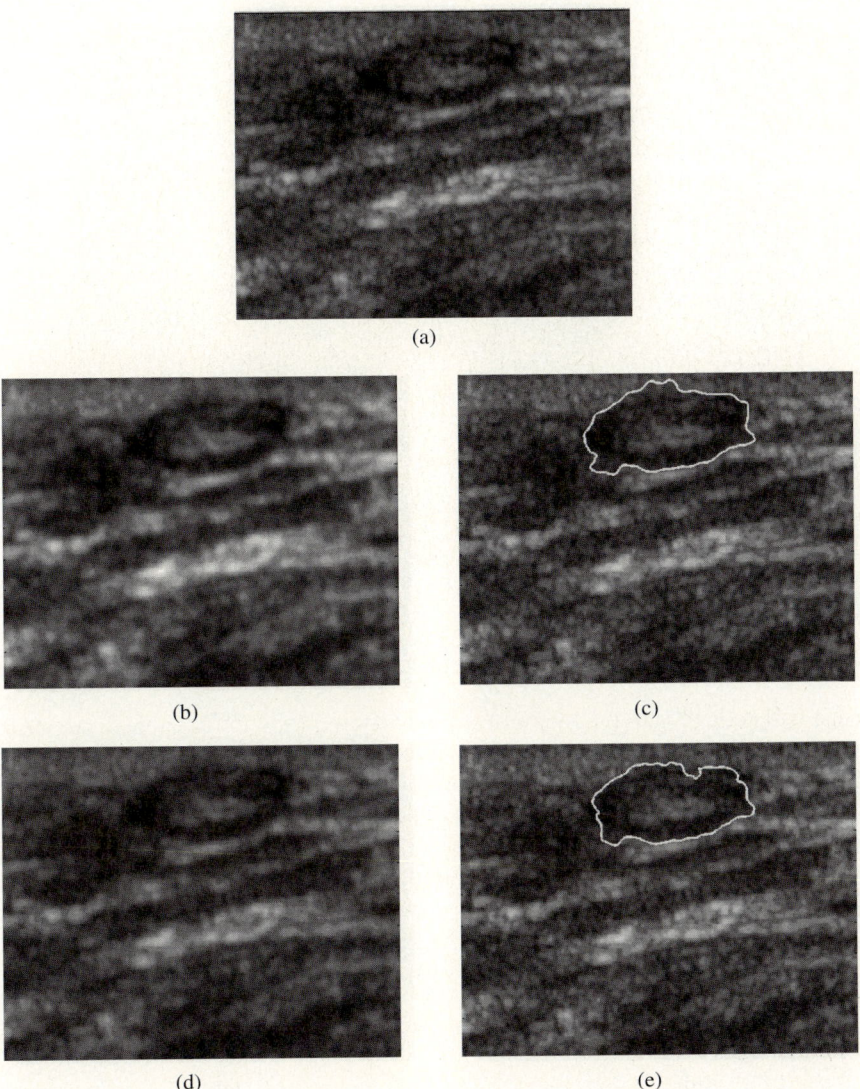

Fig. 4. Superficial venous thrombosis: (a) original image, (b)Gaussian smooth , (c) result of detection after smoothing, (d) anisotropic diffusion (e) result of contour detection after diffusion

presented in Fig. 4.e and Fig. 5.e. As we can see, the obtained accuracy is very encouraging from a clinical evaluation point of view, on healthy case as well as on pathologic case.

The problem created by missing data often encountered in ultrasound imaging is resolved thanks to fusion method, which attributes weights to spline control points proportionally to believes elaborated during the fuzzy process. Even if the

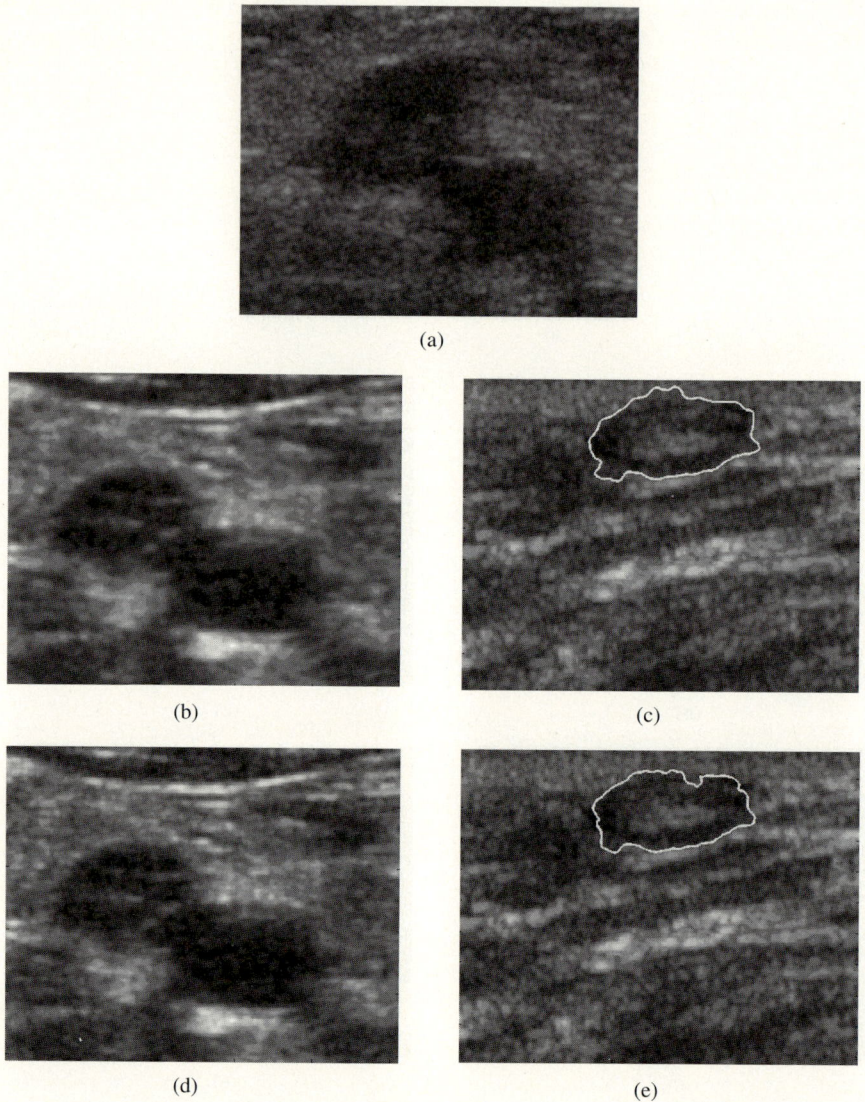

Fig. 5. Deep venous thrombosis: (a) original image, (b) Gaussian smooth, (c) result of detection after smoothing, (d) anisotropic diffusion (e) result of contour detection after diffusion

filtering depends on parameters, which must be for instance manually adjusted, we can note spectacular improvement on the results when filtering is based on anisotropic diffusion presented in Fig. 4.d and Fig. 5.d. In consequence, obtained result by this filter is better than the result provided after Gaussian one shows on Fig. 4.c and Fig. 5.c. Unfortunately, this inhomogeneous diffusion is very CPU consuming that can be a constraint for a medical application but the obtained result is very improved comparing to the result obtained with Gaussian low pass filter with $\sigma = 1.5$ (Fig. 4.b and Fig. 5.b).

4 Conclusion

In this work we proposed a new method for ultrasound image segmentation. The implemented inhomogeneous diffusion filtering gives very interesting results but has the disadvantage of being parameter dependant and time consuming for a simple pre-processing. The possibility to enhance the implementation is seriously considered. All obtained results with our method are very encouraging.

This work will be insensitively tested in order to provide the volume quantification of venous and to compare indirectly the proposed method of segmentation to other method throw the calculated volume. Finally, the same method should be extended to take into carotid plaque of bifurcation reconstruction and stenosis 3D detection.

Acknowledgments. Research was supported by CHU Brest France. Authors thanks Professor Luc Bressollette and Bruno Guias to their contributions and inspirations leading to this work.

References

1. Bressollette, L., Guias, B., Pineau, P., Oger, E., Morin, V., et al.: Three-dimensional reconstruction calibrated with ultrasonographic images. Application to the measurement of clot volume in vitro. J. Mal. Vasc. 26(2), 92–96 (2001)
2. Debon, R., Solaiman, B., Cauvin, J.M., Roux, C.: Aorta detection on endosonographic image sequence by a Hough transform based architecture. In: Proc. Conf. IEEE Eng. Med. Biol. Soc. (1999)
3. Dhibi, M., Puentes, J., Clmence, P., Bressolette, L., Guias, B., Solaiman, B.: Venous thrombosis contour detection in ultrasound image using the snake method. In: Proc. Conf. IEEE SETIT, pp. 43–46 (2005)
4. Dhibi, M., Solaiman, B., Bressolette, L.: Ellipse fitting for venous thrombosis contour detection in ultrasound imaging. In: Proc. Conf. SPIE. Medical Imaging (2007)
5. Dhibi, M., Puentes, J., Bressollette, L., Guias, B., Solaiman, S.: Volume Calculation of Venous Thrombosis Using 2D Ultrasound Images. In: Conf. Proc. IEEE Eng. Med. Biol. Soc., pp. 4002–4005 (2005)
6. Landry, A., Spence, J.D., Fenster, A.: Measurement of carotid plaque volume by 3-dimensional ultrasound. Stroke 35, 864–869 (2004)
7. Salvador, A., Maingourd, Y., Fu, S.: Optimisation of an Edge Detection Algorithm for Echocardiographic Images. In: Proc. IEEE Eng. Med. Biol. Soc., pp. 1188–1191 (2003)
8. Rabhi, A., Adel, M., Bourennane, S.: Segmentation of ultrasound images using geodesic active contours. Ultrasound in Med. Bio. 27(1), 8–18 (2006)
9. Ravhon, R., Adam, D., Zelmanovitch, L.: Validation of ultrasonic image boundary recognition in abdominal aortic aneurysm. IEEE Trans. Med. Imaging 20(8), 751–763 (2001)

10. Shankar, P.M., Forsberg, F., Lown, L.: Statistical modelling of atherosclerotic plaque in carotid B mod images- A feasibility study. Ultrasound in Med. Biol. 29(9), 1305–1309 (2003)
11. Solaiman, B., Debon, R., Roux, C.: Information fusion: application to data and model fusion for ultrasound image segmentation. Special Issue of IEEE-Trans. on Biom. Eng. 4(10), 1171–1175 (1999)
12. Weickert, J.: Anisotropic diffusion in image processing, Ph.D. thesis, Dept. of Mathematics, University of Kaiserslautern (January 1996)
13. Wilhjelm, J.E., et al.: Quantitative analysis of ultrasound B-mode images of carotid atherosclerotic plaque: Correlation with visual classification and histological examination. IEEE. Trans. Med. Imag. 17(6), 910–922 (1998)
14. Zadeh, L.A.: Fuzzy sets. Inform. Contr. 8, 338–353 (1965)

Optical Flow Based Velocity Field Control

Leonardo Fermín, Wilfredis Medina-Meléndez,
Juan C. Grieco, and Gerardo Fernández-López

Simon Bolivar University, Mechatronics Group, ELE-302
Sartenejas 1080-A Miranda, Venezuela
lfermin@usb.ve

Abstract. This work describes a velocity field controller for a mobile platform based on the estimation of its position using odometry. It is a known fact that this approach has an unbounded error attached. The approach here proposed uses a position estimation obtained by odometry, which uses the six components of the velocity calculated on the basis of the displacement vectors resulted from optical flow estimation. The control loop proposed takes the position estimation as reference and performs a velocity field control. Results obtained shows that, for closed trajectories, the tracking error is bounded.

1 Introduction

The studies of robotics platforms based on biological inspired models have demonstrated the great importance of vision for navigation. From these models, several strategies have been designed: obstacle avoidance, feature tracking, and even tracking trajectories.

The applications using visual control could be classified into two main approaches, outboard and onboard camera. With camera outboard, the position of the platform can be easily obtained and therefore controlled; whereas the onboard case needs more complex algorithms, and the most common tasks performed are the obstacle avoidance [1] [2] and the feature tracking, among others.

With the camera onboard, it is possible to obtain the velocity of the mobile platform using optical flow estimation. It consists in the detection of the velocity of the apparent movement of the objects within the image (this movement is induced by the movement of the camera) [3], which have to be integrated to obtain the position estimation.

In this work the previously mentioned approach is employed combining it with a velocity field control technique [4].

A velocity field implies a velocity vector reference for each point of the platform workspace, converging to a predefined trajectory [5]. This scheme was chosen because it can deal with the position errors obtained from the calculus of the position through the integration of the velocities.

This article is organized as follows; section 2 explains how the optic flow is calculated and some results are shown; in section 3 the calculus of the velocities based in the optic flow estimation is described; the odometer based on the velocities obtained is explained in the section 4. Section 5 is dedicated to the controller description; section 6 presents the conclusions, and future work is presented in section 7.

2 Optical Flow

Tests were made in the simulation environment Webots® which allows the fast simulation of robotics platforms as well as many of the sensors which usually are placed on them.

For the generation of the test videos a K-Team Hemisson® robot, a wheeled mobile platform with differential drive, was placed in the simulation environment. The acquisition system was a camera mounted 21 cm above the robot and with and inclination of 30°. The camera has a resolution of 320 × 240 pixels and a frame rate of 30 fps. The image of the simulation is described in Fig. 1. This configuration produces a distance profile which affects the optical flow estimation, but being known, its effect can be canceled. The effect of moving objects in the images taken is beyond the scope of this work.

There are several methods for the optical flow estimation, however due to the application studied, the main criterion employed to select them was the processing time. The methods were:

1. Differential Methods: These are based on the solution of the general equation of the flow.
   $E_x \cdot v + E_y \cdot u = -E_t$
   (a) Horn & Schunk [6].
   (b) Lucas-Kanade [7].
2. Correlation Method: These are based on block processing. A ROI is chosen in the actual image to have a pattern that will be searched in the next one through correlation comparisons.
   (a) Block Matching

Fig. 1. Simulation environment

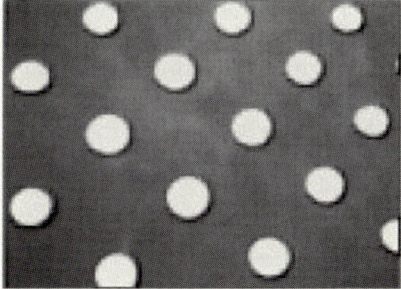

Fig. 2. Movement pattern

The methods were evaluated using the library of image processing OpenCV®
[8]; where the routines for the calculus of each of these schemes were implemented. The tests were made with the same video of the pattern movement described in Fig. 2.

2.1 Differential Methods

For each differential method, it was defined a 32×24 vector field, having one vector for each 10×10 pixels region. Both methods offered a similar response and they are described in Fig. 3. The main difference between them was the processing time.

The method of Horn & Schunk was implemented using 64 iterations, resulting in an average time of 230 ms. On the other hand, the algorithm of Lucas-Kanade resulted in 40 ms approximately which is substantially lower.

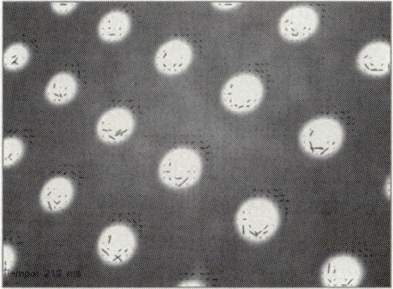

Fig. 3. Optic Flow using Horn & Schunk approach

2.2 Correlation Method

For the correlation method it was needed an image preprocessing stage which consisted on an edge enhancement improving the performance of the algorithm.

In this case, due to processing time restrictions, the image was divided into 9 regions of 100×70 pixels, besides, a discriminating criterion was used to determine vectors with the problem of the aperture, i.e. zones where the image is too homogeneous so any displacement vector cannot be found using this technique. The criterion employed was the same applied in [3] [9]: when the region with the maximum correlation surpasses the correlation of the next region, by a threshold, it is valid vector, otherwise it is ignored. The typical result is the image in Fig. 4, and as it can be observed, there are regions which don't have associated vectors.

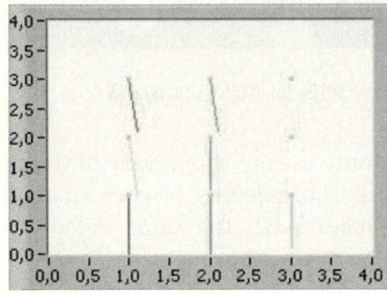

Fig. 4. Displacement vectors using a correlation approach

3 Velocity Estimations

The movements of the platform induce certain types of distribution for the vectors of the optical flow. The idea is to extract the velocities associated to a particular type, so it was designed a testbench in which an optical flow associated to velocities of the 6-DOF of the movement of a mobile platform can be generated. In the left side of Fig. 5, it can be observed the flow associated to each DOF, and in the right side, the sum of all the partial flows.

In addition, to represent the aperture problem some random vectors were removed from the final sum. To represent the noise, it was added a random variable with variable amplitude to simulate different levels of noise.

The algorithm used consisted in a square minimization of the error. The implementation is based on the matrix representation of the equations that each flow generates. As it was expected, if the level of noise is increased, there was a higher variation in the velocity detection, however, the results shown in Fig. 6 demonstrate that its average remains the same.

Table 1 shows the effects of the level of noise on the velocities estimation for rotation around the X axis and translation across Y axis).

To test the algorithm under real conditions the platform was moved at different velocities and a velocity estimates graphic was produced using each method in order to compare them. The velocities were applied to the platform according to the Fig. 7, beginning with 0 m/s and increasing 1 m/s every 14 cycles to avoid the transient effect.

The signal that comes from the optical flow algorithms is very noisy, so it has to be processed. Several filtering techniques were tested, linear as well as nonlinear ones, obtaining the best results by using a median filter.

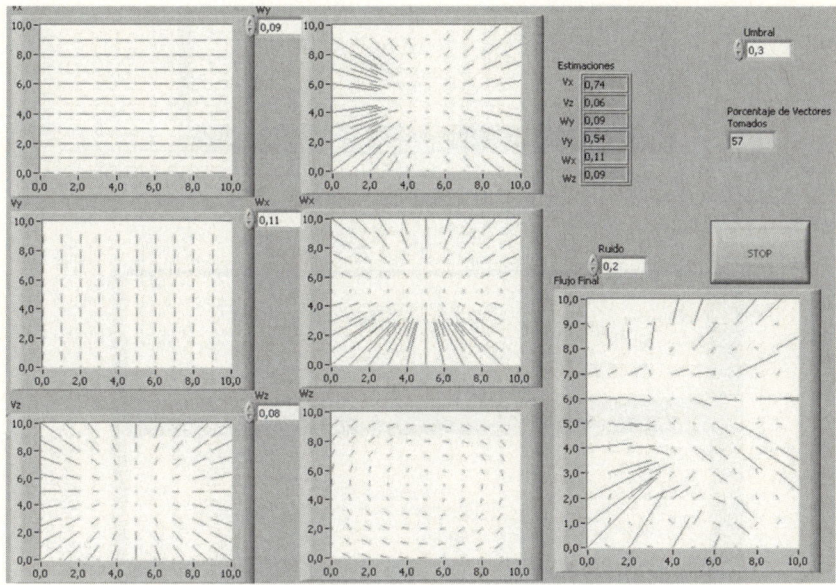

Fig. 5. Testbench for velocity detection

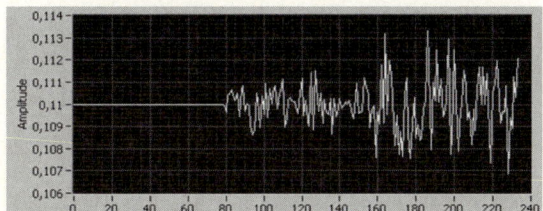

Fig. 6. Results in velocities estimation

Table 1. Effect of Noise

Movement	Noise Magnitude	Average	Standard Deviation	Error %
X Rotation	0.0	0.05	0.0000	0.0
X Rotation	0.1	0.05	0.0019	3.7
X Rotation	0.2	0.05	0.0045	8.9
Y Translation	0.0	0.70	0.0000	0.0
Y Translation	0.1	0.70	0.0228	0.7
Y Translation	0.2	0.70	0.0106	1.5

The Fig. 8 shows the result of velocities estimation from the optical flow estimation using the correlation algorithm. At high velocities, the estimation is difficult and at lower ones, two different inputs cannot be distinguished.

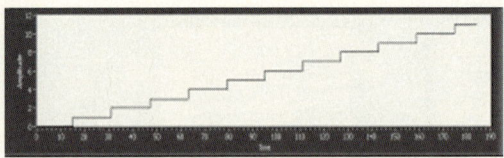

Fig. 7. Signal for velocity

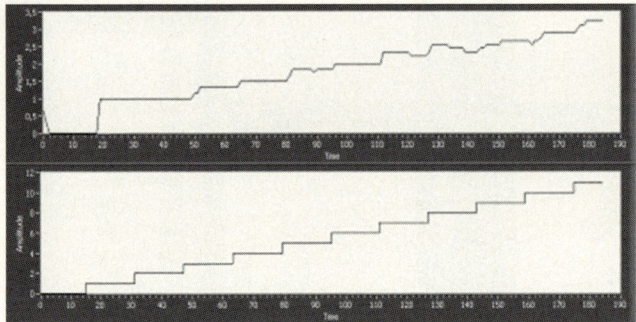

Fig. 8. Velocity estimation using correlation

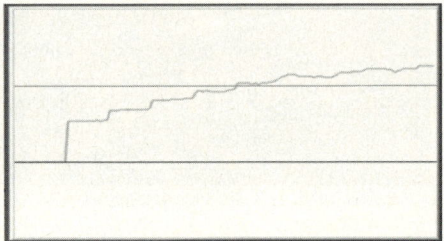

Fig. 9. Velocity detection using Horn & Schunk algorithm

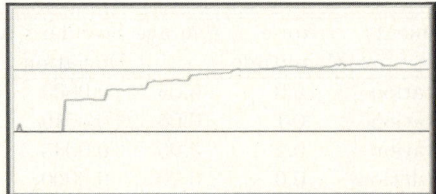

Fig. 10. Velocity detection using Lucas-Kanade algorithm

In the case of the differential algorithms the response at low velocities was better than the correlation based, but in the case of high velocities the response was also poor and difficult to extract. Fig. 9 represents the response of the algorithm using the Horn & Schunk estimate and Fig. 10 the one obtained using the Lucas-Kanade algorithm.

4 Odometry

The equations of movement of a mobile platform are represented by (1), (2) and (3).

$$x(t) = \int_0^t V(\tau) \cdot \cos(\theta(\tau)) \cdot d\tau \tag{1}$$

$$y(t) = \int_0^t V(\tau) \cdot \sin(\theta(\tau)) \cdot d\tau \tag{2}$$

$$\theta(t) = \int_0^t \omega(\tau) \cdot d\tau \tag{3}$$

From (3), the angle increases proportionally to t, as it is demonstrated by (4) and (5).

$$\theta(t) = \int_0^t \omega(\tau) \cdot d\tau + \int_0^t \Delta\omega \cdot d\tau \tag{4}$$

$$\Delta\theta(t) = t \cdot \Delta\omega \tag{5}$$

The position error, on the other hand, depends on the angle error and it increases proportional to t^2, as it is demonstrated in (6), (7) and (8).

$$\Delta x(t) = t \cdot \Delta V \cdot \cos(\theta(t)) + t \cdot V(t) \cdot \sin(\theta(t)) \cdot \Delta\theta \tag{6}$$

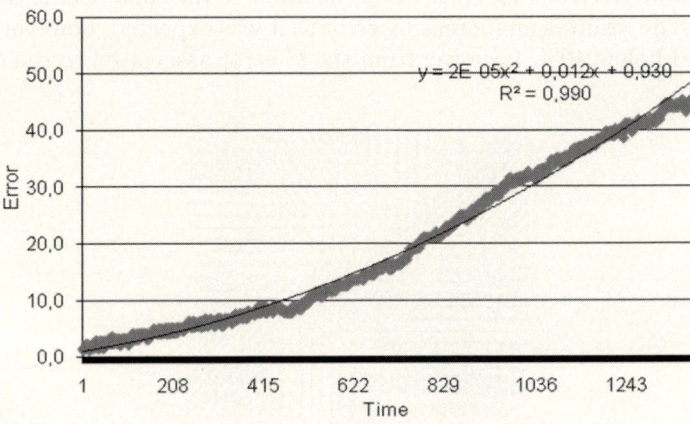

Fig. 11. Odometer Error

$$\Delta x(t) = t \cdot \Delta V \cdot \cos(\theta(t)) + t \cdot V(t) \cdot \sin(\theta(t)) \cdot (t \cdot \Delta\omega) \tag{7}$$

$$\Delta x(t) = t \cdot \Delta V \cdot \cos(\theta(t)) + t^2 \cdot \Delta\omega \cdot V(t) \cdot \sin(\theta(t)) \tag{8}$$

The quadratic behavior was proved analyzing the time variance of the position error. For the trendline based on a sequence of 1400 points, there was a correlation of 0.99, proving that the error increases by the square of time as it can be seen in Fig. 11.

5 Velocity Field Control

The main problem with the positioning using the odometer is the increasing error, in this case by the square of time. The velocity field control resulted very efficient in the trajectory tracking problem.

Now, the testbench was designed as a velocity field coding the task of following a circumference contour, as can be observed in Fig. 12.

Based on these references, an angle controller was designed, being the controlled variable the angle of the velocity vector of the platform. The control implemented was a proportional scheme with saturation in order to deal with the platform constraints. Other controllers are beyond the scope of this work.

The controller was implemented in two ways: the first one using the real position of the platform; and the second one using the position estimation obtained with the visual odometer. The results are shown in the Fig. 13 (the trajectory of the platform using its real position is the black trace).

To evaluate the performance of the complete algorithm the tracking error was measured, not the position error. The tracking error is directly the radius error because it is a circumference.

Fig. 14 shows the tracking error. This error is mainly due to the parameters of the controller that are beyond the scope of this work.

Fig. 15 shows the tracking error when the input to the controller is the position estimation. The result is an increasing error as it was expected, however this error was bounded below 40%, far away from the t^2 error associated to the odometer.

Fig. 12. Velocity field converging to a circumference contour

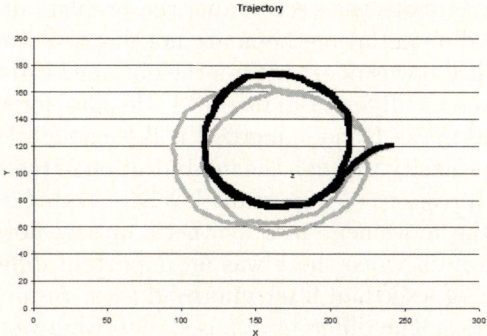

Fig. 13. Results: Trajectory

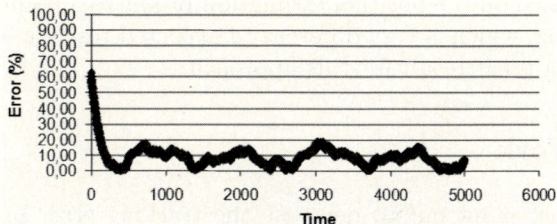

Fig. 14. Tracking error using real position

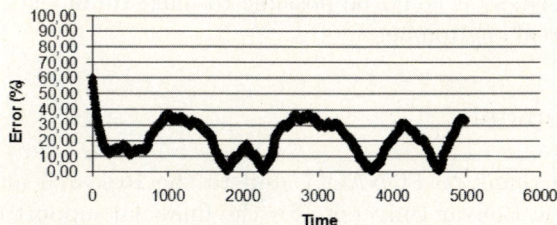

Fig. 15. Tracking error using position estimation

The mean error when the real position is used was of 8%, whereas, in the odometer based controller, the error was 20%.

6 Conclusions

This work demonstrates that it is possible to make the trajectory tracking control using as main sensor an onboard camera, and applying processing algorithm. It had a good response allowing an eventual implementation in real time.

The biggest problem encoured was the implementation of the optical flow estimation, due to the computational cost. The correlation method is well known but the need of discriminate the vectors due the problem of aperture, increase the operations. The differential methods are not employed in the industry, but they produce an optical flow smoother than the obtained through the correlation algorithm. This makes the differential methods adequate for the velocity estimation. In addition, the Lucas-Kanade method is fast enough for the application.

The estimation of velocity through the minimization of the square error proved to be robust in the presence of important levels of noise, recuperating all the six components of the movement in a synthetic optical flow. When the input was the optic flow from a video there was an important difficulty in recovering the velocity. The use of a median filter improved these results but it only works for a range of velocities. The differential based algorithm works better than the correlation based one.

It was demonstrated that the visual odometer generates a quadratic increasing error as it was predicted from the equations.

The control based on the position estimation produces a bounded error in the trajectory tracking, which is very different to expected response of an odometer. This constitutes a great result for this approach.

7 Future Work

Future works must point out to decrease the tracking error by improving the calculus of the optical flow and the velocities extraction associated, because these are the principal source of errors.

Also, it is necessary to keep on working on improving the algorithm speed and the equipment used, so it could be possible to make all of these calculations in real time over a real environment.

Acknowledgement

Authors want to thank to FONACIT and to the Research and Development Deanship of Simon Bolivar University for the financial support to this project.

References

1. Stöfler, N., Burkert, G.T.: Farber Real-Time Obstacle Avoidance Using MPEG-Processor-Based Optic Flow Sensor. In: Proceedings of the 15th International Conference on Pattern Recognition, Barcelona, Spain, pp. 161–166 (2000)
2. Hrabar, S., Sukhatme, G.S., Corke, P., Usher, K., Roberts, J.: Combined Optic-Flow and Stereo-Based Navigation of Urban Canyons for an UAV. In: Proceedings International Conference on Intelligent Robots and Systems - IROS 2005, Edmonton, Canada, pp. 3309–3316 (2005)
3. Stöfler, N.O., Schnepf, Z.: An MPEG-Processor-based Robot Vision System for Real-Time Detection of Moving Objects by a Moving Observer. In: Proceedings of the 14th International Conference on Pattern Recognition, Brisbane, Australia, pp. 477–481 (1998)

4. Kelly, R., Bugarín, E., Campa, R.: Application of Velocity Field Control to Visual Navigation of Mobile Robots. In: Proceedings of the 5th IFAC Symposium on Intelligent Autonomous Vehicles, Lisbon, Portugal (2004)
5. Medina-Meléndez, W., Fermín, L., Cappelletto, J., Murrugarra, C., Fernández-López, G., Grieco, J.C.: Vision-Based Dynamic Velocity Field Generation for Mobile Robots. In: Robot Motion Control 2007 - RoMoCo 2007. Lecture Notes in Control and Information Sciences, vol. 360, pp. 69–79 (2007)
6. Horn, B.K.P., Schunk, B.G.: Determining Optical Flow. Artificial Intelligence 17, 185–203 (1981)
7. Lucas, B.D., Kanade, T.: An Iterative Image Registration Technique with an Application to Stereo Vision. In: Proceedings DARPA Imaging Understanding Workshop, pp. 121–130 (1981)
8. http://opencvlibrary.sourceforge.net/cvreference
9. Stöfler, N.O., Farber, G.: An Image Processing Board with an MPEG Processor and Additional Confidence Calculation for Fast and Robust Optic Flow Generation in Real Environments. In: Proceedings of the 8th IEEE International Conference on Advance Robotics, Monterey, USA, pp. 845–850 (1997)

Detection of Phoneme Boundaries Using Spiking Neurons

Gábor Gosztolya and László Tóth

MTA-SZTE Research Group on Artificial Intelligence
of the Hungarian Academy of Sciences and University of Szeged
H-6720 Szeged, Aradi vértanúk tere 1., Hungary
{ggabor, tothl}@inf.u-szeged.hu

Abstract. Automatic speech recognition (ASR) is an area where the task is to assign the correct phoneme or word sequence to an utterance. The idea behind the ASR segment-based approach is to treat one phoneme as a whole unit in every respect, in contrast with the frame-based approach where it is divided into equal-sized, smaller chunks. Doing this has many advantages, but also gives rise to some new problems. One of these is the detection of potential bounds between phones, which has an effect on both the recognition accuracy and the speed of the speech recognition system. In this paper we present three ways of boundary detection: first two simple algorithms are tested, then we will concentrate on our novel method which incorporates a spiking neuron. On examining the test results we find that the latter algorithm indeed proves successful: we were able to speed up the recognition process by 35.72% while also slightly improving the recognition performance.

1 Introduction

In the problem of Automatic Speech Recognition (ASR) we have to map the correct phoneme or word sequence to a given speech signal. Most methods for this task are based on the frame-based notion, which treats the speech signal as a series of independent, equal-sized small units called frames. These frames are classified as phonemes, and then joined to make whole words or sentences, from which the most probable one is chosen. This scheme is based on the concept of combining multiple, consecutive frames into one phoneme.

The segment-based approach is founded on different principles. Here all the frames which make up a phoneme are treated together as one unit, and naturally their classification is also done together. This way we can extract more information from the context, and this leads to more precise phoneme-level identification, but doing this is not without its drawbacks. One, for instance, is that both ends of the phoneme segment become variables of the search, which raises the computational needs of the recognition process to a quadratic running time instead of the former linear one. To reduce this CPU requirement we can filter the phoneme boundary hypotheses before evaluating them. It is easy to see that this restriction has to be carried out very carefully: if we do not allow the kind of bound where there is actually a boundary between two phonemes, we will lose information, and this will probably lead to a decrease in the recognition performance. On the other hand if we allow possible boundaries where there is no need for them, we can significantly and unnecessarily slow down our speech recognition system. Hence a tradeoff should be made.

Here we will focus on the task of constructing a method for detecting probable segments in order to speed up the recognition process. To do this, we will investigate one simple method as a baseline, construct three novel segmentation algorithms, and then vary their parameters in order to make them work efficiently. In the experiments we found that we were able to not only speed up the recognition process significantly, but also improve the recognition performance, which was an unexpected bonus.

The structure of our paper is as follows. First we define the speech recognition task in a segment-based approach. Next we discuss the issue of segmentation, outlining the problem and our solutions. Then we describe the test environment and the tests we made. Lastly we present our results and draw some conclusions.

2 The Speech Recognition Problem

In speech recognition problems we have a speech signal represented by a series of observations (frames) $A = a_1 \ldots a_t$, and a set of possible phoneme sequences (words) which will be denoted by W. Our task is to find the word $\hat{w} \in W$ defined by

$$\hat{w} = arg \max_{w \in W} P(w|A), \qquad (1)$$

which, using Bayes' theorem, is equivalent to the maximization problem

$$\hat{w} = arg \max_{w \in W} \frac{P(A|w) \cdot P(w)}{P(A)}. \qquad (2)$$

Further, noting the fact that $P(A)$ is the same for all $w \in W$, we have that

$$\hat{w} = arg \max_{w \in W} P(A|w) P(w). \qquad (3)$$

Speech recognition models can be divided into two types – the discriminative and generative ones – depending on whether they use Eq. (1) or Eq. (3). In the experiments we restricted our investigations to the generative approach [1]. Speech recognition models can also be divided into two types, namely a frame-based one and a segment-based one. Here we will focus on the latter type. Note too that the factors $P(A|w)$ and $P(w)$ can be treated separately. From now on we will focus on the former one, and take $P(w)$ (the *language model*) as given.

Now let us define the word w as a phoneme-sequence $o_1, \ldots, o_n$. Next, let us divide A into non-overlapping segments $A_1, \ldots, A_n$, each A_j belonging to the corresponding phoneme o_j. (As $A = a_1 \ldots a_t$, we can define A_j as $a_{t_{j-1}} \ldots a_{t_j}$

with $1 = t_0 < t_1 < \ldots < t_n = t$.) Then, making the common assumption that the phonemes are independent, we have

$$P(A|w) = \prod_{j=1}^{n} P(A_j|o_j). \qquad (4)$$

In a segment-based model we calculate these $P(A_j|o_j)$ values by some machine-learning method using interval features (i.e. features like the mean, minimum or maximum, calculated on the whole $a_{t_{j-1}} \ldots a_{t_j}$ segment). In our system Artificial Neural Networks (ANNs) [2] are used, and their corresponding output serve as estimates for the $P(o_j|A_j)$ probability values after length normalization. From these, estimates of $P(A_j|o_j)$ can be readily obtained by a division by the priors.

2.1 Restricting the Space of Possible Segmentations

At this point we can assign a probability for a phoneme-sequence and segment-sequence pair, so in theory we can find the optimal (most probable) pairing. Unfortunately, it is practically impossible to examine all pairings, hence we have to introduce some further restrictions for technical reasons. Firstly, instead of referring to the segments as $A_1, \ldots, A_n$, we use their boundary elements, i.e. $I = [t_0, t_1, \ldots, t_n]$. This I will from now on be called a *segmentation*, and we can refer to an A_j segment with its starting and ending segmentation bounds as $[t_{j-1}, t_j]$. Next, we will limit the range of these t_j values: instead of taking any number between 1 and t, only the elements of some set T will be allowed. This set is called the set of possible segmentation bounds (or *possible segmentation* in short), while the algorithm which supplies these values for the given speech signal A is called the *segmentation algorithm*. In this paper we will focus on this issue.

It is not hard to find reasons for the existence of this T set. It is very unlikely that every value between 1 and t will be needed (partly because it would mean that all phonemes are then of identical length). On the other hand, keeping values in T which will obviously not be used in any segmentation makes speech recognition much more time-consuming without any gain. Constructing a good T also seems possible because in the ideal case it coincides with the real boundary between phones, which can hopefully be predicted by signal processing techniques. We will discuss this issue later in detail, but first we need to describe the search algorithm used.

2.2 The Search Process

The task of a search algorithm is to find the word and segmentation pair with the highest probability for the given speech signal A. Many search methods are available in the literature to do this: perhaps the most widely-used one is the Viterbi beam search [3], but here we opted for the multi-stack decoding algorithm [4].

These two methods are similar in the way they work: they handle prefixes of phoneme sequence-segmentation pairs which are called *hypotheses*. The hypotheses are stored in *priority queues* assigned to each possible phoneme boundary, which automatically discard hypotheses considered improbable relative to the other hypotheses kept there. During a search these queues are visited in increasing time order. At each step all hypotheses are taken from the actual queue and they are expanded in every possible way: a new valid phoneme is attached to the end with a new end time. Then the new hypotheses have their probabilities calculated, and are inserted into the queue belonging to their new end time. Of course these queues automatically sort the hypotheses they store, and discard the most improbable ones. In the end the result will be the most probable hypothesis belonging to the final queue which has a correct phoneme-sequence (i.e. not ending in the middle of a word).

The difference between the two search methods is how they decide whether a hypothesis in a priority queue is improbable. The Viterbi beam search discards those whose probability does not lie within a given interval relative to the best one, this threshold being called the *beam width* parameter. The multi-stack decoding algorithm, however, always keeps a fixed number of best hypotheses, and discards the rest. The corresponding constant is the *stack size*. Since we used the multi-stack decoding algorithm, we shall concentrate on it, but the following statements also apply to the Viterbi beam search method.

Choosing the right stack size parameter is of course of great importance. If it happens to be too low, the correct word-segmentation pair may be lost, because a hypothesis leading to it is discarded due to its temporary improbability. If, however, the stack size chosen is too big, it greatly slows down the search process, forcing the method to needlessly investigate and expand many non-promising hypotheses.

3 Phoneme Boundaries

In a segment-based speech recognition task it is vital that all actual phoneme boundaries (or at least frames in their immediate neighbourhood) be in the set of possible segmentation bounds T. On the other hand we should limit the size of this set, otherwise it would lead to an unacceptable loss in speed. This makes the choice of the elements of this T quite important.

A baseline phoneme boundary assignment can be simply to draw one possible phoneme boundary at every k^{th} frame. It is a brute force solution, but it makes quite satisfactory recognition scores possible. However, using this, during a search we have to consider lots of hypotheses, which naturally slows down the search process. On the other hand, this way we cannot miss any real phoneme boundary. But this method is by no means optimal, so next we will describe the algorithms we constructed for phoneme boundary detection.

3.1 Phoneme Boundary Detector Algorithms

All our algorithms will take as input a function which estimates for each point the probability of this point being a boundary position. One may think that supplying this value for all frames (all a_i values) solves the entire segmentation

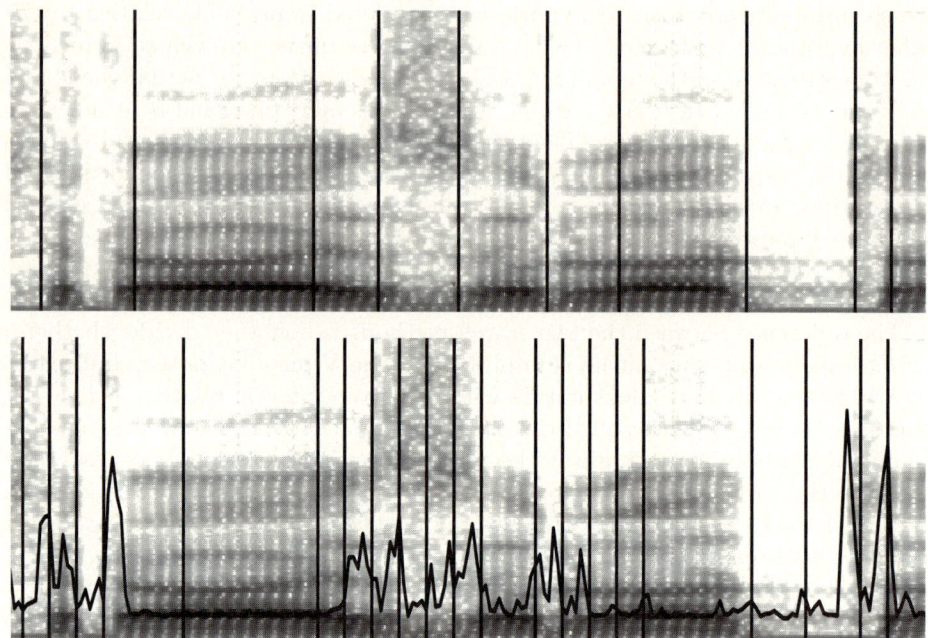

Fig. 1. Up: A spectrum with its correct segmentation. Down: the same spectrum with the b_i values and the output of the Thresholding Algorithm.

problem, but unfortunately this is not the case. While it is indeed a very important factor, in our experiments we found that determining the actual position of the segment bounds is by no means straightforward. Hence in the following we will assume that for each frame a_i we have a value b_i which tells us the probability of a phoneme boundary being at that particular place. To get these b_i values we applied a neural net, which will be described in detail in the Experiments section.

Thresholding Algorithm. This algorithm improves the default brute force segmentation method by applying a threshold. That is, we still draw a possible phoneme boundary at every k^{th} frame, but only if the corresponding b_i value is greater or equal to some minimum probability p_{min}. The lower panel of Figure 1 shows how this method works. One can see that it draws equidistant boundaries everywhere, apart from two intervals where the b_i values are permanently low. Compared to the correct segmentation shown in the upper panel, it is clear that this method draws lots of unnecessary boundaries.

Maximum Algorithm. This algorithm simply looks for positions where the b_i values take their local maxima over a given interval of neighbouring points. The advantage of this method is that the boundary positions suggested by it in many cases fall pretty close to the ones proposed by humans, as those human experts who manually process such databases also place the boundaries at

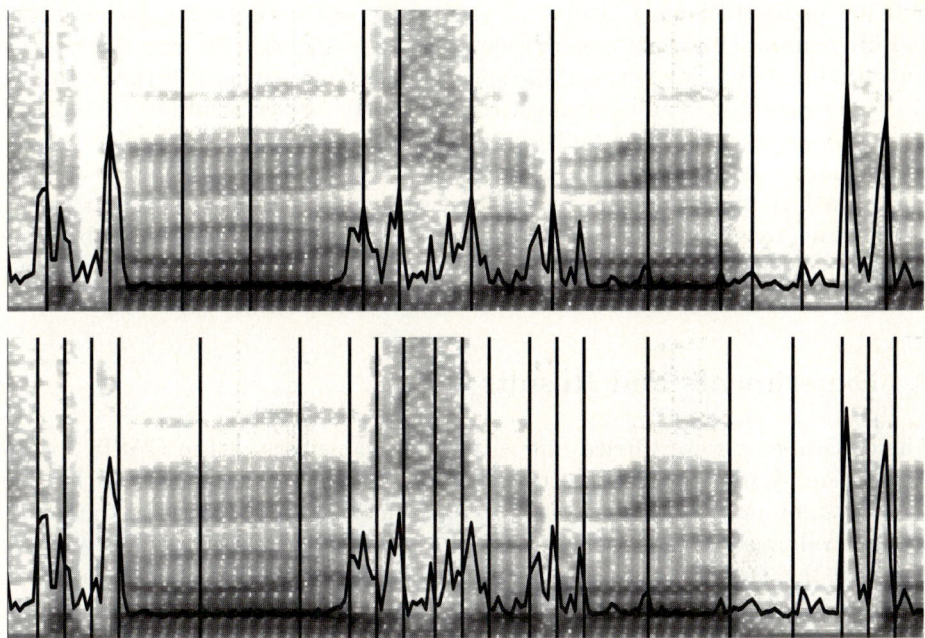

Fig. 2. Up: the same spectrum as in Figure 1 with the b_i values and the output of the Maximum Algorithm. Down: the same spectrum with the b_i values and the output of the LIF Algorithm.

the points where the spectral change is the largest. By adjusting the size k of the neighbourhood examined by the algorithm we can easily guarantee a minimal distance between the hypothesized boundaries. The algorithm can also be combined with the thresholding method described above. The behaviour of this method is illustrated in Figure 2 (upper panel).

LIF Algorithm. The problem with both the former algorithms is that we have no direct control over the density of the boundary hypotheses, so in certain cases they may not draw any boundary for a 'long' period of time. A less risky solution would be a modification of the baseline algorithm which puts boundary lines everywhere along the time axis, but dynamically adjusts the density of the markers in such a way that it is proportional to the local probability values b_i. A self-evident biologically motivated solution for this is to apply a leaky integrate-and-fire (LIF) neuron model [5]. This construct is the best-known example of the family of spiking neuron models, and its operation is very simple: the neuron integrates its input until the sum reaches a certain threshold. Then the neuron fires (emits a spike), its potential is reset, and the whole process starts again. The model can be refined by making the integration "leaky", in which case the membrane potential decays with a characteristic time constant when no input is present. The model may also incorporate an absolute refractory period, over which it is not excitable.

In our segmentation algorithm the b_i values are used as input for the integrate-and-fire neuron. Obviously, where these values are higher, their sum reaches the activation threshold sooner, and the spikes become more dense. With the tuning of the threshold one can control the density of the spikes, while by setting the refractory period a minimum distance between the neighbouring spikes can be easily guaranteed. In our experiments these parameters (e and k, respectively) were tuned manually. Obviously, we tried to set the parameters so that at the most dense areas the density of the markers were similar to those of the baseline algorithm, while at other places the boundary markers were more spaced out. The behaviour of this method is illustrated in Figure 2 (lower panel).

4 Experiments and Results

The experiments were carried out within the framework of the OASIS speech recognition system, which, due to its module-based structure and its flexible script system, provides a good basis for experimentation [6]. We used the multi-stack decoding algorithm for searching with a constant big stack size. While it is true that this parameter also affects both the running speed and the recognition scores, here we sought to test just the various segmentation algorithms.

There are many ways of measuring the correctness of a possible segmentation (and hence the algorithm which produced it). Perhaps the easiest one is to compare T with the actual phoneme bounds. To do this, first the elements of T and the actual phoneme bounds have to be mapped to each other somehow. After this step one can easily see how many boundaries are missing from T and how many of its elements are redundant. This test can be carried out in a very short time, but it is not without its drawbacks: the actual phoneme boundaries have to be known beforehand, and it is not easy to say how many missing bounds should be considered "too few". (Similarly, it is not known how many needless bounds should be judged as "too many".) Moreover, it is also not clear when two phoneme boundaries can be mapped to each other.

Since this testing method had quite a few drawbacks, we eventually chose the other option: we examined how a possible segmentation T works in practice; that is, how it affects the actual recognition scores on real data. For this we performed a thorough test involving sentence recognition, and examined how the recognition scores varied. But to do this we first have to introduce the way these two recognition measures can be calculated.

We cannot simply compare the original and the resultant sentences because this way even one badly identified word would ruin the whole sentence. We cannot compare the two sentences word for a word either, because one incorrectly inserted or omitted word would also ruin the calculated performance ratio. For this reason, usually the edit distance of the two sentences is calculated; that is, we construct the resulting sentence from the original using the following operations: inserting and deleting words, and replacing one word with another one. These operations have some cost (in our case the common values of 3, 3 and 4, respectively), then we choose an operation series with the lowest cost. Finally we can calculate the following measures:

$$Correctness = \frac{N - S - D}{N} \qquad (5)$$

and

$$Accuracy = \frac{N - S - D - I}{N}, \qquad (6)$$

where N is the total number of words in all the original sentences, S is the number of substitutions, D is the number of deletions and I is the number of insertions. We will employ these two formulas throughout our tests. As for the speed-up values, we always express the running speed of the recognition using T as the percentage of that of the baseline.

4.1 The Baseline Values

The next step is to construct a working speech recognizer setup to serve as our baseline configuration. First we have to solve the problem of phoneme recognition (see Eq. (4)), for which we will use ANNs. The features used were the typical segment-based ones (for details see [7]). We had a large, general database for training: 332 people of various ages spoke 12 sentences and 12 words each, which were recorded on different computers and sound cards via different microphones [8]. To get a segment-boundary hypothesis we used the brute force method with $k = 8$.

The tests were performed on sentences from the field of medical reports; for this purpose 150 randomly selected sentences were recorded. The language model was a simple word 3-gram; i.e. the probability of the next word depended just on the last two words spoken, and it was calculated by a statistical examination of texts in a similar field. We attained, at the word-level, correctness and accuracy scores of 96.42% and 95.34%, respectively, which then served as our baseline.

4.2 Detecting Phoneme Boundaries

For phoneme boundary detection, first a function which generates the b_i values is needed. As the phoneme boundary positions correlate well with the local spectral changes, we used the first derivates ("Δ values") of the mel-frequency cepstral coefficients (MFCC) [9] as features. For each phoneme in the train database we assigned a value of 1 to the first and last frames, a value of 0 to the middle frame, and a value between 0 and 1 to the rest of the frames depending on how close they were to the sides. Next an ANN was trained for these frames and target values in regression mode. This way evaluating this ANN on any frame a_i, say, generated the corresponding phoneme bound probability value b_i.

4.3 Test Results for the Thresholding Algorithm

Surprisingly this algorithm did not produce good results. Although with some parameters (like $k = 6$, $p_{min} = 0.065$) we were able to get a small speed-up, it also caused a decrease in the recognition scores. With some other parameters (like $k = 6$, $p_{min} = 0.070$) the recognition performance was better than that of the baseline, but this configuration was slower as well. The next table shows the interesting test results we obtained:

k	p_{min}	Correctness	Accuracy	Relative speed
6	0.065	96.06%	94.94%	90.23%
6	0.070	97.49%	95.70%	114.02%
baseline		96.42%	95.34%	100.00%

Usually the method was just too slow, or when this was not the case, the two recognition percentages fell dramatically. These results are probably due to the fact that the phoneme boundary probability estimator ANN was trained to detect the amount of change in the spectrum. Sometimes, however, this change is not abrupt, but occurs over a long time, resulting in just slightly higher b_i values over the longer period. The Thresholding Algorithm was not able to detect these changes, unlike the other two which, in contrast, could take the context of a given frame into consideration.

4.4 Test Results for the Maximum Algorithm

The overall results of this algorithm are somewhat mixed. Although visually inspecting the possible segmentations it produced should be the most promising method of the three, in practice this was not so. The explanation is that although the method usually inserted very few needless possible bounds, it also skipped some necessary ones, and in practice failing to find phoneme bounds is a very serious mistake. The results can be seen in the next table.

k	Correctness	Accuracy	Relative speed
3	93.90%	90.68%	127.73%
4	95.34%	92.83%	72.98%
5	93.19%	89.96%	48.83%
6	82.80%	78.49%	35.70%
baseline	96.42%	95.34%	100.00%

Moreover, the algorithm was tuneable only to a very limited extent: its parameter had to be a small integer, so while with the value of 3 it was slower than the basic method, even with 5 it produced a much lower accuracy value. With the only value left in between (4) the accuracy was also somewhat low (92.83% vs. the baseline 95.34%; a 53% increase in the relative error).

4.5 Test Results for the LIF Algorithm

This algorithm, unlike the others, achieved a satisfactory improvement in speed with no loss in accuracy (see Figure 3). Moreover, it was able to raise the recognition figures along with a decent gain in speed. This was probably due to the fact that the algorithm was better than the other two in some respects. While the Maximum Algorithm was very good at finding possible phoneme bounds,

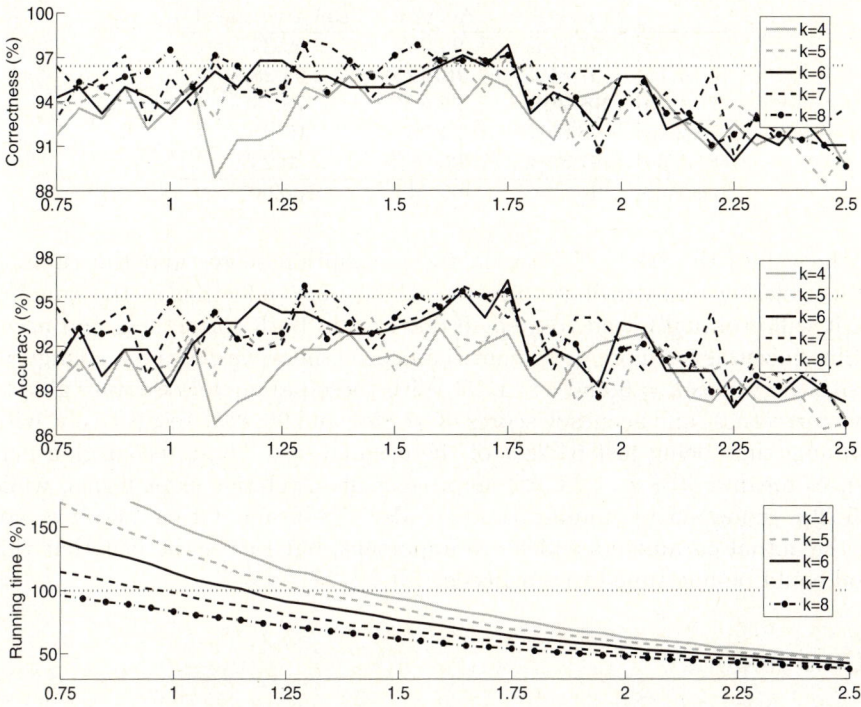

Fig. 3. The correctness, accuracy and running time results of the LIF algorithm with different k and e values; the baseline values are shown by a dotted line

it could not be tuned. On the other hand, the Thresholding Algorithm could be fine-tuned with its two parameters, but the method itself was too simple to produce really good results. The LIF algorithm, however, seems to be the golden mean between the two: it also has two parameters for tuning, and suggests phoneme bounds at much smarter places than the Thresholding Algorithm does, and it takes the context into account.

As for the increase of the recognition scores, it was a surprising result, which was probably due to the combined effect of many factors. Firstly, the phoneme bounds were placed at more accurate positions than they were with the basic method. Secondly, fewer possible phoneme bounds were generated, thus fewer unusable hypotheses were generated, which did not fill the priority queues that much (as the stack size was the same in all cases). And thirdly, due to the fewer possible phoneme bounds being used, it was possible to draw them more closely to each other (i.e. to use a smaller k value) than in the basic method. (The last statement, of course, is true for all three segmentation algorithms.) Some interesting values are shown in the next table, the best values being presented in **bold**.

k	e	Correctness	Accuracy	Relative speed
5	1.75	97.13%	94.98%	69.37%
6	1.75	97.85%	**96.42%**	64.28%
7	1.30	**98.20%**	95.69%	78.58%
7	2.20	96.05%	94.26%	**46.48%**
8	1.30	97.85%	96.05%	70.23%
baseline		96.42%	95.34%	100.00%

From the values the tradeoff between the recognition scores and the running speed can clearly be seen: if we aim for a 50% plus reduction in the running times, it will eventually lead to a small decrease in both correctness and accuracy. Since accuracy is more important than correctness, we chose the configuration with the values $k = 6$ and $e = 1.75$. With these parameters we were able to achieve correctness and accuracy scores of 97.85% and 96.42%, respectively, with the running time being just 64.28% of the original one. These recognition performances mean a 40% and 23.28% improvement in relative error terms, while the 35.72% reduction in running times is also significant. Of course, it is not really the actual parameters which are important, but rather the fact that this method could be fine-tuned to our needs.

5 Conclusions

In this paper we investigated the issue of detecting potential phoneme bounds, which is an important task in segment-based speech recognition. Since the position of the probable phoneme bounds correlate well with local spectral changes, we were able to construct a bound probability estimator function based on them. Then we introduced three novel algorithms for phoneme bound detection instead of the basic, functional-but-slow brute force method, which were all based on this estimator function. The three algorithms were all quite different: the first was a thresholding version of the basic brute force procedure, the second looked for local maxima of the probability estimator function, while the third incorporated a special spiking neuron. Out of these three the last proved to be surprisingly good for the task: not only were we able to speed up the recognition process by 35%, but we could also significantly raise the recognition scores at the same time. These points, we think, make the algorithm worthy of both further study and putting it into practice.

References

1. Jelinek, F.: Statistical Methods for Speech Recognition. MIT Press, Cambridge (1997)
2. Bishop, C.M.: Neural Networks for Pattern Recognition. Clarendon Press, Oxford (1995)
3. Hart, P.E., Nilsson, N.J., Raphael, B.: Correction to a formal basis for the heuristic determination of minimum cost paths. SIGART Newsletter (37), 28–29 (1972)

4. Bahl, L.R., Gopalakrishnan, P.S., Mercer, R.L.: Search issues in large vocabulary speech recognition. In: Proceedings of the 1993 IEEE Workshop on Automatic Speech Recognition, Snowbird, UT (1993)
5. Gerstner, W., Kistler, W.M.: Spiking Neuron Models. Cambridge University Press, Cambridge (2002)
6. Kocsor, A., Tóth, L., Kuba Jr., A.: An overview of the oasis speech recognition project. In: Proceedings of ICAI 1999, Eger-Noszvaj, Hungary (1999)
7. Glass, J., Chang, J., McCandless, M.: A probabilistic framework for feature-based speech recognition. In: Proceedings of ICSLP 1996, Philadelphia, PA, vol. 4, pp. 2277–2280 (1996)
8. Vicsi, K., Kocsor, A., Teleki, C., Tóth, L.: Hungarian speech database for computer-using environments in offices (in hungarian). In: Proceedings of MSZNY 2004, Szeged, Hungary, pp. 315–318 (2004)
9. Huang, X., Acero, A., Hon, H.-W.: Spoken Language Processing. Prentice-Hall, Englewood Cliffs (2001)

Geometric Structure Filtering Using Coupled Diffusion Process and CNN-Based Approach

Bartosz Jablonski

Institute of Computer Engineering, Control and Robotics
Wybrzeze Wyspianskiego 27, 50-370 Wroclaw, Poland
bartosz.jablonski@pwr.wroc.pl

Abstract. Image processing algorithms are being intensively researched in the last decades. One of the most influential filtering tendencies is based on partial differential equations (PDE). Different kinds of modifications of classical linear process were already proposed. Most of them are based on non-linear or anisotropic process taking into consideration local descriptor of image structure. Main goal is to remove noise and simultaneously to decrease level of blurring important features (like edges). In this paper a new approach is presented, which introduces, into non-linear diffusion process, extra knowledge about geometric structures existing on an image. Algorithm scheme is proposed and results of numerical experiments are presented. Moreover, possibilities of algorithm application within cellular neural networks paradigm will be analysed.

Keywords: image processing, diffusion filtering, geometric structures, cellular neural networks.

1 Introduction

Within image processing domain many different approaches have already been developed. Image filtering is a process, which goal is to find original, restored version of an image (e.g. noise removal added by acquisition methods) or to extract specific, important features (edges, shapes, etc.) for further analysis.

Filtering process removes noise, but also often leads to decreasing level of useful information or introducing new kinds of artifacts. Gaussian blur algorithm has high performance of noise removal, but edges on processed image are getting blurred. Gaussian blur can be computed as a simulation of linear diffusion process with homogenous parameters described by partial differential equation (PDE). Perona and Malik [15] proposed to control the process in non-linear way: speed of diffusion is decreased in areas, where existence of edges is more probable. The algorithm was initially named anisotropic, although the exact anisotropic approach was later presented by Weickert [21]. These approaches resulted in significant improvement in filtering performance and stable edges for many iterations.

Diffusion based algorithms have been studied in detail within practical and theoretical frameworks [19], [20]) also for colour images [17], robust statistics [2], higher order derivatives [23] [9] and memory terms [7].

Most approaches in diffusion filtering domain are concentrated on description of image structure based on local features like edge estimator. We propose another approach, which makes use of a priori knowledge. A new kind of algorithm will be presented, which takes into account both local edge estimation and global description of geometric structures, which are assumed to be present on an image. As a result speed of filtering process can be increased and edges can be more regularly preserved.

The rest of the paper is organised in the following way. In the second chapter state of the art PDE-based filtering methods are presented. This is the base for construction of geometric coupled filtering algorithm presented in chapter three. Chapter four consists of the results of numerical filtering experiments and examples. Analysis of cellular neural network (CNN) application is described in chapter 5. Conclusions and further works are collected in chapter six.

2   State of the Art PDE-Based Filtering Methods

Filtering processing, which is based on partial differential equations (PDEs) will be used for the construction of coupled filtering algorithm. Application of linear diffusion process leads to two effects: noise removal and simultaneously edge blurring. To remove the latter, the following modification of the process was proposed [15]:

$$\frac{\partial u}{\partial t} = div\left(c(x,y,u,t)\nabla u\right) = c(x,y,u,t)\Delta u + \nabla c \cdot \nabla u, \qquad (1)$$

where image is defined by the function of brightness $u(x,y)$ at the given time t of diffusion process. Diffusion coefficient c depends on the underlying image structure. In the cited paper the coefficient depends on the estimation of presence of edges E, which is approximated by discretization of local gradient:

$$c(x,y,z,t) = g\left(\|\nabla u(x,y,t)\|\right) \qquad (2)$$

Function g is called stopping function - it decreases diffusion speed in areas, where the presence of edges is more probable. There are many functions of this type, of which two were originally presented:

$$g_1(s) = exp\left(-\left(\frac{s}{K}\right)^2\right), g_2(s) = \frac{1}{1+(\frac{s}{K})^2}. \qquad (3)$$

where variable K plays role of contrast parameter. In the cited paper, the following approximation was proposed for the equation (1):

$$u_s^{t+1} = u_s^t + \frac{\lambda}{4}\left[c_N\nabla_N u + c_S\nabla_S u + c_W\nabla_W u + c_E\nabla_E u\right]_s^t. \qquad (4)$$

where indices N, S, W, E refer for four directions respectively of numerical local differences of the function u. Parameter λ is used for scaling purposes and assures stability of the process. This form of discretization will be also used in the next steps of our approach.

3 Introducing Extra Geometric Knowledge

The Model described by equations (1)-(4) takes into consideration only local features of the underlying image. The novelty in this paper is to combine information from two different sources: non-linear diffusion process and filtering properties of Hough transform as a general descriptor for global image geometric structure.

Gerig et al. [12] suggested a possibility to combine information from two synchronised acquisition channels by modifying diffusion coefficient. More complex approach was proposed in [16], where information about underlying structures is introduced by analytical modification of the diffusion tensor coefficient. These ideas are the starting point for our proposal.

3.1 Hough Transform Fundamentals

The original version of transformation for straight lines was given by Hough in the patent application in 1962; analytic formulation was presented by Duda and Hart in [10]. Hough transform converts two dimensional domain of image values into N-dimensional space of curve parameters. This space describes geometric structures existing on an image [1],[13]. Inverse Hough transform can also be realised [14] denoted as $H_\Psi^{-1}(u(x,y))$ or $H_\eta^{-1}(u(x,y))$. Parameter Ψ defines the threshold value (the sensitivity level) in the Hough space and parameter η defines the number of maximum values in Hough space, which are taken into account.

3.2 PDE-Based Coupled Filtering Algorithm

Coupled filtering algorithm should consider two factors: local estimator of edges and global descriptor of geometric structures. It can be obtained modifying diffusion coefficient, so it depends on both factors:

$$\frac{\partial u}{\partial t} = div\left(c_c(x,y,u,t)\nabla u\right), \qquad (5)$$

where c_c is the coupling diffusion coefficient defined as:

$$c_c = f\left(g(|\nabla u|), g_H(|\nabla H^{-1}|)\right). \qquad (6)$$

The coupling is realised in the common domain of derivatives of brightness function. The first component corresponds to standard control variable representing local edge estimator. Usage of the inverse Hough transform allows connecting two channels of information. The second component is also interpreted as an estimator of edges in the form of local gradient. However, the component results from shape analysis (carried out using Hough transform), instead of the straightforward analysis of edges (carried out using local estimation of gradient function). Hence each point in the inverse transform domain does not reflect solely local character of underlying image values, but results from global interpretation of geometric structures.

It is possible to compute the c_c in different ways depending, which factor should be enhanced. We propose the following useful additive formula:

$$c_c = g\left((1-h_f)\|\nabla u(x,y)\| + h_f\|\nabla(K_c * H_\Psi^{-1})\|\right). \tag{7}$$

The $h_f \in (0,1]$ (called Hough factor) defines the character of the coupling (more local or more global descriptor), which is used to construct the diffusion coefficient. Parameter K_c plays the role of contrast coefficient for inverse Hough transform H_Ψ^{-1}. Bounded values of c_c assure, that the overall differential process is stable and all of properties of the equation (1) are maintained.

We have also noticed, that the Hough factor can be computed adaptively, which has a great positive impact on the process:

$$h_f(t) = h_{max}\exp\left(\frac{-(N(t)-h_\mu)^2}{h_\sigma}\right), \tag{8}$$

basing on the function $N(t)$, which estimates the overall amount of noise in the image:

$$N(t) = \frac{1}{wh}\sum_{x=0}^{w-1}\sum_{y=0}^{h-1}|\nabla u(x,y,t)|, \tag{9}$$

where w, h are image dimensions width and height respectively. Parameter h_{max} is the scaling factor, which defines the maximum value of the coupling. Parameters h_μ and h_σ define shape of the kernel - the mean value and the standard deviation respectively. We noticed that the character of the process is less sensitive to the choice of these parameters, than to the choice of constant Hough factor value.

4 Filtering Performance and Examples

4.1 Experimental Methods

Experiments were carried out using two test images. Both images represent simple shapes in order to check the performance of the coupled filtering algorithm. Noise with Gaussian distribution was added to test images (see figure 1 - first pair represents test image 1, second pair - test image 2). The first sample will let us to verify filtering performance, when exact number of geometric structures is known. Unknown number of features will be considered for second image. We use standard Hough transform for straight lines and Sobel edge detection. Results are compared to the standard Perona Malik algorithm.

For numerical experiments, the following procedure was performed:

- Gaussian noise has been added to the original image.
- Noised version of the test image were filtered for the given number of iterations.
- After each iteration, signal to noise ratio (SNR) was computed to measure the similarity of filtered image to original (reference) version.
- Graph representing SNR value against iteration number has been prepared.

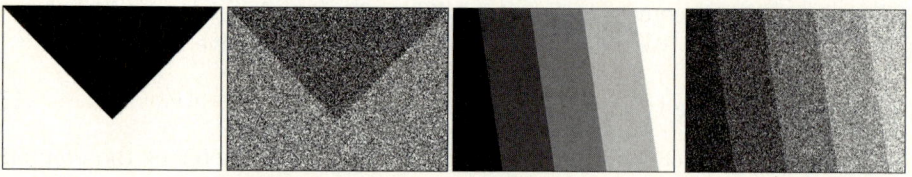

Fig. 1. Test images used in the experiments: original img. 1; Gaussian noised img. 1; original img. 2; Gaussian noised img.2

Signal to noise ratio is computed basing on mean square error value as:

$$SNR(u_0, u_1) = 10 \log_{10} \frac{\sigma_{u_0}^2}{MSE(u_0, u_1)}, \quad (10)$$

where $\sigma_{u_0}^2$ is the estimated variance of reference image function u_0. Mean square error for two images u_0 and u_1 is evaluated taking into consideration each pair of pixels:

$$MSE(u_0, u_1) = \frac{1}{wh} \sum_{i=1}^{w} \sum_{j=1}^{h} d^2(u_0(i,j), u_1(i,j)), \quad (11)$$

where w, h are image dimensions and $d(\cdot, \cdot)$ is the distance function, which in this case is computed using Euclidean distance.

4.2 Numerical Simulation

Figure 2 represents filtering performance graph, which was measured as signal to noise ratio (SNR) for coupled and standard algorithm. Coupled filtering algorithm was executed for h_σ ranging from 10 to 20 ($h_\mu = 10$). Moreover, we tested two kinds of stopping functions. SNR factor was computed after each iteration step.

The graph confirms that the usage of the coupled algorithm results in the improvement of image quality. Moreover the influence of the adaptive computation of the Hough factor is easily noticeable. After about 15 iterations the influence of geometric structure descriptor causes, that the filtering is being processed much faster than in the standard PM model. The impact of the introducing extra information does not only preserve shapes, but also accelerates the process of filtering within areas constant value without edges.

Visual comparison shows, that the quality of these aspects of an image is improved more than it would result from the simple comparison of SNR values. The measure takes into consideration global distance between two images without reflecting the quality of important meaningful features like edges. Figure 3 represents results after selected iterations of the process of the test image 1.

The standard PM process was used using stopping function g_2 (equation 3), $\lambda = 0.25$ and adaptive computation of K coefficient. Second row represents the results of the coupled process using the same basic parameters. Additive coupling was used with adaptive computation of the Hough factor ($h_\mu = 10$,

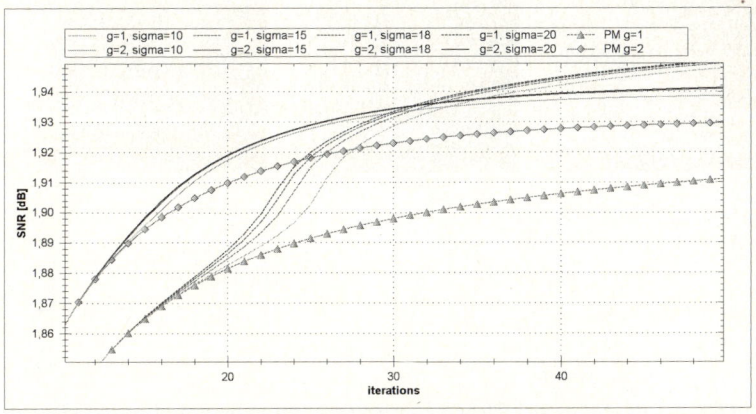

Fig. 2. Signal to noise ratio for the results of filtering processes with different parameters in comparison to the standard PM model

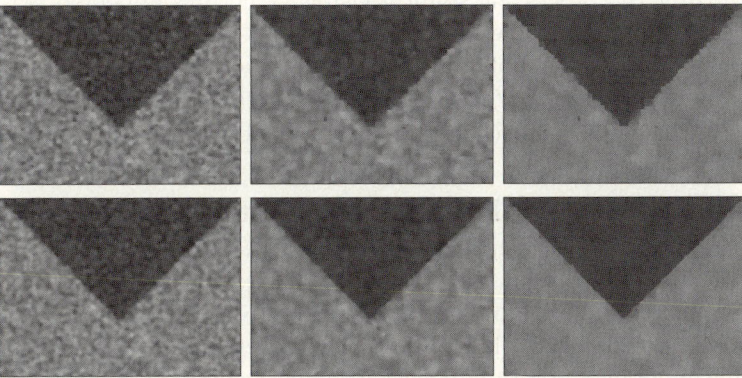

Fig. 3. The comparison of the standard PM model (first row) to the coupled filtering algorithm (second row). Columns represent results after 20, 50, 200 iterations

$h_\sigma = 20$). Inverse Hough transform used $\eta = 2$ maximum values for the geometry reconstruction and was computed every 5 iterations.

Analysing the results, several interesting effects can be distinguished: diffusion process placed away from edges is accelerated; solid regions become smoother; more artifacts are removed by coupled adaptive algorithm.

Figure 4 presents the results of filtering, when unknown number of features are being taken into consideration. Two version of coupled algorithm were used: constant Hough factor ($h_f = 0.3$, second column) and adaptive Hough factor ($h_\mu = 10$, $h_\sigma = 25$, third column).

The visual inspection of the results allows to easily notice the high level of distortion of the edges obtained using standard PM algorithm. Moreover there are areas, which in the original image (without noise) have constant brightness

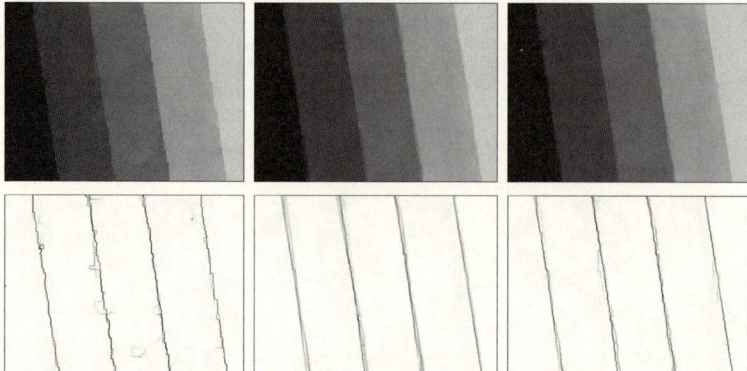

Fig. 4. The results of the image filtering (first row) and the Sobel based edge detection (second row) after 400 iterations. Standard PM algorithm - first column, level based coupling: constant Hough factor - second column, adaptive factor - third column.

values, but after the processing slightly visible structures appeared which were produced by the noise and the filtering process.

These results are the proof, that the combination of information from two channels describing image structure (both global and local) improves filtering performance in the meaning of process acceleration and the quality of output.

5 Application of Cellular Neural Network Paradigm for the Coupled Algorithm

Main disadvantage of the standard algorithm is its computational complexity. Most of image filtering algorithms process each pixel separately basing on values of neighbouring points. Solution of these problems can be approached using cellular neural networks (CNN) paradigm. Construction of the coupled algorithm, which is based on partial differential equations, is very suitable for the CNN approach.

5.1 Fundamentals of Cellular Neural Network Approach

The CNN paradigm was initially proposed by Chua et al. and developed in several consecutive papers ([4], [5], [6]). Because construction of the CNN is directed towards solving spatio-temporal problems, the approach was also applied for image filtering based on variational methods [11].

Model based on CNN is a discrete realisation of analog circuit dynamics. Network consists of n-dimensional array of identical cells, which represent dynamic systems [4]. Each cell is connected with a finite number of neighbouring cells within defined radius r. All state variables in the system are continues. State equation describing interaction of one cell can be written in the following form [4]:

$$\frac{dv_{xij}(t)}{dt} = -v_{xij}(t) + \sum_{kl \in N_r(ij)} A_{ij;kl}\left(v_{ykl}(t), v_{yij}(t)\right)$$
$$+ \sum_{kl \in N_r(ij)} B_{ij;kl}\left(v_{ukl}(t), v_{uij}(t)\right) + I_{ij}. \tag{12}$$

The output - state equation is formed as:

$$v_{yij} = f(v_{xij}). \tag{13}$$

Interaction between output variables of neighbouring cells is described by A template, which is a matrix of $(2r+1)x(2r+1)$ size. Interaction between input variables of neighbouring cells is described by the B template, which is a matrix of the same size. In general A and B templates can be nonlinear with additional delay terms. There is no direct interaction between state variables of neighbouring cells. Parameter I_{ij} represents threshold current source, which in many applications is constant $I_{ij} = I$. State - output function mapping can be any function like unity gain, threshold function (with I_{ij} used), Gaussian, etc.

Many different methods have already been developed to compute connection templates. One of the most formal approach was described by Crounse et al. [8]. Tha authors show, that different image processing tasks can be realised using cellular neural network.

5.2 Application of the CNN for Obtaining PDE Steady State Solution

Proposal of application cellular neural network for image denoising based on PDE formulation was presented in [11]. It bases on the Rudin-Osher-Fatemi model, which utilises minimisation of total variation for the following energy expression:

$$E(\Phi_0) = \iint_\Omega \left(\alpha(\Phi - \Phi_0)^2 + \lambda|\nabla\Phi_0|\right) dxdy \tag{14}$$

where Φ_0 is the original two dimensional image defined on area Ω.

In the cited paper the following CNN templates have been proposed to find a solution of the problem:

$$A = \begin{bmatrix} 0 & a & 0 \\ a & 1 & 1 \\ 0 & a & 0 \end{bmatrix}, B = \begin{bmatrix} 0 & 0 & 0 \\ 0 & b & 0 \\ 0 & 0 & 0 \end{bmatrix}, I_{ij} = I = 0, \tag{15}$$

where coefficients a, b are to be defined as:

$$a = \lambda sgn(y_{ij} - y_{kl}), b = 2\alpha(x_{ij} - u_{ij}), \tag{16}$$

where sgn is the sign function. The following notation holds: x_{ij}, u_{ij}, y_{ij} represent state, input and output of the current cell respectively and y_{kl} represents a variable of output of neighbouring cell. In the paper transfer function was linear, so output of the cell corresponds directly to current state variable.

It is possible to obtain the CNN version of the Perona-Malik process modifying coefficients of the template A in the following way:

$$\alpha_{PM} = \lambda f(y_{ij} - y_{kl})|y_{ij} - y_{kl}|, \tag{17}$$

where a function f can be computed as expressions (3). The template B could remain unchanged, because the relation can be mapped to both sides of equation (4) with the accuracy of the constant coefficient 2α.

Both models can be used for image filtering. As it was reported by Gacsadi et al. the application of the sign function (16) leads to very interesting results. Using modified version of the coefficient (17) should give comparable or even better results. The equivalent of the coupled approach will be presented in the next section.

5.3 The Proposal of Coupled Filtering Algorithm Realised by CNN

Basing on the presented approach, we can propose the application of CNN for previously presented (chapter 3) coupled algorithm. The algorithm itself (see chapter 3.2) is the realisation of specific PDE process, so it should be possible to obtain a CNN structure, which gives similar results.

We assume that ready cellular neural network solution for Hough transform will be used. Fundamentals of this idea were presented by Wu et al. [22] in the paper concerning Radon transform of a binary image.

Concept presented in [18] introduces the idea of multi-layer networks used for finding solutions of stereo vision problems. Technically the idea is similar, however the goal is completely different. In the paper of Taraglio, the solution of optimisation problem is sought. In this paper, the minimisation of image energy is obtained.

We propose to build a multi-layer cellular neural network, which consists of the following layers (see figure 5): Horizontal gradient layer (HGL) - horizontal component of the gradient; Vertical gradient layer (VGL) - vertical component; Hough transform layer (HTL) - generates geometric interpretation of the given image; Inverse Hough transform layer (IHTL) - generates geometric reconstruction of original image basing on HTL; Horizontal Hough transfer gradient layer (HHTL) - generates horizontal component of Hough transform; Vertical Hough transfer gradient layer (VHTL) - vertical component; Coupled adder layer (CAD), which combines output of previous layers (HHTL, VHTL, HGL, VGL) and generates a matrix of nonlinear coefficients; Nonlinear filtering layer (NFL), which realises the actual diffusion process modelled by partial differential equation.

Templates for the first 2 layers are very easy to construct:

$$HGL: A_1 = \begin{bmatrix} 0 & 0 & 0 \\ 0 & 1 & -1 \\ 0 & 0 & 0 \end{bmatrix}, VGL: A_2 = \begin{bmatrix} 0 & -1 & 0 \\ 0 & 1 & 0 \\ 0 & 0 & 0 \end{bmatrix}. \tag{18}$$

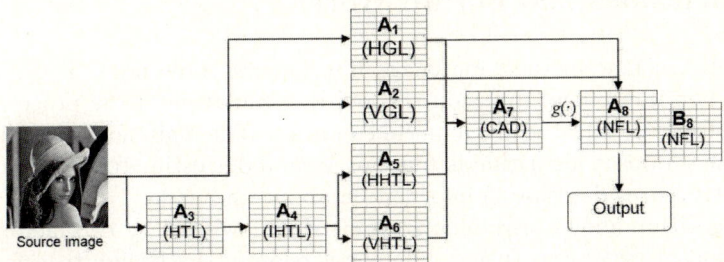

Fig. 5. The schema of the multilayer cellular neural network (CNN) realizing coupled image filtering algorithm

Definition of templates B are not necessary in this case; transfer function f is linear mapping $v_{yij} = v_{xij}$.

Layers 3rd and 4th, which are necessary to compute Hough transforms (HTL, IHTL) can be easily obtained using well known approaches (see [22] for details). Horizontal and vertical gradient of the Hough transform results are in fact counterparts of equations (18) respectively.

The most important part for the whole system of neural networks is the coupling layer (CAD), which can be modelled using the following cellular template:

$$CAD : A_7 = \begin{bmatrix} 0 & \|\alpha_V\| + \|\beta_V\| & 0 \\ -(\|\alpha_V\| + \|\beta_V\|) & 0 & \|\alpha_V\| + \|\beta_V\| \\ 0 & -(\|\alpha_V\| + \|\beta_V\|) & 0 \end{bmatrix}, \quad (19)$$

where the following notation has been assumed:

$$\alpha_H = v_{yij}(A_1), \alpha_V = v_{yij}(A_2), \beta_H = v_{yij}(A_5), \beta_V = v_{yij}(A_6). \quad (20)$$

Stopping functions (3) can be used as transfer function for the CAD layer. The final layer gives the actual result of the process due to the usage of the following templates:

$$NFL : A_8 = \begin{bmatrix} 0 & \|\alpha_V\| & 0 \\ -\|\alpha_V\| & 1 & \|\alpha_V\| \\ 0 & -\|\alpha_V\| & 0 \end{bmatrix}, B_8 = \begin{bmatrix} 0 & 0 & 0 \\ 0 & \lambda(x_{ij} - u_{ij}) & 0 \\ 0 & 0 & 0 \end{bmatrix}. \quad (21)$$

The input of the final layer is the result obtained from the CAD layer. Hence the direct mapping for the pixels in template (21). The proposed multi-layer cellular neural network can be visualised on the following figure 5. As one can notice, the main advantage of the cellular neural network approach is the possibility to compute each part of coupled algorithm independently, so the final output can be obtained only after 5 time units. Thanks to this approach it is possible to utilise the network structure for intensive streaming data processing. However the actual processing time would strongly depend on type of hardware implementation of the structure, which is out of the scope of this paper.

6 Conclusions and Future Works

In the paper a new method for image filtering was presented. The method is based on non-linear heat diffusion realised by simulation of the process for differential equation with the given initial condition. The main novelty in the paper is, that the proposed algorithm is coupling local (edge estimator) and global (geometric structures descriptor) information about an image.

The algorithm allows introducing geometric knowledge by two factors: defining the kind of structures (using different version of the Hough transform) and determining amount of structures present on an image. Additionally two methods of taking into consideration quantity of structures were proposed (exact number of structures or relative amount of features).

Presented experiments proved usefulness and properties of the proposed algorithm. The most important advantages of the algorithm include: filtering acceleration, more efficient speckles removal and better preservation of edges (more stable and regular).

Additionally we considered a possibility to implement the coupled algorithm using cellular neural network. This approach can lead to real-time filtering system for highly noised image sources. However, is still further development and testing of hardware solutions.

Future works could be also focused on the application of different kinds of inverse Hough transform [14] and better local structure descriptor [21]. Hence the proposed coupled algorithm should be considered more as a framework, rather than a single solution. It extends the approach of Snchez-Ortiz et al. [16], which considers analytical type of extra knowledge. Proposed results are the starting point for introducing extra knowledge of geometric structures into locally realised non-linear filtering process.

Acknowledgements. The author would like to gratefully acknowledge the anonymous reviewers for all valuables remarks and suggestions.

References

1. Ballard, D.: Generalized Hough transform to detect arbitrary patterns. Pattern Recognition 13(2), 111–122 (1981)
2. Black, M., Sapiro, G., Marimont, D., Heeger, D.: Robust Anisotropic Diffusion. IEEE Trans. On Image Processing 7(1), 421–432 (1998)
3. Catt, F., Lions, P.L., Morel, J.M., Coll, T.: Image selective smoothing and edge detection by nonlinear diffusion. SIAM Journal of Numerical Analysis 29, 182–193 (1992)
4. Chua, L.O., Roska, T.: The CNN Paradigm. IEEE Trans. on Circuits and Systems 40(1), 147–156 (1993)
5. Chua, L.O., Yang, L.: Cellular Neural Networks: Theory. IEEE Trans. on Circuits and Systems 35(10), 1257–1272 (1998)
6. Chua, L.O., Yang, L.: Cellular Neural Networks: Theory. IEEE Trans. on Circuits and Systems 35(10), 1273–1290 (1998)

7. Cottet, G.H., El Ayyadi, M.: Nonlinear PDE operators with memory terms for image processing. In: Proc. IEEE International Conference on Image Processing, vol. 1, pp. 481–483 (1996)
8. Crounse, K.R., Chua, L.O.: Methods for Image Processing and Pattern Formation in Cellular Neural Networks: A Tutorial. IEEE Trans. on Circuits and Systems 42(10), 583–601 (1995)
9. Didas, S., Weickert, J., Burgeth, B.: Stability and Local Feature Enhancement of Higher Order Nonlinear Diffusion Filtering. In: Kropatsch, W.G., Sablatnig, R., Hanbury, A. (eds.) DAGM 2005. LNCS, vol. 3663, pp. 451–458. Springer, Heidelberg (2005)
10. Duda, R., Hart, P.: Use of the Hough transformation to detect lines and curves in the pictures. Communications of the ACM 15(1), 11–15 (1972)
11. Gacsdi, A., Szolgay, P.: A Variational Method for Image Denoising, by using Cellular Neural Networks. In: Proc. of The 8th IEEE International Biannual Workshop on Cellular Neural Networks and their Applications, Budapest, Hungary (2004)
12. Gerig, G., Kikinis, R., Kbler, O., Jolesz, F.A.: Nonlinear Anisotropic Filtering of MRI Data. IEEE Trans. on Medical Imaging 11(2), 221–231 (1992)
13. Guil, N., Villalba, J., Zapata, E.: A Fast Hough Transform for Segment Detection. IEEE Trans. on Image Processing 4(11), 1541–1548 (1995)
14. Kesidis, A.L., Papamarkos, N.: On the Inverse Hough Transform. IEEE Trans. on Pattern Analysis and Machine Intelligence 21(12), 1329–1343 (1999)
15. Perona, P., Malik, J.: Scale-Space and Edge Detection Using Anisotropic Diffusion. IEEE Trans. On Pattern Analysis and Machine Intelligence 12(7), 629–639 (1990)
16. Snchez-Ortiz, G.I., Rueckert, D., Burger, P.: Knowledge-based tensor anisotropic diffusion of cardiac magnetic resonance images. Medical Image Analysis 3(1), 77–101 (1999)
17. Sapiro, G., Ringach, L.: Anisotropic Diffusion of Multivalued Images with Applications to Color Filtering. IEEE Trans. on Image Processing 5(11), 1582–1586 (1996)
18. Taraglio, S., Zanela, A.: A practical use of cellular neural networks: the stereovision problem as an optimisation. Machine Vision and Applications (11), 242–251 (2000)
19. Weickert, J.: Theoretical Foundations of Anisotropic Diffusion in Image Processing. Computing Supplement 11, 221–236 (1996)
20. Weickert, J.: A Review of Nonlinear Diffusion Filtering. In: ter Haar Romeny, B.M., Florack, L.M.J., Viergever, M.A. (eds.) Scale-Space 1997. LNCS, vol. 1252, pp. 3–28. Springer, Heidelberg (1997)
21. Weickert, J.: Anisotropic Diffusion in Image Processing. B.G. Teubner, Stuttgart (1998)
22. Wu, C.W., Chua, L.O., Roska, T.: A two-layer Radon transform cellular neural network. IEEE Trans. On Circuits and Systems 39(7), 488–489 (1992)
23. You, Y.L., Kaveh, M.: Image Enhancement Using Fourth Order Partial Differential Equations. IEEE Trans. on Image Processing 9(10), 1723–1730 (2000)

MARCoPlan: MultiAgent Remote Control for Robot Motion Planning

Sonia Kefi[1,2], Ines Barhoumi[1,2], Ilhem Kallel[1,2], and Adel M. Alimi[2]

[1] REGIM : REsearch Group on Intelligent Machines
Engineering School of Sfax, University of Sfax, Tunisia
[2] Department of Computer Science, High Institute of Computer Science and Management of Kairouan, University of Kairouan, Tunisia
{Adel.Alimi, Ilhem.Kallel, Sonia.Kefi}@ieee.org, Ines_isig@yahoo.fr

Abstract. A multiagent system to support a mobile robot motion planning has been presented. Baptized *MARCoPlan* (MutiAgent Remote Control motion Planning), this system deals with optimizing robot path. Considered as an agent, the robot has to optimize its motion from a start position to a final goal in a dynamic and unknown environment, on the one hand by the introduction of sub-goals, and on the other hand by the cooperation of multiagents. In fact, we propose to agentify the proximity environment (zones) of the robot; cooperation between theses zones agents will allow the selection of the best sub-goal to be reached. Therefore, the task of the planner agent to guide the robot to its destination in an optimized way will be easier. *MARCoPlan* is simulated and tested using randomly and dynamically generated problem instances with different distributions of obstacles. The tests verify some robustness of *MARCoPlan* with regard to environment changes. Moreover, the results highlight that the agentification and the cooperation improve the choice of the best path to the sub goals, then to the final goal.

Keywords: Agentification, multiagent model planning, cooperation, mobile robot, sub-goal, MARCoPlan.

1 Introduction: Environment and Planning Dilemma

The multiagent systems (SMA) are particularly adapted to propose reactive and robust solutions with complex problems whose centralized control is very penalizing. They are ideal systems to represent problems of multiple resolution methods and multiple prospects [12]. One of the complex problems is the planning tasks. Indeed, for a long time, planning has been the interest of researchers in artificial intelligence, and the first applications were founded on the assumption that the environment of the planner is stable and it is possible to create complete plans before executing them [5]. Actually, an environment is always dynamic and unknown for its navigator; it is thus a question of founding a system able to plan while having only one part of required information, the other part is formed progressively as the navigator changes its state; this is called *a continual planning* [1].

It has been proved that a distributed approach, especially the multiagent approach supports a continual planning system. In fact, multiagent approach has well known properties to find a solution through a continuous collaboration between several agents, either software or human agent [2]. There are three multiagent planning modes based on agents cooperation in order to plan a path.

The centralized planning for distributed plans is focused on the agents control and coordination in shared environments. This planning allows the generation of a partial plan, to break up the plan into sub-plans, to insert the actions of synchronization in sub-plans, to allocate sub-plans with the agents and to initiate the execution of the plan. Its main advantage is to facilitate the resolution of the conflicts and convergence towards a total solution. But, this planning is facing a major problem that is the centralization of control in only one agent [3].

The distributed planning for centralized plans is more focused on planning: its process and its extension to a distributed environment. Indeed, the process of the complex planning is distributed between various specialist agents who cooperate and communicate by sharing objectives and representations to form a coherent plan. Its major advantage is that there is no centralization of control. Yet, its disadvantage is that the cost of communication has increased and it does not guarantee convergence: the goals and the interventions incompatibility, incoherent knowledge and the partial plans have different representations [4] [9].

The distributed planning for distributed plans: The process of synthesis and execution of a multiagent plan is distributed. This planning can be task-directed where there is a total goal and the synthesis of the sub-plans is based on the results and not on the plans. Besides, it can be agent-directed where there is a mechanism allowing the execution of competitor plans, instead of a plan or a total goal [10] [11].

The mobile robot motion planning is more focused on a continual planning since the robot motion is commonly carried out from an initial position to a final goal. But there are two different types of motion method: one is directly carried out to the goal, the second by introducing sub-goals.

The former assumes that the distance from the robot to the goal decreases in case of absence of obstacles. With this method, there are two situations: the robot can navigate directly to the goal throughout an open space from where it creates a NP complete problem, or the robot can advance to the goal by avoiding obstacles while keeping an always decreasing distance to the goal (swing mode). The latter dynamically introduces intermediate goals to enable the robot to attend progressively its goal. This solution is interesting because the sub-goal is closer to the robot. Therefore, the robot can easily recognize and reach the final goal. In [7] and [13], have presented a fuzzy system of robot motion planning in a partially-known environment, based on the concept of sub goals. This system initially generates a visibility graph, and then it uses a fuzzy controller to make decisions for next step. In [8], Kim has used the concept of selection of sub-goals for the robots navigation, where the robot initially generates an environment chart, by subdividing it into sectors containing the localization information, then, if it is necessary a sub-goal is selected. It must be reliable so that the robot does not run a risk by joining it.

In this paper, we propose a multiagent model of the robot motion planning. We present a new method of multiagent planning with partial agentification of the environment and introduction of sub-goals throughout the robot motion to the final goal.

This paper is organised as follows. Section 2 proposes a multiagent architecture as well as a model based on agents to support a planning in a dynamic and a partially agentified environment. Section 3 presents our new method of the robot motion planning through the introduction of the sub goals. Section 4 justifies the contribution of the multiagent cooperation to the robot motion planning, by several simulations. To conclude, section 5 gives an idea on some fields resulting from the real world where this model can be applied.

2 Multiagent Modeling

The researchers in the multiagent field have suggested to inspire the methodologies of design human organisations, drawn from the theory of the organizations, to conceive the SMA. They have been summarized in a set of metaphors emphasizing seven essential dimensions of a multiagent cooperative organization:

- The identification of the common goal
- The nature of environment which includes the organisation of agents
- The identification of the agents and their roles
- The structure of control between agents
- The social structure of the agents
- The agents collaborative structure.
- The static structure which concerns the presentation of all classes needed to realize the simulation. Since, the paper is more focused on dynamic multiagent aspects, we have choosen not to develop this part here.

2.1 Interest of Environment Agentification

The navigation environment can be divided into elementary zones of the same size that may or may not contain obstacles. The robot has to move from one zone to another free adjacent zone. Practically, the robot motion planning in an environment towards a goal is reduced to a series of motion planning towards sub-goal in partial environment. This series ends when the sub-goal becomes the final goal. A partial environment is managed by a set of Zone Agents representing adjacent zones; we call this phenomenon *environment agentification*. Therefore, the environment agentification is done by successive steps, thing that allows to represent the robot proximity information as competitive agents when each one tries to attract the sub goal according to its direct proximity.

2.2 *MARCoPlan* Hierarchical Architecture

In *MARCoPlan*, there is an organization which structures the cohabitation rules and the agents collective work through the definition of various roles, shared resources and tasks dependency. Figure 1 presents the organisational architecture of the system.

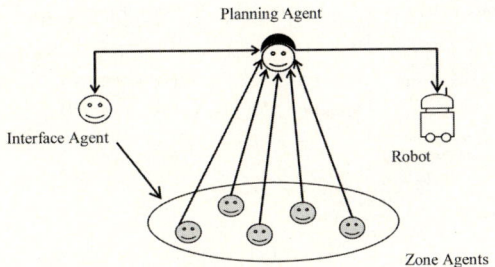

Fig. 1. MARCoPlan Architecture

All agents work simultaneously; each agent has the capacity to communicate with its acquaintances and to recognize the representation coming from others. Indeed, *MARCoPlan* manages four types of agents:

Interface Agent: It is considered as an intermediate agent between the user and the system.

Zone Agents: They are reactive agents which are dynamically created along the robot motion. They manage the proximity zones of the robot environment. Each agent determines and sends descriptive information of its zone state to the planning agent.

Planning Agent: It is a deliberative agent. It is equipped with means and mechanisms of communication to manage the interactions with other agents (cooperation and coordination). It studies the feasibility of received information from zone agents. It plans an optimal partial robot path to reach the next sub-goal, while avoiding the obstacles.

Robot Agent: Our robot is considered as a reactive agent. It can perceive, communicate and execute the plan given by the Planning Agent.

2.3 Functional and Social Structure

Nowadays, an intelligent system strives for a set of communicating and collaborating entities which are acting in an environment cannot function by itself. Thus, it is legitimate to study such systems socially. In fact, *MARCoPlan* forms a society which is made up of several groups; each group is formed of a set of agents having well defined roles. These agents will cooperate to find the shortest path guiding the robot to the goal while avoiding static and dynamic obstacles. We define the group as a primitive concept of regrouping agents. Each agent can be a member of one groups or more. We associate a set of roles with each

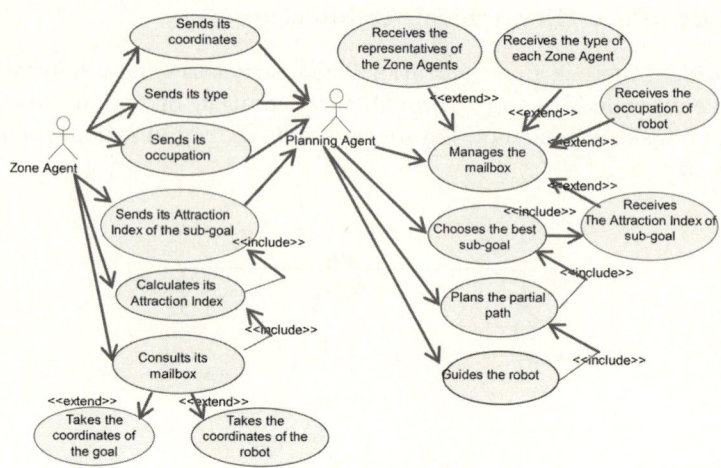

Fig. 2. Functional structure of *MARCoPlan* agents

group. A role can be a service, a function or simply an identification of another agent. For a better representation of the agents functional structures, we propose to adopt diagrams coming from UML (Unified Modeling Language). Figure 2 illustrates the functional structure of each Zone Agent and the Planning Agent.

The cooperative structure of *MARCoPlan* is based on a succession of agent actions running in parallel and exchanging synchronized messages. Figure 3 represents the associated sequence diagram which represents the dynamic aspect of agents while putting the emphasis on the messages sending chronology.

3 Multiagent Planning

The planning action implies considering an environment abstraction. Our environment is considered as a set of contiguous geometric forms (zones) having a defined size. It can contain many obstacles defined statically or dynamically. The mobile robot, occupying one zone, is managed by remote control of a multiagent system. One zone can contain one obstacle and it is possibly agentified, this means that is it can be controlled by a Zone Agent.

3.1 Planning with Introduction of Sub-goals

It is necessary to be able to prepare a path for the robot allowing it to join a sub-goal from any position. It is the role of the Planning Agent to solve this problem by remote control. Once the position of the sub-goal is known, the Planning Agent makes it possible to plan the actions to be undertaken to reach it. The robot has to move from a zone to another adjacent one. Since the robot motion planning is a very complex problem, where we obtain a NP-complete problem. Therefore, we propose to simplify it by adapting the introduction of the sub-goals method and agentification of environment; the application of an algorithm

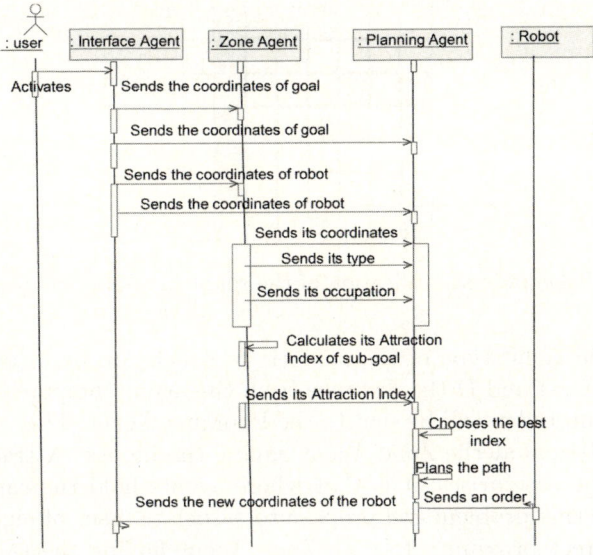

Fig. 3. Cooperative structure of *MARCoPlan*

to research the shortest path in a partial environment becomes easier. Indeed, the number of possible solutions falls considerably and the optimum path will be found in the shortest time.

Concretely, the robot motion planning to a goal, in a furnished environment, is reduced to a series of planning to sub-goals in a furnished partial environment. This series stops when the sub goal becomes the final goal. Each partial environment, which is in fact a set of adjacent zones, is controlled by a set of Zone Agent; we speak thus about the agentification of zones in a the partial environment.

As it is illustrated by figure 4, the agentification is related to the (5x5) zones of the environment, because we adopt the concept of sub-goals by using a geometrical method of planning [8], the zone of the medium hosts, the mobile robot and the other zones are likely to attract the sub-goal.

3.2 Choice of Sub-goal

In *MARCoPlan*, the sub-goal is only the projection of the final goal in a partial environment housing the robot. The sub-goal, represented by one of Zone Agent, is the one which registers the highest Attraction Index. This index is calculated by each Zone Agent and it is proportional to the distance separating the final goal of the zone in question. The expression of the Attraction Index of the zone (I, J) likely to be the sub-goal is given by the equation (1).

$$AttractionIndex = \left(\frac{T}{D+1}\right) \quad . \tag{1}$$

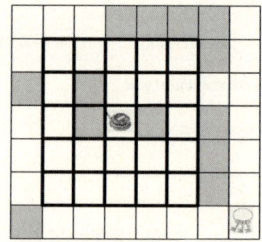

Fig. 4. Schematic agentification of the robot proximity environment

With T is the agent type of the zone (i, j), which can be 0 (occupied by an obstacle) or 1 (free) and D the distance from the agent compared to the goal.

The Attraction Index will be sent to the Planning Agent. This last will choose to lodge the sub-goal at the Zone Agent having the highest Attraction Index.

A problematic case arises; it is that where agents hold the same Attraction Index. To solve this problem, we propose to associate a set of eight agents representing its direct proximity to each Zone Agent having the same Attraction Index, in such a way that the agent is placed in the center. The agent, surrounded by less obstacles, will lodge the sub-goal.

Figure 5 illustrates the case of two Zone Agents having the same attraction index. The agent presented in the center of the discontinued zones will lodge the sub goal after computing and sending its magnet index. The expression of the magnet index is expressed by the equation (2).

$$MagnetIndex = \left(\frac{T}{(D+1) \times (Nb+1)} \right) . \qquad (2)$$

With Nb the number of Zone Agents occupied by an obstacle in the direct proximity zones.

3.3 Choice of the Best Plan Towards Sub-goal

Figure 6 presents the behavior of the Planning Agent, and particularly the path planning of the mobile robot in a partial environment going from its position towards the sub-goal. Each agent communicates with others by exchanging messages. To resolve problems related to communication coordination, we use blackboard architecture where the mailbox corresponds to a matrix (5x5). Each Zone Agent sends its coordinates, its occupation, its types (0: occupied by an obstacle, 1: free) and its Attraction Index (a number between 0 and 1) (see table 1).

The Planning Agent consults its mailbox, obtains the information sent by all Zone Agents and chooses the best sub gaol. Finally, it plans the partial path and sends a message to the robot. This message contains the direction of path.

Figures 7 and 8 present a description in pseudo code of the general behavior of our Planning Agent. In figure 7, the Planning Agent determines the adjacent

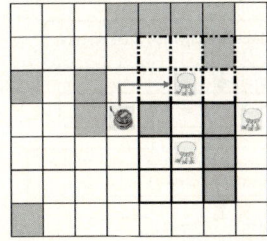

Fig. 5. Case of agents having the same Attraction Index

Table 1. Example of message sent by the Zone Agent Num 20 to the Planning Agent

$Xgoal$	$Ygoal$	$Xrobot$	$Yrobot$	$Xagent$	$Yagent$	$Type$	$Occupation$	Att_{Index}
11.0	10.0	10.0	5.0	11.0	7.0	1.0	0.0	0.25

zones accessible to the robot and initializes *TWeight* and *TTrace*. For a better understanding of the functional cycle of this agent detailed in figure 8, the following parameters are given:

Input:

- Position of the robot (R).
- Types of the agents.
- Sub-goal (SB).
- Numbers of proximity agents (N).

Output:

- The partial robot path.

Method:

- The construction of Planning Matrix (PM) from sent messages (types) from the Zone Agents. Each element of this matrix contains 1 if the horizontal or vertical passage from a zone to another is possible and 0 if not.
- The construction of *TWeight*(vector of the weights) and *TTrace* (vector storing the robot trace): the Planning Agent assigns to the first adjacent zone the weight 1 for the first passage, and for each robot motion to another zone this weight will be incremented by 1.

The Planning Agent will build a vector storing the robot path (*TPath*) according to *TTrace* . After the choice of the shortest path, the robot will move from one zone to another while going towards the sub-goal. At that moment the Zone Agents undergo a mutation and start again their tasks. Another problematic case arises; it is where there is no path of the robot to the sub-goal. Since our approach is based on the existence of a path from a starting position to the final goal, the robot will take a step towards the nearest free zone and the planning method is normally continued.

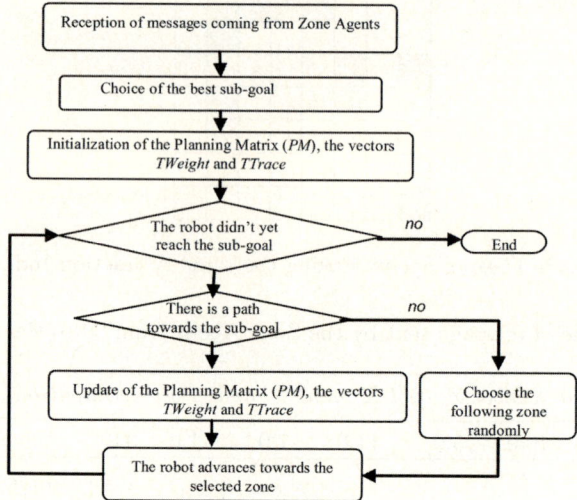

Fig. 6. Behaviour of the Planning Agent

```
For i = 1 to N do
        if (MP[N/2][i] = 1) // Initially the robot is in N/2 position
    TWeight[i] = 1
    TTrace[i] = N/2
  Endif
Endfor
```

Fig. 7. Initialization of TWeight and TTrace

```
j=1
while R <> SB do
 for i = 1 to N do
    if TWeight[i]= j then // test if the robot has passed through zone i //
        for k = 1 to N do
          if MP[i][k] = 1 // test if the passage from zone i to zone k is possible //
             if (TWeight[k] is empty) or (TWeight[k] > TWeight[i] +1)
                TWeight[k] = TWeight[i]+1 // updating TWeight and TTrace //
                TTrace[k] = i // fill TWeight and TTrace //
             Endif
          Endif
       Endfor
    Endif
 Endfor
 j : = j+1
End while
```

Fig. 8. Determination of the shortest path

4 Simulation and Results Discussion

We propose to test and validate MARCoPlan with a series of simulations. It is sufficient to present here some captions of environment of 25x25 zones (n=25). During this simulation, we show the various stages of planning by introducing sub goals. In this paper, we propose to present two experiments made out of two different examples: a first simple example with static and less furnished environment, and a second one with dynamic and more furnished environment.

Figures 9 and refimage11, illustrate the agentification of the zones as well as the robot motion carried out to achieve its final goal (the goal is indicated by the dark grey color. The figures from (9a) to (9h), and (11a) to (11p), emphasize also the sub-goal which is represented by a grey color in an agentified environment of navigation. This sub-goal changes position with the displacement of the robot (which is indicated by the black color) from one planning zone to another. The zone occupied by an obstacle is represented by the clear grey color.

To check the performance of *MARCoPlan* to find an optimal path, we have carried out several simulations where the complexity and the size of the environment have changed. We notice that the effectiveness of our method is independent on the complexity of the environment and it always provides an optimal path (see figure 10).

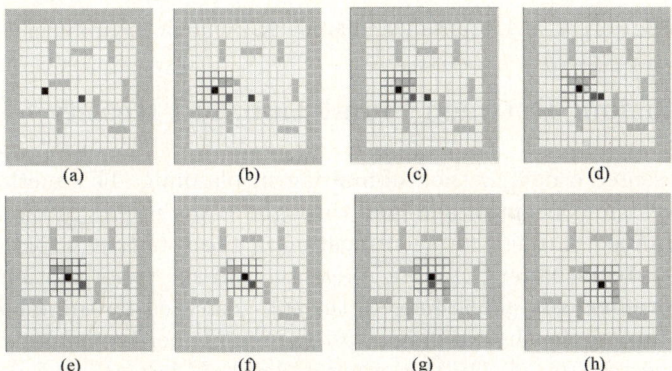

Fig. 9. A *MARCoPlan* simulation in static environment

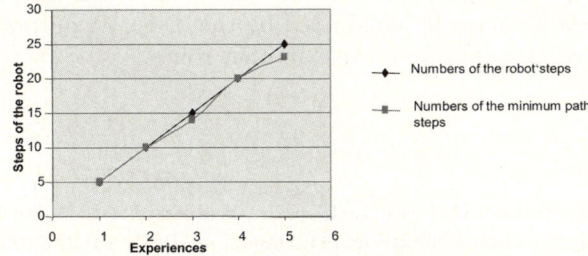

Fig. 10. *MARCoPlan* plans versus those determined by the known minimal path

Fig. 11. A *MARCoPlan* simulation in dynamic environment

5 Conclusion: Between Simulation and Reality

We have presented a new method of multiagent planning. This method consists in introducing the concept of sub-goal throughout the robots motion. The sub-goal is determined through an agentification of the robots environment. These agents cooperate with the Planning Agent to plan the robot motion towards a series of determined sub-goals. Indeed, the sub-goal is easier to be reached than a final goal. Since, the first is closer to the robot, whereas the second is more complex to be recognized. On the practical level, *MARCoPlan* can be applied to a geographically distributed system like the road supervision, especially route choice problem, thus guiding a vehicle in a road network to reach its destination with an optimal path [6] [14]. Indeed, a road is considered as a zone in which the vehicle motion is manually conducted by the driver or automatically by an intelligent navigation system to choose another route.

References

1. DesJardins, M., Durfee, E.H., Ortiz Jr., C.L., Wolverton, M.: A Survey of Research in Distributed, Continual Planning. AI Magazine 20(4), 13–22 (1999)
2. Durfee, E.H.: Distributed Continual Planning for Unmanned Ground Vehicle Teams. AI Magazine 20(4), 55–61 (1999)

3. El Fallah-Seghrouchni, A., Degirmenciyan-Cartault, I., Marc, F.: Framework for Multi-agent Planning Based on Hybrid Automata. In: Mařík, V., Müller, J.P., Pěchouček, M. (eds.) CEEMAS 2003. LNCS (LNAI), vol. 2691, pp. 226–235. Springer, Heidelberg (2003)
4. Fischer, T., Gehring, H.: Planning vehicle transhipment in a seaport automobile terminal using a multi-agent system. European Journal of Operational Research 166, 726–740 (2005)
5. Hendler, J.A., Tate, A., Drummond, M.: AI Planning: Systems and Techniques. AI Magazine 11(2), 61–77 (1990)
6. Kammoun, H.M., Kallel, I., Alimi, A.M.: RoSMAS2: Road Supervision based Multi Agent System Simulation. In: Proc of the International Conference on Machine Intelligence, Tozeur-Tunisia, pp. 203–210 (November 2005)
7. Kallel, I., Baklouti, N., Alimi, A.M.: Accuracy Preserving Interpretability with Hybrid Hierarchical Genetic Fuzzy Modeling: Case of Motion Planning Robot Controller. In: Proc. of the International Symposium on Evolving Fuzzy Systems, Lake District, UK, pp. 312–317 (September 2006)
8. Kim, J., Pearce, R.A., Amato, N.M.: Robust geometric-based localization in indoor environments using sonar range sensors. In: Proceedings of International Conference on Intelligent Robots and System. IEEE/RSJ, vol. 1, pp. 421–426 (2002)
9. Kuu-Young, Y., Chi-Haur, W.: Path feasibility and modification. Journal of robotic systems (JRS) 9(5), 613–633 (1992)
10. Meignan, D., Simonin, O., Koukam, A.: Simulation and evaluation of urban bus networks using a mutiagent approach. Simulation Modelling Practice and Theory 15, 659–671 (2007)
11. Rui, X., Ping-Yuan, C., Xiao-fei, X.: Realization of multi-agent planning system for autonomous spacecraft. Advances in Engineering Software 36, 266–272 (2005)
12. Wooldridge, M.: An introduction to multiagent systems. John Wiley and Sons, Chichester (2002)
13. Zhang, J., Knoll, A.: Integrating deliberative and reactive strategies via fuzzy modular control. In: Saffotti Drainkov, A. (ed.) Fuzzy Logic techniques for autonomous vehicle navigation, Springer, Heidelberg (1999)
14. Zhang, J., Wang, H., Li, P.: Towards the applications of multi-agent techniques in intelligent transportation systems. In: IEEE Proc. of Intelligent Transportation Systems, vol. 36, pp. 1750–1754 (2003)

A Hybrid Method of User Identification with Use Independent Speech and Facial Asymmetry

Mariusz Kubanek and Szymon Rydzek

Czestochowa University of Technology
Institute of Computer and Information Science
Center for Intelligent Multimedia Techniques
Dabrowskiego Street 73, 42-200 Czestochowa, Poland
{mariusz.kubanek,szymon.rydzek}@icis.pcz.pl
http://icis.pcz.pl

Abstract. Speaker identification is the process of identifying an unknown speaker from a set of known speakers. In a speaker identification or verification, the prime interest is not in recognizing the words but determining who is speaking the words. In systems of speaker identification, a test of signal from an unknown speaker is compared to all known speaker signals in the set. The signal that has the maximum probability is identified as the unknown speaker. In security systems based on speaker identification, faultless identification has huge meaning for safety.

In aim of increasing safety, in this work it was proposed own approach to user identification, based on independent speech and facial asymmetry. Extraction of the audio features of person's speech is done using mechanism of cepstral speech analysis.

The part of the work that deals with face recognition was based on the technique of automatic authentication of a person with assumption that the use of automatically extracted, structural characteristics of the face asymmetry (in particular within the eyes and mouth regions as the most informative parts of the face) leads to improvement of the biometrical authentication systems.

Finally, the paper will show results of user identification.

Keywords: user identification, user verification, speaker identification, speaker verification, speech signal, independent speech, speech coding, facial asymmetry.

1 Introduction

Automatic speaker identification (ASI) by machine has been an active research area for several decades, but in spite of enormous efforts, the performance of current ASI system is far from the performance achieved by humans. The problem of speaker identification is one of many research centers at present. The recognition of users is used in many areas. Interest of this discipline is a result of potential possibilities of practical applications in real world, where the audio signal is very intensely disturbed and changed [1]. The variety of applications of ASI systems, for human computer interfaces, telephony, or robotics has driven the research of a large scientific community [1,3]. The success of currently available ASI systems is however restricted to relatively controlled environments and well defined applications such as dictation or small to medium vocabulary voice-based controll commands [1,4]. Often, robust ASR systems require special positioning of the microphone with respect to the speaker resulting in a rather unnatural human-machine interface.

The most important problem in process of speaker identification is suitable coding of signal audio [5]. In general, speech coding is a procedure to represent a digitized speech signal using a few bits as possible, maintaining at the same time a reasonable level of speech quality. Speech coding has matured to the point where it now constitutes an important application area of signal processing. Due to the increasing demand for speech communication, speech coding technology has received augmenting levels of interest from the research, standardization, and business communities. Advances in microelectronics and the vast availability of low-cost programmable processors and dedicated chips have enabled rapid technology transfer from research to product development; this encourages the research community to investigate alternative schemes for speech coding, with the objectives of overcoming deficiencies and limitations. To standardization community pursues the establishment of standard speech coding methods for various applications that will be widely accepted and implemented by the industry. The business communities capitalize on the ever-increasing demand and opportunities in the consumer, corporate and network environments for speech processing products [1,5].

In [13,14,15] there was proposed approach based on holistic representation of face asymmetry characteristics to improvement of face recognition techniques as well as expression qualification. Authors used affine transformations in aim to image normalization on the basis of three landmark points – inner eyes corners and base of the nose. In the experiment it was used two asymmetry measures D-face (Density Difference) and S-face (Edge Orientation Similarity). Authors reported mean error value of 3.60/1.80% (FRR/FAR) when using both asymmetry measures in fusion with FisherFaces in task of expession qualification.

The idea of improvement of effectiveness of face recognition technique used in the hybrid method was based on processing information regarding face asymmetry in the most informative parts of the face – the eyes region. Such approach to person identification has been reported as efficient and possible to apply in real-time systems [16].

In the work, it was proposed the method to user identification based on independent speech and facial asymmetry. To extraction of the audio features of person's speech, in this work it was applied the mechanism of cepstral speech analysis. For acoustic speech recognition was used twenty dimensional MFCC (Mel Frequency Cepstral Coefficients) as the standard audio features. Fig. 1. shows scheme of analysis of audio signal.

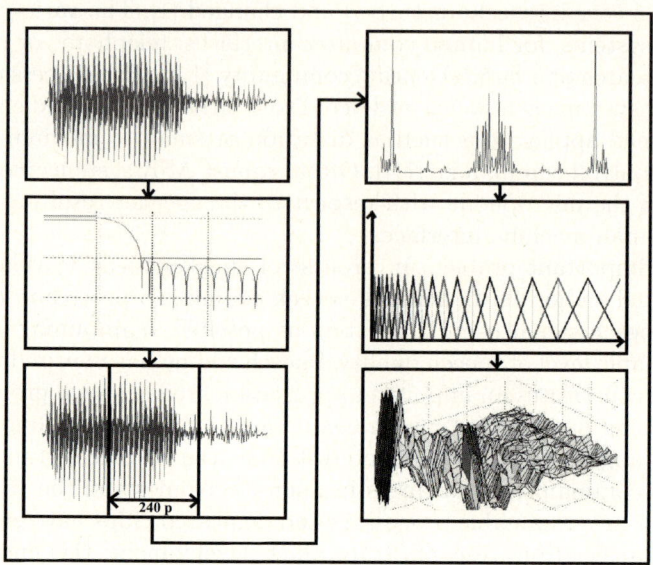

Fig. 1. Scheme of analysis of audio signal

2 Preliminary Processing of Signal

Analysis of audio channel one should to begin from filtration of signal, removing elements of signal being him disturbances. Filtration of signal permits smooth out, improvement of signal to noise ratio, limitation of width of band transfers as well as detecting of phenomena displaying oneself changes of spectrum. Additionally, making anti aliasing filtration, it prevents itself distortions as well as phenomenon of overlapping spectrum, often creating during analogue-digital processing (A/C), when upper frequencies of signal are higher than half of higher frequency of sampling of signal [6,7]. In working system in conditions

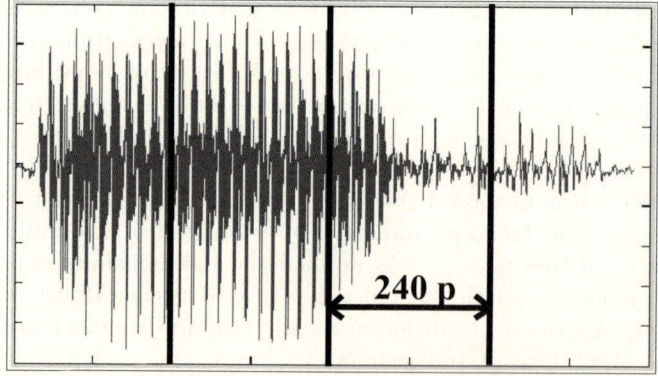

Fig. 2. Scheme of division of entrance signal onto stationary frame boxes

approximate to ideal it was been possible to skip stage of preliminary filtration in aim of acceleration of working. In real conditions of work, signal of audio speech is often considerably disturbed, therefore in work was applied preliminary filtration [2].

The non-stationary nature of the speech signal caused by dynamic proprieties of human speech result in dependence of the next stage on use of division of entrance signal onto stationary frame boxes [8]. Signal is stationary in short temporary partitions $(10 \pm 30$ ms) [9]. Every such stationary frame box was replaced by symbol of observation in process of create of vectors of observation. In created system it was assumed that length of every frame box equals 30 ms, what at given sampling of signal (8kHz) remove 240 samples. Obtained frame boxes do not overlap. Scheme of division of entrance signal onto stationary frame boxes is shown on Fig. 2.

3 The Mechanism of Cepstral Speech Analysis

Speech processing applications require specific representations of speech information. A wide range of possibilities exists for parametrically representing the speech signal. Among these the most important parametric representation of speech is short time spectral envelope [8,10]. Linear Predictive Coding (LPC) and Mel Frequency Cepstral Coefficients (MFCC) spectral analysis models have been used widely for speech recognition applications. Usually together with MFCC coefficients, first and second order derivatives are also used to take into account the dynamic evolution of the speech signal, which carries relevant information for speech recognition.

The term cepstrum was introduced by Bogert et al. in [11]. They observed that the logarithm of the power spectrum of a signal containing an echo had an additive periodic component due to echo. In general, voiced speech could be regarded as the response of the vocal track articulation equivalent filter driven by a pseudo periodic source [11].

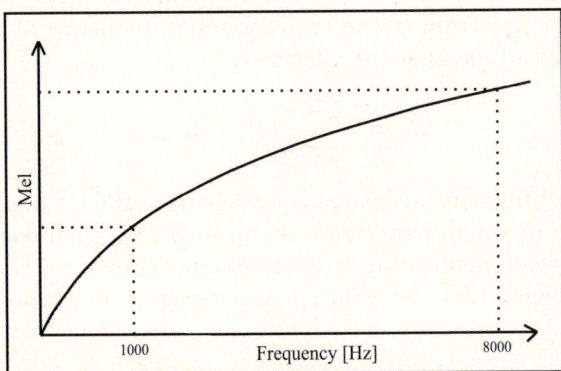

Fig. 3. Dependence between scale of frequency and mel

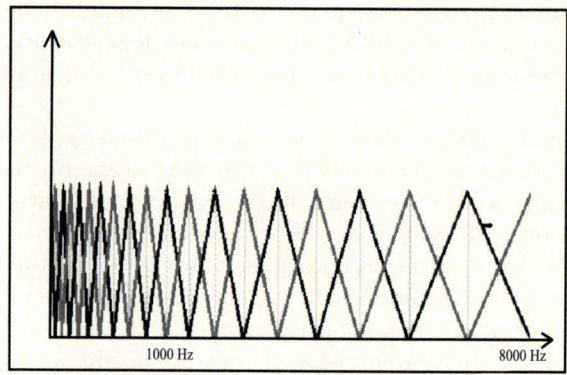

Fig. 4. Character of bank of 30 filters with widths 200 mels

The main difference between mel-cepstrum and cepstrum was that the spectrums were first passed through mel-frequency-bandpass-filters before they were transformed to the frequency domain [12].

$$f_{mel} = 2595 log_{10}(1 + f_{Hz}/700) \tag{1}$$

$$f_{Hz} = 700(10^{\frac{f_{mel}}{2595}} - 1) \tag{2}$$

The characteristics of filters followed the characteristics old human auditory system [9]. The filters had triangular bandpass frequency responses. The bands of filters were spaced linearly for bandwidth below 1000 Hz and increased logarithmically after the 1000 Hz. In the mel-frequency scaling, all the filter bands had the same width, which were equal to the intended characteristic of the filters, when they were in normal frequency scaling. Fig. 3. shows dependence between scale of frequency and mel. Character of bank of 30 filters with widths 200 mels was showed on Fig. 4.

Spectrum of signal of every frame boxes obtained by Fast Fourier Transform (FFT) comes under process of filtration by bank of filters. The next step was to calculate the members of each filter by multiplying the filter's amplitude with the average power spectrum of the corresponding frequency of the voice input. The summation of all members of a filters is:

$$S_k = \sum_{n=0}^{(N/2)-1} (P_n * A_{k,n}) \tag{3}$$

Finally, the mel-frequency cepstrum coefficients (MFCC) was derived by taking the log of the mel-power-spectrum coefficients (S_k) then convert them back to time (frequency) domain using Discrete Cosine Transform (DCT). The number of mel-coefficients (K) used, for speaker recognition purposes, was usually from 12 to 20 [12].

$$MFCC_n = \sum_{k=1}^{K} (log S_k) cos \left[n(k - 0.5) \frac{\pi}{K+1} \right] \tag{4}$$

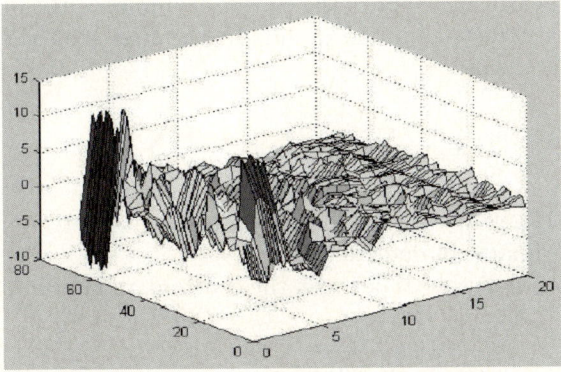

Fig. 5. Cepstral coefficients for all frame boxes

In this work, for speech coding, it was used twenty dimensional MFCC as the standard audio features. Next, obtained for all frame boxes cepstrum coefficients, shows on Fig. 5, add upped properly. In this expedient all independent statement one coded by twenty cepstral coefficients.

4 Face Asymmetry

The input information for procedure of face recognition was frontal face image taken at resolution of 768×576 pixels with 24-bit color depth. Basing on procedures of image proccessing developed for purpose of automatic face feature extraction face image was processed in order to find 8 landmark points (see table 1).

Table 1. Description of landmark points detected using automatic procedure of feature extraction

Symbol	Description
RIris	Center of right iris
LIris	Center of left iris
Rcrn.out	Outer eyelid corner of right eye
Rcrn.inn	Inner eyelid corner of right eye
Lcrn.out	Outer eyelid corner of left eye
Lcrn.inn	Inner eyelid corner of left eye
MouthRcrn	Right mouth corner
MouthLcrn	Left mouth corner

The face description vector was constructed on the basis of measure of geometrical dependencies between detected landmark points and asymmetry measures. All distance measures was done using the Muld unit which is based on the diameter of the iris. Such approach allows us to avoid necessity of image scaling in aim of size normalization. Asymmetry measurement was done in two ways.

In the case of measuring shape of the eyelids asymmetry, the input information was set of points detected on the eyelid. In the first step the coordinates of centroid was found using:

$$x_c = \frac{1}{N} \sum_{t=0}^{N-1} x(t), \quad y_c = \frac{1}{N} \sum_{t=0}^{N-1} y(t) \qquad (5)$$

where $t = 0, 1, \ldots, N - 1$, N – number of points detected on the eyelid, $(x(t), y(t))$ – coordinates of detected points.

Next, the values of distances between the centroid and eyelid points was found with use of:

$$r(t) = \sqrt{[x(t) - x_c]^2 + [y(t) - y_c]^2}, \quad t = 0, 1, \ldots, N - 1 \qquad (6)$$

Asymmetry measure was defined as set of subtracion values of corresponding distances between the eyelid and centroid calculated for right and left eye:

$$d(t) = |r_R(t) - r_L(t)|, \quad t = 0, 1, \ldots, N - 1 \qquad (7)$$

where $r_R(t)$ i $r_L(t)$ – distances between the centroid and eyelid points for right and left eye respectivelly.

In the case of measuring asymmetry of eyelid corners position, the dependence constructed on the basis of the Weber-Fechner law was used:

$$F_{Asym}(F, L) = ln\left(\frac{F}{L}\right). \qquad (8)$$

where F i L – results of the feature measurement (the distance between corresponding landmark points).

The measure of face similarity was calculated as the weighted mean value of scaled measures of features difference:

$$d\_face = \left(\sum_{i=1}^{n} w_i \cdot s_i \cdot d\_feature\right) \cdot \frac{1}{n} \qquad (9)$$

where: d_face – value of face descriptions simmilarity, s – vector of scaling values, w – weights vctor, $d\_feature$ – vector of the measures of features difference, n – number of features.

The measures of features difference was defined as euclidean distances between corresponding feature values. For features which description consist of the set of measures (eyelid shape) the difference was defined as the mean square value of the elements of the vetor.

5 Experimental Results

In our experiment, the error level of speaker identification with use only audio independent speech and coupled audio speech and facial asymmetry was tested. Research were made, using author base of independent statements and face pictures of different users. Thirty users were tested, speaking out for four long independent statements, three statements as training data and one as test data. For each user one photo was made, and vector of asymmetry was built. Research were made for different degree of disturbance of signal audio (Signal to Noise Ratio, SNR = 0, 5, 10, 20 dB). It was accept, that for SNR = 20 dB signal audio is clean. It was assumed, that face pictures is clean for different SNR of audio.

In research the recordings of independent speach were used. Samples were taken at frequency of 8 khz and 16-bit encoding. Three samples for each user were recorded, which after encoding using 20 cepstral coefficients were put in system database with corresponding person ID. In the identification process user has to read long enough sentence randomly selected by the system. Sampling of the sentence was done with the same conditions as the samples in the database. The sentence was compared with all encoded sentences in the database by calculating the euclidean distance of two vectors. Next, for the user ID of the highest probability the vector of face asymmetry stored in the database was compared with the input asymmetry vector. For audio signal and face asymmetry were asummed weights of 60% and 40% respectivelly.

Tab. 2. shows result of speaker identification with use only audio independent speech. Tab. 3. shows result of speaker identification with use audio independent speech and facial asymmetry.

The audio signal was transformed to lower quality with use of noise generator. In the experimet the level of FAR/FRR errors was invastigated. Sampling frequency was set to 8 khz because of the possibillity of easy transmission of such signals, what in the future can be used to build the system of user identification with use of the phone.

Fig. 6. and Fig. 7. shows Level of FRR and FAR of speaker identification of only audio independent speech and in connection with facial asymmetry.

Table 2. Result of speaker identification with use only audio independent speech

User	Speaker identification with use audio independent speech									
	FRR [%]					FAR [%]				
	SNR 20dB	SNR 15dB	SNR 10dB	SNR 5dB	SNR 0dB	SNR 20dB	SNR 15dB	SNR 10dB	SNR 5dB	SNR 0dB
Average for thirty users	3,33	8,88	14,44	22,22	44,44	2,22	4,44	10,00	17,77	25,55

Table 3. Result of speaker identification with use audio independent speech and facial asymmetry

User	Speaker identification with use audio independent speech and facial asymmetry									
	FRR [%]					FAR [%]				
	SNR 20dB	SNR 15dB	SNR 10dB	SNR 5dB	SNR 0dB	SNR 20dB	SNR 15dB	SNR 10dB	SNR 5dB	SNR 0dB
Average for thirty users	1,11	2,22	4,44	7,77	11,11	0,00	1,11	3,33	5,55	7,77

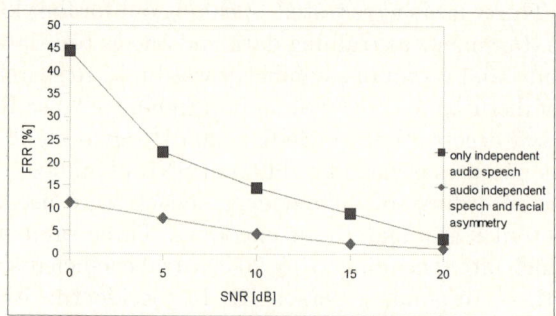

Fig. 6. Level of FRR of speaker identification of onlu audio independent speech and in connection with facial asymmetry

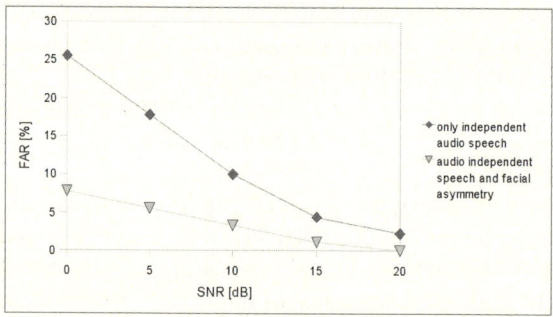

Fig. 7. Level of FAR of speaker identification of onlu audio independent speech and in connection with facial asymmetry

6 Conclusion and Future Work

In this paper, the new approach to user identification based on independent speech and facial asymmetry was presented. The basic idea is that not only speech, not only naturally looking face, but also other facial features such as facial asymmetry contain much useful information, necessary to identification. It was evaluated the robustness of the new approach. The new approach will implemented in the module of our user identification system. The method was proposed to improvements of efectivity of user identification.

An advantage of our approach is the simplicity and functionality by the proposed methods, which fuse together audio and visual signals. Obtained results of research show, that commits less of errors, if to identification we give information about tone of voice of examined user, and information about his asymmetry of face. A decisively lower level of mistakes was obtained in user identification based on independent speech and facial asymmetry, in comparison to only audio speech, particularly in facilities, where the audio signal is disrupted.

In future, we plan to built system to identification and verification identity of users, in which errors will be on level near 0 %. Also will be unrolled research of suitable selection of proportion of weights. decreasing of length of single independent statements.

Acknowledgments. This work is sponsored by the Czestochowa University of Technology under Grant No. BW-1-112-209/2007/investitage specializations.

References

1. Kubanek, M.: Analysis of Signal of Audio Speech i Process of Speech Recognition. Computing, Multimedia and Intelligent Techniques, Czestochowa 2, 55–64 (2006)
2. Kubanek, M.: Method of Speech Recognition and Speaker Identification with use Audio-Visual Polish Speech and Hidden Markov Models. In: Saeed, K., Pejas, J., Mosdorof, R. (eds.) Biometrics, Computer Security Systems and Artificial Intelligence Applications, pp. 45–55. Springer Science + Business Media, New York (2006)
3. Aydin, Y., Nakajama, H.: Realistic Articulated Character Positioning and Balance Control in Interactive Environments. In: Proceedings Computer Animation 1999, pp. 160–168 (1999)
4. Neti, C., Potamianos, G., Luttin, J., Mattews, I., Glotin, H., Vergyri, D., Sison, J., Mashari, A., Zhou, J.: Audio Visual Speech-Recognition. 2000 Final Report (2000)
5. Chu Wai, C.: Speech Coding Algorithms. Foundation and Evolution of Standardized Coders. John Wiley & Sons, New Jersey (2000)
6. Basztura, C.: Talk to Computer. Publishing House of Research Work FORMAT, Wroclaw (1992)
7. Lyons, R.G.: Input to digital processing of signal. Publishing House of Communication and Connection, Warsaw (1999)
8. Rabiner, L., Yuang, B.H.: Fundamentals of Speech Recognition. Prentice Hall Signal Processing Series (1993)
9. Wisniewski, A.M.: Hidden Markow Model in Speech Recognition. In: Bulletin IAiR WAT, Wroclaw, vol. 7 (1997)
10. Kanyak, M.N.N., Zhi, Q., Cheok, A.D., Sengupta, K., Chung, K.C.: Audio-Visual Modeling for Bimodal Speech Recognition. In: Proc. Symp. Time Series Analysis (2001)
11. Bogert, B.P., Healy, M.J.R., Tukey, J.W.: The Frequency Analysis of Time-Series for Echoes. In: Proc. 2001 International Fuzzy Systems Conference, pp. 209–243 (1963)
12. Wahab, A., See Ng, G., Dickiyanto, R.: Speaker Verification System Based on Human Auditory and Fuzzy Neural Network System. Neurocomputing Manuscript Draft, Singapore
13. Liu, Y., Schmidt, K., Cohn, J., Mitra, S.: Facial Asymmetry Quantification for Expression Invariant Human Identification. In: AFGR 2002, pp. 198–204 (2002)
14. Liu, Y., Weaver, R., Schmidt, K., Serban, N., Cohn, J.: Facial Asymmetry: A New Biometric. CMU-RI-TR (2001)
15. Mitra, S., Liu, Y.: Local Facial Asymmetry for Expression Classification. In: Proc. of the 2004 IEEE Conference on Computer Vision and Pattern Recognition (CVPR 2004) (June 2004)
16. Rydzek, S.: A Method to Automatically Authenticate a Person Based on the Asymmetry Measurements of the Eyes and/or Mouth. PhD thesis, Czestochowa University of Technology (2007)

Multilayer Perceptrons for Bio-inspired Friction Estimation

Rosana Matuk Herrera*

Department of Computer Science, Facultad de Ciencias Exactas y Naturales,
Universidad de Buenos Aires, Argentina
rmatuk@dc.uba.ar

Abstract. Few years old children lift and manipulate unfamiliar objects more dexterously than today's robots. Therefore, it has arisen an interest at the artificial intelligence community to look for inspiration on neurophysiological studies to design better models for the robots. The estimation of the friction coefficient of the object's material is a crucial information in a human dexterous manipulation. Humans estimate the friction coefficient based on the responses of their tactile mechanoreceptors. In this paper, finite element analysis was used to model a finger and an object. Simulated human afferent responses were then obtained for different friction coefficients. Multiple multilayer perceptrons that received as input simulated human afferent responses, and gave as output an estimation of the friction coefficient, were trained and tested. A performance analysis was carried out to verify the influence of the following factors: number of hidden neurons, compression ratio of the input pattern, partitions of the input pattern.

1 Introduction

Dexterous manipulation of objects requires to apply the following fingertip forces to the object of interest:

1. *Load force (LF):* Vertical force required to overcome the force of gravity.
2. *Grip force (GF):* Normal force required to prevent slips.

To avoid a slip between the fingers and the object, the grip force must be greater than the slip limit, i.e., the load force divided by the friction coefficient (μ): GF $\geq$ LF/μ. Thus, the estimation of the friction coefficient is an essential information required for the dexterous manipulation of objects.

2 Friction Estimation by Humans

There are four types of afferents in the glabrous skin of the human hand: FA-I, SA-I, SA-II and FA-II. All together, there are approximately 17,000 mechanoreceptors in the glabrous skin of each hand. The SA I and SA II units are referred to

---

* Postal address: Departamento de Computación, Pabellón 1, Ciudad Universitaria, 1428, Buenos Aires, Argentina.

as *slowly adapting afferents*, which means that they show a sustained discharge in the presence of ongoing stimuli. In contrast, the *fast adapting afferents*, FA I and FA II, fire rapidly when a stimulus is presented (or removed) and then fall silent when the stimulation cease to change [1]. The skin area where responses can be evoked from a receptor is called its receptive field (RF). The RFs of FA I and SA I units typically have sharply delineated borders and show several small zones of maximal sensitivity. In contrast, the SA II and FA II have large RFs but only one zone of maximal sensitivity and a gentle increase of threshold further and further away from this zone [2,1,3,4].

To estimate the friction coefficient of objects, humans touch the object of interest, and increment the grip force during a short time. During this time signals are generated mainly from the FA I and SA I afferents. If the object's material has a great friction coefficient, the FA I signals will have a great intensity. On the other side, if the object's material has a low friction coefficient, the FA I signals will have a low intensity. Thus, humans seem to determine the friction coefficient based on the intensity of the FA I signals [5].

3 Simulation of the Human Afferent Responses

To simulate the FA-I afferent responses, we adapted software code provided by Anna Theorin (f. Israelsson), Lars Rådman, Benoni Edin and Roland Johansson of Umeå University.

Israelsson [4] designed a computational simulator of the responses of the afferents from the glabrous skin during human manipulation. To develop her simulator program Israelsson used experimental data from a not yet published at that time study by Roland Johansson and co-workers at Umeå University. In this study subjects had their nail of their index finger glued to a support structure. A flat circular contact plate was moved against the fingertip from 0.2N up to a total of 4N. In the center of the contact plate there was a little hole with a force probe. The flat tip of the force probe was at level with the contact surface and measured the local force in a small skin area in three orthogonal directions, X, Y and Z, representing the local tangential forces and the normal force, respectively. The plate was moved in 1 mm steps across the fingertip. This way, a complete data set of local forces for each stimulus direction was obtained. In parallel experiments, afferent responses to the same type of stimuli were recorded.

Israelsson [4] generated spike trains similar to those observed in actual neurophysiological experiments in the following way:

1. For all receptor classes, except the FA II afferents, the response intensity was calculated, at specified time intervals, as a weighted sum of the forces (X, Y, Z) and first time derivates of the same three orthogonal force directions:

$$f(t) = k_1 \times F_z(t) + k_3 \times F_y(t) + k_5 \times F_x(t) + k_2 \times \frac{dF_z(t)}{dt} + k_4 \times \frac{dF_y(t)}{dt} + k_6 \times \frac{dF_x(t)}{dt}$$

where k_{1-6} represent weights that differed between the different afferents. F_z represents the force in the normal plane, F_y and F_x correspond to the

tangential forces in the Y and X directions, respectively. The weights allow to define the simulated afferents with different sensitivity to the dynamic and static components of the force development, as well as different sensitivity to forces normal and tangential to the surface.
2. To simulate the irregularity in discharges of natural spike trains, noise was added to the calculated response intensity.
3. Spike trains were generated from the response intensity signals by integration; every time the integration of the signal reached above a specified threshold a spike was generated, the integrator was reset and the process repeated for the whole simulation.

4 Method

A bio-inspired method for friction estimation was outlined in [6]. Contact between an elastic finger with a curved surface and an object with a plane surface was analyzed using the finite element method (FEM). Finite element code *MSC.Marc release 2* was used to create a 3-D model of a finger and an object. The afferent responses were obtained adapting software code of the Israelsson's human afferent simulator program. *Matlab 7* modules were programmed to process the output of the *MSC.Marc*, and artificial neural networks were programmed to process the output of the Israelsson's afferents simulator. Thus, our software contribution consists of designing the FEM model, programming the simulated experiments using the FEM model, formatting the output of the FEM model to be fed to the Israelsson's afferent simulator program, and processing the output of the Israelsson's afferent simulator program with artificial neural networks to obtain the friction coefficient (Fig. 1).

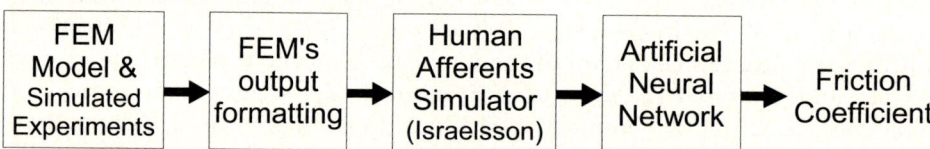

Fig. 1. Block diagram of the software modules

Maeno et al. [7] used finite element analysis to calculate the deformation of an elastic finger, contact forces, and strain distribution inside the elastic finger for various friction coefficients between the finger and the surface. Their results show that the shear strain differs when the friction coefficient differs. Then they used strain sensors to estimate the friction coefficient. In this work, we use finite element analysis as in Maeno's work to simulate a finger and an object, but we don't estimate the friction coefficient based on the shear strain. Instead, we estimate the friction coefficient based on simulated human afferent signals because we intend to follow a bio-inspired approach.

In this work, a finite element plane model of a finger with a geometry similar to the finger of Maeno et al. [7] was first generated. However, Maeno et al. only used a plane (2-D) model, and in this work we expanded the model in the Z-axis direction, and thus a 3-D model of the finger was obtained (Fig. 2). Our FEM model consisted of 288 elements and 513 nodes. The skin was modeled as a linear isotropic elastic material, for which two constants, elastic modulus and Poisson's ratio, need to be specified. Since soft tissues are generally considered incompressible owing to their predominant fluid content, the Poisson's ratio was assumed to be 0.48, close to the theoretical limit of 0.5 [8][9]. The elastic modulus was assumed to be 0.18 MPa [10]. The object was modeled as a unique block element. The Poisson's ratio of the object was set at 0.48 and the Young modulus at 0.72 MPa. Therefore the object was almost incompressible and harder than the finger.

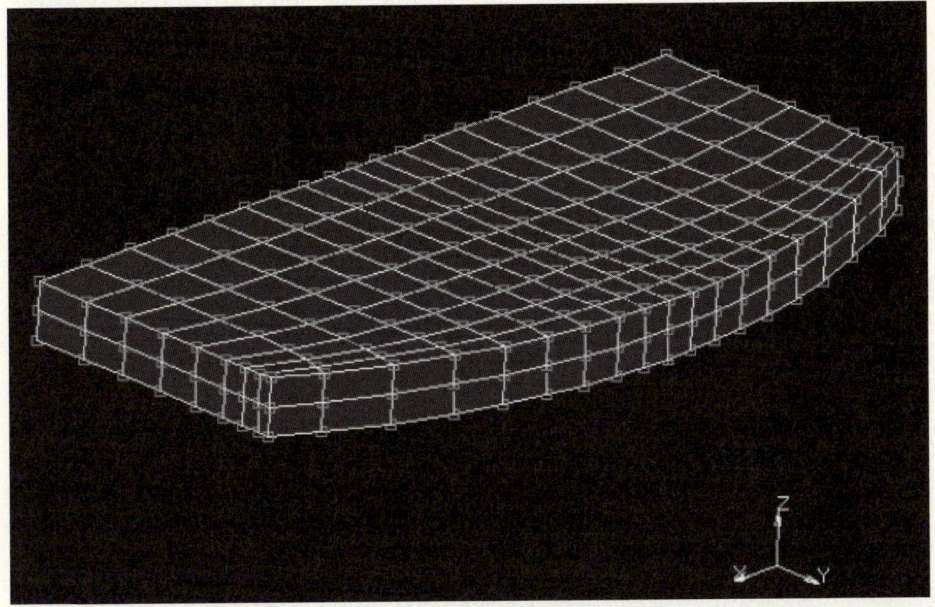

Fig. 2. 3-D FEM model of the finger

As seen in Sect. §2, the initial responses in the FA I afferents are considered responsible for the initial adjustment to a new frictional condition. As humans estimate the friction coefficient pressing their fingers against the object of interest [5,11], in this work the friction coefficient was estimated pressing the simulated finger against the surface of the object. This experiment was repeated for various friction coefficients. Thus, as the grip force of the finger against the object is raised incrementally, simulated FA I responses are obtained. A supervised artificial neural network is then used to estimate the friction coefficient, analyzing the simulated FA I responses for the different friction coefficients. The simulated

FA I responses were obtained from the values returned by the force sensors at the surface of the simulated finger as seen in Sect. §3.

The input patterns for the neural networks were built joining the simulated FA I afferent answers obtained for all the normal force increments. The friction coefficient was used as the target of the neural network. To improve the learning capacity of the neural network the patterns of the spike answers of the FA I afferents were compressed. The compression step consisted of choosing a compression factor c, dividing each one of the afferent answer patterns of size s obtained for each node of interest, where s is the number of grip force increments, in $\lceil \frac{s}{c} \rceil$ consecutive intervals, and adding the spike responses at each interval.

5 Implementation

The experiment to estimate the friction coefficient consisted of moving down the object 10 units along the y-axis, while recording the nodal stress tensors at the surface of the fingertip. The nodes of the finger at $y=0$ (i.e., the nodes situated at the finger's base) were constrained in the $x-$, $y-$ and $z-$directions, to avoid the displacement of the finger. The nodes of the object were constrained in the $x-$ and $z-$directions. Thus, normal load was increased moving the object towards the finger. The object was displaced 0.1 units along the y-direction until the total displacement reached 10 units (Fig. 3). In this way, data for 100 increments was obtained.

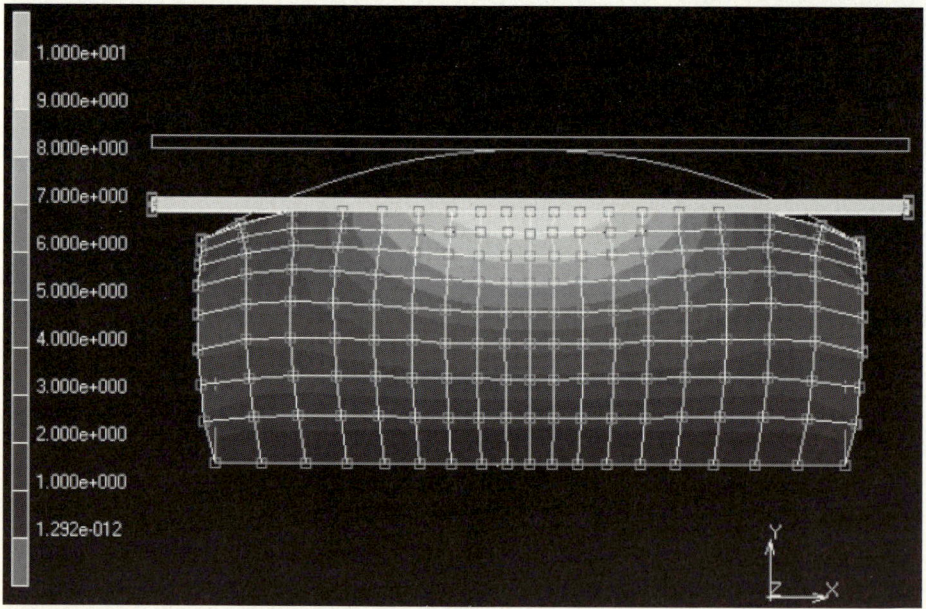

Fig. 3. Contour bands of the displacements at the final state. The image shows the original and final shapes of the finger and the object.

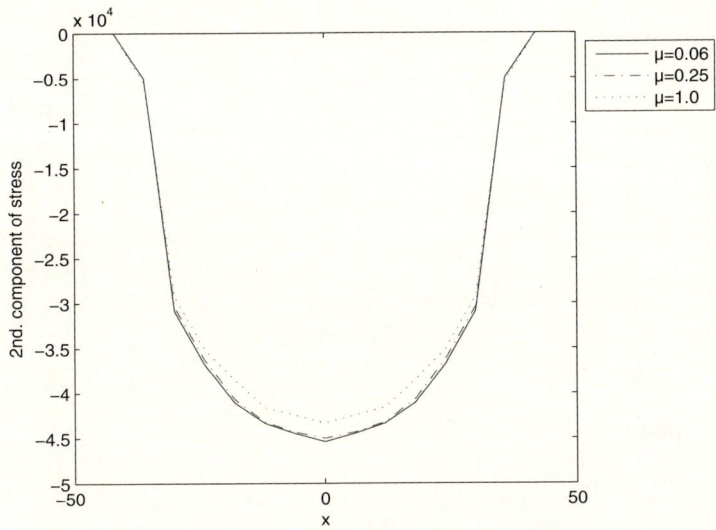

Fig. 4. Normal force F_y in the final state along the central longitudinal line of the fingertip's surface

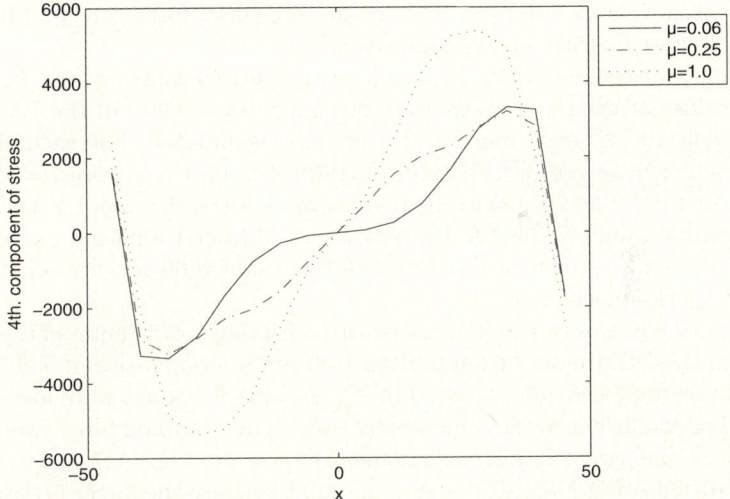

Fig. 5. Tangential force F_x in the final state along the central longitudinal line of the fingertip's surface

In the surface of the fingertip there were 3 rows of 19 nodes (Fig. 2). Figures 3, 4 and 5 reveal a symmetry of the signal patterns. Therefore, only the signals of the 10 nodes that were on the right half of the central line of the finger in contact with the object were recorded. The normal and tangential forces were obtained from the nodal stress tensors. In the designed experiment, the normal

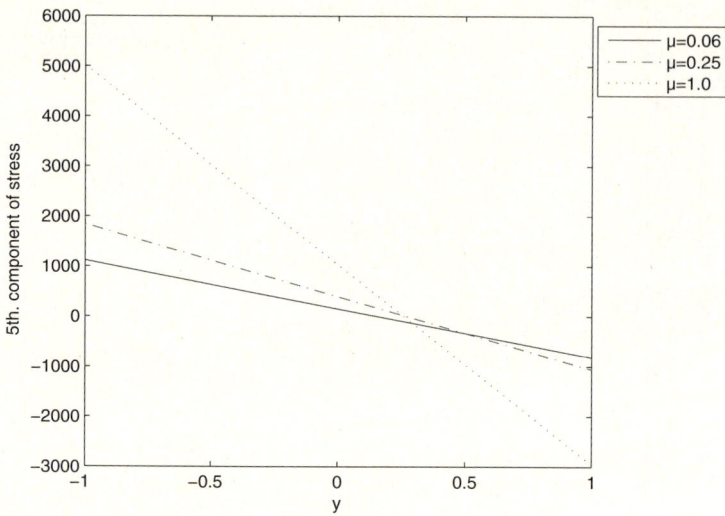

Fig. 6. Tangential force F_z in the final state along the central transversal line of the fingertip's surface

force is F_y (2nd. stress component) and the tangential forces are F_x (4th. stress component) and F_z (5th. stress component).

Therefore, for each simulated friction coefficient, measures of the F_x, F_y and F_z force values at each increment were obtained. The values of the FA I signals at each node and at each increment were then simulated. For each simulated friction coefficient, a vector of dimension 100 (because there were 100 grip force increments) with 0 and 1 values (0 if there was not spike and 1 if there was a spike) corresponding to the FA I answers was obtained for each one of the 10 nodes of interest. In this way, a pattern of dimension 1000 was obtained for each simulated friction coefficient.

The experiment was repeated increasing the friction coefficient μ of the object's material in the FEM model from 0.01 to 1.00 (with an step size of 0.005). Thus, 200 input patterns were generated. The F_x, F_y and F_z values were measured for each friction coefficient at each increment step. The obtained force values at the final state for different friction coefficients are shown in Figs. 4, 5 and 6. The normal force distribution F_y is always semi-circular because the finger is curved. The obtained curves for the normal force are similar (Fig. 4). The tangential reaction forces, i.e. the friction forces, have a local minimum and a local maximum. The sum of the tangential force F_x is always zero because no movement of the surface in the x-direction is applied (Fig. 5). However, the shapes of the curves change as the friction coefficient is increased. These results are in accordance with the results obtained by Maeno et al. [7] for their 2-D FEM model of the finger. Finally, the sum of the tangential force F_z is always zero because no movement of the surface in the z-direction is applied. The slope of the curves of the tangential force F_z decreases as the friction coefficient increases (Fig. 6).

6 Results

Multilayer perceptron is a popular model to solve supervised-learning problems. In this work, multiple multilayer perceptrons with 1 hidden layer were trained and tested. The multilayer perceptrons received as input a pattern of FA I spike signals, and gave as output an estimated friction coefficient. The activation function in the hidden layer was chosen to be a sigmoidal function in the form of a hyperbolic tangent for its convenient antisymmetric property. The output layer activation function was chosen to be linear, because the friction coefficient could be in principle any positive value. For its fast convergence, the scaled conjugate gradient algorithm developed by Moller [12] was chosen as the minimization method to train the network. This algorithm combines the model-trust region approach used in the Levenberg-Marquardt algorithm, with the conjugate gradient approach.

A random partition of the input set was done, in which the 90 percent of the patterns was chosen as training set and the remaining 10 percent as testing set. The training set was further partitioned into a subset used for estimation of the model, and a subset used for evaluation of the performance of the model. The validation subset was 10 percent of the training set. The training of the

Table 1. Mean absolute errors obtained when the FA1 spike patterns without compression were used as input of the multilayer perceptrons

# Hidden Neurons	3	4	5	10
M.a.e., first 10 incr., train set	0.121	0.131	0.138	0.108
M.a.e., first 10 incr., test set	0.140	0.148	0.145	0.149
M.a.e., first 20 incr., train set	0.105	0.095	0.092	0.077
M.a.e., first 20 incr., test set	0.120	0.131	0.123	0.135
M.a.e., first 30 incr., train set	0.065	0.059	0.056	0.072
M.a.e., first 30 incr., test set	0.162	0.159	0.179	0.187

Table 2. Mean absolute errors obtained when the FA1 spike patterns compressed with a compression ratio 5:1, were used as input of the multilayer perceptrons

# Hidden Neurons	3	4	5	10
M.a.e., first 5 incr., train set	0.140	0.137	0.138	0.130
M.a.e., first 5 incr., test set	0.159	0.158	0.156	0.159
M.a.e., first 10 incr., train set	0.133	0.130	0.133	0.117
M.a.e., first 10 incr., test set	0.144	0.149	0.147	0.165
M.a.e., first 20 incr., train set	0.118	0.109	0.113	0.105
M.a.e., first 20 incr., test set	0.135	0.142	0.142	0.142
M.a.e., first 30 incr., train set	0.114	0.114	0.108	0.087
M.a.e., first 30 incr., test set	0.139	0.137	0.138	0.138
M.a.e., first 50 incr., train set	0.090	0.102	0.089	0.081
M.a.e., first 50 incr., test set	0.135	0.148	0.139	0.148
M.a.e., all incr., train set	0.079	0.077	0.077	0.054
M.a.e., all incr., test set	0.139	0.132	0.130	0.156

Table 3. Mean absolute errors obtained when the FA1 spike patterns compressed with a compression ratio 10:1, were used as input of the multilayer perceptrons

# Hidden Neurons	3	4	5	10
M.a.e., first 10 incr., train set	0.142	0.146	0.141	0.138
M.a.e., first 10 incr., test set	0.140	0.151	0.156	0.157
M.a.e., first 20 incr., train set	0.128	0.138	0.120	0.127
M.a.e., first 20 incr., test set	0.127	0.144	0.127	0.125
M.a.e., first 30 incr., train set	0.135	0.125	0.125	0.124
M.a.e., first 30 incr., test set	0.131	0.131	0.123	0.129
M.a.e., first 50 incr., train set	0.115	0.111	0.112	0.113
M.a.e., first 50 incr., test set	0.131	0.140	0.133	0.159
M.a.e., all incr, train set	0.114	0.104	0.098	0.099
M.a.e., all incr, test set	0.110	0.107	0.117	0.117

Table 4. Mean absolute errors obtained when FA1 patterns compressed with a compression ratio 10:1, were used as input of a multilayer perceptron with 1 hidden layer of 3 neurons. The increments correspond to taking the *last* number of increments of the input pattern (i.e., "Last incr." 90 corresponds to discarding of the input pattern the first 10 increments).

Last Incr.	90	80	70	60	50	40	30	20	10
M.a.e. train set	0.139	0.154	0.160	0.170	0.168	0.175	0.183	0.217	0.234
M.a.e. test set	0.169	0.159	0.162	0.172	0.169	0.176	0.186	0.220	0.252

network on the training set was stopped at a number of epochs corresponding to the minimum point of the error-performance curve on cross-validation. The final performance of the neural network was measured as mean absolute error (i.e., the average of the absolute difference between the friction coefficient and the value returned by the neural network).

Multiple multilayer perceptrons with different number of neurons in the hidden layer were designed, trained and tested. The shown final performances correspond to the average of 10 different initializations. Multiple multilayer perceptrons that receive as input FA1 spike patterns with different compression ratios were trained and tested (tables 1, 2, and 3). As described in Sect. §5, each simulated experiment consisted of 100 grip force increments. Multiple multilayer perceptrons that receive as input a different partition of the input pattern were trained and tested (tables 1, 2, 3, and 4).

7 Discussion

As changing the friction coefficient produces different shapes for the curves of the tangential forces (Figs. 5 and 6), and the afferent answers are estimated based on values of these forces, it is expected to obtain different afferent answers for different friction coefficients. Therefore, it seems consistent a method that estimates the friction coefficient based on the afferent answers.

Multilayer perceptrons with different numbers of units in the hidden layer were trained and tested, and a multilayer perceptron with 1 hidden layer of 4 neurons and a compression ratio 1:10 using the full input pattern obtained the best performance for the testing set (table 3). The mean absolute error of the testing set for this neural network was 0.107. As the maximum value of the target test set was 1.0, this result gives a relative error for the testing set of only 10.7%. As the input patterns were generated adding noise to the simulated afferent answer, this result shows a good performance for the friction coefficient estimation. We believe that this performance can be improved enlarging the set of input patterns used to train the neural network.

Multilayer perceptrons with 3, 4 or 5 neurons in the hidden layer achieved better generalization that multilayer perceptrons with 10 hidden neurons (tables 1, 2, and 3). Thus, multilayer perceptrons with a few number of hidden neurons seems to be best suited for this problem.

As the dimension of the input patterns were too long for an efficient training of a neural network, different compression ratios of the input pattern were tested (tables 1, 2, and 3). Multilayer perceptrons using input patterns with a 10:1 compression ratio achieved the better results. Thus, the selection of a right compression ratio seems to be an important factor to achieve a good performance.

Each simulated experiment consisted of 100 grip force increments. On one hand, multiple multilayer perceptrons that receive as input only the first 20 increments (tables 1, 2, and 3) achieved a performance similar to multilayer perceptrons that received full input patterns. On the other hand, multiple multilayer perceptrons that receive input patterns without the first 40 increments showed a poor performance (table 4). Thus, the first increments of the grip force seem to be critical for the estimation of the friction coefficient.

8 Conclusions and Future Work

The estimation of the friction coefficient is a crucial information for a dexterous manipulation process in unknown environments. In this paper, a bio-inspired friction estimation method was used. This method is strongly inspired on neurophysiological studies of the human dexterous manipulation. A finite element method model was built, and simulated experiments with planar objects of different friction coefficients were done.

A performance analysis of multiple multilayer perceptrons to estimate the friction coefficient was done. The multilayer perceptrons shown a good performance in the estimation of the friction coefficient. The compression ratio of the input pattern, and the first increments of the grip force, showed to be important factors for the estimation of the friction coefficient.

The conjunction between robotics and human neurophysiology can give fruitful and interesting results. The obtained results show that a bio-inspired approach could be a promising way to explore in order to afford the dexterous manipulation of unknown objects. A future work consists in improving the obtained results using radial-basis function networks.

Acknowledgments. The author acknowledges Anna Theorin (f. Israelsson), Lars Rådman, Benoni Edin and Roland Johansson for kindly providing the software code of their afferent simulator program, and Enrique Carlos Segura for helpful discussions about this paper.

References

1. Johansson, R.: Tactile sensibility in the human hand: receptive field characteristics of mechanoreceptive units in the glabrous skin area. Journal of Physiology 281, 101–123 (1978)
2. Johansson, R.: Sensory and memory information in the control of dexterous manipulation. In: Neural Bases of Motor Behaviour, pp. 205–260. Kluwer Academic Publishers, Dordrecht (1996)
3. Johnson, K.: The roles and functions of cutaneous mechanoreceptors. Curr. Opin. Neurobiol. 11, 455–461 (2001)
4. Israelsson, A.: Simulation of responses in afferents from the glabrous skin during human manipulation. Master's thesis, Master thesis in Cognitive Science, Umeå University, Umeå, Sweden (2002)
5. Johansson, R., Westling, G.: Signals in tactile afferents from the fingers eliciting adaptive motor responses during precision grip. Exp. Brain Res. 66, 141–154 (1987)
6. Matuk Herrera, R.: A bio-inspired method for friction estimation. In: Proc. of MICAI 2007. IEEE CS Press, Los Alamitos (2007)
7. Maeno, T., Kawamura, T., Cheng, S.: Friction estimation by pressing an elastic finger-shaped sensor against a surface. IEEE Transactions on Robotics and Automation 20(2), 222–228 (2004)
8. North, J.F., Gibson, F.: Volume compressibility of human abdominal skin. J. Biomech. (11), 203–207 (1978)
9. Srinivasan, M.A., Gulati, R.J., Dandekar, K.: In vivo compressibility of the human fingertip. Adv. Bioeng. (22), 573–576 (1992)
10. Dandekar, K., Raju, B., Srinivasan, M.: 3-d finite-element models of human and monkey fingertips to investigate the mechanics of tactile sense. Journal of Biomechanical Engineering 125(5) 682–691 (2003)
11. Westling, G., Johansson, R.: Responses in glabrous skin mechanoreceptors during precision grip in humans. Exp. Brain Res. 66, 128–140 (1987)
12. Moller, M.: A scaled conjugate gradient algorithm for fast supervised learning. Neural Networks 6, 525–533 (1993)

Color Image Watermarking and Self-recovery Based on Independent Component Analysis

Hanane Mirza, Hien Thai, and Zensho Nakao

Department of Electrical and Electronics Engineering, University of the Ryukyus, Okinawa 903-0213, Japan
`{hanane,tdhien,nakao}@augusta.eee.u-ryukyu.ac.jp`

Abstract. The digital image watermarking field addresses the problem of digital image authentication and integrity. In this paper we propose a novel color image watermarking scheme based on image self-embedding and self-recovery techniques. The main idea of this algorithm is to embed a reduced content of the original image to itself, in order to be able to partially recover the deleted features from the watermarked image. Separately, the red and blue color channels are embedded, respectively in the wavelet domain by a compressed version of the original image, and in the spatial domain by binary encoded sequences generated from the original image. This allowed us, in detection stage, to prove the ownership, detect the altered blocks, and recover them. The detection and recovery bits extraction is computed using an ICA algorithm. The experimental results were satisfactory and show a high robustness against most common attacks as well as a reassuring rate of image recovery.

Keywords: Digital watermarking, color image, self-recovery, ICA.

1 Introduction

It is important to know that many court rooms around the world are, nowadays, admitting digital images as legal evidence[4], and many others are seriously considering the digital images for law enforcement[3]. Therefore, the authentication and the integrity of digital images become a serious issue[5], as it can be the key element in a judicial process. On the other hand, the powerful publicly available image processing softwares make digital forgeries very accessible. It is simple for anyone to alter the content of a digital image, by adding or deleting features from the original image without causing detectable edges[9]. In consequence, the new information marketplace, where the digital data can be a currency with two sides, addresses the need and the necessity to produce the softwares and tools necessary to protect the digital images ownership rights, their authentication, and make it possible to recover their original content in case a cutting/pasting attack was performed. Thus, we are proposing in this paper, a new content-based color image self-embedding and self-recovery scheme.

Several algorithms were previously presented regarding this matter, In one of the first techniques used for image tampering detection, Walton *et al* [1], created the theory of check-sums technique, by modifying the Least Significant Bits (LSB) of each pixel. This technique presents a high probability of tamper detecting but it is vulnerable to the block swap attacks. Fridrich *et al* [2] proposed an original fragile watermarking method of image self-embedding that consists of embedding the reduced bits of a block in a different distant block. This method could achieve the self-recovery, but if a distinct region of one image were attacked, the recovery bits would also be corrupted.

In this paper we will try to achieve both objectives, by performing two separate watermarking schemes in two different color layers of the original RGB image. As we have discussed in [6], a color RGB image can be watermarked in its three different color layers (Red,Green and Blue), and the separate watermarking process increases the overall watermarking capacity. In the current paper, we will watermark the red and blue color channels separately, using independent techniques, in order to increase security and the green layer we will keep as original in order not to degrade the quality of the image. In the blue subimage we will try to insert the necessary data to prove the image ownership and to detect its tampered regions. This can be done by embedding the robust bit extracted from a gray level version of the original image, and the embedding process was performed in the spatial domain. As for the red subimage, we inserted a compressed version of the original image, using a DCT compression technique, the embedding is performed in the wavelet domain. The objective of watermarking two color channels with different algorithms is to maximize the chances of content recovery of the image, especially after cutting/pasting attacks. The watermarked image is produced by the superposition of the watermarked red and blue subimage and the unwatermarked green subimage.

As for the extraction stage, the ICA algorithm proved to be an efficient tool[7][8] to detect and extract the mixed unknown sources. It is largely applied for the image and signal processing purposes, and we chose to implement for this experiment the FastICA algorithm for its properties, discussed later. The paper is organized as follows: the second section will discuss the proposed algorithm where we will show how we generate the watermarks to be from the original image separately for both red and blue subimages. Also we will include the embedding process illustrated by figures. In the third section using the chosen ICA algorithm we will demonstrate the detection/extraction process, and the tampered image restoration process. The last two sections will be dedicated to computer simulation results and some of our conclusions.

2 Proposed Algorithm

In the present embedding algorithm (fig.1), we first, separate the original color image $I(N \times N)$ to three color RGB channels, respectively, $R(N \times N)$, $G(N \times N)$ and $B(N \times N)$. For a high embedding capacity[10]we separately watermark the red and blue color channels with two different sets of watermark data that we call R_w, B_w. In the first step we need to generate the watermark data which is

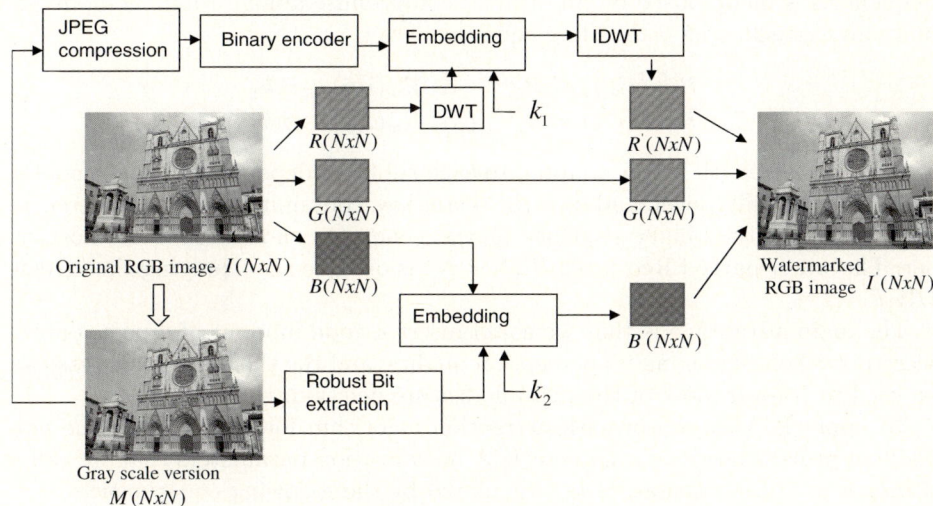

Fig. 1. Embedding process of color layers

content related to the original image, and we leave the green layer untouched, so as not to degrade the invisibility of the final watermarked image.

2.1 Recovery Bits Generation and Embedding

Since the main objective of this proposal is to recover the tampered or deleted areas of the image, we need to generate original image's content related data sets: R_w and B_w. The procedure is as follows:

First, we convert the copy of $I(N \times N)$ to gray-scale level, and we call it $M(N \times N)$.

R_w generation: As for the Red layer $R(N \times N)$ we tried to generate the R_w and perform the embedding in the wavelet domain following the next steps:
1. The Gray-scale image $M(N \times N)$ is divided into 8x8 blocks.
2. Apply *2D DCT* for each block, and divide the entire image by 8 to normalize the DCT coefficients.
3. Quantize the resulting values using the quantization matrix equivalent to the standard *50 %* quality JPEG.
4. The quantized values are further binary encoded using 64 bits only.
5. The resulting binary sequences generated in the previous step can be written as $R_w = \{R_{w1}, ..., R_{wm}\}$, where m is the number of the selected blocks to be compressed.

R_w embedding: On one hand, the red layer $R(N \times N)$ is decomposed into three levels by wavelet transform and the binary sequences R_w are inserted into midfrequency subbands by modifying wavelet coefficients belonging to two details bands at third level (R_3^{LH}, R_3^{HL}). The choice of the embedded wavelet frequencies is made based on an optimal compromise among robustness, invisibility and attack. The embedding equations are :

$$R_3^{'LH}(i,j) = R_3^{LH} + \alpha_i.R_w(i,j) + \beta.k_1 \\ R_3^{'HL}(i,j) = R_3^{HL} + \alpha_i.R_w(i,j) + \beta.k_1 \quad (1)$$

where α is a strength factor adapted to each subband depending on the smoothness and invisibility level, and k_1 is the secret key, containing the block references from which the the binary sequence R_w was generated and the subband host order. The watermarked Red layer $R'(N \times N)$ is obtained by applying the inverse DWT.

The main advantage is that we could insert a good amount of image content secretly without changing its perceptual quality, and the the embedded data can be used to recover most of the possible feature deletion.

To apply ICA for watermark extraction algorithm for $R'(N \times N)$, the embedding process needs to create an ICA initialization parameters that we call a *demix_key*, and we denote D_k_1, calculated by the following equations:

$$D\_k_1 = R_3^{*LH}(i,j) + R_3^{*HL}(i,j) + \gamma k_1(i,j) \quad (2)$$

where R_3^{*LH} and R_3^{*HL} are the subbands coefficient where the key κ is inserted and γ is a mixture strength coefficient, set to 0.5.

B_w generation: The watermark used here is designed in a way to extract a bit sequence of length L containing the robust bits of selected pixels. We divide the image $M(N \times N)$ to $(m \times m)$ block size, using the robust bit extraction algorithm detailed in [11] and illustrated in (fig.2), we extract, from each block, a binary string B_w of fixed length L; $B_w = \sum_{i=1}^{L} B_{wi}$.

B_w embedding: In the other hand, we divide the blue color subimage $B(N \times N)$ into 8x8 blocks.
1. For each block b, we denote the 64 pixels as $P_i \epsilon \{P_1, P_2...P_{64}\}$ and we define in a secret key, k_2, the information about each pixel's location in the block b and the block number ($b\#$).
2. The relation between the pixels contained in the block b from the original $B(N \times N)$ and the binary sequences B_{wi} extracted from $M(N \times N)$ is developed according to the following equation:

$$R_i = \Sigma_{i=1}^{64}.B_{wi}.C(P_i) \quad (3)$$

where $C(P_i)$ is the blue color intensity level of the pixel P_i.
3. We encrypt the binary form of the relation R_i and we embed it in the least significant bit of the pixel P_i . We can describe the embedding formula as:

$$P_i' = P_i + R_{ei} + k_2 \quad (4)$$

where R_{ei} is the binary and encrypted form of R_i and P' is the watermarked pixel if we denote a watermarked block by $b'(i,j)$, the watermarked blue layer is retrieved by the union of the watermarked blocks:

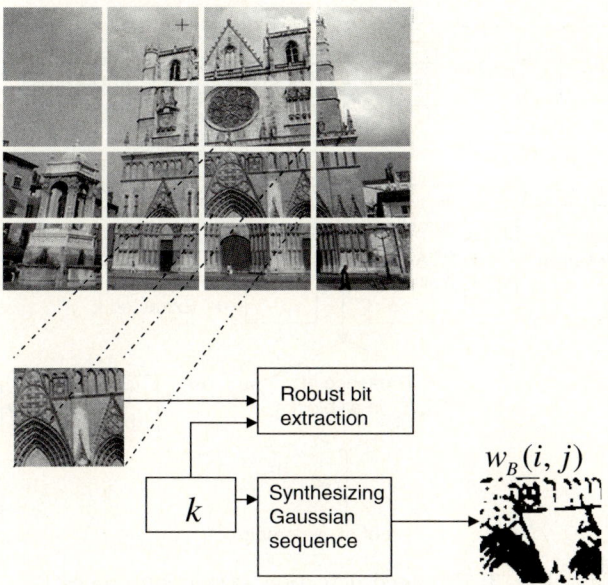

Fig. 2. Robust bits extraction

$$B^{'}(N \times N) = \bigcup_{i=1}^{\frac{N}{8}} \bigcup_{j=1}^{\frac{N}{8}} b^{'}(i,j). \tag{5}$$

The advantage of creating a relation between the extracted sequences and embedded pixels is to make alteration tamper easily detectable and the change in the quality of the image is not noticeable. The relation R_e is block dependent to avoid the loss of embedded data in case the block was removed, as it contains the block # and pixel's location information. Furthermore it is impossible, with this technique, to duplicate an entire block without making undetected damage. The main advantage of this embedding method is that the imperceptibility of the original image is not degraded as we are modifying only the LSB of selected pixels.

Similarly with the red channel, we need to create a demix_key for detection and extraction purposes of this algorithm. The demix_key can be written as:

$$D\_k_2 = P_i^* + k_2 \tag{6}$$

where P_i^* denote the pixels that contain the key k_2.

The watermarked color image $I^{'}(N \times N)$ is obtained by the superposition of the three resulting color layers, $R^{'}(N \times N)$, $B^{'}(N \times N)$ and the green $G(N \times N)$ non-watermarked layer.

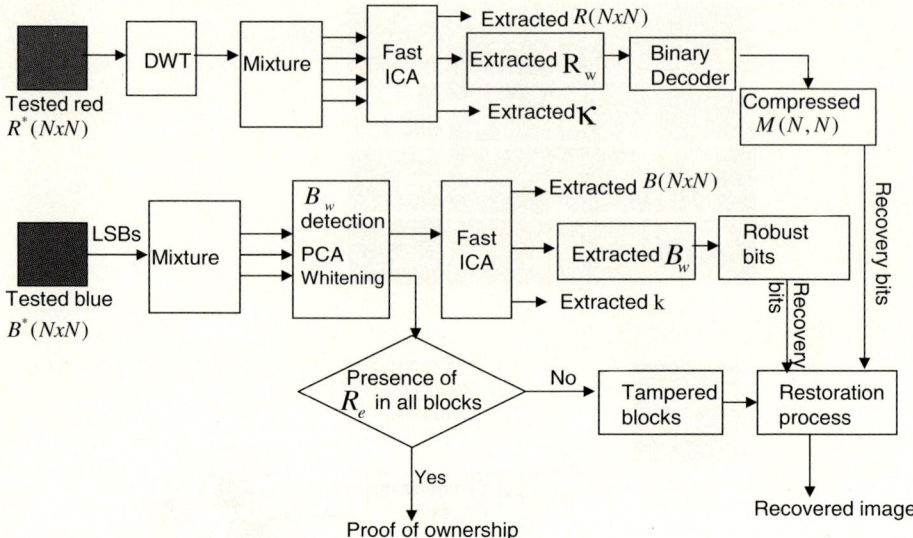

Fig. 3. Proposed detection/extraction process

2.2 Recovery Bits Extraction

Fast ICA
Independent Component Analysis (ICA) is a computing tool to extract independent sources from given mixtures of unknown sources, and this intelligent computing tool is widely applied in signal and image processing field. We consider here that the embedded recovery bits and the original image layers are unknown sources and the watermarked color layers are mixtures of those unknown sources. By creating different mixture the ICA algorithm detects and extracts the embedded recovery bits. Among the presented fixed-point algorithms, we chose in this paper to apply the FastICA [12] because it presents a certain amount of good properties mainly the fast convergence and its easy and suitable implementation for watermarking schemes.

The FastICA is based on two stages: the first is a PCA whitening of the input mixtures and the second is the FastICA by using fourth-order statistics of the signal. The extraction of the recovery bit in the red and blue color channels does not require any knowledge of original embedded recovery bits, or original image, or strength factors. We follow the next steps for extraction process(fig.3):

Step1: The watermarked color image $I^{'}(N \times N)$ is divided to the three color layers $R^{'}(N \times N), G(N \times N)$ and $B^{'}(N \times N)$, We separately apply the FastICA algorithm to both channels $R^{'}(N \times N)$ and $B^{'}(N \times N)$. respectively in step 2 and step 3.

Step2: The extraction process from the red color channel $R^{'}(N \times N)$, using the ICA algorithm is described in the following tasks:

1. The watermarked red layer R' is decomposed through DWT by three levels to obtain wavelet coefficients at $R_3^{\#LH}$ and $R_3^{\#HL}$ subbands.
2. In order to input the initialization parameter of the FastICA algorithm we create mixture signals X_1, X_2, X_3, X_4 from $R_3^{\#LH}$ and $R_3^{\#HL}$ subbands:

$$\begin{aligned} X_1 &= R_3^{\#LH} + D\_k_1 \\ X_2 &= R_3^{\#HL} + D\_k_1 \\ X_3 &= R_3^{\#LH} + R_3^{\#LH} \\ X_4 &= D\_k_1 + k_1 \end{aligned} \quad (7)$$

3. The mixture signals X_1, X_2, X_3, X_4 are also mixtures of the original wavelet transform coefficients of the original red layer (R_3^{LH}, R_3^{HL}), and the binary sequences R_w and the secret key k_1 which can be written as:

$$\begin{aligned} X_1 &= a_{11}R_3^{LH} + a_{12}R_3^{HL} + a_{13}R_w + a_{14}k_1 \\ X_2 &= a_{21}R_3^{LH} + a_{22}R_3^{HL} + a_{23}R_w + a_{24}k_1 \\ X_3 &= a_{31}R_3^{LH} + a_{32}R_3^{HL} + a_{33}R_w + a_{34}k_1 \\ X_4 &= a_{41}R_3^{LH} + a_{42}R_3^{HL} + a_{43}R_w + a_{44}k_1 \end{aligned} \quad (8)$$

where $a(i,j)\epsilon\{a_{11},...,a_{44}\}$ is an arbitrary real number.
4. Using the above described mixtures we can extract, using the fastICA algorithm, from the red layer the embedded binary sequence R'_w.

Step 3: We proceed to extract the embedded recovery bit from the $B'(N \times N)$, proceeding as follows:
1. The checking process is similar to embedding process: it consists of comparing for each block the value of Re^*, determined by the pixels of tested images with the original Re embedded in the LSB, by verifying the pixels locations and the block $b\#$.
2. This correlation process is enough to detect the altered blocks and also to claim the ownership of the original color image.
3. To extract the embedded bits from the watermarked blue layer we create the mixtures, similarly to the red layer :

$$\begin{aligned} X_1 &= P_i + D\_k_2 \\ X_2 &= P_i + e.k_2 \\ X_3 &= D\_k_2 + k_2 \end{aligned} \quad (9)$$

where $e = 0.5$ is the mixture strength factor. The same mixtures are included in the original watermarked image as:

$$\begin{aligned} X_1 &= a_{11}P + a_{12}R_e + a_{13}k_2 \\ X_2 &= a_{21}P + a_{22}R_e + a_{23}k_2 \\ X_3 &= a_{31}P + a_{32}R_e + a_{33}k_2 \end{aligned} \quad (10)$$

4. The above mixtures are used as input data for the fastICA algorithm and the the embedded sequence B_w is extracted.

The detected tampered blocks are first restored using the robust bits extracted at this stage, and we recall the extracted content from the red layer to recover each block separately.

3 Computer Simulation

The proposed algorithm was tested on three RGB color images(fig.4(a-c)) of size (512×512) (Cathedral, Airplane and Liberty). The first two images are taken by digital camera while the third one is taken by satellite. The watermarked images are shown in (fig.4(d-f)). No noticeable difference between the original and the watermarked images are detectable for the human eye, and the PSNR values are shown in table (1).

Table(1) shows the PSNR values computed between the original and the watermarked images and the recovery bits detector response in the blue and red channels. The higher is the PSNR, the better is the invisibility of the watermark, and the higher is the detection rate, the better is the robustness.

In order to test the robustness of our algorithm, the watermarked images were subjected to some common image processing attacks, including: Surrounding

(d) Watermarked Cathedral   (e) Watermarked Airplane   (f) Watermarked Liberty

Fig. 4. Original images (a-c) and Watermarked images(d-f)

Table 1. PSNR values and recovery detector response before the attacks

images	(a)	(b)	(c)
PSNR values	42.9	47.0	35.3
B_w detection rate	0.95	0.90	0.91
R_w detection rate	0.93	0.91	0.92

Table 2. Applied attacks and the resulting PSNR values and watermark detector response

Attacks	image(a)			image(b)			image(c)		
	PSNR	XB_w	XR_w	PSNR	XB_w	XR_w	PSNR	XB_w	XR_w
Surrounding Crop (94%)	52.2	0.63	0.68	27.4	0.62	0.58	59.0	0.60	0.54
Resize (448x448)	33.5	0.87	0.79	33.3	0.93	0.89	25.2	0.75	0.72
Adding Noise (power 5000)	17.3	0.81	0.78	15.7	0.87	0.73	15.2	0.76	0.71
Lowpass filtering (3x3)	29.2	0.71	0.63	31.0	0.78	0.67	22.7	0.58	0.63
Median filtering (3x3)	34.2	0.58	0.63	33.7	0.66	0.58	23.1	0.680	0.63
Jpeg (Quality=85%)	33.4	0.76	0.72	34.0	0.73	0.70	27.0	0.71	0.68
Jpeg2000 (bpp = 0.25)	32.7	0.86	0.77	33.3	0.81	0.79	23.9	0.76	0.68

Crop, Resize, Adding Noise, Lowpass filtering, Median filtering, Jpeg, and Jpeg2000. The results are shown in table(2).

After performing the attacks on the watermarked images the PSNR was re-calculated and so was the the Recovery bits extraction: we indicated the amount

$(c_1)PSNR = 21.1$ $(c_2)PSNR = 22.1$ $(c_3)PSNR = 20.5$

Fig. 5. Sample results of the proposed tamper detection and recovery algorithm

of the recovery bits detected in the watermarked tested blue channel as XB_w and the one in the red channel as XR_w.

The system showed good results for the robustness as shown in table (2). The embedded watermark was detectable enough in most tests, to prove ownership of the file at least, if not to also recover partially some of the attacked features.

As for the image restoration testing, the three watermarked images were tested by being subjected to cutting/pasting attacks, and we tried to recover the deleted feature from the extracted recovery bits. Fig.5($a_1 - a_3$) shows the modified watermarked images, the modification purpose was to delete some of the features of the watermarked images: In the Cathedral image (a_1) the middle gate was 'cut' and colored in similar color to the main color of the rest of Cathedral; in the Airplane image (a_2) the tail of the Airplane that carried all the references was deleted; in the Liberty image (a_3) the liberty statue was cut off its original stand and pasted to three different locations in same image (right, left and below the original stand). The PSNR between the tampered images and the original images was computed as well and it is shown in fig.5($a_1 - a_3$). As figure(5)($b_1 - b_3$) shows, the altered blocks of each image were detected by the algorithm. The partially recovered features are shown in fig.5 ($c_1 - c_3$). The tampered images are not perfectly recovered but enough to have an idea about the original features, which guarantee the image authentication, and make it reliable as a legal evidence.

4 Conclusions

The main contribution of this paper is to demonstrate that it is possible, through the theory and computer simulation (more results and comparisons to come), to recover the original content of a tampered color image. Watermarking two color layers is done to double the chances of recovering the embedded data and two different independent algorithms are performed to increase the security and robustness of the algorithm. The ICA extraction algorithm makes it possible to extract the embedded data after all the attacks are performed. The experimental results shows that the watermark survived all the common attacks, and the embedded content-based watermarked were still detectable, extractable and useful for tampered regions recovery.

Acknowledgments

This research was supported in part by Ministry of Internal Affairs and Communications (Japan) under Grant: SCOPE 072311002, for which the authors are grateful.

References

1. Walton, S.: Information authentication for a slippery new age. Dr. Dobbs Journal 20, 18–26 (1995)
2. Fridrich, J., Goljan, M.: Protection of digital images using self embedding. In: Symposium on Content Security and Data Hiding in Digital Media, New Jersey Institute of Technology, USA (1999)

3. Craiger, P.J., Pollitt, M., Swauger, J.: Law enforcement and digital evidence. Handbook of Information Security, New York, USA (2005)
4. Staggs, S.B.: The Admissibility of Digital Photographs in Court (2005), http://www.crime-scene-investigator.net/admissibilityofdigital.html
5. Mason, S.: Authentication of electronic evidence. Information Age, Australia (2006)
6. Miyara, K., Thai, H., Harrak, H., Nakao, Z., Nagata, Y.: Multichannel color image watermarking using PCA eigenimages. In: Advances in Soft Computing, vol. 5, pp. 287–296. Springer, Heidelberg (2006)
7. Yu, D., Sattar, F., Ma, K.-k.: Watermark Detection and Extraction Using Independent Component Analysis Method. Journal on Applied Signal Processing, EURASIP 5, 92–104 (2002)
8. Shen, M., Zhang, X., Sun, L., Beadle, P.J., Chan, F.H.Y.: A method for digital image watermarking using ICA. In: 4th International Symposium on Independent Componenet Analysis and Blind Signal Separation (ICA 2003), Japan (2003)
9. Rehmeyer, J.: Computing Photographic Forgeries. Science News, vol. 171 (2007)
10. Barni, M., Bartolini, F., De Rosa, A., Piva, A.: Color image watermarking in the KLT domain. In: SPIE, Electronic Imaging, pp. 87–95 (2002)
11. Fridrich, J.: Robust Bit Extraction from Images. In: IEEE International Conference on Multimedia Computing and Systems (ICMCS 1999), vol. 2 (1999)
12. Bingham, E., Hyvarinen, A.: A fast fixed-point algorithm for independent component analysis of complex-valued signals. Int. Journal of Neural Systems 10, 1–8 (2000)

A Fuzzy Rule-Based System with Ontology for Summarization of Multi-camera Event Sequences

Han-Saem Park and Sung-Bae Cho

Dept. of Computer Science, Yonsei University
Shinchon-dong, Seodaemun-ku, Seoul 120-749, Korea
sammy@sclab.yonsei.ac.kr, sbcho@cs.yonsei.ac.kr

Abstract. Recently, research for the summarization of video data has been studied a lot due to the proliferation of user created contents. Besides, the use of multiple cameras for the collection of the video data has been increasing, but most of them have used the multi-camera system either to cover the wide area or to track moving objects. This paper focuses on getting diverse views for a single event using multi-camera system and deals with the problem of summarizing event sequences collected in the office environment based on this perspective. Summarization includes camera view selection and event sequence summarization. View selection makes a single event sequence from multiple event sequences as selecting optimal views in each time, for which domain ontology based on the elements in an office environment and rules from questionnaire surveys have been used. Summarization generates a summarized sequence from a whole sequence, and the fuzzy rule-based system is used to approximate human decision making. The degrees of interests input by users are used in both parts. Finally, we have confirmed that the proposed method yields acceptable results using experiments of summarization.

1 Introduction

Recently, the popularization of digital cameras and advancement of data compression and storage techniques made it possible for most people to access and use the video data easily [1]. Accordingly, people can obtain the video data in various ways from using the CCTV in the public space to using a personal portable device. The video data are more useful than text documents, voices and still images because they contain much more specific and realistic information. Therefore, it has become very important to extract information that people want to search, and the studies of analyzing and summarizing video data have been investigated from all over the world [1,2].

There are various types of target video for summarization. Some videos are made by experts like movie, news, and sports, and the some other videos are collected by researchers in a preset indoor environment. The first type has clear scenes and shots divided by backgrounds. Summarization process of these videos includes content analysis, structure parsing and summarization [2]. Content analysis step maps low-level features to high-level semantic concepts. This can be conducted automatically using image processing and recognition techniques [3] or manually [4]. Structure parsing step divides the video into scenes and shots based on the result of content analysis, and summarization step provides significant parts of videos using analyzed information in previous steps. The second type videos are collected by researchers. They usually collect the data in an indoor environment and use a multi-camera system [5,6]. These videos do not have clear scenes because the backgrounds are static indoor environments, so they are divided by either the location of the users [5] or activities occurred in that domain [6]. The analyses of activities or events can provide valuable information for studies on human behavior, the design of indoor environment, etc. Besides, the summarization service can help people to remember what they did in the past [5].

This paper targets the video collected in an office environment with the multi-camera system. The proposed method subdivides the target videos into event sequences, selects the optimal camera views, and summarizes event sequence considering the degree of interests input by users. For view selection, we have made the domain ontology describing the elements in the office and the rules from questionnaire surveys. For summarization, we have used the fuzzy rule-based system to approximate a human decision making process in events evaluation [8]. The users should input the degrees of interests to each event, person, and object so that the system can provide personalized summary of target video.

2 Multi-camera System

Most conventional multi-camera systems used the multiple cameras to cover wide area and to track moving persons and objects. J. Black et al. developed the multi-camera system that tracked and extracted the moving object in an outdoor environment [9], and they attempted to apply the system to an indoor environment. Another main purpose of using the multi-camera system is in a security. F. Porikli set several cameras in a building to track the moving objects and summarize that information [10]. Recently, studies for summarization and retrieval or studies for human activity analyses have been investigated. Y. Sumi et al. analyzed the subjects using the sensors including multiple cameras and provided a simple summarization [6], G. C. Silva et al. exploited the multi-camera system to summarize and retrieve information of persons living in a home-like ubiquitous environment [5].

However, the multi-camera systems have another use where they can provide diverse views using the multiple cameras. Because a single view for a certain activity or event might not provide the exact information, we can obtain this correct and diverse information through multiple views. C. Zhang et al. extracted the event information using IR tags and four cameras, and provided the retrieval system that searched for event the user wanted with a simple query [7]. They focused on an advantage of diverse views by the multi-camera systems. We also focus on this possibility, and also we used the users' degrees of interests to events, persons, and objects to provide the personalized summary of event sequences.

3 Multi-camera Office Environment

3.1 Setting Multi-camera System in Office Environment

To collect the office event sequence, we set eight cameras in the lab, shown in the left figure of Figure 1. We set all cameras to focus on the same area so that the system can provide diverse views for a single event. The right figure illustrates a captured example of different views.

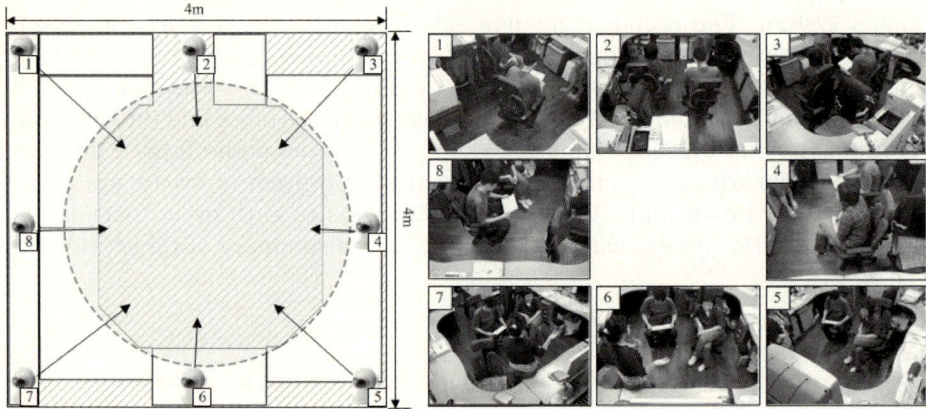

Fig. 1. Cameras set in an office environment and the target area (left) and an example of event captured by the multi-camera system (right)

3.2 Annotation of Events, Persons, and Object

This paper regards the collected video as an event sequence and provides the summarization service with that sequence. All basic information including events, persons, objects, and their positions have been annotated manually, and these works have been performed based on the event definition as follows.

- Entry (A),
  if stand (A, entrance-area) and face (A, in)
- Leaving (A),
  if stand (A, entrance-area) and face (A, out)
- Calling (A),
  if hold (A, phone) and speak (A)
- Vacuuming (A),
  if hold (A, vacuum cleaner) and stand (A, center-area)
- Eating (A),
  if hold (A, food)

- Resting (A),
  if rest (A, corner-area) or stretch (A, corner-area)
- Work (A),
  if sit (A, corner-area) and {use (A, computer) or hold (A, document)}
- Printing (A),
  if exist (A, printer-area) and hold (A, printout)
- Conversation (A, B),
  if {exist (A, x) and exist (B, y) and close (x, y)} and
  {speak (A) or speak (B)}
- Meeting (A, B, C),
  if {exist (A, x) and exist (B, y) and exist (C, z) and close (x, y, z)} and
  {speak (A) or speak (B) or speak (C)} and
  {hold (A, document) or hold (B, document) or hold (C, document)}
- Seminar (A, B),
  if {stand (A, screen-area) and sit (B, center-area)} and speak (A)

These eleven events are normal ones that could happen in the office environment. We have decided these events according to the related works dealing with human activities or events. Some of these works classified the events based on objects such as phone, table, chair, book, keyboard [11], and some other works classified the events based on persons and their activities [12]. Many of the previous video summarization studies have performed the annotation automatically or semi-automatically using image processing and pattern recognition techniques [1,3]. The current version of the system relies on manual annotation, and it will be replaced by automatic annotation in the future work.

4 The Proposed Method

Figure 2 provides an overview of the proposed summarization system. As mentioned before, summarization process is divided into a view selection module and a summarization module. View selection, which selects an optimal view for a single point in time, is performed using rules based on domain ontology. Summarization is performed using fuzzy rule-based system. In summarization, fuzzy system evaluates each event in event sequence, and then events with high evaluation score are selected as important ones.

4.1 User Input

To summarize an office event sequence considering users' personal interests, we attempt to use user input of the degree of interest (DOI) to each event, person, and object. User interest is an important factor to design an interface or to interact with users. A user input has a form of integer between 0 (not interested) to 3 (interested a lot), and it can be used for view selection and the personalized summarization.

We have segmented the event sequence into shots. Shot, a basic unit for the video data, is defined as a set of consecutive frames with the same event, person,

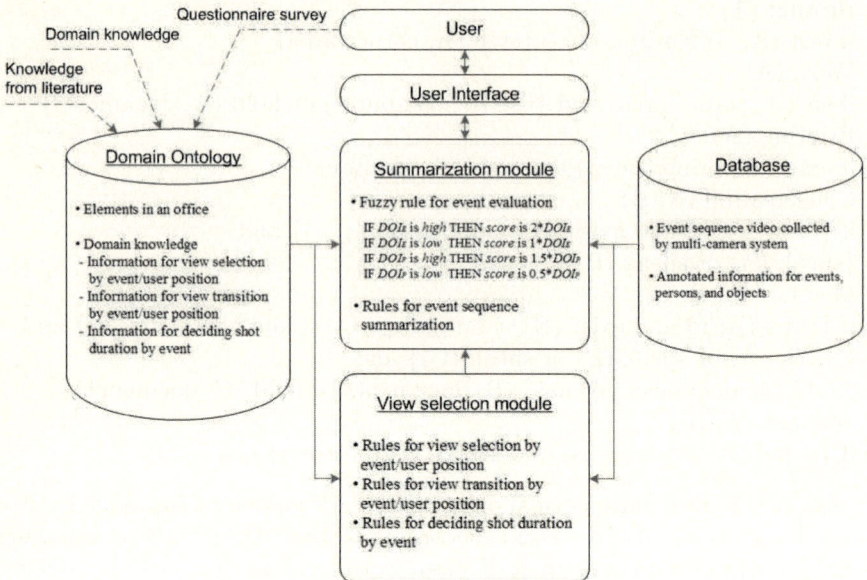

Fig. 2. An overview of the proposed summarization system

and object. DOI value has been calculated by shot, and DOI those one shot Si are defined as follows.

$$DOI^2_{E_j,S_i} = \sum_{frame f \in S_i} DOI_{E_j}(f) \qquad (1)$$

$$DOI^2_{P_k,S_i} = \sum_{frame f \in S_i} DOI_{P_k}(f) \qquad (2)$$

$$DOI^2_{O_l,S_i} = \sum_{frame f \in S_i} DOI_{O_l}(f) \qquad (3)$$

Equations (1) to (3) show the DOI value for one shot S_i to event E_j, person P_k, and object O_l. Adding DOI values of all frames in a shot, its square root is used as a final DOI value so that the duration does not have a significant influence.

4.2 Domain Ontology

Domain ontology comprises elements in an office and domain knowledge for view selection. The former describes each element in office environment and their relationships, and the latter describes information required to design rules for view selection. Figure 3 shows the element description of "Place" and "Meeting". Domain knowledge includes information for camera view selection and view

```
<owlr:Rule>
    <owlr:antecedent>
        <owlr:individualPropertyAtom owlr:property="locate">
            <owlr:variable owlr:name="A">
            <owlr:variable owlr:name="2">
        </owlr: individualPropertyAtom>
        <owlr:individualPropertyAtom owlr:property="happen">
            <owlr:variable owlr:name="Work">
        </owlr: individualPropertyAtom>
    </owlr:antecedent>
    <owlr:consequent>
        <owlr:individualPropertyAtom owlr:property="view">
            <owlr:variable owlr:name="3">
        </owlr: individualPropertyAtom>
    </owlr:consequent>
</owlr:Rule>
```

Fig. 3. Domain knowledge description: An information description for view selection

transition by event and user position, and information for deciding shot duration by event. Most of this information is based on the questionnaire surveys and their analyzed results. Questionnaires were surveyed by ten graduate students. Figure 4 illustrates an example of information description for view selection, which means "If person A locates in area 2 and event Working happens, and then view #2 should be selected." Representation format has referred the syntax for ORL (Owl Rule Language) [13].

4.3 View Selection and Event Sequence Summarization

View selection generates a single event sequence from multi-camera event sequences. It includes a selection of an optimal event at the same point in time and a selection of an optimal view among views showing the same event. Previous works exploited simple rules to select one sensor among many of them. Y. Sumi et al. selected a sensor that had a high priority based on predefined priority [7]. To select an optimal camera view considering all variables including

```
Procedure ViewSelection
    var      N: the number of events in given time point
             M: the number of views for a given event
    function SelectEvent(Ei): a function that select an event with the highest DOIE
             SelectView(Ei, Vj): a function that select a view for an event Ei based on rules
                                 in domain ontology
    begin
             for i=1 to N
                     SelectEvent(Ei)
             for j=1 to M
                     SelectView(Ei, Vj)
    end
```

Fig. 4. A pseudo-code for a view selection

IF (DOI_E is high and DOI_P is high and DOI_O is high) THEN score = $2 \times DOI_E + 2 \times DOI_P + 1 \times DOI_O$
IF (DOI_E is high and DOI_P is high and DOI_O is low) THEN score = $2 \times DOI_E + 2 \times DOI_P + 0.25 \times DOI_O$
IF (DOI_E is high and DOI_P is low and DOI_O is high) THEN score = $2 \times DOI_E + 0.5 \times DOI_P + 1 \times DOI_O$
IF (DOI_E is high and DOI_P is low and DOI_O is low) THEN score = $2 \times DOI_E + 0.5 \times DOI_P + 0.25 \times DOI_O$
IF (DOI_E is low and DOI_P is high and DOI_O is high) THEN score = $0.5 \times DOI_E + 2 \times DOI_P + 1 \times DOI_O$
IF (DOI_E is low and DOI_P is high and DOI_O is low) THEN score = $0.5 \times DOI_E + 2 \times DOI_P + 0.25 \times DOI_O$
IF (DOI_E is low and DOI_P is low and DOI_O is high) THEN score = $0.5 \times DOI_E + 0.5 \times DOI_P + 1 \times DOI_O$
IF (DOI_E is low and DOI_P is low and DOI_O is low) THEN score = $0.5 \times DOI_E + 0.5 \times DOI_P + 0.25 \times DOI_O$

Fig. 5. TSK fuzzy rules for event evaluation

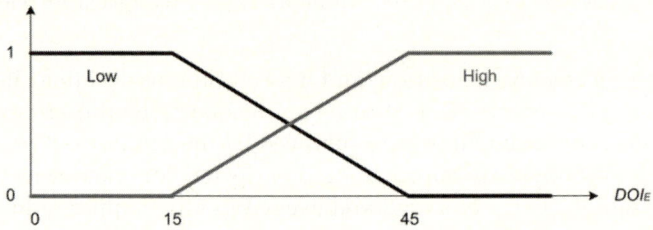

Fig. 6. A fuzzy membership function for DOI_E

user input, the proposed system constructed the domain ontology with the domain knowledge through questionnaire survey. Figure 4 shows a pseudo-code explaining a view selection process.

Next, the system evaluated each event in single event sequence generated by view selection to summarize it. TSK fuzzy system has been utilized in this paper. Fuzzy rules designed by domain knowledge in domain ontology are as follows. Getting users' DOI values as an input, the system calculates the final score for each event.

Before the fuzzy inference with these fuzzy rules, a fuzzy membership function shown in Figure 6 has been used to fuzzify the user inputs, DOI values, so that they could be used as input variables of the fuzzy system. It depicts a fuzzy membership function for DOI_E, and functions for DOI_P and DOI_O also have the trapezoidal function. This type of membership function is very simple, but widely used [14].

In summarization step, important events are selected by the rank based on the evaluated score. Here, shots of one long event, which splitted into several events due to an event happening in-between, cannot be selected more than twice, and events with low evaluated score were excluded from summary. Duration of each event is also based on the domain ontology, and central frames were selected for summary.

5 Experiments

5.1 Scenario and Data Collection

Experimental data were collected in the office environment presented in section 3. We designed a realistic scenario that could happen in an office assuming three persons in one day (from 9:00 a.m. to 6:00 p.m.). Figure 7 illustrates this scenario. Here, EN, PR, CONVERS, CA, and LE stand for entry, printing, conversation, calling and leaving, respectively.

For data collection, Sony network camera (SNC-P5) were used, and video were saved with the resolution of 320 240 and frame rate of 15 fps using MPEG video format.

5.2 Result of Summarization

Based on the scenario in Figure 7, we have made a single event sequence with view selection, and user inputs were assumed as Table 1 and 2. DOI values to objects were assumed as 0.

First, we have selected a single event sequence with view selection process. Most events in selected view were related to person C and event Vacuuming, Printing, Meeting, and Seminar, which had high DOI values. Subsequently, experiments for summarization have been conducted using this selected event sequence. This sequence contains 29 shots, and user inputs were assumed as in section 5.2. Figure 10 provides finally evaluated score by shots using fuzzy rules from DOI_E and DOI_P values.

In Figure 8, shots with the highest scores are shot #20 and shot #23 #26. The common characteristics of these shots are as follows. They are related to

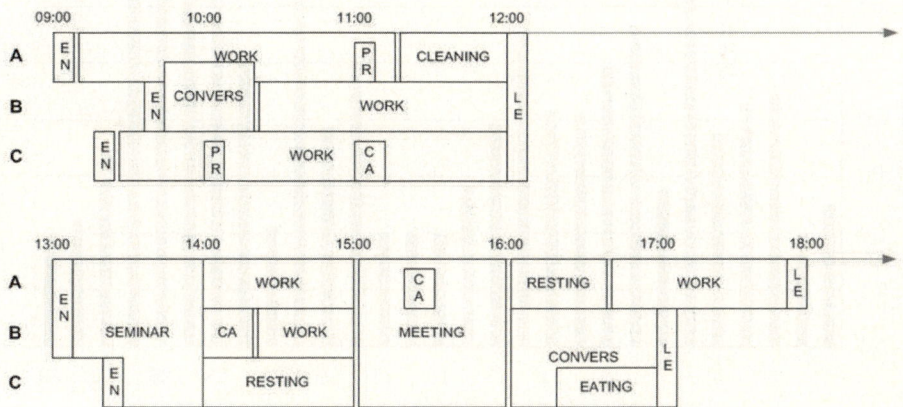

Fig. 7. A scenario of office events in one day

Table 1. User input (DOI value to event)

Event	DOI_E
Entry	2
Leaving	1
Calling	1
Vacuuming	3
Eating	0
Resting	0
Work	1
Printing	3
Conversation	2
Meeting	3
Seminar	3

Table 2. User input (DOI value to person)

Person	DOI_P
A	1
B	1
C	3

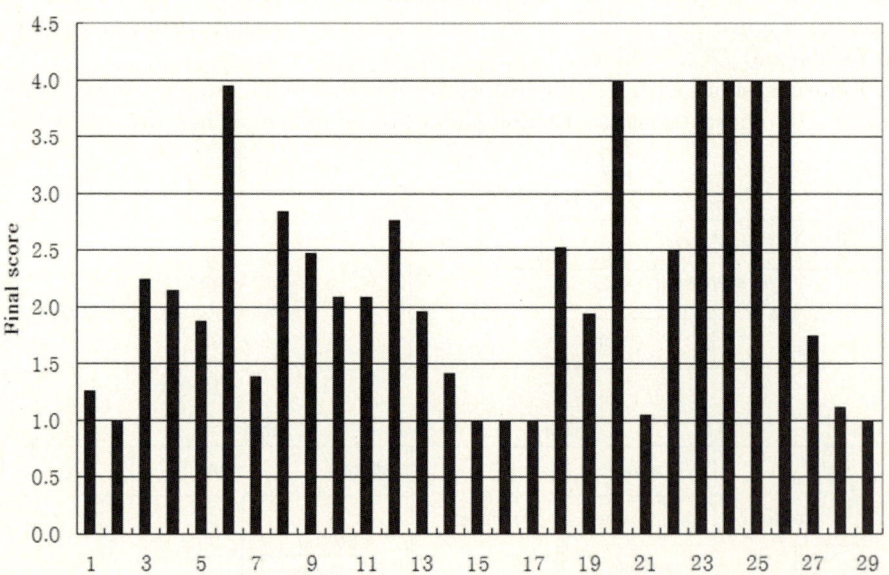

Fig. 8. Change of finally evaluated score by shot number

Table 3. A summarized event

S#	C#	F#(start)	F#(end)	Person	Event	Score
3	6→1	400	570	C	Entry	2.2
4	1	616	690	C	Work	2.1
6	1→6→1	950	1230	C	**Printing**	3.9
9	3→6→3	1810	2050	A	**Printing**	2.5
10	1	2065	2140	C	Calling	2.1
12	6→3→1→6	2395	2750	A	**Vacuuming**	2.8
13	1→6	2750	2900	C	Leaving	2
20	2→3	4216	4365	A, B, C	**Seminar**	4
22	1	5081	5155	C	Resting	2.5
25	2	6441	6515	A, B, C	**Meeting**	4
26	1	6731	6805	B, C	**Meeting**	4
27	6	7241	7335	C	Leaving	1.8

events with high DOI_E values. They are also related to person C, which has a high DOI_P value, and they have long duration. Shot #6 is exceptionally short, but it is about Printing event, which also has a high DOI_E value, and person C.

Table 3 shows a summarized event sequence. Here, S# means the shot number, C# means the camera view number, and F# means the frame number, and more than two views have been selected for C# in case of events with movement. Score represents a finally evaluated score by fuzzy rules. View selection in view transition has been performed using domain knowledge in domain ontology as described in section 4.2.

6 Conclusions

We collected the video with office events in a multi-camera environment and proposed the summarization system. The system generated a single event sequence from manually annotated event sequences and summarized the sequence using fuzzy rule-based system. For view selection, the domain ontology based on questionnaire surveys and literatures was used. Also, users' degrees of interest were used for personalized summarization. With experiments of view selection and summarization, we confirmed that the summarized event sequence was acceptable.

Future work will focus on the automatic annotation of office events. The design and implementation of user friendly interface to use the proposed method is required to help easy use and effective presentation. Also, subjective test to evaluate the experimental result will be performed.

Acknowledgement. This research was supported by MIC, Korea under ITRC IITA-2008-(C1090-0801-0011).

References

1. Zhu, X., et al.: Hierarchical video content description and summarization using unified semantic and visual similarity. Multimedia Systems 9(1), 31–53 (2003)
2. Li, Y., et al.: Techniques for movie content analysis and skimming. IEEE Signal Processing Magazine 23(2), 79–89 (2006)
3. Tseng, B.L., et al.: Using MPEG-7 and MPEG-21 for personalizing video. IEEE Multimedia 11(1), 42–53 (2004)
4. Petkovic, M., Jonker, W.: An overview of data models and query languages for content-based video retrieval. In: Int. Conf. on Advances in Infrastructure for E-Business, Science, and Education on the Internet. l'Aquila, Italy (2000)
5. Silva, G.C., et al.: Evaluation of video summarization for a large number of cameras in ubiquitous home. In: Proc. of the 13th ACM Int. Conf. on Multimedia, pp. 820–828 (2005)
6. Farwer, Sumi, Y., Mase, K., et al.: Collaborative capturing and interpretation of interactions. In: Pervasive 2004 Workshop on Memory and Sharing of Experiences, pp. 1–7 (2004)
7. Zhang, C.C., et al.: MyView: Personalized event retrieval and video compositing from multi-camera video images. In: Smith, M.J., Salvendy, G. (eds.) HCII 2007. LNCS, vol. 4557, pp. 549–558. Springer, Heidelberg (2007)
8. Dorado, A., et al.: A rule-based video annotation system. IEEE Trans. on Circuits and Systems for Video Technology 14(5), 622–633 (2004)
9. Black, J., Ellis, T.: Multi-camera image tracking. Image and Vision Computing 24, 1256–1267 (2006)
10. Porikli, F.M., Divakaran, A.: Multi-camera calibration, object tracking, and query generation. In: IEEE Int. Conf. on Multimedia and Expo, vol. 1, pp. 653–656 (2003)
11. Fogarty, J., et al.: Predicting human interruptibility with sensors. ACM Trans. on Computer-Human Interaction 12(1), 119–146 (2005)
12. Oliver, N., et al.: Layered representations for learning and inferring office activity from multiple sensory channels. Computer Vision and Image Understanding 96(2), 163–180 (2004)
13. Horrocks, I., Patel-Schneider, P.F.: A proposal for an OWL rules language. In: Proc. of the 13th Int. World Wide Web Conf., pp. 723–731 (2004)
14. Lertworasirikul, S., et al.: Fuzzy data envelopment analysis (DEA): A possibility approach. Fuzzy Sets & Systems 139(2), 3–29 (1998)

# A New Approach to Interactive Visual Search with RBF Networks Based on Preference Modelling

Paweł Rotter[1,2] and Andrzej M.J. Skulimowski[2]

[1] European Commission, DG Joint Research Centre, Institute for Prospective Technological Studies, Seville, Spain, on leave from ([2])
[2] AGH-University of Science and Technology, Chair of Automatic Control, Krakow, Poland
`{rotter, ams}@agh.edu.pl`

Abstract. In this paper we propose a new method for image retrieval with relevance feedback based on eliciting preferences from the decision-maker acquiring visual information from an image database. The proposed extension of the common approach to image retrieval with relevance feedback allows it to be applied to objects with non-homogenous colour and texture. This has been accomplished by the algorithms, which model user queries by an RBF neural network. As an example of application of this approach, we have used a content-based search in an atlas of species. An experimental comparison with the commonly used content-based image retrieval approach is presented.

Keywords: Content-based image retrieval, relevance feedback, interactive multimedia retrieval, RBF networks, distance-based matching, preference modelling

1 Introduction

Low and still decreasing costs of digital image processing devices, availability of high storage capacities and fast access to Internet contribute to the rapid growth of importance of computer-assisted multimedia search. Specifically, search in databases containing graphical documents without using the annotations, often referred to as Content-Based Image Retrieval (CBIR), raises the need for effective algorithms. Up till now (February 2008), the content-based approaches are still at the prototype stage, or are used within a specific restricted field, while all popular Internet image search engines are based on textual information only (description of images) or use quantitative or normative information describing low-level image properties such as resolution, file format, size etc., without analysing the image contents. In spite of intensive research in the field of content-based image retrieval in last years 1, the performance of CBIR systems which have been developed until now is still unsatisfactory and most of potential application areas (like Internet search engines, e-commerce) cannot fully benefit from content-based search.

The most popular content-based image retrieval systems, which are based on interaction with the user during the search process, like MARS 2, MindReader 3, VisualSeek 4 or QBIC 5, were developed about ten or more years ago. Except the systems which are designed for a specific class of images, "semantic gap" 1 (between low-level features extracted from an image and high-level, semantic features that users have in mind) make search result often disappointing.

The method here presented, like a large part of CBIR systems, uses the relevance feedback, i.e. adaptation of parameters during interactive search process, based on the user's assessment of the relevance of images presented by the system in consecutive iterations of the search process. Our method is designed for a specific class of images: we assume that we deal with objects which can be decomposed to sub-objects such that the main object can be classified based on shape, texture and colour distribution of sub-objects and mutual spatial relations between them in 2-dimensional image. We also assume that translation, scaling and 2D rotation do not change the class of the object, but 3D transformations are not considered in this paper. Therefore e.g. photos of the same 3D object made from different positions may be considered as objects belonging to different classes. We assume that images are preliminarily processed in such a way that objects appear on homogeneous background.

Since the proposed approach requires some preliminary processing or/and selection of images in a database, it can be used to perform a search for an object based on its memorized appearance, without using any additional information about the object, such as e.g. the name of a species searched in an atlas. In general, methods based on RBF networks (Sec. 4) can be also suitable for search in heterogeneous databases and for applications, where content-based search is motivated by lack of textual information associated with images, like in approach proposed in 6. The search process can be based on global image features rather than on understanding image semantics. An example of application where the method presented in this paper can be successfully applied is an interactive atlas of species, which we have developed based on the method described hereby.

The paper is organized as follows. In Sec.2, we briefly describe existing image retrieval methods and point out their limitations. In Sec.3, we present an adaptation of the existing method for the recognition of non-homogeneous objects. We assume that each object can be broken down into sub-objects with a specified shape (unique for each class), homogenous texture and colour distribution. Similarity of two objects is calculated based on similarity of shape, texture and colour of corresponding sub-objects and on similarity of spatial relations between them. This adaptation enables application for broader class of images and allows for experimental comparison with the approach which we propose in Sec. 4. Our method is based on an approximation of user's preferences in feature space by the RBF neural network 7. We propose the network which is composed from RBF sub-networks which correspond to representations of images (like colour, texture, shape). The output of each sub-network is a function of a distance between input feature vector and network coefficients which are modified through the learning process and correspond to system's guess about the object which the user is looking for. The distance measures vary for sub-networks and are adjusted to specific representation. In Sec. 5 we compare both methods using as an example an interactive atlas of species, developed in Matlab environment.

2 State of the Art in Interactive CBIR Methods

A comprehensive review of research on relevance feedback in image retrieval can be found in 8. As presented in 9, relevance feedback methods modify either a query, represented as a single point or a set of points in feature space, or a similarity measure used to calculate the distance between the query and images in database, or both. The method proposed by Rui, Huang and Mehrotra 10 can be regarded as a typical approach to image retrieval with relevance feedback. In the next section we will propose modifications of this method, which aim to extend its area of applications to objects with non-homogenous colour and texture. Then we will discuss the performance of methods based on the approach described below comparing them with the method based on neural networks proposed in Sec. 4.

For the interpretation of performance and results of algorithms we will use the notion of *utility*, coming from decision theory. Specifically, the *deterministic utility function* v is a quantitative measure of satisfaction of the user defined on the set of feasible decisions (a more precise definition and properties can be found e.g. in 11). In our approach, having defined a measure of proximity ρ between a model M and images in a database, which satisfies all axioms of the metrics except, perhaps, the condition that $\rho(M,I) = 0 \Rightarrow M=I$, one can assign the deterministic utility value to the images retrieved so that for every fixed M it is monotonic (decreasing) with respect to ρ. This principle will be used in this paper, for images represented in the feature space F and a utility function v constructed in such a way that the image which corresponds to the maximum satisfaction of the user's preferences irrespectively whether unique or not, maximizes simultaneously v on the feature space F. Let us remark that another approach, where features themselves (but not their distance to the model, because the latter is not defined) are monotonically increasing with respect to the user's preference, is presented e.g. in 11 and 12.

The selection of object features to be used for calculation of similarity between images plays a crucial role in image recognition and retrieval systems. Much work has been done on finding the features which are most useful for recognition, i.e. those that give high similarity of objects from the same class and low similarity of objects belonging to different classes. The methodology which consists of a choice of a specific similarity measure and a scalarization method (usually by means of a distance measure or a weighted sum of several similarity measures; see e.g. 13 for more information about distance scalarization) before recognition (retrieval) process is referred to as the *isolated approach* 14. In image retrieval, interaction with the user is possible and even desired to much greater extent than in image recognition. It is assumed that users do not have any specialised knowledge on image analysis so, in interactive feedback, they only need to provide evaluation of individual images in the form of grades which express the *relevance* of images. In each iteration the system presents to the user several images, and the relevance information given by the user is a starting point for upgrading similarity function parameters. Therefore images presented by the system in the next iteration should better correspond to the user's preferences. Besides the parameters of similarity function, descriptors of a query object can also be modified. Starting values are calculated based on an image provided by the user (who wants to find other images similar to the one/ones he already has) or randomly chosen in the first iteration, if a query image was not provided. The term *virtual query* refers to a point in feature space which corresponds to the image the user is looking for. The concept described above is referred to as *relevance feedback*. Rui, Huang and Mehrotra 10 proposed an approach, where functions describing the similarity of objects are defined at three levels:

1° object – an area with homogenous colour and texture,
2° feature – e.g. colour or texture,
3° feature representation – e.g. a colour histogram or average value of Gabor transform for a given area.

They assume that the user's utility function is a linear combination of preferences concerning image features (like colour, texture or shape) – for example, if shape is k_1 times more relevant than texture, that means that the ratio of corresponding coefficients in linear combination of these features is equal to k_1. Moreover, they assume that preferences for a specific image feature are a linear combination of similarities of feature representations Ψ_i. For example, in this approach the representation Ψ_1 of shape (e.g. similarity of shape coefficients) can be assumed k_2 times more relevant than the representation Ψ_2 (e.g. similarity of Fourier descriptors). Coefficients describing the preferences in the feature space, in our case k_1 and k_2, can be modified in every iteration of algorithm based on *relevance feedback*, provided by the user.

Based on the assumptions given above, the distance between a query object q and the model m can be expressed as a linear combination of functions Ψ_i, which defines the distance for feature representation i:

$$d(q,m) = \sum_{i=1}^{I} u_i \Psi_i(q_i, m_i, P_i), \tag{1}$$

where:
q and m denote the query object and model,
q_i and m_i are representations (vectors, with different dimensions for different i),
$u_i > 0$ are weight coefficients and
P_i is a set of parameters of metric in space of representation i.

In the above presented theoretical framework, the calculation of parameters of similarity functions is formulated as a minimization problem:

$$\sum_k \sum_{i=1}^{I} \pi_k u_i \Psi_i(q_i, m_i^{(k)}, P_i) \longrightarrow \min_{u_i, q_i, P_i}, \tag{2}$$

where π_k defines the degree of k-th image relevance for the user and is positive for *relevant*, zero for *indifferent* and negative for *non-relevant* images (i.e. images with negative relevance, which are examples of what the user is not looking for).

When optimal parameters u_i^*, q_i^*, P_i^* are selected based on (2), the object sought is a solution to the optimization problem:

$$\sum_{i=1}^{I} u_i^* \Psi_i(q_i^*, m_i^{(k)}, P_i^*) \longrightarrow \min_k. \tag{3}$$

K objects with the smallest value of (3) are presented to the user, who can again assign to them a degree of relevance in order to recalculate optimal search parameters according to (2) and perform the next iteration of the algorithm. Many authors use a formula proposed by Rocchio 16, cf. also 15 for adjustment of the virtual query to user's preferences. This idea is based on moving a virtual query towards the centre of gravity of *relevant* objects (in the descriptors' space) and in the opposite direction to the centre of gravity of *non-relevant* objects. Further heuristics have been proposed to find parameters of similarity function, cf. 2. Ishikawa, Subramanya and Faloutsos in 3 gave analytical solution of (2) but only for a specific class of similarity function.

The methods described above are based on the assumption that the user is looking for an object with pre-specified values of descriptors, and the associated utility function is monotonically decreasing with the distance between the vector of descriptors of a query and retrieved object. Selection of a metric is the only way to influence utility function and it limits the indifferent sets to be spheres in the selected metric. However, the assumption that they have such a shape is not justified, they may be non-convex and even disconnected – if the utility function is multimodal. For example, if the user wants to find one of several objects, for every alternative component of the associated query there is a corresponding local maximum of utility function.

Above we have presented a typical approach to the image retrieval problem with relevance feedback. This methodology may be, in our view, inconsistent because:

- the assumption of linearity of user preferences is not justified. It seems that, in most cases, these preferences are nonlinear,
- in the hitherto approach, search results not only depend on the ordinal structure of ranks assigned by the user to objects, but also on their values. This is not coherent with the basic assumptions of utility theory,
- the assumption that any object sought can be represented by a single point in feature space does not always correspond to real-life situations.

3   An Extension of Rui, Huang and Mehrotra – Type Method for Non-homogeneous Objects

The main drawback of the image retrieval approach described in the previous Section is a non-satisfactory performance when applied to complex objects. In this section, we propose an extension of the above methods allowing to apply them to complex objects, consisting of several sub-objects with different colours and textures.

The methods presented in Sec. 2 are based on the assumption that values of descriptors are similar for the whole object. This means that objects must have homogeneous colour and texture – or at least that classification can be done based on global distribution of these features in the image. Here we assume that we deal with non-homogeneous objects, which are composed of a certain number of homogeneous sub-objects. We also assume that pre-specified classes of transformations (e.g. isometry) do not change the classification of an object.

Finding similarity between a pair of objects in our adaptation of methods described in the previous section consists of:

- finding transformation of one of compared objects respect to the other which gives the best match
- segmentation of both objects into sub-objects with homogeneous colour and texture
- finding similarity of corresponding pairs of sub-objects
- scalarization of sub-objects similarities

Therefore, the search algorithm can be presented as follows:

Algorithm 1. Interactive retrieval of non-homogeneous objects based on distance from virtual query

Phase 1. Calculation/upgrade of the virtual query

Step 1. Presentation of K images to the user, who can assign to them grades (positive or negative). Value of π_k ranks user's perception of similarity between a given image and the sought image q^*. The image with the highest value of π_k is denoted by m^*

Step 2. Segmentation of object m^*. Here, we used EdgeFlow segmentation algorithm 17, combining local features of colour, texture and phase.

Step 3. For each object $m^{(k)}$ with a defined value π_k finding parameters of transformation (i.e. translation, rotation and scaling) which gives the best match to the model m^*, by minimization of the Hausdorff distance.

Step 4. Matching of objects m^* and $m^{(k)}$ according to transformation found in Step 3 and calculation of local descriptors of colour and texture for areas of object $m^{(k)}$ corresponding to sub-objects of m^*.

Step 5. Calculation of new virtual query descriptors, as sum of descriptors for K objects, weighted with π_k:

$$q_i^r = \sum_{k=1}^{K} \pi_k m_i^{(k,r)} \Big/ \sum_{k=1}^{K} |\pi_k|, \qquad (4)$$

where r is an index of sub-object of the object m*, i – index of representation and q_i^r, $m_i^{(k,r)}$ are i-th representations of k-th object in area corresponding to r-th sub-object of m^*. Correspondence is determined by the best matching of m^* and $m^{(k)}$.

Phase 2. Calculation of ranking of images from the database

Step 6. Calculation of similarities of all images in the database to the virtual query with descriptors given by (4), as the sum of homogeneous sub-objects, weighted by their areas.

Step 7. If the set of top K images is the same set as in the previous iteration, STOP. Otherwise, return to Step 1.

Observe that the purpose of matching in Step 4 is to find a correspondence between sub-objects of objects: m^* and $m^{(k)}$.

The extension presented above allows using RHM method for retrieval of non-homogeneous objects, composed of sub-objects with different colours and textures. However, the extension still has the weaknesses of this type of method, which we have listed at the end of Section 2. In the next section we propose an alternative approach to the problem, where an RBF network is used for approximation of user's preferences. In Sec. 5 both approaches are compared.

4 Image Retrieval Based on RBF Networks

In Sec. 2 we have mentioned the shortcomings of the common approach to image retrieval. Here, we will propose the application of neural networks with *Radial Basis Function* (RBF NN, cf. 7) to approximate user's preferences. In contrast to methods presented in Secs. 2 and 3, we do not impose a distance to a query (i.e. a point of the feature space) as a measure of preferences, but assume that user's preferences can be expressed by an arbitrary utility function, to be approximated with the neural network (cf. Fig. 1).

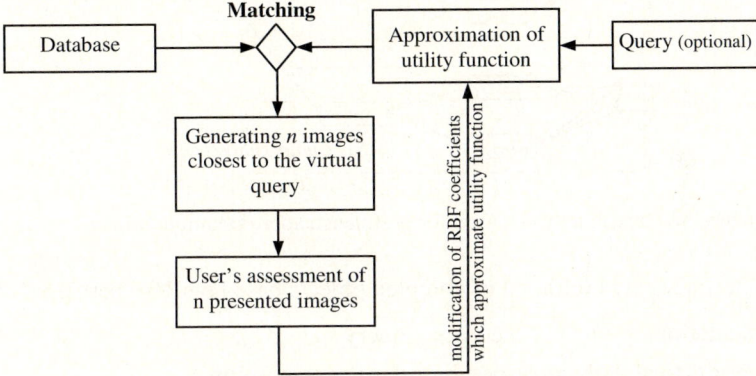

Fig. 1. A scheme of image retrieval system using RBF NN approximation of user preferences

Usually, in neural networks output of a neuron is a weighted sum of elements of input vector, with weighting coefficients w that can be modified in the learning process. In RBF networks coefficients w are not weights: the output of a neuron depends on the *distance* between the vector w and an input vector. Neuron's input can be interpreted as a set of descriptors of the graphical object: output is high when the object's features represented by input vector is similar to object's features represented by the neuron (vector w). In the training process, the neuron is taught to represent any feature desired by the user. Vector w corresponds to a local maximum of utility function in a subspace of a feature space and for a simplified model of preferences, as assumed for methods described in Secs. 2 and 3, w contains coordinates of the virtual query in the feature space.

The here proposed network used for object retrieval differs from classical (as described in 7) RBF network in that the distance measure need not be homogeneous across the network. Different parts of the network are designed for different *representations* (image features, such as shape, texture, colour) and local distances are based on similarity measures specific to these features (e.g. Hausdorff distance, similarity of Gabor coefficients, histogram intersection). The network structure used for preference elicitation is depicted in Fig. 2. For every representation m_i there is associated a separate sub-network S_i with a specific metric. The number of neurons for i-th representation is chosen automatically in such a way that the sub-network S_i can be trained, so that the margin of error is below an associated threshold. Therefore the network is able to approximate utility functions regardless of their shape. The output neuron, which combines outputs of sub-networks (cf. Fig. 2) is trained separately by linear regression.

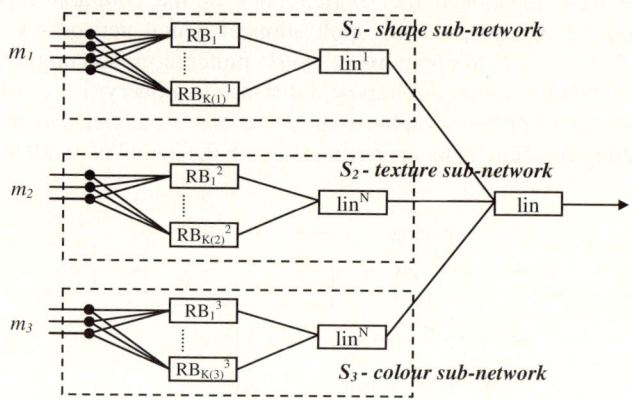

Fig. 2. Structure of a neural network used for preference approximation in image retrieval 12

Algorithm 2. Interactive retrieval of complex objects based on RBF network

Phase 1. Calculation/upgrade of the virtual query

Steps 1-4 are identical with corresponding steps of Algorithm 1.

Step 5. Training of the network. Three sub-networks (for: shape, texture and colour) are trained separately. Training set is a set of images presented to the user. For each of these images:
    a. Input vector of the i-th sub-network, $i=\{1$ for shape, 2 for texture, 3 for colour$\}$ is a subset of features $m_i^{(k,r)}$
    b. Desired output of sub-network is relevance π_k
Weights of the output neuron are found using linear regression in order to minimize total error of the network.

Phase 2. Calculation of ranking of images from the database

Step 6. Simulation of the network for all images in the database and ranking them according to value of the network output.

Step 7. If the set of top K images is the same set as in the previous iteration, STOP. Otherwise, return to Step 1.

5 An Experimental Comparison of Common and RBF NN-Based Methods

The methods described in Secs. 3 and 4 have been implemented as a prototype image retrieval system in Matlab environment (cf. Fig. 4). They were then tested and

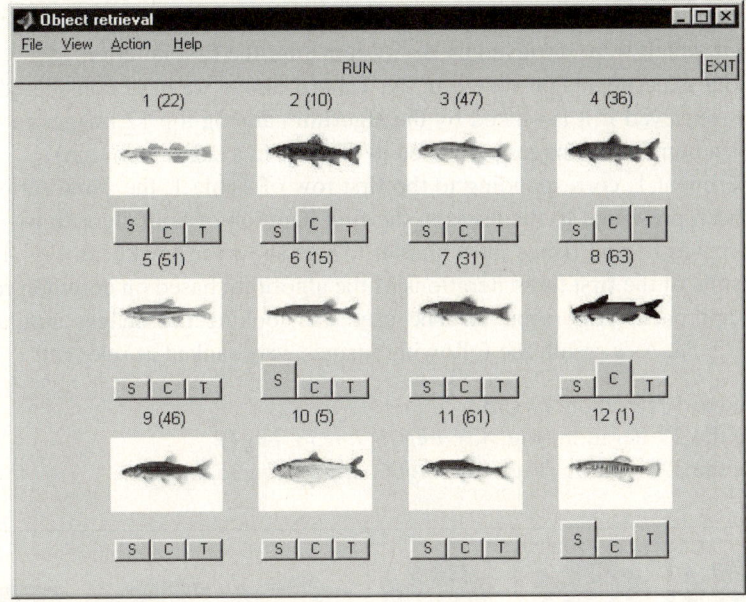

Fig. 3. A screenshot of our image retrieval system applied to a Fish Dataset. The system allows the user to either mark selected images as relevant (to train the output neuron with linear regression) or to use bars (S,C,T) to indicate relevant feature (Shape, Colour or Texture).

compared using a sample database of 75 fish images. Small size of testing set allows the user to mark all relevant images manually and compare with set of images retrieved by the system. The set of testing images will be referred to in the sequel as Fish Dataset (http://home.agh.edu.pl/~rotter/CBIR/Fishes75).

In Table 1 we present the results of experimental comparison of both algorithms. In each experiment the users selected in the whole images from the Fish Dataset relevant to his/her query (identifiers of these images are listed in column 2) and then tried to retrieve them by pointing relevant images (numbers in bold) among those presented

Table 1. Results of experimental comparison of Algorithm 1 based on classical methodology and Algorithm 2 based on RBF neural networks

Experiment No.	Images in Fish Dataset pointed out as relevant	Relevant images retrieved* after one iteration		Percentage of relevant images found in one iteration	
		Algorithm 1	Algorithm 2	Algorithm 1	Algorithm 2
1	**06**, 19, 26, 38, 52, 59, 65, 71, 73	06, 19, 26, 38, 65	06, 19, 26, 38, 52,59,65,71,73	56 %	100 %
2	**01**, 21, 23, 28, 37, 47, 53	23, 53	1, 21, 23, 47, 53	28 %	72 %
3	**02, 10, 11**, 18, 29, 36, 31, 46	2, 10, 11, 29, 36, 31, 46	2, 10, 11, 29, 36, 31, 46	88 %	88 %
4	**05**, 14, 27	05, 14	05, 14, 27	67 %	100 %

by the system. Performance of algorithms was assessed based on percentage of relevant images which were retrieved by the system after the first iteration.

In all four experiments presented in the Table 1 above we have considered that the image was retrieved if it is ranked by the algorithm among top 12 images in the database (12 is a number of images presented to the user in a single iteration)

In Experiment 1, corresponding to the first row of Table 1, the same set of 12 images has been presented to the users in the first iteration of both algorithms. The task to the users was to find fishes most similar to that shown in img06. In Fig. 4, we can see the results of the first three iterations of the algorithm based on common relevance feedback approach (Algorithm 1). The user was looking for images similar to the *img06* in Fig 5a. As a result, the following images were ranked as relevant:

- Before the 1st iteration: *img06*
- Before the 2nd iteration (Fig. 4a): *img06, img38, img19*
- Before the 3rd iteration (Fig. 4b): *img06, img19, img38, img65*

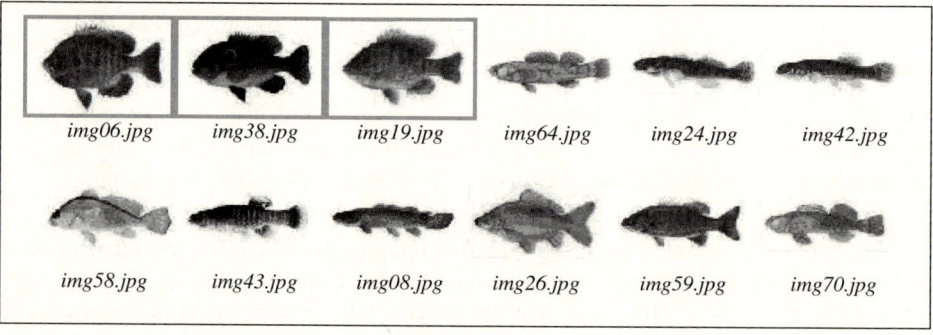

(a)

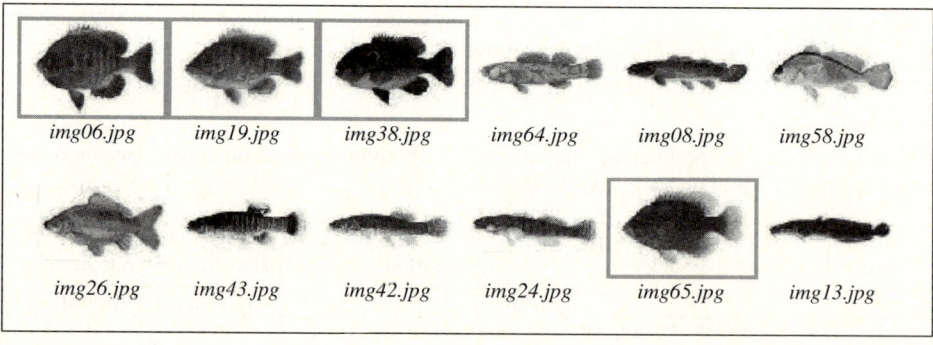

(b)

Fig. 4. The output of the first and second iterations of the algorithm based on methods described in Sec. 2. The frames in figures (a) and (b) show which objects were marked as relevant.

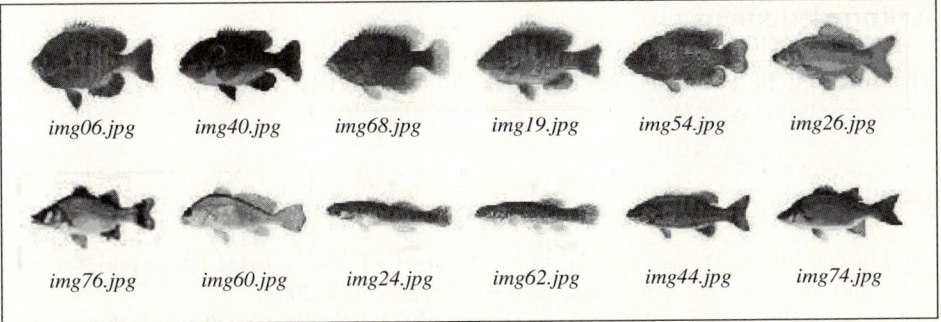

Fig. 5. The images retrieved in a single iteration of the RBF algorithm

Even the results of one iteration of the RBF-NN algorithm can be satisfactory: the first five images in Fig. 5 are most similar to the query image in the whole database. Another observation is that the order of images retrieved can vary, but the set of individual elements remains stable.

Referring to the results of experiments summed up in Tab. 1, one can conclude that, in terms of the number of relevant images retrieved by each algorithm, the performance of the proposed algorithm in three of four experiments was better and in one case the same as results of the classical retrieval algorithms.

6 Discussion and Conclusions

In this paper we have proposed a new method for Content-based Image Retrieval with relevance feedback based on neural networks with Radial Basis Function (RBF) and compared this method to the common relevance feedback approach, described in Secs. 2 and 3. Both methods were implemented in Matlab and tested on an interactive atlas of species. Experiments on the selected dataset show persistently higher performance of the RBF-based method. Therefore the proposed method can be useful for some type of applications specified by the assumptions presented in Sec. 1 and allow for effective retrieval of information based on a memorised image. As examples of applications one can mention interactive atlases, multimedia encyclopaedias, search in trademark databases and in e-commerce catalogues. The common feature of these applications is that the user knows the appearance of the object searched for, but this knowledge is insufficient to formulate a clear narrative description of desired object's features. Such situation may happen particularly when the classification of images depends on low-level features, which are not directly perceived, or are difficult to articulate. An interesting direction of further research is to include 3D invariants in description of objects, what could allow for retrieval of 3-dimensional objects based on 2D projections. Another direction of potential expansion of the approach here presented is interactive visual search in web databases (cf. e.g. 18), where the semantic information and image annotation can be combined with the feature-space-based methods.

Acknowledgment

The research of P. Rotter has been partially supported by the project CHORUS (www.ist-chorus.org).

References

1. Liu, Y., Zhang, D., Lu, G., Ma, W.Y.: A survey of content-based image retrieval with high-level semantics. Pattern Recognition 40(1), 262–282 (2007)
2. Rui, Y., Huang, T.S., Mehrotra, S., Ortega, M.: A Relevance Feedback Architecture in Content-based Multimedia Information Retrieval Systems. In: Proc. of IEEE Workshop on Content-based Access of Image and Video Libraries, in conjunction with CVPR 1997, Puerto Rico, June 20, 1997, pp. 82–89 (1997)
3. Ishikawa, Y., Subramanya, R., Faloutsos, C.: MindReader: Querying databases through multiple examples. In: 24th VLDB Conference, New York (1998)
4. Smith, J.R., Chang, S.F.: VisualSEEk: a fully automated content-based image query system. In: Proc. ACM Intern. Conf. Multimedia, Boston, MA (May 1996)
5. Faloutsos, C., et al.: Efficient and Effective Querying by Image Content. Journal of Intelligent Information Systems 3(3-4) (July 1994)
6. Chuang, S.C., et al.: A Multiple-Instance Neural Networks Based Image Content Retrieval System. In: Proceedings of the First International Conference on Innovative Computing, Information and Control (ICICIC) (2006)
7. Park, J., Sandberg, I.W.: Universal approximation using radial-basis-function networks. Neural Computation 3, 246–257 (1991)
8. Cheng, P.C., Chien, B.C., Ke, H.R., Yang, W.P.: A two-level relevance feedback mechanism for image retrieval. In: Expert Systems with Applications (to appear, 2008)
9. Doulamis, N., Doulamis, A.: Evaluation of relevance feedback schemes in content-based in retrieval systems. Signal Processing: Image Communication 21(4), 334–357 (2006)
10. Rui, Y., Huang, T.S., Mehrotra, S.: Relevance Feedback Techniques in Interactive Content-Based Image Retrieval. In: Proc. of IS&T and SPIE Storage and Retrieval of Image and Video Databases VI, San Jose, CA, January 24-30, 1998, pp. 25–36 (1998)
11. Skulimowski, A.M.J., Rotter, P.: Preference elicitation, reference sets and relevance feedback. In: 65th Meeting of the European Working Group Multiple Criteria Decision Aid, Poznan, Poland, April 12-14 (2007)
12. Rotter, P.: Application of multicriteria optimization methods in image interpretation (in Polish). PhD Thesis, AGH-University of Science and Technology, Krakow (2004)
13. Skulimowski, A.M.J.: Methods of Multicriteria Decision Support Based on Reference Sets. In: Caballero, R., Ruiz, F., Steuer, R. (eds.) Advances in Multiple Objective and Goal Programming. Lecture Notes in Econ. and Math. Systems, vol. 455, pp. 282–290. Springer, Heidelberg (1997)
14. Lew, M.S.: Principles of Visual Information Retrieval. Springer, London (2001)
15. Müller, H., Müller, W., Marchand-Maillet, S., Squire, D.M.: Strategies for positive and negative relevance feedback in image retrieval. In: Proc. of the International Conference on Pattern Recognition (ICPR 2000), Barcelona, Spain, September 3-8, 2000. Computer Vision and Image Analysis, vol. 1, pp. 1043–1046 (2000)
16. Rocchio, J.J.: Relevance Feedback in Information Retrieval. In: Salton, G. (ed.) The SMART Retrieval System – Experiments in Automatic Document Processing, pp. 313–323. Prentice Hall, Englewood Cliffs (1971)

17. Ma, W.Y., Manjunath, B.S.: Edge flow: a framework of boundary detection and image segmentation. In: Proc. IEEE International Conference on Computer Vision and Pattern Recognition, San Juan, Puerto Rico, June 1997, pp. 744–749 (1997)
18. Tadeusiewicz, R.: Intelligent Web mining for semantically adequate images. In: Węgrzyn-Wolska, K.M., Szczepaniak, P.S. (eds.) Advances in intelligent Web mastering : Proceedings of the 5th Atlantic Web Intelligent Conference – AWIC 2007, Fontainebleau, France, June 25–27, 2007. Advances in Soft Computing, vol. 43, pp. 3–10. Springer, Heidelberg (2007)

An Adaptive Fast Transform Based Image Compression

Kamil Stokfiszewski[1] and Piotr S. Szczepaniak[1,2]

[1] Institute of Computer Science,
Technical University of Lodz
ul. Wolczanska 215, 93-005 Lodz, Poland
[2] Systems Research Institute,
Polish Academy of Sciences
Newelska 6, 01-447 Warsaw, Poland
stokfi@ics.p.lodz.pl

Abstract. The paper deals with image compression performed using an adaptive fast transform-based method. The point of departure is a base scheme for fast computation of certain discrete transforms. The scheme can be interpreted in terms of the neural architecture whose parameters (neurons' weights) can be adjusted during learning on set data, here images. The same basic network topology enables realization of diverse transformations. The results obtained for the task of image compression are presented and evaluated.

1 Introduction

Due to their useful properties, discrete trigonometric orthogonal transforms are in the center of interest in digital signal processing, the image compression inclusive. The most commonly used transforms of this kind are: the discrete Fourier transform (DFT), discrete Hartley transform (DHT), as well as discrete sine and cosine transforms of the second type (DST, DCT, respectively), with DCT being part of the JPEG[1] image compression method [1,2,3]. In their four basic variants, the discrete cosine DCT_N^X and sine DST_N^X transformations [8] are successfully employed in the signal processing practice. For data compression, the second (DCT_N^{II}, DST_N^{II}), the third (DCT_N^{III}, DST_N^{III}), as well as the fourth (DCT_N^{IV}, DST_N^{IV}), of the variants are most frequently used [8]. In the present work, only the DFT for real sequences, DHT, DCT and DST are considered:

$$DFT_N(n,k) = exp\left(-j2\pi nk/N\right) \quad (1)$$

$$DHT_N(n,k) = cos\left(2\pi nk/N\right) + sin\left(2\pi nk/N\right) \quad (2)$$

$$DCT_N^{II}(n,k) = cos\left(\pi(2n+1)k/(2N)\right) \quad (3)$$

$$DST_N^{II}(n,k) = sin\left(\pi(2n+1)(k+1)/(2N)\right) \quad (4)$$

$$n,k = 0,1,\ldots,N-1 \text{ where } n \neq 0, n \neq N-1.$$

---
[1] The latest JPEG standard uses also the wavelet transform.

Several fast algorithms have been developed to calculate discrete transforms; the respective literature is abundant, cf. [1,2,3,4,11,12,13]. DFT for complex sequences, and DHT for real ones are often used as departure points for determination of other transforms, cosine and sine in particular [7]. Another procedure, which consists in taking the DCT as the basis [4], is also possible. Due to these solutions, the computational complexity in comparison with the original problem becomes usually reduced from $\mathcal{O}(N^2)$ to $\mathcal{O}(N log_2 N)$, which brings the improvement in time consumption that is crucial for practical applications. In [5], neural realization of a uniform fast algorithm for calculation of orthogonal transformations and its application to linear filtering are presented. In this paper, the application of this neural approach to the image compression is described and the results of its use in practice are evaluated.

2 Generalized Computational Scheme

The general definition of the linear transform in the matrix form

$$\mathbf{y}_N = \mathbf{A}_N \mathbf{x}_N \tag{5}$$

involves orthogonal discrete transformations, like those of Fourier, Hartley, Walsh, Hadamard, cosine and sine [6,7,8]; for all of them $\mathbf{y}_N$ has the spectral character. Here $\mathbf{x}_N^T = [\, x_0 \; x_1 \; \ldots \; x_{N-1}\,]$ and $\mathbf{y}_N^T = [\, y_0 \; y_1 \; \ldots \; y_{N-1}\,]$ are samples of real input and transformed signals, respectively; $\mathbf{A}_N$ denotes the matrix of transformation; T stands for transposition. Generalization of the results presented in [4] leads to the sequence of operations shown in Fig. 1; they allow calculation of the direct DCT and DST, DHT and DFT.

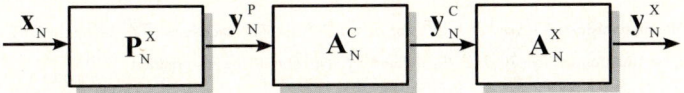

Fig. 1. The sequence of operations for calculation of direct transforms

In block $\mathbf{P}_N^X$ that depends on the kind of transformation, the rearrangement of data is performed. Block $\mathbf{A}_N^C$ is common to all the transformations DCT and DST of the second kind, as well as to DHT and DFT when real sequences are transformed. Matrix $\mathbf{A}_N^C$ performs an intermediate calculation and its result $\mathbf{y}_N^C$ is identical for all the considered transformations. In block $\mathbf{A}_N^X$ the transition from the intermediate transformation $\mathbf{y}_N^C$ to the required $\mathbf{y}_N^{CII}$, $\mathbf{y}_N^{SII}$, $\mathbf{y}_N^F$ or $\mathbf{y}_N^H$ is carried out. It is worth mentioning that blocks $\mathbf{A}_N^C$ and $\mathbf{A}_N^X$ apply fast algorithms. Figures from 2 to 5 present in detail the above scheme of unified, fast calculation of four considered orthogonal transforms. In Fig. 6 the flow diagram of the fast algorithm for determination of DCT_N^{II} for $N = 8$ is presented (which is a special case of Algorithm 1), while in Fig. 7 its base operations are explained. For inverse transformations (6) a similar scheme, shown in Fig. 8, is valid.

Algorithm 1. Fast cosine transformation of the second kind (FCTII).

Let $\vec{x}_N$, be the input sequence, $N = 2^m$, and $m = 1,2,3,\ldots$. The sequence $\vec{y}_N^{CII}$ to be calculated is the DCT of the second kind, i.e. transformation (5) in which matrix A_N is determined by formula (3).

Step 1. $J = N$. Calculate the auxiliary sequences $a(n)$ and $b(n)$

$$a(n) = x_N(2n) + x_N(2n+1), \quad b(n) = (-1)^n (x_N(2n) - x_N(2n+1)), \quad \text{(A1)}$$
$$n = 0,1,\ldots,N/2-1$$

Step 2. $J = J/2$. For each sequence obtained in the previous step, the Step 1 is repeated until $J > 1$. (The result is: N one-point sequences $l_1^{(j)}, j = 0,1,\ldots,N-1$).

Step 3. $K = 1$. Under formula (A2) two consecutive K – point sequences are united into one $2K$ – point sequence

$$\left.\begin{array}{l} l_{2K}^{(j)}(0) = l_K^{(2j)}(0),\ l_{2K}^{(j)}(K) = \sqrt{2}/2 \cdot l_K^{(2j+1)}(0); \\ l_{2K}^{(j)}(k) = C_{4K}^k \cdot l_K^{(2j)}(k) + S_{4K}^k \cdot l_K^{(2j+1)}(K-k), \\ l_{2K}^{(j)}(2K-k) = C_{4K}^k \cdot l_K^{(2j+1)}(K-k) - S_{4K}^k \cdot l_K^{(2j)}(k), \\ k = 0,1,\ldots,K-1, \ j = 0,1,\ldots,N/K-1, \end{array}\right\} \quad \text{(A2)}$$

where $C_M^m = \cos(2\pi m/M)$, $S_M^m = \sin(2\pi m/M)$.

Step 4. Repeat Step 3 for the sequences obtained in the previous step while $K=2,4,\ldots N/2$. (For $K = 1$ only the first two equations of the set of formulae (A2) are used).

Step 5. Set $\vec{y}_N^{CII} \equiv \vec{l}_N^{(0)}$, which gives the desired sequence. Stop.

Fig. 2. Algorithm for fast computation of the N-point discrete cosine transform

---

Algorithm 2. Fast sine transformation of the second kind (FSTII).

Let $\vec{x}_N$ be the input sequence where $N = 2^m$ and $m = 1,2,3,\ldots$. The sequence $\vec{y}_N^{SII}$ being the DST of the second kind, i.e. transformation (5) with A_N defined by (4) should be determined.

Step 1. Form the auxiliary sequence $\vec{d}_N$: $d_N(n) = (-1)^n x_N(n)$, $n = 0,1,\ldots,N-1$.

Step 2. Using Algorithm 1 calculate the DCT of the sequence $\vec{d}_N$. The result is the sequence $\vec{d}_N^{CII}$.

Step 3. Because of the identity $y_N^{SII}(k-1) \equiv d_N^{CII}(N-k)$, the desired result has been achieved. Stop.

Fig. 3. Algorithm for fast computation of the N-point discrete sine transform

$$\mathbf{x}_N = \mathbf{A}_N^{-1} \mathbf{y}_N \qquad (6)$$

Here block $(\mathbf{A}_N^X)^{-1}$ executes transition from various transformations to the intermediate transformation $\mathbf{y}_N^C$. Block $(\mathbf{A}_N^C)^{-1}$ has the same structure for the DCT and DST of the third kind, as well as for the DHT and the inverse DFT of

Algorithm 3. Fast Hartley transformation (FHT).

Let $\vec{x}_N$ be the input sequence, $N = 2^m$ and $m = 1,2,3,\ldots$. The sequence $\vec{y}_N^H$, i.e. the DHT expressed by (5) with A_N defined by (2) should be determined.

Step 1. Form the auxiliary sequence $\vec{z}_N$: $z_N(2n) = x_N(n)$, $z_N(2n+1) = x_N(N-1-n)$, $n = 0,1,\ldots,N/2-1$.

Step 2. Using Algorithm 1 calculate the DCT of $\vec{z}_N$. In the last run (with $K=N/2$) of Step 3 of Algorithm 3 replace formula (A2) by

$$\left.\begin{aligned}
&l_N^{(0)}(0) = l_{N/2}^{(0)}(0),\ l_N^{(0)}(N/2) = l_{N/2}^{(1)}(0); \\
&l_N^{(0)}(k) = (C_{2N}^k - S_{2N}^k)/2 \cdot l_{N/2}^{(0)}(k) + (C_{2N}^k + S_{2N}^k)/2 \cdot l_{N/2}^{(1)}(N/2-k), \\
&l_N^{(0)}(N-k) = (C_{2N}^k - S_{2N}^k)/2 \cdot l_K^{(1)}(N/2-k) - (C_{2N}^k + S_{2N}^k)/2 \cdot l_{N/2}^{(0)}(k), \\
&k = 0,1,\ldots,N/2-1.
\end{aligned}\right\} \quad (A3)$$

Step 3. Set $\vec{y}_N^H \equiv \vec{l}_N^{(0)}$. Stop.

Fig. 4. Algorithm for fast computation of the N-point discrete Hartley transform

Algorithm 4. Fast Fourier transformation of real sequence (FFT).

Let $\vec{x}_N$ be the input sequence where $N = 2^m$ and $m = 1,2,3,\ldots$. The sequence $\vec{y}_N^F$ being the DFT, i.e. transformation (5) with A_N defined by (1), should be determined.

Step 1. Form the auxiliary sequence $\vec{z}_N$: $z_N(2n) = x_N(n)$, $z_N(2n+1) = x_N(N-1-n)$, $n = 0,1,\ldots,N/2-1$.

Step 2. Using Algorithm 1 calculate the DCT of $\vec{z}_N$. In the last run (with $K = N/2$) of Step 3 of Algorithm 1 replace (A2) by the formulae

$$\left.\begin{aligned}
&l_N^{(0)}(0) = l_{N/2}^{(0)}(0),\ l_N^{(0)}(N/2) = l_{N/2}^{(1)}(0); \\
&l_N^{(0)}(k) = C_{2N}^k \cdot l_{N/2}^{(0)}(k) + S_{2N}^k \cdot l_{N/2}^{(1)}(N/2-k), \\
&l_N^{(0)}(N-k) = -(C_{2N}^k \cdot l_{N/2}^{(1)}(N/2-k) - S_{2N}^k \cdot l_K^{(0)}(k)), \\
&k = 0,1,\ldots,N/2-1.
\end{aligned}\right\} \quad (A4)$$

Step 3. The required sequence is $\vec{y}_N^F \equiv \vec{l}_N^{(0)}$. Stop.

Fig. 5. Algorithm for fast computation of the N-point discrete Fourier transform

the real sequence. It is the inversion of block $\mathbf{A}_N^C$. In block $(\mathbf{P}_N^X)^{-1}$ the rearrangement of data, inverse to the rearrangement of $\mathbf{P}_N^X$, is carried out. Note that the difference between the pairs $\{(\mathbf{A}_N^X)^{-1}, (\mathbf{A}_N^C)^{-1}\}$ and $\{\mathbf{A}_N^X, \mathbf{A}_N^C\}$ is determined only by the order of their elements and the values of coefficients in the vector rotation formulae. In other words, when looking at the scheme in Fig. 6 from right to left, one obtains the structure realizing the sequence $(\mathbf{A}_8^X)^{-1}$, $(\mathbf{A}_8^C)^{-1}$.

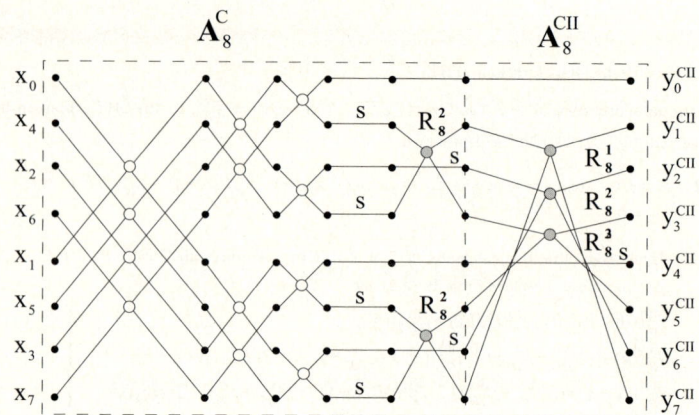

Fig. 6. Diagram of 8-point blocks $\mathbf{A}_N^C$ and $\mathbf{A}_N^X$ for DCT of the second kind

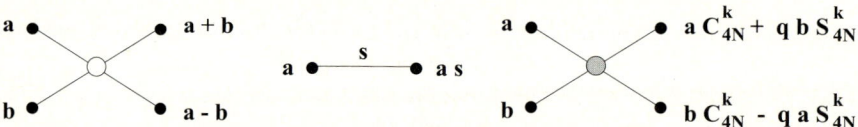

Fig. 7. Base operations: a) summation/subtraction, b) multiplication c) vector rotation

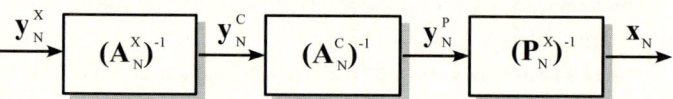

Fig. 8. Generalized scheme of inverse DCT and DST, DHT and DFT calculation

It is well known (cf. [11,12,13]) that if the class of real images is considered a set of statistical realizations of the ideal, probabilistic first-order stationary Markov field image model then, form the statistical point of view, the best pair of transforms approximating such images from a truncated set of transform coefficients is the Karhunen-Loéve (KLT) transform pair. It can also be shown that under certain conditions, cf. [12,13], many known orthogonal transforms, including DCT and DST family, DFT and DHT, constitute special cases of the mentioned Karhunen-Loéve transform for a various types of Markov fields which, according to the last argument, makes them theoretically optimal in compression of images comprising such models. The last fact and the fact that the proposed algorithms 1 – 4 are proved analitically to unify computational structure for at least four of the known transform types suggests that the same structure might be capable of realizing other discrete (orthogonal) transforms, optimal in image compression of a wider set of classes of real images comprising Markov field image model. This capability might be achieved, for example, by a statistical adaptation of the sparsely structured matrices constituting the proposed computational procedure. This, in turn, may be realized through a neural network learning, which has an additional advantage, namely the ability of employing parallel calculations, typical of neural methods, which potentially makes such adaptive approach even more efficient. To sum up, the discussion reveals so far that the proposed procedure of determination of orthogonal transformations is general enough to include several types of known transforms in a single structure, yet it's still fast, since its computational complexity can easily be shown, cf. [9], to be of order $\mathcal{O}(2\,Nlog_2N)$. In addition, aforementioned generality property of the proposed unified structure brings hope to achieve new types of discrete transforms adopted to image compression tasks for various classes of real images exhibiting different statistical properties, through optimization process. This process can be efficiently realized with neural methods, since the proposed computational scheme may be easily transferred to a unified, fast neural network architecture. This architecture will be presented in the foregoing sections of the paper.

3 Neural Realization

As known (cf. [6]), both of the transformations (5) and (6) can be realized directly in a linear neuron network with one layer consisting of N neurons with N inputs and one output. Training of such a network consists in a selection of elements in $\mathbf{A}_N$ or $\mathbf{A}_N^{-1}$. For example, the network task in data coding is to choose M coefficients, where M is essentially smaller than N. Generally, network training requires then determination of $2N^2$ elements and proves to be a time consuming process. To eliminate this imperfection, a new approach to the design of neural networks for signal processing has been proposed in [5] and presented in the previous paragraph. The calculation scheme shown in Fig. 1 can be closed in the following formula:

$$\mathbf{y}_N^X = \mathbf{A}_N^X \mathbf{A}_N^C \mathbf{P}_N^X \mathbf{x}_N \tag{7}$$

In linear neuron networks, multiplication on matrices $\mathbf{A}_N^X$ or $\mathbf{A}_N^C$ is performed with the use of one layer of neurons. Therefore, at first sight, the replacement of one matrix in transformation (5) or (6) with three in transformation (7) is neither of theoretical nor practical importance unless the association of two new layers substantially enriches the network capabilities. In general, anything realizable in a network with two, three or more layers, can be realized in one layer with the same success. However, when the sparseness property of matrices $\mathbf{A}_N^X$ and $\mathbf{A}_N^C$ is taken into account, such a replacement appears quite justified. Firstly, the neuron structure becomes essentially simpler. Namely, it is possible to replace the vector rotation by two neurons with two inputs and one output, see Fig. 9, where v_1^i and v_2^i denote weights of the i-th neuron e_i, $i = 1, 2$.

Obviously, the training of such a neuron becomes more effective than that of a standard neuron with N inputs and one output. Secondly, considering the best case, matrices $\mathbf{A}_N^C$ and $\mathbf{P}_N^X$ do not need any training, which means that the learning process can be reduced to the determination of matrix $\mathbf{A}_N^X$, most

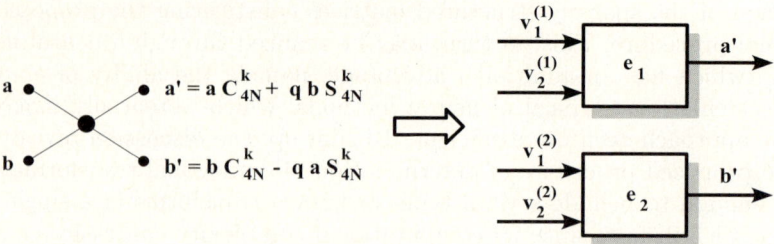

Fig. 9. Replacement of vector rotation by two neurons e_1 and e_2

elements of which are equal to zero. Consequently, the training has to be applied only to N two-input neurons, instead of to N neurons with N inputs, which results in a considerable saving of computation time. According to algorithms 1 - 4 presented in the previous section, in the worst case half of the weights (which correspond to rotations) of the single neural network have to undergo the process of training. Even then computational complexity of the proposed architecture outperforms that of a standard network by a ratio of $Nlog_2 N$ to N^2 accordingly. So far, it has been shown that the four considered transformations, namely DCT and DST of the second kind, as well as DHT and DFT of real sequences, can be realized on the basis of the simplified neural architecture. Similarly, with the use of the scheme in Fig. 8 and the replacement method (see Fig. 9) applied to inversion $(\mathbf{A}_N^X)^{-1}$ one arrives at a neural network that enables fast calculation of inverse transformations.

4 Application to Image Compression

To evaluate the developed neural architectures, image compression with the use of orthogonal transformations has been performed. In general, three main operations need to be realized, i.e. a) direct transformation T; b) selection and quantization; c) inverse transformation T^{-1} [6]. The architecture of the network is shown in Fig. 10. It is a two-dimensional realization of the scheme shown in Figures 1 and 8. Layers are numbered from 0 to $2Q - 1$ (from left to right). The search for transformation is realized through learning; the method is a standard backpropagation, simplified due to the linearity of transformations performed by neurons. Input to the network are (here randomly) chosen image fragments having 8×8 pixels while on the output the reconstruction of the input is expected. The part of the network subjected to learning (scc below for variants of learning) is initialized by random values of weights taken from the interval $(-1, 1)$; moreover, a part of the weights of the quantization and dequantization layers are set to zero. Neither during the forward input signal propagation nor during the error backpropagation are the signals of the neurons with weight coefficients set initially to zero let through by the quantization layer when they appear at the input (propagation) and output (backpropagation). Consequently, the learning process makes the weights of the remaining neurons of the quantization layers as well as the weights of the neurons of other network layers associated with

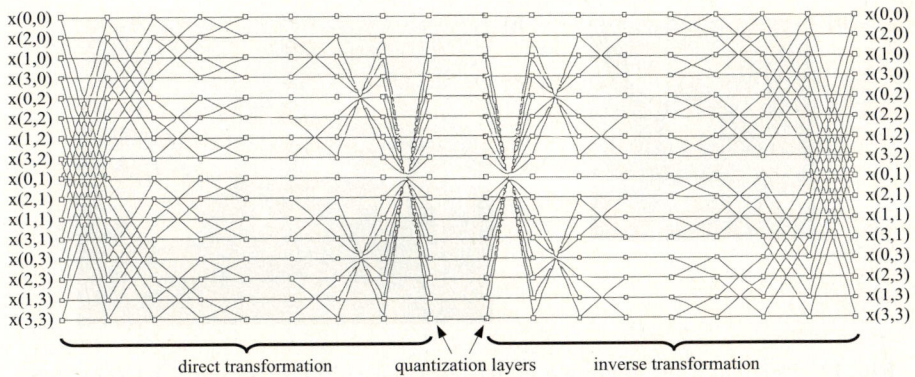

Fig. 10. Two-dimensional 4×4 - point network ($Q = 9$)

them undergo adaptation so that, with a prescribed compression degree, for each learning matrix given as input its possibly most faithful copy could be obtained at the output, despite incomplete connection scheme in the quantization layers. The learning process constructed in this way enables adaptive acquisition of the direct and inverse transformations with a prescribed compression degree and resulting from the respective two parts of the network. These two parts are then used separately during the compression and decompression of images.

In the experiments carried out to test the potential capabilities of the applied network architecture and the weight adaptation method, the networks subjected to learning have been of size 8×8 points, while the input learning matrices have been randomly chosen fragments of 8×8 pixels of the learning image in the 8-bit scale of grayness - an example of the image is shown in Fig. 11a. The testing comprises the image compression and decompression with the two parts of the already learned network, i.e. the direct and inverse ones, used separately, and with the number of the direct transformation coefficients set to zero being the same as during the learning process. Analogous learning and testing procedures were applied, and results were reported, in [10] for a standard three-layer perceptron counterpart of the considered network scheme, thus enabling direct qualitative comparison between two architectures in various experimental setups. Those results are gathered in Tables 1, 2 and 3, along with qualitative performance of the standard DCT coding schemes carried out under identical experimental conditions. Subjective visual performance of all three mentioned coding methods for a chosen set of tests can be verified by looking at Fig. 11 and 12, while the objective quality is checked in a standard way by calculation of PSNR:

$$PSNR = 10 \, log_{10} \left(\frac{255^2}{MSE} \right) \quad (8)$$

where MSE denotes the mean square error and is defined as:

$$MSE = \frac{1}{m \cdot n} \sum_{x=1}^{m} \sum_{y=1}^{n} \left[f(x,y) - f_c(x,y) \right]^2 \quad (9)$$

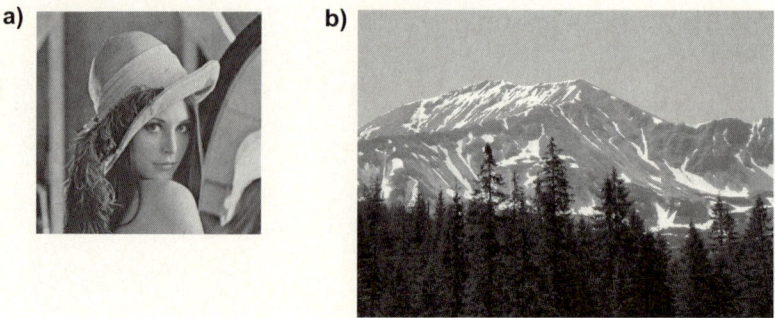

Fig. 11. Original 8-bit grayscale images of sizes a) 512 × 512 and b) 1024 × 768

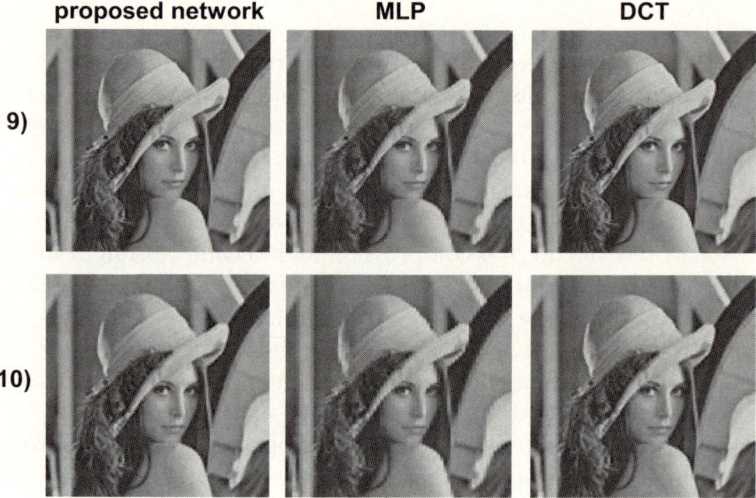

Fig. 12. Experiments 9 and 10 for the proposed network, perceptron and the DCT

where m, n is the dimension of the test image, and $f(x,y)$, $f_c(x,y)$ are gray levels of pixels of the original and the compressed images, respectively.

The experiment covers two variants that differ in the number of the network layers undergoing learning in the direct and inverse transformations. The number of the layers in the network part corresponding to the direct transformation is expressed by the formula $Q = 2\,log_2 N + 1$, where N denotes the number of inputs (and outputs) of the network. In view of the symmetry of the network architecture an identical formula is valid for the part that realizes the inverse transformation. Let the network layers be numbered from 0 up to $2Q - 1$. In the first experimental variant, adaptation concerns the weights in the last layer of the direct network (of number $Q - 1$), in the quantization layer (of number Q), and in the first layer of the inverse network (of number $Q - 1$). In the second variant, half the number of weights of the joint direct and inverse network scheme, from

Table 1. Results obtained for image shown in Fig. 11a

experiment no	1	2	3	4	5	6
compression coefficient	4	8	4	8	12.8	21.3
learning rate $\cdot\, 10\,E-8$	1	2	7	7	8	7
no of layers undergoing learning	13	13	3	3	3	3
no of training images for network	128	128	128	128	128	128
PSNR for multilayer perceptron	–	–	–	–	26.1	25.2
PSNR for proposed network	32.6	32.1	39.1	34.6	30.5	28.9

the layer of number $(Q-1)/2+1$ to the one numbered $(3Q-1)/2$, are subjected to learning. In both cases the values of the weights not supposed to undergo adaptation are set to the magnitudes known from the fast universal algorithm for calculation of transforms DCT, DST, DFT, and DHT, presented earlier in this paper. In both the discussed variants the applied values of the compression coefficient, defined here as the ratio of the number of input (output) neurons to the number of the neurons with non-zero weight coefficients in the quantization layer, are assumed to be given in advance and equal to 4, 8, 12.8 and 21.3, respectively. In the case of neural networks i.e. the proposed network and the perceptron, Tables 1, 2 and 3 present the best results that have been achieved throughout the series of the tests performed. Table 1 contains parameters of the experiment and the quantitative evaluation of the decompression quality determined for the same image (Fig.11a).

It is seen that in terms of the PSNR values the results are evaluated very highly. Let us check whether the network that has already undergone one image processing learning (Fig. 11b) is useful for compression of another (Fig. 11a). Table 2 includes additionally the value of PSNR obtained on application of a usual DCT transform. Another network learning that has been performed involves ten different images, while the tests have been carried out using the image in Fig. 11a not included in the learning set. For comparison, Table 3 contains the results of a usual DCT obtained on the testing image. For DCT, the compression coefficient is equal to the quotient of the number of all elements of the transformation matrix (here 8×8) and the number of its non-zero elements.

It is noticeable that the network results are a little better than those received through the application of DCT and a lot better than for the perceptron network.

Taking into consideration computational complexity of the proposed network architecture i.e. $\mathcal{O}(2\,Nlog_2N)$ as opposed to standard perceptron network - $\mathcal{O}(N^2)$, it is obvious that time consumption gain of both the learning process and latter forward signal propagation in the trained network should be in favour of the proposed architecture over the standard perceptron, even for high compression ratios. However it is interesting to verify this intuition practically, especially in the case of the learning process. In order to do that, a few experiments were carried out. In each of them, the forward stages of both the proposed network and the perceptron were supposed to be trained for the ability of one dimensional discrete cosine transform computation. In the case of the proposed network only

Table 2. Results: learning on image in Fig. 11b; testing on image in Fig. 11a

experiment no	7	8	9	10
compression coefficient	4	8	12.8	21.3
learning rate · 10 E − 8	2	2.5	2.5	2
no of learning epochs	107	43	57	35
no of training images for network	640	640	640	640
PSNR for DCT - test image	39.162	34.873	30.544	28.880
PSNR for mlt. perceptron - test image	–	–	27.800	26.500
PSNR for prop. network - test image	39.510	34.959	30.549	28.884

Table 3. Results for ten learning images and one testing image (Fig. 11a)

experiment no	11	12	13	14
compression coefficient	4	8	12.8	21.3
learning rate · 10 E − 8	7	7.5	7	8
no of training images for network	320	640	1280	640
PSNR for DCT - test image	39.162	34.873	30.544	28.880
PSNR for prop. network - test image	39.325	34.925	30.551	28.889

the rotation layers were trained. The same fixed accuracy of computation of the DCT was chosen for both networks as a stop criterion. As shown by the experiments, the proposed architecture has a huge advantage over the standard perceptron, far greater than the one resulting from theoretical considerations. For a 64-point transform (one dimensional counterpart of the two dimensional 8×8 - point transform), the learning time of the proposed network has been statistically 420 times smaller than the perceptron learning time, and accordingly, for a 16-point transform the network was 16 times faster. As for a 256-point transform, and above (up to 1024-point), the proposed network learning time did not exceed 40 minutes (on the computer equipped with the Celeron 700 MHz processor), while the perceptron did not succeed in reaching the stop criterion, despite multiple attempts. It is also worth mentioning that the computational complexity of the proposed architecture is about 2 times greater than the fastest known DCT implementation, [11,12,13], which might be considered a price for the architecture's flexibility in adaptaion to various discrete transforms.

5 Conclusions

The presented neural architecture is based on a unified procedure for fast computation of various orthogonal transformations. The method assures effective learning of the network because the original task of the complexity of order $\mathcal{O}(N^2)$ has been replaced by the task with the complexity of order $\mathcal{O}(Nlog_2 N)$. The (supervised) learning may be performed with the use of any standard method suitable for adjusting weights in linear networks assuming that the layer network organization is taken into consideration, which is not the usual situation in the networks of this type. Moreover, the network is able to realize a wide class of known transformations, and consequently, it is flexible in adaptation to any new task of the form [4]. The application to image compression shows that the results are satisfactory in the aspects of both time consumption and quality.

Acknowledgments. The authors are grateful to Professor M.M.Jacymirski from the Technical University of Lodz, Poland, for many fruitful discussions.

References

1. Nelson, M., Gailly, J.: The Data Compression Book. M&A Books, New York (1996)
2. Pennebacker, W.B., Mitchell, J.L.: JPEG Still Image Data Compression Standard. Van Nostrand Reinhold, New York (1993)
3. Sayood, K.: Introduction to Data Compression. Morgan Kaufmann, San Francisco (2000)
4. Jacymirski, M.M.: Fast Algorithms for Orthogonal Trigonometric Transformations. Academic Express, Lviv, no. 219, Ukraine (in Ukrainian) (1997)
5. Jacymirski, M., Szczepaniak, P.S.: Neural Realization of Fast Linear Filters. In: Proc. of 4th EURASIP-IEEE Region & International Symposium on Video/Image Processing and Multimedia Communications, Zadar, Croatia, pp. 153–157 (2002)
6. Lu, F.-L., Unbehauen, R.: Applied Neural Networks for Signal Processing. Cambridge University Press, Cambridge (1998)
7. Bracewell, R.N.: The Hartley Transform. Oxford University Press, New York (1986)
8. Wang, Z.: Fast Algorithms for the Discrete W Transform and for the Discrete Fourier Transform. IEEE Trans. ASSP 32(4), 803–816 (1984)
9. Jacymirski, M., Wiechno, T.: A Novel Method of Building Fast Fourier Transform Algorithms. In: Proceedings of the International Conference on Signals and Electronic Systems (ICSES 2001), Lodz, Poland, September 18-21, 2001, pp. 415–422 (2001)
10. Osowski, S:: Algorithmic Approach to Neural Networks, Wydawnictwa Naukowo-Techniczne, Warszawa (in Polish) (1996) ISBN 83-204-2197-7
11. Ahmed, U.N., Rao, K.R.: Orthogonal Transforms for Digital Signal Processing. Springer, New York (1975)
12. Rao, K.R., Yip, P.: Discrete cosine transform: algorithms, advantages, applications. Academic Press Professional, San Diego (1990)
13. Ahmed, N., Natarajan, T., Rao, K.R.: Discrete Cosine Transform. IEEE Transactions on Computers, 90–93 (1974)

Emotion Recognition with Poincare Mapping of Voiced-Speech Segments of Utterances

Krzysztof Ślot, Jaroslaw Cichosz, and Lukasz Bronakowski

Institute of Electronics, Technical University of Lodz, Wolczanska 211/215, 90-924
Lodz, Poland
{kslot,jaroslaw.cichosz,lukasz.bronakowski}@p.lodz.pl
http://www.eletel.p.lodz.pl

Abstract. The following paper introduces a set of novel descriptors of emotional speech, which allows for a significant increase in emotion classification performance. The proposed characteristics - statistical properties of Poincare Maps, derived for voiced-speech segments of utterances - are used in recognition in combinations with a variety of both commonly used and some other, original descriptors of emotional speech. The introduced features proved to provide useful information into a classification process. Emotion recognition is performed using binary decision trees, which perform extraction of different emotions at consecutive decision levels. Classification rates for the considered six-category problem, which involved anger, boredom, joy, fear, neutral and sadness, are at the level up to 79% for both speaker-dependent and speaker-independent cases.

Keywords: Emotional Speech Recognition, Feature Selection, Poincaré Maps.

1 Introduction

Emotion recognition becomes an increasingly important research direction in speech analysis. A successful solution to this challenging problem would enable a wide range of important applications. A correct assessment of emotional state of a user can significantly improve a quality of emerging, natural language-based human-computer interfaces. It can be applied in monitoring of psycho-physiological states of individuals in several demanding work environments to assess a level of stress or fatigue. Emotional speech can provide significant clues for forensic data analysis. Finally, correct assessment of emotional load can improve performance of speech and speaker recognition systems.

Despite an enormous amount of research that has been done in the field of emotion recognition [1],[2],[3],[4],[5],[6],[7] the problem is still far from its satisfactory solution. There are numerous difficulties encountered in an analysis of emotional load that is embedded in speech signals. For example, it is hard to clearly identify specific properties of a speech signal that could play a dominant role in conveying particular emotions. It seems that emotions are expressed through mutual interactions of a variety of features from various speech-formation levels, from speech generation, through articulation and prosody. Another fundamental problem of the domain is a difficulty in collecting a reliable experimental material, so that different recognition approaches are tested on databases of speech uttered by actors and actresses, which are not guaranteed to correctly reflect the real emotional processes. Therefore, a performance of the proposed methods is relatively low, although some of the reported recognition rates (at the level of 60-70% for four-five category problems) are comparable to evaluation capabilities of humans.

Speech signal-based assessment of emotions is a pattern recognition problem, so standard pattern recognition methodology, which involves feature space derivation and feature classification, is applied to do the task. The main effort in research on emotion recognition is put on the first of the two elements. A search for novel, appropriately tailored descriptors of emotional speech is a focus of majority of the proposed methods and several features have been identified as important indicators of emotional state. They can be broadly classified into three major categories [4], [8], [9], [10]: energy-related (e.g. energies of voiced or unvoiced speech or various spectral components), frequency-related (pitch and its various descendants) and time-related (speaking rate, pause or voiced-speech percentage and others). On the contrary, emotion classification is performed using standard techniques that analyze either single feature vectors (SVM [11], Neural Networks [4], k-NN [12] etc.) or feature vector sequences (HMM) [9], [13], [10].

The following paper introduces a set of new descriptors of emotional speech - statistical characteristics of Poincare Maps, derived for voiced segments of a speech signal. The proposed features summarize variability in utterance production, which is most likely weakly correlated to any of the aforementioned categories of commonly used emotion descriptors. Therefore, it has been expected that they can introduce meaningful information into emotion classification process. Indeed, experiments have been shown that introduction of the proposed features results in a significant increase in correct recognition rates, both in speaker-dependent (from 72% to 76% for [14] database) and speaker-independent (from 72% to 79%) emotion classification rates for six-category experiments (joy, anger, boredom, sadness, fear and neutral). A binary desicion tree, designed to extract emotions at each of its levels, have been used for emotion classification throughout the experiments.

A structure of the paper is the following. The proposed Poincare-mapping based descriptors of emotional speech are introduced in Section 2. A procedure for selection of features that are applied for emotion classification at consecutive nodes of the decision tree is presented in Section 3. Experimental evaluation of the proposed method, based on two different emotional speech databases - for German and Polish languages - is given in Section 4.

2 Vowel Pronunciation Variability Assessment Using Poincare Maps

Voiced-speech segment pronunciation variability is a property that has not been considered so far by research community in emotional speech analysis, although

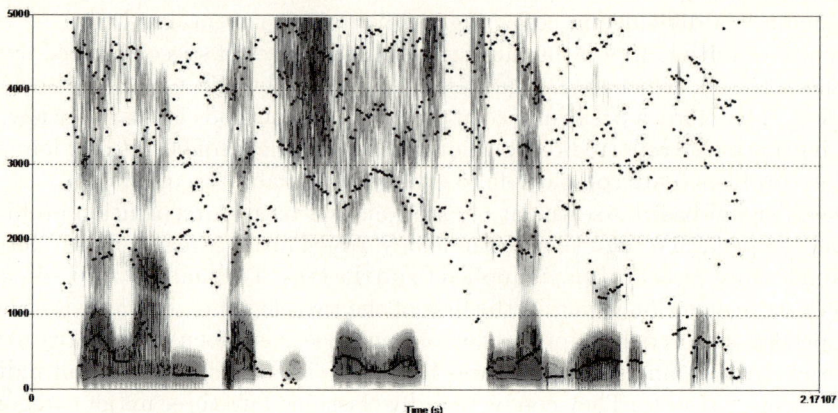

Fig. 1. Evolution of the first four formants (dotted lines) shown on a spectrogram of sample utterance

non-stationary nature of production of voiced-speech portions is a well-known phenomenon [15]. An objective of our research was to examine, whether one can extract from this variability any emotion-specific pattern and to exploit it as an additional clue for emotion assessment. Therefore, we decided to construct an appropriate, quantitative representation of the considered property and apply it as an additional feature to be considered in emotion recognition [16]. A possible way to characterize dynamic changes in pronunciation of voiced-speech is to use Poincare mapping of selected speech signal characteristics, which is a common approach used in nonlinear system analysis.

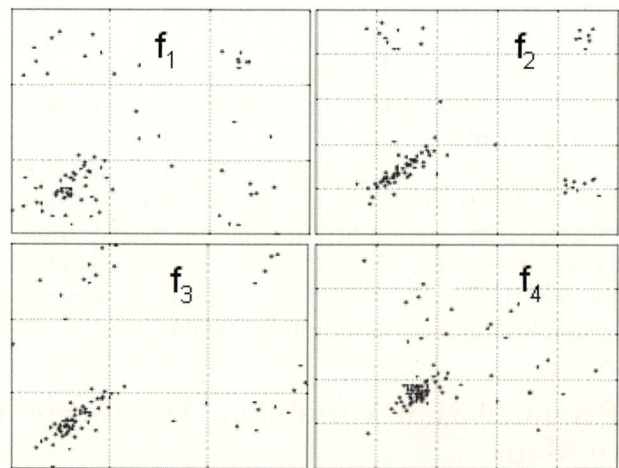

Fig. 2. Poincare Maps produced for four formant frequencies for voiced speech segments of a sentence uttered with an emotion 'fear'

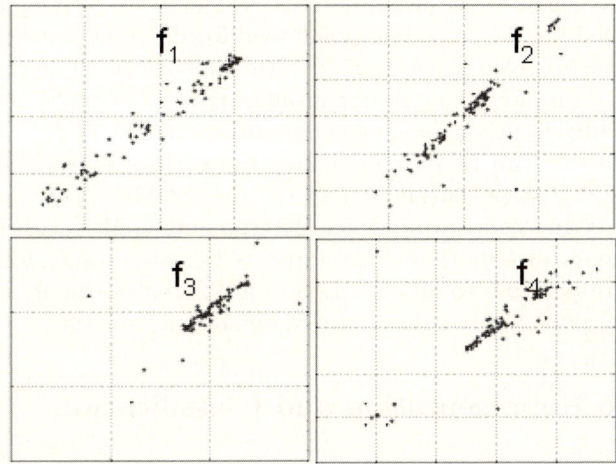

Fig. 3. Poincare Maps produced for four formant frequencies for voiced speech segments of a sentence uttered with an emotion 'joy'

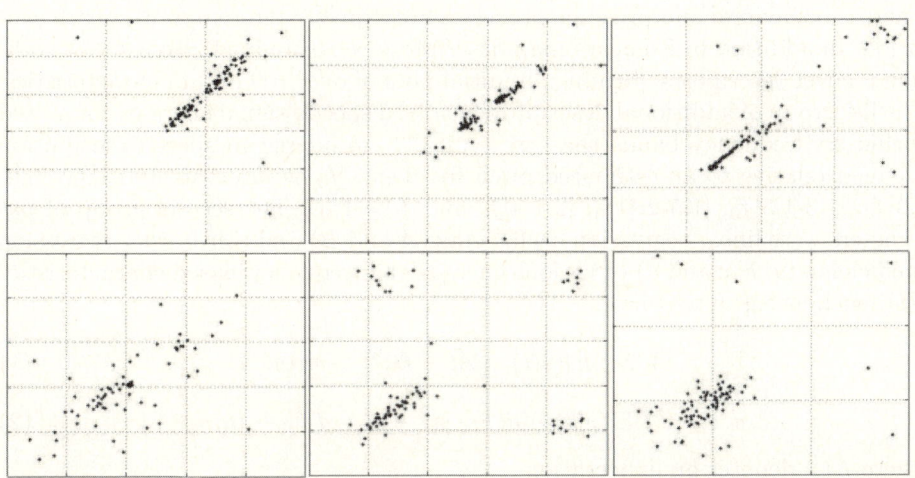

Fig. 4. Poincare Maps produced for the second formant evolution for the same sentence uttered by three different speakers with two emotional loads: 'fear' (top row) and 'joy' (bottom row)

Poincare mapping that is adopted for the purpose of speech-signal analysis, assumes sampling of some voiced-speech segment properties at regular time intervals. Four consecutive formant frequencies have been considered as properties of interest. Sample evolution of these formant frequencies (derived using the PRAAT software [17]) on a spectrogram of some utterance, is shown in Fig. 1).

Sample Poincare maps derived for a sentence, which has been uttered by an actor in two different ways, involving two different emotional loads: 'fear' (Fig. 2) and 'joy' (Fig. 3) show substantial differences. Moreover, although exact forms of maps produced for the same emotional load for different speakers vary, their overall appearance is clearly category-dependent (Fig. 4). Therefore, Poincare Maps seem to be an interesting source of information for emotion classification.

A clear difficulty in using the maps is a complex nature of data distributions. Since no simple rule can be extracted from maps that correspond to different emotions, basic statistical parameters of the distributions were considered, including moments and central moments, determinants and traces of co-variance matrices, boundary distribution characteristics etc. As a result, four sets of features, each corresponding to an evolution of a different formant, have been derived, forming a pool of novel descriptors of emotional speech .

3 Emotion Representation and Classification

Speech characteristics that are commonly used in emotion recognition can be grouped into three aforementioned categories, which cover energy, frequency and temporal domains. Spectral speech characteristics result from vocal tract morphology and vocal fold operation, signal energy reflects vocal tract excitation, whereas behavioral and prosodic processes are expressed in temporal speech evolution. In addition to Poincare map descriptors, we examined also a set of additional novel descriptors, forming an initial pool of over 100 input characteristics. The fist group of additional descriptors involved speech-signal energies computed within six frequency bands that are typically considered in speech recognition (defined relative to an estimated pitch frequency f_0 as intervals $[0-0.8]f_0$, $[0.8-1.3]f_0$, $[1.3-1.5]f_0$, $[1.5-2.1]f_0$, $[2.1-4]f_0$ and $f > 4f_0$). The second group of parameters were linear regression coefficients (A and B) and third-order regression coefficients (a, b, c and d) of various energy- and frequency-based characteristics $s(t)$ (such as e.g. pitch) i.e.:

$$A, B : E[(s(t) - At - B)^2] \to min; \qquad (1)$$

$$a, b, c, d : E[(s(t) - at^3 - bt^2 - ct - d)^2] \to min, \qquad (2)$$

where $E(.)$ denotes an expectation.

Since the presented three groups of descriptors correspond to different phenomena, it has been assumed that three-dimensional feature spaces can be an appropriate input domain for the purpose of emotion characterization. A choice of specific feature triplets that perform the best in emotion discrimination has been done using a feature selection procedure, where results of training data classification are used as a criterion of triplets' performance.

Of many possible classification strategies, a binary decision-tree has been selected. At each node of a tree a particular emotion is 'extracted' from a mixture of the remaining classes, i.e. a two-category classification problems are to be performed. Recognition of six different types of emotional speech - 'anger', 'fear', 'sadness', 'boredom', 'joy' and 'neutral' - was an objective of the research, so corresponding decision trees were composed of five decision levels (Fig. 5a).

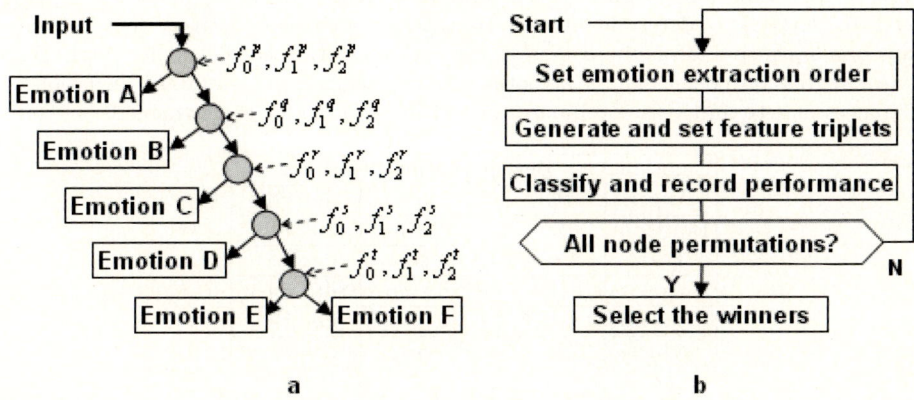

Fig. 5. A structure of the adopted binary decision tree (a) and tree derivation procedure (b)

An adopted triplet selection procedure was combined with selection of an optimum tree structure in the following way (Fig. 5b). Given a specific permutation of emotion extraction order, five triplets are formed from an initial pool of over 100 speech signal characteristics. To ensure good discrimination power, only triplets composed of weakly correlated features are accepted for further analysis. The triplets define three-dimensional feature spaces that are used at five consecutive levels of the decision tree. Classification at each level is performed using k-NN classifier (a value of k=3 has been adopted) and classification scores are recorded for triplet performance evaluation. The presented procedure is computationally expensive - it is repeated for all 120 node permutations and for all possible feature triplets - yet, it is performed off-line, so its efficiency is not critical. As a result, every emotion is assigned a triplet that performed the best at its extraction. Sample results of the procedure have been presented in Table 1, where three best performing feature sets for each of the considered emotions are shown.

Table 1. Emotions and corresponding, best-performing feature triplets (MP, min_P - mean and mean minimum local pitch; A_P, A_{Pmax}, A_{Pmin} - linear regression coefficient for temporal evolution of pitch and its local maxima / minima; $E_{1.3-1.5}$ - an energy of a frequency band between 1.3 and 1.5 of a mean pitch; min_E, ME, MEV - minimum and mean energy and mean energy of voiced speech; std(E) - standard deviations of energy; ND - normalized duration of an utterance; VP - percentage of voiced speech

Emotions	Feature sets		
Anger	min_P, A_{Pmin}, MEV	ND, ME, $E_{1.3-1.5}$	MP, stdE, A_{Pmin}
Boredom	MEV, MP, A_P	ND, A_P, MP	ND, A_P, min_P
Fear	VP, MP, A_{Pmin}	A_P, ME, $E_{1.3-1.5}$	std(E), A_P, MP
Joy	min_E, $E_{1.3-1.5}$, A_{Pmin}	ND, $E_{1.3-1.5}$, , A_{Pmin}	std(E), $E_{1.3-1.5}$, A_P
Neutral	std(E), $E_{1.3-1.5}$, A_{Pmax}	std(E), $E_{1.3-1.5}$, A_P	ND, $E_{1.3-1.5}$, A_P
Sadness	ND, min_E, $E_{1.3-1.5}$	std(E), MP, A_P	ND, $E_{1.3-1.5}$, A_{Pmin}

Table 2. Emotions and corresponding, best-performing feature quads (det(F0)...det(F3) - determinants of co-variance matrices derived for Poincare Maps of consecutive formants; d_P - a third-order regression coefficient for pitch; A_{sP} - linear regression coefficient for smoothed pitch; std(F_i,x-y) - standard deviation of differences of x-th and y-th coordinates of a Poincare map for the formant F_i)

Emotions	Best performing features			
Anger	ME	VP	A_{Pmin}	det(F1)
Boredom	ND	A_P	d_P	det(F0)
Fear	A_P	ME	A_{sP}	det(F2)
Joy	std(E)	VP	$E_{1.3-1.5}$	std(F_1,x-y)
Neutral	std(E)	A_P	$E_{1.3-1.5}$	det(F1)
Sadness	ME	min_P	A_{Pmin}	std(F_0,x-y)

Derived feature triplets have been used as a basis of the proposed emotion representation and emotion recognition procedure. The proposed Poincare maps express non-stationarity of voiced-speech production and are related to a different phenomenon than the ones addressed by 'conventional' characteristics of emotional speech. Therefore, we expect Poincare map descriptors to be weakly correlated with such characteristics and to provide additional, meaningful information for the purpose of emotion classification. To verify this hypothesis, we decided to produce four-element feature vectors, which extend the winning triplets by a single Poincare map descriptor. This has been done by taking sixty triplets (ten winners per each emotion), and testing a performance of the extended, four-element sets. Again, an exhaustive search has been performed, yielding a set of feature quadruples, ordered by their score (sample sets are shown in Table 2).

4 Experimental Evaluation of the Proposed Method

Two different databases of emotional speech were used to evaluate the proposed approach. The first one [14], contains over five hundred sentences, uttered in German by five males and five females, and is one of the benchmarks for the considered domain. The database includes seven different categories of emotional speech: 'anger', 'fear', 'sadness', 'boredom', 'joy', 'disgust' as well as 'no emotion'. The second one is a database of 240 sentences uttered in Polish, which has been constructed in a similar fashion and includes six types of emotions (excluding 'disgust' category). To enable cross-language comparisons of emotion recognition performance, we focused on classification of six emotions that are common for both databases.

Emotion classification procedure has been performed using binary decision trees, derived in a way described in the previous section. Datasets used for classifier derivation and emotion classification were disjoint. The only restriction that has been made was a separate processing of emotional speech uttered by males and females.

Emotion classification procedure, depicted in Fig. 6, begins with an evaluation of features that are required by decision trees. The first pool of features (including

Table 3. Emotion recognition results for speaker-dependent recognition and feature triplets

	Speaker-dependent		Speaker-independent	
	Triplets	Quadruples	Triplets	Quadruples
German database	72.4%	75.6%	72%	79%
Polish database	76.3%	79.3%	64%	76.2%

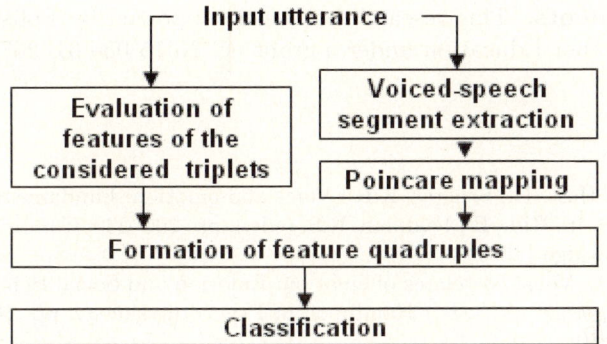

Fig. 6. Block diagram of emotion classification procedure

pitch, energies etc.) is derived based on a full utterance, whereas Poincare Map descriptors are formed only for voiced-speech segments. Therefore, such segments are automatically extracted from input utterances and used as a domain for Poincare mapping. Five best-performing decision trees, derived during the classifier training phase, are used in parallel for sample utterance recognition. Classification results are combined using majority voting mechanism - possible draws are considered as classification errors.

Emotion classification results are summarized in Table 3. As it can be seen, even in case of triplet-based classification, results are very high and they are comparable for both languages used in the experiments. Introduction of Poincare Map-based features improves a performance, especially for the speaker-independent emotion recognition case. Considering a large (six-element) set of target emotions, these results are better than any of the reported so far.

5 Conclusion

The presented paper shows a new family of speech signal characteristics that have proved useful in emotion recognition. Variability in uttering of voiced-speech segments seems to be a source of useful information on emotional state of a speaker. There are several alternative properties of a speech signal that evolve in time in a non-stationary manner, so that they can be used as a basis for Poincare mapping, instead of formant frequencies. Also, one can come out with several other descriptors of Poincare maps, which could provide better discriminative power.

Feature selection has been adopted as a strategy for feature space derivation. Supervised feature extraction methods, such as linear / nonlinear discriminant analysis or supervised principal component analysis can be another way to proceed, which potentially could produce even better emotion modeling. Also, other classification strategies could be considered instead of the proposed binary decision trees, however, no simple clues seem to exist for their choice for the considered, hard problem of multi-class emotion recognition.

Acknowledgments. This research has been supported by Polish Ministry of Science and Higher Education under a grant no. N515 058 31/2677.

References

1. Kappas, A., Hess, U., Scherer, K.R.: Voice and emotion: Fundamentals of Nonverbal Behavior. In: Rim, B., Feldman, R.S. (eds.), pp. 200–238. Cambridge University Press, Cambridge (1991)
2. Scherer, K.R.: Vocal correlates of emotion: Emotion and Social Behavior. In: Wagner, H., Manstead, A. (eds.) Handbook of Psychophysiology, pp. 165–197. Wiley, London (1989)
3. Scherer, K.R.: Vocal affect expression: A review and a model for future research. Psychol. Bull. 99(2), 143–165 (1986)
4. Batliner, A., Huber, R., Spilker, J.: The recognition of Emotion. In: Int. Conf. on Spoken Language Processing, pp. 122–130 (2000)
5. Scherer, K.R.: Vocal communication of emotion: A review of research paradigms. Speech Communication 40, 227–256 (2003)
6. Cowie, R., et al.: Emotion recognition in human-computer interaction. IEEE Signal Processing magazine 18(1), 32–80 (2001)
7. Douglas-Cowie, E., Cowie, R., Schroder, M.: Speech and emotion. Speech Communication 40, 1–257 (2003)
8. Scherer, K.R.: Expression of emotion in voice and music. J. Voice 9(3), 235–248 (1995)
9. Zhou, G., Hansen, J.H.L., Kaiser, J.F.: Nonlinear feature based classification of speech under stress. IEEE Transactions on Speech and Audio Processing 9(3), 201–216 (2001)
10. Nwe, T.L., Foo, S.W., De Silva, L.C.: Speech emotion recognition using hidden Markov models. Speech Communication 41, 603–623 (2003)
11. Kwon, O.K., Chan, K., Hao, J., Lee, T.W.: Emotion recognition by speech signals. In: Int. Conf. EUROSPEECH 2003, Geneva, Switzerland, pp. 125–128 (2003)
12. Dellaert, F., Polzin, T., Waibel, A.: Recognizing emotion in speech. In: International Conference on Spoken Language Processing (ICSLP), pp. 1970–1973 (1996)
13. Nogueiras, A., Moreno, A., Bonafonte, A., Marino, J.B.: Speech emotion recognition using Hidden Markov Models. In: Proceedings of Eurospeech, Aalborg, Denmark, pp. 2679–2682 (2001)
14. Burkhardt, F., Paeschke, A., Rolfes, M., Sendlmeier, W., Weiss, B.: A Database of German Emotional Speech. In: Proc. of Int. Conf. Interspeech 2005, Lisbon, pp. 1517–1520 (2005)

15. Skijarov, O., Bortnik, B.: Chaos and speech rhythm. In: International Joint Conference on Neural Network, vol. 4, pp. 2070–2075 (2005)
16. Bronakowski, lot, K., Cichosz, J., Kim, J.: Application of Poincare Map-Based Description of Vowel Pronunciation Variability for Emotion Assessment in Speech Signal. In: Int. Symp. on Information Technology Convergence (ISITC 2007), Korea, pp. 175–178 (2007)
17. Boersma, P., Weenink, D.: PRAAT, a system for doing phonetics by computer. Glot International 5(9/10), 341–345 (2005)

Effectiveness of Simultaneous Behavior by Interactive Robot

Masahiko Taguchi[1], Kentaro Ishii[1], and Michita Imai[2]

[1] Graduate School of Science and Technology, Keio University
[2] Faculty of Science and Technology, Keio University
3-14-1 Hiyoshi, Yokohama, Kanagawa 223-8522, Japan
{taguchi,kenta,michita}@ayu.ics.keio.ac.jp

Abstract. The ability of a robot to do gaze-drawing and to gesture have become recognized as essential elements in achieving joint attention of the real world during human-robot interaction. However, the ability of a robot using such non-verbal actions to interrupt human action is unclear. We have conducted an experiment in which a robot tries to interrupt human action to demonstrate the effectiveness of an interactive robot acting simultaneously with a person. Using the results of this experiment, we developed a system that a robot can use to predict human motion so that the robot can automatically perform simultaneous behavior.

Keywords: Simultaneous behavior, Human-robot interaction.

1 Introduction

Many recent studies of human-robot interaction have looked at communication using real world information.

For a robot and person to share real world information, they must establish joint attention. This entails the use of non-verbal communication such as gaze-drawing and gesturing [3,4,5].

Kawakami et al. focused on the closeness motion of a handshake and developed a handshake robot system that has a closeness motion acceptable to people [1]. This system has an arm robot based on a closeness motion model that incorporates the two-dimensional delay elements of human action. The characteristic frequency of the two-dimensional delay motion was calculated using the results of human-human handshake motion analysis. In an experiment evaluating the system, they used three characteristic frequencies and determined a handshake timing acceptable to people. Sugiyama et al. developed a simultaneous behavior system in which a robot predicts a real world object to which a person will point within 0.3 seconds and initiates gaze-drawing before the person has finished pointing. The person's actions are measured using a Vicon motion capture system. In their experiment, the robot acted using simultaneous behavior in the experimental group and using a gaze-drawing action after the person completed the action in the control group. When the person pointed to an object, he or she used the deictic or number label attached to the object. The results showed that simultaneous behavior of the robot made it easier for the person to use deictic labels, which promoted expression sharing between the person and robot.

Both Kawakami et al. [1] and Sugiyama et al. [2] did not question whether robot gaze-drawing and gesturing effectively interrupt human actions. That is, whether the means people use to naturally interrupt communication works in human-robot interaction has not been verified. We have experimentally verified that it does and have used our findings to implement a system that enables a robot to automatically behave simultaneously with the actions of a person, a system that is small enough to run on a humanoid robot.

Section 2 formulates a model for a robot to interrupt human actions, and Section 3 explains simultaneous behavior. Section 4 describes our experiment and presents the results. Section 5 discusses the results. Section 6 describes a system in which a robot automatically acts using simultaneous behavior. Section 7 describes future work, and Section 8 concludes the paper with a brief summary.

2 Model for Interrupting Human Action

We define "robots interrupt human action" to mean that robots behave in such a way as to express their intention to a person and the person interrupts his or her action as a result. We formulate this as shown in Eq. (1), where R represents the robot, and $Act(I, P, B)$ means that the robot acts in accordance with behavior B to express intention I to person P.

$$R.Act(I, P, B) \tag{1}$$

$$R = Robot$$
$$Act(I, P, B) = to\ communicate\ I\ to\ P\ using\ B$$
$$I = Intention\ of\ Robot$$
$$P = Person$$
$$B = Behavior$$

A change in the person's action is formulated as Eq. (2). An interruption of the person's action is formulated as Eq. (3). The P in both equations is the same P as in Eq. (1). $ChangeAction(X)$ means that person changes his or her action to action X, the New action. If it set New on $Int(This)$ as Eq. (3), the person interrupts the current action.

$$P.ChangeAction(New) \tag{2}$$

$$ChangeAction(X) = Change\ Action\ to\ X$$
$$New = New\ Action$$

$$P.ChangeAction(Int(This)) \tag{3}$$

$$Int(X) = Interrupt\ X$$
$$This = Current\ action$$

A relation in which behavior B occurred because of action A is formulated as Eq. (4).

$$CA \to RA \qquad (4)$$

$$CA = Causative\ Action$$
$$RA = Resulting\ Action$$

For the final formulation, we use Eqs. (1) (2) (3) and (4) to formulate Eq. (5), which represents a robot performing behavior B to express intention I to a person and the person interrupting his or her action as a result. The formulation of Eq. (5) is addressed in this paper.

$$R.Act(I, P, B) \to P.ChangeAction(Int(This)) \qquad (5)$$

3   Use of Simultaneous Behavior

We use a simultaneous behavior approach to realize a robot that uses some behavior to express its intention to a person so that the person interrupts his or her current action. Simultaneous behaviors are a natural means of communication between people. They include gaze-drawing and gesturing, both of which are used to stop another person's action. For example, a mother might call out "No!" to stop a child from taking a snack from a plate. She would also watch the child's actions with her eyes. In this example, the mother's watching action would start before the child's stopping action started. Such overlapping actions are call *simultaneousbehavior*. Simultaneous behavior for a robot means that the robot starts an action before a person's corresponding action finishes. That is, the person and robot act simultaneously.

In this research, our goal is to formulate Eq. (6) that the simultaneous behavior sets B of Eq. (5), and what robot's action interrupts human action.

$$R.Act(I, P, SB) \to P.ChangeAction(Int(This)) \qquad (6)$$

$$SB = Simultaneous\ Behavior$$

4   Experimental Testing of Hypothesis Validity

We experimentally tested the validity of Eq. (6), i.e., whether the simultaneous behavior of a robot effectively interrupt a person's action.

4.1   Hypothesis

Robots using simultaneous behavior are better able to interrupt human action than robots not using simultaneous behavior.

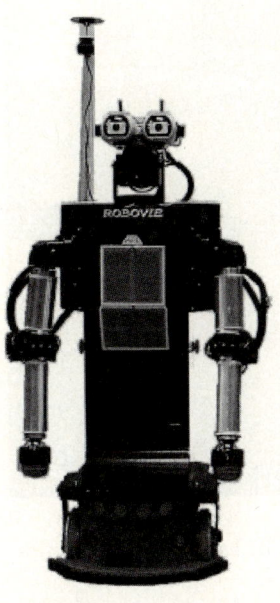

Fig. 1. Robovie-II

Table 1. Specifications of Robovie-II

Size	1140 (H)× 520 (W) × 500 (D) mm
Weight	39 kg
Movement	Power Wheeled Steering, 1 Caster
Degrees of Freedom	Eyes: 2DOF × 2 Head: 3DOF Arms: 4DOF × 2
Speed	1.6 m/s (max)
Sensors	Omni-Directional Camera × 1 Color CCD Camera × 2 Ultrasonic Range Sensor × 24 Bumper Sensor × 10 Touch Sensor × 2 Tactile Sensor (Pressure Conductive Elastomer) × 16 Microphone × 1 Joint Angle Sensor (potentiometer) × 11 Encoder × 2
Controller	Intel Pentium III, 933 MHz HDD, 20 GB OS, Linux
Communication	Wireless LAN (IEEE801.11b)
Power Supply	DC 24V
Operating Time	4 hours

4.2 Method

Overview. The participants in this experiment were presented with three plates containing three different types of snacks and were told to sequentially take one from each plate in random order. The robot called out "dame!" (which means "No!" in Japanese) when the person started to take the third snack. The robot's intention, I in Eq. (6), was to stop the person from taking the third snack. The robot used gaze-drawing before the person tried to take the third snack (using simultaneous behavior) in the experimental group and gaze-drawing after the person took the third snack (not using simultaneous behavior) in the control group. The robot was controlled using the Wizard of Oz method.

Equipment. We used the Robovie-II communication robot [6]. Fig. 1 shows an image of Robovie-II, and Table 1 lists its specification. "Robovie" was developed for natural human-robot communication.

Participants. The 27 participants were in their early 20s and were science and technological undergraduate or graduate students. We assigned 13 to the experimental group and 14 to the control group.

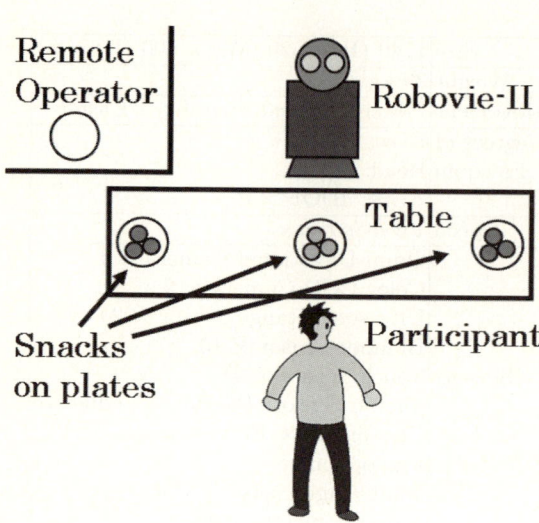

Fig. 2. Experimental Room   **Fig. 3.** Experimental Scenes

Procedure. Each participant interacted with Robovie in a one-to-one relationship in the experimental environment, a room with a large conference table. The three plates of snacks were on the table, at the far left, in the center, and at the far right. A diagram of the room is shown in Fig. 2, and two typical scenes are shown in Fig. 3. The upper photo is for the experimental group, and the lower one is for the control group. The robot's head is at the top-left in both photos. In Fig. 2, the robot is on the opposite side of the table. In Fig. 3, it is at the end of the table.

Before entering the room, the participants were given three instructions that disguised the actual purpose of the experiment.

1. Take one of each kind of snack and taste it.
2. Tell the robot which one tasted best in order to adjust the robot's utterances.
3. Move in front of the robot after tasting the snacks in order to clarify how the robot acts when someone takes a snack and to optimize the Wizard of Oz remote operation.

Thirty seconds after the person tried to take the third snack, we administered a short questionnaire. One question (yes or no) was whether the person heard the robot say "dame!" A second question asked what he or she thought when the robot said "dame!"

Unknown to the participant, an operator in separated area of the room observed the participant's actions through a video camera mounted in the root. The operator controlled the robot's movements, including the calling out of "dame!".

Table 2. Results for Experimental Group

	Did not hear robot's utterance	Did not understand robot's intention	Understood robot's intention	Total
Did not stop action	0	2	1	3
Stopped once	3	1	1	5
Pulled hand back or Stepped back and faced robot	0	0	0	0
Never took third snack	0	0	5	5
Total	3	3	7	13

(units: participants)

Table 3. Results for Control Group

	Did not hear robot's utterance	Did not understand robot's intention	Understood robot's intention	Total
Did not stop action	2	0	1	3
Stopped once	1	1	1	3
Pulled hand back or Stepped back and faced robot	1	2	4	7
Never took third snack	0	0	1	1
Total	4	3	7	14

(units: participants)

Evaluation Method. We conducted a χ^2 test for two conditions. The independent variable (SB in Eq. (6)) was whether the robot acted on the basis of simultaneous behavior, and the dependent variable ($P.ChangeAction\ (Int(This))$ of Eq.(6)) was whether the subject took the third snack. One test condition was that the person could hear the robot utter "dame!" (condition 1). The other was that the person understood the robot's intention "stop the person from taking the third snack" (condition 2).

4.3 Results

The results for the experimental group are summarized in Table 2, and those for the control group are summarized in Table 3. The first row in each table means "the person did not stop the action of taking the third snack." The second row means "the person stopped the action once but then continued." The third row means "the person pulled his or her hand back and/or stepped back and faced

Table 4. Results for Condition 1

	Experimental Group	Control Group	Total
Took third snack	5	9	14
Did not take third snack	5	1	6
Total	10	10	20

(units: participants)

Table 5. Results for Condition 2

	Experimental Group	Control Group	Total
Took third snack	2	6	8
Did not take third snack	5	1	6
Total	7	7	14

(units: participants)

Robovie but then continued." The fourth row means "the person never took the third snack." All the participants eventually took the third snack, except for those represented by the results in the fourth row. The column headings are clear enough without further explanation. If you can, make the row headings equally clear so that you don't have to explain them either.

Table 4 summarizes the results for condition 1. The result of the χ^2 test was p = 0.051<0.10, which shows there was a significant difference for the robot's simultaneous behavior causing the person not to take the third snack. Table 5 summarizes the results for condition 2. The result of the χ^2 test was p = 0.031<0.05, which confirms there was a significant difference for the robot's simultaneous behavior causing the person not to take the third snack.

These results confirm our hypothesis (Eq. 6): robots using simultaneous behavior interrupt human action more effectively than robots not using simultaneous behavior.

5 Discussion

5.1 Hearing Robot's Utterance

The participants who could not hear the robot's utterance answered the second question in several ways. Two (Two) of the three (four) experimental (control) group participants said, "I heard something but did not understand it." One of the remaining two in the control group who heard something said, "I heard as UE." (3) The remaining three said they heard nothing. Although there is no relationship between whether a robot behaves simultaneously and whether the person hears the robot's utterance, one participant in the control group who understood the robot's intention said, "The utterance was a little hard to hear, but I heard maybe "dame." Sometimes, the robot's utterance was split into two distinct sounds: "da-me." We thus need to improve the speaker or utterance module.

5.2 Understanding Robot's Intention

The participants who could hear the robot's utterance but did not understand its intention answered the second question in various ways. One such participant in the experimental group said, "The robot recognized fourth snack when I ate the third snack." Another said, "I took an unwanted action." A third said, "The robot did not recognize eating or my action was wrong." One participant in the control group said, "I did not take the next snack because the robot did not recognize eating." Another said, "My action was not still, so the robot could not recognize me." A third said, "I was taking the third snack before I finished eating the second one; I was taking snacks faster than the robot could recognize my actions." All of these answers are related to misrecognition by the robot. These participants were apparently more concerned about the experiment than the interaction. The logical flow here is confusing. The results just presented seem negative, yet the following conclusion is positive. It would appear that the robot expressed its intention easily by encouraging immersion in interaction [7].

5.3 Participants Who Did Not Take Third Snack

Five participants in the experimental group and one in the control group apparently understood the robot's intention (This is not true for all five participants according to what you say below.) as they never took the third snack. Those in the experimental group answered the second question as follows: "The robot identified the end of the experiment," "the third snack was the robot's favorite," "I did not understand, so I asked the robot why did you stop my action," "the third snack was reserved for the robot," and "I did not understand." The one in the control group said, "It was the ending time of the experiment." The experimental group participants apparently thought the interruption and utterance were robot controlled, while the control group one apparently thought they were experimenter controlled.

6 Automatic Simultaneous Behavior System

Using the results described above, we implemented a system that enables a robot to automatically behave simultaneously.

The target environment consists of a table on which there are two cups. A person and a robot face each other across the table. This is the same simplified environment described in Section 4, the one we used for evaluating the effectiveness of simultaneous robot behavior.

The person's action, i.e., acceleration of the hand, was measured using a Bluetooth-based, light-weight, triaxial accelerometer sensor, a "B-Pack" [8]. A photograph of the B-Pack sensor is shown in Fig. 4. It was secured to the back of a glove worn by the participant, as shown in Fig. 5. It sent acceleration readings to a computer through a Bluetooth connection.

Fig. 4. B-Pack Sensor

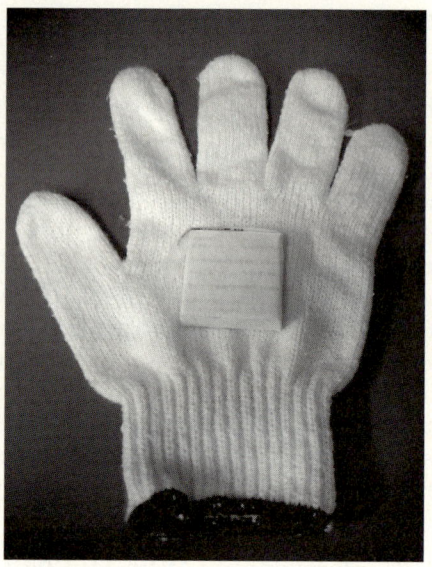

Fig. 5. Glove with B-Pack Sensor Attached

There are five basic steps in this system.

1. The B-Pack sensor sends triaxial acceleration readings to a laptop PC through a Bluetooth connection every 10 ms.
2. Predictor software running on the PC receives the readings and calculates whether the hand has moved by using the data for the last ten readings (100 ms).
3. If the hand has moved, the predictor software uses the data for the last 20 readings (200 ms) and calculates the direction of movement (left or right). Using these results, it predicts which cup the person intends to take.

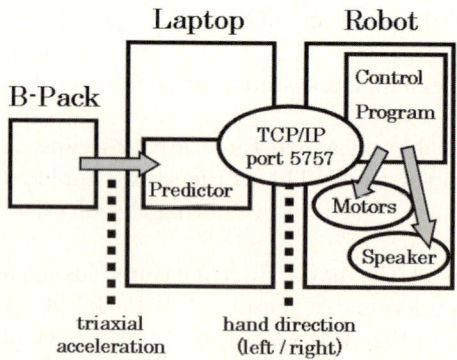

Fig. 6. Operation of Simultaneous Behavior System

Fig. 7. Demonstration of Simultaneous Behavior System

4. The predictor sends the prediction results to the robot through the robot's TCP/IP 5757 port.
5. When the robot's control program receives the prediction results, it operates its motors and speaker in accordance with defined rules in order to direct its gaze toward the predicted cup and say, "left" or "right."

The operation of the system is diagramed in Fig. 6, and a photograph of it being demonstrated is shown in Fig. 7. Testing demonstrated that this system enables a robot to exhibit simultaneous behavior.

7 Future Work

Our finding that half the participants in the experimental group took the third snack is attributed to the instructions they received being inadequate. Apparently, teaching is a higher priority than the robot's intention. We plan to investigate this more thoroughly using another robot and a similar experiment.

8 Conclusion

We have experimentally investigated whether simultaneous behavior by a robot effectively interrupts a person's action and have implemented a system that enables a robot to automatically initiate simultaneous behavior.

References

1. Kawakami, Y., Jindai, M., Watanabe, T.: Development of a handshake robot system based on human handshake motions. In: Proceedings of Human Interface Symposium 2006, Kurashiki, Okayama, Japan, September 2006, pp. 63–68 (2006) (in Japanese)
2. Sugiyama, O., Kanda, T., Imai, M., Ishiguro, T., Hagita, N.: Natural deictic communication with humanoid robots. In: Proceedings of IEEE/RSJ International Conference on Intelligent Robots and Systems (IROS 2007), San Diego, CA, USA, October 2007, pp. 1441–1448 (2007)
3. Imai, M., Ono, T., Ishiguro, H.: Physical relation and expression: Joint attention for human-robot interaction. IEEE Transaction on Industrial Electronics (ITIED 6) 50(4), 636–643 (2003)
4. Kozima, H., Nakagawa, C., Yasuda, Y.: Interactive robots for communication-care: A case-study in autism therapy. In: Proceedings of 14th IEEE International Workshop on Robot and Human Interactive Communication (RO-MAN 2005), Nashville, TN, USA, August 2005, pp. 341–346 (2005)
5. Kanda, T., Kamashima, M., Imai, M., Ono, T., Sakamoto, D., Ishiguro, H., Anzai, Y.: Embodied Cooperative Behavior for a Humanoid Robot that Communicates with Humans. Journal of the Robotics Society of Japan 23(7), 132–143 (2005) (in Japanese)
6. Kanda, T., Ishiguro, H., Ono, T., Imai, M., Maeda, T., Nakatsu, R.: Development of "Robovie" as Platform of Everyday-Robot Research. The Institute of Electronics, Information and Communication Engineers (IEICE) Transactions J85-D-I(4), 380–389 (2002) (in Japanese)

7. Imai, M., Narumi, M.: Robot Behavior for Encouraging Immersion in Interaction. In: Proceedings of Complex Systems Intelligence and Modern Technological Applications (CSIMTA 2004), Cherbourg, France, September 2004, pp. 591–598 (2004)
8. Ohmura, R., Naya, F., Noma, H., Kogure, K.: B-Pack: A Bluetooth-based Wearable Sensing Device for Nursing Activity Recognition. In: Proceedings of 1st International Symposium on Wireless Pervasive Computing 2006 (ISWPC 2006), Phuket, Thailand (January 2006) (CD-ROM)

Part VI

Bioinformatics and Medical Applications

Quality-Driven Continuous Adaptiation of ECG Interpretation in a Distributed Surveillance System

Piotr Augustyniak

AGH University of Science and Technology, 30 Mickiewicza Ave. 30-059 Krakow
Poland
august@agh.edu.pl

Abstract. Principal rules defining the adaptation of ECG interpretation software in a distributed surveillance network are presented in this paper. Thanks to the pervasive access to wireless digital communication services, the intelligent monitoring networks automatically solve difficult medical cases thanks to the auto-adaptation of data interpretation and transmission to the variable patient status and technical constrains. The foundation of this innovative approach is the use of selected diagnostic parameters in a loopback modifying the running interpretive software. The auto adaptive process maximizes the general estimate of patient description quality aggregating the divergence values of particular parameters modulated by the medical relevance factor dependent on the status of patient. Our approach is motivated by the outcomes from the research on human experts behavior, statistics of the procedures reliability and usage as well as tests in a prototype client-server application. The tests yielded very promising results: the convergence of the remotely computed diagnostic outcome was achieved in over 80% of software adaptation attempts. Comparing to the rigid reporting mode, avoiding unnecessary computation extends the autonomy time by 65% and the transmission channel occupation was reduced by 3,1 to 5,6 times

1 Introduction

Thanks to the wide accessibility of wearable computers and wireless digital communication, distant diagnostic tools for cardiology are already commercialized with success [6] [14]. The advantages of this technique are usually stressed in context of cardiovascular out-hospital patients [9] [11] [13]. Recent reports show that it has a much larger impact on the quality of live [18] [23]. Very similar from a technological viewpoint are applications for a seamless monitoring of vital signs from sportsmen, elderly people living on their own or members of services exposed to danger. The artificial intelligence embedded in wearable recorders simulates well the continuous assistance of medical experts without imposing limits on the everyday activity or mobility range.

  A conventional star-shaped topology is usually applied in the surveillance networks. The management is performed by the central server which cooperates with several remote wearable recording devices over a digital data link. Two approaches to the automatic signal interpretation, currently represented in marketed systems, employ either the transmission of raw signal to the center or the signal processing embedded in the remote device. The first method involves high cost of telecommunication service paid for the amount of the data sent. In the case of the second method, the limitation of the quality and reliability is implied by a compromise between the computational power and the energy consumption.

Our approach postulates the interpretation task sharing between the remote recorder and the central server. This process is flexible and automatically controlled by the central server via bi-directional digital wireless transmission channel. The interpretation of the recorded ECG signal and the adaptivity of diagnostic report content are driven by multicriterial optimization including following aspects:

- Assessment of the reliability of remote interpretation with consideration of the received diagnostic data consistency and with the reference to redundant interpretation results computed by the central server from a copy of raw ECG signal occasionally pulled from the remote recorder.
- The evaluation of the patient status and projection of the most probable further diagnostic goal made on a background of human experts guidelines, preference factors and behavior statistic studies.
- The estimation of the remote resources availability such as battery status, CPU workload, memory usage and wireless link quality.

The distributed process of ECG interpretation is asymmetric. It is always initiated in the remote device in consequence of signal acquisition and continued as far as necessary and possible. Technical limitations of the resources available from a remote recorder, causing certain interpretation tasks to be too difficult for such a small battery-operated wearable device, are compensated by the complementary interpretation thread running in parallel on the server. The server collects medical data at various possible processing stages ranging from a raw ECG signal to a full diagnostic outcome. Consequently, it continues the interpretation towards the final verdict and helps to clarify ambiguous or remotely unresolved records. The adaptation of remote interpretation process is achieved in real time with the use of diagnostically-oriented libraries of subroutines. The selected functions are uploaded from the repository managed by the supervising central server [26] and dynamically linked with the running software. The adaptation is performed automatically, but occasionally, in critical cases, it is supervised by a human expert.

Three innovative concepts were a background for our interdisciplinary research:

- The application of medical results and statistics concerning the human experts behavior to the modification of the ECG interpretation process.
- The use of wireless feedback and dynamically linked libraries for modification of the interpretation software in the run. The feedback transmits control data, executable code of software libraries as well as messages to the patient.

– The use of seamless optimization of the processing, which implies the flexibility of data transmission format.

The idea of a distributed ECG interpretation assumes the application of human experts skills in the signal interpretation and rules of knowledge exchange. Studies carried out within the framework of presented research included:

- analysis of human experts behavior and their mutual relations,
- research on visual ECG trace conspicuity and scanpath-based identification of interpretation strategies in human [3], [4],
- procedure reliability and usage statistics, processing error propagation studies [24], [25] and others.

This paper presents first the principal aspects of adaptivity: interpretation procedures optimization, flexible reporting format and non-uniform sampling of the ECG signal. In the second part, main adaptation management procedures are explained: implementation of expert-derived diagnosis priority, automatic assessment of result reliability and estimation of available remote resources. Results of the first prototype adaptivity test, the discussion and considerations for further research close the paper.

2 Principal Adaptivity Aspects

2.1 Modifications of Processing Parameters

The ECG interpretation algorithms performs detection and recognition processes based on heuristically determined values of factors and thresholds. In a rigid architecture software they are often referred as constants, because their values are not modified in course of processing [16]. Typical examples are normal-abnormal limits depending on sex, race or medication. The modification of ECG interpretation software is possible by setting conditional rules of adjustment for these "constants" values in particular procedures. The advantage of this method is a very little amount of data necessary to be sent from the supervising center to the remote recorder, so changes in the software are applied immediately. Due to limited influence on the signal processing, the modification of parameters is applicable rather to the recorder customization than to the radical adaptation of its functionality.

2.2 Rearrangement of the Processing Chain

The architecture of a typical ECG interpreting software is designed with the aim of issuing an universal diagnostic outcome. Specialized recorders for exercise test or Holter monitoring often use an alternative application-specific processing chain [15], [12], [20], [22], [8], [10], [19]. Considering the constraints of computational resources in a remote wearable recorder [21], the universal processing has to be based on average quality procedures. Such universal solutions are outperformed by task-oriented routines in a dynamically modified processing chain.

Consequently, our first step towards the agile implementation was the detailed analysis of data flow, in particular data stream reduction at subsequent stages of the interpretation process and cumulation of the processing error. This analysis yielded a new software processing architecture (fig. 1) composed of mandatory basic procedures and facultative specialized libraries of functions.

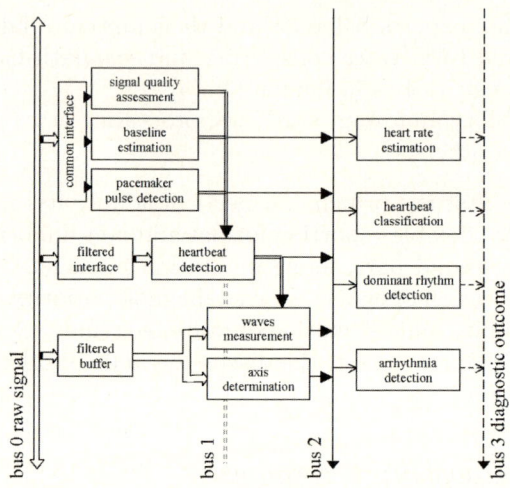

Fig. 1. The architecture of remote ECG recorder interpretation optimized for the data flow and processing error propagation

The proposed architecture offers following advantages:

– Defines mandatory and optional procedures and their reciprocal dependences,
– Minimizes processing error propagation through the interpretation chain,
– Allows considerable reduction of data stream on early stages of the interpretation process.

All the ECG interpretation procedures were classified to one of the following categories:

– Recorder-only procedures.
– Recorder defaults (preloaded dll's)
– Recorder optionals (downloadable dll's)
– Server-only procedures

The range of ECG interpretation modifications is determined by the contents of two middle categories. All these subroutines are compiled as dynamically linked libraries for two platforms: the remote recorder and the server. Requirements concerning simultaneously linked libraries and intermediate results that must be available in a compatible format for both platforms for correct processing are defined by a global cross-dependence table.

2.3 Report Contents and Frequency Adaptation

Inside the ECG interpretation process, the medical information is represented in various data forms including signals, meta-data and final diagnostic parameters. Transmission of the data at various processing stages is therefore an intrinsic necessity of the distributed software architecture with adaptive task sharing [5]. The flexible communication format contains mandatory data description fields and optional data containers of variable size (fig. 2). For future extensions, it also provides a support for various data types (e.g. patient communications, global positioning coordinates etc.). The modifiable communication protocol is very useful for optimization of wireless channel use and helps maintaining the monitoring costs at a commonly acceptable level. Basic interpretation results are usually delivered for all the monitoring time and more detailed reports are occasionally issued for short time intervals. The delivery of detailed reports can be initiated by the server or by the remote recorder in case of occurrence or suspicion of any event. The report includes interpretation results and all the data required for their verification. It may also include a corresponding strip of raw ECG signal. The processing and reporting modifications are based on the implementation of rules derived in results of cardiologist's behavior analysis.

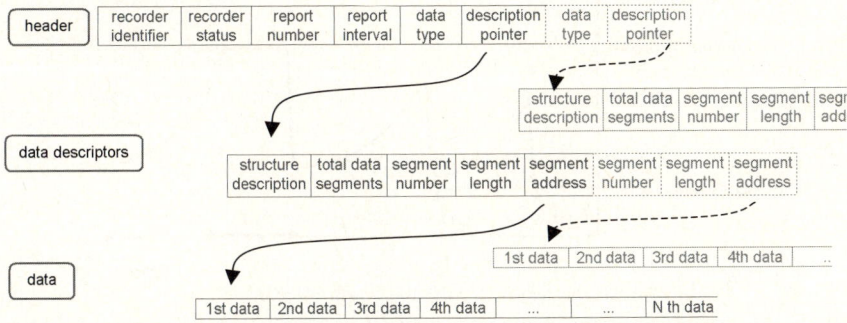

Fig. 2. Data communication format; mandatory fields are bordered by the solid line, optional fields are bordered by the dash line

The temporal variability significantly differs for particular diagnostic parameters. The reporting frequency, considered as a minimum update rate satisfying the Shannon theorem for cardiac series, ranges from 3 Hz (once per a heart beat) for the heart rate (HR) to 0,0033 Hz (once per a 5-minutes strip) for the ST-segment depression. The variability statistics studied in a short-time context of abnormal events show increased probability of further variations. Consequently, the frequency of patient status sampling is determined individually considering the past and present values of each parameter. The report format flexibility supports the transmission of each diagnostic parameter with its appropriate sampling interval. In the case the mid-point samples need to be estimated (e.g. for comparizon with another data set), irregularly sampled data streams are interpolated by the recipient. The supervising server controls the reporting frequency independently for each co-operating remote recorder. Diagnostic reports are delivered with the frequency determined by the server in the context of previous diagnostic data. Reports accidentaly missing (e.g. due to communication failures) are easily detected by the recipient and re-transmitted. In the case of difficult signals, whose interpretation exceeds the capacity of the remote recorder, the raw signal is appended to the report. This allows the complementary processing thread running on the server to complete the ECG interpretation. Once the server determines the patient status, new remote interpretation tasks and reporting interval are accordingly imposed to the remote recorder over the wireless feedback. In very rare cases, the interpretation requirements exceed the availabile resources in the remote recorder. In such case the recorder calculates only basic diagnostic parameters (e.g. heart rate) in real time. The report is sent continuously and contains the raw signal, so the whole interpretation is performed by the server thread. The diagnostic outcome is validated in order to determine whether the patient status allows for returning to the remote interpretation mode. Figure 3 summarizes all modes of reporting frequency control.

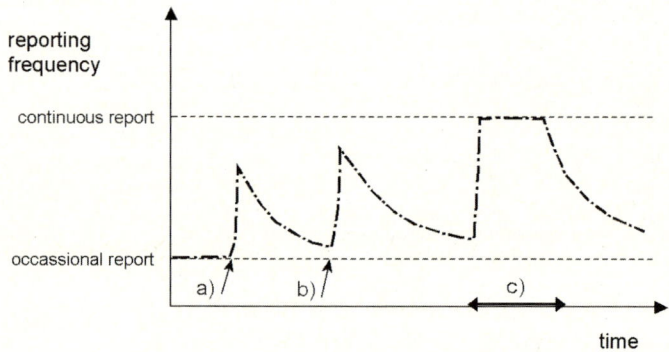

Fig. 3. Examples of reporting frequency control a) and b) - pathological ECG interpreted remotely, c) - unexpected event causing raw signal transmission and server-side interpretation

2.4 Expert-Derived Diagnosis Priority

The automatic management of interpretation task sharing directly influences the monitoring costs and diagnostic reliability. Implementation-dependent approaches specify the processes running on the remote recorder, on the central server or shared between these devices. For the last category, task sharing rules define the medically-derived conditions for processing adaptation and algorithms for automatic data verification. Diagnostic interpretation flow charts, specified in standards and professional guidelines, usually have a form of path with conditional branches. These expresses the dependency of subsequent ECG analysis steps of precedent stages' results. Studying different approaches found in the literature, we found useful to transform a common diagnostic path to a binary tree. Its nodes define medically-justified conditions where an automatic choice is made for the most appropriate way of further ECG interpretation.

Thanks to a functional modification of the prototyped ECG interpretation software we carried out experimental studies concerning doctors preferences about the contents of ECG diagnostic report. In the selection window, displaying the proposal of final report contents, the default layout was replaced by randomly ordered and randomly pre-selected items. Once the interpretation is completed, the human expert had to select or deselect results he or she wish to include in the printed and saved report. The selected items and the order in which a doctor made the choice were memorized with the diagnostic outcome. Finally, the statistical review of these data in context of 14 most common diseases revealed the knowledge about expected doctors preferences. These studies confirm and quantify the common, but poorly documented belief, that depending on the suspected disease, some diagnostic results are more important than others for a human expert. Moreover, assuming that several common diseases may be reliably diagnosed in course of a fully automated process, our studies result in attributing each disease a hierarchical list of most expected diagnostic parameters and their maximal update interval. This relation allows for disease-dependent report modification, issued by the server as a command to adapt the remote interpretation and to update the data priority list.

3 Results

The prototype scale-limited implementation of a wireless monitoring network used the PDA-based adaptive ECG monitor and the stationary PC-based server with 100Mb static IP Internet connection and Linux OS. The tests were focused on the benefits expected from the processing and transmission adaptivity, and on the correctness of automatic decisions about the adaptivity and interpretation task sharing. The software provided options for selective disabling the adaptivity features. In all experiments we used 58 artificial signals originating from looped normal CSE records with inserts of pathological CSE records [27]. A multi-channel programmable generator reproduced signals of the total duration of 1-1.5 hour. The correct medical findings were known from the test signal database and used in the test result processing as references for diagnoses computed in the distributed architecture. The first group of tests was aimed at the comparizon of agile and rigid procedures implemented on the same hardware platform and at quantifying the advantages of the adaptive interpretation software over the methods being in use today. Two principal benefits are:

- Significant reduction of digital communication costs achieved with content-adaptive signal and data representation.
- Extension of the autonomy time due to avoiding unnecessary computation and data transmission.

With respect to the uniform regular reporting, our tests show significant reduction of transmitted data volume. This is the result of report content management (5.6 times in average), non-uniform signal sampling (3.1 times) and irregular reporting (2.6 times) driven by individual values of expected diagnostic parameters variability. The resulting benefit does not accumulate all these optimization aspects, because the opportunity for using of the last two appears occasionally. Besides the savings on transmitted data volume, the report content management offers an advantage of lower power dissipation achieved by avoiding unnecessary computation and data transmission. In consequence, the remote recorder battery life was extended in average by 65% comparing to the software with all optimization options disabled running on the same PDA. The total of 2751 12-leads ECG records were processed in the prototyped system. Among them, 1808 records were correctly interpreted by the remote recorder 857 records (31,2%) needed software adaptation for the correct remote interpretation, and 86 records (3,1%) were assessed as too complicated for the remote interpretation. Among the software adaptation attempts 768 (89,6%) cases were correct, while the remaining 10,4% failed due to incorrect estimation of available remote resources. Resources overestimation resulting in the operating system crash and thus monitoring discontinuity occurred in 27 (1%) cases.

4 Discussion

A significant progress has been made with formulating ECG interpretation task sharing rules, and their implementation and validation as a software controlling the distributed monitoring system. The test results show considerable improvement of diagnosis quality in comparison with the rigid software, however for further development the detailed analysis of outliers is highly recommended. The newly proposed adaptive distributed processing combines advantages of methods widely spread today:

- The load of transmission channel is close to the values of remotely interpreting applications,
- The interpretation reliability is close to the values achieved in centralized interpretation architectures with an occasional human expert assistance.

The software adjustment and test process not described in this paper lasted for 14 months. Several unexpected problems emerged so far have to be considered before the common application in clinical practice. The most problematic issue is the compatibility of the interpretive software designed for two different platforms. The points where the interpretation could be taken over from the remote recorder and continued by the server require the server interpretation thread to follow the architecture of remote recorder software. The use of diagnostic results as arguments for processing-modifying functions looks interesting, but it should be more thoroughly investigated to reveal any unexpected system behavior and to predict the possible medical risk.

Acknowledgment

Scientific work supported by the Polish State Committee for Scientific Research resources in years 2004-2007 as a research project No. 3 T11E 00127.

References

1. Aldroubi, A., Feichtinger, H.: Exact iterative reconstruction algorithm for multivariate irregularly sampled functions in spline-like spaces: the Lp theory. Proc. Amer. Math. Soc. 126(9), 2677–2686 (1998)
2. Augustyniak, P.: Adaptive Discrete ECG Representation - Comparing Variable Depth Decimation and Continuous Non-Uniform Sampling. Computers in Cardiology 29, 165–168 (2002)
3. Augustyniak, P.: How a Human Perceives the Electrocardiogram: The Pursuit of Information Distribution through Scanpath Analysis. Computers in Cardiology 30, 601–604 (2003)
4. Augustyniak, P., Tadeusiewicz, R.: Investigation of Human Interpretation Process Based on Eyetrack Features of Biosignal Visual Inspection. In: IEEE 27-th Annual IEEE-EMBS Conference, paper nr 89 (2005)
5. Augustyniak, P.: Content-Adaptive Signal and Data in Pervasive Cardiac Monitoring. Proc. Computers in Cardiology 32, 825–828 (2005a)
6. Balasz, G., Kozmann, G., Vassanyi, I.: Intelligent Cardiac Telemonitoring System. Computers in Cardiology 31, 745–748 (2004)
7. Banitsas, K.A., Georgiadis, P., Tachakra, S., Cavouras, D.: Using handheld devices for real-time wireless Teleconsultation. In: Proc. 26th Conf. IEEE EMBS, pp. 3105–3108 (2004)
8. Bar-Or, A., Healey, J., Kontothanassis, L., Van Thong, J.M.: BioStream: A system architecture for real-time processing of physiological signals. In: Proc. 26th Conf. IEEE EMBS, pp. 3101–3104 (2004)
9. Bousseljot, R., et al.: Telemetric ECG Diagnosis Follow-Up. Computers in Cardiology 30, 121–124 (2003)
10. CardioSoft, Version 6.0 Operator's Manual. GE Medical Systems Information Technologies, Inc. Milwaukee (2005)
11. Chiarugi, F., et al.: Continuous ECG Monitoring in the Management of Pre-Hospital Health Emergencies. Computers in Cardiology 30, 205–208 (2003)
12. DRG, MediArc Premier IV Operator's Manual version 2.2 (1995)
13. Fayn, J., et al.: Towards New Integrated Information and Communication Infrastructures in E-Health. Examples from Cardiology. Computers in Cardiology 30, 113–116 (2003)
14. Gonzalez, R., Jimenez, D., Vargas, O.: WalkECG: A Mobile Cardiac Care Device. Computers in Cardiology 32, 371–374 (2005)
15. HP, M1700A Interpretive Cardiograph Physician's Guide ed. 4. Hewlett-Packard (1994)
16. IBM, Electrocardiogram Analysis Program Physician's Guide (5736-H15) 2nd edn. (1974)
17. IEC, 60601-2-47. Medical electrical equipment: Particular requirements for the safety, including essential performance, of ambulatory electrocardiographic systems (2001)

18. Korsakas, S., et al.: Electrocardiosignals and Motion Signals Telemonitoring and Analysis System for Sportsmen. Computers in Cardiology 32, 363–366 (2005)
19. Macfarlane, P.W., Devine, B., Clark, E.: The University of Glasgow (Uni-G) ECG Analysis Program. Computers in Cardiology 32, 451–454 (2005)
20. Nihon Kohden, ECAPS-12C User Guide: Interpretation Standard revision A (2001)
21. Paoletti, M., Marchesi, C.: Low computational cost algorithms for portable ECG monitoring units IFMBE Proc. Medicon paper 231 (2004)
22. Pinna, G.D., Maestri, R., Gobbi, E., La Rovere, M.T., Scanferlato, J.L.: Home Telemonitoring of Chronic Heart Failure Patients: Novel System Architecture of the Home or Hospital in Heart Failure Study. Computers in Cardiology 30, 105–108 (2003)
23. Puzzuoli, S.: Remote Transmission and Analysis of Signals from Wearable Devices in Sleep Disorders Evaluation. Computers in Cardiology 32, 53–56 (2005)
24. Straszecka, E., Straszecka, J.: Uncertainty and imprecision representation in medical diagnostic rules, IFMBE Proc, Medicon, paper 172 (2004)
25. Tadeusiewicz, R., Augustyniak, P.: Information Flow and Data Reduction in the ECG Interpretation Process. In: IEEE 27 Annual EMBS Conf., paper 88 (2005)
26. Telisson, D., Fayn, J., Rubel, P.: Design of a Tele-Expertise Architecture Adapted to Pervasive Multi-Actor Environments. Application to eCardiology. Computers in Cardiology 31, 749–752 (2004)
27. Willems, J.L.: Common Standards for Quantitative Electrocardiography 10-th CSE Progress Report, ACCO publ., Leuven (1990)

Detection of Eyes Position Based on Electrooculography Signal Analysis

Robert Czabański, Tomasz Przybyła, and Tomasz Pander

Silesian University of Technology,
Institute of Electronics,
ul. Akademicka 16, 44-101 Gliwice, Poland
{robert.czabanski, tomasz.przybyla, tomasz.pander}@polsl.pl

Abstract. In this paper we reported initial work at development of a human-machine interface system for people with severe disabilities based on measurement of electrooculography signal (EOG). We proposed a system for detecting and predicting eyes position using EOG. We applied a zero-order Takagi-Sugeno-Kang model with modified inference procedure for this task. To calculate values of parameters of fuzzy system we used results of EOG signal segmentation. Experimentation shows the usefulness of the presented method for improving functionality of interface systems that can assist people with limited upper body mobility.

1 Introduction

Communication between human beings seem to be much more simple than between a man and computer machine. This difficulty increases when a person is disable. Eye movement can be applied as a mean to communicate with computer. The idea of such interface is the following. The human-machine interface device records the electrooculogram (EOG) signal and then EOG is digitally processed to generate steering signals for a computer.

The EOG signal is based on electrical measurement of the potential difference between the cornea and the retina. The cornea-retinal potential creates an electrical field in the front of a head. This field changes in orientation as the eyeballs rotate. The electrical changes can be detected by electrodes which are placed near eyes. It is possible to obtain independent measurements from each of the one pair of eyes. For a healthy man, the movement of eyes is coupled. Thus it is adequate to measure the vertical motion of only single eye together with the horizontal motion of a pair of eyes. The amplitude of EOG signal varies from 50 to 3500 μV with a frequency range of about DC-100 Hz. Its behavior is practically linear for gaze angles of $\pm 30^o$ [1]. It should be pointed out here that the variables measured in the human body (any biopotentials) are almost always recorded with a noise and often have a non-stationary features. Their magnitude varies with time, even when all possible variables are under control. This means that the variability of the EOG signals depends on many factors that are difficult to determine [1].

EOG signal can be recorded for the horizontal and the vertical direction of eye movement. On the basis of EOG signal it is possible to detect where one person is looking. The exact detection of the final eyes position is not a trivial task because of the saccadic eye movements. They are characterized by a rapid shift from one point of fixation to another. The saccadic movements are extremely variable, with wide variation in latent period, time to peak velocity, peak velocity and saccade duration [1]. The saccade is the fastest movement of an external part of the human body. The peak angular speed of the eye during a saccade reaches up 1000 degrees per second and it last from about 20 to 200 milliseconds.

In this paper we reported initial work at the development of the human-machine interface system for people with severe disabilities. At that point our aim is the detection (prediction) of eye position using Takagi-Sugeno-Kang (TSK) fuzzy model.

The rest of the paper is divided into the following sections: Section 2 describes the segmentation method we used to calculate values of parameters of fuzzy rules for TSK fuzzy system. In Section 3 the proposed fuzzy model of eye movement is presented. Section 4 shows the experimental results followed by the conclusion in Section 5.

2 Segmentation of the EOG Signal

The EOG signal represents an electric activity of the eyeball muscles. An example of the electrooculography signal is presented on Fig. 1. Looking at the example of the EOG signal, it can be found the rapid changes of amplitude that correspond with consciously eye movement to another part of the scene. Also small and rather fast changes of signal amplitude can be observed in between. These parts of analyzed signal are called saccadic eye motion. Let us define the gradient of the EOG signal as

$$dx(n) = |x(n+1) - x(n)|, \tag{1}$$

where $x(n)$ is the $n-th$ sample of the EOG signal, and $1 \leq n \leq N-1$. Analyzing the gradient signal $dx(n)$ it can be found the moment, when *large* amplitudes occur. It matches the change of observed object or part of the scene. Hence, when the gradient values are *small* then we deal with saccade movements. An example of the EOG gradient signal is presented on Fig. 1.

The notions *large* amplitude and *small* amplitude can be examined from the fuzzy set theory point of view. So, we can state that the interesting parts of analyzed signal occur when amplitudes of gradient signal are small.

For the estimation of the membership values one of many fuzzy clustering methods can be utilized. Mainly, as a clustering results we obtain the membership grades for the clustered data and the cluster prototypes. Unfortunately, after clustering process only for the samples from the input data set the membership grades are known. When the input data set would be changed, for some samples from the data set we could not estimate the membership values. So, in this

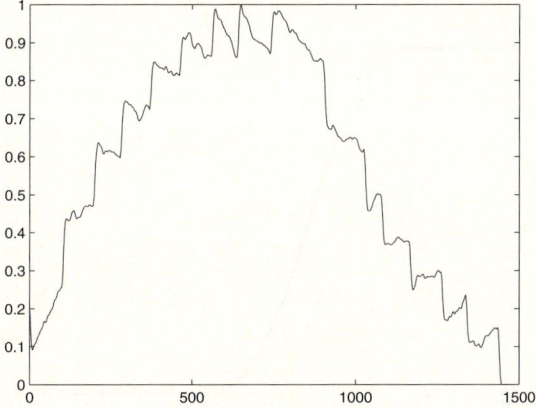

Fig. 1. An example of the EOG signal

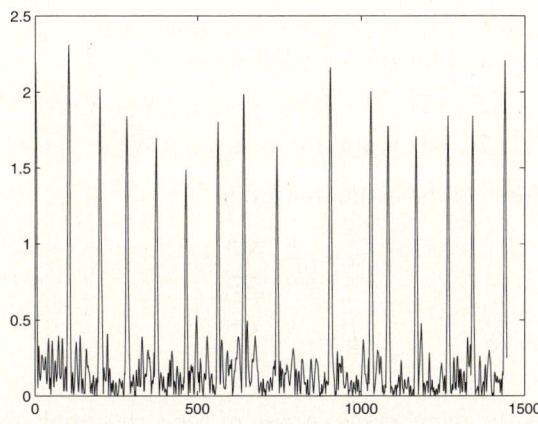

Fig. 2. The gradient of EOG signal. The amplitude has been 100 times increased due to picture readability.

work, instead of clustering data for each analyzed signal, we proposed a $\mathcal{Z}(x)$ membership function in the following form:

$$\mathcal{Z}(x) = \begin{cases} 1, & \text{if } x < a, \\ 1 - 2\left(\frac{x-a}{b-a}\right)^2, & \text{if } a \leq x \leq \frac{a+b}{2}, \\ 2\left(\frac{b-x}{b-a}\right)^2, & \text{if } \frac{a+b}{2} \leq x \leq b, \\ 0, & \text{if } x > b. \end{cases} \quad (2)$$

An example of the $\mathcal{Z}$ membership function has been presented on Fig. 3.

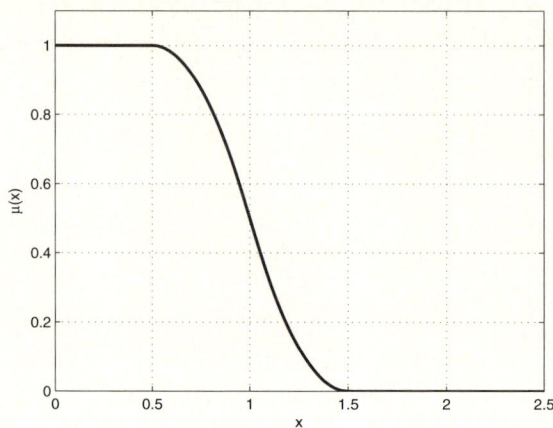

Fig. 3. $\mathcal{Z}$ membership function, where $a = 0.5$ and $b = 1.5$

The two parameters a and b can be computed in the following way: for the clustered data set $\mathcal{X}$ and for the sets defined as:

$$\begin{aligned}\mathcal{I}_1 &= \{i: \ 1 - \mu(x_i) < \varepsilon, \ 1 \leq i \leq N\}, \\ \mathcal{I}_0 &= \{i: \ \mu(x_i) < \varepsilon, \ 1 \leq i \leq N\},\end{aligned} \quad (3)$$

the parameter values can be estimated from

$$\begin{aligned} a &= \frac{1}{|\mathcal{I}_1|} \sum_{i \in \mathcal{I}_1} x_i, \\ b &= \frac{1}{|\mathcal{I}_0|} \sum_{i \in \mathcal{I}_0} x_i. \end{aligned} \quad (4)$$

The $\mu(x_i)$ denotes the membership grade of the i-th sample from the data set, $x_i \in \mathcal{X}$ and $1 \leq i \leq N$, the ε describes the tolerance limit explained further in this section. The notations $|\mathcal{I}_1|$ and $|\mathcal{I}_0|$ describe the cardinal numbers of $\mathcal{I}_1$ and $\mathcal{I}_0$ sets, respectively.

After the clustering process, the membership values are equal to one, only for these samples from the input data set that are equal to the cluster prototype (i.e. the distance between the samples and the prototype is equal to zero) [2], [3]. Hence, as the a parameter value, the obtained cluster prototype value that corresponds to *small* cluster can be applied. For the methods that utilize the kernel functions a problem of the cluster prototype value estimation occurs. Therefore, to avoid such kind of problems, we have proposed another way for estimation the a parameter value.

The presented equations (3) and (4) can be interpreted as a mean of these samples, that have the membership grades not smaller than $(1 - \varepsilon)$ for the problem of the a parameter value estimation, or the membership grades not higher than ε for the calculation of b parameter.

The segments (the saccade parts of the EOG signal) are determined as the fragments of the signal, where the membership values of the signal gradient to the *small* set are higher than 0.5 (generally, when the membership vales are higher than $1/c$, where c is the number of clusters).

3 Eye Movement Model

To build a model of eye movement we used a TSK fuzzy model [4], [5]. TSK fuzzy system generates inference results based on fuzzy if-then rules. Every fuzzy conditional statement from a rule base may be written as follows:

$$\underset{i=1,2,\ldots,I}{\forall} \quad R^{(i)} : \text{if } \underset{j=1}{\overset{t}{\text{and}}} \left(x_{0j} \text{ is } A_j^{(i)} \right) \text{ then } y^{(i)}(\boldsymbol{x}_0) = p_0^{(i)} + p_1^{(i)} x_{01} + \ldots + p_t^{(i)} x_{0t}, \tag{5}$$

where I denotes the number of fuzzy if-then rules, t is the number of inputs, x_{0j} is the j-th element of the input vector $\boldsymbol{x}_0 = [x_{01}, x_{02}, \ldots, x_{0t}]^T$, $A_j^{(i)}$ are linguistic values of fuzzy sets in antecedents, $y^{(i)}(\boldsymbol{x}_0)$ is the output of each rule and $\forall_{j=0,1,\ldots t}\ p_j^{(i)}$ are parameters of the linear function in consequents.

We perform the detection of eyes position using single samples of EOG signal and zero-order TSK model. For the assumptions made, the above formula can be rewritten in the form:

$$R^{(i)} : \text{if } \left(x_0(n) \text{ is } A^{(i)} \right) \text{ then } y^{(i)} = p_0^{(i)}, \tag{6}$$

where $x_0(n)$ is an EOG sample.

If we assume also that fuzzy sets of linguistic values in rule antecedents have Gaussian membership functions, then we can evaluate the grade of membership (the firing strength) for the i-th rule using the following formula:

$$A^{(i)}(x_0(n)) = \exp\left[-\frac{1}{2} \left(\frac{x_0(n) - c^{(i)}}{s^{(i)}} \right)^2 \right], \tag{7}$$

where $c^{(i)}$ and $s^{(i)}$, for $i = 1, 2, \ldots, I$, are membership function parameters, center and dispersion, respectively.

In the TSK system the weighted average aggregation of fuzzy rules is applied. The output of the fuzzy model y_0 is then inferred as follows:

$$y_0 = \frac{\sum_{i=1}^{I} A^{(i)}(x_0(n))\, y^{(i)}}{\sum_{i=1}^{I} A^{(i)}(x_0(n))}. \tag{8}$$

To increase the prediction quality, we modified the original TSK inference procedure: we calculate the output value using only rules for which the normalized firing strength is grater than a predefined level:

$$A_\varepsilon^{(i)}\left(x_0\left(n\right)\right) = \begin{cases} A^{(i)}\left(x_0\left(n\right)\right), & \text{if } \frac{A^{(i)}(x_0(n))}{\max\limits_{i=1}^{I}\left(A^{(i)}(x_0(n))\right)} > \varepsilon, \\ 0, & \text{otherwise}. \end{cases} \quad (9)$$

where $\varepsilon \in [0,1]$ is a pre-set value.

Finally, the output of the fuzzy system is given as:

$$y_0 = \frac{\sum\limits_{i=1}^{I} A_\varepsilon^{(i)}\left(x_0\left(n\right)\right) p_0^{(i)}}{\sum\limits_{i=1}^{I} A_\varepsilon^{(i)}\left(x_0\left(n\right)\right)}. \quad (10)$$

For $\varepsilon = 0$ we get the original TSK inference method, for $\varepsilon = 1$ only one rule, with maximum premise membership grade is taken for the output calculation.

To avoid rapid changes of the fuzzy system output value which come from the EOG signal variation being a result of saccadic eye movements we add two rules of thumbs in the inference process:

1. we allow the fuzzy system output to change only if ℓ successive EOG samples decrease $(x_0(n) < x_0(n-1) < x_0(n-2) < \ldots < x_0(n-\ell))$ or increase $(x_0(n) > x_0(n-1) > x_0(n-2) > \ldots > x_0(n-\ell))$,
2. we allow the fuzzy system output to increase (decrease) only if ℓ successive EOG samples increase (decrease).

Too little value of the ℓ parameter may lead to rapid switching of the system output, on the other hand, too high value of ℓ results in decrease of eye position detection quality. For values of ℓ in a range $[3,8]$ we did not get the significant change of the quality of eyes position detection. In our experiments we used $\ell = 3$.

To establish values of the model parameters we used the results of EOG signal segmentation. The number of the rules equals the number of segments detected in EOG signal. Values of the rules output are given by EOG signal level measured at the point when eyes reach the final position. Centers of membership functions of fuzzy sets in rules premises are given as:

$$c^{(i)} = x_0^{(i)}(f), \quad (11)$$

where $x_0^{(i)}(f)$ is the first sample of the i-th saccade part of the EOG signal, and dispersions $s^{(i)}$ are defined as a standard deviation of data in a range $\left[x_0^{(i)}(f) : x_0^{(i)}(l)\right]$, where $x_0^{(i)}(l)$ is the last sample of the i-th EOG segment.

4 Numerical Experiments

In our investigations, the real, *in vivo* acquired signal was analyzed. As an data acquisition unit we used the Biopac system hardware. All analyzed signals were not preprocessed. For registration of eye movements we constructed a segment

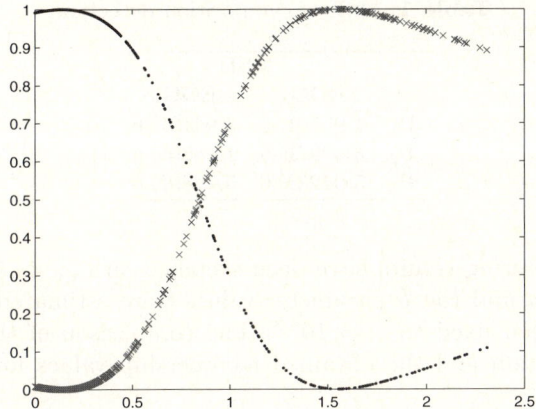

Fig. 4. The membership values for the *small* (dots), and the *large* fuzzy set (crosses)

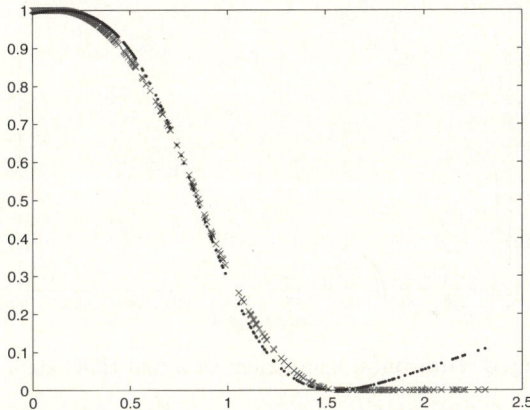

Fig. 5. The comparison between the $\mathcal{Z}$ membership (crosses) and clustering results (dots)

of LEDs which were displayed sequentially. In our experiments we used EOG signals obtained from three different persons (P_1-P_3). We recorded two separate EOG signals for the horizontal and the vertical movement of eyes.

At the first stage we applied a method of the EOG signal segmentation. Firstly, gradient signal based on (1) was estimated. Next, gradient values were clustered into two groups corresponding to the *small* and the *large* amplitude fuzzy sets. As the clustering method, we used the familiar fuzzy c–means clustering method (FCM) proposed by Bezdek [2]. The following parameters were fixed:

- the number of clusters $c = 2$,
- the fuzzyfier $m = 2$,
- the tolerance for the FCM method $\epsilon = 10^{-5}$.

Table 1. MSE of eye position detection

	MSE	
	EOG_x	EOG_y
P_1	1.9556E-6	3.9419E-6
P_2	3.9827E-6	1.7860E-6
P_3	5.0423E-6	5.7189E-6

The gradient clustering results have been sketched on Fig. 4. After the clustering process, the a and the b parameter values were estimated. Value of the ε parameter has been fixed to $\varepsilon = 10^{-3}$. The comparison of the estimated (z) membership function and the obtained membership values for the given data

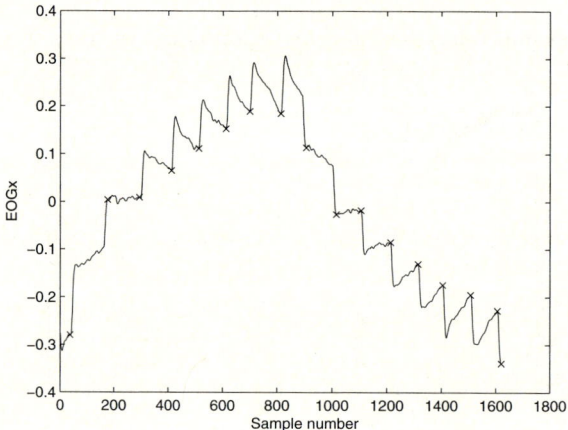

Fig. 6. Horizontal component of a real EOG signal

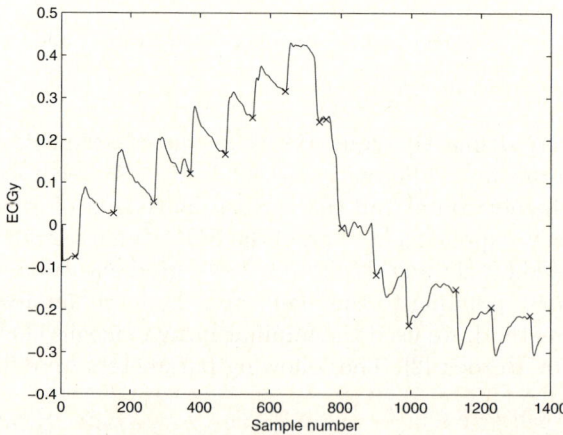

Fig. 7. Vertical component of a real EOG signal

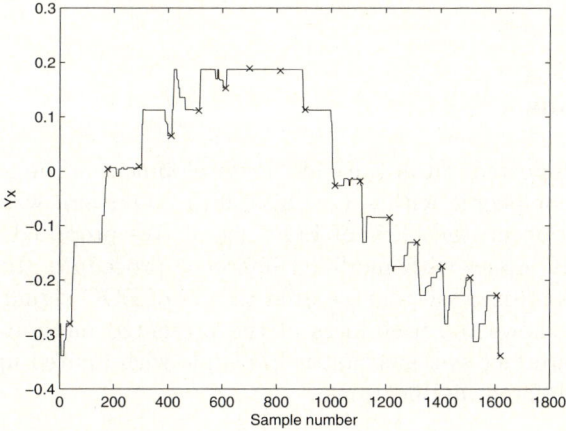

Fig. 8. Fuzzy system outputs corresponding to horizontal EOG

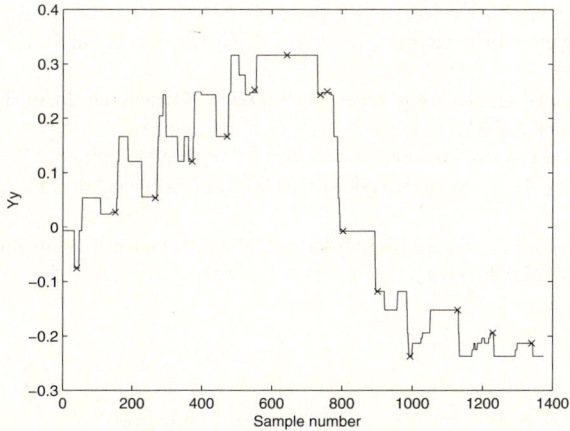

Fig. 9. Fuzzy system outputs corresponding to vertical EOG

set, are plotted on Fig. 5. On the basis of the segmentation results we established parameters' values of TSK fuzzy system. Then we applied the fuzzy system in a process of eyes position detection. We changed value of ε in a range $[0.75, 0.95]$. The lowest values of positioning errors (mean square errors) were obtained for $\varepsilon = 0.95$. Table 1 shows MSE values of the detection of eye position measured at the destination points. Fig. 6 and 7 show the horizontal (EOG_x) and vertical (EOG_y) components of a real EOG signal (P_1) respectively. We marked with symbol 'x' EOG samples at the point where an eye reached the final position. Fig. 8 and 9 show outputs of the fuzzy system correspondingly to horizontal (y_x) and vertical (y_y) EOG signal components. The results of our experiments confirm the ability of application of the proposed system for prediction of eyes position on the basis of EOG signal. The presented solution enables to reduce the negative influence of saccadic eye movements on quality of the detection of eye position.

5 Conclusions

In this paper we reported initial work at the development of the human-machine interface system for people with severe disabilities. Our aim was to detect and predict eyes position on the basis of EOG signal. We proposed an application of zero-order TSK model with modified inference procedure. To calculate values of parameters of fuzzy system we used results of EOG signal segmentation. Experimentation shows the usefulness of the presented method for improving functionality of interface systems that help people with limited upper body mobility to control different applications.

References

1. Barea, R., Boquete, L., Mazo, M., Lpoez, E., Bergasa, L.M.: E.O.G. guidance of wheelchair using neural networks. In: Proc. 15th Int. Conf. on Pattern Recognition, Barcelona (2000)
2. Bezdek, J.C.: Pattern Recognition with Fuzzy Objective Function Algorithms. Plenum, New York (1981)
3. Pedrycz, W.: Konwledge–Based Clustering. Wiley–Interscience, Chichester (2005)
4. Sugeno, M., Kang, G.T.: Structure identification of fuzzy model. Fuzzy Sets Syst. 28, 15–33 (1988)
5. Takagi, T., Sugeno, M.: Fuzzy identification of systems and its applications to modeling and control. IEEE Trans. Syst., Man Cybern. 15, 116–132 (1985)

An NLP-Based 3D Scene Generation System for Children with Autism or Mental Retardation

Yılmaz Kılıçaslan, Özlem Uçar, and Edip Serdar Güner

Department of Computer Engineering, University of Trakya,
Ahmet Karadeniz Yerleşkesi, 22030 Edirne, Turkey
{yilmazkilicaslan, ozlemucar, eserdarguner}@trakya.edu.tr
http://tbbt.trakya.edu.tr/

Abstract. It is well-known that people with autism or mental retardation experience crucial problems in thinking and communicating using linguistic structures. Thus, we foresee the emergence of text-to-image conversion systems to let such people establish a bridge between linguistic expressions and the concepts these expressions refer to via relevant images. S2S is such a system for converting Turkish sentences into representative 3D scenes via the mediation of an HPSG-based NLP module. A precursor to S2S, a non-3D version, has been tested with a group of students with autism and mental retardation in a special education center and has provided promising results motivating the work presented in this paper.

1 Introduction

The use of educational technologies for supporting the education of disabled children continues to increase both in quantity and quality. Particularly with personal computers that have advanced by leaps and bounds and become cheap enough to be ubiquitous in the last thirty years, software technology offers great opportunities for disabled people to communicate and socialize.[1] Griswold et al. [7] list 'poor comprehension of abstract concepts' among other cognitive problems exhibited by individuals with autism. It is also observed that some people with mental retardation experience the same sort of problems: as they have difficulties in grasping abstract concepts, they tend to think in terms of concrete visual images rather than linguistic expressions [14]. A crucial claim bearing particular relevance to the matter under discussion is that computer-based technologies have a potential to significantly alleviate educational and social and/or communicative handicaps that individuals with mental retardation or autism face ([8], [15]).

To this effect, we have developed a series of software programs as part of a research project[2] to assist the education and training of children with autism

---
[1] See [13] for a comprehensive and detailed account of computer-based technologies that can serve to include people with disabilities into the mainstream of society.
[2] The University of Trakya Scientific Research Project Office supported this work with grant number TUBAP-760.

and mental retardation.[3] In the first and second cycles of the project, two modules were consecutively developed to map words to images and sentences to 2D pictures ([11]). Several experiments have been performed using these modules with 88 children in Yağmur Çocuklar Psychological Counseling and Special Education Center, Istanbul. An improvement between 20 to 25% was observed in the learning performances of the children when they were assisted with the modules.[4] These results and suggestions by the trainers encouraged us to move to the third cycle of project in order to incorporate a 3D scene generator to the system. This latest version of the software awaits to be tested with children.

Except for two previous works of ours, to the best of our knowledge, there is no work based on a notion of conversion from natural language sentences to scenes developed with the aim of assisting the education of children suffering from autism or mental retardation. However, there is an abundance of work reported to be about scene generation based on natural language. Even though it did not have a graphics component, the SHRDLU system [16] was one of the early AI programs that successfully used natural language to move various objects around in a closed virtual world. The system by Adorni et al [1] was intended to imagine a static scene described by means of a sequence of simple phrases, focusing particularly on the principles of equilibrium, support and object positioning. The Put system by Clay & Wilhelms [3] allowed spatial arrangements of existing objects on the basis of an artificial subset of English consisting of expressions of the form *Put(X P Y)*, where X and Y are objects, and P is a spatial preposition. The WordsEye system by Coyne & Sproat [4], which is currently under active development, is a major improvement over the restrictions of the preceding systems. It is not confined to a closed world but provides a blank slate where the user can paint a picture with a text describing not only spatial relations but also actions performed by objects in the scene. S2S comes close to Put in limiting object configuration to spatial relations and it is like WordsEye in allowing entirely natural linguistic expressions as input. However, S2S generates considerably less complicated scenes compared to WordsEye. This is partially a shortcoming of a system which is still under development but also a requirement imposed by the current field of application, namely assisting the education of disabled children not capable of grasping complicated configurations.

2   Architectural Design

S2S is composed of four components: *a Lexical Preprocessor, a Natural Language Processor, a Scene Generator,* and *a Renderer.* The following diagram shows the interaction of these components in terms of data flow between them:

---

[3] The precursory work was disseminated in a special issue of the IEEE Journal of Pervasive Computing dedicated to works in progress in healthcare systems and other applications ([5]).

[4] See [15] for a detailed discussion of the experimental results.

An NLP-Based 3D Scene Generation System for Children

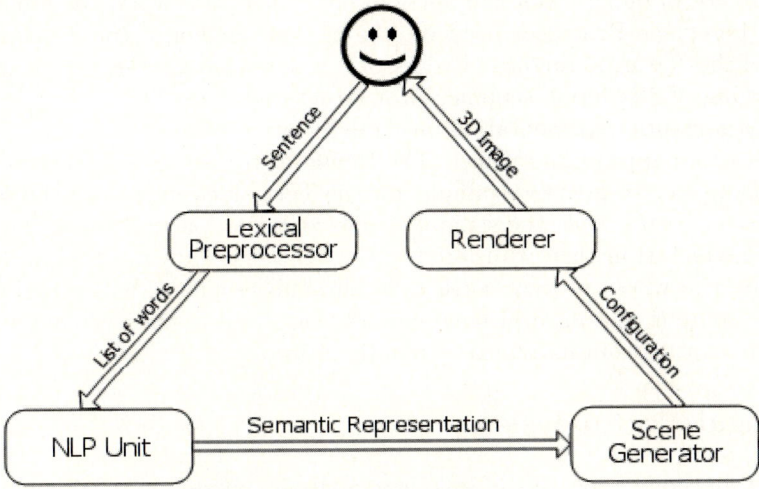

Fig. 1. The architectural design of S2S

$$\begin{bmatrix} \text{ENTITY\_1} & \text{chair} \\ \text{ENTITY\_2} & \text{table} \\ \text{RELATION} & \text{on} \end{bmatrix}$$

Fig. 2. Semantic representation for *The chair is on the table*

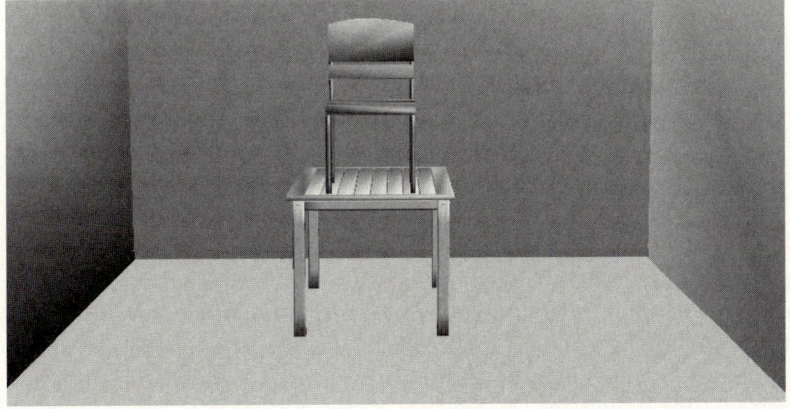

Fig. 3. The scene generated for *The chair is on the table*

The Lexical Preprocessor transforms the given natural language sentence into a list containing the words of the sentence with any capital letters and punctuation removed away. As the aspects of interpretation encoded by capital letters and any kinds of punctuation fall outside the scope of this work, the input to the Natural Language Processor need be free of these orthographic elements. The output of the Natural Language Processor is a semantic representation encoding the meaning of the input sentence in a structured way. The Scene Generator decodes a semantic representation into a description of scene formulated by parameters set to appropriate values. The Renderer translates a scene description into a 3D image. To give an example, for the Turkish equivalent of the sentence *The chair is on the table* the semantic representation partially shown in Fig. 2 is yielded which is in turn translated to the 3D scene shown in Fig. 3.

In sum, the whole process consists of an analysis phase, which extracts the semantic content of a natural language sentence, and a synthesis phase, which builds up a scene from individual semantic objects.

3 Linguistic Analysis

The main burden of the task of linguistic analysis, parsing natural language expressions in order to extract their semantic content, is carried by the Natural Language Processor. The internal structure of this component is as shown in Fig. 4.

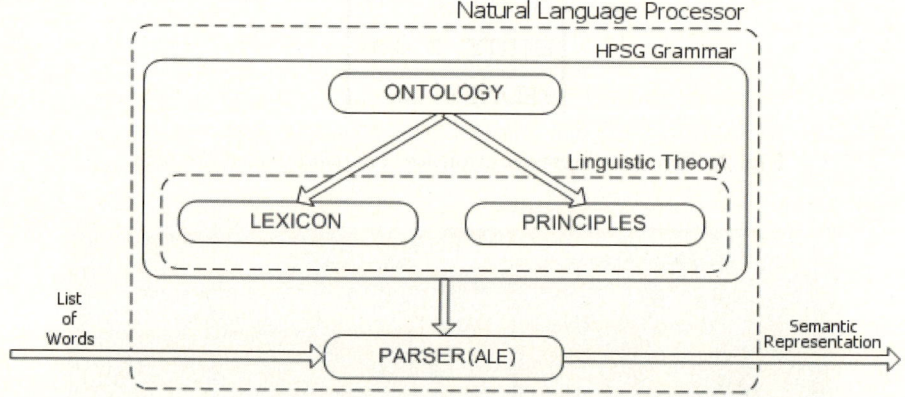

Fig. 4. The internal structure of the Natural Language Processor

The Natural Language Processor consists of two main parts: a parser and a grammar. The sub-component responsible for the parsing process is the Attribute Logic Engine (ALE) (version 3.2.1), which is an integrated phrase structure parsing and definite clause logic programming system in which the data structures are typed feature structures [2]. Feature structures serve as the main representational device in the framework adopted in this study. A feature structure consists of two pieces of information: a type (which every feature structure must have) and a finite set of feature-value pairs (which can possibly be empty) (see Fig. 2). A feature-value pair is defined recursively, where the value itself is a feature structure which can also be an atomic object. An important operation defined over pairs of feature structures is unification, which refers to an operation which has gained widespread recognition as a general tool in computational linguistics since Kay's [9] seminal work. If two feature structures are consistent, unifying them results in a feature structure subsuming the information contained in both of them; otherwise, the operation fails.[5]

The parsing process is driven by a Head-driven Phrase Structure Grammar (HPSG) [12] which we have designed and implemented in order to handle a fairly large fragment of Turkish.[6] An HPSG grammar can be split into three units: an ontology, a lexicon and a set of principles. The ontology is a hierarchically organized inventory of universally available types of linguistic entities, together with a specification of their appropriate features and their value types. The lexicon is a system of lexical entries and lexical rules. The principles include universal and language specific constraints which every linguistic structure to be generated by the system must obey. Fig. 5 shows a description of the feature structure which the system assigns to the sentence *The chair is on the table*:

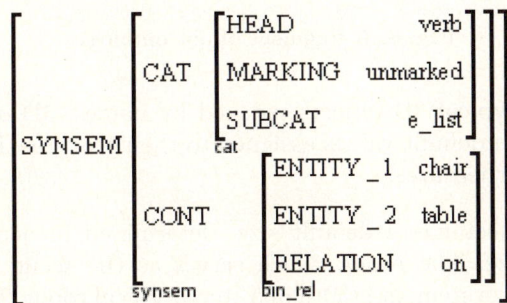

Fig. 5. The feature structure assigned to *The chair is on the table*

The SYNSEM (SYNTAX-SEMANTICS) feature includes a complex of syntactic and semantic information about the modeled linguistic sign. The CAT (CATEGORY) feature encodes syntactic information whereas the CONT (CONTENT) value constitutes the sign's contribution to context-independent aspects of the semantic interpretation. Every semantic value must be of a type subsumed by *sem_obj* (*semantic_object*). A fragment of the semantic type hierarchy utilized in our grammar is given in Fig. 6.

4 Scene Synthesis

The Scene Generator extracts out of the CONT value of a sentence a configuration of possibly underspecified 3D entities together with the attributes they bear and the relations they stand to each other. Each entity is searched in a

---

[5] See [2] for a detailed discussion of strong typing, feature structures and unification as implemented in the Attribute Logic Engine (ALE).
[6] See [10] for a semantico-pragmatically oriented grammar of a fragment of Turkish developed within a modified version of the HPSG formalism.

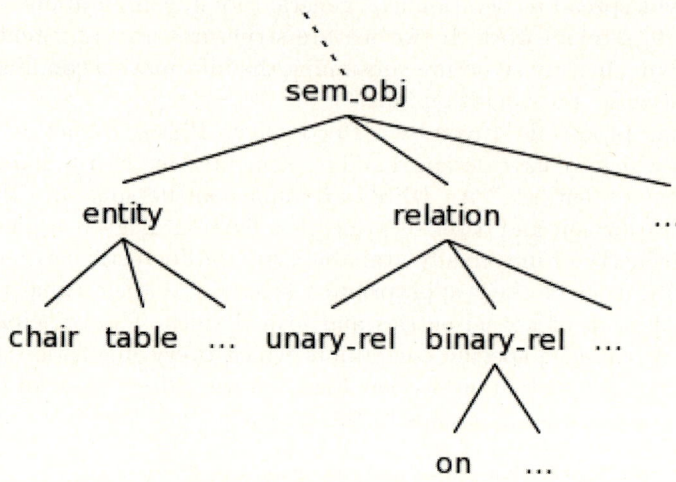

Fig. 6. A fragment of our ontology

database of stereotypical 3D objects indexed by names. 3D objects are stored in the database with default values assigned to their features. Each 3D object is associated with four features:

- **Size:** Each object has a default size, determined in accordance with its stereotypical size. The room, which serves as the scene for all situations described in the system, is 1/50th of a stereotypical room. The default size of each object is scaled accordingly. In addition to the default size, each object is also associated with the ratios by which it is to be scaled when its size is specified with adjectives, such as *big* and *small*. However, a lower bound is placed on the size of each object in order to preclude counterintuitive cases that might, for instance, arise with the use of comparative adjectives like *smaller* and *much smaller*.
- **Color:** Each object comes with a default color, which is encoded in the corresponding RGB format. Objects are allowed to change color depending on relevant specifications in input sentences. To the extent that it is available in the conventionally used list of colors (e.g. red, blue, yellow, lilac), the color feature can take any value.
- **Texture:** Each object is covered with a default texture, which is stored as a bitmap image. However, like the other features, the texture of an object can be re-specified in the input expression (e.g. *tartan ball*, *striped ball*).
- **Spatial tags:** The exact depiction of spatial relations requires the shapes of the objects in question to be known. To this effect all objects in the database are associated with spatial tags such as *canopy area, top surface, back surface, front surface, base,* and *cup*. Each tag is further associated with a size feature to encode the information concerning its dimensional measures.

An NLP-Based 3D Scene Generation System for Children   935

Fig. 7. The chair is to the left of the table

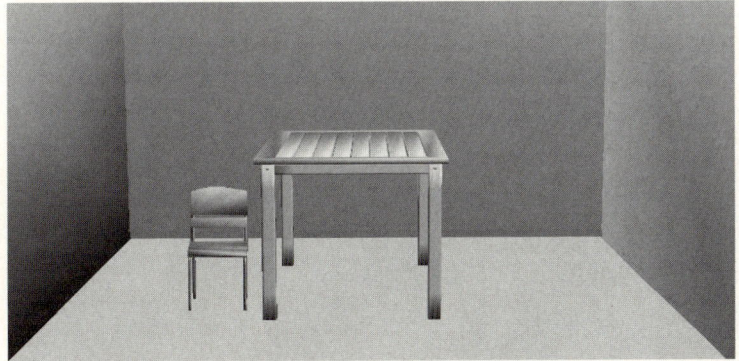

Fig. 8. The small chair is to the left of the table

The default values of objects are kept unchanged if not specified in the input sentence; otherwise, they are overridden in favor of the specified value. Fig. 7 and 8 show the scenes corresponding to the sentences *The chair is to the left of the table* and *The small chair is to the left of the table*, respectively.

The configuration of objects is restricted to spatial relations. These are the relations denoted by prepositions like *in, on, under, left to, right to, in front of, behind,* and *next to*.[7] In fact, as discussed by [6], some reverse spatial relations can be defined in one integral relationship function. The *on-under, left-right* and *in front of-behind* pairs are defined as such relations in our framework.

Another important fact is that the exact positioning of objects relative to each other requires their shapes and surfaces to be taken into consideration. For instance, the ball in the scene referred to by the sentence *The ball is on the chair* will occupy the top surface area of the chair rather than the top of its back. (See Fig. 9)

---

[7] The Turkish equivalents of English prepositions are right adjoined to their associated noun phrases, i.e. they are postpositions.

Fig. 9. The ball is on the chair

Fig. 10. The chair is under the table

Moreover, the Scene Generator shows the flexibility of resizing an object if it does not fit in a specified area. For example, if a chair is instructed to be placed in the canopy area of a table (e.g. in the intended reading of the sentence *The chair is under the table*), an appropriately sized version of the chair is employed. (See Fig. 10)

However, resizing objects in this way should not lead to scenes conflicting with the common–sense understanding of the world. As the system is intended to assist the education and training of children with autism or mental retardation, equipping it with a capacity to generate non–common–sensical situations (e.g. a situation described by the sentence *The house is under the table*) would not be in agreement with this intention. Therefore, when the size of an object falls below the lower bound specified in the database for this object, the system will prompt an appropriate warning message, rather than producing the scene.

The outcome of the process described above is a configuration data ready to be rendered. The Renderer (which runs in 3D Developer Studio v.8.0 by

3DSTATE) depicts object configurations starting from a default position in a room environment.

5 Conclusion

Many children with autism or mental retardation experience difficulties in thinking using abstract concepts. It is widely acknowledged that with the aid of methods of visual education, the abstractness of linguistic structures can be circumvented, and learning and understanding with all senses can be promoted. We believe S2S is a new and promising approach to assisting the thinking and learning process of the autistic or mentally retarded with visually concrete representations. It should be emphasized that S2S is not intended to replace 3D software tools, but rather to augment such technology in two dimensions: 1) incorporating Turkish as the natural language to be processed and 2) using the software in the field of special education.

Acknowledgments

We are grateful to the University of Trakya Scientific Research Project Office for providing all sorts of support for the realization of the work. We are indebted to Algı Special Education and Rehabilitation Center, Yağmur Çocuklar Psychological Counseling and Special Education Center and Armağan Dönertaş Education, Rehabilitation and Research Center for Disabled Children for very valuable comments and co-operation at different stages of this work. We are also thankful to the Scientific and Technological Research Council of Turkey (TUBITAK) for supporting one of the authors through scholarship in conducting his MSc research covering the NLP part of the work.

References

1. Adorni, G., Di Manzo, M., Giunchiglia, F.: Natural Language Driven Image Generation. In: COLING 1984, pp. 495–500 (1984)
2. Carpenter, B., Penn, G.: The Attribute Logic Engine User's Guide, Version 3.2.1. University of Toronto (2001), http://www.cs.toronto.edu/~gpenn/ale.html
3. Clay, S.R., Wilhelms, J.: Put: Language-Based Interactive Manipulation of Objects. IEEE Computer Graphics and Applications, 31–39 (1996)
4. Coyne, B., Sproat, R.: WordsEye: An Automatic Text-to-Scene Conversion System. In: Siggraph Proceedings (2001)
5. Dröes, R.M., Mulvenna, M., Nugent, C., Finlay, D., Donnelly, M., Mikalsen, M., Walderhaug, S., van Kasteren, T., Kröse, B., Puglia, S., Scanu, F., Migliori, M.O., Uçar, E., Atlig, C., Kılıçaslan, Y., Uçar, Ö., Hou, J.: Healthcare Systems and Other Applications. IEEE Pervasive Computing 6(1), 59–63 (2007)
6. Durupınar, F., Kahramankaptan, U., Cicekli, I.: Intelligent indexing, retrieval and construction of crime scene photographs. In: Proceedings of the 13th Turkish Symposium on Artificial Intelligence and Neural Networks, pp. 297–306 (2004)

7. Griswold, D.E., Barnhill, G.P., Myles, B.S., Hagiwara, T., Simpson, R.L.: Asperger syndrome and academic achievement. Focus on Autism and Other Developmental Disabilities 17(2), 94–102 (2002)
8. Jacklin, A., Farr, W.: The computer in the classroom: how useful is the computer as a medium for enhancing social interaction with young people with autistic spectrum disorders? British Journal of Special Education (research section) 32(4), 209–217 (2005)
9. Kay, M.: Functional Grammar. In: Proceedings of the Fifth Annual Meeting of the Berkeley Linguistic Society, pp. 142–158 (1979)
10. Kılıçaslan, Y.: A Form-Meaning Interface for Turkish. Ph.D. Dissertation. University of Edinburgh (1998)
11. Kılıçaslan, Y., Uçar, Ö., Güner, E.S., Bal, K.: An NLP-Based Assistive Tool for Autistic and Mentally Retarded Children: an Initial Attempt. Trakya University Journal of Science 7(2), 101–108 (2006)
12. Pollard, C., Sag, I.A.: Head-Driven Phrase Structure Grammar. University of Chicago Press and CSLI Publications, Chicago, London (1994)
13. Poole, J.B., Skymcilvain, E., Jackson, L., Singer, Y.: Education for an Information Age: Teaching in the Computerized Classroom, 5th edn. (2005)
14. Turkish Foundation of Support and Education for Autistics Periodical (TODEV), vol. 2 (2002)
15. Uçar, Ö.: Development of Artificial Intelligence-Based Assistive Tools for the Education of Disabled Children. Unpublished PhD Thesis, University of Trakya (In Turkish) (2007)
16. Winograd, T.: Understanding Natural Language. PhD thesis, Massachusetts Institute of Technology (1972)

On Using Energy Signatures in Protein Structure Similarity Searching

Bożena Małysiak, Alina Momot, Stanisław Kozielski, and Dariusz Mrozek

Silesian University of Technology, Department of Computer Science,
Akademicka 16, 44-100 Gliwice, Poland
bozena.malysiak@polsl.pl, alina.momot@polsl.pl,
stanislaw.kozielski@polsl.pl, dariusz.mrozek@polsl.pl

Abstract. The analysis of small molecular substructures (like enzyme active sites) in the whole protein structure can be supported by using methods of similarity searching. These methods allow to search the 3D structural patterns in a database of protein structures. However, the well-known methods of fold similarity searching like VAST or DALI are not appropriate for this task. Methods that benefit from a dependency between a spatial conformation and potential energy of protein structure seem to be more supportive. In the paper, we present a new version of the EAST (Energy Alignment Search Tool) algorithm that uses energy signatures in the process of similarity searching. This makes the algorithm not only more sensitive, but also eliminates disadvantages of previous implementations of our EAST method.

1 Introduction

Proteins are an important class of biological macromolecules that play a key role in all biological reactions in living cells. They are involved in many cellular processes, e.g.: reaction catalysis, energy storage, signal transmission, maintaining of cell structure, immune response, transport of small biomolecules, regulation of cell growth and division [1], [2].

The appropriate activity of proteins in cellular reactions usually depends on their spatial conformations that determine physical, chemical and biological properties of reacting molecules. The conformation of a molecule is the arrangement of its atoms in space, which can be changed by a rotation around single bonds [3]. Therefore, the knowledge of the general protein structure and how proteins fold and obtain their stable states allows to understand and to predict the protein function in organisms. Moreover, the further detailed analysis of protein structures and possible conformations allows to model the protein function and behavior, which is important in drug design processes, and plan intended modifications of the structure to influence the function and activity, e.g. in protein engineering. For all these reasons, the analysis of protein structures became very important to study the entire chains of complex processes proteins are involved in [1]. For example, the activity of enzymes in catalytic reactions depends on an exposition of some typical parts of their 3D structures, called active sites [1], [4].

Conformation and chemical features of active sites allow to recognize and to bind substrates during the catalysis [2], [5]. For these reasons, the study of active sites and spatial arrangement of their atoms is essential while analyzing the activity of proteins in particular reactions [4]. One of the key components of the analysis is a comparison of one protein structure to other structures (e.g. stored in a database). This can be supported by methods of structural similarity searching. Having a group of proteins indicating a strong similarity of selected structural parts, one can explore the atomic arrangement of these fragments that take part in respective reactions. Techniques of similarity searching allow to seek the 3D structural patterns in the database of protein structures. Unfortunately, this is a very complicated task because of three reasons:

1. proteins are very complex, usually composed of thousands of atoms;
2. the searching process is usually carried out through the comparison of a given structure to all structures in the database;
3. the number of protein structures in databases, like PDB (Protein Data Bank) [6] rises exponentially every year and now amounts to 46 377 (October 9, 2007).

The first problem is usually solved by decreasing the protein structures complexity in the search process. The most popular methodologies developed so far are based on various representations of protein structures in order to reduce the search space. A variety of structure representations was proposed so far, e.g. secondary structure elements (SSE) in VAST [7], locations of the C_α atoms of a protein body and intermolecular distances in DALI [8], aligned fragment pairs (AFPs) in CE [9], or 3D curves in CTSS [10], and many others. These methods are appropriate for homology modeling or function identification. During the analysis of small parts of protein structures that can be active sites in cellular reactions it is required to use more precise methods of comparison and searching. There is also a group of algorithms of similarity searching based on the atomic potentials. They use Molecular Interaction Potentials (MIPs) or Molecular Interaction Fields (MIFs), e.g. [11], [12], [13]. MIPs/ MIFs are results of interaction energies between the considered compounds and relevant probes [12]. This group of algorithms is usually used in the process of drug design.

Relations between the chemical variety of the protein sequence, the structure and the shape of energy are described by the theory of molecular mechanics [14]. We designed and implemented the EAST (Energy Alignment Search Tool) algorithm, which benefits from the dependency between the protein structure and the conformational energy of the structure. The algorithm was presented in our previous works [15], [16]. In our research, we calculate so called *energy profiles* (EPs) that are distributions of various potential energies along amino acid chains of proteins. Energy profiles reduce the structure complexity in a similarity search process. In the paper, we extend the definition of energy profiles by introducing the *energy signature* term. We further developed our EAST algorithm incorporating energy signatures and thus, eliminating weaknesses of the previous implementations. In the paper, we briefly describe the idea of energy profiles and energy signatures (section 2), and a new version of the EAST algorithm (section 3). In section 4 we give an example of the system usage and a short discussion on basic advantages of the new EAST in comparison to its previous implementation.

2   Protein Structure Energy Signatures

In the section we describe the terms of energy characteristics, energy profiles and energy signatures. Since energy profiles represent protein structures all calculations of energy profiles are based on Cartesian coordinates of small groups of atoms that constitute each amino acid in a protein molecule polypeptide chain.

Let the $R = (r_1, r_2, \ldots, r_m)$ be a sequence of m residues in the polypeptide chain, and $X_i^{n_i}$ be a set of atomic coordinates building the ith residue r_i (n_i is a number of atoms of the the ith residue r_i depending on the type of the residue), then the simplified protein structure can be expressed as a sequence of small groups of atoms $X = (X_1^{n_1}, X_2^{n_2}, \ldots, X_m^{n_m})$.

The *energy characteristics* or *energy distribution* E^t, where t is a type of energy, is a sequence of energy values - *energy points* e^t calculated for groups of atoms $X_i^{n_i}$ constructing the consecutive residues r_i in the protein polypeptide chain R:

$$E^t = (e_1^t, e_2^t, \ldots, e_m^t), \tag{1}$$

where e_i^t is an energy point of the ith residue, t is a type of energy. Since we have several types of potential energy we have many energy characteristics that contribute to the energy profile.

We define the single protein *energy profile EP* as a sequence of energy characteristics of various types of energy, determined for a given protein structure:

$$EP = (E^{st}, E^{ben}, E^{tor}, E^{vdw}, E^{el}, E^{tot})^T, \tag{2}$$

where E^{st} denotes a distribution of the bond stretching energy, E^{ben} is a distribution of the angle bending component energy, E^{tor} is a distribution of the torsional angle energy, E^{vdw} is a distribution of the van der Waals energy, E^{el} denotes a distribution of the electrostatic energy. E^{tot} is a distribution of the total energy which is a summary of all component energies in each residue.

Thus, the energy profile for a single protein structure can be presented in a form of the matrix:

$$EP = \begin{bmatrix} e_1^{st} & e_2^{st} & \ldots & e_m^{st} \\ e_1^{ben} & e_2^{ben} & \ldots & e_m^{ben} \\ \ldots & & & \\ e_1^{tot} & e_2^{tot} & \ldots & e_m^{tot} \end{bmatrix}, \tag{3}$$

where each row contains energy characteristics of a single type of energy, while the column of the matrix will be called *energy signature* of a single residue. Therefore, the energy signature of ith residue r_i described as e_i^S is a sequence of consecutive energy types for the residue:

$$e_i^S = (e_i^{st}, e_i^{ben}, \ldots, e_i^{tot})^T. \tag{4}$$

One energy signature represents all energy features of one residue in the protein structure and can be treated as a vector in the 6-dimensional space. One energy characteristics represents one type of energy for all residues in a the protein structure. One energy profile represents all energy features for the whole molecule. It is worth noting that the single protein *energy profile EP* can be also treated as a sequence of energy signatures of consecutive residues (E^S):

$$EP = E^S = (e_1^S, e_2^S, \ldots, e_m^S). \tag{5}$$

This representation reduces the search space during a comparison of two protein structures. In our approach, we compute energy profiles based on the protein atomic coordinates retrieved from the public, macromolecular structure database Protein Data Bank (PDB) [6]. During the calculations we use TINKER [17] application of molecular mechanics and Amber [18] force field which is a set of physical-chemical parameters. We had computed complete energy profiles for about 5 000 protein structures from the PDB and we stored them in a special database. For this purpose, we designed and developed the Energy Distribution Data Bank (EDB). The EDB is a foundation for the similarity search task. With the use of the EDB protein structures can be compared to each other on the basis of their energy profiles, in order to find strong structural similarities, places of discrepancies, or possible mutations.

3 Searching with the Use of Energy Signatures

We can use energy profiles to search structurally similar proteins or search just some particular parts of their structures. The search process is performed on the energy level with the use of a new version of the EAST (Energy Alignment Search Tool) algorithm and profiles stored in the EDB. The similarity searching is realized through the comparison and *alignment* of a given query protein energy profile and each candidate protein profile from the EDB. In the old version of the EAST algorithm an energy profile was represented by only one chosen energy characteristics. In contrast with the old version in the new one we consider energy signatures. Therefore, the **alignment** can be thought as a juxtaposition of two sequences of energy signatures that gives the highest number of identical or similar signatures (corresponding to residues). The word *similar* means that energy signatures do not have to be identical but they should indicate some similarity with the given range of tolerance. As a consequence of a suitable juxtaposition on the basis of appropriate similarity measures, it is possible to evaluate a similarity degree of compared molecules (profiles) and optionally find regions of dissimilarity. The EAST algorithm (Fig. 1) examines identities, similarities (both are qualified as matches) and disagreements (mismatches) of compared energy signatures. Analogically to the nucleotide or amino acid sequences alignment, some mismatching positions and gaps can appear in the best alignment of energy profiles. These mismatches and gaps take into account possible evolutionary changes in different organisms.

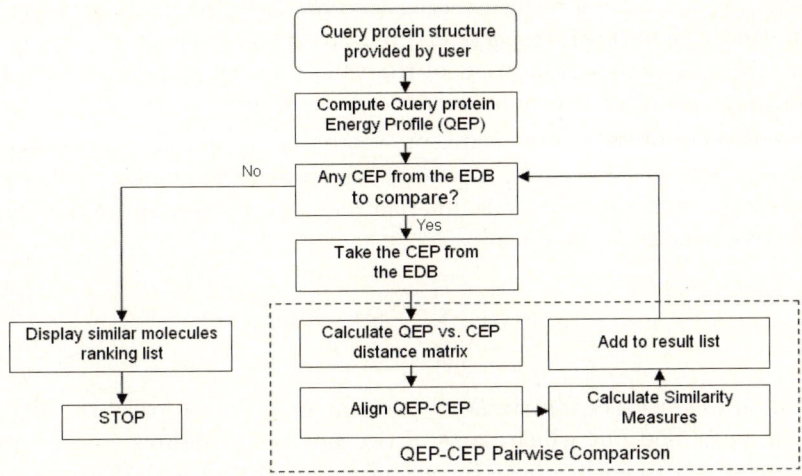

Fig. 1. Overview of the similarity search process with the EAST algorithm

During the search process with the use of our EAST algorithm a user specifies a query protein (or a part of a molecule) with a known structure (Fig. 1). This query-structure is then transformed into the energy profile (EP).

Query-molecule EP (QEP) can be compared with other energy profiles stored in the EDB. This is a pairwise comparison - each pair is constituted by the query energy profile QEP and the energy profile of one molecule from the EDB (candidate energy profile, CEP). The pairwise comparison is repeated for each possible pair QEP-CEP (a number of candidate CEPs can be narrowed by the Preselection phase [15]). In the pairwise comparison phase, both QEP and CEP are seen as sequences of energy signatures $E_Q^S = (e_{Q,1}^S, e_{Q,2}^S, \ldots, e_{Q,n}^S)$ and $E_C^S = (e_{C,1}^S, e_{C,2}^S, \ldots, e_{C,n}^S)$. This super phase consists of several sub-phases.

During the pairwise comparison we build a distance matrix D similar to that presented in [19]. However, on the contrary to the previous version of the EAST, the matrix is used to check the distances between energy signatures (not energy characteristics). The energy signature distance matrix allows to compare all energy signatures of the query molecule $E_Q^S = (e_{Q,1}^S, e_{Q,2}^S, \ldots, e_{Q,n}^S)$ (QEP) to all energy signatures of the candidate molecule $E_C^S = (e_{C,1}^S, e_{C,2}^S, \ldots, e_{C,n}^S)$ (CEP). In the energy signature distance matrix D, a single cell entry $d_{s,ij}^{QC}$ denotes the distance between the energy signature of the ith residue of query protein Q and the energy signature of the jth residue of candidate protein C:

$$d_{s,ij}^{QC} = \left(\frac{(e_{Q,i}^{str} - e_{C,j}^{str})^2}{p_{str}} + \frac{(e_{Q,i}^{bend} - e_{C,j}^{bend})^2}{p_{bend}} + \frac{(e_{Q,i}^{tor} - e_{C,j}^{tor})^2}{p_{tor}} + \frac{(e_{Q,i}^{vdw} - e_{C,j}^{vdw})^2}{p_{vdw}} + \frac{(e_{Q,i}^{coul} - e_{C,j}^{coul})^2}{p_{coul}} \right)^{\frac{1}{2}}, \quad (6)$$

where $e_{Q,i}^t$, $e_{C,j}^t$ are component energies (of type t) in the query and candidate protein energy signatures, respectively; parameters p_{str}, p_{bend}, p_{tor}, p_{vdw}, p_{coul} are divisors computed on the basis of the values described later in the section. Actually, $e_{Q,i}^t$, $e_{C,j}^t$, are coordinates of energy signature vectors and the value of $d_{s,ij}^{QC}$ is a distance of two energy signature vectors in the 5-dimensional space (the e^{tot} term is not considered). In a general case, when the number of component energies in the energy signature is different (equal to k) we can use the following formula to evaluate the distance:

$$d_{s,ij}^{QC} = \sqrt{\sum_{t=1}^{k} \frac{(e_{Q,i}^t - e_{C,j}^t)^2}{p_t}}. \qquad (7)$$

Based on the distance matrix D we perform an optimization of the alignment path. We must find out which parts of two protein structures (represented by EPs) correspond to each other. To optimize the alignment path we use modified, energy-adapted Smith-Waterman method (originally published in [20]) that produces the similarity matrix S according to the following rules:

$$\begin{aligned}
S_{i0} &= S_{0j} = 0, \\
S_{ij}^{(1)} &= S_{i-1,j-1} + \vartheta(d_{s,ij}^{QC}), \\
S_{ij}^{(2)} &= \max_{1 \leq k \leq n} \{S_{i-k,j} - \omega_k\}, \\
S_{ij}^{(3)} &= \max_{1 \leq l \leq m} \{S_{i,j-l} - \omega_l\}, \\
S_{ij}^{(4)} &= 0, \\
S_{ij} &= \max_{v=1,2,3,4} \{S_{i,j}^v\},
\end{aligned} \qquad (8)$$

for $1 \leq j \leq m$ and $1 \leq i \leq n$, where ω_k and ω_l are gap penalties for horizontal and vertical gaps of length k and l, respectively, and $\vartheta(d)$ is a function which takes a form of similarity award $\vartheta^+(d_{s,ij}^{QC})$ for matching energy signatures ($e_{Q,i}^S$ and $e_{C,j}^S$) or a form of mismatch penalty $\vartheta^-(d_{s,ij}^{QC})$ in the case of mismatch.

Default settings for the energy-adapted Smith-Waterman method are: similarity award $\vartheta^+(d_{s,ij}^{QC}) = 1$ for the Smith-Waterman method in standard version [16] and $\vartheta^+(d_{s,ij}^{QC}) \in \langle 0, 1 \rangle$ for the Smith-Waterman method in fuzzy version [16], mismatch penalty $\vartheta^-(d_{s,ij}^{QC}) = -1/3$, affine [21] gap penalty $\omega_k = 1.2 + k/3$, where k is a number of gaps and $d_{s,ij}^{QC}$ is a single cell value of the distance matrix D. The similarity award $\vartheta^+(d_{ij}^{AB})$ for a match depends on distance $d_{s,ij}^{QC}$.

The match/mismatch can be resolved on the basis of the distance $d_{s,ij}^{QC}$ between the considered signatures and the additional parameter d_0^S called cutoff value. In the implementation of the EAST algorithm, the **cutoff value** d_0^S means the highest possible difference between energy signatures $e_{Q,i}^S$ and $e_{C,j}^S$ when we treat them as similar. The cutoff value determines a range of tolerance for energy discrepancies. If $d_{s,ij}^{QC} \leq d_0^S$, then two energy signatures match to each other. If $d_{s,ij}^{QC} > d_0^S$, then two energy signatures do not match (mismatch).

In the previous version of the EAST algorithm [15], [19] we used only one chosen energy type in the similarity searching. Therefore, we compared and calculated a distance between energy points from energy characteristics of two molecules (not energy signatures and entire EPs). Consequently, cutoff values d_0^t for various types of energy t were different. Recommended defaults of cutoff values d_0^t for different energies are presented in table 1 [19].

Table 1. Recommended ranges of cutoff values d_0^t and default values of the cutoff d_0^t for searching with the use of each type of energy characteristics of the EP

Component energy of EP	Cutoff value d_0^t [Kcal/mole] from	to	Default value of cutoff d_0^t [Kcal/mole]
bond stretching	0.6	1.6	1.0
angle bending	0.8	4.0	1.6
torsional angle	0.8	1.9	1.4
van der Waals	3.1	7.1	5.1
electrostatic	2.4	8.4	6.1

The cutoff value d_0^S for energy signatures and values of divisors p_{str}, p_{bend}, p_{tor}, p_{vdw}, p_{coul} are computed on the basis of cutoff values d_0^t for the particular energy type t and participation weights predetermined individually for each component energy in the energy profile, according to the following deduction.

Let $d_{t,ij}^{AB}$ be the difference between the energy value of ith energy point $e_{A,i}^t$ (of an energy characteristic of type t of a molecule A) and the energy value of jth energy point $e_{B,j}^t$ (of an energy characteristic of type t of a molecule B), computed according to the expression $d_{t,ij}^{AB} = |e_{A,i}^t - e_{B,j}^t|$, then

$$\left(d_{t,ij}^{AB}\right)^2 < \left(d_0^t\right)^2, \qquad (9)$$

where d_0^t is cutoff value for energy characteristic of type t. Since d_0^t is positive

$$\frac{\left(d_{t,ij}^{AB}\right)^2}{\left(d_0^t\right)^2} < 1 \qquad (10)$$

for $t \in \{1, 2, \ldots, k\}$, thus

$$\sum_{t=1}^{k} \left(\frac{d_{t,ij}^{AB}}{d_0^t}\right)^2 < k, \qquad (11)$$

where k is a number of energy signature components (e.g. 6). Dividing both sides of the inequality (11) by k we obtain

$$\sum_{t=1}^{k} v_t \left(\frac{d_{t,ij}^{AB}}{d_0^t}\right)^2 < 1, \qquad (12)$$

where $v_t = 1/k$ (for $t \in \{1, 2, \ldots, k\}$) is a participation weight of each component in a process of computation of the energy signature distance matrix. Since in the searching process using the EAST algorithm some of the components may be adjudged to be more relevant than others, thus participation weights of the energy profile components may be different. Generally, v_t (for $t \in \{1, 2, \ldots, k\}$) are arbitrarily chosen positive values, which sum up to 1.

Then we transform formula (12) into

$$\sum_{t=1}^{k} \frac{\left(d_{t,ij}^{AB}\right)^2}{\left(d_0^t\right)^2 v_t^{-1}} < 1. \tag{13}$$

Both sides of the inequality are nonnegative, therefore

$$\sqrt{\sum_{t=1}^{k} \frac{\left(d_{t,ij}^{AB}\right)^2}{\left(d_0^t\right)^2 v_t^{-1}}} < 1. \tag{14}$$

Since the distance between energy signatures i and j of molecules A and B should satisfy inequality

$$d_{t,ij}^{AB} = \sqrt{\sum_{t=1}^{k} \frac{(e_{A,i}^t - e_{B,j}^t)^2}{p_t}} < d_0^S, \tag{15}$$

where d_0^S is cutoff value for the energy signature distance matrix, therefore comparing formulae (14) and (15) we obtain the values of divisors:

$$p_t = \left(d_0^t\right)^2 v_t^{-1}. \tag{16}$$

and $d_0^S = 1$ for $d_{t,ij}^{AB} = |e_{A,i}^t - e_{B,j}^t|$. The values of parameters p_t and participation weights for each component of the energy profile are presented in table 2.

One of the results of the Alignment phase is the *Smith-Waterman Score*, which quantifies a similarity between two energy profiles. The *Smith-Waterman Score*

Table 2. Recommended values of participation weights v_t and parameters p_t for particular components of the energy profile while searching with the use of energy signatures

Component energy of EP	Participation weights v_t	Assumed values of w_t	Default value of cutoff d_0^t[Kcal/mole]	Values of parameters p_t
bond stretching	0.1560	0.5	1.0	6.4
angle bending	0.1560	0.5	1.6	16.4
torsional angle	0.3125	1.0	1.4	6.3
van der Waals	0.0625	0.2	5.1	416.2
electrostatic	0.3125	1.0	6.1	119.1

is a function of all awards for matches, penalties for mismatches and gaps. It is calculated for the optimal alignment path. In the next phase, we calculate additional similarity measures: *RMSD (Root Mean Square Deviation)* and *Score* [15]. Finally, the identifier of the candidate molecule (PDB ID), similarity measures and additional information are appended to a result list. The single pairwise comparison is finished and it is repeated for the query molecule and the next candidate molecule until there is no more molecules to compare (Fig. 1).

At the end of the process, a user obtains a list of n molecules sorted from the most to the least similar, where n is specified by a user at the beginning of the searching.

4 Discussion on the New EAST Algorithm

We performed a set of tests of the new implementation of the EAST algorithm. We concentrated on molecules representing different enzymes, especially those taking part in signal transduction processes [22]. During the experiments we used the EDB database containing about 5 000 EPs of protein structures. Beneath, we present an example of the EAST search process using energy signatures and the case study. The example shows the advantages of the new version of the EAST. The search process was carried out for the molecule 1TB7, which represents catalytic domain of Human Phosphodiesterase 4D in complex with AMP. Phosphodiesterase 4D is an enzyme supporting the transformation of the cAMP molecules. cAMP molecules play an important role in signal transduction processes through the activation of protein kinases. They also regulate the passage of Ca^{2+} ion channels. Therefore, both phosphodiesterases and cAMPs are important elements of signal pathways [23].

The results of the EAST similarity search process that uses energy signatures are presented in Fig. 2. The process was run with default settings of the EAST

```
Best results for job: 2007-08-19 12:58:44
Cut-off: 1.0; id threshold: 0.3; energy type: Energy signature
S-W type: Fuzzy; mismatch: -0.3334; gap open: 1.0; gap ext.: 0.3334
```

PDB ID	Chain	Length	Matches	Match%	RMSD	Score	S-W Score
1TBB	A	314	292	92	0.509	573.68	250.77
1Q9M	A	314	261	83	0.942	277.22	192.91
1OYN	A	314	262	83	0.826	317.06	192.32
1PTW	A	314	262	83	0.815	321.57	185.12
1ROR	A	308	214	69	1.843	116.10	136.99
1XMU	A	300	210	70	1.864	112.66	126.81
1RO9	A	290	209	72	1.685	124.06	126.39
1RO6	A	307	201	65	1.990	101.01	124.13
1TB5	A	306	208	67	1.812	114.79	118.63

Fig. 2. Results of the similarity search process with the use of the EAST algorithm for the molecule 1TB7 (Catalytic Domain of Human Phosphodiesterase 4D in Complex with AMP). Energy type: energy signatures.

```
Best results for job: 2006-08-11 16:03:02
Cut-off: 6.1; id threshold: 0.5; energy type: Charge-charge
S-W type: Fuzzy; mismatch: -0.3334; gap open: 1.0; gap ext.: 0.3334

PDB ID  Chain  Length  Matches  Match%  RMSD      Score       S-W Score
------  -----  ------  -------  ------  --------  ----------  ---------
1TBB    A      298     271      90      2.795     96.97       219.70
1Q9M    A      278     261      93      2.511     103.94      198.21
1PTW    A      284     267      94      2.821     94.65       197.53
1OYN    A      279     255      91      2.924     87.20       192.52
1ROR    A      232     210      90      3.125     67.19       129.08
1RO9    A      230     211      91      3.307     63.80       129.00
1TB5    A      233     211      90      3.355     62.89       124.72
1XMU    A      233     211      90      3.205     65.84       124.57
1RO6    A      217     201      92      2.921     68.82       117.28
```

Fig. 3. Results of the similarity search process (old version of the EAST algorithm) for the molecule 1TB7 (Catalytic Domain of Human Phosphodiesterase 4D in Complex with AMP). Energy type: electrostatic.

algorithm (table 2). All molecules (*PDB ID* and *Chain*) in the result list represent proteins from the Phosphodiesterase family in various conformations or in complexes with other molecular compounds. The results were verified on the basis of descriptions and comments of molecules.

We can compare the results presented in Fig. 2 with the results of the searching carried out using a previous version of the EAST algorithm [15] presented in Fig. 3. In the situation, the old version of the EAST algorithm used only one component energy type (electrostatic) from the protein energy profile.

The results can be interpreted based on similarity measures: *Score* and *Smith-Waterman Score* (*SW-Score*) - the higher value the higher similarity between a presented molecule and a given query molecule; *RMSD* - the lower value the better quality of the alignment. We can also look at additional output parameters (for all, the higher value the better similarity): *Length* - an alignment frame, *Matches* - a number of aligned positions, *Match%* - a percentage of matching positions in the alignment frame.

Both result lists (Fig. 2 and Fig. 3) contain the same set of molecules. However, similarity measures are different because we used different component energies in the similarity searching. For the new version of the EAST algorithm, values of the *Score* similarity measure are higher, values of the *RMSD* lower, and alignment frames are wider than for the old version of the EAST. In other words, the new version is more sensitive than the old one.

The search process that uses energy signatures has an additional interesting feature, which is a self-compensation. If the distance between energy values of one component energy exceeds its cutoff value (d_0^t), then the excess can be compensated by other components of the energy profile. However, the excess cannot be large and the distance of energy values of other components must be proportionally low.

The comparison of spatial structures of a given query molecule 1TB7 and one of the resultant molecules 1OYN is presented in Fig. 4.

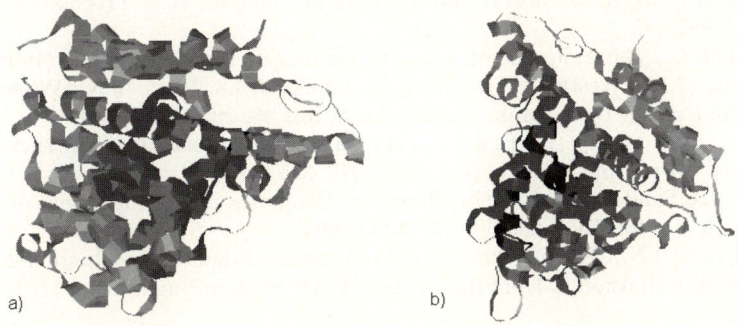

Fig. 4. Spatial structures (chain A only): a) molecule 1TB7 (Catalytic Domain of Human Phosphodiesterase 4D in Complex with AMP), b) molecule 1OYN (Crystal Structure of Phosphodiesterase PDE4D2 in Complex with (R,S)-Rolipram). Ribbon representation in the RasMol [24].

5 Concluding Remarks

The EAST algorithm has a general purpose. It can be applied in searching proteins or parts of proteins, which possess a defined structural conformation against a database of protein structures. For example, it can be used to find molecules having a specific catalytic region, which is important for a particular intracellular reaction. Molecules found in the search process may indicate similar biological behavior of the given region. This is exactly the case when we use the EAST. Furthermore, the similarity searching with the EAST algorithm can be supportive during the analysis of signal pathways, where conformational changes are significant for the protein activity and entire reactions of the pathways. The EAST can be also used to find molecules having similar biological behavior so that a given primary molecule can be replaced by one of the molecules returned as a result.

We developed the new version of the EAST algorithm, which compares energy signatures during the search process. The new implementation of the EAST algorithm benefits from the self-compensation of small energy discrepancies and in consequence, it is much more sensitive than the old version. After eliminating the weaknesses of the previous version the new EAST is still fault-tolerant and as fast as its predecessor.

References

1. Dickerson, R.E., Geis, I.: The Structure and Action of Proteins, 2nd edn., Benjamin/Cummings, Redwood City, Calif. Concise (1981)
2. Lodish, H., Berk, A., Zipursky, S.L., et al.: Molecular Cell Biology, 4th edn. W. H. Freeman and Company, New York (2001)
3. Leach, A.R.: Molecular Modelling. Principles and Applications, 2nd edn. Pearson Education Ltd, London (2001)

4. Fersht, A.: Enzyme Structure and Mechanism, 2nd edn. W.H. Freeman & Co., New York (1985)
5. Branden, C., Tooze, J.: Introduction to Protein Structure. Garland (1991)
6. Berman, H.M., Westbrook, J., Feng, Z., Gilliland, G., Bhat, T.N., Weissig, H., et al.: The Protein Data Bank. Nucleic Acids Res 28, 235–242 (2000)
7. Gibrat, J.F., Madej, T., Bryant, S.H.: Surprising Similarities in Structure Comparison. Curr. Opin. Struct. Biol. 6(3), 377–385 (1996)
8. Holm, L., Sander, C.: Protein Structure Comparison by Alignment of Distance Matrices. J. Mol. Biol. 233(1), 123–138 (1993)
9. Shindyalov, I.N., Bourne, P.E.: Protein Structure Alignment by Incremental Combinatorial Extension (CE) of the Optimal Path. Protein Engineering 11(9), 739–747 (1998)
10. Can, T., Wang, Y.F.: CTSS: A Robust and Efficient Method for Protein Structure Alignment Based on Local Geometrical and Biological Features. In: Proceedings of the 2003 IEEE Bioinformatics Conference (CSB), pp. 169–179 (2003)
11. Goodford, P.J.: Computational Procedure for Determining Energetically Favourable Binding Sites on Biologically Important Macromolecules. J. Med. Chem. 28, 849–857 (1985)
12. Rodrigo, J., Barbany, M., et al.: Comparison of Biomolecules on the Basis of Molecular Interaction Potentials. J. Braz. Chem. Soc. 13(6), 795–799 (2002)
13. Ji, H., Li, H., Flinspach, M., Poulos, T.L., Silverman, R.B.: Computer Modeling of Selective Regions in the Active Site of Nitric Oxide Synthases: Implication for the Design of Isoform-Selective Inhibitors. J. Med. Chem., 5700–5711 (2003)
14. Burkert, U., Allinger, N.L.: Molecular Mechanics. American Chemical Society, Washington D.C. (1980)
15. Mrozek, D., Małysiak, B., Kozielski, S.: EAST: Energy Alignment Search Tool. In: Wang, L., Jiao, L., Shi, G., Li, X., Liu, J. (eds.) FSKD 2006. LNCS (LNAI), vol. 4223, pp. 696–705. Springer, Heidelberg (2006)
16. Mrozek, D., Małysiak, B., Kozielski, S.: An Optimal Alignment of Proteins Energy Characteristics with Crisp and Fuzzy Similarity Awards. In: Proc. of the IEEE International Conference on Fuzzy Systems (FUZZ-IEEE), pp. 1508–1513 (2007)
17. Ponder, J.: TINKER - Software Tools for Molecular Design, Dept. of Biochemistry & Molecular Biophysics, Washington University, School of Medicine, St. Louis (June 2001)
18. Cornell, W.D., Cieplak, P., et al.: A Second Generation Force Field for the Simulation of Proteins, Nucleic Acids, and Organic Molecules. J.Am. Chem. Soc. 117, 5179–5197 (1995)
19. Mrozek, D., Małysiak, B.: Searching for Strong Structural Protein Similarities with EAST. Journal of Computer Assisted Mechanics and Engineering Sciences 14, 681–693 (2007)
20. Smith, T.F., Waterman, M.S.: Identification of Common Molecular Subsequences. J. Mol. Biol. 147, 195–197 (1981)
21. Altschul, S.F., Erickson, B.W.: Optimal Sequence Alignment Using Affine Gap Costs. Bull. Math. Biol. 48(5-6), 603–616 (1986)
22. Ray, L.B.: The Science of Signal Transduction. Science 284, 755–756 (1999)
23. Zhang, K.Y.J., Card, G.L., Suzuki, Y., Artis, D.R., et al.: A Glutamine Switch Mechanism for Nucleotide Selectivity by Phosphodiesterases. Mol.Cell 15, 279–286 (2004)
24. Sayle, R., Milner-White, E.J.: RasMol: Biomolecular Graphics for All. Trends in Biochemical Sciences (TIBS) 20(9), 374 (1995)

SYMBIOS: A Semantic Pervasive Services Platform for Biomedical Information Integration

Myriam Mencke[1], Ismael Rivera[1], Juan Miguel Gómez[1],
Giner Alor-Hernandez[2], Rubén Posada-Gómez[2], and Ying Liu[3]

[1] Universidad Carlos III de Madrid
myriam.mencke@uc3m.es, ismael.rivera@uc3m.es, juanmiguel.gomez@uc3m.es
[2] Division of Research and Postgraduate Studies,
Instituto Tecnologico de Orizaba, Mexico
gineralor@computacion.cs.cinvestav.mx, rposada@itorizaba.edu.mx
[3] University of Texas at Dallas, United States of America
ying.liu@utdallas.edu

Abstract. Applying semantic pervasive services to Biomedical research is providing a new breed of intelligent applications which can tackle with the heterogeneity and intrinsic complexity of biomedical information integration. Using semantics leverages the potential of enabling cross-interoperability among a variety of storage and data formats widely distributed both across the Internet and within individual organizations. In this paper, we present SYMBIOS, a fully-fledged biomedical information integration solution based on semantic pervasive services that combine a Service Oriented Architecture (SOA) and semantically-empowered techniques to ascertain biomedical information intelligent integration. We discuss our approach with a proof-of-concept implementation where the breakthroughs and efficiency of integrating the biomedical publications database MEDLine, the Database of Interacting Proteins (DIP) and the Munich Information Center for Protein Sequences (MIPS) has been tested.

1 Introduction

Integration and exchange of data within and among organizations is a universally recognized need in bioinformatics and genomics research. By far the most obvious frustration of a life scientist today is the extreme difficulty in putting together information available from multiple distinct sources. A commonly noted obstacle for integration efforts in bioinformatics is that relevant information is widely distributed, both across the Internet and within individual organizations, and is found in a variety of storage formats, both traditional relational databases and non-traditional sources (e.g. text data sources in semi-structured text files or XML, and the result of analytic applications such as gene-finding application or homology searches).

Arguably the most critical need in biomedical data integration is to overcome semantic heterogeneity i.e. to identify objects in different databases that represent the same or related biological objects (genes, proteins, etc) and to resolve the differences in database structures or schemas, among the related objects. Such data integration is technically difficult for several reasons. First, the technologies on which different databases are based may differ and do not inter-operate smoothly. Standards for cross-database communication allow the databases (and their users) to exchange information. Secondly, the precise naming conventions for many scientific concepts (such as individual genes, proteins or drugs) in fast developing fields are often inconsistent, and so mappings are required between different vocabularies. Third, the precise underlying biological model for the data may be different (scientists view things differently) and so to integrate these data requires a common model of the concepts that are relevant and their allowable relations. This reason is particularly crucial because unstated assumptions may lead to improper use of information, on the surface, appears to be valid. Fourth, as our understanding of a particular domain improves, not only will data change, but even database structures will evolve. Any users of the data source, including in particular any data integrators must be able to manage such data source evolution.

Since the current Web is an environment primarily developed for human users, the need of adding semantics to the Web becomes more critical as organizations rely on the service-oriented architecture paradigms to expose functionality and data sources by means of Pervasive Services or Web Services. The Semantic Web is about adding machine-understandable and machine-processable metadata to Web resources through its key-enabling technology: ontologies [1]. Ontologies are a formal, explicit and shared specification of a conceptualization [2]. The breakthrough of adding semantics to Web Services leads to the Semantic Web Services (SWS) paradigm [3], which offers the possibility of ascertain which services could best fit the wishes and fulfill the goals of the user. SWS can be discovered, located and accessed since they provide formal means of leveraging different vocabularies and terminologies and foster mediation. However, the problem of bridging the gap between the current Web, primarily designed for human users whose intentions are expressed in natural language, and the formalization of those wishes remains. Potential users might deter from using semantic pervasive services, since its underlying formalization and unease of use hampers its use from rich user-interaction perspective.

Hence, we present in this paper our work on a Semantic Pervasive Services Platform for Biomedical Information Integration (SYMBIOS) platform, which fosters the intelligent interaction between pervasiveness and the Semantic Technologies available. The remainder of this paper is organized as follows. Section 2 describes the Biomedical Information and Integration Discovery with Semantic Web Services (SYMBIOS) platform for Semantic Web services. Section 3 presents the proof-of-concept implementation based in a real world scenario in which the integration of the biomedical publications database PubMed, the Database of Interacting Proteins (DIP) and the Munich Information Center for Protein Sequences (MIPS) has been tested. Finally, conclusions and related work are discussed in Section 5.

2 Problem Statement and Motivating Scenario

This section describes the problem of heterogeneous data addressed by SYMBIOS, in the context of a motivating scenario in which the problems faced by a user are displayed.

2.1 Problem Statement

The objective of the SYMBIOS platform is to interpret a user's input query in terms of semantic goals, and employ a semantically-powered architecture to access integrated data which returns specific results to the user. The problem faced is that the references to biological data, such as genes and proteins, are not homogeneous in each database. The SYMBIOS platform uses semantic descriptions of these objects in order to generate integrated information which spans multiple databases, to provide accurate results to a user in a user-oriented approach.

2.2 Motivating Scenario

We present a real-world scenario which shows the information required by a user, Alice, and illustrates the problems faced by the user when searching for particular data. Alice is a biochemist and wants to find information about the gene ASB1[1] and how it affects the anthranilate protein. The first issue she is confronted with is that the gene with the name asb1 occurs in many species: *Arabidopsis thaliana* (a plant from the Brassicaceae family), *Danio rerio* (the zebrafish), and *Mus musculus* (the common house mouse). In all species it is referred to as asb1. Although Alice knows that she requires information relating to the *Arabidopsis thaliana* species, she cannot uniquely use the search term asb1. Secondly, the name of the protein of this gene has multiple synonyms, for example, for the species Arabidopsis thaliana it is referred to as: ANTHRANILATE SYNTHASE BETA SUBUNIT, ANTHRANILATE SYNTHASE BETA SUBUNIT 1, F4F7_39, F4F7.39, TRP4, TRYPTOPHAN BIOSYNTHESIS 4, WEAK ETHYLENE INSENSITIVE7, among others. Therefore, if, for example, Alice searches the PubMed database, she will have enter a number of search terms and attempt the search a number of times, impeding efficiency.[2] For each search, she will have to manually examine the articles returned in the search results until she finds an article relevant to her particular query.

The third issue is that not only does the name of the gene have multiple synonyms, the protein name for the gene also has multiple descriptors depending on which database is searched. If Alice consults the DIP database and searches for

---

[1] The gene ASB1 has many functions in the *Arabidopsis thaliana* species, however its effective functioning depends on its interaction with other compounds. For example, both the gene ASA1 and ASB1 are key elements in the regulation of auxin production in *Arabidopsis thaliana*.

[2] However, fortunately, in some cases it is possible to define species when performing a gene-related search in web-bases, for example, in PubMed.

the gene using the "by sequence" field, by entering the gene's peptide sequence into the search query, this query returns labels for the gene's protein according to multiple descriptors. The descriptors include DIP's identifier for proteins, of the form <DIP:nnnN>, its E-VAL (a numeric coding value for the gene), and its label according to each database cross-referenced by DIP: the protein database PIR (Protein Information Resource), SWISSPROT, and GENBANK. The protein name/description is also provided.

In fact the user does not require multiple labels for a single gene. She requires one description of the gene and the interactions it is involved in, which are generated by the semantic capabilities of SYMBIOS. When Alice uses SYMBIOS and enters the query "Give me information about gene ASB1 and its effects on the anthranilate protein", SYMBIOS provides a solution to the heterogeneous descriptors of gene objects usually returned to the user in traditional gene databases. The SYMBIOS platform integrates the information in PubMed, DIP, and MIPS, by filtering the user's natural language input, and annotating the relevant input terms with semantic identifiers from the Gene Ontology. The Gene Ontology links all of the different labels from various databases, by annotating the name of the gene with a single term. Thus the problem of heterogeneous inconsistent data sources is addressed. Subsequently, the user's goal, in terms of semantic descriptors, is used by the architecture to provide integrated results to the user. The architecture will be described in the next section.

3 SYMBIOS: Biomedical Information Integration Discovery

SYMBIOS is a two-faced software agent designed to interact with human beings as a gateway or a man-in-the-middle to access Semantic Web Services. The main goal of the system is to help users express their needs in terms of information retrieval and achieve both biological data and analytical functionality integration by means of SWS. SYMBIOS allows users to either state their needs via natural language or go through a list of the most important terms, extracted from the Gene Ontology[3] (GO). For this, SYMBIOS makes use of ontology-driven data mining. This implies that it firstly captures and gathers the terms for which the user would like to search (e.g. Gene ASB1, Protein anthranilate) by using as a reference the aforementioned terms of the GO. Secondly, it builds up a "lightweight ontology" i.e. a very simple graph made of the relationships of those terms. The lightweight ontology is then managed as the formal "goal" the system has to achieve on behalf of its user. A "goal" in SWS technology refers to the aim a user expects to fulfill by the use of the service. When SYMBIOS has inferred the goal derived from the users' wishes, it starts looking for the necessary services to achieve the objective. With that purpose, SYMBIOS queries the repository of services semantic descriptions. Once the appropriate services have been located, SYMBIOS orchestrates them and starts their execution. Finally,

---

[3] See http://www.geneontology.org/

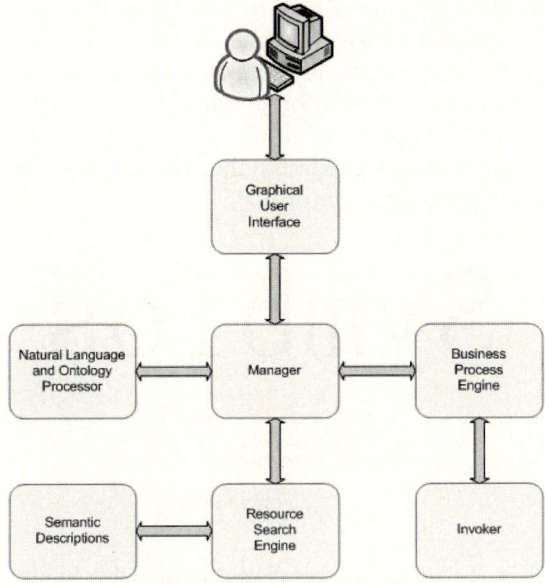

Fig. 1. The SYMBIOS Architecture

SYMBIOS retrieves the outcome resulting from the integration of the applications being accessed (e.g. all the biomedical, biological publications and medical databases) and presents it back to the user.

Fig. 1 illustrates the architecture of the SYMBIOS platform and the communications of its main components. The component who interacts with the user is the GUI, placed between the user itself and the Manager. The core component is the Manager, which supervises all the process and act as intermediary among the other components. The Manager uses the Natural Language and Ontology Processor component to get the "lightweight ontology" of the query. On the other hand, both the Resource Search Engine and the Semantic Descriptions are responsible for finding and selecting the most suitable services. Then, the Business Process Engine determines how the orchestration should take place and the Invoker invokes the services one after the other. In the following subsections, a concise description of each of these components will be presented.

A more in-depth description of the case study is provided in the next section, where the different components of the SYMBIOS architecture fit in the context of an added-value scientific case study.

3.1 Graphical User Interface

This is the component which allows people to interact with SYMBIOS. The users have two possibilities: they can either introduce a text in natural language or use the ontology-guided tool (but this option would require further work, which is not envisaged in this section), which assists them in expressing their goals. Then, the Manager component assumes the control of the execution.

As shown in Fig. 2, for example, a biomedical researcher could write in plain natural language: "Give me information about the gene ASB1 and its effects on the anthranilate protein". This would result in the invoking of Natural Language Processing (NLP) techniques to express the most important concepts of the query to configure a "lightweight ontology".

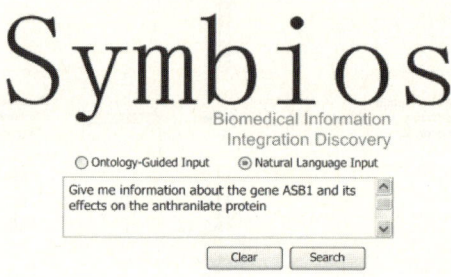

Fig. 2. An example of Natural Language Input

The other possibility of interaction is shown in Fig. 3. The user writes the query, selecting the options displayed by the interface. In the example, once detected the word "gene", the interface displays a list of genes to be selected.

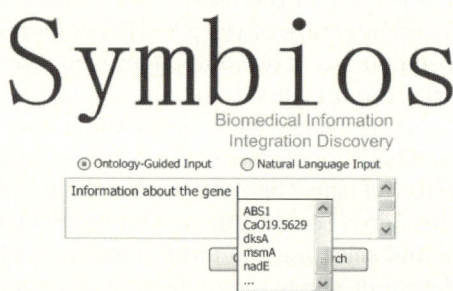

Fig. 3. An example of Ontology Guided Input

3.2 Manager

As the main component of our architecture, the Manager coordinates all processes among the various components of the architecture. It follows the execution semantics explained in the previous paragraph and ends the SYMBIOS execution once the invoker has returned its results, and the corresponding information is forwarded to the end-user.

3.3 Natural Language and Ontology Processor

This component performs the natural language processing of the query made by the user. It creates a very simple graph made of the relationships of the terms of the query as shown in Fig. 4. This "lightweight ontology" can then be managed as a formal goal. The challenges faced by the application stem from the highly ambiguous nature of natural language, which the Natural Language and Ontology Processor attempts to disambiguate.

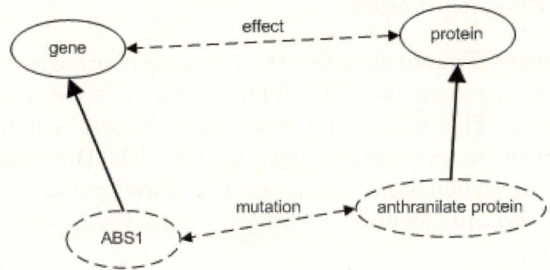

Fig. 4. The lightweight ontology created from the query

3.4 Invoker

This component simply uses the grounding of the Web Services technology to invoke and interact with the particular Semantic Web Service, no matter how or where the services are provided, selected by the Resource Search Engine.

The Invoker allows completely dynamic invocation of Web services, based upon examination of the meta-data about the service at runtime.

3.5 Semantic Descriptions

Semantic descriptions of the information resources connected to SYMBIOS are currently paving the way to provide the automation and dynamic discovery of information resources. In our particular case, these semantic descriptions are SWS descriptions.

In our architecture, the user can use a two-fold approach to access these services: they can be described and annotated either with a semantic web language, such as OWL (WEB ONTOLOGY LANGUAGE), or these descriptions can be found in a SWS goal repository. Particularly, RDF (RESOURCE DESCRIPTION FRAMEWORK) use is highly advisable given the good results of the SPARQL language in its querying.

3.6 Resource Search Engine

The Resource Search Engine component searches and locates the most accurate biomedical resource to be accessed. A pivotal concept that enables the location of these ERP (Enterprise Resource Planning) systems is usually known as discovery, which implies locating a semantic description of a resource that may have been previously unknown, and that meets certain functional criteria.

In SYMBIOS, discovery takes place querying the RDF descriptions of the multiple biomedical information systems, which are interconnected. This query is performed using the SPARQL (SPARQL Protocol and RDF Query Language)[4] language. The results of the query are a given number of SWS. These SWS will be invoked by the Invoker.

3.7 Business Process Engine

The Business Process Engine specifies the order or choreography of the invocations of SYMBIOS, regarding how the different sources to be accessed or queried, should be addressed. This implies an order for temporal constraints and also a particular "workflow" or execution of such queries. The Business Process Engine also raises a challenge when a particular choreography provides more added-value than a given one and optimization techniques must be applied.

4 Using SYMBIOS for Biomedical Information Integration

In this section, we present a real-world based use case scenario in order to validate the appropriateness of the architecture presented above and show the advantages provided by SYMBIOS from the user perspective. First, the three data sources that are being integrated are described. Then, some insights on the ontology utilized for integrating the disparate data sources are discussed. Thirdly, Web Services developed that are capable of displaying some parts of the functionality available in the biomedical Web portals, along with their semantic annotation are depicted. Finally, a typical example of how a common user might use the application is shown.

4.1 Data and Functionality Sources

Firstly, the PubMed[5] database, a free search engine offered by the United States National Library of Medicine (as part of the Entrez information retrieval system). The inclusion of an article in PubMed does not endorse that article contents and the service allows searching the MEDLINE database. MEDLINE covers over 4,800 journals published and also offers access to citations to articles that are out-of-scope (e.g., covering plate tectonics or astrophysics) from certain MEDLINE journals, primarily general science and general chemistry journals, for which the life sciences articles are indexed for MEDLINE, in-process citations which provide a record for an article before it is indexed and added to MEDLINE or converted to out-of-scope status and citations that precede the date that a

---

[4] See http://www.w3.org/TR/rdf-sparql-query/
[5] See http://pubmed.gov/

journal was selected for MEDLINE indexing (when supplied electronically by the publisher).

Secondly, the Database of Interacting Proteins (DIP)[6] database catalogs experimentally determined interactions between proteins. It combines information from a variety of sources to create a single, consistent set of protein-protein interactions. The data stored within the DIP database were curated, both, manually by expert curators and also automatically using computational approaches that utilize the knowledge about the protein-protein interaction networks extracted from the most reliable, core subset of the DIP data.

Finally, the Munich Information Center for Protein Sequences (MIPS)[7] database was selected, through its endpoint MPact. MPact[8] provides a common access point to interaction resources at MIPS. It is designed to support for both downloading and uploading data related to protein interaction. It provides the user with intuitive query forms to quickly retrieve the interactions of interest. Graphical representations allow an easy navigation through the protein interaction networks.

4.2 The Integrating Ontology

An ontology of biological terminology provides a model of biological concepts that can be used to form a semantic framework for data storage, retrieval and analysis tasks. As it was aforementioned, the Gene Ontology (GO) [4] is the common reference model employed in SYMBIOS.

GO is an ontology that describes attributes of gene products. GO offers a way to resolve the semantic heterogeneity of the annotations of genetic products in diverse databases: annotations of different databases are linked to the same GO term. The main GO components are the terms and their relations. Each term has a sole identifier apart from the name of the term. GO is divided into three independent ontologies: molecular functions, biological processes, and cellular components. Molecular functions describe the basic molecular role of gene products. Each biological process is comprised of different molecular functions and describes its role at conceptual level. The cellular component ontology represents the structure of the eukaryotic cell.

4.3 Pervasive Services for Integration

In order to test SYMBIOS, we have developed several Web Services that expose some of the functions provided by the three biomedical sources accessed. Each of these services has been subsequently annotated with semantic content using the OWL-S approach. For this purpose the OWL-S Editor [5], a plugin for Protégé that allows users to semantically annotate services, has been applied. So, for example, the service that provides global access to the MIPS database consist of two methods, namely, getDescription and getInteraction, which return the

---

[6] See http://dip.doe-mbi.ucla.edu/
[7] See http://mips.gsf.de/services/ppi
[8] See http://mips.gsf.de/genre/proj/mpact/

```
- <wsdl:definitions targetNamespace="http://www.inf.uc3m.es/">
    <wsdl:documentation>MPact Service</wsdl:documentation>
  + <wsdl:types></wsdl:types>
  + <wsdl:message name="getDescriptionMessage"></wsdl:message>
  + <wsdl:message name="getDescriptionResponseMessage"></wsdl:message>
  + <wsdl:message name="getInteractionMessage"></wsdl:message>
  + <wsdl:message name="getInteractionResponseMessage"></wsdl:message>
  + <wsdl:portType name="MPact_WebServicePortType"></wsdl:portType>
  - <wsdl:binding name="MPact_WebServiceSOAP11Binding" type="axis2:MPact_WebServicePortType">
      <soap:binding transport="http://schemas.xmlsoap.org/soap/http" style="document"/>
    + <wsdl:operation name="getDescription"></wsdl:operation>
    + <wsdl:operation name="getInteraction"></wsdl:operation>
    </wsdl:binding>
  + <wsdl:binding name="MPact_WebServiceSOAP12Binding" type="axis2:MPact_WebServicePortType"></wsdl:binding>
  + <wsdl:binding name="MPact_WebServiceHttpBinding" type="axis2:MPact_WebServicePortType"></wsdl:binding>
  + <wsdl:service name="MPact_WebService"></wsdl:service>
  </wsdl:definitions>
```

Fig. 5. MPact Service WSDL file

```
- <j.1:Service rdf:ID="getDescriptionService">
    <j.1:describedBy rdf:resource="#getDescriptionProcess"/>
  - <j.1:presents>
    - <j.0:Profile rdf:ID="getDescriptionProfile">
        <j.1:presentedBy rdf:resource="#getDescriptionService"/>
      - <j.0:hasInput>
        + <j.2:Input rdf:ID="term"></j.2:Input>
        </j.0:hasInput>
        <j.0:hasOutput rdf:resource="#description"/>
      + <j.0:textDescription rdf:datatype="http://www.w3.org/2001/XMLSchema#string"></j.0:textDescription>
        <j.0:serviceName rdf:datatype="http://www.w3.org/2001/XMLSchema#string">getDescription</j.0:serviceName>
      </j.0:Profile>
    </j.1:presents>
  + <j.1:supports></j.1:supports>
  </j.1:Service>
```

Fig. 6. A portion of the OWL-S file for the MPact service

description and the interactions of a protein respectively given its name. An extract of the WSDL file that describes the capability of the service is shown in Fig. 5. Finally, a piece of the OWL-S semantic description of the service is presented in Fig. 6

4.4 Use Case Scenario

In this section we present a use case scenario (see Fig. 7) in which a user, hereafter Alice, aims to get some information about a particular gene. The first thing Alice must do is to type the application URL on her standard Web browser. Then, she expresses her query through a sentence in natural language as shown in Fig. 2. Once she had clearly stated her goal, she presses the "Search" button and SYMBIOS starts to process the query. Firstly, it translates the natural language sentence into a lightweight ontology by means of a NLP tool. In the second place, SYMBIOS searches for those services that can achieve the goal. This is done by matching the goal lightweight ontology with the semantic description of the services' capabilities. Next, the services are ordered by the Business

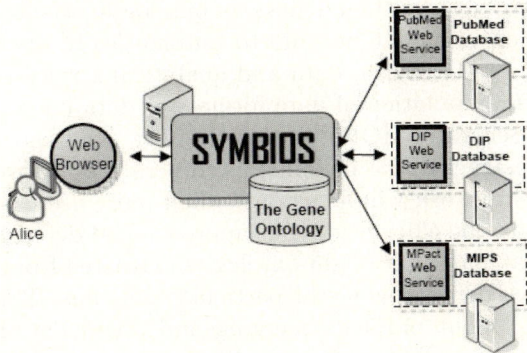

Fig. 7. Use Case Scenario

Process Engine and invoked. Finally, the results are integrated and returned to the user.

Our experience shows that users prefer guided typing when they begin using SYMBIOS. This is because free typing becomes frustrating for novel users when SYMBIOS is not able to match user' goals with goals in the repository. According to our experience, when users acquire enough experience they move to free typing. However, guided typing is not so easy when the number of options increase. Critical mass in goals repositories and more available services hamper the efficiency of the selection criteria. A potential solution is dividing the list in a hierarchical manner in order to avoid hoarding of an excessive number of elements to the user.

5 Conclusions and Related Work

The increasing volume and diversity of information in biomedical research is demanding new approaches for data integration in this domain. Semantic Web Services along with Semantic Web technologies can leverage the potential of biomedical information integration and discovery, facing the problem of semantic heterogeneity of biomedical information sources. In this paper we have proposed a solution and a proof-of-concept implementation to integrate both biological data and analytical functionality from disparate sources. Semantic Web Services provide access to these sources and SYMBIOS executes them according to the user wishes. The forthcomings of our approach are mainly two, namely: striving for a rich and ease-of-use interaction to understand the user goals and the ability to interact with the existing Semantic Web Services.

Several projects are running at the moment aiming at developing a reference architecture for data and functionality integration in this application field. This is the purpose of the BioMOBY project[9] . A system for providing interoperability between biological data hosts and analytical services is being built within the

---

[9] See http://biomoby.org

scope of this project. This system defines an ontology-based messaging standard through which a client might be able to automatically discover and interact with task-appropriate biological data and analytical service providers, without requiring manual manipulation of data formats as data flows from one provider to the next. The Semantic MOBY[10] project stems from BioMOBY. It intends to provide an infrastructure for enabling interoperability between Web Services.

Finally, our future work will focus on finding more use cases and real-world scenarios to validate the efficiency of our approach and determine the feasibility of the semantic match of lightweight ontologies extracted from natural language text and ontologies defining goals in particular domains. This work is related to existing efforts about ontology merging and alignment. Future version of SYMBIOS will be oriented towards that direction.

References

1. Berners-Lee, T., Hendler, J., Lassila, O.: The semantic web. Scientific American 284(5), 28–37 (2001)
2. Borst, W.: Construction of Engineering Ontologies for Knowledge Sharing and Reuse. Centre for Telematics and Information Technology (1997)
3. Fensel, D., Bussler, C.: The Web Service Modeling Framework WSMF. Electronic Commerce Research and Applications 1(2), 113–137 (2002)
4. Ashburner, M., Ball, C., Blake, J., Botstein, D., Butler, H., Cherry, J., Davis, A., Dolinski, K., Dwight, S., Eppig, J., et al.: Gene ontology: tool for the unication of biology. The Gene Ontology Consortium. Nat Genet 25(1), 25–29 (2000)
5. Elenius, D., Denker, G., Martin, D., Gilham, F., Khouri, J., Sadaati, S., Senanayake, R.: The OWL-S Editor – A Development Tool for Semantic Web Services. In: Gómez-Pérez, A., Euzenat, J. (eds.) ESWC 2005. LNCS, vol. 3532, pp. 78–92. Springer, Heidelberg (2005)

---

[10] See http://semanticmoby.org

The PCR Primer Design as a Metaheuristic Search Process

L. Montera[1] and M.C. Nicoletti[2]

[1] PPG-Biotchnology, UFSCar – Brazil & The University of Newcastle – Australia
[2] Departament of Computer Science, Universidade Federal de S. Carlos – Brazil
{montera,carmo}@dc.ufscar.br

Abstract. The Polymerase Chain Reaction process is a well-known technique for the *in vitro* amplification of a DNA sequence. The success of a PCR depends on several parameters particularly the primer sequences used. Since the design of a suitable pair of primer involves a reasonable number of variables, which can have a range of different values, computer programs are commonly used to assist this task. This paper approaches the design of a pair of primer sequences as a search process throughout the space defined by all possible primer sequence pairs, directed by an evaluation function that combines the many variables involved in a primer design; an experiment and its results are discussed.

Keywords: Primer design, Simulated Annealing, PCR, Bioinformatics.

1 Introduction

In order to carry on studies and analysis of the structure and functionalities of certain proteins, it is necessary to have available enough amount of their corresponding DNA. Until very recently, microorganisms were used for generating multiple copies of a DNA sequence, a time-consuming and labor-intensive task. The discovery of the DNA polymerase enzyme and the development of a simple method named PCR (Polymerase Chain Reaction) [1] were the keys for establishing a laboratorial procedure for synthesizing many copies of a DNA molecule, without the use of microorganisms.

Although the main goal of the PCR is to amplify (i.e., to generate copies) a specific DNA fragment (named target sequence), PCR is also used in many other applications, such as cloning and mutation detection. There are two essential requirements for conducting a PCR reaction: a DNA template and a primer sequence pair. The DNA template is the DNA to be amplified. Primers are an appropriate short nucleotide sequences that are added to the reaction in order to mark the limits of the target region, i.e., the region of the template to be amplified. In general, two primers are used: the forward, for flanking the beginning of the target region and the reverse, for flanking its end. The primer pair is one of the most important parameters for a successful PCR. Its design involves several variables; determining a suitable value of some of them may require heavy calculations. Although many different types of software are available for assisting a primer design, the process is still not well defined, mainly due to the number of variables involved and to the lack of consensus in relation to their values. This paper approaches the design of a primer sequence pair as a search process implemented as an interactive software named SAPrimer, based on the simulated annealing algorithm.

Following this introductory section, Section 2 briefly describes the PCR, focusing on the importance of a well designed primer pair for its success. Section 3 details the main variables to be taken into account when designing primers, as well as the procedures used for determining their values. Section 4 discusses the proposed search based approach as well as presents details of the SAPrimer software. Finally, Section 5 presents the final considerations.

2 The PCR and the Importance of Primers

As defined in [2], "PCR is an elegant but simple technique for the *in vitro* amplification of target DNA utilizing DNA polymerase and two specific oligonucleotide or primer sequences flanking the region of interest". The DNA polymerase synthesizes new strands of DNA in 5' → 3' direction from a single-stranded template. So that, to make a complementary strand from a template in 5' → 3' direction, it is necessary a 3' → 5' primer (reverse) and to make a complementary strand from a template in direction 3' → 5' it is necessary a 5' → 3' primer (forward). The three steps of the PCR are:

(1) *Denaturing*: double-stranded DNA molecules are heated so that the double-stranded DNA molecules are separated completely into two single-stranded sequences;
(2) *Annealing*: the temperature is lowered such that primers anneal to the single-stranded sequences;
(3) *Extension*: the temperature is raised again up to the temperature that is optimum for the polymerase to react. The DNA polymerases use the single-stranded sequences as templates to extend the primers that have been annealed to the templates.

The success of a PCR is highly dependent on the primer pair chosen and the experimental conditions in which the reaction occurs, such as the number of cycles, the temperature and time involved in each individual step as well as the quality and the volume of primers used in the annealing step. The denaturation and annealing temperatures are directly dependent on the primers used. An exhaustive list of variables and parameters that interfere with a PCR as well as a discussion about the PCR, its results and limitations can be found in [2] and [3].

3 Primers and Their Main Characteristics

The primer pair that can be used in the *Annealing* step of a PCR is not unique, since forward and reverse primers with different sizes are possible. The pair of primers that promotes the best results of a PCR (i.e., the pair that optimizes the amount and the specificity of the product) is named *optimum pair*. Finding the optimum pair involves the simultaneous analysis of many parameters like the primer size, the primer contents of cytosine and guanine bases, the melting temperature of the DNA strands, the 3' end, the specificity, the formation of secondary structure between forward and reverse primer or between two forward and/or two reverse primers, etc. In spite of the non-existence of a consensus for the exact values of these parameters several studies can be found in the literature establishing values, or intervals of values for them. The following subsections discuss the main parameters to be considered when designing an adequate pair of prime sequences for a PCR.

3.1 Repeats, Runs and Secondary Structures

The repetition of nucleotide sequences (named *repeat*) inside of a primer sequence should be avoided since the occurrence of repeats can favor the occurrence of *misprimes*, i.e., the annealing between the template and the primer sequence that causes the amplification of a region different from the target region. The occurrence of a long repetition of one single base in the primer sequence is named *run*. Runs should be avoided because they can favor misprimes.

Another important characteristic that should be avoided during the primer design is their *self-complementarity*, which promotes the primer-primer annealing. Self-complementary primer sequences can affect the PCR efficiency by reducing the concentration of single-stranded primers since some annealed primer-primer could be extended by the polymerase, resulting in an unwanted non-specific product. Three distinct primer-primer annealing situations can occur, resulting the formation of secondary structures named *self-dimer* (annealing between two forward or two reverse primer sequences), *hetero-dimer* (annealing between one forward and one reverse primer sequence), and *hairpins* (annealing of a primer sequence (forward or reverse), to itself).

3.2 Specificity and Primer Length

A forward primer is considered to be specific if it anneals to the template just at the beginning of the target region. A reverse primer is considered to be specific if it anneals to the template just at the ending of the target region. The specificity of a primer is highly important to assure that the PCR product will correspond exactly to the amplification of the target region. One way to evaluate the specificity of a primer sequence is by 'sliding' it along the length of the template, trying to detect alternative priming sites, other than the target region. For obvious reasons, primers that promote alternative annealing sites are not a good choice. The specificity is closely related to the primer length.

The choice of primer lengths, however, involves at least three parameters: specificity, annealing stability and cost. The longer the primer length is, the smaller are the chances of existing alternative priming sites; also, the longer the primers, the more specific they are. Longer primers are more stable (due to the greater number of hydrogen bonds that they form with the template). Longer primers, however, are more biased to the formation of secondary structures and financially more expensive to be produced. Shorter primers, in spite of their lower cost, are prone to anneal outside the target region, resulting in non-specific product, lowering the quality of the PCR product. There is no single optimum length for a primer. A rule-of-thumb suggested in [4] is "primers of 18–30 nucleotides in length are the best".

3.3 The %CG Content and the 3' end

The percentage of cytosine (C) and guanine (G) bases (%CG) in a primer sequence is very important because these numbers provide information about the annealing stability/strength. A ligation between a thymine (T) and an adenine (A) base occurs due to the formation of two hydrogen bonds; a ligation between a cytosine (C) and a guanine (G) base occurs due to the formation of three hydrogen bonds, making the latter more stable and more difficult to be formed and broken. As a consequence, the CG content of a primer directly influences the temperature in which the annealing between the primer and the template will occur. In general, primers with a CG content varying between 40% and 60% are preferred.

Mismatches can occur during the annealing between a primer and a template. They can be located anywhere (inside or at the end of the primer–template complex) and can affect the stability of the complex, provoking undesirable side effects as far as the efficiency of the polymerase extension process is concerned. A mismatch located at (or near) the 3' end of a primer (where the extension by polymerase starts) is more damaging effective than those located at other positions [5]. Based on this information, it can be inferred that the 3' end of a primer should be well "stuck" to the template (i.e., no mismatches), so that the polymerase can start and conduct the extension process efficiently. Due to the strong ligation between the C and G bases, the presence of either, at the 3´end of a primer should be preferred (over the occurrence of a T or A) since this will (potentially) assure more stability to the primer-template complex.

3.4 Melting and Annealing Temperatures

The melting temperature (T_m) is the temperature at which 50% of the DNA molecules are in the duplex form and the other 50% are in the denaturated form. In a PCR, it is expected that while the template molecules denature, the primer molecules anneal to the single-stranded resulting sequences (templates). The temperature at which the annealing between the primer and the template occurs is defined as the annealing temperature (T_a). The T_m can be defined in relation to both, the product (amplified templates) and the primers; the T_a calculation is particularly dependent on both. There are several different methods to estimate the T_m value, which can be broadly classified according to the methodology they use: *Basic* (only considers the %CG content), *Salt Adjusted* (takes into account the salt concentration at the solution) and *Thermodynamic* (uses the Nearest Neighbor model).

The most basic formula for the T_m calculation (eq. (1) in Table 1) was given in [6], where |C|, |G|, |A| and |T| represent, respectively, the number of cytosine, guanine, adenine and thymine bases present in the DNA sequence. Eq. (1) establishes that the value of T_m is directly related to the length and contents of a DNA sequence. Another basic formulation, proposed in [7], is given by eq. (2) in Table 1, which assumes that the reaction occurs at the standard conditions of pH = 7.0, 50mM Na^+ and 50nM of primer concentration. The salt adjusted formulation proposed in [8], considering the same values for the pH and sequence concentration as in [7], is given by eq. (3), in Table 1.

Table 1. Basic and salt adjusted T_m formulas

$$T_m = 2*(|A|+|T|) + 4*(|C|+|G|) \qquad (1)$$

$$T_m = \frac{64.9 + 41*(|C|+|G|-16.4)}{(|A|+|T|+|C|+|G|)} \qquad (2)$$

$$T_m = 100.5 + 41*\frac{|C|+|G|-6.4}{|A|+|T|+|C|+|G|} - \frac{820}{|A|+|T|+|C|+|G|} + 16.6*\log[Na^+] \qquad (3)$$

Formulations dependent on the Nearest Neighbor (NN) model are widely used; one of the reasons for that is because "the stability of a DNA duplex appears to depend primarily on the identity of the nearest neighbor bases", as stated in [9]. Considering the four bases, there are sixteen different pairwise nearest neighbor possibilities that can be used to predict the stability and the T_m of a duplex. The NN model establishes values for the enthalpy and entropy variation (symbolically represented as ΔH and ΔS, respectively) to each one of the sixteen pairs. Several studies propose different values for ΔH and ΔS, as those ones described in [9], [10] and [11]. The ΔH and ΔS values for a DNA sequence $X = x_1 x_2 \ldots x_n$ are calculated by eqs (4).

$$\Delta H = \sum_{i=1}^{n-1} \Delta H(x_i, x_{i+1}) \qquad \Delta S = \sum_{i=1}^{n-1} \Delta S(x_i, x_{i+1}) \qquad (4)$$

A commonly used formulation to calculate the T_m value considering the contribution of the NN model was proposed in [12] and is given by eq. (5), where $R = 1.987$ cal/°C*mol is the molar gas constant, γ is the primer concentration in the solution, $[Na^+]$ is the salt concentration, and ΔH and ΔS the enthalpy and entropy variation of the primer sequences, respectively.

$$T_m = \frac{\Delta H}{\Delta S + R*\ln\frac{\gamma}{4}} - 273.15 + 16.6*\log[Na^+] \qquad (5)$$

A few other proposals to calculate T_m can be found in the literature. It is important to mention, however, that all the attempts to define a proper value for T_m are only an approximation of the real melting temperature since, as commented in [13], "a proper computation of the primer melting temperature does not appear to exist". Reference [14] presents a comparative study of different melting temperature calculation methods as well as the influence of the different NN interaction values available on the T_m.

Although there have been some attempts to estimate T_a (such as in [12]), it seems there is a consensus in the literature that the T_a value should be empirically determined (see [3]).

4 A Simulated Annealing Based Algorithm for the PCR Primer Design – Approaching the Design of a Primer as an Optimization Problem

Generally speaking, the simulating annealing [15] is a probabilistic algorithm suitable for finding a good approximation to a global optimum of a given function in a large search space. It is based on successive steps, which depend on an arbitrarily set parameter named temperature (T). In this paper, the design of a primer pair has been approached as an optimization problem, using the simulating annealing to conduct a search process throughout the space of all possible primer pairs, trying to find an optimal solution (i.e., a primer pair) to a function. The SA technique is heavily dependent on an appropriate choice of the function to be optimized. For this particular domain, the function was constructed based on the primer relevant characteristics for a successful PCR when amplifying a given DNA target, as described in Section 3.

Before presenting and discussing the function used in the experiment, a set of basic metrics that have been implemented for evaluating primer characteristics are described in Table 2, where *fp* and *rp* represent the forward and reverse primer in a primer pair, respectively.

In order to 'measure' how 'good' a primer pair (*fp*, *rp*) is (in relation to the probability of a successful PCR), it is mandatory to evaluate its conformity to the pre-established (user-defined) range of values, as well as to check for the occurrence of any of the unwanted characteristics, such as runs, repeats, secondary structures and non-specificity. The pre-established range of parameter values that should be defined by the user, are listed in Table 3.

In Table 3, the parameter MAX_DIF establishes the maximum allowed T_m difference between the forward and the reverse primer. The T_m, as well as the T_m difference, are measured in degree Celsius (°C). The 3'_END is a boolean parameter

Table 2. Basic metrics to evaluate a primer

Metric	Output	Description
len(*fp*), len(*rp*)	integer	Gives the length of the argument sequence
CG(*fp*), CG(*rp*)	real	Gives the % of C and G bases in the argument sequence
T_m(*fp*), T_m(*rp*)	real	Gives the T_m of the argument sequence
$T_m$dif(*fp*, *rp*)	real	Gives the T_m difference between the argument sequences
3'_end(*fp*), 3'_end(*rp*)	boolean	Checks the existence of a C or G base at the 3' end of the argument sequence
run(*fp*), run(*rp*)	boolean	Checks the existence of runs in the argument sequence
repeat(*fp*), repeat(*rp*)	boolean	Checks the existence of repeats in the argument sequence
spec(*fp*), spec(*rp*)	boolean	Checks the specificity of the argument sequence
sec(*fp*, *rp*)	boolean	Checks the existence of secondary structures in the argument sequences

Table 3. User-defined parameter values

Parameter	Range of values
LENGTH_INTERVAL	[MIN_LEN, MAX_LEN]
%CG_CONTENT	[MIN_CG, MAX_CG]
Tm	[MIN_Tm, MAX_Tm]
MAX_DIF	
3'_END	

that specifies the user preference (or not) for the occurrence of a base C or G at the 3' end of the primer sequences.

As there is no agreement about the best T_m formula to be used, the T_m value is estimated by the average of all T_m values calculated by using all distinct formulas described in the Section 3.4. In case of the T_m that use the enthalpy and entropy contribution, one calculation is made for each distinct NN interaction values proposed in [9], [10] and [11].

The proposed function for evaluating a primer pair (*fp*, *rp*) used for implementing the SA algorithm is given by eq. (6). It 'measures' how well the argument pair fits the pre-established range of values (for the characteristics given in Table 3) and how 'good' it is, concerning others (e.q. absence of repeats, runs, etc.), by associating a 'cost' to the unwanted values the characteristics can have. The highest the function value is, the less suitable is the primer pair given as its argument.

$$\begin{aligned}\text{tot\_cost}(fp, rp) = &\text{ len\_cost}(fp) + \text{len\_cost}(rp) + \%CG\_cost(fp) + \\ &+ \%CG\_cost(rp) + 3*(T_m\_cost(fp) + T_m\_cost(rp)) + \\ &+ T_m\text{dif\_cost}(fp, rp) + 3'\_end\_cost(fp) + \\ &+ 3'\_end\_cost(rp) + \text{run\_cost}(fp) + \text{run\_cost}(rp) + \\ &+ \text{repeat\_cost}(fp) + \text{repeat\_cost}(rp) + \text{spec\_cost}(fp) + \\ &+ \text{spec\_cost}(rp) + \text{sec\_struc\_cost}(fp, rp)\end{aligned} \quad (6)$$

Letting sq represent *fp* or *rp*, each individual cost function is defined as:

$$\text{len\_cost}(sq) = \begin{cases} 0 \text{ if } MIN\_LEN \leq \text{len}(sq) \leq MAX\_LEN \\ MIN\_LEN - \text{len}(sq) \text{ if } \text{len}(sq) < MIN\_LEN \\ \text{len}(sq) - MAX\_LEN \text{ if } \text{len}(sq) > MAX\_LEN \end{cases}$$

$$\%CG\_cost(sq) = \begin{cases} 0 \text{ if } MIN\_CG \leq CG(sq) \leq MAX\_CG \\ MIN\_CG - CG(sq) \text{ if } CG(sq) < MIN\_CG \\ CG(sq) - MAX\_CG \text{ if } CG(sq) > MAX\_CG \end{cases}$$

$$T_m\_cost(sq) = \begin{cases} 0 \text{ if } MIN\_T_m \leq T_m(sq) \leq MAX\_T_m \\ MIN\_T_m - T_m(sq) \text{ if } T_m(sq) < MIN\_T_m \\ T_m(sq) - MAX\_T_m \text{ if } T_m(sq) > MAX\_T_m \end{cases}$$

$$T_m\text{dif\_cost}(fp,rp) = \begin{cases} 0 & \text{if } T_m\text{dif}(fp,rp) \leq \text{MAX\_DIF} \\ T_m\text{dif}(fp,rp) - \text{MAX\_DIF} & \text{otherwise} \end{cases}$$

$$3'\_\text{end\_cost}(sq) = \begin{cases} 0 & \text{if } 3'\_\text{end}(sq) = \text{true} \\ 5 & \text{otherwise} \end{cases}$$

$$\text{run\_cost}(sq) = \begin{cases} 0 & \text{if } \text{run}(sq) = \text{false} \\ 5 * \text{number of runs} & \text{otherwise} \end{cases}$$

$$\text{reapeat\_cost}(sq) = \begin{cases} 0 & \text{if } \text{reapet}(sq) = \text{false} \\ 5 * \text{number of repeats} & \text{otherwise} \end{cases}$$

$$\text{spec\_cost}(sq) = \begin{cases} 0 & \text{if } \text{spec}(sq) = \text{true} \\ 5 * \text{number of alternative priming sites} & \text{otherwise} \end{cases}$$

$$\text{sec\_struc\_cost}(fp,rp) = \begin{cases} 0 & \text{if } \text{sec}(fp,rp) = \text{false} \\ \sum_{i=1}^{5} \text{highest\_cost}(G_i) & \text{otherwise} \end{cases}$$

The first three cost functions assign a cost to a primer (forward or reverse) whose values are outside the pre-established limits given in Table 3. The fourth cost function ($T_m dif\_cost$) assigns a cost to a primer pair depending on its temperature difference be higher than the user-defined parameter MAX_DIF.

The inclusion of the fifth cost function for the calculation of the value of *tot_cost* is dependent on an information given by the user (his/her preference (or not) for a base C or G at the 3' end of the primers). If the user has no preference, both *3'_end_cost(fp)* and *3'_end_cost(rp)* are not included on the *tot_cost* calculation. Otherwise, the non-existence of a C or G base at the 3' end of a primer, adds the arbitrary cost of 5 to the *tot_cost* value. The cost associated to the presence of runs and repeats is given by the functions *run_cost* and *repeat_cost*, respectively which add a cost of 5 to each time a run or a repeat is found. The *spec_cost* function is similar to the two previous cost functions.

The last cost function assigns a cost to the possible secondary structures that may be formed in each of the following five groups (G_i, i = 1,...,5): hetero-dimer, self-dimer (forward), self-dimer (reverse), hairpin (forward) and hairpin (reverse). In each group, different annealing situations can happen. The *sec_struc_cost* function takes into consideration only the annealing situation with the highest cost per group (*highest_cost(G_i)*). The cost of any annealing situation is given as the sum of the numbers of A-T matches (cost 2 each) and C-G matches (cost 4 each), as suggested in [13].

The pseudocode of the SA algorithm implemented is given in Figure 1. The algorithm starts by randomly choosing a pair of primers (referred to as current),

```
procedure SAPrimer
  begin
    fp_cur = find_primer(MIN_LEN, MAX_LEN)
    rp_cur = find_primer(MIN_LEN, MAX_LEN)
    cost_cur = tot_cost(fp_cur, rp_cur)
    T = 200
    decreasing_factor = 0.999
    while (T > 0.01)
      fp_new = find_primer_neighbor (len(fp_cur))
      rp_new = find_primer_neighbor (len(rp_cur))
      cost_new = tot_cost (fp_new, rp_new)
      if (cost_new < cost_cur) then
        // change the current primer pair
        fp_cur = fp_new
        rp_cur = rp_new
        cost_cur = cost_new
      else
        num = random( )
        if (num < exp$\left(\frac{-\Delta E}{T}\right)$) then
          fp_cur = fp_new
          rp_cur = rp_new
          cost_cur = cost_new
      T = T * decreasing_factor
  end
```

Fig. 1. Pseudocode of the SAPrimer procedure

i.e. (*fp_cur*, *rp_cur*) such that |*fp_cur*| = *m* and |*rp_cur*| = *n* with MIN_LEN ≤ *m*, *n* ≤ MAX_LEN; the first *m* bases and the last *n* complementary bases of the target DNA sequence are the *fp_cur* and *rp_cur* respectively and their cost is evaluated.

At each step, the algorithm randomly chooses a new candidate-pair (*fp_new*, *rp_new*) in the neighborhood of the current pair, as any primer pair such that |*fp_cur*| − 3 ≤ |*fp_new*| ≤ |*fp_cur*| + 3 and |*rp_cur*| − 3 ≤ |*rp_new*| ≤ |*rp_cur*| + 3; its cost is them evaluated. The cost values of both solutions are then compared and the primer pair that has the smaller cost becomes the current pair. Even if the current pair has not changed, there is still a chance of the new candidate pair to become the current pair, depending on a probability function based on both, the T parameter and the ΔE parameter (where ΔE is the difference between the new and current solution cost). The acceptance of a solution with a higher cost is an attempt to prevent local minima. The value of 200 assigned to the parameter T was empirically determined as well the decreasing_factor of 0.999.

Figure 2 exemplifies the use of the SAPrimer when looking for a suitable primer pair to amplify a codant region of the gene ADAM9 (GenBank accession NM_003816). The parameters in Figure 2 have their default values. The SAPrimer prompts the best primer pair found, showing its length, %CG and T_m. The restriction %CG was not satisfied by the reverse primer. The %CG of 32.14, however, is not too far from the user-defined minimum of 40%. Inspecting the 3' end of the ADAM9

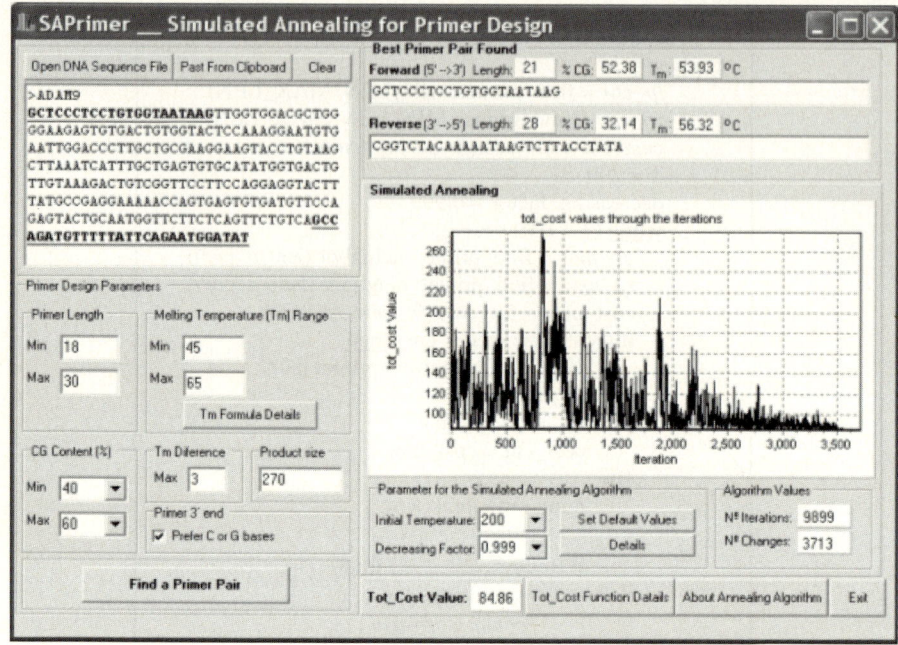

Fig. 2. SAPrimer interface

sequence in Figure 2, it can be noticed that the number of C/G bases is low. Notice that, even if the reverse primer chosen was length 35 (bigger than the user-defined maximum), the %CG would be 34.3, the T_m would be 65.48 °C and the T_m difference (between forward and reverse) equal to 11.55 °C, which would be a worst pair as far as the user-requirements concern.

5 Conclusions

The efficiency of a PCR reaction is highly dependent on the efficiency of the primer used [16]. As primer design involves the optimization of many parameters, computer programs are commonly used in this task. Some programs, however, do not find a solution when some restrictions are not satisfied. The SAPrimer software always finds the best possible primer pair to amplify a specific DNA sequence, even when some restrictions can not be satisfied (for example, when the given DNA sequence does not have an appropriated %CG or when the T_m value of a primer (forward or reverse) does not respect user-defined range of values). The work will proceed by implementing other heuristic search strategies (starting with genetic algorithm) in an attempt to identify the most suitable type of search for dealing with the problem of primer design.

Acknowledgments. The first author would like to thank CAPES for the scholarship granted and the opportunity to be financially supported while spending six months as a research visitor at The University of Newcastle, Australia, working in collaboration with The Newcastle Bioinformatics Initiative group.

References

1. Mullis, K.B., Faloona, F.: Specific synthesis of DNA in vitro via a polymerase-catalyzed chain reaction. Methods Enzymology 155, 335–350 (1987)
2. Metzker, M.L., Caskey, T.C.: Polymerase Chain Reaction (PCR), Encyclopedia of Life Science. Nature Publishing Group (2001)
3. Innis, M.A., Gelfand, D.H.: Optimization of PCRs. In: Innis, Gelfand, Sninsky, White (eds.) PCR Protocols. Academic Press, New York (1990)
4. Kamel, A., Abd-Elsalam: Bioinformatic tools and guideline for PCR primer design. African Journal of Biotechnology 2(5), 91–95 (2003)
5. Kwok, S., Kellogg, D.E., McKinney, N., Spasic, D., Goda, L., Levenson, C., Sninsky, J.: Effects of primer-template mismatches on the polymerase chain reaction: Human Immunodeficiency Virus 1 model studies. Nucleic Acids Research 18, 999–1005 (1990)
6. Wallace, R.B., Shaffer, J., Murphy, R.F., Bonner, J., Hirose, T., Itakura, K.: Hybridization of synthetic oligodeoxyribonucleotides to phi chi 174 DNA: the effect of single base pair mismatch. Nuclic Acids Research 6, 3543–3557 (1979)
7. Marmur, J., Doty, P.: Determination of the base composition of deoxyribonucleic acid from its thermal denaturation temperature. Journal of Molecular Biology 5, 109–118 (1962)
8. Howley, P.M., Israel, M.F., Law, M.-F., Martin, M.A.: A rapid method for detecting and mapping homology between heterologous DNAs. Journal of Biological Chemistry 254, 4876–4883 (1979)
9. Breslauer, K.J., Frank, R., Blocker, H., Marky, L.A.: Predicting DNA duplex stability from the base sequence. Proc. Natl. Acad. Sci. USA 83, 3746–3750 (1986)
10. Sugimoto, N., Nakano, S., Yoneyama, M., Honda, K.: Improved thermodynamic parameters and helix initiation factor to predict stability of DNA duplexes. Nucleic Acids Research 24, 4501–4505 (1996)
11. SantaLucia Jr., J., Allawi, H.T., Seneviratne, P.A.: Improved Nearest-Neighbor parameters for predicting DNA duplex stability. Biochemistry 35, 3555–3562 (1996)
12. Rychlik, W., Spencer, W.J., Rhoads, R.E.: Optimization of the annealing temperature for DNA amplification in vitro. Nucleic Acids Research 18, 6409–6412 (1990)
13. Kämpke, T., Kieninger, M., Mecklenbug, M.: Efficient primer design algorithms. Bioinformatics 17(3), 214–225 (2001)
14. Panjkovich, A., Melo, F.: Comparison of different melting temperature calculation methods for short DNA sequences. Bioinformatics 21(6), 711–722 (2005)
15. Kirkpatrick, S., Gelatt, C.D., Vecchi, M.P.: Optimization by simulated annealing. Science 220, 671–680 (1983)
16. He, Q., Marjamäki, M., Soini, H., Mertsola, J., Viljanen, M.K.: Primers are decisive for sensitivity of PCR. BioTechniques 17(1), 82–87 (1994)

On Differential Stroke Diagnosis by Neuro-fuzzy Structures

Krzysztof Cpałka[1,2], Olga Rebrova[3], Tomasz Gałkowski[1], and Leszek Rutkowski[1,2]

[1] Czestochowa University of Technology, Poland,
Department of Computer Engineering
[2] Academy of Humanities and Economics, Poland,
Department of Artificial Intelligence
[3] Russian Academy of Medical Sciences, Russia,
Institute of Neurology
cpalka@kik.pcz.czest.pl, olga@neurology.ru,
lrutko@kik.pcz.czest.pl, tom@kik.pcz.czest.pl

Abstract. In this paper we develop a neuro-fuzzy system for stroke diagnosis. A novel concept of weights describing importance of antecedents and rules will be incorporated into construction of such systems. Simulation results based on 298 real stroke data will be presented.

1 Introduction

Both fuzzy systems and neural networks, along with probabilistic methods, evolutionary algorithms, rough sets and uncertain variables, constitute a consortium of soft computing techniques (see e.g. [1], [10], [11]). These techniques are often used in combination. For example, fuzzy inference systems are frequently converted into connectionist structures called neuro-fuzzy systems which exhibit advantages of neural networks and fuzzy systems. We present various soft computing methods for differential stroke diagnosis.

One of the most actively developing directions of computational intelligence [18] and soft computing [20] is concerned with applications of neural networks and fuzzy systems [6] in medicine. Both techniques allow overcoming problems of medical data analysis, e.g. their mainly qualitative character, incompleteness of the information and non-linearity of formal descriptions. In this paper we will develop a neuro-fuzzy system for stroke diagnosis. We will present an alternative method to that studied in [16] and based on neural networks. Our interest in differential stroke diagnosis is justified by the following circumstances. The death rate caused by vascular diseases in Russia is one of the highest in the world. Among all death cases caused by cerebral-vascular pathology the stroke diagnosis with the pointing of its type is fixed only in 21% of the cases, non-differential diagnosis like "poor brain blood circulation" - in 39% of cases. It is registered about 400 000 stroke annually and the rate of diseases among the population of able-bodied age is permanently increasing. Death rate because of stroke reaches

40% and 62% of the persons who survived after a stroke become invalids. The prevention has crucial importance in death reduction and invalidity reduction caused by the stroke. The essential effect gives the optimization of the patient aid system based on heterogeneity concept. That's why it is of great importance to provide comprehensive support to medical staff in making the fast and exact diagnostic decision that defines further tactics of treatment. It is important for stroke type diagnosis as well as for pathogenetic subtype of ischemic stroke diagnosis. It is also very important to make quality differential clinical diagnosis of stroke type in short time. Usually rate of mistakes in the diagnosis is in the range of 20-45%. So the problem of decision support in making a competent diagnosis is of a great interest for medical doctors. In the paper we will present various types on neuro-fuzzy systems for differential stroke diagnosis.

2 Various Types on Neuro-fuzzy Systems for Differential Stroke Diagnosis

In this paper we use various neuro-fuzzy systems for differential stroke diagnosis. In particular, we consider a multi-input, single-output neuro-fuzzy systems mapping $\mathbf{X} \to \mathbf{Y}$, where $\mathbf{X} \subset \mathbf{R}^n$ and $\mathbf{Y} \subset \mathbf{R}$.

The fuzzifier performs a mapping from the observed crisp input space $\mathbf{X} \subset \mathbf{R}^n$ to the fuzzy sets defined in $\mathbf{X}$. The most commonly used fuzzifier is the singleton fuzzifier which maps $\bar{\mathbf{x}} = [\bar{x}_1, \ldots, \bar{x}_n] \in \mathbf{X}$ into a fuzzy set $A' \subseteq \mathbf{X}$ characterized by the membership function

$$\mu_{A'}(\mathbf{x}) = \begin{cases} 1 \text{ if } \mathbf{x} = \bar{\mathbf{x}} \\ 0 \text{ if } \mathbf{x} \neq \bar{\mathbf{x}} \end{cases}. \tag{1}$$

The fuzzy rule base consists of a collection of N fuzzy IF-THEN rules in the form

$$R^{(k)} : \left[\text{IF } x_1 \text{ is } A_1^k \text{ AND} \ldots \text{AND } x_n \text{ is } A_n^k \text{ THEN } y \text{ is } B^k \right], \tag{2}$$

where $\mathbf{x} = [x_1, \ldots, x_n] \in \mathbf{X}$, $y \in \mathbf{Y}$, $A_1^k, A_2^k, \ldots, A_n^k$ are fuzzy sets characterized by membership functions $\mu_{A_i^k}(x_i)$, whereas B^k are fuzzy sets characterized by membership functions $\mu_{B^k}(y)$, respectively, $k = 1, \ldots, N$. The firing strength of rules is given by

$$\mu_{\mathbf{A}^k}(\bar{\mathbf{x}}) = \underset{i=1}{\overset{n}{T}} \left\{ \mu_{A_i^k}(\bar{x}_i) \right\}. \tag{3}$$

The fuzzy inference determines a mapping from the fuzzy sets in the input space $\mathbf{X}$ to the fuzzy sets in the output space $\mathbf{Y}$. Each of N rules (2) determines a fuzzy set $\bar{B}^k \subset \mathbf{Y}$ given by the compositional rule of inference

$$\bar{B}^k = A' \circ \left(\mathbf{A}^k \to B^k \right), \tag{4}$$

where $\mathbf{A}^k = A_1^k \times A_2^k \times \ldots \times A_n^k$. Fuzzy sets $\bar{B}^k$, according to the formula (4), are characterized by membership function expressed by the sup-star composition

$$\mu_{\bar{B}^k}(y) = \sup_{\mathbf{x} \in \mathbf{X}} \left\{ T \left\{ \mu_{A'}(\mathbf{x}), \mu_{\mathbf{A}^k \to B^k}(\mathbf{x}, y) \right\} \right\}, \tag{5}$$

where T can be any operator in the class of t-norms. It is easily seen that for a crisp input $\bar{\mathbf{x}} \in \mathbf{X}$, i.e. a singleton fuzzifier (1), formula (5) becomes

$$\mu_{\bar{B}^k}(y) = \mu_{\mathbf{A}^k \to B^k}(\bar{\mathbf{x}}, y) = I\left(\mu_{\mathbf{A}^k}(\bar{\mathbf{x}}), \mu_{B^k}(y)\right), \tag{6}$$

where $I(\cdot)$ is a t-norm (Mamdani approach) or fuzzy implication [7] (logical approach). The aggregation operator, applied in order to obtain the fuzzy set B' based on fuzzy sets $\bar{B}^k$, is the t-norm or t-conorm operator, depending on the type of fuzzy inference.

The defuzzifier performs a mapping from a fuzzy set B' to a crisp point $\bar{y}$ in $\mathbf{Y} \subset \mathbf{R}$. The COA (centre of area) method is defined by the following formula

$$\bar{y} = \frac{\int\limits_{\mathbf{Y}} y \mu_{B'}(y) \, dy}{\int\limits_{\mathbf{Y}} \mu_{B'}(y) \, dy}, \tag{7}$$

or by

$$\bar{y} = \frac{\sum\limits_{r=1}^{N} \bar{y}^r \mu_{B'}(\bar{y}^r)}{\sum\limits_{r=1}^{N} \mu_{B'}(\bar{y}^r)} \tag{8}$$

in the discrete form, where $\bar{y}^r$ denotes centres of the membership functions $\mu_{B^r}(y)$, i.e. for $r = 1, \ldots, N$

$$\mu_{B^r}(\bar{y}^r) = \max_{y \in \mathbf{Y}} \{\mu_{B^r}(y)\}. \tag{9}$$

2.1 Mamdani Approach with Non-parameterized Triangular Norms

In this section we present neuro-fuzzy system of the Mamdani type with non-parameterized triangular norms. In Mamdani approach function (6) is given by

$$\mu_{\bar{B}^k}(y) = T\left\{\mu_{\mathbf{A}^k}(\bar{\mathbf{x}}), \mu_{B^k}(y)\right\}, \tag{10}$$

and the aggregation is carried out by

$$B' = \bigcup_{k=1}^{N} \bar{B}^k. \tag{11}$$

The membership function of B' is computed by the use of a t-conorm, that is

$$\mu_{B'}(y) = \mathop{S}_{k=1}^{N} \{\mu_{\bar{B}^k}(y)\}. \tag{12}$$

In our study we use Zadeh families triangular norms. The Zadeh t-norm, described by min operator, is given as follows

$$T\{\mathbf{a}\} = \min_{i=1,\ldots,n} \{a_i\}, \tag{13}$$

and the Zadeh t-conorm, described by max operator, is given by

$$S\{\mathbf{a}\} = \max_{i=1,\dots,n}\{a_i\}. \tag{14}$$

Consequently, in Mamdani approach with non-parameterized triangular norms formula (8) takes the form

$$\bar{y} = \frac{\sum_{r=1}^{N} \bar{y}^r \, \underset{k=1}{\overset{N}{S}} \left\{ T \left\{ \begin{array}{c} T\left\{\mu_{A_1^k}(\bar{x}_1),\dots,\mu_{A_N^k}(\bar{x}_N)\right\}, \\ \mu_{B^k}(\bar{y}^r) \end{array} \right\} \right\}}{\sum_{r=1}^{N} \underset{k=1}{\overset{N}{S}} \left\{ T \left\{ \begin{array}{c} T\left\{\mu_{A_1^k}(\bar{x}_1),\dots,\mu_{A_N^k}(\bar{x}_N)\right\}, \\ \mu_{B^k}(\bar{y}^r) \end{array} \right\} \right\}}. \tag{15}$$

2.2 Mamdani Approach with Parameterized Triangular Norms

In this section we present neuro-fuzzy system of the Mamdani type with adjustable triangular norms. The hyperplanes corresponding to parameterized families of t-norms and t-conorms can be adjusted in the process of learning of an appropriate parameter. We use Dombi families triangular norms. The Dombi t-norm is given as follows

$$\vec{T}\{\mathbf{a}; p\} = \left(1 + \left(\sum_{i=1}^{n}\left(\frac{1-a_i}{a_i}\right)^p\right)^{\frac{1}{p}}\right)^{-1}, \tag{16}$$

and the Dombi t-conorm is given by

$$\vec{S}\{\mathbf{a}; p\} = 1 - \left(1 + \left(\sum_{i=1}^{n}\left(\frac{a_i}{1-a_i}\right)^p\right)^{\frac{1}{p}}\right)^{-1}, \tag{17}$$

where $\vec{T}$ and $\vec{S}$ stand for a t-norm and t-conorm of the Dombi family parameterized by $p \in (0, \infty)$.

Consequently, in Mamdani approach with parameterized triangular norms formula (15) takes the form

$$\bar{y} = \frac{\sum_{r=1}^{N} \bar{y}^r \, \underset{k=1}{\overset{N}{\vec{S}}} \left\{ \vec{T} \left\{ \begin{array}{c} \vec{T}\left\{\mu_{A_1^k}(\bar{x}_1),\dots,\mu_{A_N^k}(\bar{x}_N); p_k^\tau\right\}, \\ \mu_{B^k}(\bar{y}^r); p_k^I \end{array} \right\}; p^{\mathrm{agr}} \right\}}{\sum_{r=1}^{N} \underset{k=1}{\overset{N}{\vec{S}}} \left\{ \vec{T} \left\{ \begin{array}{c} \vec{T}\left\{\mu_{A_1^k}(\bar{x}_1),\dots,\mu_{A_N^k}(\bar{x}_N); p_k^\tau\right\}, \\ \mu_{B^k}(\bar{y}^r); p_k^I \end{array} \right\}; p^{\mathrm{agr}} \right\}}. \tag{18}$$

2.3 Logical Approach with Non-parameterized Triangular Norms

In this section we present neuro-fuzzy system of the logical type with Zadeh non-parameterized triangular norms described by formulas (13) and (14). When we use logical approach with an S-implication, function (6) is defined as follows

$$\mu_{\bar{B}^k}(y) = S\left\{1 - \mu_{\mathbf{A}^k}(\bar{\mathbf{x}}), \mu_{B^k}(y)\right\}, \tag{19}$$

and the aggregation is carried out by

$$B' = \bigcap_{k=1}^{N} \bar{B}^k. \tag{20}$$

The membership function of B' is determined by the use of a t-norm, i.e.

$$\mu_{B'}(y) = \underset{k=1}{\overset{N}{T}}\left\{\mu_{\bar{B}^k}(y)\right\}. \tag{21}$$

Consequently, in Logical approach with non-parameterized triangular norms formula (8) takes the form

$$\bar{y} = \frac{\sum_{r=1}^{N} \bar{y}^r \underset{k=1}{\overset{N}{T}} \left\{ S \left\{ \begin{array}{c} 1 - T\left\{\mu_{A_1^k}(\bar{x}_1), \ldots, \mu_{A_N^k}(\bar{x}_N)\right\}, \\ \mu_{B^k}(\bar{y}^r) \end{array} \right\} \right\}}{\sum_{r=1}^{N} \underset{k=1}{\overset{N}{T}} \left\{ S \left\{ \begin{array}{c} 1 - T\left\{\mu_{A_1^k}(\bar{x}_1), \ldots, \mu_{A_N^k}(\bar{x}_N)\right\}, \\ \mu_{B^k}(\bar{y}^r) \end{array} \right\} \right\}}. \tag{22}$$

2.4 Logical Approach with Parameterized Triangular Norms

In this section we present neuro-fuzzy system of the logical type with Dombi adjustable triangular norms described by formulas (16) and (17). Consequently, formula (22) takes the form

$$\bar{y} = \frac{\sum_{r=1}^{N} \bar{y}^r \underset{k=1}{\overset{N}{\vec{T}}} \left\{ \vec{S} \left\{ \begin{array}{c} 1 - \vec{T}\left\{\mu_{A_1^k}(\bar{x}_1), \ldots, \mu_{A_N^k}(\bar{x}_N); p_k^\tau\right\}, \\ \mu_{B^k}(\bar{y}^r); p_k^I \end{array} \right\}; p^{\mathrm{agr}} \right\}}{\sum_{r=1}^{N} \underset{k=1}{\overset{N}{\vec{T}}} \left\{ \vec{S} \left\{ \begin{array}{c} 1 - \vec{T}\left\{\mu_{A_1^k}(\bar{x}_1), \ldots, \mu_{A_N^k}(\bar{x}_N); p_k^\tau\right\}, \\ \mu_{B^k}(\bar{y}^r); p_k^I \end{array} \right\}; p^{\mathrm{agr}} \right\}}. \tag{23}$$

3 Simulation Results

The stroke data contains 298 instances and each instance is described by 30 attributes. In our experiments all sets are divided into a learning sequence (268 sets) and a testing sequence (30 sets). Out of 298 data samples, 211 cases represent ischemic stroke (IS), 73 cases represent hemorrhagic stroke (HS) and 14 cases describe subarachnoid hemorrhage (SAH). We used the Mamdani type neuro-fuzzy system (15) with non-parameterized triangular norms, Mamdani type neuro-fuzzy system (18) with parameterized triangular norms, logical type neuro-fuzzy system (22) with non-parameterized triangular norms and logical type neuro-fuzzy system (23) with parameterized triangular norms. The experimental results for the stroke problem are depicted in Table 1.

Table 1. Simulation results for stroke diagnosis

No	Type of the system	Learning accuracy [%]	Testing accuracy [%]
1	Mamdani type neuro-fuzzy system (15) with Zadeh triangular norms ($N=6$, Gaussian membership function)	6.71	16.00
2	Mamdani type neuro-fuzzy system (18) with Dombi triangular norms ($N=6$, Gaussian membership function)	7.83	6.66
3	logical type neuro-fuzzy system (23) with Zadeh triangular norms ($N=6$, Gaussian membership function)	5.97	10.00
4	Logical type neuro-fuzzy system (22) with Dombi triangular norms ($N=6$, Gaussian membership function)	6.34	6.66

4 Conclusions

Various types on neuro-fuzzy systems for differential stroke diagnosis for differential diagnostics of three types of stroke were developed. A high accuracy of our approach is demonstrated in simulation results depicted in Table 1. The logical type neuro-fuzzy system (22) with Dombi triangular norms seems to be superior over other structures.

Acknowledgment

This work was partly supported by the Foundation for Polish Science (Professorial Grant 2005-2008) and Polish Ministry of Science and Higher Education (Habilitation Project 2007-2010, Special Research Project 2006-2009, Polish-Singapore Research Project 2008-2010 and Research Project 2008-2010).

References

1. Aliev, R.A., Aliev, R.R.: Soft Computing and its Applications. World Scientific Publishing, Singapore (2001)
2. Alonso, J.M., Cordon, O., Guillaume, S., Magdalena, L.: Highly Interpretable Linguistic Knowledge Bases Optimization: Genetic Tuning versus Solis-Wetts. Looking for a good interpretability-accuracy trade-off. In: Proc. of the 2007 IEEE Int. Conf. on Fuzzy Systems, pp. 1–6 (2007)
3. Amaral, T.G., Crisostomo, M.M.: An Approach to Improve the Interpretability of Neuro-Fuzzy Systems. In: Proc. of the 2006 IEEE Int. Conf. on Fuzzy Systems, pp. 1843–1850 (2006)
4. Casillas, J., Cordon, O., Herrera, F., Magdalena, L. (eds.): Interpretability Issues in Fuzzy Modeling. Springer, Heidelberg (2003)

5. Czabanski, R.: Neuro-Fuzzy Modelling Based on a Deterministic Annealing Approach. Int. J. Appl. Math. Comput. Sci. 15(4), 561–576 (2005)
6. Czogała&, E., Łęski, J.: Fuzzy and Neuro-Fuzzy Intelligent Systems. Physica-Verlag, Heidelberg (2000)
7. Fodor, C.: On fuzzy implication. Fuzzy Sets and Systems 42, 293–300 (1991)
8. Gorzałczany, M.: Computational Intelligence Systems and Applications: Neuro-Fuzzy and Fuzzy Neural Synergisms. Springer, Heidelberg (2002)
9. Guillaume, S.: Designing fuzzy inference systems from data: An interpretability-oriented review. IEEE Trans. Fuzzy Syst. 9(3), 426–443 (2001)
10. Kasabov, N.: Foundations of Neural Networks, Fuzzy Systems and Knowledge Engineering. The MIT Press, CA (1996)
11. Kecman, V.: Learning and Soft Computing. MIT, Cambridge (2001)
12. Kumar, M., Stoll, R., Stoll, N.: A robust design criterion for interpretable fuzzy models with uncertain data. IEEE Trans. Fuzzy Syst. 14(2), 314–328 (2006)
13. Łęski, J., Henzel, N.: A Neuro-Fuzzy System Based on Logical Interpretation of If-then Rules. Int. J. Appl. Math. Comput. Sci. 10(4), 703–722 (2000)
14. Łęski, J.: A Fuzzy If-Then Rule-Based Nonlinear Classifier. Int. J. Appl. Math. Comput. Sci. 13(2), 215–223 (2003)
15. Manley-Cooke, P., Razaz, M.: An efficient approach for reduction of membership functions and rules in fuzzy systems. In: Proc. of the 2007 IEEE Int. Conf. on Fuzzy Systems, pp. 1–6 (2007)
16. Rebrova, O., et al.: Expert system and neural network for stroke diagnosis. International Journal of Information Technology and Intelligent Computing 1(2), 441–453 (2006)
17. Riid, A., Rustern, E.: Interpretability of Fuzzy Systems and Its Application to Process Control. In: Proc. of the 2007 IEEE Int. Conf. on Fuzzy Systems, pp. 1–6 (2007)
18. Rutkowski, L.: Computational Intelligence. Springer, Heidelberg (2007)
19. Rutkowski, L.: Flexible Neuro-Fuzzy Systems. Kluwer Academic Publishers, Dordrecht (2004)
20. Rutkowski, L.: New Soft Computing Techniques for System Modeling, Pattern Classification and Image Processing. Springer, Heidelberg (2004)
21. Rutkowski, L., Cpałka, K.: Flexible neuro-fuzzy systems. IEEE Trans. Neural Networks 14(3), 554–574 (2003)
22. Yager, R.R., Filev, D.P.: Essentials of fuzzy modelling and control. John Wiley & Sons, Chichester (1994)

Novel Quantitative Method for Spleen's Morphometry in Splenomegally

Tomasz Sołtysiński

Institute of Precision and Biomedical Engineering, Department of Mechatronics,
Warsaw University of Technology, Św. Andrzeja Boboli 8, 02-525 Warsaw, Poland
solek@mchtr.pw.edu.pl

Abstract. Novel method for spleen's semiautomatic accurate quantitative morphometry adaptable in diagnosis of splenomegally is described. The method is based on multiscale wavelet image decomposition, Bayesian inference that reveals the most probable structure delineation in the image and spectral method to smooth and approximate the most probable representation of a real contour hidden in a noisy or fuzzy data.

1 Introduction

The imaging of spleen by means of ultrasonography (as shown on fig. 2) is the most straightforward way to confirm the diagnosis based on manual inspection of a patient with splenomegally. The noninvasive technics of medical imaging [30] allow for the qualitative visual inspection of size, shape and structure of internal organ [11, 12, 13, 14, 15, 16, 17] and for quantitative measurement of particular property. Such a measurement may be performed by a manual placement of mark points on the image or by semi- or fully automatic procedure. It is usually hard to perform in the case of spleen, as its image is inhomogenous, disturbed or partially missing due to the lack of echo from the region beneath the ribs. Even if the manual placement of the mark points allows for precise determination of spleen's boundary, it is done locally. On the other hand, full automatic delineation is hard to perform due to the lack of prior knowledge on missing spleen data in the image or high level of speckle noise as well as fuzzy nature of ultrasonic imaging technics. Delineation of the fuzzy boundaries on the image is thus challenging to determine [2, 3] and when it is done in automatic manner the inaccuracy of edge detection may be not acceptable. Hence, throughout this study, the novel method is presented that has semiautomatic nature and allows the investigator to place the control mark points as well as enables the computer to find the best approximation of the required organ's shape in terms of maximization of probabilities of edge detection.

The proposed method is a new approach to image segmentation that combines the speed of spectral method in contour detection [7] and the Bayesian inference [18, 19, 20, 21] that allows for the most probable estimation of initial edge map. Additionally multiscale data decomposition [23] is adapted. Further in the study the method is refereed as Bayesian constrained spectral method (BCSM). The method has been successfully applied to ultrasonic data [10, 24, 28], CT brain data with aneurysm described in details in [9] as well as to multimodal PET-CT, MRI-CT data as indicated in [10]. Some other applications of the method, like supporting the analysis of molecular imaging by automatic cell's morphometry has also been tested [29].

In presented approach the real contour of investigated and segmented spleen is extracted in the iterative process of solving the nonlinear partial differential equation (PDE). This step is realized in Fourier space by fast spectral method. PDE is approaching the function, that reveals the real edge of the object and starts from an initially guessed edge map. This iterative process is controlled by two parameters that describe fidelity of reconstruction: one, μ is steering the PDE and the other is responsible for slight smoothing of the resolved subsequent partial approximations of the final solution [10, 24]. The contour detection depends very much on *a priori* knowledge which, in this particular case, is given by an examinator who manually determines control mark points. These points are constraining an enhanced prior that takes part in Bayesian inference in addition to the prior obtained from multiscale decomposition. All further steps are fully automatic and the final morphometric parameters like diameteres and circumference lengths may be revealed in required units. The final shape or circumference is described by a 1D radial function that may be further analysed.

2 Methods

The method is characterized by a flow chart shown on figure 1. The left branch describes wavelet multiscale decomposition, noise analysis and reduction as well as the construction of priors. The right branch is showing the path of Bayesian inference and spectral method driven by initially infered edge map. Detailed description of the method is available in recently published work [9, 10, 24, 25, 28, 29]. The most general framework necessary for the spleen's morphometry is further explained. In opposite to previously shown fully automatic approach the method presented here is semiautomatic in nature as explained in the subsequent sections.

2.1 Application of PDE's in Shape Reconstruction

Contouring may be expressed as solving the partial differential equation (PDE). This is the approach commonly found in methods based on active contours [13, 14, 15] that are iteratively approaching the final contour. Following the work presented in [7] the final contour f may be found by solving the elliptic equation of Helmholtz type

$$\nabla^2 f - \mu(f - g) = 0 \quad (1)$$

This equation uses known variable which is initially guessed edge map g. It is solved in spherical coordinates. Moving the g term to right side and relating it to

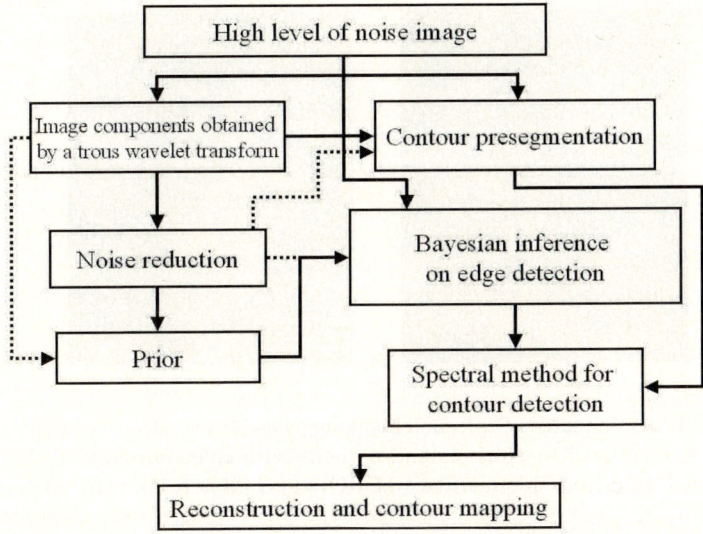

Fig. 1. Flow chart showing the main paths one needs to follow when applying BCSM for spleen morphometry

a previously found value of f, called f_n the PDE can be expressed in linearized form

$$\alpha \nabla^2 f_{n+1} - f_{n+1} = g_{f_n} \qquad (2)$$

Such an equation is further easily solved by the fast spectral method. Applying $\alpha = 1/\mu$ the solution may be controlled by value of μ.

2.2 Bayesian Edge Map Determination

The edge map determination [10] is based on Bayesian approach [6, 18, 19, 20, 21] and provides the best choice among all others taking into account the risk associated with each one and mutual relationship between them.

The edge map g is determined by Bayesian inference on edge placement in image data. Let $P(E_i/I)$ denote the required probability of the most appropriate edge in a data set. This is the conditional probability as it depends on the contents of I. $P(E_i/I)$ is the probability of the fact that the pixel in I belongs to the edge class E_i, knowing the value of intensity of this point. Let $P(I/E_i)$ be a probability of how much the value or intensity of a point is depending on edge class E_i. This term serves as a kernel. $P(E_i)$ is simply the probability of existence of the edge class E_i among all other detected edge classes. Edge class is a set of some subsequent pixel intensity values. Then the required probability can be found by solving the Bayes rule:

$$P(E_i/I) = \frac{P(I/E_i)P(E_i)}{P(I)} = \frac{P(I/E_i)P(E_i)}{\sum_i P(I/E_i)P(E_i)} \qquad (3)$$

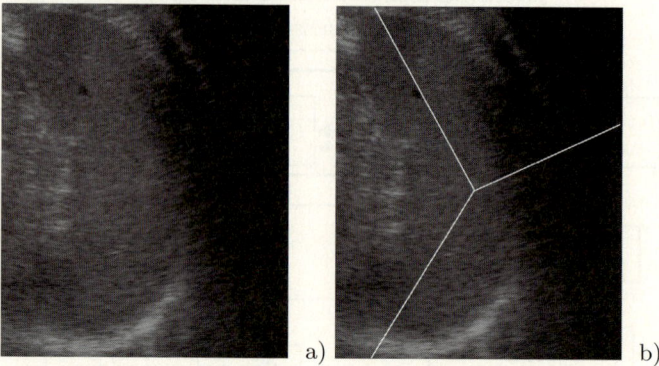

Fig. 2. a) Ultrasonic image of spleen. This image was selected from a sample of a dozen different spleens of healthy subjects and patients with splenomegally. b) The scanning radii. For each point on circumference of ROI the radius from centroid to this point is taken. Data for inference are collected along the radius. Three example radii are shown, for clockwise scanning.

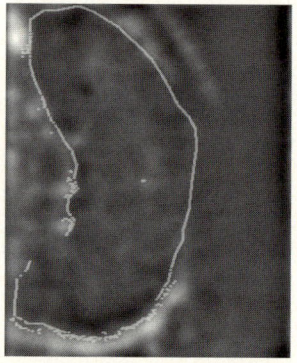

Fig. 3. Raw edge map infered in Bayesian way

$P(I)$ is a sum of all probabilities $P(I/E_i)$ weighted by $P(E_i)$ and thus remaining constant. $P(I)$ is only a normalizing term and can be excluded from further analysis. The standard way of solving Bayes equation is to maximize of the right side over the parameter E_i (maximum likelihood, ML step) and then to maximize of the found solution over all accessible data (maximum-a-posteriori, MAP step). The $P(E_i)$ is a prior and contains *a priori* knowledge.

In practice the $P(I/E_i)$ is estimated from the histogram of I along given radius (2) from centroid to a point of circumference of some bounding box selecting the region of interest (ROI). The histogram [10] is shrank in such a way that each bin is equal to the edge size assuming that each expected edge covers the same number of intensity levels. Calculating the normalized probability over all edges E_i [10], ML step is performed and the most probable edge in I is estimated.

Then the MAP is done by searching for maximum over the data itself, and usually the first maximum in $P(I/E_i)$ is detected as an edge. $P(E_i)$ may be a constant value if we assume all edge classes as equally probable or may be distributed uniquely according to the prior knowledge. From $P(E/I)$ the polar edge map, $g(\theta)$ is derived (shown on fig. 3). The θ is an angle between the first and the current scanning radius. A centroid is automatically determined from a uniquely processed data. Its processing is done by a technic of multiscale decomposition and denoising done by the $\grave{a}$ trous algorithm [5, 10, 25, 27] and its simple modification known as multiresolution support [27].

2.3 Speckles Removal and Priors Construction by Application of $\grave{a}$ trous Algorithm

The noise common in ultrasonic imaging has Rayleigh distribution and speckle nature. When ultrasonic data is rescaled logarithmically the noise becomes Gaussian. This makes the application of Gaussian noise model [26] suitable for removal the speckles. Gaussian noise model can also be adapted as the first level of approximation [26] in original data and used to initially determine the rate of noise reduction.

In presented approach the noise is removed by first decomposing the data by $\grave{a}$ trous transform (fig. 4), cancellation of nonsiginificant wavelet coefficients [22] on each scale (component) according to the noise model, and subsequent synthesis. All details of the technic has been explained in [10, 27].

Multiscale components obtained from wavelet decomposition are excellent source of data for prior construction. The prior is used in inference to maximize the probability of correct edge detection in the image data and is constructed by summation of the components containing the most grain details only. The prior used in discussed study is shown on figure 6.

The same components are good source of knowledge to determine the centroid of the structure hidden in the image data. It is generally expected that in the region of structure there is less pixels of high intensities than in a surrounding

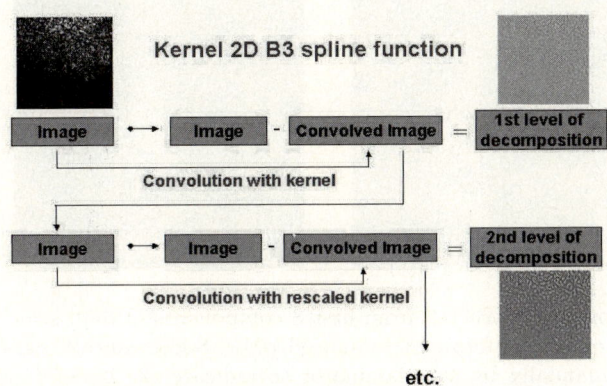

Fig. 4. Scheme describing $\grave{A}$ trous algorithm

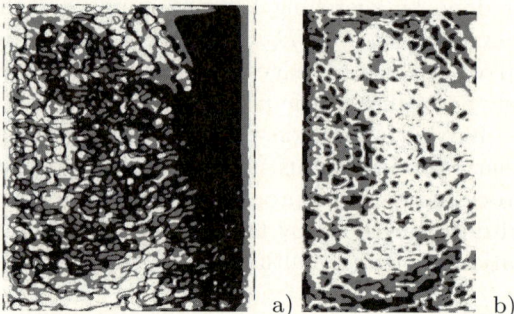

Fig. 5. a) Result of multiresolution support technic [27] applied to input image of spleen. b) The inverted image of data.

area. When multiresolution support (MRS) [27] technic is applied with the low threshold (fig. 5) and the resulting image is inverted the data takes the form shown on figure 5. Taking such an image data as the input for centroid calculation, the latter is easily and automatically determined. MRS is a technic that combines the thresholded components of multiscale *à trous* decomposition after the noise at each level has been automatically determined by median filtering of each component.

2.4 Manually Incorporated Enhanced Prior

Ultrasonic imaging of spleen is often highly disturbed by an underskin fat or the lack of echo from beneath the ribs. Automatic analysis by Bayesian inference

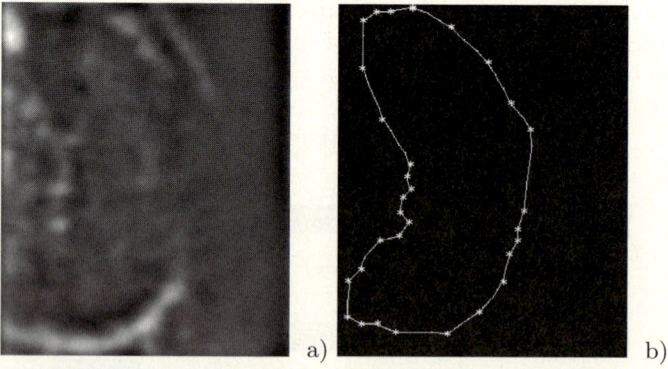

Fig. 6. a) The prior constructed from first 3 components of multiscale decomposition obtained by *à trous* transform. b) Enhanced prior. Some control mark points (stars) have been put manually by an examinator to indicate the guessed position of local edge. This is *a priori* knowledge, in this case the knowledge of an expert. In practice, around 10 mark points may be enough.

usually fails in such a case leading to undetected or misclassified edges. A good solution to overcome such limitations is the incorporation of an additional prior. For the purpose of this study another prior has been put manually to indicate the position of invisible edge in the structure. When all mark points are settled the lines connecting them along the suspected structure's boundaries are automatically found and the initial shape estimation is prepared, as shown on figure 6. This is *a priori* subjective knowledge of examinator and can't be determined automatically, however it affects the Bayesian inference significantly where no other data is available. The missing part of the image substitutes only its minor fraction and the few mark points are analysed together with all the automatically determined data what minimizes the making of wrong decision on mark point position's and edge estimation. The spectral method is further applied to smooth the initial rough edge map globally. This step makes the manual incorporation of prior knowledge safe and jusitified.

2.5 Fast Spectral Method

The spectral methods are widely used for all kinds of problems that can be expanded into Fourier series and solved in Fourier space. For the purpose of this study, following [7] Cheong's method [1] is adapted what results in the equation given in final form:

$$\mathbf{B}_e \hat{f}_e = \mathbf{A}_e \hat{g}_e, \quad \mathbf{B}_o \hat{f}_o = \mathbf{A}_o \hat{g}_o \tag{4}$$

with subscripts e for even and o for odd n, $\hat{f}$ and $\hat{g}$ denote the column vector of expansion coefficients of $f(\theta)$ and $g(\theta)$, respectively. $\mathbf{B}$ is a tridiagonal matrix containing the left hand side of the above equation and $\mathbf{A}$ is the tridiagonal matrix with constant coefficients along each diagonal corresponding to right side of shown equation.

Symbols $\mathbf{B}$ and $\mathbf{A}$ denote two dimensional matrices of $J/2 \times J/2$ size which contain the tridiagonal components only. The f and g denote here the column vectors with cosinus transform's coefficients, $f_m(\theta)$ and $g_m(\theta)$, respectively. Symbol ˆ denotes vector's transposition.

For odd indexes (B_o, A_o and f_o oraz g_o) the above shortcut equation is fully revealed in the following way:

$$\begin{pmatrix} b_1 & c_1 & & & \\ a_3 & b_3 & c_3 & & \\ & \ddots & \ddots & \ddots & \\ & & a_{J-3} & b_{J-3} & c_{J-3} \\ & & & a_{J-1} & b_{J-1} \end{pmatrix} \begin{pmatrix} f_1 \\ f_3 \\ \vdots \\ f_{J-3} \\ f_{J-1} \end{pmatrix} = \begin{pmatrix} 2 & -1 & & & \\ -1 & 2 & -1 & & \\ & \ddots & \ddots & \ddots & \\ & & -1 & 2 & -1 \\ & & & -1 & 2 \end{pmatrix} \begin{pmatrix} g_1 \\ g_3 \\ \vdots \\ g_{J-3} \\ g_{J-1} \end{pmatrix} \tag{5}$$

where coefficients a_n, b_n, c_n have the form described in [9, 10, 24]. The above equation looks similar for even components.

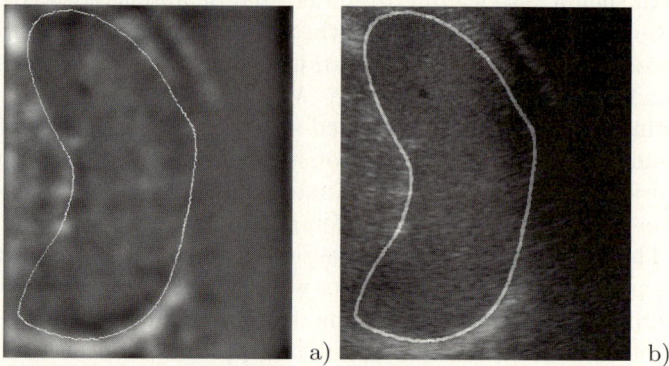

Fig. 7. a) Structure's boundary overlaid on the prior and b) on the original input image

3 Results - The Spleen's Morphometry

After BCSM's performance the delineated contours are overlaid on the original image as shown on figure 7.

In effect of application of the presented method a radial function that describes the boundary of the structure is revealed. As it is given in functional form, its further processing is much easier. Most descriptive parameters like diameteres are easy to determine. For instance, to derive the maximum distance between any points belonging to the radial function, what corresponds to the maximum diameter the following formulae may be applied when the x and y dimensions of a pixel in a certain units are known:

$$D = max(\sqrt{(x_k - x_j)^2 + (y_k - y_j)^2}) \quad for\ all\ k\ and\ j \qquad (6)$$

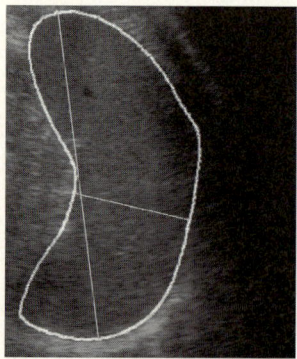

Fig. 8. The diameteres derived from contour radial function

where x_k, y_k, x_j, y_j denotes the points on the image also belonging to radial function f at position k and j.

Similar formulae can be applied for the minimum diameter, however some additional condition is necessary that eliminates all minimal distances that are not diameters. This is equivalent to search for a local minimum in a data.

For the presented case of spleen the following values have been revealed: D_{min}=4.20 cm, D_{max}=10.04 cm. The positions of diameters are shown on figure 8.

4 Discussion and Final Conclusions

The method shown in the study is robust and well suited to the noisy data of speckle type. Having discussed a single case the method has also been proved to be valid for more than a dozen ultrasonic acquisitions of spleens both in healthy and disordered subjects. It is expected to support particularly the diagnostics of splenomegally. The method provides the functional form of the contour what simplifies further morphometric analysis and makes its quantification straightforward. The calculation of length of object's boundary is executable when x and y dimensions of a pixel in certain units are given. This opens a number of possible applications for future work like quantitative comparative analysis of spleens among groups of healthy and disordered subjects when their spleens are transformed into common coordinate space. Such space is easy to construct with the help of other imaging modalities and proper design of acquisition protocol. The presented study is a form of general prescription of the method suitable to modify and different tuning depending on the quality of the data, acquisition technic and scanner properties. The main and unique advantage of the method comes from the opportunity to provide the exact length of spleen's boundary and maximal or minimal diameters instead of the ones guessed to be maximal after visual estimation by an examiner.

Acknowledgments

The author would like to thank Dr Janusz Hałka for medical management, comments and his kind support in ultrasonic data acquisition. This scientific work has been funded by resources of Science and High Education Ministry of Poland for years 2006-2007, in the frame of research grant number N518 023 31/1387.

References

1. Cheong, H.: Double Fourier series on a sphere: Applications to elliptic and vorticity equations. Jour. of Comp. Physics 157(1), 327–349 (2000)
2. Dydenko, I., Friboulet, D., Gorce, J.M., D'hooge, J., Bijnens, B., Magnin, I.E.: Towards ultrasound cardiac image segmentation based on the radiofrequency signal. Medical Image Analysis 7, 353–367 (2003)

3. Hammoude, A.: Endocardial border identification in two-dimensional echocardiografic images: review of methods. Comp. Med. Imag. Graph. 32, 181–193 (1998)
4. Heimdal, A., Stoylen, A., Torp, H., Skjaerpe, T.: Real-time strain rate imaging of the left ventricle by ultrasound. J. Am. Soc. Echocard. 11, 1013–1019 (1998)
5. Holschneider, M., Kronland-Martinet, R., Morlet, J., Tchamitchian, P.: A real time algorithm for signal analysis with the help of the wavelet transform, Wavelets: Time-Frequency Methods and Phase-Space, pp. 286–297. Springer, Heidelberg (1989)
6. Laidlaw, D.H., Fleischer, K.W., Barr, A.H.: Partial volume segmentation with voxel histograms. In: Bankman, I.N. (ed.) Handbook of Medical Imaging, Processing and Analysis, pp. 195–214. Academic Press, London (2000)
7. Li, J., Hero, A.O.: A Fast Spectral Method for Active 3D Shape Reconstruction. Jour. of Math. Imaging and Vision 20, 73–87 (2004)
8. Mitchell, S.C., Bosch, J.G., Lelieveldt, B.P.F., van der Geest, R.J., Reiber, J.H.C., Sonka, M.: 3-D Active Appearrance Models: Segmentation of Cardiac MR and Ultrasound Images. IEEE Trans. on Medical Imaging 21(9), 1167–1178 (2002)
9. Soltysinski, T.: Novel algorithm for active surface reconstruction by Bayesian constrained spectral method. In: Hozman, J., Kneppo, P. (eds.) IFMBE Proceedings, Prague: IFMBE, Proceedings of the 3rd European Medical & Biological Engineering Conference - EMBEC05. Prague, Czech Republic, 20-25.11.2005, vol. 11, pp. 4191–4195 (2005) ISSN 1727-1983
10. Soltysinski, T.: Bayesian constrained spectral method for segmentation of noisy medical images. Theory and applications. In: Sordo, M., Sachin, V., Jain, L.C. (eds.) Advanced Computational Intelligence Paradigms in Healthcare - 3, Studies in Computational Intelligence, vol. 107, Springer, Heidelberg (2008)
11. Sahoo, P.K., Soltani, S., Wong, A.K.C.: A survey of thresholding techniques. Computer Vision, Graphics, and Image Processing 41, 233–260 (1988)
12. Beucher, S., Lantuejoul, C.: Use of watersheds in contour detection. In: International Workshop on image processing, real-time edge and motion detection/estimation, Rennes, France (September 1979)
13. Kass, M., Witkin, A., Terzopoulos, D.: Snakes - Active Contour Models. International Journal of Computer Vision 1(4), 321–331 (1987)
14. Osher, S., Sethian, J.A.: Fronts propagating with curvature dependent speed: Algorithms based on Hamilton-Jacobi Formulations. Journal of Computational Physics 79, 12–49 (1988)
15. McInerney, T., Terzopulos, D.: Deformable models in medical image analysis: A survey. Med. Img. Analysis (2) (1996)
16. Malladi, R., Sethian, J.A., Vemuri, B.C.: Shape modeling with front propagation: A level set approach. IEEE Transactions on Pattern Analysis and Machine Intelligence 17(2), 158–175 (1995)
17. Xu, C., Prince, J.L.: Snakes, Shapes, and Gradient Vector Flow. IEEE Transactions on Image Processing 7(3), 359–369 (1998)
18. Bernardo, J.M., Smith, A.F.M.: Bayesian Theory. John Wiley & Sons, New York (1994)
19. Berry D.A.: Statistics: A Bayesian Perspective, Duxbury, Belmont (1996)
20. Berger, J.O.: Statistical Decision Theory and Bayesian Analysis. Springer, New York (1985)
21. Box, G.E.P., Tiao, G.C.: Bayesian Inference in Statistical Analysis. John Wiley & Sons, New York (1973)

22. de Stefano, A., White, P.R.: Selection of thresholding scheme for image noise reduction on wavelet components using bayesian estimation. Jour. of Math.Imaging and Vision 21, 225–233 (2004)
23. Sharon, E., Brandt, A., Basri, R.: Segmentation and boundary detection using multiscale intensity measurements. In: IEEE Computer Society Conference on Computer Vision and Pattern Recognition, vol. 1, p. 469 (2001)
24. Soltysinski, T., Kałużyński, K., Pałko, T.: Cardiac ventricle contour reconstruction in ultrasonographic images using Bayesian constrained spectral method. In: Rutkowski, L., Tadeusiewicz, R., Zadeh, L.A., Żurada, J.M. (eds.) ICAISC 2006. LNCS (LNAI), vol. 4029, pp. 988–997. Springer, Heidelberg (2006)
25. Soltysinski, T.: Speckle Noise Removal By *A Trous* Decomposition And Threshold-limited Synthesis In Ultrasonic Cardiac Images. In: Proceedings of the conference Biosignal 2006, Brno Technical University, Czech Republic, vol. 18 (2006)
26. Gravel, P., Beaudoin, G., De Guise, J.A.: A method for modeling noise in medical images. IEEE Transactions on medical imaging 23(10), 1221–1232 (2004)
27. Starck, J.-L., Murtagh, F.: Astronomical image and data analysis. Springer, Berlin (2002)
28. Soltysinski, T.: Influence of multiscale denoising on Bayesian constrained spectral method in segmentation of noisy medical images. IFMBE Proceedings 14 (2006)
29. Sołtysiński, T.: Novel Bayesian constrained spectral method for cell morphometry in multimodal molecular imaging. Molecular Imaging 6(5), 366 (2007)
30. Bankman, I.N. (ed.): Handbook of Medical Imaging, Processing and Analysis. Academic Press, London (2000)

Part VII

Various Problems of Artificial Intelligence

Part II

Parallel Single-Thread Strategies in Scheduling

Wojciech Bożejko, Jarosław Pempera, and Adam Smutnicki

Wrocław University of Technology, Institute of Computer Engineering, Control and Robotics, Janiszewskiego 11-17, 50-372 Wrocław, Poland
{wojciech.bozejko,jaroslaw.pempera}@pwr.wroc.pl,
adam.smutnicki@student.pwr

Abstract. This paper, as well as coupled paper [2], deal with various aspects of scheduling algorithms dedicated for processing in parallel computing environments. In this paper, for the exemplary problem, namely the flow-shop scheduling problem with makespan criterion, there are proposed original methods for parallel analysis of a solution as well as a group of concentrated and/or distributed solutions, recommended for the use in metaheuristic approaches with single-thread trajectory. Such methods examine in parallel consecutive local sub-areas of the solution space, or a set of distributed solutions called population, located along the *single* trajectory passed through the space. Supplementary multi-thread search techniques applied in metaheuristics have been discussed in complementary our paper [2].

1 Introduction

Most of scheduling cases derived from production practice belong to the class of strongly NP-hard combinatorial optimization problems. While known exact algorithms own, necessarily, exponential computational complexity, then even the significant increase of computers power will result incomparably small increase of the size of instances we can solved. Thus, there exist two, not conflicted mutually, approaches, which allow one to solve large-size instances in the acceptable time: (1) approximate methods (chiefly metaheuristics), (2) parallel methods. The best hybrid combination of both is the one which we really have needed.

Quality of the best solutions generated by search algorithms strongly depends on the number of analyzed solution, and thus on the running time. Time and quality have opposable tendency in that sense, that finding better solution requires significant growth of computation time. Through the parallel processing one can increase the number of checked solutions (per time unit). In the paper there are proposed several solutions algorithms, dedicated for analysis single solution as well as group of solutions employed in widely used metaheuristics.

In the scope of single-thread search, dedicated fundamentally for uniform multiprocessor system of small granularity, there are proposed a few original solution methods, taking into account various design technology and different needs applied by modern discrete optimization algorithms, namely: (a) single solution analysis (dedicated for simulated annealing SA, simulated jumping SJ, random search RS), (b) local neighborhood analysis (for tabu search TS, adaptive memory search AMS, descending search DS), (c) analysis of population of distributed solutions (for genetic approach GA, scatter search SS). Special attention has been paid to efficiency, cost and speedup of methods depending on the used parallel computing environment. For each algorithm there has been shown theoretical evaluation of its numerical properties as well as comparative analysis of potential benefits from proposed approaches. The multi-thread search, dedicated for uniform as well as non-uniform multiprocessor systems of large granularity (such as mainframe, clusters, distributed systems linked through the network), have been have been discussed in the supplementary paper [2].

We assume that the reader knows the basic notions, see e.g. [5]: theoretical parallel architectures, theoretical models of parallel computations, granularity, threads, cooperation, speed up, efficiency, cost, cost optimality, computational complexity, real parallel architectures and parallel programming languages.

2 The Problem

We consider, as the test case, the well-known in the scheduling theory, strongly NP-hard problem, called the permutation flow-shop problem with the makespan criterion and denoted by $F||C_{max}$. Skipping consciously the long list of papers dealing with this subject, we only refer the reader to recent reviews and best up-to-now algorithms [4,6,7].

The problem has been introduced as follows. There is n jobs from a set $J = \{1, 2, \ldots, n\}$ to be processed in a production system having m machines, indexed by $1, 2, \ldots, m$, organized in the line (sequential structure). Single job reflects one final product (or sub product) manufacturing. Every job is performed in m subsequent stages, in common way for all tasks. Stage i is performed by machine i, $i = 1, \ldots, m$. Every job $j \in J$ is split into sequence of m operations $O_{1j}, O_{2j}, \ldots, O_{mj}$ performed on machines in turn. Operation O_{ij} reflects processing of job j on machine i with processing time $p_{ij} > 0$. Once started job cannot be interrupted. Each machine can execute at most one job at a time, each job can be processed on at most one machine at a time.

The sequence of loading jobs into system is represented by a permutation $\pi = (\pi(1), \ldots, \pi(n))$ on the set J. The optimization problem is to find the optimal sequence π^* so that

$$C_{max}(\pi^*) = \min_{\pi \in \Pi} C_{max}(\pi). \qquad (1)$$

where $C_{max}(\pi)$ is the makespan for permutation π and Π is the set of all permutations. Denoting by C_{ij} the completion time of job j on machine i we have $C_{max}(\pi) = C_{m,\pi(n)}$. Values C_{ij} can be found by using either recursive formula

$$C_{i\pi(j)} = \max\{C_{i-1,\pi(j)}, C_{i,\pi(j-1)}\} + p_{i\pi(j)}, \quad i = 1, 2, \ldots, m, \quad j = 1, \ldots, n, \qquad (2)$$

with initial conditions $C_{i\pi(0)} = 0$, $i = 1, 2, \ldots, m$, $C_{0\pi(j)} = 0$, $j = 1, 2, \ldots, n$, or non-recursive one

$$C_{i\pi(j)} = \max_{1=j_0 \leq j_1 \leq \ldots \leq j_i = j} \sum_{s=1}^{i} \sum_{k=j_{i-1}}^{j_i} p_{s\pi(k)}. \qquad (3)$$

Computational complexity of (2) is $O(mn)$, whereas for (3) is $O(\binom{j+i-2}{i-1}(j+i-1))$ $= O(\frac{(n+m)^{n-1}}{(n-1)!})$. In practice the former formula has been commonly used.

Note that the problem of transforming sequential algorithm for scheduling problems into parallel one is nontrivial, because of strongly sequential character of computations carried out by (2) and by other known scheduling algorithms.

3 Single-Thread Search

We consider a parallel algorithm which uses single (master) process to guide the search. Thread performs in the cyclic way (iteratively) two leading tasks: (A) goal function evaluation for single solution or a set of solutions, (B) management, e.g. solution filtering and selection, collection of history, updating. Part (B) takes statistically 1-3% total iteration time, then its acceleration is useless. Part (A) can be accelerated in parallel environment in various manner – our aim is to find either *cost optimal* method or non-optimal in the cost sense but offering *the shortest running time*.

Single solution. In each iteration we have to find a goal function value for *single* fixed π. Calculations can be spread into parallel processors in a few ways.

Theorem 1. *For a fixed π, $C_{max}(\pi)$ in the problem $F||C_{max}$ can be found on CREW PRAM machine in the time $O(n+m)$ by using m processors.*

Proof. Without the loss of generality one can assume that $\pi = (1, 2, \ldots, n)$. Calculations of $C_{i,j}$ by using (2) have been clustered. Cluster k contains values C_{ij} such that $i + j - 1 = k$, $k = 1, 2, \ldots, n + m - 1$ and requires at most m processors. Clusters are processed in order $k = 1, 2, \ldots, n + m - 1$. ○

Fact 1. *Speedup of method from Theorem 1 is $O(\frac{nm}{n+m})$, efficiency is $O(\frac{n}{n+m})$.*

Presented method is not cost optimal. Its efficiency slowly decreases with increasing m. For example for $n = 10$, $m = 3$ efficiency is about 77%, whereas for $m = 10$ is 50% [1]. Clearly, for $n \gg m$ efficiency tends to 100%, for $n \ll m$ quickly decreases to 0. Proposed method requires fixed a priori number of processors $p = m$, which seems to be not troublesome since usually in practice $m \leq 20$. Emulation of calculations by using $p < m$ processors increases computational complexity to $O((n+m)m/p)$, although construction of proper algorithm remains open. If $p \geq m$, then $(p - m)$ processors will be unloaded.

---
[1] Evaluation is true with certain constant multiplier.

Theorem 2. *For a fixed π, $C_{max}(\pi)$ in the problem $F||C_{max}$ can be found on CREW PRAM machine in the time $O(n+m)$ by using $O(\frac{nm}{n+m})$ processors.*

Proof. Without the loss of generality one can assume that $\pi = (1, 2, \ldots, n)$. Let $p \leq m$ be the number of used processors. Calculation process will be carried out for *levels* $k = 1, 2, \ldots, d$, $d = n + m - 1$ in that order. On the level k we perform calculation of n_k values $C_{i,j}$ such that $i + j - 1 = k$, $\sum_{k=1}^{d} n_k = nm$.

We cluster n_k elements on the level k into $\left\lceil \frac{n_k}{p} \right\rceil$ groups; first $\left\lfloor \frac{n_k}{p} \right\rfloor$ groups contains p elements each, whereas remain elements (at most p) belong to the last group. Parallel computations on level k are performed in the time $O(\left\lceil \frac{n_k}{p} \right\rceil)$. Total calculation time is equal the sum along all levels and is of order

$$\sum_{k=1}^{d} \left\lceil \frac{n_k}{p} \right\rceil \leq \sum_{k=1}^{d} \left(\frac{n_k}{p} + 1 \right) = \frac{nm}{p} + d = \frac{nm}{p} + n + m - 1. \quad (4)$$

We are seeking for the number of processors p, $1 \leq p \leq m$, for which efficiency of parallel algorithm is $O(1)$ – this ensures cost optimality of the method. Value p can be found from the following condition

$$\frac{1}{p} \frac{nm}{\frac{nm}{p} + n + m - 1} = c = O(1) \quad (5)$$

for some constant $c < 1$. After a few simple transformations of (5) we have

$$p = \frac{nm}{n + m - 1} \left(\frac{1}{c} - 1 \right) = O(\frac{nm}{n+m}). \quad (6)$$

Setting $p = O(\frac{nm}{n+m})$, we get the total calculation time of C_{ij} values equal to

$$O(\frac{nm}{p} + n + m - 1) = O(\frac{nm}{\frac{nm}{n+m}} + n + m) = O(n + m). \quad (7)$$

□

Fact 2. *Speedup of method based on Theorem 2 is $O(\frac{nm}{n+m})$, cost is $O(nm)$.*

The method is cost optimal and allows one to control efficiency as well as speed of calculations by choosing the number of processors and adjusting the parameters of calculations to the real number of parallel processors existed in the system. Besides this, Theorem 2 provides the "optimal" number of processors that ensures cost optimality of the method. This number can be set by flexible adaptation of the number of processors to both sizes of the problem, namely n and m simultaneously. For example, for $n \gg m$ we have $p \approx m$, for $n \ll m$ we have $p \approx n \ll m$, whereas for $n \approx m$ we have $p \approx n/2$.

Observe that both Theorems 1 and 2 own the same bound $O(n + m)$ on the computational complexity, being the natural consequence of sequential structure of formula (2). In order to obtain higher speedup we need to give up the scheme (2). On the current state of knowledge, there remains only non-recursive scheme (3) of very high computational complexity.

Table 1. Times of 100.000 cost's function calculations due to the method from the Theorem 2 ($n = 150$, $m = 150$)

processors	wall time (sec.)	CPUs time (sec.)	memory used (kB)
1	10	9	2 040
2	10	9	1 188
4	11	9	2 144
16	13	18	61 020

Table 1 presents results of calculations of the method based on Th. 2 on the cluster of 8 dual-core Intel Xeon 2.4 GHz processors connected by Gigabit Ethernet installed in the Wrocław Center of Networking and Supercomputing. As we can see the cost of calculations, measured as a CPUs time (the sum of computations time for all the processors), is constant for 1,2 and 4 processors. For 16 processors cost raises due to the communication in the cluster, which has not got a shared memory. Usage of the memory is raising because of copying the data into each processor's local memory.

Theorem 3. *For a fixed π, C_{max} in the problem $F||C_{max}$ can be found on CREW PRAM machine in the time $O(m + \log n)$ by using $O(\frac{(n+m)^{n-1}}{m(n-1)!})$ processors.*

Proof. Without the loss of generality one can assume that $\pi = (1, 2, \ldots, n)$. We use the formula (3), which can be re-written in the form of

$$C_{ij} = \max_{1 = j_0 \leq j_1 \leq \ldots \leq j_i = j} \sum_{s=1}^{i} \left(P_{s,j_s} - P_{s,j_{s-1}-1} \right), \qquad (8)$$

where $P_{s,t} = \sum_{k=1}^{t} p_{sk}$ is the prefix sum, $t = 1, 2, \ldots, n$. For a fixed s the value of $P_{s,t}$ can be found in the time $O(\log n)$ on $O(n/\log n)$ processors, $t = 1, 2, \ldots, n$. Thus, all $P_{s,t}$ for $t = 1, 2, \ldots, n$, $s = 1, 2, \ldots, m$ can be found by using $O(mn/\log n)$ processors in time $O(\log n)$ once at the begin. For the goal function we need C_{mn}. The number of all subsequences $(j_0, j_1, \ldots, j_m)$ satisfying condition $1 = j_0 \leq j_1 \leq \ldots \leq j_m = n$ corresponds one-to-one to the number of combinations of $m - 1$ elements with repetitions on the $(n-2)$-th element set, and is equal $\binom{n+m-2}{m-1}$. Original method of generating such subsequences in the time $O(m)$ by using $\binom{n+m-2}{m-1}$ processors one can find in [1]. Next, by using $\binom{n+m-2}{m-1}$ processors one can find sequentially all sums $\sum_{s=1}^{m} \left(P_{s,j_s} - P_{s,j_{s-1}-1} \right)$ from the formula (8) for all subsequences in the time $O(m)$. To find value C_{mn} we have to find maximum among $\binom{n+m-2}{m-1}$ calculated sums sum, which can be found in the time $O(\log \binom{n+m-2}{m-1})$ by using $O(\binom{n+m-2}{m-1}/\log \binom{n+m-2}{m-1})$ processors.

Computational complexity of this step, through the inequality,

$$\binom{n+m-2}{m-1} = \frac{m(m+1)\ldots(n+m-2)}{(n-1)!} \leq \frac{(n+m)^{n-1}}{(n-1)!} \qquad (9)$$

and well-known equation $\log(n!) = \Theta(n \log n)$ is equal

$$O(\log \frac{(n+m)^{n-1}}{(n-1)!}) = O(n \log(1 + \frac{m}{n})). \tag{10}$$

Note, that (10) implies

$$n \log(1+\frac{m}{n}) \leq \lim_{n \to \infty} (n \log(1+\frac{m}{n})) = \log \lim_{n \to \infty} (1+\frac{m}{n})^n = \log e^m = m \log e \tag{11}$$

which is $O(m)$. Since the computational complexity of remain algorithm steps is not greater than $O(\max(m, \log n)) = O(m + \log n)$, this is also the final computational complexity of the method. The number of processor used is

$$O(\max(\frac{mn}{\log n}, \frac{\binom{n+m-2}{m-1}}{\log \binom{n+m-2}{m-1}})) = O(\frac{(n+m)^{n-1}}{m(n-1)!}).$$

Fact 3. Speedup of the method from Theorem 3 is $(\frac{mn}{m+\log n})$.

The number of processors growths exponentially with increasing n, decreasing simultaneously efficiency very quickly. Thus, the result has rather theoretical than practical meaning.

Neighborhood API. Neighborhood based on adjacent pairwise interchange (API) of elements in permutation is the simplest one and commonly used. Sequential algorithms that searches API uses so called *accelerator* to speed up the run by suitable decomposition and aggregation of computations for relative solutions, see [8]; this can be applied only for the problem $F||C_{max}$. Since some further theorems refer to this concept, we will introduce it briefly.

Let π be the permutation that generates neighborhood API and $v = (a, a+1)$ be the pair of adjacent positions, such that their interchange in π lead us to new solution π_v. At first for permutation π we calculate

$$r_{st} = \max\{r_{s-1,t}, r_{s,t-1} + p_{s\pi(t)}\}, \quad t = 1, 2, \ldots, n, \quad s = 1, 2, \ldots, m, \tag{12}$$

$$q_{st} = \max\{q_{s+1,t}, q_{s,t+1} + p_{s\pi(t)}\}, \quad t = n, \ldots, 2, 1, \quad s = m, \ldots, 2, 1, \tag{13}$$

where $r_{0t} = 0 = q_{m+1,t}$, $t = 1, 2, \ldots, n$, $r_{s0} = 0 = q_{s,n+1}$, $s = 1, 2, \ldots, m$. $C_{max}(\pi_v)$ for single interchange $v = (a, a+1)$ can be found in the time $O(m)$ from equations

$$C_{max}(\pi_v) = \max_{1 \leq s \leq m} (e_s + q_{s,a+2}), \tag{14}$$

$$e_s = \max\{e_{s-1}, d_s\} + p_{s,\pi(a)}, \quad s = 1, 2, \ldots, m, \tag{15}$$

$$d_s = \max\{d_{s-1}, r_{s,a-1}\} + p_{s\pi(a+1)}, \quad s = 1, 2, \ldots, m. \tag{16}$$

Initial conditions are as follows : $e_0 = 0 = d_0$, $r_{s0} = 0 = q_{s,n+2}$, $s = 1, 2, \ldots, m$. Neighborhood API contains $n-1$ solutions π_v, $v = (a, a+1)$, $a = 1, 2, \ldots, n-1$, and is searched conventionally in the time $O(n^2 m)$; by using *sequential accelerator* for API we can do it in the time $O(nm)$.

Theorem 4. *For a fixed π, neighborhood API for $F||C_{max}$ problem can be searched on CREW PRAM machine in the time $O(n+m)$ by using $O(\frac{n^2 m}{n+m})$ processors.*

Proof. Skipping solution affinity, we allocate for each π_v the number $O(\frac{nm}{n+m})$ processors, which allows us to find all $C_{max}(\pi_v)$ in the time $O(n+m)$, see Theorem 2. The best solution in the neighborhood can be found in the time $O(n)$ by using single processor. □

There is a dilemma to which version of sequential algorithm should be compared the parallel method - with or without sequential accelerator? If we take the best one (with accelerator), then we have the following evaluation.

Fact 4. *Speedup of the method from Theorem 4 is $O(\frac{nm}{n+m})$, efficiency is $O(\frac{1}{n})$.*

Presented method is not cost optimal, its efficiency quickly decreases with growing n. Note, that if sequential accelerator cannot be applied (as an example for $F||\sum C_i$ problem), the presented method is cost optimal with efficiency $O(1)$. Employing knowledge about relationship among solutions in the neighborhood one can prove significantly stronger result.

Theorem 5. *For a fixed π, neighborhood API for $F||C_{max}$ problem can be searched on CREW PRAM machine in the time $O(n+m)$ by using $O(\frac{nm}{n+m})$ processors.*

Proof. Let $v = (a, a+1)$. We design parallel algorithm using sequential accelerator for API. Values r_{st}, q_{st} are generated once, at the begin of the search, in the time $O(n+m)$ using $O(\frac{nm}{n+m})$ processors, in the way analogous to that from proof of Theorem 2. This is cost optimal method. The process of overlooking of API neighborhood has been split into groups of cardinality $\left\lceil \frac{n}{p} \right\rceil$ each, where $p = \left\lceil \frac{nm}{n+m} \right\rceil$ is the number of processors used. Computations in each group is performed independently. Processor $k = 1, 2, ..., p$ serves the group defined by v

$$v = ((k-1)\left\lceil \frac{n}{p} \right\rceil + a, (k-1)\left\lceil \frac{n}{p} \right\rceil + a + 1), \quad a = 1, 2, \ldots, \left\lceil \frac{n}{p} \right\rceil \quad (17)$$

for $k = 1, 2, ..., p-1$, and

$$v = ((p-1)\left\lceil \frac{n}{p} \right\rceil + a, (p-1)\left\lceil \frac{n}{p} \right\rceil + a + 1), \quad a = 1, 2, \ldots, n - (p-1)\left\lceil \frac{n}{p} \right\rceil - 1 \quad (18)$$

for $k = p$. The last group can be incomplete. Since the computational complexity of finding $C_{max}(\pi_v)$ in single group equals to

$$\left\lceil \frac{n}{p} \right\rceil O(m) = O(\frac{nm}{p}) = O(\frac{nm}{\frac{nm}{n+m}}) = O(n+m),$$

then all $C_{max}(\pi_v)$ can be found in the same time. Each processor, while calculating sequentially its portion of $C_{max}(\pi_v)$ values, can simultaneously store the best solution in this group. To this aim, it additionally performs

$$\left\lceil \frac{n}{p} \right\rceil - 1 = O(\frac{n}{p}) = O(\frac{n}{\frac{nm}{n+m}}) = O(\frac{n+m}{m}), \qquad (19)$$

comparisons to the best solution, which has no influence on the provided earlier computational complexity. Choosing the best solution among the whole API neighborhood requires p comparisons of best values found for all groups. This can be done in the time $O(\log p)$, by using $p = O(\frac{nm}{n+m})$ processors. The last fact follows from the following sequence of inequalities

$$\log p = \log \left\lceil \frac{nm}{n+m} \right\rceil < \log(\frac{nm}{n+m} + 1) =$$

$$= \log(\frac{nm+n+m}{n+m}) = \log(\frac{(n+1)(m+1)-1}{n+m}) =$$

$$= (\log((n+1)(m+1)-1) - \log(n+m) < \log((n+1)(m+1)) =$$

$$= \log(n+1) + \log(m+1) < n+1+m+1 \qquad (20)$$

Fact 5. Speedup of the method from Theorem 5 is $O(\frac{nm}{n+m})$, efficiency is $O(1)$.

Neighborhood INS. Neighborhood INS based on insertions of elements in permutation, has the computational complexity $O(n^3 m)$ for the searching. For INS and problem $F||C_{max}$ there is known *sequential accelerator*, see e.g. [6,8], which reduces this complexity to $O(n^2 m)$. We will shown stronger result for parallel algorithm.

Theorem 6. *For a fixed π, neighborhood INS for $F||C_{max}$ problem can be searched on CREW PRAM machine in the time $O(n + m)$ by using $O(\frac{n^2 m}{n+m})$ processors.*

Proof. Let $v = (a, b)$ defines INS neighborhood for permutation π as follows: job $\pi(a)$ has been removed from its position and then is inserted so that its new position in resulting permutation π_v becomes b; $a, b \in \{1, \ldots, n\}$, $a \neq b$. Let r_{st}, q_{st}, $s = 1, 2, \ldots, m$, $t = 1, 2, \ldots, n-1$, be values found by (12)–(13) for permutation obtained from π by removing job $\pi(a)$. For each fixed $a = 1,2,\ldots,n$ values r_{st}, q_{st} can be found in the time $O(n + m)$ by using $O(\frac{nm}{n+m})$ processors in the way analogous to Theorem 2. Employing $O(\frac{n^2 m}{n+m})$ processors we can perform such calculations in the time $O(n+m)$ for all permutations obtained from π by removing job $\pi(a)$, $a = 1, 2, \ldots, n$. For each fixed a values $C_{max}(\pi_{(a,b)})$, $b = 1,2,\ldots,n$, $b \neq a$ can be found using (14) in the time $O(m)$. We split the whole computation process on $p = \left\lceil \frac{n^2 m}{n+m} \right\rceil$ groups, each of which is assigned to separate processor. Since INS neighborhood contains $(n-1)^2 = O(n^2)$ solutions, all $C_{max}(\pi_v)$ can be found in the time $\left\lceil \frac{(n-1)^2}{p} \right\rceil O(m) = O(n+m)$. The best solution in the neighborhood can be found in the time $O(\log(n^2)) = O(2 \log n) = O(\log n)$ by using n processors. The whole method has complexity $O(n + m + \log n) = O(n + m)$ and employ $O(\frac{n^2 m}{n+m})$ processors. □

Fact 6. Speedup of the method from Theorem 6 is $O(\frac{n^2 m}{n+m})$, efficiency is $O(1)$.

Neighborhood NPI. The neighborhood is generated swapping any pair of jobs $\pi(a)$, $\pi(b)$, for $a, b \in \{1, 2, \ldots, n\}$, $a \neq b$. We start from the description of sequential accelerator, [8], used in the parallel version presented below. Direct method of searching the neighborhood NPI has the computational complexity $O(n^3 m)$. Sequential accelerator for NPI reduces this complexity to $O(n^2 m)$.

Let $v = (a, b)$, $a \neq b$ defines the move that generates a new permutation π_v. Without the loss of generality we can assume that $a < b$, due to symmetry. Next, let r_{st}, q_{st}, $s = 1, 2, \ldots, m$, $t = 1, 2, \ldots, n$ be values found by (12)–(13) for π. Denote by D_{st}^{xy} the length of the longest path between nodes (s, t) and (x, y) in the grid graph $G(\pi)$, [6]. The method of calculating $C_{\max}(\pi_v)$ can be decomposed into following steps. At the begin we calculate the length of the longest path which going to the node (s, a), joining the job $\pi(b)$, put by v on the position a

$$d_s = \max\{d_{s-1}, r_{s,a-1}\} + p_{s,\pi(b)}, \quad s = 1, 2, \ldots, m, \tag{21}$$

where $d_0 = 0$. Then we calculate the length of the longest path going to node $(s, b-1)$, joining the part of $G(\pi)$ located between jobs on positions from $a+1$ to $b-1$, invariant for $G(\pi)$

$$e_s = \max_{1 \leq w \leq s} (d_w + D_{w,a+1}^{s,b-1}), \quad s = 1, 2, \ldots, m. \tag{22}$$

In the successive step we calculate the the length of the longest path going to node (s, b), joining job $\pi(a)$, put by v on position b

$$f_s = \max\{f_{s-1}, e_s\} + p_{s,\pi(a)}, \quad s = 1, 2, \ldots, m, \tag{23}$$

where $f_0 = 0$. Finally we obtain

$$C_{\max}(\pi_v) = \max_{1 \leq s \leq m} (f_s + q_{s,b+1}). \tag{24}$$

The value of $C_{\max}(\pi_v)$ can be found if we have suitable D_{st}^{xy}. These values can be calculated recursively, for fixed t and $y = t+1, t+2, \ldots, n$, by using equality

$$D_{st}^{x,y+1} = \max_{s \leq k \leq x} (D_{st}^{ky} + \sum_{i=k}^{x} p_{i\pi(y+1)}) \tag{25}$$

where $D_{st}^{xt} = \sum_{i=s}^{x} p_{i\pi(t)}$. The formula (25) can be re-written in the form of

$$D_{st}^{s,t+1} = D_{st}^{st} + p_{s,\pi(t+1)}, \quad D_{st}^{x0} = D_{st}^{0y} = 0, \tag{26}$$

$$D_{st}^{x,y+1} = \max\{D_{st}^{xy}, D_{st}^{x-1,y}\} + p_{x,\pi(y+1)}, \tag{27}$$

$x = 1, 2, \ldots, m$, $y = 1, 2, \ldots, n$, which allows to find all D_{st}^{xy}, $x = 1, 2, \ldots, m$, $y = 1, 2, \ldots, n$ for fixed (s, t) in the time $O(nm)$. Finally, sequential calculation of all $O(n^2)$ values $C_{\max}(\pi_v)$ (including all D_{st}^{xy}, $x, s = 1, 2, \ldots, m$, $y, t = 1, 2, \ldots, n$) can be processed in the time $O(n^2 m^2)$.

Theorem 7. *For a fixed π, neighborhood NPI for $F||C_{max}$ problem can be searched on CREW PRAM machine in the time $O(nm)$ by using $O(n^2m)$ processors.*

Proof. We employ the parallel counterpart of sequential accelerator. Let each of $\frac{n(n-1)}{2}$ elements of the neighborhood will be associated with stakes of $O(m)$ processors. For fixed (s,t), and all $x = 1, 2, ..., m$, $y = 1, 2, ..., n$, values D_{st}^{xy} can be find sequentially in the time $O(nm)$. Using $O(nm)$ processors we can calculate D_{st}^{xy} for all $x, s = 1, 2, ..., m$, $y, t = 1, 2, ..., n$ in the time $O(nm)$ once at the begin. Let us analyze calculation of $C_{\max}(\pi_v)$ for fixed v. Values d_s in (21) we have to perform sequentially in the time $O(m)$. Maximum among m values in (22) can be performed in parallel, for all s, by using $O(m)$ processors in the time $O(m)$. Formula (23) we calculate sequentially, for each s, in the time $O(m)$. Single value $C_{\max}(\pi_v)$ in (24) requires m independent adding operations and then finding maximum among m numbers. We do this sequentially in the time $O(m)$. Finally, parallel computations of $O(n^2)$ values $C_{\max}(\pi_v)$ can be performed in the time $O(m)$ by using $O(n^2m)$ processors. Since process of generating D_{st}^{xy} had complexity $O(nm)$, this is the final complexity of the whole method. ○

Fact 7. *Speedup of the method from Theorem 7 is $O(nm)$, efficiency is $O(\frac{1}{n})$.*

A set of distributed solutions. Evolutionary algorithms (e.g. GA, SS) work on a small set of distributed solutions called *population* $\mathcal{P} \subset \mathcal{X}$. Its processing requires, among others, calculation of adaptation function for solutions from $\mathcal{P}$ - frequently this is the goal function of the scheduling problem. Skipping affinity among solutions from $\mathcal{P}$, one can disperse calculations into $|\mathcal{P}|$ sub-processes calculated independently. So has been done in Theorem 4 and can be applied also here. In this section we examine possibility of employing similarity between dispersed solutions in $\mathcal{P}$ to speed up single iteration of GA by using parallel processing. The design process consists of two phases: (1) detecting relationship in $\mathcal{P}$, and (2) using relationship to reduction of the calculation cost.

Let $\mathcal{P}$ be the population of cardinality $k = |\mathcal{P}|$. Permutation $\pi \in \mathcal{P}$ can be interpreted as ancestral line $\pi(1) \to \pi(2) \to \ldots \to \pi(n)$ containing descendants of this line in successive generations $i = 1, 2, \ldots, n$. Descendant i-th in ancestral line π is defined by triple (a, i, π) so that $\pi(i) = a$. Line π can be represented by graph $H(\pi) = (V(\pi), E(\pi))$, with the set of nodes

$$V(\pi) = \{(a, i, \pi) : \ a = \pi(i), \ i = 1, 2, \ldots, n\} \quad (28)$$

representing descendants in successive generations, and set of arcs

$$E(\pi) = \{((a, i, \pi), (b, i+1, \pi)) : \ a = \pi(i), \ b = \pi(i+1), \ i = 1, 2, \ldots, n\}, \quad (29)$$

representing chronological order of descendants; graph $H(\pi)$ is the chain.

To compare ancestral line having common ancestor we introduce identity relation ($\equiv$). For two lines $\pi, \sigma \in \mathcal{P}$ nodes (a, i, π) and (a, i, σ) are identical (refer to the same person), if $\pi(j) = \sigma(j)$, $j = 1, 2, \ldots, i$; this fact will be denoted by

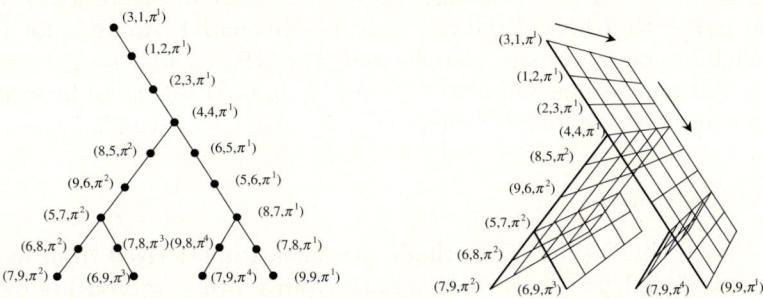

Fig. 1. Genealogy tree $H(\mathcal{P})$ (left). Corresponding 3D computing net (right).

$(a,i,\pi) \equiv (a,i,\sigma)$. Note that if $(a,i,\pi) \equiv (a,i,\sigma)$, permutations π and σ own common prefix $\pi(1), \pi(2), \ldots, \pi(i)$.

Genealogy tree of population $\mathcal{P}$ is the graph with nodes $V(\mathcal{P})$ and arcs $E(\mathcal{P})$

$$H(\mathcal{P}) = (V(\mathcal{P}), E(\mathcal{P})), \quad V(\mathcal{P}) = \bigcup_{\pi \in \mathcal{P}} V(\pi), \quad E(\mathcal{P}) = \bigcup_{\pi \in \mathcal{P}} E(\pi) \qquad (30)$$

and identity relation on the set $V(\mathcal{P})$. Genealogy tree for $\mathcal{P} = (\pi^1, \pi^2, \pi^3, \pi^4)$, $\pi^1 = (3,1,2,4,6,5,8,7,9)$, $\pi^2 = (3,1,2,4,8,9,5,6,7)$, $\pi^3 = (3,1,2,4,8,9,5,7,6)$, $\pi^4 = (3,1,2,4,6,5,8,9,7)$ is shown in Figure 1.

We define the *concentration* of population $\mathcal{P}$ as follows

$$c(\mathcal{P}) = \frac{kn}{|V(\mathcal{P})|}. \qquad (31)$$

Clearly, $1 \leq c(\mathcal{P}) \leq k$; value $c(\mathcal{P}) = 1$ corresponds to k completely different solutions. Value $c(\mathcal{P}) = k$ corresponds to k identical solutions. Our fundamental aim is to avoid repetitive calculations for common parts of ancestral lines. To find goal function values for solutions from $\mathcal{P}$ we need C_{ij}, $i = 1, 2, \ldots, m$ for each j. The proposed computing scheme in 3D net is shown in Figure 1. It is obtained by copying in depth a genealogy tree. This type of calculation constitutes new quality. We can save on calculations skipping the number of nodes equal

$$mkn - m|V(\mathcal{P})| = (1 - \frac{1}{c(\mathcal{P})})mkn = \alpha(\mathcal{P})mkn. \qquad (32)$$

To evaluate potential profits for this approach, one needs to find a coefficient $\alpha(\mathcal{P})$, having sense of a fraction of skipped nodes. For $c(\mathcal{P}) = 1$ there is no saves. For regular $\mathcal{P}$ one can evaluate $\alpha(\mathcal{P})$ analytically. For API, $k = n$ and

$$\alpha(\mathcal{P}) = \frac{(n-1)(n-2)}{2n^2}. \qquad (33)$$

Since $\alpha(\mathcal{P})$ is nondecreasing with n, then

$$\alpha(\mathcal{P}) \leq \lim_{n \to \infty} \alpha(\mathcal{P}) = \frac{1}{2}, \qquad (34)$$

which implies $c(\mathcal{P}) \leq 2$, independently on n. The maximal theoretically obtainable profit is less that half of the cost paid in Theorem 4. Analysis for remain regular neighborhoods is more complicated. For NPI as well as INS we have $\alpha(\mathcal{P}) \to \frac{1}{3}$. For non-regular populations (such as in GA) $\alpha(\mathcal{P})$ can be evaluated only experimentally. We found it equal approximately 5% to 10%.

4 Conclusions

In single-thread single-solution methods, parallelization derived from basic recursive formula lead us to cost optimal algorithms, other approaches own low efficiency although offers high speed. The latter results can be perceived as the first evaluation of lower bound on calculation speed under number of processor tending to infinity. Some obtained in this section results can be extended to EREW PRAM model. This observation follows from the fact that problem data can be copied n times (this can be done in time $O(\log n)$ using $O(n/\log n)$ processors in the initial phase), therefore it is easy to modify algorithms in proofs of theorems to obtain versions of the theorems for the EREW model.

In single-thread neighborhood-search methods, the whole neighborhood can be searched in the time of the same order that for single solution under sufficiently increased number of processors. That fact seems to be quite natural, and this computational complexity appears to be obvious bound on the neighborhood analysis time. Also these results can be extended to EREW PRAM model.

Relationship in single-thread distributed-solution search does not propose anything new with respect to approach ignoring similarity between solutions. In fact, relationship has no influence on the computational complexity, but offers reduction by a constant multiplier the number of processor engaged. Finally, profits are small comparing to complexity range of implementation.

References

1. Bożejko, W.: Parallel job scheduling algorithms (in Polish). Report PRE 29/2003, Ph.D. dissertation, Wroclaw University of Technology (2003)
2. Bożejko, W., Pempera, J., Smutnicki, A.: Multi-thread parallel metaheuristics in scheduling. In: Proceedings of ICAISC (accepted, 2008)
3. Crainic, T.G., Toulouse, M.: Parallel metaheuristics. In: Crainic, T.G., Laporte, G. (eds.) Fleet management and logistics, pp. 205–251. Kluwer, Dordrecht (1998)
4. Grabowski, J., Pempera, J.: New block properties for the permutation flow shop problem with application in tabu search. Journal of Operational Research Society 52, 210–220 (2000)
5. Grama, A., Kumar, V.: State of the Art in Parallel Search Techniques for Discrete Optimization Problems. IEEE Transactions on Knowledge and Data Engineering 11, 28–35 (1999)
6. Nowicki, E., Smutnicki, C.: A fast tabu search algorithm for the permutation flow shop problem. European Journal of Operational Research 91, 160–175 (1996)
7. Nowicki, E., Smutnicki, C.: Some aspects of scatter search in the flow-shop problem. European Journal of Operational Research 169, 654–666 (2006)
8. Smutnicki, C.: Scheduling algorithms (in Polish), EXIT, Warszawa (2002)

Artificial Immune System for Short-Term Electric Load Forecasting

Grzegorz Dudek

Department of Electrical Engineering, Czestochowa University of Technology,
Al. Armii Krajowej 17, 42-200 Czestochowa, Poland
dudek@el.pcz.czest.pl

Abstract. This paper proposes a novel model, based on the artificial immune system, to solve the problem of short-term load forecasting. An artificial immune system is trained to recognize antigens which encode sequences of load time series. The created immune memory is a representation of these sequences. In the forecast procedure a new incomplete antigen, containing only the first part of the sequence, is presented to the model. The second forecasted part of the sequence is reconstructed from activated antibodies. The model was verified using several real data examples of the short-term load forecast.

Keywords: artificial immune system, short-term electric load forecasting, similarity-based method.

1 Introduction

The load demand on an electrical power system varies depending on such factors as seasonal effects, work cycles of industrial plants, meteorological conditions, legal and religious holidays, failures of networks and devices etc. Some of these factors are random. A basic requirement in the operation of power systems is to balance the system load by the system generation at all times. Load forecasting is a very important task for electricity companies in order to manage the production, transmission and distribution of electricity in a secure and efficient way. Accurate load forecasts are essential to optimize unit commitment, economic dispatch, hydro scheduling, hydro-thermal coordination, spinning reserve allocation and interchange evaluation. Moreover, the electricity markets could not function without load forecasts. An accurate load forecast allow a lot of money to be saved, e.g. an increase of only 1% in forecast error caused an increase of 10 million pounds in operating cost per year for one electric utility in United Kingdom [1].

Short-term load forecasting (STLF) is defined as forecasting system load demand from one hour to one week ahead. Many techniques have been investigated to solve the STLF problem in the last tree decades. Conventional STLF methods use smoothing techniques, regression methods and statistical analysis. Regression methods are usually used to model the relationship of load consumption and other factors (weather, day type, customer class) [2]. ARMA and related models are very popular (also known as Box-Jenkins, time series, or transfer function models) [3], where the load is modeled by an autoregressive moving average difference equation.

These models are based on the assumption that the data have an internal structure, such as autocorrelation, trend and seasonal variation.

In recent years, artificial intelligence methods (AI) have been widely applied to STLF [4]. AI methods for forecasting have shown an ability to give better performance in dealing with non-linearity and other difficulties in modeling the time series. They do not require any complex mathematical formulations or quantitative correlation between inputs and outputs. The AI methods most often used to STLF can be divided as follows:

- Neural networks (NN) – multilayer perceptron [5], RBF NN [6], Kohonen NN [7], counterpropagation NN [8], recurrent NN [9];
- Fuzzy systems [10], [11];
- Expert systems [12], [13].

Expert systems are heuristics models, which are usually able to take both quantitative and qualitative factors into account. A typical approach is to try to imitate the reasoning of a human operator. The idea is to reduce the analogical thinking behind the intuitive forecasting to formal steps of logic. Neural networks, on the other hand, do not rely on human experience but attempt to learn by themselves the functional relationship between system inputs and outputs. Fuzzy logic models map a set of input variables to a set of output variables. These variables need not be numerical and may be expressed in natural language. Most commonly, a fuzzy logic model includes the mapping of input values to output values using IF-THEN logic statements.

In order to overcome some of the limitations of individual methods, hybrid AI models have been constructed, such as neural networks combined with fuzzy systems [14], [15] or neural network-fuzzy expert systems [16], [17].

New STLF methods are still being created. Some of them are based on machine learning and pattern recognition techniques, e.g. regression trees [18], cluster analysis methods [19] and support vector machines [20]. Other original approaches have also been developed, such as a method using fractal geometry [21], the point function method [22] and a canonical distribution of the random vector method [23].

This paper presents an artificial immune system (AIS) as a way of modeling to STLF. The merits of AIS lie in its pattern recognition and memorization capabilities. AIS are being used in many applications such as [24], [25], [26] anomaly detection, pattern recognition, data mining, computer security, adaptive control, and fault detection. Antigen recognition, self-organizing memory, immune response shaping, learning from examples, and generalization capability are valuable properties of immune systems which can be brought to potential forecasting models. In AIS learning occurs through modification of the number and affinities of the antibodies. The cross-reactivity threshold is the parameter which determines the model generalization level. In the proposed method, sequences of the load time series are encoded in antigens. Immune memory after learning is a representation of a set of antigens. When a new incomplete, composed only of the first part of time series sequence, antigen is presented, it is recognized by some antibodies. The second (forecasted) part of the sequence is reconstructed from these antibodies.

2 Forecasting Model Based on the Artificial Immune System

The problem of STLF, considered in this work, is the one-day ahead power system daily load curve forecasting. The daily load curve is represented by the 24-component vector P, whose components are the following hourly loads. The input variables are 24 hourly loads of the day preceding the day of forecast. It is assumed that the information about the future realization of the load time series is included in the time series preceding the forecast moment. This assumption for the load time series, which are characterized by annual, weekly and daily cycles due to the changes in industrial activities and climatic conditions, was confirmed by statistical tests [27]. Other factors influencing load (atmospheric temperature, humidity, wind speed, precipitation and cloud cover) are not employed in the proposed model. They are important in the power systems in which electrical heating and air-conditioning are common.

Let P_x be a vector of hourly power system loads in the following hours of the day preceding the day of forecast $P_x = [P_x(1), P_x(2), ..., P_x(24)]$, and let P_y be a vector of hourly loads of the day of forecast $P_y = [P_y(1), P_y(2), ..., P_y(24)]$. These vectors are preprocessed in order to get rid of the time series trend and seasonality, and simplify the model. The load patterns are introduced: input $x = [x(1), x(2), ..., x(24)]$ and output $y = [y(1), y(2), ..., y(24)]$, which are vectors with components defined as follows:

$$x(i) = \frac{P_x(i)}{\overline{P}_x} \quad i = 1, 2, ..., 24 \tag{1}$$

$$y(i) = \frac{P_y(i)}{\overline{P}_x} \quad i = 1, 2, ..., 24 \tag{2}$$

where $\overline{P}_x$ is the daily mean load for the day preceding the day of forecast.

The model learns to map $x \rightarrow y$. After learning the input pattern x is presented to the model and the pattern y is obtained as a model output. Formula (2) is used to receive the forecasted load curve P_y.

The model is based on the artificial immune system. Concatenated patterns x and y form antigens. Thus each antigen is composed of 48 amino acids (Fig. 1) which are real numbers. It is assumed that the only components of the immune system are antibodies built analogously to antigens. Each antibody has two chains – x to detect the x-chain of the antigens and y to memorize the y-chain of detected antigens.

The task of the immune system is to learn to map the set of antigens into the set of antibodies. The immune memory is an effect of learning. For each day of the week (Monday, ..., Sunday) the separate immune memory is created using antigens representing only this day (e.g. for forecasting the Sunday load curve, system learns from antigens which x-chain represents the Saturday pattern and y-chain represents the Sunday pattern). The quality criterion of the immune memory is the forecast error. The forecasting procedure applying the learned immune memory runs in the following order. The new antigen consisting only of the x-chain is presented. It is detected by the antibodies with similar x-chains and the y-chain of the antigen is reconstructed from y-chains of these antibodies.

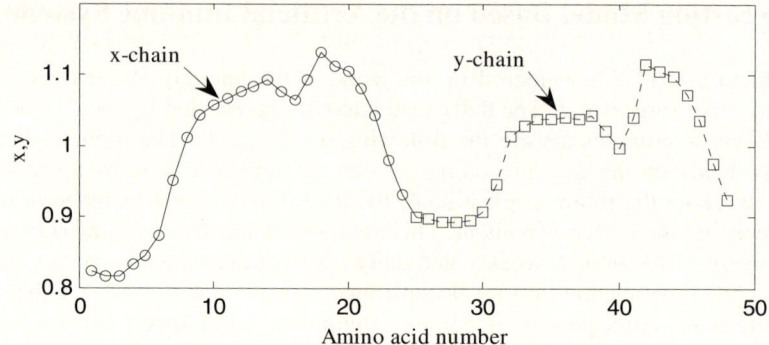

Fig. 1. The antigen and antibody structure

The detailed algorithm of the immune memory creation in the proposed STLF model is described below.

The immune memory creation algorithm in the STLF model

```
1. Loading of the training set of antigens
2. Generation of the initial antibody population
3. Calculation of the affinity of antibodies for antigens
4. Activated antibody detection and evaluation
5. Do until the stop criterion is reached
   5.1. Clonal selection
   5.2. Clone hypermutation
   5.3. Antibody affinity calculation
   5.4. Activated antibody detection and evaluation
   5.5. Selection of the best antibodies
```

Ad. 1. The whole dataset is divided into two subsets – training one and test one. The first sequences of the time series (typically two thirds of the whole time series) are included in the training set and the latest sequences are included in the test set. Immune memory is trained using the training set, and after learning the model is tested using the test set.

Ad. 2. An initial antibody population is created by copping all the antigens from the training set (antibodies and antigens have the same structure). This way of initialization prevents inserting antibodies in empty regions without antigens.

Ad. 3 and 5.3. The affinity measure is based on the distance between x-chains of antigens and antibodies. The Euclidean distance is used:

$$d = \sqrt{\sum_{i=1}^{24}[x_{Ab}(i) - x_{Ag}(i)]^2} \qquad (3)$$

where x_{Ab} and x_{Ag} are the x-chains of the antibody and antigen, respectively.

Ad. 4 and 5.4. If the affinity of the antibody for the antigen is smaller than or equal to the cross-reactivity threshold r, it means that the antigen lies in the antibody recognition region (the antibody is activated by the antigen). For this antibody the forecast error (MAPE, which is traditionally used in STLF models) is calculated:

$$\delta = \frac{1}{24}\sum_{i=1}^{24}\left|\frac{y_{Ab}(i) - y_{Ag}(i)}{y_{Ag}(i)}\right| \cdot 100\% \qquad (4)$$

where y_{Ab} and y_{Ag} are the y-chains of the antibody and antigen activating this antibody.

If several antigens lie in the antibody recognition region, the error is calculated for each of them. The mean error $\bar{\delta}$ is applied to evaluate the antibody and is minimized in the following iterations of the algorithm.

Ad. 5. The algorithm stops if the maximum number of iteration L is reached.

Ad. 5.1. Each antibody cases secreting as many clones as many antigens are in its recognition region. Thus most clones are generated in the dense clusters of antigens.

Ad. 5.2. The main goal of hypermutation is to improve the diversity of the immune system in order to effectively recognize new antigens. The hypermutation is realized as follows. Each clone of the antibody is shifted towards different antigen lying in the recognition region of this antibody. The bigger the error δ for the given antigen is, the bigger shift toward this antigen is. The shift is calculated according to the formulae:

$$x_{Ab}(i) = x_{Ab}(i) + \eta(i)[x_{Ag}(i) - x_{Ab}(i)] \quad i = 1,2,...,24 \qquad (5)$$

$$y_{Ab}(i) = y_{Ab}(i) + \eta(i+24)[y_{Ag}(i) - y_{Ab}(i)] \quad i = 1,2,...,24 \qquad (6)$$

where $\eta \in (0, 1)$ is a learning coefficient calculated from the hyperbolic tangent sigmoid function as follows:

$$\eta(i) = \frac{2}{1 + \exp[-\beta \delta N_i (1,0.1)]} - 1 \quad i = 1,2,...,48 \qquad (7)$$

where β is the shape parameter and $N_i(1, 0.1)$ are the independent normally distributed random numbers with mean 1 and standard deviation 0.1.

Random factor in formula (7) is introduced to avoid stagnation of the learning process caused by getting into local minimum trap of the error function. For the shape parameter $\beta = 0.04$ and error $\delta = 1\%$ the value of the learning coefficient is minor – $\eta \cong 0.02$, and consequently the shift is minor too. For the higher errors – 2%, 5%, 10%, 100% the η-value is higher – about 0.04, 0.1, 0.2 and 0.96, respectively. This type of hypermutation produces new antibodies only in the regions covered by antigens.

Ad. 5.5. For each antigen from the training set, the set of antibodies activated by this antigen is determined. Only one antibody from this set, with the best evaluation $\bar{\delta}$, is selected to the next population. So the clonal expansion, unnecessary in this model, is halted. The maximum number of antibodies in the next population is equal to the number of antigens, but the real number of antibodies is usually smaller because the same antibody could be selected by the several antigens (it depends on the value of the cross-reactivity threshold r). Outlier, i.e. antigen lying away from other antigens, is represented by the separate antibody.

Forecast procedure. After learning the antibodies represent overlapping clusters of similar antigens. In the forecast procedure new antigen having only x-chain is presented. The Ω set of antibodies, activated by this antigen, is determined. The y-chains of these antibodies storage average y-chains of antigens from the training set with similar x-chains. The y-chain of the input antigen is reconstructed from the y-chains of the antibodies contained in the Ω set (denoted by $\mathbf{y}_{Ab}^{j}$):

$$\hat{y}_{Ag}(i) = \frac{\sum_{j=1}^{|\Omega|} w_j y_{Ab}^j(i)}{\sum_{j=1}^{|\Omega|} w_j} \quad i = 1, 2, \ldots, 24 \qquad (8)$$

where $w_j \in (0, 1)$ is the weight which value is dependent on the distance d_j between the input antigen and the j-th antibody from the Ω set:

$$w_j = 1 - \frac{d_j}{r} \quad j = 1, 2, \ldots, |\Omega| \qquad (9)$$

Antibodies, closer to the antigen, have the higher influence on the forecast forming. If an antigen is not recognized by antibodies, it means that it represents a new shape of the load curve, not contained in the training set. In this case the cross-reactivity threshold r is consistently being increased until the antigen is recognized by one or more antibodies. The level of confidence in the forecast in such a case is low and the forecast should be verified.

3 Application Examples

The described above artificial immune system for STLF was implemented in Matlab and was applied to five real STLF problems. Data are described in Table 1. Usually the smaller power system is, the more irregular and harder to forecasting load time series is. The measure of the load time series regularity could be the forecast error (MAPE) determined by the naïve method. The forecast rule in this case is as follows: the load curve of the day of forecast is the same as seven days ago. The mean forecast errors, calculated according to this naïve rule, are presented in Table 1.

The model parameters – the cross-reactivity threshold r, the shape parameter β and the maximum number of iterations L were determined after the preliminary tests. An increase in the r-value causes an increase in the training set error, but the test set error behavior is rather irregular, especially in the case of irregular time series. So the choice of the r-value is not obvious. The β parameter is not critical. The similar results were received for different values of this parameter. The L-parameter value

Table 1. Description of data used in experiments

Data symbol	Data description	Forecast error of the naïve method, %
A	Time series of the hourly loads of the Polish power system from the period 2002-2006, mean load of the system ~16 GW	4,25
B	Time series of the hourly loads of the Polish power system from the period 1997-2000, mean load of the system ~15,5 GW	4,38
C	Time series of the hourly loads of the local power system from the period July 2001-January 2003, mean load of the system ~1,2 GW	6,59
D	Time series of the hourly loads of the local power system from the period June 1998-July 2002, mean load of the system ~300 MW	7,45
E	Time series of the hourly load demands of the chemical plant from the period 1999-2001, mean load demand of the plant ~80 MW	17,46

should ensure stabilization of the training error at the fixed level. However, the test error is often not stabilized varying in value up and down. The parameter values used in experiments were: $\beta = 0.04$, $L = 50$, r – half of the mean distance between antigens and initial population of antibodies.

Results of forecasting – mean errors for training ($MAPE_{trn}$) and test sets – are presented in Table 2. Results of test sets include two cases – one where unrecognized antigens are not taken into account (percentage of these antigens is shown in Table 2) and second where the cross-reactivity threshold r increases until recognition of these antigens. The forecast calculated for these untypical antigens is not very reliable and accurate, so the mean errors in the second case ($MAPE_{tst2}$) are higher than in the first case ($MAPE_{tst1}$).

More detailed results for the test parts of A (most regular) and E (most irregular) time series are presented in figures. Fragments of the A and E time series and their forecasts – in Fig. 2, $MAPE_{tst1}$ for each day type and hour – in Fig. 3 and percentage error (PE_{tst1}) histograms – in Fig. 4.

For comparison, forecast using the simple nearest neighbor method was calculated. The method applies the following rule: y-chain paired with the input x-chain is the same as the y-chain paired with nearest neighbor (found in the training set) of the input

Table 2. Forecast errors

Data symbol	$MAPE_{trn}$	$MAPE_{tst1}$	Percent of unrecognized antigens	$MAPE_{tst2}$	$MAPE_{tst3}$	$MAPE_{tst4}$	$MAPE_{tst5}$
A	1.56	1.77	5.80	1.88	2.05	-	-
B	1.62	1.82	8.90	2.29	2.63	2.24	2.11
C	2.05	3.16	16.48	4.46	4.76	4.89	4.07
D	2.79	3.55	8.85	4.00	4.17	3.71	3.52
E	3.77	6.41	22.74	8.60	9.47	8.32	8.06

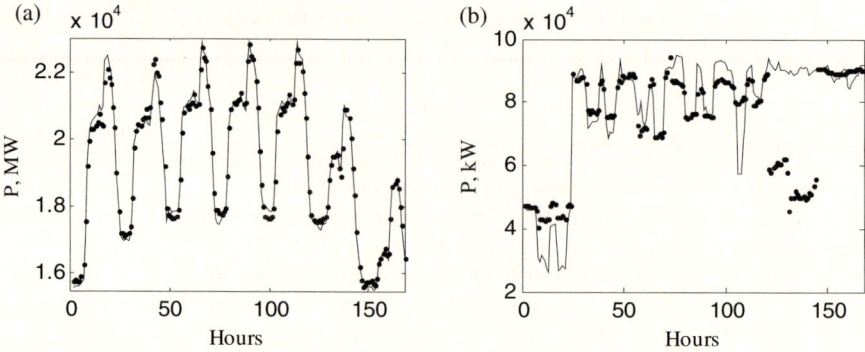

Fig. 2. Fragments (one week) of the test A (a) and E (b) time series (*solid lines*) and their forecasts (*dots*)

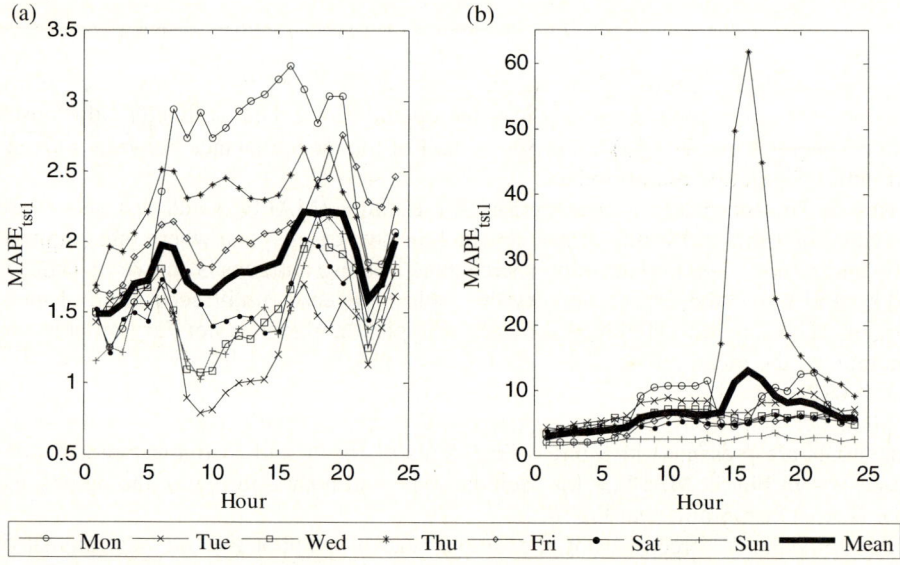

Fig. 3. MAPE$_{tst1}$ for each day type and hour for A (a) and E (b) time series

x-chain. The forecast results for this method (MAPE$_{tst3}$) are presented in Table 2. In this table the results (taken from [27]) for two other STLF models are also presented: for model based on the neural network (MAPE$_{tst4}$) and for model based on the fuzzy clustering (MAPE$_{tst5}$). These models are described in [27] and [11].

An example of the antibody and antigens, which activated this antibody, is shown in Fig. 5. The antigens, recognized by the same antibody, have the similar y-chains in case of the regular time series (Fig. 5(a)). Thus forecasts are more accurate. It is different in case of irregular series (Fig. 5(b)) – large y-chain dispersion causes high errors of forecast.

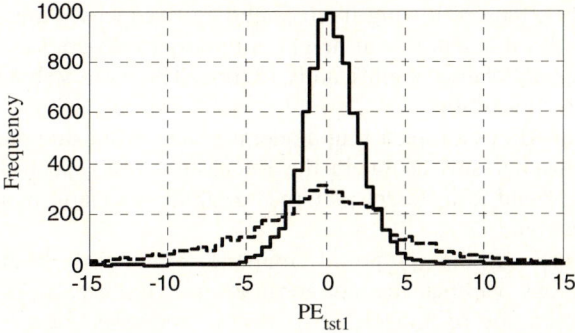

Fig. 4. PE_{tst1} histograms for A (*solid line*) and E (*dashed line*) time series

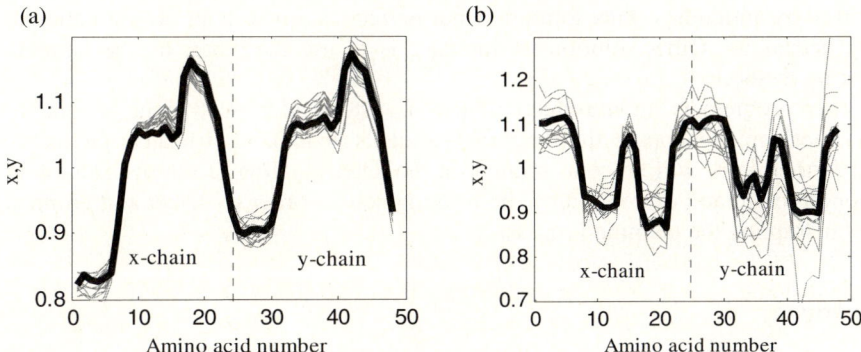

Fig. 5. A group of similar antigens and activated by them antibody (*thick line*) for A (a) and E (b) time series

The peak observed in Fig. 3(a) is a result of the drastic change in load in one day, possibly caused by a failure (load forecast errors of this day reached 2300%; $MAPE_{tst1}$ for the E time series without this day decreased to 5.07%). The prediction of this event was impossible because probable lack of information about this event in the time series sequence before its existence and it was not represented in the training set. For irregular time series, like E series, surely there were many similar situations, e.g. in Fig. 2(b) the forecast for Saturday is completely wrong.

Empirical distributions of errors PE_{tst1} (Fig. 4) are rather symmetrical (skewness close to 0), similar in shape to the normal distribution, but steeper (kurtosis higher than 3).

The proposed AIS has a lot of interesting properties as the forecasting model, but yet it is not developed enough to compete with other AI models such as models using neural networks and fuzzy logic (Table 2), which have been finishing up by many researchers for many years.

4 Conclusions

The proposed STLF model belongs to the class of similarity-based models. These models are based on the assumption that, if patterns of the time series sequences are similar to each other, then following them patterns of sequences are similar to each other as well. It means that patterns of neighboring sequences are staying in a certain relation, which does not change significantly in time. The more stable this relation is, the more accurate forecasts are.

The idea of using AIS as a forecasting model is a very promising one. The immune system has some mechanisms useful in the forecasting tasks, such as an ability to recognize and to respond to different patterns, an ability to learn, memorize, encode and decode information.

The disadvantage of the proposed immune system is limited ability to extrapolation. Regions without the antigens are not represented in the immune memory. However, a lot of models, e.g. neural networks, have problems with extrapolation. But the AIS has a detection mechanism of outliers, i.e. antigens laying outside the recognition regions of antibodies. In the proposed approach for such antigens the cross-reactivity threshold is increasing, and finally these antigens are recognized by antibodies. This solution is not perfect as forecasts in such situation are usually inaccurate. Other solution to this case is to use the other, maybe heuristic, forecasting method.

Another problem is an introduction to the system additional input information which is not homogeneous with time series elements (loads), e.g. wheatear factors.

The further work will be concentrated on the determination of the better antibody receptor structure and on rebuilding the training sets in order to detect and eliminate outliers disrupting the learning process.

References

1. Bunn, D., Farmer, E.: Economic and Operational Context of Electric Load Prediction. In: Bunn, D., Farmer, E. (eds.) Comparative Models for Electrical Load Forecasting, pp. 3–11. Wiley, Chichester (1985)
2. Engle, R.F., Mustafa, C., Rice, J.: Modeling Peak Electricity Demand. Journal of Forecasting 11, 241–251 (1992)
3. Gross, G., Galiana, F.D.: Short-Term Load Forecasting. Proc. IEEE 75(12), 1558–1573 (1987)
4. Metaxiotis, K., Kagiannas, A., Askounis, D., Psarras, J.: Artificial Intelligence in Short Term Load Forecasting: A State-of-the-Art Survey for the Researcher. Energy Conversion and Management 44, 1525–1534 (2003)
5. Khotanzad, A., Afkhami-Rohani, R., Lu, T.L., Davis, M.H., Abaye, A., Maratukulam, D.: ANNSTLF - A Neural Network Based Electric Load Forecasting System. IEEE Trans. on Neural Networks 8(4), 835–846 (1997)
6. Dudek, G.: Short Term Load Forecasting using Radial Basis Neural Networks. In: 5th National Conference on Forecasting in Power Engineering PE 2000 (in Polish), Wydawnictwo Politechniki Częstochowskiej, Częstochowa, pp. 59–68 (2000)
7. Osowski, S., Siwek, K., Kądzielawa, A.: Neural Network Approach to Load Forecasting. In: 2nd Conference on Neural Networks and Their Applications, Wydawnictwo Politechniki Częstochowskiej, Częstochowa, pp. 355–360 (1996)
8. Dudek, G.: Application of the Hecht-Nielsen Neural Network to Short Term Load Forecasting. In: International Conference on Present-Day Problems of Power Engineering APE 1997 (in Polish), Gdańsk-Jurata, vol. 4, pp. 65–72 (1997)

9. Vermaak, J., Botha, E.C.: Recurrent Neural Networks for Short-Term Load Forecasting. IEEE Trans. on Power Systems 13(1), 126–132 (1998)
10. Papadakis, S.E., Theocharis, J.B., Bakirtzis, A.G.: A Load Curve Based Fuzzy Modeling Technique for Short-Term Load Forecasting. Fuzzy Sets and Systems 135, 279–303 (2003)
11. Dudek, G.: Short-Term Electrical Load Forecasting using Fuzzy Clustering Method. Przegląd Elektrotechniczny (in Polish) 9, 26–28 (2006)
12. Rahman, S., Hazim, O.: A Generalized Knowledge-Based Short Term Load Forecasting Technique. IEEE Trans. on Power Systems 8(2), 508–514 (1993)
13. Markovic, M., Fraissler, W.: Short-Term Load Forecast by Plausibility Checking of Announced Demand: An Expert System Approach. Eur. Trans. Electr. Power Eng/ETEP 3(5), 353–358 (1993)
14. Kodogiannis, V.S., Anagnostakis, E.M.: Soft Computing Based Techniques for Short-Term Load Forecasting. Fuzzy Sets and Systems 128, 413–426 (2002)
15. Liang, R.-H., Cheng, C.-C.: Short-Term Load Forecasting by a Neuro-Fuzzy Based Approach. Electrical Power and Energy Systems 24, 103–111 (2003)
16. Dash, P.K., Liew, A.C., Rahman, S., Dash, S.: Fuzzy and Neuro-Fuzzy Computing Models for Electric Load Forecasting. Engng Applic. Artif. Intell. 8(4), 423–433 (1995)
17. Tamimi, M., Egbert, R.: Short Term Electric Load Forecasting via Fuzzy Neural Collaboration. Electric Power System Research 56, 243–248 (2000)
18. Dudek, G.: Regression Tree as a Forecasting Tool. In: National Conference on Forecasting in Power Engineering PE 2004 (in Polish), Wydawnictwo Politechniki Częstochowskiej, Częstochowa, pp. 99–105 (2004)
19. Dudek, G.: Application of Hierarchical Clustering Methods to the Daily Electrical Load Profile Forecasting. Przegląd Elektrotechniczny (in Polish) 9, 9–11 (2006)
20. Pai, P.-F., Hong, W.-C.: Support Vector Machines with Simulated Annealing Algorithms in Electricity Load Forecasting. Energy Conversion and Management 46, 2669–2688 (2005)
21. Dobrzańska, I.: Hausdorff Dimension as a Tool of Step Prediction. Badania Operacyjne i Decyzje (in Polish) 2, 31–62 (1991)
22. Łyp, J.: Methodology of the Electric Load Analysis and Forecasting for the Local Power Systems. Ph.D. Dissertation (in Polish), Dept. Elect. Eng., Częstochowa University of Technology (2003)
23. Popławski, T., Dąsal, K.: Canonical Distribution in Short-Term Load Forecasting. In: Szkutnik, J., Kolcun, M. (eds.) Technical and Economic Aspect of Modern Technology Transfer in Context of Integration with European Union, Mercury-Smekal Pub House, Kosice, Slovak Republic, pp. 147–153 (2004)
24. Dasgupta, D., Ji, Z., Gonzalez, F.: Artificial Immune System (AIS) Research in the Last Five Years. In: Congress on Evolutionary Computation, Canberra, Australia, pp. 123–130 (2003)
25. Timmis, J., Knight, T., De Castro, L.N., Hart, E.: An Overview of Artificial Immune Systems. In: Paton, R., Bolouri, H., Holcombe, M., Parish, J.H., Tateson, R. (eds.) Computation in Cells and Tissues: Perspectives and Tools for Thought. Natural Computation Series, pp. 51–86. Springer, Heidelberg (2004)
26. Wierzchoń, S.: Artificial Immune Systems. Theory and Applications. Akademicka Oficyna Wydawnicza EXIT, Warsaw (in Polish) (2001)
27. Dudek, G.: Short-Term Load Forecasting using Fuzzy Clustering and Genetic Algorithms. Final report of the Polish State Committee for Scientific Research founded grant no. 3T10B02329. Dept. Elect. Eng., Częstochowa University of Technology (unpublished, in Polish) (2006)

Ant Focused Crawling Algorithm[*]

Piotr Dziwiński and Danuta Rutkowska

Department of Computer Engineering
Czestochowa University of Technology, Poland
dziwinski@kik.pcz.czest.pl,
drutko@kik.pcz.czest.pl

Abstract. This paper presents a new algorithm for hypertext graph crawling. Using an ant as an agent in a hypertext graph significantly limits amount of irrelevant hypertext documents which must be downloaded in order to download a given number of relevant documents. Moreover, during all time of the crawling, artificial ants do not need a queue to central control crawling process. The proposed algorithm, called the Focused Ant Crawling Algorithm, for hypertext graph crawling, is better than the Shark-Search crawling algorithm and the algorithm with best-first search strategy utilizing a queue for the central control of the crawling process.

1 Introduction

Enormous growth of the Internet and easy access to the network enable huge amount of users to access WWW resources through search engines. Changeability and large amount of WWW pages is a big challenge for modern crawlers and search engines. They should reflect WWW resources as accurately as possible and also hold information about the resources as fresh as possible. Complete crawling entire Web is impossible in reasonable time, no matter which technology is available at the site where the search engines operate. An ideal crawler should be able to recognize relevance and importance of Web pages. The crawlers can order new links extracted from downloaded WWW pages by use of different methods. Some of them are measurements of similarity between pages and a current query, amount of links to point out WWW pages or the most popular Page Rank.

Most crawler algorithms use a queue that globally control the process of crawling. The first crawler algorithm was the Simple-Crawler method mentioned in [1,2]. In the Simple-Crawler algorithm, a crawler extracts URL addresses from documents and includes them at the end of the queue without ordering. Another type of the crawlers is called selective crawlers [1]. The selective crawlers select a next download page with respect to some criterions (relevance or importance of the page). Relevance of the documents can be calculated using a

---

[*] This work was partly supported by the Foundation for Polish Science (Professorial Grant 2005-2008) and the Polish State Committee for Scientific Research (Grant N516 020 31/1977), Special Research Project 2006-2009, Polish-Singapore Research Project 2008-2010, Research Project 2008-2010.

classifier. In order to classify WWW documents, different methods can be employed, such as: k-nearest neighbors algorithm, Naive Bayes, support vector machines, decision trees, neural networks, fuzzy rules or neuro-fuzzy systems; see e.g. [1, 3, 4, 5, 6, 7, 8, 9, 10, 11, 12]

If a crawler inserts a new URL in the queue, in the order depending on the relevance, we obtain the best-first search strategy [13, 14, 15]. In many cases, crawling on entire web is not required. Instead, it can perform crawling only a relevant part of the web. In this way, the focused crawling algorithm [16, 17] has been obtained. Relevant areas of the web are small enough for the focused crawler to operate on such areas in finite time. Locality of the subject in the web are studied by Davison [18]. Rungsawang et al. [19] proposed the consecutive crawling to take advantage of experience from earlier crawling processes. They built a knowledge-base employed to produce better results for the next crawling. The first crawling algorithm inspired by the nature – behavior of some animals or insects – is the Fish-Search algorithm [20]. There is one of the first dynamic search heuristics based on intuition that relevant documents often have relevant neighbors [18]. The best version of the Fish-Search algorithm is the Shark-Search algorithm introduced in [21]. The Shark-Search algorithm bases on the same intuition but introduces a real measure of relevance of anchors extracted from new documents. The relevance of the anchors is based on relevance of a current document, content and context of the anchor. However in this case, the proposed Shark-Search algorithm exploits a queue for central control of the crawling process.

How we could dispose the queue for the central control of crawling is a question for which the answer we get from the nature – from real ants that solved this problem long time ago. There is a lot of algorithms developed based on behavior of real ants. Most of them are known as ant colony optimization (ACO). The ACO has been applied to the Traveling Salesman Problem (TSP) [22, 23, 24], graph coloring [25, 26], dynamic shortest path problems arising in telecommunication networks, dynamic cleaning problem [27]. Wagner I. A. et al. [28] adopt artificial ants to consider the problem of deciding whether graph $G(V, E)$, $V-$ set of vertices, E – set of edges, is Hamiltonian. This problem is a special case of the TSP. Moving rules of the artificial ants were employed to control robots in multi agent systems to solve the distributed covering problem [28]. In [29], artificial ants are used to network covering in the Vertex Ant Walk (VAW) algorithm.

The goal of this article is to introduce the Ant Focused Crawling Algorithm which does not require the queue and decreases the number of downloaded irrelevant documents. Artificial ants leave pheromone trails on the ground in order to mark some favorable paths that should be followed by other members of the colony. The ants, using the pheromone trails, share information about the problem to be solved. In this way, all members of the colony have access to the information about relevance of the vertices that can be visited in next steps in the hypertext graph while progressing the crawling process. This crawling process performed by artificial ants can be running simultaneously by many artificial ants (crawling robots) without central control by the queue.

Section 1 provides background information on the selective crawling, focused crawling, ant colony optimization, and the proposed algorithm. Section 2 surveys the Ant Focused Crawling Algorithm. Section 3 highlights experimental results for the first crawling process compared with the Shark-Search algorithm and the best-first search strategy. Moreover, the crawling process performed by the proposed algorithm is analyzed with regard to the number of repetitions. Section 4 concludes the article.

2 Ant Focused Crawling Algorithm

Effective crawling the Internet by crawlers is a main issue concerning search engines. There are two types of crawling: passive crawling and active crawling. In the active crawling, crawlers are controlled by utilize a queue to central control of the crawling process. The passive crawling consists of crawling without any central control. In this article, new Ant Focused Crawling Algorithm which does not use the central control in the form of the queue, is proposed. This algorithm saves system and memory resources of the hardware.

The ant colony optimization (ACO) takes inspiration from the foraging behavior of some ant species. Ants moves from their nests to food and leave pheromone trails on the ground, in order to mark favorable path that should be followed by other members of the colony. In this way, individual ants discover the shortest path between the nest and the source of food using undirected communication in the form of the pheromone trail. In the similar way, we adopt the ant behavior with regard to the focused crawling hypertext graph $\mathbf{G}$. Hypertext graph $\mathbf{G}(\mathbf{V}, \mathbf{E})$ is a directed graph in which vertice $v \in V$ and edge $e \in E$ correspond to WWW documents and links in documents, respectively. Artificial ants move in the hypertext graph. The pheromone in vertices contain information about relevance, number of visits, time of visits. Artificial ants selects next vertice v according to the rule as follows [30]

$$v = \begin{cases} \arg\max_{\omega \in J(u)} \left\{ [\tilde{\tau}(\omega)]^\alpha \cdot [\eta(\omega)]^\beta \right\} & \text{if} \quad q \leq q_0 \\ p_{u \to v} & \text{if} \quad q > q_0 \end{cases} \quad (1)$$

where

$$p_{u \to v} = \begin{cases} \dfrac{[\tilde{\tau}(v)]^\alpha \cdot [\eta(v)]^\beta}{\sum_{\omega \in J(u)} [\tilde{\tau}(\omega)]^\alpha \cdot [\eta(\omega)]^\beta} & \text{if} \quad v \in J(u) \\ 0 & \text{if} \quad v \notin J(u) \end{cases} \quad (2)$$

$J(u)$ – set of vertices connected with vertice u by use of edges $e \in \mathbf{E}$ in the hypertext graph $\mathbf{G}$,
q_0 – parameter, $q_0 \in [0, 1]$,
q – random number belonging to $[0, 1]$,
$p_{u \to v}$ – probability of the movement from vertex u to vertex v,
α – importance of the pheromone trail,
β – importance of the heuristic information,
$\tilde{\tau}(u)$ – value of the pheromone smell perceived by artificial ants.

Equations (1) and (2) are similar to the rule used in the ant colony system [22, 23, 31].

Artificial ants move in the hypertext graph **G** from vertex u to vertex v, where $u, v \in \mathbf{G}$, using the set of edges E, and leave pheromone trails $\tau_l(v)$, $\tau_r(v)$, $\tau_{\Delta t}(v)$.

Pheromone $\tau_l(v)$ guarantees that the searching process is similar to the Hamiltonian cycle in a graph [32]. Pheromone $\tau_r(v)$ directs the ants to move in the relevant area of the hypertext graph. Pheromone $\tau_{\Delta t}(v)$ causes refreshment of the visited vertices after specific time which depends on the frequency change of the documents in the hypertext graph or the WWW network.

It is proposed that values of pheromone smell perceived by artificial ants are calculated as follows

$$\widetilde{\tau}(v) = [\tau_r(v)]^{\alpha_r} \cdot \left[\frac{1}{1+\tau_l(v))}\right]^{\alpha_l} \cdot [\tau_{\Delta t}(v)]^{\alpha_{\Delta t}} \quad (3)$$

where:
$\alpha_r, \alpha_l, \alpha_{\Delta t}$ – parameters which weight the relative importance of the pheromones $\tau_r(v)$, $\tau_l(v)$, $\tau_{\Delta t}(v)$, respectively.

Pheromone $\tau_r(v)$ is refreshed by all ants in each step from vertex u to vertex v in the hypertext graph G, similarly as in the ant system presented in [23], according to the equation

$$\tau_r(u) \leftarrow (1 - \varphi_r) \cdot \tau_r(u) + \varphi_r \cdot \Delta\tau_r^k(u) \quad (4)$$

where:
ϕ_r – evaporation rate of pheromone $\tau_r(v)$,
$\Delta\tau_r^k(u)$ – quality of the pheromone $\tau_r(v)$, dependent on the relevance of the vertices available from vertex v and the next ones – $ch[v]$ (children of v) through ant k.

Every ant has a small memory that contains a specified number of visited vertices. This memory is used for local control of the crawling process (but is functioning differently as the queue), and avoids cycles that occur frequently in the hypertext graph. Quality of the pheromone $\Delta\tau_r^k(u)$ is calculated based on memory of the ants. It is dependent on relevance at further vertices. This is done with a specified delay and is associated with reinforcement learning. In this way, each ant contains small path $\mathbf{Q}^k$ which includes relevance of the visited vertices. Each ant uses path $\mathbf{Q}^k$, and leaves pheromone $\Delta\tau_r^k(u)$ calculated by the equation

$$\Delta\tau_r^k(u) = \sum_{i=1}^{L} \delta^i q_i^k \quad (5)$$

where:
δ – coefficient reducing influence of the relevance of the following vertices; $\delta \in [0, 1]$,
q_i^k – relevance of vertice i remembered for ant k; $q_i^k \in \mathbf{Q}^k$,
L – length of path $\mathbf{Q}^k$.

Equation (5) is effective only in the first crawling process performed by the Ant Focused Crawling Algorithm. We develop another equation for successive crawling process, in the form:

$$\Delta \tau_r^k(u) = \frac{\sum_{i=1}^{L} \delta^i \cdot q_i^k}{\sum_{j=1}^{L} \delta^j} \qquad (6)$$

Pheromone $\tau_l(v)$ is increased by all ants during visiting each vertice. This strategy warrants convergence of the algorithm to the path approximate to the Hamiltonian path, like in the VAW algorithm [29].

Pheromone $\tau_{\Delta t}(u)$ is calculated as follows

$$\tau_{\Delta t}(u) = \frac{t_a(u) - t(u)}{t_{od}(u)} \qquad (7)$$

where:
$t_a(u)$ – current time in vertice u,
$t(u)$ – visit time in vertice u by an artificial ant,
$t_{od}(u)$ – reference time (referenced to speed of the change in vertice u).

During the crawling process performed by artificial ants, members of the colony select the next vertice according to the value of the pheromone in the vertice and heuristic information $\eta(u)$, see Equations (1),(2). The heuristic information in ant algorithms evaluates a local solution of the problem. In the case of the TSP, the heuristic information is calculated as follows:

$$\eta(u, v) = \frac{1}{d(u, v)} \qquad (8)$$

where $d(u, v)$ – distance measure between cites.

In the hypertext graph, the heuristic information should be related to potential relevance $r_p(v)$ of the following vertices or relevance $R(v)$ of the document with the context. It is proposed that the heuristic information is calculated as follows:

$$\eta(v) = \begin{cases} R(v) & \text{if} \quad d(v) \neq \emptyset \\ r_p(v) & \text{if} \quad d(v) = \emptyset \end{cases} \qquad (9)$$

where:
$r_p(v)$ – potential relevance of the next vertice [20]; see Equation (11),
$d(v)$ – document for vertice v.

The relevance of a downloaded document with a context of the hypertext graph depends on the relevance of the document for vertice v and the arithmetic average of the heuristic information about the descendant vertices $ch[v]$, and is calculated as follows:

$$R(v) = r(d(v)) + \frac{1}{|ch[v]|} \sum_{\omega \in ch[v]} \eta(\omega) \qquad (10)$$

where:
$ch[v]$ – set of descendant vertices of vertice v (children of v),
$|ch[v]|$ – number of descendant vertices of vertice v.

Equation (10) is similar to that presented by Mark et al. [1,33]; this involves the context of the graph.

The potential relevance of vertice v depends on the inherited relevance of the document $r_d(v)$ for vertice v, relevance of the content of the anchor $r_a(u,v)$ and relevance of the neighborhood of the anchor $r_{ac}(u,v)$. Potential relevance $r_p(v)$ is calculated as follows [20]:

$$r_p(v) = \gamma \cdot r_d(v) + (1-\gamma) \cdot r_n(v) \qquad (11)$$

where:
γ – coefficient of the inherited relevance of the document $r_d(v)$ for vertice v; see Equation (13),
$r_n(v)$ – anchor relevance, given by Equation (15).

Inherited relevance $r_d(v)$ for vertice v depends on relevance of vertice u, which contains document $d(u)$, $u \in pa[v]$, is calculated as follows:

$$r_d(v) = \begin{cases} r(d(u)) \cdot \delta & \text{if} \quad r(d(u)) > \epsilon;\ u \in pa[v] \\ r_d(u) \cdot \delta & \text{if} \quad r(d(u)) \leq \epsilon;\ u \in pa[v] \end{cases} \qquad (12)$$

where:
ϵ – threshold relevance value, like that in the Shark-Search algorithm [21],
δ – coefficient of reduction of the relevance for the inherited relevance of parent documents $pa[u]$.

If we have more than one vertice $u \in pa[v]$, the inherited relevance is given by:

$$r_d(v) = \frac{1}{|pa[v]|} \sum_{u \in pa[v]} \begin{cases} r(d(u)) \cdot \delta & \text{if} \quad r(d(u)) > \epsilon \\ r_d(u) \cdot \delta & \text{if} \quad r(d(u)) \leq \epsilon \end{cases} \qquad (13)$$

where $|pa[v]|$ – number of parent vertices of vertice v.

The relevance of the anchor in vertice v consists of relevance of the anchor text $r_a(u,v)$ of document in vertice u and relevance of the context of the anchor $r_{ac}(u,v)$ of document in vertice u, and is calculated similarly as in [20]:

$$r_n(v) = \beta \cdot r_a(u,v) + (1-\beta) \cdot r_{ac}(u,v) \qquad (14)$$

where β – coefficient of the influence of the content relevance of the anchor $r_a(u,v)$ and the context of the anchor $r_{ac}(u,v)$.

If we have more than one parent vertice $u \in pa[v]$, the relevance of the anchor is given by:

$$r_n(v) = \beta \cdot \frac{1}{|pa[v]|} \sum_{u \in pa[v]} r_a(u,v) + (1-\beta) \cdot \frac{1}{|pa[v]|} \sum_{u \in pa[v]} r_{ac}(u,v) \qquad (15)$$

The relevance of the anchor text $r_a(u,v)$ is defined as [20]:

$$r_a(u,v) = \text{sim}(q, \text{anchor}(u,v)) \qquad (16)$$

The relevance of the anchor context is calculated by [20]:

$$r_{ac}(u,v) = \begin{cases} \text{sim}(q, \text{ConAn}(u,v)) & \text{if} \quad r_a(u,v) \leq 0 \\ 1 & \text{if} \quad r_a(u,v) > 0 \end{cases} \quad (17)$$

where:
anchor(u,v) – anchor text for edge $u \to v$,
$\text{ConAn}(u,v)$ – text in the specified neighborhood of the anchor, for the edge $u \to v$.

3 Experiments and Results

The experiments were performed for JAVA documentation [34], initially parsed, and indexed. The parsed documentation was saved as a compressed XML file, which after reading and decompressing becomes the indexed hypertext graph **G**. All experiments were performed for the same simple query q and for the same starting address. The experiments for a specified parameter are repeated some number of times.

In the experiments, we obtained satisfied results which overcome comparable algorithms. Figure 1a shows the number of relevant documents as a function of the number of all downloaded documents during entire crawling process. Figure 1b illustrates total amount of relevant information as a function of the number of all downloaded documents. The total amount of relevant information is calculated as follows [20, 21]:

$$Sum\_Inf(\mathbf{D^r}, q) = \sum_{d(u) \in \mathbf{D^r}} d(d(u), q) \quad (18)$$

where $\mathbf{D}^r$ – set of downloaded relevant documents.

The proposed new Ant Focused Crawling Algorithm, for hypertext graph crawling, overcomes compared algorithm – achieves results 33% better those obtained from the Schark-Search algorithm. Moreover, it skip an irrelevant area of the graph **G** and is focused only on relevant areas. In addition better stabilization over period of the crawling process is observed. The proposed algorithm does not use the queue to central control of the crawling process. Effectiveness of the algorithm [1] is calculated as follows:

$$e = \frac{r_t}{t} \quad (19)$$

where:
t – number of downloaded documents in specific period of time;
r_t – number of relevant downloaded documents for which the relevance is better than ϵ, where ϵ – threshold of the relevance.

An ideal crawling algorithm should obtain effectiveness equal to 1. The proposed algorithm achieves effectiveness better than two compared algorithms. The efficiency of co-operation of the artificial ants in the process of crawling

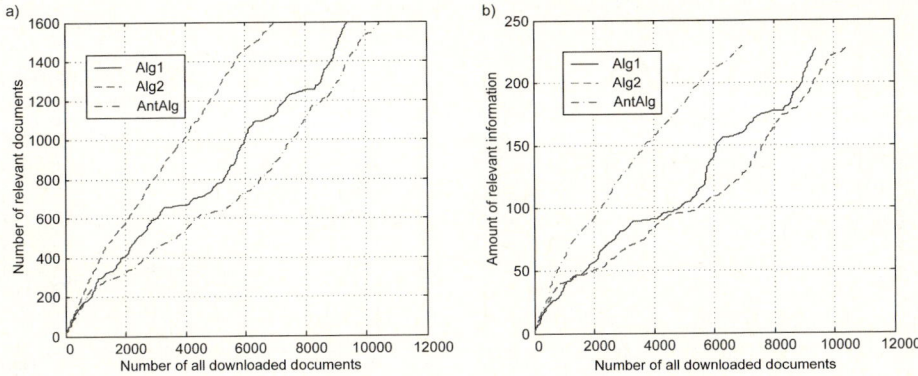

Fig. 1. Compared crawling algorithm with respect to (a) – number of relevant documents as a function of the number of all downloaded documents; (b) – amount of relevant information as a function of the number of all downloaded documents; Alg1 – focused crawling algorithm with the best-first search strategy; Alg2 – Shark-Search crawling algorithm; AntAlg – Ant Focused Crawling Algorithm

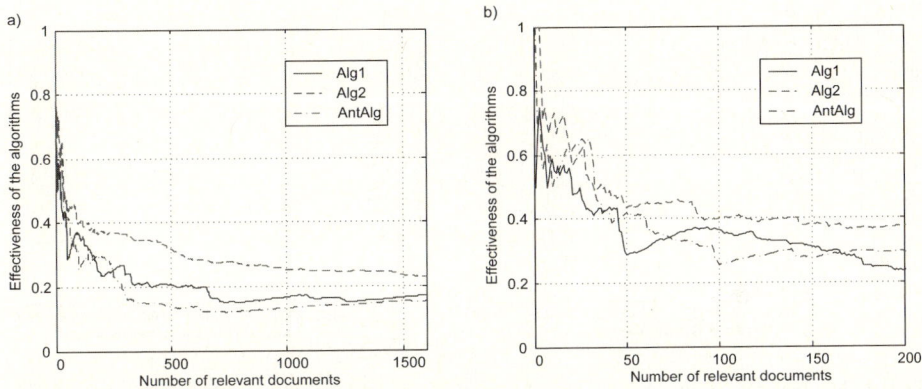

Fig. 2. Compared crawling algorithm with respect to effectiveness (a) – for number of relevant documents; (b) – for first 200 relevant documents; Alg1 – focused crawling algorithm with the best-first search strategy; Alg2 – Shark-Search crawling algorithm; AntAlg – Ant Focused Crawling Algorithm

is shown in Fig. 3. For more ants, the proposed algorithm obtains the same or better results. This feature of the proposed algorithm can be useful for crawling the web by many agents without using the queue for central control.

For successive crawling, an ant colony possesses experience about location of relevant information in the hypertext graph, saved in the form of pheromone. In successive crawling process, individual ants use information from earlier processes, and the Ant Focused Crawling Algorithm obtains better results, what is shown in Fig. 4. The proposed algorithm is useful in order to maintain freshness

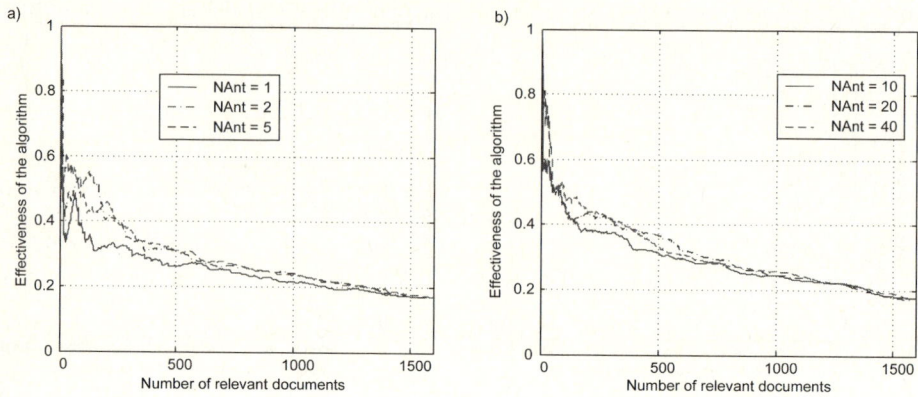

Fig. 3. Effectiveness of the Ant Focused Crawling Algorithm as a function of the number of relevant documents, depending on the number of ants; (a) – for the number of ants: 1,3,5; (b) – for the number of ants: 10,20,40

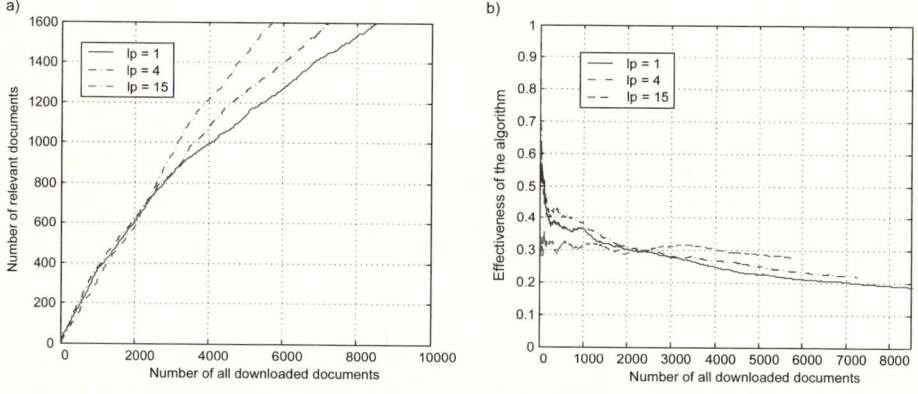

Fig. 4. Evaluation of the Ant Focused Crawling Algorithm depending on the number of repetitions: $lp = 1, 4, 15$ of the crawling process ; (a) – number of relevant documents as a function of the number of all downloaded documents ; (b) – effectiveness as a function of the number of all downloaded documents

of information in a local database about a part of the web. Moreover, the use of pheromone $\tau_{\Delta t}(v)$ makes possible focusing the ants in the changeable area of the hypertext graph.

4 Conclusion

The proposed new Ant Focused Crawling Algorithm, by applying indirected communication similar to that observed in some ant species, enables a crawling process performed by many agents (artificial ants) without any central control.

In this way, the proposed algorithm saves memory and hardware requirements. Moreover, it produces better results than two other compared algorithms.

References

1. Baldi, P., Frasconi, P., Smyth, P.: Modeling the Internet and the Web, Probabilistic Methods and Algorithms. Wiley, Chichester (2003)
2. Cormen, T.H., Leiserson, C.E., Rivest, R.L., Stein, C.: Introduction to Algorithms, 2nd edn. MIT Press, Cambridge (2001)
3. Cortez, C., Vapnik, V.N.: The hybrid application of an inductive learning method and a neural network for intelligent information retrieval. Machine Learning 20, 1–25 (1995)
4. Kłopotek, A.M.: Intelligent Search Engines. EXIT (in polish) (2001)
5. Duch, W., Adamczak, R., Diercksen, G.H.F.: Classification, association and pattern completion using neural similarity based methods. International Journal of Applied Mathematic and Computer Science 10(4), 101–120 (2000)
6. Bilski, J.: The UD RLS algorithm for training feedforward neural networks. International Journal of Applied Mathematic and Computer Science 15(1), 115–123 (2005)
7. Łęski, J., Henzel, N.: A neuro-fuzzy system based on logical interpretation of if-then rules. International Journal of Applied Mathematic and Computer Science 10(4), 703–722 (2000)
8. Łęski, J.: A fuzzy if-then rule-based nonlinear classifier. International Journal of Applied Mathematic and Computer Science 13(2), 215–223 (2003)
9. Piegat, A.: Fuzzy Modeling and Control. Physica-Verlag (2001)
10. Rutkowska, D., Nowicki, R.: Implication-based neuro-fuzzy architectures. International Journal of Applied Mathematic and Computer Science 10(4), 675–701 (2000)
11. Dziwiński, P., Rutkowska, D.: Algorithm for generating fuzzy rules for WWW document classification. In: Rutkowski, L., Tadeusiewicz, R., Zadeh, L.A., Żurada, J.M. (eds.) ICAISC 2006. LNCS (LNAI), vol. 4029, pp. 1111–1119. Springer, Heidelberg (2006)
12. Dziwiński, P., Rutkowska, D.: Hybrid algorithm for constructing DR-FIS to classification www documents. In: Some Aspects of Computer Science, EXIT Academic Publishing House, Warsaw (2007)
13. Russell, S., Norvig, P.: Artificial Intelligence: A Modern Approach. Prentice-Hall, Englewood Cliffs (1995)
14. Cho, J., Garcia-Molina, H., Page, L.: Efficient crawling through URL ordering. Computer Networks and ISDN Systems 30, 161–172 (1998)
15. Baeza-Yates, R., Castillo, C., Marin, M., Rodriguez, A.: Crawling a country: Better strategies than breadth-first for web page ordering. In: International Word Wide Web Conference (2005)
16. Chakrabarti, S., van den Berg, M., Dom, B.: Focused crawling: a new approach to topic-specific web resource discovery. Computer Networks (31), 1623–1640 (1999)
17. Diligenti, M., Coetzee, F.M., Lawrence, S., Giles, C.L., Gori, M.: Focused crawling using context graphs. In: 26th International Conference on Very Large Data Bases, pp. 527–534 (2000)
18. Davison, B.D.: Topical locality in the web. In: 23rd Ann. Int. ACM SIGIR Conference on Research and Development in Information Retrieval, pp. 272–279 (2000)

19. Rungsawang, A., Angkawattanawit, N.: Learnable topic-specific web crawler. Computer Applications 28, 97–114 (2005)
20. Hersovici, M., Jacovi, M., Maarek, Y., Pelleg, D., Shtalhaim, M., Ur, S.: The shark-search algorithm – an application: tailored web site mapping. In: 7th International World-Wide-Web Conference on Computer Networks, pp. 317–326 (1998)
21. De Bra, P., Post, R.: Information retrieval in the world wide web: making client-based searching feasible. Computer Networks and ISDN Systems 27(2), 183–192 (1994)
22. Dorigo, M., Gambardella, L.M.: Ant colony system: A cooperative learning approach to the traveling salesman problem. IEEE Transactions on Evolutionary Computation 1(1), 53–66 (1997)
23. Dorigo, M., Birattari, M., Stützle, T.: Ant colony optimization, artificial ants as a computational intelligence technique. IEEE Computational Intelligence Magazine, 28–39 (November 2006)
24. Pintea, C.M., Pop, P.C., Dumitrescu, D.: An ant-based technique for the dynamic generalized traveling salesman problem. In: 7th WSEAS International Conference on Systems Theory and Scientific Computation, vol. 7 (2007)
25. Vesel, A., Zerovnik, J.: How good can ants color graphs? Journal of Computing and Information Technology - CIT 8, 131–136 (2000)
26. Dowsland, K.A., Thompson, J.M.: An improved ant colony optimisation heuristic for graph coloring, vol. 156, pp. 313–324. Elsevier Science Publishers B. V (2008)
27. Altshuler, Y., Bruckstein, A., Wagner, I.: Swarm robotics for a dynamic cleaning problem. In: Swarm Intelligence Symposium, SIS 2005, pp. 209–216 (2005)
28. Wagner, I.A., Lindenbaum, M., Bruckstein, A.M.: Distributed covering by ant-robots using evaporating traces. IEEE Transactions on Robotics and Automation 15(5) (1999)
29. Wagner, I.A., Lindenbaum, M., Bruckstein, A.M.: Efficiently searching a graph by a smell-oriented vertex process. Annals of Mathematics and Artificial Intelligence 24, 211–223 (1998)
30. Birattari, M., Pellegrini, P., Dorigo, M.: On the invariance of ant colony optimization. IEEE Transactions on Evolutionary Computation 11(6) (2007)
31. Dorigo, M., Maniezzo, V., Colorni, A.: Ant system: Optimization by a colony of cooperating agents. IEEE Transactions on Systems, Man, and Cybernetics – Part B 26(1), 29–41 (1996)
32. Yanowski, V., Wagner, I.A., Lindenbaum, M., Bruckstein, A.: A distributed ant algorithm for efficiently patrolling a network. Algorithmica 37, 165–186 (2003)
33. Mark, E.: Searching for information in a hypertext medical handbook. Communications of the ACM (31), 880–886 (1988)
34. Documentation for the Java Platform, Standard Edition (2008),
    http://java.sun.com/javase/reference/index.jsp

Towards Refinement of Clinical Evidence Using General Logics

Patrik Eklund, Robert Helgesson, and Helena Lindgren

Umeå University, Department of Computing Science
SE-90187 Umeå Sweden
{peklund,c03rhn,helena}@cs.umu.se

Abstract. Clinical knowledge building upon evidence-based medicine is typically represented in textual guidelines, thus providing a rather informal description from a logical point of view. Further, the context which provides utility of these guidelines is not specified in any detail with respect to workflow and underlying motivations for decision-making. In addition, the level of detail is mostly static in the sense that measurements and decision values are fixed and intended for specific user groups. There is thus a lack of flexibility which disables knowledge to be shifted coherently between user levels in the entire workflow and decision process. In this paper, we will discuss formalizations of the underlying logical structures of guidelines from the viewpoint of being represented appropriately at each user level. Further, to establish a formal correctness criterion, the shift from one level of representation to another is required to be morphic in the categorical sense. General logics [7] is the selected generalized, and categorical, framework for our approach to flexible guideline representation. Our medical scope is dementia differential diagnosis based on consensus guidelines [2], and we focus on types of cognitive disorders as a prerequisite for further diagnostic tasks.

1 Introduction

Diagnosis of cognitive disorder differentiates between cognitive impairment, delirium and dementia. Similarly, in the case of dementia, differentiation is required in further steps when diagnosing dementia of Alzheimer's type and various non-Alzheimer dementia types such as vascular dementia. In these further steps neurological functions e.g. as determined by radiological signs become important parts of the diagnosing process. It is also also important to note how dementia may connect with a number of general medical conditions, and dementia is generally also classified with respect to its severity.

In this view we should note that diagnosis is a prerequisite for intervention which may be based e.g. on organic, personal and social aspects. In an organic view, interventions are mostly pharmacological with improvement of neurotransmissions or avoidance of neuronal degeneration as typical goals. At a personal level objectives are to maintain or even improve capabilities with respect to daily living, thus enabling elderly to remain in their own homes. Personal and social levels of intervention are performed by different professional groups as compared to corresponding interventions at organic level. Nevertheless, the overall resources of information is the same albeit used at different levels with different degrees of specificity. In this holistic view rule base representation is quite challenging as the requirement is that underlying knowledge and well-foundedness of information for diagnosis and treatment must be concise and sufficient also for smooth transfer between professional domains.

Diagnosis of cognitive disorder is initially based on observations concerning memory functions such as progressing difficulties with episodic and semantic memory. Representation of these 'measurements' can be done with different levels of specificity and confidence. At each representation level rules must be 'correct' is some sense with respect to the given evidence-based guidelines. However, even more important is correctness preservation when shifting from one representation level to another. Concerning this correctness preservation, general logics provides the appropriate formal framework for management transformations between logics and logical representations.

General logics [7] provides an appropriate framework for representing these refinements of information. A logic described within its categorical framework contains suitable building blocks, e.g. in form of functors and natural transformations, that enables representation of information at various user levels. General logics has previously not been used for representation of clinical guidelines, and therefore we need some further categorical developments to support our developments. A typical basic question in this respect is whether to analyze rules within one and the same entailment system involving a category of signatures or to split the entailment system into different entailment systems based on separating signatures of interest. We adopt the latter and establish the necessary theoretical framework for this approach.

Throughout this paper we assume the reader to be familiar with basic concepts in category theory, e.g. as in [1], and universal algebra, e.g. as in [6]. The paper is organized as follows. Section 2 provides in summary the necessary categorical tools we need for our purposes. Some categorical properties necessary for our treatment are established in Section 3. In Section 4 we describe our selected logics and we represent our cognitive disorder differentiating diagnosis rules in these logical systems. Section 5 concludes the paper.

2 General Logics

Definitions, concepts and notations used in this section are adopted from [7] and [5].

Definition 1. *An* entailment system *is a triple* $\mathscr{E} = (\mathtt{Sign}, Sen, \vdash)$ *where*

- $\mathtt{Sign}$ *is a category of signatures,*
- *Sen is a functor* $Sen : \mathtt{Sign} \to \mathtt{Set}$*, and*
- $\vdash$ *is a family of binary relations consisting of*

$$\vdash_{\Sigma} \subseteq \mathcal{P}Sen(\Sigma) \times Sen(\Sigma)$$

for each signature $\Sigma \in \mathrm{Ob}(\mathtt{Sign})$ where $\vdash_{\Sigma}$ is called a Σ-entailment subject to the condition that each $\vdash_{\Sigma}$

- is reflexive, that is, $\{\varphi\} \vdash_{\Sigma} \varphi$ for each $\varphi \in Sen(\Sigma)$;
- is monotone, that is, $\Gamma' \vdash_{\Sigma} \varphi$, for all $\Gamma' \subseteq \Gamma$ such that $\Gamma \vdash_{\Sigma} \varphi$;
- is transitive, that is, given sentences φ_i, $i \in I$, such that $\Gamma \vdash_{\Sigma} \varphi_i$ and, additionally, $\Gamma \cup \{\varphi_i \mid i \in I\} \vdash_{\Sigma} \psi$, then $\Gamma \vdash_{\Sigma} \psi$; and
- is an $\vdash$-translation, meaning that, if $\Gamma \vdash_{\Sigma} \varphi$ then for all $\mu \in \mathrm{Hom}_{\mathtt{Sign}}(\Sigma, \Sigma')$, it is the case that $\mathcal{P}Sen(\mu)(\Gamma) \vdash_{\Sigma'} Sen(\mu)(\varphi)$.

If $\Gamma \vdash_{\Sigma} \varphi$, then we say that Γ is the set of axioms and φ is derivable from Γ or, alternatively, that φ is a logical consequence of Γ. Rules are presented in theories.

Definition 2. *A ($\mathcal{E}$-)theory for an entailment system $\mathcal{E} = (\mathtt{Sign}, Sen, \vdash)$ is a pair $T = (\Sigma, \Gamma)$ such that*

- *$\Sigma \in \mathrm{Ob}(\mathtt{Sign})$ is a signature, and*
- *$\Gamma \subseteq Sen(\Sigma)$ is a set of axioms.*

The category of theories is as follows.

Definition 3. *Let $\mathcal{E} = (\mathtt{Sign}, Sen, \vdash)$ be an entailment system. Then the category of $\mathcal{E}$-theories, denoted $\mathtt{Th}_{\mathcal{E}}$, or $\mathtt{Th}$ for short, is such that $\mathrm{Ob}(\mathtt{Th})$ is the set of all $\mathcal{E}$-theories and each theory morphism $(\Sigma, \Gamma) \xrightarrow{\mu_{\mathtt{Th}}} (\Sigma', \Gamma')$ with $\Sigma, \Sigma' \in \mathrm{Ob}(\mathtt{Sign})$, $\Gamma \subseteq Sen(\Sigma)$, $\Gamma' \subseteq Sen(\Sigma')$, consists of a signature morphism $\Sigma \xrightarrow{\mu} \Sigma'$ such that*

$$\Gamma' \vdash_{\Sigma'} Sen(\mu)(\varphi)$$

holds for each $\varphi \in \Gamma$. If $\mu_{\mathtt{Th}}$ also fulfills the condition that $\mathcal{P}Sen(\mu)(\Gamma) \subseteq \Gamma'$, that is, axioms are mapped to axioms, then it is said to be axiom-preserving. We let $\mathtt{Th}_0$ denote the subcategory of $\mathtt{Th}$ containing the same objects but only axiom-preserving morphisms.

Mappings between entailment systems can now be introduced. Let $\mathcal{S} : \mathtt{Th} \to \mathtt{Sign}$ and $\mathcal{A} : \mathtt{Th} \to \mathtt{Set}$ be the forgetful functors taking a theory to its underlying signature and set of axioms, respectively. That is, for some theory, $T = (\Sigma, \Gamma) \in \mathrm{Ob}(\mathtt{Th})$, we have $\mathcal{S}(T) = \Sigma$ and $\mathcal{A}(T) = \Gamma$. Further, let $\mathcal{T} : \mathtt{Sign} \to \mathtt{Th}$ be the functor which for some signature Σ is such that $\mathcal{T}(\Sigma) = (\Sigma, \varnothing)$, i.e. given a signature, $\mathcal{T}$ yields the theory containing that signature but holds no axioms.

Definition 4. *Let*

$$\mathcal{E} = (\mathtt{Sign}, Sen, \vdash) \text{ and } \mathcal{E}' = (\mathtt{Sign}', Sen', \vdash')$$

be two entailment systems. Then a map from $\mathscr{E}$ to $\mathscr{E}'$ is a pair $(\Phi, \alpha) : \mathscr{E} \to \mathscr{E}'$ where
$$\mathrm{Th}_0 \xrightarrow{\Phi} \mathrm{Th}'_0$$
is a functor from the category of $\mathscr{E}$-theories to the category of $\mathscr{E}'$-theories, and
$$Sen \xrightarrow{\alpha} Sen' \circ \mathcal{S}' \circ \Phi \circ \mathcal{T}$$
is a natural transformation such that

(i) Φ maps theory signatures with no regard to axioms, that is, $\mathcal{S}'\Phi = \mathcal{S}'\Phi\mathcal{T}\mathcal{S}$;
(ii) the theories $\Phi(\Sigma, \Gamma)$ and $\Phi(\Sigma, \varnothing)$ are such that
$$(\mathcal{A}\Phi(\Sigma, \Gamma))^\bullet = (\mathcal{A}'\Phi(\Sigma, \varnothing) \cup \mathcal{P}\alpha_\Sigma(\Gamma))^\bullet;\ \text{and}$$
(iii) for each $\Sigma \in \mathrm{Ob}(\mathrm{Sign})$,
$$\Gamma \vdash_\Sigma \varphi \implies \mathcal{P}\alpha_\Sigma(\Gamma) \cup \mathcal{A}'\Phi(\Sigma, \varnothing) \vdash'_{\mathcal{S}'\Phi(\Sigma, \varnothing)} \alpha_\Sigma(\varphi).$$

We say that the map (Φ, α) is conservative if instead of (iii), the stronger condition
$$\Gamma \vdash_\Sigma \varphi \iff \mathcal{P}\alpha_\Sigma(\Gamma) \cup \mathcal{A}\Phi(\Sigma, \varnothing) \vdash'_{\mathcal{S}\Phi(\Sigma, \varnothing)} \alpha_\Sigma(\varphi)$$
holds.

3 Splitting Entailment Systems

We may in an intuitive sense split an entailment system into two separate entailment systems by choosing appropriate signature morphisms. This split will also give rise to a map between the two newly constructed entailment systems. The following proposition makes this claim precise.

Proposition 1. *Let $\mathscr{E} = (\mathrm{Sign}, sen, \vdash)$ be an entailment system. Additionally, let $\mathscr{E}_1 = (\mathrm{Sign}_1, Sen_1, \vdash^1)$ and $\mathscr{E}_2 = (\mathrm{Sign}_2, Sen_2, \vdash^2)$ be such that Sign_1 and Sign_2 are subcategories of Sign and $Sen_1, Sen_2, \vdash^1,$ and $\vdash^2$ are the restrictions of sen and $\vdash$ to the signatures of their respective signature categories. Further, place on Sign_1 and Sign_2 the restriction that they may contain only identity morphisms and, finally, for each $\Sigma_1 \in \mathrm{Ob}(\mathrm{Sign}_1)$, choose a signature morphism $\mu \in \mathrm{Hom}_{\mathrm{Sign}}(\Sigma_1, \Sigma_2),\ \Sigma_2 \in \mathrm{Ob}(\mathrm{Sign}_2)$.*

Under these conditions $\mathscr{E}_1$ and $\mathscr{E}_2$ are entailment systems and $(\Phi, \alpha) : \mathscr{E}_1 \to \mathscr{E}_2$ as given below is a map between them. The functor $\Phi : \mathrm{Th}_0^1 \to \mathrm{Th}_0^2$ takes $\mathscr{E}_1$-theories to $\mathscr{E}_2$-theories and is defined for each $\mathscr{E}_1$-theory (Σ, Γ) by
$$\Phi(\Sigma, \Gamma) = (\mu(\Sigma), \mathcal{P}Sen(\mu)(\Gamma))$$
$$\Phi(\mathrm{id}_{(\Sigma, \Gamma)}) = id_{\Phi(\Sigma, \Gamma)}$$
and $\alpha : Sen_1 \to Sen_2 \circ \mathcal{S}_2 \circ \Phi \circ \mathcal{T}_1$ is a natural transformation similarly defined
$$\alpha_\Sigma = Sen(\mu).$$

Proof. First we observe that since a known entailment system has been restricted to a subset of its signatures and the signature morphisms have at the same time been limited to only the identity morphisms, it is trivially true that $\mathcal{E}_1$ and $\mathcal{E}_2$ are entailment systems. It remains to show that (Φ, α) is a map from $\mathcal{E}_1$ to $\mathcal{E}_2$.

Second, since $\mathtt{Sign}_1$ and $\mathtt{Sign}_2$ only contain identity morphisms, Φ is by construction a functor and likewise α is a natural transformation since

$$\alpha_\Sigma \circ Sen_1(\mathrm{id}_\Sigma) = (Sen_2 \circ \mathcal{S}_2 \circ \Phi \circ \mathcal{T}_1)(\mathrm{id}_\Sigma) \circ \alpha_\Sigma$$

for all $\Sigma \in \mathrm{Ob}(\mathtt{Sign}_1)$. We now have that condition (i) holds since for each $\Sigma \in \mathrm{Ob}(\mathtt{Sign}_1)$ and $\Gamma, \Gamma' \in \mathcal{P}sen_1(\Sigma)$ it is the case that

$$\mathcal{S}_2\Phi(\Sigma, \Gamma) = \mu(\Sigma) = \mathcal{S}_2\Phi\mathcal{T}_1\mathcal{S}_1(\Sigma, \Gamma').$$

Condition (ii) holds since

$$\begin{aligned}(\mathcal{A}_2\Phi(\Sigma, \Gamma))^\bullet &= (\mathcal{P}Sen(\mu)(\Gamma))^\bullet \\ &= (\mathcal{P}Sen(\mu)(\varnothing) \cup \mathcal{P}Sen(\mu)(\Gamma))^\bullet \\ &= (\mathcal{A}_2\Phi(\Sigma, \varnothing) \cup \mathcal{P}\alpha_\Sigma(\Gamma))^\bullet.\end{aligned}$$

Finally, condition (iii) holds since by the definition of entailment systems we know that

$$\Gamma \vdash_\Sigma \varphi \implies \mathcal{P}Sen(\mu)(\Gamma) \vdash_{\Sigma'} Sen(\mu)(\varphi)$$

for each $\Sigma' \in \mathrm{Ob}(\mathtt{Sign})$ such that $\mu \in \mathrm{Hom}_{\mathtt{Sign}}(\Sigma, \Sigma')$. This is generalization of the situation given by the definition of maps of entailment systems applied to (Φ, α), i.e., we have

$$\begin{aligned}\Gamma \vdash^1_\Sigma \varphi &\implies \mathcal{P}\alpha_\Sigma(\Gamma) \cup \mathcal{A}_2\Phi(\Sigma, \varnothing) \vdash^2_{\mathcal{S}_2\Phi(\Sigma, \varnothing)} \alpha_\Sigma(\varphi) \\ &\implies \mathcal{P}Sen(\mu)(\Gamma) \vdash^2_{\mu(\Sigma)} Sen(\mu)(\varphi).\end{aligned}$$

4 Cognitive Disorder Differential Diagnosis

In this section we will consider mappings between entailment systems, where the object sets for the categories of signatures are one-pointed. It is then natural to require that the signature categories contain only identity morphisms for their respective signatures. The justification for this split of signature categories and entailment systems is given by Proposition 1.

The mapping situation is based on the following scenarios.

Scenario 1 (*Granularity of knowledge*) . *Successful pharmacological treatment of type Alzheimer's disease requires an early detection of cognitive decline. An early detection happens typically during home care, i.e. when professionals from the social and nursing area are in direct and frequent contact with the elderly. Diagnosis cannot be performed in this environment, but initial and initiating observations of symptoms and signs used in guidelines for diagnosis of dementia can be (systematically using different validated screening tools) collected and brought forward to* primary care physicians. These medical experts in turn carry out preliminary diagnostic tasks before referring the patient to a regular geriatric investigation concerning the suspected state of dementia. In the process different screening-tools and guidelines are used which capture knowledge at different levels of granularity.

What are the respective information types and rule representations for these professional groups? Can we guarantee consistency when information and knowledge is mapped between ontological domains as understood and used by these professional groups?

Scenario 2 (*Many-valuedness of truth*). During the course of diagnosis it is common that information viewed is considered not accurate enough or knowledge entailed is understood as just vaguely true. Nevertheless, clinicians may have to reach conclusions and make decisions as a basis for further intervention steps based on these premises. This means that qualifications and truth values may have to be sharpened to becoming binary.

When moving from vagueness, and even accurately estimated vagueness, to presenting decisions and rules based thereof in a more crisp fashion, how should these transformations be specified so that information loss is minimized and the remaining crisp situation is an optimal representation of the vague situation?

Let $\mathscr{E}_i = (\text{Sign}_i, sen_i, \vdash_i)$, $i = 1, 2$, be entailment systems, where $\text{Ob}(\text{Sign}_i) = \{(S_i, \Omega_i)\}$. We will picture $\mathscr{E}_1$ as an equational logic basically using boolean terms. In order to do this let $S_1 = \{\text{BOOL}\}$ and $\Sigma_1 = \{\text{false}, \text{true}, \text{OR}\}$ with $\text{false}, \text{true} :\to \text{BOOL}$ and $\text{OR} : \text{BOOL} \times \text{BOOL} \to \text{BOOL}$.

The sen_i functors are specified as for equational logic according to

Definition 5. *The Sen^{Eq} functor for equational logic is given by*

$$Sen^{Eq}(\Sigma) = \{(X, t, t') \mid t, t' \in T_{\Sigma, s} X, s \in S, \text{ and } X \text{ a family of variables for } \Sigma\}$$

for any signature $\Sigma = (S, \Omega) \in \text{Ob}(\text{Sign}^{Eq})$, where

$$Sen^{Eq}(\mu)((X, t, t')) = (\mu(X), \mu(t), \mu(t'))$$

for some signature morphism $\mu : \Sigma \to \Sigma'$.

The equations for OR are as usual. Part of the theory $T_1 = (\Sigma_1, \Gamma_1)$, where $\Sigma_1 = (S_1, \Omega_1)$, for classifying delirium according to DSM-IV guidelines [2] is given as

```
op false true : -> BOOL .
op OR : BOOL BOOL -> BOOL .
var ... X_episodic X_semantic X_shortterm ... : BOOL .
eqn OR ... .
eqn DEO_CogDisDSMdelirium =
    ...
    OR
```

```
        {
          X_episodic
          X_semantic
          X_shortterm
        }
        ... .
  eqn X_episodic = false .
  eqn X_semantic = true .
  eqn X_shortterm = true .
```

In this view, we should see

```
  eqn OR ... .
  eqn DEO_CogDisDSMdelirium =
    ...
    OR
      {
        X_episodic
        X_semantic
        X_shortterm
      }
      ... .
```

as the set of axioms, and

```
    X_episodic = false .
    X_semantic = true .
    X_shortterm = true .
```

as specific patient data, in the end stored and retrieved from the electronic patient record.

This arrangement could in Scenario 1 be seen as the rule base used within home care and could in Scenario 2 be seen as part of the guideline needed for crisp decisions.

Entailment system $\mathscr{E}_2$ will also be equational but with more sorts and operators. We will include the sort qualICF to enable representation of qualification values according to ICF [9]. Let now $S_2 = \{\text{BOOL}, \text{qualICF}\}$ and $\Omega_2 = \{\text{false}, \text{true}, \text{OR}, 0, 1, 2, 3, 4, \text{a}\}$ with false, true $:\to$ BOOL and OR : BOOL $\times$ BOOL $\to$ BOOL, together with 0, 1, 2, 3, 4 $:\to$ qualICF and a : qualICF $\to$ BOOL. The operator a thus converts qualification values to boolean values. (Part of) theory $T_2 = (\Sigma_2, \Gamma_2)$, where $\Sigma_2 = (S_2, \Omega_2)$, becomes[1]

```
  op false true : -> BOOL .
  op OR : BOOL BOOL -> BOOL .
```

---

[1] Granularity of memory types follows [8], i.e. a distinction is made between semantic and episodic memory, thus extending the ICF codes [9] in the example with an additional level of granularity.

```
op 0 1 2 3 4 : qualICF .
op a : qualICF -> BOOL .
var ... X_b14411 X_b14412 X_b1440 ... : BOOL .
eqn OR ... .
eqn a 0 = false .
eqn a 1 = true .
eqn a 2 = true .
eqn a 3 = true .
eqn a 4 = true .
eqn DEO_CogDisDSMdelirium =
  ...
  OR
    {
      a X_b14411    *** ICF code for longterm episodic memory
      a X_b14412    *** ICF code for longterm semantic memory
      a X_b1440     *** ICF code for shortterm memory
    }
  ...
eqn X_b14411 = 0 .
eqn X_b14412 = 1 .
eqn X_b1440 = 3 .
```

Thus e.g. a 0 = false $\in sen_2((S_2, \Omega_2))$.

The signature morphism $\mu : \Sigma_1 \to \Sigma_2$ is given by $\mu(\text{BOOL}) = \text{BOOL}$, and $\mu(\text{false}) = \text{false}, \mu(\text{true}) = \text{true}, \mu(\text{OR}) = \text{OR}$. The entailment morphism $\mathcal{E}_1 \to \mathcal{E}_2$ is then trivial as it basically embeds knowledge without changing granularity.

The embedding $\mathcal{E}_2 \to \mathcal{E}_1$, on the other hand, is non-trivial. In this case we need an 'identity' i : BOOL $\to$ BOOL as an operator in Ω_1, i.e. having

```
eqn i false = false .
eqn i true = true .
```

as its equations in Γ_1.

Remark 1. Using negation instead of i obviously changes the rule base completely. However, by using i, misinterpretations due to cultural differences in the intuitive mappings can be handled, such as situations when negations are confirmed with a positive response can be captured here, which else requires extensive context information (e.g. "affected memory function? - yes" vs. "memory function? - affected").

Let the signature morphism $\mu : \Sigma_2 \to \Sigma_1$ be given by $\mu(\text{BOOL}) = \text{BOOL}$ and $\mu(\text{qualICF}) = \text{BOOL}$. Further, $\mu(\text{false}) = \text{false}, \mu(\text{true}) = \text{true}, \mu(\text{OR}) = \text{OR}$, $\mu(0) = \text{false}, \mu(1) = \text{true}, \mu(2) = \text{true}, \mu(3) = \text{true}, \mu(4) = \text{true}$ and $\mu(a) = i$.

We now have e.g.

$$sen(\mu)(X\_b14411 = 0) = (X\_episodic = \text{false})$$

$$sen(\mu)(aX\_b14411 = \text{false}) = (X\_episodic = \text{false})$$

Qualification values could also have been converted within a many-valued logic framework, and in this case `BOOL` would be extended to capture many-valuedness. Converters must be defined accordingly. In future papers we will explore these examples in more detail, and including more complete versions of rule bases. More elaborate ingredients are then necessary, such as provided by formal descriptions in [3,4].

5 Conclusions

We have addressed the problem of sustaining correctness when transforming clinical evidence between different levels of care and contexts of interpretation, with focus on granularity of knowledge. Instead of trying to capture the whole care process in one (static) formalisation, i.e. one entailment system, we capture different phases using appropriate entailment systems for each. This is accomplished by splitting entailment systems in the framework of general logics, which also provides the transformational properties for sustaining correctness in refinements of evidence.

References

1. Adámek, J., Herrlich, H., Strecker, G.: Abstract and concrete categories. Wiley-Interscience, New York (1990)
2. American Psychiatric Association, Diagnostic and statistical manual of mental disorders, 4th edn. (DSM-IV-TR), Text Revisions, American Psychiatric Association (2000)
3. Gottwald, S.: A Treatise on Many-Valued Logics. Studies in Logic and Computation, vol. 9. Research Studies Press, Baldock (2001)
4. Hájek, P.: Metamathematics of Fuzzy Logic. Kluwer Academic Publishers, Dordrecht (1998)
5. Helgesson, R.: A categorical approach to logics and logic homomorphisms, UMNAD 676/07, Umeå University, Department of Computing Science (2007)
6. Loeckx, J., Ehrich, H.-D., Wolf, M.: Specification of Abstract Data Types. Wiley, Chichester (1996)
7. Meseguer, J.: General logics. In: Ebbinghaus, H.-D., et al. (eds.) Logic Colloquium 1987, pp. 275–329. Elsevier, North-Holland (1989)
8. Tulving, E.: Episodic and semantic memory. In: Tulving, E., Donaldson, W. (eds.) Organization of Memory, pp. 382–402. Academic Press, Inc., New York (1972)
9. WHO, International classification of functioning, disability and health: ICF (accessed 2007-04-17), http://www3.who.int/icf/icftemplate.cfm

Appendix

The rulebase for cognitive disorder differential diagnosis is presented in a (partly informal) syntax for an equational logic. The rulebase tries to evaluate the presence of dementia based on evidence concerning memory functions. However, the available evidence in the example is not sufficient for diagnosis.

```
begin module CogDisDSM
  sort BOOL qualICF qualDSM dgDSM dgICD10 .
  op false true : -> BOOL .
  op NOT : BOOL -> BOOL .
  op AND OR : BOOL BOOL -> BOOL .
  op 0 1 2 3 4 : qualICF .
  op normal mild significant : qualDSM .
  op 294.8;1 294.8;2 780.09 : dgDSM .
  op F03 F05.9 : dgICD10 .
  op DTO_CogDisDSM : dgDSM -> BOOL .
  op DEO_CogDisDSMdelirium DEO_CogDisDSMdementia
      DEO_CogDisDSMamnestic : dgDSM -> BOOL .
  op a : qualICF -> BOOL .
  op b : qualDSM -> BOOL .
  op f : qualICF -> qualDSM .
  op g : dgDSM -> dgICD10 .
  var X_BOOL : BOOL .
  var X_rapidOnset X_DSMfluct X_insidiousOnset X_progressive: BOOL .
  var X_dgDSM : dgDSM .
  var X_dgICD10 : dgICD10 .
  var X_ICFb110 ... X_ICFb164 : qualICF
  eqn NOT true = false .
  eqn NOT false = true .
  eqn AND true X_BOOL = X_BOOL .
  eqn AND false X_BOOL = false .
  eqn OR true X_BOOL = true .
  eqn OR false X_BOOL = X_BOOL .
  eqn f 0 = normal .
  eqn f 1 = mild .
  eqn f 2 = significant .
  eqn f 3 = significant .
  eqn f 4 = significant .
  eqn g 294.8;1 = F03 .
  eqn g 294.8;2 = F03 .
  eqn DTO_CogDisICD10 g X_dgDSM = DTO_CogDisDSM X_dgDSM .
  eqn DTO_CogDisDSM X_dgDSM =
    OR
      {
        DEO_CogDisDSMdelirium X_dgDSM
        DEO_CogdisDSMamnesticState X_dgDSM
        DEO_CogDisDSMdementia X_dgDSM
      } .
  eqn DEO_CogDisDSMdelirium X_dgDSM =
    AND
```

```
      {
    NOT a X_ICFb110        *** consciousness
    a X_ICFb140            *** attention
    X_rapidOnset
    X_DSMfluct
    OR
       {
         a X_b14411
         a X_b14412
         a X_b1440
         a X_b156          *** agnosia
         a X_b167          *** aphasia
         a X_b176          *** apraxia
         a X_b164          *** executive
       }
   } .
eqn DEO_CogDisDSMamnestic X_dgDSM =
   AND
     {
       NOT a X_ICFb110
       NOT DEO_DSMdementia X_dgDSM
       a X_b156
       a X_b167
       a X_b176
       a X_b164
       OR
         {
           a X_b14411
           a X_b14412
           a X_b1440
         }
     } .
eqn DEO_CogDisDSMdementia X_dgDSM
   AND
     {
       NOT a X_ICFb110
       X_insidiousOnset
       X_progressive
       OR
         {
           a X_b14411
           a X_b14412
           a X_b1440
         }
       OR
```

```
              {
                a X_b156
                a X_b167
                a X_b176
                a X_b164
              }
          } .
    eqn X_b14411 = 0 .
    eqn X_b14412 = 1 .
    eqn X_b1440 = 3 .
    eqn X_dgDSM = 294.8;2 .
*** eqn X_dgDSM = 294.8;1 .
*** eqn X_dgDSM = 780.09 .
end module
```

An Empirical Analysis of the Impact of Prioritised Sweeping on the DynaQ's Performance

Marek Grześ and Daniel Kudenko

Department of Computer Science
University of York
York, YO10 5DD, UK
{grzes, kudenko}@cs.york.ac.uk

Abstract. Reinforcement learning tackles the problem of how to act optimally given observations of the current world state. Agents that learn from reinforcements execute actions in an environment and receive feedback (reward) that can be used to guide the learning process. The distinguishing feature of reinforcement learning is that the model of the environment (i.e., effects of actions or the reward function) are not known in advance. Model-based approaches represent a class of reinforcement learning algorithms which learn the model of dynamics. This model can be used by the learning agent to simulate interactions with the environment. DynaQ and its extended version with prioritised sweeping are the most popular examples of model-based approaches. This paper shows that, contrary to common belief, DynaQ with prioritised sweeping may perform worse than pure DynaQ in domains where the agent can be easily misled by a sub-optimal solution.

1 Introduction

Reinforcement learning (RL) is concerned with the design of intelligent agents that can learn how to act given observations of the world. In contrast to supervised learning, RL agents are not given instructive feedback on what the best decision in a particular situation is. Rather, they execute actions which have some impact on the environment and the environment subsequently provides numerical feedback for some states (usually only for goal states), which can be used to guide the learning process. The agent uses this information to find a policy which maximises the accumulated reward. A policy defines which action should be taken in a given state and is usually represented as a value function $Q(s, a)$ which estimates "how good" it is to execute action a in state s. This can be expressed in terms of the expected reward which can be obtained when action a will be chosen in state s and the given policy followed thereafter [10].

There are two broad categories of RL algorithms. Model-free solutions learn the value function in a direct way. They update the value function after each interaction with the environment. Model-based algorithms [5,6,9] learn additionally how the world responds to its actions (transition probabilities) and what reward is given (reward function) and use this model to make further updates to their value function. In this way, knowledge about responses from the environment (encoded in the model) can be used many times by the learning agent, especially several time steps after a particular interaction. There may be also other applications of the model. It can be, for example, used to predict future states in order to expedite learning.

In this paper the analysis of the two most popular model-based RL algorithms (DynaQ [9] and DynaQ with prioritised sweeping [5]) is conducted. Prioritised sweeping organises the updates of the value function according to a measure of their urgency, which should lead to improved performance. This paper shows that contrary to common belief and results published in the literature [5,6,7] DynaQ with prioritised sweeping (PS) may behave worse than DynaQ when the problem is difficult in terms of exploration, that is when the agent can be easily misled by a sub-optimal solution. The paper also proposes an explanation of this behaviour.

The rest of the paper is organised as follows. In the next section DynaQ and PS with different measures of priority are briefly discussed. Various algorithms and problem domains and their characteristics are presented in section 3. Experimental results with the analysed algorithms and a deeper analysis to explain their behaviour are presented in section 4. Section 5 shows possible avenues of future research and section 6 concludes the paper.

2 DynaQ and Prioritised Sweeping

RL algorithms which learn their policy via a value function apply mostly temporal-difference updates to propagate information about values of states or state-action pairs. These updates are based on the difference of two temporally different estimates of a particular state or state-action value. Model-free Q-learning is such a method [10]. It updates state-action values by the formula:

$$Q(s,a) \leftarrow Q(s,a) + \alpha[r + \gamma \max_{a'} Q(s',a') - Q(s,a)]. \tag{1}$$

It modifies the value of taking action a in state s, when after executing this action the environment returned reward r and moved to a new state s'. The value of the next state-action pair $Q(s',a')$ is selected according to the max over actions in the next state s'. α is a learning rate and γ the discount factor. Because this formula propagates information in the backward direction such updates are named also backups.

The Q-learning algorithm applies its update rule (Equation 1) to pair (s,a) only after each actual transition $(s,a) \rightarrow (s',r)$. The idea of model-based learning is to compactly encode such transitions in the model and later use the model for simulated updates. The most popular approaches to model-based RL are DynaQ (see Figure 1) and prioritised sweeping. These algorithms perform additional simulated backups based on the model which they learn (Figure 1, lines 5 − 9 in DynaQ).

1: **loop**
2:   Execute action a in current state s; observe next state s' and reward r
3:   $Q(s,a) \leftarrow Q(s,a) + \alpha[r + \gamma \max_{a'} Q(s',a') - Q(s,a)]$
4:   $Model(s,a) \leftarrow (s',r)$
5:   Repeat N times
6:     $s \leftarrow$ random previously observed state
7:     $a \leftarrow$ random action previously taken in state s
8:     $(s',r) \leftarrow Model(s,a)$
9:     $Q(s,a) \leftarrow Q(s,a) + \alpha[r + \gamma \max_{a'} Q(s',a') - Q(s,a)]$
10: **end loop**

Fig. 1. The DynaQ algorithm [10]

The distinguishing feature of DynaQ is that state-action pairs for simulated backups are chosen randomly. It has been noted [5,6] that focusing backups on particular state-action pairs can be more profitable. The motivation for this is that if simulated transitions are generated uniformly like in DynaQ, then many unnecessary backups can be made before the useful ones will be chosen. For this reason PS applies a priority queue to prioritise backups according to some priority measure. After action execution the model is updated, the priority for the current state-action pair is evaluated, and the pair is added to a priority queue. Before the next action is called in the "real world", the N most urgent pairs are taken from the priority queue for simulated backups.

Previous results on mainly shortest path problems show that PS leads to better convergence in terms of the number of performed backups [5,6,7].

2.1 Evaluating Priority

The priority p for the update of a state-action pair (s,a) is usually evaluated using the prediction difference (temporal-difference in RL terminology which is used for backups) [6]:

$$p \leftarrow |r + \gamma \max_{a'} Q(s',a') - Q(s,a)|. \qquad (2)$$

The reward r and the next sate s' can be obtained as a real response from the environment or simulated using the model.

An alternative and similar approach to evaluate potential is applicable in domains with a well-defined start state [6]. It is based on the distance $d(s)$ of the current state s to the start state. In our experiment the following version was used:

$$p \leftarrow \gamma^{d(s)} |r + \gamma \max_{a'} Q(s',a') - Q(s,a)|. \qquad (3)$$

2.2 Managing the Priority Queue

From the point of view of the discussion in the latter part of the paper it is worth considering how the predefined number of simulated backups N is related to the real number of simulated backups which can be performed.

```
ADD2QUEUE(s,a,p)
if p > δ and ¬In(queue, s, a) then
   Insert(queue, s, a, p)
end if
```

Fig. 2. Inserting state-action pair (s, a) with priority p to a priority queue with a tiny threshold δ [5]

Basic DynaQ always performs N backups. It unconditionally chooses N pairs uniformly and because of this it usually looses (in terms of the number of backups during the learning period) in comparisons with PS. When a priority queue is used there can be less elements in the queue than the corresponding value of N in DynaQ and hence less backups can be performed. It is supposed to lead to a more rational performance of backups (i.e., states which have already converged do not need to be updated). But it can be also noted that if the priority queue becomes empty while the agent still follows a sub-optimal path in its interaction with the environment, it may be difficult to update other states which still require it and which are necessary for reaching the optimal solution.

The content of the priority queue is influenced by the way how the queue is managed. The schema considered in this paper (shown in Figure 2) comes from the standard version of PS [5]. Each state-action pair is added to the queue only if its priority is higher than the defined threshold δ. Because small values of δ can lead to the rapid growth of the size of the priority queue an additional condition is checked: a given state-action pair can appear only once in the priority queue.

3 Experimental Design

In this section, the evaluated algorithms with their parameters and the problem domains are discussed. Some characteristics of these domains are also highlighted.

3.1 Evaluated Algorithms and Parameters

DynaQ and its extended version with prioritised sweeping are used in our experiments in a form presented in Sutton and Barto's [10] book. They are used in an unchanged form unless it is explicitly indicated in the paper. The following values of common parameters were applied: $\alpha = 0.1$, $\gamma = 0.99$, the number of episodes per experiment 10^4. In all experiments ϵ-greedy exploration strategy was used where ϵ was decreased linearly from 0.3 in the first episode to 0.01 in the last episode. In all cases the number of simulated backups N was 50.

3.2 Experimental Domains

A family of maze-based navigation problems is investigated. The agent is placed in a maze and has to learn how to navigate to a goal position G from a start

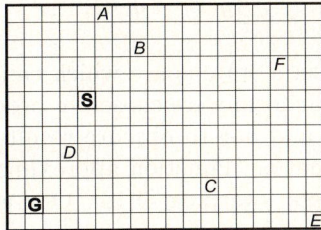

Fig. 3. The maze problem with flags

position S. It has eight actions available which deterministically lead to one of eight neighbouring cells.

The first task is to check the properties of tested algorithms on the standard version of the maze on which DynaQ and PS have been mostly tested [9]. A 20x20 maze (400 states) without obstacles is used in which $(1,1)$ is a starting position and $(20, 20)$ the goal position.

The aim of the second task is to provide our experimental setting with a more difficult (in terms of exploration) maze problem (see Figure 3). In this case the agent has to collect 6 flags $(A - F)$ before reaching the goal position G (14976 states). Because the episode ends when the agent reaches the goal position regardless of the number of collected flags, this kind of maze has been used to evaluate sophisticated exploration strategies [3,8]. The reward is zero in all except the goal position. Upon reaching the goal position the reward proportional to the number of collected flags is given.

In the first task there is exactly one goal state. The improvement of the current policy during the learning process is based on shortening the current path. The policy which is obtained can differ only slightly from its predecessor (updates with only small changes can lead to the best result). In contrast, the second maze has a set of goal states, i.e., all states at position G with all combinations of collected flags. The improvement of the policy in this case can be seen as a more radical activity when the agent has to switch to the policy which collects more flags or collects them in a different sequence. Smaller changes like updates with a constant number of flags can operate only on a family of sub-optimal solutions.

Because the action cost is uniformly zero in these domains the following characteristic will support our discussion. The optimal solution in the first maze is the shortest one in terms of the number of steps. If we define Π as a set of policies $\pi \in \Pi$ which represent solutions to a given sequential decision problem and define $g(\pi)$ as the expected number of steps needed to execute given policy π, the following formula can express this relationship:

$$\forall_\pi g(\pi) \geq g(\pi^*), \qquad (4)$$

where π^* is the policy which yields the highest expected long-term outcome. In the second maze the optimal solution does not necessary have to be the shortest one in terms of the number of steps:

$$\exists_\pi g(\pi) < g(\pi^*). \tag{5}$$

These artificial tasks with a state space which can be explicitly enumerated and represented in a tabular form were chosen in order to have understandable properties of the domains and not to obscure results by other factors like for example function approximation. Function approximation can be a problem in itself for analysed prioritised sweeping [4,10]. Our goal is to show particular properties of the algorithms, which raise awareness when more complex domains are considered.

4 Experimental Results

In this section experimental results are presented. Firstly the tested algorithms are evaluated on the simple maze problem. Evaluation on a more complicated version of the maze poses some questions, and more experiments to approach these questions are performed and discussed.

4.1 Maze Problem

The experiment on the first maze intended to check whether our implementation of the tested algorithms confirms the expected trend that PS is more efficient in terms of the number of backups. The reward function as a function of the number of backups is given in Figure 4. As expected, it shows that both tested prioritised versions of DynaQ are more efficient in terms of the number of backups.

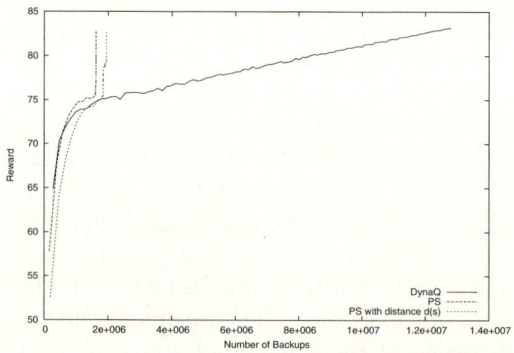

Fig. 4. Reward on the simple maze problem as a function of the number of backups. The y axis shows the average reward for 10 runs of the algorithm.

4.2 Maze Problem with Flags

Contrary to common belief, results on the maze with flags presented in Figure 5 show superiority of pure DynaQ over its prioritised versions. DynaQ is able to find the best solution, i.e., to bring all six flags to the goal position. All

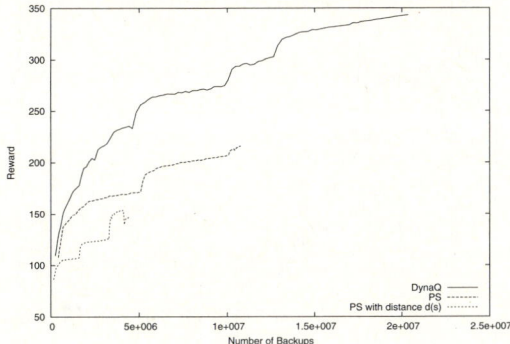

Fig. 5. Reward on the maze problem with flags as a function of the number of backups. The y axis shows the average reward for 10 runs of the algorithm.

algorithms were given the same number of episodes (10^4). However the real number of backups depends heavily on at least the two following factors: the number of steps within one episode and the number of elements in the priority queue. In practice these factors influenced prioritised algorithms and curves for both versions of prioritised sweeping end after smaller number of backups after reaching sub-optimal solutions. It is worth noting the properties of the domain used here. Sub-optimal solutions are relatively short in terms of the number of steps. Because they are shorter, they converge quicker (this effect is accelerated by prioritised sweeping). More experiments to investigate this behaviour were conducted and results are discussed below.

In Figure 5 one can observe that DynaQ also converges at the beginning to a sub-optimal solution which does not collect all flags but in latter episodes is able to improve it. During the first episodes, the algorithm saves information (in the model) about highly rewarded states and by propagating this information via simulated updates is able to switch to a better solution in latter episodes. There can be a problem with DynaQ with a priority queue when states with a high value are visited only during the initial random exploration (a high value of ϵ in first episodes) and are not visited during latter episodes. It should be noted that simulated backups have to reach such states many times. The initial conclusions can be drawn from the size of the priority queue. In Figure 6 the size of the priority queue at the beginning of each episode is shown. It can be noted that at some point in time (in this particular case after 5000 episodes) the priority queue remains almost empty (this size can change within the episode, because this graph is for the moment when the previous episode has just finished and the next one is about to start). An empty priority queue means that there are no backups. The algorithm assumes that it has converged. In fact it has converged, but only to the sub-optimal solution. To look more deeply at this problem, Figure 7a shows how many times states with a certain number of collected flags were removed from the priority queue (successive `top` and `pop` operations which take and remove the first element from the priority queue).

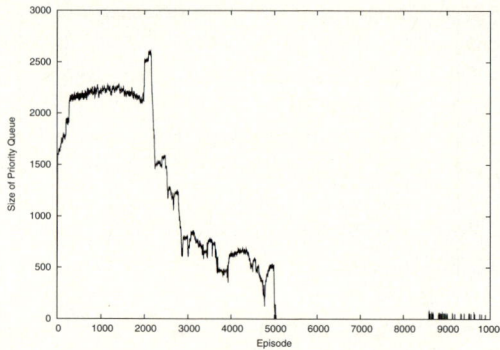

Fig. 6. The size of the priority queue at the beginning of each episode

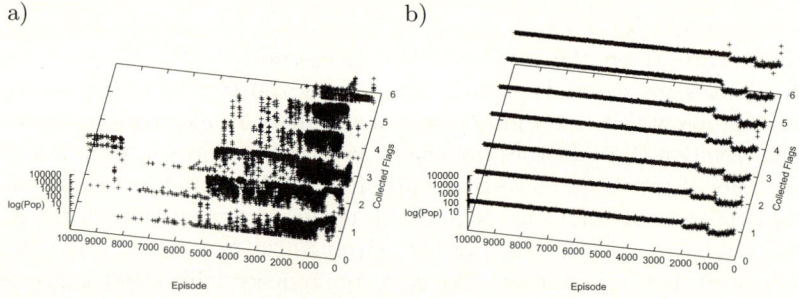

Fig. 7. The number of times states with 0 to 6 collected flags were: a) removed from the priory queue in the PS algorithm, b) drawn by DynaQ

The final result of this experiment led to a solution which collects two flags. This is reflected in Figure 7a. States which were most often removed from the priority queue and only states which were backed up in the last part of the experiment are those with zero, one and two collected flags. After random exploration at the beginning of learning those states with a higher number of collected flags appeared in the priority queue and there were some updates of these states. But this graph shows that because the algorithm followed mostly a sub-optimal path in the "real" experience with the environment, states with the higher numbers of flags did not manage to keep being backed up. A corresponding graph for DynaQ, Figure 7b, shows that states with all numbers of flags were updated within entire period of learning. In the experiment presented in Figure 7b one flag was being collected during the first 1000 episodes, three flags between episodes 1000 and 2000 and mostly six flags thereafter. Updates of states with the higher numbers of flags especially during the first 1000 episodes were essential to make it possible to switch to better (longer) solutions. These longer solutions in latter episodes caused the higher numbers of backups noticeable in the graph. The minimal priority for states to be put into the queue was 10^{-4} in our experiments so this parameter should not obscure the presented results.

These further analysis led us to conclude that the observed drop in the performance can be explained by some characteristics of the evaluation domains. The maze domain with flags has many sub-optimal goals. Their solutions may be much shorter in the number of steps (see Equation 5) and they usually converge quicker especially when misleadingly accelerated by prioritised sweeping. Our conclusions drawn from experiments on the maze problems are of interest when also bigger domains are considered.

Model-based RL with a priority queue can lead to problems when there are sub-optimal solutions which dominate especially during the early stages of learning. DynaQ is able to propagate information about highly assessed states, and prioritised sweeping can miss these states (shown in Figure 7a). DynaQ overcomes these problems because it always chooses states randomly for simulated backups and states from those areas which do not appear any more in the real experience have the same chances of being updated.

The priority queue is managed in a slightly different way in Rayner's et al. [7] Prioritised-LRTA*. In this case the size of the priority queue is limited, but pairs with even small priority can be added to the queue. When the queue is full and the priority of the new element is lower than the priority of the minimal element in the queue, the new state-action pair is discarded. Otherwise it replaces the minimum in the queue. The advantage of this approach is that the maximal queue size guarantees a hard limit on the memory used and elements with even small priority can be added to the queue when it is not full. For a broader analysis of the concept of prioritised sweeping this scheme was adapted (see Figure 8) to the RL algorithms used in this paper. This procedure adds another user defined parameter: the maximal size of the priority queue. To have a deeper insight into the results some values of the maximal size of the priority queue in range $10^2 \div 10^5$ were tested. Results in this setting combined with results of DynaQ and PS are shown in Figure 9. The lowest performance was obtained for the smallest sizes of the priority queue (10^2 and $5*10^2$). Increasing the size of the priority queue leads to results comparable with PS, but still lower than

```
ADD2QUEUE(s,a,p)
if p > 0 then
    if ¬ Full(queue) and ¬ In(queue, s, a) then
        Add(queue, s, a, p)
    else
        (s_min, a_min) ← state-action pair with smallest priority in queue
        if Priority(s_min, a_min) < p then
            Remove(queue, s_min, a_min)
            Add(queue, s, a, p)
        end if
    end if
end if
```

Fig. 8. Inserting state-action pair (s, a) with priority p to the priority queue with limited size. Adapted from Prioritised-LRTA* [7].

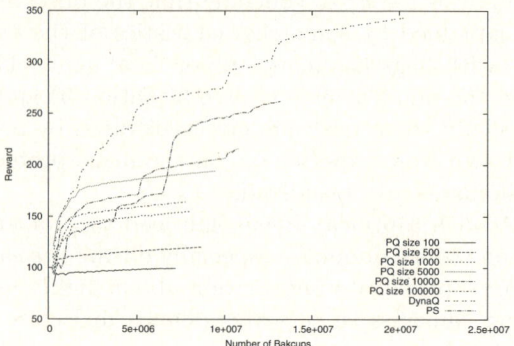

Fig. 9. The performance with different maximal sizes of the priority queue

DynaQ. It shows that the concept of the limited size of the priority queue with the modified procedure for inserting elements to the queue does not overcome the problems of prioritised sweeping on the tested domain. The best results are for large sizes of the priority queue when such configuration becomes equivalent with the basic version of prioritised sweeping.

Andre *et al.* [1] introduced a general framework for prioritised sweeping (GenPS) in which states are prioritised in a representation-specific manner. This approach can be used, e.g., with the factored representation [2] which allows for a more rapid learning and model acquisition. According to [1], one way to improve the performance of PS on the presented class of problems could be a factored representation. In our experiments, a look-up table to represent the value function was used with which PS becomes a special case of the GenPS approach [1] with an explicit state-based representation.

5 Future Work

This paper indicates some problems of prioritised sweeping in RL setting on a class of learning tasks which are difficult in terms of exploration. A particular avenue towards better performance of PS on such tasks can be explored. In our experiments the existing measures of priority were applied (see Equations 2 and 3). Hence a first line of research to improve performance of PS on this class of problems can be the design and application of a new specialised measure. Such a measure should take into account the fact that there may be many sub-optimal solutions and should preserve good performance on (stochastic) shortest path problems on which current versions work well. An engineering approach based on the knowledge about the domain and the exploration of an underlying structure of the task can be adopted.

6 Conclusion

Contrary to common belief from previous research, prioritised sweeping can hurt the performance of the DynaQ algorithm. The domain on which DynaQ with prioritised sweeping performs worse was presented. It was hypothesised that reasons for this drop in the performance are some characteristics of the domain: it has many potential goals and sub-optimal solutions can be much shorter in the number of steps and converge quicker especially when misleadingly accelerated by prioritised sweeping.

The practical conclusions from this work relate to a wider scope of domains. The problem presented in the paper can occur in tasks which have properties discussed in Section 3.2 (Equation 5) and at least special care needs to be taken when deploying RL solutions for such problems.

Acknowledgments

This research was sponsored by the United Kingdom Ministry of Defence Research Programme.

References

1. Andre, D., Friedman, N., Parr, R.: Generalized prioritized sweeping. In: Proceedings of the 1997 conference on Advances in Neural Information Processing Systems, pp. 1001–1007 (1997)
2. Boutilier, C., Dean, T., Hanks, S.: Decision-theoretic planning: Structural assumptions and computational leverage. Journal of Artificial Intelligence Research 11, 1–94 (1999)
3. Dearden, R., Friedman, N., Russell, S.J.: Bayesian Q-learning. In: Proceedings of the Fifteenth National Conference on Artificial Intelligence, pp. 761–768. AAAI Press (1998)
4. Kalyanakrishnan, S., Stone, P., Liu, Y.: Model-based reinforcement learning in a complex domain. In: RoboCup-2007: Robot Soccer World Cup XI. Springer, Berlin (2008)
5. Moore, A.W., Atkenson, C.G.: Prioritized sweeping: Reinforcement learning with less data and less time. Machine Learning 13, 103–130 (1993)
6. Peng, J., Williams, R.J.: Efficient learning and planning within the dyna framework. In: Proceedings of the 1993 IEEE International Conference on Neural Networks, pp. 168–174 (1993)
7. Rayner, D.C., Davison, K., Bulitko, V., Anderson, K., Lu, J.: Real-time heuristic search with a priority queue. In: Proceedings of the 2007 International Joint Conference on Artificial Intelligence, pp. 2372–2377 (2007)
8. Strens, M.J.A.: A bayesian framework for reinforcement learning. In: Proceedings of the 17th International Conference on Machine Learning, pp. 943–950 (2000)
9. Sutton, R.S.: Integrated architectures for learning, planning, and reacting based on approximating dynamic programming. In: Proceedings of the Seventh International Conference on Machine Learning, pp. 216–224 (1990)
10. Sutton, R.S., Barto, A.G.: Reinforcement Learning: An Introduction (Adaptive Computation and Machine Learning). MIT Press, Cambridge (1998)

Heuristic Algorithms for Solving Uncertain Routing-Scheduling Problem

Jerzy Józefczyk and Michał Markowski

Institute of Information Science and Engineering, Wrocław University of Technology,
Wybrzeże Wyspiańskiego 27, 50–370 Wrocław, Poland
`Jerzy.Jozefczyk@pwr.wroc.pl, Michal.Markowski@student.pwr.wroc.pl`

Abstract. A combined routing-scheduling problem is considered in the paper. It consists in scheduling of tasks on moving executors. The case with non-preemptive and independent tasks, unrelated executors as well as interval execution times to minimize the makespan is investigated. The worst-case scheduling problem based on an absolute regret is formulated. Solution algorithms of polynomial computational complexity, which use simulated annealing as well as tabu search approaches, are presented. The results of numerical experiments are given.

Keywords: manufacturing systems, scheduling, movement, decision making, uncertainty, robustness, optimization problems.

1 Introduction

Scheduling of tasks in discrete manufacturing systems involves new complex problems and so new assumptions. Considering moving different manufacturing system elements is an example of such a new assumption, e.g. [11], [9]. The movement in discrete manufacturing systems may apply to different objects and may be understood and considered in different ways. The obvious is to move products to be manufactured, but one can also consider to move executors performing manufacturing tasks on products in manufacture located at stationary workstations.

1.1 Problem of Routing-Scheduling

A model with moving executors, referred to as routing-scheduling or task scheduling on moving executors, is considered in this paper. It is mainly manufacturing systems where such complex scheduling problems occur in cases when the movement of products to be manufactured is impossible or too expensive and it is necessary to move executors. This may occur, for instance, when products in manufacture (such as large engine casings, railway carriage components, etc.) are too large or too heavy to move them among executors (machines). Another example are maintenance tasks on stationary machines or transportation tasks performed by mobile robots or automated guided vehicles in flexible manufacturing systems. Also automated service systems and computer systems, where the times of the

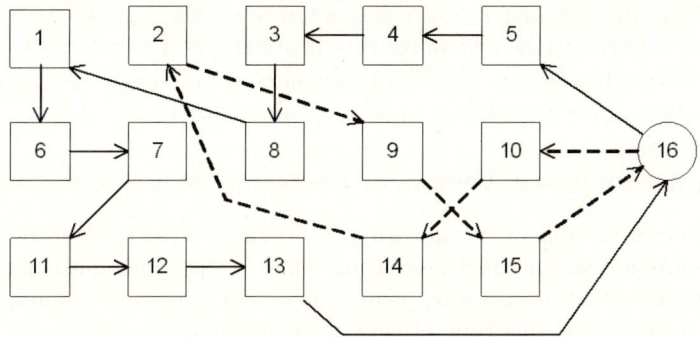

Fig. 1. Example of a layout of workstations and executor routes

movement of executors (agents) are not negligible, can be the prospective areas for the application of routing-scheduling problem under consideration.

In order to define more precisely the scope of this paper, the main terms will be explained now. A *task* is understood as in scheduling theory, but it takes on a new meaning since the movement of executors is taken into account. An *executor* is a subject (usually a technological device) performing tasks. It performs a task at a place called a *workstation* which is located on a plane or in a space. Also a *depot* – a workstation where each executor starts and ends its work – is distinguished. The considered problem's central idea is: to perform a task, the executor should drive up to the workstation. Hence, each task consists of two parts: driving up to the workstation and performing an *activity* at the workstation (no activity is executed at the depot). Consequently, *execution time* (the main datum for each task scheduling problem) is the sum of the *driving up time* and the *activity performing time*. This generalization defines a new task scheduling problem in which not only a subset of tasks for each executor must be derived but also routes should be determined for the executors. The determination of routes is necessary since the order in which tasks are performed by each executor affects the quality of scheduling for the whole set of tasks. An example of executor routes for fifteen tasks and two executors is shown in Fig. 1. It is assumed in the deterministic version of the routing-scheduling problem that all the problem parameters (in particular, all the execution times) can be specified precisely in advance of routing-scheduling whereby the outcome (in the form of an optimal schedule) can be determined exactly.

1.2 Related Problems

Similar problems, but for flow-shop task scheduling, are considered in, for instance, [2]. Directly related to the problem of routing-scheduling are the traveling salesman problems with time windows stated in [5] and further developed in, for example, [19] where the makespan problem with time windows and additional resources taken into account is considered, see [20] and [6] for a survey. The movement of subjects performing tasks also occurs in hoist scheduling problems where, however, the routes of the hoists being in charge of transporting products in manufacture among machines are restricted to tracks, e.g. [4]. Another group of problems, referred to as vehicle routing problems, is also related to the problem under consideration, e.g. [7]. The problem considered can also be treated as a generalization of the scheduling of tasks with setup times, e.g. [24], [25].

1.3 Non-deterministic Version of Problem – Worst-Case Approach

This paper deals with the generalization of previous investigations, leading to a non-deterministic case in which execution times – the routing-scheduling problem's most important parameters – are uncertain. Problem formulations and solution algorithms for this type of routing problem are given in, e.g., [10]. The stochastic approach is the most popular (and widely investigated) way to model uncertainty in decision making problems see, for instance, surveys: [23], [21], [22], [3]. It is assumed that a certain probability distribution exists over the space of all the possible realizations (scenarios) of all the random parameters of the problem and the objective is to determine a solution that fulfils a selected probabilistic performance index. The main drawbacks of the stochastic approach are discussed in [18].

They have motivated many researchers to develop other approaches which would cope with the uncertainty of parameters more adequately without using any probability distribution as the description of parameter uncertainty. An approach based on an absolute regret ([27]) is proposed in the paper. When the maximum function (used as the operator of a substantiation) is applied to the absolute regret, a minimax decision making problem, being a special case of worst-case decision making problems, is obtained. The corresponding criterion (decision) are referred to as the worst-case absolute regret criterion (decision). The worst-case approaches produce decisions which have reasonable criterion values under any likely realization of the uncertain parameters. They lead, however, to conservative solutions based on an anticipation that the worst might happen. Additionally, the worst-case solutions perform well across all scenarios and they hedge against the worst of all possible scenarios. Worst-case decision making formulations were introduced by Gupta and Rosenhead in [8] and Rosenhead *et al.* in [26] and were developed by Kouvelis and Yu in [18] as well as by Averbakh in [1].

The investigation of a non-deterministic version of the routing-scheduling is well grounded. In the problem considered, the uncertainty is associated mainly with execution times, i.e. both driving up times and activity performing times. It is obvious that in manufacturing systems both above times as routing-scheduling problem parameters may be imprecise, plausible or unpredictable. It is possible, however, to specify the bounds of the times and consequently, to determine their intervals instead of their crisp values as in the deterministic version of the routing-scheduling problem. This is possible, for instance, when both times are the results of other decision making problems. Such a case is discussed in [12], [13], [14] where an idea of a two-level decision making system with task scheduling for moving executors located at the upper level and collision-free motion control of a group of executors at the lower level is presented.

This paper is organized as follows. In Section 2, the routing-scheduling problem is introduced and formulated as a discrete optimization problem for both deterministic and interval execution times. A simple version with unrelated executors performing a set of non-preemptive, independent tasks, where the aim is to minimize makespan, is addressed. The worst-case absolute regret version of the uncertain problem is solved. It is obvious that it is NP-hard optimization problem because its deterministic version is NP-hard one. In the consequence, it is important to have computationally effective solution algorithms. Two metaheuristics were applied and the corresponding solution algorithms were determined. The first solution algorithm is based on simulated annealing approach. The early results have been presented in [16] and in [15] for the other version of the uncertain problem with the relative regret instead of the absolute regret. Now, this heuristic algorithm is compared with the second solution algorithm founded on tabu search metaheuristics. Such an approach was presented at first in [17]. In the paper, it is developed. In particular, the different definition of moves was applied. Both heuristic algorithms are presented in Section 3 together with the procedure of calculating the values of the performance index which is very important for the performance index considered. The results of numerical experiments as well as conclusions complete the paper.

2 Problem Formulation

2.1 Deterministic Case

Let us formulate the routing-scheduling problem in the deterministic case. Only one task, being a sequence of driving up and performing an activity by the same executor, is carried out at every workstation, or exactly speaking, on a product in manufacture located at the workstation. The sets of tasks and workstations are identically denoted as $\boldsymbol{H} = \{1, 2, \ldots, H\}$, where H is the number of tasks or workstations. A depot for executors, being the place where no activity is performed and which is the beginning and end of each executor's route, is distinguished among the workstations. The depot is denoted as $h = H + 1$. Then $\overline{\boldsymbol{H}} = \boldsymbol{H} \cup \{H+1\}$ is a set of workstations with the depot. Similarly, $\boldsymbol{R}$ and R are respectively a set of executors and a number of executors. Task execution times $\tau_{r,g,h}$ have now a more complex form than for a version without moving executors, depending not only on the executors but also on the workstations from which the executors set off to perform the tasks. Each execution time $\tau_{r,g,h}$ is sum $\tau_{r,g,h} = \hat{\tau}_{r,h} + \check{\tau}_{r,g,h}$, where $\hat{\tau}_{r,h}$ and $\check{\tau}_{r,g,h}$ are respectively the time for which activity h is performed by executor r at workstation h and the time for which the executor drives up to workstation h from workstation g. Assumption $\check{\tau}_{r,h,h} = +\infty, h = 1, 2, \ldots, H+1$ means that it is not possible for executors to return to the same workstation without performing any activity. Times $\tau_{r,g,h}$ form three-dimensional matrix $\boldsymbol{\tau} = [\tau_{r,g,h}]_{r=1,2,\ldots,R;\ g,h=1,2,\ldots,H+1}$. In order to formulate the corresponding optimization problem, let us define a decision variable as matrix $\boldsymbol{\gamma} = [\gamma_{r,g,h}]_{g,h=1,2,\ldots,H+1;\ r=1,2,\ldots,R}$ where $\gamma_{r,g,h} = 1(0)$ if executor r performs task h after driving up from workstation g (otherwise). The following constraints imposed on decision matrix γ ensure the feasibility of the solutions

$$\gamma_{r,g,h} \in \{0,1\}, r \in \boldsymbol{R}, g, h \in \overline{\boldsymbol{H}}, \tag{1}$$

$$\gamma_{r,h,h} = 0, r \in \boldsymbol{R}, h \in \overline{\boldsymbol{H}}, \tag{2}$$

$$\sum_{r=1}^{R} \sum_{g=1}^{H+1} \gamma_{r,g,h} = 1, h \in \boldsymbol{H}, \tag{3}$$

$$\sum_{g=1}^{H+1} \gamma_{r,g,p} = \sum_{h=1}^{H+1} \gamma_{r,p,h}, r \in \boldsymbol{R}, p \in \overline{\boldsymbol{H}}, \tag{4}$$

$$\gamma \in \boldsymbol{S} \tag{5}$$

where $\boldsymbol{S} = \{ \gamma: \sum_{g \in \overline{\boldsymbol{H}}_S} \sum_{h \in \overline{\boldsymbol{H}}_S} \gamma_{r,g,h} \leq \overline{H}_S - 1,$

$\overline{\boldsymbol{H}}_S$ is any non-empty subset of $\overline{\boldsymbol{H}}$, $r \in \boldsymbol{R}$ $\}$,

$$\sum_{h=1}^{H} \gamma_{r,H+1,h} = 1, r \in \boldsymbol{R}. \tag{6}$$

Apart from obvious constraints (1) and (2), the other ones can be explained as follows. Because of (3), each task has to be performed by the executor after completing the previous task or after driving up from the depot. The continuity of the routes for executors results from (4). Conditions (5) and (6) ensure that the routes do not contain sub-cycles and each executor sets off once from a depot, respectively. The constraints form set $\boldsymbol{\Gamma}$ of feasible matrices γ i.e. $\boldsymbol{\Gamma} = \{\gamma: (1) \wedge (2) \wedge (3) \wedge (4) \wedge (5) \wedge (6) \text{ hold}\}$. The makespan

$$Q(\boldsymbol{\tau},\boldsymbol{\gamma}) = \max_{r=1,2,\ldots,R} \sum_{h=1}^{H+1} \sum_{g=1}^{H+1} \gamma_{r,g,h} \tau_{r,g,h} \tag{7}$$

is used as the performance index.

The routing-scheduling problem can be formulated as discrete optimization problem $\mathbf{P}(\boldsymbol{\tau})$: Find the minimum of $Q(\boldsymbol{\tau},\boldsymbol{\gamma})$ with respect to $\boldsymbol{\gamma} \in \boldsymbol{\Gamma}$ for fixed value of $\boldsymbol{\tau}$. Let us also denote $Q^*(\boldsymbol{\tau}) \triangleq \min_{\gamma \in \Gamma} Q(\boldsymbol{\tau},\boldsymbol{\gamma})$.

2.2 The Case of Interval Execution Times

The exact values of execution times $\tau_{r,g,h}$ or their components are often not known. Let us assume that $\tau_{r,g,h} \in [\underline{\tau}_{r,g,h}, \overline{\tau}_{r,g,h}]$, $0 \leq \underline{\tau}_{r,g,h} \leq \overline{\tau}_{r,g,h}$ where values of $\underline{\tau}_{r,g,h}, \overline{\tau}_{r,g,h}$ are known and given. Then each realization of matrix $\boldsymbol{\tau}$ of execution times $\tau_{r,g,h}$ as a parameter of routing-scheduling problem $\mathbf{P}(\boldsymbol{\tau})$ is called a scenario. Set $\boldsymbol{D}_\mathrm{T}$ of all scenarios is the Cartesian product of all intervals $[\underline{\tau}_{r,g,h}, \overline{\tau}_{r,g,h}]$. The worst-case approach is proposed to cope with this uncertainty. The objective is to find matrix $\boldsymbol{\gamma}$ that performs well for all the scenarios in the sense of makespan (7). The worst-case optimization approach can be used for this purpose, e.g. [18], [1]. The uncertainty in decisions due to the uncertainty in parameters $\boldsymbol{\tau}$ can be evaluated in different ways. In further considerations, the absolute regret, being the difference $Q(\boldsymbol{\tau},\boldsymbol{\gamma}) - Q^*(\boldsymbol{\tau})$ is used. Then, the uncertain optimization problem, referred to as **UP** which corresponds to optimization problem $\mathbf{P}(\boldsymbol{\tau})$, consists in the minimization with respect to $\boldsymbol{\gamma} \in \boldsymbol{\Gamma}$ of the term

$$z(\boldsymbol{\gamma}) = \max_{\boldsymbol{\tau} \in \boldsymbol{D}_\mathrm{T}} [Q(\boldsymbol{\tau},\boldsymbol{\gamma}) - Q^*(\boldsymbol{\tau})] \qquad (8)$$

called the worst-case absolute regret criterion, i.e.

$$\min_{\boldsymbol{\gamma} \in \boldsymbol{\Gamma}} z(\boldsymbol{\gamma}) = \min_{\boldsymbol{\gamma} \in \boldsymbol{\Gamma}} \max_{\boldsymbol{\tau} \in \boldsymbol{D}_\mathrm{T}} [Q(\boldsymbol{\tau},\boldsymbol{\gamma}) - Q^*(\boldsymbol{\tau})] = z(\boldsymbol{\gamma}''). \qquad (9)$$

As the result, matrix $\boldsymbol{\gamma}''$ (a worst-case optimal in the sense of (8) and expressing the worst-case absolute regret schedule) and value $z(\boldsymbol{\gamma}'')$ of (8) are obtained. The optimization problem (9) is NP-hard because its deterministic version is NP-hard. Two heuristic solution algorithms are presented in the next section.

3 Worst-Case Absolute Regret Routing-Scheduling Algorithm

Two different well-known metaheuristics were used for solving the routing-scheduling problem investigated: simulated annealing (SA) and tabu search (TS).

3.1 Simulated Annealing Algorithm

Let us introduce the basic notation:
N – number of iterations of the algorithm which is used also as the stop condition,
$n, n = 1, 2, \ldots, N$ – index of the current iteration of the algorithm,
$\theta, \theta \in [\theta_{\min}, \theta_{\max}]$ – parameter of the algorithm called temperature,
$\boldsymbol{\gamma}_n$ – solution in the nth step of the algorithm given in the form of matrix $\boldsymbol{\gamma}$
$\boldsymbol{\gamma}'$ – the best solution determined by the algorithm.

The values of parameters $N, \theta_{\min}, \theta_{\max}$ as well as randomly generated initial solution $\boldsymbol{\gamma}_0$ are the data for the solution algorithm that can be presented in seven steps:

1. Set $n = 0$, $\boldsymbol{\gamma}' = \boldsymbol{\gamma}_0$ and calculate $z(\boldsymbol{\gamma}_0)$.
2. Find a new solution $\boldsymbol{\gamma}_{n+1}$ in the neighborhood of $\boldsymbol{\gamma}_n$ and calculate $z(\boldsymbol{\gamma}_{n+1})$.
3. If $z(\boldsymbol{\gamma}_{n+1}) \leq z(\boldsymbol{\gamma}_n)$ set $\boldsymbol{\gamma}' = \boldsymbol{\gamma}_{n+1}$ and go to Step 6, otherwise go to the next step.
4. Generate randomly the number d from the interval $[0, 1]$ according to the rectangular distribution.
5. If $\exp(\frac{z(\boldsymbol{\gamma}_n) - z(\boldsymbol{\gamma}_{n+1})}{\theta}) > d$ set $\boldsymbol{\gamma}_{n+1} = \boldsymbol{\gamma}_n$, otherwise do not change $\boldsymbol{\gamma}_n$.

6. Set $\theta = \frac{\theta}{1+\lambda\theta}$ where $\lambda = \frac{\theta_{max}-\theta_{min}}{N\theta_{max}\theta_{min}}$.
7. If $n < N$ set $n = n+1$ and go to Step 2, otherwise stop the algorithm with the solution γ'.

The following procedure has been applied to generate the neighborhood. The task is taken randomly from the longest working executor, and it is inserted into any possible position for other executors.

3.2 Tabu Search Algorithm

The second solution algorithm referred to as TS is based on the tabu search metaheuristics. Let us denote additionally (with respect to the notation introduced for SA):
TL, L – tabu list and its length, respectively,
$\boldsymbol{R}, K$ – parameters of the algorithm, i.e. the set of executors finishing their work as the earliest, the number of moves, respectively,
$\gamma^r = [\gamma_{r,g,h}]_{g,h=1,2,\ldots,H+1}$ – part of matrix γ concerning executor r,
H^r – number of tasks performed by executor r.

1. Obtain γ_0 as the solution of deterministic problem $\mathbf{P}(\overline{\tau})$ as well as $z' \triangleq z(\gamma_0)$.
2. Set TL $= \emptyset, n = 1, \gamma' = \gamma_0$.
3. Define the attribute of solution γ' as the set $A(\gamma') = \{(m_r(i), m_r(j)) : r \in R, i, j = 1, 2, \ldots, H^r, i < j\}$.
4. For $\tau = \overline{\tau}$ determine executor $\overline{r}$ which finishes its work as the latest as well as set $\boldsymbol{R}$.
5. For $k = 1$ to K determine the move by performing the following activities:
   (a) Select randomly tasks h and $\tilde{h}$, respectively performed by executors $\overline{r}$ and s where $s = \overline{r}$ or $s \in \boldsymbol{R}$.
   (b) Obtain new sub-matrices $\tilde{\gamma}^{\overline{r}}$ and $\tilde{\gamma}^s$ and so new solution $\tilde{\gamma}$ after performing the move, i.e. after changing positions of tasks h and $\tilde{h}$.
   (c) Define sets of attributes $A(\tilde{\gamma})$ and $A(\gamma', \tilde{\gamma}) = A(\gamma') \backslash A(\tilde{\gamma})$.
6. For all K moves select such a solution $\tilde{\gamma}$ referred to as $\overline{\gamma}$ that $(A(\gamma', \tilde{\gamma}) \notin$ TL $\wedge z(\overline{\gamma}) = \min z(\tilde{\gamma}))$. Then set $\gamma' = \overline{\gamma}, z' = z(\gamma')$
7. Place $A(\gamma', \overline{\gamma})$ at the end of TL and fit the length of TL to L by eliminating the appropriate number of its first elements.
8. If $n < N$ set $n = n+1$ and go to Step 4, otherwise stop the algorithm with γ' and z' as the result.

3.3 Calculation of the Performance Index

The effective calculation of the performance index value is the crucial issue for both algorithms. The following Lemma shows how to simplify the maximization of the absolute regret with respect to the uncertain matrix τ.

Lemma. For any $\gamma \in \boldsymbol{\Gamma}$

$$\max_{\tau \in D_T} [Q(\tau, \gamma) - Q^*(\tau)] = \max_{\tau \in D'_T} [Q(\tau, \gamma) - Q^*(\tau)] \quad (10)$$

where D'_T is a Cartesian product of all 2-element sets $\{\underline{\tau}_{r,g,h}, \overline{\tau}_{r,g,h}\}$.

Proof. Let γ^* denote the optimal solution of $\mathbf{P}(\tau)$ for fixed matrix τ. Then

$$z(\gamma) = \max_{\tau \in D_T} [\max_{r=1,2,\ldots,R} \sum_{h=1}^{H+1} \sum_{g=1}^{H+1} \tau_{r,g,h} \gamma_{r,g,h} - \max_{r=1,2,\ldots,R} \sum_{h=1}^{H+1} \sum_{g=1}^{H+1} \tau_{r,g,h} \gamma^*_{r,g,h}]. \quad (11)$$

Let j and j^* be indices of executors for which the sums in (11) are the greatest, i.e.

$$j = \arg \max_{r=1,2,\ldots,R} \sum_{h=1}^{H+1} \sum_{g=1}^{H+1} \tau_{r,g,h} \gamma_{r,g,h} \text{ and } j^* = \arg \max_{r=1,2,\ldots,R} \sum_{h=1}^{H+1} \sum_{g=1}^{H+1} \tau_{r,g,h} \gamma^*_{r,g,h}.$$

Thus, (11) can be rewritten as

$$z(\gamma) = \max_{\tau \in D_T} [\sum_{h=1}^{H+1} \sum_{g=1}^{H+1} \tau_{j,g,h} \gamma_{j,g,h} - \sum_{h=1}^{H+1} \sum_{g=1}^{H+1} \tau_{j^*,g,h} \gamma^*_{j^*,g,h}] \quad (12)$$

and in the consequence as

$$z(\gamma) = \max_{\tau \in D_T} [\sum_{h=1}^{H+1} \sum_{g=1}^{H+1} (\tau_{j,g,h} \gamma_{j,g,h} - \tau_{j^*,g,h} \gamma^*_{j^*,g,h})]. \quad (13)$$

Let us notice that values of times $\tau_{j,g,h}$ and $\tau_{j^*,g,h}$ are only important with respect to calculation of $z(\gamma)$. The values of other times can be set as arbitrary, for example they can obtain values of the bounds of corresponding intervals. Hence, it is enough to fix $\tau_{r,g,h} = \overline{\tau}_{r,g,h}$ and $\tau_{r^*,g,h} = \underline{\tau}_{r,g,h}$ to maximize the difference in (13). ∎

So, it is sufficient to consider the bounds of intervals instead of all intervals. Let $Q^r(\tau_r, \gamma)$ denote the time executor r is engaged, i.e. $Q(\tau,\gamma) = \max_{r=1,2,\ldots,R} Q^r(\tau_r, \gamma)$ where τ_r is a part of matrix τ with the execution times corresponding to executor r.

The following procedure, which calculates $z(\gamma)$, results from Lemma.

1. Calculate $Q^r(\tau_r, \gamma)$ for $r = 1, 2, \ldots, R$ and find the index of the executor that works the longest time, i.e.

$$j = \arg \max_{r=1,2,\ldots,R} Q^r(\overline{\tau}_r, \gamma) = \arg \max_{r=1,2,\ldots,R} \sum_{h=1}^{H+1} \sum_{g=1}^{H+1} \overline{\tau}_{r,g,h} \gamma_{r,g,h}.$$

2. Obtain the deterministic matrix $\hat{\tau}$ with the following elements

$$\hat{\tau}_{r,g,h} = \begin{cases} \overline{\tau}_{r,g,h}, & \text{for } r = j \text{ and } \gamma_{r,g,h} = 1, \\ \underline{\tau}_{r,g,h}, & \text{otherwise}. \end{cases}$$

3. Obtain $\gamma^*, Q^{*j^*}(\hat{\tau})$ as the solution of problem $\mathbf{P}(\hat{\tau})$ and determine j^*.
4. Calculate $z(\gamma) = Q^j(\overline{\tau}_j \gamma) - Q^{*,j^*}(\hat{\tau})$.

4 Numerical Experiments

The empirical evaluation of heuristic solution algorithms is an important part of their determination. The dependencies of empirical performance indices on different parameters characterizing the problem as well as the algorithm are investigated during such a research. The value of criterion and the time of computation serve as empirical performance indices for the problem considered. The former can be treated as the measure of uncertainty, the importance of the latter is evident. The following parameters of the solution algorithms were chosen as the most important for the evaluation conducted: $N, \theta_{max}, \theta_{min}$ for SA and N, L, K for TS. The following values have been obtained as the result of the experiments: $N = 200$ for SA and 100 for TS, $\theta_{max} = 100000, \theta_{min} = 1, L = 3H, K = 20$. The example of the results is given in Fig. 2 and Fig. 3. It turned out that the increase of the number of iterations improves the solution algorithm SA in a moderate way. This dependence for TS is more expressive. The investigations concerning TS shows that the dependence of the quality of the algorithm on L is not crucial.

Table 1. Dependence of T and δ on H for all solution algorithms investigated

H	SA		SA(A)		TS		TS(A)	
	$T[s]$	δ	$T[s]$	δ	$T[s]$	δ	$T[s]$	δ
5	23	21	1	25.8	4	18	1	18.7
10	153	12.8	1	18.3	80	12.8	4	13.1
15	263	14.3	1	19.4	375	14.6	18	14.6
20	439	14	1	19	384	13.4	20	13.2
25	646	14.3	1	19.8	1122	14	46	13.6
30	970	14.5	1	18.8	1078	13.8	51	13.3
35	1387	15	1	19.1	1737	14	78	13.5
40	1812	14.9	1	18.1	1987	14	88	13.1
45	2346	15	2	19	2760	14.4	136	13.3

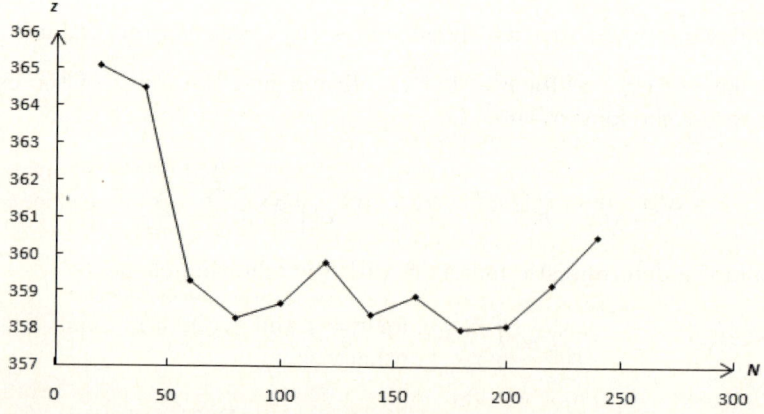

Fig. 2. Dependence of z on N for algorithm SA

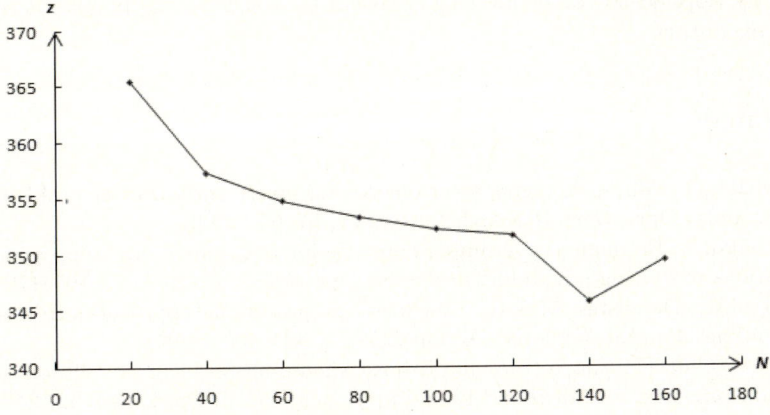

Fig. 3. Dependence of z on N for algorithm TS

Both heuristic algorithms were compared for the case when $R = 2, H = 25$ and the execution times have been randomly generated according to the rectangular distribution: activity performing times (assumed as deterministic in the numerical experiments) from interval $[10, 30]$ as well as lower and upper bounds of the driving up times from intervals $[40, 70]$ and $[70, 90]$, respectively. The versions of both algorithms referred to as TS(A) and SA(A) have been launched when the minimum value of $z(\gamma)$ has been calculated in the approximate way. The dependence of T and δ on H is presented in Table 1 where index $\delta = \frac{z}{H}$ is the average value of z with respect to H. The initial experiments show that TS is better than SA in terms of δ (the times of computation are similar). It is worth to notice that applying the approximate versions of the algorithms enables to shorten substantially the times of computation, and, first of all, it does not decrease the quality of the solutions with respect to δ.

5 Conclusions

The uncertain version of the routing-scheduling problem is investigated in the paper. The uncertainty is understood in terms of interval execution times. The simple task scheduling (assignment) problem is addressed. The worst-case approach based on the absolute regret was used as the description of the uncertainty. Because of the NP-hardness of the basic deterministic problem, the heuristic solution algorithm, using simulated annealing and tabu search approaches, were employed for solving the uncertain version. The solution algorithms use the procedure which makes easier the calculation of the performance index. The slight difference between the accuracy of TS and TS(A) is rather surprising and require additional investigations.

The approach can be easily applied to other routing-scheduling problems. It is also worth to notice that the solution algorithm presented in this paper can be directly applied to corresponding classical task scheduling problems without moving executors.

References

1. Averbakh, I.: Minimax regret solutions for minimax optimization problems with uncertainty. Operations Research Letters 27, 57–65 (2000)
2. Averbakh, I., Berman, O.: A simple heuristic for m-machine flow-shop and its applications in routing-scheduling problems. Operations Research 47, 165–170 (1999)
3. Birge, J.R., Dempster, M.A.H.: Stochastic programming approaches to stochastic scheduling. Journal of Global Optimization 9, 383–409 (1996)
4. Che, A., Chu, C.: Single-track multi-hoist shcduling problem: a collision-free resolution based on a branch-and-bound approach. Int. J. Prod. Res. 42, 2435–2456 (2004)
5. Christofides, N., Mingozzi, A., Toth, P.: State-space relaxation procedures for the computation bounds to routing problems. Networks 11, 145–164 (1981)
6. Desrosiers, J., Dumas, Y., Solomon, M., Soumis, F.: Time constrained routing and scheduling. Report G-92-42, GERAD, Montreal (1992)
7. Fisher, M.L.: Vehicle Routing. In: Ball, M.O., Magnanti, T.L., Monma, C.L., Nemhauser, G.L. (eds.) Handbooks in Operations Research and Management Science. Network Routing, vol. 8, pp. 1–33. North-Holland, Amsterdam (1995)
8. Gupta, S.K., Rosenhead, J.: Robustness Approach to International Sourcing. Annals of Operations Research 59, 165–193 (1972)
9. Józefczyk, J.: An algorithm for scheduling tasks on moving executors in complex operation systems. In: Proceedings of 1st IFAC Workshop on Manufacturing Systems MIM 1997, Vienna, Austria, pp. 139–144 (1997)
10. Józefczyk, J.: Scheduling of manufacturing tasks with uncertain processing times. In: Proceedings of 14th International Conference on Systems Engineering, Coventry, UK, vol. 1, pp. 294–299 (2000)
11. Józefczyk, J.: Scheduling tasks on moving executors to minimise the maximum lateness. European Journal of Operational Research 131, 171–187 (2001)
12. Józefczyk, J.: Application of knowledge-based pattern recognition to movement control of a group of vehicles. International Journal of Knowledge Based Intelligent Engineering Systems 6, 192–198 (2002)
13. Józefczyk, J.: Knowledge-based two-level control of a group of mobile executors. Integrated Computer-Aided Engineering 10, 191–202 (2003)
14. Józefczyk, J.: Decision-making algorithms in two-level complex operation system. Decision Support Systems 38, 171–182 (2004)
15. Józefczyk, J.: Robust algorithm for the routing-scheduling problem with uncertain execution times. In: Proc. of 19th Int. Conf. on Production Research, Valparaiso, Chile (2007) [cd-rom]
16. Józefczyk, J.: Robust decision-making algorithm for two-level operational system. In: Proc. of 4th IFAC International Conference on Management and Control of Production and Logistics, Sibiu, Romania, pp. 611–616 (2007)
17. Józefczyk, J., Salwach, T.: An uncertain routing-scheduling problem in a class of manufacturing systems. In: Proceedings of the 8th International Conference on The Modern Information Technology in the Innovation Processes of the Industrial Enterprises. MITIP 2006, Budapest, Hungary, pp. 547–552 (2006)

18. Kouvelis, P., Yu, G.: Robust Discrete Optimization and Its Applications. Kluver Academic Publishers, Boston (1997)
19. Langevin, A., Desrochers, M., Desrosiers, J., Gelinas, S., Soumis, F.: A two-commodity flow formulation for the traveling salesman and the makespan problems with time windows. Networks 23, 631–640 (1993)
20. Mingozzi, A., Bianco, L., Ricciardelli, S.: Dynamic programming strategies for the traveling salesman problem with time window and precedence constraints. Operations Research 45, 365–377 (1997)
21. Möhring, R.H., Radermacher, F.J., Weiss, G.: Stochastic scheduling problems I. General strategies, Zeitschrift für Oper. Res. 28, 193–260 (1984)
22. Möhring, R.H., Radermacher, F.J., Weiss, G.: Stochastic scheduling problems II. Set strategies, Zeitschrift für Oper. Res. 29, 65–104 (1985)
23. Pinedo, M.L., Schrage, L.: Stochastic shop scheduling: a survey. In: Dempster, M.A.H., Lenstra, J.K., Rinooy Kann, A.H.G. (eds.) Deterministic and Stochastic Scheduling, Reidel, Dordrecht (1982)
24. Potts, C.N., Kovalyov, M.Y.: Scheduling with batching: a review. European Journal of Operational Research 120, 228–249 (2000)
25. Rajendran, C.H., Ziegler, H.: Scheduling to minimize the sum of weighted flow time and weighted tardiness of jobs in a flowshop with sequence-dependent setup times. European Journal of Operational Research 149, 513–522 (2003)
26. Rosenhead, M.J., Elton, M., Gupta, S.K.: Robustness and optimality as criteria for strategic decisions. Oper. Res. Quarterly 23, 413–430 (1972)
27. Savage, L.J.: The theory of statistical decision. Journal of American Statist. Assoc. 46, 55–67 (1951)

Life Story Generation Using Mobile Context and Petri Net

Young-seol Lee and Sung-Bae Cho

Dept. of Computer Science, Yonsei University
Shinchon-dong, Seodaemun-ku, Seoul 120-749, Korea
tiras@sclab.yonsei.ac.kr, sbcho@cs.yonsei.ac.kr

Abstract. People mainly organize their experience as a kind of narratives. Story generated from user's information in mobile environment can help share his experience with other people and recall his meaningful memory. In this paper, we propose a method that generates a story with Petri net and user contexts. In order to verify the usefulness of the proposed method, we show an example of generating user's experience to story with user context in mobile environment. Comparison of user's report and generated story confirms the validity of the automatic story generation.

1 Introduction

People mainly organize their experience as a form of narratives such as stories, excuses, myths, and so on [1]. They tends to relate their meaningful memory with events in a story. They can easily understand causality between events by organizing experience into familiar narratives. It is human ability that David Blair and Tom Meyer call "Narrative Intelligence"[2]. As people that have mobile device with diverse functions increase, it is getting stored more and more information in it. If the information in mobile device is organized to story, it can be used to share his experience with other people, and recall his meaningful memory.

Generally, there are three kinds of approaches for story generation: character based approach, script based approach, and intermediate approach [4]. In character based approach, characters in a story are implemented as autonomous, intelligent agents. Characters in story interact with other characters or external environment. Story is naturally generated from the interactions. Characters can choose their own action with their goals, emotions and perceptions of environment,which are continually updated by their own and other agents' actions. In other words, the story is based on actions that are autonomously performed by the characters in a bottom up fashion. In script based story generation, characters have no autonomy and intelligence. They also cannot affect story directly. It is designed by story designer or author. A general disadvantage of script based story is lacking diversity of story. Branching scripts can solve this problem. The choice of branches is made by the user. However, the diversity of story is still limited. Intermediate approach has basic scripted plot, and characters have limited autonomy in the plot. This means that there are some pre-specified branches in the story and characters must select one branch among them.

Among the three kinds of approaches, script based approach is selected in this paper. In real life, user's log in a mobile device is massive and difficult to understand. Also, there are too many variables, such as user's goal, emotion, state and other people or extra environment, to consider story generation using character based approaches. Therefore, it is advantageous to use pre-described scripts for creating coherent, causal story from massive but limited data. We use Petri net for script representation, which is appropriate solution for modeling real world and particularly such systems that are of interest for computer science, physics, and sociology [5]. Petri net is used to represent non-linear story line that changes as a user's intervention [6]. Petri net is appropriate for expressing changeable story from user context, and modeling the effect of events to other events causally.

In this paper, we propose a method that generates a story with Petri net and user contexts. In order to show the probability of the proposed method, we make the example of generating user's experience to story with user context from mobile device.

2 Related Works

2.1 Story Generation

Story graph is a branching story structure: a directed graph of nodes connected by arcs that represent user choices [7]. A node in story graph is a story fragment, which is an event sequence, and coonected to the other node by arc which represents story branch. User's action in the branch decides next story fragment. This process is looped until no more story branch exists. Every possible path through the graph represents a story. In story graph, user should select limited action among story branches. If user does not select the candidates of story branch, the process of generating story stops and abnormally terminates.

Narrative mediation is to generate a linear story structure that represents the most appropriate story [7]. The linear story structure includes actions that the user should perform. Narrative mediation has expressive power as much as acyclic story graph and merits for expressing detailed story branches. First of all, narrative mediation designs the ideal story structure. It is possible that user performs an exceptional action that interferes with the ideal story structure. When an exception occurs, the system can prevent the exceptional action from interfering with the story structure or incorporate the exceptional action into the story and generating a new linear story structure that is not threatened by the exception. In narrative mediation, ideal story always exists and it is difficult to design many different stories beginning from one ideal story.

2.2 Petri Net

Petri net is a proper method for modeling a system that contains asynchronous elements [8]. The elements in Petri net are causally connected, which represent relationships of events and conditions. Petri net contains the set of place (P), set of transition (T), input function (I) and output function (O). Input function (I) means relationships between transition t_j and the set of places, which is represented as $I(t_j)$. Output function (O) means relationships between transition t_j and the set of places, which is represented as $O(t_j)$.

$$PetriNet = (P, T, I, O) \tag{1}$$

$$P = (p_1, p_2, p_3, \ldots, p_n), n \geq 0 \tag{2}$$

$$T = t_1, t_2, t_3, \ldots, t_m, m \geq 0 \tag{3}$$

$$I : P \to T^\infty \tag{4}$$

$$O : T \to P^\infty \tag{5}$$

Table 1 shows the names and functions of the elements in Petri net. Place (P) is a form of circle, transition (T) is a form of black rectangle and I and O are a form of line with arrows. Finally, token is expressed as points in place (P), which represents the current status of the system. Figure 1 shows the example of

Table 1. Name and description of the Petri net elements

Name	Description
Place (P)	Status or condition of event
Transition (T)	Happening of event
Edge (E)	Relationship between P and T
Token	Current status of the system

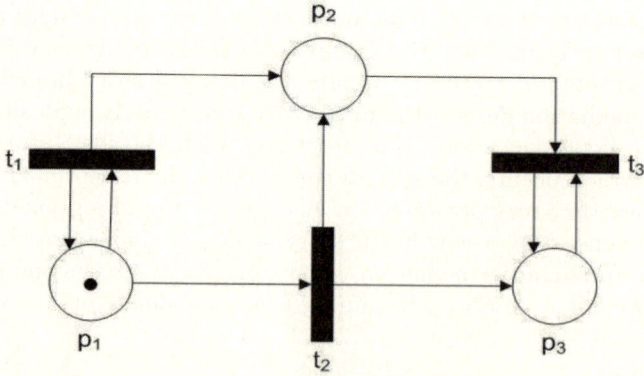

Fig. 1. An example of Petri net

the system using Petri net. If all input places connected with a transition have tokens, it can be fired. When the transition is fired, the tokens are disappeared in input places and generated in the output places. For an example, in Figure 1, transition t_1 can be fired. When t_1 is fired, a token in place p_1 is disappeared. The output places of t_1, and places p_1 and p_2 have tokens. As a result, after t_1 is fired, p_1 and p_2 get to have tokens respectively.

3 Proposed Method

In order to generate a story about human, we first collect his raw data from mobile device. Second, the collected raw data are converted to meaningful information. In this paper, raw log data from mobile devices are converted to user contexts, which are user's actions or emotions extracted by processing raw data [3]. Finally, story should be generated from the contexts.

In this paper, user contexts are user's actions or emotions derived from various log using mobile devices such as GPS log, SMS, Call history, Photos taken, music listened, weather, and so on. We extract the user contexts according to the probabilities using modular Bayesian networks [3].

3.1 Basic Definitions

In this paper, event, user's action or experience, is basic component of plot. A plot is a set of events arranged temporarily or causally and has a main event

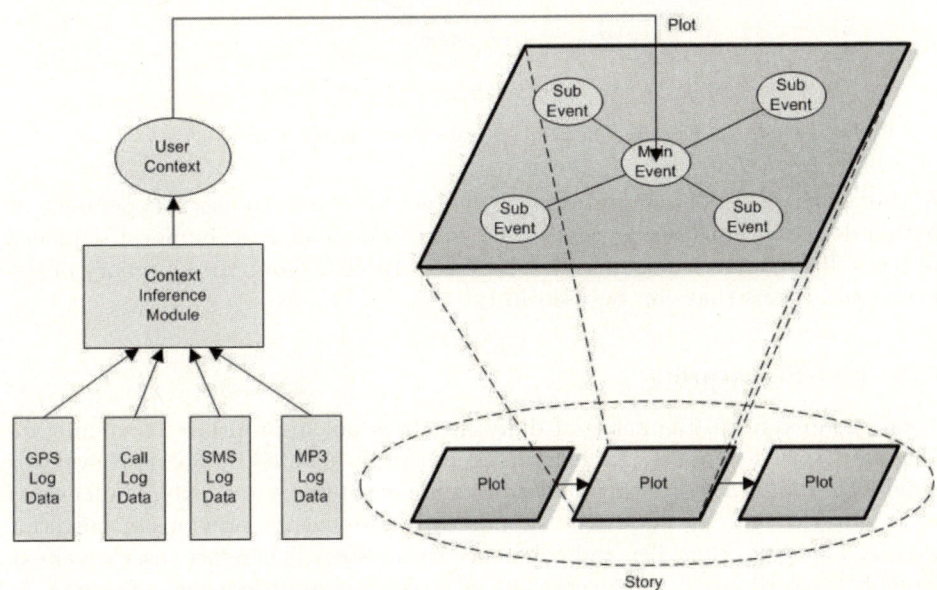

Fig. 2. System overview

Table 2. Classification and examples of event

Classification	Examples
leisure/traveling/viewing	sports, traveling, playing, hunting, dancing, drama, playing instruments, etc
work/school/army	retirement, vacation, absence, entrance, graduation, business, enter the army, discharge from the military service
living/household/trade	decoration, meal, bath, shopping, rest, cleaning, washing up, etc
association/negotiation/promise /quarrel/social intercourse	visit, farewell, reconciliation, break of friendship, competition, request, invitation, etc
religion	worship, a prayer, etc
ceremonies	marriage, divorce, a funeral, etc

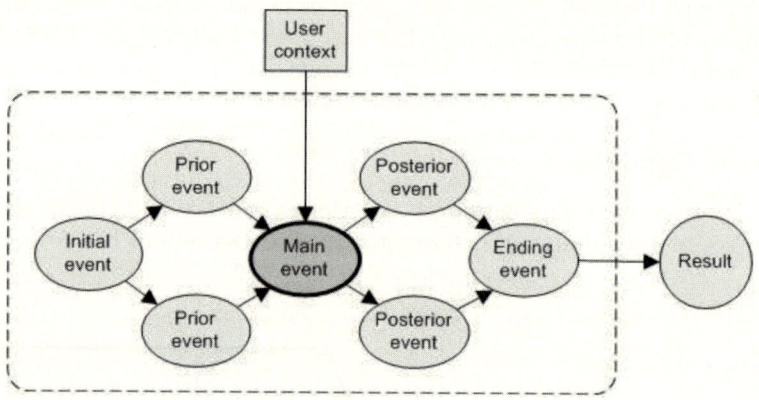

Fig. 3. The conceptual structure of a plot

in the center of it. The main event of the plot is related to user's experience or action derived from user contexts. The story is defined as a temporal sequence of plots. It is the abstraction of the daily life. Table 2 shows the summary of the events or actions that can be used in the story.

3.2 Plot Structure

User's contexts are fragments of daily life. It is not appropriate to enumerate them for story generation. In general, event can be affected by previous event or condition. Plot is a short story that has a main event and sub-events. Main events related to the contexts should be supported by sub-events, previous or posterior events. This gives causality and coherence to the story. Therefore, user's context cannot be expressed as one event, but as a plot, a causal sequence of events. In some cases, the result of a plot can affect the other plots. The system should be able to pass the result of a plot to other plots. Figure 3 shows the conceptual structure of the plot.

Table 3. Elements of Petri net for story generation

Picture	Name	Function
○	place	Condition causally required by action or event Result or status after the event
▬	transition	User's action or event
→	arc (edge)	Link between places and transitions Representation of input or output functions which connect user's action and status
●	token	Representation of current status of story It used to pass result of current plot to posterior plots

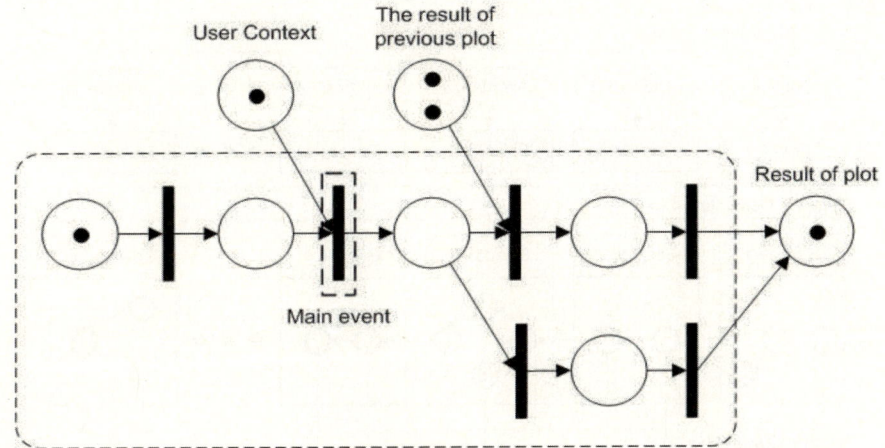

Fig. 4. A conceptual structure of the plot

3.3 Representation of Plot Using Petri Net

Petri net is used to represent the plot, which can clearly represent temporal and causal relationship of events and the connection between user's context and the main event. It is also naturally represented to pass the result of plot by transferring tokens from a plot to other. Table 3 shows the functions of component of Petri net in plot representation.

Figure 4 is an example of plot represented by Petri net. In Figure 4, generating story starts with inserting a token at the initial place in the plot. Main event needs user's context. The result of previous plot changes the flow of story in current plot. Because the result of the previous plot changes the current flow of story, the whole story can be changed by user's previous experience. Finally, the result of current plot is represented by inserting tokens in result place.

Here, place can be departmentalized to user context, result of event, condition of event, result of plot and condition of selecting an event. Typical Petri net treats these various roles as place, but we create additional elements for them as shown in Table 4.

Table 4. Additional elements of Petri net for story generation

Name	Role	Explanation
Core	user context	User context connected to main event of a plot
Branch	condition of selecting an event	Branch determines which event can run by the result of previous plot
Status	result of plot	Status stores the result of a plot and delivers it to the next plot
Place	result and condition of event	Represents result and condition of events

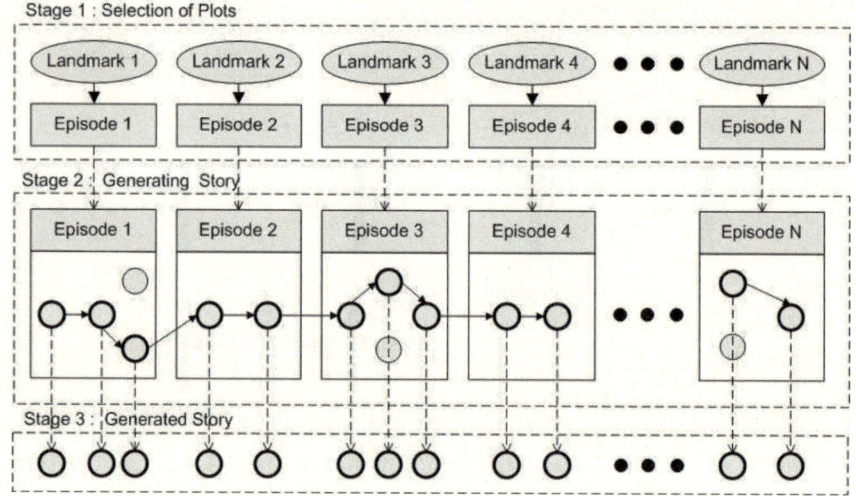

Fig. 5. The selection of plots and story generation

3.4 Plot Selection and Story Generation

The whole process of story generation consists of 3 stages. First stage is selection of plots, which depends on user contexts. Selected plots are arranged according to temporal sequence. Second stage is to process plots. Each selected plot extracts events related to user contexts and previous plots. Third stage is to generate story. The sequence of events is a story that represents user's daily life. After all, story generation is done by selection, arrangement and processing of plots. Figure 5 shows the whole process of story generation related to user's context.

Figure 6 shows the process of event extraction in a plot. It starts at checking the current state of the plot. The state is expressed by allocation of tokens in places of the plot. Current state represents the condition of running events and determines which event can occur in the plot. If several events can run at the same time, one event is selected by the priority of events which is determined

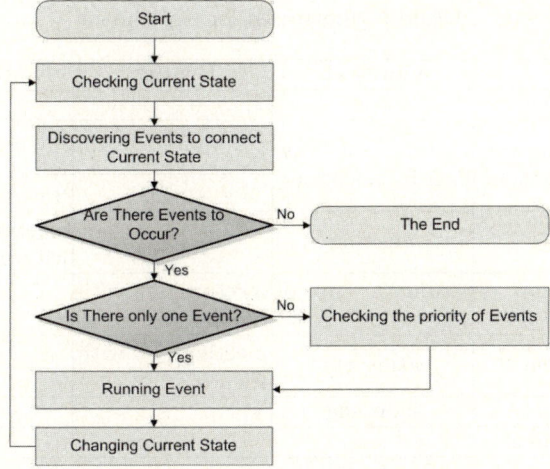

Fig. 6. Flow of Plot Processing

beforehand. After running an event, the state is changed by the effect of the event. This process is repeated until any event cannot run.

4 Experimental Results

4.1 Experimental Environment

In this paper, mobile logs are used to select plots and generate a story. They are collected from three undergraduate students for two weeks [1]. The whole system is developed by Visual C++ 6.0 in Windows XP Professional.

4.2 Experimental Setup

We make some plots to generate a story from user's contexts. Table 5 shows the kinds of plots for experiment and the requirements of selection of plots, the result of the previous plots for change of current story, and the result of the current plots. In this experiment, 'going out' and 'homecoming' are always selected plots for ease of experiment. Story starts at 'going out' and ends at 'homecoming'.

4.3 Results

Figure 7 is a generated story from a day for one user. The story is stored in XML file format. The story in Figure 7 is the result of arrangement with 'going out,' 'meeting,' 'eating out,' 'shopping,' 'taking a cup of tea,' and 'homecoming' plots. The result of 'meeting' is 'meeting with friends', which affects 'eating out' and 'taking a cup of tea' plots. The system generated the stories, 'eating out with friends' and 'taking a cup of tea with friends'. However, 'shopping' plot has

Table 5. Plots used for experiment

Plot	Context	Branch	Result
going out	-	-	-
cheerful conversation by telephone	cheerful conversation by telephone	-	talking over the telephone with a friend
meeting	meeting	-	meeting with friends
taking a cup of tea	taking a cup of tea	meeting with friends	-
eating out	eating out	meeting with friends	-
shopping	shopping	-	-
homecoming	-	-	-

```
<landmark name="get up" />
<landmark name="washing up" />
<landmark name="going out" />
<landmark name="meeting with friends" />
<landmark name="go to restaurant" />
<landmark name="eating out with friends" />
<landmark name="go to shopping center" />
<landmark name="shopping" />
<landmark name="go to cafe" />
<landmark name="taking a cup of tea with friends" />
<landmark name="homecoming" />
```

Fig. 7. The result XML file of story generation

Table 6. Comparison with user's experience and selected plot

Time	Place	Experience	Plot
07:30	home	get up	going out
08:45	bus stop	take a bus	going out
09:15	engineering college	attending class	-
11:00	cafeteria	meal	-
12:00	eating house	treating a junior a meal	-
13:30	auditorium	club activity	meeting
20:00	restaurant	a dinner with club members	eating out
21:10	Baskin Robbins 31	eating icecream	taking a cup of tea
22:00	home	homecoming	homecoming

no story branch that could be affected by 'meeting friends', and 'shopping with friends' cannot be generated.

Table 6 is the comparison of the generated story and user's real life. 'taking lessons' or 'common eating' are not candidates of plot in the system. 'treating

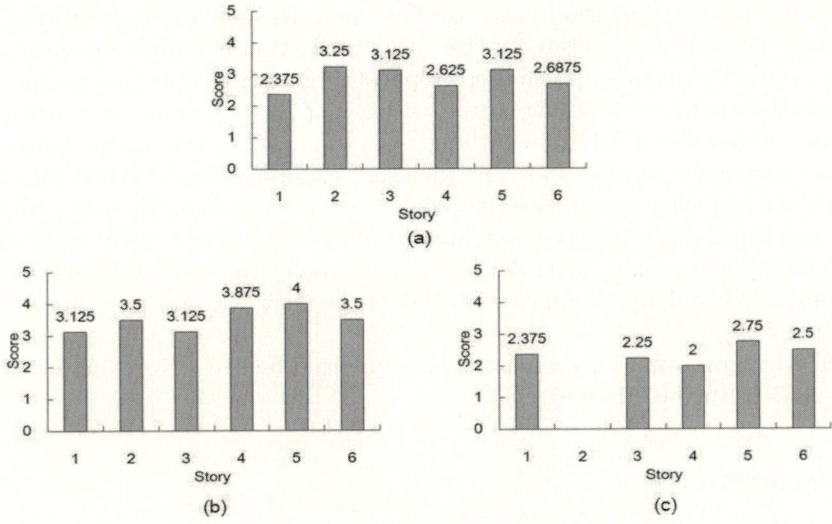

Fig. 8. The results for subjective test

a junior a meal' is not extracted from user's context. It is granted that they are not discovered in the generated story. In the case of 'shopping', it is contained in the generated story, but user did not experience it. User's contexts related to 'shopping' is extracted incorrectly.

4.4 Evaluation

We perform subjective test to evaluate the proposed method. After 6 generated stories and collected log data are shown to 9 subjects, following questions are requested.

- Does story show his life well?
- Does story coincide with his visited places?
- Does story coincide with his photographs?

Score of evaluation is between 1(very bad) and 5(very good). Figure 8(a) shows the result of first question. It represents the fitness between story and log data. Figure 8(b) shows second evaluation of the coincidence between story and visited places. Final evaluation shows the fitness between story and photographs in Figure 8(c).

5 Conclusion and Future Work

In this paper, we have attempted to generate a story from user contexts. Here, story is defined as a causal sequence of user experienced events or actions. In order to make plots from user's context collected by mobile device, we have proposed a modeling method based on Petri net. To show the possibility of the proposed method, we generated a story from collection of mobile log data.

The result of comparison with user's report and generated story shows many coincidences except for extracting user's context inaccurately or having no scripted plot for user's context. Addition of the sensor device and increasing the number of kinds of contexts is required to reflect the user's experience in the generated story. The increase of accuracy of the contexts is needed for the coincidence with plot and experience. Also, adding many plots is required to reflect many real events, which might help to show more detailed story. Finally, the evaluation of the story is required to evaluate the appropriateness of the story with user.

Acknowledgement. This research was supported by MIC, Korea under ITRC IITA-2007-(C1090-0701-0046).

References

1. Mateas, M., Sengers, P.: Narrative Intelligence. The AAAI Press, Menlo Park, Technical Report FS-99-01, Fall (1999)
2. Bruner, J.: The Narrative Construction of Reality. In: Mateas, M., Sengers, P. (eds.) Narrative Intelligence, ch. 3, pp. 41–62 (2003)
3. Hwang, K.-S., Cho, S.-B.: Modular Bayesian Networks for Inferring Landmarks on Mobile Daily Life. In: The 19th Australian Joint Conference on Artificial Intelligence, pp. 929–933 (2006)
4. Theune, M., Faas, S., Nijholt, A., Heylen, D.: The Virtual Storyteller: Story Creation by Intelligent Agents. In: Proceedings Technologies for Interactive Digital Storytelling and Entertainment TIDSE, pp. 204–215 (2003)
5. Farwer, B.: Recovery and Reset in Object Petri Nets with Process Markings. In: Proceedings of CS&P 2006 - Concurrency, Specification and Programming, pp. 47–57 (2006)
6. Brom, C., Abonyi, A.: Petri Nets for Game Plot. In: Proceedings of AISB Artificial Intelligence and Simulation Behaviour Convention, Bristol, vol. 3, pp. 6–13 (2006)
7. Riedl, O.M., Young, M.R.: From Linear Story Generation to Branching Story Graphs. IEEE Journal of Computer Graphics and Animation 26(3), 23–31 (2006)
8. Peterson, L.J.: Petri Net Theory and The Modeling of Systems. Prentice Hall (1981)
9. Murata, T.: Petri Nets: Properties, Analysis and Applications. Proceedings Of the IEEE 77(4) (1989)
10. Riedl, O.M., Saretto, J.C., Young, M.R.: Managing Interaction between Users and Agents in a Multi Agent Storytelling Environment. In: Proceedings of the 2nd International Conference on Autonomous Agents and Multi Agent Systems, Melbourne, Australia (July 2003)

On the Minima of Bethe Free Energy in Gaussian Distributions

Yu Nishiyama[1] and Sumio Watanabe[2]

[1] Department of Computational Intelligence and Systems Science,
Tokyo Institute of Technology,
4259, Nagatuta, Midori-ku, Yokohama, 226-8503 Japan
nishiyudesu@cs.pi.titech.ac.jp
[2] Precision and Intelligence Laboratory, Tokyo Institute of Technology,
4259, Nagatsuta, Midori-ku, Yokohama, 226-8503 Japan
swatanab@pi.titech.ac.jp

Abstract. Belief propagation (BP) is effective for computing marginal probabilities of a high dimensional probability distribution. Loopy belief propagation (LBP) is known not to compute precise marginal probabilities and not to guarantee its convergence. The fixed points of LBP are known to accord with the extrema of Bethe free energy. Hence, the fixed points are analyzed by minimizing the Bethe free energy.

In this paper, we consider the Bethe free energy in Gaussian distributions and analytically clarify the extrema, equivalently, the fixed points of LBP for some particular cases. The analytical results tell us a necessary condition for LBP convergence and the quantities which determine the accuracy of LBP in Gaussian distributions. Based on the analytical results, we perform numerical experiments of LBP and compare the results with analytical solutions.

1 Introduction

The algorithm called belief propagation (BP) is known to be effective for computing marginal probabilities of a high dimensional probability distribution. Since BP is a fundamental algorithm, it has been studied in various fields such as probabilistic inference for artificial intelligence, error correcting codes (e.g., Turbo, LDPC codes), code division multiple access (CDMA) systems, and probabilistic image processing.

Though BP was originally derived and is valid under the condition that the target distribution expresses a tree graph, the algorithm is applicable even to the graph with loops. Then, the BP which is applied to the graph with loops is called loopy belief propagation (LBP). LBP is known not to compute precise marginal probabilities and not to guarantee its convergence since LBP lacks mathematical foundations. To examine and validate the performance of LBP, there are studies of clarifying the conditions for LBP convergence [1] or the accuracy of the approximate marginal probabilities [2][3][4]. The accuracy has also been theoretically studied from the viewpoint of information geometry [5][6].

On the other hand, the fixed points of LBP are known to accord with the extrema of a functional called Bethe free energy [7][8], which originates in statistical physics. From the equivalence, the computation of marginal probabilities is changed to the optimization problem of minimizing Bethe free energy.

In this paper, we apply Bethe approximation to multi-dimensional Gaussian distributions and consider the Bethe free energy in Gaussian distributions. Then, we analytically clarify the extrema of the Bethe free energy, equivalently, the fixed points of LBP for some particular cases. After that, on the basis of the analytical results, we perform numerical experiments of LBP and examine how the results agree with analytical solutions. This study helps us to reveal fundamentally the performance of LBP.

The accuracy of Bethe approximation in Gaussian distributions has numerically been studied in probabilistic image processing [9][10]. In those papers, they used LBP for restoring degraded images and assessed the accuracy of LBP by an estimated hyperparameter that maximizes the marginal likelihood of Gaussian graphical models. The results of this paper correspond to performing an analytical approach to the studies.

Specifically, in this paper, we analytically give the extrema of Bethe free energy when the inverse covariance matrix of Gaussian distributions forms a single loop (section **3.1**). Then, we give a necessary condition for LBP convergence and show what quantity determines the accuracy of LBP (section **3.1**). Besides, in the case of the single loop, we mention the necessary and sufficient condition for LBP convergence (section **4**). Subsequently, when the inverse covariance matrix forms an arbitrary graph containing multiloops, we analytically give the expansions at small interactions (section **3.2**). Based on the analytical results, we perform numerical experiments of LBP for each case (section **5**). Finally, conclusion and some future works follow (section **6**).

2 Bethe Approximation

Here, we review the mathematical framework of Bethe free energy.

2.1 Bethe Free Energy

Let the target distribution whose marginal distributions we wish to compute be the distribution of

$$p(\boldsymbol{x}) = \frac{1}{Z} \prod_{\{ij\} \in B} W_{ij}(x_i, x_j), \tag{1}$$

where $\boldsymbol{x} \in R^d$, the function $W_{ij}(x_i, x_j)$ represents the strength of interaction among variables x_i and x_j, the set B includes all pairs of two variables among which there exists interaction, and Z is the normalization constant. When we approximate the target distribution by a trial distribution $r(\boldsymbol{x})$, the minimization of Kullback Leibler divergence from $p(\boldsymbol{x})$ to $r(\boldsymbol{x})$ is equivalent to that of the functional.

$$F(r) = -S(r) - \sum_{\{ij\} \in B} \int r_{ij} \log W_{ij}(x_i, x_j) dx_i dx_j,$$

$$S(r) \equiv -\int r(\boldsymbol{x}) \log r(\boldsymbol{x}) d\boldsymbol{x}$$

with respect to $r(\boldsymbol{x})$. In Bethe approximation, the entropy term $S(r)$ is approximated by $S_{Bethe}(\{r_i\}, \{r_{ij}\})$ as

$$S(r) \simeq S_{Bethe}(\{r_i\}, \{r_{ij}\}) \equiv \sum_{i=1}^{d} S_i(r_i) + \sum_{\{ij\} \in B} \{S_{ij}(r_{ij}) - S_i(r_i) - S_j(r_j)\}$$

$$= \sum_{\{ij\} \in B} S_{ij}(r_{ij}) - \sum_{i=1}^{d} (|\mathcal{N}_i| - 1) S_i(r_i),$$

where r_i and r_{ij} are marginal densities $r_i(x_i)$, $r_{ij}(x_i, x_j)$ of $r(\boldsymbol{x})$ and entropies $S_i(r_i)$, $S_{ij}(r_{ij})$ are defined by $-\int r_i \log r_i dx_i$ and $-\int r_{ij} \log r_{ij} dx_i dx_j$, respectively. The set $\mathcal{N}_i$ includes all variables $\{x_j\}$ which directly interact with x_i and $|\mathcal{N}_i|$ is the number of elements of the set $\mathcal{N}_i$. Under the approximation, Bethe free energy is defined by a functional of marginal densities $\{r_i\}$, $\{r_{ij}\}$ as

$$F_{Bethe}(\{r_i\}, \{r_{ij}\}) \equiv -S_{Bethe}(\{r_i\}, \{r_{ij}\}) - \sum_{\{ij\} \in B} \int r_{ij} \log W_{ij} dx_i dx_j.$$

The optimal marginal densities $\{r_i^*\}$, $\{r_{ij}^*\}$ that give the extrema of Bethe free energy are equivalent to those computed by LBP. By considering the variational condition of Bethe free energy, we obtain the extrema as follows:

$$r_i^* = \frac{1}{Z_i} \prod_{k \in \mathcal{N}_i} \exp\left\{-\frac{\eta_{ik,i}(x_i)}{|\mathcal{N}_i| - 1}\right\}, \qquad i \in \{1, \cdots, d\},$$

$$r_{ij}^* = \frac{1}{Z_{ij}} W_{ij} \exp\left\{-\eta_{ij,i}(x_i) - \eta_{ij,j}(x_j)\right\}, \qquad \{ij\} \in B.$$

Here, $\eta_{ij,i}(x_i)$, $\eta_{ij,j}(x_j)$, $(\{ij\} \in B)$ are Lagrange multipliers in order for functions r_i^* and r_{ij}^* to satisfy marginal properties of each other. When we change the multipliers as $\exp\{-\eta_{ik,i}(x_i)\} = \prod_{l \in \mathcal{N}_i \setminus \{k\}} \mathcal{M}_{l \to i}(x_i)$, the extrema of Bethe free energy are expressed by

$$r_i^* = \frac{1}{Z_i} \prod_{k \in \mathcal{N}_i} \mathcal{M}_{k \to i}(x_i), \quad r_{ij}^* = \frac{W_{ij}}{Z_{ij}} \left(\prod_{k \in \mathcal{N}_i \setminus \{j\}} \mathcal{M}_{k \to i}(x_i)\right) \left(\prod_{k \in \mathcal{N}_j \setminus \{i\}} \mathcal{M}_{k \to j}(x_j)\right). \quad (2)$$

The expression is equivalent to those which appear in LBP. The functions $\mathcal{M}_{i \to j}$, $\mathcal{M}_{j \to i}$, $(\{ij\} \in B)$ are called messages in LBP. Messages which are necessary for computing $\{r_i^*\}$ and $\{r_{ij}^*\}$ are determined by the fixed points $\mathcal{M}_{i\to j}^*, \mathcal{M}_{j\to i}^*$, $(\{ij\} \in B)$ which are the results of the update equations:

$$\mathcal{M}_{i\to j}^{(t+1)}(x_j) = \frac{1}{\tilde{Z}_{ij}} \int_{-\infty}^{\infty} W_{ij}(x_i, x_j) \prod_{k \in \mathcal{N}_i \setminus \{j\}} \mathcal{M}_{k\to i}^{(t)}(x_i) dx_i,$$

$$\mathcal{M}_{j\to i}^{(t+1)}(x_i) = \frac{1}{\tilde{Z}_{ji}} \int_{-\infty}^{\infty} W_{ij}(x_i, x_j) \prod_{k \in \mathcal{N}_j \setminus \{i\}} \mathcal{M}_{k\to j}^{(t)}(x_j) dx_j, \quad \{ij\} \in B. \quad (3)$$

2.2 Bethe Free Energy in Gaussian Distributions

Since we focus on the Bethe free energy in Gaussian distributions, We regard the target distribution (1) as multi-dimensional Gaussian distributions with mean vector $\mathbf{0}$ and inverse covariance matrix S, that is,

$$p(\mathbf{x}) = \sqrt{\frac{\det S}{(2\pi)^d}} \exp\left\{-\frac{1}{2}\mathbf{x}^T S \mathbf{x}\right\}.$$

Then, we write $p(\mathbf{x}) \sim N(\mathbf{0}, S)$ and components of the matrix S as $(S)_{ij} = s_{i,j}$. Then, the function W_{ij} can be expressed by a form of

$$W_{ij}(x_i, x_j) = \exp\left\{-\frac{1}{2}(\tilde{s}_{i,i} x_i^2 + 2 s_{i,j} x_i x_j + \tilde{s}_{j,j} x_j^2)\right\},$$

where $\tilde{s}_{i,i}$ denotes $\tilde{s}_{i,i} \equiv \frac{s_{i,i}}{|\mathcal{N}_i|}$ and $|\mathcal{N}_i| > 0$ for $\forall i$.

We set the probability densities of messages as Gaussian distributions given by $\mathcal{M}_{i\to j} \sim N(0, \lambda_{i\to j})$, $\mathcal{M}_{j\to i} \sim N(0, \lambda_{j\to i})$, $(\{ij\} \in B)$. Then, update equations of messages (3) reduce to those of inverse variances $\lambda_{i\to j}, \lambda_{j\to i}, (\{ij\} \in B)$ as follows:

$$\bar{\lambda}_{i\to j}^{(t+1)} = -\frac{\bar{s}_{ij}^2}{1 + \sum_{k \in \mathcal{N}_i \setminus \{j\}} \bar{\lambda}_{k\to i}^{(t)}}, \quad \bar{\lambda}_{j\to i}^{(t+1)} = -\frac{\bar{s}_{ij}^2}{1 + \sum_{k \in \mathcal{N}_j \setminus \{i\}} \bar{\lambda}_{k\to j}^{(t)}}, \quad \{ij\} \in B. \quad (4)$$

Here, $\bar{\lambda}_{i\to j}$ and $\bar{s}_{ij}$ denote $\bar{\lambda}_{i\to j} \equiv \frac{\lambda_{i\to j}}{s_{jj}} - \frac{1}{|\mathcal{N}_j|}$ and $\bar{s}_{ij} \equiv s_{ij}/\sqrt{s_{ii}}\sqrt{s_{jj}}$, respectively. After computing the fixed points of inverse variances $\lambda_{i\to j}^*, \lambda_{j\to i}^*$, $(\{ij\} \in B)$, equivalently, the fixed points of messages $\mathcal{M}_{i\to j}^*, \mathcal{M}_{i\to j}^*, (\{ij\} \in B)$, we obtain the optimal marginal densities $\{r_i^*\}$ and $\{r_{ij}^*\}$ by substituting those messages to eqs.(2). The marginal densities $\{r_i^*\}, \{r_{ij}^*\}$ give the extrema of Bethe free energy and the fixed points of LBP in Gaussian distributions.

3 Analytical Solutions

Here, we analytically give the optimal marginal densities $\{r_i^*\}$ and $\{r_{ij}^*\}$ in two particular cases of Gaussian distributions. All the proofs in this section are written in [11].

3.1 When Matrix S Forms a Single Loop

At first, we consider the case that inverse covariance matrix S forms a single loop. Then, without loss of generality, the set B can be restricted to $B \in \{\{12\}, \{23\}, \cdots, \{d-1\,d\}, \{d1\}\}$ and we can obtain the following results.

Theorem 1. *When the inverse covariance matrix S forms a single loop, the optimal messages $\mathcal{M}^*_{i \to j}$, $\mathcal{M}^*_{j \to i}$, ($\{ij\} \in B$) are given by*

$$\mathcal{M}^*_{i \to i+1}(x_{i+1}) \sim N(0, \frac{s_{i,i+1}\Delta_{i,i+1} - s_{i+1,i+2}\Delta_{i+1,i+2} \pm \sqrt{\mathcal{D}}}{2\Delta_{i+1,i+1}}),$$

$$\mathcal{M}^*_{i \to i-1}(x_{i-1}) \sim N(0, \frac{s_{i,i-1}\Delta_{i,i-1} - s_{i-1,i-2}\Delta_{i-1,i-2} \pm \sqrt{\mathcal{D}}}{2\Delta_{i-1,i-1}}),$$

$$\mathcal{D} \equiv (\det S)^2 + (-1)^d 4 s_{1,2} s_{2,3} \cdots s_{d-1,d} s_{d,1} \det S$$

for $i \in \{1, \cdots, d\}$. Here, $\{\Delta_{i,j}\}$ are cofactors of the matrix S and periodic boundary conditions hold, i.e., $s_{d,d+1} \equiv s_{d,1}$, $s_{1,0} \equiv s_{1,d}$, $\Delta_{d,d+1} \equiv \Delta_{d,1}$, $\Delta_{1,0} \equiv \Delta_{1,d}$.

The above theorem immediately tells us the optimal marginal densities $\{r_i^*\}$ and $\{r_{ij}^*\}$ as follows.

Theorem 2. *When the inverse covariance matrix S forms a single loop, the optimal marginal densities $\{r_i^*\}$ and $\{r_{ij}^*\}$ are given by*

$$r_i^* \sim N(0, \frac{\det S}{\Delta_{i,i}}\sqrt{1+\epsilon}), \quad r_{i,i+1}^* \sim N\left(\mathbf{0}, \begin{bmatrix} \frac{E_{i,i+1}}{\Delta_{i,i}} & s_{i,i+1} \\ s_{i,i+1} & \frac{E_{i,i+1}}{\Delta_{i+1,i+1}} \end{bmatrix}\right),$$

where ϵ and $E_{i,i+1}$ are given by

$$\epsilon \equiv \frac{(-1)^d 4 s_{1,2} \cdots s_{d,1}}{\det S}, \quad E_{i,i+1} \equiv \frac{\det S + \sqrt{\mathcal{D}}}{2} - s_{i,i+1}\Delta_{i,i+1},$$

respectively.

While the analytical result in theorem 2 gives the minimum of Bethe free energy, the true marginal densities are given by $p_i(x_i) \sim N(0, \det S/\Delta_{i,i})$. By comparing both inverse variances, we know that the difference is characterized by the value of ϵ. In other words, the value ϵ decides the accuracy of Bethe approximation. Bethe approximation offers high performance if ϵ is small enough but not if otherwise. In order for marginal densities $\{r_i^*\}$, $\{r_{ij}^*\}$ to be equal to the true marginal densities, the value ϵ has to be 0, equivalently, one of the values from $s_{1,2}$ to $s_{d,1}$ must be 0. This condition means that a single loop turns to a tree graph. Hence, the condition explains the fact that BP applied to a tree graph computes precise marginal probabilities. In addition, in order for the fixed point $\{r_i^*\}$ and $\{r_{ij}^*\}$ to exist, the value ϵ must satisfy $\epsilon \geq -1$. Formally, we obtain the following necessary condition for LBP convergence.

Condition 1. *When the inverse covariance matrix S forms a single loop, $\epsilon \geq -1$ must be satisfied for LBP convergence.*

The results in this section can easily be extended to more general graphs that have a single loop and arbitrary tree structures [12]. Then, the value ϵ also plays a crucial role in determining the accuracy of Bethe approximation in the Gaussian distributions.

3.2 When Matrix S Forms an Arbitrary Graph

When the inverse covariance matrix S forms an arbitrary graph, we give the expansions of the optimal marginal densities $\{r_i^*\}$ at small interactions. Specifically, we introduce a new parameter s and change the matrix S as $S = S_d + sS_o$, where matrix S_d is the diagonal matrix which has the same diagonal components as the original matrix S and matrix S_o has the same off-diagonal components as the matrix S and all diagonal components are 0. Parameter s can adjust the strength of interactions among variables. Then, we give the expansions of optimal marginal densities $\{r_i^*\}$ with respect to s up to third order. We obtain the following results.

Theorem 3. *When the inverse covariance matrix S forms an arbitrary graph, the optimal marginal densities $\{r_i^*\}$ are expanded as follows with respect to parameter s.*

$$r_i^* \sim N(0, \Lambda_i^*(s)), \quad \Lambda_i^*(s) = s_{i,i} - \sum_{j=1(\neq i)}^{d} \frac{s_{j,i}^2}{s_{j,j}} s^2 + O(s^4). \tag{5}$$

While the analytical result in theorem 3 gives the expansions of an extremum of Bethe free energy, the true marginal densities p_i are expanded as follows.

$$p_i \sim N(0, \det S/\Delta_{i,i}), \quad \frac{\det S}{\Delta_{i,i}} = s_{i,i} - \sum_{j=1(\neq i)}^{d} \frac{s_{j,i}^2}{s_{j,j}} s^2 + \delta_i s^3 + O(s^4),$$

$$\delta_i \equiv \frac{s_{i,i}}{3}[\text{tr}(S_d^{-1}S_o)^3 - \text{tr}\{(S_d)_{i,i}^{-1}(S_o)_{i,i}\}^3], \tag{6}$$

where $(S_d)_{i,i}, (S_o)_{i,i}$ are minor matrices of S_d and S_o with the i-th row and column deleted, respectively. By comparing the approximate and the true expansions, we know that the coefficients accord with each other up to second order but differ from third order. Hence, the difference is mainly characterized by the values $\{\delta_i\}$ at small interactions ($s \ll 1$) and the values determine the accuracy of Bethe approximation in the Gaussian distributions.

4 LBP Convergence

Since the fixed points of LBP are clarified for the graph of a single loop in section **3.1**, we mention the LBP convergence towards the fixed points.

4.1 When Matrix S Forms a Single Loop

To examine the LBP convergence, we expand the update equations (4) around the fixed points $\{\bar{\lambda}^*_{i \to j}\}$ and make the update equations (4) linear. Under we define the displacement of the inverse variances $\{\bar{\lambda}^{(t)}_{i \to j}\}$ as

$$\Delta\bar{\lambda}^{(t)}_{i \to i+1} \equiv \bar{\lambda}^{(t)}_{i \to i+1} - \bar{\lambda}^*_{i \to i+1}, \quad \Delta\bar{\lambda}^{(t)}_{i+1 \to i} \equiv \bar{\lambda}^{(t)}_{i+1 \to i} - \bar{\lambda}^*_{i+1 \to i}, \quad i \in \{1, 2, \cdots, d\}$$

and also the displacement vector $\Delta\bar{\boldsymbol{\lambda}}^{(t)}$ as

$$\Delta\bar{\boldsymbol{\lambda}}^{(t)} \equiv (\Delta\bar{\lambda}^{(t)}_{1 \to 2}, \cdots, \Delta\bar{\lambda}^{(t)}_{d \to 1}, \Delta\bar{\lambda}^{(t)}_{2 \to 1}, \cdots, \Delta\bar{\lambda}^{(t)}_{1 \to d})^T,$$

the update equations (4) are linearized near the fixed points as

$$\Delta\bar{\boldsymbol{\lambda}}^{(t+1)} = A_{BP}\Delta\bar{\boldsymbol{\lambda}}^{(t)}, \quad A_{BP} \equiv \begin{pmatrix} \overline{A}_{BP} & O \\ O & \underline{A}_{BP} \end{pmatrix},$$

$$\overline{A}_{BP} \equiv \begin{pmatrix} 0 & & & a_{1 \to 2} \\ a_{2 \to 3} & 0 & & \\ & \ddots & \ddots & \\ & & a_{d \to 1} & 0 \end{pmatrix}, \quad \underline{A}_{BP} \equiv \begin{pmatrix} 0 & a_{2 \to 1} & & \\ & 0 & \ddots & \\ & & \ddots & a_{d \to d-1} \\ a_{1 \to d} & & & 0 \end{pmatrix},$$

where components $\{a_{i \to j}\}$ of matrices $\overline{A}_{BP}, \underline{A}_{BP}$ are given by

$$a_{i \to i+1} \equiv \frac{\bar{s}^2_{i,i+1}}{(1 + \bar{\lambda}^*_{i-1 \to i})^2}, \quad a_{i \to i-1} \equiv \frac{\bar{s}^2_{i,i-1}}{(1 + \bar{\lambda}^*_{i+1 \to i})^2}, \quad i \in \{1, 2, \cdots, d\}.$$

The eigenvalues $\bar{\eta}, \underline{\eta}$ of matrices $\overline{A}_{BP}, \underline{A}_{BP}$ are given by

$$\bar{\eta} = \left(\prod_{j=1}^{d} a_{j \to j+1}\right)^{\frac{1}{d}} \exp\left\{i\frac{\pi(d+1)}{d} + i\frac{2\pi}{d}m\right\},$$

$$\underline{\eta} = \left(\prod_{j=1}^{d} a_{j \to j-1}\right)^{\frac{1}{d}} \exp\left\{i\frac{\pi(d+1)}{d} + i\frac{2\pi}{d}m\right\}, \quad m \in \{0, \cdots, d-1\},$$

respectively[1]. From the eigenvalues $\bar{\eta}, \underline{\eta}$, the necessary and sufficient condition for LBP convergence, when the vector $\bar{\boldsymbol{\lambda}}^{(t)}$ is close enough to the fixed points $\bar{\boldsymbol{\lambda}}^*$, is given by

$$\left(\prod_{j=1}^{d} a_{j \to j+1}\right)^{\frac{1}{d}} < 1 \quad \text{and} \quad \left(\prod_{j=1}^{d} a_{j \to j-1}\right)^{\frac{1}{d}} < 1.$$

Thus, the LBP convergence is characterized by the geometric mean of $a_{i \to j}$.

---

[1] Here i denotes the imaginary unit.

5 Numerical Experiments

Here, we perform some numerical experiments of LBP in Gaussian distributions and compare the results with the analytical solutions in section **3**.

5.1 When Matrix S Forms a Single Loop

At first, we perform the numerical experiments of LBP when matrix S forms a single loop. We show examples of the matrix S where the marginal probabilities computed by LBP are comparatively different from the true values. When we consider the matrices of

$$S_1 \equiv \begin{pmatrix} 10 & -9.9 & 1 \\ -9.9 & 10 & 0.4 \\ 1 & 0.4 & 10 \end{pmatrix}, S_2 \equiv \begin{pmatrix} 10 & 9 & 0 & 1 \\ 9 & 10 & 4 & 0 \\ 0 & 4 & 10 & 1.9 \\ 1 & 0 & 1.9 & 10 \end{pmatrix},$$

the experiments of LBP result in figures 1,2, respectively. Each matrix has high correlations among the variables. In figures 1,2, dots indicate the inverse variances Λ_3, Λ_4 of marginal densities r_3, r_4 computed by LBP depending on the iteration counts of message update equations (4), respectively. Dashed lines indicate the analytical inverse variances which we give in theorem 2. The other straight lines indicate the true inverse variances. From figures 1,2, both results of LBP converge to the analytical solutions in theorem 2 and yield discrepancies from the true values. The discrepancies are caused by the values ϵ for matrices S_1, S_2. The values ϵ are calculated as $\epsilon = 41.684211$ and $\epsilon = 25.786993$, respectively. Since the values ϵ are large, LBP does not promise the high performances.

Next, we consider the values ϵ depending on dimension d. Practically, we consider the following $d \times d$ matrices

$$S_3(d) = \begin{pmatrix} 10 & 4.9 & 0 & \cdots & 0 & 4.9 \\ 4.9 & 10 & 4.9 & 0 & \cdots & 0 \\ 0 & 4.9 & 10 & 4.9 & 0 & \cdots \\ \vdots & \ddots & \ddots & \ddots & \ddots & \vdots \\ 0 & \cdots & 0 & 4.9 & 10 & 4.9 \\ 4.9 & 0 & \cdots & 0 & 4.9 & 10 \end{pmatrix}, S_4(d) = \begin{pmatrix} 10 & -4.9 & 0 & \cdots & 0 & -4.9 \\ -4.9 & 10 & -4.9 & 0 & \cdots & 0 \\ 0 & -4.9 & 10 & -4.9 & 0 & \cdots \\ \vdots & \ddots & \ddots & \ddots & \ddots & \vdots \\ 0 & \cdots & 0 & -4.9 & 10 & -4.9 \\ -4.9 & 0 & \cdots & 0 & -4.9 & 10 \end{pmatrix}.$$

Then, values ϵ depending on dimension d result in figures 3,4. For each figure 3,4, the value ϵ approaches 0 exponentially with the increases of dimension d. Hence, the larger dimension d becomes, the more accurate LBP becomes for the graph of a single loop. It is interesting from figure 3 that the value ϵ, which decides the accuracy of LBP, fluctuates depending on whether dimension d is even or odd.

5.2 When Matrix S Forms an Arbitrary Graph

Here, we consider the inverse covariance matrix S that forms an arbitrary graph but the correlations among the variables are small. Then, from theorem 3, we

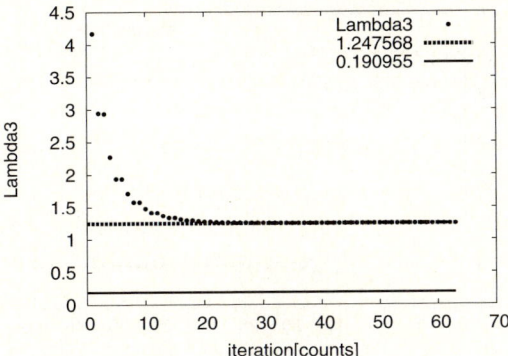

Fig. 1. Convergence of inverse variance Λ_3 of marginal density r_3 in matrix S_1. Vertical axis shows Λ_3. Horizontal axis shows iteration counts of eq.(4). Dashed line and the other straight line show analytical solution in theorem 2 and the true inverse variance, respectively.

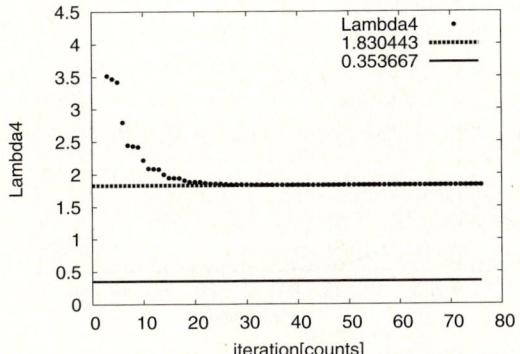

Fig. 2. Convergence of inverse variance Λ_4 of marginal density r_4 in matrix S_2. Vertical axis shows Λ_4. Horizontal axis shows iteration counts of eq.(4).

know that the values $\{\delta_i\}$ play a major role in determining the accuracy of Bethe approximation. In order to grasp the properties of $\{\delta_i\}$, we examine the values $\{\delta_i\}$ for some particular matrices S. When matrix S forms either a tree graph or a single loop, $\{\delta_i\}$ satisfy $\delta_i = 0$ for $i \in \{1, \cdots, d\}$. Hence, the values $\{\delta_i\}$ involve more complex information than that of a tree graph or a single loop. Subsequently, when matrix S forms an uniformly and fully connected graph where the diagonal and off-diagonal components are uniformly composed of s_0 and s, respectively, the values $\{\delta_i\}$ are calculated as $\delta_i = (d-1)(d-2)\frac{1}{s_0^2}$ for $i \in \{1, \cdots, d\}$. We denote the matrix as $S(s_0, s, d)$. Since the values $\{\delta_i\}$ increase quadratically with dimension d, LBP degrades the accuracy with the increases of dimension d.

Practically, we consider the series of matrices $S(10, 0.1, d)$, which express an uniformly and fully connected graph. Then, we depicted several quantities of inverse variances depending on dimension d in figure 5. Black square dots indicate

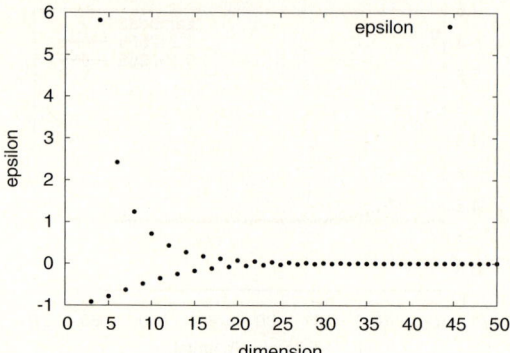

Fig. 3. The value ϵ depending on dimension d in the series of matrices $S_3(d)$. Vertical axis is ϵ. Horizontal axis is d.

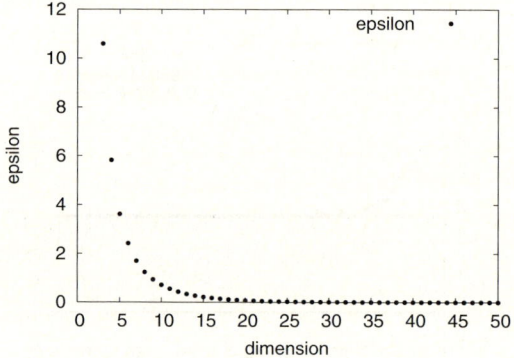

Fig. 4. The value ϵ depending on dimension d in the series of matrices $S_4(d)$. Vertical axis is ϵ. Horizontal axis is d.

the true inverse variances and open square dots indicate the numerical results of LBP. We know that LBP produces discrepancies from the true values with the increases of dimension d. Open triangle dots indicate the expansions of inverse variances that we give in theorem 3. Since the triangle dots trace the open square dots very well, we know that the expansions in theorem 3 can sufficiently approximate the marginal probabilities computed by LBP. Black triangle dots indicate the results obtained by adding the values $\delta_i s^3$ to the numerical results of LBP. This correction is intended to improve LBP so as to be more accurate and easily suggested by comparing eqs.(5) with eqs.(6). From figure 5, though the corrected results surpass original results, the correction still produces discrepancies from the true values. The discrepancies are caused by higher order terms with respect to s and imply that we need more information of higher order terms when dimension d is large.

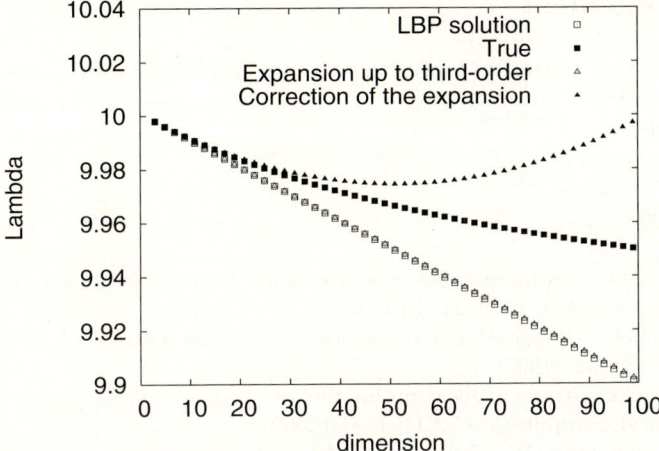

Fig. 5. Several inverse variances depending on dimension d in the matrices $S(10, 0.1, d)$. Black square dots show true inverse variances. Open square dots show the results of LBP. Open triangle dots show the expansion in theorem 3. Black triangle dots show the results of a correction of LBP.

6 Conclusion and Future Works

In this paper, we analyzed the Bethe free energy in Gaussian distributions and analytically clarified the extremum. Especially, when the inverse covariance matrix of Gaussian distribution forms a single loop, we gave a necessary condition for LBP convergence and revealed that the value ϵ determines the accuracy of Bethe approximation. When the inverse covariance matrix forms an arbitrary graph but the interactions among the variables are small, we revealed that the values $\{\delta_i\}$ mainly determine the accuracy of Bethe approximation. Based on the analytical results, we performed numerical experiments of LBP and verified that the numerical results agree with analytical results very well. This paper contributes to clarifying the theoretical properties underlying LBP.

Some of future works are as follows. In association with LBP, Convex concave procedure (CCCP) is known to be a convergent algorithm for minimizing Bethe free energy [13]. To compare CCCP with LBP enables us to specify the difference of their dynamics towards the fixed points. In probabilistic image processing, all the accuracies of mean field approximation (MFA), Bethe approximation, generalized BP (GBP), and a method based on TAP free energy have numerically been studied [14]. To evaluate the accuracies analytically and compare those with this paper enables us to know the differences of their properties and their performances. Another is to establish correction algorithms for LBP based on the analytical results in this paper.

Acknowledgement

This research was partially supported by the Ministry of Education, Science, Sports and Culture, Grant-in-Aid for JSPS Fellows and for Scientific Research 18079007, 2007.

References

1. Heskes, T.: On the uniqueness of loopy belief propagation fixed points. Neural Computation 16(11), 2379–2414 (2004)
2. Weiss, Y.: Belief propagation and revision in networks with loops. Technical report 1616, MIT AI lab (1997)
3. Weiss, Y.: Correctness of local probability propagation in graphical models with loops. Neural Computation 12(1), 1–41 (2000)
4. Weiss, Y., Freeman, W.: Correctness of belief propagation in gaussian graphical models of arbitrary topology. Neural Computation 13(10), 2173–2200 (2001)
5. Ikeda, S., Tanaka, T., Amari, S.: Stochastic reasoning, free energy, and information geometry. Neural Computation 16(9), 1779–1810 (2004)
6. Ikeda, S., Tanaka, T., Amari, S.: Information geometry of turbo and low-density parity-check codes. IEEE Trans. Inf. Theory 50(6), 1097–1114 (2004)
7. Kabashima, Y., Saad, D.: The TAP approach to intensive and extensive connectivity systems. In: Opper, M., Saad, D. (eds.) Advanced Mean Field Methods -Theory and Practice, pp. 65–84. MIT Press, Cambridge (2001)
8. Yedidia, J., Freeman, W., Weiss, Y.: Bethe free energy, kikuchi approximations, and belief propagation algorithms. Technical Report TR2001-16, Mitsubishi Electric Research Laboratories (2001)
9. Tanaka, K., Shouno, H., Okada, M.: Accuracy of the bethe approximation for hyperparameter estimation in probabilistic image processing. J.Phys. A, Math. Gen. 37(36), 8675–8696 (2004)
10. Tanaka, K., Inoue, J., Titterington, D.M.: Loopy belief propagation and probabilistic image processing. In: proceedings of 2003 IEEE International workshop on Neural Networks for Signal Processing, vol. 13, pp. 383–392 (2003)
11. Nishiyama, Y., Watanabe, S.: Theoretical Analysis of Accuracy of Belief Propagation in Gaussian Models. IEICE Technical Report 107(50), 23–28 (2007)
12. Nishiyama, Y., Watanabe, S.: Theoretical Analysis of Accuracy of Gaussian Belief Propagation. In: Proceedings of International Conference on Artificial Neural Networks 2007, pp. 29–38 (2007)
13. Yuille, A.L.: CCCP algorithms to minimize the Bethe and Kikuchi free energies: convergent alternatives to belief propagation. Neural Computation 14(7), 1691–1722 (2002)
14. Tanaka, K.: Generalized belief propagation formula in probabilistic information processing based on gaussian graphical model. IEICE D-II, Vol. J88-D-II, No. 12, pp. 2368–2379 (in Japanese) (2005)

An Application of Causality for Representing and Providing Formal Explanations about the Behavior of the Threshold Accepting Algorithm

Joaquín Pérez[1], Laura Cruz[2], Rodolfo Pazos[1], Vanesa Landero[1],
Gerardo Reyes[1], Héctor Fraire[1], and Juan Frausto[3]

[1] Centro Nacional de Investigación y Desarrollo Tecnolgico (CENIDET)
Departamento de Ciencias Computacionales
AP 5-164, Cuernavaca 62490. México
[2] Instituto Tecnológico de Ciudad Madero (ITCM)
División de Estudios de Posgrado e Investigación. Cd. Madero México
[3] Instituto Tecnológico de Estudios Superiores de Monterrey (ITESM)
Departamento de Ciencias Computacionales. Cuernavaca, 62589, México
{jperez,pazos,greyes}@cenidet.edu.mx

Abstract. The problem of algorithm selection for solving NP problems arises with the appearance of a variety of heuristic algorithms. The first works claimed the supremacy of some algorithm for a given problem. Subsequent works revealed the supremacy of algorithms only applied to a subset of instances. However, it was not explained why an algorithm solved better a subset of instances. In this respect, this work approaches the problem of explaining through causal model the interrelations between instances characteristics and the inner workings of algorithms. For validating the results of the proposed approach, a set of experiments was carried out in a study case of the Threshold Accepting algorithm to solve the Bin Packing problem. Finally, the proposed approach can be useful for redesigning the logic of heuristic algorithms and for justifying the use of an algorithm to solve an instance subset. This information could contribute to algorithm selection for NP problems.

1 Introduction

In many organizations there exist a large number of complex optimization problems. Such problems are known to the scientific community in computer science as combinatorial optimization problems and belong to the NP-hard class. Heuristic algorithms have been proposed as a good alternative for solving very large instances [1,2]. Unfortunately, in real-life situations, there is usually no algorithm that outperforms all the other algorithms for all instances [3], and therefore, the problem of selecting the best algorithm arises. The works related to this topic have tried to analyze the behavior of heuristic algorithms for solving the selection problem.

Cohen comments in his book [4] that the explanation of an observed behavior evolves through three stages: description (D), prediction (P) and causality (C).

Therefore, in the case of behavior analysis of heuristic algorithms, the evolution through these stages has taken place through the following approaches. Initially, the works reported by the scientific community focused on showing the superiority of an algorithm for some specific problem [5]. Subsequent works found domination regions for algorithms; i.e., instances subsets of a problem where each algorithm outperforms the others. Additionally, they used this knowledge for predicting the best algorithm for a new instance of the problem [6]. Currently, there exist some few works that provide informal explanations for algorithms behavior [7]. However, up to now the problem of explaining formally why an algorithm is the best for solving an instances set had remained open.

1.1 A Comparative Table of Related Works

Some related works are presented in Table 1. Column 2 indicates for each work the analysis type: descriptive (D), predictive (P) and causal (C). Columns 3 and 4 indicate whether the algorithm analysis includes information from the description of the problem instance (PD) or from a sample of the solution space of a problem instance (PS). Columns 5, 6 and 7 indicate whether the analysis considers the information from: algorithm behavior (AB), search trajectory (ST) and algorithm structure (AS). Column 8 indicates if the works present formal explanations of the algorithm performance.

Table 1. Related works

Work	Analysis Type	Problem Indicators		Algorithm Indicators		AS	Formal Explanation
		PD	PS	AB	ST		
Hoos 1998 [5]	D			✓			
Soares 2003 [8]	P	✓					
Pérez 2004 [6]	P	✓					
Hoos 2004 [9]	C	✓			✓		
Lemeire 2004 [7]	C			✓			✓
Konak 2005 [10]	C					✓	
Pérez 2007 [11]	C		✓	✓			✓
This Paper	C	✓	✓	✓	✓	✓	✓

The works presented in [6,8,12,13] used decision tree or k-nearest neighborhood algorithms for predicting the best algorithm for a new problem instance. The works reported in [9,10] provide informal explanations through a tabular analysis of experimental results of algorithm behavior on instances of the problem. Other works [7,11] carry out a causal analysis using structure and parameter learning algorithms, which provide a formal model of algorithm behavior. The work in [11] explains formally why an algorithm performs better in solving an instance set; however, its explanation is limited since it does not include other indicators of the problem and algorithm. A survey of the specialized literature, revealed the inexistence of a formal model that explains the association between indicators of problem instances and indicators of the characteristics of an algorithm that solves the instances successfully. The solution to this problem is important, since it may provide a solid foundation for the selection of algorithms for solving given instances of NP-hard problems. Therefore, the problem of explaining why an algorithm dominates in an instance region is approached in this paper. The causal analysis seems to be promising for providing a solution to this problem (see Table 1). The solution approach presented permits systematically finding relations between influencing indicators (columns 3-6 of Table 1) of dominance of an algorithm and the inner workings of the algorithm (column 7), in order to provide formal explanations through causal analysis.

2 General Solution Approach

Figure 1 shows the principal components of solution approach. First, information on the instances that constitute the study case is characterized by indicators of the parameters of the problem instance (PD) and indicators of a sample of the problem solutions space (PS). All indicators are explained in section 3.1. After that, two heuristic algorithms of interest are characterized (it is explained in section 3.2) and executed on the instances.

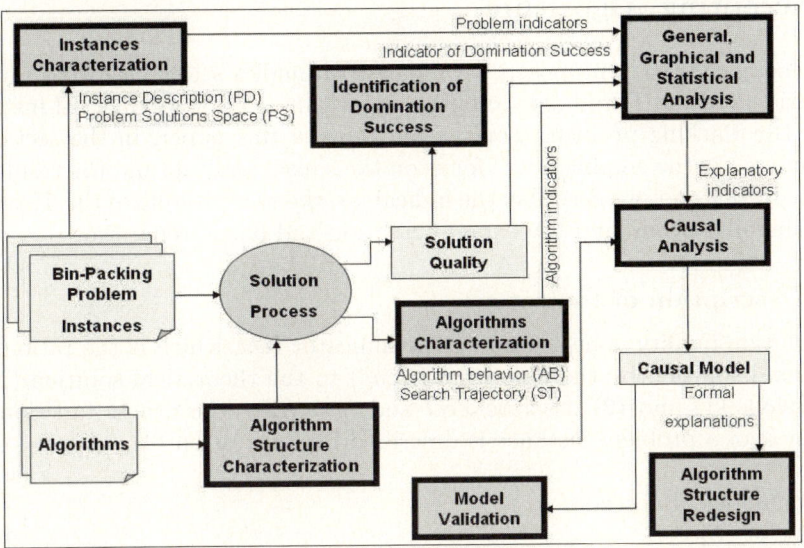

Fig. 1. A general solution approach

The algorithm information is collected during the solution process and characterized by indicators, which constitute: the problem solutions space during execution, the internal behavior of the algorithm (AB), and the search trajectory (ST). Once the solution process ends, the quality indicator of the solution found is calculated for each algorithm in solving each problem instance; with this information the domination success of each algorithm is identified (It is explained in section 3.3). One algorithm is chosen to analyze its performance and explain its behavior. The information represented by the indicators of the chosen algorithm, solution quality, domination success and the problem indicators, are used for a general, graphical and statistical analysis, aiming at finding a minimal set of explanatory indicators (this process is explained in our previous work [11]). Afterwards, a causal analysis is performed using the explanatory indicators, generating a causal model. This model constitutes the relations between the explanatory indicators (that represent instances and algorithm behavior) and the superiority or inferiority of the algorithm to solve them (domination success). From these relations, conditional probability functions are obtained for finding the formal explanations on why an algorithm outperforms another on an instance subset. Finally, we used the information of the internal structure of the analyzed algorithm and the interpretation of these explanations to formulate the redesign in order to improve its performance; and we validated the final causal model to predict if the analyzed algorithm will outperform another when solving a new instance.

3 A Study Case: Analyzing the Performance of Threshold Accepting Algorithm

The general approach proposed is validated through a study case that involves two variants of the Threshold Accepting algorithm [14] for solving 324 instances of the Bin Packing problem. For the purposes of this paper, in the section on the study case, the emphasis will focus on the causal analysis and the results obtained. To this end, we describe: the indicators, the two variants of the Threshold Accepting algorithm, and the experimentation and data preparation.

3.1 Description of Indicators

The solution quality is measured by the indicator $rcal$, which is the ratio of the best found solution by the algorithm ($Zenc$) to the theoretical solution ($Zcal$). Expressions (1) and (2) describe $Zcal$ and $rcal$. Also, the fitness function of a solution x of a problem instance is described by Expression (3), where:

$n = $ number of objects
$w_i = $ size of object i
$c = $ capacity of the container(bin)
$F_j = $ sum of objects in bin j
$numbin = $ number of bins
$x = \{\{F_1, F_2, \ldots, F_{numbin}\}, numbin\}$

$$Zcal = \frac{\sum_{i=1}^{n} w_i}{c} \qquad (1)$$

$$rcal = \frac{Zenc}{Zcal} \qquad (2)$$

$$f(x) = \frac{\left(\sum_{j=1}^{numbin}\left(\frac{F_j}{c}\right)\right)^2}{numbin} \qquad (3)$$

Problem description (PD). Each problem instance is characterized by the indicator cv. It uses the problem parameters to calculate the variation coefficient of the object sizes and it is described by Expression (4).

$$cv = \frac{\sqrt{\frac{1}{n}\sum_{i=1}^{n}\left(w_i - \bar{w}\right)^2}}{\bar{w}} \qquad (4)$$

Problem solutions space (PS). We obtained a sample of the solutions space of each problem instance prior to algorithms experimentation. This sample is built by generating 100 random solutions, where each generated solution x is evaluated by the fitness function $f(x)$. The variability vo of these values is calculated.

Algorithm behavior(AB). The behavior of the algorithm during execution is observed and recorded through indicator pa, which is the number of solutions accepted by the algorithm during the search process.

Search trajectory (ST). The trajectory traced by the algorithm during the search process (solutions generated during execution) is characterized using one of two alternatives. In the first, we used two known indicators: the autocorrelation coefficient ($coef$) and autocorrelation length ($long$), described in [2,9]. In the second, we proposed to characterize the trajectory using three indicators: number nc of inflexion points, number nv of valleys and the average size tm of the valleys. Figure 2 shows an example of the algorithm search trajectory. Axis x represents the steps traced by the algorithm (solutions generated during execution) and axis $f(x)$ represents the values of the fitness function for each solution.

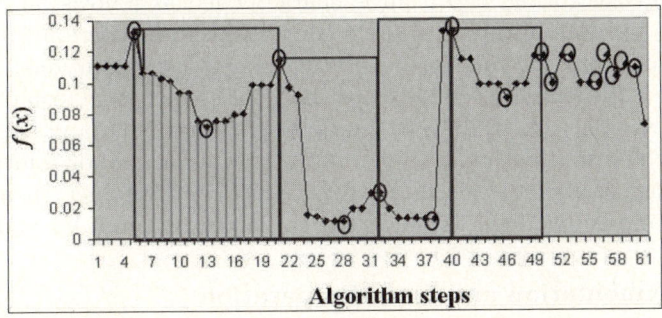

Fig. 2. Identified inflexion points and valleys

The inflexion points are marked by a circle and the identified valleys are marked by a rectangular. A valley is considered by means of Expression (5). The indicator nv is the average of number identified valleys.

$$f(x_{i+1}) \leq f(x_i), 1 \leq i \leq (i+vx), vx \geq 6$$
$$f(x_{vx+(j+1)}) \geq f(x_{vx+j}), 0 \leq j \leq vy, vy \geq 1 \qquad (5)$$

The area under the curve of a valley defined by the finite interval $[i, vy]$ is divided into $(vy\text{-}i)$ subintervals, which are shown in Figure 2. Each rectangular area that corresponds to a subinterval is calculated by Expression (6). The area of an identified valley is calculated by Expression (7), where:

B = base of rectangular area that is located below valley
H = high of rectangular area that is located below valley
b = base of triangular area that is located above valley
h = high of triangular area that is located above valley

$$tm_{11} = (B \times H) - \left(\frac{b \times h}{2}\right) = (1 \times f(x_i)) - \left(\frac{1 \times (f(x_i) - f(x_{i+1}))}{2}\right) \qquad (6)$$

$$tm_1 = \sum_{i=1}^{vy} tm_{1i} \qquad (7)$$

The indicator (tm) is the average of the sizes of identified valleys.

3.2 Description of Variants of Threshold Accepting Algorithm

The internal structure of heuristic algorithms may be characterized by four main aspects: parameter tuning, which can be static or dynamic considering information from the problem instance; initial solution, which can be generated randomly or by some deterministic procedure; neighborhood generation method; and finally, the stop criterion, which may happen after a given number of iterations (DI) or when there is no improvement in the solution (CO). An example of characterization is illustrated in Figure 3. The internal structure of variant V1 of the Threshold Accepting algorithm is characterized as follows: an initial temperature 1, a random initial solution, a convergent stop criterion (until freezing is reached), and one method for generating neighboring solutions: swap (1, 0) [15]. Variant V2 is similar to V1, except that it uses several alternative methods (swap(1, 0), swap(1, 1), swap(1, 2), swap(2, 2), swap(0, 1), and swap(2, 1)), it tries initially the first method, if it can not generate a neighbor solution because the interchange of objects exceeds the capacity of one container; then it tries the second one, and so on. Table 2 shows the characterization of V1 and V2.

3.3 Experimentation and Data Preparation

We randomly selected 324 instances of the Bin Packing Problem defined by [16,17]. Variants V1 and V2 were executed 15 times for each problem instance.

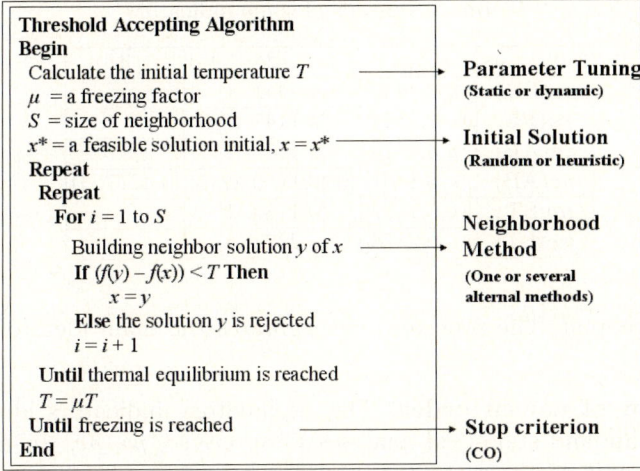

Fig. 3. Main parts of the Threshold Accepting algorithm

Table 2. Variants of the Threshold Accepting algorithm

Variant	Tabu List		Initial Solution		Neighborhood		Stop Criterion	
	Static	Dynamic	Random	Heuristic	One	Several	DI	CO
V1	✓		✓		✓		✓	✓
V2	✓		✓			✓	✓	✓

The indicator of domination success (*success*) of each variant was established using two criteria: First, the variant with the best *rcal* from 15 executions on an instance is considered superior for solving this instance (*success*=1); second, if the variants have the same *rcal*, the variant with the smallest number of evaluations of the fitness function is considered (*success*=1); otherwise (*success*=0). The number of dominant cases for each variant were V1=141 and V2=183. Therefore, we choose the variant V2 to analyzed and explain why it was superior with respect to V1.

Data Preparation. The data were normalized using the min-max method, thus rendering all the values in the interval [0, 1]. Also, the data were discretized by the MDL method [18]. Table 3 shows the intervals (1, 2 and 3) of the indicators. In the following sections, the value ranges for the indicators will be denoted according to the columns of Table 3; for example, the value range [0, 0.4440] that corresponds to column 1 of the row $cv(\mathrm{PD})$ will be denoted by $cv=1$.

3.4 Causal Analysis

The procedure used to build a causal model for variant V2 of the Threshold Accepting algorithm, applied to the solution of the Bin Packing problem, is

Table 3. Intevals of some indicators

Indicators	1	2	3
cv(PD)	[0, 0.4440]	[0.4441, 1]	
vo(PS)	[0, 0.1140]	[0.1141, 1]	
pa(AB)	[0, 0.4510]	[0.4511, 1]	
nc(AB)	[0, 0.1402]	[0.1403, 0.4276]	[0.4277, 1]
nv(ST)	[0, 0.1182]	[0.1183, 1]	
tm(ST)	[0, 0.2378]	[0.2379, 1]	

presented hereupon. The procedure incorporates the main ideas of Cohen and Spirtes [4,19].

Specification of causal order. The explanatory indicators identified from general, graphic and statistical analyses were: cv, vo, pa, nc, nv, tm, $coef$ and $long$. An example of these analyses can be showed in our previous work [11]. The two kinds of indicators for the trajectory traced by the algorithm ($coef$, $long$ and nc, nv, tm) provide the same information about the search trajectory (ST). Therefore, we built two different causal models: the first with information cv, vo, pa, $coef$ and $long$; the second with information cv, vo, pa, nc, nv and tm. The construction of causal models was carried out using the causal inference software TETRAD (www.phil.cmu.edu/projects/tetrad_download/) and the structure learning algorithm PC [19] with a confidence level of 95%, it is shown in Figure 4. PC algorithm starts by generating a complete undirected graph (Line 3), then thins that graph by removing edges with zero order conditional independence relations, thins again with first order, and so on (Lines 4-11). After that, it orients the edges with the evidence found (Lines 12-13). The remaining undirected edges are oriented making sure that no directed cycles occur (Lines 14-17). It is evident from Figure 5 that the first causal model (a) did not yield relevant information about direct causes of the algorithm performance, in terms of the algorithm search trajectory ($coef$, $long$). Conversely, the second causal model (b) shows that the indicators cv, pa, nc and tm are direct causes. For validating this model, we used the causal inference software HUGIN (Hugin Expert, www.hugin.com) and generated another model, using the same information and the PC algorithm (see Figure 6).

Figure 6 confirms that the indicators cv, vo, pa, nc and tm are direct causes of superiority ($success$=1) or inferiority ($success$=0) of variant V2. Therefore, this model is considered for the next analyses presented in this paper.

Estimation of the model. Tables of conditional probability (CPT) of the indicators were calculated using the parameter learning algorithm Maximum Expectation (EM) [20]. We focused on the most important probabilities of direct causes of node $success$. These probabilities and their experience (Exp) are presented in Table 4.

Model interpretation. Causal relations which showed the largest values of conditional probability and experience were interpreted; so that the following

PC Algorithm	
1	**Begin**
2	V = a vertex set, $ADJ(x)$ = a set of vertices adjacent to x, $i=0$
3	Generate the complete undirected graph G on the vertex set V
4	**Repeat**
5	For each $x \in V$
6	For each $y \in ADJ(x)$
7	**If** there is a subset $W \subseteq ADJ(x) - \{y\}$ where
8	$\|W\|=i$, x and y are d-separated [19] given subset W
9	**Then** $Wxy=W$, delete edge $x\text{—}y$
10	$i=i+1$
11	**Until** $\|ADJ(x)\| \leq i, \forall\, x$
12	For each triple of vertices x, y, z
13	**If** $(x\text{—}y\text{—}z)$ and $(y \notin Wxz)$ **Then** orient $x\text{—}y\text{—}z$ as $x{\to}y{\leftarrow}z$
14	**Repeat**
15	**If** $x{\to}z\text{—}y$ **Then** orient $z\text{—}y$ as $z{\to}y$
16	**If** there is a directed path from x to z and an edge **Then** orient $x\text{—}z$ as $x{\to}z$
17	**Until** no more edges can be oriented
18	**End**

Fig. 4. PC Algorithm

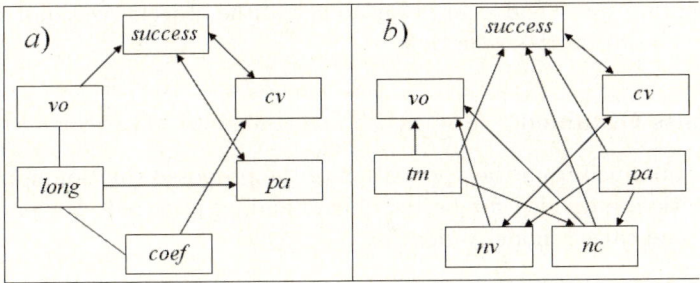

Fig. 5. Causal models generated by Tetrad

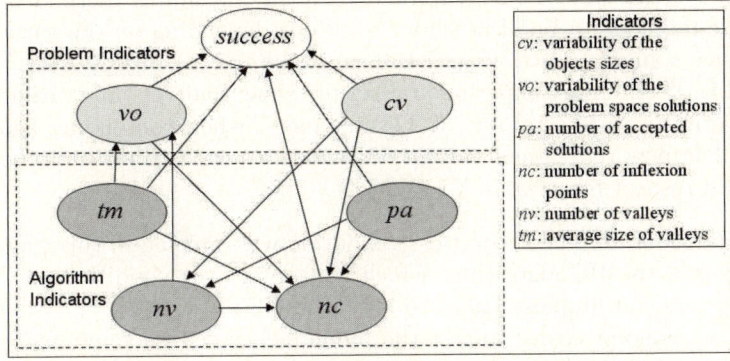

Fig. 6. Causal model generated by Hugin

Table 4. Causal Relations

Causal Relations	%	Exp
P(success=1 |cv=1, vo=2, pa=2, nc=2, tm=2)	89.47	19
P(success=1 |cv=1, vo=2, pa=2, nc=2, tm=2)	100	12
P(success=0 |cv=2, vo=1, pa=1, nc=1, tm=1)	100	17

explanations were inferred from these relations. The explanations obtained reveal that a particular combination of values for the influential indicators (*cv*, *vo*, *pa*, *nc*, *tm*) and its relation with the algorithm structure (V2), characterize the performance in terms of superiority or inferiority for solving problem instances. These quantitative data hinted at concluding that the use of several alternate methods for constructing neighbor solutions by V2 supports a large number of accepted solutions when the variability of the instance solution space is large and the variability of the objects sizes is small. This situation permits the algorithm to intensify its search in the solution space. Therefore, the search trajectory is better suited to the problem solutions space (large number of inflexion points and large valleys), since it allows V2 to enter and get out of large valleys. Conversely, variant V2 is at disadvantage when there is a small variability of the instance solution space and a large variability of the objects sizes and therefore, V1 performs better in this situation.

3.5 Results Obtained

The results obtained from the application of the proposed solution approach for the study case are divided for better understanding into: redesign proposal for variant V2 and causal model validation.

Redesign of algorithm structure. The Threshold Accepting algorithm is described in Figure 7. Variant V2 uses several methods for generating neighboring solutions (only lines 8, 9). The conclusions in section 3.4 allow us to redesign the V2 (V3) algorithm. Thus, V3 could perform one simple method or several alternative methods to build neighbor solutions depending on the variability of the instance solution space *vo* and the variability of the objects sizes *cv* (only lines 10-14). Preliminary experimental results show that the new variant V3 attains superiority in 62.35% out of 324 instances, when contending against V1 and V2. The new redesigned variant V3 yields a performance improvement of 24.7% with respect to variants V1 and V2.

Model Validation. The theoretical validation was omitted in this example because we used the PC algorithm, which guarantees the conditions of Markov: minimality and faithfulness [19]. We used the NETICA software (Norsys Corporation, www.norsys.com) to test the model generated on some instances, for which we ignored whether variant V2 will be superior or not. We obtained a prediction success of 86.29%, which surpasses the values obtained by the classifiers C4.5 (84.67%) and Nave Bayes (81.04%).

	Threshold Accepting Algorithm
1	**Begin**
2	$T = 1$ (initial temperature); $\mu = 0.85$ (a freezing factor)
3	$S = 100$ (size of neighborhood)
4	$x^* =$ an initial feasible solution; $x = x^*$
5	**Repeat**
6	**Repeat**
7	**For** $i = 1$ to S
8	Generate neighbor solution y of x using swap (1, 0) or swap (1, 1) } V2
9	or swap (1, 2) or swap (2, 2) or swap (0, 1) or swap (2, 1)
10	Calculate cv, vo from parameters of problem instance
11	**If** ($cv=1$ and $vo=2$) **Then**
12	Generate neighbor solution y of x using swap (1, 0) or swap (1, 1) } V3
13	or swap (1, 2) or swap (2, 2) or swap (0, 1) or swap (2, 1)
14	**Else** Generate neighbor solution y of x using swap (1, 0)
15	**If** ($f(y) - f(x)) < T$ **Then** $x = y$
16	**Else** the solution y is rejected
17	$i = i + 1$
18	**Until** thermal equilibrium is reached
19	$T = \mu T$
20	**Until** freezing is reached
21	**End**

Fig. 7. Threshold Accepting Algorithm Redesign

4 Conclusions and Future Work

This work presents a new approach for solving the problem of explaining why an algorithm outperforms another on a set of instances through a causal analysis. For validating the proposed approach a set of experiments were carried out, for generating a causal model that showed the interrelation of instances indicators of the Bin Packing problem and structure particularities of the Threshold Accepting algorithm. The most important indicators for problem and algorithm were proposed to build the final causal model. We obtained a prediction percentage of 86.29%. The formal explanations found permitted to devise an improvement to the logic of variant V2 (V3) of the Threshold Accepting algorithm. Preliminary experimental results show that the new redesigned version V3 yields a performance improvement of 24.7% with respect to variants V1 and V2.

Finally, the main contributions of the proposed approach are: advancement of insight on algorithm behavior, permitting to obtain formal explanations on the reasons for an algorithm to outperform another on a set of problem instances; and the formulation of redesign proposals for the internal logic of algorithms for improving their performance. Additionally, we consider that this approach can be useful for justifying the use of an algorithm for solving an instances subset of the Bin Packing problem, and it can contribute to the general problem of algorithms selection for NP problems. As future work we are interested in extending this study to other variants of the Threshold Accepting algorithm and other heuristic algorithms.

References

1. Garey, M.R., Jhonson, D.S.: Computers and Intractability, a Guide to the Theory of NP-completeness. W. H. Freeman and Company, New York (1979)
2. Merz, P., Freisleben, B.: Fitness Landscapes and Memetic Algorithm Design. New Ideas in Optimization, pp. 245–260. McGraw-Hill Ltd. (1999)
3. Wolpert, D.H., Macready, W.G.: No Free Lunch Theorems for Optimization. IEEE Transactions on Evolutionary Computation 1, 67–82 (1997)
4. Cohen, P.: Empirical Methods for Artificial Intelligence. The MIT Press, Cambridge (1995)
5. Hoos, H.: Stochastic Local Search Methods, Models, Applications, PhD Thesis, Department of Computer Science from Darmstadt University of Technology (1998)
6. Pérez, O., Pazos, R.: A Statistical Approach for Algorithm Selection. In: Ribeiro, C.C., Martins, S.L. (eds.) WEA 2004. LNCS, vol. 3059, pp. 417–431. Springer, Heidelberg (2004)
7. Lemeire, J., Dirkx, E.: Causal Models for Parallel Performance Analysis. In: 4th PA3CT Symposium, Edegem, Belgium (2004)
8. Soares, C., Pinto, J.: Ranking Learning Algorithms: Using IBL and Meta-Learning on Accuracy and Time Results. Journal of Machine Learning 50(3), 251–277 (2003)
9. Hoos, H.: Stochastic Local Search Methods, Models, Applications, PhD Thesis, Department of Computer Science from Darmstadt University of Technology (1998)
10. Konak, A.: Simulation Optimization Using Tabu Search: An Empirical Study. In: Steiger, M., Armstrong, F., Joines, B.,, J.A. (eds.) Proceedings of the 37th Conference on Winter Simulation, pp. 2686–2692 (2005)
11. Pérez, J., Cruz, L., Landero, V., Pazos, R.: Explaining Performance of the Threshold Accepting Algorithm for the Bin Packing Problem: A Causal Approach. In: Proceedings of 14th International Multi-conference, Advanced Computer Systems, Polland (2007)
12. Pérez, J., Pazos, R.: Comparison and Selection of Exact and Heuristic Algorithms. In: Laganá, A., Gavrilova, M.L., Kumar, V., Mun, Y., Tan, C.J.K., Gervasi, O. (eds.) ICCSA 2004. LNCS, vol. 3045, pp. 415–424. Springer, Heidelberg (2004)
13. Pérez, J., Pazos, R.: A Machine Learning Approach for Modeling Algorithm Performance Predictors. In: Torra, V., Narukawa, Y. (eds.) MDAI 2004. LNCS (LNAI), vol. 3131, pp. 70–80. Springer, Heidelberg (2004)
14. Sanvicente, H., Frausto, J.: A Method to Establish the Cooling Scheme in Simulated Annealing Like Algorithms. In: International Conference on Computational Science and Applications. LNCS, vol. 3, pp. 755–763. Springer, Heidelberg (2004)
15. Fleszar, K., Hindi, K.S.: New Heuristics for One-dimensional Bin Packing. In: Computers and Operations Research, vol. 29, pp. 821–839. Elsevier Science Ltd., Amsterdam (2002)
16. Beasley, J.E.: OR-Library. Brunel University (2006), http://people.brunel.ac.uk/~mastjjb/jeb/orlib/binpackinfo.html
17. Scholl, A., Klein, R.: (2003), http://www.wiwi.uni-jena.de/Entscheidung/binpp/
18. Fayyad, U.M., Irani, K.B.: Multi-Interval Discretization of Continuous-Valued Attributes for Classification Learning. In: 13th International Joint Conference of Artificial Intelligence, pp. 1022–1029 (1993)
19. Spirtes, P., Glymour, C., Scheines, R.: Causation, Prediction, and Search, 2nd edn. MIT Press, Cambridge (2001)
20. Lauritzen, S.: The EM algorithm for Graphical Association Models with Missing Data. In: Computational Statistics Data Analysis, vol. 19, pp. 191–201. Elsevier Science, Amsterdam (1995)

A Method for Evaluation of Compromise in Multiple Criteria Problems

Henryk Piech and Pawel Figat

Institute of Mathem. and Inform., Technical University of Czestochowa,
Dabrowskiego 73, 42-200 Czestochowa, Poland
h.piech@adm.pcz.czest.pl

Abstract. A graphical method for the modeling of compromise in multiple criteria problems solution is proposed. The method is based on the analysis of the strategic games characteristics and takes into account both the players cooperation for the compromise solution searching and the influence of rejection of the better solutions in favor of the compromise ones. The visualization of the considered problem is based on triangle type representation of the local criteria and can be realized both in deterministic and interval or fuzzy versions. The methodology for building the models of compromise based on the comparative analysis of possible solutions is proposed. In comparison with the approaches based the on the polygon method in games theory [32], our proposition is evidently less algorithmic complex and seems as more suitable for the comparative analysis. Its usefulness becomes especially apparent is the case of small number of local criteria.

1 Introduction

The solution of multiple criteria problems especially when we deal with the conflicting local criteria is a complex problem [8,16,17,28,35]. In such cases, the final stage of multiple criteria optimization is the searching of compromise. It is the most neuralgic phase of the solution process since according to the games theory (here we treat the local criteria as the players), revealing a single predominant solution for the group of criteria is not usually the case [2,26,29]. Similarly as in case of risk there are many different ways to define a compromise depending on the studied situation [3,6,15,30,31]. The most of such cases we can treat as a strategic game, even if we play only with ourselves, with nature or with time [3,5,25]. If all players (all criteria) achieve the largest payment (the optimum value) for concrete set of parameters, the value of compromise is equal to zero since a compromise is unnecessary. On the other hand, if all players are not satisfied with their payments, the value of compromise is equal to zero too. In situation when only a single player is satisfied, and the others are not, we can assume that the value of compromise is equal to zero as yet [8,11,12]. Expression "I agree to this condition, but I would prefer the others " represents verbally the sense of partial satisfaction. In the methods based on polygons (convex) games [2], the participation in creating the compromise is based the on defining two person game value. In the case of k local criteria there are $\binom{k}{2}$ of such possible combinations. We define here the value of game using graphic (or geometrical) methods [26]. We define the weight of zones of compromise as follows:

$$w(i,k) = f_k(xd(i)) - wm(k),$$

where $w(i,k)$ is the weight of i-th zone of compromise for k-th criterion, $f_k(xd(i))$ is the value of k-th criterion at the bottom of i-th zone of compromise, $wm(k)$ is the value of game for k-th criterion.

The known combination methods for calculations used in the framework of this method increase the complexity of algorithm, and additionally it does not eliminate the subjective character of the weights assignment procedure. In the current paper we propose a method which allows to operate with the subjective experts' opinions concerned to the weights assignment in the two - person game strategy. In the multicriterial problem, we create the compositions of such opinions and evaluate the level of compromise using graphical representation of the problem.

2 The General Approach to Defining and Modeling of Compromise Using Graphical Methods

There are different approaches to the classification and definition of partial satisfaction proposed in the literature [20,23]. One of them is the acceptance level graduating type approach based on the consideration of the acceptance, admissibility, anxiety and critical states (Fig.1,2). The example of dependence (it can be dependence in the statistical sense) of the degree of general compromise in the case of five criteria on the number of accepted criteria (or in the terms of game theory, on the number of satisfied players') is shown in Fig.1. On the vertical axis we have a level of general compromise, which is aggregated with the help of sum of degrees of local criteria acceptance. It this example, the values of general compromise rises linearly or nonlinearly (the five variants of dependence are showed by dashed lines) with increase of the number of the accepted criteria. It is seen in Fig.1 that the greater is the angle ($90^0 - alpha$), where alpha is the angle between horizontal axis and tangent to function of compromise, the more easily is the improving of the compromise value by increasing the number of accepted criteria. Approximation of the discrete compromise evaluations cm by continuous function allows to estimate the degree of skepticism dsc in respect to the belief in the effectiveness of achieved compromise defined as follows:

$$dsc = cm'(lu) = tan(90^0 - alpha(lu)),$$

where lu is the number of satisfied players or the number of accepted criteria.

So the skepticism expresses the small belief in effectiveness of compromise searching which grows along with the increasing of the satisfied players number Any multiple criteria problem solution or an approach to such a solution in context of compromise searching can be considered from two points of view:

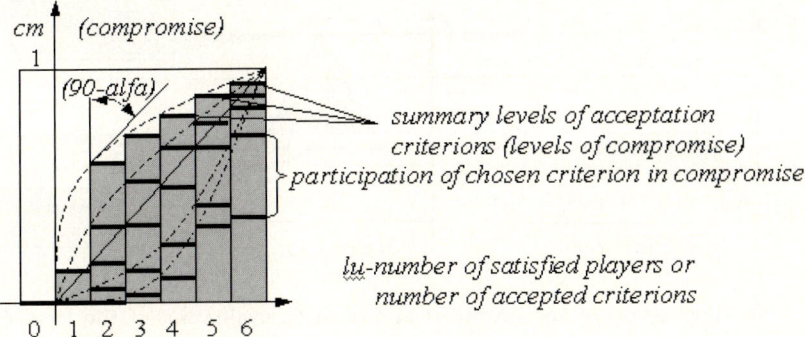

Fig. 1. The five variants (dashed lines) of the discrete compromise evaluations approximation

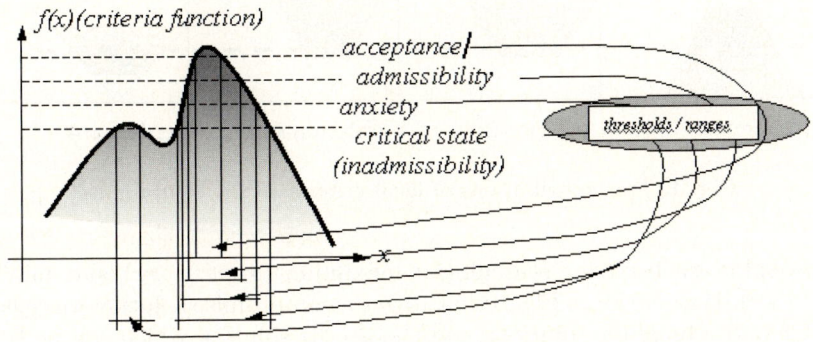

Fig. 2. The acceptance levels for the maximized local criterion

taking into account the participation of the players (local criteria) in compromise or rejection of the "best" solutions in favor of the compromise ones. That is why, we propose to evaluate the participation (contribution) of the local criteria in the creation of compromise as follows:

$$part = |xr - xs| \,/\, |xopt - xs| \quad \textit{in continuous version,}$$

$$part = |xp - xs| \,/\, |xopt - xs| \quad \textit{in discrete or linguistic version,}$$

where *part* is the contribution to compromise,

$$xp = \{w = xa \text{ "'or"'} \; xd \text{ "'or"'} \; xo \text{ "'or"'} \; xs;$$

$$((min\,|xr - w| + w) \geq xr \; for \; xopt \geq xr) \vee$$

$$((min\,|xr - w| + w) \leq xr \; for \; xopt \leq xr)\}$$

In above expressions, xp is the minimal distance between the thresholds of considered zones of acceptance xa, xd, xo and xs (see Fig.3) and the values of local

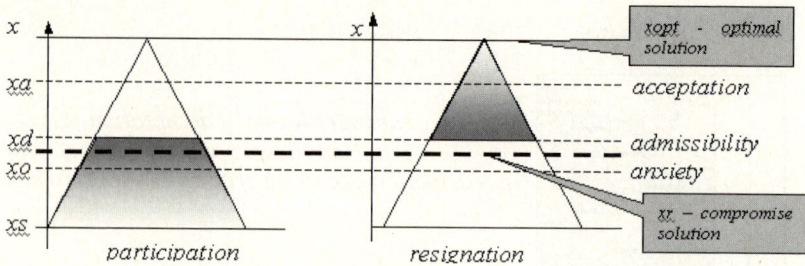

Fig. 3. The illustration of the proposed approach to evaluation of the local criteria participation in the compromise

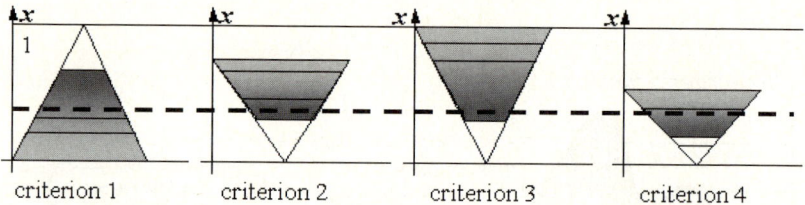

Fig. 4. The contributions of local criteria to the compromise

criteria value xr both for searching a maximum ($xopt \geq xr$) and minimum ($xopt \leq xr$). It is shown in Fig. 3 that to get a compromise solution we reject acceptable and admissible solutions and choose the solution which can be treated as "anxiety" one.

The structure presented in Fig. 3 we call "triangle of compromise" for the considered criterion (or for the player). To build the "triangle of compromise" the consequences of withdrawing from global optimum should be analyzed. So to create the "triangle of compromise" using the local criteria function and considered zones of acceptance we build the mapping of acceptance zones on the axis of so called "conventional arguments", x, (representing their group) analyzing the zone in which the solution will be located. The structure of the " triangles of compromise" should be scaled taking into account the bounds xs, xo, xd, xa and $xopt$. This procedure allows to represent the problem in form that is not dependent on the number of local extrema and the number of arguments. Let's consider the example of the compromise evaluation in the case of several criteria (Fig. 4).

Let us assume for simplicity the weights of acceptance zones as follows: acceptance- 4, admissibility -3, anxiety -2, critical state -1 . Then in the case shown in Fig. 4, we get the next value of compromise:

$w\_cm = (part(1) + part(2) + part(3) + part(4))/(m*n) = (3+3+3+2)/16 = 11/16 = 0,6875,$

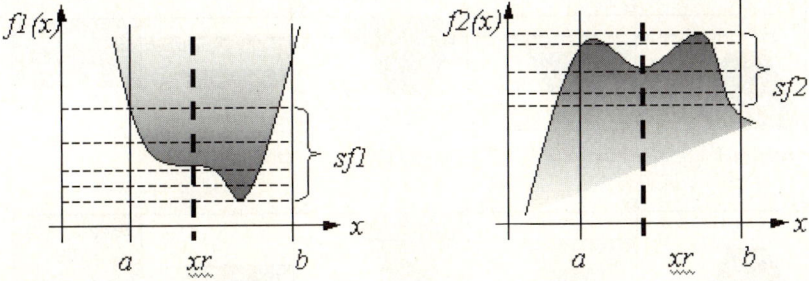

Fig. 5. Two criteria functions with zones of treatment of results sf_1, sf_2. The scales of $f_1(x)$ and $f_2(x)$ are chosen in such a way that they allow to compare the values of local criteria.

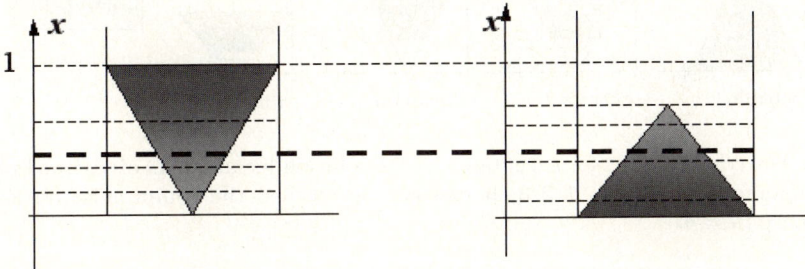

Fig. 6. The mapping of the defined zones of treatment on the axis of conventional arguments and creating the triangles of compromise

where m is the number of zones, n is the number of criteria (in our case, we have $n=4$).

The general algorithm for the creating of triangles of compromise is as follows:

1. Preliminary mathematical formalization of local criteria and corresponding zones of the result treatment (see Fig.5).

2. The normalizing of $f_1(x)$ and $f_2(x)$ to the interval [0,1]: the value of $max\{sf_1, sf_2, ...sf_n\}$ corresponds to "1".

3. The mapping of the defined set of zones of treatment on the axis of conventional arguments (see Fig. 6).

Turning back to compromise evaluation depending on the number of satisfied players we can represent a compromise on the level of acceptance for chosen number of criteria (see Fig.7).

The approached to treatment the solution can be different, i.e., admissibility, anxiety, critical zone. However for the comparison and creating the dependence of the compromise level on the number of satisfied players they should be the same for all criteria. The graphical illustration of the method for compromise evaluation on the level of acceptance is shown in Fig.8.

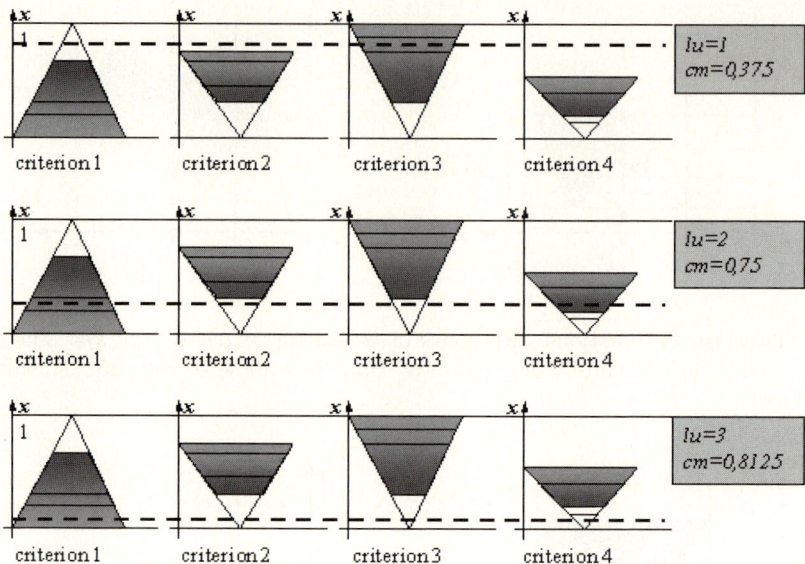

Fig. 7. The compositions for acceptance of the solution for numbers of satisfied players on acceptance levels $lu = 1, 2, 3$. It can not be reached the compromise on level of acceptance $lu = 4$.

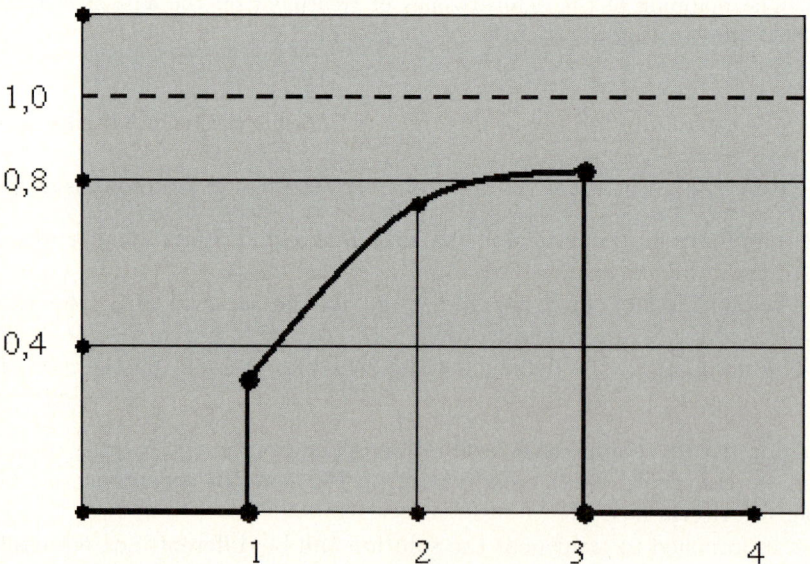

Fig. 8. Dependence of the compromise level on the number of satisfied players

3 The Model of Compromise under Uncertainty

Creating the zones of treatment of the solution is not easy problem. Usually it is difficult to use the traditional deterministic methods and real valued thresholds of solution treatment to solve this problem [10,20] since the statistical analysis and subjective experts' opinions lead inevitable to the structure of initial data in form of interval or fuzzy values. The influence of uncertainty on the compromise evaluation depends on intervals widths which directly refer to the thresholds of compromise zones location. The interval form of the thresholds representation is illustrated in Fig. 9.

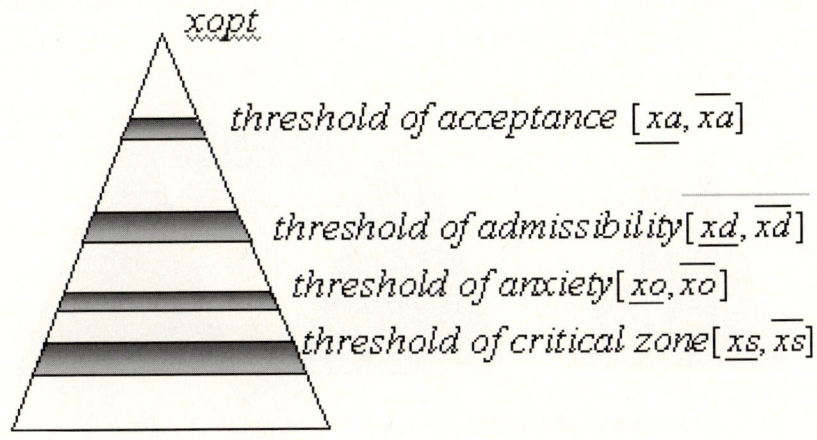

Fig. 9. Threshold intervals in the triangle of compromise

The use of interval analysis for the compromise evaluation leads to the interval solutions. Obviously, the use of interval thresholds leads to the enlargement of the range of neighboring zones (see Fig.10).

Let's consider example presented in Fig. 10. To define the bottom level of compromise evaluation for the number of satisfied players $lu = 1$ we should consider at least three compromise solutions xr_1, xr_2, xr_3. At first, we choose the maximal estimations among minimal and possible ones. Similarly we estimate the upper levels of compromise(choosing the sets of solutions xr), but we choose the maximal estimations among maximal and possible ones as follows:

$\overline{cm}(lu = 1) = max\{(4+0+0)/12; (3+4+1)/12; (1+1+4)/12\} = 0,667$
$\underline{cm}(lu = 1) = max(4+0+1)/12; (3+4+1)/12; (2+2+4)/12 = 0,667$
$\overline{cm}(lu = 2) = \overline{cm}(lu = 2) = (4+4+1)/12 = 0,75$
$\underline{cm}(lu = 3) = \overline{cm}(lu = 3) = 0$

In this example, the bottom level of compromise evaluation, $\underline{cm}$, is caused by the shift of thresholds between zones to the upper bounds in the case when we are looking for the maxi- mum or to the bottom bounds in the case when we

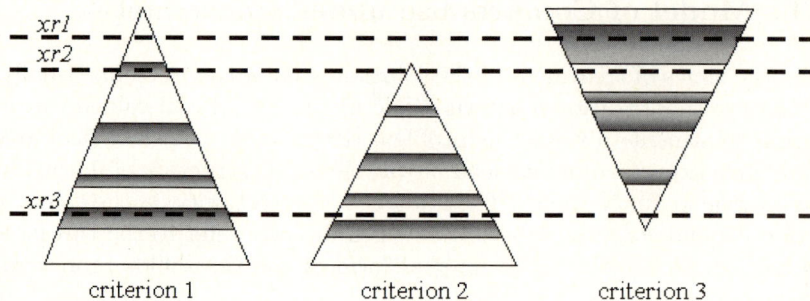

Fig. 10. The examples - compromising solutions xr_1, xr_2, xr_3 and their location in interval zones of classifying of solution

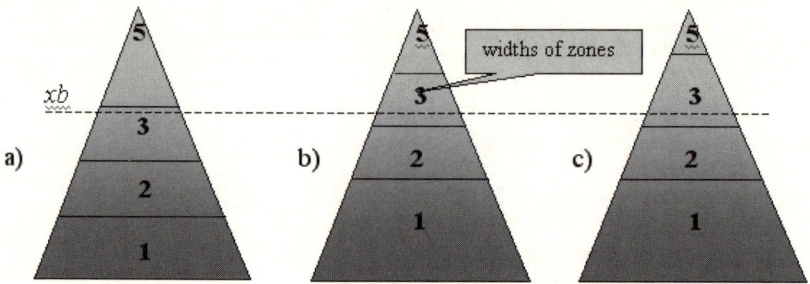

Fig. 11. The comparison of the degrees of skepticism at point xb : $dsc(and) < dsc(b) < dsc(c)$

are looking for the minimum (this reduces the statistical (e.g., average) value of compromise). The upper level of compromise evaluation, $\overline{cm}$, is connected with the shift of thresholds between zones to the bottom bounds in the case when we are looking for the maximum or to upper bounds in the case when we are looking for the minimum (this enlarges the statistical value of compromise). The degree of skepticism in compromise evaluation is rising with decreasing of the weights w assigned to the compromise zones ($w \downarrow \Rightarrow dsc \uparrow$) and the widths of zones wtx ($wtx \downarrow \Rightarrow dsc \uparrow$). The degree of skepticism dsc is increasing to distance between the solution and optimum when the distribution of weights, w, is in the "opposite trend", i.e., with rising of the speed of weights incrementing in relation to the incrementing of zones widths speed ($\Delta w \downarrow > \Delta wtx \downarrow \Longrightarrow dsc \uparrow$, where $\Delta w \downarrow$ - speed of reducing the weights, $\Delta wrx \downarrow$- speed of reducing the width of zones)). Above analysis is illustrated in Fig.11.

4 Conclusions

The searching of compromise solution is the final phase of multiple criteria optimization. The compromise evaluation can be carried out both on base of deterministic information or using uncertain one. In the last case, the solution referred to the level of compromise should be obtained in interval or fuzzy form. The compromise level evaluations based on the participation or resignation (Fig. 3) are usually obtained with different dose of skepticism that is essential for analysis of compromise stability and the real role of participants. Decreasing of the widths of solution treating zones reduces the statistical value of compromise and enlarges the degree of skepticism of this value. An important role in compromise evaluation and its characteristics play the number of zones as well as the distribution of the weights and widths of zones (Fig. 11). The use of the triangle of compromise facilitates the searching of the compromise optimal solutions and allows to quick comparative analysis of the different variants of optimum solution.

References

1. Antonson, E.K., Otto, K.N.: Improving Engineering Design with Fuzzy Sets Department of Mechanical Engineering. MIT Press, Cambridge (1995)
2. Aumann, R.: Survey of repeated games, Easy in game theory and Mathematical Economics in Honor of Oscar Morgenstern, Institute Mannheim, Vein (1981)
3. Baas, S.M., Kwakernaak, H.: Rating and Raking of Multiple-Aspects Alternatives Using Fuzzy Sets. Automatica 1, 47–58 (1977)
4. Baebera, S., Jackson, M.O.: Choosing of Barbera Jackson model-Stable Majority Rules. In: V Conference SAET, Iscia (2001)
5. Bawa, V.S.: Optimal Rules for Ordering Uncertain Prospect. Journal of Economic 2, 95–121 (1975)
6. Best, M.: Fisible Conjugate Direction Method to Solve Linearly Constrained Optimization Problems. Journal of Optimization Theory and Applications 16 (1975)
7. Brans, J.P., Mareshal, B.: The PromCalc and GAIA decision support system for multicriteria decision aid. Decision Support Systems 12, 297–310 (1994)
8. Buckley, J.J.: The multiple - judge multiple criteria ranking problem. A fuzzy set approach. Fuzzy Sets and Systems 13, 23–37 (1984)
9. Bustnce, H., Burillo, P.: Interval valued fuzzy relations in a set structures. J. Fuzzy Math. 4, 765–785 (1996)
10. Caprani, O., Madsen, K.: Mean Value Forms in Interval Analysis. Computing 25, 147–154 (1980)
11. Choi, D.Y., Oh, K.W.: Asa and its Application to Multi- criteria Decision Making. Fuzzy sets and Systems 114, 89–102 (2000)
12. Chu, A., Kalaba, R., Springarn, R.: A comparition of Two Methods for Determining the Weights Belonging to Fuzzy Sets. Journal of Optimization (1996)
13. Czogala, E., Perdycz, W.: Elementy i metody teorii zbiorów rozmytych, PWN, Warszawa (1985)
14. Deluca, A., Termini, S.: A Definition of a nonprobabilistic entropy the of Fuzzy sets theory. Information and Control 20, 301–312 (1972)
15. Dubois, D., Koenig, J.L.: Social choise axioms for fuzzy set aggregation. Fuzzy sets and Systems 43, 257–274 (2003)
16. Eschenauer, H., Kolski, J., Osyczka, A.: Multicriteria Design Optimalization. Springer, Berlin (1990)
17. Fuller, R., Carlson, C.: Fuzzy Multiple Criteria Decision Making. Fuzzy Sets and Systems 78, 139–153 (1996)

18. Hauke, W.: Using Jager's t-noms for Aggregation of Fuzzy Intervals. Fuzzy sets and Systems 101, 59–65 (1999)
19. Hofbauer, J.: Stability for Best Response Dynamic. University of Viena (1995)
20. Jaulin, L., Kieffer, M., Didrit, O., Walter, E.: Applied Interval Analysiss. Springer, London (2001)
21. Kacprzyk, J.: Wieloetapowe sterowanie rozmyte, WNT, Warszawa (2001)
22. Kalynmoy, D., Lother, T., Laumans, M., Zitzler, E.: Scalable Multi-objective Opimization Test Problems (2002)
23. Kaufman, A., Gupta, M.: Introduction to Fuzzy Arithmetic-Theory and applications, p. 349. Van Nostrand Reinhold, New York (1985)
24. Klir, G.J., Folger, T.A.: Fuzzy Sets, Uncerteinty and Information. Prentice-Hall, Englewood Cliffs (1988)
25. Lachwa, A.: Fuzzy world of files, numbers, relations, facts, rules and decisions, Akademicka Oficyna Wydawnicza Exit, Warsaw (in polish) (2001)
26. Papadimitriou, C.H.: Games against nature. J. Comp.System Sci. 31, 288–301 (1985)
27. Piegat, A.: Fuzzy modelling and controlling. Akademicka Oficyna Wydawnicza Exit, Warsaw (2003) (in polish)
28. Roy, B.: Wielokryterialne wspomaganie decyzji, WNT, Warszawa (1990)
29. Straffin, P.D., Gier, T.: Wydawnictwo Naukowe Scholar, Warszawa (2001)
30. Syslo, M., Deo, N., Kowalik, J.: Discrete optimisation algorithms. PWN, Warsaw (1995) (in polish)
31. Vavock, V.: Aggregating Strategies. In: Conference on Computational Learning Theory (1990)
32. Watson, J.: Strategia. Wprowadzenie do teorii gier. WNT, Warszawa (2005)
33. Wrather, C., Yu, P.L.: Probability Dominance. Journal of Optimization Theory and Application 36 (1982)
34. Yager, R.: Modeling Uncertainty Using Partial Information. Information Sciences 121, 233–294 (1999)
35. Yager, R.: Non Numeric Multicriteria, Multiperson Decision Making. Group Decision and Negotiations 2, 81–93 (1993)
36. Yager, R.: On the Measure of Fuzziness and Negation, Membership in the Unit Interval. Int. J. Gen. Systems 5, 221–229 (1979)
37. Zadeh, L.A.: Fuzzy limitations calculus; design and systems, methodological problems. Ossolineum (1980) (in polish)
38. Zeleny, M.: Multiple Criteria Decision Making. In: Business Venturing, vol. 7, pp. 505–518. Mc-Grow-Hill, New York (1987)
39. Zorychta, K., Ogryczak, W.: Programowanie liniowe i cakowitoliczbowe. WNT, Warszawa (1981)

Neural Networks as Prediction Models for Water Intake in Water Supply System*

Izabela Rojek

Kazimierz Wielki University in Bydgoszcz,
Institute of Environmental Mechanics and Applied Computer Science,
Chodkiewicza 30, 85-064 Bydgoszcz, Poland
izarojek@ukw.edu.pl
http://www.imsis.ukw.edu.pl

Abstract. The paper presents neural networks as models for prediction of the water intake. For construction of prediction models three types of neural networks were used: linear network, multi-layer network with error backpropagation and Radial Basis Function network (RBF).

The prediction models were compared for obtaining optima quality prognosis. Prediction models were done for working days, Saturdays and Sundays. The research was done for selected nodes of water supply system: detached house node and nodes for 4 hydrophore stations from different pressure areas of water supply system. Models for Sundays were presented in detail. Further research concerning the creation of prognosis models should be directed towards constructing models not only for particular days, but also for the complete week, four seasons of the year: spring, summer, autumn and winter, and finally the entire year.

Keywords: prediction model, neural network, water supply system.

1 Introduction

Water network is central to the municipal system of water supply and sewage system. The load of water network influences the work of pumps in the water intake and treatment station, the load of sewage system and the load of sewage treatment plant [6].

Accurate forecasting of load and operation of the water supply system will allow energy efficient operation of pumps at the water intake, as well as effective management of the technological process in the sewage treatment plant by preparing it for a certain load of sewage and pollution.

That is why the most important issue is the prognosis of the load of water supply system in the long run.

In the contemporary world, people as well as enterprises are flooded by data and information from a multitude of sources. Unfortunately this valuable data, the accumulation and processing of which costs millions, are usually lost in

---

* The paper is issued in range of the project no R11 001 01 financed by the Ministry of Higher Education and Sciences.

databases and data warehouses. The problem stems form the lack of qualified analysts who could convert data into knowledge. Detailed analysis of information might prove very useful in establishing a relation with a client, and tailoring an offer for the client's special needs. In this case traditional forecasting methods are insufficient. That is why data mining is nowadays essential [1],[3],[5].

Data Mining is based on three principles: data, techniques and modelling. These processes cannot be automated, they need to be operated by a user. Data Mining techniques - decision trees, neural networks, and regression are implemented onto different tools called Enterprise Miner. Data Mining focuses on making a client aware of the complexity of data they have at their disposal and the possibilities of using his data effectively. For the purposes of Data Mining, data warehouses are constructed which collect data from different, sometimes globally dispersed sources.

The practical side of data mining can be put down to two areas:

- prediction,
- description.

Prediction consists in using the known values of variables with a view to forecasting values of these or other variables in the future. Description consists in creating a clear knowledge representation in the form of graphs, formulas, rules and tables. Such descriptions, in the form of description models, are frequently used to support decision processes.

In the paper different type of neural networks were applied to forecast the load of water network. The experiments had two aims.

Firstly the paper presents and compares prediction models in the form of selected types of neural network: multi-layer network with backward error propagation, linear network, and the network with radial basic functions. After comparison, we can say which type of neural network is the best for prediction of load of water network.

Secondly the research was done for selected nodes of water supply system: detached house node and nodes for 4 hydrophore stations from different pressure areas of water supply system. Specific nodes were selected for checking whether different networks nodes require different prognosis models, or whether one shall suffice.

2 Prediction as a Task of Data Mining

Data mining is not an easy process. Six basic steps which shall render the process effective can be identified here [1],[3]:

- Understand and properly define the problem/task which is the object of mining. Moreover, the background or surroundings in which the problem occurs should also be properly examined.
- Select a set of data which shall be subject to mining. The set has to be a substantial sample of the entire amount of data. The selection refers to objects, their attributes (variables), span of time, geographical area, the size of sample etc.

- Decide how to prepare data for processing. For example: should age be represented as a range (for example 40-45 years) or as a figure (for example 40 years).
- Select an algorithm (or combinations) of data mining and write a program processing this algorithm on the basis of the prepared data. Very often there is a need to go back to item 3 or even 2 if the results are unsatisfactory.
- Analyse results of the program execution and choose the ones which are the result of work. At this stage, a close cooperation between an analyst and specialist in the explored area is needed. The results should be presented in the form commonly accepted in the organization in question.
- Submit the results to the management of the organization and suggest ways of using them.

Forecasting is different from other data mining methods in that it refers to the future. Methods and techniques used for forecasting comprise traditional methods of estimating the value of a point and trusting interval, basic line regression and correlation and multiple regression, as well as data mining methods such as neural networks, decision trees and the method of k-closest neighbours.

3 Neural Network as a Forecasting Model

Neural networks return continuous value at the exit, hence they are excellent for estimating and forecasting. Such networks may analyze many variables at a time. It is possible to create a model even if the solution is very complex.

The drawbacks of neural networks are the difficulties in setting architectural parameters, falling into local minima, long learning process and lack of clear interpretation [2],[4].

Three types of neural networks were used for construction of the forecasting model.

Linear network - represented by a network with no hidden layers, however the neurons present in the output layer are fully linear (i.e. these are neurons the collective stimulation of which is determined as a linear combination of input values and which have a linear activation function). During the operation of the network, the inputs are multiplied by weight matrix, which collectively forms a vector of output signals.

In *multi-layer network with error backpropagation*, the signals flow from entrance to exit. Multi-layer networks are formed with many layers of neurons. The entrance of each neuron from a given layer is linked with exits of all neurons in the preceding layer. In the model of a multi-layer network, the first layer is called the input layer, the last - output layer, and the rest - hidden layers. The number of layers and neurons is at random.

RBF - Radial Basis Function network usually has one hidden layer with radial neurons, each of which models a Gauss' surface of answers. Since these functions are strongly non-linear, one hidden layer is enough to model a function of any shape. Yet, for an RBF to form a successful model of any function, the network structure needs to dispose of many radial neurons. If there are a sufficient number of radial neurons, each important detail of a modeled function can have the needed radial neuron attached, which guarantees, that the obtained solution shall genuinely reproducer the given function.

4 Neural Model for Prediction of Water Intake in a Water Supply System Node

Experiments were conducted for selected water supply system nodes: detached house node, nodes for 4 hydrophore stations from different pressure areas of water supply system.

For each node, three models were drawn: working day, Saturday, Sunday, using three types of neural networks: linear network, multi-layer network with error backpropagation and RBF network.

For each type of network, a research examining which network produced best results was done. Analysis and comparison of prediction models for water intake was carried out.

The paper was illustrated with graphs presenting data from a Sunday.

4.1 Data Preparation

Measuring data presents state of water supply system in selected measuring nodes. Measurement was taken every hour. The data is transferred to database by used PlusGSM mobile communications.

The data is used as learning files for prediction models. Data was processed through clearing and conversion. Data was divided into periods: working day, Saturday and Sunday. Missing values were put.

Figure 1 gives an example of a fragment of a file with measurement data. The file contained data for the flow during 3 months.

Data were divided into the learning set (80% of data) and the testing set (20% of data).

Figure 2 gives a fragment of a learning file. Data is grouped according to year, month, day and hour. Knowing the model of the facility, the reaction to various input violations should be analysed.

Year	Month	Day	Hour	Flow
2007	1	14	0	6
2007	1	14	1	3

Fig. 1. Fragment of file containing node data - Sunday

-	-	-	-	input	input	output
Year	Month	Day	Hour	Flow t-1	Flow	Flow t+1
2007	1	14	0	0	6	3
2007	1	14	1	6	3	3
2007	1	14	2	3	3	3

Fig. 2. Learning file fragment

It is interesting to defining the future state of the facility for the time $t+n$, where n is the prediction horizon, t contains the input changes history up to the present. In order to construct time sequences which are later used in forecasting model, the values of flow before the moment t (Flow t-1) and after the moment t (Flow t+1) were added.

4.2 Drawing a Neural Network Model for Prediction

Model of a Multi-Layer Neural Network with Error Backpropagation.
A model of a multi-layer neural network with backward error propagation (MLP) was drawn. The model is equipped with two inputs (Flow t-1, Flow) and one output (Flow t+1) and one hidden layer containing 5, 10 or 15 neurons.

Figure 3 illustrates example of values of RMS error during learning for different structures of MLP neural network. In the same manner other parameters (learning quality, testing quality, testing error) were obtained, too.

These structures were taught with different conditions of ending the process, i.e. the end after reaching the number of periods equal to 1000, 10000 or 100000. To every combination, a RMS error was compared.

As a result the neural network of 2-10-1 proved to be the most accurate model, when the learning processed was stopped at 10000 periods because values of RMS error were smallest [7]. Results obtained for the new input values proved the point. The parameters of 2-10-1 network are the following:

- learning quality (0,420672),
- testing quality (0,473670),
- learning error (0,123733),
- testing error (0,130733),
- number of inputs (2),
- number of hidden layer neurons (10).

Having drawn a model of a neural network, it was assessed as a prediction model. The stages of prognosis procedure are as follows:

- observation of time sequence process,
- creating a model of the observed sequence,

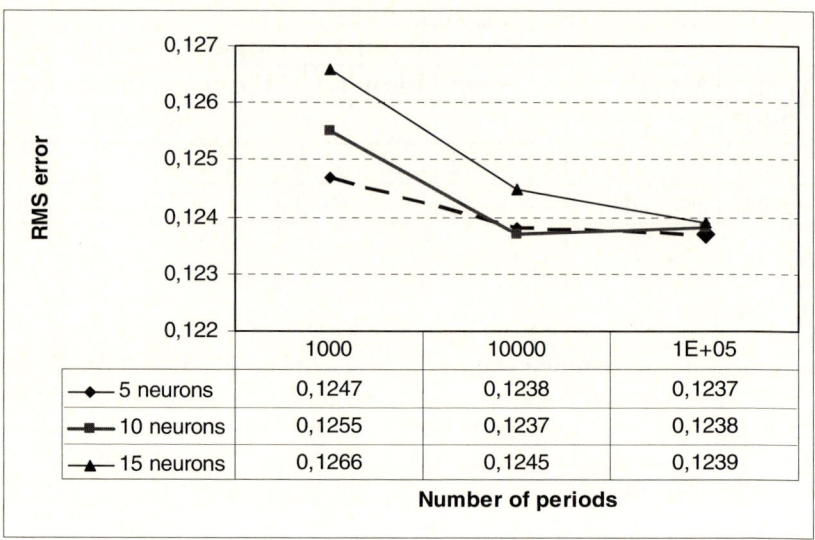

Fig. 3. Analysis of the RMS error results for the number of neurons equal to 5,10,15

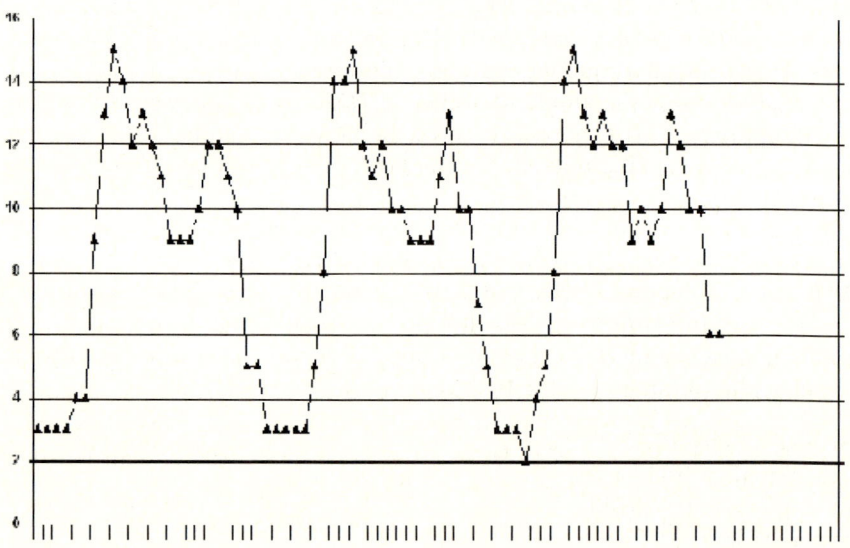

Fig. 4. Time sequence containing real data on water flow in a node

- conversion of the time sequence model into a shape which enables prognosis (construction of forecasting algorithm).

For the prediction of water intake in a node for Sunday, one-series prognoses for time sequence models were used.

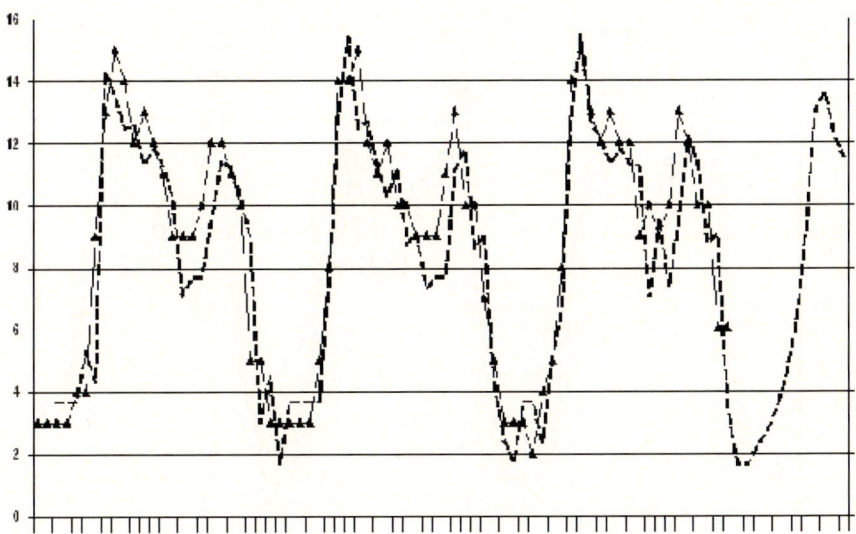

Fig. 5. Neural network prediction - broken line

The first stage is the presentation of a time sequence containing real data on water flow in a node (Fig. 4).

Then a neural network model was used for prediction of water network load. Black lines with triangles constitute real values of water flow, broken line - prognosis done by a neural network (Fig. 5).

Assessing the accuracy of prognosis is possible while testing during which process a part of input time sequence is out of the set used for identifying model parameters.

Testing showed relevance of the model to historical data, on the basis of which the prognosis is done.

Linear Network Model. Similar research was done for linear network model. The model is equipped with two inputs (Flow t-1, Flow) and one output (Flow t+1). These structures were taught with different conditions of ending the process, i.e. the end after reaching the number of periods equal to 1000, 10000 or 100000. An RMS error was compared with every combination.

The neural network of 2-1 proved to be the most accurate model, when the learning processed was stopped at 10000 periods because values of RMS error were smallest.

Results obtained for the new input values proved the point. The parameters of the best linear network are the following:

- learning quality (0,499988),
- testing quality (0,513962),
- learning error (0,135304),
- testing error (0,187106),

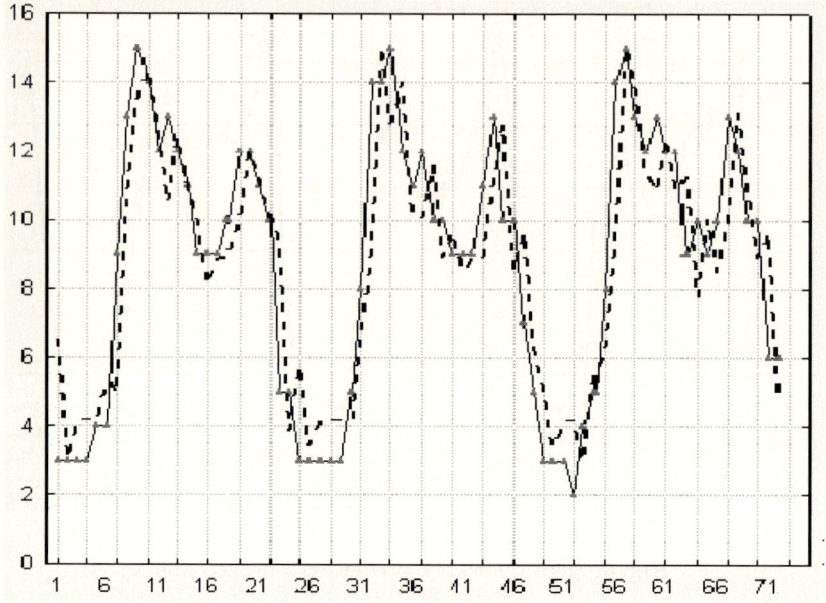

Fig. 6. Prediction model with the use of linear neural network - broken line

- number of inputs (2),
- number of hidden layer neurons (0).

Figure 6 shows a prognosis model using linear neural network. Continuous line shows real values of water flow, broken line- prognosis done by the neural network.

RBF Network Model. A model of a RBF neural network was drawn. The model is equipped with two inputs (Flow t-1, Flow) and one output (Flow t+1) and one hidden layer containing 2, 3, 4, 5 or 9 neurons. These structures were taught with different conditions of ending the process, i.e. the end after reaching the number of periods equal to 1000, 10000 or 100000. To every combination, a RMS error was compared. Figure 7 shows parameters describing experiments for different structures of RBF networks.

The neural network of 2-9-1 proved to be the most accurate model. Results obtained for the new input values proved the point because values of RMS error were smallest.

The parameters of the RBF network are the following:

- learning quality (0,313510),
- testing quality (0,428445),
- learning error (0,079589),

Type of Network Structure	Learning Quality	Testing Quality	Learning Error	Testing Error	Number of Inputs	Number of Hidden Layer Neurons
RBF 2-5-1	0,399129	0,456325	0,101466	0,124039	2	5
RBF 2-2-1	0,538986	0,877917	0,136829	0,216179	2	2
RBF 2-9-1	**0,313510**	**0,428445**	**0,079589**	**0,120050**	**2**	**9**
RBF 2-4-1	0,434599	0,730911	0,110329	0,194728	2	4
RBF 2-3-1	0,384227	0,616717	0,097541	0,169967	2	3

Fig. 7. Parameters describing experiments for different structures of RBF networks

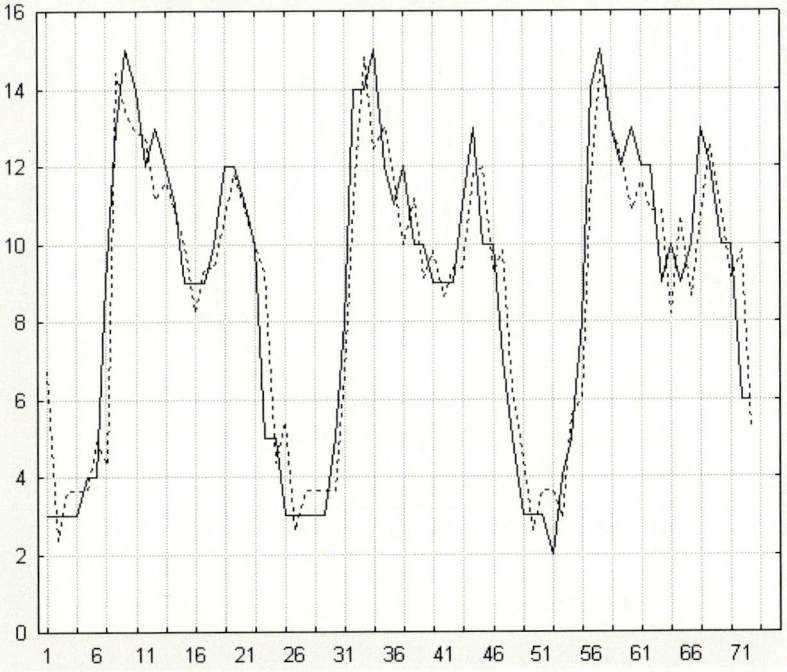

Fig. 8. Prediction model with the use of RBF network - broken line

- testing error (0,120050),
- number of inputs (2),
- number of hidden layer neurons (9).

Figure 8 shows prediction model with the use of RBF network. Continuous line presents real values of water flow, broken line means prognosis done by a neural network of RBF type.

Type of Network Structure	Learning Quality	Testing Quality	Learning Error	Testing Error	Number of Inputs	Number of Hidden Layer Neurons
MLP 2-10-1	0,420672	0,473670	0,0825	0,130733	2	10
Linear 2-1	0,499988	0,513962	0,135304	0,187106	2	0
RBF 2-9-1	**0,313510**	**0,428445**	**0,079589**	**0,120050**	**2**	**9**

Fig. 9. Parameters describing experiments for the best MLP, Linear and RBF networks

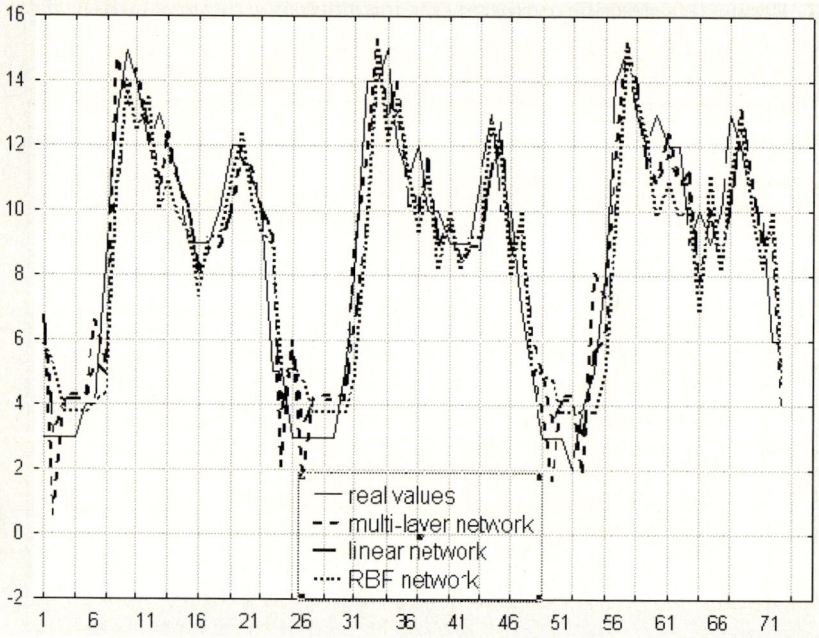

Fig. 10. Comparison of prognosis models

Comparison of Prognosis Models. All these models forecast the load of water supply system very well. These models were tested on selected network nodes. Radial Basic Function model proved to be the most accurate. Figure 9 shows that parameters of RBF network are smallest.

Figure 10 presents a comparison between prognosis models using linear neural networks, multi-layer network with error backpropagation, and radial basic function network.

5 Conclusion

The general principle of science states that if there is a possibility to choose between a complex and a basic model, the basic model should always be given priority - unless of course the more complex one is significantly better than the basic one for the particular data.

This principle should also be applied to neural networks. Having analyzed various prediction models, it was the RBF network model which proved the most effective, and then the multi-layer network with error backpropagation should be listed.

Since these networks are similar in adjustment while choosing network to create the prognosis model, the complexity of a model should also be taken into account - the learning time, calculation time and accuracy. We have selected the RBF network model for further research.

Further research concerning the creation of prognosis models should be directed towards constructing models not only for particular days, but also for the complete week, four seasons of the year: spring, summer, autumn and winter, and finally the entire year.

When acquiring new models, these models should be taught in any moment in time. Further research into when and how these models should be taught.

Having constructed all these models, a full model for prediction the load of water supply system at any time shall be obtained.

References

1. Hand, D., Mannila, H., Smith, P.: Data Mining. WNT, Warsaw (2005) (in polish)
2. Krawiec, K., Stefanowski, J.: Machine Learning and Neural Networks. Publishing House of Poznan University of Technology, Poznan (2004) (in polish)
3. Larose, D.T.: Discovering Knowledge in Data. An Introduction to Data Mining. PWN, Warsaw (2006) (in polish)
4. Rutkowska, D., Pilinski, M., Rutkowski, L.: Neural Networks, Genetic Algorithms and Fuzzy Logic. PWN, Warsaw (1997) (in polish)
5. Michalski, R.S., Bratko, I., Kubat, M.: Machine Learning and Data Mining. John Wiley & Sons, Chichester (1998)
6. Studzinski, J., Bogdan, L.: Computer Aided Decisions Making System for Management, Control and Planning Water and Wastewater Systems. In: Applications of Informatics in Science, Engineering and Management, System Research Institute, Polish Academy of Sciences. Systems Research, Warsaw, vol. 49, pp. 149–157 (2006)
7. Rojek, I.: Neural Model for Prediction of Water Consumption in Water Network. In: Studies&Proceedings Polish Association for Knowledge Management, Bydgoszcz, vol. 10, pp. 173–180 (2007) (in polish)

Financial Prediction with Neuro-fuzzy Systems

Agata Pokropińska[1,3] and Rafał Scherer[2,3]

[1] Institute of Mathematics and Computer Science
Jan Długosz University
al. Armii Krajowej 13/15, 42-200 Częstochowa, Poland
http://imi.ajd.czest.pl/
[2] Department of Computer Engineering, Częstochowa University of Technology
al. Armii Krajowej 36, 42-200 Częstochowa, Poland
http://kik.pcz.pl
[3] Academy of Humanities and Economics in Łódź, ul. Rewolucji 1905 nr 64,
Łódź, Poland
a.pokropinska@ajd.czest.pl, rafal@ieee.org
http://www.wshe.lodz.pl

Abstract. An application of neuro-fuzzy systems to supporting trading decisions is presented. The system has the ability to use expert knowledge and to be fitted to the learning data by various machine learning techniques. The proposed approach uses the backpropagation algorithm to determine system parameters on the basis of several indices. Experiments were made on past data showing relatively good performance of the proposed approach.

1 Introduction

Along with conventional methods, soft computing techniques are used for predicting stock market. These methods do not need the information about the probability model. Most often neural networks [1] are used [4][6]. Fuzzy systems [2][8] are also sometimes used in financial applications [3][11]. They are sometimes a better choice thanks to their ability to use knowledge in the form of intelligible IF-THEN fuzzy rules. They fall into one of the following categories [8], depending on the type connective between the antecedent and the consequent in fuzzy rules: Takagi-Sugeno (consequents are functions of inputs), Mamdani-type (consequents and antecedents are related by the min operator or generally by a t-norm) and (Logical-type reasoning method - consequents and antecedents are related by fuzzy implications). Neuro-fuzzy systems work well in tasks with underlying nonlinear patterns or complex relationships thus they fit perfectly to the environment of stock markets. Compared to traditional methods, neuro-fuzzy systems process imprecise information, which helps to make decisions in real world applications. Examples of neuro-fuzzy systems applications include

- optimal portfolio selection,
- forecasting the future direction of prices or stock market indicators,

- determining trend,
- generating entry or exit signals.

Usually investors use technical and / or fundamental analysis to help making their investment decisions. Technical analysis is based on three premises

- market action discounts everything,
- prices move in trends,
- history tends to repeat itself.

Supporters of the technical analysis claim that all factors influencing the price are included in it. It results from conviction that price behaviors reflect changes in supply and demand relationship. It does not mean that technical analysts reject claim that economics factors cause stock market trend changes. They claim only that the market is easier to understand and predict when we analyze its history in the form of charts. By drawing price charts, analysts try to find out a trend i.e. the direction towards which the prices go. Trend recognition in its early stage let investors make transaction decision, which should generate the profit in the future. Analysts assume that the probability of continuing the trend is greater than of its reversing. Analyzing the price charts allows discover repeating patterns (formations) by which prices are moving. It arises from repeatability of human behavior in given situations. Technical analysts, knowing repeating patterns (formations), try to find them in current quoting and make predictions about the future. While technical analysis concentrates on researching market behavior, fundamental analysis deals with economics factors of supply and demand. All factors influencing the price are researched to determine the real value. The value depends on supply and demand law. If real value is lower than the current market price, it means that the good is over priced and should be sold. Conversely if the determined price is greater that its current market price the good should be bought because it is underestimated. In both approaches to forecasting the market the same problem is trying to be solved - determine probable direction of price change but it is done from two different perspectives. The knowledge of these methods is not required when we use neuro-fuzzy systems but the lack of it can be an obstacle to designing the effective transaction system.

In the paper we use Mamdani-type neuro-fuzzy systems to generate sell or buy signals on the basis on the WIG20 index of the Warsaw Stock Exchange. The system is described in Section 2 and the experiments in Section 3.

2 Neuro-fuzzy System

In the paper we use Mamdani neuro-fuzzy systems designed using the Larsen inference rule. Output of such systems is given by

$$y = \frac{\sum_{r=1}^{N} \bar{y}^r \cdot \mu_{\bar{B}^r}(\bar{y}^r)}{\sum_{r=1}^{N} \mu_{\bar{B}^r}(\bar{y}^r)}, \quad (1)$$

where membership functions of fuzzy sets $\bar{B}^r$, $r = 1, \ldots, N$, are given by

$$\mu_{\bar{B}^r}(y) = \sup_{x \in X} \left[\mu_{A'}(x) \overset{T}{*} \mu_{A^r \to B^r}(x, y) \right]. \tag{2}$$

If we assume that we use the singleton defuzzification, above formula takes the following form

$$\mu_{\bar{B}^r}(y) = \mu_{A^r \to B^r}(\bar{x}, y) = T(\mu_{A^r}(\bar{x}), \mu_{B^r}(y)). \tag{3}$$

Let us observe that

$$\mu_{A^r}(\bar{x}) = \underset{i=1}{\overset{n}{T}} \left(\mu_{A_i^r}(\bar{x}_i) \right), \tag{4}$$

and assume that

$$\mu_{B^r}(\bar{y}^r) = 1. \tag{5}$$

Then, using a t-norm property $(T(a, 1) = a)$, we obtain a formula describing a single Mamdani fuzzy system

$$\bar{y} = \frac{\sum_{r=1}^{N} \bar{y}^r \cdot \underset{i=1}{\overset{n}{T}} \left(\mu_{A_i^r}(\bar{x}_i) \right)}{\sum_{r=1}^{N} \underset{i=1}{\overset{n}{T}} \left(\mu_{A_i^r}(\bar{x}_i) \right)}. \tag{6}$$

Linguistic variables are described by the Gaussian membership function

$$\mu_{A_i^r}(x_i) = \exp\left[-\left(\frac{x_i - \bar{x}_i^r}{\sigma_i^r} \right)^2 \right]. \tag{7}$$

If we choose product t-norm, then the neuro-fuzzy system with the center average defuzzification and Mamdani-type relation is described by

$$\bar{y} = \frac{\sum_{r=1}^{N} \bar{y}^r \left[\prod_{i=1}^{n} \exp\left(-\left(\frac{x_i - \bar{x}_i^r}{\sigma_i^r} \right)^2 \right) \right]}{\sum_{r=1}^{N} \left[\prod_{i=1}^{n} \exp\left(-\left(\frac{x_i - \bar{x}_i^r}{\sigma_i^r} \right)^2 \right) \right]}, \tag{8}$$

and the network structure of the system is depicted in Fig. 1.

Learning of the considered system is carried out by a gradient method. Let us denote a pair of input vector and output signal by $\mathbf{x}^q$ and d^q, respectively, where q denotes vector number, $q = 1, \ldots, M$. Let us define the error

$$Q^q = \frac{1}{2} [f(\mathbf{x}^q) - d^q]^2. \tag{9}$$

During learning we tune parameters $\bar{y}^r, \bar{x}_i^r, \sigma_i^r$ to minimize (9) for all learning vectors. The most common approach is the use of gradient algorithms. Value of $\bar{y}^r$ parameter is computed by

$$\bar{y}^r(l+1) = \bar{y}^r(l) - \eta \frac{\partial Q}{\partial \bar{y}^r} |l, \tag{10}$$

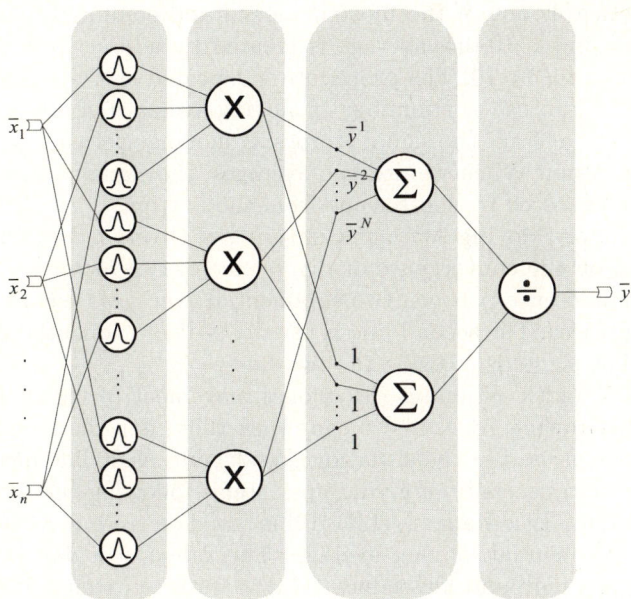

Fig. 1. Single Mamdani neuro-fuzzy system

where $l = 0, 1, 2...$ is a step number of the gradient algorithm, and η is a learning coefficient. Output signal value depends on $\bar{y}^r$ only by the numerator of (8). Let us assume that $a = \sum_{r=1}^{N} \bar{y}^r z^r$, $b = \sum_{r=1}^{N} z^r$ and $z^r = \prod_{i=1}^{n} \exp\left(-\left(\frac{x_i - \bar{x}_i^j}{\sigma_i^j}\right)^2\right)$. Then we can write modification of the parameter $\bar{y}^r$ for consecutive steps as follows

$$\bar{y}^r(l+1) = \bar{y}^r(l) - \eta \frac{f-d}{b} z^r . \qquad (11)$$

An analogical deliberation for parameters $\bar{x}_i^r$ and σ_i^r yields

$$\bar{x}_i^r(l+1) = \bar{x}_i^r(l) - \eta \frac{f-d}{b}(\bar{y}^r - f) z^r \frac{2(x_i^q - \bar{x}_i^r(l))}{\sigma_i^{r2}(l)} , \qquad (12)$$

$$\sigma_i^r(l+1) = \sigma_i^r(l) - \eta \frac{f-d}{b}(\bar{y}^r - f) z^r \frac{2(x_i^q - \bar{x}_i^r(l))^2}{\sigma_i^{r3}(l)} . \qquad (13)$$

Formulas can be treated as a special case of the backpropagation algorithm.

3 Numerical Simulations

The neuro-fuzzy system was learned with the past data of the WIG20 index from the Warsaw Stock Exchange. The input data is composed of different WIG20 values and several indices.

Moving Average is one of the most universal and commonly used technical indicators. The simple 10-day average is created by adding last closing prices and dividing the sum by 10. The procedure is repeated everyday. To obtain sell or buy signals we can use any number of moving averages of any length. When we use one moving average the signal is crossing moving average line by the price line from the bottom. When we use two averages, the signal is generated when longer average is crossed by shorter one from the bottom.

MACD oscillator (Moving Average Convergence Divergence) is the system of two lines based on the convergence or divergence of two moving averages. The first line is the difference between two exponential averages (usually 12- and 26-day) of closing prices. The second line is usually 9-day exponential average from the first one. The signal is crossing the two lines.

Wilder's ADX index estimates directional movement of trends on the scale from 0 to 100. Growing ADX line means that there is a distinct trend and a suitable system can used in the situation. Weakening ADX line means that the lack of a trend, and necessity of giving up systems following the trend.

Wilder's Average True Range (ATR) [12] is based on trading ranges smoothed by an N-day exponential moving average. The range of a day trading is the difference between high and low values. The true range extends it to yesterday closing price if it was outside of today range. The average true range is an N-day exponential moving average of the true range values.

In the numerical experiment, there were used 881 daily quotations (631 learning vectors and 250 testing ones). Data comes from the period from Jan. 01, 2004 and Jun. 29, 2007. Input data include

- WIG20 O - WIG20 value at opening,
- WIG20 Z - WIG20 at session closing,
- WIG20 max - maximal WIG20 value during session,
- WIG20 min - minimal WIG20 value during session,
- WIG O - WIG index at opening,
- WIG Z - WIG index at session closing,
- Volume,
- ADX,
- ATR,
- MACD,
- MA 4, MA 9, MA 18 - closing WIG20 value from 4, 9, 18 consecutive sessions.

There are two output states possible, which are desired value during learning:

- long position - buying at the session opening,
- short position - selling at the session opening.

Given position is kept after session closing (it allows minimizing commission value) till generating opposite signal to the actual one. Desired learning signal was created by subjective analyzing WIG20 index charts. It is a theoretical approach, assuming that there exists an instrument similar to futures contract

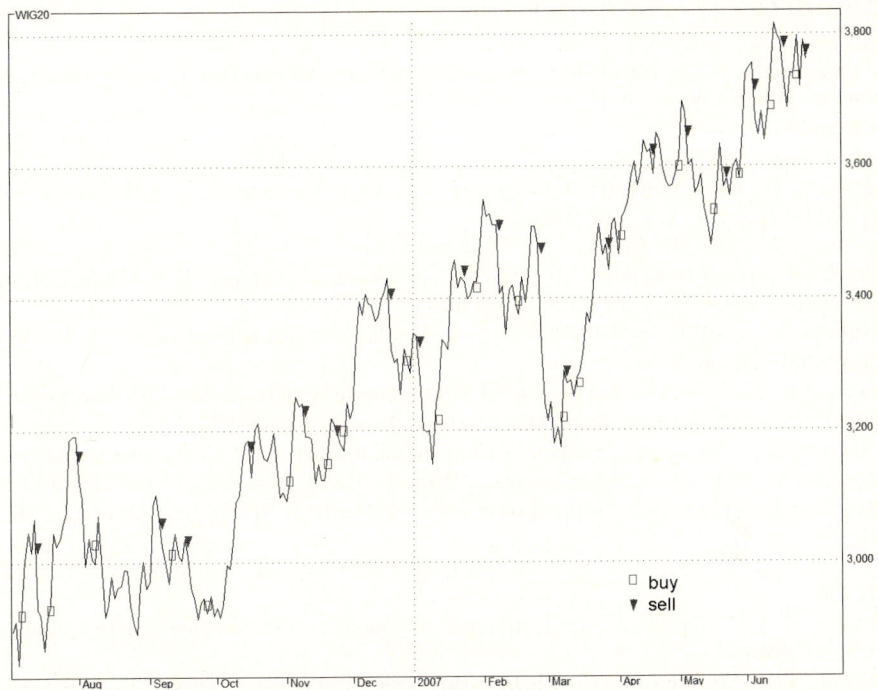

Fig. 2. Buy/sell signals generated by the neuro-fuzzy systems

but with the value correlated with the index value. The outcome from WIG20 investment has no practical value but only researching purposes. Preparing the investment strategy using the index can have two interpretations: trading portfolio of indexed shares, investing in one instrument with similar characteristic to the index or buying/selling futures contracts on the WIG20 index. All neuro-fuzzy system parameters were tuned by the backpropagation algorithm [10]. The system generated signals resulting in 41% profit for the whole period being considered. It should be noted that in the period the WIG20 index value risen by 30% thus the real profit above the WIG20 change is 11%. This is a gross profit without commissions.

4 Conclusions

Investing using neuro-fuzzy systems can provide better results than the one based on traditional indices. It should be noted that the experiments were conducted using past data and the results were compared with real values. Finance markets are unpredictable, so neuro-fuzzy system learning on the past data from the past can predict with worse accuracy in the future. Neuro-fuzzy systems base on the interpretable knowledge so they can be initialized with some expert knowledge.

References

1. Domaradzki, R.: Application of Neural Networks to Generating Decision Strategies on the Warsaw Stock Exchange. AGH, Krakow (2007)
2. Czogała, E., Lęski, J.: Fuzzy and Neuro Fuzzy Intelligent Systems. Physica Verlag, Heidelberg (2000)
3. Dourra, H., Siy, P.: Investment using technical analysis and fuzzy logic. Fuzzy Sets and Systems 127, 221–240 (2002)
4. Gencay, R., Qi, M.: Pricing and Hedging Derivative Securities with Neural Networks: Bayesian Regularization. Early Stopping, and Bagging, IEEE Transactions on Neural Networks 12(4), 726–734 (2001)
5. Jajuga, K.: Foundations of Investing on the Warsaw Stock Exchange. GPW, Warsaw (2006) (in polish)
6. Kwon, Y.-K., Moon, B.-R.: A Hybrid Neurogenetic Approach for Stock Forecasting. IEEE Transactions on Neural Networks 18(3), 851–864 (2007)
7. Murphy, J.J.: Technical Analysis of the Financial Markets: A Comprehensive Guide to Trading Methods and Applications. Prentice Hall Press, Englewood Cliffs (1999)
8. Rutkowski, R.: Flexible Neuro-Fuzzy Systems. Kluwer Academic Publishers, Dordrecht (2004)
9. Van Tharp, K.: Trade Your Way to Financial Freedom, 2nd edn. McGraw-Hill (2006)
10. Wang, L.-X.: Adaptive Fuzzy Systems And Control. PTR Prentice Hall, Englewood Cliffs (1994)
11. Wang, Y.-F.: Predicting stock price using grey prediction system. Expert Systems with Applications 22, 33–39 (2002)
12. Wilder, J.W.: New Concepts in Technical Trading Systems, Trend Research (1978)

Selected Cognitive Categorization Systems

Ryszard Tadeusiewicz[1] and Lidia Ogiela[2]

[1] AGH University of Science and Technology, Institute of Automatics,
Al. Mickiewicza 30, PL-30-059 Kraków, Poland
rtad@agh.edu.pl

[2] AGH University of Science and Technology, Faculty of Management,
Al. Mickiewicza 30, PL-30-059 Kraków, Poland
logiela@agh.edu.pl

Abstract. This paper demonstrates that AI methods can be applied to the development of intelligent IT systems. They also facilitate an in-depth analysis of the meaning presented in cognitive categorization information systems - in particular UBIAS systems (*Understanding Based Image Analysis Systems*). This paper also presents the IT mechanisms of object meaning description on selected examples of long bone fractures image analysis. The procedures for such semantic reasoning are based on the model of cognitive resonance and cognitive analysis. These have been applied to the task of interpreting the meaning of selected diagnostic images from the long bone fractures system as an intelligent analysis module in IT systems. The application presented in this paper is of a research character and it serves the preparation of efficient lesion detection methods applied to a dataset originating from of the long bone fractures structures.

1 Introduction

All fields of science have progressed for years, but one of the fastest to develop is computer science, as it is applied in all other scientific disciplines, from humanities to technical sciences. The rapid progress of computer science has also influenced economics, sociology and philosophy. It was on the scientific foundations from these fields that the method of analysis, understanding and reasoning was built. That method is the starting point and the reason for the operation of intelligent information system. Cognitive analysis methods based on the principles of understanding the researched phenomena using the model of analysis conducted by humans have been mentioned more and more and with increasing understanding recently. Every one of us, when analysing a phenomenon in depth, tries to learn the reasons for its occurrence, the whole course of it, but also tries to draw conclusions as to those reasons and the course of the phenomenon. In addition, reasoning about the effects that a given phenomenon may cause seems extremely important. This method of analysis is not only the way to analyze the phenomenon correctly, but it also guarantees that the analysis will be right, the phenomenon will be classified and understood correctly and the reasoning about the future, meaning the possible effects of the phenomenon, will be true. This process is most usually the prerogative of humans, but there are situations where it can be transferred and applied for computer (machine) analysis. A properly designed system using step by step analysis processes based on how humans think is referred to as a cognitive system operating on the basis of the cognitive analysis conducted for a given phenomenon. Such systems are currently used with great success for understanding and analyzing various types of medical images.

Intelligent, cognitive information systems used to analyze varied, often extremely complex medical images have been developing extremely fast for many years as scientists and researchers try to answer the question how much the efficiency of this type of systems will allow humans to be replaced in making the final decision and whether this type of process is at all practicable. The whole class of computer systems designed for analyzing various types of images as well as the whole class of diagnostic support systems have overstepped their originally set functional limits which restricted the operation of such systems to visualizing and classifying patterns. At first it was thought that those systems would be used only for diagnostic jobs, so their operations would boil down to making simple statements without the practical possibility of verifying those. This type of IT systems were not sufficient for meaning interpretation jobs and for analyzing complex medical data, which would require imitating the thought processes of diagnosticians and taking steps towards understanding the semantics of the analyzed images. Consequently, within the broad class of IT systems, a subclass of systems was developed which were oriented towards jobs of analyzing various medical patterns, with the capability of conducting semantic reasoning based on the meaning information contained in the analyzed image. This is the UBIAS (Understanding Based Image Analysis Systems) class, the functional structure of which the authors have defined [11,13].

UBIAS cognitive information systems were thus developed on the basis of intelligent IT systems whose purpose was not just the simple analysis of data by storing, processing and interpreting it, but mainly an analysis based on understanding and reasoning of an about the semantic contents of the processed data. This is a significant extension of the capabilities of previous information systems.

Every information system which analyzes a selected image or information based on certain characteristic features of it contains in its database the knowledge indispensable for performing the correct analysis or reasoning, which forms the basis for generating the system's expectations of the analysis conducted. Combining the actual features of the analyzed image with the expectations of the semantic contents of the image generated based on the knowledge (about the pattern studied), brings about a phenomenon called the cognitive resonance. This phenomenon has been described more broadly in the publication [18,23], but the notion behind it will also be presented in the next subsection.

UBIAS cognitive information systems are based on methods which lay down structural reasoning techniques to fit patterns [14,19,24]. Consequently, the structure of the image being analyzed is compared during the analysis process to the structure of the image representing such a pattern. The comparison is conducted using sequences of derivation rules which allow this pattern to be generated unanimously. These rules, sometimes called productions, are defined in a specially introduced grammar, which in turn defines a certain formal language or a so-called image language. The image (information) recognized in this way is assigned to the class which contains the pattern representing it. The analysis and reasoning process is conducted using the phenomenon of cognitive analysis, whose main element and also one of its foundations is the cognitive resonance phenomenon.

2 Cognitive Aspects of the Operation of Intelligent Information Systems

Every information system operates on the basis of various external and internal conditions. One of the internal conditions is the degree of improvement of the analysis conducted by the system and the ability of its possible extension and streamlining. The improvement of the process of analyzing various forms of data should be based on building functionalities that conduct the cognitive analysis to ensure its correct course and the best reasoning. The cognitive analysis process runs based on two fundamental though processes that occur in the human brain:

– understanding processes,
– reasoning processes.

Both processes are of the same importance in the analysis process because they complement each other. Conducting the understanding process is necessary to make an in-depth analysis of a given phenomenon, but to analyze data completely, the system must conduct a reasoning analysis.

In the case of data analysis, the first stage in the whole understanding process is to input into the system the analyzed image (it may have the form of image-type data, but also of feature vectors) which is then subjected to a process that analyses its basic features, including its form, shape, structure and the content of the analyzed data (fig. 1).

The analysis stage isolates characteristic features of the analyzed data present only in this particular case. Here it is worth stressing that an intelligent information system, while analyzing data, also performs many other operations aimed at generating certain significant expectations as to the contents of the analyzed data.

The process of generating expectations is based on the knowledge gained by the system about a given phenomenon, so the more knowledge the system has, the greater the number of generated expectations as to the contents of the data analyzed. Combining both processes and results achieved by completing them leads to cognitive resonance, i.e. the isolation of common features obtained both by the analysis conducted and generated using the knowledge implemented in the system.

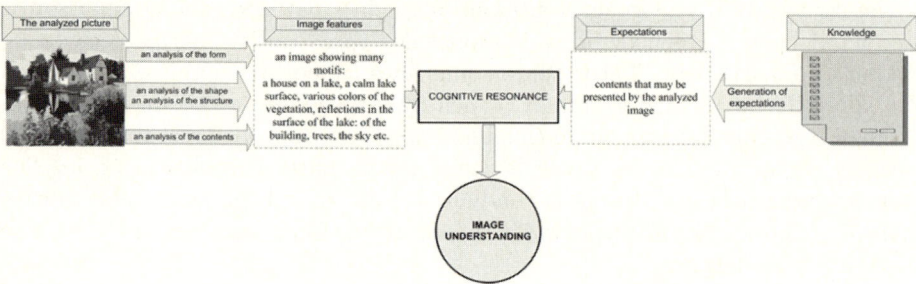

Fig. 1. Cognitive resonance in data analysis

Cognitive resonance becomes the starting point for the data understanding process which consists not just in the simple analysis of data, but uses its semantic meaning which allows it to reason.

The reasoning process which follows the understanding process is necessary for the correct data analysis, because if it is not conducted, there is no way of forecasting and drawing conclusions as to the future of the phenomenon in question. Thus, conducting an analysis without the reasoning process would impoverish the entire analysis process by limiting it to understanding only the reasons why the phenomenon occurred, without any chance of determining its further development.

3 Cognitive Categorization Systems as an Opportunity to Develop Intelligent Information Systems

A cognitive analysis based on processes of learning about and understanding the phenomenon considered offers an opportunity for developing intelligent information systems in almost every field in which data analysis is currently used. The general class of information systems that use cognitive analysis for semantic interpretation and reasoning are referred to as cognitive categorization systems, among which systems for understanding various forms of information as well as data analysis and interpretation systems feature significantly (fig. 2).

UBCCS cognitive categorization systems are divided into decision-making systems (UBDSS) designed to make decisions and draw conclusions based on decisions taken by the system. Management support systems (UBMSS) developed for taking strategic corporate decisions represent a subclass of cognitive systems. Apart from UBDSS and UBMSS systems, there are also UBIAS cognitive systems for analyzing image data which are used to analyze various types of medical images [12,13,14,15,16,17,18] and which represent the main subject of this publications. UBPAS systems are person authentication systems which include biometric identification systems, for example. Another subclass of UBCCS are UBDAS systems for various signal analysis and UBACS for automatic control.

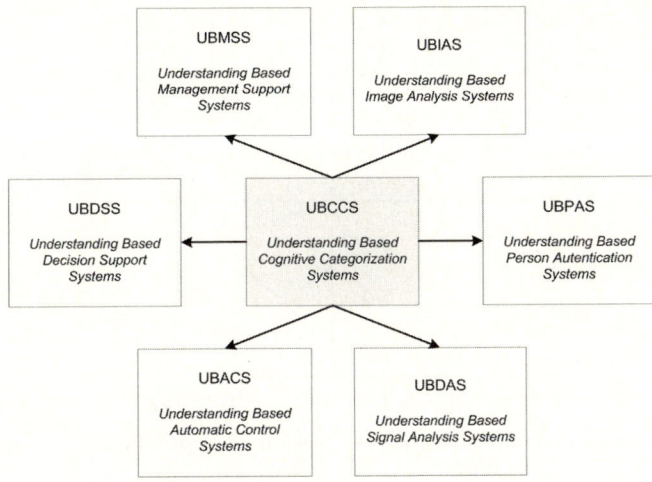

Fig. 2. The classification of cognitive categorization systems

All the presented cognitive categorization system types are widely applied in fields ranging from medicine and economics to technical or natural sciences and the military.

The operation of all the above cognitive categorization systems is based on using cognitive analysis methods for classical data analysis and for reasoning on the basis of the semantics (meaning) of the information analyzed. A novel approach to data analysis in cognitive categorization systems uses cognitive resonance, which plays a significant role in understanding the data analyzed [10,18]. Attempts to conduct analyses in UBIAS systems seem particularly important due to the extremely complex type of data subjected to system analysis.

4 UBIAS Systems as an Example of Using Cognitive Categorization for Interpreting Long Bone Fractures

Example of application of structural formalism for semantic categorization of medical images is lesion analysis in case of leg bones abnormalities or injuries interpretation. Such analysis is possible both for arm and leg bones, but further will be presented example of interpretation of various types (shapes) of leg bones fracture, and stages of theirs recovery.

In order to describe the shape of the analyzed bone in the form of linguistic representation, it is necessary to approximate the bone edges by sections, and introduce a terminal symbol labels, depending on theirs orientation (figure 3). Such operation allow to obtain a string of symbols which represent the outer edges of analyzed structures. The way of assignment terminal symbols to the spatial orientation angels was presented in fig. 3.

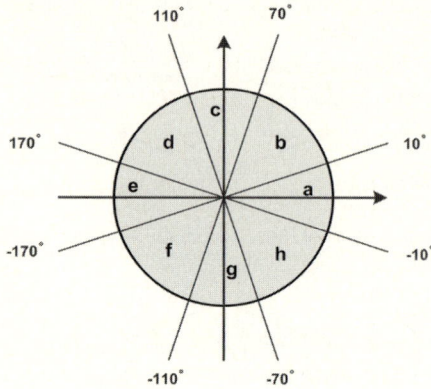

Fig. 3. Terminal symbols and theirs relation to approximate section angels

Fig. 3 presents the way of making linguistic description for approximating section describing the upper bone edges. For similar description of bottom edges the angels must be changed of 180°.

For detection of the most common leg bone fractures the following attributed grammar has been proposed (fig. 4):

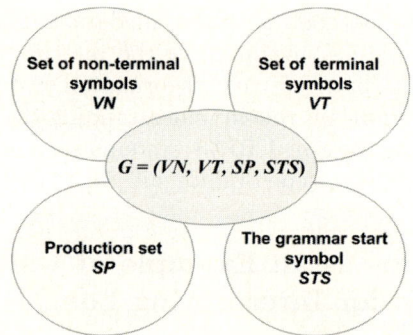

Fig. 4. Attributed grammar G

where:
VN = {RESULT, FRACTURE, FISSURE, TRANSVERSE, SPIRAL, ADHESION, DELAYED_UNION, DISPLACED_M1, DISPLACED_M2, DISPLACED1, DISPLACED2, LONGITUDINAL, A, B, C, D, E, F, G, H}
VT = {'a', 'b', 'c', 'd', 'e', 'f', 'g', 'h'} where symbols defined as follows:
a∈[-10°, 10°], b∈(10°, 70°], c∈(70°, 110°], d∈(110°, 170°], e∈(170°, -170°), f∈(-110°, -170°], g∈(-70°, -110°], h∈(-10°, -70°].
STS = RESULT. A production set SP is presented in table 1.

Table 1. Set of grammar rules defining types of fractures

Lesion	Grammar rules
Fissure fracture	RESULT → FRACTURE → A FISSURE A FISSURE → H B | H A B
Transverse fracture	RESULT → FRACTURE → A H E | A TRANSVERSE E | A G F E | A TRANSVERSE H E TRANSVERSE → H G | H F
Spiral fracture	RESULT → FRACTURE → A SPIRAL A SPIRAL → ADHESION F | ADHESION G F | ADHESION F E | ADHESION F G | H F | G F | F G | F H | F ADHESION → B A H | B H
Displaced fracture	RESULT → FRACTURE → DISPLACED_M1 F | DISPLACED1 F | DISPLACED_M2 D | DISPLACED2 D DISPLACED_M1 → B A | B G | B H DISPLACED_M2 → H G | H F | H E DISPLACED1 → B A H G | B A H | B A G | B A G H DISPLACED2 → H G F E | H G E
Delayed union fracture	RESULT → FRACTURE → A DELAYED_UNION A DELAYED_UNION → ADHESION ADHESION | ADHESION A ADHESION | ADHESION G ADHESION | ADHESION C ADHESION | ADHESION G A ADHESION | ADHESION G C ADHESION | ADHESION G A C ADHESION | ADHESION A C ADHESION | ADHESION C ADHESION | ADHESION B C ADHESION ADHESION → B A H | B H
Longitudinal fracture	RESULT → FRACTURE → A LONGITUDINAL E LONGITUDINAL → TRANSVERSE TRANSVERSE | TRANSVERSE E TRANSVERSE | TRANSVERSE E H | H E TRANSVERSE | H E H
Adhesion	RESULT → FRACTURE → A ADHESIONE ADHESION → B A H | B H
Elements of the detected lesions	A→'A'A|'A' B→'b'B|'b' C→'c'C|'c' D→'d'D|'d' E→'e'E|'e' F→'f'F|'f' G→'g'G|'g' H→'h'H|'h'

The proposed grammar is designed not just for simple image analysis, but also becomes the starting point for conducting a semantic reasoning about the analyzed fractures. The examples of bone fractures described by the authors have been subjected to a descriptive analysis leading to making a medical diagnosis, but an attempt has also been made to reason out the substantive and semantic content of the analyzed image presenting a long bone fracture.

An example result of bone fracture detection and analysis is presented in fig. 5. Figure 5a presents a limb fracture after a certain time of union, together with the periosteum growing around it. The UBIAS system has recognised this lesion as a bone fracture at the phase of hard bone matter growth. The fracture is visible in the image, but the analyzed area is partially filled with periosteum. Figure 5b presents the bone fracture with the oblong fracture automatically detected by the UBIAS system.

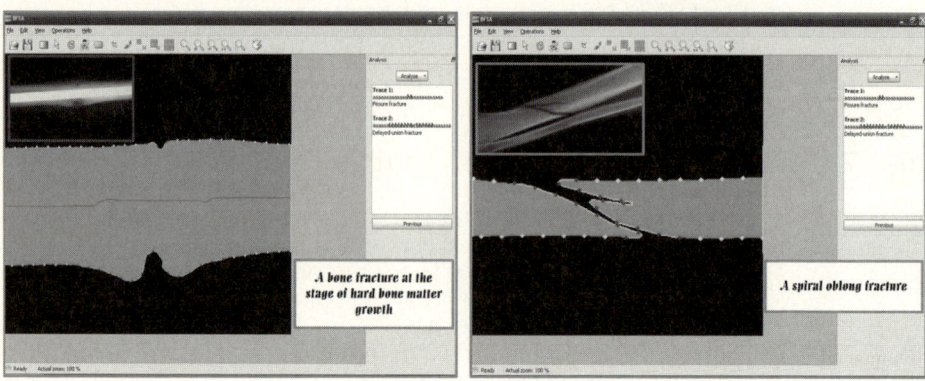

Fig. 5. a) A description of a bone fracture; the marked fragment shows the bone fracture at the phase of hard bone matter growth detected by the system b) A description of a bone fracture with a oblong fracture detected by the system marked

The UBIAS systems presented here, designed for analyzing images of lower and upper limb fractures, serve to analyze various types of fractures, including spiral, longitudinal, displaced ones, factures at the phase of bone reconstruction and at the phase of hard bone matter growth. The jobs of analyzing the lesions and pathologies in the field of long bone fractures make use of cognitive analysis on the basis of which the system for cognitive analysis and interpretation of image-type data conducts reasoning and analysis using semantic information contained in the image under consideration.

5 Conclusions

Accepting cognitive analysis as an indispensable element of data analysis processes offers the opportunity to conduct an in-depth semantic analysis of the phenomenon being researched. The semantic analysis based on the learning and understanding processes taking place in the human mind offers an in-depth and "broadband" analysis.

Acknowledgement

This work has been supported by the Ministry of Science and Higher Education Republic of Poland, under project number N516 025 32/2881.

References

1. Albus, J.S., Meystel, A.M.: Engineering of Mind - An Introduction to the Science of Intelligent Systems. A Wiley-Interscience Publication, John Wiley & Sons Inc. (2001)
2. Branquinho, J. (ed.): The Foundations of Cognitive Science. Clarendon Press, Oxford (2001)

3. Brejl, M., Sonka, M.: Medical image segmentation: Automated design of border detection criteria from examples. Journal of Electronic Imaging 8(1), 54–64 (1999)
4. Burgener, F.A., Meyers, S.P., Tan, R.K., Zaunbauer, W.: Differential Diagnosis in Magnetic Resonance Imaging. Georg Thieme Verlag (2002)
5. Chomsky, N.: Language and Problems of Knowledge: The Managua Lectures. MIT Press, Cambridge (1988)
6. Davis, L.S. (ed.): Foundations of Image Understanding. Kluwer Academic Publishers, Dordrecht (2001)
7. Duda, R.O., Hart, P.E., Stork, D.G.: Pattern Classification, 2nd edn. A Wiley-Interscience Publication, John Wiley & Sons Inc. (2001)
8. Klopotek, M.A., Wierzchoń, S.T., Trojanowski, K. (eds.): Intelligent Information Processing and Web Mining Proceedings of the International IIS: IIP WM 2004 Conference Held in Zakopane, May 17-20, 2004. Springer, Poland (2004)
9. Meystel, A.M., Albus, J.S.: Intelligent Systems - Architecture, Design, and Control. A Wiley-Interscience Publication, John Wiley & Sons Inc., Canada (2002)
10. Ogiela, L.: Usefulness assessment of cognitive analysis methods in selected IT systems. Ph. D. Thesis. AGH, Kraków (2005) (in polish)
11. Ogiela, L.: Cognitive Understanding Based Image Analysis Systems (UBIAS) of the Diagnostic Type. In: IEEE International Workshop on Imaging Systems and Techniques - IST Krakow, Poland, May 4-5, CD-ROM (2007)
12. Ogiela, L., Tadeusiewicz, R., Ogiela, M.: Cognitive Analysis In Diagnostic DSS-Type IT Systems. In: Rutkowski, L., Tadeusiewicz, R., Zadeh, L.A., Żurada, J.M. (eds.) ICAISC 2006. LNCS (LNAI), vol. 4029, pp. 962–971. Springer, Heidelberg (2006)
13. Ogiela, L., Tadeusiewicz, R., Ogiela, M.R.: Cognitive Approach to Visual Data Interpretation in Medical Information and Recognition Systems. In: Zheng, N., Jiang, X., Lan, X. (eds.) IWICPAS 2006. LNCS, vol. 4153, pp. 244–250. Springer, Heidelberg (2006)
14. Ogiela, M.R., Tadeusiewicz, R., Ogiela, L.: Inteligent Semantic Information Retrieval In Medical Pattern Cognitive Analysis. In: Gervasi, O., Gavrilova, M.L., Kumar, V., Laganá, A., Lee, H.P., Mun, Y., Taniar, D., Tan, C.J.K. (eds.) ICCSA 2005. LNCS, vol. 3483, pp. 852–857. Springer, Heidelberg (2005)
15. Ogiela, M.R., Tadeusiewicz, R., Ogiela, L.: Image languages in intelligent radiological palm diagnostics Pattern Recognition 39, pp. 2157–2165. Elsevier Ltd, Amsterdam (2006)
16. Ogiela, M.R., Tadeusiewicz, R., Ogiela, L.: Graph image language techniques supporting radiological, hand image interpretations. In: Computer Vision and Image Understanding 103, pp. 112–120. Elsevier Inc., Amsterdam (2006)
17. Tadeusiewicz, R., Ogiela, L., Ogiela, M.R.: Cognitive Analysis Techniques in Business Planning and Decision Support Systems. In: Rutkowski, L., Tadeusiewicz, R., Zadeh, L.A., Żurada, J.M. (eds.) ICAISC 2006. LNCS (LNAI), vol. 4029, pp. 1027–1039. Springer, Heidelberg (2006)
18. Tadeusiewicz, R., Ogiela, M.R.: Medical Image Understanding Technology, Artificial Intelligence and Soft-Computing for Image Understanding. Springer, Berlin (2004)

19. Tadeusiewicz, R., Ogiela, M.R.: New Proposition for Intelligent Systems Design: Artificial Understanding of the Images as the Next Step of Advanced Data Analysis after Automatic Classification and Pattern Recognition. In: Kwasnicka, H., Paprzycki, M. (eds.) Intelligent Systems Design and Applications, Proceedings 5th International Conference on Intelligent Systems Design and Application, pp. 297-300. IEEE Computer Society Press. Los Alamitos (2005); Proceedings 5th International Conference on Intelligent Systems Design and Application - ISDA 2005, September 8-10, Wrocaw (2005)
20. Tadeusiewicz, R., Ogiela, M.R.: Automatic Image Understanding - A New Paradigm for Intelligent Medical Image Analysis (opening article). Bio.-Algorithms And Med.-Systems 2(3), 5–11 (2006)
21. Tadeusiewicz, R., Ogiela, M.R.: Why Automatic Understanding? In: Beliczynski, B., Dzielinski, A., Iwanowski, M., Ribeiro, B. (eds.) ICANNGA 2007. LNCS, vol. 4432, pp. 477–491. Springer, Heidelberg (2007)
22. Tadeusiewicz, R., Ogiela, M.R., Ogiela, L.: A New Approach to the Computer Support of Strategic Decision Making in Enterprises by Means of a New Class of Understanding Based Management Support Systems. In: Saeed, K., Abraham, A., Mosdorf, R. (eds.) CISIM 2007 - IEEE 6th International Conference on Computer Information Systems and Industrial Management Applications Elk, Poland, June 28-30, pp. 9–13. IEEE Computer Society Press, Los Alamitos (2007)
23. Tadeusiewicz, R., Ogiela, M.R.: Automatic Understanding of Images. In: Adipranata, R., Rostianingsih, S. (eds.) Proceedings of International Conference on Soft Computing Intelligent System and Information Technology (ICSIIT 2007), Bali - Indonesia, July 26-27, 2007. Special Book for Keynote Talks Informatics Engineering Department Petra Christian University Surabaya, pp. 13–38 (2007)
24. Tanaka, E.: Theoretical aspects of syntactic pattern recognition. Pattern Recognition 28, 1053–1061 (1995)
25. Zhong, N., Raś, Z.W., Tsumoto, S., Suzuki, E. (eds.): Foundations of Intelligent Systems 14th International Symposium ISMIS 2003, Maebashi City, Japan (2003)

Predictive Control for Artificial Intelligence in Computer Games

Paweł Wawrzyński[1], Jarosław Arabas[2], and Paweł Cichosz[2]

[1] Institute of Control and Computation Engineering
[2] Institute of Electronic Systems,
Warsaw University of Technology,
Nowowiejska 15/19, 00-665 Warsaw, Poland
{p.wawrzynski,j.arabas,p.cichosz}@elka.pw.edu.pl

Abstract. The subject of this paper is artificial intelligence (AI) of non-player characters in computer games, i.e. bots. We develop an idea of game AI based on predictive control. Bot's activity is defined by a currently realized plan. This plan results from an optimization process in which random plans are continuously generated and reselected.

We apply our idea to implement a bot for the game Half-Life. Our bot, Randomly Planning Fighter (RPF), defeats the bot earlier designed for Half-Life with the use of behavior-based techniques. The experiments prove that on-line planning can be feasible in rapidly changing environment of modern computer games.

1 Introduction

Computer video games constitute a rapidly growing industry, which has catched up with more mature entertainment industries during the last decade [1]. The increasing computing power available for home users, as well as technological achievements in computer graphics hardware and software, have allowed an outstanding progress in the complexity and reality of game environments. It is one major factor permitting vastly improved enjoyability of recent computer games. There is another possible factor that could lead to even further and similarly spectacular improvement of gaming experience still awaiting a comparable progress. This is the improvement of the behavior exhibited by non-player game characters (NPCs or bots), constituting opponents or partners for human players. Most existing approaches to NPC behavior control are based on relatively simple sets of rules or finite-state machines representing pre-programmed patterns of a bot's reaction to particular environment conditions [2]. Such bots are either too primitive and easily predictable to provide sufficient challenge to human players, or require overly complex code and substantial programming effort to implement and test (whilst still remaining prone to nonsense behavior in some conditions due to possible programmer overlookings). It is more than likely that human-level behavior cannot be achieved using pre-programmed reaction patterns of manageable complexity. A promising alternative approach is to employ techniques developed in the area of artificial intelligence (AI).

The requirement of beeing competitive to humans is very closely related to the long-term goal of AI, which consists in creating artificial systems exhibiting human-like intelligent behavior [11]. Whereas in most current successful applications of intelligent computational techniques only specific narrowly defined tasks have to be performed [9,12], the domain of computer games provides a good testbed for a more general intelligent behavior task [8]. Techniques that prove successful when applied to computer games may turn out extremely useful also outside the entertainment industry, e.g., for providing high-quality opponent simulation for military training.

The paper presents a novel approach to applying inteligent computation to creating characters for action games. In such games the human player controls a character in a virtual world, interacting (usually fighting) with other characters, either controlled by other human players or other bots. The most common genre of such games is known as FPS (first person shooter), with Doom, Quake, or Half-Life beeing some well-known examples. Our bot control algorithm combines high computational efficiency, permitting application in complex game environments, with excellent exhibited skill level, which is essential for providing good gaming experience to human players.

2 Computer Game Artificial Intelligence

The possibility of applying AI for creating computer game characters is quite obvious and probably has been pointed out independently by several rearches and game developers. Systematic work in this direction has been initialized in the late 90s. Despite relatively little progress made so far in bringing intelligent computational techniques to the game industry, the term 'game AI' has become quite popular and constitutes a desirable selling point for computer games. The major goals of game AI in a technical rather than marketing sense can be summarized as creating bots behaving reasonably in a variety of different conditions, hardly predictable, capable of observing and using the environment, and self-improving.

The idea of applying AI techniques to create control algorithms for game bots was convincingly argued for by Laird and van Lent [8], who were among major originators of game AI within the academic AI community. They demonstrated its potential by creating a *Quakebot* [7], a non-player character for the popular FPS game Quake II. Their bot maintains a hierarchy of operators and a corresponding hierarchy of abstract goals to be achieved using these operators. The control algorithm performs a main loop in which operators matching the current state are proposed, selected, and applied. This is essentially a search algorithm (as used by AI problem solving techniques) and has been implemented within a well-known general problem solving architecture, Soar [6]. Subsequently Laird [5] proposed introducing a form of planning to this framework, consisting in the anticipation of opponent actions (assuming a set of possible opponent tactics).

Orkin [10] presents a somewhat different approach, which is entirely based on real-time planning, where possible actions (or operators) are specified through a set of preconditions and effects and an A*-based planning algorithm is used to create sequences of actions to achieve particular goals. Significant changes in the environment can trigger plan revision.

Spronck in his thesis [13] examines the possible use of machine learning techniques for making inteligent bots adaptive. He considers the usefulness of both offline learning (during game development) and online learning (during actual performance). The solution proposed in the thesis is based on an evolutionary learning method and a *dynamic scripting* technique. There is no search or planning involved in this approach. Instead, scripts (rule bases) representing the bot's control policy are being refined using their observed performance as feedback.

Khoo et al. [3] point out severe computational efficiency problems with bot control algorithms based on extensive search and planning and argue that the complexity involved in such approaches may not be neccessary to ensure a good level of game playability. They suggest that a successful and seemingly humanlike bot can be achieved using much simpler and more efficient techniques borrowed from behavior-based robotics. They describe a bot called *LedgeWalker* for the *Half-Life* game environment which maintains a prioritized set of simple behaviors.

Our approach described in the remainder of this paper is motivated by looking for a most reasonable compromise among several different views and research directions expressed in the prior work. Whereas we agree that full-fledged search or planning may be both overly expensive and unnecessary, we find the simplistic behavior-based approach insufficient to match the challange of complex games and good human players. Instead of employing general purpose search and planning techniques we therefore come out with a tailored real-time planning algorithm, with plans consisting of a predefined set of tasks. Unlike for operators in traditional planning, their effects are expressed numerically in terms of expected costs or profits rather than postconditions that should hold upon their completion. This permits a highly efficient planning algorithm and creates a natural possibility of employing offline or online learning to achieve adaptation and self-improvement capabilities.

3 The General Idea: Planning and Executing Tasks

Designing a bot control system, we intend to utilize capabilities of planning and problem solving. However, we also want to exploit the fact that certain behaviors leading the bot to nontrivial objectives are quite easy to implement by game developers. Therefore, we organize the bot control in two level hierarchy. At the lower level, the bot realizes one of preprogrammed *tasks*. A task is a complex activity that leads to a certain tactical objective, like shooting the player to its death or running toward a wall corner to hide behind it. The task may be parameterized: e.g. the running toward a wall is parameterized by the specific wall corner to which the bot is to run. At the upper control level, a sequence of tasks, *a plan*, is optimized with respect to a numeric criterion.

A definition of a task encompasses, among others, two elements. On of them is the task's *feasibility condition* that characterizes game states that enable the begin of the task. The other one is the final state that defines a prognosis of the game state at the moment of the task completion. If one task's final state satisfies the feasibility condition of another task, these tasks are *logically connected*.

A sequence of logically connected tasks is a *plan*. On the higher level of control, random plans are generated, the best one among them is chosen to be realized. Each plan is characterized by a numeric *income* and *cost*; their difference, a *profit*, defines the quality of the plan that allows to determine which one is the best. The incom, cost, as well as the final state of the task is computed on the basis of the initial game state of the task.

3.1 Tasks

A task is defined by six entities: a procedure that determines activietis of the bot in each state, completion and feasibility conditions, income and cost, and a final state. All of them are defined below.

A procedure that determines activities of the bot in each state. E.g. "running toward a wall corner in order to hide behind it" requires determining the direction to run to, running, and probably shooting the chasing player.

A completion condition. The task can not be realized any longer when the game state meets this condition. E.g. "running toward a wall corner..." is completed when the bot is already hidden behind the wall corner.

While the above entities are necessary to define the bot's current activity, the below ones are necessary to make plans.

A feasibility condition. It is a condition that the game state satisfies when a begin of the task makes sense. Usually, it is negation of the completion condition: if it makes sense to begin a certain task, its continuation makes sense as well. However, there may be some differences: a badly armed bot may want to escape from a visible player and hide behind a wall corner. However, during the escape, the bot actually does not have to see the player.

The lasts three elements are defined by a prognosis of the course of a task to its completion. The prognosis is based on a given game state.

An income. It is the most natural to measure income in the units of player's health. There may be several sources of task's income and the most important are: (i) a prognosis of losses that the bot is going to inflict on the player and (ii) weapons and ammunition that the bot is going to collect.

A cost. As above, income is naturally measured in the units of the player's health. There are several elements that may define the cost of the task: (i) a prognosis of losses that the player inflicts on the bot and (ii) the ammunition that the bot is going to waste shooting the player.

A final state. It is a prognosis of the game state when the task is completed.

Notice the important feature of our concept of task: it is accompanied by a deterministic prediction, based on a given game state, of the course of the task until its completion. Within our design, only one possible behavior of bot's surrounding is analyzed. On one hand, it is a weakness, because it prevents the bot from being ready to various possible actions of the player. On the other hand, it is strength, becuase it enables fast planning in a rapidly changing environment.

3.2 Planning

The general idea of planning is to develop, maintain, and actualize a set of plans such that each plan is a sequence of tasks. The plan currently realized is the one with the highest profit. Within each plan maintained in the memory, the tasks have to be logically connected and the current game state does not satisfy the completion condition of the first task of any plan. The profit of the plan is calculated as the sum of the profits of the contained task and a heuristic quality assessment of the plan's final state.

Let us discuss our idea of planning in reference to the well known minimax algorithm usually applied in computer programs that play board games [12,9]. In comparison to this method, our planning mechanism has two important features. First, various moves of the other opponent (i.e. the player) are not analyzed. The opponent is assumed to move in the known way in response to our move (i.e. to performing a given task by the bot). Second, a whole game tree is not build. There is no computational power for that since the moves are to be determined fast in real time. Only some paths in the game tree are generated, the plans, and the best of these is chosen.

Because of the severe computational power limitations, the planning has to be a continuous process yielding good plans through all the time. For a radical change of the game state (e.g. the player suddenly appears), the planning process should be efficient enough to determine a sensible, not necessary optimal, plan at once (e.g. to determine whether to attack or escape the player). However, when the game state changes slowly according to the current plan, the process should approximate the optimal plan.

In order to meet the requirements mentioned above, the planning process maintains a small set of plans. They are the best ones from among those analyzed so far. Once a certain number of game cycles (or every game cycle) these plans are exposed to the following operations:

Verification. For each plan it is checked whether the present game state satisfies the completion condition of its first task. If so, the first task is deleted and it is checked whether the present game state satisfies the completion condition of the new first task. If so, the plan is deleted.

Extension. New plans are generated and existing plans are extended, i.e. randomly generated (albeit logically connected) tasks are joined to their ends.

Evaluation. Profits of all plans are computed. A certain number of the most profitable plans remain and others are deleted.

Selection. The most profitable plan is chosen as the one executed. Its first task is preformed in the following game cycle.

Let us now compare our bot controller to the *Simple Predictive Controller* (SPC). The idea of SPC is to determine a control stimulus within each step as follows. A sequence of control stimuli on a fixed horizon starting from the present moment is optimized assuming a deterministic response of the controlled system to control.

The first stimulus of the best sequence becomes the current control action. In principle, what we propose here follows the idea of SPC, only the optimization process is specific: There is no way to perform an exhaustive optimization of the sequence of control stimuli, and thus we suggest to pick the best one from among randomly generated few.

4 Tasks and Plans in Randomly Planning Fighter

Veryfying the idea presented in the previous section, we implment a bot for the game Half-Life. The bot is called *Randomly Planning Fighter* (RPF). It is coded in C++ with the use of FlexBot, the game-programmer interface developed at Northwestern University [4]. FlexBot enables the bot to "sense" the surrounding world by sensors and to execute actions by actuators. It is a freeware provided by the authors with an exhaustive documentation.

4.1 Perception, Memory, and Consciousness of RPF

All the information available to the bot come from its sensors. The information is processed and stored in three structures that support planning and realizing tasks: (i) geometrical consciousness collects and stores information on walls around the bot, (ii) extended state stores relevant information previously read from the sensors, and (iii) some information is directly read from sensors.

Geometrical consciousness. Unfortunately, through sensors, the bot can not identify walls which are a key tool for strategic behavior like hiding behind them or attacking from around their corners. However, FlexBot enables the bot to measure distance to the nearest obstacle in a given direction. The bot can thus identify walls based on the following property: three obstacles on the same line usually make a wall. The bot maintains a map of hypothetical walls around it. Periodically, once every 5 game cycles, the bot measures distance in 10 random directions to gather data and use it to adjust this map.

Extended state stores all the information necessary for planning that is not available in geometrical consciousness. Within the planning process the predictions are made on changes of the extended state. It includes (i) the position and health of the bot, (ii) the weapon and the armament of the bot, (iii) objects available to collect, (iv) *level of danger* defined as follows; 4 - the bot is under fire and does not see the player, 3 - the bot can see the player, 2 - the bot has recently seen the player, 1 - the bot has heard the player, 0 - the bot has not seen nor heard the player recently; levels 4, 2, and 1 are maintained for 4 seconds from the time their reason occurred; this time is registered, (v) the position of the player (relevant for 3rd level of danger), (vi) the ealth level of the player and the path to the point the player has been seen for the last time.

Perception. There is some information that is not kept in the extended state that is needed for performing tasks, like bot orientation in the space. This is directly read from the sensors.

4.2 Incomes, Costs, and Final State Prognosis

Incomes and costs are measured by means of the same units as health level which, for players as well as for bots, belongs to the interval $[0, 100]$. Typically, incomes and costs are governed by the following rules. (i) A cost of the bot's health loss is equal to the loss. (ii) An income from the player's health loss is 3 times greater than the loss. (iii) Elimination of the player yields the income of 100. (iv) Elimination of the bot costs 100. (v) Ammunition usage costs as much as health loss caused by the ammunition when applied in the optimal way (from the distance that is optimal for this weapon). (vi) Income induced by the ammunition collecting is equal to health loss caused by the ammunition when applied optimally. (vii) Income induced by the weapon collecting is equal to health loss caused by one magazine of this weapon when applied optimally.

In presence of the player, the typical prognosis of the income, cost, and final state is implied by an assessment of health losses incured by the bot and the player before the completion of the task. The assessment is based on the assumption that the bot uses its weapon reasonably and the player uses the *typical* weapon. The bot can not find out what kind of weapon the player has, and thus it assumes typical player's weapon defined as one that causes 1.5 times larger damages than a pistol.

4.3 Tasks

Our RPF bot realizes a number of tasks. We will present all of them, albeit most of them very briefly.

Permanent attack. The bot changes weapon to the best for the current distance to the player. Then, it shoots the player and moves such that the distance between the bot and the player optimal for the currently applied weapon is stabilized. When the current ammunition is exhausted, the bot either reloads the magazine or changes the weapon. Concurrently, the bot moves chaotically in order to make difficult to take a bearing on it, but does not allow the player to hide behind any wall. *Completion condition.* The player is not visible (or alive) any more. *Feasibility condition.* Danger level equal to 3. *Income and cost:* Both are typical, but it is not considered that both the player and the bot survive. *The final state:* Either the bot or the player survives, depending on the calculation of the power of the bot fire, the power of the player fire (assumed typical), the bot health, and the player health. In case the bot is predicted to survive, its prognosed final position is equal to its initial position.

Wild attack. Like "permanent attack", but the bot all the time gets closer to the player. If necessary, the bot changes weapon suitably to the distance to the player. If the distance is small enough, the bot uses the bar.

Attack with collecting weapon. The bot shoots the player moving toward weapons lying on the floor. *Feasibility condition:* The player is visible and the bot is aware of weapons that are possible to collect.

Attack from around a corner. The bot is close to a wall corner. It recurrently hides behind the corner and moves from around it to shoot the player.

Escape from a visible player. The bot shoots the player moving toward a wall corner to hide behind it.

Escape from an invisible player. The bot moves toward a wall corner watching the point that the player is expected to appear.

Armaments in presence of an invisible player. The player has been visible, but now he is not. The bot is watching the point of possible emergence of the player and goes toward a weapon lying on the floor.

Lying in wait for a player. The bot knows where the player may emerge. It is standing and watching this point with the weapon ready to fire.

Collecting a weapon. The bot is coming to a weapon lying on the ground while it is watching dangerous places around it.

Territory exploration. The bot is walking toward a randomly chosen point and watching dangerous places around it.

Looking around. The bot is turning, looking, and identifying walls around it (this task is useful when the bot has just been "born").

Escape to the closest wall. The bot is turning around and moving toward the closest wall. *Feasibility condition:* Danger level equal to 4, i.e. the bot has just got shoot from an invisible player.

4.4 Plans

Let us call a plan *complete* if it satisfies at least one of the following conditions:

a) Its last task ends up with death of the bot.
b) It contains more than 1 task and ends up with danger level equal to 0 or 1.
c) It consists of the maximal number of tasks.

Consequently, if the bot attacks according to a complete plan, it either beats the player, suffers defeat, or the attack is very long. Also, we call a plan *c-complete* if it is complete but does not satisfy the conditions (a) nor (b) and only satisfies the condition (c).

All the time the bot keeps in memory a certain number of complete and incomplete plans. Periodically a property is restored that the complete plans contain in sum no more than n_1 tasks and there are no more than 3 incomplete plans. Furthermore, the maximal number of tasks within a plan is $n_1/3$. n_1 is a constant parameter.

Core Planning Procedure. The planning is performed in portions of computational complexity $O(n_1)$. A portion is a single execution of the core planning procedure:

1. For each incomplete plan, draw an extending task. If there are fewer than 3 such plans, draw missing single-task plans.
2. If Step 1. turned a certain incomplete task into complete, make the complete plans containing in sum no more than n_1 tasks. To this end, sort the complete plans with respect to their profit and delete the least profitable plans.

Active plan selection. The plan that is being executed is called *active*. If there is more than one complete plan, the most profitable form those complete is selected to be active. Otherwise, the most profitable from all the plans becomes active.

Each game cycle planning. In each game cycle it is checked whether the completion condition of the first task in the active plan is satisfied. If so, this task is realized. Otherwise, the first tasks of all plans are deleted. All the plans that were initially c-complete and not active, are deleted. If the first task of the active plan is feasible, it is executed. Otherwise, all the plans are deleted, Core Planning Procedure is executed n_2 times, the active plan is selected and executed.

Recurrently executed planning procedure. Once every 5 game cycles, a more exhaustive planning is performed. The following steps are then executed:

1. For each plan it is verified whether its first task is feasible and all its tasks are logically connected. The plan that does not comply is deleted.
2. Profit predictions are calculated for all the plans on the basis of the current game state. The prediction is equal to the sum of profits of tasks increased by the heuristic profit estimate of the final state.
3. n_3 steps of Core Planning Procedure are executed.
4. The active plan is selected.

5 Experimental Study

It seems that some of the RPF bot behavior can be regarded humanlike. The bot avoids empty spaces, where there is no place for hiding. While wandering, the bot tries to pick up as good weapon as possible. After enemy is observed, the bot attacks him. When the enemy succeeds to hide, the bot initially takes a safe position behind some terrain obstacle, and then begins to wander again in a hope to observe the enemy again. The only behaviour which cannot be classified as humanlike is a frequent looking around performed by the bot, i.e. the bot makes a 360 degrees turn about its vertical axis to observe the whole space in which it acts. This effect helps the bot to attain good efficiency in planning its actions (i.e. surviving in combats and killing its enemies), but can be recognized by human player as an unpleasant feature.

The above qualitative characteristcs is insufficient in comparing different bot algorithms, therefore we attempted to make a quantitative analysis. This attempt is discussed in the next subsection.

Numerical Results

The primary objective of the computer game is to entertain the human player. Unfortunately, this kind of objective can be neither directly expressed nor measured. Therefore we decided to express the bot "quality" as its ability to survive and to kill many opponents.

Definitely, human players differ significantly in their game playing skills, so the reliable human evaluation of the bot quality should involve multiple human players and will be cost- and time-consuming. Before such tests start, it seems adviceable to apply preliminary comparative testing of bots developed according to different approaches. We were able to make comparison to the *LedgeWalker* [3].

We performed tests under the following conditions. Only two bots act in a specific, relatively small and simplified space. The space contains of a large room with a small room in one corner, and a collections of cubic obstacles in another corner. The small room has two exits, and the obstacles are large enough for a bot to hide behind. A view at the game space is given in Fig.1.

Fig. 1. Picture of the game space used for tests

In the game space, two bots take part in the tournament; let us number them. Each bot is created in a random place of the game space, and under creation it is assigned a certain health level. During the tournament, its health level changes (e.g., when the bot is hit by the opponent, the health level decreases), and when health level falls to zero, the bot dies off, and a new bot appears in the game space. This new bot is assigned the same number, and is driven by the same algorithm. Thus it is possible to perform a long-lasting tournament to compare different algorithms to implement the bot behavior.

We have made preliminary nonsystematic tests and found that the RPF approach works quite well with the parameters $n_1 = 9$, $n_2 = 1$, and $n_3 = 1$. Therefore we used this values in the numerical experiments described in the further text.

We have performed five experiments (see Table 1 for the results). In the first experiment (labeled RPF), we attempted to verify if it is possible for the RPF bot to win over another RPF bot. If such effect were observed, this would indicate that the long-lasting tournament could not lead to reliable comparison of two bot algorithms. Then we performed two long-lasting tournaments when the opponents were: RPF and Ledgewalker. The experiments (labeled with LW1 and LW2) differed in the allocation of bot numbers to the bot algorithms, in order to indicate possible influence of the number assignment on the numerical results. Finally we made two very long-lasting tournaments of RPF vs. Ledgewalker (labeled LW3 and LW4).

For each experiment, we report how many opponents were killed in the single life period of the RPF bot, and how frequently such observation was made during the whole long-lasting tournament. We also indicate how many life periods of the RPF bot were observed during the experiment, and the fraction of the number of RPF bot enemies killed by the number of RPF bots killed. The last number is expected to exceed one and is the indicator of the RPF method relative superiority over the counterpart method.

Table 1. Results of experiments with RPF

opponent	RPF	LW1	LW2	LW3	LW4
#bots	1094	128	130	1523	2045
#kills		freq. kills			
0	0.36	0.3	0.31	0.15	0.27
1	0.42	0.31	0.32	0.52	0.38
2	0.13	0.21	0.16	0.22	0.19
3	0.043	0.062	0.11	0.059	0.080
4	0.021	0.055	0.038	0.031	0.042
5	0.010	0.031	0.0077	0.0092	0.023
6	0.0055	0.0078	0.023	0.012	0.013
7	0	0.0078	0.0077	0.0026	0.0044
8	0.00091	0	0	0.0026	0.0077
9	0.00091	0.0078	0	0.00066	0.00054
10	0.00091	0	0.0077	0.00066	0.00054
11	0	0.0078	0	0	0
12	0	0	0	0	0
13	0	0	0	0	0.0005
exp. kills	1.00	1.55	1.52	1.42	1.41

From first three columns of Table 1 we read that the assignment of numbers to bots does not influence the result, so the game provides a ,,fair play" environment. When playing against Ledgewalker, RPF is superior, since the RPF bots are able to kill approximately 1.4 – 1.5 enemies before being killed. The tournament length does not seem to have substantial influence on the evaluation of the RPF algorithm quality.

6 Conclusions

In this study, we searched for a design of artificial intelligence controlling activity of non-player characters (bots) in computer games. We developed a hierarchical AI based on preprogrammed behaviors at the lower level and predictive control at the upper level. The preprogrammed behaviors, tasks, aim at certain tactical objectives. A complete definition of a tasks encompasses several entities that enable joining tasks into sequences i,e., plans. At the upper control level the plan is optimized to maximize the expected profit of its execution. The optimization process consists in repeatedly performed random generation and selection of the best plan.

In the experimental study, a bot designed on the basis of the presented schema, Randomly Planning Fighter, significantly outperforms a bot designed with the use of behavior-based techniques. In the next step of this research, we are going to introduce self-improvement into bot's performance.

References

1. Crandall, R.W., Sidak, J.G.: Video games: Series business for America's economy. Technical report, Entertainment Software Association (2006)
2. Houlette, R., Fu, D.: The Ultimate Guide to FSMs in Games. Charles River Media (2003)
3. Khoo, A., Dunham, G., Trienensm, N., Sood, S.: Efficient, realistic npc control systems using behavior-based techniques. In: Proceedings of the AAAI 2002 Spring Symposium Series: Artificial Intelligence and Interactive Entertainment, Menlo Park, CA (2002)
4. Khoo, A., Zubek, R.: Applying inexpensive ai techniques to computer games. IEEE Intelligent Systems 17(4), 48–53 (2002)
5. Laird, J.E.: It knows what you're going to do: adding anticipation to a quakebot. In: Müller, J.P., Andre, E., Sen, S., Frasson, C. (eds.) Proceedings of the Fifth International Conference on Autonomous Agents, Montreal, Canada, pp. 385–392. ACM Press (2001)
6. Laird, J.E., Newell, A., Rosenbloom, P.S.: SOAR: An architecture for general intelligence. Artificial Intelligence 33(1), 1–64 (1987)
7. Laird, J.E., van Lent, M.: Developing an artificial intelligence engine. In: Proceedings of the Game Developers Conference, San Jose, CA, pp. 577–588 (1999)
8. Laird, J.E., van Lent, M.: Human-level ai's killer application: Interactive computer games. In: AAAI Fall Symposium Technical Report, North Falmouth, MA, pp. 80–97 (2000)
9. Luger, G.F.: Artificial Intelligence: Structures and Strategies for Complex Problem-Solving. Addison-Wesley (2004)
10. Orkin, J.: Agent architecture considerations for real-time planning in games. In: AIIDE, pp. 105–110 (2005)
11. Russell, S.J.: Rationality and intelligence. Artificial Intelligence 94, 57–77 (1997)
12. Russell, S.J., Norvig, P.: Artificial Intelligence: A Modern Approach, 2nd edn. Prentice-Hall (2002)
13. Spronck, P.H.M.: Adaptive Game AI. PhD thesis, Maastricht University, The Netherlands (2005)

Building a Model for Time Reduction of Steel Scrap Meltdown in the Electric Arc Furnace (EAF): General Strategy with a Comparison of Feature Selection Methods

Tadeusz Wieczorek[1], Marcin Blachnik[1,2], and Krystian Mączka[1]

[1] Silesian University of Technology, Electrotechnology Department,
Katowice Krasińskiego 8, Poland
{tadeusz.wieczorek,marcin.blachnik,krystian.maczka}@polsl.pl
[2] Helsinki University of Technology, Adaptive Informatics Research Centre,
P.O. Box 5400, 02015 Espoo, Finland

Abstract. Time reduction of steel scraps meltdown during the electic arc process is really a challenging problem. Typically the EAF process is stochastic without any determinism and only simple and naive rules are currently used to manage such processes. The goal of the paper is to present the way, which have been considered, to build an accurate model concerning different feature selection methods that would be helpful in predicting the end of the meltdown and maximum energy needed by the furnace.

1 Introduction

The arc-electric steelmaking process is usually carried out in three stages: the melt process (EAF), the ladle furnace process (LHF or VAD) and the casting process (CCS) - the second and the third stage of the process will not be considered in the paper. Steel scrap, alloys and fluxes are melted in electric arc furnaces, which serve as melting unit only because the liquefying must be done in as short time as possible. The liquefying of the charged materials is performed by electric energy, but this stage of the process can be supported by the use of wall and door burners in order to deliver energy to areas which cannot be directly reached by the electric arc.

Controlling of the EAF process is carried out by determination of electrical working points (set points). It takes place in step-wise adjustment of the output voltage of the furnace transformer and in an adjustment of the control value. The determination of the set points are based on experience with modeling of the EAF processes. This method of determining the set points is never completely accurate and has got a static character. A number of non-linear influences and the temporary change in the process must be taken into account. The operation of the furnace is unstable, especially during the initial meltdown period that can result in significant stochastic electrical disturbances. The EAF characteristics are essentially nonlinear, and the fluctuations of the voltage amplitude depend on the kind and quality of the charge, the usage of the furnace and its mode of operation. The amount of electrical energy transferred to the melting steel should be optimally maintained during the entire melting and refining process, with the optimal distribution of radiated heat within the furnace. The known conventional control and optimization methods (based on differential or integral equations, equilibrium chemistry, and material and energy balances) are not able to solve tasks of this kind. Due to the variability and complexity of the EAF process, accurate optimization must be based on actual operating data, which are often distorted, and requiring significant pre-processing. The steelmaking process is continually changing, as the furnace is being used, procedures and raw materials are being changed. This requires such a data-based optimization system which is adaptive. Computational intelligence (CI) methods can take over both tasks, modeling and adaptation at the same time. [7]

Research on the intelligent control of the electric-arc steelmaking process is done by the authors within the research project [1]. This paper describes one of the detailed subproblems that researchers faced, that was the energy and time reduction of steel scrap meltdown in the electric arc furnace. This problem is a typical approximation task, however there is a wide range of unexpected breaks and mistakes in the industrial data series. It makes it difficult for classical methods of analysis.

Another real problem, faced by researches, was the size of the dataset. Each data series describing each individual meltdown consist of 100.000 data samples of currents, voltages and powers for each of three electrodes, extended by other process parameters, and over 1000 different meltdowns have to be taken into consideration in this problem. Such a huge amount of data requires well suited preprocessing steps where the attributes were manually selected by some experts, and only 18 input attributes were finally chosen describing the dataset. Independently of the manually feature extraction, there was a problem of finding possibly the best regression model among a large group of all possible models. The goal is to find the one which accurately fits the data preserving generalization abilities. Testing all possible model is, of course, to complex, to be performed. However trying to search for the most promising one, one has to consider different groups of algorithms. In presented paper we have compared simple linear model with multilayer perceptron neural network (MLP), support vector machine for regression based on ϵ insensitive loss function with various kernels, and one sparse Bayesian method called relevance vector machine (RVM) with Gaussian kernels.

Except for the neural network, which automatically do the feature weighting, all other methods generally require feature selection, feature extraction or feature weighting during the data preprocessing. This leads to simple tree step procedure which can be used for searching a good model.

---

[1] Project partially sponsored by the grant No 6ZR9 2006 C/06742 from the Polish Ministry of Education and Science (MNiSZW).

During the calculations we also did a simple comparison and discussion of different approaches to feature selection. Obtained results were very surprising, where simple and fast combination of filters and wrappers approach overcome search based methods.

All these problems are considered in this paper. In section 2 the dataset is described including also preprocessing and manual feature extraction methods. Section 3 presents the way of finding a good model, and description three different steps of data analysis, and final section 4 includes discussion of obtained results for different feature selection methods and conclusions about building accurate models.

2 Data Description

Data used to computations was collected in the real process of steel production in one of Polish steelworks during EAF process. This factory uses electronic recording of process parameters. For controling of metallurgical devices, programmable logic controllers are used (Simatic-5 and Simatic-7). Signals from the controllers are written and stored in Industrial SQL Server database. There are over four hundred signals stored in the database which describe actual process of metal scrap meltdown in the electric arc furnace. There are, amongst other things: primary and secondary voltage, primary and secondary current, transformer tap position, electricity consumption, power coefficient, positions of electrodes etc. Each signal brings minimum 1 record of data per second. Due to such a great amount of data, proper analysis and preprocessing of data have to be done. The first step was to analyze time series to estimate duration of particular phases of the process i.e. time of melting the first, second and third basket of metal scrap, time of refining, time of the particular load of baskets. As it turned out, the task was not easy because melting in the EAF is a very unstable process. Breaks occur very often which makes this process very stochastic. The cause of these breaks can be: damaging of an electrode, overheating of the conditioning system, necessity of repairing of furnace walls and cover, which are hardly able to predict.

In the figure (1) chart of the primary current of the first electrode during EAF process is presented (a) Correct process - it is easy to identify particular phases of a process: first three parts - melting of baskets and the last two parts - two steps of refining. And (b) - incorrect process in which many breaks occur and it is difficult to recognize phases of process. Because of the fact that the process is very unstable, the values of signals are too stochastic and each melting process is different from the previous one, so untypical methods of analysis have to be used.

Estimation of duration of particular process phases required an analysis of overlapping signals: i.e. primary current of all three electrodes and active electric energy and finally calculation of particular phases duration. The final set was created as a combination of information about particular phases of process computed on

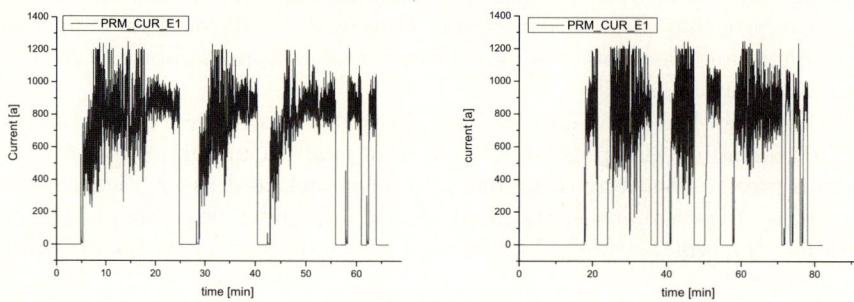

(a) Correct process: first three parts - melting of baskets. The last two parts - two steps of refining.

(b) Incorrect process: difficult identification of process phases

Fig. 1. Chart of the primary current of the first electrode during EAF process

the basis of time series analysis and global information about processes. In the end, for analysis the following features were chosen: weight of each of three baskets (respectively BM_1, BM_2, BM_3), sum of melting time of particular baskets, sum of charge time of particular baskets, the amount of used oxygen, carbon, and gas, the copper content. Output variables, which should be predicted, were electric energy used in the refining phase (total RE_1 and energy used per one ton of product RE_2).

The most important information for a furnace operator is the time when one should finish the melting process. Due to the fact that the scheme of melting of all three baskets is always the same, correction of the process is made only during the last phase - refining phase. It is the last phase of the steel production process in the EAF (two last parts of chart at figure 1 subsection (a)). In this phase carbon and oxygen are inserted too. The last task in this phase is to heat the metal bath to the proper temperature, which guarantees safe steel casting. During the refining phase electric energy consumption per time is constant, therefore while predicting energy that should be consumed, the time, in which the process should be finished, is estimated automatically .

3 Building a Model

In general, the approach of building a regression model can be divided into a few steps. In the beginning ,a brief look on the dataset is required, and in this task, in our problem the output variables required to be analyzed in detail, while they were not directly measured but obtained after manual data processing. In the next step, the most promising model or models is selected. These models (or model) are used in further steps. In the last stage attribute selection and extraction is used to increase generalization abilities.

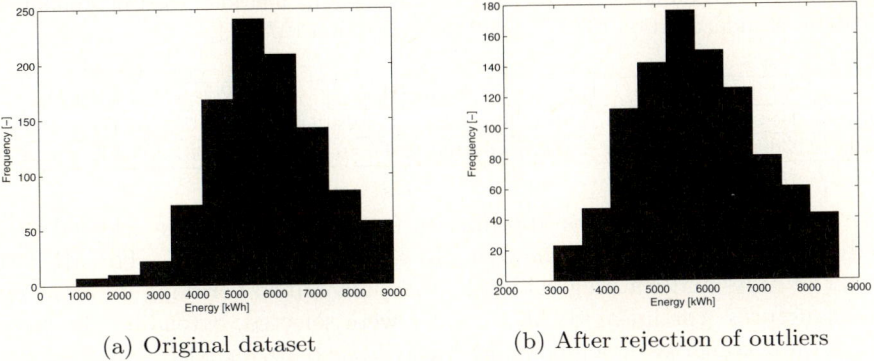

(a) Original dataset   (b) After rejection of outliers

Fig. 2. Histogram of the output variables before and after removing outliers

3.1 General Look of the Dataset

In the dataset, there were two different output variables and the goal was to optimize just one of them RE_1 $[kWh]$ or RE_2 $[kWh/T]$. These two variables were simply related to themselves according to the equation

$$RE_2 = \frac{RE_1}{BM_1 + BM_2 + BM_3} \qquad (1)$$

So only one of them was necessary to be predicted.

The course of physical process and the analysis of the histogram of RE_1 variable proved the existence of outliers vectors in the dataset. In this task, simple algorithm based on one dimensional statistical analysis using quartile rules was used.

All the vectors which did not fulfill inequality (2) were marked as outliers and removed from dataset.

$$y_i = \begin{cases} > Q_1 - \alpha \cdot IQR \\ < Q_3 + \alpha \cdot IQR \end{cases} \qquad (2)$$

Where Q_1 and Q_3 are first and third quartile, and IQR is an interquartile range $(Q_3 - Q_1)$, α is a factor, which in our calculations was determined equal 1.

After removing outliers size of the dataset reduce form 1014 to 960 vectors. Histogram of these two distributions are presented in figure (2)

3.2 Regression Models

From the wide range of different regression models, we have considered well-known methods like multilayer perceptron (MLP), liner model (LR) such as support vector machine (SVM) with linear kernel, but also nonlinear SVM with RBF and polynomial kernels and one sparse Bayesian method called relevance vector machine (RVM)[6].

Table 1. 5xCV mean square error (MSE) for different methods used in simulations after input standardization (MSE_s) and normalization (MSN_n)

	LR	MLP	SVM RBF	SVM Poly	RVM
MSE_n	0.0383±0.0012	0.0377 ± 0.0016	0.0349 ± 0.0014	0.0358 ± 0.0013	0.0361 ± 0.0014
MSE_s	0.0386±0.0015	0.0378 ± 0.0015	0.0400 ± 0.0007	0.0385 ± 0.0015	0.0405 ± 0.0030

For all of these algorithms appropriate free parameters were selected using greedy search procedure and the five-fold cross validation (CV). For MLP the number of units and learning rate were manually tuned verifying different network structures. For linear SVM, C and ϵ were selected, extending the search space with a kernel parameter for nonlinear kernels (σ for RBF kernel or a for polynomial kernel), and similarly for RVM where the only hyperparameter was the kernel parameter.

During the first step, all these models were applied, with only outliers rejection without any data preprocessing, except for data normalization and standardization, therefore all values of each input feature were in range of [0, 1] after normalization or they were of $\mu = 0$ mean and standard deviation $\sigma = 1$ after standarization. Obtained results are presented in table (1).

Results obtained in this table prove that normalization (MSE_n) is a better solution than standardization (MSE_s). None other feature weighting methods have been considered.

The best results were obtained with the SVM with RBF kernel for $C = 1$, $\epsilon = 0.13$, and kernel width $\gamma = 1$. It has to be noticed that the hyperparameter space imposes limitation on SVM. SVM has 3 hyperparameters, which requires searching in cubic space, which is very complicated and time-consuming. However, this process can be sped up because for a given kernel matrix (especially when it can be stored in memory) different models can be built with different quadratic optimization parameters like C and ϵ reducing the overall computational complexity. In other words, in step one - for given kernel parameter kernel matrix is built only once. In step two, all possible C and ϵ values should be considered and tested, and finally these steps should be repeated for each kernel matrix. Less complicated with lower computational cost is RVM. It requires only one dimensional simple search. In the primary calculations, it gave a bit poorer results then the leading SVM did. However, preliminary calculations in the third stage pointed its limitations in this application. Finally, SVM with RBF kernel was selected as the reference model and considered in further stages.

3.3 Feature Selection and Extraction

Feature selection. Feature selection is one of the most important steps in building an accurate model. Various approaches to this problem can be found exp. in [2]. Generally, as it is described in [8], there can be three different groups of feature selection methods defined: filters, wrappers and embedded methods. The first group uses the external measure to find the most informative features, independent of the regression algorithm (however they could be irrelevant to particular regression algorithm, for details see [3]). In this approach, usually mutual information is used or other methods like nonparametric noise estimation (NNE) [9]. On the other hand, wrappers use final regression or classification model to judge the selected feature subset. In this approach, different searching algorithms like forward selection or backward elimination and genetic algorithms are used. The last group of methods uses attribute selection which is directly built in, or dedicated to particular model.

Another approach to feature selection based on joining filters and wrappers advantages, sometimes called frappers (as abbreviation of filter-wrappers)[2], giving very good results, have been used in classification task (ex. during NIPS'03 Feature Selection Challenge). Various feature ranking algorithms can be used as fast filters [8]. A typical ranking is determined by coefficient $f_j(\boldsymbol{x_j}, \boldsymbol{y})$ where j indicates number of features ($j = 1 \ldots m$). $f(\boldsymbol{x_j}, y)$ provides relevance of feature $\boldsymbol{x_j}$ describing $\boldsymbol{y}$. In our Calculations two different rankers have been used:

– simple linear correlation coefficient:

$$CC(\boldsymbol{x_j}, \boldsymbol{y}) = \frac{\mathrm{cov}(\boldsymbol{x_j}, \boldsymbol{y})}{\sqrt{E(\boldsymbol{x_j}^2) - E^2(\boldsymbol{x_j})}\sqrt{E(\boldsymbol{y}^2) - E^2(\boldsymbol{y})}} \qquad (3)$$

– mutual information

$$MI(\boldsymbol{x_j}, \boldsymbol{y}) = \sum_{i=1}^{n}\sum_{k=1}^{n} p(x_{ji}, y_k) \log\left(\frac{p(x_{ji}, y_k)}{p(x_{ji})p(y_k)}\right) \qquad (4)$$

In frappers, coefficient vector $\boldsymbol{f} = [f_1, f_2, \ldots m]$ is sorted and ordered from the most relevant features to the least ones. Afterwards, regression model is used to determine the best feature subset adding one by one feature according to the relevance order. In case of large feature set instead of adding each feature the subset may be determined by using some threshold value (Θ) optimized using regression algorithm (ex. see [2], [4]).

$$\boldsymbol{F}_R = \bigvee_{j=1\ldots m} f_j > \Theta \qquad (5)$$

where F_R is the new feature subset.

Obtained results for different feature selection methods with SVM, used as regression algorithm, are presented in table (2). For all these variable selection methods InfoSel++[2] library were used. The same table presents a number of selected features for each validation in CV test, and a mean number of features with their standard deviation. For comparison, forward selection and backward elimination were used. The best results were obtained for frappers feature selection method with MI used to determine ranking values. Other methods, especially based on simple searching, obtained even worse results then without feature selection.

---

[2] Project partially sponsored by the grant No N N519 1506 33 from the Polish Ministry of Education and Science (MNiSZW).

Table 2. 5xCV mean square error (MSE) for different feature selection methods: FS - forward selection, BE - backward elimination, FC - CC frapper, FM - MI frapper, number of selected features in each fold ($cv_1 \ldots cv_5$) mean number of features μ and standard deviation of selected features (σ^2)

	MSE	cv_1	cv_2	cv_3	cv_4	cv_5	μ	σ^2
FS	0.0363±0.001	6	11	6	8	7	7.6	2.07
BE	0.0353±0.0013	16	15	12	11	13	13.4	2.07
FC	0.0354±0.0012	17	16	12	15	16	15.2	1.92
FM	0.0348±0.0014	17	14	15	13	13	14.4	1.67

Table 3. 5xCV mean square error (MSE) for different feature selection methods used after linear transformation of the dataset and with concatenated datasets: FS - forward selection, BE - backward elimination, FC - CC frapper, FM - MI frapper

	MSE	cv_1	cv_2	cv_3	cv_4	cv_5	μ	σ^2
Linear transformation								
FS	0.0358±0.0012	7	13	13	12	12	11.4	2.5
BE	0.0353±0.0015	16	16	15	14	14	15	1
FC	0.0351±0.0013	17	17	16	17	14	16.2	1.3
FM	0.0350±0.0014	17	17	17	17	16	16.8	0.4
Concatenated datasets								
FS	0.0374±0.0014	8	16	7	5	10	9.2	4.2
BE	0.0356±0.0019	30	28	23	32	29	28.4	3.4
FC	0.0358±0.0018	23	29	25	32	19	25.6	5.1
FM	0.0356±0.0018	25	31	23	29	31	27.8	3.6

Feature extraction. Feature extraction such as PCA, ICA may be useful if correlated features exist in the dataset. For example, when meltdown additives are related to steel scrap weight or when steel scrap weight is correlated with the duration of furnace charge. If such relation exists after transformation, new features will be obtained. This new set of features consists of more informative features, while the useless ones (less informative features) are automatically rejected.

In our experiments, we have only considered simple feature extraction method based on linear transformation of the dataset using orthogonal-triangular decomposition (QR factorization) [1] used to rotate the coordinate system in the directions of highest variance of dataset (QR factorization is often used for PCA analysis, however we have not used the eigenvalues). In the basic approach, after transformation of all features, feature selection algorithms were used to reject useless features from the new dataset. Unfortunately, this approach did not help us to build a better model. The results are presented in table (3)

Our last choice, concerning building the model, was concatenation of both datasets: before and after the QR transformation, so the number of features got doubled. The idea of proposed approach is motivated by our observation that after the QR transformation some of relevant features may become destroyed, what may cause decreasing the overall model accuracy, especially in frappers methods. This may occur when linear transformation is not supervised.

4 Results Discussion and Conclusions

4.1 Conclusions on the Feature Selection Algorithms

The searching method, used for building the model, uses a simple search technique - 'best first', determining the size of feature set. It usually adds or removes (respectively for forward and backward selection) one by one single feature searching among current dataset. This approach is computationally very expensive and in the worst case requires $t(t-1)/2$ repetitions of model training. Pudil et al. [5] did a comparison among different search techniques in classification task.

Frapper algorithms require much less computational effort determining feature subset in m (size of the feature set) repetitions of training regression algorithm. Presented results pointed out that frappers which use mutual information as the ranking value overcome other methods, even the search based ones. This observation was very surprising, however the possible reason for that is high variance of obtained feature set in each validation. This problem can be observed in the results in the tables (2,3), where the frapper algorithm, based on mutual information, is also the most stable. Only in concatenated dataset, backward elimination overcame a bit frappers, obtaining smaller variance.

However, frapper based on Pearson's correlation coefficient (CC), did not give satisfactory results. The reason for that, is assumption of linear dependences between both random variables and in general, this assumption is not fulfilled in regression problems. Mutual information is not restricted to these conditions and simply could be understood as an amount of information, by the use of which, knowing one of the variables, the other one can be easily described.

In general, CC is not a valid ranking builder for nonlinear regression and it should not be used in this type of problems. Statistical correlation metrics based on rank tests (like Spearman's rank correlation coefficient) could be a possible extension to our work but they have not been tested yet. This type of tests are usually insensitive to correlation type.

Another observation arising due to the results of the tables (2) and (3) is the problem known as 'variance in feature selection', which is especially important in understanding of obtained results. This high variance of selected number of attributes in each fold may be explained by small representativity of particular input variables. During the cross validation test, each fold include reduced sample size, which may impact on rejecting attributes.

Feature extraction methods have a positive influence on this variance (as those based on QR decomposition). In all tested methods the variance of number of selected attributes in each validation fold sharply decreases and all attribute selection methods become much more stable, selecting similar number of features.

4.2 Conclusions on Building Regression Model

The method, used for building regression model, can be divided into three main steps: analysis of outliers and the distribution of the variables, including experts consulting; selecting the most promising model form the large group representing different approaches to the regression problem; and building the final model extending its abilities by feature selection and extraction. These three steps usually provide good results.

The first step done in cooperation with domain experts allows pruning the dataset from outliers vectors and choosing the most important features, what leads to building more accurate model.

In typical agnostic methods of model selection, without any special prior knowledge of the dataset, the model space is becoming very large, so a possible solution is primary model selection basing on the entire dataset and further calculations using the most promising one. The problem of model accuracy estimation came up in this step . Typical hold-out test can not be used on this type of datasets (too small number and to noisy samples), but other methods like cross-validation (CV), bootstrap or 'leave on out' are much more accurate, providing good accuracy estimation. Our results, based on CV estimation, selected the SVM model with Gaussian kernels.

The last step including feature selection should be also done very carefully, while many different methods require different computational effort, and give different results. Our results prove high quality of frapper approaches, reducing computational cost, and giving very good overall generalization of the system.

Currently described in this paper system require empirical verification. In recent work most attention will be payed to apply it into the industrial environment where it will be used by EAF engineers to check its real quality, and further to support their work.

Acknowledgment. Authors are grateful for the support by the Polish Ministry of Education and Science (MNiSZW), research grants No 6ZR9 2006 C/06742 and N N519 1506 33.

References

1. Anderson, E., et al.: LAPACK User's Guide. SIAM (1999)
2. Guyon, I., Gunn, S., Nikravesh, M., Zadeh, L.: Feature Extraction: Foundations and Applications. Physica Verlag, Springer (2006)
3. Kohavi, R., John, G.: Wrappers for feature subset selection. Artificial Intelligence, special issue on relevance 97(1-2), 273–324 (1997)
4. Blachnik, M., Duch, W., Wieczorek, T.: Threshold rules decision list. In: Moczulski, W., Burczynski, T., Cholewa, W. (eds.) Methods of artificial intelligence, Gliwice. AI-METH Series, pp. 23–24 (2005)
5. Novovivcov'a, J., Pudil, P., Kittler, J.: Floating search methods in feature selection. Pattern Recognition Letters 15, 1119–1125 (1994)
6. Tipping, M.: The relevance vector machine. In: Advances in Neural Information Processing Systems, Morgan Kaufmann (2000)

7. Wieczorek, T.: Intelligent control of the electric-arc steelmaking process using artificial neural networks. Comp. Methods in Material Science 6, 9–14 (2006)
8. Duch, W., Wieczorek, T., Biesiada, J., Blachnik, M.: Comparision of feature ranking methods based on information entropy. In: Proc. of International Joint Conference on Neural Networks, Budapest, Hungary, pp. 1415–1420. IEEE Press (2004)
9. Yu1, Q., Severin, E., Lendasse, A.: Variable selection for financial modeling. In: 13th International Conference on Computing in Economics and Finance, Montreal, Quebec, Canada (2007)

Epoch-Incremental Queue-Dyna Algorithm

Roman Zajdel

Faculty of Electrical and Computer Engineering,
Rzeszow University of Technology, W. Pola 2, 35-959 Rzeszow, Poland
rzajdel@prz-rzeszow.pl

Abstract. The basic reinforcement learning algorithm, as Q-learning, is characterized by short time-consuming single learning step, however, the number of epochs necessary to achieve the optimal policy is not satisfactory. There are many methods that reduce the number of necessary epochs, like TD($\lambda > 0$), Dyna or prioritized sweeping, but their learning time is considerable. This paper proposes a combination of Q-learning algorithm performed in incremental mode with executed in epoch mode method of acceleration based on environment model and distance to terminal state. This approach ensures the maintenance of short time of a single learning step and high efficiency comparable with Dyna or prioritized sweeping. Proposed algorithm is compared with Q(λ)-learning, Dyna-Q and prioritized sweeping in the experiments on three maze tasks. The time-consuming learning process and number of epochs necessary to reach the terminal state is used to evaluate the efficiency of compared algorithms.

Keywords: reinforcement learning, Dyna-Q, prioritized sweeping.

1 Introduction

The efficiency of basic reinforcement learning algorithm, e.g. Q-learning [19], AHC [2] or Sarsa [12], measured in number of epochs to obtain the optimal policy is relatively small. Hence, the number of practical implementations of these algorithms to more complex problems is rather small. However, the unquestionable advantage of these algorithms is a small computational complexity, which implies the extremely short learning time of each epoch. The short learning time is essential for application of the reinforcement learning algorithms to the on-line control problems. Thus, the acceleration methods of reinforcement learning algorithms should ensure both, relatively small computational complexity, and high efficiency. Unfortunately, the acceleration methods used up to date, that reduce the number of epochs needed to obtain the optimal policy, require significantly more learning time. For example, one of the most often used acceleration method, like temporary-differences mechanism TD($\lambda > 0$) requires the additional memory elements known as eligibility traces. The learning time of TD($\lambda > 0$) algorithm significantly grows, because the updates of the policy function Q are made for all states, not for only one (actual) state, like in the basic reinforcement learning TD(0) [16], [19].

Much more efficient learning methods in comparison to TD($\lambda > 0$), e.g. Dyna [14], [15] or prioritized sweeping [9], [10] belong to the class of memory based methods as well. The basic idea of these methods is the usage of the adaptive environment model to reinforcement learning. Their efficiency is much higher than in case of TD($\lambda > 0$) [16], but it is paid off by the increase of learning time. These algorithms perform a fixed number of updates of the policy function Q for these pairs of state-action which were activated in the past. Furthermore, in Dyna algorithm these update states are chosen at random, and in the case of prioritized sweeping the update states are prioritized by the value of adequate temporary differences error.

In this paper the epoch-incremental reinforcement learning algorithm is proposed in order to obtain both, the high efficient learning method, and very short learning time. The main idea of this algorithm is to combine the simplest form of Q-learning algorithm, i.e. 1-step Q-learning, which is performed in incremental mode, with the acceleration method performed in epoch mode. The acceleration method proposed here is based on the Queue-Dyna method [10], which is performed not in incremental mode but in epoch one. The queue procedure is also changed, because it is oriented to determine the shortest path from terminal state backwards to all visited states during the episode. Proposed epoch-incremental Queue-Dyna algorithm is compared to three often used reinforcement learning methods, i.e. Q(λ)-learning, Dyna-Q and prioritized sweeping. These algorithms are used to solve several grid worlds control problem.

2 Reinforcement Learning

Reinforcement learning addresses the problem of the agent that must learn to perform a task through trial and error interaction with an unknown environment [16]. The agent and the environment interact continuously until the terminal state is not reached. The agent senses the environment through its sensors and, based on its current sensory inputs, selects an action to perform in the environment. Depending on the effect of its action, the agent obtains a reward. Its goal is to maximize the discounted sum of future reinforcements r_t received in long run, what is usually formalized as $\sum_{t=0}^{\infty} \gamma^t r_t$, where $\gamma \in [0, 1]$ is the agent's discount rate. The application of nonlinear models is influenced by the fact of the complexity of training data. The models presented below seem to be most suitable for this particular problem.

2.1 Q-Learning

There exist different types of reinforcement learning algorithms. The Q-learning proposed by Watkins [19] is one of the most often used. This algorithm computes the table of all values $Q(s, a)$ (called Q-table), by successive approximations. $Q(s, a)$ represents the expected payoff that agent can obtain in state s after it performs action a. The Q-table is updated according to the following formula:

$$Q(s, a) \leftarrow Q(s, a) + \beta \left(r + \gamma \max_{a'} Q(s', a') - Q(s, a) \right) \qquad (1)$$

where the maximization operator refers to the action value a' which may be performed in next state s' and $\beta \in (0, 1]$ is the learning rate. The basic Q-learning algorithm (1-step Q-learning) can be significantly improved considering the history of state activation represented by eligibility traces. The eligibility trace is parametrized by recency factor λ, therefore this enriched learning method is called Q(λ)-learning [16], [19]. The eligibility trace of each state becomes large after state activation and then it decreases exponentially until the state is not visited again [4].

2.2 Dyna-Q and Prioritized Sweeping

Dyna and prioritized sweeping require the model of the environment in order to improve the policy represented by Q-table. Given a state and an action, a model produces a prediction of the next state and the next reward.

Dyna-Q is a Q-learning based algorithm which simultaneously uses experience to build the model and to adjust the policy. Additionally, it uses a model to adjust the policy. Dyna-Q operates in a loop of interaction with the environment as follows. After each transition $(s, a) \rightarrow (s', r)$, the model is recorded in its table entry for (s, a) which predicts that (s', r) will appear. It is followed by the update of the Q value at state s and at action a using rule (1). Then, some number (denoted further as L) of additional updates is performed on the basis of the model. Each of these updates consists of random choice of the state action pair (s, a) that has been experienced before and the query to the model with this pair. Model returns the last observed next state s' and reward r as its prediction. Finally, the update according to (1) is performed. A reasonable value of L can be determined on the basis of the relative speeds of computation and of taking action [6].

The prioritized sweeping algorithm is similar to the Dyna-Q, except that its updates are not chosen at random but depending on priority p which is the absolute value of temporal differences error $|r + \gamma \max_{a'} Q(s', a') - Q(s, a)|$. If the priority p is greater than the arbitrary defined threshold, then the pair (s, a) is stored into queue (called $PQueue$) with priority. Finally, only L states, with the highest priority are updated according to (1) using the model, similarly to Dyna-Q [16].

3 Navigation Task

The exemplary maze shown in the Fig. 1 is a typical navigation task. The maze is a 6 by 9 grid of states, one of which is marked as starting state 'S', and one of which is the goal state 'G'. The shaded states are barriers and cannot be entered. Agent can move in one of four possible directions (actions): LEFT, RIGHT, UP, and DOWN.

The reward is zero for all transitions except for the ones leading to the goal state 'G', for which it is 1. If the goal state is reached, the learning episode is finished and the agent is transported to the start state to begin the next trial [14].

Fig. 1. Sutton's 6x9 grid world with 46 reachable states [14]

4 Epoch-Incremental Queue-Dyna

The proposed epoch-incremental Queue-Dyna algorithm described below combines short learning time of the 1-step Q-learning, and the efficiency of model based learning method derived from Dyna-Q and prioritized sweeping. Furthermore, this proposition is based on the following four observations.

If the prioritized sweeping algorithm is applied to grid world problem (Fig. 1), during the first learning episode only the state action pair leading directly into the goal has greater than zero priority p. This means, that it is not necessary to do all the updates, except for the last one, that leads to the goal state. This suggests, that the updates could be performed backwards from the goal state, which is the greatest source of the reinforcement signal. From this state the learning process of action-value function Q begins.

The second observation refers to the time, when the update process of Q-table is performed. Namely, if the L updates are performed incrementally, then the learning time is increased significantly, which is essential for use of this algorithm to control real objects, e.g. cart-pole [2], ball-beam [20], or mobile robot [1]. Thus, the adaptive control algorithm, based on reinforcement method should ensure short learning time. The solution, resulting from reduction of the number of updates L, can decrease the efficiency of learning [14].

Thirdly, Dyna and prioritized sweeping algorithm are based on the adaptive model of environment which at the beginning of learning process is unreliable, because it is built on the basis of too few environment observation. Therefore, one suggests to defer the model based learning process until some number of states is visited (as more as possible), and the model is more credible. As one can notice, the best moment to perform the updates is when the terminal state is reached what, in general, occurs if a large number of states is visited.

The last observation is related to the real-world application of the reinforcement learning algorithms. Namely, ones each learning episode is finished, the system is restored to the starting state no instantly, but in nonzero time. This time can be utilized to perform the updates, which in Dyna and prioritized sweeping algorithms are performed in incremental mode so far, i.e. before ending the learning episodes.

The pseudocode of complete algorithm proposed in this work is presented in Fig. 2. All four observations mentioned above are taken into consideration here. At the beginning, the Q-table and the tables necessary for the model are

1. Initialize $Q(s,a), N(s,a), N(s,a,s')$ and $Model(s,a)$ arbitrarily for all $s \in S$ and $a \in A$
2. Repeat (for each episode):
   2.1. Initialize s, a
   2.2. Repeat (for each step of episode):
      2.2.1. Choose action a in s using policy derived from Q (e.g. ϵ-greedy)
      2.2.2. Execute action a, observe reward r and resultant state s'
      2.2.3. $Q(s,a) \leftarrow Q(s,a) + \beta(r + \gamma \max_{a'} Q(s',a') - Q(s,a))$
      2.2.4. $Model(s,a) \leftarrow s', r$
      2.2.5. $N(s,a) \leftarrow N(s,a) + 1; N(s,a,s') \leftarrow N(s,a,s') + 1$
      2.2.6. $p(s,a,s') \leftarrow N(s,a,s')/N(s,a)$
      2.2.7. $s \leftarrow s'$
      until s is TERMINAL
   2.3. $PQueue \leftarrow (s = TERMINAL, a = 0, d = 0)$
   2.4. while $PQueue_d$ is not empty
      2.4.1. $d \leftarrow d + 1$
      2.4.2. for each $s' \in PQueue_{d-1}$
         2.4.2.1. for each $a \in A$, for each $s \in preds(s') \cap$ (not in PQueue)
            2.4.2.1.1. if $p(s,a,s') > 0$, $PQueue \leftarrow s,a,d$
   2.5. Repeat for all (except first) elements of $PQueue$
      2.5.1. $s,a \leftarrow PQueue$
      2.5.2. $s',r \leftarrow Model(s,a)$
      2.5.3. $Q(s,a) \leftarrow Q(s,a) + \beta(r + \gamma \max_{a'} Q(s',a') - Q(s,a))$

Fig. 2. Epoch-incremental Queue-Dyna algorithm with priority determined by estimated value of the distance to terminal state

initialized arbitrary (step 1). The learning process consists of some number of the episodes. Each of them starts from initial state and action (step 2.1), that are often the same in each episode. Then, the episode loop is begun, in which the 1-step Q-learning algorithm is performed first (steps 2.2.1-2.2.3). Then, in step 2.2.4, the observed next state s' and obtained reward r are saved to the table $Model$ for state-action pair (s, a). The information about the possible transition to state s' within executing action a in state s is stored in table $p(s,a,s')$, where $N(s,a)$ denotes the number of time in which action a is executed in s, and $N(s,a,s')$ is the number of times resulting in transition to state s' [3], [9], [10], [17] (steps 2.2.5 and 2.2.6). The incremental part of learning algorithm is performed until the terminal state (the goal state 'G' in grid world) is reached. Then, coming back from terminal state, the queue ($PQueue$) of every state-action pair visited (if $p(s,a,s') > 0$) is maintained, prioritized by the minimum number of time steps d to the terminal state (steps 2.3 and 2.4). The $preds(s')$ is the set of all states, which are observed as immediate predecessors of state s'. Because the queue loop (step 2.4) is performed for all visited states, and moreover, the update procedure is executed in epoch mode, there is no problem in choice of updates number L here, as in case of Dyna and prioritized sweeping.

The exemplary effect of queuing procedure (step 2.4) performed after first episode is shown in Fig. 3. In the top-left corner of each cell, the number of the state is placed, and the distance d is located in the bottom-right corner. The

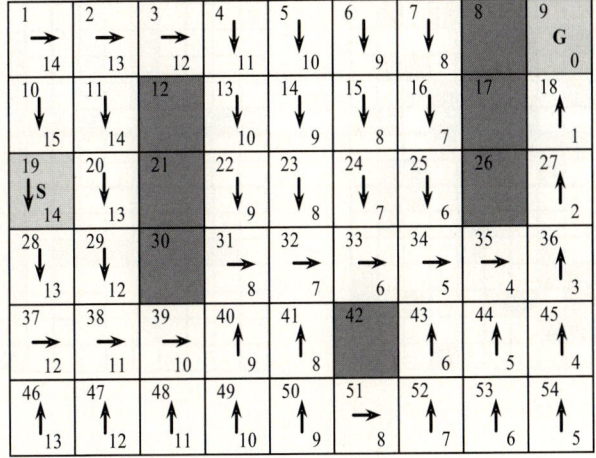

Fig. 3. The policy and the distance to the terminal state 'G' represented by the queue *PQueue* after first episode

arrows represent the last action executed. One can also notice, that *PQueue* visualized this way is the optimal policy. Hence, the updates of Q-table based on *PQueue* (step 2.5) should ensure the improvement of the policy.

5 Empirical Results

Grid worlds are often used to compare how quickly different reinforcement learning methods converge to a stable solution [5], [7], [8], [11], [13], [18]. Three different grid worlds are used here: Sutton's 6x9 (Fig. 1) [14], More and Atkeson's 14x14 (Fig. 4. (a)) [9] and 14x14 (Fig. 4. (b)). The minimal number of agent's moves from 'S' to 'G' is 14, 25 and 70 respectively, and the number of reachable states is 46, 130 and 161.

The proposed epoch-incremental Queue-Dyna algorithm is compared with three well-known methods: $Q(\lambda)$-learning with replacing eligibility traces, Dyna-Q, and prioritized sweeping. The initial values of $Q(s,a)$, $N(s,a)$, $N(s,a,s')$ and $Model(s,a)$ are set to zero, the step size $\beta = 0.1$, the discount rate $\gamma = 0.95$ and the exploration parameter $\epsilon = 0.1$. For Dyna-Q and prioritized sweeping, the number of updates (L) for 6x9 grid world is set to 50 [2]. In case of both 14x14 grid worlds, the number of updates equals 100 for these algorithms. The learning curves show the number of steps taken by the agent in each episode, averaged over 30 repetitions of the experiments (Fig. 5-7). The first step per episode for 6x9 grid world, first two steps per episodes for Moore and Atkeson 14x14 grid world and first three steps per episodes for the 14x14 grid world with 161 states are significantly bigger than the remaining ones, therefore they are omitted in the Fig. 5-7. Since the action selection ϵ-greedy method is used

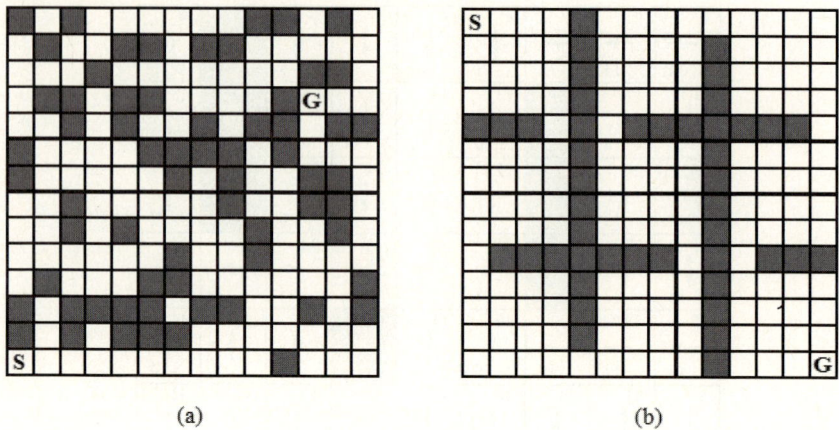

Fig. 4. The 14x14 grid worlds with 130 (a) and 161 (b) reachable states

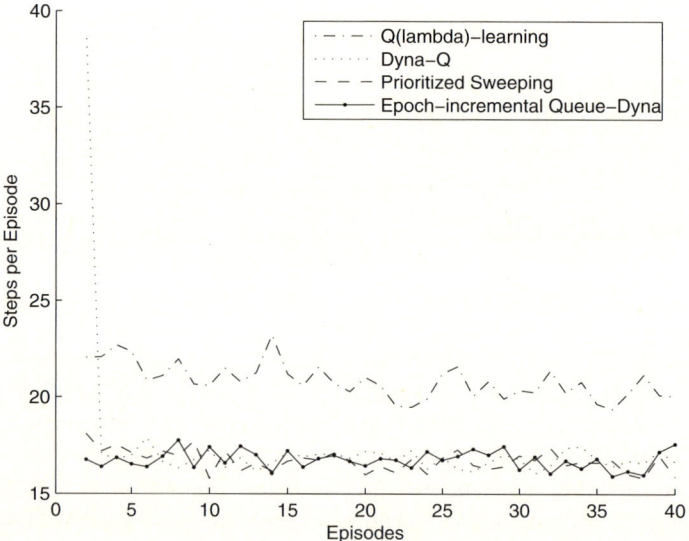

Fig. 5. The average learning curves from second episode for grid world with 46 reachable states

($\epsilon = 0.1$), the minimal numbers of achieved moves from 'S' to 'G' are at last 10% greater than the perfect result, as shown in the left column in Tab. 1. The $Q(\lambda)$-learning algorithm appears the worst for 46 and 161 states grid worlds. The 130 states grid world is different from the others, because from the start state to the goal state there are two main paths with 25 and 51 moves. Therefore, such big deviation from established value, which shown in Fig. 6, are caused

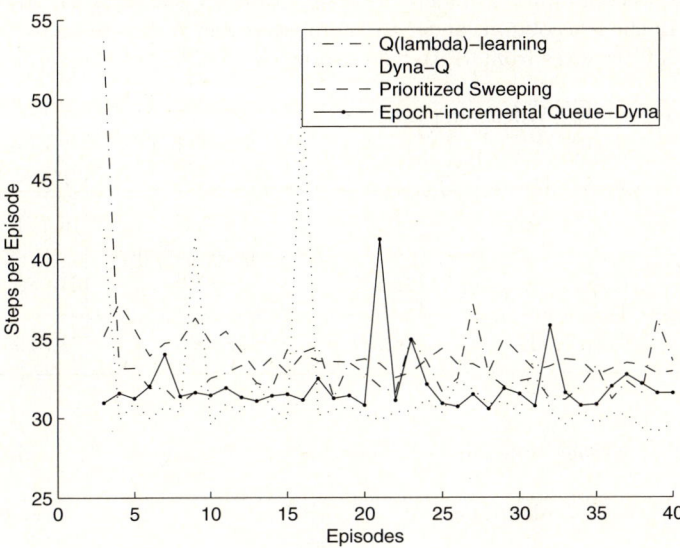

Fig. 6. The average learning curves from third episode for grid world with 130 reachable states

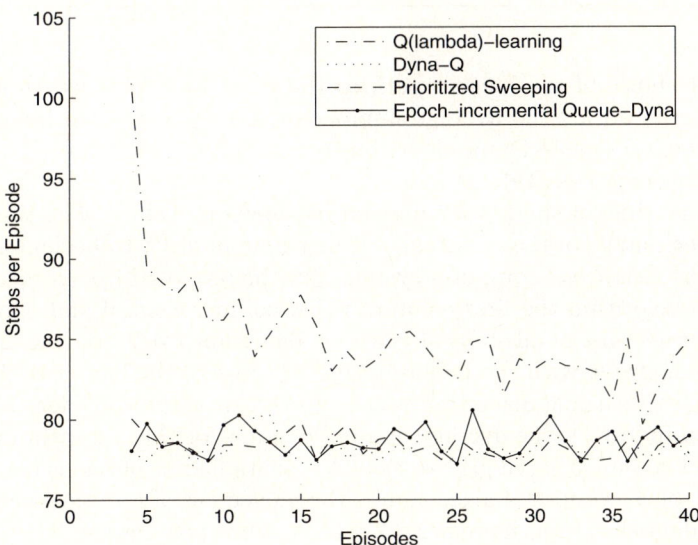

Fig. 7. The average learning curves from fourth episode for grid world with 161 reachable states

Table 1. The average number of steps necessary to reach the terminal state 'G' from start one 'S'. In the left column, under the number of grid worlds reachable states, the minimal number of moves from 'S' to 'G' is shown.

algorithm environment	$Q(\lambda)$-learning	Dyna-Q	Prioritized Sweeping	Epoch-incremental Queue-Dyna
46 states min 14	20.861	17.379	16.722	16.771
130 states min 25	32.720	31.348	33.832	31.980
161 states min 70	85.069	78.271	78.283	78.616

Table 2. The average time runs calculated as a percentage of the maximal time

algorithm environment	$Q(\lambda)$-learning	Dyna-Q	Prioritized Sweeping	Epoch-incremental Queue-Dyna	
				incremental and epoch	incremental
46 states	4.1%	14.3%	100.0%	17.5%	2.3%
130 states	1.4%	4.5%	100.0%	11.3%	0.8%
161 states	2.7%	6.8%	100.0%	7.6%	0.6%

by the agent choice of the longest path to the goal state. It is worth to notice that for all grid worlds the average number of steps per episode for proposed epoch-incremental Queue-Dyna algorithm is comparable with the Dyna-Q and prioritized sweeping (see Tab. 1).

The average time of single learning step is shown in Tab. 2. The time is calculated as percentage of the maximal learning time in order to become independent from performance of computer system. For the proposed Epoch-incremental Queue-Dyna algorithm the first column, titled as 'incremental and epoch' contains the average time of all steps of learning algorithm (step 2 on Fig. 2). These times are comparable with the Dyna-Q time. In spite of the fact that the learning time of $Q(\lambda)$-learning algorithm is the shortest, it is not considered because, as it is shown in Tab. 1, the average number of steps for this algorithm is unsatisfactory for two out of three grid worlds. As mentioned in previous section, the main feature of the proposed algorithm is the movement of the acceleration part of learning algorithm from incremental mode to the epoch mode. Therefore the learning time of incremental part (step 2.2 on Fig. 2) is presented in separated column, named 'incremental' in Tab. 2. This time for all three grid worlds is significantly shorter, not only for Dyna-Q and prioritized sweeping, but also for $Q(\lambda)$-learning.

6   Conclusions

The highly efficient methods of reinforcement learning, as Dyna or prioritized sweeping are based on a model, which is improved and used to build a policy in incremental mode. These methods ensure that the optimal policy is obtained in small number of epochs, but unfortunately in much longer learning time.

In this paper the new method of acceleration for reinforcement learning algorithm is proposed. The main idea of this method is to extract the part of algorithm, which is responsible for the acceleration of learning process. This part is executed after ending the incremental learning stage, that is in epoch mode. In this way, the total (incremental and epoch) learning time is comparable with Dyna learning time admittedly, but the incremental learning time is significantly shortened. Furthermore, the efficiency of proposed algorithm, measured in number of epochs to obtain the optimal path from start state to the goal state remains comparable with the efficiency of Dyna and prioritized sweeping methods. Receiving of such a high efficiency algorithm with short incremental learning time is essential for planned implementation of this algorithm to more complex and realistic environments.

References

1. Asadpour, M., Siegwart, R.: Compact Q-learning optimized for micro-robots with processing and memory constraints. In: Robotics and Autonomous Systems. European Conference on Mobile Robots, vol. 48(1), pp. 49–61 (2004)
2. Barto, A.G., Sutton, R.S., Anderson, C.W.: Neuronlike adaptive elements that can solve difficult learning problem. IEEE Trans. SMC 13, 834–847 (1983)
3. Barto, A.G., Bradtke, S.J., Singh, S.P.: Learning to Act using Real-Time Dynamic Programming. Artificial Intelligence. Special Vol. on Computational Research on Interaction and Agency 72(1), 81–138 (1995)
4. Cichosz, P.: Learning systems. WNT, Warsaw (2000) (in polish)
5. Crook, P., Hayes, G.: Learning in a State of Confusion: Perceptual Aliasing in Grid World Navigation. In: Proc. of Towards Intelligent Mobile Robots (2003)
6. Kaelbing, L.P., Litman, M.L., Moore, A.W.: Reinforcement Learning: A Survey. Journal of Artificial Intelligence Research 4, 237–285 (1996)
7. Lanzi, P.L.: Adaptive Agents with Reinforcement Learning and Internal Memory. In: Proc. of the Sixth International Conference on the Simulation of Adaptive Behavior, pp. 333–342. The MIT Press, Cambridge (2000)
8. Loch, J., Singh, S.: Using eligibility traces to find the best memoryless policy in partially observable Markov decision processes. In: ICML, pp. 323–331 (1998)
9. Moore, A.W., Atkeson, C.G.: Prioritized sweeping: Reinforcement learning with less data and less time. Machine Learning 13, 103–130 (1993)
10. Peng, J., Williams, R.J.: Efficient learning and planning within the Dyna framework. In: Meyer, J., Roitblat, H., Wilson, S. (eds.) From Animals to Animats 2, USA, pp. 281–290 (1993)
11. Pickett, M., Barto, A.G.: PolicyBlocks: An Algorithm for Creating Useful Macro-Actions in Reinforcement Learning. In: Proc. of the International Conference on Machine Learning, vol. 19, pp. 506–513 (2002)

12. Rummery, G.A., Niranjan, M.: On line q-learning using connectionist systems. Technical Report CUED/F-INFENG/TR 166, Cambridge University Engineering Department (1994)
13. Sherstov, A.A., Stone, P.: Improving Action Selection in MDP's via Knowledge Transfer. In: Proc 20th the Nation Conference on Artificial Intelligence, vol. 20(2), pp. 1024–1029 (2005)
14. Sutton, R.S.: Integrated Architectures for Learning, Planning, and Reacting Based on Approximating Dynamic Programming. In: Proc. of Seventh Int. Conf. on Machine Learning, pp. 216–224 (1990)
15. Sutton, R.S.: Planning by incremental dynamic programming. In: Proc. of the Ninth Conference on Machine Learning, pp. 353–357 (1991)
16. Sutton, R.S., Barto, A.G.: Reinforcement learning: An Introduction. MIT Press, Cambridge (1998)
17. Tadepalli, P., Ok, D.: Model–Based Average Reward Reinforcement Learning. Artificial Intelligence 100, 177–224 (1998)
18. Tanner, B., Sutton, R.S.: Temporal-Difference Networks with History. In: Proc. of the 2005 International Joint Conference on Artificial Intelligence, pp. 865–870 (2005)
19. Watkins, C.J.C.H.: Learning from delayed Rewards. PhD thesis, Cambridge University, Cambridge, England (1989)
20. Wellstead, P.E.: Introduction to Physical System Modelling, Control System Principles (2000)

Part VIII

Agent Systems

Utilizing Open Travel Alliance-Based Ontology of Golf in an Agent-Based Travel Support System

Agnieszka Cieślik[1], Maria Ganzha[2], and Marcin Paprzycki[2]

[1] Department of Mathematics and Information Technology,
Warsaw University of Technology, Warsaw, Poland
`agnieszka.cieslik@gmail.com`
[2] Systems Research Institute, Polish Academy of Science, Warsaw, Poland
`{maria.ganzha,marcin.paprzycki}@ibspan.waw.pl`

Abstract. Currently, we are developing an agent-based travel support system, in which ontologically demarcated data is used to facilitate personalized information provisioning. Recently we have shown how Open Travel Alliance golf-related messages can be reverse-engineered to create an *OTA ontology of golf*. The aim of this paper is to illustrate how these ontologies are going to be used in the system. In addition to the general scenario, details concerning implementation of needed translators will be discussed.

1 Introduction

Our current work is devoted to developing a comprehensive agent-based traveler support system, and involves a number of sub-projects. First we have been developing a model agent-based e-commerce system (see [2,1], and references to our earlier work contained there). There we study utilization of agent-facilitated autonomous price negotiations in a general e-commerce scenario. This work was extended to facilitate possibility of airline ticket auctioning [25,23,24,20,26]. Second, we work on creation of an agent based *Travel Support System* (*TSS*) [18,7,10]. In the *TSS*, travelers are to obtain personalized information related to their travel needs (e.g. favorite hotels, restaurants, etc.). The main idea of the *TSS* is to utilize a central repository of semantically demarcated travel data, and operate on it to deliver personalized information [7,8]. While these two projects (the airline ticket auctioning system and the Travel Support System) are being developed separately, in [25] we have discussed issues involved in their possible merger. Since the ontologically demarcated data is the central component of the *TSS*, the two projects were conceptually merged through development of a common travel ontology. Within the *TSS*, we have initially developed an ontology of hotels and restaurants [6,9]. In the airline ticket auctioning system we have utilized the fact that the Open Travel Alliance (OTA) [15] has proposed a set of *messages* designed to facilitate meaningful communication about travel-related activities such as flights or golf course reservations. Here, it has to be stressed that, while on the way to become industry standard, these messages *do not* explicitly define an ontology. Therefore, in [26] we have proposed how OTA air-travel-related messages can be used as a basis for development of an ontology of air-travel. We have proceeded with development of such ontology and later merged it with the existing ontology of restaurant and hotel (results—a complete ontology of restaurants, hotels and air travel—can be found within [22]). Most recently, following the example set forward in [26] we have shown how the *OTA golf messaging* can be turned into an ontology of golf [3]. The aim of this paper is to describe how systems utilizing the *OTA ontology of golf* can collaborate with entities that utilize *OTA golf messaging* (and, for instance, for one reason or another, do not work with ontologically demarcated data).

To this effect we proceed as follows. In the next section we briefly describe OTA golf messages as well as the proposed *OTA ontology of golf*. Next we provide context in which *OTA golf messages* and the *OTA ontology of golf* can interact. Finally, we discuss *how* these interactions can be implemented using currently existing technologies.

2 OTA Golf Messages and Ontology

As all OTA messages concerning various "areas of travel," OTA golf-related messages come in pairs [17]. There is a request (RQ) message (a query) and, corresponding to it, a response (RS) message. As what concerns this paper, the OTA standard identifies three pairs of golf-related messages (detailed description can be found in [17] and [3], Table 1):

- *OTA_GolfCourseSearchRQ*—request for course information; used to find golf courses that satisfy a given set of criteria,
- *OTA_GolfCourseSearchRS*—list of courses that meet the requested criteria,
- *OTA_GolfCourseAvailRQ*—requests information about course availability,
- *OTA_GolfCourseAvailRS*—provides information about course availability,
- *OTA_GolfCourseResRQ*—requests a reservation of a given golf course,
- *OTA_GolfCourseResRS*—confirms (or denies) reservation of a given course.

These messages allow interested party: (1) to find a golf course with specific characteristics (claimed to include all features that any golfer could think off), (2) to check if a course of interest is available at a specific time and under a specified set of conditions (e.g. start time, or price), and (3) to make an actual reservation.

To illustrate the form of OTA messages, in Figure 1 we present an example on an *OTA_GolfCourseAvailRQ* message (see, [17]). In this message four friends specify that they would like to play golf on June 22nd, and the requested tee-off time is to be between 14:00 and 15:30. They are interested in playing at a specific golf course with the identifier *PL4321* (it is assumed that through an earlier query-message they have established that *PL4321* is the course that they are interested in). The maximum price that they are willing to pay for 18 holes is $75.00 per person. The aim of this message is to find if the *PL4321* course is available at a given time and if the price condition is satisfied.

```xml
<?xml version="1.0" encoding="UTF-8" ?>
<OTA\_GolfCourseAvailRQ xmlns=
    "http://www.opentravel.org/OTA/2003/05"
xmlns:xsi="http://www.w3.org/2001/XMLSchema-instance"
xsi:schemaLocation=
''http://www.opentravel.org/OTA/2003/05
            OTA_GolfCourseAvailRQ.xsd''
EchoToken="12345"
        TimeStamp="1003-05-31T13:20:00-05:00"
        Target="Production" Version="1.001"
        SequenceNmbr="123456">
<GolfCourseTeeTimes CourseID="PL4321">
<GolfCourseTeeTime Start="2003-10-31T14:00:00"
        End="2003-10-31T15:30:00"
        NumberOfGolfers="4"
        NumberOfHoles="18"
        NumberOfTimes="1"
        MaxPrice="75.00"
        CurrencyCode="USD">
</GolfCourseTeeTime>
</GolfCourseTeeTimes>
</OTA\_GolfCourseAvailRQ>
```

Fig. 1. Example of an OTA golf course availability query message

Analysis of OTA golf-related messages (see, [3] for a complete description) revealed that two core concepts have to be defined: (a) *Golf Course*—specifying static features of a golf course (e.g. Course ID, Address, Architect, etc.) and (b) *Golf Course Tee Time*—defining (dynamic) information necessary for a reservation of a golf course (e.g. Start date and time, Price, Number of golfers, etc.). After identifying these two core concepts, taking into account the fact that we are developing and extending an existing ontology of travel (the *TSS ontology*), we have analyzed which of its parts should be re-used. As a result, in [3] we have introduced the resulting *OTA golf ontology* and indicated how it is going to be integrated with the *TSS ontology*.

3 Utilizing OTA Golf Messages and OTA Golf Ontology

Let us now consider how the *OTA golf messages* and the *OTA golf ontology* can be utilized in the general context of "Internet-travel" and in the context of our *Travel Support System*. Here, we have to bring forward a slightly bigger picture of the "world of travel" that can provide a canvas for what we are doing. First, let us assume that the OTA messaging becomes an industry-wide standard for travel-related communication. The second assumption is that the idea of the Semantic Web takes off as expected by its proponents, and utilization of ontologically demarcated data starts to become a standard ([4]). Finally, let us also assume that software agents start to play an important role in the computational fabric (see for instance arguments put forward in [13]). Note that we do not claim that these assumptions are uncontroversial; rather, we simply accept them as the departure point and focus on developing a system that is going to work if they are to be true. However, our work is also geared toward assessing feasibility of these assumptions by attempting at implementing a system that utilizes them as its foundation. From the three assumptions follows naturally a vision of a situation in which at least the following stakeholders participate in travel-related activities:

- *Users* represented by their *Personal Agents* (*PA*); here the notion of the *Personal Agent* follows the concept of "agent as a personal assistant" put forward in [14]; to support travel needs of *Users*, their *PAs* may contact either *Travel Service Providers* to obtain a specific information / reservation, or *Travel Agencies* which can provide, for instance, complete travel packages (e.g. airline ticket + car rental + hotel + golf). Obviously, in the first case content personalization will be facilitated by the *PA* alone. In the second case, it is possible that initial content personalization will take place within the *Travel Agency* (note that the *TSS*, can be viewed as such agency), which will utilize past interactions with a representative of a given *User* and, for instance, data clustering-based stereotypes, to select and rank offers out of existing possibilities.
- *Travel Service Providers* that provide information about, and facilitate reservations of, specific travel entities (e.g. hotel chains, individual hotels, restaurants, golf course operators), as well as global reservation systems (e.g. Amadeus). Their role is most likely going to be limited to content delivery. The only exception may be related to existing loyalty programs, which will allow some of such entities (e.g. Mariott Hotels) to acquire, store and utilize customer data.
- *Travel agencies*, which may play the same role as *Travel Service Providers*; here we assume that it is unlikely that "anyone" will have access to global reservation systems (e.g. for security reasons). They also provide integrated services (e.g. a vacation package to Milan, consisting of: airline reservation + hotel + opera tickets). Their profit may come, for instance, from selling extra services on the basis of knowledge of habits of their users (similarly to Amazon.com suggesting additional items based on similarities of behavior of their customers).

Let us now discuss how these stakeholders may want to store the necessary data and in this context consider the question: will there be a single ontology of travel. While the ideal situation is promoted by project CYC [16], where a single ontology of "everything" is to be developed, this vision is unlikely to materialize for a variety of reasons (e.g. multilinguality of the world, pragmatic needs of individual players etc.). Instead, we can expect that (a) some entities will move toward ontologies very slowly, e.g. old players such as global reservation systems, (b) some entities will utilize domain and business specific ontologies, e.g. hotel chains may use a combination of a "hotel as a tourist entity" ontology and "hotel as a business entity" ontology, while have no use and knowledge of ontologies of other travel entities, (c) *Personal Agents* may use simplified ontologies, that are large enough to support their users, e.g. in such ontology concepts related to "hotel as a place for a conference" (including capacity of meeting rooms) may be omitted. Therefore, we can expect that different stakeholders of the "world of travel" will utilize different data representation (ontologically demarcated, or not). Furthermore, even if data will be stored in an ontologically demarcated fashion, different players will use different ontologies.

These considerations point back to one of main reasons of creation of the OTA messaging system. While each travel entity may use different data storage, they all should be able to communicate utilizing OTA messaging. Obviously this means that each time messages are to be exchanged, a number of translations needs to take place:

- In the case of *Travel Service Providers*, incoming OTA requests have to be translated into queries matching their internal data representation. Resulting responses have to be translated "back" into the OTA response messages and send to requesters.
- We should assume that communication between the *User* and its *Personal Agent* does not involve OTA messages. Rather, the *User* fills-in a form (e.g. an HTML template) and the resulting querystring is send to the *PA* (see, [5] for more details about non-agent entities communicating with software agents). The *Personal Agent* takes the *User*-query and translates it into an OTA request message, which can be send either to *Travel Service Providers*, or to *Travel Agencies*. Obtained OTA response has to be translated into the local ontology, as this is the data representation that is used by the *PA* to filter and order responses (later translated into user readable form and displayed on the user device; for more details see, [5]).
- The *Travel Agency* (e.g. the *TSS*) receives OTA requests from the *User*. Some of them can be answered directly by the *TSS*. For instance, since in the *TSS* we gather data, and keep it fresh by systematic updates, static elements such characteristics of the golf course (represented by the *Golf Course* concept) can be found by querying the internal database of teh *TSS*. Specifically, in the current design of the *TSS*, ontologically demarcated travel data is kept in the Jena repository [12]. Therefore, the OTA request message is translated into the SPARQL query [19] and executed. The result may then either be translated into an OTA response message and send to the *PA*, or further processed (e.g. to propose other travel related items that a given *User* may be interested in and in this way to maximize its profit [10]). The second possibility is that the original request requires access to *Travel Service Providers* (e.g. a request to check availability of a given golf course). Such message can be forwarded to an appropriate *Travel Service Provider* to obtain the necessary data (see above). The response is then treated as if it was obtained from the local database.

The scenario involving *Travel Service Providers* is uninteresting, as we cannot speculate what is their internal data representation. Furthermore, currently the *Personal Agent* is an internal part of the *TSS* (see [10]). Therefore, to illustrate

Table 1. TA actions depending on received messages

Message	TA Actions
message *TA_translate_from_OTAGolf-CourseSearchRQ*	TA translates the *OTAGolfCourseSearchRQ* XML message to the structure *Conditions*
message *TA_translate_from_OTAGolf-CourseSearchRS*	TA translates the *OTA_GolfCourseSeachRS* XML message to the list of instances of the *GolfCourse* ontology.
message *TA_translate_from_OTAGolf-CourseAvailRS*	TA translates the *OTA-GolfCourseAvailRS* XML message to the list of instances of the *GolfCourseTeeTime* ontology
message *TA_translate_to_OTAGolf-CourseSearchRS*	TA translates the instances of the *GolfCourse* ontology to the *OTAGolfCourseSearchRS* XML message.
message *TA_translate_to_OTAGolf-CourseAvailRQ*	TA translates the structure *Map* to the *OTAGolfCourseAvailRQ* XML message
message *Close_system_action*	TA finishes its activity

how to implement necessary translations we will focus on golf messages and the needs of the *Travel Support System*. However, at this stage we have not implemented golf-related functionalities directly within the *TSS*. Instead, for testing purposes, we have implemented it as a separate sub-system and introduced a number of auxiliary agents, out of which the most important one is the *Translation Agent* (*TA*). Actions undertaken by the *TA* depend on received messages and have been summarized in Table 1 (it should be obvious that the *TA*, or its functions could be used directly by (or within as a sub-agent of) the *Personal Agent* to fulfill its role in *User* support):

Here, the *Conditions* structure contains list of objects of the class *Condition* and has the form:

```
class Condition implements jade.content.Concept
{
    String name_;       /*name of the feature (e.g. ``Architect'')*/
    boolean required_;  /*is given criterion is required?*/
    String valueString; /* value (e.g. ``Jan Kowalski'')*/
    String operation_;  /*operation*/
}
```

Class *Condition* is used to specify criteria of a requested golf course (criteria based on the *OTA_GolfCourseSearchRQ* message). This structure is used to generate the SPARQL query to be executed on the Jena repository.

The *Map* is a structure from the *TSS*. In the Golf sub-system it is used to specify details of the question regarding golf course availability. *Map* contains the list of objects of the class *MapEntry* and has the form:

```
class MapEntry implements jade.content.Concept
{
    private String key;    /*name of parameter (e.g. "golfCourseId")*/
    private String value;  /*value of parameter(e.g. "AW313")*/
}
```

Classes *Conditions*, *Condition*, *Map* and *MapEntry* extend class *jade.content.Concept* and are part of the *GolfCourseOntology*.

3.1 Implementing Message Translations

Agent *TA*, during the above summarized translations of messages utilizes classes generated by the Castor and the Jastor software [21,11]. Castor is an Open Source data binding framework for Java. Castor's Source Code Generator creates a set of Java classes which represents an object model for an *XMLSchema*. The input file for the source code generator is an *XSD* file. We used Castor to generate classes for all six OTA messages (*OTA_GolfCourseSearchRQ*, *OTA_GolfCourseSearchRS*, *OTA_GolfCourseAvailRQ*, *OTA_GolfCourseAvailRS*, *OTA_GolfCourseResRQ*, *OTA_GolfCourseResRS*). Castor generates classes, not only for messages but also for all types of their attributes. For instance let us consider a snippet of the the XMLSchema file for the *OTA_GolfCourseSearchRQ* message:

```
elementFormDefault="qualified" version="1.005" id="OTA2006A">
<xs:include schemaLocation="OTA_GolfCourseTypes.xsd"/>
<!--xs:include schemaLocation="OTA_GolfCourseSearchRQTypes.xsd"/-->
<!--xs:include schemaLocation="OTA_GolfCommonTypes.xsd"/>
<xs:include schemaLocation="OTA_CommonTypes.xsd"/>
<xs:include schemaLocation="OTA_AirCommonTypes.xsd"/>
<xs:include schemaLocation="OTA_SimpleTypes.xsd"/-->

<?xml version="1.0" encoding="UTF-8"?>
<xs:schema xmlns:xs="http://www.w3.org/2001/XMLSchema"
    xmlns="http://www.opentravel.org/OTA/2003/05"
    targetNamespace="http://www.opentravel.org/OTA/2003/05">
<...appropriate headers come here...>
    <xs:annotation>
        <xs:documentation xml:lang="en"> </xs:documentation>
    </xs:annotation>
    <xs:element name="OTA_GolfCourseSearchRQ">
        <xs:annotation>
            <xs:documentation xml:lang="en"> </xs:documentation>
        </xs:annotation>
        <xs:complexType>
            <xs:sequence>
                <xs:element name="Criteria">
                    <xs:complexType>
                        <xs:sequence>
                            <xs:element name="Criterion" maxOccurs="99">
                                <xs:complexType>
                                    <xs:attributeGroup ref="CriteriaGroup"/>
                                </xs:complexType>
                            </xs:element>
                        </xs:sequence>
                    </xs:complexType>
                </xs:element>
                <...>
</xs:schema>
```

We can see that within the *OTA_GolfCourseSearch* message there is a list of *Criterion*, which is an attribute that has reference to the *CriteriaGroup*. Now, part of the *XMLSchema* file for the *CriteriaGroup* has the form:

```
<xs:annotation>
    <xs:documentation xml:lang="en">
    </xs:documentation>
</xs:annotation>
```

```xml
<xs:attribute name="Name" type="StringLength1to32" use="required">
    <xs:annotation>
        <xs:documentation xml:lang="en">
        </xs:documentation>
    </xs:annotation>
<xs:attributeGroup name="CriteriaGroup">
<... appropriate headers come here ...>
    </xs:attribute>
<xs:attribute name="Value" type="StringLength1to16" use="required">
    <xs:annotation>
        <xs:documentation xml:lang="en"></xs:documentation>
    </xs:annotation>
</xs:attribute>
<xs:attribute name="Required" type="xs:boolean" use="required">
    <xs:annotation>
        <xs:documentation xml:lang="en"></xs:documentation>
    </xs:annotation>
</xs:attribute>
<xs:attribute name="Operation" type="StringLength1to16" use="optional">
    <xs:annotation>
        <xs:documentation xml:lang="en"></xs:documentation>
    </xs:annotation>
</xs:attribute>
</xs:attributeGroup>
```

Criterion has attributes: *Name* (type *String*; attribute *required*), *Value* (type *String*; attribute *required*), *Required* (type *Boolean*; attribute *required*), *Operation* (type *String*; attribute *optional*). Castor generates a class for the *Criterion* with methods *get* and *set*. Let us present a fragment of such resulting class:

```
public class Criterion implements java.io.Serializable {
    /** A code representing the criterion on which to filter */
    private java.lang.String _name;
    /** The value of the criterion */
    private java.lang.String _value;
    /** A flag establishing if this criterion
        must be met (value \textit{Yes}) */
    private boolean _required;
    /* keeps track of state for field: _required */
    private boolean _has_required;
    /*Other operations to be used as the filter (e.g. GT, LT, etc.). */
    private java.lang.String _operation;
    //- Constructors -/
    public Criterion() {
        super();
    } //-- golfCourse.translations.castor.Criterion()
    //- Methods -/
    /*@return the value of field 'name'. */
    public java.lang.String getName()
    {
        return this._name;
    } //--- java.lang.String getName()
    /*@return the value of field 'operation'. */
    public java.lang.String getOperation()
    {
        return this._operation;
    } //--- java.lang.String getOperation()
    /* @return the value of field 'required'. */
    public boolean getRequired()
    {
        return this._required;
    } //--- boolean getRequired()
    /*@return the value of field 'value'. */
    public java.lang.String getValue()
    {
```

```
            return this._value;
        } //-- java.lang.String getValue()
    /*Method hasRequired */
    public boolean hasRequired()
        {
            return this._has_required;
        } //-- boolean hasRequired()
<...continued...>
```

In class generated for OTA_GolfCourseSearchRQ are methods to get and set list of Criteria:

```
public class OTA_GolfCourseSearchRQ implements java.io.Serializable {
...
    /** Field _criteria          */
    private golfCourse.translations.castor.Criteria _criteria;
...
    /** Returns the value of field 'criteria'.
     * @return the value of field 'criteria'.        */
    public golfCourse.translations.castor.Criteria getCriteria()
        {
            return this._criteria;
        } //-- golfCourse.translations.castor.Criteria getCriteria()

    /** Sets the value of field 'criteria'.
     * @param criteria the value of field 'criteria'.*/
    public void setCriteria(
            golfCourse.translations.castor.Criteria criteria)
        {
            this._criteria = criteria;
        } //-- void setCriteria(golfCourse.translations.castor.Criteria)
```

All requested classes generated by Castor have method *marshal* and static method *unmarshal*. These methods are used to convert Java classes to XML and to transform that XML back into the Java code. Method *marshal* converts an instance of a class to XML. Note using the method *marshal* we can transform only instances of a class, not the class itself. We instantiate (or obtain from a factory or from another instance-producing mechanism) that class to give it a specific form. Then, we populate fields of that instance with the actual data. Obviously that instance is unique; it bears the same structure as any other instances of the same class, but the data is separate. For instance, when we want to create the XML file from the *OTA_GolfCourseSearchRQ* message, we have two classes: *TA_GolfCourseSearchRQ* and *Criterion*. We must create instances of these classes and insert data into them. Here we will present only an example of utilization of the *marshall* method.

```
//create instance of OTA_GolfCourseSearchRQ class
OTA_GolfCourseSearchRQ ota = new OTA_GolfCourseSearchRQ();
// set data to this instance
...
// create instance of Criteria
Criteria criteria = new Criteria();
//put data from list of structure Condtion to Criteria
for(Iterator iter = conditions.getAllConditions(); iter.hasNext();)
{
    Condition condition = (Condition)iter.next();
//create instance of class Criterion
    Criterion criterion = new Criterion();
    criterion.setName(condition.getName_());
```

```
        criterion.setOperation(condition.getOperation_());
        criterion.setRequired(condition.getRequired_());
        criterion.setValue(condition.getValueString());
        criteria.addCriterion(criterion);
}
//put instance of Criteria to instance of class OTA_GolfCourseSearchRQ;
ota.setCriteria(criteria);
}
```

After that, we can convert these instances to XML:

```
/*put values to OTA (object of class OTA_GolfCourseSearchRQ)*/
...
Writer writer = new StringWriter();
try {    /* convert object to stream (XML text)*/
    ota.marshal(writer);
    }
 catch(MarshalException e)  {...}
 catch(ValidationException e)  {...}
```

And we get XML:

```
<?xml version="1.0" encoding="UTF-8"?>
<OTA_GolfCourseSearchRQ xmlns="http://www.opentravel.org/OTA/2003/05"
xmlns:xsi="http://www.w3.org/2001/XMLSchema-instance"
xsi:schemaLocation=
   "http://www.opentravel.org/OTA/2003/05_OTA_GolfCourseSearchRQ.xsd"
      EchoToken="54321"
      TimeStamp="2003-11-12T10:30:00"
      Target="Production" Version="1.001"
      SequenceNmbr="2432"
      PrimaryLangID="en" ID="FL4902"
      DetailResponse="true">
<Criteria>
<Criterion Name="Architect" Value=''Robert Jones'' Required="false"/>
<Criterion Name="Slope" Value="110"
        Required="true" Operation="LessThan"/>
</Criteria>
</OTA_GolfCourseSearchRQ>
```

The second generator Jastor is used for work with ontologies. It generates classes for them (like Castor for the XMLSchema). Next, we can use Jastor to convert instances of these classes to instances of ontologies and transform back instances of ontologies to objects of these classes. Jastor generates Java interfaces, implementations, factories and listeners for ontologies. For instance, for the ontology *GolfCourse*, Jastor has generated four files:

- interface *GolfCourse* extends *com.ibm.adtech.jastor.Thing*
- interface *GolfCourseListener* extends *com.ibm.adtech.jastor.ThingListener*
- class *GolfCourseImpl* extends *com.ibm.adtech.jastor.ThingImpl*
- class *GolfCourseFactory* extends *com.ibm.adtech.jastor.ThingFactory*

We used Jastor to generate classes for all ontologies needed in the system: *GolfCourse, GolfCourseTeeTime, Contacts, Description, Price, Fee, AddressRecord,* and *OutdoorLocation.* For instance ontology *GolfCourseTeeTime* has parameters: *golfCourseId (String), amount (float), currencyCode (String), startDate (String), endDate (String), maxPrice (float), numberOfHoles (integer), numberOfTimes (integer), list of fees (Fee).* For this ontology Jastor generates the interface *GolfCourseTeeTime* with methods *get/set* for properties and class *GolfCourseTeeTimeImpl* that implements this interface. Let us see a snippet of this interface for the *golfCourseId*

```
public interface GolfCourseTeeTime extends com.ibm.adtech.jastor.Thing {
    ...
    /** Gets the 'golfCourseID' property value
     * @return      {@link java.lang.String}
     * @see         #golfCourseIDProperty       */
    public java.lang.String getGolfCourseID()
        throws com.ibm.adtech.jastor.JastorException;

    /**Sets the 'golfCourseID' property value
     * @param       {@link java.lang.String}
     * @see         #golfCourseIDProperty */
    public void setGolfCourseID(java.lang.String golfCourseID)
        throws com.ibm.adtech.jastor.JastorException;
<... continued for remaining parameters...>
```

Interfaces generated by Jastor for the ontology extend the interface *com.ibm.adtech.jastor.Thing*. Classes generated by Jastor extend the class *com.ibm.adtech.jastor.ThingImpl* that implements the interface *com.ibm.adtech.jastor.Thing*.

Work with Jastor is very similar to work with Castor. First Jastor generate classes for the ontologies (like Castor for XMLSchema) and then we work only with instances of these classes. We can convert instance of a class generated by Jastor to instance of an ontology (like instances of a class generated by Castor to XML). We can also transform back instances of an ontology to instances of a class generated by Jastor (like converting XML to instances of a class generated by Castor). During translation Agent *TA* uses classes generated by Castor and Jastor. So the *TA* has only to take values from the object of one class and put it to the object of another class.

4 Concluding Remarks

In this paper we have discussed how OTA messages can be used to connect systems that utilize various forms of internal data representation (ontological or not). We have identified three groups of main stakeholders of the "world of travel," i.e. *Users*, *Service providers* and *Intermediaries* (e.g. *Travel Agencies*, or our own *Travel Support System*. Next, we have discussed scenarios that lead to communication between these three groups of players and specified what kind of translations between OTA messages and the *OTA ontology of golf* have to take place within our *TSS*. Finally we have discussed and illustrated on examples how Castor and Jastor software can be used to implement necessary translations. Our current work is devoted to merging the *OTA ontology of golf* and the translation mechanisms with the existing *TSS* and its *ontology of travel*.

References

1. Bădică, C., Bădiţă, A., Ganzha, M., Paprzycki, M.: E-Service Intelligence—Methodologies, Technologies and Applications. In: Developing a Model Agent-based E-commerce System, pp. 555–578. Springer, Berlin (2007)
2. Bădică, C., Ganzha, M., Paprzycki, M.: Journal of Universal Computer Science. Implementing Rule-Based Automated Price Negotiation in an Agent System 13, 244–266 (2007)

3. Cieślik, A., Ganzha, M., Paprzycki, M.: Developing open travel alliance-based ontology of golf. In: Proceedings of the 2008 WEBIST conference (to appear, 2008)
4. Fensel, D.: Ontologies: A Silver Bullet for Knowledge Management and Electronic Commerce. Springer, New York (2003)
5. Gawinecki, M., Gordon, M., Kaczmarek, P., Paprzycki, M.: The problem of agent-client communication on the internet. Scalable Computing: Practice and Experience 6(1), 111–123 (2005)
6. Gawinecki, M., Gordon, M., Nguyen, N.T., Paprzycki, M., Szymczak, M.: Rdf demarcated resources in an agent based travel support system. In: G.M. et al. (eds.) Informatics and Effectiveness of Systems, Katowice, pp. 303–310. PTI Press (2005)
7. Gawinecki, M., Gordon, M., Nguyen, N.T., Paprzycki, M., Vetulani, Z.: Ontologically Demarcated Resources in an Agent Based Travel Support System. In: Advanced Knowledge International, Adelaide, Australia, pp. 219–240 (2005)
8. Gawinecki, M., Kruszyk, M., Paprzycki, M.: Ontology-based stereotyping in a travel support system. In: Proceedings of the XXI Fall Meeting of Polish Information Processing Society, pp. 73–85. PTI Press (2005)
9. Gordon, M., Kowalski, A., Paprzycki, M., Pełech, T., Szymczak, M., Wasowicz, T.: Internet 2005. In: Ontologies in a Travel Support System, pp. 285–300. Technical University of Wroclaw Press (2005)
10. Gordon, M., Paprzycki, M.: Designing agent based travel support system. In: ISPDC 2005: Proceedings of the ISPDC 2005 Conference, pp. 207–214. IEEE Computer Society Press, Los Alamitos (2005)
11. http://jastor.sourceforge.net/
12. http://jena.sourceforge.net/
13. Jennings, N.R.: An agent-based approach for building complex software systems. Commun. ACM 44(4), 35–41 (2001)
14. Maes, P.: Agents that reduce work and information overload. Commun. ACM 37(7), 30–40 (1994)
15. http://www.opentravel.org
16. http://www.cyc.com/
17. OTA MessageUserGuide2006V1.0 (2006)
18. Salam, A.F., Stevens, J. (eds.): Utilizing Semantic Web and Software Agents in a Travel Support System, pp. 325–359. Idea Publishing Group, Hershey
19. http://www.w3.org/TR/rdf-sparql-query/
20. Szymczak, M., Gawinecki, M., Vukmirovic, M., Paprzycki, M.: Ontological Reusability in State-of-the-art Semantic Languages. Knowledge Management Systems, pp. 129–142. PTI Press,
21. http://www.castor.org/
22. http://www.e/-travel.sourceforge
23. Vukmirovic, M., Ganzha, M., Paprzycki, M.: Developing a Model Agent-based Airline Ticket Auctioning System, pp. 297–306. Springer, Berlin (2006)
24. Vukmirovic, M., Paprzycki, M., Szymczak, M.: Designing ontology for the open travel alliance airline messaging specification. In: Bohanec, M., et al. (eds.) Proceedings of the 2006 Information Society Multiconference (2006)
25. Vukmirovic, M., MichałSzymczak,, Ganzha, M., Paprzycki, M.: Utilizing ontologies in an agent-based airline ticket auctioning system. In: Luznar, V., et al. (eds.) Proceedings of the 28th ITI Conference, Piscatawy, NJ, pp. 385–390. IEEE (2006)
26. Vukmirovic, M., Szymczak, M., Gawinecki, M., Ganzha, M., Paprzycki, M.: Designing new ways for selling airline tickets. Informatica 31(3), 93–104 (2007)

Collaborative Recommendations Using Bayesian Networks and Linguistic Modelling

Luis M. de Campos, Juan M. Fernández-Luna, and Juan F. Huete

Department of Computer Science and Artificial Intelligence
Technical School of Computer Science, University of Granada, 18071–Granada, Spain
{lci,jmfluna,jhg}@decsai.ugr.es

Abstract. This paper presents a model designed under the formalism of Bayesian Networks to deal with the problem of collaborative recommendation. It has been designed to perform efficient and effective recommendations. We also consider the fact that the user can usually use vague ratings for the products, which might be represented as fuzzy labels. The complete proposal is evaluated with MovieLens.

1 Introduction

Over the last ten years, there has been a massive increase in the amount of information available on Internet, and it is often hard for users to access this information, possibly because they are unaware it exists. This offers an attractive framework for research into new, accurate and efficient techniques for accessing this information. In this framework, *Recommender Systems (RS)* have emerged to help people deal with this overload of information. Broadly speaking, an RS provides specific suggestions about items (or actions) within a given domain, which may be considered of interest to the user [1]. There are several types of RS, classified according to the information used when recommending. This paper focuses on the variant called *collaborative filtering*, which attempts to identify groups of people with similar tastes to the user's and to recommend items that they have liked. In this case, the objective is usually to predict the utility of an unseen item for an active user based on items previously rated by other users.

In this paper, we propose to model an RS by using the *Bayesian network (BN)* [2] formalism. Given an unobserved item, we will therefore be able to obtain the most probable vote that a user would give to that item. It must be pointed out that fuzziness exists in the rating process. Rating an item usually implies the selection of a vote from a set of labels. However, there is often no meaningful way to set the boundary between two consecutive labels. For this reason, we shall also explore the advantages of considering the rating alternatives as vague concepts. Fuzzy sets formalize the idea of graded membership of an element to a given set. Although users might think in terms of vague concepts when rating, it is quite common for an RS to eventually store their ratings using crisp values with the consequent loss of information. What we offer in this paper, therefore, is a linguistic modelling in the system output so that the crisp votes generated by the BNs are converted into fuzzy labels which are closer to the users.

Our approach for modelling RSs therefore involves processing two different types of uncertainty: probability arising from a lack of knowledge of how the different users are related, and fuzziness concerned with the ambiguity or vagueness in the description of the ratings. We shall show how the combination of both theories leads to an improvement when modelling a collaborative RS.

In order to describe our model, this paper is organised into the following parts: Section 2 will briefly review Bayesian networks and Recommender Systems; Section 3 will describe the model, explaining its topology and how it is learnt, the probability estimation, the inference mechanism and finally, how the linguistic modelling is carried out; Section 4 will show the experiments and results; and finally, Section 5 will discuss the main conclusions and further work.

2 Background

2.1 Bayesian Networks

In recent decades, Bayesian networks [2] have become one of the most consolidated methodologies for probabilistic inference. These graphical models are capable of efficiently representing and manipulating n-dimensional probability distributions by combining a qualitative and a quantitative representation of the problem by means of, on the one hand, a *directed acyclic graph* (DAG), $G = (V, E)$, where the nodes in V represent the random variables from the problem we want to solve, and the topology of the graph (the arcs in E) encodes dependence relationships between the variables, with the absence of an arc between any two nodes representing an independence relationship between the variables; and, on the other, a set of conditional probability distributions drawn from the graph structure: for each variable $X_i \in V$ we have a family of conditional probability distributions $Pr(X_i \mid pa(X_i))$, where $pa(X_i)$ represents any combination of the values in the parent set of X_i in G, $Pa(X_i)$.

Once the BN is completed, it specifies a complete joint probability distribution over all the variables, i.e. given a configuration $c = (x_1, x_2, \ldots, x_n)$ over the set of variables $X_1, \ldots, X_n$, with x_i being the value that variable X_i takes in c, then $Pr(c) = \prod_{i=1}^{n} Pr(x_i|pa(x_i))$, where $pa(x_i)$ are the values taken by the parent set of X_i in c. This decomposition of the joint distribution results in important savings in storage requirements.

In a probabilistic context, the user usually has some evidence of the state that a variable (or a set of variables) takes. The problem is to compute the posterior probability distribution for a variable given the evidence, $Pr(X_i|ev)$ [2].

2.2 Recommender Systems

The usual formulation of the recommendation problem is to predict the vote or rating that an active user would give to an unseen item. This estimation can be used to recommend those items with the highest estimated ratings to the user. RSs are usually classified into the three main categories based on how the recommendations are made. The first type are *content-based RSs*, which store content information about each item to be recommended. This information will be used to recommend similar items to those favoured by the user in the past, based on how similar certain items are to each other or the similarity with respect to user preferences. The second kind are *collaborative filtering RSs*, which attempt to identify groups of people with similar tastes to those of the user and recommend items that they have liked (predicting the vote for a given user depends on the votes of people with similar tastes or preferences). In order to do so, they use some kind of aggregation measure considering the ratings of other (most similar) users for the same item. Alternatively, predictions may be made by building (offline) an explicit model of the relationships between items. This model is then used (on-line) to finally recommend the product to the users. In this approach, the predictions are not therefore based on any ad hoc heuristic but rather on a model learnt from the underlying data using statistical and machine learning techniques. Finally, *Hybrid RSs* combine both previous approaches.

Considering collaborative RSs, we can distinguish between two approaches. The first approach uses BN learning algorithms to learn a full joint probability distribution about items and then uses this distribution to make on-line predictions [3,4]. BN-based classifiers [5,6,7] have also been applied. The second approach builds several conditional models and predicts the likelihood of an individual item given a combination of the observed votes for other users [8].

3 The Collaborative Bayesian Network-Based Model

We shall consider a large number m of items $\mathcal{I} = \{I_1, I_2, \ldots, I_m\}$, a large set of n users, $\mathcal{U} = \{U_1, U_2, \ldots, U_n\}$. For each user, a set of ratings about the quality of certain observed items in $\mathcal{I}$. The user's ratings (preferences) are values in the set $\mathcal{S}$. The set of observed data can then be viewed as a very sparse $n \times m$ matrix, $\mathbf{R}$ (users only rate a very small proportion of items). In the matrix, $\mathbf{R}[a][j]$ represents the rate of user U_a for the item I_j and will also be denoted as $s_{a,j}$, assuming 0 when the item has not been rated by the user.

We are interested in representing the relation $\mathcal{I} \longrightarrow \mathcal{U}$ in a BN, modelling the database of user votes for the set of observed items, as well as the relation $\mathcal{U} \longrightarrow \mathcal{U}$, modelling the relationships between users. We shall therefore consider the set of items $\mathcal{I}$ and users $\mathcal{U}$ as variables in the BN (nodes in the graph).

It is clear that the voting pattern of each user (U_a) will depend directly on the vote given to each observed item. We shall include an arc from each item, I_j, voted by user U_a to the node representing that user. Each item $I_j \in \mathcal{I}$ will have an associated random binary variable, taking values from the sets $\{i_{j,0}, i_{j,1}\}$ (not relevant, relevant, respectively, to the user's interest). In the case of user variables, $U_a \in \{u_{a,1} \ldots, u_{a,r}\} \cup \{u_{a,0}\}$ (a new state ($u_{a,0}$) is added to represent the fact that the user has no interest in voting).

Our model might be able to represent relations between users, $\mathcal{U} \longrightarrow \mathcal{U}$. These should be modeled in the BN by the inclusion of arcs between any two similar users. As the similarities between two users tend to be symmetric, we

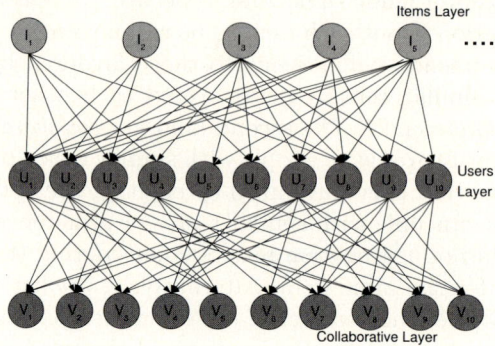

Fig. 1. Collaborative Recommending System Topology

would be including a cycle in the BN, something which is forbidden. To solve this problem, we propose that a new set of nodes $\mathcal{V}$ be considered to denote collaborative votes. There is therefore one collaborative node for each user in the system, i.e. $\mathcal{V} = \{V_1, V_2, \ldots, V_n\}$, which will also be used to predict the vote that the active user could give to an unseen item and they will therefore take their values in the set of valid rating labels, i.e. $\{v_{a,1}, v_{a,2}, \ldots, v_{a,r}\}$, omitting $v_{a,0}$ as an alternative state (see Figure 1).

3.1 Learning User Relationships

We shall now describe how the selection of similar users for a given one, V_a, is performed from the database of votes, in order to form its set of parents in the graph, $Pa(V_a)$ (those user variables, $U_b \in \mathcal{U}$, with U_a and U_b having the greatest similarity between their tastes). Thus, given a similarity measure, $Pa(V_a)$ is obtained using a threshold or p variables with the highest similarity.

A first measure is the *Cosine Measure* [9], based on the computation of the cosine of the angle formed by two vectors (any two rows of matrix **R**). In the range $[0, 1]$, the greater the similarity between the vectors, the greater the cosine:

$$Cosine(U_a, U_b) = \frac{\sum_{I_j \in \mathcal{I}} r_{a,j} \cdot r_{b,j}}{\sqrt{\sum_{I_j \in \mathcal{I}} r_{a,j}^2} \sqrt{\sum_{I_j \in \mathcal{I}} r_{b,j}^2}} \quad (1)$$

A second alternative is *Pearson's Correlation Coefficient*, which determines whether there is a linear relationship between two variables (domain $[-1, 1]$). A 0 valoue means that there is absolutely no correlation; 1 means that there is an exact and positive correlation, and -1 that the correlation is exact but negative.

$$Pearson(U_a, U_b) = \frac{\sum_{I_j \in P(U_a) \cap P(U_b)} (r_{a,j} - \overline{r}_a)(r_{b,j} - \overline{r}_b)}{\sqrt{\sum_{I_j \in P(U_a) \cap P(U_b)} (r_{a,j} - \overline{r}_a)^2 \sum_{I_j \in P(U_a) \cap P(U_b)} (r_{b,j} - \overline{r}_b)^2}} \quad (2)$$

In this case, the summations over I_j are over those items for which both users U_a and U_b have recorded votes. In addition, $\bar{r}_a$ is the mean vote for user U_a.

A common problem due to the sparsity of the data set R arises: let us consider two users rating a common item. Then, $Pearson(U_a, U_b) = 1$, so U_b will be set as the parent of V_a and also U_a is a parent of V_b, resulting in low quality parent sets. In order to avoid this situation, we also propose taking into account the number of items that both U_a and U_b rated simultaneously, i.e. their overlap degree. The criterion can therefore be implemented with two alternatives [1]:

The *Jaccard Coefficient* [9] measures the overlap degree between two sets by dividing the numbers of items observed by both users (intersection) and the number of different items from both sets of rated products (union):

$$Jaccard(U_a, U_b) = \frac{|\{I_j \in \mathcal{I}/r_{a,j} \neq 0\} \cap \{I_j \in \mathcal{I}/r_{b,j} \neq 0\}|}{|\{I_j \in \mathcal{I}/r_{a,j} \neq 0\} \cup \{I_j \in \mathcal{I}/r_{b,j} \neq 0\}|} \quad (3)$$

A second choice is the *Overlap Coefficient*, which substitutes the denominator of Equation 3 for the number of products rated by one of the two users:

$$Overlap(U_a, U_b) = \frac{|\{I_j \in \mathcal{I}/r_{a,j} \neq 0\} \cap \{I_j \in \mathcal{I}/r_{b,j} \neq 0\}|}{|\{I_j \in \mathcal{I}/r_{a,j} \neq 0\}|}. \quad (4)$$

The final similarity measures that we propose are combinations of both criteria: vote correlation between common items and the overlap degree, i.e.

$$\begin{aligned} PearsonJC(U_a, U_b) &= abs(Pearson(U_a, U_b)) \times Jaccard(U_a, U_b), \\ PearsonOC(U_a, U_b) &= abs(Pearson(U_a, U_b)) \times Overlap(U_a, U_b). \end{aligned} \quad (5)$$

where *abs* denotes the absolute value, as we take into account those users with a high positive correlation value (very similar tastes), and those with very low positive correlations (same items but totally opposite votes).

3.2 Probability Estimation

In order to complete the model's specification, the numerical values for the conditional probabilities must be estimated from the data sets, but prior to this, we shall introduce some notation. $x_{i,j}$ denotes the fact that variable X_i takes the jth-value. We write $Pr(x_{i,j}|pa(X_i))$ for $Pr(X_i = x_{i,j}|pa(X_i))$, with $pa(X_i)$ denoting a configuration of the parent set of X_i, $Pa(X_i)$.

With respect to the set of items $\mathcal{I}$, since they are root nodes in the graph, they store marginal probability distributions which are linear in size to the number of states. Variables $\mathcal{U}$ and $\mathcal{V}$ must store a set of conditional probability distributions with an exponential size to the number of parents. Since a user can rate a large number of items and a collaborative node might be related to a large number of users, assessing and storing these probability values can be quite complex. We therefore propose a weighted-sum canonical model to represent these probabilities, enabling us to design a very efficient inference procedure.

---

[1] U_a is the user for which we are learning its set of parents.

Thus, for a given X_i, we define $Pr(x_{i,j}|pa(X_i)) = \sum_{Y_k \in Pa(X_i)} w(y_{k,l}, x_{i,j})$, where $w(y_{k,l}, x_{i,j})$ is a weight measuring how this l^{th} value of variable Y_k describes the jth state of X_i. For every item, I_j, a priori probability distributions are estimated: $Pr(i_{j,1}) = \alpha$ and $Pr(i_{j,0}) = 1 - \alpha, \forall I_j \in \mathcal{I}$, α being a constant.

For every user node U_k, we need to assess a set of conditional probability distributions, one for each possible configuration of the set of items rated by U_k, representing its rating pattern. Considering the above restrictions, these will be computed using a canonical model: assuming that the user U_k rated an item I_j with the label l, their weights could be defined by means of: $w(i_{j,0}, u_{k,s}) = 0, \forall s \neq 0$; $w(i_{j,0}, u_{k,0}) = 1/|Pa(U_k)|$; $w(i_{j,1}, u_{k,s}) = \frac{\phi(s|l)}{|Pa(U_k)|}, \forall s \neq 0$, $w(i_{j,1}, u_{k,0}) = 0$.

Focusing on collaborative nodes $\mathcal{V}$, for each node V_a we must compute those weights $w(u_{b,\bullet}, v_{a,\bullet})$ given by users U_b with similar tastes, i.e. $U_b \in Pa(V_a)$. On one side, and in view of the fact that user ratings are related statistically, these weights might be considered to depend on the frequency that user U_a votes with value s given that user U_b has the state t, i.e. $freq(u_{a,s}|u_{b,t})$, and on the other, considering that the highest weights are assigned to the most similar users, it seems natural that these weights will also depend on the similarity degree between users. The way the weight associated to user U_b is distributed is therefore defined by means of the following equation:

$$w(u_{b,t}, v_{a,s}) = \frac{freq(u_{a,s}|u_{b,t}) \times sim(U_a, U_b)}{|Pa(V_a)| \sum_{U_b \in Pa(V_a)} sim(U_a, U_b)}, \text{ with } s \in \mathcal{S}, t \in \mathcal{S} \cup \{0\} \quad (6)$$

With respect to the estimation of $freq(u_{a,s}|u_{b,t})$: if user U_b has no interest in voting ($U_b = u_{b,0}$) then $Pr(u_{a,s}|u_{b,0}) = 1/r$, for all $s \in \mathcal{S}$ (the weight associated with the 'no interest in voting' situation will be distributed uniformly among the different candidate rates at collaborative nodes). If user U_b voted, then $freq(u_{a,s}|u_{b,t})$ with $t \neq 0$ can be estimated by means of:

$$freq(u_{a,s}|u_{b,t}) = \frac{N^*(u_{b,t}, v_{a,s}) + \beta q_s}{N^*(u_{b,t}) + \beta}, 1 \leq t, s \leq r.$$

$N^*(u_{b,t}, v_{a,s})$ is the number of items from $I(U_a) \cap I(U_b)$ that have been rated with value t by user U_b and with s by user U_a. $N^*(u_{b,t})$ is the number of items in $I(U_a) \cap I(U_b)$ voted with t by user U_b. Values β and q_s are the parameters of a Dirichlet prior over user ratings with $\sum_{i=1}^{r} q_i = 1$.

3.3 Recommending: Inference in the Bayesian Network

In order to predict the satisfaction degree that a user would give to a new item acting as evidence we shall compute the a posteriori probability distribution for the collaborative node V_a, $Pr(V_a = s|ev)$ for all $s \in \mathcal{S}$. Although general purpose algorithms do exist, they take exponential time with the number of parents when applied to a BN with the proposed topology [2]. Nevertheless, considering that the evidence only affects user nodes and the conditional independence statements represented in the network, the a posteriori probabilities for the collaborative nodes can be computed efficiently by using the advantages of the canonical weighted-sum representation in Section 3.2.

Theorem 1: Let l_{X_a} denote the number of states that X_a takes in the collaborative BN network and let Y_j be a node in $Pa(X_a)$. Let us assume that the set of conditional probability distributions over X_a are expressed using a canonical weighting scheme, i.e. $Pr(x_{a,s}|pa(X_a)) = \sum_{Y_j \in Pa(X_a)} w(y_{j,t}, x_{a,s})$, where $y_{j,t}$ is the value that variable Y_j takes in the configuration $pa(X_a)$ and $w(\cdot, \cdot)$ are a set of non-negative weights verifying that $\sum_{s=1}^{l_{X_a}} \sum_{Y_j \in Pa(X_a)} w(y_{j,t}, x_{a,s}) = 1, \forall\ pa(X_a)$. If the evidence, ev, is only on ancestors of X_a, the exact a posteriori probabilities can then be computed with the following formula:

$$Pr(x_{a,s}|ev) = \sum_{Y_j \in Pa(X_a)} \sum_{t=1}^{l_{Y_j}} w(y_{j,t}, x_{a,s}) \cdot Pr(y_{j,t}|ev).$$

Propagation would therefore comprise two steps: computation of posterior probability in those user nodes, U_a, affected by the evidence, and with this information, the computation of the a posteriori distributions in the corresponding nodes, V_a. The next step would be vote selection, with two alternatives: the most probable vote, $MaxPostP$ ($r_{V_a} = \arg\max_l Pr(V_{a,l}|ev)$, and the vote with the largest difference between the a priori and a posteriori distributions, $MaxDifPostPPriP$ ($r_{V_a} = \arg\max_l (Pr(V_{a,l}|ev) - Pr(V_{a,l}))$. This would involve rewarding those states with the greatest increase in probability.

3.4 Vote Modelling

We assume that it is more difficult for a user to rate a product with an exact value from the set $\mathcal{S}$ of possible alternatives than to say 'this item is good'. Additionally, in this case he/she is not ruling out the fact that the item could be 'bad' or 'excellent', the previous and subsequent grades on the scale. Since the judgements are not usually strict and have a certain degree of flexibility, it seems appropriate to use the fuzzy-set formalism to describe the degree of satisfaction of a user evaluating an item. Following Zadeh's [10] definition, a fuzzy set A of a reference set Ω is identified by its membership function, $\mu_A : \Omega \longrightarrow [0, 1]$, where $\mu_A(x)$ is the membership degree of element $x \in A$, $\forall x \in \Omega$.

User ratings will then be considered as fuzzy observations from $\mathcal{S}$, i.e. each particular vote could be seen as a fuzzy set from $\mathcal{S}$, i.e. $\mathcal{S}_L = \{l_1, l_2, \ldots, l_r\}$[2].

As mentioned before, we are focusing on linguistic modelling at the output, i.e. instead of offering a value from $\mathcal{S}$ as the prediction, it would return a fuzzy label in $\mathcal{S}_L$ which is closer to the user's interest. Having computed $Pr(V_a|ev)$, we would then select the fuzzy label $l \in \mathcal{S}_L$ that best predicts the user's vote.

One alternative for carrying out this selection is to compute the a posteriori probability for each fuzzy event, $l \in \mathcal{S}_L$, and then return the most probable fuzzy

---

[2] For instance, with the MovieLens collection, $\mathcal{S}_L = \{$ 1 = Awful, 2 = Fairly Bad, 3 = OK, 4 = Enjoyable, 5 = Must see $\}$, while $\mathcal{S} = \{1, 2, 3, 4, 5\}$.

set (*ProbFL*): vote $= \arg\max_l \{\sum_{s=1}^{r} \mu_l(s) Pr(V_a = s|ev)\}$. A second alternative is to use a similarity measure between the a posteriori probabilities and the fuzzy labels, giving as output the fuzzy label which is most similar to the a posteriori values. A direct similarity measure cannot be applied since we are not only talking about fuzzy labels and probabilities. It is therefore necessary to make transformations so as to allow both fuzzy labels and probability values to be compared with the same language: Possibility Theory [11].

We shall use $\Pi_{ev}(V_a)$ to denote the possibility distribution over the ratings obtained after transforming the a posteriori probability distribution $Pr(V_a|ev)$ and $\Pi_l(V_a)$ to denote the possibility distribution representing the fuzzy label l. We can then use a similarity measure between them in order to select the best rate, returning the label which best matches the (a posteriori) possibility values.

We propose to use a similarity measure based on a geometric distance model, the idea being that the smaller the distance between Π_A and Π_B, the greater the similarity between them. Given two possibility distributions, a one parameter class of distance functions can be defined as [12]:

$$d_z(\Pi_A, \Pi_B) = \left[\sum_{x=1}^{n} abs(\pi_A(x) - \pi_B(x))^z\right]^{\frac{1}{z}} \tag{7}$$

In this paper, we propose to use the parameter $z = 2$, and therefore the predicted vote will be the one with the lowest d_2, i.e. vote $= \arg\min_l \{d_2(\Pi_{ev}(V_a), \Pi_l(V_a))\}$. This approach will be called *DistPossM*.

In order to transform $Pr(V_a|ev)$ into $\Pi_{ev}(V_a)$, all that needs to be done is to normalize it by using the value of maximum probability [13], i.e. $\pi(x_i) = \frac{Pr(x_i)}{\max_{j=1}^{n} Pr(x_j)}$. For $l \in S_L \longrightarrow \Pi_l(V_a)$, it is satisfied as directly as $\pi_l(s) = \mu_l(s)$. This technique will be noted as *MaxPoss*.

An alternative to *MaxPoss* is the so-called *AcumPoss*. The probabilities are sorted increasingly. The associated possibility would be the sum of the probabilities of the events which are below it in the ranking [13]: let σ be an increasingly sorted ranking of the events considering the associated posterior probabilities, and $\sigma(j)$ the event ranked j^{th}, then $\pi(x_i) = \sum_{j=1}^{i} Pr(x_{\sigma j} | ev), i = 1, \ldots, n$.

We must remember that with *MaxPostP*, we can apply both *ProbFL* and *DistPossM*. With this last one, it is possible to apply *MaxPoss* and *AcumPoss*, but with *MaxDifPostPPriP*, we are only able to apply *DistPossM* with *MaxPoss* (differences between probabilities are not probability distributions).

4 Experimentation

The most widely used experimental data set in the recommendation field, and therefore the one that we have selected for our experiments, is currently *Movie-Lens* [14], consisting of 100,000 ratings (1= Awful, 2= Fairly bad, 3= It's Ok, 4= Will enjoy, 5= Must see) for 1682 movies by 943 users. The data set is divided into 5 training and test sets (disjoint 80% – 20%) for 5-fold cross validation.

In order to test the performance of our model, we shall measure its capability to predict a user's true rating or preferences (system accuracy). Following [15],

Table 1. MAE for Cosine and Pearson variants per number of parents

#P	MaxPostP			MaxDifPostPPriP		
	Cosine	PearsonJC	PearsonOC	Cosine	PearsonJC	PearsonOC
5	0.8495 ± 0.0063	0.8513 ± 0.0085	0.8127 ± 0.0059	1.1268 ± 0.0042	1.3398 ± 0.0126	0.9888 ± 0.0049
10	0.8316 ± 0.0056	0.8263 ± 0.0069	0.7918 ± 0.0041	0.9427 ± 0.0041	1.0734 ± 0.0098	0.8550 ± 0.0013
20	0.8224 ± 0.0061	0.8117 ± 0.0068	**0.7861 ± 0.0037**	0.8592 ± 0.0046	0.9246 ± 0.0074	0.8073 ± 0.0032
30	**0.8191 ± 0.0061**	0.8075 ± 0.0059	0.7872 ± 0.0041	0.8350 ± 0.0045	0.8714 ± 0.0063	0.7939 ± 0.0036
50	0.8198 ± 0.0056	**0.8029 ± 0.0055**	0.7886 ± 0.0051	0.8193 ± 0.0044	0.8301 ± 0.0057	0.7848 ± 0.0042
75	0.8238 ± 0.0062	0.8047 ± 0.0045	0.7938 ± 0.0045	0.8123 ± 0.0047	0.8105 ± 0.0043	0.7837 ± 0.0038
100	0.8249 ± 0.0056	0.8056 ± 0.0049	0.7966 ± 0.0047	**0.8092 ± 0.0048**	**0.8020 ± 0.0038**	**0.7817 ± 0.0037**

we adopt the mean absolute error (MAE) which measures how close the system predictions are to the user's rating for each movie by considering the average absolute deviation between a predicted rating and the user's true rating: $MAE = \sum_{i=1}^{N} abs(p_i - r_i)/N$, with N being the number of cases in the test set, p_i the vote predicted for a movie, and r_i the true rating. In each result table presented, we shall show the average MAE obtained after repeating the experiment for each training and test sets and the standard deviation for the 5 experiments.

The main objective of the experimentation that we have designed is to measure the general performance of the model and compare it with other models. Specific objectives are: 1) to investigate the best vote selection method; 2) to determine the ideal size of the sets of parents of collaborative nodes; 3) to discover which similarity measure performs best; and 4) to observe whether linguistic modelling is a useful technique and how it should be carried out.

The first step is to design a battery of experiments without linguistic modelling in an attempt to find answers to objectives 1) to 3). From previous experimentation, the values of the parameters are: $\alpha = 0$, $\beta = 1$ and $q_i = 1/5$.

We need to conduct various experiments, considering a different number of parents (#P) fixed for all the collaborative node (5, 10, 20, 30, 50, 75 and 100), selected using Cosine, PearsonJC and PearsonOJ similarities. In terms of the technique for selecting the favourite vote, we shall test *MaxPostP* and *MaxDifPostPPriP*.

The results of the experiments are shown in Table 1. The tendency is for the best performance to be reached systematically by the Pearson Coefficient corrected with the overlap over the other two, independently of the prediction method used. *MaxDifPostPPriP* is a slight improvement on *MaxPostP*. Regarding the suitable number of parents, it seems that an intermediate number (20, 30 or 50) is the most appropriate when using the *MaxPostP* method, and a large number of them with *MaxDifPostPPriP* (100). The behaviour of the first method seems to be as expected: with a low number there is not enough information, and with a large number, noise is introduced, leading to bad recommendations in both cases. We were, however, surprised by the behaviour of the second method.

One reason why this situation occurs is that when an item is instantiated to perform a recommendation for a given user, if the item has not been evaluated by either of its parents there is no change in its a posteriori probability distribution with respect to the a priori since the collaborative node does not receive any influence by any other node. When the number of parents is low, this situation is more probable, so there will be more collaborative users with identical

Table 2. MAE obtained using the a priori distribution of U_a

#P	MaxPostP			MaxDifPostPPriP		
	Cosine	PearsonJC	PearsonOC	Cosine	PearsonJC	PearsonOC
5	0.8286 ± 0.0052	0.8252 ± 0.0056	0.8043 ± 0.0047	0.8430 ± 0.0045	0.8487 ± 0.0051	0.8136 ± 0.0056
10	0.8201 ± 0.0050	0.8105 ± 0.0047	0.7871 ± 0.0038	0.8250 ± 0.0042	0.8285 ± 0.0049	0.7899 ± 0.0040
20	0.8154 ± 0.0057	0.8025 ± 0.0057	0.7832 ± 0.0037	0.8112 ± 0.0049	0.8109 ± 0.0058	0.7798 ± 0.0036
30	**0.8148 ± 0.0057**	0.8008 ± 0.0054	**0.7850 ± 0.0041**	**0.8062 ± 0.0048**	0.8012 ± 0.0052	0.7772 ± 0.0038
50	0.8167 ± 0.0054	**0.7984 ± 0.0051**	0.7870 ± 0.0047	0.8029 ± 0.0050	**0.7908 ± 0.0043**	**0.7745 ± 0.0043**
75	0.8215 ± 0.0060	0.8018 ± 0.0043	0.7926 ± 0.0043	0.8031 ± 0.0052	0.7872 ± 0.0039	0.7767 ± 0.0040
100	0.8232 ± 0.0055	0.8033 ± 0.0046	0.7956 ± 0.0045	0.8032 ± 0.0051	0.7860 ± 0.0037	0.7764 ± 0.0040

probability distributions, and therefore, the difference is 0, and the selection of the vote is basically random. This implies that lots of prediction mistakes are made. As the number of parent increases, it is more likely that someone has rated the instantiated items, so the distributions will be different, and fewer cases presented. In order to confirm this, we have counted the number of times that this situation arises, finding the following summarised results (for 5 and 100 parents): Cosine (4464.4, 175.4), PearsonJC (6839.8, 332.2) and PearsonOC (2939.0, 144.2). When this number decreases, the number of parents increases.

Having detected the problem in a node V_a, we have adopted the solution of predicting the vote with the maximum a priori probability computed in its clone node U_a. Table 2 shows the results considering the 'parent help'.

In this case, the overlap variant performs better in both prediction methods. With respect to the number of parents, we can see how in both approaches, intermediate values show the best results. The values shown in this table are better than those presented in Table 2, so the use of this technique is suitable. In terms of which prediction method is preferable, *MaxDifPostPPriP* obtains the best results although the difference is not entirely significant.

Focusing on the application of the linguistic modelling to the output of the system, we have used the following set of linguistic labels in this second stage of the experimentation (with triangular membership functions):

$FL1$: $\mu_{FL1,l_1} = \{1/1, 0.5/2, 0/3, 0/4, 0/5\}$, $\mu_{FL1,l_2} = \{0.5/1, 1/2, 0.5/3, 0/4, 0/5\}$,
$\mu_{FL1,l_3} = \{0/1, 0.5/2, 1/3, 0.5/4, 0/5\}$, $\mu_{FL1,l_4} = \{0/1, 0/2, 0.5/3, 1/4, 0.5/5\}$,
$\mu_{FL1,l_5} = \{0/1, 0/2, 0/3, 0.5/4, 1/5\}$.

$FL2$: $\mu_{FL2,l_1} = \{1/1, 0.5/2, 0.25/3, 0/4, 0/5\}$, $\mu_{FL2,l_2} = \{0.5/1, 1/2, 0.5/3, 0.25/4, 0/5\}$,
$\mu_{FL2,l_3} = \{0.25/1, 0.5/2, 1/3, 0.5/4, 0.25/5\}$, $\mu_{FL2,l_4} = \{0/1, 0.25/2, 0.5/3, 1/4, 0.5/5\}$,
$\mu_{FL2,l_5} = \{0/1, 0/2, 0.25/3, 0.5/4, 1/5\}$.

$FL3$: $\mu_{FL3,l_1} = \{1/1, 0.25/2, 0/3, 0/4, 0/5\}$, $\mu_{FL3,l_2} = \{0.5/1, 1/2, 0.5/3, 0/4, 0/5\}$,
$\mu_{FL3,l_3} = \{0.25/1, 0.5/2, 1/3, 0.5/4, 0.25/5\}$, $\mu_{FL3,l_4} = \{0/1, 0/2, 0.5/3, 1/4, 0.5/5\}$
$\mu_{FL3,l_5} = \{0/1, 0/2, 0/3, 0.25/4, 1/5\}$.

$FL4$: $\mu_{FL4,l_1} = \{1/1, 0.5/2, 0.25/3, 0/4, 0/5\}$, $\mu_{FL4,l_2} = \{0.25/1, 1/2, 0.25/3, 0/4, 0/5\}$,
$\mu_{FL4,l_3} = \{0/1, 0.25/2, 1/3, 0.25/4, 0/5\}$, $\mu_{FL4,l_4} = \{0/1, 0/2, 0.25/3, 1/4, 0.25/5\}$,
$\mu_{FL4,l_5} = \{0/1, 0/2, 0.25/3, 0.5/4, 1/5\}$.

Our aim here is not to tune the model by experimenting with a wide set of fuzzy labels, but rather to determine whether there is an improvement in the recommending ability of our model when linguistic modelling is applied. We have therefore generated the fuzzy labels by changing the granularity degree between them, without their design being too exhaustive.

Table 3 shows the result of the experimentation when linguistic modelling is applied to *MaxPostP* and *MaxDiffPPostPPri* but only for 20, 30 and 50 parents

Table 3. MAE values for various experiments related to linguistic modelling

#P	FL1	FL2	FL3	FL4
	PostP-MaxPoss-FL?-DistPossM			
20	**0.7384 ± 0.0053**	0.7401 ± 0.0048	0.7735 ± 0.0040	0.9247 ± 0.0049
30	0.7400 ± 0.0045	0.7427 ± 0.0047	0.7787 ± 0.0042	0.9347 ± 0.0059
50	0.7432 ± 0.0046	0.7470 ± 0.0040	0.7895 ± 0.0049	0.9558 ± 0.0068
	PostP-PriorP-MaxPoss-FL?-DistPossM			
20	**0.7357 ± 0.0047**	0.7368 ± 0.0039	0.7690 ± 0.0030	0.9044 ± 0.0057
30	0.7374 ± 0.0042	0.7395 ± 0.0039	0.7749 ± 0.0035	0.9211 ± 0.0066
50	0.7412 ± 0.0043	0.7442 ± 0.0035	0.7863 ± 0.0044	0.9473 ± 0.0073
	PostP-AcumPoss-FL?-DistPossM			
20	**0.7666 ± 0.0049**	0.7795 ± 0.0047	0.7407 ± 0.0054	0.7893 ± 0.0040
30	0.7678 ± 0.0048	0.7805 ± 0.0050	0.7396 ± 0.0054	0.7905 ± 0.0044
50	0.7672 ± 0.0055	0.7823 ± 0.0054	0.7383 ± 0.0054	0.7916 ± 0.0051
	PostP-PriorP-AcumPoss-FL?-DistPossM			
20	**0.8007 ± 0.0058**	0.8126 ± 0.0053	0.7757 ± 0.0061	0.7977 ± 0.0047
30	0.7924 ± 0.0053	0.8045 ± 0.0054	0.7649 ± 0.0055	0.7976 ± 0.0049
50	0.7834 ± 0.0059	0.7982 ± 0.0057	0.7549 ± 0.0056	0.7973 ± 0.0055
	PostP-FL?-ProbFL			
20	**0.7412 ± 0.0049**	0.7512 ± 0.0042	0.8054 ± 0.0049	1,03 ± 0.0048
30	0.7434 ± 0.0047	0.7539 ± 0.0035	0.8131 ± 0.0048	1,05 ± 0.0047
50	0.7477 ± 0.0049	0.7610 ± 0.0039	0.8277 ± 0.0054	1,08 ± 0.0054
	PostP-PriorP-FL?-ProbFL			
20	**0.7381 ± 0.0045**	0.7470 ± 0.0032	0.8001 ± 0.0038	1,01 ± 0.0056
30	0.7405 ± 0.0043	0.7502 ± 0.0027	0.8087 ± 0.0039	1,03 ± 0.0054
50	0.7457 ± 0.0046	0.7579 ± 0.0033	0.8241 ± 0.0047	1,07 ± 0.0059
	MaxDifPostPPriP-MaxPoss-FL?-DistPossM			
20	**0.8098 ± 0.0040**	0.8527 ± 0.0042	0.8198 ± 0.0038	0.7957 ± 0.0042
30	0.7960 ± 0.0052	0.8415 ± 0.0054	0.8073 ± 0.0051	0.7890 ± 0.0045
50	0.7891 ± 0.0054	0.8355 ± 0.0054	0.8009 ± 0.0046	0.7862 ± 0.0048

(the other sizes show worse results). From this, the main conclusion that we can draw is that the linguistic modelling usually helps improve the performance of the model when an appropriate selection of the parameters is carried out. More specifically, and focusing initially on *MaxPoss*, the best linguistic label is *FL1*, with the remaining labels obtaining worse values. The use of the information provided by the parents is a useful technique when the probability distributions in the nodes from $\mathcal{V}$ are equal. If we deal with the *AcumPoss* method, the tendency changes substantially as this method does not improve *MaxPostP*. In this case, the best values are obtained with *FL3* and the technique of using the parent distribution does not improve the basic approach. Finally, with *ProbFL*, the behaviour is similar to *MaxPoss*. With respect to the linguistic modelling applied to *MaxDiffPPostPPri*, the *MaxPoss* method works best with *FL4*.

More specifically, we can draw the following general conclusions from this detailed experimentation: 1) the Jaccard coefficient modified with the overlap is the best measure for selecting user parents; 2) it is better to select an intermediate number of parents; 3) we think that it is better to make recommendations by using *MaxPostP* with and without linguistic modelling; 4) the application of linguistic modelling is interesting. The use of *MaxPoss* and *FL1* recommended.

We should compare our model's performance with other published RSs, experimenting with MovieLens and MAE, in order to discover its potential as a recommender system. The best results published so far are around $0.72 - 0.73$ of the MAE metric: Kim and Yum [16] with 0.70, Li and Kim [17] got 0.735, Sarwar et al. [18], 0.72; Chen and Yin [19], 0.732 and Mobasher et al. [20] got 0.73. With these data to hand, we can conclude that our model, obtaining a better MAE of 0.7357, competes well with the best published standards.

5 Conclusions and Further Research

In this paper, we have proposed a general BN-based model for collaborative recommendation, which is both effective and efficient. We have also studied the possibility of considering the set of ratings as vague concepts. Schematically, our system consists of two components: the first uses probabilistic reasoning to compute the probability distribution over the expected vote by means of a Bayesian network, and efficient methods for learning the topology, estimating the probability distributions and propagating; and the second computes the user's vote (the fuzzy set), thereby better representing this probability distribution.

By way of future work, we are planning to study mechanisms to incorporate better specifications of the products into the system and new methods for estimating the weights stored in the nodes of the BN. We do, however, wonder, like [15], whether *"users are sensitive to a change in the mean absolute error of 0.01.* This observation suggests that we might explore different directions instead of merely continuing to improve the MAE metric. In the future, we therefore plan to study problems such as how our system can communicate its reasoning to the users, the minimum amount of data (ratings) required for us to yield accurate recommendations, or how to include item information when recommending.

Acknowledgment. Work supported by the Spanish 'Ministerio de Educación y Ciencia' (TIN2005-02516), 'Consejería de Innovación, Ciencia y Empresa de la Junta de Andalucía' (TIC-276), and Spanish research programme Consolider Ingenio 2010: MIPRCV (CSD2007-00018).

References

1. Resnick, P., Varian, H.R.: Recommender systems. CACM 40(3), 56–58 (1997)
2. Pearl, J.: Probabilistic reasoning in intelligent systems: networks of plausible inference. Morgan Kaufmann Publishers Inc., San Francisco (1988)
3. Schiaffino, S.N., Amandi, A.: User profiling with case-based reasoning and bayesian networks. In: IBERAMIA-SBIA, Open Discussion Track (2000) pp. 12–21 (2000)
4. Butz, C.: Exploiting contextual independencies in web search and user profiling. In: Proc. of World Congress on Computational Intelligence, pp. 1051–1056 (2002)
5. Breese, J.S., Heckerman, D., Kadie, C.: Empirical analysis of predictive algorithms for collaborative filtering. In: 14th UAI Conference, pp. 43–52 (1998)
6. Miyahara, K., Pazzani, M.J.: Collaborative filtering with the simple bayesian classifier. In: Pacific Rim Int. Conf. on Artificial Intelligence, pp. 679–689 (2000)
7. Robles, V., Larrañaga, P., Peña, J., Marbán, O., Crespo, J., Pérez, M.: Collaborative filtering using interval estimation naive bayes. LNCS (LNAI), pp. 46–53. Springer (2003)
8. Heckerman, D., Chickering, D.M., Meek, C., Rounthwaite, R., Kadie, C.: Dependency networks for inference, collaborative filtering, and data visualization. J. Mach. Learn. Res. 1, 49–75 (2001)
9. van Rijsbergen, C.J.: Information retrieval. Butter Worths, London (1979)
10. Zadeh, L.A.: Probability measures from fuzzy events. Math. Anal. Applications. 23, 421–427 (1968)

11. Zadeh, L.A.: Fuzzy sets as a basis for a theory of possibility. Fuzzy Sets and Systems 1, 3–28 (1978)
12. Zwick, R., Carlstein, E., Budescu, D.: Measures of similarity among fuzzy concepts: a comparative analysis. Internat. J. Approximate Reasoning 1, 221–242 (1987)
13. Klir, G., Parviz, B.: Probability-possibility transformations: A comparison. Int. Journal of General Systems 21, 291–310 (1992)
14. Miller, B., Albert, I., Lam, S., Konstan, J., Rield, J.: Movielens unplugged: Experiences with an occasionally connected recommender systems. In: Proc. of Int'l Conf. Intelligent User Interfaces, pp. 263–266 (2002)
15. Herlocker, J.L., Konstan, J.A., Terveen, L.G., Riedl, J.T.: Evaluating collaborative filtering recommender systems. ACM Trans. Inf. Syst. 22(1), 5–53 (2004)
16. Kim, D., Yum, B.: Collaborative filtering based on iterative principal component analysis. Expert Systems with Applications 28(4), 823–830 (2005)
17. Li, Q., Kim, B.: Clustering approach for hybrid recommender system. In: IEEE/WIC Proc. of the Int. Conf. on Web Intelligence, pp. 33–38 (2003)
18. Sarwar, B., Karypis, G., Konstan, J., Riedl., J.: Application of dimensionality reduction in recommender system – a case study. In: Proc. ACM WebKDD (2000)
19. Chen, J., Yin., J.: Recommendation based on influence sets. In: Proc. of the Workshop on Web Mining and Web Usage Analysis (2006)
20. Mobasher, B., Jin, Y.Z.X.: Semantically enhanced collaborative filtering on the web. In: Berendt, B., Hotho, A., Mladenič, D., van Someren, M., Spiliopoulou, M., Stumme, G. (eds.) EWMF 2003. LNCS (LNAI), vol. 3209, pp. 57–76. Springer, Heidelberg (2004)

Autonomous Parsing of Behavior in a Multi-agent Setting

Dieter Vanderelst* and Emilia Barakova

Designed Intelligence Group, Eindhoven University of Technology, The Netherlands
`dieter.vanderelst@emailengine.org`

Abstract. Imitation learning is a promising route to instruct robotic multi-agent systems. However, imitating agents should be able to decide autonomously what behavior, observed in others, is interesting to copy. Here we investigate whether a simple recurrent network (Elman Net) can be used to extract meaningful chunks from a continuous sequence of observed actions. Results suggest that, even in spite of the high level of task specific noise, Elman nets can be used for isolating re-occurring action patterns in robots. Limitations and future directions are discussed.

1 Introduction

According to Thorndike [1] an organism is able to learn by imitation if it can acquire new behavioral skills by directly copying them from others. Imitation, like other forms of social learning [2,3], has, potentially, an enormous ecological advantage [4]. It allows animals to be flexible learners while avoiding the dangers associated with individual learning [5]. The behavior of others has often already been shaped by its consequences and can therefore be assumed to be save and rewarding to imitate [2]. Humans and some primates have been found to imitate [6,2,7,8,9].

Another property of imitation, together with its ecological value, is that it can support the spread of behavior through a population of individuals [10,11,12]. Several observational studies [6] have yielded evidence for this in groups of primates but recently also experimental evidence has been reported. For example, Bonnie [8] thought individual chimpanzees to deposit tokens in a box to receive a reward. Subsequently, these individuals were introduced into a population of naive animals. After some time the rewarding behavior was copied by the other animals and its frequency in the population increased. Similar findings have since then been reported by Whiten [13].

The ecological advantages and the capacity to support the spread of behavior make imitation learning a potentially interesting mechanism to support learning in robotic multi-agent systems [14]. In a multi-agent setting, agents could search simultaneously for a solution for a given problem (e.g. how to pick up food). Once a single agent has found a solution, this innovation could be imitated by others and could propagate through the population. In this way, learning by

---
* Corresponding author.

imitation could drastically reduce the total number of learning trials needed for a population of agents to solve a problem [14].

2 Problem Statement

Imitating agents in a multi-agent setting face a number of fundamental problems [15,16]. One of the most important questions in the multi-agent context is how agents can autonomously select the behavior that should be copied[1]. Great apes and humans seem to be very good at determining what behavior should be imitated when they observe a demonstrator [17]. However, for robots in the multi-agent scenario sketched above, determining what they should imitate is no simple task[2].

To see why this is a problem, imagine a number of agents exploring an artificial world in which different types of food and supplies are available (like in [11]). In this artificial world different resources need to be approached differently to use them. For example, nuts need to be gathered from the ground and smashed against a rock before they can be eaten while a banana must be picked from a tree and carefully peeled. It is assumed that originally all agents are naive concerning the rules governing the world. When the agents are released in the world they start off generating sequences of behavior in the hope of finding a sequence that gives access to some rewarding fruit. Agents that find such a sequence will remember it for future use. This means that after a while agents will alternate between generating new behavior (exploration) and exploiting gathered knowledge (when a fruit for which a known sequence is stored is encountered [19]. Exploitation behavior will consist of fairly fixed action sequences like picking up a nut, smashing it onto a rock and eating it.

An agent that tries to observe and imitate the action sequences of the others sees a continuous sequence of various kinds of actions. The actions which are, from the viewpoint of the observed agents, meaningful and can be parsed into exploring and exploiting parts, are unordered from the viewpoint of a learning agent. There is no a priori way for the learning agent to parse the actions of its colleagues. It can not know which sequences it should copy and which are to be ignored. It should copy the exploiting action sequences and ignore the exploring action sequences. However, it can not know where the exploiting sequences start or end. Even if it is assumed that the imitating agent can detect when an other agent is rewarded (i.e. the end of an exploiting sequence), it can not know where the sequence of events that lead to the reward started.

Parsing a continuous stream of actions is reminiscent of another problem in cognition: that of segmenting a continuous input stream of sounds into words. In a set of seminal papers Elman [20,21] presented a simple recurrent neural network (figure 1a) that was able to segment a sequence of letters into words. His network was constructed to take an input letter n and predict the next

---

[1] This question has been termed as "What To Imitate?" by Dautenhahn [15].
[2] Even in a setting where humans act as explicit teachers, determining what to imitate is not easy for a robot [18].

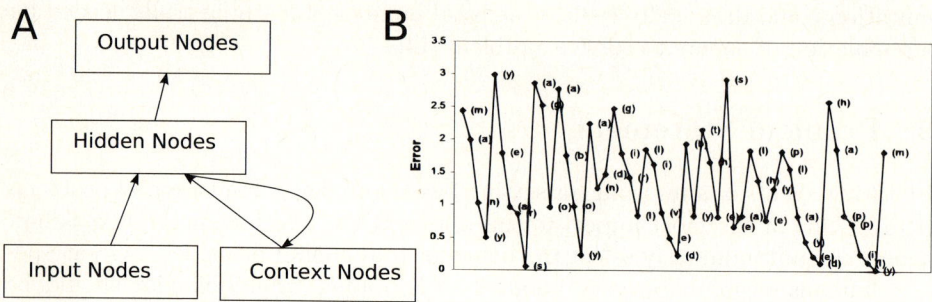

Fig. 1. (a) Schematic representation of an Elman net. On each feedforward sweep, the activation of the Hidden Nodes is copied to the Context Nodes. The Context Nodes are used as an additional set of input nodes on the next feedforward sweep.(b)The error curve produced by an Elman network that was trained to predict the next letter in a sequence from the previous one. Here the sequence "many-years-ago-a-boy-and-a-girl-lived-by-the-sea-they-played-happily" is given to the network.

letter $n+1$ in the sequence from n. After some training the network was capable of making good predictions. The network made its predictions based on the co-occurrence statistics available in the data. Furthermore, segmenting the sequence into words was possible using the error signal produced by the network while executing this task. Figure 1b shows an error curve that was produced while the network processed an input stream of letters. Inspecting the error curve, it can be seen that the error is high at the boundaries between words while it drops over the course of a word. This is caused by the fact that while a word unfolds, the next letter becomes more and more predictable with each new letter. On the other hand, at the boundaries between words, the next letter is very hard to predict since it is not determined by the previous one (in the dataset provided to the net). Therefore, the error provides a good clue as to what are recurring sequences in the input, and these correlate highly with words [20]. The network learned which parts of the input should be regarded as meaningful chunks.

A similar solution might be used to let agents autonomously decide what to imitate (see [22] for a related suggestion). An imitating agent could notice that some sequences of actions are consistently executed in the same order. Rewarding action sequences will be repeated often by the demonstrating robots to reap the fruits of their explorations. Using an Elman-net, an observing agent could try to predict the actions of its fellow agents. After a given amount of training the observer could use the error curve to isolate the segments of the input that are interesting to imitate (or a least to evaluate before attempting imitation). Exploiting sequences will be characterized by being predicable (low predicting error).

The current study focuses on a simplified version of the multi-agent setting sketched in the previous paragraphs. In this study it will be assumed that the demonstrating agent has already discovered several rewarding sequences at the start of the experiment and that it does not learn any new behavior in the course of the experiment. A second simplification is that only a single demonstrating robot will be considered. This setting mimics the situation in the cited experiments of Bonnie and Whiten [8,13] where a single, well-trained, animal is observed by others.

The number of studies related to the current one is very limited. While imitation in multi-agent settings seems to be a promising learning mechanism, until now little or no research has been done in this area[3]. Most research on imitation in artificial agents focuses on human-machine imitation (see [18] for an overview) where the human is a teacher that clearly marks the boundaries of the to-be-imitated behavior. In such a setting there is no need for an agent to detect the boundaries between meaningful chunks of actions since they are marked by the teacher [16]. To the best of our knowledge only [16] has investigated the use of imitation in the context of embodied autonomous multi-agent systems where no explicit teacher is present.

3 Methods

3.1 Experimental Scenario: Selecting Action Sequences

The experimental setting is designed to test whether an Elman network can be used by an observing agent to select interesting sequences in a continuous stream of actions. Therefore, an autonomous robot will be observed while it alternates between exploring (random action sequences) and exploiting (predetermined action sequences). It will be tested whether the network can be used to isolate the predetermined action sequences.

3.2 Data Collection and Preprocessing

All experiments reported in this paper were conducted using the e-puck robot platform (http://www.e-puck.org). A single robot was used during the experiments. The e-puck is a small mobile robot measuring 70 mm in diameter and 55 mm in height. The robot is equipped with infrared distance sensors that are located around the body at $10°, 45°, 90°, 270°, 315°$ and $350°$ with respect to the heading direction of the robot. Two sensors located at the back of the robot were not used in the reported experiments. The robot was controlled by a personal computer through a Bluetooth interface. A rectangular arena was constructed for the robot which measured about 100 cm × 70 cm. The arena was fenced by cardboard walls which were about 10 cm high. The robots movements were filmed by a Logitech QuickCam camera (http://www.logitech.com) suspended about 160 cm above the floor of the arena. The camera captured the entire arena using 320 × 240 pixels at 10 Hz. The floor of the arena was white. The robot was fitted with a black cap for maximal contrast so that tracking the robot was easy. All image processing and tracking of the robot was done using RoboRealm

---

[3] See the recent ECAL workshop on social learning in robot for some explorative papers: http://laral.istc.cnr.it/slea/.

software (http://roborealm.com). Processing the images of the camera included correcting for radial distortion. The tracking software provided the approximate location of the center of the robot in each camera frame and its speed.

In this study the robot could only drive straight on and to turn in place. The robot was allowed to drive 4 fixed driving distances: 32 mm, 64 mm, 96 mm and 128 mm. Also 10 turning angles were fixed: -120°, -90°, -60°, -30°, 0°, 30°, 60°, 90°, 120° and 180°. Negative values denote counterclockwise turns. Given the constraints imposed on the movements, all movement of the robot was an alternation between turning in place conform to one of the fixed angles, followed by driving one of the fixed distances. The robot iteratively selected a turning angle and a traveling distance to execute.

The e-puck executed two different kinds of behavior. First, the robot could execute exploration behavior. In this mode a turning angle and a traveling distance were selected at random on each iteration. Second, after each turn and drive action the robot could, with a probability of 0.3, select at random one of 3 patterns to execute. The patterns are depicted in figure 2. Detailed information on the patterns can be found in table 1. While the robot was driving, the distance sensors were probed each 200 ms to determine whether it was about to hit the walls of the arena. If the robot detected a wall, it aborted its current action and moved away from the wall. In case the robot was executing one of the predetermined patterns of action, the pattern was aborted and a random move was initiated after the avoidance maneuver.

In the experiment reported here, the robot executed 1500 moves consisting of turning and driving. This amounted to about 120 minutes and 74702 image frames.

The robot path captured by the camera was preprocessed using R-software [23]. Preprocessing aimed at reconstructing the actions of the robot from the camera images as an observing agent could do. Figure 3 illustrates the preprocessing steps. Note that only a small subset of data have been used for these plots. Plotting an entire dataset would result in graphs that are too cluttered.

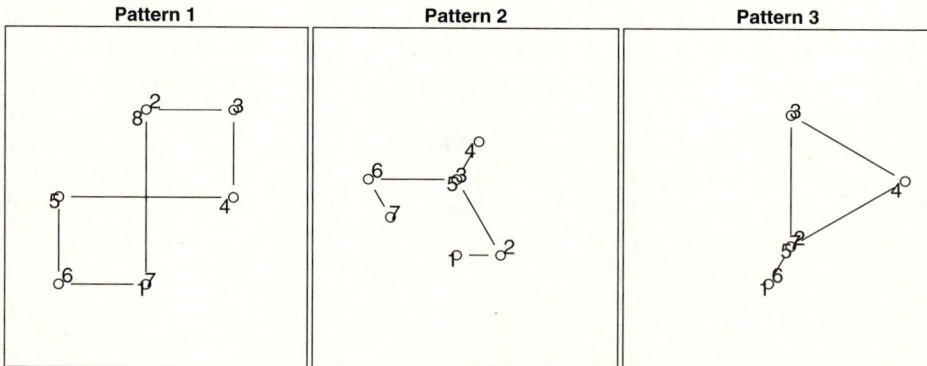

Fig. 2. These are the three predefined patterns that could be driven by the robot. Numbers signify the order of execution. See table 1 for details about these patterns.

Table 1. This table lists the turning angles and the driving distances that made up the 3 predefined patterns plotted in figure 2

Step	Pattern 1		Pattern 2		Pattern 3	
	Angle (Degrees)	Distance (mm)	Angle (Degrees)	Distance (mm)	Angle (Degrees)	Distance (mm)
1	0	128	90	32	30	32
2	90	64	-120	64	-30	96
3	90	64	60	32	120	96
4	90	128	180	32	120	96
5	-90	64	60	64	-30	32
6	-90	64	-120	32	180	32
7	-90	128				

As can be seen in plot 3a-c, the path of the robot consists of random sequences interwoven with a number of predefined patterns. A cubic smoothing spline was fitted to the raw robot path. The smoothed path is plotted in figure 3b. The smoothed track was segmented in order to reconstruct the moves the robot executed. Segmenting the track was done based on the detected speed of the robot. Because the top of the e-puck robot is perfectly round, it looks as if it stands completely still, from the viewpoint of the overhead camera, when it turns in place. Therefore, local minima in the speed curve signify points in time when the robot was (probably) executing a turn. The smoothed robot path was segmented at these points in time. Figure 3c depicts the segmented version of the robot path. Next, the segmentation points were connected by straight lines because the robot could only drive straight on between two turning points. From the segmented path it was trivial to calculate the sequence of the approximate angles the robot turned and the distances it drove.

The final result of the preprocessing step is an approximate reconstruction of the program that has been executed by the robot. This is a record of the actions of the robot as perceived by the camera (or any other onlooker).

Figure 4 depicts the distribution of the angles and distances that were detected by the camera. As can be seen in this graph, the execution and the perception of the moves of the robots was liable to noise. The distribution of the detected distances shows five clusters labeled as *a-e*. Cluster *b-e* correspond to the 4 distances the robot could drive in the experiment. Cluster *a* contains very short traveling distances. These are caused by over-segmentation of the robot path and by instances in which the driving of the robot was aborted due to the detection of the walls. The distribution of the turning angles also shows overlapping clusters. The fact that both distributions are characterized as a number of overlapping clusters indicates that the level of noise was relatively high.

Because of the noise, some further processing of the perceived moves was necessary before the data could be fed to an Elman network. The turning angles and the traveling distances were discretized. Turning angles were mapped onto the nearest 30°. After this, the data contained 12 different turning angles

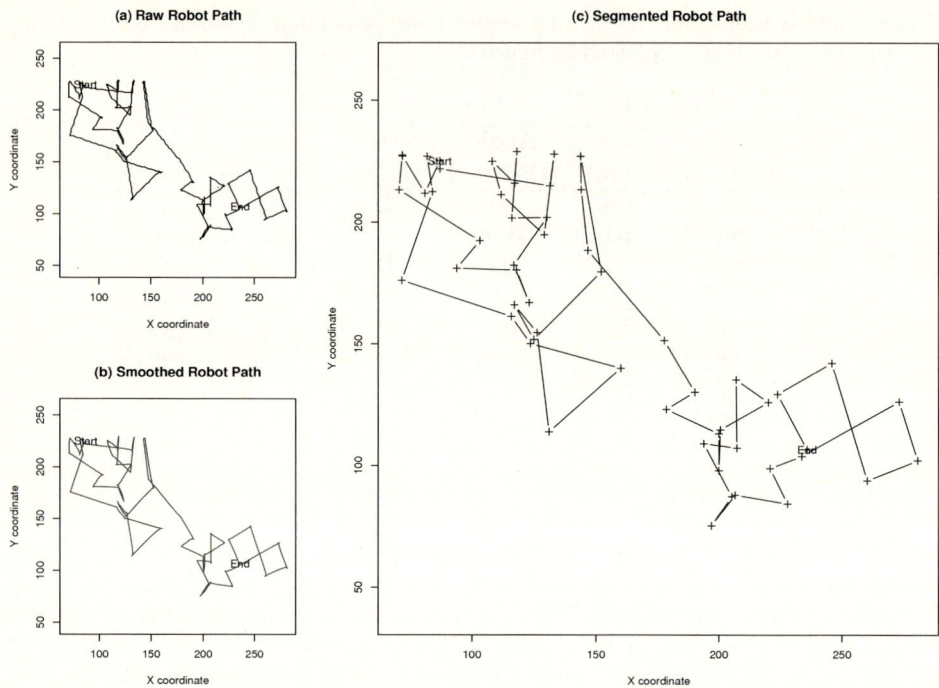

Fig. 3. These plots depict the raw (a), smoothed (b) and segmented (c) path the robot drove in experiment 1. for reasons of clarity only a small subset moves of the robot have been plotted.

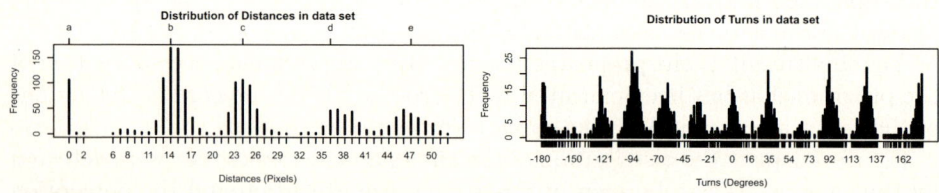

Fig. 4. (a) The distribution of traveling distances detected by the camera in the dataset reported in the results section. Letters $a-e$ refer to five clusters present in this distribution. (b) The distribution of the turning angles in the same data.

(360/12) instead of the 10 used by the robot. Traveling distances were mapped onto the nearest cluster center (0, 14, 24, 36 or 47). Moves that were classified as belonging to the first cluster (i.e. 0), were discarded from further processing. These final data cleaning steps can be considered as reflecting categorical perception.

3.3 Training the Elman Network

A generic Elman network was implemented (see [20,21] for more details about the structure of Elman networks). The network had 12 input nodes coding the turning angles and 4 inputs that coded each of the 4 traveling distances present in the data. The hidden layer of the network consisted of 30 nodes. All neurons had a sigmoid activation function. The network was trained by presenting it with a turning angle and a traveling distance by setting the corresponding input nodes to 1 (other nodes were assigned an activation value of 0). So, an input vector consisted of 16 values of which 2 were set to 1 to signify the current angle and driving distance. Importantly, each turning angle and traveling distance was assigned a coding input neuron at random. In this way, the coding of the moves was completely abstract. Thus, although the predefined patterns were visually symmetrical (see figure 2), from the viewpoint of the network they were not. The visual symmetry could not be exploited by the network to learn to recognize the patterns. After the presentation of an angle and distance, the network predicted the next turning angle and driving distance the robot would execute. Simple gradient descent (Error Backpropagation) was used to adapt the connection weights of the network after each presented input. After updating the network connections, the next turning angle and traveling distance was presented to the network. In this way the whole data set was presented 10 times to the network (i.e. 10 epochs of training). This amounted to about 15000 training trials.

4 Results

The experiment and the training of the network were replicated several times using slightly different parameter settings. However, qualitatively the results were always similar to the ones reported in this section.

Data about the moves executed by the robot can be found in table 2. These data show that a large proportion of the patterns were not completed. This introduced additional noise into the training data.

Figure 5 depicts the most important training results. Plot 5a shows the change in the prediction error by the network in the form of a density plot. One can see that after some initial training, there is a bifurcation in the error. After about 2000 trials, most moves of the robot are well predicted (low error) while others are not (high error). This binomial distribution of the error is also clearly visible in plot 5b. A Gaussian Mixture Model [24] with 2 components was fitted to the error distribution across all trianing trials to obtain an objective threshold to separate trials for which the prediction error was low and trials for the error was high. At about a value of 0.9, an error value had an equal probability of belonging to either of the two clusters (assuming equal priors for both components). This value is indicated by an arrow in figure 5b. This value was used as a cut-off to identify trials in which the network had made a good prediction. Trials in which the network predicted the next step with an error lower than 0.9, were considered as trials with a low error. Subsequently, sequences of trials longer than 2 steps, in which the prediction was better than the cut-off, were identified

Table 2. This table lists the number of patterns the robot drove while collecting the data reported in the results section. Also the number of random moves is listed. The number of patterns and random moves that were terminated because the robot detected a wall are listed separately.

	Termination		
	Normal	Stopped	Total
Pattern 1	39	19	58
Pattern 2	47	21	68
Pattern 3	37	17	54
Total	123	57	180
Random	274	46	320

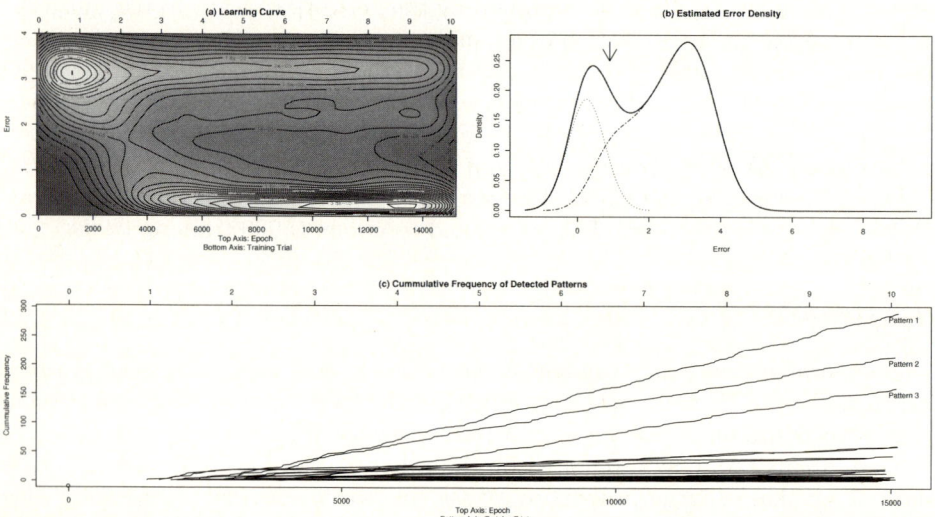

Fig. 5. These plots depict the training results of the experiment. (a) A density estimation of the prediction error of the Elman network in individual learning trials in function of the number of trials. (b) A density estimation of the prediction error collapsed across all learning trials. (c) The cumulative frequency of all different sequences of actions that were well predicted by the network (see text for details).

in the data. The cumulative frequency of all identified sequences is plotted in figure 5c as a function of the number of training trials. Many different sequences were identified ($n = 141$). However, most of these are identified only a few times. Only 10 patterns were recognized 20 times or more while 54 were encountered only once. As can be seen in plot 5c, three patterns are clearly encountered more often than all others. These are the patterns the network was supposed to learn (see labels in plot 5c). So, by analyzing the error curve produced by the Elman

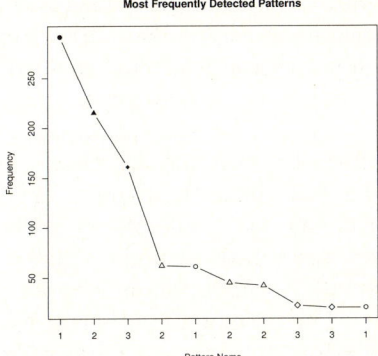

Fig. 6. Total frequency of detection of the most frequently detected patterns of actions. ●; Pattern 1, ▲; pattern 2, ◆; Pattern 3. Opened markers signify subpatterns of the patterns denoted with a similar shape.

network during training, the three predefined patterns could be isolated as being high frequent sequences the network could predict very well.

The fact that the network reliably isolated the three predefined patterns is further demonstrated in figure 6. This plot depicts the frequency of the patterns that were detected more than 19 times in the course of the training. As stated before, the most frequently observed patterns were the three target patterns. Of equal interest is the fact that the other detected patterns were parts of the goal patterns.

5 Discussion

The reported experiment investigated whether an Elman network could be used to reliably isolate re-occurring action sequences. The results show that Elman nets can indeed be used for this task. In contrast to the original simulation studies of Elman [25,20,21], the data collected by observing the robot contained substantial amounts of task-specific noise. Nevertheless, given enough learning trials, the error curve could be used to reconstruct the three goal patterns. Therefore it is possible for an observer to extract the interesting parts of the behavior of a demonstrator by using a simple recurrent network.

However, while the proposition of the study was confirmed by the data, their main valor might be to suggest new lines of research and improvements of the current approach. To this end some issues of the study will be discussed.

A first issue is the learning speed demonstrated in the current study. The Elman network needs many trials before a reliable extraction of the patterns is possible (something that was also experienced by Elman in his original studies). This makes the mechanism, exactly as it is implemented in this paper, an unlikely candidate to be used in a multi-agent setting. Adaptations to speed up learning will be necessary.

The current study is still far away from having a collection of robots learning from each other through imitation. The current experiment modeled one robot (the camera) watching another one continuously (from a favorable perspective). When implementing the current setup with different demonstrators and observers, the noise levels in the perception of each and every robot will rise. Even if robots are allowed to watch each other from the viewpoint used in this paper. A robot will have to choose between a number of agents to observe. Some of these will perform very well while others might be still learning themselves. If the robot chooses to imitate an under trained co-agent, it will not be able to learn the task. Instead, it will extract any re-occurring sequences that are coincidently demonstrated. Furthermore, in absence of any re-occurring pattern in the demonstrators behavior, nothing will be learned. In short, because any learner has access to the behavior of trained as well as untrained individuals, the noise in the perception will increase and the chance of mastering the task decrease. To avoid such a scenario, more flexible learning mechanisms, adapted to multi-agent settings, must be researched. One obvious way to extend the current mechanism is to add reinforcement (or a similar mechanism) - a source of information also used by animals and humans.

Adding reinforcement to the learning mechanism will also reduce the number of learning trials needed. A robot could try out each (partial) sequence of actions it discovers in the behavior of others. If the action sequence is successful according to some measure, the behavior should be consolidated. If not, it should be discarded until further training changes it in some respect. Such a mechanism will speed up learning drastically since a target sequence has to be isolated only once. Furthermore, this way the number of demonstrations of unadaptive behavior in a set of robots is kept at bay which reduces the noise level [10].

We are currently carrying out simulation studies to investigate different options for extending the current research.

References

1. Thorndike, E.: Animal intelligence: an experimental study of the associative process in animals. Psychological Review, Monograph Supplements 2, 551–553 (1889)
2. Zentall, T.R.: Imitation: definitions, evidence, and mechanisms. Animal Cognition 9(4), 335–353 (2006)
3. Noble, J., Todd, P.M.: Imitation or something simpler? modelling simple mechanisms for social information processing. In: Imitation in Animals and Artifacts, MIT Press, Cambridge (2002)
4. Leadbeater, E., Chittka, L.: Social learning in insects–from miniature brains to consensus building. Current Biology 17(16), R703–R713 (2007)
5. Boyd, R., Richardson, P.J.: An evolutionary model of social learning: the effect of spatial and temporal variation. In: Zentall, R.R., Galef, B.j. (eds.) Social Learning: Psychological and Biological Perspectives, pp. 29–48. Erlbaum, Hillsdale (1988)
6. Whiten, A.: The second inheritance system of chimpanzees and humans. Nature 437(7055), 52–55 (2005)

7. Stoinski, T.S., Whiten, A.: Social learning by orangutans (pongo abelii and pongo pygmaeus) in a simulated food-processing task. Journal of Comparative Psychology 117(3), 272–282 (2003)
8. Bonnie, K.E., Horner, V., Whiten, A., de Waal, F.B.M.: Spread of arbitrary conventions among chimpanzees: a controlled experiment. Proceedings in Biological Science 274(1608), 367–372 (2006)
9. Price, E., Caldwell, C.A.: Artificially generated cultural variation between two groups of captive monkeys, colobus guereza kikuyenis. Behavioral processes 27(1), 13–20 (2007)
10. Parisi, D.: Cultural evolution in neural networks. IEEE Expert: Intelligent Systems and Their Applications 12(4), 9–11 (1997)
11. Noble, J., Franks, D.W.: Social learning mechanisms compared in a simple environment. In: Proceedings of the Eighth International Conference on Artificial Life (2002)
12. Nakamaru, M., Levin, S.A.: Spread of two linked social norms on complex interaction networks. Journal of Theoretical Biology 230(1), 57–64 (2004)
13. Whiten, A., Spiteri, A., Horner, V., Bonnie, K.E., Lambeth, S.P., Schapiro, S.J., de Waal, F.B.M.: Transmission of multiple traditions within and between chimpanzee groups. Current Biology 17(12), 1038–1043 (2007)
14. Pini, G., Tuci, E., Dorigo, M.: Evolution of social and individual learning in autonomous robots. In: Ecal Workshop: Social Learning in Embodied Agents (2007)
15. Dautenhahn, K., Nehaniv, C.: An agent-based per- spective on imitation. In: Dautenhahn, K., Nehaniv, C.L. (eds.) Imitation in Animals and Artifacts, MIT Press (2007)
16. Belpaeme, T., de Boer, B., Jansen, B.: The dynamic emergence of categories trough imitation. In: Dautenhahn, K., Nehaniv, C.L. (eds.) Imitation in Animals and Artifacts, MIT Press (2007)
17. Carpenter, M., Call, J.: The question of what to imitate: inferring goals and intentions from demonstrations. In: Dautenhahn, K., Nehaniv, C.L. (eds.) Imitation in Animals and Artifacts, MIT Press (2007)
18. Breazeal, C., Scassellati, B.: Robots that imitate humans. Trends in Cognitive Science 6(11), 481–487 (2002)
19. Thrun, S.B.: Efficient exploration in reinforcement learning. Technical report, Pittsburgh, PA, USA (1992)
20. Elman, J.L.: Finding structure in time. Cognitive Science 14(2), 179–211 (1990)
21. Elman, J.L.: Learning and development in neural networks: the importance of starting small. Cognition 48(1), 71–99 (1993)
22. Moga, S., Gaussier, P.: Artificial neural network for sequence learning. In: International Joint Conference on Artificial Intelligence (2003)
23. Team, R.D.C.: R: A language and environment for statistical computing (2007) ISBN 3-900051-07-0
24. Fraley, C., Raftery, A.E.: MCLUST version 3 for R: Normal mixture modeling and model-based clustering. Technical Report 504, University of Washington, Department of Statistics (September 2006)
25. Elman, J.L.: Distributed representations, simple recurrent networks, and grammatical structure. Machine Learning 7, 195–225 (1991)

On Resource Profiling and Matching in an Agent-Based Virtual Organization[*]

Grzegorz Frąckowiak[1], Maria Ganzha[1], Maciej Gawinecki[1], Marcin Paprzycki[1], Michał Szymczak[1], Myon-Woong Park[2], and Yo-Sub Han[2]

[1] Systems Research Institute, Polish Academy of Science, Warsaw, Poland
{maria.ganzha,marcin.paprzycki}@ibspan.waw.pl
[2] Korea Institute of Science and Technology, Seoul, Korea
{myon,emmous}@kist.kist.re.kr

Abstract. In our work we are designing agent based support for workers in a virtual organization. In the system under development, all resources are to be ontologically demarcated and utilized through semantically-driven information processing techniques. This paper contains preliminary considerations concerning resource profiling and ontological matching methods which we intend to utilize within the system. Two applications that we are currently developing are used to illustrate the proposed approach.

1 Introduction

Our recent work is devoted to personalized worker support in an agent-based virtual organization. The basic assumption underlying our work is that emergent technologies such as software agents [25] and ontologies [19] should be utilized as a foundation around which the proposed system should be conceptualized. In particular: (i) organizational structure, consisting of "roles" played by various entities within the organization and interactions between them, should be represented by software agents and their interactions, and (ii) domain knowledge, resource profiles and resource matching should be based on ontologies and associated with them reasoning machinery. Thus far, in [20] we have outlined processes involved when a "project" is introduced into an organization (approached from the point of view of resource management). Later, in [4] we have conceptualized roles played by various entities identified in [20]. This allowed us to establish roles to be played by (a) software agent(s) alone, (b) by human(s), and (c) by human-agent team(s). We have also discussed agent interactions and introduced a specific application, a *Duty Trip Support* (*DTS*), to illustrate how the proposed top level design of the system can be utilized in practice. Separately, in [3] we have proposed how e-learning can be introduced into the system in support of adaptability of human resources. Finally, in [21] we introduced basic ontologies to be used in the system (including the concept of a generic resource).

---

[*] Work was partially sponsored by the KIST-SRI PAS "Agent Technology for Adaptive Information Provisioning" grant.

The aim of the current work is to discuss how we plan to introduce resource profiles and use them to perform simple matchmaking operations. The context of our work is provided by two applications that have been selected as initial test cases for our system. The first one is a rather simple *Grant Assistant System* (*GAS*), while the second one is the above mentioned *DTS*. Implementation and experiments with these two subsystems will provide us with experiences that will be generalized in two directions: (a) to support more involved reasoning, and as a result (b) to support a more general case of personalized information provisioning in an organization. Having said that, we start with providing a brief background of main features of the system under development.

2 System Overview

Our system is conceived as an agent-based virtual organization, which as its core provides support for project-oriented collaborative work [3]. Here, the notion of the project is very broad an includes installation of a satellite TV antenna as well as design and implementation of an intranet-based information system for a corporation. Structure of the organization and interactions between its participants are functionalized using software agents and their interactions. Domain knowledge and profiles of resources (human and non-human) are to be represented using ontologies [3,4,21] and overlay-based profiles (see below). Out of agents existing in the system, the *Personal Agent* (*PA*) and the *Organization Provisioning Manager* (*OPM*) play key roles in the context of this paper. The *Personal Agent* is associated with each human member of the organization. In addition to being the interface between its *User* and the system, the *PA* can support her in any role required by the organization (see [4] for an extended discussion). The *OPM*, on the other hand, is an entity responsible for organizational resource management. In this capacity the *OPM* has access to profiles of all resources present in the organization.

To focus our attention let us introduce two scenarios depicted in Figure 1. In the *Grant Assistant System* (*GAS*), the *OPM* of a university (or a research institute) receives a grant announcement and its role is to deliver it to these and only these *PA*s that represent *Users* that may be interested in exactly this announcement. In other words, the *OPM* has to decide who (which *PA*) should receive a given announcement, based on ontologically demarcated profiles describing faculty in the university (researchers in the institute). Here, we assume that the announcement is a resource that has already an assigned profile based on the internal domain ontology (note that specifying entity inside, or outside, of the system that performs profile demarcation is of no importance here).

In the *Duty Trip Support* (*DTS*) scenario, the *OPM* undertakes the role of a *Travel Assistant*. As can be seen in [4] (specifically, see the sequence diagram presented there) the *OPM* is contacted two times while the *PA* helps its *User* during the Duty Trip related activities. First, when the traveler requests a list of other institutions/persons she can visit during her trip. Second, when a specific advice is sought as far as accommodations are concerned (e.g. hotel and/or

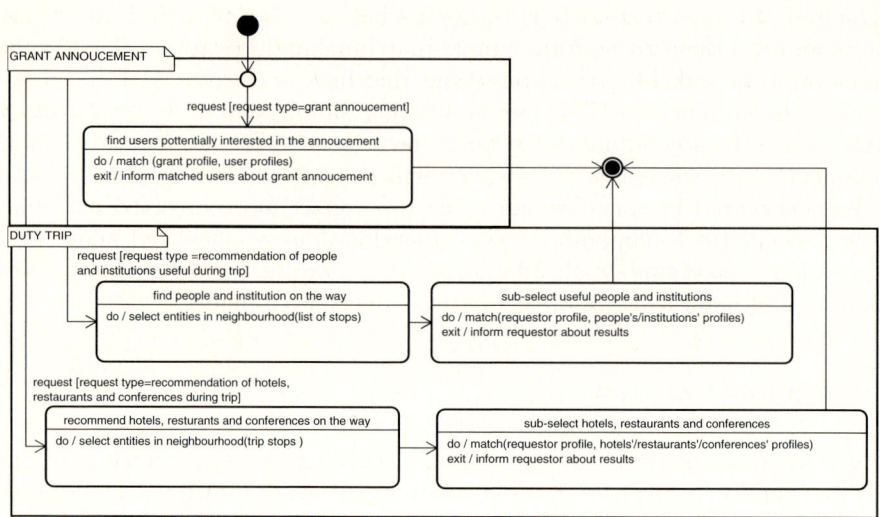

Fig. 1. Matching scenarios in the proposed system

restaurant advice). In both cases the *OPM* has to search the (local) database of ontologically demarcated data to find the pertinent information. For instance, let us assume that a researcher is to travel to a conference that is going to take place in Poznan, Poland. In this case the first question is: which universities/individuals located "close-enough" to Poznan, she may want to visit during her trip. The second question is: where should she stay and eat while in Poznan.

Let us now describe how the desired results are to be obtained in both scenarios. Before we proceed let us stress that we assume that all data *within* a given organization is demarcated utilizing a common (for that organization) ontology. Therefore, in what follows we do not have to deal with matching differing and potentially incompatible (external) ontologies. All that we are interested in is: how to establish "distances between resources" within a single ontology and how to use this information in the above described scenarios.

3 Defining Profile of a Resource

In the proposed system every entity is conceptualized as a *resource*; regardless of it being an employee of the organization, or a laptop computer. Since we use two ontologies (organizational and domain specific), resource profile is very likely going to utilize both of them. The organizational ontology allows to define structure of the organization and place a resource within this structure, while domain specific ontologies allow to place the resource within the "area of expertise" of the organization [21]. For instance, a mass spectrometer, may be assigned to a given research laboratory (with a specific list of human resources that have access to it), but also be demarcated utilizing an "ontology of chemistry." Let us note that most profiles defined in our system are likely not only to be demarcated in terms of organizational and domain ontologies, but also to have an "access specification." Here, consider the fact that while for most (grant) announcements anyone should be able to read them, access to modify (write or delete) grant announcements should be restricted to authorized entities.

When considering the resource profile we also follow our work in the design of the Travel Support System (TSS) [7]. There, a profile of a human was defined as a set of data about a given individual and his opinions which reflect his attitude toward certain features of objects, or toward objects demarcated within the system [5]. This approach allowed us to deduce the potential interest of a person in a Korean restaurant knowing that she likes meat dishes and dislikes vegetarian cuisine. While, here, we consider both human and non-human resources, for the time being we opt for conceptualizing profiles similarly to the way it was done in the TSS. However, it is still possible that our approach will become readjusted as we continue to analyze project requirements and various development options.

3.1 Defining Human Resource Profile

Let us now consider a sample profile of a human resource. Specifically, a professional experience related fragment of a profile of an employee of a Research Institute in East Asia could be defined as (note that names of disciplines and their codes are standardized for a given Organization and Country, e.g. considering a South Korean Institute–as in the example below, it is likely to be the standard of the Korea Science and Engineering Foundation [12]):

```
:SamplePersonalProfile  a  onto:ISTPersonalProfile;
   onto:belongsTo  resourceNamespace:someHumanResource;
   onto:id  "1234567890"^^xsd:string;
   person:name  "John_Doe"^^xsd:string;
   person:gender  person:Male;
   person:birthday  "1982-01-01T00:00:00"^^xsd:dateTime.
:SampleProfile  a  onto:ISTExperienceProfile;
   onto:belongsTo  resourceNamespace:someHumanResource;
   onto:doesResearchInFields
      scienceNamespace:Aerodynamics-30501,
      scienceNamespace:Marine_Hydrodynamics-30505,
      scienceNamespace:Turbulence-30412;
   onto:knowsFields
      [a  onto:Knowledge;
       onto:knowledgeObject  scienceNamespace:Aerodynamics-30501;
       onto:knowledgeLevel  "0.75"^^xsd:float],
      [a  onto:Knowledge;
       onto:knowledgeObject  scienceNamespace:Marine_Hydrodynamics-30505;
       onto:knowledgeLevel  "0.40"^^xsd:float],
      [a  onto:Knowledge;
       onto:knowledgeObject  scienceNamespace:Turbulence-30412;
       onto:knowledgeLevel  "0.90"^^xsd:float];
   onto:managesProject  :project1.
:project1  a  onto:ISTProject;
   onto:managedBy  resourceNamespace:someHumanResource;
   onto:period
      [a  onto:Period;
       onto:from  "2008-06-01T00:00:00"^^xsd:dateTime;
       onto:to  "2009-05-31T00:00:00"^^xsd:dateTime];
   onto:fieldsRef  scienceNamespace:Aerodynamics-30501;
   onto:projectTitle ''Very Interesting Aerodynamics Project''^^xsd:string.
```

From the example we can tell that a person identified as *someHumanResource* specializes in *Aerodynamics* and his *level of knowledge* is identified as 0.75 (for more info about assigning levels of skills, or more generally "temperature" to a feature, see [5,7,9]), *Marine Hydrodynamics* (level of knowledge identified as 0.4), and *Turbulence* (level of knowledge 0.9). Additionally, this person is scheduled to manage a project entitled: "Very Interesting Aerodynamics Project," which starts on June 1st, 2008 and ends on May 31st, 2009.

Let us note, that the scientific interests of a given employee (*onto:knowsFields* in the above example) can be replaced by professional skills describing worker in any discipline. For instance, they could as well be used to specify that a given programmer has knowledge of Java (level 0.7), C++ (level 0.5), Oracle (level 0.65), etc. In this way the proposed approach is both robust and flexible.

3.2 Defining Non-human Resource Profile

To illustrate how we plan to introduce and utilize profiles of non-human resources, we present a fragment of a profile of a Duty Trip Report utilized within the *DTS* scenario (note that, as any document in the organization, the Duty Trip Report is also a resource with its own profile).

```
:SampleDTProfile a onto:DTProfile;
        onto:belongsTo resourceNamespace:SomeDT;
        onto:employee resourceNamespace:someHumanResource;
        onto:status dtStatusNamespace:Application
        onto:destination geo:AmesdaleCity;
        onto:period
            [a onto:Period;
             onto:from "2008-02-01T00:00:00"^^xsd:dateTime;
             onto:to "2008-02-12T00:00:00"^^xsd:dateTime].
```

In the listing above *SampleDTProfile* is a profile of resource represented by *SomeDT*. In our example the latter is a Duty Trip Report resource. The employee who this profile directly refers to is represented by the *someHumanResource*. Hence, we can tell that a person represented in the system as *someHumanResource* applied for a duty trip (*SomeDT*). The current status of that Duty Trip Report is *Application* and the trips destination is Amesdale, Canada. The researcher intends to stay there for twelve days (from February 1st till February 12th, 2008). Again, properties of the *DTProfile* refer to the system schemas (organization and domain ontologies) and fulfill the data model requirements set by the Duty Trip Support System which we develop (see also [21]). Please note that, in order not to overly complicate the example, the snippet above does not cover all properties of the *DTProfile* class defined in [21]. Following listing introduces context of the *SampleDTProfile* instance and serves as geospatial information data for the purpose of our example:

```
geo:CanadaCountry a onto:Country;
        onto:name "Canada"^^xsd:string.
geo:QuebecArea a onto:Area;
        onto:name "Quebec"^^xsd:string;
        onto:isInCountry :CanadaCountry.
geo:OntarioArea a onto:Area;
        onto:name "Ontario"^^xsd:string;
        onto:isInCountry :CanadaCountry.
```

```
geo:AnnavilleCity   a   onto:City;
        onto:name  "Annaville"^^xsd:string;
        onto:long  "-72.433"^^xsd:float;
        onto:lat   "46.217"^^xsd:float;
        onto:isInCountry  :CanadaCountry;
        onto:isInArea     :QuebecArea.
geo:AnjouCity   a   onto:City;
        onto:name  "Anjou"^^xsd:string;
        onto:long  "-73.533"^^xsd:float;
        onto:lat   "45.6"^^xsd:float;
        onto:isInCountry  :CanadaCountry;
        onto:isInArea     :QuebecArea.
geo:AmesdaleCity   a   onto:City;
        onto:name  "Amesdale"^^xsd:string;
        onto:long  "-92.933"^^xsd:float;
        onto:lat   "50.017"^^xsd:float;
        onto:isInCountry  :CanadaCountry;
        onto:isInArea     :OntarioArea.
```

Here we defined *CanadaCountry* as an instance of the *Country* class, *QuebecArea* and *OntarioArea* as an instances of the *Area* class; both of them are regions of the *CanadaCountry*. Additionally, three instances of the class *City* were defined: *AnnavilleCity*, *AnjouCity* and *AmesdaleCity*. The first two are located in the *QuebecArea* whereas the last one is in the *OntarioArea*. Please note that only coordinates of cities are given. On the other hand, countries are represented in our system as sets of regions (areas) and these are in turn represented as sets of cities and are located in a particular country. Let us note that we do not claim that such a representation of data in our geospatial information support subsystem is the most efficient solution to the problem, but we may assume that it is sufficient enough for the purpose of our information provisioning system and the Duty Trip Support application.

In order to illustrate the way that our system is going to work, let us introduce one more example of a non-human resource, which is an information about a conference.

```
:SampleConferenceProfile  a  onto:ConferenceProfile;
    onto:belongsTo  resourceNamespace:SomeConference;
    onto:primaryFields
            scienceNamespace:Aerospace_Ship_and_Ocean_Engineering-30500,
            scienceNamespace:Mechanical_Engineering-30400;
    onto:specialSessionField
            scienceNamespace:Acoustics_and_Noise-30405;
    onto:takesPlaceAt  geo:AnnavilleCity;
    onto:period
            [a onto:Period;
             onto:from  "2008-02-07T00:00:00"^^xsd:dateTime;
             onto:to    "2008-02-09T00:00:00"^^xsd:dateTime. ].
```

The conference introduced here, is going to be held in the second week of February 2008 in Annaville, Canada. Here, the location of the conference (object of the *takesPlaceAt* property) references the geospatial information which was introduced in the previous listing. As far as the conference is concerned, its main topics are *Aerospace*, *Ship and Ocean Engineering* and *Mechanical Engineering*. Furthermore, a special session is planned to cover *Acoustics and Noise*.

Let us now propose how the above introduced resource profiles can be utilized in our system.

4 Calculating Distances between Resources

From the scenarios described above and summarized in Figure 1, we can easily see the need for resource matching (finding distances between two or more resources). To focus our attention, let us present a few examples of types of resource matching operations that have to be implemented in our system (this list is not intended to be exhaustive, but rather to point to some classes of needed resource matching and/or distance calculations):

1. computing distance between two geographical locations, to be able to establish if a given location is close-enough to the place where the employee is to travel,
2. matching a non-human resource (e.g. a grant, hotel, restaurant, conference) with a human-resource, to find if a person who is planning a trip could be interested in a given nearby located conference, or if an employee is potentially interested in a grant announcement,
3. matching two human resources to find out who are the researchers that a person planning a trip may be interested in visiting.

Upon further reflection it is easy to notice that the way the distance between resources should be calculated depends on types of objects which are arguments of calculations. For example, the distance between value of *onto:destination* property of the *SampleDTProfile* and the value of the *onto:takesPlaceAt* of the *ConferenceProfile* class instance will be calculated in a different way than the distance between values of *onto:period* properties of the *SampleDTProfile* and the *ConferenceProfile* class instances. The following object types that appear in our work can be distinguished, based on different approach to calculate their distance (examples of specific calculations involve the above presented ontology snippets):

1. Objects which represent geographical locations—distance between the *onto:destination* property range of the *onto:DTProfile* class and the *onto:takesPlaceAt* of the *ConferenceProfile* class.
2. Numeric objects—distance between *onto:long* property values.
3. Date objects—distance between *onto:from* and the *onto:from* (or the *onto:to*) property values.
4. Enumerable objects—distance between *onto:primaryFields* property values/range of the *ConferenceProfile* and the range of *onto:doesResearchInFields* property of the *HRProfile* class.

Let us now discuss proposed approaches to distance calculations/resource matching for the four distinguished classes of properties.

4.1 Location Based Calculations

City, *country* and *area* are classes designed to represent geographical locations which may be visited by the *User*. These classes have properties which allow to build a tree structure of countries, areas and cities. For instance, *CanadaCountry*, *QuebecArea*, *AnnavilleCity*, *AmesdaleCity* and *AnjouCity* were samples of geo-locations introduced above. They represent a part of an administrative division of Canada. First level in our structure is a country, the second level is an area and finally city (Annaville and Anjou) is the third one. Available properties allow to query for neighbor (adjacent) instances of the same class. This approach requires access to administrative divisions of the world data, otherwise it may be of little value in terms of facilitating a location based advice. Apart from the administrative division tree, these classes allow to describe actual geo-coordinates of objects. Location based advising can be performed by calculating object's distances using the general formula (*long* - longitude, *lat* - latitude, *alt* - altitude):

$$\sqrt{(long_0 - long_1)^2 + (lat_0 - lat_1)^2 + (alt_0 - alt_1)^2}$$

Note that in most business travel scenarios the altitude (alt_0 and alt_1) is of little relevance and can be omitted. In the case of searching for a nearby conference or an organization which may be visited during a Duty Trip, performance issues have to be considered. Calculating distance according to the above formula from data in a semantic (Jena-based [11]) repository is currently very likely to be inefficient. However, we have chosen the PostgreSQL database for the persistence layer of the GIS information [16]. This database engine supports GIS processing and allows to perform geospatial calculations with a designated component. Instead of calculations involving directly the semantically described data, we propose to create a link between (a) semantically described objects which represent geographical locations and have been described using their coordinates, and (b) the copy of the object's geo-information represented within, and calculated using, the PostgreSQL GIS engine. Obviously, similar calculations can be performed not only for conferences and/or institutions, but also for all other geo-objects (e.g. restaurants and hotels) as their coordinates are described in the same way as cities (hotel, restaurant and city are subclasses of the *Spatial Thing* class in our travel ontology; see [7,22]). Therefore, the *DTS* system will be able to provide at least the following geo-info-based advice:

1. Location notes and tips (textual information about a location which was added to Duty Trip Reports - class in the ontology: *Location Specific Notes*),
2. Organizations and people that can be visited (objects of *Organization Contact* and *Contact Person* classes, these objects are created by the employees during the Duty Trip Report's creation),
3. Information about nearby conferences of possible interest (based on location of the trip and the conference as well as on the personal interest and conference topics),
4. Hotels and restaurants (based on the *Hotel* and *Restaurant* TSS ontology classes),
5. Car rental and golf courses (ontology extensions based on the OTA specification [15]).

4.2 Numeric and Date Object Calculations

Computing distance between numeric and date object is rather obvious. The distance will be represented by the result of difference operation on these objects. In the first case, the result will be a number, in the latter case the result will be a time period (e.g. of a stay in a given place). Note that currently most major programming languages provide date calculation support hence we believe this issue should not be discussed in more detail (assuming there are no problems with date representation and deserialization).

4.3 Enumerable Object Calculations

In case of ontology, enumerable values can be more complex than *enums* known from popular programming languages. In an ontology, class instances can also be enumerable values. In that case complex structures can be constructed, representing relations between objects. For instance, presented above *ScientificField* class falls under the OWL *oneOf* restriction, however each instance of that class has property values which refer to other instances of that class. This results in a graph-like structure of enumerable values.

To calculate distance between two object of enumerable type, let us note first that if the structure of *enum* values is flat (plain list with no relations between objects) it can be assumed that the distance is 0 if the values match, otherwise it equals to 1. An example of such simple enumerable is the *Gender* property, which is utilized in the human resource profile. Here we have two values: *Male* and *Female* and if they match the distance is 1, and 0 otherwise.

Let us now present a method for calculating distance between class instances which involve transitive, non-symmetric properties, which is a simplified case of method introduced in [17]. Here, a path in a directed graph is calculated for all relations. Let us assume that R is such a transitive, not symmetric relation (*property* in the OWL notation). Then the distance between two vertices of a graph of relation R: v_0 and v_k ($dist_R(v_0, v_k)$) is calculated according to the following algorithm:

1. If there exists $path_R(v_0, v_k)$ in the graph of relation R, then the shortest one can be found and

$$dist_R(v_0, v_k) = length(shortestPath_R(v_o, v_k));$$

   otherwise go to 2nd step.
2. Let $X = \{x : path_R(x, v_0) \text{ and } path_R(x, v_k) \text{ exist}\}$. Find such $y \in X$, that $length(path_R(y, v_0))$ is minimal among all vertices belonging to X (i.e. this is the shortest path):

$$dist_R(v_0, v_k) = 10^{length(path_R(y, v_0))} + length(shortestPath_R(y, v_k))$$

Let us now describe calculation of distance between research interests of a human resource and conference topics, while utilizing examples introduced above. According to the proposed algorithm the following distance values can be found (here we calculate all-against-all distance values):

```
$dist_SF = path_isSubfieldOf$
$dist_SF$(Aerodynamics-30501, Aerospace_Ship_and_Ocean_Engineering-30500)=10
$dist_SF$(Aerodynamics-30501, Mechanical_Engineering-30400)=101
$dist_SF$(Aerodynamics-30501, Acoustics_and_Noise-30405)=102
$dist_SF$(Marine_Hydrodynamics-30505, Aerospace_Ship_and_Ocean_Engineering-30500)=10
$dist_SF$(Marine_Hydrodynamics-30505, Mechanical_Engineering-30400)=101
$dist_SF$(Marine_Hydrodynamics-30505, Acoustics_and_Noise-30405)=102
$dist_SF$(Turbulence-30412, Aerospace_Ship_and_Ocean_Engineering-30500)=101
$dist_SF$(Turbulence-30412, Mechanical_Engineering-30400)=10
$dist_SF$(Turbulence-30412, Acoustics_and_Noise-30405)=11
```

These values allow us to utilize a number of strategies to establish "closeness" of two resources. The simplest one would be, if for any two properties the distance is below a certain threshold, then a conference should be recommended as potentially interesting. Note that distance could be also scaled by the level of knowledge of the specialist in the field. However more involved considerations are also possible. In this context let us note that values of the $dist_R(v_0, v_k)$ function allow us to specify how far are the graph nodes located from each other in terms of a transitive, not symmetric relation R. In the case of research specialization modeling relation we can assume that the maximum length of $path_R(v_0, v_k)$ is 9. In our ontology an example of such relation is the *isSubfieldOf* property of the *ScientificField* class, where the maximum length of $path_{SF}(v_0, v_k)$ is 2. Additionally, infinite distance is not considered. With such assumptions we are able to distinguish following groups of conclusions which can be drawn from the function values:

1. If $dist_R(v_0, v_k) = 0$, then $v_0 = v_k$
2. If $dist_R(v_0, v_k) = n$ and $0 < n < 10$, then $R(v_0, v_k) = true$ and v_k is n-deep specialization of v_0
3. If $dist_R(v_0, v_k) = n$ and $n = 10^k, k > 0$, then $R(v_0, v_k) = false$ and v_0 is k-deep specialization of v_k
4. If $dist_R(v_0, v_k) = n$ and $10^k < n < 10^(k+1), k > 0$, then $R(v_0, v_k) = false$ and v_0 is $n - 10^k$-deep specialization of k-deep specialization of v_k

For instance, if

$$dist_{SF}(Turbulence - 30412, Acoustics\_and\_Noise - 30405) = 11,$$

we may say that there is a node of which both *Turbulence* and *Acoustics and Noise* are direct specializations.

These observations allow us to develop a number of reasoning scenarios that utilize not only information about numerical distance, but also following form it knowledge about the structure of relations. Developing a reasoning engine utilizing this information is next step in our work.

Let us note that the proposed algorithm is a specific case of the method proposed in [17]. Basic difference between them is as follows. The method proposed in [17] assumes existence of multiple relations linking any class instance (node) from a single node and merging edges which represent relations of the same direction between the same nodes; thus, the distance is computed including all properties (relations) of classes (concepts). The algorithm presented above, on the other hand, is restricted to one selected property (relation) of a class (concept) and an inverse of the selected property. This pair represent generalization and specialization relations between concepts. We claim that the algorithm presented here can be substituted for the algorithm of [17] by adjusting appropriate weights to concepts relations. Specifically, used here weirths of 1 for specialization and 10 for generalization.

5 Concluding Remarks

In this paper we have presented preliminary results concerning resource matching in a virtual organization. Specifically, we have considered comparisons between specific values within ontologically demarcated resources (geographical distance, date and numeric objects, as well as enumerated types). In the next phases of our work we plan to apply these methods in automatic resource matching. Specifically, we will develop a matching engine that will use rules of resource relevance and as a result will be able to calculate the distance between them depending on the matching query context (i.e. scenario and types of compared resources).

An important problem which has to be taken into account is the performance of proposed operations. This is especially the case as we are already aware of deficiencies of the *Jena* persistence and query engine. This is one of the reasons for the planned mixed environment in which the PostgreSQL GIS engine is to be used. However, the final determination of the implementation details will have to be made on the basis of experiments performed with realistic datasets.

References

1. Biesalski, E., Abecker, A.: Human Resource Management with Ontologies. In: Althoff, K.-D., Dengel, A., Bergmann, R., Nick, M., Roth-Berghofer, T.R. (eds.) WM 2005. LNCS (LNAI), vol. 3782, pp. 499–507. Springer, Heidelberg (2005)
2. Bizer, C., Heese, R., Mochol, M., Oldakowski, R., Tolksdorf, R., Eckstein, R.: The Impact of Semantic Web Technologies on Job Recruitment Processes. In: Proc. International Conference Wirtschaftsinformatik (WI 2005), Bamberg, Germany (2005)
3. Ganzha, M., Paprzycki, M., Popescu, E., Bădică, C., Gawinecki, M.: Agent-Based Adaptive Learning Provisioning in a Virtual Organization. In: Advances in Intelligent Web Mastering. Proc. AWIC 2007, Fontainebleu, France. Advances in Soft Computing, vol. 43, pp. 25–40. Springer (2007)
4. Ganzha, M., Paprzycki, M., Gawinecki, M., Szymczak, M., Frackowiak, G., Bădică, C., Popescu, E., Park, M.-W.: Adaptive Information Provisioning in an Agent-Based Virtual Organization—Preliminary Considerations. In: Proceedings of the SYNASC Conference, IEEE CS Press (in press)
5. Gawinecki, M.: Modelling User on the Basis of Interactions with a WWW Based System, Master Thesis, Adam Mickiewicz University (2005)
6. Gawinecki, M., Gordon, M., Paprzycki, M., Vetulani, Z.: Representing Users in a Travel Support System. In: Kwasnicka, H., et al. (eds.) Proceedings of the ISDA 2005 Conference, pp. 393–398. IEEE Press (2005)

7. Gawinecki, M., Gordon, M., Nguyen, N.T., Paprzycki, M., Szymczak, M.: RDF Demarcated Resources in an Agent Based Travel Support System. In: Golinski, M., et al. (eds.) Informatics and Effectiveness of Systems, pp. 303–310. PTI Press, Katowice (2005)
8. Gawinecki, M., Gordon, M., Nguyen, N.T., Paprzycki, M., Zygmunt Vetulani, Z.: Ontologically Demarcated Resources in an Agent Based Travel Support System. In: Katarzyniak, R.K. (ed.) Ontologies and Soft Methods in Knowledge Management, Advanced Knowledge International, Adelaide, Australia, pp. 219–240 (2005)
9. Greer, J., McCalla, G.: Student modeling: the key to individualized knowledge based instruction. NATO ASI Series F, vol. 125 (1993)
10. HR-XML Consortium, http://www.hr-xml.org/
11. Jena Semantic Web Framework, http://jena.sourceforge.net/
12. Korea Science and Engineering Foundation, http://www.kosef.re.kr/english_new/index.html
13. Mochol, M., Wache, H., Nixon, L.: Improving the Accuracy of Job Search with Semantic Techniques. In: Abramowicz, W. (ed.) BIS 2007. LNCS, vol. 4439, pp. 301–313. Springer, Heidelberg (2007)
14. Montaner, M., López, B., de la Rosa, J.L.: A taxonomy of recommender agents on the Internet. Artif. Intell. Rev. 19(4), 285–330 (2003)
15. Open Travel Alliance, http://www.opentravel.org/
16. PostGIS: HOME, http://postgis.refractions.net/
17. Rhee, S.K., Lee, J., Park, M.W.: Ontology-based Semantic Relevance Measure. In: Kim, S., et al. (eds.) Proceedings of the The First International Workshop on Semantic Web and Web 2.0 in Architectural, Product and Engineering Design, 6 pages (2007), http://ftp.informatik.rwth-aachen.de/Publications/CEUR-WS/Vol-294/
18. Schmidt, A., Kunzmann, C.: Towards a Human Resource Development Ontology for Combining Competence Management and Technology-Enhanced Workplace Learning. In: Meersman, R., Tari, Z., Herrero, P. (eds.) OTM 2006 Workshops. LNCS, vol. 4278, pp. 1078–1087. Springer, Heidelberg (2006)
19. Semantic Web, http://www.w3.org/2001/sw/
20. Szymczak, M., Frackowiak, G., Ganzha, M., Gawinecki, M., Paprzycki, M., Park, M.-W.: Resource Management in an Agent-based Virtual Organization—Introducing a Task Into the System. In: Proc. of the WI/IAT 2007 Workshops, IEEE CS Press, Los Alamitos (in press, 2007)
21. Szymczak, M., Frackowiak, G., Ganzha, M., Gawinecki, M., Paprzycki, M., Park, M.-W.: Adaptive Information Provisioning in an Agent-Based Virtual Organization—Ontologies in the System (submitted for publication)
22. Szymczak, M., Gawinecki, M., Vukmirovic, M., Paprzycki, M.: Ontological Reusability in State-of-the-art Semantic Languages. In: Olszak, C., et al. (eds.) Knowledge Management Systems, pp. 129–142. PTI Press, Katowice (2006)
23. Training, http://en.wikipedia.org/wiki/Training
24. Tzelepis, S., Stephanides, G.: A conceptual model for developing a personalized adaptive elearning system in a business environment. In: Current Developments in Technology-Assisted Education, vol. III, pp. 2003–2006. Formatex Publishing House (2006)
25. Wooldridge, M.: An Introduction to MultiAgent Systems. John Wiley & Sons (2002)

Knowledge Technologies-Based Multi-Agent System for Semantic Web Services Environments

Francisco García-Sánchez, Rodrigo Martínez-Béjar, Rafael Valencia-García, and Jesualdo T. Fernández-Breis

Faculty of Computer Science, University of Murcia,
30100 Espinardo (Murcia), Spain
{frgarcia,rodrigo,valencia,jfernand}@um.es

Abstract. Intelligent Agents and Semantic Web Services are two technologies with great potential in the future. Striking new applications can be developed by using them. However, in order for these technologies to be fully applicable to real settings, several problems remain unsolved. In particular, Semantic Web Services need for an upper software entity able to deal with them, while Intelligent Agents have historically suffered from a number of drawbacks not yet resolved. In this paper, we claim that integrating these technologies into a joint environment can overcome part of their respective problems while strengthening their advantages. The necessity for integrating these technologies and its benefits are analyzed in this work. Based on this study, we present SEMMAS, a framework for seamlessly integrating Intelligent Agents and Semantic Web Services. The fundamentals of the framework and its architecture are explained and a proof-of-concept implementation is described.

1 Introduction

The Intelligent Agents (IAs) topic has been broadly studied over the last 30 years and currently the referred topic is being revisited due to its relation to the Semantic Web (SW) and the potential benefits that can be reached from their integration. An IA can be defined as a computer system situated in some environment and capable of autonomous action in this environment in order to meet its design objectives [1]. IAs are characterized by a set of basic properties including reactivity (i.e. the ability to perceive its environment and respond to changes in it in a timely fashion), pro-activeness (i.e. the ability to exhibit goal-directed behavior by taking the initiative), and social ability (i.e. the ability to interact with other agents). Before the emergence of the SW, agents had to face the problems derived from the lack of structure in the information published on the Web. The SW [2] involves the addition of machine-readable, semantic annotations to Web resources by using ontologies [3] as the backbone technology. Thus, in the so called Web 3.0, agents will be able to automatically process the machine-readable, semantic contents of the Web and new powerful opportunities will open up for both application developers and users.

On the other hand, Web Services (WSs) arose as the best solution for remote execution of functionality. This was partly due to properties such as operating system and programming language-independence, interoperability, ubiquity and the possibility to develop loosely-coupled systems. Thus, WS technology transforms the Web from a mere repository of information into a distributed source of functionality. However, as the Web grows in both size and diversity, there is an increased need to automate aspects of WSs such as discovery, execution, selection, composition and interoperation [4]. Semantic Web Service (SWS) technology [5], that is, the semantic annotation of services' capabilities, has been the solution proposed. The W3C is currently examining various approaches with the purpose of reaching a standard for the SWS technology including OWL-S[1] (OWL Web Ontology Language for Services), WSMO[2] (Web Service Modeling Ontology), SWSF[3] (Semantic Web Services Framework), WSDL-S[4] (Web Service Semantics), and SAWSDL[5] (Semantic Annotations for WSDL). OWL-S, WSMO and SWSF consist of ontologies that contain the concepts and properties necessary to describe a WS. In contrast, both WSLD-S and SAWSDL propose to associate semantic annotations to WSDL documents by means of WSDL extensibility elements.

The IA and WS paradigms are often viewed as similar and competing. However, several research studies have demonstrated that, despite having overlapping functionalities, the cooperative interaction between IAs and SWSs can lead to the development of new more powerful applications (see [6,7,8]). In this paper, we list some research projects carried out with the purpose of combining IA and WS technologies in an integrated environment, and point out the factors that hamper their applicability. Then, based on this previous analysis, we present the Semantic Web Services and Multi-Agent System (SEMMAS) framework, which overcomes the shortcomings of the current proposed solutions. SEMMAS is based on a loosely-coupled infrastructure and makes use of ontologies to facilitate the agent-service interoperation. With this approach, applications benefit from the autonomy, pro-activeness, dynamism and goal-oriented behavior IAs provide, and the high degree of interoperability across platforms WSs advocate.

The rest of the paper is organized as follows. In Section 2, different tools and research studies related to the integration of Intelligent Agents and (Semantic) Web Services are analyzed. The framework for effectively integrating these two major technologies is formulated in Section 3. In Section 4, a proof-of-concept implementation of the SEMMAS framework is described. Finally, conclusions and future work are put forward in Section 5.

2 Related Work

Different research projects have explored how to support the cooperative interaction between IAs and SWSs. Blacoe and Portabella [9] point out that, in order to

---

[1] http://www.w3.org/Submission/OWL-S/
[2] http://www.w3.org/Submission/WSMO/
[3] http://www.w3.org/Submission/SWSF/
[4] http://www.w3.org/Submission/WSDL-S/
[5] W3C Recommendation from August 28th, 2007; http://www.w3.org/TR/sawsdl/

integrate what they called 'agent-based services' and 'web-based services', three main scenarios are possible: (1) WSs provide the more basic level functionality and agents provide higher-level functions by using, combining and choreographing WSs, so achieving added-value functions; (2) communication in WSs and agents becomes equivalent, so the distinction between them disappears ('agents in web service wrappers'); and (3) both types remain separate creating a heterogeneous service space and interoperating through gateways and translation processes. Next, some of the approaches developed so far are discussed.

It was Hendler [10] who first described how the ontology languages of the SW could lead to more powerful agent-based approaches to using services offered on the Web. In this work, the author presented the foundations of what is now called SWSs. He proposed a method for describing the way the invocation of services should be done by agents by means of an ontology language such as DAML+OIL. Once the invocation characteristics of a service are semantically described, agents would be able to determine the specific information needed for invoking the service.

In latter studies, the integration of agents and (semantic) WSs has been successfully addressed for different purposes. The Semantic Web FRED project (SWF) combines agent technology, ontologies, and SWSs in order to develop a system for automated cooperation [6]. In this system, software agents ('Freds') perform tasks on behalf of their owners and interact among them if they have to. In order to resolve a task, agents make use of Services, computational resources that allows automated resolution of tasks. The authors distinguish three types of Services: plans (Java programs), processes (complex and nested services), and external WSs (through WSDL). A major problem of the SWF is that it seems to be tightly bound to WSMO. In fact, the SWF is supposed to be a "WSMO Implementation". The framework we present here includes mechanisms to support all the current SWS approaches. It is even possible to dynamically incorporate support for new solutions.

Another example is the GODO (Goal Oriented DiscOvery) system [7], which is an agent located between users and the WSMX environment. When users want to send goals to WSMX they have to write them down in WSML, a formal language that can be hard to understand to a common user (actually there exist some adapters that receive goals in other syntax, e.g. SOAP messages, and transform them into WSML syntax). GODO is able to transform user requests in natural language into WSMX goals in WSML. With this purpose, it incorporates a language analyzer that determines the ontology-relevant elements (i.e., concepts, attributes, attribute values, and relationships) within a sentence, thus producing a lightweight ontology. Once these elements have been obtained, GODO uses the ontology to generate the goals to be executed and sends them to WSMX. GODO's main drawback is that it interacts with an execution environment and not with SWSs as such. Consequently, few of the advantages of agents systems are exploited (e.g. agent-based negotiation can not take place between parties and agents techniques for tasks such as discovery, composition and invocation have to be implemented ad-hoc within the SWSs infrastructure).

A further related solution is the one provided by the "Agents and Web Services Interoperability Working Group (AWSI WG)"[6], which is part of the IEEE FIPA Standards Committee. The idea of the working group is to create a middleware able to handle the fundamental differences between Agent Technology and Web Services, that is, the use of different communication protocols (ACL vs. SOAP), service description languages (DF-Agent-Description vs. WSDL) and service registration mechanisms (DF vs. UDDI). With this approach, the so called AgentWeb Gateway middleware [8] facilitates the required integration without changing existing specifications and implementations of both technologies. This solution complies with the third category of Blacoe and Portabella's classification. We believe that having both IAs and WSs working at the same level is not conceptually appropriate and does not take into account the fundamental differences between these paradigms.

The approaches presented so far suffer from a number of shortcomings mainly due to their inability to completely benefit from the advantages of the argued combination. In fact, they restrain the possibilities by considering only some of the properties of either IAs or SWSs. The aim of the framework proposed here is to fully exploit the benefits of this integration. This is achieved by taking advantage of the positive aspects of both technologies separately while combining them so as to avoid their drawbacks. For this, it is important to notice that agents' original conception differs from that of WSs, so that their strengths and weaknesses are different.

3 SEMMAS: A Framework for Intelligent Agents and Semantic Web Services Integration

As it was stated before, while (Semantic) Web Service infrastructures provide for a high degree of interoperability across platforms and operating systems, they do not posses either enough degree of autonomy or ability to automatically adapt to changing situations. In these settings, agents can contribute to make systems more autonomous and dynamic thus maximizing their perceived utility. In line with this, the framework presented here comprises both IAs and SWSs working cooperatively in the same environment. This solution stems from a basic underlying hypothesis: IAs and WSs must lie on two different layers of abstraction due to the conceptual differences between these technologies from their very conception. The main idea behind agent technology was not for IAs to be able to provide services, but to become autonomous entities that incorporate intelligence, being capable of exhibiting pro-active, goal-oriented behavior, and interacting (either competitively or cooperatively) with other entities in order to satisfy their design objectives. In contrast, WSs were conceived with the purpose of providing globally accessible software components with a particular functionality.

Next the foundations of the referred framework are presented and the main elements of the architecture enumerated.

---

[6] http://www.fipa.org/subgroups/AWSI-WG.html

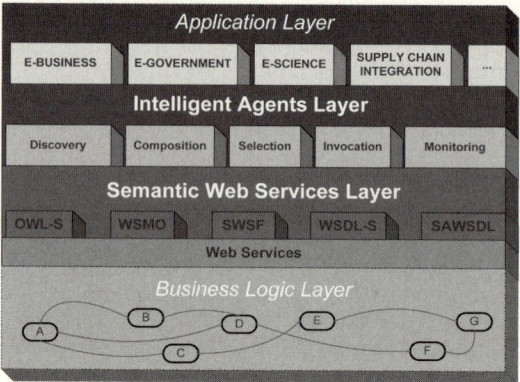

Fig. 1. The framework layout

3.1 SEMMAS Foundations

Ontologies are the paramount technology of the approach introduced in this paper as they act as the 'glue' that binds the components of the framework architecture. In this work, an ontology is seen as "a formal and explicit specification of a shared conceptualization" [3]. Firstly, ontologies function as domain vocabularies so that Web Services and agents share the same interpretation of the terms contained in the messages that they exchange. Secondly, ontologies are useful to semantically describe Web Services' capabilities and processes. This semantic description can then be automatically processed by software entities, so that Web Service discovery, composition, selection, execution and monitoring can be done without human intervention. Finally, from the agents' perspective, each agent's local domain-related knowledge may be extracted from, or built upon, the application domain ontology. Moreover, inter-agent communication may be carried out by means of a common vocabulary based on an agreed ontology.

The framework presented here is based on a multi-tier architecture that is composed of four different layers (see Fig. 1). The lower layer, namely, the Business Logic Layer provides for the most specific operations. It comprises the internal business processes within companies. Upon this layer, WSs are deployed that show off some parts of the internal business process and make that functionality publicly available. These services along with the semantic description of their capabilities lie on the second layer, namely the Semantic Web Services Layer. Adding semantic annotations to WSs capabilities can help software entities to (semi-)automatically interact with them in a dynamic way. In particular, new services can emerge and others may change their functionality or even disappear at run-time, but the system would keep on working and the changes would be reflected on the application instantly. These sophisticated software entities (i.e. IAs) that interact with, and take advantage of, basic services are located in the Intelligent Agents Layer. IAs make use of the semantic annotation of services capabilities to automatically discover, compose, invoke and monitor WSs. They are also able to dynamically exhibit and propagate the changing functionality

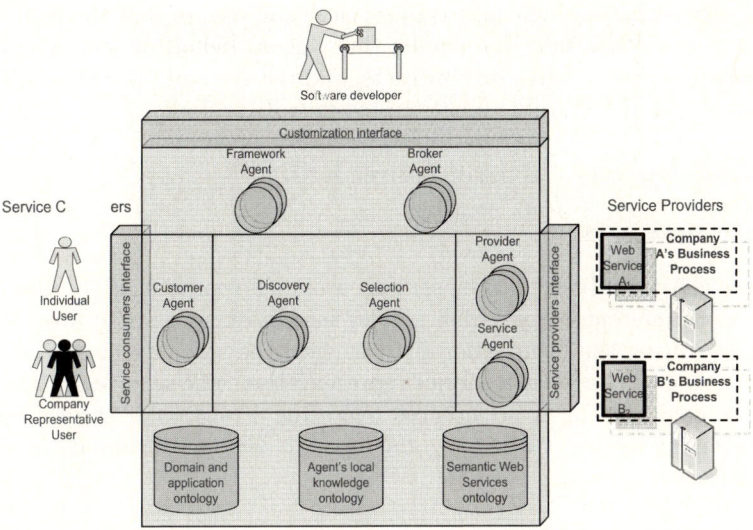

Fig. 2. The SEMMAS Architecture

provided in lower layers. Finally, the Application Layer is responsible for organizing (i.e. orchestrating and coordinating) agents to actually perform useful activities for users. In this way, depending on the agents available in the system and the way they inter-operate, different user-tailored applications can be obtained.

3.2 SEMMAS Architecture

The SEMMAS (SEMantic web service and MultiAgent System) framework comprises two of the layers identified above, the Intelligent Agents and the Semantic Web Services layers. As a result, the framework becomes independent of both the application domain and the actual applications to be developed. In order to create an application, programmers only have to set the appropriate domain ontologies and decide on which agents to instantiate and which services to access. Thus, the framework can be considered as a reference architecture for several scenarios and complex environments such as e-commerce, e-science or e-government.

The architecture that constitutes SEMMAS is composed of three main components (see Fig. 2): a set of IAs that constitute a MAS, three ontology repositories, and three different interfaces for interacting with the external actors that have been identified (i.e., service providers, service requesters and software developers).

In the platform proposed to run the system, three main groups of agents are distinguished: agents that act on behalf of service owners ('Provider Agent' and 'Service Agent'), agents that act on behalf of service consumers ('Customer Agent', 'Discovery Agent', and 'Selection Agent'), and agents that perform management tasks ('Framework Agent' and 'Broker Agent'). Those acting on behalf of service owners manage the access to services and ensure that the contracts are fulfilled. On the other side, the agents that act on behalf of service consumers have to locate services, agree on contracts, and receive and present results. Management agents have a double function: to control and balance the system workload, and to help in solving the interoperability mismatches. A description of the seven different types of agents is given next:

Service owners agents

– Provider Agent: It acts as a service provider representative. The entities set their preferences regarding service execution and these are taken into account during the negotiation process with the service consumers.
– Service Agent: It acts as a service representative. The service provider establishes a concrete set of preferences regarding a particular service and these are taken into account when negotiating with service consumers (in this case, the Selection Agent).

Service consumers agents

– Customer Agent: It acts as a user representative. First, (individual or collective) users indicate their preferences and specify the goal to be achieved. Then, the goal is carried out and the results given back to the user. The intermediate process happens transparently to the user.
– Discovery Agent: It is in charge of searching in the Semantic Web Services repository for the service or set of services (i.e. composition) that satisfy the requisites established by the users.
– Provider representative role: It interacts with service providers. At a high level of abstraction, it must be able to enforce the conditions present in the company's business strategy
– Selection Agent: It is in charge of selecting the most appropriate (single or compound) service from the set of services found by the discoverer according to the users' preferences. A negotiation process with the different service representatives (i.e. Service Agent) is carried out for that purpose.

Framework management agents

– Broker Agent: It is responsible for solving interoperability issues. Three different levels are considered: data mediation, process mediation and functional interoperability.
– Framework Agent: It is responsible for monitoring and ensuring a correct functioning of the platform. This type of agent also controls and balances the workload.

Only abstract, high-level description of the tasks that each agent must carry out is given. Nothing is said about the actual implementation of these tasks. The framework has been designed so that there is no a fixed way each task should be performed. In fact, it is at run-time when it is determined which implementation is actually used. For this purpose, the 'role' concept is introduced. Roles are encapsulations of dynamic behavior and properties that can be played by agents. The use of roles presents a number of benefits that can be summarized as follows [11]: (1) roles are dynamic and flexible; (2) roles are responsibility-driven; (3) roles are context-sensitive. We distinguish between roles dealing with service-related issues from those related to the framework management. Both roles categories are enumerated next:

Service-related Roles

- Broker role: It represents the functionality needed for solving all kind of interoperability problems (data, process and functional mediation)
- Composer role: It allows the achievement of a goal by means of several composed services
- Invoker role: It invokes a Web Service once the operation to be executed and the parameters are known
- Matchmaker role: It finds the services whose semantic descriptions match the goal that was sent by the user
- Monitor role: It ensures that the contracts established for the execution of the services by both service owners and service consumers are fulfilled
- Ontology manager role: It includes functionality associated with the access and processing of ontologies
- Selector role: It provides the functionality necessary for the selection of a service from a list of services according to a set of preferences

Platform Management Roles

- Negotiator role: It enacts the actual negotiation process between the parties establishing the basis for the system execution
- Platform manager role: It incorporates functions to control and balance the system workload
- Provider representative role: It interacts with service providers. At a high level of abstraction, it must be able to enforce the conditions present in the company's business strategy
- Service representative role: It interacts with service consumers by, firstly, determining their wishes and, then, returning the expected results
- Consumer representative role: It interacts with service consumers by, firstly, determining their wishes and, then, returning the expected results
- Global monitor role: It monitors the events in the application, detects possible problems, and defines the actions to take in case of error

At run-time each agent decides on what roles to play depending on the goal it pursues. Nevertheless, some of the roles are mandatory for some agents, as they characterize the type of agent the agents belong to. Thus, for example, the framework agents should take over both the 'Global Monitor' and the 'Platform Manager' roles. Both factors, the actual implementation of these roles and the agent election of the roles to take, eventually determine the agent behavior.

Regarding ontologies, three different kinds have been identified: application and domain ontology, agent local knowledge ontology, and Semantic Web Services ontologies. The application ontology contains the knowledge entities (i.e. concepts, attributes, relationships, and axioms) to model the application in which the framework is to be employed. The domain ontology, on the other hand, represents a conceptualization of the specific domain the framework is going to be applied in. This ontology supports the communication among the components in the framework without misinterpretations. The agent local knowledge ontology contains, for each agent, the knowledge about the environment that agent has. This ontology generally includes knowledge about the assigned tasks, and the mechanisms and resources available to achieve those tasks. Finally, the ontologies that contain the semantic description of Web Services are kept in the Semantic Web Services repository.

At last, three different interfaces have been included within the framework architecture. They aim at enabling the interaction with the actors that are external with respect to the framework: service consumers, service providers, and software developers. Software developers can, by means of their interface, customize the application by setting up the specific ontologies to be used. They also have to instantiate and configure the core agents necessary for the proper functioning of the system (customer, provider and service agents will be launched as needed at run-time).

Once the application has been properly set up, both service consumers and service providers can register in the system and use it as a meeting point. Through their interface, service providers can modify the list of services they provide and set the conditions under which a service they provide must be executed. Service consumers, on the other hand, can, by means of their interface, query the system and trigger the execution of one or several Web Services in order to fulfill a particular goal.

4 Proof-of-Concept Implementation

Based on the above described framework, a prototype of a Semantic Web Services execution environment has been implemented. This tool comprises a Multi-Agent System (MAS) that complies with the requirements imposed by the framework, and a Web application that constitutes the user interfaces. For the design of this complex MAS the INGENIAS methodology [12] was applied. One of the main reasons for choosing INGENIAS was that, while most of the other methodologies do not have any reference implementation for developing the models and diagrams proposed by the methodology, INGENIAS comes along with a toolkit, the INGENIAS Development Kit (IDK)[7]. It provides the means to create most of the diagrams and models required by the methodology making it more convenient for software designers to carry out the system specification. Besides, it incorporates a function to automatically generate JADE files from the developed

---

[7] http://sourceforge.net/projects/ingenias/

diagrams. JADE (Java Agent Development Framework)[8] is the agent platform that has been used for implementing the system. JADE is the most widely used agent platform for research projects worldwide and it seems to be at a quite mature stage of development.

Apart from JADE, various tools and libraries has been used to develop the framework. The latest Web programming techniques based on JavaServer Pages (JSP) and Servlets have been applied for the user interface. In order to communicate the Web interface with the MAS, the JadeGateway[9] was employed. Exploiting the semantic content of ontologies has been possible through the use of the Jena API[10]. KAText[13], a natural language processing tool, has been utilised to support the interaction of non-expert users with the system. Finally, similarly to other research studies (e.g. [14]), we have followed a planning approach for service composition. In particular, we have made use of a library that implements a STRIPS-like planner[11]. This mechanism enables the automatic design of simple sequential workflows.

As it was stated before, SEMMAS is an application- and domain-independent framework that can be useful in various environments. In order for software developers to adapt the framework to a particular application domain they must carry out four main tasks:

1. Design the application and domain ontologies: the knowledge about the relevant entities in the application environment and their relationships, and the tasks the application should be able to perform must be represented by means of ontologies. OWL (Web Ontology Language) is the ontology language used in this work.
2. Develop the Web Services that provide the most basic functionality: services must be exhibited as Web Services. In general, it is the service providers who are responsible for implementing the services they aim to offer. Apache Axis2 has been chosen as the WS implementation framework.
3. Semantically annotate the Web Services' capabilities: once the services have been properly implemented and their WSDL syntactic descriptions are available, they must be described semantically. This is also a task service providers will usually carry out. As for software developers, they must set up the URLs of the repositories the platform will have access to.
4. Implement the roles and instantiate the necessary agents: ad-hoc or currently existing techniques for service discovery, selection, composition and invocation must be included as roles implementations. Then, agents must be instantiated and assigned a initial set of roles to play[12].

The scope of this prototypic implementation includes the basic interactions between agents, the access of a single SWSs repository, and the (sequential)

---

[8] http://jade.tilab.com/
[9] http://jade.tilab.com/doc/tutorials/JadeGateway.pdf
[10] http://jena.sourceforge.net/
[11] http://www.dcs.shef.ac.uk/~pdg/com1080/java/strips/
[12] For more details concerning the proof-of-concept implementation visit http://www.semmas.com

execution of WSs. Thus, the full life cycle for the provision of Semantic Web Services is covered to some extent.

5 Conclusions and Future Work

In this paper, we first introduce both Intelligent Agent and Semantic Web Service technologies and point out the potential benefits that can be reached from their combination. Then, we analyze the plausibility of integrating Web Services and Multi-Agent Systems into a combined architecture and the advantages derived from it through a comprehensive list of approaches. Based on the benefits envisioned of such integration and the drawbacks of the current solutions, we present SEMMAS, a new framework for Intelligent Agents and Semantic Web Services integration. This framework aims to exploit the striking potential of the two technologies (concepts) under question while overcoming their deficiencies. In SEMMAS, the basic underlying hypothesis is that Intelligent Agents should be placed on a layer above Web Services, and be responsible for higher-level tasks. We believe this is the most suitable approach due to the conceptual differences between these technologies from their very conception. Particularly, the main idea behind agent technology was not for intelligent agents to be able to provide services, but to become autonomous entities that incorporate intelligence, are capable of exhibiting pro-active, goal-oriented behavior, and can interact (either competitively or cooperatively) with other entities in order to satisfy their design objectives. In contrast, Web Services were conceived with the purpose of providing globally accessible software components that exhibit particular functionalities.

One of the major contributions of our proposal is that it enables the integration of Agent and Web Service Technologies without changing their existing specifications. Besides, with this approach applications can benefit from the autonomy, pro-activeness, dynamism and goal-oriented behavior agents provide, and the high degree of interoperability across platforms and operating systems Web Services advocate. Based on the above described framework, a prototype of a Semantic Web Services execution environment has been implemented. This tool comprises a Multi-Agent System that complies with the requirements imposed by the framework, and a Web application that constitutes the user interfaces.

As further work, we plan to evaluate the framework in terms of its performance and usability in domains such as Bioinformatics, eCommerce, and eGovernment. In addition, more sophisticated composition mechanisms must be investigated in order to make the framework useful in real settings. Security concerns have not yet been addressed, and will be part of future upgrades. Also for the future it is left the integration of Grid services within the framework.

Acknowledgements

This work has been possible thanks to the Spanish Ministry for Science and Education through projects TIN2006-14780 and TSI2007-66575-C02-02, the Spanish

Ministry for Public Works through project PT-2006-055-241-ICPP and the Murcian Government under projects TIC-INF 06/01-0002 and BIO-TEC 06/01-005.

References

1. Wooldridge, M.: An introduction to MultiAgent Systems. John Wiley & Sons Ltd. (2002)
2. Berners-Lee, T., Hendler, J., Lassila, O.: The Semantic Web. Scientific American, 34–43 (May 2001)
3. Studer, R., Benjamins, R., Fensel, D.: Knowledge Engineering: Principles and Methods. Data and Knowledge Engineering 25(1-2), 161–197 (1998)
4. Fensel, D., Bussler, C.: The Web Service Modeling Framework WSMF. Electronic Commerce Research and Applications 1(2), 113–137 (2002)
5. McIlraith, S., Son, T.C., Zeng, H.: Semantic Web Services. IEEE Intelligent Systems 16(2), 46–53 (2001)
6. Stollberg, M., Roman, D., Toma, I., Keller, U., Herzog, R., Zugmann, P., Fensel, D.: Semantic Web Fred - Automated Goal Resolution on the Semantic Web. In: Proc. of the 38th Hawaii International Conference on System Sciences (2005)
7. Gómez, J.M., Rico-Almodóvar, M., García-Sánchez, F., Toma, I., Han, S.: GODO: Goal Oriented Discovery for Semantic Web Services. In: Discovery on the WWW Workshop (SDISCO 2006), Beijin, China (2006)
8. Shafiq, O., Suguri, H., Ali, A., Fensel, D.: A first step towards enabling Interoperability between Software Agents and Semantic Web Services: Multi Agent Systems adapting Web Services Standards. IBIS - Interoperability in Business Information Systems 2(2), 97–117 (2006)
9. Blacoe, I., Portabella, D.: Guidelines for the integration of agent-based services and web-based services. Deliverable D2.4.4 (WP2.4), Knowledge Web project (2005)
10. Hendler, J.: Agents and the Semantic Web. IEEE Intelligent Systems 16(2), 30–37 (2001)
11. Zhao, L., Mehandjiev, N., Macaulay, L.: Agent Roles and Patterns for Supporting Dynamic Behavior of Web Services Applications. In: Proc. of the 3rd International Conference on Autonomous Agents and Multi-Agent Systems (AAMAS), New York, USA (2004)
12. Pavón, J., Gómez-Sanz, J., Fuentes, R.: The INGENIAS Methodology and Tools. In: Henderson-Sellers, B., Giorgini, P. (eds.) Agent-Oriented Methodologies, pp. 236–276. Idea Group Publishing (2005)
13. Valencia-García, R., Castellanos-Nieves, D., Fernández-Breis, J.T., Vivancos-Vicente, P.J.: A methodology for extracting ontological knowledge from spanish documents. In: Gelbukh, A. (ed.) CICLing 2006. LNCS, vol. 3878, pp. 71–80. Springer, Heidelberg (2006)
14. Kuter, U., Sirin, E., Parsia, B., Nau, D., Hendler, J.: Information gathering during planning for web service composition. Journal of Web Semantics 3(2-3), 183–205 (2003)

Distributed Graphs Transformed by Multiagent System

Leszek Kotulski

Department of Automatics, AGH Univeristy of Science and Technology
kotulski@agh.edu.pl

Abstract. Graph transformations are a powerful notation formally describing different aspects of the software systems. Multiagent systems are one of the most promising ways to introduction of the parallel computation. The reason of difficulties in joining these approaches was centralized way of the graph representation. The GRADIS framework offers the possibility of splitting the graph describing the problem onto a few partial graphs, that can be maintained in different places. Moreover, the GRADIS distributed environment makes the application of old rules possible, in order to modify these set of partial graphs. Basing on this framework, we show how to introduce the multiagent system and we show an example estimation of improvement time complexity of the multiagent system in comparison to the centralized one.

1 Introduction

Agents based system technology has generated a lot of excitement in recent years because of its promise as a new paradigm for conceptualizing, designing, and implementing software systems. The characteristics of the Multiagent Systems are [15]:

1. assumption that each agent has incomplete information or capabilities for solving the problem and, thus, a limited point of view;
2. there is no global system control;
3. data are decentralized;
4. computations are asynchronous.

Graph transformations are a powerful notation formally describing different aspects of the software systems. This area has been continuously and systematically developed for over 40 years [14,6,7]. Most of the propositions are based on the centralized graph representation. This put multiagent systems and graph transformations into opposite sides. It seams to be interested to fill the gap between them.

The GRADIS framework [13] offers the possibility of splitting the graph, describing the problem into a few partial graphs, that can be maintained in different places. Moreover, the GRADIS distributed environment makes the application of old rules (defined for the centralized approach) possible, in order to modify these

set of partial graphs. Let us notice that the set of Local Graph Transformation Systems fulfill all the aspects of the Multiagent System characteristic.

The scope of the paper is following. In section two - the form of the distributed complementary graph proposed by the GRADIS framework is introduced for labelled graphs. In section three the concept of applying by agents the graph transformation rules to these environment is considered. Finally the usefulness of multiagent system solution is discussed, especially in the context of an improvement of the computational efficiency.

2 Distributed Graph Complementary Form

In this paper we will consider labelled (attributed) graphs. Let Σ^v and Σ^e be a sets; the elements of Σ^v are used as node labels and the elements of Σ^e are used as edge labels. The graph structure is defined as follows:

Definition 1. *A (Σ^v,Σ^e)-graph is a triple (V, E ,v-lab) where V is nonempty set nodes, $E \subseteq V \times \Sigma^e \times V$ is a set of edges, and v-lab:$V \to \Sigma^v$ is a node labelling function.*

One can extend this graph definition e.g. by introduction attributing functions for both nodes and edges, but these extensions will not influence on the rules of the centralized graph distribution and their transformation, because of that they will not be considered here.

Our intention is to split a given graph G into a few parts and to distribute them to different locations. Transformation of each subgraph G_i will be controlled by its Local Graph Transformation System ($LGTS_i$); each of LGTS's can be treated as an agent that cooperates with multiagent environment to achieve a common goal using Node Label Controlled graph grammars [12] .

To maintain the compatibility between the centralized graph G and the set of distributed graphs, some nodes (called border nodes) should be replicated and placed in the proper partial graphs. Graphically, we will mark a border node by a double circle; we also introduce the set Border(G_i) of all border nodes of the given graph G_i. During the splitting of the graph we check whether an edge connecting two nodes crosses a border among the subgraphs.

The function PathS(G,v,w) returns all sets of the nodes belonging to the edges creating any acyclic connection in G between v and w. For example, for the graph G presented in figure 1A PathS(G,a,c)={{a,c},{a,b,c}, {a,d,e,c}, {a,d,f,g,e,c}}. GRADIS framework offers the way of splitting the graph G onto the set of partial graphs (that are in the complementary form – see definition 2). Let's note that this splitting is not dependent from the type of graph grammar supporting graph transformations. The split graphs can be autonomously transformed by a graph transformation system defined for the support of the centralized graph transformations according to the rules (1 and 2) introduced in section 3.

Definition 2. *Set of partial graphs $G_i=(V_i,E_i,v\text{-}lab_i)$, for i=1..k, is the complementary form of graph G iff there exists a set of injective homomorphisms s_i from G_i to G such as:*

1. $\bigcup_{i=1..k} s_i(G_i) = G$
2. $\forall i,j = 1..k \ \ (s_i(V_i) \cap S_j(V_j)) = (s_i(Border(G_i)) \cap s_j(Border(G_j)))$
3. $\forall w \in V_i \ \forall v \in V_j : \exists p \in PathS(G,w,v) \Rightarrow \exists b \in Border(G_i) : s_i(b) \in p$
4. $\forall j = 1..k \ v \in Border(G_j) \iff (\exists w \in G_j : w \ is\_connected\_with \ v)$
   or $G_j = \{v\}$

Algorithm 1
Let H be a subgraph of G, two partial graphs H' and H" are created in the following steps:

1. initially H'=H and H"=G-H;
2. for every v∈H such that there exists a node w∈G-H which is a neighbor of v in G: we replicate v (denoting a copy of v as v') and :
   (a) we keep v∈H' and mark it as a border node
   (b) we attach a node v' to H" with all edges connecting v with the nodes belonging to G-H and we mark it as a border node,
   (c) proper references are added to v and v', they allows us to designate the set of all replicas of v, (iff v has been a border node before this operation these references are also updated in all its previous replicas);
3. optionally, some reindexation of $V_{H'}$ and $V_{H''}$ can be made for optimizing local transformations.

The unique addressing of nodes in the glued graph (after 3-rd step of the algorithm 2.1 or any sequence of local graph transformations) is guaranteed by the remembering their indices as the pairs of the form (local_graph_id, local_index). We also assume that marking the node as a border one is associated with designation for it of a unique index in the border nodes set (border_id is equal to 0). Splitting graph the G into a few new partial graphs can be accomplished by the recursive execution of the algorithm 2.1 on the already obtained partial graphs H' or H" and so on. An example of G and the set of two distributed partial graphs are presented on figures 1B and 1C respectively.

Algorithm 2
The reconstruction of G from the set of partial graphs, $\{G_i\}$ is made in the following way:

1. for border nodes, one of the replicas, indexed as (0,gbni), where gbni is anacronim of the global border node index, is added to V,
2. all normal (not border) nodes are added to V, with their local indexation,
3. edges in E are inherited from the local graphs (if one node of an edge is a border node in the final edge its global representative appears).
4. labelling function *lab* is the union of lab_i.

Fig. 1A represents the graph that is reconstruction from the graphs presented in fig. 1B, with the indexation introduced by the algorithm 2.

For any border node v in graph G_i we can move boundary in such a way, that all nodes (located inside another partial graph), that are connected with v are

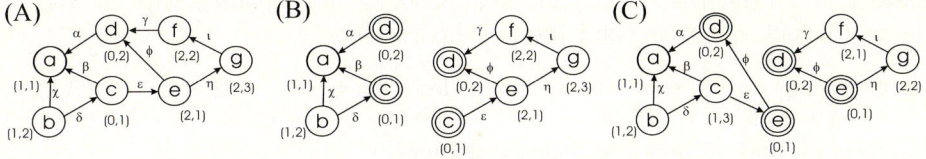

Fig. 1. Centralized and complementary partial graphs

incorporated to G_i as a border nodes and replicas of v are removed from another graphs (i.e. v becomes a normal node); we will do it using incorporate(v,i) operation. For graphs presented in fig. 1B an incorporate((0,1),1) operation leads us to the graphs presented in fig. 1C.

Graph grammars provide a mechanism in which the local transformations on a graph can be modelled in a mathematically precise way. A graph grammar consist of a finite set of productions having a form of triples (L,R,E) where L and R are the graphs (left- and right-hand side graph of production, respectively) and E is some embedding mechanism. The left-hand side graphs should be matched to some subgraph of G, in such a way that the graphical structure and labels are preserved; formally we will describe this matching by (iso)morphism m. Such a production can be applied to graph G whenever there is an m occurrence of L in G. It is applied by removing m(L) from G, replacing it by (an isomorphic copy) of R, and finally using embedding mechanism E to attach R to the remainder G-m(L).

3 Multiagent Cooperation

Graph grammars provide a mechanism in which the local transformations on a graph can be modelled in a mathematically precise way. A graph grammar consist of a finite set of productions having a form of triples (L,R,E) where L and R are the graphs (left- and right-hand side graph of production, respectively) and E is some embedding mechanism. The left-hand side graphs should be matched to some subgraph of G, in such a way that the graphical structure and labels are preserved; formally we will describe this matching by (iso)morphism m. Such a production can be applied to graph G whenever there is an m occurrence of L in G. It is applied by removing m(L) from G, replacing it by (an isomorphic copy) of R, and finally using embedding mechanism E to attach R to the remainder G-m(L).

The proposed idea of applying a production p:L→ (R,E) to a distributed graph can be described by two rules:

Rule 1. each agent autonomously decides to apply this production when three conditions are fulfilled:

1. an m-th occurrence of L inside G_i can be found;
2. none of border nodes are removed;
3. all removed edges belong to G_i.

Rule 2. when the first rule cannot be applied, agent cooperates with the rest of the multiagent system in the following three steps:

1. gathering an information: for all nodes v such that $v \in \text{Border}(G_i) \cap V_{m(L)}$ agent is looking for such minimal k that the graph B =k- neighborhood(v) covers L and all removed nodes and edges.
2. preparing an environment: all nodes belonging to B and not belonging to G_i are incorporated to the graph G_i, as a result of a sequential execution of incorporate(...) operations, made in the transactional mode.
3. applying a production: a production L→(R,E) can be applied in a newly created graph G_i' according to the rule 1 (local derivation).

In the algorithmic approach (often called the connecting one) in an embedding transformation we consider edges between nodes belonging to the m(L) subgraph and the reminder G-m(L) graph. This type of embedding transformation bases on set of these edges and describes how to generate a new set of edges, that will connect the G-m(L) graph with the graph R of right hand side of production p.

Formally, an embedding transformation in edge labelled directed graph with Neighborhood Controlled Embedding (edNCE) grammars over the Σ and Γ is defined by a connection relation $E \subset \Sigma \times \Gamma \times \Gamma \times V_G \times \{in, out\}$. A connection instruction $(\sigma, \alpha, \beta, v, d) \in E$ will be interpreted, in a context of the production L→R, as follows: for any node labelled as σ such that it was connected with any node belonging to m(L) (before m(L) has been removed) by an edge labelled by α a new edge connecting this node with $v \in V_R$ is created; this edge is labelled by β, and d points this edge direction.

Applying a production p:L→(R,E) in a distributed environment depends on the type of a grammar. In most of applicable graph grammars we assume that L consists of a single node u (see [5] for details). That simplifies application of such productions in a distributed environment of the partial graphs. For any G_i if u belongs to Border(G_i) then G_i'=incorporate(u,i) covers B= 1- neighborhood(u) so the production p can be applied locally.

Let us consider an example of application of the production presented in figure 2A with the connection relation

E= { $(a,\beta,\beta,1,in), (b,\delta, \delta,1,out),(b,\delta,\kappa,2,out),(e,\varepsilon,\kappa,2,in)$ }

to the graph presented in figure 1B. Node labelled by c is a border node in the graph G_1 so we have to incorporate it to G_1 before applying p– new form of G_1 is presented in fig. 2B.

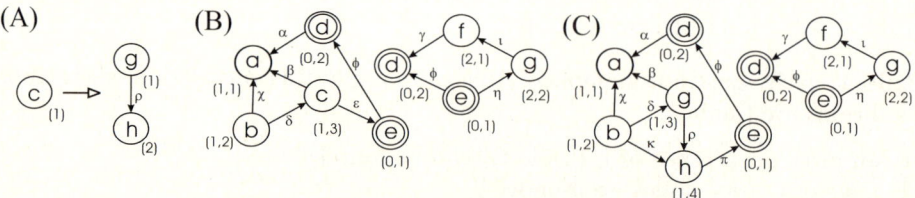

Fig. 2. Production p and graph G before and after its application

Now the graph consisting of node (1,3), that is m isomorphic to left-hand side of the production, is removed from G_1 and three dangling edges appear. The edge labelled β coming to the node labelled by a is replaced by the same edge but attached to the node 1 of the right-hand side of production; the edge labelled δ coming from the node labelled by b is replaced by two edges: one that is attached to the node 1 with the same labelling and the second that is attached to node 2 with κ labelling; the edge labelled ε coming to the node labelled by e is replaced by an edge attached to the node 2 with π labelling.

Let us consider the complexity of the presented algorithm; each of these incorporate operations needs at most cooperation with all distributed environments ($\sharp d$); thus the algorithm complexity is $O(\sharp d)$.

One of the most significant features of the Multiagent Systems is their ability to solve the problems that are too large for the centralized agent to solve, because of resource limitation or sheer the risk of having one centralized system that could be a performance bottleneck or could fail at critical times. Its computational complexity grows, because the concurrency of computation is exploited.

Let us consider the problem of finding given subpattern inside a pattern. Following the Flasinski [9,10] we assume that the pictorial pattern recognition can be described with the help of hierarchical structural representation. The proposed ETPL(k) graph grammars are able to solve a membership problem with $O(n^2)$ time complexity, where n is the number nodes in the considered graph.

The simplest (but probably not more efficient idea) is to check all subgraphs of the same dimension as the pattern (k) in their correspondence to this patter. In this case we have to check graphs; we are still in polynomial complexity (because $\binom{n}{k}$ is less then n^k) but in case on graph size measured in hundreds of nodes it is impossible to make a computation.

However we can decentralize our solution by splitting the graph into set of partial graphs that will be maintained by the agents; each agent maintains a graph of about 50 nodes so it makes this computation in ten seconds. The problem appears in finding patterns belonging to the two or more subgraphs (i.e. divided by subgraphs borders). We solve this problem by creating for each border node (pointed as RNR) a subgraph with replicas of nodes, that distance from it is less than the maximal distance between any two nodes in the pattern generated by the given grammar (see [11] for details).

The table 1 presents the estimation of the run time of both of the centralized and multiagent solution on the following assumptions:

- computations are made on the computers with the 1000 GHz processor,
- $O(n^2)$ computational complexity has been estimated by $10*n^2$ processor operations,
- patterns represented by a graph grammar are limited to 10 nodes.
- number of border nodes has been estimated on assumption that the border nodes are near at the same distance from the starting point, so the proportion of the border nodes to the internal nodes is the same as in the circle.

Table 1. Estimation of the subpattern searching time

Graph dimension	Centralized solution	Agent solutions - parallel					
		Subgraph dimension=50		Subgraph dimension=40		Subgraph dimension=30	
		Agent comput.	Number of agents	Agent comput.	Number of agents	Agent comput.	Number of agents
200	7 months	10 sec.	4+25	0.9 sec	5+31	0.03 sec.	7+40
250	7 years	10 sec.	5+35	0.9 sec	7+42	0.03 sec.	9+53
300	45 years	10 sec.	6+40	0.9 sec	8+53	0.03 sec.	10+77

– we skip the evaluation of the initial partial graph designation, that does not depends on the graph size (for one thousand nodes is less the one second) but depends more on the speed of network connection (time of the allocation of partial graphs to the distributed agents).

Lets note, that two first assumptions, can change this evaluation only in a linear way, so they do not change the proportions between centralized and agents computation presented in the table 1. The case of the border node estimation is also not critical, because their number is limited by the number of the nodes in the centralized graph.

To sum up, even when agent calculations will be performed sequentially on one computer, finding all patterns in 300 nodes graph by 46 agents (6 agents maintaining 50 elements graphs + 40 agents checking replicas created for RNR element, that should be less then 50 nodes) takes 460 seconds instead of 45 years.

4   Conclusions

It is obviously known that multiagent environment due to parallel agents processing improves the speed of the problem solving. Sometimes in the spectacular way, like in the finding subpatterns problem presented above, that makes this computation sensible in practice. However, always the key problem is the way of the problem distribution and designation of the rules of the agent's cooperation. GRADIS framework solves these problems in the case when systems are represented by graphs and they are modified by agents with help of graph transformations. The centralized graph can be split using algorithm 2.1, and each agent behavior is represented by incorporate operation and the transformation rules of the centralized system restricted to the local partial graph maintained by this agent. It seems to be promising idea for searching the new area of agents solutions, based on the algorithmic graph transformation approach.

The concept of distribution (to specify a concurrent and distributed system) is considered in the algebraic approach. The distributed graph transformation [14] was developed to naturally express computations in a system composed of interacting parts. In general, there is assumed the existence of some global state, represented as a graph, that is used to synchronize execution of the amalgamated productions or synchronized productions. In [2], a distributed attributed graph (Distr(AGr)) has two levels: a network and a local level. Network nodes are assigned to the local graphs, which represent their state. These graphs are transformed via distributed rules. They consist of a network rule and a set of local rules, one for each node of network rule. In the double push-out approach, *dangling* and *identification* conditions, two additional *locality* conditions should be verified [8,16]. The satisfying of the *locality* conditions by some local graph morphism assures that the push-out can be constructed component wise. There are a few examples of usefulness of Distr(AGr) in visual design of distributed system [8], specifying integrated refactoring [1], modelling and analysis of a distributed simulation protocols [3]. The problem of evaluation of distributed rules and graphs in a real distributed environment was not considered until now (the last implementation [13] is made in a centralized environment with help of AToM tool [4]).

These concepts seems to be less applicable to multiagent solutions than the GRADIS approach.

References

1. Bottoni, P., Parisi-Presicce, F., Taentzer, G.: Specifying integrated refactoring with distributed graph transformations. In: Pfaltz, J.L., Nagl, M., Böhlen, B. (eds.) AGTIVE 2003. LNCS, vol. 3062, pp. 220–235. Springer, Heidelberg (2004)
2. Corradini, A., Montanari, U., Rossi, F., Ehrig, H., Heckel, R., Löwe, M.: Algebraic approaches to graph transformation - part i: Basic concepts and double pushout approach. In: Rozenberg (ed.) [14], pp. 163–246.
3. de Lara, J., Taentzer, G.: Modelling and analysis of distributed simulation protocols with distributed graph transformation. In: ACSD, pp. 144–153. IEEE Computer Society (2005)
4. de Lara, J., Vangheluwe, H.: Atom^3: A tool for multi-formalism and meta-modelling. In: Kutsche, R.-D., Weber, H. (eds.) ETAPS 2002 and FASE 2002. LNCS, vol. 2306, pp. 174–188. Springer, Heidelberg (2002)
5. Ehrenfeucht, A., Harju, T., Rozenberg, G.: 2-structures - a framework for decomposition and transformation of graphs. In: Rozenberg [14], pp. 401–478.
6. Ehrig, H., Engels, G., Kreowski, H.-J., Rozenberg, G. (eds.): Handbook of graph grammars and computing by graph transformation. applications, languages, and tools, vol. 2. World Scientific Publishing Co., Inc., River Edge (1999)
7. Ehrig, H., Kreowski, H.-J., Montanari, U., Rozenberg, G. (eds.): Handbook of graph grammars and computing by graph transformation. concurrency, parallelism, and distribution, vol. 3. World Scientific Publishing Co., Inc., River Edge (1999)
8. Fischer, I., Koch, M., Taentzer, G., Volle, V.: Distributed graph transformation with application to visual design of distributed systems. [7], 269–340
9. Flasiński, M.: Distorted pattern analysis with the help of node label controlled graph languages. Pattern Recogn. 23(7), 765–774 (1990)
10. Flasinski, M.: On the parsing of deterministic graph languages for syntactic pattern recognition. pattern recognition 26(1), 1–16 (1993)

11. Kotulski, L.: On the efficient finding subpatterns with the help of etpl(k) graphs grammar (in preparation)
12. Kotulski, L.: Supporting software agents by the graph transformation systems. In: Alexandrov, V.N., van Albada, G.D., Sloot, P.M.A., Dongarra, J. (eds.) ICCS 2006. LNCS, vol. 3993, pp. 887–890. Springer, Heidelberg (2006)
13. Kotulski, L.: On the distribution graph transformations. Reports of Department Automatics AGH, 2/2007
14. Rozenberg, G. (ed.): Handbook of graph grammars and computing by graph transformation. foundations, vol. I. k World Scientific Publishing Co., Inc., River Edge (1997)
15. Sycyra, K.P.: Multiagent system. AI MAgazine, 79–92 (1998)
16. Taentzer, G.: Distributed graphs and graphs transformation. In: Applied Categorical Structures, Special Issues on Graph Transformations, vol. 7 (1999)

Multi-agent Logics with Interacting Agents Based on Linear Temporal Logic: Deciding Algorithms

Vladimir Rybakov

Department of Computing and Mathematics,
Manchester Metropolitan University,
John Dalton Building, Chester Street, Manchester M1 5GD, U.K.
Siberian Federal University, Krasnoyarsk, Russia
V.Rybakov@mmu.ac.uk

Abstract. We introduce[1] a multi-agent logic $\mathcal{MA}_{\mathcal{LTL}}$ – a variant of the linear temporal logic LTL with embedded multi-agent knowledge with interacting agents. The logic is motivated by semantics based on potentially infinite runs with time points represented by clusters of states with distributed knowledge of the agents. We address properties of local and global knowledge modeled in this framework, consider modeling of interaction between agents by possibility to puss information from one agent to others via possible transitions within time clusters of states. Main question we are focused on is the satisfiability problem and decidability of the logic $\mathcal{MA}_{\mathcal{LTL}}$. Key result is proposed algorithm which recognizes theorems of $\mathcal{MA}_{\mathcal{LTL}}$ (so we show that $\mathcal{MA}_{\mathcal{LTL}}$ is decidable). It is based on verification of validity for special normal reduced forms of rules in models with at most triple exponential size in the testing rules. In the final part we discuss possible variations of the proposed logic.

Keywords: multi-agent logic, hybrid logics, knowledge based reasoning, relational Kripke/Hintikka models, decidability algorithms.

Multi agent logic nowadays is one of active areas in AI and Knowledge Representation (cf. Bordini et.al.[3], Dix et. al. [6], Hoek et.al. [14], Fisher [9], Hendler [16], Kacprzak, [18] Wooldridge [41,42]). Agent is an autonomous entity that acts in the world, interacting with its environment and with other agents. Multi-Agent Systems - systems that may include many such entities - are becoming more and more popular in Computer Science and Artificial Intelligence. First, they offer a paradigm for modeling the reality around us as a world inhabited by such autonomous, active, possibly intelligent elements. Second, they can be used as a methodology that enables design and implementation of large systems in a really modular way. In a sense, multi agent logics came from a particular application of multi-modal logics to reasoning about knowledge (cf. Fagin et.al [8,7], Halpern and Shore [15]), where modal-like operations K_i responsible for knowledge of

---

[1] This research is supported by Engineering and Physical Sciences Research Council (EPSRC), U.K., grant EP/F014406/1.

individual agents were embedded to the language of propositional logic. These operations were intended to model effects and properties of agents knowledge in changing environment and logical inference, tools devoted to derivation of new knowledge from known facts. The exact nature of the agents is a matter of some controversy. They are sometimes claimed to be autonomous. For example a household floor cleaning robot can be autonomous in that it is dependent on a human operator only to start it up. On the other hand, in practice, all agents are under active human supervision. Furthermore, the more important the activities of the agent are to humans, the more supervision that they receive. In fact, autonomy is seldom desired; instead interdependent systems are needed.

Therefore it seems to be relevant to work out efficient tools to work with interacting agents. Besides, it is an open question which basic logic can be put in the background to develop multi-agent systems. There are many approaches, but, if the logic used as the basis in very expressive - the undecidability phenomenon can happen (cf. Kacprzak [18] with reduction the decidability to the domino problem). If the basic logic is just boolean logic, and agent are autonomous, decidability for standard systems usually follows reasonably easy (cf. [8,7]). If we wont decidability then, still, the background logic must me expressive enough to handle properties interesting for AI and CS community. We choose as background the linear temporal logic LTL.

Temporal logics are currently the most widely used specification formalism for reactive systems. They were first suggested to be used for specifying properties of programs in late 1970's (cf. Pnueli [21]). The temporal framework most used is linear-time propositional temporal logic LTL, which has been studied from various viewpoints of its application (cf. e.g. Manna and Pnueli [19,20], Clark E. et al., [4]). Temporal logic has numerous applications to safety, liveness and fairness, to various problems arising in computing (cf. Eds. Barringer, Fisher, Gabbay and Gough, [1]). Model checking for LTL formed a direction in logic in computer science, which uses, in particular, applications of automata theory (cf. Vardi [5,39]). Temporal logics itself can be considered as a special case of hybrid logics, e.g. as a bimodal logic with some laws posed on interaction of modalities to imitate the flow of time. Mathematical theory devoted to study of various aspects of interaction for temporal operations (e.g. axiomatizations of temporal logics) and to construction of effective semantic theory based on Kripke/Hintikka-like models and temporal Boolean algebras, formed a highly technical branch in non-classical logics (cf. e.g. van Benthem [38,37], Goldblatt [11], Gabbay and Hodkinson [13], Hodkinson [17]).

In this paper we make an attempt to develop technique for implementing multi-agent knowledge logics with interacting agents to linear temporal logic LTL. We propose a multi-agent logic $\mathcal{MA}_{\mathcal{LTL}}$ (described briefly above in the abstract) and address the problems of satisfiability and decidability for $\mathcal{MA}_{\mathcal{LTL}}$. Our paper suggests an algorithm which recognizes theorems of $\mathcal{MA}_{\mathcal{LTL}}$ (so we show that $\mathcal{MA}_{\mathcal{LTL}}$ is decidable). This algorithm is based at reduction of arbitrary formulas in the language of $\mathcal{MA}_{\mathcal{LTL}}$, first, to inference rules, then to special normal reduced forms of this rules, and then at checking of validity of these rules at (special) bent models with size efficiently bounded from the size of the rules. We show also that some possible cases of definition for local knowledge and global knowledge (in this framework) are expressible via time operations of kind Until, and discuss possible variations of the implemented logical language.

1 Preliminaries, Definitions, Notation

Base of the language for the multi-agent logic we consider is the Linear Temporal Logic LTL. Linear temporal logics differ from typical modal logics by presence of specific temporal operations Until and Next which cannot be expressed in standard modal language. We start by introduction of semantic objects which upon our logic will be based. They are the following Kripe/Hintikka-like models with linear discrete time. The frame $\mathcal{N}_C := \langle \bigcup_{i \in N} C(i), R, R_1, \ldots R_m, Next \rangle$ is a tuple, where N is the set of natural numbers, C(i) are some nonempty sets, $R, R_1, \ldots R_m$ are binary accessibility relations imitating possible agents transitions. For all elements a and b from $\bigcup_{i \in N} C(i)$, $aRb \Leftrightarrow [a \in C(i)$ and $b \in C(j)$ and $i < j]$ or $[a, b \in C(i)$ for some $i]$; any R_j is a reflexive, transitive and symmetric relation, and $\forall a, b \in \bigcup_{i \in N} C(i)$, $aR_j b \Rightarrow [a, b \in C(i)$ for some i); $a \, Next \, b \Leftrightarrow [\exists i((a \in C(i)) \& (b \in C(i+1)))]$. These frames are intended to model the reasoning (or computation) in discrete time, so each $i \in N$ (any natural number i) is the time index for the cluster of states arising after the step in current computation. Any $C(i)$ is a finite set of all possible states in the time point i, and R models discrete current of time. Relations R_j are intended to model knowledge of agents for any current time point i in the cluster of states $C(i)$. So, as usually, any R_j is supposed to be $S5$-like relation, i.e an equivalence relation at any $C(i)$. We suppose the reasonong/computation to be simultaneous and parallel - after a step, new cluster of possible states appears, and agents will be given new access rules to the information in this time cluster of states. However, the agents cannon predict, which access rules they will have (that is why we, in particular, do not use *nominals*). The *Next* relation is the standard one – it describers all states available in the next time point cluster.

For description of reasonong/computation modeled in structures $\mathcal{N}_C$ we propose the following language (based on extended language of LTL and a language for agents knowledge logic with interaction). It uses the language of the standard LTL (which extends the language of Boolean logic by operations **N** (next), **U** (until)) and the new operations $\mathbf{U}_w$ (weak until) and $\mathbf{U}_s$ (strong until). We also employ the language of the agents knowledge logic, which extends the language of Boolean logic by modal-like unary operations $\mathbf{K}_j$, $1 \leq j \leq m$ and additional unary operations $\mathbf{CK_L}$, $\mathbf{CK_G}$ and $\mathbf{IntK}$. Formation rules for formulas are as usual. The intended meaning of the operations is as follows.

$\mathbf{K}_j \varphi$ means the agent j *knows* φ in the current state of a time cluster.
$\mathbf{CK_L} \varphi$ means that φ is a *local common knowledge* in the in the current state of a time cluster, that all agents knows φ.
$\mathbf{CK_G} \varphi$ means φ is a *global common knowledge* in the in the current state of a time cluster (i.e. that since now all agents will know it always).

IntKφ means that in the current state φ may be known by
interaction between agents.
Nφ has meaning φ holds in the *next time cluster* of states (state);
$\varphi \mathbf{U} \psi$ can be read: φ holds until ψ will hold;
$\varphi \mathbf{U}_w \psi$ has meaning φ weakly holds until ψ will hold;
$\varphi \mathbf{U}_s \psi$ has meaning φ strongly holds until ψ will hold;

As in standard definition of Kripke/Hintikka models on frames, for any collection of propositional letters $Prop$ and any frame $\mathcal{N}_C$, a valuation in $\mathcal{N}_C$ is a mapping which assigns truth values to elements of $Prop$ in $\mathcal{N}_C$. Thus, for any $p \in Prop$, $V(p) \subseteq \mathcal{N}_C$. We will call $\langle \mathcal{N}_C, V \rangle$ a model (a Kripke/Hinikka model).

For any such model $\mathcal{M}$, the truth values are extended from propositions of $Prop$ to arbitrary formulas as follows (for $a \in \mathcal{N}_C$, we denote $(\mathcal{N}_C, a) \Vdash_V \varphi$ to say that the formula φ is true at a in $\mathcal{N}_C$ w.r.t. V). The rules for computation truth values are:

$\forall p \in Prop$, $(\mathcal{M}, a) \Vdash_V p \Leftrightarrow a \in V(p)$;

$(\mathcal{M}, a) \Vdash_V \varphi \wedge \psi \Leftrightarrow (\mathcal{M}, a) \Vdash_V \varphi \wedge (\mathcal{M}, a) \Vdash_V \psi$;

$(\mathcal{M}, a) \Vdash_V \neg \varphi \Leftrightarrow not[(\mathcal{M}, a) \Vdash_V \varphi]$;

$(\mathcal{M}, a) \Vdash_V \mathbf{K}_j \varphi \Leftrightarrow \forall b[(a \ R_j \ b) \Rightarrow (\mathcal{M}, b) \Vdash_V \varphi]$.

So, $\mathbf{K}_j \varphi$ says that φ holds in all states available for the agent j.

$(\mathcal{M}, a) \Vdash_V \mathbf{IntK} \varphi \Leftrightarrow \exists a_{i1}, a_{i2}, \ldots, a_{ik} \in \mathcal{M}$

$$[aR_{i1}a_1 R_{i2}a_2 \ldots R_{ik}a_k \ \& \ (\mathcal{M}, a_k) \Vdash_V \varphi].$$

Thus, φ may be known by interaction between the agents if there is a path of transitions by agents accessibility relations which leads to a state where φ holds.

$(\mathcal{M}, a) \Vdash_V \mathbf{CK_L} \varphi \Leftrightarrow \forall j \forall b[(a \ R_j \ b) \Rightarrow (\mathcal{M}, b) \Vdash_V \varphi]$.

So, φ is of *local common knowledge* if it holds in all states which are accessible in the *current time point* for any agent.

$(\mathcal{M}, a) \Vdash_V \mathbf{CK_G} \varphi \Leftrightarrow \forall b[(a \ R \ b) \Rightarrow (\mathcal{M}, b) \Vdash_V \varphi]$.

Thus, φ is of *global common knowledge* if it holds in all states in all future (and current) time clusters. This definition differs from standard meaning of just common knowledge (it does not refer to agents) but is seemed to be meaningful and plausibly justified.

$(\mathcal{M}, a) \Vdash_V \mathbf{N} \varphi \Leftrightarrow \forall b[(a \ Next \ b) \Rightarrow (\mathcal{M}, b) \Vdash_V \varphi]$;

$(\mathcal{M}, a) \Vdash_V \varphi \mathbf{U} \psi \Leftrightarrow \exists b[(aRb) \wedge ((\mathcal{M}, b) \Vdash_V \psi) \wedge$

$$\forall c[(aRcRb)\&\neg(bRc){\Rightarrow}(\mathcal{M},c)\Vdash_V\varphi]];$$

$$(\mathcal{M},a)\Vdash_V\varphi\mathbf{U}_w\psi \Leftrightarrow \exists b[(aRb) \wedge ((\mathcal{M},b)\Vdash_V\psi) \wedge \forall c[(aRcRb)\&\neg(bRc)\&$$

$$\&(c \in C(i)) \Rightarrow \exists d \in C(i)(\mathcal{M},d)\Vdash_V\varphi]];$$

$$(\mathcal{M},a)\Vdash_V\varphi\mathbf{U}_s\psi \Leftrightarrow \exists b[(aRb) \wedge b \in C(i) \wedge \forall c \in C(i)((\mathcal{M},c)\Vdash_V\psi)\wedge$$

$$\forall c[(aRcRb)\&\neg(bRc){\Rightarrow}(\mathcal{M},c)\Vdash_V\varphi]];$$

Definition 1. *For a Kripke structure $\mathcal{M} := \langle \mathcal{N}_C, V\rangle$ and a formula φ, we say that (i) φ is satisfiable in $\mathcal{M}$ (denotation – $\mathcal{M}\Vdash_{Sat}\varphi$) if there is a state b of $\mathcal{M}$ ($b \in \mathcal{N}_C$) where φ is true: $(\mathcal{M},b)\Vdash_V\varphi$. φ is valid in $\mathcal{M}$ (denotation – $\mathcal{M}\Vdash\varphi$) if, for any b of $\mathcal{M}$ ($b \in \mathcal{N}_C$), the formula φ is true at b ($(\mathcal{M},b)\Vdash_V\varphi$).*

Definition 2. *For a Kripke frame $\mathcal{N}_C$ and a formula φ, we say that φ is satisfiable in $\mathcal{N}_C$ (denotation $\mathcal{N}_C\Vdash_{Sat}\varphi$) if there is a valuation V in the frame $\mathcal{N}_C$ such that $\langle\mathcal{N}_C,V\rangle\Vdash_{Sat}\varphi$. φ is valid in $\mathcal{N}_C$ (i.e. $\mathcal{N}_C\Vdash\varphi$) if not($\mathcal{N}_C\Vdash_{Sat}\neg\varphi$).*

Definition 3. *The logic $\mathcal{MA}_{\mathcal{LTL}}$ is the set of all formulas which are valid in all frames $\mathcal{N}_C$.*

Thus, a formula φ in the language of $\mathcal{MA}_{\mathcal{LTL}}$ is satisfiable iff there is a valuation V in the Kripke frame $\mathcal{N}_C$ which makes φ satisfiable: $\langle\mathcal{N}_C,V\rangle\Vdash_{Sat}\varphi$. Obviously that a formula φ is satisfiable iff $\neg\varphi$ is not a theorem of $\mathcal{MA}_{\mathcal{LTL}}$: $\neg\varphi \notin \mathcal{MA}_{\mathcal{LTL}}$, and vise versa, φ is a theorem of $\mathcal{MA}_{\mathcal{LTL}}$ ($\varphi \in \mathcal{MA}_{\mathcal{LTL}}$) if $\neg\varphi$ is not satisfiable. We may observe that, in the rules above, the treatment of $\mathbf{U}$ is slightly different from standard one – it is sufficient for ψ to be true at least in one state of the achieved current time cluster. The operation $\mathbf{U}_w$ more drastically differs from the standard $\mathbf{U}$, – it is sufficient for φ to be true only in a certain state of all time clusters before ψ will true at a state. And the strong until – $\varphi\mathbf{U}_s\psi$ – means that there is a time point i, where the formula ψ is true at all states in the time cluster $C(i)$, and φ holds in all states of all time points j proceeding i. Using operations $\mathbf{U}$ and $\mathbf{N}$ we can define all standard temporal and modal operations. For instance, modal operations may be trivially defined:

Lemma 1. *The following holds:*

(i) $\Diamond\varphi \equiv true\mathbf{U}\varphi \in \mathcal{MA}_{\mathcal{LTL}}$;

(ii) $\Box\varphi \equiv \neg(true\mathbf{U}\neg\varphi) \in \mathcal{MA}_{\mathcal{LTL}}$.

The temporal operation $\mathbf{F}\varphi$ (φ *holds eventually*, which, in terms of modal logic, means φ is possible (denotation $\Diamond\varphi$)), can be described as $true\mathbf{U}\varphi$. The temporal operation $\mathbf{G}$, where $\mathbf{G}\varphi$ means φ *holds henceforth*, can be defined as $\neg\mathbf{F}\neg\varphi$. Using these derived logical operations we can easy describe common knowledge operations accepted in the current framework:

Lemma 2. *The following holds:*

(i) $\mathbf{CK}_G \varphi \equiv \Box \varphi \in \mathcal{MA_{LTL}}$;

(ii) $\mathbf{CK}_L \varphi \equiv \bigwedge_{1 \leq i \leq n} (\mathbf{K}_i \varphi) \in \mathcal{MA_{LTL}}$.

Thus the initially specified language for $\mathcal{MA_{LTL}}$ is a bit *superfluous* and we can omit operations for local and global common knowledge because they are expressible via others. The standard temporal operations together with knowledge operations add more expressive power to the language. For instance, the formula $\Box \neg K_1 \neg \varphi$ says that, for any future time cluster and for any state a of this cluster the knowledge φ is *discoverable* for agent 1, it has access to a state b where φ holds. The new temporal operations $\mathbf{U}_s$ and $\mathbf{U}_s$ brings new unique features in the language. For instance the formula $\Box_w \varphi := \neg(\top \mathbf{U}_s \neg \varphi)$ codes the *weak necessity*, it says that in any time cluster $C(i)$ there is a state where φ is true. The formula $\neg(\varphi \mathbf{U}_w \Box \varphi) \wedge \Diamond \Box \varphi$ says that, there is an earliest future time cluster $C(i)$, since which φ will be true in all states of all time clusters, but, before $C(i)$ φ fails in a state of any time cluster. Such properties are problematic to be expressed in standard modal or temporal operations. So, the logic expresses subtle second order properties of systems, and we devote the rest of the paper to finding an efficient algorithm to check satisfiability in $\mathcal{MA_{LTL}}$ (to show that $\mathcal{MA_{LTL}}$ is decidable).

2 Key Results, Decidability Algorithm

We will implicitly model the universal modality, which will be useful instrument (but we did not put the universal modality in the language to avoid overcomplicated language, which may bring difficulties to show decidability). This will allow to make shorter proofs and to avoid any reasoning by induction of the length of formulas, and even rid of formulas with nested operations. This approach is based on our previous techniques to handle inference rules (cf. [23] – [34]). Actually here we extend our research at Logic of Discovery in Uncertain Situations presented in [34] and research at a hybrid of LTL and knowledge logic without interaction for agents (autonomous one) reported recently at the workshop on Hybrid Logics (2007, Dublin). We will represent formulas by rules and then convert such rules in a special normal reduced form. Recall, a (sequential) (inference) rule is an expression

$$\mathbf{r} := \frac{\varphi_1(x_1, \ldots, x_n), \ldots, \varphi_m(x_1, \ldots, x_n)}{\psi(x_1, \ldots, x_n)},$$

where $\varphi_1(x_1, \ldots, x_n), \ldots, \varphi_m(x_1, \ldots, x_n)$ and $\psi(x_1, \ldots, x_n)$ are some formulas constructed out of letters $x_1, \ldots, x_n$. Letters $x_1, \ldots, x_n$ are variables of $\mathbf{r}$, we use notation $x_i \in Var(\mathbf{r})$.

Definition 4. *A rule* **r** *is said to be* **valid** *in a Kripke model* $\langle \mathcal{N}_C, V \rangle$ *with the valuation V (we will use notation* $\mathcal{N}_C \Vdash_V \mathbf{r}$*) if*

$$[\forall a \, ((\mathcal{N}_C, a) \Vdash_V \bigwedge_{1 \leq i \leq m} \varphi_i)] \Rightarrow \forall a \, ((\mathcal{N}_C, a) \Vdash_V \psi).$$

Otherwise we say **r** *is* **refuted** *in* $\mathcal{N}_C$*, or* **refuted** *in* $\mathcal{N}_C$ *by V, and write* $\mathcal{N}_C \not\Vdash_V \mathbf{r}$. *A rule* **r** *is* **valid** *in a frame* $\mathcal{N}_C$ *(notation* $\mathcal{N}_C \Vdash \mathbf{r}$*) if, for any valuation V of* $Var(\mathbf{r})$, $\mathcal{N}_C \Vdash_V \mathbf{r}$

For any formula φ we can convert it in the rule $x \to x/\varphi$ and employ the technique of reduced normal forms for inference rules as follows. First, it is trivial that

Lemma 3. *A formula φ is a theorem of* $\mathcal{MA}_{\mathcal{LTL}}$ *iff the rule* $(x \to x/\varphi)$ *is valid in any frame* $\mathcal{N}_C$.

A rule **r** is said to have the *reduced normal form* if $\mathbf{r} = \varepsilon_c / x_1$ where

$$\varepsilon_c := \bigvee_{1 \leq j \leq m} (\bigwedge_{1 \leq i,k \leq n, i \neq k} [x_i^{t(j,i,0)} \wedge (\mathbf{N}x_i)^{t(j,i,1)} \wedge (x_i \mathbf{U} x_k)^{t(j,i,k,0)} \wedge$$

$$(x_i \mathbf{U}_w x_k)^{t(j,i,k,1)} \wedge (x_i \mathbf{U}_s x_k)^{t(j,i,k,2)} \wedge$$

$$\bigwedge_{1 \leq s \leq m} (\neg \mathbf{K}_s \neg x_i)^{t(j,i,m,3)} \wedge \mathbf{Int}\mathbf{K} x_i^{t(j,i,4)}]),$$

and all x_s are certain letters (variables), $t(j,i,z), t(j,i,k,z) \in \{0,1\}$ and, for any formula α above, $\alpha^0 := \alpha$, $\alpha^1 := \neg \alpha$.

Definition 5. *Given a rule* $\mathbf{r_{nf}}$ *in the reduced normal form,* $\mathbf{r_{nf}}$ *is said to be a normal reduced form for a rule* **r** *iff, for any frame* $\mathcal{N}_C$, $\mathcal{N}_C \Vdash \mathbf{r} \Leftrightarrow \mathcal{N}_C \Vdash \mathbf{r_{nf}}$.

Similar to Lemma 3.1.3 and Theorem 3.1.11 from [26], it follows

Theorem 1. *There exists an algorithm running in (single) exponential time, which, for any given rule* **r**, *constructs its normal reduced form* $\mathbf{r_{nf}}$.

Decidability of $\mathcal{MA}_{\mathcal{LTL}}$ will follow (Lemma 3) if we will find an algorithm recognizing rules in reduced normal form which are valid in all frames $\mathcal{N}_C$. The next lemma does not follow by standard filtration technique (because interaction of agents – potentially unbounded paths of combinations of agent's accessibility relations), but it may be proved by a some bit refined one. Actually it is the starting point to handle interactions of agents.

Lemma 4. *A rule* $\mathbf{r_{nf}}$ *in the reduced normal form is refuted in a frame* $\mathcal{N}_C$ *if and only if* $\mathbf{r_{nf}}$ *can be refuted in a frame of kind* $\mathcal{N}_C$ *with time clusters* $C(i)$ *of size at most square polynomial in size of* $\mathbf{r_{nf}}$.

To describe our algorithm further we need the following (special) *bent* finite Kripke models. Take any frame $\mathcal{N}_C$ and some numbers k, m, where $m > k > 1$ and modify $\mathcal{N}_C$ as follows. The frame $\mathcal{N}_C(k, m)$ has the structure: $\mathcal{N}_C(k, m) := \langle \bigcup_{1 \le i \le m} C(i), R, R_1, \ldots R_n Next \rangle$, where R is the accessibility relation from $\mathcal{N}_C$ extended by pairs (x, y), where $x \in C(i), y \in C(j)$) and $i, j \in [n, m]$, so xRy holds for all such pairs. Any relation R_j is simply transferred from $\mathcal{N}_C$, and $Next$ is merely the relation from $\mathcal{N}_C$ extended by

$$\forall a \in C(m) \forall b \in C(k)(a \ Next \ b = true).$$

If given a valuation V of letters from a formula φ in $\mathcal{N}_C(k, m)$, the truth values of φ can be defined at elements of $\mathcal{N}_C(k, m)$ by the modified rules to the ones for frames $\mathcal{N}_C$ above (actually just in accordance with standard meaning of truth values for time operations and knowledge modalities). We describe below steps for time operations. For a cluster $C(i)$, $NxtC(i)$ is the next (by the operation $Next$ for worlds) for $C(i)$ cluster $C(j)$. The operation Nxt is a function. Let $a \in C(i) \subseteq \mathcal{N}_C(k, m)$.

$$(\mathcal{N}_C(k,m), a) \Vdash_V \varphi \mathbf{U} \psi \Leftrightarrow \exists t \in N(t \ge 0) \& \exists b \in Nxt^t C(i)[(\mathcal{N}_C(k,m), b) \Vdash_V \psi$$

$$\& \forall t_1 ((t_1 \in [0, t-1] \Rightarrow \forall c \in Nxt^{t_1} C(i)((\mathcal{N}_C(k,m), c) \Vdash_V \varphi)].$$

$$(\mathcal{N}_C(k,m), a) \Vdash_V \varphi \mathbf{U}_w \psi \Leftrightarrow \exists t \in N(t \ge 0) \& \exists b \in Nxt^t C(i)$$

$$[(\mathcal{N}_C(k,m), b) \Vdash_V \psi \& \forall t_1 ((t_1 \in [0, t-1] \Rightarrow$$

$$\exists c \in Nxt^{t_1} C(i)((\mathcal{N}_C(k,m), c) \Vdash_V \varphi)].$$

$$(\mathcal{N}_C(k,m), a) \Vdash_V \mathbf{N}^{-1} \mathbf{U}_s \psi \Leftrightarrow \exists t \in N(t \ge 0) \& \forall b \in Nxt^t C(i)$$

$$[(\mathcal{N}_C(k,m), b) \Vdash_V \psi \& \forall t_1 ((t_1 \in [0, t-1] \Rightarrow$$

$$\forall c \in Nxt^{t_1} C(i)((\mathcal{N}_C(k,m), c) \Vdash_V \varphi)].$$

Using Lemma 4 as the basis, we get

Lemma 5. *A rule $\mathbf{r_{nf}}$ in the reduced normal form is refuted in a frame $\mathcal{N}_C$ if and only if $\mathbf{r_{nf}}$ can be refuted in a frame $\mathcal{N}_C(k, m)$ by a valuation V with some special properties, where the size of the frame $\mathcal{N}_C(n, m)$ is exponential from a square polynomial on size of $\mathbf{r_{nf}}$.*

From Theorem 1, Lemma 3 and Lemma 5 we immediately derive

Theorem 2. *The logic $\mathcal{MA}_{\mathcal{LTL}}$ is decidable. The algorithm for testing a formula to be a theorem in $\mathcal{MA}_{\mathcal{LTL}}$ consists of validity verification (w.r.t. special valuations) of rules in reduced normal form at frames $\mathcal{N}_C(n, m)$ of size exponential from a square polynomial on size of $\mathbf{r_{nf}}$.*

Some other operations allowing to model $\mathbf{U}_s$ and $\mathbf{U}_w$ may be suggested. For instance, consider the following new relation R_s on frames $\mathcal{N}_C$: $\forall i \in N, \forall a, b \in C(i)(aR_sb)$. The relation R_s plays especial role to model the knowledge of a supervisor (omniscient agent) who knows the information in all states of the current time point. Let $\Box_s := K_s$, $\Diamond_s := \neg K_s \neg$. We use notation $\equiv_{sem}$ to say that the truth values of formulas in frames $\mathcal{N}_C$ coincide. It is easy to see that (i) $\varphi \mathbf{U}_w \psi \equiv_{sem} \Diamond_s \varphi \mathbf{U} \Diamond_s \psi$; (ii) $\varphi \mathbf{U}_s \psi \equiv_{sem} \Box_s \mathbf{U} \Box_s \psi$. So, having in disposal an supervisor agent, we can obtain weak and strong *Until*. The logic $\mathcal{MA}_{\mathcal{LTL}}^S$ in the language with K_s and without $\mathbf{U}_s$ and $\mathbf{U}_w$ obeys the technique presented in this paper for $\mathcal{MA}_{\mathcal{LTL}}$, and we can get the decidability with the same estimation of complexity. Another way to vary/extend the language is to add variants of the operation $\mathbf{N}$. For instance, we could consider the operation $\mathbf{N}_w$ – weak next with interpretation $(\mathcal{M}, a) \Vdash_V \mathbf{N}_w \varphi \Leftrightarrow \exists b [(a \text{ Next } b) \wedge (\mathcal{M}, b) \Vdash_V \varphi]$, and the logic with this new operation again will be decidable. Moving in this direction further, we can consider a new specific operation $Next_w$ on frames $\mathcal{N}_C$ being a restriction of $Next$, say $\forall a, b \in \bigcup_{i \in N} C(i), a \text{ } Next_w \text{ } b \Rightarrow [a \in C(i) \text{ for some } i \text{ and } b \in C(i+1)]$; $\forall a \in \bigcup_{i \in N} C(i)[a \in C(i) \Rightarrow \exists b \in C(i+1)(a \text{ } Next_w b) \wedge \forall c \in C(i) \forall d \in C(i+1)((cNext_wd) \Leftrightarrow (a \text{ } Next_w \text{ } d))]$. The technique for satisfiability and decidability for $\mathcal{MA}_{\mathcal{LTL}}$ from this paper is seemed to work for this case again. Also some restrictions for agent's accessibility relations R_i may be considered by introduction hierarchy between these relations R_i (but not only the presence of the omniscient supervisor s with R_s as above). This hierarchy may be arbitrary desirable one (of kind $R_i \subseteq R_j$ for supervision), and the suggested above technique works again.

3 Conclusion, Future Work

The paper presents technique to show decidability of $\mathcal{MA}_{\mathcal{LTL}}$ and similar logics. Instruments suggested seem to be reasonably flexible and may work for other various logics with background from AI and CS. There are many avenues for research on logic $\mathcal{MA}_{\mathcal{LTL}}$ and its variants. Axiomatizability and research on complexity are interesting. Logics obtained from $\mathcal{MA}_{\mathcal{LTL}}$ by refining via introduction operations *Since* and *Previous* based on $C(i)$ with $i \in N$, by considering indexing time based on integers (but not only natural numbers as above) are actual and interesting. And it seems these questions may be approached by evolving suggested technique.

References

1. Barringer, H., Fisher, M., Gabbay, D., Gough, G.: Advances in Temporal Logic. Applied logic series, vol. 16. Kluwer Academic Publishers, Dordrecht (1999)
2. Blackburn, P., Marx, M.: Constructive Interpolation in Hybrid Logic. Journal of Symbolic Logic 68(2), 463–480 (2003)
3. Bordini, R.H., Fisher, M., Visser, W., Wooldridge, M.: Model Checking Rational Agents. IEEE Intelligent Systems 19(5), 46–52 (2004)

4. Clarke, E., Grumberg, O., Hamaguchi, K.P.: Another look at LTL Model Checking. In: Dill, D.L. (ed.) CAV 1994. LNCS, vol. 818, Springer, Heidelberg (1994)
5. Daniele, M., Giunchiglia, F., Vardi, M.: Improved Automata Generation for Linear Temporal Logic. In: Halbwachs, N., Peled, D.A. (eds.) CAV 1999. LNCS, vol. 1633, Springer, Heidelberg (1999)
6. Dix, J., Fisher, M., Levesque, H., Sterling, L.: Special Issue on Logic-Based Agent Implementation – Editorial. Annals of Mathematics and Artificial Intelligence 41(2-4), 131–133 (2004)
7. Geanakoplos, F.J., Halpern, J., Vardi, M.Y.: The hierarchical approach to modeling knowledge and common knowledge. International Journal of Game Theory 28(3), 331–365 (1999)
8. Fagin, R., Halpern, J., Moses, Y., Vardi, M.: Reasoning About Knowledge. The MIT Press (1995) ISBN-10: 0-262-06162-7 ISBN-13: 978-0-262-06162-9
9. Fisher, M.: Temporal Development Methods for Agent-Based Systems. Journal of Autonomous Agents and Multi-Agent Systems 10(1), 41–66 (2005)
10. Gabbay, D.M., Hodkinson, I.M.: An axiomatisation of the temporal logic with Until and Since over the real numbers. Journal of Logic and Computation 1, 229–260 (1990)
11. Goldblatt, R.: Logics of Time and Computation. CSLI Lecture Notes, vol. 7 (1992)
12. Goldblatt, R.: Mathematical Modal Logic: A View of its Evolution. J. Applied Logic 1(5-6), 309–392 (2003)
13. Gabbay, D.M., Hodkinson, I.M.: An axiomatisation of the temporal logic with Until and Since over the real numbers. Journal of Logic and Computation 1, 229–260 (1990)
14. van der Hoek, W., Wooldridge., M.: Towards a Logic of Rational Agency. Logic Journal of the IGPL 11(2), 133–157 (2003)
15. Halpern, J., Shore, R.: Reasoning about common knowledge with infinitely many agents. Information and Computation 191(1), 1–40 (2004)
16. Hendler, J.: Agents and the semantic web. IEEE Intelligent Systems 16(2), 30–37 (2001)
17. Hodkinson, I.: Temporal Logic and Automata. In: Gabbay, D.M., Reynolds, M.A., Finger, M. (eds.) Temporal Logic: *Mathematical Foundations and Computational Aspects*, vol. 2, pp. 30–72. Clarendon Press, Oxford (2000)
18. Kacprzak, M.: Undecidability of a multi-agent logic. In: Fundamenta Informaticae, vol. 45(2-3), pp. 213–220. IOS Press, Amsterdam (2003)
19. Manna, Z., Pnueli, A.: Temporal Verification of Reactive Systems: Safety. Springer (1995)
20. Manna, Z.,, P.: The Temporal Logic of Reactive and Concurrent Systems: Specification. Springer (1992)
21. Pnueli, A.: The Temporal Logic of Programs. In: Proc. of the 18th Annual Symp. on Foundations of Computer Science, pp. 46–57. IEEE (1977)
22. Pnueli, A., Kesten, Y.: A deductive proof system for CTL^*. In: Brim, L., Jančar, P., Křetínský, M., Kucera, A. (eds.) CONCUR 2002. LNCS, vol. 2421, pp. 24–40. Springer, Heidelberg (2002)
23. Rybakov, V.V.: A Criterion for Admissibility of Rules in the Modal System $S4$ and the Intuitionistic Logic. Algebra and Logic 23(5), 369–384 (1984) (Engl. Translation)
24. Rybakov, V.V.: Rules of Inference with Parameters for Intuitionistic logic. Journal of Symbolic Logic 57(3), 912–923 (1992)
25. Rybakov, V.V.: Hereditarily Structurally Complete Modal Logics. Journal of Symbolic Logic 60(1), 266–288 (1995)

26. Rybakov, V.V.: Admissible Logical Inference Rules. Series: Studies in Logic and the Foundations of Mathematics, vol. 136. Elsevier Sci. Publ., North-Holland (1997)
27. Rybakov, V.V., Kiyatkin, V.R., Oner, T.: On Finite Model Property For Admissible Rules. Mathematical Logic Quarterly 45(4), 505–520 (1999)
28. Rybakov, V.V.: Construction of an Explicit Basis for Rules Admissible in Modal System S4. Mathematical Logic Quarterly 47(4), 441–451 (2001)
29. Rybakov, V.V.: Logical Consecusions in Intransitive Temporal Linear Logic of Finite Intervals. Journal of Logic Computation 15(5), 633–657 (2005)
30. Rybakov, V.V.: Logical Consecusions in Discrete Linear Temporal Logic. Journal of Symbolic Logic 70(4), 1137–1149 (2005)
31. Rybakov, V.V.: Logical Consecusions in Intransitive Temporal Linear Logic of Finite Intervals. Journal of Logic Computation 15(5), 633–657 (2005)
32. Rybakov, V.V.: Linear Temporal Logic with Until and Before on Integer Numbers, Deciding Algorithms. In: Grigoriev, D., Harrison, J., Hirsch, E.A. (eds.) CSR 2006. LNCS, vol. 3967, pp. 322–334. Springer, Heidelberg (2006)
33. Rybakov, V.V.: Until-Since Temporal logic Based on Parallel Time with Common Past. In: Artemov, S.N., Nerode, A. (eds.) LFCS 2007. LNCS, vol. 4514, pp. 486–497. Springer, Heidelberg (2007)
34. Rybakov, V.V.: Logic of Discovery in Uncertain Situations – Deciding Algorithms. In: Apolloni, B., Howlett, R.J., Jain, L. (eds.) KES 2007, Part II. LNCS (LNAI), vol. 4693, pp. 950–958. Springer, Heidelberg (2007)
35. Thomason, S.K.: Semantic Analysis of Tense Logic. Journal of Symbolic Logic 37(1) (1972)
36. van Benthem, J.: The Logic of Time. Reidel, Dordrecht, Synthese Library 156 (1983)
37. van Benthem, J., Bergstra, J.A.: Logic of Transition Systems. Journal of Logic, Language and Information 3(4), 247–283 (1994)
38. van Benthem, J.: The Logic of Time. Kluwer (1991)
39. Vardi, M.: An automata-theoretic approach to linear temporal logic. In: Proceedings of the Banff Workshop on Knowledge Acquisition (Banff 1994) (1994)
40. Vardi, M.: Reasoning about the past with two-way automata. In: Larsen, K.G., Skyum, S., Winskel, G. (eds.) ICALP 1998. LNCS, vol. 1443, pp. 628–641. Springer, Heidelberg (1998)
41. Wooldridge, M.J., Weiß, G., Ciancarini, P. (eds.): AOSE 2001. LNCS, vol. 2222. Springer, Heidelberg (2002)
42. Wooldridge, M., Dunne, P.E.: The Computational Complexity of Agent Verification. In: Meyer, J.-J., Tambe, M. (eds.) ATAL 2001. LNCS (LNAI), vol. 2333. Springer, Heidelberg (2002)

On Multi Agent Coordination in the Presence of Incomplete Information

Krzysztof Skrzypczyk

Department of Automatic Control, Silesian University of Technology, Akademicka 16,
44-100 Gliwice, Poland
`krzysztof.skrzypczyk@polsl.pl`

Abstract. This paper presents the problem of the coordination of actions in a multi-agent system. The main difficulty in resolving this problem is the limited information. Each agent is provided only with partial information about the state of the team of agents. This paper presents a hybrid technique that combines the game theory tool and the voting schema, which is applied to create the method of coordination that deals with this problem. Appropriate simulation results of the proposed techniques are presented.

1 Introduction

The key issue in a problem of controlling a team of robots is the coordination of the movements of individual team mates. In a mobile robotics domain the basic problem that is often considered is the simple point-to-point navigation task, the execution of which requires motion planning. The mutual interactions between individual robots sharing common workspace, even if the workspace is free of obstacles make the problem of motion planning non trivial. Another problem that is considered in multi-robot domain is controlling a formation of mobile robots. There are many variations of this problem that have been reported in literature. One of them consists in moving the formation satisfying the geometrical constraints that are put on the movements of the agents. A number of solutions to this type of problem have been presented in literature. In the work by Belta and Kumar [2] the heuristic approach based on geometrical dependencies was applied to control a group of mobile robots. In another work by Gerkey [4] an application of auction methods to multi robot coordination was proposed. In the paper by Sequiera [7] negotiation method was applied to provide multiple robot coordination. Another interesting approach to the solution of aforementioned problem was presented in the work by Lawton [6]. Dynamic, decentralized methodology was applied to maneuver the formation of robots under hard geometrical constraints. Another branch of coordination methods is a group of methods and techniques based on game theory [1]. The game theory is a convenient tool dedicated to solve problems of coordination in a multi robot domain. In the work by Skrzypczyk [8] a problem of tracking a moving target by the formation of robots is presented. Another interesting work by LaValle [5] addresses the problem of coordinated task planning for multiple robots with independent goals. Unfortunately an application of the previously highlighted works is limited to centralized control architecture where complete information of the process is provided. In the work by Chaimowicz [3] a quite effective solution of the problem of coordinating multiple robots transporting an object is presented and a very interesting approach of synchronization of the distributed system was discussed. Despite a number of successful solutions to the problem of incomplete information in the multi agent systems, the problem has still been relevant. In this paper the method of controlling a team of mobile robots in case of incomplete information is presented. The method is a hybrid one that utilizes noncooperative game theory and voting schema. As an illustration of the method a problem of moving the formation of mobile robots is presented. What makes the problem complex to solve is the limited information about the state of the team the given team mate is provided with. The fusion of game theory results and voting technique allows us to create the control law that provides the completion of the team task.

1.1 Overview of the System

Let us consider the multi-agent system architecture the diagram of which is presented in fig.1. The system consists of N robotic agents. Each agent is defined by the three features [9]:

- Perception;
- Reasoning;
- Action.

The team of robotic agents is intended to perform the navigational task called a mission. The aforementioned features define competencies of the given robotic agent. The perception determines how extensive the knowledge of both the performed task and the state of the other team-mates the given robotic agent is provided with. Another aspect covered by this feature is the communication ability of the agent. The more extensive perception the better abilities of the agent to interact with the other team-mates. Also the mission itself can be performed more efficiently in case of more extensive perception. Another feature that defines the agent is its reasoning abilities. The reasoning determines the way that the input data that comes from the sensing and communication mechanism are processed into an action the agent effects the environment. The last feature named *action* defines the variety of ways the agent can effect the environment. Here in the paper we consider the multi-agent system under the following assumptions:

- The system works in a synchronic way;
- All the agents have common knowledge about the mission;
- Each agent has limited both communication and sensing abilities.

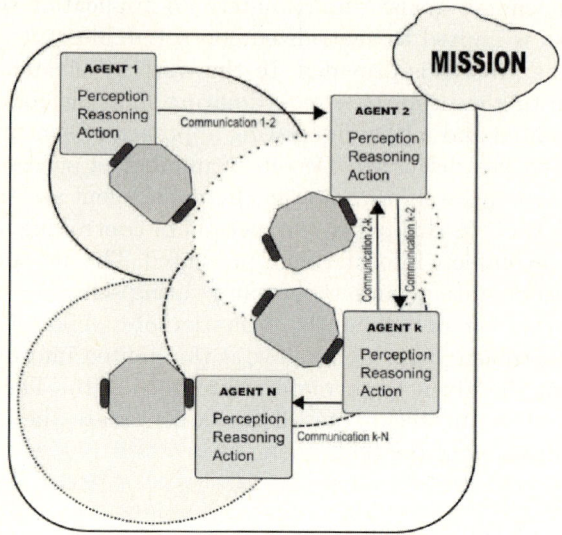

Fig. 1. The multi-agent system structure with incomplete information

2 The Problem Statement

Without a loss of generality let us consider a multi-robot system. In such a case each agent is intended to control a mobile robot. The mobile robot and the agent controlling the movement of the robot will be called robotic-agent hereafter for simplicity. In order to clearly address the problem let us define a few notions that shall be used in this paper.

2.1 The Team of Robots and the Mission

First let us denote the team of robotic-agents as:

$$R = \{R_1, R_2, ...R_N\} \qquad (1)$$

The mission consists in moving the team in the way that the centre of the team C_{Rm} moves from the initial position S to the desired one T. Additional constraints are also put on the formation shape. In this case the constraint is the limited space the team is expected to occupy. The space is defined by a circle of the radius r_N. The circle is fixed to the center point C_{Rm} of the team. In this paper we consider the motion in a 2D space therefore $S, T \in \Re^2$. The pose of the given ith robotic agent in the moment of discrete time n is defined by a vector:

$$X_i(n) = [x_{R_i}, y_{R_i}, \Theta_{R_i}]^T \qquad (2)$$

where x_{Ri}, y_{Ri} elements denote location of the centre of the ith robot platform and Θ_{Ri} denotes its heading. Both are related to a cartesian coordinates frame fixed to the workspace. In this work the given robotic agent is assigned to control a differential driven mobile platform. The platform is controlled by changing a velocity of its right and left wheel. In this paper the control of the platform in the moment of time n is defined by a vector:

$$u(n) = [v, \omega]^T \text{ where } \omega \in \langle -\omega_{max}, \omega_{max} \rangle \ , \ v \in \langle 0, v_{max} \rangle \quad (3)$$

where ω, v denote angular and linear velocity of the platform and ω_{max}, v_{max} are the maximal possible velocities. The control defined by (3) can be easily transformed to the velocity of the left and the right wheel by:

$$\begin{bmatrix} v_L \\ v_R \end{bmatrix} = \begin{bmatrix} 1 & 0.5L \\ 1 & -0.5L \end{bmatrix} \begin{bmatrix} v \\ \omega \end{bmatrix} \quad (4)$$

where v_L, v_R are the velocities of the left and the right wheel and L is the distance between the wheels. In this paper we assume that each agent can effect the environment with finite number of actions. The finite number of controls are obtained by discretizing the control space (3).Therefore the set of possible actions of ith agent is defined by:

$$U_{R_i} = \{u_{i,1}, ..., u_{i,K_i}\} \quad (5)$$

where

$$u_l = [v_l, \omega_l]^T : \omega_l \in \{\omega_{l,1}, \omega_{l,2}, ..., \omega_{l,\Omega_{R_l}}\}, v_l \in \{v_{l,1}, v_{l,2}, ..., v_{l,V_{R_l}}\} \quad (6)$$

where Ω_{R_i} and V_{R_i} are quantities of levels of distrecization, correspondingly, of angular and linear velocity. The number of possible actions of ith agent is equal $K_i = \Omega_{R_i} V_{R_i}$.

2.2 Limited Perception

There are three sources of informational limitations in the considered system. The first one is caused by the limited range of sensors and communication devices the given agent is equipped with. Therefore ith agent of the team (1) is able to perceive only a part of the team. So let us define the group of agents perceived by ith one in the moment of time n using the set of indices:

$$\Phi_{R_i}(n) = \{i\} \cup \{i_1, ..., i_l\} \quad \text{where} \quad i_1, ..., i_l \in \{1, ..., N\} \quad (7)$$

The cardinal of the set (7) is denoted by $N_{R_i} = \overline{\overline{\Phi}}_{R_i}$. Another limitation concerns information about the location of the other agents ith agent is provided with. Namely the given agent knows only the position of the other one. The heading of the other agent is estimated on the basis of its past locations. Therefore let us denote the pose of the kth agent perceived by ith agent in the moment of time n:

$$X_i^k(n) = \begin{cases} x_{R_i}, y_{R_i}, \Theta_{R_i} & \text{for} \quad k = i \\ x_{R_k}, y_{R_k}, \hat{\Theta}_{R_k}^i & \text{for} \quad k \neq i \end{cases} \quad (8)$$

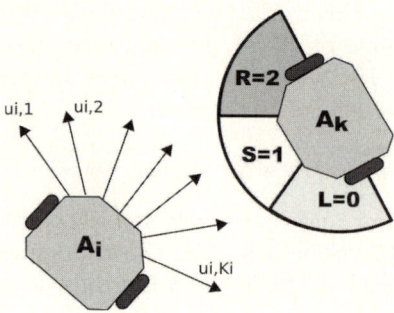

Fig. 2. An interpretation of the limitation concerning the knowledge of the actions of the other team-mates

where $\hat{\Theta}^i_{R_k}$ denotes the estimation of the heading of kth agent made by ith one. The assumptions imposed on the architecture of the presented system imply that also each agent has the limited knowledge about possible actions of the other team-mates. Therefore a set of possible actions of kth agent from the point of view of ith agent is defined as:

$$U^k_i = \{L = 0, S = 1, R = 2\} \text{ for } k \neq i \wedge k \in \Phi_{R_i} \qquad (9)$$

It is clear that the control system of ith agent has the complete knowledge about all the actions the agent can take. But considering the possible actions of the other team-mates the knowledge is limited to possible three actions the team-mate is assumed to be able to take. Figure 2 illustrates the discussed limitation. So the aim of the design process is to create a method of control of the individual team-mates that will meet all the previously discussed limitations and will allow the team to perform the mission defined in section 2.1.

3 The Method of the Coordination

The problem of the coordination of multiple agents that share the common workspace can be perceived as a conflicting situation between the individual agents. When the knowledge is complete and common to all of the agents the problem can be relatively easily solved using the game theory framework [1]. The matters are getting more complicated when the aforementioned assumption can not be satisfied. In this paper we propose to combine properties of the two commonly used techniques: the game theory and the voting technique. Figure 3 illustrates an idea of the proposed technique of coordination.

The technique can be described as follows. Let us consider the problem of coordination from the point of view of ith agent. Due to the limited perception (look section 2.2) this agent is only able to detect and communicate with agents that belong to the group defined by (7). Therefore the sub-game between them is defined. The game results in determining an optimal strategy (action) for

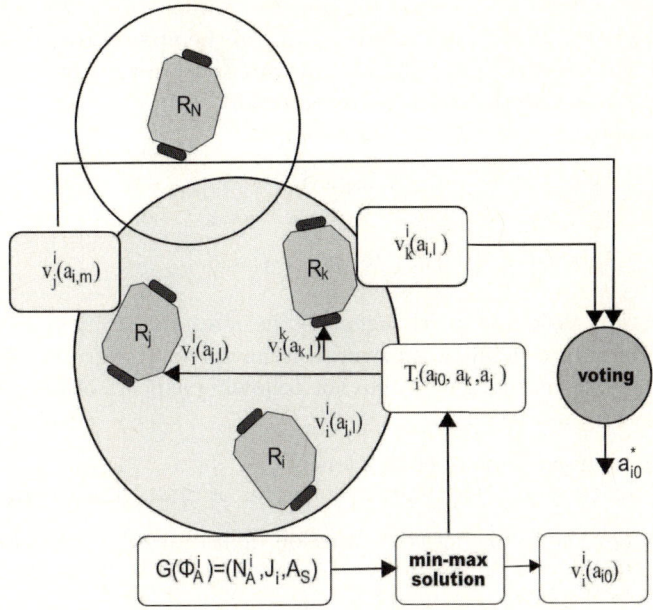

Fig. 3. A diagram illustrating the proposed technique

ith agent. It should be emphasized that the action is determined for ith agent on the basis of information about the state of the group Φ_{R_i} and the common knowledge about the mission. Simultaneously on the basis of the result of the game ith agent votes for actions of the rest of the Φ_{R_i} team-mates. The meaning of these votes is that ith agent will keep its best result if the rest of the Φ_{R_i} team-mates apply the actions that the ith agent voted for. Of course the same line of thought can be done for each agent from the (1). That results in that simultaneously the other agents can vote the action of ith. Finally an action that ith agent applies is the one that is the result of the voting process.

4 The Model of the Game

In this section the model of the game is presented. The game is a model of a decision making process that happens between the agents that belong to the given sub-team. Analyzing the game from the point of view of ith agent the result of the game is the solution that determines an optimal policy for the player. Simultaneously it is the basis for voting for actions that other team mates are supposed to take. Let us denote the game played between the gents of the sub-team Φ_A^i by:

$$G_i(\Phi_{R_i}) = (N_{R_i}, J_{R_i}, A_{S,R_i}) \quad i = 1, ..., N \tag{10}$$

where $N_{R,i}$ is a number of agents playing the game. J_{R_i} is the cost function related to ith player. It determines an amount of the virtual cost that the player pays when all the players adopt the given strategy from the action space A_{S,R_i}. The action space of the given sub-team perceived from the point of view of ith agent is defined by:

$$A_{S,R_i} = (A_i \times A_{i_1} \times, ..., \times A_{i_l}), \quad i, i_1, ..., i_l \in \Phi_{R_i}$$
$$A_i = U_{R_i} \quad (11)$$
$$A_p = U_i^p = \{0, 1, 2\} \text{ for } p \in \{i_1, ..., i_l\}$$

The cost function assigned to ith agent is the mapping $J_{R_i} : A_{S,R_i} \to \Re$. The cost function is very crucial to an overall behavior of the group of the agents. Therefore during designing the system the following aspects of the group behavior must be taken into account:

- Collision free movement of each agent;
- Tracking the target by the team (the center of mass of the team);

Here in the paper the following cost function definition was proposed:

$$J_{R_i} = k_{i,1} \hat{L}_{C,T}^i + k_{i,2} \frac{1}{\min_{j \in \Phi_{R_i}} \hat{L}_{i,j}} \quad (12)$$

where $\hat{L}_{C,T}^i$ is the predicted distance of the centre of the sub-team to the target location T. The distance is computed using the equation:

$$\hat{L}_{C_{R_i},T}(n) = \sqrt{\left(x_T - \hat{x}_{C_{R_i}}(n)\right)^2 + \left(y_T - \hat{y}_{C_{R_i}}(n)\right)^2} \quad (13)$$

where

$$\hat{C}(\Phi_{R_i}) = (\hat{x}_{C_{R_i}}, \hat{y}_{C_{R_i}}) = \left(\frac{1}{N_{R_i}} \sum_{k \in \Phi_{R_i}} \hat{x}_{R_k}(a_k), \frac{1}{N_{R_i}} \sum_{k \in \Phi_{R_i}} \hat{y}_{R_k}(a_k) \right) \quad (14)$$

The $(\hat{x}_{C_{R_i}}, \hat{y}_{C_{R_i}})$ denotes the predicted centre of the team defined by Φ_{R_i} whereas $(\hat{x}_{R_k}(a_k)), \hat{y}_{R_k}(a_k))$ is the predicted position of kth agent that took the action $a_k = a_k(n)$ in the moment of time n. The action a_k is defined by:

$$a_k(n) = \begin{cases} U_{R_i} & \text{for } k = i \\ U_i^k & \text{otherwise} \end{cases} \quad (15)$$

The first component in (12) is responsible for an accurate tracking the moving target by the sub-team. The second one is to prevent all the team-mates from collisions. The $L_{i,j}$ is the predicted distance between ith and jth agent:

$$\hat{L}(a_i(n), a_j(n)) = \sqrt{\left(\hat{x}_{R_i}(a_i) - \hat{x}_{R_j}(a_j)\right)^2 + \left(\hat{y}_{R_i}(a_i) - \hat{y}_{R_j}(a_j)\right)^2} \quad (16)$$

In (13) and (16) $\hat{x}_{R_{i,j}}(a_{i,j}), \hat{y}_{R_{i,j}}(a_{i,j})$ denote the predicted location of the kth agent after applying the strategy a_k estimated by ith agent. It is determined using the following formulae:

$$\begin{aligned}
\hat{x}_{R_k} &= x_{R_k} + v_k(a_k)\Delta t \cos\left(\Theta_{R_k} + \omega_k(a_k)\Delta t\right) \\
\hat{y}_{R_k} &= y_{R_k} + v_k(a_k)\Delta t \sin\left(\Theta_{R_k} + \omega_k(a_k)\Delta t\right) \quad \text{for} \quad k = i \\
\hat{x}_{R_k} &= x_{R_k} + \hat{v}_k \Delta t \cos\left(\hat{\Theta}_{R_k} + \tfrac{\pi}{3}\delta a_k\right) \\
\hat{y}_{R_k} &= y_{R_k} + \hat{v}_k \Delta t \sin\left(\hat{\Theta}_{R_k} + \tfrac{\pi}{3}\delta a_k\right) \quad \text{otherwise} \\
\delta &\in \langle 0, 1 \rangle
\end{aligned} \quad (17)$$

where δ is a uniformly distributed random value. The ($v_k(a_k)$ and $\omega(a_k)$) denote linear and angular velocity the robot moves with as a result of applying action a_k. Whereas $\hat{\Theta}_{R_k}$ is an estimation of the heading of the kth agent made by agent ith The coefficients $k_{i,1}, k_{i,2}$ in (12) adjust an influence of the components of the cost function. They are adjusted in an experimental way.

4.1 The Solution of the Game

In the presented problem each agent is not sure of decisions of the other agents and has incomplete knowledge about possible actions of the other team mates. Therefore the reasonable solution of the game (10) can be *min-max* strategy. This strategy is commonly used to solve problems that have the aforementioned property. So the solution of (10) is given by:

$$a_{i0} = \min_{a_i} \max_{i_1,\ldots,i_l \in \Phi_{R_i}/i} J_{R_i}(a_{i0}, a_{i_1}, \ldots, a_{i_l}) \quad (18)$$

Taking the action (18) the ith agent secures his possible loss at a level called the security level. The solution is further used in a voting process for actions of the team-mates.

5 The Voting

Repeating the reasoning discussed in section 4 for all the sub-teams and for all the agents we obtain "optimal" strategies for all the agents. But it has to be stressed that they are "optimal" only for each particular agent. Moreover they are determined on the basis of only partial information about the team. Therefore a technique that allows to propagate the local information to the whole team is necessary. In this paper a voting technique is used to provide the aforementioned propagation. Figure 4 illustrates general idea of the voting technique.

Let us assume the process is to be controlled by a team of agents. Also there is a set of the actions the agents can act on the process. The voting idea is that each agent votes for a chosen action. The action selected by the agent provides

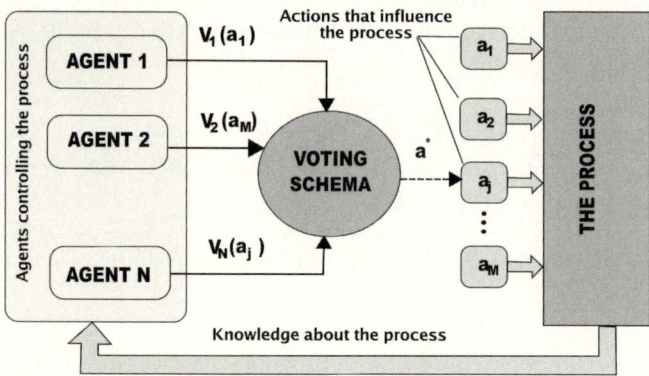

Fig. 4. The voting technique illustration

optimal control of the process. The optimality of course is only valid from the point of view of the given agent. If all the agents vote then according to applied voting schema the given action will be selected. The selected action satisfies expectations of the majority of the agents. The player adopting the solution (18) selects "the best from the worst" cases that may occur. Simultaneously the agent votes with the highest priority for his own action:

$$v_i^i(a_{i0}) = 1 \tag{19}$$

The agent playing (18) determines a set of actions of the other players it gives the votes weighted w_0 (the value of the weight is tuned experimentally). The set A_0 satisfies the following:

$$A_0 = \{a_{i0}, a_{i_10}, ..., a_{i_l0}\} = \min_{i_1,...,i_l \in \Phi_{R_i}/i} T_{R_i}(a_{i0}, a_{i_1}..., a_{i_l}) \tag{20}$$

The performance index T_i that is to be minimized, reflects the limitation put on the formation boundaries. Therefore the agents selects the actions that both satisfy (18) and guarantee the lowest as possible size of the formation. The index is calculated according to:

$$T_{R_i}(a_{i0}, a_{i_1}, ..., a_{i_l}) = \sum_{k \in \Phi_{R_i}} \hat{L}_{C,R_k}(a_k) \tag{21}$$

Where $\hat{L}_{C,R_k}(a_k)$ denotes predicted distance of the kth agent to the center of the mass of the team (14) after taking the action a_k. As a result of the aforementioned procedure the set of actions is determined. The set defines actions of the other team-mates $v_i^l(a_{l0}) = w_0$ the ith player wants to vote for.

5.1 Solution of the Voting

If the aforementioned reasoning is repeated for all the agents the set of votes on an action of ith agent will be obtained. Therefore the final and optimal action in the sense of this method is obtained from:

$$a_{i0}^* = \max_a V_i(a) \qquad (22)$$

The solution (22) corresponds to the action that obtains maximum number of points as a result of the voting process. Therefore the V_i denotes:

$$V_i(a) = \sum V_i^k(a) w_k \quad a \in U_{R_i} w_k = \begin{cases} 1 & \text{for } k = i \\ w_0 & \text{otherwise} \end{cases} \qquad (23)$$

If the aforementioned modelling process is applied for each agent the result will be a set of actions $A^* = a_{i0}^*$ for $i = 1, 2, ..., N$. These actions if applied would cause a change of the state of the formation of agents (1). This change allows the formation to realise a part of the mission.

6 Simulation

In this section an evaluation of the proposed method in the simulation is presented. The simulation is made for a team of three robots. The team is intended to realize the mission defined in the section 2.1. The studies were done for a group of three robots.

Let us assume the initial configuration (2) of the group of robots: X=(10, 10, 45), (50, 50, 10), (10, 80, -45)[cm, cm, deg]. The size of the robots is determined by a radius of their platforms and in the case of the robot used in the experiment is equal 2.8 [cm]. Constraints are put on the formation shape. In this case the constraint is defined by the limited space the team is expected to occupy. The space is defined by a circle of a radius $r_N = 15$[cm]. The goal the center of the team is intended to reach is determined by the point T(100, 90)[cm, cm]. Sampling time $\Delta t = 0.1$[s]. The set of possible actions of robots is set to the values $U_1, U_2, U_3 = \{(v_i, -50), (v_i, -25), (v_i, 0), (v_i, 25), (v_i, 50)\}$ [cm/s,deg/ s] $v_i = 10, 15$[cm/s]. In the first experiment the perceptual and communication limitations are set to the distance (radius) equal 2.8 [cm]. That means there is not both communication and perception between individual agents. Figure 5 shows the result of the simulation. It can be seen that due to lack of communication and information about the rest of the team each agent tries only to reach the target. There is no cooperation in moving in the formation.

Next experiment proves efficiency of the proposed method. The informational and communication range was increased to 20 [cm]. Now the agents can "see" each other and communicate to each other if they are close enough to each other. The mechanism of voting now can work and the cooperative behavior of the group can be seen in fig.6. The team acts the way that realizes the mission. The center of mass of the group moves toward the target and the formation size limitation is satisfied.

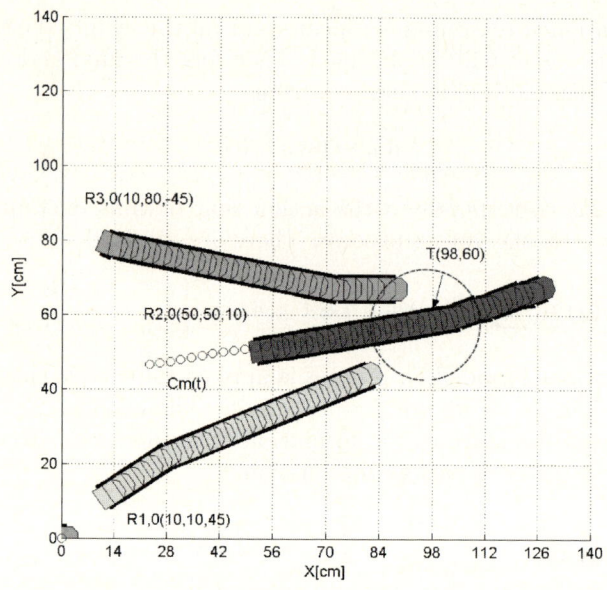

Fig. 5. Result of the first experiment showing the effects of lack of information

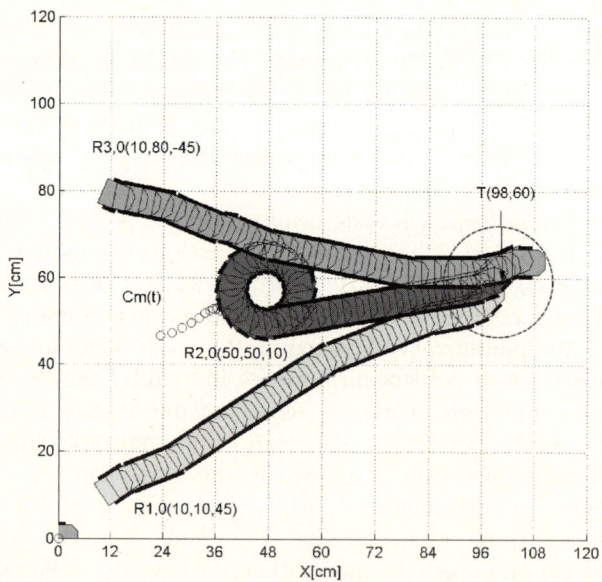

Fig. 6. Result of the second experiment showing the cooperative behavior of the team

7 Conclusion

In the paper a novel method of coordination in a multi-agent system was presented. The method takes into account limitations of perception and communication that can occur in real multi agent systems. The hybrid method utilizing advantages of game theory and voting technique was presented. The simulation results seem to be very promising. Therefore more research are planned that will allow the correction an verification of potential weak points within the method. Finally it is planned to apply this method to a real mobile multi robot system.

Acknowledgements

This work has been supported by Polish Ministry of Science funds in the years 2007-2008.

References

1. Basar, M., Olsder, G.J.: Dynamic Noncooperative Game Theory. Mathematics in Science and Engeneering. Academic Press Inc., London (1982)
2. Belta, C., Kumar, V.: Optimal motion generation for groups of robots: a geometric approach. ASME Journal of Mechanical Design 126, 63–70 (2004)
3. Chaimowicz, L., Kumar, V., Campos, M.: A Paradigm for Dynamic Coordination of Multiple Robots. Special issue on Advances in Multi-Robot Systems 18(5), 758–786 (2004)
4. Gerkey, B., Mataric, M.J.: Sold!: Auction methods for multi-robot coordination. IEEE Transactions on Robotics and Automation, special issue on Advances in Multi-Robot Systems 18(5), 758–786 (2002)
5. LaValle, S., Hutchinson, S.: Optimal Motion Planning for Multiple Robots Having Independent Goals. IEEE Trans. On Robotics and Automation 14(6), 912–925 (1998)
6. Lawton, J.R., et al.: A Decentralized Approach to Formation Maneuvers. IEEE Transactions on Robotics and Automation 19(6), 933–941 (2003)
7. Sequiera, J., Ribeiro, M.I.: A Negotiation Model for Cooperation Among Robots. In: Proc. of European Control Conference, Porto, Portugal (2001)
8. Skrzypczyk, K.: Game theory based target following by a team of robots. In: Proc. of the 4-th International Workshop on Robot Motion and Control (RoMoCo 2004), Puszczykowo, Poland, June 17-20, 2004, pp. 91–96 (2004)
9. Stone, P., Veloso, M.: Multiagent systems: A survey from a machine learning perspective. Autonomous Robots 3(8), 345–383 (2000)

Author Index

Alimi, Adel M. 806
Alor-Hernandez, Giner 951
Arabas, Jarosław 1137
Augustyniak, Piotr 909

Barakova, Emilia 741, 1198
Barhoumi, Ines 806
Bartczuk, Łukasz 197
Białko, Michał 479
Bielecka, Marzena 3
Bielecki, Andrzej 3
Bilski, Jarosław 11
Blachnik, Marcin 1149
Bousmalis, Konstantinos 365
Bożejko, Wojciech 995
Bronakowski, Lukasz 886
Brzostowski, Krzysztof 17
Bundzel, Marek 513

Chmielowiec, Anna 3
Cho, Sung-Bae 850, 1064
Chonnaparamutt, Winai 741
Cichosz, Jaroslaw 886
Cichosz, Paweł 1137
Cieślik, Agnieszka 1173
Cierniak, Robert 752
Cpalka, Krzysztof 207, 974
Cruz, Laura 1087
Cruz-Chávez, Marco A. 376
Cyganek, Bogusław 217
Czabański, Robert 919
Czarnowski, Ireneusz 522
Czibula, István-Gergely 693

de Campos, Luis M. 1185
Debon, Renaud 762
De Giusti, Armando 438
Dembczyński, Krzysztof 533
Dhibi, Mounir 762
Díaz-Parra, Ocotlán 376
Drapała, Jarosław 17
Dreżewski, Rafał 388
Drozda, Stanislaw 405
Duch, Włodzisław 566, 643
Dudek, Grzegorz 1007

Dudek-Dyduch, Ewa 28
Dymova, Ludmila 229
Dziwiński, Piotr 1018

Ech-cherif, A. 717
Eklund, Patrik 1029

Fermín, Leonardo 771
Fernández-Breis, Jesualdo T. 1222
Fernández-López, Gerardo 771
Fernández-Luna, Juan M. 1185
Figat, Pawel 1099
Frąckowiak, Grzegorz 1210
Fraire, Héctor 1087
Frausto, Juan 1087

Gabryel, Marcin 398, 405
Ganzha, Maria 1173, 1210
García-Sánchez, Francisco 1222
Gawinecki, Maciej 1210
Gałkowski, Tomasz 974
Gómez, Juan Miguel 951
Gorzałczany, Marian B. 40
Gosztolya, Gábor 782
Grąbczewski, Krzysztof 545
Greco, Salvatore 556
Grieco, Juan C. 771
Grochowski, Marek 566
Grudziński, Karol 578
Grześ, Marek 1041
Güner, Edip Serdar 929

Hájek, Petr 52
Halawa, Krzysztof 62
Han, Yo-Sub 1210
Hayes, Gillian M. 365
Helgesson, Robert 1029
Horzyk, Adrian 28
Hrebień, Maciej 588
Huete, Juan F. 1185

Imai, Michita 896
Ishii, Kentaro 896

Jablonski, Bartosz   794
Jalali-Farahani, Farhang   469
Jankowski, Norbert   545
Jędrzejowicz, Piotr   522
Jelonkiewicz, Jerzy   71
Józefczyk, Jerzy   1052
Juárez-Romero, David   376

Kacprzyk, Janusz   241, 351
Kallel, Ilhem   806
Kefi, Sonia   806
Kılıçaslan, Yılmaz   929
Klęsk, Przemysław   597
Kluska, Jacek   253
Korbicz, Józef   588
Korytkowski, Marcin   265, 405
Korzeń, Marcin   597
Kosiński, Witold   414, 489
Kotowski, Stefan   414
Kotulski, Leszek   1234
Kotłowski, Wojciech   533
Kozak, Karol   609
Kozielski, Stanisław   939
Krętowska, Małgorzata   78
Krętowski, Marek   426
Kryzhanovsky, Boris   89
Kryzhanovsky, Vladimir   89
Krzyżak, Adam   101
Kubanek, Mariusz   818
Kudenko, Daniel   1041
Kurzynski, Marek   620
Kwasnicka, Halina   450

Landero, Vanesa   1087
Landowski, Marek   661
Lanzarini, Laura   438
Ławryńczuk, Maciej   111
Lecornu, L.   717
Lee, Young-seol   1064
Leza, Victoria   438
Lindgren, Helena   1029
Liu, Ying   951

Markowska-Kaczmar, Urszula   450
Markowski, Michał   1052
Martínez-Béjar, Rodrigo   1222
Martínez-Rangel, Martín G.   376
Martyna, Jerzy   632
Marusak, Piotr M.   273
Maszczyk, Tomasz   643

Matuk Herrera, Rosana   828
Mazurkiewicz, Jacek   175
Małysiak, Bożena   939
Medina-Meléndez, Wilfredis   771
Mencke, Myriam   951
Michalewicz, Zbigniew   414
Mirza, Hanane   839
Momot, Alina   939
Montera, L.   963
Morkowski, Maciej   652
Mrozek, Dariusz   939
Mączka, Krystian   1149

Nakao, Zensho   839
Nicoletti, M.C.   963
Niewiadomski, Adam   285
Nishiyama, Yu   1075
Nowicki, Jakub   414
Nowicki, Robert   652

Obuchowicz, Andrzej   460
Ogiela, Lidia   1127
Olej, Vladimír   52

Pander, Tomasz   919
Paprzycki, Marcin   1173, 1210
Park, Han-Saem   850
Park, Myon-Woong   1210
Pazos, Rodolfo   1087
Pempera, Jarosław   995
Pérez, Joaquín   1087
Pfaffmann, Jeffrey O.   365
Piech, Henryk   1099
Pieczyński, Andrzej   295
Piegat, Andrzej   661
Pokropińska, Agata   1120
Popczyński, Piotr   426
Posada-Gómez, Rubén   951
Prętki, Przemysław   460
Prokopowicz, Piotr   306
Przepiórkiewicz, Bartosz   414
Przybył, Andrzej   71
Przybyła, Tomasz   919
Przystałka, Piotr   123
Puchala, Dariusz   135

Rafajłowicz, Ewaryst   146, 671
Razavi, Fatemeh   469
Rebrova, Olga   974
Rewak, Aleksander   620

Author Index

Reyes, Gerardo   1087
Rivera, Ismael   951
Robak, Silva   295
Róg, Izabela   229
Rojek, Izabela   1109
Roozmand, Omid   681
Rotter, Paweł   861
Rudziński, Filip   40
Rusiecki, Andrzej   154
Rutkowska, Danuta   197, 1018
Rutkowski, Leszek   207, 265, 398, 405, 974
Rybakov, Vladimir   1243
Rydzek, Szymon   818

Sas, Jerzy   729
Schäfer, Dominik   101
Scherer, Rafał   265, 317, 1120
Şerban, Gabriela   693
Seredynski, Franciszek   164
Sevastjanov, Pavel   229
Sinčák, Peter   513
Siwik, Leszek   388
Słoński, Marek   186
Słowik, Adam   479
Słowiński, Roman   533, 556
Skaruz, Jaroslaw   164
Skrzypczyk, Krzysztof   1254
Skubalska-Rafajłowicz, Ewa   671
Skulimowski, Andrzej M.J.   861
Ślot, Krzysztof   886
Smoląg, Jacek   11
Smutnicki, Adam   995
Socała, Jolanta   489
Solaiman, B.   717
Sołtysiński, Tomasz   981
Stapor, Katarzyna   609

Starczewski, Artur   324
Starczewski, Janusz T.   333
Stokfiszewski, Kamil   874
Straszecka, Ewa   341
Świątek, Jerzy   17
Szczęch, Izabela   556
Szczepaniak, Piotr S.   874
Szczepkowski, Marcin   450
Szmidt, Eulalia   351
Szymczak, Michał   1210

Tadeusiewicz, Ryszard   1127
Taguchi, Masahiko   896
Tatjewski, Piotr   111
Thai, Hien   839
Tóth, László   782
Trojanowski, Krzysztof   499

Uçar, Özlem   929

Valencia-García, Rafael   1222
Vanderelst, Dieter   1198

Walkowiak, Tomasz   175
Waszczyszyn, Zenon   186
Watanabe, Sumio   1075
Wawrzyński, Paweł   1137
Wieczorek, Tadeusz   1149
Wilbik, Anna   241

Yatsymirskyy, Mykhaylo   135

Zajdel, Roman   1160
Zamanifar, Kamran   681
Zdunek, Rafal   705
Zemirline, A.   717
Zolnierek, Andrzej   729

Printing: Mercedes-Druck, Berlin
Binding: Stein+Lehmann, Berlin

Lecture Notes in Artificial Intelligence (LNAI)

Vol. 5097: L. Rutkowski, R. Tadeusiewicz, L.A. Zadeh, J.M. Zurada (Eds.), Artificial Intelligence and Soft Computing – ICAISC 2008. XVI, 1269 pages. 2008.

Vol. 5078: E. André, L. Dybkjær, W. Minker, H. Neumann, R. Pieraccini, M. Weber (Eds.), Perception in Multimodal Dialogue Systems. X, 311 pages. 2008.

Vol. 5032: S. Bergler (Ed.), Advances in Artificial Intelligence. XI, 382 pages. 2008.

Vol. 5027: N.T. Nguyen, L. Borzemski, A. Grzech, M. Ali (Eds.), New Frontiers in Applied Artificial Intelligence. XVIII, 879 pages. 2008.

Vol. 5012: T. Washio, E. Suzuki, K.M. Ting, A. Inokuchi (Eds.), Advances in Knowledge Discovery and Data Mining. XXIV, 1102 pages. 2008.

Vol. 5009: G. Wang, T. Li, J.W. Grzymala-Busse, D. Miao, A. Skowron, Y. Yao (Eds.), Rough Sets and Knowledge Technology. XVIII, 765 pages. 2008.

Vol. 4994: A. An, S. Matwin, Z.W. Raś, D. Ślęzak (Eds.), Foundations of Intelligent Systems. XVII, 653 pages. 2008.

Vol. 4953: N.T. Nguyen, G.S. Jo, R.J. Howlett, L.C. Jain (Eds.), Agent and Multi-Agent Systems: Technologies and Applications. XX, 909 pages. 2008.

Vol. 4946: I. Rahwan, S. Parsons, C. Reed (Eds.), Argumentation in Multi-Agent Systems. X, 235 pages. 2008.

Vol. 4944: Z.W. Raś, S. Tsumoto, D. Zighed (Eds.), Mining Complex Data. X, 265 pages. 2008.

Vol. 4938: T. Tokunaga, A. Ortega (Eds.), Large-Scale Knowledge Resources. IX, 367 pages. 2008.

Vol. 4933: R. Medina, S. Obiedkov (Eds.), Formal Concept Analysis. XII, 325 pages. 2008.

Vol. 4930: I. Wachsmuth, G. Knoblich (Eds.), Modeling Communication with Robots and Virtual Humans. X, 337 pages. 2008.

Vol. 4929: M. Helmert, Understanding Planning Tasks. XIV, 270 pages. 2008.

Vol. 4924: D. Riaño (Ed.), Knowledge Management for Health Care Procedures. X, 161 pages. 2008.

Vol. 4923: S.B. Yahia, E.M. Nguifo, R. Belohlavek (Eds.), Concept Lattices and Their Applications. XII, 283 pages. 2008.

Vol. 4914: K. Satoh, A. Inokuchi, K. Nagao, T. Kawamura (Eds.), New Frontiers in Artificial Intelligence. X, 404 pages. 2008.

Vol. 4911: L. De Raedt, P. Frasconi, K. Kersting, S. Muggleton (Eds.), Probabilistic Inductive Logic Programming. VIII, 341 pages. 2008.

Vol. 4908: M. Dastani, A. El Fallah Seghrouchni, A. Ricci, M. Winikoff (Eds.), Programming Multi-Agent Systems. XII, 267 pages. 2008.

Vol. 4898: M. Kolp, B. Henderson-Sellers, H. Mouratidis, A. Garcia, A.K. Ghose, P. Bresciani (Eds.), Agent-Oriented Information Systems IV. X, 292 pages. 2008.

Vol. 4897: M. Baldoni, T.C. Son, M.B. van Riemsdijk, M. Winikoff (Eds.), Declarative Agent Languages and Technologies V. X, 245 pages. 2008.

Vol. 4894: H. Blockeel, J. Ramon, J. Shavlik, P. Tadepalli (Eds.), Inductive Logic Programming. XI, 307 pages. 2008.

Vol. 4885: M. Chetouani, A. Hussain, B. Gas, M. Milgram, J.-L. Zarader (Eds.), Advances in Nonlinear Speech Processing. XI, 284 pages. 2007.

Vol. 4874: J. Neves, M.F. Santos, J.M. Machado (Eds.), Progress in Artificial Intelligence. XVIII, 704 pages. 2007.

Vol. 4870: J.S. Sichman, J. Padget, S. Ossowski, P. Noriega (Eds.), Coordination, Organizations, Institutions, and Norms in Agent Systems III. XII, 331 pages. 2008.

Vol. 4869: F. Botana, T. Recio (Eds.), Automated Deduction in Geometry. X, 213 pages. 2007.

Vol. 4865: K. Tuyls, A. Nowe, Z. Guessoum, D. Kudenko (Eds.), Adaptive Agents and Multi-Agent Systems III. VIII, 255 pages. 2008.

Vol. 4850: M. Lungarella, F. Iida, J.C. Bongard, R. Pfeifer (Eds.), 50 Years of Artificial Intelligence. X, 399 pages. 2007.

Vol. 4845: N. Zhong, J. Liu, Y. Yao, J. Wu, S. Lu, K. Li (Eds.), Web Intelligence Meets Brain Informatics. XI, 516 pages. 2007.

Vol. 4840: L. Paletta, E. Rome (Eds.), Attention in Cognitive Systems. XI, 497 pages. 2007.

Vol. 4830: M.A. Orgun, J. Thornton (Eds.), AI 2007: Advances in Artificial Intelligence. XIX, 841 pages. 2007.

Vol. 4828: M. Randall, H.A. Abbass, J. Wiles (Eds.), Progress in Artificial Life. XII, 402 pages. 2007.

Vol. 4827: A. Gelbukh, Á.F. Kuri Morales (Eds.), MICAI 2007: Advances in Artificial Intelligence. XXIV, 1234 pages. 2007.

Vol. 4826: P. Perner, O. Salvetti (Eds.), Advances in Mass Data Analysis of Signals and Images in Medicine, Biotechnology and Chemistry. X, 183 pages. 2007.

Vol. 4819: T. Washio, Z.-H. Zhou, J.Z. Huang, X. Hu, J. Li, C. Xie, J. He, D. Zou, K.-C. Li, M.M. Freire (Eds.), Emerging Technologies in Knowledge Discovery and Data Mining. XIV, 675 pages. 2007.

Vol. 4811: O. Nasraoui, M. Spiliopoulou, J. Srivastava, B. Mobasher, B. Masand (Eds.), Advances in Web Mining and Web Usage Analysis. XII, 247 pages. 2007.

Vol. 4798: Z. Zhang, J.H. Siekmann (Eds.), Knowledge Science, Engineering and Management. XVI, 669 pages. 2007.

Vol. 4795: F. Schilder, G. Katz, J. Pustejovsky (Eds.), Annotating, Extracting and Reasoning about Time and Events. VII, 141 pages. 2007.

Vol. 4790: N. Dershowitz, A. Voronkov (Eds.), Logic for Programming, Artificial Intelligence, and Reasoning. XIII, 562 pages. 2007.

Vol. 4788: D. Borrajo, L. Castillo, J.M. Corchado (Eds.), Current Topics in Artificial Intelligence. XI, 280 pages. 2007.

Vol. 4775: A. Esposito, M. Faundez-Zanuy, E. Keller, M. Marinaro (Eds.), Verbal and Nonverbal Communication Behaviours. XII, 325 pages. 2007.

Vol. 4772: H. Prade, V.S. Subrahmanian (Eds.), Scalable Uncertainty Management. X, 277 pages. 2007.

Vol. 4766: N. Maudet, S. Parsons, I. Rahwan (Eds.), Argumentation in Multi-Agent Systems. XII, 211 pages. 2007.

Vol. 4760: E. Rome, J. Hertzberg, G. Dorffner (Eds.), Towards Affordance-Based Robot Control. IX, 211 pages. 2008.

Vol. 4755: V. Corruble, M. Takeda, E. Suzuki (Eds.), Discovery Science. XI, 298 pages. 2007.

Vol. 4754: M. Hutter, R.A. Servedio, E. Takimoto (Eds.), Algorithmic Learning Theory. XI, 403 pages. 2007.

Vol. 4737: B. Berendt, A. Hotho, D. Mladenic, G. Semeraro (Eds.), From Web to Social Web: Discovering and Deploying User and Content Profiles. XI, 161 pages. 2007.

Vol. 4733: R. Basili, M.T. Pazienza (Eds.), AI*IA 2007: Artificial Intelligence and Human-Oriented Computing. XVII, 858 pages. 2007.

Vol. 4724: K. Mellouli (Ed.), Symbolic and Quantitative Approaches to Reasoning with Uncertainty. XV, 914 pages. 2007.

Vol. 4722: C. Pelachaud, J.-C. Martin, E. André, G. Chollet, K. Karpouzis, D. Pelé (Eds.), Intelligent Virtual Agents. XV, 425 pages. 2007.

Vol. 4720: B. Konev, F. Wolter (Eds.), Frontiers of Combining Systems. X, 283 pages. 2007.

Vol. 4702: J.N. Kok, J. Koronacki, R. Lopez de Mantaras, S. Matwin, D. Mladenič, A. Skowron (Eds.), Knowledge Discovery in Databases: PKDD 2007. XXIV, 640 pages. 2007.

Vol. 4701: J.N. Kok, J. Koronacki, R. Lopez de Mantaras, S. Matwin, D. Mladenič, A. Skowron (Eds.), Machine Learning: ECML 2007. XXII, 809 pages. 2007.

Vol. 4696: H.-D. Burkhard, G. Lindemann, R. Verbrugge, L.Z. Varga (Eds.), Multi-Agent Systems and Applications V. XIII, 350 pages. 2007.

Vol. 4694: B. Apolloni, R.J. Howlett, L. Jain (Eds.), Knowledge-Based Intelligent Information and Engineering Systems, Part III. XXIX, 1126 pages. 2007.

Vol. 4693: B. Apolloni, R.J. Howlett, L. Jain (Eds.), Knowledge-Based Intelligent Information and Engineering Systems, Part II. XXXII, 1380 pages. 2007.

Vol. 4692: B. Apolloni, R.J. Howlett, L. Jain (Eds.), Knowledge-Based Intelligent Information and Engineering Systems, Part I. LV, 882 pages. 2007.

Vol. 4687: P. Petta, J.P. Müller, M. Klusch, M. Georgeff (Eds.), Multiagent System Technologies. X, 207 pages. 2007.

Vol. 4682: D.-S. Huang, L. Heutte, M. Loog (Eds.), Advanced Intelligent Computing Theories and Applications. XXVII, 1373 pages. 2007.

Vol. 4676: M. Klusch, K.V. Hindriks, M.P. Papazoglou, L. Sterling (Eds.), Cooperative Information Agents XI. XI, 361 pages. 2007.

Vol. 4667: J. Hertzberg, M. Beetz, R. Englert (Eds.), KI 2007: Advances in Artificial Intelligence. IX, 516 pages. 2007.

Vol. 4660: S. Džeroski, L. Todorovski (Eds.), Computational Discovery of Scientific Knowledge. X, 327 pages. 2007.

Vol. 4659: V. Mařík, V. Vyatkin, A.W. Colombo (Eds.), Holonic and Multi-Agent Systems for Manufacturing. VIII, 456 pages. 2007.

Vol. 4651: F. Azevedo, P. Barahona, F. Fages, F. Rossi (Eds.), Recent Advances in Constraints. VIII, 185 pages. 2007.

Vol. 4648: F. Almeida e Costa, L.M. Rocha, E. Costa, I. Harvey, A. Coutinho (Eds.), Advances in Artificial Life. XVIII, 1215 pages. 2007.

Vol. 4635: B. Kokinov, D.C. Richardson, T.R. Roth-Berghofer, L. Vieu (Eds.), Modeling and Using Context. XIV, 574 pages. 2007.

Vol. 4632: R. Alhajj, H. Gao, X. Li, J. Li, O.R. Zaïane (Eds.), Advanced Data Mining and Applications. XV, 634 pages. 2007.

Vol. 4629: V. Matoušek, P. Mautner (Eds.), Text, Speech and Dialogue. XVII, 663 pages. 2007.

Vol. 4626: R.O. Weber, M.M. Richter (Eds.), Case-Based Reasoning Research and Development. XIII, 534 pages. 2007.

Vol. 4617: V. Torra, Y. Narukawa, Y. Yoshida (Eds.), Modeling Decisions for Artificial Intelligence. XII, 502 pages. 2007.

Vol. 4612: I. Miguel, W. Ruml (Eds.), Abstraction, Reformulation, and Approximation. XI, 418 pages. 2007.

Vol. 4604: U. Priss, S. Polovina, R. Hill (Eds.), Conceptual Structures: Knowledge Architectures for Smart Applications. XII, 514 pages. 2007.

Vol. 4603: F. Pfenning (Ed.), Automated Deduction – CADE-21. XII, 522 pages. 2007.

Vol. 4597: P. Perner (Ed.), Advances in Data Mining. XI, 353 pages. 2007.

Vol. 4594: R. Bellazzi, A. Abu-Hanna, J. Hunter (Eds.), Artificial Intelligence in Medicine. XVI, 509 pages. 2007.

Vol. 4585: M. Kryszkiewicz, J.F. Peters, H. Rybinski, A. Skowron (Eds.), Rough Sets and Intelligent Systems Paradigms. XIX, 836 pages. 2007.